List of the elements with their symbols and atomic masses

	Symbol	Atomic number	Atomic mass
Actinium	Ac	89	(227)ᵃ
Aluminum	Al	13	26.9815
Americium	Am	95	(243)
Antimony	Sb	51	121.75
Argon	Ar	18	39.948
Arsenic	As	33	74.9216
Astatine	At	85	(210)
Barium	Ba	56	137.33
Berkelium	Bk	97	(247)
Beryllium	Be	4	9.0122
Bismuth	Bi	83	208.980
Boron	B	5	10.811
Bromine	Br	35	79.904
Cadmium	Cd	48	112.41
Calcium	Ca	20	40.08
Californium	Cf	98	(249)
Carbon	C	6	12.01115
Cerium	Ce	58	140.12
Cesium	Cs	55	132.905
Chlorine	Cl	17	35.453
Chromium	Cr	24	51.996
Cobalt	Co	27	58.9332
Copper	Cu	29	63.54
Curium	Cm	96	(247)
Dysprosium	Dy	66	162.50
Einsteinium	Es	99	(254)
Erbium	Er	68	167.26
Europium	Eu	63	151.96
Fermium	Fm	100	(253)
Fluorine	F	9	18.9984
Francium	Fr	87	(223)
Gadolinium	Gd	64	157.25
Gallium	Ga	31	69.72
Germanium	Ge	32	72.59
Gold	Au	79	196.967

	Symbol	Atomic number	Atomic mass
Hafnium	Hf	72	178.49
Hahnium	Ha	105	(260)
Helium	He	2	4.0026
Holmium	Ho	67	164.930
Hydrogen	H	1	1.00797
Indium	In	49	114.82
Iodine	I	53	126.9045
Iridium	Ir	77	192.2
Iron	Fe	26	55.847
Krypton	Kr	36	83.80
Lanthanum	La	57	138.91
Lawrencium	Lr	103	(257)
Lead	Pb	82	207.19
Lithium	Li	3	6.941
Lutetium	Lu	71	174.97
Magnesium	Mg	12	24.305
Manganese	Mn	25	54.9380
Mendelevium	Md	101	(256)
Mercury	Hg	80	200.59
Molybdenum	Mo	42	95.94
Neodymium	Nd	60	144.24
Neon	Ne	10	20.179
Neptunium	Np	93	237.0482
Nickel	Ni	28	58.70
Niobium	Nb	41	92.906
Nitrogen	N	7	14.0067
Nobelium	No	102	(253)
Osmium	Os	76	190.2
Oxygen	O	8	15.9994
Palladium	Pd	46	106.4
Phosphorus	P	15	30.9738
Platinum	Pt	78	195.09
Plutonium	Pu	94	(242)
Polonium	Po	84	(210)
Potassium	K	19	39.098

	Symbol	Atomic number	Atomic mass
Praseodymium	Pr	59	140.907
Promethium	Pm	61	(147)
Protactinium	Pa	91	231.0359
Radium	Ra	88	226.0254
Radon	Rn	86	(222)
Rhenium	Re	75	186.2
Rhodium	Rh	45	102.905
Rubidium	Rb	37	85.47
Ruthenium	Ru	44	101.07
Rutherfordium	Rf	104	(257)
Samarium	Sm	62	150.35
Scandium	Sc	21	44.956
Selenium	Se	34	78.96
Silicon	Si	14	28.086
Silver	Ag	47	107.868
Sodium	Na	11	22.9898
Strontium	Sr	38	87.62
Sulfur	S	16	32.064
Tantalum	Ta	73	180.948
Technetium	Tc	43	(99)
Tellurium	Te	52	127.60
Terbium	Tb	65	158.925
Thallium	Tl	81	204.37
Thorium	Th	90	232.038
Thulium	Tm	69	168.934
Tin	Sn	50	118.69
Titanium	Ti	22	47.90
Tungsten	W	74	183.85
Uranium	U	92	238.03
Vanadium	V	23	50.942
Wolfram	W	74	183.85
Xenon	Xe	54	131.30
Ytterbium	Yb	70	173.04
Yttrium	Y	39	88.905
Zinc	Zn	30	65.38
Zirconium	Zr	40	91.22

ᵃApproximate values for radioactive elements are listed in parentheses.

Chemistry

The central science

The cover photo shows the surface of a high-purity crystal of arsenic (magnified about 1100 times) after it has been subjected to a vacuum at 282°C for 4 hours. The pyramidlike objects are actually pits in the surface caused by the evaporation of arsenic molecules. This photo was taken in connection with a study on the molecular processes by which molecules evaporate from crystalline solids. Photos such as this show that the most rapid evaporation occurs where imperfections are found in the crystal. (Photo courtesy of Gerd M. Rosenblatt, Carl A. Hultman, and Michael B. Dowell, Department of Chemistry, The Pennsylvania State University)

THEODORE L. BROWN
University of Illinois

H. EUGENE LeMAY, JR.
University of Nevada

2ND EDITION

Chemistry

The central science

Prentice-Hall, Inc., Englewood Cliffs, New Jersey 07632

Chemistry: The central science, **2nd edition**
THEODORE L. BROWN ▲ H. EUGENE LEMAY, JR.

10 9 8 7 6 5 4 3 2 1

Printed in the United States of America

PRENTICE-HALL INTERNATIONAL, INC., *London*
PRENTICE-HALL OF AUSTRALIA PTY. LIMITED, *Sydney*
PRENTICE-HALL OF CANADA, LTD., *Toronto*
PRENTICE-HALL OF INDIA PRIVATE LIMITED, *New Delhi*
PRENTICE-HALL OF JAPAN, INC., *Tokyo*
PRENTICE-HALL OF SOUTHEAST ASIA PTE. LTD., *Singapore*
WHITEHALL BOOKS LIMITED, *Wellington, New Zealand*

Library of Congress Cataloging in Publication Data
Brown, Theodore L
    Chemistry: the central science.

    Includes index.
    1. Chemistry. I. LeMay, Harold Eugene, 1940–
joint author. II. Title.
QD31.2.B78 1981   540   80-16771
ISBN 0-13-128504-1

*This book was composed on film in Baskerville and Helvetica, with
Serif Gothic display type. Development editor: Raymond Mullaney;
acquiring editor: Elizabeth G. Perry; designer: Lorraine Mullaney;
production assistant: Richard D. Kilmartin; manufacturing buyer:
Raymond Keating; art preparation: Vantage Art, Inc.*

To those from whom we have learned,
and to our wives and children

Contents

Appendices

Preface

The preface is almost always the last part of a textbook to be written. Only when the book has assumed its final form can the author declare to the reader what he thinks he has accomplished. Thus, what is for students the beginning is, in one sense, for authors the culmination of a long, often difficult journey. In this preface we want to give you some feeling for the ideas and concepts that have informed our writing of the text. We also wish to suggest how you can best make use of this book in your study of chemistry.

You will notice that this is the second edition of *Chemistry: The Central Science*. The first edition has been a very successful text; apparently many teachers of chemistry (and many students as well, to judge from letters we have received) felt that we had developed a good book. But, as with any first attempt, there is bound to be room for improvement. During the past few years, with the help of our publisher we have asked a great many teachers and students for suggestions as to how the book might be improved. In addition, we have watched carefully for the need to make changes to keep the book up to date. This second edition represents the end result of more than 2 years of work. The changes we have made are numerous, and some of them are important, but they need not concern you. What you should know is that we have adhered to the general approach and style that distinguished the first edition. Our aim has been to present chemistry to you in a clear, readable fashion. We have tried continually to keep in mind the audience for whom the book is intended—you, the student.

Most of you are studying chemistry because it has been declared an essential part of the curriculum in which you are enrolled. That curriculum may be agriculture, dental hygiene, electrical engineering, geology, microbiology, metallurgy, paleontology, or one of many other related areas of study. It is fair to ask why it is that so many diverse areas of study should all relate in an essential way to chemistry. The answer is that chemistry is, by its nature, the *central* science. In any area of human activity that deals with some aspect of the material world, there must inevitably be a concern for the fundamental character of the materials involved—their endurance, their interactions with other materials, and their changes under a given set of conditions. This is true whether the materials involved are a polymer used to coat electronics components, the color used by a Renaissance painter, or the blood cells of a child born with sickle-cell anemia. It is very likely that chemistry plays an important role in the profession to which you now aspire, or may decide later to pursue. You will be a better professional, a more creative and knowledgeable person, if you understand the chemical concepts applicable to your work and are able to apply these concepts as needed.

The relationship of chemistry to professional goals is important, and this factor provides reason enough for you to study chemistry. There is, however, an even more important reason. Because chemistry is so central and so intimately involved in almost every aspect of our contact with the material world, this science is an integral part of our culture. The involvement of chemistry in our lives goes much deeper than the well-known advertising slogan, "Better things for better living through chemistry." In addition to all the obvious ways in which we use the products of chemical research and production—plastic bags, children's toys, counter tops, weed and insect killers, photographic films—we indirectly use thousands of chemical products via the foods we eat, the cars we drive, the medical care we receive, and so forth. During the past several years, we have become increasingly aware that our use of chemicals has had a profound and frightening effect on our environment. Indeed, many scientists are convinced that we have so intensely polluted this planet and so unthinkingly sowed the seeds of future pollution that the fate of civilization is all but sealed. Whether this is so remains to be seen; however, if you are to be a responsible citizen, you will surely need to be informed on many complex issues involving chemistry and the use of chemicals. Because vested interests have a powerful stake in public policy, the public often is presented with conflicting information and claims. You can more fully appreciate and analyze the complex issues put before you if you understand the fundamental principles involved and keep them in mind during your reading and study.

With all of these considerations in mind, you should now be impatient and eager to begin your study of chemistry. Now that you are ready to go, we should say something about how this book can best help you. You might first take a few minutes to glance through the table of contents. The particular sequence of chapters that we have chosen is one that we feel promotes a natural unfolding of the science of chemistry. However, the order in which the chapters of the book are covered in the classroom will be determined by your instructor. You should not be disturbed if the order is not the same as the order in the book. The book has been written so as to make allowance for alternative chapter orders and, in some instances, for the complete omission of certain chapters. Notice that several chapters interspersed throughout the book deal with the chemical aspects of the world in which we live: the air, the earth, and the waters on the earth's surface. In these chapters we have attempted to connect the chemical facts and principles introduced in other, usually earlier, chapters to the familiar (and sometimes not so familiar) aspects of our surroundings on earth. Your instructor, the person who will guide you through this book, may not feel that there is sufficient time to cover some or all of the materials in these chapters. We suggest that you read them anyway; they will help you appreciate the many ways in which chemical concepts and observations are related to contemporary life.

If you should at some point encounter a term or concept you are expected to know but can't remember, use the index at the back of the book. A good index is a rarity; we have worked hard to make your index in this book as complete and accurate as possible. Use it often. (Remember the index also when you later use the book as a reference, after having finished the course. It can help you find what you want more quickly than any other means.)

The difficulties that many chemistry students experience often can be traced to faulty exposition and confusing explanations in their text. This book has been worked on very thoroughly by many people to ensure that it is as clear, concise, and free of confusion as possible. However, you may find that a single reading of a chapter will not suffice if you are to use the book effectively as a learning tool. We suggest that you read every assigned chapter as early as possible, preferably before the material is covered in lecture. This will make you aware of important concepts and terms even before they are treated by the lecturer. Later, you will need to go through the assigned sections of the book much more carefully, making sure that you understand the new terms and problems put before you. We have inserted a great many *sample exercises* into the text, so that you might have clearly worked-out examples of problem solving of various types. You should study these exercises carefully, noting every aspect of them, especially if numerical problem solving is involved.

The review section at the end of each chapter is an integrated package designed to help you determine whether you have in fact learned all the material assigned you in each chapter. The *summary* points out the highlights of the chapter; sometimes we say things a little differently in the summary in order to add an extra element of understanding to what you have gotten from the chapter itself. The *key terms* that you should know are also collected for your convenience. The *learning goals* are placed at the end of the chapter to enable you to test yourself. You should make sure that you can meet each learning goal. This can best be done if you state a definition and then check it, write a formula and then check it, or solve a problem and then check it. It may happen, of course, that your instructor will not have covered part of the material in a chapter. You can then skip over the learning goals for this material, but you should still read the complete summary and learn all of the key terms. By learning even nonrequired terms and concepts you can expand your chemical vocabulary with little effort.

The *exercises* at the back of each chapter are designed to test your understanding of the materials covered in the chapter. They are grouped according to topic, except for a number of additional exercises. The purpose of the additional exercises is to test your ability to solve a problem when it is not clearly identified as to topic. Also, some of the questions in this category require the application of material from more than one topic area. Problems marked with brackets are, in general, a little more difficult to solve than the others. We have prepared a solutions manual that contains detailed answers to all the end-of-chapter exercises; you should consult this manual only after working out problems on your own.

Finally, you should note that there are several appendices following Chapter 25. These are designed to aid you in various ways. You should get acquainted with what is there by glancing through them before the course gets under way. In particular, note that answers are provided to many of the end-of-chapter exercises. Color question numbers in the text indicate that the answer to the question is in the answer section following the appendices.

Your instructor may have elected to have you purchase the *Student's Guide* designed for use with the text. This guide, written by Professor James C. Hill, of California State University, Sacramento, is a nicely organized and well-written supplement to the text. You will find it filled with helpful ideas, problem-solving techniques, and fresh insights into the materials presented in the text. We are very happy that Jim has agreed to write the study guide; we feel that it is a valuable learning aid for use with the text.

Most general chemistry courses involve laboratory as well as classroom work. There is a very good reason for this. Chemistry is an experimental science; the entire theoretical structure of chemistry is based on the results of laboratory experiments. As you study chemistry, you should try to relate what you learn in the classroom and from the text to operations and observations made in the course of your laboratory work. A very fine laboratory manual for use with this text has been written by Professors John H. Nelson and Kenneth C. Kemp of the Department of Chemistry, University of Nevada, Reno. We believe that it is also an important learning tool in your study of chemistry.

During the many years that we have been practicing chemists, we have found chemistry to be an exciting intellectual challenge and an extraordinarily rich and varied part of our human cultural heritage. We hope that all the hassles you must face regarding course grades will not keep you from sharing with us some of that enthusiasm and appreciation. We have, in effect, been engaged by your instructor to help you learn chemistry. We are confident that we've done that job well. In any case, we would appreciate your writing us, either to tell us of the book's shortcomings, so that we might do better, or of its virtues, so that we'll know where we have helped you most.

Theodore L. Brown
School of Chemical Sciences
University of Illinois
Urbana 61801

H. Eugene LeMay, Jr.
Department of Chemistry
University of Nevada
Reno 89557

Acknowledgments

This book owes its final shape and form to the assistance and hard work of many people. Several colleagues reviewed the manuscript and helped us immensely by sharing their insights and criticizing our initial writing efforts. We would like especially to thank the following:

David L. Adams, North Shore Community College
Charles Anderson, Pacific Lutheran University
O. T. Beachley, Jr., State University of New York at Buffalo
Jon M. Bellama, University of Maryland
Jesse S. Binford, University of South Florida
Robert Brasted, University of Minnesota
Rachel J. Britton, Purdue University
Thomas Cassen, University of North Carolina at Charlotte
Ronald J. Clark, Florida State University
A. J. Crossfield, American River College
Geoffrey Davies, Northeastern University
Leland S. Endres, California State Polytechnic College
Wade Freeman, University of Illinois at Chicago Circle
Roy G. Garvey, North Dakota State University
Thomas Gilmore, Delaware Technical and Community College
Dorothy Goldish, California State University, Long Beach
Henry Griffin, University of Michigan
Arnulf P. Hagen, University of Oklahoma
Albert W. Herlinger, Loyola University
James C. Hill, California State University, Sacramento
Mary O. Hillis, Vassar College
Richard Hunt, California State University, Long Beach
Joseph Kanamueller, Western Michigan University
Edwin M. Larsen, University of Wisconsin, Madison
Gerard A. L'Heureux, Holyoke Community College
C. R. Matthews, Pennsylvania State University
Catherine Middlecamp, Hobart & William Smith Colleges

R. E. Mitchell, Texas Technological University
Robert Niedzielski, University of Toledo
Stanley J. Opella, University of Pennsylvania
J. Gary Pruett, University of Pennsylvania
Patricia Ann Redden, St. Peter's College
Russell Trimble, Southern Illinois University at Carbondale
James W. Viers, Virginia Polytechnic Institute
Kenneth W. Watkins, Colorado State University
Philip K. Welty, Miami University
Charles M. Wheeler, University of New Hampshire
Robert Whitaker, University of South Florida
Steven S. Zumdahl, University of Illinois

We deeply appreciate the assistance of the following members of Prentice-Hall's College Division: David R. Esner, Director, Product Development Department, Elizabeth G. Perry, Chemistry Editor, Fred Henry, former Chemistry Editor, and Robin Bartlett, Marketing Manager, who provided guidance, support, and counsel throughout the project; Dudley R. Kay, Associate Director of Product Marketing Management, whose hard and imaginative work helped to make the first edition such a success; Lorraine Mullaney, Associate Art Director, whose design for the first edition was difficult to improve upon and whose talents are again evident in this edition; and, most particularly, Raymond Mullaney, Assistant Director, Product Development Department, who has been with this project from its beginnings. His sharp eye, impeccable taste, and good judgment have guided the transformation from rough manuscript to a finished book of which we are all proud.

Finally, special thanks are due to Becky Lazaro and Mary Kaylor, who so ably typed manuscript copy from our rough drafts.

Introduction: some basic concepts

Perhaps the only thing permanent about our world is change. All around us are numerous examples of change in ourselves and our environment. Trees change color in autumn, iron rusts, snow melts, paint peels, seeds become flowers, and logs burn. We grow up, we grow old. Living plants and animals undergo continual change, and even dead plants and animals continue to change as they decay. Such changes have long fascinated people and have prompted them to look more closely at nature's working in hopes of better understanding themselves and their environment.

Understanding change is closely tied to understanding the nature and composition of matter. Matter is the physical material of the universe; it is anything that occupies space and has mass. Chemistry is the science that is concerned primarily with matter and the changes that it undergoes. Therefore, as we begin our study of chemistry, our primary focus will be on matter. First, however, let's sketch a somewhat broader picture of chemistry.

Chemistry is a changing science. Therefore, the questions that chemists seek to answer are constantly changing also. Because of this, we might define chemistry as what chemists do. In many regards this definition is unsatisfactory. Nevertheless, it does suggest that chemistry itself changes as chemists absorb new information from other fields, tackle new problems, or reexamine old ones in new ways. One of the important activities of chemists is the synthesis of new materials or the improvement in the ways of making old ones. This aspect of chemistry has had great impact on our lives; chemists have synthesized new fibers, medicines, fertilizers, pesticides, and structural materials. Many new chemicals never find any commercial use but are nevertheless important to chemists in answering subtle questions about matter and its changes. In

1 Introduction:
some basic concepts

1.1 The emergence of
chemistry: a historical
perspective

designing ways to synthesize new materials, it is useful to know the factors that determine how fast and to what extent the required changes proceed. Such knowledge allows chemists to improve, avoid, or control many changes of matter. This knowledge is necessary, for example, in devising ways to clean up automobile exhaust or to make fertilizers at lower cost. Chemists are also interested in determining the identity and concentration of substances. Such analysis may involve determining the quality of a soap in a manufacturing operation, the concentration of a pollutant in the air, the amount of gold in a potential ore, the amount of mercury in a lake, the identity of the substances in some physiologically active mixture, or the chemicals resulting from the utilization of a drug in the body. Chemists are interested not only in determining what things are made of, but also in discovering the ways their composition and structure are related to properties. For example, what makes a particular substance poisonous or sweet or hard or explosive?

The intent of this text is to introduce you to basic chemical facts and theories, not as ends in themselves, but as means to help you understand the material world and to recognize the constraints and opportunities it provides. We hope that this text will provide not only a firm foundation for further scientific studies, whether they be in chemistry or some other field, but also a background to enable you to evaluate scientific information found in news media and periodicals. In the remainder of this chapter we will consider some background material useful to your studies— the metric system, uncertainty in measurement, and problem solving in chemistry. We will also briefly explore the historical and philosophical background of chemistry and the scientific approach to problems.

1.1 The emergence of chemistry: a historical perspective

Chemistry has two roots. First, it is rooted in the craft traditions such as metallurgy, brewing, tanning, and dyeing, which provided a practical understanding of how matter behaves. Second, it can also be traced to the philosophers of ancient Greece who concerned themselves with questions of the basic nature of matter. The growth of chemistry is in some respects a reflection of the practical problems overcome in the course of human society's cultural and technical development. In addition, however, chemistry is an expression of humankind's innate curiosity and desire to understand its surroundings without regard to the practical application of that understanding.

Metallurgy, the science and art of obtaining and working with metals, exemplifies the development of chemical knowledge through the craft traditions. This craft developed for many years and achieved considerable sophistication without any theoretical framework that would explain metallurgical operations and guide their development. Developments were made largely through trial and error and through accidental discoveries. For the most part, the early pattern of discovery of metals followed their ease of recovery from ores, the earthy mixtures that are mined as sources of metals. Gold was one of the first metals used because it is found in nature in an uncombined, metallic state, for example, as gold nuggets. Copper, which is more abundant, was not used until about 3500 B.C. At that time processes for obtaining copper metal from its ores were discovered. This discovery was undoubtedly accidental. It could have occurred when some copper ore was dropped or thrown into the coals of a fire. Copper metal can be obtained by heating copper ore with charcoal. Methods for obtaining iron, which is much more abundant than copper, did not develop until about 1500 B.C.

People probably attempted to understand changes in matter even before they began to use these changes to their advantage. We know that many early explanations presumed the existence of supernatural powers. In contrast to this approach, the early Greeks sought to understand matter and its changes purely on the basis of logic. However, they were not especially concerned with using these ideas as guides to improve their crafts. Modern science differs from the approach of the Greeks; it depends not only on logic but also on the systematic gathering of facts and on careful observations. Furthermore, the ideas of science are widely used to guide the development of our technologies.

The ideas of the early Greek philosophers such as Anaximander (sixth century B.C.) and Empedocles (fifth century B.C.) found their most lasting expression in the teachings and writings of Plato (427–347 B.C.) and Aristotle (384–322 B.C.). The early Greek philosophers had proposed that all of nature was composed of four elements: fire, earth, air, and water. This idea provides a logical explanation for many observations. For example, a green log can burn, producing smoke (air) and flame (fire), leaving behind ashes (earth), and perhaps even giving a fleeting glimpse of sap (water). The idea of four elements was extended by emphasizing the basic properties of the elements. These were coldness, hotness, dryness, and wetness. Each element, in its ideal form, had two associated properties as shown schematically in Figure 1.1. For example, water was the wet, cold element. The Greeks believed that an element could be changed into any other element by altering its properties. For example, water could be changed to earth by replacing the property of wetness with the property of dryness.

This concept and its associated logic persisted for over 1000 years, influencing thought through the Middle Ages. However, by the time of the Middle Ages the list of elements and their associated properties had grown. Operating in this framework of logic the alchemists sought to convert common or ordinary metals into gold (Figure 1.2). Their attempts were based on the idea that addition of the essential quality or property of "nobility" to these metals would cause them to "grow" into gold. In the course of their futile attempts to bring about such changes the alchemists discovered new chemicals and developed new ways for working with them.

The Greek ideas of elements, as further developed during the Middle Ages, spawned the idea of a combustible property associated with flammable materials. The element associated with this property was known as phlogiston after the Greek word for fire. At this point, the idea of basic or elementary properties of matter had become quite confused with that of elements. The concept of phlogiston was used to explain many observations. It was believed that objects could burn only so long as they still contained phlogiston. After phlogiston had escaped from an object it was no longer flammable.

Early scientists noted that when metals are heated in air they lose their metallic properties such as their luster and are converted to pow-

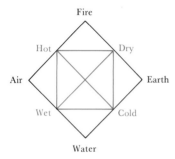

FIGURE 1.1 A schematic representation of the four elements of the Greeks and the four properties associated with these elements.

1 Introduction:
some basic concepts

1.1 The emergence of
chemistry: a historical
perspective

FIGURE 1.2 *The Alchemist,* a seventeenth-century painting by the Flemish painter David Teniers (1582–1649). (*Bettmann Archive*)

FIGURE 1.3 Joseph Priestley (1733–1804); Priestley became interested in chemistry at the age of 39, perhaps through his personal acquaintance with Benjamin Franklin. Because he lived next door to a brewery where he could obtain carbon dioxide, his initial studies involved this gas and were later extended to other gases. Because he was suspected of sympathizing with the American and French Revolutions, his church, home, and laboratory in Birmingham were burned by a mob in 1791. Priestley had to flee in disguise. He eventually emigrated to the United States in 1794 where he lived his remaining years in relative seclusion in Pennsylvania. Although his discovery of "dephlogisticated air" (oxygen) eventually led to the downfall of the phlogiston theory, Priestley stubbornly continued to support this theory even after strong evidence had brought it into serious question. Priestley was a scientific conservative, although he was very liberal in his religious and political views. (*Library of Congress*)

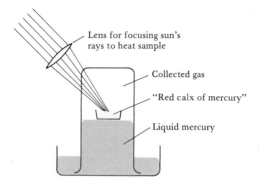

Lens for focusing sun's
rays to heat sample

Collected gas

"Red calx of mercury"

Liquid mercury

FIGURE 1.4 Priestley's experiment showing how he prepared "dephlogisticated air" (oxygen).

dery materials with much different properties that were known as calxes.* For example, under appropriate conditions, iron can be converted to rust by heating it in air. Thus, rust is a calx of iron. According to phlogiston theory, a metal is converted to its calx when it loses phlogiston. In the eighteenth century Joseph Priestley (Figure 1.3), an English clergyman and self-trained amateur scientist, heated the red calx of mercury and thereby generated mercury metal together with a gas (oxygen). His experiment is shown in Figure 1.4. Priestley found that objects burn more vigorously in this gas than in air. He called the gas "dephlogisticated air" because its properties suggested that the gas had a particularly large capacity for phlogiston.

THE BIRTH OF MODERN CHEMISTRY

In 1772, a wealthy French nobleman named Antoine Lavoisier (Figure 1.5) began experimenting with combustion, the act of burning. By weighing objects before and then after combustion, Lavoisier observed that burning objects gain weight. This effect is easily observed when metals are converted to their calxes; Lavoisier showed that it is also true when nonmetals like sulfur are burned. Furthermore, he observed that when combustion is carried out in a closed container, there is no change in weight; the weight of all of the substances in the container is the same before and after combustion even though they change form. There was no evidence for the loss of phlogiston. Instead, the experiments indicated that when a substance burns it gains something from the air. The weight gained by the burning sample was the same as the weight lost by the air. This is why there is no net change in weight when combustion is carried out in a closed container. These quantitative measurements could not be explained using the phlogiston idea. Lavoisier also found that combustion in dephlogisticated air, which he renamed oxygen, gives the same products as combustion in air. It was concluded that air contains oxygen. Thus, the role of air in combustion is not to carry off phlogiston but to provide oxygen.

On the basis of these observations, Lavoisier rejected the phlogiston theory. In its place he proposed that when an object burns, oxygen is removed from air and becomes incorporated into the burning object. In this view, the combustion of coal, which contains carbon, involves a

---

*The word calx is no longer used in chemistry, but we will use it in this brief account of chemical history to illustrate the reasoning followed by the early chemists.

FIGURE 1.5 Antoine Lavoisier, after a painting by Louis David. Lavoisier (1743–1794) conducted studies that led to the downfall of the phlogiston theory and the birth of modern chemistry. Unfortunately, Lavoisier's career was cut short by the French Revolution. He was not only a member of the French nobility but also a tax collector. He was guillotined in 1794 during the final months of the Reign of Terror. He is now generally considered to be the father of modern chemistry because of his reliance on carefully controlled experiments and his use of quantitative measurements. The tribunal that sentenced him to death, however, declared that France had no need for scientists. The great mathematician Lagrange, then living in Paris, remarked: "It took but a moment to cut off that head, though a hundred years perhaps will be required to produce another like it." (*Courtesy Burndy Library*)

reaction between oxygen and carbon. The product of this reaction is a gaseous substance called carbon dioxide. The reaction can be summarized as

Carbon + oxygen $\longrightarrow$ carbon dioxide

where the arrow may be read as "produces" or "yields." Because combustion could be understood better without reference to phlogiston, that idea slowly died. Today we define combustion as a rapid reaction accompanied by heat and light; we find that most common combustion reactions do indeed involve oxygen as Lavoisier suggested.

Air is only 20 percent oxygen; consequently, combustion is much more vigorous in pure oxygen than in air. The fire that broke out on the *Apollo 1* rocket in January 1967 and resulted in the deaths of astronauts Grissom, White, and Chaffee burned so very fast because the space capsule was filled with pure oxygen. Because of that incident, a mixture of 60 percent oxygen and 40 percent nitrogen has been used in space capsules ever since. A fire may be extinguished by placing a blanket over it because the blanket excludes the oxygen. People in a closed room can be suffocated by a fire because it removes oxygen from air.

Lavoisier is generally considered the father of modern chemistry because of his reliance on carefully controlled experiments and his use of quantitative measurements and not merely qualitative observations. In his studies Lavoisier also abandoned the ancient idea that if a material contains a particular element it must have the properties of that element. He adopted an idea proposed in 1661 by Robert Boyle. Boyle proposed that elements are the basic substances out of which all other substances can be made and into which all other substances can be decomposed or resolved. We will pursue the idea of elements more fully in Chapter 2.

1.2 The scientific approach

The fundamental activity of science is making careful observations. These may be of both a qualitative and quantitative nature and often involve controlled experiments. Scientists seek general relations that will unify their observations. A concise verbal statement or a mathematical equation that summarizes a broad variety of observation and experience is known as a scientific law. A familiar example is the law of gravity; it summarizes our experience that what goes up must come down. We also seek to understand our laws. A tentative explanation is called a hypothesis. A hypothesis is useful only if it can be used to make predictions that can be tested by further experiments and thereby verified or refuted. A hypothesis that continually withstands such tests is called a theory. A theory may serve to unify a broad area and may provide a basis for explaining many laws. Such is the case with the atomic theory of matter, which we will begin to examine in Chapter 2.

There is no fail-proof, step-by-step scientific method that scientists use. The approaches of various scientists depend on their temperament, circumstances, and training. Rarely will two scientists approach the same problem in the same way. The scientific approach involves doing one's utmost with one's mind to understand the workings of nature. Just because we can spell out the results of science so concisely or neatly in textbooks does not mean that scientific progress is smooth, certain, and predictable. The path of any scientific study is likely to be irregular and uncertain; progress is often slow and many promising leads turn out to be dead ends. Through the course of our studies we will see that serendipity (fortunate accidental discovery) has played an important role in the development of science. What we will often miss discussing are the doubts, conflicts, clashes of personalities, and revolutions of perception that have led to our present ideas. We should also remember that our theories are not chiseled into stone; they are tentative.

SELECTION OF THEORIES

No hypothesis or theory can ever be exposed to all the possible tests necessary for absolute verification. However, a hypothesis or theory can be disproved by obtaining experimental results inconsistent with it. Scientific advance depends on such disproofs to eliminate faulty hypotheses and theories. In the absence of such disproofs, scientists often choose between hypotheses or theories by comparing the ability of each to explain the evidence at hand; the one that explains the facts better is chosen.

It has been suggested that scientific progress is more rapid when scientists are open and imaginative enough to formulate several alternative hypotheses to explain their observations. Experiments can then be designed to test these alternatives, thereby excluding some of them. This approach involves a continual search for alternative hypotheses. A person who works with only a single hypothesis can become strongly attached to it. Research can then become a strenuous and devoted attempt to force nature into the conceptual boxes supplied by that hypothesis.

In the end it is the collective judgment of the scientific community that effectively decides among theories. Therefore, one of the most important activities of a scientist is public disclosure of scientific results through publication. The authority of science does not rest ultimately on the individual who has done the work, but rather on whether others can repeat the work and obtain the same results or extend the work in a self-consistent fashion.

THE DEVELOPMENT OF SCIENCE

In the traditional view science progresses by the gradual accumulation of factual knowledge and of ever more encompassing and useful laws and theories to unify it. Indeed, this may serve as an explanation of what is meant by scientific progress. The term "useful" can be taken to mean better problem-solving ability. It might also suggest that progress involves movement toward greater technological benefits. If this is true, we must ask whether the benefits involve only an increase in the quantity of "things" or whether they reflect a consideration for the quality of life. Similarly, progress is often thought to involve our increased ability to manipulate nature. Perhaps we are beginning to realize that this ability also means learning how to live in harmony with nature. The idea that our goal is to "conquer nature" suggests exploitation and ignores the fact that we are part of nature.

The philosopher and historian Thomas S. Kuhn has suggested that there are discontinuities in the development of science that involve revolutionary changes in the ways scientists perceive and approach nature. Kuhn suggests that at any given time scientists operate under a set of ideas about what nature is like that is rather universally accepted as general truth. This intellectual matrix of ideas and beliefs, which have been called **paradigms,** gives direction to experimental efforts and influences what scientists are able to perceive. "Facts" may be distorted by our expectations or totally overlooked because they were not anticipated. In designing experiments, the types of questions asked are conditioned by the types of answers expected. All people must make judgments, and scientists are not immune to the frailties and possibilities of prejudice that beset all persons. A scientific revolution occurs when a paradigm such as the phlogiston theory undergoes change.

Persons involved in such revolutions often refer to a flash of insight or a new perception, indicating a new way of seeing things. These revolutions often involve young scientists or persons who are new to a particular field of science and therefore are not deeply committed to its prevalent paradigms. These revolutions are often triggered by observations that fail to yield to explanations in terms of the theories of the day. Kuhn suggests the following:

> Sometimes a normal problem, one that ought to be solvable by known rules and procedures, resists the reiterated onslaught of the ablest members of the group within whose competence it falls. On other occasions a piece of equipment designed and constructed for the purpose of normal research fails to perform in the anticipated manner, revealing an anomaly that cannot, despite repeated effort, be aligned with professional expectation. In these and other ways besides, normal science repeatedly goes astray. And when it does—when, that is, the professional can no longer evade the anomalies that subvert the existing tradition of scientific practice—then begin the extraordinary investigations that lead the practice of science. . . . They are the tradition-shattering complements to the tradition-bound activity of normal science.*

1.3 Measurement and the metric system

Lavoisier's studies of combustion (Section 1.1) should impress upon us the importance of quantitative measurements. This idea is simply common sense to us now, although it has not always been so. Consider a person who is ill. Using only sense perception we may conclude that this person is running a fever, but we are not certain, and another person may disagree with our conclusion. To tell accurately, we use a thermometer,

*Thomas S. Kuhn, *The Structure of Scientific Revolutions* (Chicago: The University of Chicago Press, 1962), p. 5.

and much can depend upon the result of our measurement—for example, whether the person's temperature is 98.6°F or 102°F. This example suggests three points about measurement. First, our five senses are the most important tools we have, but they are limited. We therefore resort to various instruments to extend and quantify our sense perceptions. Second, an advantage of quantitative data is that they allow different people to obtain the same results, thus avoiding many arguments based on opinions. Third, measurements depend on a standard of reference. For instance, there is a considerable difference between 102°F and 102°C. (The meaning of °F and °C will be discussed shortly.)

The standards used in science are those of the metric system. This is the system of weights and measures used throughout most of the world; the United States is also moving toward adopting it in many facets of society. The weights of most canned products in the grocery store are now given in grams as well as in ounces; there are even a few highway signs that show distance in both miles and kilometers (Figure 1.6).

FIGURE 1.6 A road sign along an interstate highway in Michigan, showing distance in metric and English-system units. (*Michigan Department of State Highways and Transportation*)

TABLE 1.1 Basic SI units

Physical quantity	Name of unit	Abbreviation
Mass	Kilogram	kg
Length	Meter	m
Time	Second	s or sec
Electric current	Ampere	A
Temperature	Kelvin	K
Luminous intensity	Candela	cd
Amount of substance	Mole	mol

According to international agreement reached in 1960, certain basic metric units and units derived from them are to be preferred in scientific use. The preferred units are known as International System units (commonly called SI units, from the French, *Système International*). The basic units of the SI system are given in Table 1.1. Non-SI units are to be progressively discouraged and with time phased out. Adoption of SI units is an attempt to further systematize the metric system. However, until SI units are fully adopted by practicing scientists, it is necessary to be aware of both SI units and the non-SI units that are still in use. Whenever we first encounter a non-SI unit in the text, the proper SI unit will also be given.

The primary standards used for the basic SI units are selected on the basis of their being reproducible, unchanging, and capable of use for precise measurement. Basically, however, they are otherwise arbitrarily selected. For example, the kilogram is defined as the mass of a standard platinum-iridium cylinder that is stored at the International Bureau of Weights and Measures at Sèvres, France.

The metric system employs a series of prefixes to indicate decimal fractions of the various basic measurements. The most common of these are given in Table 1.2.* In using the metric system and in working problems throughout this text, it is important to have a comfortable familiarity with exponential notation. If you are unfamiliar with exponential notation or want to review it, refer to Appendix A.1.

*It is interesting that the monetary system of the United States is decimal: 0.01 of a dollar is a cent (centidollar) while 0.001 of a dollar (a tenth of a cent) is a mill (millidollar). Extending this usage, we could theoretically refer to $1000 as a kilodollar or kilobuck.

TABLE 1.2 Selected prefixes used in the metric system

Prefix	Abbreviation	Meaning	Example
Mega-	M	10^6	1 megameter (Mm) = 1×10^6 m
Kilo-	k	10^3	1 kilometer (km) = 1×10^3 m
Deci-	d	10^{-1}	1 decimeter (dm) = 0.1 m
Centi-	c	10^{-2}	1 centimeter (cm) = 0.01 m
Milli-	m	10^{-3}	1 millimeter (mm) = 0.001 m
Micro-	μ[a]	10^{-6}	1 micrometer (μm) = 1×10^{-6} m
Nano-	n	10^{-9}	1 nanometer (nm) = 1×10^{-9} m
Pico-	p	10^{-12}	1 picometer (pm) = 1×10^{-12} m

[a] This is the Greek letter mu (pronounced "mew").

TABLE 1.3 Metric-English system equivalents

Length	Mass	Volume
1 meter = 1.094 yards	1 kilogram = 2.205 pounds	1 liter = 1.06 quarts
2.54 centimeters = 1 inch	453.6 grams = 1 pound	1 cubic foot = 28.32 liters

LENGTH

The basic SI unit of length is the meter (m). From the comparisons between metric and English-system measurements given in Table 1.3 we can see that the meter is only slightly longer than a yard. A diagrammatic comparison between metric and English-system measures of length is made in Figure 1.7. We will consider interconversion of English and metric system measures more closely in Section 1.5. For the moment it is more important that we clearly understand the use of the prefixes that are given in Table 1.2.

VOLUME

The measure for volume is a derived unit based on the fundamental SI unit of length cubed, m^3. The cubic meter, m^3, is the volume of a cube that is 1 m on each edge. Related units such as the cubic centimeter, cm^3 (sometimes written cc), or cubic decimeter, dm^3, are also used. Another common measure of volume is the liter (L), a volume roughly the size of a quart (refer to Table 1.3 and to Figure 1.8). A liter is the volume occupied by 1 cubic decimeter, dm^3. There are 1000 mL in a liter, and each milliliter is the same volume as a cubic centimeter. Thus milliliter and cubic centimeter are commonly used interchangeably in expressing volume. The liter is the first metric unit that we have encountered that is not an SI unit.

The devices most frequently used in chemistry to measure volume are illustrated in Figure 1.9. Pipets and burets allow delivery of liquids with more accurately known volumes than do graduated cylinders. Volumetric flasks are used to prepare accurately a designated volume of solution.

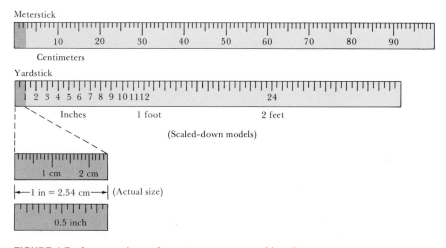

FIGURE 1.7 A comparison of common measures of length.

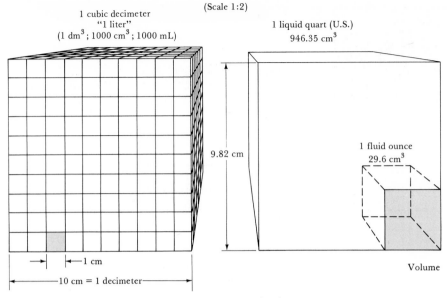

FIGURE 1.8 A comparison of common measures of volume.

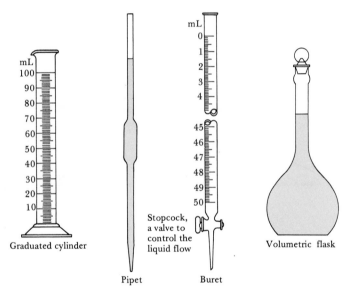

FIGURE 1.9 Common devices used in chemistry laboratories to measure volume.

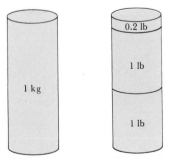

FIGURE 1.10 A comparison of common measures of mass.

MASS

The basic SI unit of mass* is the kilogram (kg). As shown in Table 1.3 and Figure 1.10, a kilogram is equal to 2.2 lb. The unit of mass used most frequently in chemistry is the gram (g), which is $\frac{1}{1000}$ of a kilogram. The

*Mass and weight are often incorrectly thought to be the same. Mass is a measure of the amount of material in an object; the weight of that object, however, depends not only on its mass, but also on the attractive force of gravity. In outer space, where gravitational forces are very weak, an astronaut may be weightless, but he is not massless. In fact, he has the *same* mass as he has on earth. Nevertheless, it is common practice to use the terms mass and weight interchangeably.

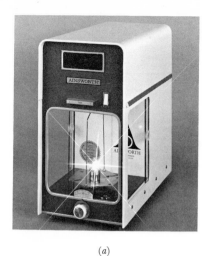

(a)

(b)   (c)

FIGURE 1.11 Three common types of laboratory balances: (a) analytical balance; (b) triple-beam balance of the stirrup type; (c) triple-beam platform balance. (a, *Denver Instrument Company, Ainsworth Division; b and c, Ohaus Scale Corp.*)

mass of an object is determined by balancing it against a set of known masses using a device known as a balance. Several types of common laboratory balances are shown in Figure 1.11.

TABLE 1.4 Densities of some selected substances

Substance	Density (g/cm³)
Air	0.001
Balsa wood	0.16
Water	1.00
Table salt	2.16
Iron	7.9
Gold	19.32

DENSITY

Density is a quantity widely employed by chemists to identify substances. It is defined as the amount of mass in a unit volume of the substance:

$$\text{Density} = \text{mass/volume} \qquad [1.1]$$

Density is commonly expressed in units of grams per cubic centimeter (g/cm^3 or $g\ cm^{-3}$). The densities of some common substances are listed in Table 1.4.

SAMPLE EXERCISE 1.1

Calculate the density of mercury if 1.00×10^2 g occupies a volume of $7.36\ cm^3$.

Solution:

$$\text{Density} = \frac{\text{mass}}{\text{volume}} = \frac{1.00 \times 10^2\ g}{7.36\ cm^3} = 13.6\ g/cm^3$$

The terms *density* and *weight* are sometimes confused. A person who says that iron weighs more than air generally means that iron has a higher density than air; 1 kg of air has the same mass as 1 kg of iron.

TEMPERATURE

The temperature scales commonly employed in scientific studies are the Celsius (or centigrade) and Kelvin scales; as previously noted, the Kelvin is the SI unit of temperature. The Celsius scale is based on assignment of

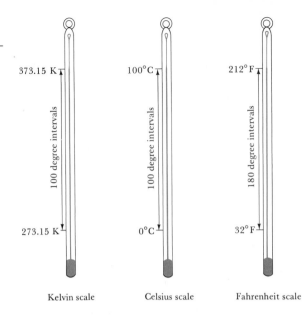

373.15 K

273.15 K

100 degree intervals

Kelvin scale

100°C

0°C

100 degree intervals

Celsius scale

212°F

32°F

180 degree intervals

Fahrenheit scale

FIGURE 1.12 A comparison of the Fahrenheit, Celsius, and Kelvin temperature scales.

0°C to the freezing point of water and 100°C to its boiling point at sea level. The corresponding temperatures in the Fahrenheit scale are 32°F and 212°F. There are 100° between the freezing point and boiling point of water in the Celsius scale, whereas there are 180° between these points on the Fahrenheit scale. Consequently, the Celsius and Fahrenheit scales are related as follows:*

$$°C = \frac{100}{180}(°F - 32°) = \frac{5}{9}(°F - 32°) \qquad [1.2]$$

The Kelvin scale is based on the properties of gases, and its origins will be considered more fully in Chapter 9. Zero on this scale corresponds to −273.15°C, and the size of a Kelvin is the same as a degree Celsius. The Kelvin and Celsius scales are therefore related by Equation [1.3].

$$K = °C + 273.15 \qquad [1.3]$$

[According to the SI convention, a degree sign (°) is not used with the Kelvin scale. Thus we write 273 K and not 273°K.] Further comparisons between the Celsius, Kelvin, and Fahrenheit scales are made in Figure 1.12 and Table 1.5.

*Using Equation [1.2] it can be shown that −40°C = −40°F. This fact permits another simple method of converting between Celsius and Fahrenheit scales. To convert a temperature from Fahrenheit to Celsius add 40°, multiply the result by $\frac{5}{9}$, and then subtract 40°. To convert from Celsius to Fahrenheit add 40°, multiply the result by $\frac{9}{5}$, and then subtract 40°

TABLE 1.5 Some comparisons of Fahrenheit, Celsius, and Kelvin temperatures

Absolute zero	−460°F	−273°C	0 K
Freezing point of water	32°F	0°C	273 K
Average room temperature	68°F	20°C	293 K
Normal body temperature	98.6°F	37°C	310 K
Boiling point of water	212°F	100°C	373 K

SAMPLE EXERCISE 1.2

If a weather forecaster predicts that the temperature for the day will reach 30°C, what is the predicted temperature (a) in K; (b) in °F?

Solution:

(a) $K = 30 + 273 = 303$

(b)
$$°C = \frac{5}{9}(°F - 32°)$$

$$30° = \frac{5}{9}(°F - 32°)$$

$$\left(\frac{9}{5}\right)(30°) = °F - 32°$$

$$54° + 32° = °F$$

$$86° = °F$$

We sense temperature as a measure of the hotness or coldness of an object. Indeed, temperature determines the direction of heat flow; heat always flows spontaneously from a substance at high temperature to one at low temperature. Thus we feel the influx of energy when we touch a hot stove, and we know that the stove is at a higher temperature than our hand. Temperature is referred to as an *intensive property,* meaning that its value does not depend on the amount of material chosen. Thus, two samples of a liquid may have the same temperature though one has a volume of one cup and the other sample fills a bathtub. Density is another example of an intensive property; the density of mercury, as calculated in Sample Exercise 1.1, is 13.6 g/cm³ whether one has 100 g or 1000 g.

By contrast, both volume and mass are *extensive* properties, because they depend on the amount of material. Similarly, the *heat content* of a sample is an extensive property. In some types of solar heating systems for homes, air is heated by passage through solar panels exposed to the sun. The heated air is then passed through a large bed of stones. Because the heat content of the stones is large, the heat stored in them during the day is sufficient to heat the house during the night. Note the important distinction between heat content, an extensive property, and temperature, an intensive property.

SAMPLE EXERCISE 1.3

Consider the following description, labeling each property or characteristic as intensive or extensive: "The yellow sample is solid at 25°C. It weighs 6.0 g and has a density of 2.3 g/cm³."

Solution: Mass is an extensive property; color, physical state (that is, solid), temperature, and density are intensive properties.

1.4 Uncertainty in measurement

All measurements have some degree of uncertainty; how great the uncertainty is depends on both the accuracy of the measuring device and the skill of its operator. For example, the magnitude of the uncertainty associated with weighing an object will depend on the type of balance employed. On a triple-beam platform balance (Figure 1.11), the mass of a sample substance can be measured to the nearest 0.1 g; mass differences less than this cannot be detected on this balance. We might therefore indicate the mass of a dime measured on this balance as 2.2 ± 0.1 g; the ± 0.1 (read plus or minus 0.1) is a measure of the accuracy of the measurement. It is important to have some indication of how accurately any measurement is made; the $\pm$ notation is one way to accomplish this. We could measure the mass of the dime more accurately on an analytical balance, to the nearest 0.0001 g if we are careful. We might therefore report the mass as 2.2405 ± 0.0001 g. It is common to drop the $\pm$ notation with the understanding that *there is uncertainty of at least one unit in the last digit of the measured quantity;* that is, measured quantities are often reported in such a way that only the last digit is uncertain. All of the digits, including the uncertain one, are called significant digits or, more commonly, significant figures. The number 2.2 has two significant figures, while the number 2.2405 has five significant figures.

SAMPLE EXERCISE 1.4

What is the difference between 4.0 g and 4.00 g?

Solution: Many people would say there is no difference, but a scientist would note the difference in the number of significant figures between the two measurements. The number 4.0 has two significant figures while 4.00 has three. This implies that the second measurement has been made more accurately. A mass of 4.0 g indicates that the mass of the sample must be between 3.95 g and 4.05 g, closer to 4.0 g than to 3.9 g or 4.1 g. A mass of 4.00 g means that the sample must have a mass between 3.995 and 4.005 g, closer to 4.00 g than to 3.99 g or 4.01 g.

The following rules apply to determining the number of significant figures in a measured quantity:

1. All nonzero digits are significant—457 cm (three significant figures); 0.25 g (two significant figures).
2. Zeros between nonzero digits are significant—1005 kg (four significant figures); 1.03 cm (three significant figures).
3. Zeros to the left of the first nonzero digit in a number are not significant; they merely indicate the position of the decimal point—0.02 g (one significant figure); 0.0026 cm (two significant figures).
4. When a number ends in zeros that are to the right of the decimal point, they are significant—0.0200 g (three significant figures); 3.0 cm (two significant figures).
5. When a number ends in zeros that are not to the right of a decimal point, the zeros are not necessarily significant—130 cm (two or three significant figures); 10,300 g (three, four, or five significant figures). The way to remove this ambiguity is described below.

Use of standard exponential notation avoids the potential ambiguity of whether the zeros at the end of a number are significant (rule 5). For example, a mass of 10,300 g can be written in exponential notation showing three, four, or five significant figures:

1.03×10^4 g	(three significant figures)
1.030×10^4 g	(four significant figures)
1.0300×10^4 g	(five significant figures)

In these numbers all the zeros to the right of the decimal point are significant (rules 2 and 4).

The rules we have stated apply to nonintegral measured quantities; for these cases the number of significant figures is indicative of the associated uncertainty of the measurement. It is important to distinguish these nonintegral numbers from exact, integral ones, those that are defined or that result from counting. For example, there are exactly 3 ft in a yard, exactly four people in my immediate family, exactly 1000 g in a kilogram, and exactly 12 eggs in a dozen eggs. These are examples of exact numbers, numbers with no associated uncertainty. They can be considered to have an infinite number of significant figures. The number 1 in the metric-English system equivalents given in Table 1.3 is an exact number.

In carrying measured quantities through calculations the rule used is that the accuracy of the result is limited by the least accurate measurement. *In multiplication and division the result must be reported as having no more significant figures than the measurement with the fewest significant figures.* When the result contains more than the correct number of significant figures it must be rounded off.

For example, the area of a rectangle whose edge lengths are 6.221 cm and 5.2 cm should be reported as 32 cm^2:

$$\text{Area} = (6.221 \text{ cm})(5.2 \text{ cm}) = 32.3492 \text{ cm}^2 \longrightarrow \text{round off to } 32 \text{ cm}^3$$

We round off to two significant figures because 5.2 cm has only two significant figures.

In rounding off numbers, the following rules are followed (each example is rounded to two digits):

1 If the leftmost digit to be removed is more than 5, the preceding number is increased by 1; 2.376 rounds to 2.4.
2 If the leftmost digit to be removed is less than 5, the preceding number is left unchanged; 7.248 rounds to 7.2.
3 If the leftmost digit to be removed is 5, the preceding number is not changed if it is even and is increased by 1 if it is odd; 2.25 rounds to 2.2; 4.35 rounds to 4.4.

The rule used to determine the number of significant figures in multiplication and division cannot be used for *addition and subtraction*. For these operations, *the result should be reported to the same number of decimal places as that of the term with the least number of decimal places.* The following is an example:

This number limits	20.4	⟵ one decimal place
the number of significant	1.322	⟵ three decimal places
figures in the result ⟶	83	⟵ zero decimal places
	104.722	⟶ round off to 105

SAMPLE EXERCISE 1.5

A gas at 25°C exactly fills a container previously determined to have a volume of 1.05×10^3 cm^3. The container plus gas are weighed and found to have a mass of 837.6 g. The container, when emptied of all gas, has a mass of 836.2 g. What is the density of the gas at 25°C?

Solution: The mass of the gas is just the difference in the two masses: $(837.6 - 836.2)$ g $= 1.4$ g. Notice that this figure has only two significant figures, even though the individual masses each have four. Note also that there is one significant figure to the right of the decimal place. This means that there is an uncertainty of 0.1 g in the measured 1.4 g mass of the gas. Since there are 14 tenths of a gram in 1.4 g, we can say that the mass of gas is known to within approximately 1 part in 14. On the other hand, the volume is known to about one part in a hundred, since the uncertainty in this quantity is in the second decimal place in the number 1.05. That is, the volume is in the range from 1.04×10^3 cm^3 to 1.06×10^3 cm^3. We have that

$$\text{Density} = \frac{\text{mass}}{\text{volume}}$$

The density is thus given by 1.4 g/1.05×10^3 cm^3. It can be expressed as 0.0013 g/cm^3 or 1.3 mg/cm^3. There are two significant figures in this quantity, corresponding to the smaller number of significant figures in the two numbers that form the ratio.

This example illustrates an important point about significant figures; they are always related to the uncertainties in our ability to measure or estimate something. Whenever you do an experiment or evaluate someone else's experimental results, the uncertainty in the measured values is as important as the number itself.

In conducting laboratory experiments, it is important to choose the apparatus you employ so as to achieve the accuracy your experiment requires. For example, suppose you need to measure out 10 mL of a liquid to an accuracy of about one part in fifty, for use as a reagent in a reaction. One-fiftieth of 10 is $10/50 = 0.2$. This means you must measure out 10.0 ± 0.2 mL of the liquid. It would be a mistake to employ a 100 mL graduated cylinder as illustrated in Figure 1.9. The markings on this cylinder correspond to 1 mL intervals; it would be impossible to estimate between them to a precision of 0.2 mL. On the other hand, a 10 mL graduated cylinder with markings every 0.1 or 0.2 mL would be sufficiently accurate.

It is also important to have a feeling for significant figures when using a calculator, because they ordinarily display more digits than are significant. As an example, suppose that a student accurately weighs a nickel, and finds it to have a mass of 4.9566 g. He looks up the density of the alloy from which the coin is made and finds it listed as 8.8 g/cm^3. The volume of the nickel he weighed is then:

$$\text{Volume} = \frac{\text{mass}}{\text{density}} = \frac{4.9566 \text{ g}}{8.8 \text{ g/cm}^3}$$

Carrying out this division on a calculator, the student obtains 0.56325. This must be rounded off to two significant figures: 0.56 cm^3. In future calculations, answers will be given to the proper number of significant figures; the round-off process will not be shown.

SAMPLE EXERCISE 1.6

How many significant figures are there in each of the following numbers? (a) 4.003; (b) 6.023×10^{23}; (c) 5000; (d) the sum $15.3 + 0.2334$; and (e) the product (16)(5.7793).

Solution: (a) Four; the zeros are significant figures. (b) Four; the exponential term does not add to the number of significant figures. (c) One, two, three, or four. In this case, the ambiguity could have been avoided by using standard exponential notation. Thus, 5×10^3 has only one significant figure; 5.00×10^3 has three. (d) Three; the sum, expressed to the proper three significant figures, is 15.5. (e) Two; the product, expressed to the proper two significant figures, is 92.

1.5 Dimensional analysis—an approach to problem solving

Before we go on, perhaps a word of caution is in order. Sometimes students have little difficulty reading their chemistry text or following the lecture and yet have difficulty on exams. In some instances the problem is lack of familiarity with terms. Often the problem is that the students have a passive but not an active understanding of the material. They can see how someone else has worked a problem, but they are unable to work any on their own. An active understanding involves being able to use the material in new situations, including especially working problems that are not identical to those used as sample exercises in the text. It is important to use the problems at the end of each chapter to test yourself to determine how well you are able to use the material in the chapter. Colored numbers indicate problems whose answers can be found at the back of the book. Bracketed numbers indicate problems of above-average difficulty. Wherever possible we have used an approach that can be referred to as **dimensional analysis** in solving problems. If you develop a facility with this approach, which is illustrated in the following discussion, your work will be much easier. If you need a review of basic mathematics, refer to Appendix A.

A number reported for a measured quantity is meaningless unless its units are specified. If units are treated as algebraic quantities, they can be carried through all calculations and will indicate whether the calculation has been performed correctly. This approach is illustrated in the following examples and in the sample exercises that follow.

Consider the conversion of mass from pounds to kilograms. If a man weighs 175 lb, what is his mass in kilograms? From Table 1.3 we have the following relationship: 1 kg = 2.205 lb. We can therefore write the following equalities:

$$1 = \frac{1 \text{ kg}}{2.205 \text{ lb}}; \quad 1 = \frac{2.205 \text{ lb}}{1 \text{ kg}}$$

These equalities, which can be read as 1 kg per 2.205 lb and 2.205 lb per kilogram, are referred to as **unit conversion factors.** Multiplication of a quantity by these factors changes the units in which the quantity is expressed but not its value. To convert pounds to kilograms we choose the unit conversion factor that cancels pounds:

$$? \text{ kg} = (175 \text{ lb})\left(\frac{1 \text{ kg}}{2.205 \text{ lb}}\right) = 79.4 \text{ kg}$$

Now consider a more complex conversion of units, the calculation of the number of inches in 3.00 km. We can begin by writing the equality we are working toward:

? in. = 3.00 km

From the relations shown in Tables 1.2 and 1.3 and from our basic knowledge of the English system we can write the following equalities:

1 km = 1000 m; 1 m = 1.094 yd; 1 yd = 36 in.

Therefore,

$$\frac{1000 \text{ m}}{1 \text{ km}} = 1; \quad \frac{1.094 \text{ yd}}{1 \text{ m}} = 1; \quad \frac{36 \text{ in.}}{1 \text{ yd}} = 1$$

If we multiply 3.00 km by these factors we have

$$? \text{ in.} = (3.00 \text{ km})\left(\frac{1000 \text{ m}}{1 \text{ km}}\right)\left(\frac{1.094 \text{ yd}}{1 \text{ m}}\right)\left(\frac{36 \text{ in.}}{1 \text{ yd}}\right)$$

$$= 1.18 \times 10^5 \text{ in.}$$

Each conversion factor is applied so as to cancel the units of the preceding factor. This converts kilometers successively to meters to yards to inches. Because we are left with the proper units, we know that the problem has been correctly set up. (There are exactly 1000 m in a kilometer and exactly 36 in. in a yard. The number of significant figures in the result, in this case three, is thus determined by the number of significant figures in the quantity being converted.)

Because 1 m = 1.094 yd, it is also true that 1 m/1.094 yd = 1. If this factor is applied instead of its reciprocal as above, the result is

$$? \text{ in.} = (3.00 \text{ km})\left(\frac{1000 \text{ m}}{1 \text{ km}}\right)\left(\frac{1 \text{ m}}{1.094 \text{ yd}}\right)\left(\frac{36 \text{ in.}}{1 \text{ yd}}\right)$$

$$= 9.88 \times 10^4 \frac{\text{m}^2 \text{ in.}}{\text{yd}^2}$$

Clearly the units do not cancel to give the desired units, so we know that we have not obtained the conversion we want.

SAMPLE EXERCISE 1.7

You have to pour 2.0 cubic yards (yd³) of concrete for a patio. What is this volume in cubic meters (m³)?

Solution:

$$? \text{ m}^3 = (2.0 \text{ yd}^3)\left(\frac{1 \text{ m}}{1.094 \text{ yd}}\right)\left(\frac{1 \text{ m}}{1.094 \text{ yd}}\right)\left(\frac{1 \text{ m}}{1.094 \text{ yd}}\right)$$

$$= (2.0 \text{ yd}^3)\left(\frac{1 \text{ m}}{1.094 \text{ yd}}\right)^3$$

$$= 1.5 \text{ m}^3$$

SAMPLE EXERCISE 1.8

You are approaching a city and see a sign indicating a speed limit of 40 km/hr. What is the corresponding speed in miles per hour?

Solution:

$$? \frac{mi}{hr} = \left(40 \frac{km}{hr}\right)\left(\frac{1000\ m}{1\ km}\right)\left(\frac{1.094\ yd}{1\ m}\right)\left(\frac{1\ mi}{1760\ yd}\right)$$

$$= 25 \frac{mi}{hr}$$

SAMPLE EXERCISE 1.9

The acid in an automobile battery (a solution of sulfuric acid) has a density of 1.2 g/cm³. What is the mass (in grams) of 200 mL (2.00 × 10² mL) of this acid?

Solution:

$$? \ g = (2.00 \times 10^2\ mL)\left(\frac{1\ cm^3}{1\ mL}\right)\left(1.2 \frac{g}{cm^3}\right)$$

$$= 2.4 \times 10^2\ g$$

Notice that density can be thought of as a unit conversion factor for converting volume to mass or vice versa.

SAMPLE EXERCISE 1.10

In Sample Exercise 1.1, the density of mercury was found to be 13.6 g/cm³. Convert this to the basic SI units of kg/m³.

Solution: Because the units in both numerator and denominator are to be changed, we need to employ two different unit conversion factors. These are:

$$1 = \frac{1\ kg}{1000\ g}; \ 1 = \frac{100\ cm}{1\ m}$$

$$\text{Density} = \left(\frac{13.6\ g}{1\ cm^3}\right)\left(\frac{100\ cm}{1\ m}\right)^3\left(\frac{1\ kg}{1000\ g}\right)$$

$$= 1.36 \times 10^4\ kg/m^3$$

Note that the first unit conversion factor is cubed to provide the correct dimension of volume.

Dimensional analysis cannot be used on all problems that we will work, and you should feel free to abandon it and work problems stepwise if you wish. However, you should *always* carry units throughout all of your calculations, making sure that they cancel properly. Whenever you finish a calculation, look at both the units and magnitude of your answer and ask yourself whether your answer makes any sense. This will help you to avoid making some embarrassingly simple errors.

FOR REVIEW

Summary

We have defined chemistry as the study of the properties, composition, and changes of matter. Chemistry has two origins: (1) the craft traditions such as metallurgy and (2) the more philosophical search for basic understanding of matter. Modern chemistry rests upon certain scientific laws arrived at through both qualitative observations and quantitative measurements. Hypotheses are devised to provide a tentative explanation for the laws. If the hypotheses are successful they become theories.

Measurements are made using the metric system,

which is based on the decimal system. Uncertainties associated with measurements can be expressed by use of significant figures. We have seen that when measured quantities are carried through calculations, it is important to keep track of units.

Learning goals

Having read and studied this chapter, you should be able to:

1 Use the metric system and list the basic metric units and the common prefixes.
2 Interconvert metric and English-system measurements using dimensional analysis.
3 Convert temperatures between the Fahrenheit, Celsius, and Kelvin scales.
4 Perform calculations involving density.
5 Determine the number of significant figures in a derived quantity.

Key terms

Among the more important terms and expressions used for the first time in this chapter are the following:

Combustion (Section 1.1) is a process that pro-

ceeds so rapidly that flames or light are produced.
Density (Section 1.3) is mass per unit volume.
A hypothesis (Section 1.2) is a trial idea or explanation; a tentative theory.
An intensive property (Section 1.3) is one that is independent of the amount of material under consideration; an extensive property (Section 1.3) depends on the amount of material being considered.
A scientific law (Section 1.2) is a concise verbal or mathematical statement that summarizes one or more observed relationships between physical quantities.
Mass (Section 1.3) is a measure of the amount of "stuff" in an object. It measures the resistance of a stationary object to be moved. In SI units, mass is measured in kilograms.
Matter (introduction) is the physical material (the "stuff") of the universe; it is anything that occupies space and has mass.
Paradigms (Section 1.2) are sets of basic ideas and beliefs that are rather universally believed.
Significant figures (Section 1.4) are the digits that indicate the precision with which a measurement has been made—those digits of a measured number that have uncertainty only in the last digit.
A theory (Section 1.2) is an explanation of a set of related observations.

EXERCISES

Introductory concepts

1.1 List some of the ways in which chemistry differs from the practice of law, both as an intellectual discipline and in practice.

1.2 It has been said that in experimental science, the most progress is made when there is more than one hypothesis that could reasonably explain a set of observations. Why is this so?

1.3* When two charges Q_1 and Q_2 are separated by a distance d, the force between them is given by $F = Q_1Q_2/d^2$. This equation is due to the French physicist Charles Augustin de Coulomb (1736–1806). Should this be called Coulomb's law or Coulomb's theory? Explain.

1.4 In the early history of science, it was almost universally believed that the substances that make up living matter are inherently different from those that make up inanimate matter, in that the former were formed through the agency of some vital principle, present only in living matter. In 1828, Friedrich Wöhler synthesized the compound urea (a constituent of urine) from ammonium cyanate, a purely inorganic substance. Discuss the importance of Wöhler's experiment in the light of Thomas Kuhn's view of the development of science.

1.5 Distinguish between the following: (a) a theory and a hypothesis; (b) a paradigm and a law; (c) experiment and theory; (d) quantitative measurement and qualitative observation.

*Color indicates answer in "Answers to Selected Exercises."

Metric system; SI units

1.6 What are the basic SI units appropriate to express the following: (a) the length of a racetrack; (b) the volume of a swimming pool; (c) the mass of a bar of gold; (d) the area of the state of Rhode Island; (e) the length in time of a lunar cycle?

1.7 What word prefixes indicate the following multipliers: (a) 1×10^3; (b) 0.1; (c) 1×10^6; (d) 1×10^{-6}; (e) 1×10^{-12}?

1.8 If 1.00 g of gold occupies a volume of 0.0518 cm^3 at 20°C, what is the volume in cubic centimeters of: (a) 1.00 kg gold; (b) 1.00 dg gold; (c) 1.00 mg gold; (d) 10.0 g gold; (e) 1.00×10^2 ng gold; (f) 0.100 mg gold?

1.9 (a) 6.50 m is _____ cm
(b) 33 kg is _____ mg
(c) 12 sec is _____ msec
(d) 43 mg is _____ g
(e) 2.35 kg is _____ g
(f) 22.5 mm is _____ μm

1.10 (a) 5×10^2 km is _____ m
(b) 326 g is _____ kg
(c) 476 nm is _____ cm
(d) 82 μsec is _____ msec

1.11 Indicate whether the following units measure length, area, volume, or mass: (a) cm^3; (b) km^2; (c) mg; (d) dm.

1.12 The parsec, a unit of distance used in astronomy, is 3.08×10^{13} km. How many meters is this? How many meters are there in a microparsec? How many centimeters in a nanoparsec?

1.13 The equivalent of 1 kg is _____ lb; the equivalent of 1 m is _____ in.; the equivalent of 1 oz is _____ g; the equivalent of 1 pint is _____ L; the equivalent of 55 cm is _____ ft.

Significant figures

1.14 North Dakota with a population of 649,000 (1977 estimate) has an area of 70,664 mi². What is the average area in square meters per state resident? Explain the reason for the number of significant figures your answer contains.

1.15 Indicate the number of significant figures in each of the following: (a) 43.55; (b) 5.67×10^2; (c) 0.00346; (d) 3×10^6; (e) 2.7×10^{-3}; (f) 1200; (g) 1300.41.

1.16 How many significant figures are there in each of the following numbers: (a) 5.44×10^{23}; (b) 0.114; (c) 128; (d) 60,000; (e) 0.00582; (f) 0.34×10^6?

1.17 Round off each of the following numbers to four significant figures: (a) 4,567,985; (b) 2.3581×10^3; (c) 42,560; (d) 0.00238866; (e) 9.87553.

1.18 Express the following numbers in appropriate exponential notation: (a) 1245; (b) 65,000 to show three significant figures; (c) 59,750 to show four significant figures; (d) 0.00456.

1.19 Carry out the following operations and round off the answers to the appropriate number of significant figures: (a) 3.22×0.17; (b) $4568/1.3$; (c) $1.987/(3.46 \times 10^8)$; (d) $0.0003/162$; (e) $(12.3 + 0.092)/8.3$; (f) $328 \times (0.125 + 5.43)$.

1.20 Carry out the following mathematical operations and round off to the appropriate number of significant figures: (a) $(16.788 - 15.990)/118.9$; (b) $3.4 \times 2.668/1012$; (c) $15.67 + 0.8896 + 2.0 + 1.2 \times 10^{-4}$; (d) $(4.43 \times 1.254) + 0.18$.

Conversions of units; dimensional analysis

1.21 (a) A man of height 5 ft 10 in. is _____ m tall. (b) A car weighing 4.85×10^3 lb has a weight of _____ kg. (c) A kilometer is _____ ft in length. (d) A gas tank that holds 11.5 U.S. gal will hold _____ L of gas.

1.22 (a) A fathom, used as a measure of water depths, is defined as 1.8288 m, exactly. What is the depth in fathoms of Lake Superior, which is 1302 ft at maximum depth? (b) A furlong, a measure of distance used in horse racing, is defined as 201.168 m. What is the distance in miles of the Kentucky Derby, which is a 10.000-furlong race?

1.23 (a) A wine barrel containing 31 gal contains _____ L of wine. (b) One grain is defined as 64.799 mg. One gram thus contains _____ grains. (c) A pennyweight is 1.555 g. There are _____ pennyweight in 2.52 kg. (d) The so-called "long ton" is exactly 2240 lb. This corresponds to _____ kg.

1.24 (a) A standard cranberry barrel contains 5286 in.³ Its volume is _____ dm³. (b) An engine piston displace-

ment of 320 in.³ corresponds to _____ L. (c) 100 hectares = 1 km² = _____ m² = _____ mi².

1.25 A graduated cylinder on a balance has a mass of 57.832 g. An organic liquid, toluene, with a density of 0.866 g/cm³ is added until the combined mass reads 87.127 g. What is the volume of the liquid in the graduated cylinder?

1.26 Convert the measures in the following statements to an appropriate metric unit. (a) The longest road tunnel is St. Gotthard, Switzerland-Italy: 10 mi, 120 yd. (b) The biggest dam is at New Cornelia Tailings, United States: 274,026,000 yd³ of material used. (c) The deepest mine is the Western Deep Levels Gold Mine, South Africa: 11,647 ft. (d) The great pyramid of Cheops is 481 ft high and 755 ft on each side at the base and is built of 2,300,000 blocks weighing 2.5 tons each. (e) The 1928 Indianapolis 500-mi race was won by Lou Meyer, with an average speed of 99.482 mph.

1.27 Consider the following conversion table found in a handbook titled *Surveyor's Chain Measure:*

7.92 in. = 1 link

100 links = 1 chain

1 chain = 4 rods

Using this table and other conversion factors, answer the following questions. (a) How many chains in a mile? (b) How many meters per rod? (c) How many square chains in 1 mi². (d) How many cubic rods in 1 yd³?

1.28 (a) An auto traveling at 62 mph has a speed of _____ km/hr. (b) The same auto has a speed of _____ m/sec. (c) Fish sell for $2.45/lb. This is equivalent to $_____/kg. In West Germany, where the Mark exchanges for $0.46, the fish would sell for _____ Mark/kg. (d) A 12-fluid-oz bottle of soft drink contains _____ L. (Recall that there are 32 fluid oz in a quart.)

1.29 Convert the quantities in the following account to metric units (indicate how many significant figures each quantity possesses). The moon is at a distance of 238,850 miles from earth and revolves around it with a period of 27.32 days. The moon's radius is 1081 miles, and its mass is estimated to be 1.62×10^{23} lb.

1.30 Convert the quantities in the following account to the basic SI units of kg-m-sec (indicate how many significant figures each quantity possesses). Jupiter, the largest of the planets in our solar system, is approximately 4.84×10^8 miles from the sun and revolves around it with a period of 11.86 years. The average density of the planet is 1.330 g/cm³.

1.31 A metal block has dimensions of 4.5 cm × 12.54 cm × 1.25 cm. Calculate its volume in cubic centimeters; in cubic meters.

1.32 Acceleration due to gravity on earth, g, has a magnitude of 980.66 cm/sec². What is the magnitude of g in basic SI units?

1.33 A cylindrical container of radius r and height h has a volume of $\pi r^2 h$. (a) Calculate the volume in cubic centimeters of a cylinder with radius 6.5 cm and a height 28.6 cm. (b) Calculate the volume in cubic meters of a cylinder that is 8.0 ft high and 20.0 in. in diameter. Calculate the mass in kilograms of water required to fill the cylinder of part (b) if the dimensions listed are the inside dimensions. The density of water is 1.00 g/cm³.

1.34 What is the temperature in °C of an animal with a body temperature of 100.6°F?

1.35 Convert each of the following temperatures to the scale indicated: (a) 67°F to °C; (b) −8°F to °C; (c) 355°C to K; (d) 150 K to °C; (e) 120°F to K; (f) −50°F to K.

1.36 What is the mass of 1 dm³ of mercury, density 13.6 g/cm³?

1.37 An organic liquid that occupies a volume of 3.47 L has a weight of 4.268 kg. What is the density of this liquid in kilograms/liter?

1.38 A 600-mL bottle of methanol is sold in a discount store as a gasoline additive to prevent freezing, for $0.39. The density of methanol is 0.79 g/cm³ at room temperature. What is the price of the methanol per kilogram?

1.39 A graduated cylinder weighs 204.58 g. Some liquid is added, and the volume is read as 58.3 mL. The mass of cylinder and liquid is 251.65 g. Calculate the density of the liquid.

1.40 The density of air at ordinary atmospheric pressure and 25°C is 1.19 g/L. What is the weight in kilograms of the air in a room that measures 8.2 × 13.5 × 2.75 m?

[1.41]* Suppose you had a glass marble, a graduated cylinder into which the marble fit comfortably, some water, and a balance of the type shown in Figure 1.11(b) that is calibrated in units of 0.02 g. Describe how you would determine the density of the marble.

Additional exercises

1.42 List each of the following as an intensive or extensive property: weight; color; density; volume; temperature; melting point; ease of corrosion.

1.43 Is the use of significant figures in each of the following statements appropriate? Why or why not? (a) The

1976 circulation of *Reader's Digest* magazine was 17,887,299. (b) In the United States, 1.4 million persons have the surname Brown. (c) The average annual rainfall in San Diego, California, is 20.54 in. (d) The population of East Lansing, Michigan, in 1979 was 51,237.

1.44 On August 24, 1960, the temperature at the Vostok Station in Antarctica was recorded as −127°F. What is this in °C? In K?

1.45 Convert the quantities in the following statements to metric units. (a) Women of medium frame, those 5 ft, 3 in. in height, should weigh in the range 110–122 lb. (b) A man 5 ft, 9 in. in height, body weight about 154 lb, should consume about 2.5 oz of protein each day (1 oz = 28.35 g).

1.46 The maximum allowable concentration of carbon monoxide in urban air is 10 mg/m³ over an 8-hour period. At this level, what mass of carbon monoxide is present in a room that is 2.5 × 15 × 40 m in dimensions?

1.47 What is the ratio of the size of 1°C to 1°F?

1.48 Convert the following quantities to the basic SI units: (a) 4.5 g/cm³; (b) 248 cm²; (c) 289 g cm/sec²; (d) 2230 dm³/hr.

1.49 Describe the most important characteristics of scientific activity. In what respects does a theory about the properties of a set of chemicals differ from a theory of the origins of the characters in James Joyce's *Ulysses*?

1.50 A few years ago, a cartoon pictured a thief making his getaway, gun in one hand and a bucket of gold dust in the other. If the bucket had a volume of 8 qt and was full of gold whose density is 19.3 g/cm³, what was its mass? Comment on the thief's strength.

1.51 If an automobile is able to travel 22.4 miles on a gallon of gasoline, what is the gas mileage in kilometers/liter?

1.52 In 1978 the world's record for the 100-m dash was 9.9 sec. To cover 100 m in 9.9 sec, what must the average speed be in miles per hour?

1.53 What is the radius, in centimeters, of a metal sphere whose mass is 2.00×10^2 g if the metal is: (a) iron, density = 7.86 g/cm³; (b) aluminum, density = 2.70 g/cm³; (c) lead, density = 11.3 g/cm³. The volume of a sphere is given by the equation $V = 4\pi r^3/3$.

---

*Brackets indicate difficult questions.

2

Our chemical world: atoms, molecules, and ions

Our present understanding of the changes we see around us—such as the melting of ice and the burning of wood—is intimately tied to our understanding of the nature and composition of matter. For example, before we can hope to understand what is happening when ice melts we must know what ice is—what it is composed of. It is possible to resolve or separate matter into a great variety of different pure substances. These are materials or portions of matter whose composition and intrinsic properties are uniform throughout. For example, seawater can be separated into several different pure substances, the most abundant being water and ordinary table salt (sodium chloride).

In this chapter we will examine the composition of matter. We will attempt to answer many fundamental questions: What types of pure substances are there? How can matter be separated into pure substances? Can pure substances be broken into simpler components? How do substances differ at the microscopic or atomic level? How do we represent the compositions of substances and how do we name them?

There is much more in the answers to these questions than can be contained in a single chapter. You should regard this chapter as, in part, an introduction. At this point you need to master some basic ideas and concepts and acquire a chemical vocabulary to facilitate laboratory work and further study of the text.

| 2.1 | States of matter |

As a starting point in answering the questions we have posed, it is useful to note that matter exists in three states: gas (also known as vapor), liquid, or solid. A gas has neither a shape of its own nor a fixed volume. It takes the shape and volume of any container into which it is introduced. It can be compressed to fit a small container; it will expand to occupy a large one. Air is a gas.

A liquid has no specific shape; it assumes the shape of the portion of any container that it occupies. It does not expand to fill the entire container; it has a specific volume. Furthermore, a liquid is only slightly compressible. Water and gasoline are common liquids.

A solid has a firmness that is not associated with either gases or liquids. It has a fixed volume and shape. Like liquids, solids are not compressible to any appreciable degree. Numerous objects around us are solids—nails, coins, salt, and sugar to name a few.

The state of a substance depends on temperature and pressure. Above 100°C, water exists as a gas, known as steam. Between 0°C and 100°C, it exists as a liquid. Below 0°C, it exists as a solid—ice.

Changes of state, such as the change of ice to liquid water, are examples of physical changes. Physical changes are ones that do not involve creation of new substances; they involve no change in the composition of the specimen of matter under consideration. Chemical changes, also called chemical reactions, involve conversion of one substance into another. Every pure substance has a unique set of properties or characteristics that allows us to recognize it and distinguish it from other substances. Chemical properties are those properties that refer to the way a substance is able to change into other substances (its reactivity, how it "reacts"). The physical properties of a substance are those that do not involve a change in the chemical identity of the substance.

SAMPLE EXERCISE 2.1

Chlorine is a greenish-yellow gas with a density of 3.21 g/L. It can be changed to a liquid by cooling to −34.6°C; it reacts explosively with sodium to form sodium chloride (table salt). Which of these properties are physical properties and which are chemical?

Solution: The color and density of chlorine and the temperature at which it changes state, from a gas to a liquid, are all physical properties. They do not involve a change of chlorine into any other substance. The ability of chlorine to react explosively with sodium is a chemical property. In reacting with sodium, chlorine is changed into a different substance, sodium chloride.

| 2.2 | The classification of matter |

All specimens of matter can be classified either as pure substances or as mixtures of two or more substances. Most matter around us consists of mixtures. Mixtures are characterized by variable composition and by the fact that they can be separated by physical means. That is, mixtures can be separated by taking advantage of differences in physical properties such as boiling points. For example, we can recognize that blood is a mixture because its composition may vary in many ways, such as in its iron content. Furthermore, blood can be separated into two components, packed cells and plasma, by centrifugation, a physical method of separation. Some common methods of separating mixtures are discussed in Section 2.3.

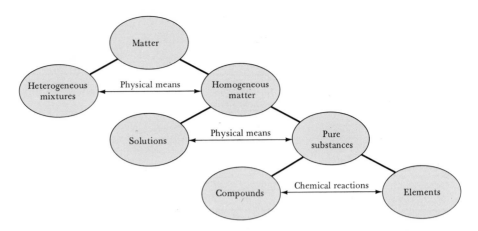

FIGURE 2.1 The classification of matter.

Any portion of matter that is uniform throughout is said to be *homogeneous*. Homogeneous mixtures are known as solutions. For example, when salt and water are mixed, a homogeneous mixture, or solution, forms. The salt is said to dissolve in the water. Mixtures that are not homogeneous are said to be *heterogeneous*. When water and clay are mixed, no solution forms; the resultant mixture is heterogeneous. A heterogeneous mixture can generally be separated by some physical means into homogeneous components.

We have seen that solutions are homogeneous mixtures. They are of uniform properties throughout, but these properties will vary from one solution to another, depending on the relative amounts of the components from which the solution is formed. By contrast, a pure substance is a homogeneous material with a constant, invariable composition and a distinct set of intrinsic properties. Water is an example of a pure substance. It is a particular kind of pure substance, because it is composed of more than one element; in this case two, hydrogen and oxygen. Such a pure substance, called a compound, can be decomposed by chemical reaction into elements. Thus water can be decomposed into hydrogen and oxygen. These two substances are termed elements to signify that they cannot be decomposed by chemical means into still simpler substances. (We will consider a more sophisticated criterion for defining an element later in this chapter.) Elements are the simplest substances. There are 106 known elements, but most of the millions of known compounds are formed from only about two dozen elements.

The classification of matter into mixtures, substances, compounds, and elements is summarized in Figure 2.1. Before we examine compounds and elements more closely, let's consider some of the ways a mixture can be separated into its component parts.

2.3 Separation of mixtures

Chemists often need to separate mixtures into their component substances. For example, separation procedures are used both to determine the composition of mixtures and to purify substances. A large number of separation procedures have been developed. Three of the most common of these are filtration, distillation, and chromatography.

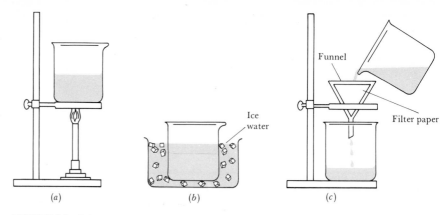

FIGURE 2.2 Schematic representation of laboratory recrystallization and filtration. (*a*) The solids are dissolved in a minimum amount of hot water. (*b*) The mixture is cooled. (*c*) The mixture is filtered after solids crystallize from the solution.

FILTRATION

Solids are readily separated from liquids by passing the mixture through a filter, a barrier with many small openings. This method is called filtration. This is a simple method for separating sediment from water in the course of the treatment of water for drinking purposes, because the particles of sand and clay making up the sediment do not pass through the filter.

Filtration is often used in conjunction with procedures that take advantage of different solubilities of substances (that is, differences in the abilities of substances to mix to form solutions). For example, a mixture

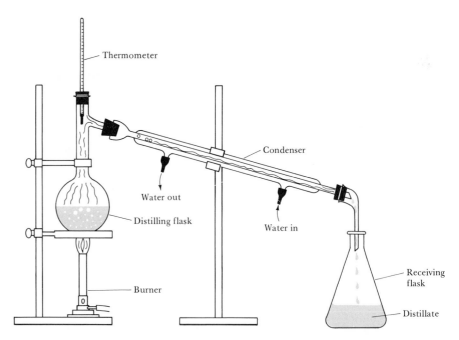

FIGURE 2.3 A simple laboratory distillation setup. Cool water circulating through the jacket of the condenser causes the liquid to condense.

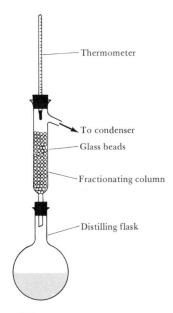

Thermometer

To condenser

Glass beads

Fractionating column

Distilling flask

FIGURE 2.4 A distilling flask with a fractionating column attached.

of 10 g of table salt and 10 g of baking soda can be partially separated by dissolving it in 100 mL of water at about 70°C and then lowering the temperature of the solution to 0°C. The table salt exhibits little change in solubility as its temperature is lowered, and it remains in solution. The baking soda, however, is much less soluble at 0°C than at 70°C, and about 9 g of this substance will separate from solution at the lower temperature. The solution is filtered to remove the baking soda. The procedure is summarized in Figure 2.2.

DISTILLATION

In distillation, differences in the volatilities of substances (that is, differences in the ease with which substances form gases) are utilized. Imagine that we wish to remove salt from seawater so that the water can be used for drinking purposes. The seawater could be heated in an apparatus like that shown in Figure 2.3. Water vaporizes at a much lower temperature than salt, and so the water boils off, leaving a residue of salt in the distilling flask. The water is condensed by cooling elsewhere in the system and collected. The liquid obtained by condensation of vapor in a distillation is known as the distillate.

If a mixture of several volatile substances is distilled, the vapor will be richer in the more volatile component. For example, the distillate obtained from the distillation of wine has a higher alcohol content than the wine, because alcohol is more volatile than water. Small percentages of minor components that impart flavor and aroma also are found in the distillate, which is known as brandy. However, a single simple distillation does not effect complete separation of the components of wine. If several components of a mixture have similar volatilities, repeated distillations may be necessary for complete separation.

A fractionating column effects in a single operation what may require several simple distillations. This procedure is called **fractional distillation.** The column, shown in Figure 2.4, has a packing such as glass beads that provides cooling space where part of the vapor condenses as it moves upward from the distilling flask. The condensed liquid is richer in the least volatile component. As the condensed liquid trickles down the beads toward the distillation flask, it comes in contact with fresh vapor moving upward from the flask. Because the vapor is hotter than the liquid on the beads, heat interchange occurs. As a result, the more volatile part of the liquid vaporizes and the less volatile part of the vapor condenses. Thus the vapor becomes further enriched in the more volatile component. Because many such heat interchanges occur along the column, only the most volatile component or components of the mixture are finally able to reach the condenser and escape.

Fractional distillation is used to separate crude oil into fractions. Commercially this is done using equipment of very sophisticated design. The fractions referred to as gasoline, kerosene, and lubricating oil differ in volatilities. The more volatile component, gasoline, boils in the approximate range of 60–150°C, whereas kerosene and lubricating oil boil in the 150–250°C and 250–350°C ranges, respectively.

CHROMATOGRAPHY

In chromatography, separation is achieved by utilization of differences in the degree to which various substances are adsorbed onto the surface of an inert material. (An inert material is one that does not undergo a chemical change.) The difference between adsorption and absorption should be noted; as illustrated in Figure 2.5, adsorption is a surface

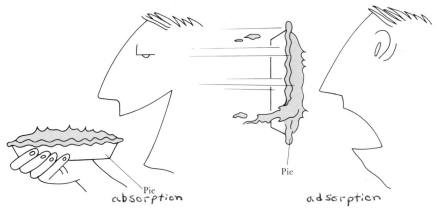

FIGURE 2.5 Adsorption is a surface phenomenon.

phenomenon. The name chromatography arose from its use in separating pigments; it means literally "graphing of colors." For example, a solution containing the colored pigments of a leaf may be washed through a column packed with alumina (aluminum oxide). The various components will move through the column at different speeds due to differences in the degree to which they are adsorbed. This type of separation, illustrated in Figure 2.6, is known as column chromatography. By using various instruments instead of visual inspection to determine the location of the components, it is not necessary to restrict the technique to colored materials.

When the adsorbent material is paper and the solution containing the mixture moves upward through the paper, the technique is called paper chromatography. As the liquid moves upward on the paper, it carries the mixture along. The components that are adsorbed most strongly to the paper move most slowly. The technique is illustrated in Figure 2.7.

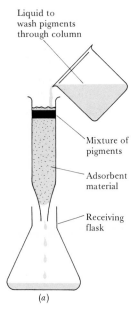

Liquid to
wash pigments
through column

Mixture of
pigments

Adsorbent
material

Receiving
flask

(a)

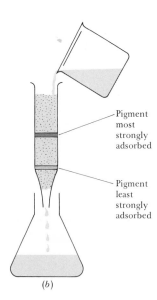

Pigment
most
strongly
adsorbed

Pigment
least
strongly
adsorbed

(b)

FIGURE 2.6 Separation by column chromatography: (a) initial stage; (b) after some time has elapsed.

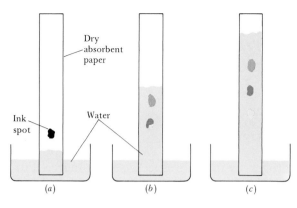

FIGURE 2.7 Separation of ink into components by paper chromatography. (*a*) Water begins to move up the paper. (*b*) Water moves past the ink spot, lifting different components of the ink at different rates. (*c*) Water has separated the ink into three different components.

2.4 Pure substances

Elements are the basic substances out of which all matter is composed. In light of the seemingly endless variety in our world, it is perhaps surprising that there are only 106 known elements. Not all of these are of equal importance or abundance. Ninety percent, by weight, of the portion of the earth to which we have access for raw materials is composed of only five elements: oxygen, silicon, aluminum, iron, and calcium. Over 90 percent of the human body is composed of just three elements: oxygen, carbon, and hydrogen. At the other extreme, about 20 elements are either found in nature in only minute traces or are prepared in the laboratory and are therefore available in only very small quantities.

Some of the more familiar elements are listed in Table 2.1, together with the chemical symbols used to denote them. All of the known elements are listed on the front inside cover of this text. It can be seen that the abbreviation or symbol for an element consists of one or two letters with the first letter capitalized. These symbols are usually derived from the English name (first and second columns of Table 2.1), but sometimes they are derived instead from a foreign name (third column). You will need to know these symbols and to learn others as we encounter them in the text.

COMPOUNDS

As was mentioned earlier, compounds are substances composed of two or more elements united chemically in definite proportions by mass. We can gain clearer insight into these substances by examining a common example, water. With the discovery of methods of generating electricity, it was found that water could be decomposed into the elements hydrogen and oxygen, as shown in Figure 2.8. Pure water consists of 89 percent

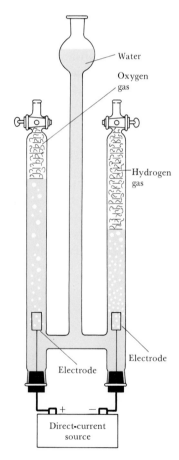

FIGURE 2.8 Decomposition of the compound water into the elements hydrogen and oxygen by passing a direct electrical current through it (electrolysis).

TABLE 2.1 Some common elements and their symbols

Carbon (C)	Aluminum (Al)	Copper (Cu, from *cuprum*)
Fluorine (F)	Barium (Ba)	Iron (Fe, from *ferrum*)
Hydrogen (H)	Calcium (Ca)	Lead (Pb, from *plumbum*)
Iodine (I)	Chlorine (Cl)	Mercury (Hg, from *hydrargyrum*)
Nitrogen (N)	Helium (He)	Potassium (K), from *kalium*
Oxygen (O)	Magnesium (Mg)	Silver (Ag, from *argentum*)
Phosphorus (P)	Platinum (Pt)	Sodium (Na, from *natrium*)
Sulfur (S)	Silicon (Si)	Tin (Sn, from *stannum*)

TABLE 2.2 Comparison of water, hydrogen, and oxygen

	Water	Hydrogen	Oxygen
Physical state[a]	Liquid	Gas	Gas
Normal boiling point	100°C	−253°C	−183°C
Density[a]	1.00 g/mL	0.090 g/L	1.43 g/L
Combustible?	No	Yes	No

[a]At room temperature and atmospheric pressure.

oxygen and 11 percent hydrogen by mass, irrespective of its source. Furthermore, the properties of water are clearly unique and much different from those of its constituent elements as seen in Table 2.2. In forming compounds the elements lose their characteristic properties.

The observation that the elemental composition of a pure compound is always the same is known both as the law of constant composition and the law of definite proportions. Although this law has been known for over 150 years, the general belief persists among some people that there is a fundamental difference between compounds prepared in the laboratory and the corresponding compounds found in nature. However, a pure compound has the same composition and properties regardless of source. Both chemists and nature must use the same elements and operate under the same natural laws. Differences in composition and properties between substances indicate that the compounds are not the same or that at least one is impure. Harmful chemicals such as strychnine are made by nature as well as by people. Nature is capable of making many compounds, especially those found in living systems, that chemists are not yet able to prepare. Chemists, on the other hand, have succeeded in forming many substances, for example, synthetic fibers and certain pesticides, not found in nature.

SAMPLE EXERCISE 2.2

Identify the following as element, compound, or mixture: milk, gold, table salt, ink.

Solution: Gold (Au) is an element (refer to the table of elements on the front inside cover of the text).

Table salt is a compound that we have now mentioned several times. It is composed of the elements sodium (Na) and chlorine (Cl). Both milk and ink are mixtures and are recognized as such by their variable compositions.

2.5 The atomic theory

The classification of matter as mixtures and substances, compounds and elements, is central to modern chemistry. It helps us to systematize many chemical facts. However, it also raises a number of questions. Why is one element different from another? Why is a compound different from a mixture? Why do elements combine to form compounds? These questions are connected with the facts that we have already discussed. We need a theory to help us explain these facts. We therefore shift our attention now to a discussion of our present theory of matter, the atomic theory. We will find that this theory provides answers to the questions we have raised.

The seeds of the atomic theory go back at least to the time of the ancient Greeks. The Greeks pondered a seemingly abstract question: Can matter be divided endlessly into smaller and smaller pieces, or is it composed of some ultimate particle that cannot be further divided? The main line of Greek thought, following the views of Plato and Aristotle, was that matter is continuous. However, some Greek philosophers, notably Democritus, disagreed with this view and argued that matter was composed of small indivisible particles that Democritus called *atomos,* meaning indivisible. This atomic concept was also central to the natural philosophy of the Roman poet and philosopher Lucretius, who lived in the first century B.C. He wrote a famous poem, *De Rerum Natura* (On the Nature of Things), in which he elaborated at length on the atomic view of matter.

Even if it were granted that matter is atomic in nature, the question arises how the atoms of different substances differ from one another. Lucretius suggested that the atoms of substances that have a bitter taste have barbs on their surfaces that scrape the tongue, whereas the atoms of substances with a bland taste must have a smooth surface. Not much improvement in the atomic view of matter occurred in the 18 centuries following Lucretius. The philosophical ideas of Plato and Aristotle, neither of whom accepted the atomistic view of matter, held sway in European thought for many centuries. Even though the atomic idea was occasionally revived, early proponents of the particulate theory of matter relied largely on intuition to support their views. During this long period, however, there was a thin, intermittent stream of experimental work. Much of it was prompted by erroneous notions, such as the alchemical belief that common metals such as lead might be transformed into precious metals. Nevertheless, experience of how chemical substances react with one another accumulated, and more quantitative methods of studying chemical reactions were developed. The way was prepared for a new and more meaningful statement of an atomic theory. It came, in the early years of the nineteenth century, from John Dalton, an English schoolteacher (Figure 2.9). Dalton's atomic theory, published in the period 1803–1807, was strongly tied to experimental observation. His efforts were so successful that his theory has dominated our thinking since his time and has had to undergo little revision.

The basic postulates of Dalton's theory were as follows:

1 Each element is composed of extremely small particles called atoms.
2 All atoms of a given element are identical.
3 Atoms of different elements have different properties (including different masses).
4 Atoms of an element are not changed into different types of atoms by chemical reactions; atoms are neither created nor destroyed in chemical reactions.
5 Compounds are formed when atoms of more than one element combine.
6 In a given compound, the relative number and kind of atoms are constant.

FIGURE 2.9 John Dalton (1766–1844) was the son of a poor English weaver. Because he was a Quaker, Dalton's quiet life-style stands in contrast to the life-styles of Priestley and Lavoisier. Dalton began teaching at the age of 12; he spent most of his years in Manchester, where he taught both grammar school and college. His life-long interest in meteorology led him to study gases and hence to chemistry and eventually to the atomic theory. It was perhaps because Dalton's training was not in chemistry that he was able to approach problems with a viewpoint different from that of chemists of the time. (*Library of Congress*)

This theory provides us with a mental picture of matter. As represented schematically in Figure 2.10, we visualize an element as being composed of tiny particles called atoms. Atoms are the basic building blocks of matter; they are the smallest units of an element that can combine with other elements. Compounds involve atoms of two or more elements combined in definite arrangements. Mixtures do not involve the intimate interactions between atoms that are found in compounds.

Dalton's theory embodies several simple laws of chemical combination that were known at the time. Because atoms are neither created nor destroyed in the course of chemical reactions (postulate 4), it is readily evident that matter is neither created nor destroyed in such reactions. Thus we have the law of conservation of matter, which was discovered by Lavoisier. This law is one of the principal topics of Chapter 3. The law of constant composition, which was cited in Section 2.4, is explained by postulate 6: In a given compound the relative number and kind of

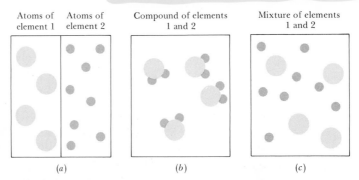

FIGURE 2.10 Difference between elements, compounds, and mixtures as visualized through Dalton's atomic theory. (*a*) Elements are composed of small particles called atoms. All atoms of a given element are identical; atoms of different elements are different. (*b*) Compounds involve atoms of two or more elements combined in definite arrangements. (*c*) Mixtures have variable compositions. There is no restriction on the relative numbers of atoms of elements 1 and 2.

atoms is constant. A third law discovered by Dalton and consistent with his theory is the law of multiple proportions: When two elements form more than one compound, for a fixed mass of one element the masses of the second element are related to each other by small whole numbers. For example, the substances water and hydrogen peroxide both consist of the elements hydrogen and oxygen. In water there are 8.0 g of oxygen for each gram of hydrogen, whereas in hydrogen peroxide there are 16.0 g of oxygen for each gram of hydrogen. The ratio of the masses of oxygen that combine with a gram of hydrogen in these compounds is in the ratio of the small whole number two: Hydrogen peroxide has twice as much oxygen per unit mass of hydrogen as water does. Using the atomic theory, we understand this to mean that hydrogen peroxide contains twice as many oxygen atoms per hydrogen atom as does water. We now know that water contains one oxygen atom for each two hydrogen atoms, whereas hydrogen peroxide contains two oxygen atoms for each two hydrogen atoms.

Thus we see that the atomic theory ties together many observations and helps us explain them. To our earlier question of what makes one element different from another we can now answer that they have different types of atoms. However, this explanation only begs a further question: How are the atoms of different elements different from each other? We need to consider the structure of the atom to answer this question. This topic is taken up in the next section. We will see that as we begin to understand the structure of the atom, we will begin to understand many more aspects of matter.

2.6 The structure of the atom

Our present understanding of atomic structure is very precise. Chemists and physicists are able to use a wide range of sophisticated instruments to measure in great detail all the properties of individual atoms. It is therefore difficult to imagine that only 100 years ago scientists knew very little about atoms beyond what was contained in Dalton's atomic theory. Dalton and his contemporaries viewed the atom as an indivisible object. However, data slowly accumulated to indicate that the atom has a substructure of smaller particles. We will consider here just a few of the most important experiments that led to our present model of atomic structure.

CATHODE RAYS AND ELECTRONS

In the mid-1800s a number of investigators began to study electrical discharge through evacuated tubes. Radiation is produced within such tubes when voltages become high enough to permit current flow (about 1000 volts). This radiation became known as cathode rays because it emanated from the negative electrode, or cathode. This and a number of additional facts suggested that the radiation consisted of a stream of negatively charged particles that were named "electrons." For example, the rays travel in straight lines in the absence of magnetic or electric fields. However, they are deflected by magnetic and electric fields in a manner expected for negatively charged particles. The behavior of a negatively charged particle in a magnetic field and in an electric field is shown in Figure 2.11. The movement of the rays can be determined

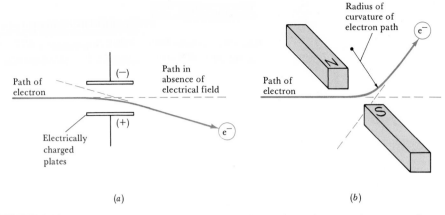

FIGURE 2.11 The behavior of a negatively charged particle such as an electron moving through an electric field (*a*) and a magnetic field (*b*). The path of the electron in (*a*) is bent because the negatively charged electron is attracted toward the plate of opposite charge and repelled by the plate carrying the same charge. Magnetic fields, as in (*b*), cause charged particles moving at right angles to the magnetic field to follow a circular path.

because of their ability to cause certain materials, including glass, to give off light, or fluoresce. In fact, a television picture tube is a cathode-ray tube; the television picture results from fluorescence from the television screen. Because the rays (electrons) were found to be independent of the nature of the cathode material, it was deduced that they are a basic component of all matter.

In 1897 the British physicist J. J. Thomson (Figure 2.12) was able to measure the ratio of the electrical charge to the mass of the electron using a cathode-ray tube such as that shown schematically in Figure 2.13. When only the magnetic field is turned on, the electron strikes point *A* of the tube. When the magnetic field is off and the electric field is on, the electron strikes point *C*. When both the magnetic and electric fields are

FIGURE 2.12 J. J. Thomson (1856–1940). (*Copyright Cavendish Laboratory, University of Cambridge*)

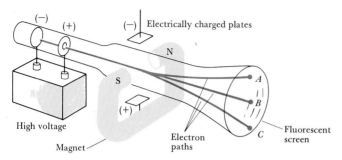

FIGURE 2.13 Cathode-ray tube with perpendicular magnetic and electric fields.

off or when they are balanced so as to cancel each other's effects, the electron strikes point B. By carefully and quantitatively determining the effects of magnetic and electric fields on the motion of the cathode rays, Thomson was able to determine the charge-to-mass ratio of 1.76×10^8 coul/g.*

The external force exerted on an electron moving in a magnetic field is given by Hev, where H is the strength of the magnetic field, e is the charge on the electron, and v is its velocity. The motion of the electron is determined by the balance of this force and the tendency of the electron to continue its straight-line motion (the centrifugal force on the electron). This latter force is given by the relation mv^2/r, where m is the mass of the electron, v is its velocity, and r is the radius of the curved path taken by the electron as it moves through the magnetic field (refer to Figure 2.11). Thus the path taken by the electron must be consistent with the equation

$$Hev = \frac{mv^2}{r}$$

By rearranging this equation, we see that the radius of the path of the electron moving through the magnetic field is given by $r = mv^2/Hev = mv/He$. The larger the radius, the smaller the deflection of the particle from the straight-line path that it would take in the absence of the magnetic field. Thus, the more massive the particle, the greater its velocity, the smaller the magnetic field, and/or the smaller the

charge on the particle, the greater the tendency of the particle to continue its straight-line motion.

The relation given above can also be arranged into the form $e/m = v/rH$. The velocity of the electron can be determined by balancing the effects of the magnetic field against a perpendicular electric field so that the electron moves in a straight line. The magnitude of the force exerted by the electric field on the electron is given by Ee where E is the strength of the electric field. For the electron moving in a straight line through the balanced magnetic and electric fields

$$Hev = Ee$$

so that

$$v = \frac{E}{H}$$

Thus, from the first relation, $e/m = v/rH = E/H^2r$, where all of the quantities on the right side of the equation can be determined experimentally. Thus, Thomson's experiment permitted determination of the ratio e/m for the electron.

In 1909, Robert Millikan of the University of Chicago determined the charge on the electron by measuring the effect of an electric field on the rate at which charged oil droplets fall under the influence of gravity. The apparatus that he used is shown schematically in Figure 2.14. The rate at which the droplets fall in air is determined by their size and mass. By watching a particular droplet, Millikan could measure its rate of fall and calculate from this its mass. The experiment was arranged so that a source of radioactivity was near the droplets. The radioactive source

*The coulomb (coul or C) is the SI unit for electrical charge.

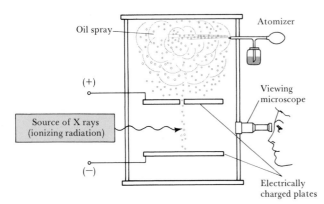

FIGURE 2.14 Schematic representation of Millikan's apparatus for studying the rate of fall of oil droplets.

caused charges to form. The charged particles floating around in the air would often become attached to an oil droplet. When an electrical charge was applied to the plates, charged oil droplets in the region between the plate would be acted upon by the electric field. Their fall could be accelerated, retarded, or even reversed, depending on the charge on the droplet and the polarity of the voltage applied to the plates. By carefully measuring the effects of the electrical field on the movements of many droplets, Millikan found that the charge on the oil drops was always an integral multiple of 1.60×10^{-19} coul, which he deduced was the charge of the electron. The mass of the electron, 9.11×10^{-28} g, was then calculated by combining Millikan's value of the charge with Thomson's charge-to-mass ratio:

$$\text{Mass} = \frac{1.60 \times 10^{-19} \text{ coul}}{1.76 \times 10^8 \text{ coul/g}} = 9.11 \times 10^{-28} \text{ g}$$

RADIOACTIVITY

The discovery of radioactivity by the French scientist Henri Becquerel in 1896 provided additional evidence for the complexity of the atom. Becquerel's imagination had been captured by W. C. Roentgen's discovery of X rays, which had been reported in January 1896. Roentgen had been quick to grasp the practical importance of his discovery, and within a short time X rays had been used in medicine. Members of the international scientific community also sensed that this was something big. Becquerel was well aware that certain substances, upon exposure to sunlight, become luminous, a phenomenon referred to as fluorescence. He sought to determine whether such fluorescent substances gave off X rays. In his initial experiments, Becquerel chose to work with a fluorescent uranium mineral. He placed this in the sunlight over a photographic plate that had been carefully wrapped to protect it from the direct radiation of the sun. When the plate was developed, he found the image of the mineral on the plate. Toward the end of February 1896, Becquerel incorrectly reported that penetrating rays, presumably X rays, could be induced by sunlight and emitted as part of fluorescence. However, the weather turned bad, and Becquerel had to postpone further studies. While the sun stayed behind the clouds, Becquerel kept the mineral and the wrapped photographic plate in a desk drawer. On March 1, 1896, he

decided to develop the plate, not expecting to find any images. He was surprised to find very intense silhouettes. Becquerel concluded correctly this time that the mineral was producing a spontaneous radiation and referred to this phenomenon as radioactivity. At Becquerel's suggestion, Marie Sklodowska Curie (Figure 2.15) and her husband, Pierre, began their famous experiments to isolate the radioactive components of the mineral, called pitchblende.

Further study of the nature of radioactivity, principally by the British scientist Ernest Rutherford, revealed three types of radiation—alpha (α), beta (β), and gamma (γ) radiation. Each type differed in electrical behavior and penetrating ability. The behavior of these three types of radiation in an electric field is shown in Figure 2.16.

It was possible to show that the β radiation was identical to a stream of high-speed electrons. The particles of this radiation were called β

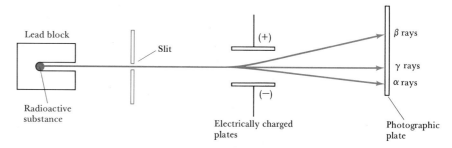

FIGURE 2.16 Behavior of alpha (α), beta (β), and gamma (γ) rays in an electric field.

TABLE 2.3 Summary of the properties of alpha, beta, and gamma rays

	Type of radiation		
	α	β	γ
Charge	$2+$	$1-$	0
Mass	6.64×10^{-24} g	9.11×10^{-28} g	0
Relative penetrating power	1	100	1000
Identity	$^4_2$He nuclei	Electrons	High-energy radiation

particles. In units of the charge of the electron, each β particle has a charge of $1-$. The α rays were found to consist of positively charged particles with a charge of $2+$. The α particles are comparatively much more massive than the β particles. Rutherford was able to show that the α particles combined with electrons in the surroundings to form atoms of helium. He thus concluded that the α rays consist of the nuclei of helium atoms; that is, of the positively charged core of the helium atom. The γ rays are high-energy radiation like X rays; they do not consist of particles. The α rays have low penetrating power; they are stopped by paper. The β particles have about 100 times greater penetrating ability, and the γ rays about 1000 times greater penetrating ability than α rays. The comparative properties of the three types of radiation are summarized in Table 2.3.

RUTHERFORD AND THE NUCLEAR ATOM

By 1909 Rutherford had firmly established that α rays consisted of helium nuclei with a $2+$ charge. Once he had unraveled the nature of α rays, he began to use them to study the structure of the atom. By this time it was well accepted that the atom was electrical in nature and contained electrons. The prevalent model of the atom, as developed by J. J. Thomson, pictured the atom as a cloud of positive charge in which negatively charged electrons were embedded like seeds in a watermelon.

In 1910 Rutherford and his co-workers performed an experiment that led to the downfall of Thomson's model. Rutherford was studying the manner of scattering of a narrow beam of α particles as they passed through a thin gold foil. He had found slight scattering, on the order of 1 degree, which was consistent with Thomson's model. One day Hans Geiger, an associate of Rutherford's, suggested that Ernest Marsden, a 20-year-old undergraduate working in their laboratory, get some experience in conducting such experiments. Rutherford suggested that Marsden see if α particles were scattered through large angles. In his own words:

> I may tell you in confidence that I did not believe they would be since we knew that the α particle was a very massive particle with a great deal of energy. . . . Then I remember two or three days later Geiger coming to me in great excitement and saying, "We have been able to get some α particles coming backwards." . . . It was quite the most incredible event that has ever happened to me in my life. It was almost as if you fired a 15-inch shell into a piece of tissue paper and it came back and hit you.

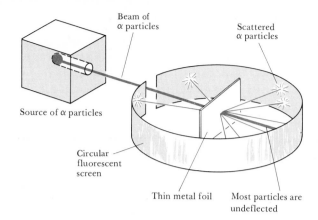

Beam of
α particles

Scattered
α particles

Source of α particles

Circular
fluorescent
screen

Thin metal foil   Most particles are
undeflected

FIGURE 2.17 Rutherford's experiment on the scattering of α particles.

What Rutherford and his co-workers had observed was that the vast majority of α particles passed directly through the foil without deflection. Only a few underwent deflection, some even bouncing back in the direction from which they had come, as shown in Figure 2.17.

By 1911 Rutherford was able to explain these observations as follows: Most α particles pass directly through the foil because most of the atom is space containing only electrons, which have a very small mass. The few α particles that undergo deflection do so because they have come close to the small nucleus of the gold atom in which the positive charge of the atom resides. The gold atom is known to be much more massive than the helium atom. Furthermore, Rutherford could assume that the gold nucleus contains a much higher positive charge than the helium nucleus. The repulsion between the highly charged gold nucleus and the α particle coming directly at it is strong enough to deflect the less massive α particle backward in space. This is represented schematically in Figure 2.18. It was in this way that the concept of the nuclear atom was born.

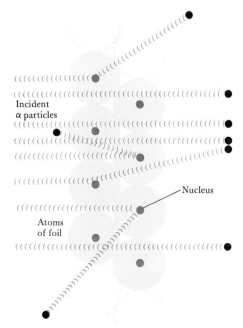

Incident
α particles

Nucleus

Atoms
of foil

FIGURE 2.18 Rutherford's model explaining his experiment with scattering of α particles. The gold foil is actually several thousand atoms thick. The α particles are deflected backward only when they collide directly with, or pass very close to, the much heavier, positively charged gold nuclei. According to Coulomb's law, like charges repel one another. The α particles, with 2+ charge, and the gold nuclei, with 79+ charge, undergo a strongly repulsive interaction when the α particle closely approaches a gold nucleus. The less massive α particle is deflected from its path by the repulsive interaction. Only a small fraction of the particles are strongly deflected, because the gold nucleus has a small volume.

THE MODERN VIEW OF ATOMIC STRUCTURE

Since the time of Rutherford, physicists have learned a great deal about the detailed structure of atomic nuclei. In the course of these discoveries the list of subnuclear particles has grown long and continues to increase. As chemists we can take a very simple view of the atom, because only three subatomic particles, the proton, neutron, and electron, have a bearing on chemical behavior.

Protons have a positive charge; neutrons are uncharged; electrons have a negative charge. The charges on the proton and electron are equal in magnitude. Because atoms have no net electrical charge, there are equal numbers of electrons and protons in an atom. The protons and neutrons reside together in a very small volume within the atom known as the nucleus. Most of the rest of the atom is space in which the electrons move. The electrons are attracted to the nucleus and kept from flying off completely free in space by the attraction that exists between particles of unlike electrical charge (coulombic or electrostatic attraction). The charges and masses of the subatomic particles are summarized in Table 2.4.

Two bodies of the same charge repel each other. A positively charged body and a negatively charged one will be attracted to each other. The force of the interaction (F) between two charged bodies is given by Coulomb's law: $F = Q_1 Q_2 / d^2$ where Q_1 and Q_2 are the magnitudes of the charge on bodies 1 and 2, and d is the distance between them. The formula indicates that doubling the magnitude of one of the charges will double the force of attraction or repulsion. Doubling the distance of separation between the charges will reduce the force to one-fourth its previous value.

Atoms have diameters on the order of 1 Å. The angstrom (Å), which is 10^{-10} m, is widely used to indicate dimensions on the atomic scale. For example, the diameter of a chlorine atom is 1.8 Å. In time the angstrom may be replaced by a more acceptable SI unit such as the picometer or nanometer. In these units the diameter of a chlorine atom is 180 pm or 0.18 nm. For the most part we will employ angstroms throughout the text to indicate atomic and molecular dimensions.

SAMPLE EXERCISE 2.3

The diameter of a U.S. penny is 19 mm. How many chlorine atoms would fit side-by-side along this diameter?

Solution: The diameter of a single chlorine atom is 1.8 Å. We convert units as follows (the symbol for chlorine is Cl):

$$\frac{\text{No. of Cl atoms}}{\text{penny}} = \left(\frac{1 \text{ Cl atom}}{1.8 \text{ Å}}\right)\left(\frac{10^{10} \text{ Å}}{1 \text{ m}}\right)$$

$$\times \left(\frac{1 \text{ m}}{10^3 \text{ mm}}\right)\left(\frac{19 \text{ mm}}{1 \text{ penny}}\right)$$

$$= 1.1 \times 10^8 \frac{\text{Cl atoms}}{\text{penny}}$$

This exercise helps to illustrate how very small atoms are in relationship to ordinary dimensions.

TABLE 2.4 Comparison of the proton, neutron, and electron

Particle	Charge	Mass
Proton	Positive $(1+)$	1.67×10^{-24} g
Neutron	None (neutral)	1.67×10^{-24} g
Electron	Negative $(1-)$	9.11×10^{-28} g

The diameters of atomic nuclei are on the order of 10^{-4} Å, only a small fraction of the diameter of the atom as a whole. If the atom were scaled upward in size so that the nucleus were 2 cm in diameter (about the diameter of a penny), the atom would have a diameter of 200 m (about twice the length of a football field.) The proton and neutron have approximately equal mass, but the electron weighs only about $\frac{1}{1835}$ as much as either of these. Therefore the tiny nucleus carries most of the mass of the atom. Indeed, the density of the nucleus is on the order of 10^{13}–10^{14} g/cm^3. A matchbox full of material of such density would weigh over $2\frac{1}{2}$ billion tons. Astrophysicists have suggested that the matter in the interior of a collapsed star may reach approximately this density.

An illustration of the atom that incorporates the features we have just discussed is shown in Figure 2.19. The electrons, which take up most of the volume of the atom, play the major role in chemical reactions. The significance of representing the region containing the electrons as an indistinct cloud will become clear in later chapters when we consider the energies and spatial arrangements of the electrons.

The identity of an element depends on the number of protons in the nucleus of an atom of that element. In fact we may define an element as a substance whose atoms all have the same number of protons. Thus all atoms of the element carbon have six protons and six electrons. Most also have six neutrons although some have more. Atoms of a given element that differ in number of neutrons, and consequently in mass, are called isotopes. The symbol $^{12}_{6}\text{C}$ or simply ^{12}C (read "carbon twelve," carbon-12) is used to represent the carbon atom with six protons and six neutrons. The number of protons, which is called the atomic number, is shown by the subscript. Since all atoms of a given element have the same atomic number, this subscript is redundant and hence often omitted. The superscript is called the mass number and is the total number of protons plus neutrons in the atom. Some carbon atoms contain six protons and eight neutrons and are consequently represented as ^{14}C (read "carbon fourteen"). Normally, subscripts and superscripts are used with the symbol for an element only when reference is made to a particlular

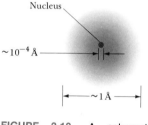

FIGURE 2.19 A schematic cross-sectional view through the center of an atom. The nucleus, which contains positive protons and neutral neutrons, is the location of virtually all the mass of the atom. The rest of the atom is mainly the space in which the light, negatively charged electrons move.

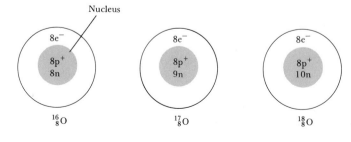

FIGURE 2.20 The distribution of subatomic particles in three isotopes of oxygen (p^+ = proton, n = neutron, e^- = electron). The shaded circle represents the nucleus; the eight electrons are in space surrounding the nucleus, as illustrated in Figure 2.19.

isotope of that element. Three isotopes of oxygen and their chemical symbols are shown schematically in Figure 2.20. The term nuclide is applied in a general way to a nucleus with a specified number of protons and neutrons. For example, the nucleus of $^{16}_{8}O$ is referred to as the $^{16}_{8}O$ nuclide. (We will have more to say about the isotopic compositions of the elements in Section 3.7.)

SAMPLE EXERCISE 2.4

How many protons, neutrons, and electrons are there in ^{197}Au?

Solution: According to the list of elements given in the front inside cover of this text, gold has an atomic number of 79. Consequently ^{197}Au has 79 protons, 79 electrons, and $197 - 79 = 118$ neutrons.

SAMPLE EXERCISE 2.5

Write the nuclear isotope symbols for the three isotopes of hydrogen, with mass numbers of 1, 2, and 3.

Solution: Because all three of the hydrogen isotopes must have the same number of protons, 1, the three symbols are: $^{1}_{1}H$; $^{2}_{1}H$; $^{3}_{1}H$.

On the atomic level gold, oxygen, and carbon differ in terms of the number of protons, neutrons, and electrons their respective atoms contain. These subatomic particles, however, are common to all substances. We can therefore state that an atom is the smallest representative sample of an element, because breaking the atom into subatomic particles destroys its identity.

In order to change a base or common metal like lead, atomic number 82, to gold, atomic number 79, requires removal of 3 protons from the nucleus of the lead atom. Because the nucleus is extremely small and buried in the heart of the atom, and because of the very strong binding forces between particles in the nucleus, this removal is exceedingly difficult. The energies required to cause changes in the nucleus are enormously greater than the energies associated with even the most vigorous chemical reactions. Thus, we still agree with Dalton that atoms of an element are not changed into different types of atoms by chemical reactions. Therein lies the futility of the alchemists' attempts to change base metals to gold.

2.7 The periodic table: a preview

Dalton's atomic theory, and the various empirical laws that it helped to explain (Section 2.5), set the stage for a vigorous growth in chemical experimentation during the early part of the nineteenth century. As the body of chemical observations grew, and the list of known elements expanded, attempts were made to find regularities in chemical behavior. These efforts culminated in the development of the periodic table in 1869. We will have much to say about the periodic table in later chapters, but it is so important and useful that you should become acquainted with it now.

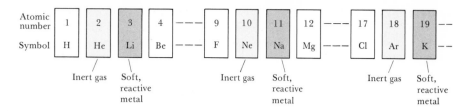

FIGURE 2.21 Arrangement of elements by atomic number to illustrate the periodic or repeating pattern in properties that is the basis of the periodic table.

Many elements show very strong similarities to each other. For example, lithium (Li), sodium (Na), and potassium (K) are all soft, very reactive metals. The elements helium (He), neon (Ne), and argon (Ar) are very nonreactive gases. If the elements are arranged in order of increasing atomic number, their chemical and physical properties are found to show a repeating or periodic pattern. For example, each of the soft, reactive metals, lithium, sodium, and potassium, comes immediately after one of the nonreactive gases, helium, neon, and argon, as shown in Figure 2.21. The arrangement of elements in order of increasing atomic number, with elements having similar properties placed in vertical columns, is known as the periodic table. The periodic table is given on the front inside cover of your text for easy reference. In most classrooms where chemistry is taught, large periodic tables are hung on the walls—a testimony to their usefulness. You may notice slight variations in periodic tables from one book to another, or between the lecture hall and the text. These are matters of style or the particular information included; there are no fundamental differences.

The elements in a column of the periodic table are known as a family or group. They are identified as group 1A, 2A, and so forth, as shown at the top of the periodic table. For example, three familiar elements that have similar properties are copper (Cu), silver (Ag), and gold (Au), which occur together in group 1B. Some groups are also described by a family name. The members of group 1A—lithium, sodium, potassium, rubidium (Rb), cesium (Cs), and francium (Fr)—are known as the alkali metals. The members of group 2A—berylium (Be), magnesium (Mg), calcium (Ca), strontium (Sr), barium (Ba), and radium (Ra)—are known as the alkaline earth metals. The members of group 7A—fluorine (F), chlorine (Cl), bromine (Br), iodine (I), and astatine (At)—are known as the halogens. The members of group 8A—helium (He), neon (Ne), argon (Ar), krypton (Kr), xenon (Xe), and radon (Rn)—are known as the noble gases, inert gases, or rare gases.

We will learn in Chapters 5 and 6 that the elements in a family of the periodic table have similar properties because they have the same type of arrangement of electrons at the periphery of their atoms. However, we need not wait until then to make good use of the periodic table; after all, the table in pretty much its modern form was invented by chemists who knew nothing of the electronic structures of atoms! We can use the table, as they intended, to correlate the behaviors of elements and to aid in remembering many facts. You will find it helpful to refer to the periodic table frequently in studying the remainder of this chapter.

Which of the following elements would you expect to show the greatest similarity in chemical and physical properties: Li, Be, F, S, Cl?

Solution: The elements F and Cl should be most alike because they are in the same family (group 7A, the halogen family).

One pattern that is evident when elements are arranged in the periodic table is the grouping together of the **metallic elements.** These elements, which are grouped together on the left side of the periodic table, share many characteristic properties, such as luster and high electrical and heat conductivity. The metallic elements are separated from the **nonmetallic elements** by the diagonal line that runs across the right side of the periodic table from boron (B) to astatine (At). Note that the majority of the elements are metallic. The nonmetals, which occupy the upper right side of the periodic table, are gases, liquids, or crystalline solids. They lack those physical characteristics that distinguish the metallic elements. Many of the elements that lie along the line that separates metals from nonmetals, such as antimony (Sb), possess properties intermediate between those of metals and nonmetals. These elements are often referred to as **semimetals** or **metalloids.**

2.8 Molecules and ions

We have seen that the atom is the smallest representative sample of an element. However, only the noble gas elements are normally found in nature as isolated atoms. Most matter is composed of molecules or ions, which are formed from atoms.

MOLECULES

Molecules are composed of combinations of tightly bound atoms. The resultant assembly or package of atoms behaves in many ways as a single object, just as a television set composed of many parts can be recognized as a single object. The nature of the forces (bonds) that bind the atoms together will be examined in Chapter 7.

When elements exist in molecular form, they contain only one type of atom. For instance, the element oxygen, as it is normally found in air, consists of molecules composed of pairs of oxygen atoms. This molecular form of oxygen is represented by the **chemical formula** O_2 ("oh two"). The subscript in this formula indicates that two oxygen atoms are present in each molecule. The molecule is said to be **diatomic.** Oxygen also exists in a form known as ozone, which consists of three bound oxygen atoms. Correspondingly, the chemical formula for ozone is O_3. Ozone and "normal" oxygen exhibit quite different chemical properties.

The elements that normally occur as diatomic molecules are hydrogen, oxygen, nitrogen, and the halogens. Their location in the periodic table is shown in Figure 2.22. When we speak of the substance hydrogen, we mean H_2 unless we indicate explicitly otherwise. Likewise, when we speak of oxygen, nitrogen, or any of the halogens, we are referring to O_2, N_2, F_2, Cl_2, Br_2, or I_2. Thus the properties of oxygen and hydrogen listed earlier in Table 2.2 are those of O_2 and H_2. In other forms, these elements behave much differently.

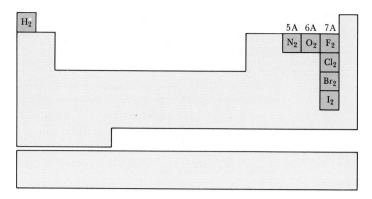

FIGURE 2.22 Elements that exist as diatomic molecules at room temperature.

Molecules of compounds contain more than one type of atom. For example, a molecule of water consists of two hydrogen atoms and one oxygen atom. It is therefore represented by the chemical formula H_2O (read "aitch two oh"). Another compound composed of these same elements is hydrogen peroxide, H_2O_2. These two compounds have quite different properties. A number of common molecules are shown in Figure 2.23. Pay close attention to how the chemical formula of each molecule reflects its composition.

SAMPLE EXERCISE 2.7

Which of the molecules shown in Figure 2.23 are compounds and which are elements?

Solution: The compounds are H_2O, CO_2, CO, H_2O_2, CH_4, and C_2H_4. The elements are O_2, O_3, and H_2 (one type of atom).

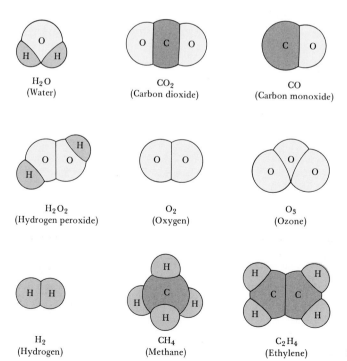

H_2O
(Water)

CO_2
(Carbon dioxide)

CO
(Carbon monoxide)

H_2O_2
(Hydrogen peroxide)

O_2
(Oxygen)

O_3
(Ozone)

H_2
(Hydrogen)

CH_4
(Methane)

C_2H_4
(Ethylene)

FIGURE 2.23 Representations of some simple, common molecules.

Chemical formulas such as those shown in Figure 2.23, which indicate the actual numbers and types of atoms within a molecule, are known as molecular formulas. It is easier to determine the relative number of each type of atom in a molecule than the actual number of atoms. A chemical formula that gives only the relative number of atoms of each type in a molecule is known as an empirical or simplest formula. The subscripts in such a formula are smallest whole-number ratios. For example, hydrogen peroxide, whose molecular formula is H_2O_2, has an empirical formula of HO; ethylene, whose molecular formula is C_2H_4, has an empirical formula of CH_2. The molecular formula may be identical with the empirical formula as in the case of H_2O.

SAMPLE EXERCISE 2.8

Glucose, a substance also known as blood sugar and as dextrose, has the chemical formula $C_6H_{12}O_6$. Is this the empirical formula for glucose?

Solution: $C_6H_{12}O_6$ is not the empirical formula, because the empirical formula has subscripts that are smallest whole-number ratios. The smallest ratios are obtained by dividing each subscript by the largest common factor, in this case 6. The empirical formula for glucose is CH_2O. Some formulas, such as H_2O and CO_2, are both empirical and molecular formulas.

In experimentally determining the molecular formula for a substance, its empirical formula is determined first. The total mass of the molecule is then determined. From the empirical formula and the mass of the molecule, its molecular formula can be determined. But now we're getting ahead of ourselves; we'll consider how empirical and molecular formulas are determined when we get to Section 3.7.

It sometimes happens that a substance exists as a three-dimensional structure in which there are no discrete atoms or molecules. Examples include the elements carbon and iron. In such cases the substance is represented by the symbol for the element, or by the empirical formula if it is a compound. As an example, boron nitride consists of a three-dimensional array of boron and nitrogen atoms. It is represented by the empirical formula BN.

Often the chemical formulas of molecules are written so as to show the relative arrangements of the atoms. For example, the formulas for water and hydrogen peroxide can be written as follows:

Water   Hydrogen peroxide

Such formulas are known as structural formulas. The lines between the symbols for the elements represent the bonds that hold the respective atoms together. These formulas indicate which atoms are attached to which; however, they do not necessarily tell anything about the shapes of molecules.

IONS

Whereas the nucleus of an atom is unchanged by ordinary chemical processes, atoms readily gain or lose electrons. These transactions take place at the periphery of the atom. If electrons are removed or added to a neutral atom, a charged particle called an ion is formed. For example, the sodium atom, which has 11 protons and 11 electrons, can lose an electron. The resulting ion has 11 protons and 10 electrons; its net charge is consequently $1+$. This ion is symbolically represented as Na^+. The net charge on an ion is represented by a superscript; $+, 2+$, and $3+$ mean a net charge resulting from loss of one, two, or three electrons, respectively. The superscripts $-, 2-$, and $3-$ represent net charges resulting from gain of one, two, or three electrons, respectively. The formation of the Na^+ ion from a Na atom is shown schematically below:

$$11p^+ \ 11e^- \quad \xrightarrow[\text{electron}]{\text{Lose one}} \quad 11p^+ \ 10e^-$$

$$\text{Na atom} \qquad\qquad \text{Na}^+ \text{ ion}$$

Chlorine, with 17 protons and 17 electrons, can gain an electron in chemical reactions, producing the Cl^- ion:

$$17p^+ \ 17e^- \quad \xrightarrow[\text{electron}]{\text{Gain one}} \quad 17p^+ \ 18e^-$$

$$\text{Cl atom} \qquad\qquad \text{Cl}^- \text{ ion}$$

In general, metal atoms lose electrons most readily, whereas nonmetal atoms tend to gain electrons.

Groups of atoms joined together as in a molecule but having a net positive or negative charge are known as polyatomic ions. An example is NO_3^-, known as the nitrate ion. We will consider further examples in the next section.

The chemical properties of ions are greatly different from those of the atoms from which they are derived. The change of an atom or molecule to an ion is like that from Dr. Jekyll to Mr. Hyde: Although the body may be essentially the same (plus or minus a few electrons), the personality is much different.

Many atoms gain or lose electrons so as to end up with the same number of electrons as the closest noble gas. The members of the noble gas family are chemically very nonreactive and form very few compounds. We might deduce that this is because they have very stable electron arrangements. Nearby elements can obtain these same stable arrangements by losing or gaining electrons. For example, loss of one electron from an atom of sodium leaves it with the same number of electrons as the neutral neon atom (atomic number 10). Similarly, when chlorine gains an electron it ends up with 18, the same as argon (atomic number 18). We will content ourselves with this simple observation in explaining the formation of ions until later chapters in which we consider chemical bonding.

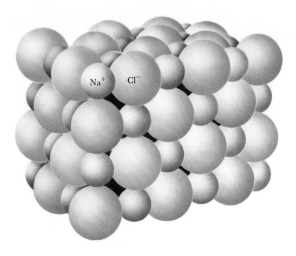

FIGURE 2.24 The arrangement of ions in sodium chloride (NaCl).

SAMPLE EXERCISE 2.9

Predict the charges expected for the most stable ions of barium and oxygen.

Solution: Refer to the periodic table. Barium has atomic number 56. The nearest noble gas is xenon, atomic number 54. Barium can obtain the stable arrangement of 54 electrons by losing two of its electrons, thereby forming the Ba^{2+} ion.

Oxygen has atomic number 8. The nearest noble gas is neon, atomic number 10. Oxygen can obtain this stable electron arrangement by gaining two electrons, thereby forming an ion of $2-$ charge, O^{2-}.

The gain and loss of electrons occur in transactions between different kinds of atoms. The compounds that result from these electron transfers are composed of positively and negatively charged ions instead of molecules. For example, NaCl, ordinary table salt, is such an ionic compound; it consists of equal numbers of Na^+ and Cl^- ions.

These ions are arranged in a three-dimensional array as shown in Figure 2.24. Because there is an overall ratio of one Na^+ ion for each Cl^- ion, the overall structure has no net charge; the compound is neutral. Because there is no discrete molecule of NaCl, we are able to write only an empirical formula. This same situation is found with other compounds that are composed of ions.

2.9 Naming of inorganic compounds

As you proceed in your study of this text, and in chemistry laboratory work, you will need to refer to specific chemical substances by name. We present here some of the basic rules for naming simple compounds. You may not have immediate use for some of the rules, but they are gathered here in one place for your convenience whenever a question of nomenclature, or naming of substances, arises.

There are now about 4 million known chemical substances. Naming them all would be a hopelessly complicated task if each had a special name independent of all the others. Many important substances that have been known for a long time, such as water (H_2O) and ammonia (NH_3), do have individual, traditional names. For most substances, however, we rely upon a set of rules that lead to an informative, systematic name for each substance.

One of the earliest classification schemes in chemistry was the distinction between inorganic and organic compounds. Organic compounds contain carbon, usually in combination with hydrogen, oxygen, nitrogen, and sulfur. Organic compounds were first associated only with plants and animals. However, a great number of organic compounds have now been prepared that do not occur in nature. We will discuss the chemistry and naming of organic compounds in Chapter 24; however we will have many occasions throughout the text to illustrate chemical principles with organic compounds as examples. In this section we will consider the basic rules for naming inorganic compounds, those that the early chemists associated with the nonliving portion of our world.

For the purpose of naming them at least, most compounds can be considered to be composed of two parts, one that is "positive" in character, the other that is "negative." In ionic compounds there really are atoms or groups of atoms that carry positive and negative charges. In molecular compounds, it is possible to associate a more positive nature with one portion of the molecule than with another. We will get to the rules for doing this in a later chapter. For now, it is important only to remember that, with very few exceptions, the positive part of the molecule is named first and listed first in writing the formula for the compound. The negative part is named and written last. To see how the rule applies, let us first consider the naming of ionic compounds.

IONIC COMPOUNDS

Ions may be monatomic, that is, composed of a single atom, or polyatomic, formed from two or more atoms.* Consider first the naming of the positive ions, also called cations. Monatomic cations are most commonly formed from metallic elements. They take the name of the element itself:

Na^+ sodium ion $\qquad$ Zn^{2+} zinc ion

If an element can form more than one positive ion, the positive charge of the ion is indicated by a Roman numeral in parentheses following the name of the metal:

Fe^{2+} iron(II) ion $\qquad$ Cu^+ copper(I) ion

Fe^{3+} iron(III) ion $\qquad$ Cu^{2+} copper(II) ion

At this stage you have no way of knowing which elements commonly exist in more than one charge state. This need not be a source of difficulty; if there is any doubt in your mind, use the Roman numeral designation of charge as part of the name. It is never wrong to use this form of charge designation, even though it may sometimes be unnecessary.

An older method still widely used for distinguishing between two differently charged ions of a metal is to use the endings *-ous* or *-ic*; these endings represent the lower and higher charged ions, respectively. They are used together with the root of the Latin name of the element:

*The polyatomic ions are sometimes referred to in elementary chemistry texts as radicals. This is an incorrect use of that word, and if you learned it in this context, forget it.

Fe^{2+} ferrous ion Cu^+ cuprous ion

Fe^{3+} ferric ion Cu^{2+} cupric ion

The only common polyatomic cations are those given below:

NH_4^+ ammonium ion Hg_2^{2+} mercury(I) or mercurous ion

The name mercury(I) ion is given to Hg_2^{2+} because it can be considered to consist of two Hg^+ ions. Mercury also occurs as the monatomic Hg^{2+} ion, which is known as the mercury(II) or mercuric ion.

Negative ions are called anions. Monatomic anions are most commonly formed from atoms of the nonmetallic elements. They are named by dropping the ending of the name of the element and adding the ending *-ide*:

H^- hydride ion O^{2-} oxide ion N^{3-} nitride ion

F^- fluoride ion S^{2-} sulfide ion P^{3-} phosphide ion

Only a few common polyatomic ions end in *-ide*:

OH^- hydroxide ion CN^- cyanide ion O_2^{2-} peroxide ion

Table 2.5 lists the most common cations and anions. Notice that there are many polyatomic anions containing oxygen. Anions of this kind are referred to as oxyanions. A particular element such as sulfur may form

TABLE 2.5 Common ions

Positive ions (cations)	Negative ions (anions)
Ammonium (NH_4^+)	Acetate ($C_2H_3O_2^-$)
Copper(I) or cuprous (Cu^+)	Bromide (Br^-)
Hydrogen (H^+)	Chloride (Cl^-)
Silver (Ag^+)	Chlorate (ClO_3^-)
Sodium (Na^+)	Cyanide (CN^-)
Potassium (K^+)	Fluoride (F^-)
	Hydrogen carbonate (HCO_3^-)
Barium (Ba^{2+})	or bicarbonate
Calcium (Ca^{2+})	Hydrogen sulfate (HSO_4^-)
Chromium(II) or chromous (Cr^{2+})	or bisulfate
Cobalt(II) or cobaltous (Co^{2+})	Hydroxide (OH^-)
Copper(II) or cupric (Cu^{2+})	Iodide (I^-)
Iron(II) or ferrous (Fe^{2+})	Nitrate (NO_3^-)
Lead(II) or plumbous (Pb^{2+})	Perchlorate (ClO_4^-)
Magnesium (Mg^{2+})	Permanganate (MnO_4^-)
Manganese(II) or manganous (Mn^{2+})	
Mercury(II) or mercuric (Hg^{2+})	Carbonate (CO_3^{2-})
Tin(II) or stannous (Sn^{2+})	Chromate (CrO_4^{2-})
Zinc (Zn^{2+})	Oxide (O^{2-})
	Peroxide (O_2^{2-})
Aluminum (Al^{3+})	Sulfate (SO_4^{2-})
Chromium(III) or chromic (Cr^{3+})	Sulfide (S^{2-})
Iron(III) or ferric (Fe^{3+})	Sulfite (SO_3^{2-})
	Phosphate (PO_4^{3-})

more than one oxyanion. When this occurs, there are rules for indicating the relative numbers of oxygen atoms in the anion. When an element has two oxyanions, the name of the one that contains more oxygen ends in -*ate*; the name of the one with less oxygen ends in -*ite*:

NO_2^-	nitrite ion	SO_3^{2-}	sulfite ion
NO_3^-	nitrate ion	SO_4^{2-}	sulfate ion

When the series of anions of a given element extends to three or four members, as with the oxyanions of the halogens, prefixes are also employed. The prefix *hypo-* indicates less oxygen, whereas the prefix *per-* indicates more oxygen:

ClO^- hypochlorite ion (less oxygen than chlorite)

ClO_2^- chlorite ion

ClO_3^- chlorate ion

ClO_4^- perchlorate ion (more oxygen than chlorate)

Notice that if you memorize the rules just indicated, you need only know the name for one oxyanion in a series to deduce the names for the other members.

SAMPLE EXERCISE 2.10

The formula for the selenate ion is SeO_4^{2-}. Write the formula for the selenite ion.

Solution: The selenite ion should have one less oxygen than the selenate ion; hence, SeO_3^{2-}.

Because many names of ions predate the establishment of systematic rules, there are many exceptions to the rules. For example, the permanganate ion is MnO_4^-; we thus expect that the manganate ion should be MnO_3^-, but this ion is unknown. The name manganate is given to the species MnO_4^{2-}.

Many polyatomic anions that have high charges readily add one or more hydrogen ions to form anions of lower charge. These ions are named by prefixing the word hydrogen or dihydrogen, as appropriate, to the name of the hydrogen-free anion. An older method, which is still used, is to use the prefix *bi-*:

HCO_3^- hydrogen carbonate (or bicarbonate) ion

HSO_4^- hydrogen sulfate (or bisulfate) ion

$H_2PO_4^-$ dihydrogen phosphate ion

We are now in a position to combine the names of cations and anions to name and write the formulas for ionic compounds. The following examples illustrate what is involved:

sodium chloride   NaCl

barium bromide	$BaBr_2$
copper(II) nitrate	$Cu(NO_3)_2$
mercurous chloride	Hg_2Cl_2
aluminum oxide	Al_2O_3

Notice that an overall zero charge is provided for by adjusting the ratios of cations and anions, depending on their charges. Thus, in the second example, two Br^- ions are required to balance the charge of the single Ba^{2+} cation. In the third example, two nitrate ions, NO_3^-, are required to balance the charge of the Cu^{2+} ion. The formula for the entire anion must be enclosed in parentheses so that it is clear that the subscript 2 applies to all the atoms of the anion. The final example, aluminum oxide, is a little more complicated in that more than one of both cation and anion are needed to achieve charge balance. Two Al^{3+} cations are needed to balance the total charge of three O^{2-} anions.

SAMPLE EXERCISE 2.11

Name the following compounds: (a) K_2SO_4; (b) $Ba(OH)_2$.

Solution: (a) Because K^+ is the potassium ion and SO_4^{2-} is the sulfate ion, the name of this compound is potassium sulfate. (b) Because Ba^{2+} is the barium ion and OH^- is the hydroxide ion, the name of this compound is barium hydroxide.

SAMPLE EXERCISE 2.12

Many toothpastes contain stannous fluoride, and foods can be enriched in iron by addition of iron(II) carbonate (ferrous carbonate). Write the chemical formulas for these two compounds.

Solution: The stannous ion, also known as the tin(II) ion, is Sn^{2+}. The fluoride ion is F^-. Two F^- ions are needed to balance the positive charge of Sn^{2+}, forming a neutral compound. The formula is therefore SnF_2.

The iron(II) ion is Fe^{2+}, while the carbonate ion is CO_3^{2-}. The compound is $FeCO_3$, one Fe^{2+} ion balancing the charge of one CO_3^{2-} ion.

These examples should indicate the importance of remembering the charges of the common ions.

ACIDS

There is an important class of compounds known as acids that are named in a special way. These compounds will be discussed further in Section 3.3 and then extensively considered in Chapter 15. For purposes of naming, the acids may be thought of as formed from hydrogen ions and an anion. When the anion is a simple monatomic species, the name of the acid has a prefix *hydro-* and an ending, *-ic*, as in these examples:

from Cl^- chloride ion—HCl hydrochloric acid

from S^{2-} sulfide ion—H_2S hydrosulfuric acid

Many of the most important acids are derived from oxyanions. The name of the acid is related to the name of the anion; when the name of the anion ends in -*ate*, the name of the acid ends in -*ic*. Anions whose names end in -*ite* have associated acids whose names end in -*ous*. Prefixes in the name of the anion are retained in the name of the acid. These rules are illustrated by the oxyacids of chlorine:

ClO^- hypochlorite ion—$HClO$ hypochlorous acid

ClO_2^- chlorite ion—$HClO_2$ chlorous acid

ClO_3^- chlorate ion—$HClO_3$ chloric acid

ClO_4^- perchlorate ion—$HClO_4$ perchloric acid

SAMPLE EXERCISE 2.13

Name the following acids: (a) HNO_3; (b) H_2SO_4; (c) H_2SO_3.

Solution: (a) Because NO_3^- is the nitrate ion, HNO_3 is nitric acid. (b) Because SO_4^{2-} is the sulfate ion, H_2SO_4 is sulfuric acid. (c) Because SO_3^{2-} is the sulfite ion, H_2SO_3 is sulfurous acid.

TABLE 2.6 Prefixes used in naming multiple compounds formed between nonmetals

Prefix	Meaning
Mono-	1
Di-	2
Tri-	3
Tetra-	4
Penta-	5
Hexa-	6

OTHER COMPOUNDS

We will consider most other rules for naming compounds as the need arises. We will also occasionally encounter common names that are still widely used for certain compounds, especially in applied science. When we do, we will give their systematic names also. However, before we end this discussion, we will consider the systematic names of compounds that are nonionic. These are named with the more positively charged element first. (In Chapter 6 we will discuss the assignments of relative charges to atoms.) The relative numbers of atoms of each element are indicated by prefixes, as listed in Table 2.6. We use the same suffixes in naming the more negative element as are used in naming ionic compounds. Some examples illustrate the rules:

NF_3 nitrogen trifluoride

N_2F_4 dinitrogen tetrafluoride

CO carbon monoxide

CO_2 carbon dioxide

SO_2 sulfur dioxide

SO_3 sulfur trioxide

FOR REVIEW

Summary

Matter exists in three states: **gas, liquid,** and **solid.** Most matter consists of a **mixture** of substances. Mixtures can be either **homogeneous** or **heterogeneous;** homogeneous mixtures are called **solutions.** Mixtures can be resolved into two types of **pure substances: compounds** and **elements.** A variety of techniques can be used to separate mixtures into their components. We considered three of these: **filtration,** distillation, and **chromatography.** Each substance has a unique set of **chemical** and **physical properties** that can be used to identify it.

Atoms are the basic building blocks of matter; they are the smallest units of an element that can combine with other elements. Atoms are composed of a **nucleus** (containing **protons** and **neutrons**) and **electrons** that move around the nucleus. We considered some of the historically significant experiments that led to this model of the atom: Thomson's experiments on the behavior of cathode rays (a stream of electrons) in magnetic and electric fields; Millikan's oil-drop experiment; Becquerel's and Rutherford's studies of radioactivity; and Rutherford's studies of the scattering of α particles by thin metal foils.

Elements can be classified by **atomic number**, the number of protons in the nucleus of an atom. All atoms of a given element have the same atomic number. The **mass number** of an atom is the sum of the number of protons and neutrons. Atoms of the same element that differ in mass number are known as **isotopes.**

The **periodic table** is an arrangement of elements in order of increasing atomic number with elements with similar properties placed in vertical columns. The elements in a vertical column are known as a **periodic family** or **group**. The **metallic elements**, which comprise the majority of the elements, are on the left side of the table; the **nonmetallic elements** are located on the right side.

Atoms can combine to form **molecules.** Atoms can also either gain or lose electrons, thereby forming charged particles called **ions.**

Each element has a one- or two-letter symbol. These symbols are combined to represent compounds, as in the formula H_2O for water. Three types of formulas were considered: **simplest** (or empirical) **formulas, molecular formulas,** and **structural formulas.** We also discussed the basic rules for naming inorganic compounds.

Learning goals

Having read and studied this chapter, you should be able to:

1 Differentiate between the three states of matter.
2 Distinguish between elements, compounds, and mixtures.
3 Give the chemical symbols for the elements discussed in this chapter.
4 Distinguish between the physical and chemical properties of a substance.
5 Describe the separation of mixtures by filtration, distillation, and chromatography.
6 Describe the composition of an atom in terms of protons, neutrons, and electrons.
7 Give the approximate size, relative mass, and charge of an atom, proton, neutron, and electron.
8 Write the chemical symbol for an element (for example, $^{12}_{6}C$), having been given its mass number and atomic number, and perform the reverse operation.
9 Write the symbol and charge for an atom or ion, having been given the number of protons, neutrons, and electrons, and perform the reverse operation.
10 Describe the properties of the electron as seen in cathode rays. Describe the means by which J. J. Thompson determined the ratio e/m for the electron.
11 Describe Millikan's oil-drop experiment and indicate what property of the electron he was able to measure.
12 Cite the evidence from studies of radioactivity for the existence of subatomic particles.
13 Describe the experimental evidence for the nuclear nature of the atom.
14 Use the periodic table to predict the charges of monatomic ions.
15 Write the simplest formula for a compound, having been given the charges of the ions from which it is made.
16 Write the name of a simple inorganic compound, having been given its chemical formula, and perform the reverse operation.

Key terms

Among the more important terms and expression used for the first time in this chapter are the following:

Adsorption (Section 2.3) is the adhesion of a thin layer of atoms, molecules, or ions to the surface of a substance.

Alpha (α) particles (Section 2.6) are helium nuclei (consisting of two protons and two neutrons) emitted from the nuclei of certain radioactive atoms.

The **atomic number** (Section 2.6) of an element is the number of protons in the nucleus of one of its atoms.

Beta (β) particles (Section 2.6) are electrons emitted from the nuclei of certain radioactive atoms.

A **chemical change** (Section 2.1) is a process in which one or more substances are converted into other substances.

Chemical properties (Section 2.1) are those properties of a substance that describe its composition and reactivity—how a substance "reacts" or changes into other substances.

A **compound** (Section 2.2) is a substance composed of two or more elements united chemically in definite proportions by mass. A compound can consist of either molecules or of ions.

An **electron** (Section 2.6) is a negatively charged subatomic particle found outside the atomic nucleus. It forms a part of all atoms. It has a mass about $\frac{1}{1835}$ times that of a proton.

An element (Section 2.2) is a substance whose atoms all have the same atomic number; there are 106 known elements. Elements cannot be separated into simpler substances by ordinary chemical means.

An empirical formula or simplest formula (Section 2.8) is a chemical formula that shows the kinds of atoms and their relative numbers in a substance.

Gamma (γ) radiation (Section 2.6), or gamma rays, is a form of radiation similar to X rays that is emitted by radioactive atoms.

Ions (Section 2.8) are electrically charged atoms or groups of atoms (polyatomic ions). Ions can be positively or negatively charged, depending on whether electrons are lost (positive) or gained (negative) from the atoms.

Isotopes (Section 2.6) are atoms of the same element containing different numbers of neutrons and therefore having different masses.

The mass number (Section 2.6) is the sum of the number of protons and neutrons in the nucleus of a particular atom.

A molecular formula (Section 2.8) is a chemical formula that indicates the actual number of atoms of each element in one molecule of a substance.

A molecule (Section 2.8) is a chemical combination of two or more atoms.

A neutron (Section 2.6) is a neutral particle found in the nucleus of an atom; it has approximately the same mass as a proton.

A physical change (Section 2.1) is a change (such as a phase change) that occurs with no change in chemical properties.

Physical properties (Section 2.1) are those properties (such as color and freezing point) that can be measured without changing the composition of a substance.

A proton (Section 2.6) is a positively charged subatomic particle found in the nucleus of any atom.

Radioactivity (Section 2.6) is the spontaneous disintegration of an unstable atomic nucleus with accompanying emission of alpha, beta, or gamma radiation.

A solution (Section 2.2) is a homogeneous mixture.

A structural formula (Section 2.8) is a formula that shows not only the number and kind of atoms in a molecule, but also the arrangement of the atoms.

EXERCISES

States of matter: separations

2.1 Give an example of: (a) a liquid metal; (b) a compound other than water that exists in either solid or liquid form, depending on temperature; (c) a compound that changes state directly from solid to gaseous.

2.2 List some properties of table salt that would distinguish it from refined sugar and indicate whether each property is physical or chemical.

2.3 Which of the following are heterogeneous mixtures, which are pure substances, and which are solutions: (a) battery fluid; (b) gold bars in a Swiss bank; (c) sand; (d) Seven-up; (e) a Bufferin or Anacin tablet?

2.4 Which of the following are chemical and which physical processes: (a) the burning of magnesium ribbon in a camera flash cube; (b) forming copper wire from a bar of copper; (c) heating the filament of an incandescent lamp to provide illumination; (d) distillation of crude petroleum to obtain gasoline; (e) crystallization of sodium bicarbonate from cold water; (f) tarnishing of silver?

2.5 Characterize the following as chemical or physical changes: (a) melting of ice; (b) hard-boiling of an egg; (c) magnetizing iron filings; (d) combustion of coal; (e) dissolving salt in water; (f) dissolving an Alka-Seltzer tablet in water; (g) evaporation of water from a lake.

2.6 Describe the procedure you might employ to: (a) separate methyl alcohol, boiling point 65°C, from butyl alcohol, boiling point 118°C (the two alcohols form a solution when mixed); (b) separate solid silver chloride, insoluble in water, from sodium chloride; (c) separate the colored pigments found in a section of a plant leaf; (d) separate the components present in diesel fuel.

2.7 Consider the following description of the element sodium: "Sodium is silver-white and soft. It is a good conductor of electricity. It can be prepared by electrolysis of molten sodium chloride. The metal boils at 883°C; the vapor is violet-colored. Sodium metal tarnishes rapidly in air. It burns on heating in air or in an atmosphere of bromine vapor." Indicate which of the properties in this description are physical and which are chemical.

2.8 It is not uncommon to find strata or pockets of nearly pure substances in geological formations (for example, quartz or copper metal). Considering that these formations developed by deposition from molten material, discuss the similarities in these geological processes and the separation procedure in Figure 2.2.

2.9 In 1807 Humphrey Davy passed an electric current through molten potassium hydroxide and isolated a bright, shiny, very reactive substance. He claimed the discovery of a new element, which he named potassium. In those days, before the advent of modern instruments, what was the basis on which one could claim that a substance was an element?

2.10 Consider the separation of the components of ink, using paper chromatography, as illustrated in Figure 2.7. Are the components absorbed or adsorbed onto the paper? Which of the three components shown is absorbed or adsorbed most strongly? Explain.

Dalton's atomic theory; atomic structure

2.11 When the elements hydrogen and bromine are caused to react, a gas is formed. The composition of this product gas is the same, no matter what the relative amounts of hydrogen and bromine from which it is formed. How is this observation explained in terms of Dalton's atomic theory? What "law" does it illustrate?

2.12 Ascorbic acid, vitamin C, can be isolated from the juices of citrus fruits. It can also be synthesized in the laboratory in a series of reactions beginning with sorbose, a readily obtained sugar. In what respects should the two types of vitamin C differ in their nutritional value? Bearing in mind the material of Section 2.3, if the two materials did differ, in what respects would they be different?

[2.13] What evidence or line of argument can you cite in support of postulate 4 of Dalton's atomic theory? (This is *not* an easy question; one approach to an answer might be found by reflecting on the process shown in Figure 2.8.)

2.14 A chemistry student prepared a series of compounds containing only nitrogen and oxygen:

Compound	Mass of nitrogen	Mass of oxygen
A	16.8	19.2
B	17.1	39.0
C	33.6	57.3

Show how these data obey the law of multiple proportions.

2.15 A solid white substance A is heated strongly. It decomposes to form a new white substance B and a gas. The gas has exactly the same properties as the product obtained when a carbon rod is burnt in an excess of oxygen. What can we say about whether solids A and B and the gas are elements or compounds?

2.16 Using the type of illustration shown in Figure 2.20 of the text, make an illustration for each of the following atoms: (a) ^{23}Na; (b) ^{25}Mg; (c) ^{7}Li.

2.17 The tracks of energetic particles such as α and β rays through gases can be followed in a device called the Wilson cloud chamber. The tracks appear as streaks. The tracks of a few α particles through air are shown in Figure 2.25. Explain why the paths of the α particles are mostly long and account for the sudden change in direction of one particle, as shown.

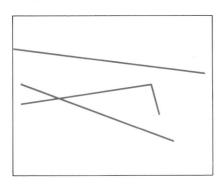

FIGURE 2.25

2.18 In a physics laboratory a student carried out the Millikan oil-drop experiment, using several oil droplets for her measurements, and calculated the charges on the drops. She obtained the following data:

Droplet	Calculated charge (coul)
A	1.60×10^{-19}
B	3.15×10^{-19}
C	4.81×10^{-19}
D	6.31×10^{-19}

What is the significance of the fact that the droplets carried different charges? What conclusion can the student draw from these data regarding the charge of the electron? What value (and to how many significant figures) should she report for the electronic charge?

2.19 Assuming that the mass of an atom is simply the sum of the masses of all the particles in the nucleus and the surrounding electrons (this isn't exactly true, as we'll see later, but it is close), calculate what fractions of the mass of a ^{7}Li atom are due to protons, neutrons, and electrons.

2.20 The diameter of the iridium (Ir) atom is estimated to be 2.7 Å. Express this in picometers (pm); in nanometers (nm). What number of Ir atoms end-to-end would form a line 1 mm long?

2.21 Astronomers have made certain observations on the galaxy M87 that appear to be consistent with the presence of a black hole (a collapsed star in which the gravitational forces are so strong that no matter or radiation can escape from it). They estimate the density of this object to be equal to the density our sun would have if it were compressed to a sphere of 1.5-km diameter. The mass of our sun is estimated to be 2×10^{30} kg. What density would the sun have if it were so compressed? Compare this with the density of ordinary liquid hydrogen, 0.07 g/cm^3.

2.22 What restrictions on the numbers of protons, neutrons, and electrons apply to each of the following: (a) atoms of a particular isotope of an element; (b) different isotopes of the same element; (c) nuclides with the same mass number?

2.23 Indicate the number of protons, neutrons, and electrons present in each of the following atoms: (a) ^{13}C; (b) ^{55}Mn; (c) ^{97}Mo.

2.24 Determine the number of protons, neutrons, and electrons present in each of the following nuclei: (a) ^{6}Li; (b) ^{57}Fe; (c) ^{27}Al; (d) ^{19}F.

2.25 Write the correct symbol, with both superscript and subscript, for each of the following (use the list of elements on the front inside cover): (a) the nuclide of potassium with mass 39; (b) the nuclide of chlorine with mass 35; (c) the nuclide of silicon with mass 29; (d) two isotopes of sulfur, one with 16 neutrons, the other with 18 neutrons.

The periodic table

2.26 Using the table on the front inside cover as a guide, write the symbol for each of the following elements;

locate the element on the periodic table and indicate whether it is a metal, metalloid, or nonmetal: (a) manganese; (b) selenium; (c) bromine; (d) zinc; (e) chromium; (f) germanium; (g) sulfur.

2.27 Locate each of the following elements on the periodic table. Indicate whether it is a metal, metalloid, or nonmetal; give the name of the element: (a) H; (b) K; (c) N; (d) Sb; (e) F; (f) Ba; (g) Cd; (h) Cs; (i) Se.

2.28 Which pair in each of the following groups of elements would you expect to be most similar in chemical and physical properties? Explain. (a) Ca, P, Si, I, Sr, Ag, Ni; (b) Nb, Mo, As, Br, Sb, N.

2.29 Suppose you have a glass flask containing equal numbers of hydrogen and chlorine molecules. How do the contents of this flask differ from those of a flask containing an equal total weight of hydrogen chloride molecules?

2.30 Write the empirical formulas corresponding to each of the following molecular formulas: (a) P_4O_{10}; (b) C_4H_8; (c) C_6H_{14}; (d) SiO_2; (e) $Na_2B_4O_7$; (f) $C_4O_4H_2$.

2.31 Each of the following elements is capable of forming an ion in chemical reactions. Refer to the periodic table and predict the charge found on the most stable ionic form: (a) K; (b) F; (c) Sc; (d) Ca; (e) O; (f) At.

2.32 Predict the formula for the ionic compound formed between each of the following pairs of elements: (a) Li, O; (b) Sr, S; (c) Cl, Mg; (d) Sc, F; (e) Cs, Br.

2.33 Identify each of the following: (a) an element of group 2A that has 20 electrons in the neutral atom; (b) an inert element with 36 protons in the nucleus; (c) an element with atomic weight between 70 and 90 that tends to form ions of 1− charge; (d) a metallic element that reacts with oxygen to form a compound of formula MO. The 2+ ion, M^{2+}, has 28 electrons.

2.34 From the group of elements F, Kr, Ca, Cs, W, Sb, and S, select the one that: (a) should form an ion with a 2− negative charge; (b) is least reactive chemically; (c) should form an ion of 1+ charge; (d) is a metalloid.

Naming inorganic compounds

2.35 Write the empirical formula for each of the following substances: (a) acetic acid, $H_4C_2O_2$; (b) benzene, C_6H_6; (c) dichloroethylene, $C_2H_2Cl_2$; (d) nicotine, $C_{10}H_{14}N_2$; (e) tetraphosphorus decasulfide, P_4S_{10}; (f) mercurous chloride, Hg_2Cl_2. Which of these compounds are organic, which are inorganic?

2.36 Name the following ions: (a) Cr^{2+}; (b) Sr^{2+}; (c) Br^-; (d) S^{2-}; (e) Cs^+; (f) Ti^{3+}.

2.37 Name the following compounds: (a) $NaClO_3$; (b) K_3PO_4; (c) $TiBr_3$; (d) $Ba(HSO_3)_2$; (e) $Zn(CN)_2$.

2.38 Write the formulas for each of the following compound: (a) calcium chloride; (b) potassium cyanide; (c) sodium bicarbonate; (d) aluminum bisulfate; (e) copper(I) bromide; (f) iron(III) chloride; (g) cupric carbonate; (h) ammonium sulfate; (i) nickel(IV) oxide; (j) potassium permanganate; (k) barium peroxide; (l) zinc phosphate.

2.39 Write the chemical formula for each substance mentioned in the following word descriptions (use the front inside cover to find the symbols for the elements you don't know). (a) Vanadium(III) bromide is a colored solid. (b) Zinc bicarbonate can be heated to form zinc oxide, carbon dioxide, and water. (c) On treatment with hydrofluoric acid, silicon dioxide forms silicon tetrafluoride; water is also formed. (d) Sulfur dioxide reacts with water to form sulfurous acid; this acid can be oxidized to sulfuric acid. (e) The substance hydrogen phosphide is commonly called phosphine. (f) Perchloric acid reacts with cadmium to form cadmium(II) perchlorate.

2.40 Write the correct formula for each of the following oxyacids or salt derived from an oxyacid: (a) sodium nitrite; (b) calcium hypochlorite; (c) phosphorous acid; (d) potassium biselenate; (e) lithium chlorate; (f) sodium perbromate; (g) periodic acid; (h) lead(II) arsenite.

2.41 Assume that you encounter the following phrases in your reading. What is the chemical formula for each compound mentioned? (a) Potassium chlorate is used as a laboratory source of oxygen. (b) Sodium hypochlorite is used as a household bleach. (c) Ammonia is important in the synthesis of fertilizers such as ammonium nitrate. (d) Hydrofluoric acid is used to etch glass. (e) The smell of rotten eggs is due to hydrogen sulfide. (f) When hydrochloric acid is added to sodium bicarbonate (baking powder), carbon dioxide gas forms.

2.42 Give the chemical name for each of the following compounds: (a) $Sr(NO_3)_2$; (b) $SnBr_4$; (c) $AgNO_3$; (d) KCN; (e) $Al(OH)_3$; (f) $NaHSO_4$; (g) $(NH_4)_2SO_4$; (h) HIO_3; (i) CuBr.

2.43 Give the chemical name for each of the following compounds: (a) $Ca(HCO_3)_2$; (b) Na_2CO_3; (c) $CoCl_2$; (d) $Ba(OH)_2$; (e) Na_2O_2; (f) Na_2O; (g) K_2S; (h) HI; (i) H_2SeO_4.

Additional exercises

2.44 Distinguish between the members of the following pairs: (a) homogeneous and heterogeneous; (b) gas and liquid; (c) atomic number and mass number; (d) chemical and physical properties; (e) Ca and Ca^{2+}; (f) hydrochloric acid and chloric acid; (g) iron(II) and iron(III); (h) sodium carbonate and sodium bicarbonate; (i) H_2O and H_2O_2; (j) metal and nonmetal; (k) H and He; (l) chloride and chlorate.

2.45 Distinguish between the members of the following pairs: (a) alpha and beta particles; (b) proton and neutron; (c) ion and atom; (d) molecular formula and empirical formula; (e) chemical change and physical change; (f) solid and liquid; (g) O and O^{2-}; (h) C and Ca; (i) sodium sulfate and sodium sulfite; (j) sodium hydrogen phosphate and sodium dihydrogen phosphate; (k) cuprous chloride and cupric chloride; (l) O_2 and O_3.

2.46 Identify each of the following substances as gases, liquids, or solids under ordinary conditions: (a) mercury; (b) iron; (c) aluminum; (d) oxygen; (e) alcohol; (f) water; (g) hydrogen; (h) helium; (i) chlorine.

2.47 List as many properties as possible that would allow you to distinguish between a solid, a liquid, and a gas.

2.48 Identify each of the following elements if a neutral atom of each (a) contains 3 protons; (b) has an atomic number of 13; (c) contains 10 electrons.

2.49 Fill in the gaps in the following table:

Symbol	$^{12}_{6}C$	$^{17}_{8}O^{2-}$			
Protons	6		12		8
Neutrons	6		13	12	10
Electrons	6	10		10	10
Net charge	0	2−	0	1+	

2.50 Fill in the gaps in the following table:

Symbol	$^{37}Cl^-$	^{40}Ar	
Protons			21
Neutrons			23
Electrons			
Net charge			3+

2.51 Fill in the gaps in the following table:

Symbol	$^{9}_{4}Be^{2+}$	
Atomic number		35
Mass number		80
Net charge		0

2.52 (a) Explain why ^{18}O and ^{16}O have essentially identical chemical properties. (b) Explain why the symbols for an alpha particle and a beta particle could be given as $^{4}_{2}\alpha$ and $^{0}_{-1}\beta$.

2.53 The formula of the compound formed between cadmium and oxygen is CdO. Predict the formula of the compound formed between (a) cadmium and chlorine; (b) cadmium and sulfur; (c) zinc and oxygen.

2.54 Describe the contributions to atomic theory made by the following persons: (a) Dalton; (b) Rutherford; (c) Thomson; (d) Millikan; (e) Becquerel.

2.55 List as many properties as possible that would allow you to distinguish between (a) silver and aluminum; (b) water and alcohol.

2.56 Which of the following are elements and which are compounds: (a) I_2; (b) CO_2; (c) S_8; (d) NH_3; (e) H_2O_2; (f) O_3?

2.57 The formula of the compound formed by aluminum and oxygen is Al_2O_3. Predict the formula of the compound formed between: (a) gallium and oxygen; (b) aluminum and sulfur; (c) gallium and sulfur.

2.58 Describe what is meant by the term "family" in speaking of the elements and the periodic table.

2.59 The elements K, Ca, and Sc are arranged in a horizontal sequence in the periodic table. The elements Mg, Ca, and Sr are arranged in a vertical sequence. We refer to the latter trio as a "family" of elements. By contrast, K, Ca, and Sc are not members of a common family. What is the meaning of the term "family" as used here? Why are Mg, Ca, and Sr members of the same family, whereas K, Ca, and Sc are not?

3

Conservation of mass; stoichiometry

Antoine Lavoisier (Section 1.1) was among the first to draw his conclusions about chemical processes from careful, quantitative observations. His work laid the basis for the law of conservation of mass, one of the most fundamental laws of chemistry. In this chapter, we will consider many practical problems based on the law of conservation of mass. These problems involve the quantitative relationships between substances undergoing chemical changes. The study of these quantitative relationships is known as stoichiometry (pronounced stoy-key-AHM-uh-tree), a word derived from the Greek words *stoicheion* ("element") and *metron* ("measure").

3.1 Law of conservation of mass

Studies of countless chemical reactions have shown that the total mass of all substances present after a chemical reaction is the same as the total mass before the reaction. This observation is embodied in the law of conservation of mass: There are no *detectable* changes in mass in any chemical reaction.* More precisely, *atoms are neither created nor destroyed during a chemical reaction;* instead, they merely exchange partners or become otherwise rearranged. The simplicity with which this law can be stated should not mask its significance. As with many other scientific laws, this law has implications far beyond the walls of the scientific laboratory.

*In Chapter 20, we will discuss the relationship between mass and energy summarized by the equation $E = mc^2$ (E is energy, m is mass, and c is the speed of light). We will find that whenever an object loses energy it loses mass, and whenever it gains energy it gains mass. These changes in mass are too small to detect in chemical reactions. However, for nuclear reactions, such as those involved in a nuclear reactor or in a hydrogen bomb, the energy changes are enormously larger; in these reactions there are detectable changes in mass.

FIGURE 3.1 Untreated domestic sewage entering streams and lakes is symptomatic of the way society ignores the law of conservation of mass. Here, raw sewage is shown entering the Mississippi River. (*USDA-SCS photo by C. H. Aubol*)

The law of conservation of mass reminds us that we really can't throw anything away. If we discharge wastes into a lake to get rid of them, they are diluted and seem to disappear. However, they are part of the environment. They may undergo chemical changes or remain inactive; they may reappear as toxic contaminants in fish or in water supplies or lie on the bottom unnoticed. Whatever their fates, the total mass of wastes is unchanged (see Figure 3.1).

The law of conservation of mass suggests that we are converters, not consumers. In drawing upon nature's storehouse of iron ore to build the myriad iron-containing objects used in modern society we are not reducing the number of iron atoms on the planet. We may, however, be converting the iron to less useful, less available forms from which it will not be practical to recover it later. For example, consider the millions of old washing machines that lie buried in dumps. Of course, if we expend enough energy, we can bring off almost any chemical conversions we choose. We have learned in recent years, however, that energy itself is a limited resource. Whether we like it or not, we must learn to conserve all our energy and material resources.

3.2 Chemical equations

We have seen (in Sections 2.4 and 2.8) that chemical substances can be represented by symbols and formulas. These chemical symbols and formulas can be combined to form a kind of statement, called a chemical equation, that represents or describes a chemical reaction. For example, the combustion of carbon in coal involves a reaction with oxygen (O_2) in the air to form gaseous carbon dioxide (CO_2). This reaction is represented as:

$$C + O_2 \longrightarrow CO_2 \qquad\qquad [3.1]$$

We read the + sign to mean "reacts with" and the arrow as "produces." Carbon and oxygen are referred to as reactants and carbon dioxide as the product of the reaction.

It is important to keep in mind that a chemical equation is a description of a chemical process. Before you can write a complete equation you must know what happens in the reaction or be prepared to predict the products. In this sense, a chemical equation has qualitative significance; it identifies the reactants and products in a chemical process. In addition, a chemical equation is a quantitative statement; it must be consistent with the law of conservation of mass. This means that the equation must contain equal numbers of each type of atom on each side of the equation. When this condition is met the equation is said to be *balanced*. For example, Equation [3.1] is balanced, because there are equal numbers of carbon and oxygen atoms on each side.

A slightly more complicated situation is encountered when methane (CH_4), the principal component of natural gas, burns and produces carbon dioxide (CO_2) and water (H_2O). The combustion is "supported by" oxygen (O_2), meaning that oxygen is involved as a reactant. The unbalanced equation is

$$CH_4 + O_2 \longrightarrow CO_2 + H_2O \qquad\qquad [3.2]$$

The reactants are shown to the left of the arrow, the products to the right. Notice that the reactants and products both contain one carbon atom. However, the reactants contain more hydrogen atoms (four) than the products (two). If we place a coefficient 2 in front of H_2O, indicating formation of two molecules of water, there will be four hydrogens on each side of the equation:

$$CH_4 + O_2 \longrightarrow CO_2 + 2H_2O \qquad\qquad [3.3]$$

Before we continue to balance this equation, let's make sure that we clearly understand the distinction between a coefficient in front of a formula and a subscript in a formula. Refer to Figure 3.2. Notice that

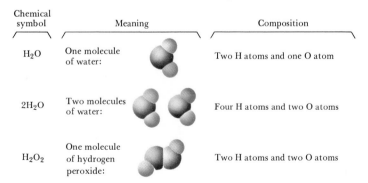

Chemical symbol	Meaning		Composition
H_2O	One molecule of water:		Two H atoms and one O atom
$2H_2O$	Two molecules of water:		Four H atoms and two O atoms
H_2O_2	One molecule of hydrogen peroxide:		Two H atoms and two O atoms

FIGURE 3.2 Illustration of the difference in meaning between a subscript in a chemical formula and a coefficient in front of the formula.

changing a subscript in a formula, such as from H_2O to H_2O_2, changes the identity of the chemical involved. The substance H_2O_2, hydrogen peroxide, is quite different from water. *The subscripts in the chemical formulas should never be changed in balancing an equation.* On the other hand, placing a coefficient in front of a formula merely changes the amount and not the identity of the substance; $2H_2O$ means two molecules of water, $3H_2O$ means three molecules of water, and so forth. Now let's continue balancing Equation [3.3]. There are equal numbers of carbon and hydrogen atoms on both sides of this equation; however, there are more oxygen atoms among the products (four) than among the reactants (two). If we place a coefficient 2 in front of O_2 there will be equal numbers of oxygen atoms on both sides of the equation:

$$CH_4 + 2O_2 \longrightarrow CO_2 + 2H_2O \qquad [3.4]$$

The equation is now balanced. There are four oxygen atoms, four hydrogen atoms, and one carbon atom on each side of the equation. The balanced equation is shown schematically in Figure 3.3.

Now, let's look at a slightly more complicated example, analyzing stepwise what we are doing as we balance the equation. Combustion of octane (C_8H_{18}), a component of gasoline, produces CO_2 and H_2O. The balanced chemical equation for this reaction can be determined by using the following four steps.

First, the reactants and products are written in the unbalanced equation:

$$C_8H_{18} + O_2 \longrightarrow CO_2 + H_2O \qquad [3.5]$$

Before a chemical equation can be written the identities of the reactants and products must be determined. In the present example this information was given to us in the verbal description of the reaction.

Second, the number of atoms of each type on each side of the equation is determined. In the reaction above there are 8C, 18H, and 2O among the reactants, and 1C, 2H, and 3O among the products; clearly, the equation is not balanced, because the number of atoms of each type differs from one side of the equation to the other.

Third, to balance the equation, coefficients are placed in front of the chemical formulas to indicate different quantities of reactants and prod-

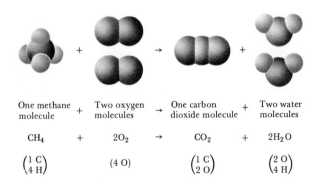

One methane molecule	+	Two oxygen molecules	→	One carbon dioxide molecule	+	Two water molecules
CH_4	+	$2O_2$	→	CO_2	+	$2H_2O$
$\begin{pmatrix}1\text{ C}\\4\text{ H}\end{pmatrix}$		(4 O)		$\begin{pmatrix}1\text{ C}\\2\text{ O}\end{pmatrix}$		$\begin{pmatrix}2\text{ O}\\4\text{ H}\end{pmatrix}$

FIGURE 3.3 The balanced chemical equation for the combustion of CH_4. The drawings of the molecules involved call attention to the conservation of atoms through the reaction.

ucts, so that the same number of atoms of each type appears on both sides of the equation. To decide what coefficients to try first, it is often convenient to focus attention on the molecule with the most atoms, in this case C_8H_{18}. This molecule contains 8C, all of which must end up in CO_2 molecules. Therefore, we place a coefficient 8 in front of CO_2. Similarly, the 18H end up as $9H_2O$. At this stage the equation reads

$$C_8H_{18} + O_2 \longrightarrow 8CO_2 + 9H_2O \qquad [3.6]$$

Although the C and H atoms are now balanced, the O atoms are not; there are 25O atoms among the products but only 2 among the reactants. It takes $12.5O_2$ to produce 25O atoms among the reactants:

$$C_8H_{18} + 12.5O_2 \longrightarrow 8CO_2 + 9H_2O \qquad [3.7]$$

However, this equation is not in its most conventional form, because it indicates half an O_2 molecule among the reactants, and half an O_2 molecule does not exist as such. Therefore, we must go on to the next step.

Fourth, for most purposes a balanced equation should contain the smallest possible whole-number coefficients. Therefore, we multiply each side of the equation above by two, removing the fraction and achieving the following balanced equation:

$$2C_8H_{18} + 25O_2 \longrightarrow 16CO_2 + 18H_2O \qquad [3.8]$$

16C, 36H, 50O	16C, 36H, 50O
Reactants	Products

The atoms are inventoried below the equation to show graphically that the equation is indeed balanced. You might note that although atoms are conserved, molecules are not—the reactants contain 27 molecules while the products contain 34. All in all, this approach to balancing equations is largely trial and error. It is much easier to verify that an equation is balanced than actually to balance one, so practice in balancing equations is essential.

It should also be noted that the physical state of each chemical in a chemical equation is often indicated parenthetically using the symbols (g), (l), (s), and (aq) to indicate gas, liquid, solid, and aqueous (water) solution, respectively. Thus the balanced equation above can be written:

$$2C_8H_{18}(l) + 25O_2(g) \longrightarrow 16CO_2(g) + 18H_2O(l) \qquad [3.9]$$

Sometimes an upward arrow ($\uparrow$) is employed to indicate the escape of a gaseous product, whereas a downward arrow ($\downarrow$) indicates a precipitating solid (that is, a solid that separates from solution during the reaction). Often the conditions under which the reaction proceeds are indicated above the arrow between the two sides of the equation. For example, the temperature or pressure at which the reaction occurs could be so indicated. The symbol Δ is often placed above the arrow to indicate the addition of heat.

SAMPLE EXERCISE 3.1

Balance the following equation:

$$Na(s) + H_2O(l) \longrightarrow NaOH(aq) + H_2(g)$$

Solution: A quick inventory of atoms reveals that there are equal numbers of Na and O atoms on both sides of the equation, but that there are two H atoms among reactants and three H atoms among products. To increase the number of H atoms among reactants, we might place a coefficient 2 in front of H_2O:

$$Na(s) + 2H_2O(l) \longrightarrow NaOH(aq) + H_2(g)$$

Now we have four H atoms among reactants but only three H atoms among the products. The H atoms can be balanced with a coefficient 2 in front of NaOH:

$$Na(s) + 2H_2O(l) \longrightarrow 2NaOH(aq) + H_2(g)$$

If we again inventory the atoms on each side of the equation, we find that the H atoms and O atoms are balanced but not the Na atoms. However, a coefficient 2 in front of Na gives two Na atoms on each side of the equation:

$$2Na(s) + 2H_2O(l) \longrightarrow 2NaOH(aq) + H_2(g)$$

If the atoms are inventoried once more we find two Na atoms, four H atoms, and two O atoms on each side of the equation. The equation is therefore balanced.

3.3 Chemical reactions

Our discussion in Section 3.2 focused on how to balance chemical equations given the reactants and products for the reactions. You were not asked to predict the products for a reaction. Students sometimes ask how the products are determined. For example, how do we know that sodium metal (Na) reacts with water (H_2O) to form H_2 and NaOH as shown in Sample Exercise 3.1? These products are identified by experiment. As the reaction proceeds, there is a fizzing or bubbling where the sodium is in contact with the water (if too much sodium is used the reaction is quite violent, and so small quantities would be used in our experiment). If the gas is captured, it can be identified as H_2 from its chemical and physical properties. After the reaction is complete, a clear solution remains. If this is evaporated to dryness, a white solid will remain. From its properties this solid can be identified as NaOH. However, it is not necessary to perform an experiment every time we wish to write a reaction. We can predict what will happen if we have seen the reaction or a similar one before. So far we have seen too little chemistry to predict the products for many reactions. Nevertheless, even now you should be able to make some predictions. For example, what would you expect to happen when potassium metal is added to water? We have just discussed the reaction of sodium metal with water, for which the balanced chemical equation is

$$2Na(s) + 2H_2O(l) \longrightarrow 2NaOH(aq) + H_2(g) \qquad [3.10]$$

Because sodium and potassium are in the same family of the periodic table (the alkali metal family, family 1A), we would expect them to behave similarly, producing the same types of products. Indeed, this prediction is correct, and the reaction of potassium metal with water is

$$2K(s) + 2H_2O(l) \longrightarrow 2KOH(aq) + H_2(g) \qquad [3.11]$$

You can readily see that it will be helpful in your study of chemistry if you are able to classify chemical reactions into certain types. We have just considered two examples of a type we might call reaction of an active metal with water. Let's briefly consider here a few of the more important and common types you will be encountering in your laboratory work and in the chapters ahead.

COMBUSTION IN OXYGEN

We have already encountered three examples of combustion reactions: the combustion of carbon, Equation [3.1]; of methane, Equation [3.4]; and of octane (C_8H_{18}), Equation [3.8]. Remember that combustion is a rapid reaction that usually produces a flame. Most of the combustions we observe involve O_2 as a reactant. From the examples we have already seen it should be easy to predict the products of the combustion of propane C_3H_8. We expect that combustion of this compound would lead to carbon dioxide and water as products, by analogy with our previous examples. That expectation is correct; propane is the major ingredient in LP (liquid propane) gas, used for cooking and home heating. It burns in air as described by the balanced equation

$$C_3H_8(g) + 5O_2(g) \longrightarrow 3CO_2(g) + 4H_2O(l) \qquad [3.12]$$

If we looked at further examples, we would find that combustion of compounds containing oxygen atoms as well as carbon and hydrogen (for example, CH_3OH) also produces CO_2 and H_2O.

ACIDS, BASES, AND NEUTRALIZATION

Table 3.1 lists several acids and bases and the amount of each compound produced in the United States each year. You can see that these substances are produced in enormous quantities; they are among the most important chemicals in industry and in the chemical laboratory.

You will nearly always encounter acids and bases in the form of solutions of these substances in water. Acids are substances whose water solutions contain an excess of H^+ ions. Although we often represent hydrochloric acid as $HCl(aq)$, it exists in water as $H^+(aq)$ and $Cl^-(aq)$ ions. Thus the process of dissolving hydrogen chloride gas in water to form hydrochloric acid can be represented as follows:

$$HCl(g) \xrightarrow{\text{H}_2\text{O}} HCl(aq)$$

or

$$HCl(g) \xrightarrow{\text{H}_2\text{O}} H^+(aq) + Cl^-(aq)$$

The H_2O given above the arrows in these equations is to remind us that the reaction medium is water. Pure sulfuric acid is a liquid; when it dissolves in water it releases H^+ ions in two successive steps:

$$H_2SO_4(l) \xrightarrow{\text{H}_2\text{O}} H^+(aq) + HSO_4^-(aq)$$

$$HSO_4^-(aq) \xrightarrow{\text{H}_2\text{O}} H^+(aq) + SO_4^{2-}(aq)$$

TABLE 3.1 U.S. production of some acids and bases, 1977

Compound	Formula	Annual production (kg)
Acids:		
Sulfuric	H_2SO_4	3.1×10^{10}
Phosphoric	H_3PO_4	7.1×10^9
Nitric	HNO_3	6.7×10^9
Hydrochloric	HCl	2.3×10^9
Bases:		
Sodium hydroxide	$NaOH$	9.5×10^9
Ammonia	NH_3	1.5×10^{10}

Thus, although we frequently represent aqueous solutions of sulfuric acid as $H_2SO_4(aq)$, these solutions actually contain a mixture of $H^+(aq)$, $HSO_4^-(aq)$, and $SO_4^{2-}(aq)$.

Bases are compounds that produce an excess of hydroxide ions, OH^-, in water. A base such as sodium hydroxide does this because it is an ionic substance composed of Na^+ and OH^- ions. When NaOH dissolves in water, the cations and anions simply separate in the solution:

$$NaOH(s) \xrightarrow{H_2O} Na^+(aq) + OH^-(aq)$$

Thus, although aqueous solutions of sodium hydroxide might be written as $NaOH(aq)$, it exists as $Na^+(aq)$ and $OH^-(aq)$ ions. Many other bases such as $Ca(OH)_2$ are also ionic hydroxide compounds. However, NH_3 is a base although it is not a compound of this sort.

You may have noticed in Table 3.1 that ammonia, NH_3, is produced in very large quantities. In fact, after sulfuric acid and lime (CaO), it is the third largest in production of U.S. chemicals. This important substance is used in fertilizers and as a primary chemical in other chemical processes. It may seem odd at first glance that ammonia is a base, because it contains no hydroxide ions. However, we must remember that the definition of a base is that it *produces* an excess of hydroxide ions in water. Ammonia does this by a reaction with water. We can represent the dissolving of ammonia gas in water as follows:

$$NH_3(g) + H_2O(l) \longrightarrow NH_4^+(aq) + OH^-(aq) \qquad [3.17]$$

Solutions of ammonia in water are often labeled ammonium hydroxide, NH_4OH, to remind us that ammonia solutions are basic. (Ammonia is referred to as a weak base, which means that not all the NH_3 that dissolves in water goes on to form NH_4^+ and OH^- ions; but that is a matter for Chapter 15, and it need not concern us here.)

Solutions of acids and bases have very different properties. Acids have a sour taste, whereas bases have a bitter taste.* Acids can change the

---

*Tasting chemical solutions is, of course, not a good practice. However, we have all had acids such as ascorbic acid (vitamin C), acetylsalicylic acid (aspirin), and citric acid (in citrus fruits) in our mouths, and we are familiar with the characteristic sour taste. It differs from the taste of soaps, which are mostly basic.

colors of certain dyes in a specific way that differs from the effect of a base. For example, the dye known as litmus is changed from blue to red by an acid, and from red to blue by a base. In addition, acidic and basic solutions differ in chemical properties in several important ways. When a solution of an acid is mixed with a solution of a base, a neutralization reaction occurs. The products of the reaction have none of the characteristic properties of either the acid or base. For example, when a solution of hydrochloric acid is mixed with precisely the correct quantity of a sodium hydroxide solution, the result is a solution of sodium chloride, a simple ionic compound possessing neither acidic nor basic properties. (In general, such ionic products are referred to as salts.) The neutralization reaction can be written as follows:

$$HCl(aq) + NaOH(aq) \longrightarrow H_2O(l) + NaCl(aq) \qquad [3.18]$$

When we write the reaction as we have here, it is important to keep in mind that the substances shown as (aq) are present in the form of the separated ions, as discussed above. Notice that the acid and base in Equation [3.18] have combined to form water as a product. The general description of an acid-base neutralization reaction in aqueous solution, then, is that an acid and base react to form a salt and water. Using this general description we can predict the products formed in any acid-base neutralization reaction. Sample Exercise 3.2 illustrates how we might go about predicting the products and writing a balanced equation for a neutralization reaction.

SAMPLE EXERCISE 3.2

Write a balanced equation for the reaction of hydrobromic acid, HBr, with barium hydroxide, Ba(OH)$_2$.

Solution: The products of any acid-base reaction are a salt and water. The salt is that formed from the cation of the base, Ba(OH)$_2$, and the anion of the acid, HBr. The charge on the barium ion is 2+ (see Table 2.5), and that on the bromide ion is 1−. Therefore, to maintain electrical neutrality, the formula for the salt must be BaBr$_2$. The unbalanced equation for the neutralization reaction is therefore

$$HBr(aq) + Ba(OH)_2(aq) \longrightarrow H_2O(l) + BaBr_2(aq)$$

To balance the equation we must provide two molecules of HBr to furnish the two Br$^-$ ions and to supply the two H$^+$ ions needed to combine with the two OH$^-$ ions of the base. The balanced equation is thus

$$2HBr(aq) + Ba(OH)_2(aq) \longrightarrow$$
$$2H_2O(l) + BaBr_2(aq)$$

PRECIPITATION REACTIONS

One very important class of reactions occurring in solution is the precipitation reaction, in which one of the reaction products is insoluble. We will concern ourselves in this brief introduction with reactions between acids, bases, or salts in aqueous solution. As a simple example, consider the reaction between hydrochloric acid solution and a solution of the salt silver nitrate, AgNO$_3$. When the two solutions are mixed, a finely divided white solid forms. Upon analysis this solid proves to be silver chloride, AgCl, a salt that has a very low solubility* in water. The reaction as just described can be represented by the equation

$$HCl(aq) + AgNO_3(aq) \longrightarrow AgCl(s) + HNO_3(aq) \qquad [3.19]$$

The formation of a precipitate in a chemical equation may be represented by a following (s), by a downward arrow following the formula for the solid, or by underlining the formula for the solid. You are reminded once again that substances indicated by (aq) may be present in the solution as separated ions.

The chemical equations that represent a precipitation reaction must be balanced to take account of the charges on the ions involved. The commonly encountered ions are listed in Table 2.5. The following examples and Sample Exercise 3.3 should make it clear how such equations are balanced.

$$Pb(NO_3)_2(aq) + Na_2CrO_4(aq) \longrightarrow PbCrO_4(s) + 2NaNO_3(aq) \qquad [3.20]$$

$$CaCl_2(aq) + 2NaOH(aq) \longrightarrow Ca(OH)_2(s) + 2NaCl(aq) \qquad [3.21]$$

SAMPLE EXERCISE 3.3

When iron(III) bromide in aqueous solution is mixed with an aqueous solution of sodium sulfide, a precipitate of iron(III) sulfide forms. Write a balanced equation to describe the reaction.

Solution: Our first task is to determine the formulas for the reactants and products. The reactants are $FeBr_3$ and Na_2S. It will require three sulfide ions, S^{2-}, to balance the charge of two Fe^{3+} ions. Thus the formula for the solid product must be Fe_2S_3. The remaining products are the Na^+ and Br^- ions that remain in solution. Together these form NaBr(aq).

The *unbalanced* equation for the precipitation reaction is thus

$$FeBr_3(aq) + Na_2S(aq) \longrightarrow Fe_2S_3(s) + NaBr(aq)$$

We need to have two $FeBr_3$ units and three Na_2S units on the left to furnish the required number of ions on the right. This in turn requires that we have six NaBr units on the right to achieve a mass balance. The balanced equation is

$$2FeBr_3(aq) + 3Na_2S(aq) \longrightarrow Fe_2S_3(s) + 6NaBr(aq)$$

A balanced equation implies a quantitative relation between the reactants and the products involved in a chemical reaction. Thus complete combustion of a molecule of C_3H_8 requires exactly five molecules of O_2, no more and no less, as shown in Equation [3.12]. Although it is not possible to count directly the number of molecules of each type in any reaction, this count can be made indirectly if the mass of each molecule is known. Indeed, this indirect approach is the one taken to obtain quantitative information about the amounts of substances involved in any chemical transformation. Therefore, before we can pursue the quantitative aspects of chemical reactions further, we must explore the concept of atomic and molecular weights.

3.4 Atomic and molecular weights

Dalton's atomic theory led him and other scientists of his time to a new problem. If it is true that atoms combine with one another in the ratios of

---

*Solubility will be considered in some detail in Chapter 12. It is a measure of the amount of substance that can be dissolved in a given quantity of solvent (see Section 3.9).

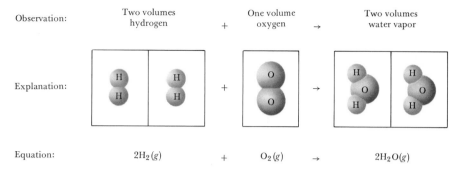

Observation:	Two volumes hydrogen	+	One volume oxygen	→	Two volumes water vapor
Explanation:		+		→	
Equation:	$2H_2(g)$	+	$O_2(g)$	→	$2H_2O(g)$

FIGURE 3.4 Gay-Lussac's experimental observation of combining volumes shown together with Avogadro's explanation of this phenomenon.

small whole numbers to form compounds, what *are* the ratios with which they combine? Atoms are too small to be measured individually by any means available in the early nineteenth century. However, if one knew the *relative* masses of the atoms, then by measuring out convenient quantities in the laboratory, one could determine the relative numbers of atoms in a sample. Consider a simple analogy: Suppose that oranges are on the average four times heavier than plums; the number of oranges in 48 kg of oranges will then be the same as the number of plums in 12 kg of plums. Similarly, if you knew that oxygen atoms were on the average 16 times more massive than hydrogen atoms, then you could weigh out 16 g of oxygen gas to obtain the same number of oxygen atoms as there are hydrogen atoms in 1 g of hydrogen gas. Thus, the problem of determining combining ratios becomes one of determining the relative masses of the atoms of the elements.

This is all very well, but there was great difficulty in getting started. Since atoms and molecules can't be seen, there was no simple way to be sure about the relative numbers of atoms in *any* compound. Dalton thought that the formula for water was HO. However, the French scientist Gay-Lussac showed in a brilliant set of measurements that it required two volumes of hydrogen gas to react with one volume of oxygen to form two volumes of water vapor. This observation was inconsistent with Dalton's formula for water. Furthermore, if oxygen were assumed to be a monatomic gas, as Dalton did, one could obtain two volumes of water vapor only by splitting the oxygen atoms in half, which of course violates the concept of the atom as indivisible in chemical reactions.

The Italian physicist Amedeo Avogadro suggested in 1811 that Gay-Lussac's results could be explained if it were assumed that both hydrogen and oxygen exist in the gas phase as diatomic molecules H_2 and O_2. Avogadro also proposed that equal volumes of gases, if measured at the same temperature and pressure, contain equal numbers of molecules. (When we talk about gases in Chapter 9 we will see the basis for Avogadro's hypothesis, which is entirely correct.) The way in which Avogadro's hypothesis explains Gay-Lussac's observations is illustrated in Figure 3.4. The key here is in assuming that the reacting gases are diatomic. Thus, it is possible to obtain two volumes of water vapor from one volume of oxygen gas. Because the combining volume of hydrogen is twice that of oxygen, the formula for water must be H_2O, not HO as Dalton had suggested.

How many liters of NH_3 will be produced when 2 L of N_2 combine with 6 L of H_2 (all volumes measured at the same temperature and pressure)?

Solution: The balanced equation is

$$N_2(g) + 3H_2(g) \longrightarrow 2NH_3(g)$$

From the coefficients it can be deduced that each liter of N_2 combines with 3 L of H_2 to produce 2 L of NH_3. Thus 4 L of NH_3 will form from 2 L of N_2 and 6 L of H_2.

It required several years for Avogadro's ideas to gain acceptance, but by about 1860 matters had pretty well straightened themselves out. In the meantime chemists such as Berzelius (Figure 3.5) had been making painstaking measurements of the masses of the elements that combined with one another and had established the atomic weights* of many elements with quite good precision.

Because different types of atoms contain different numbers of subatomic particles, they differ in mass. Atomic weights were originally assigned on a relative scale with hydrogen given a value of 1. Modern atomic weights are based on the assignment of an atomic mass of exactly 12 for the ^{12}C isotope of carbon (Section 2.6). We refer to the units of these atomic weights as atomic mass units (amu). Thus, ^{12}C has an atomic weight of exactly 12 amu. The atomic weights of elements are reported as average values, reflecting the relative abundances of each isotope of each element. Naturally occurring chlorine is 75.53 percent ^{35}Cl, which has a mass of 34.969 amu, and 24.47 percent ^{37}Cl, which has a mass of 36.966 amu. The average atomic weight for chlorine can be calculated to four significant figures from this information:

*The term "atomic weight" is a bit of a misnomer; what is really meant is atomic mass. However, the term atomic weight has become so commonly used that there is little point in attempting to change. We will use the expressions atomic or molecular weight but will refer to individual atoms or molecules in terms of mass.

FIGURE 3.5 Jons Jakob Berzelius (1779–1848), Swedish chemist. Berzelius is one of the truly great figures in nineteenth-century science. He developed the modern system of symbols and formulas in chemistry and discovered several elements. Berzelius was an exceptionally gifted experimentalist. His determinations of the atomic weights of several elements were the most accurate known for many years. (*Bettman Archive*)

$$\text{Av AW}^* = (75.53\%)(34.969 \text{ amu}) + (24.47\%)(36.966 \text{ amu})$$
$$= 26.41 \text{ amu} + 9.05 \text{ amu}$$
$$= 35.46 \text{ amu}$$

The last digit in this calculation is uncertain; the accepted value for the atomic weight of chlorine, to five significant figures, is 35.453 amu. Two methods of determining atomic weights are considered in the next two sections of this chapter. The atomic weights of the elements are listed below the symbol for the element on the periodic table found on the front inside cover of this text. They are also listed in the table of elements on the front inside cover.

The formula weight of a substance is merely the sum of the atomic weights of each atom it contains. For example, H_2SO_4, sulfuric acid, has a formula weight of 98.0 amu:

$$\text{FW} = 2(\text{AW of H}) + \text{AW of S} + 4(\text{AW of O})$$
$$= 2(1.0 \text{ amu}) + 32.0 \text{ amu} + 4(16.0 \text{ amu})$$
$$= 98.0 \text{ amu}$$

Here we have rounded off the atomic weights so that our result has three significant figures. We will round off the atomic weights in this way for most problems.

If the chemical formula of a substance is its molecular formula, then the formula weight is called the molecular weight. For example, the molecular formula for glucose (the sugar transported by the blood to body tissues to satisfy energy requirements) is $C_6H_{12}O_6$. The molecular weight of glucose is therefore

$$6(12.0 \text{ amu}) + 12(1.0 \text{ amu}) + 6(16.0 \text{ amu}) = 180.0 \text{ amu}$$

With ionic substances such as NaCl, which exist as three-dimensional arrays of ions (Figure 2.24), it is really inappropriate to speak of molecules. Thus we cannot write molecular formulas and molecular weights for such substances. The formula weight of NaCl is

$$23.0 \text{ amu} + 35.5 \text{ amu} = 58.5 \text{ amu}.$$

*The abbreviation AW is used for atomic weight, MW for molecular weight, and FW for formula weight. The SI abbreviation for amu is simply u, as in 35.46 u. We will not be using this abbreviation in the text.

SAMPLE EXERCISE 3.5

(a) Calculate the formula weight of sucrose, $C_{12}H_{22}O_{11}$ (table sugar); (b) calculate the percentage of carbon in this compound.

Solution: (a) The formula weight of $C_{12}H_{22}O_{11}$ is calculated as follows:

$$\text{FW} = 12(12.0 \text{ amu}) + 22(1.0 \text{ amu}) + 11(16.0 \text{ amu})$$
$$= 342 \text{ amu}$$

(b) The total mass of carbon in $C_{12}H_{22}O_{11}$ is $12(12.0 \text{ amu}) = 144$ amu. To obtain the percentage of carbon we calculate the fraction of total mass that is carbon, then multiply by 100:

$$\% \text{ C} = \frac{\text{total mass of C in } C_{12}H_{22}O_{11}}{\text{mass of } C_{12}H_{22}O_{11}} \times 100$$

$$= \frac{12(12.0 \text{ amu})}{342 \text{ amu}} \times 100 = 42.1\%$$

The most accurate means presently available for determining atomic weights is provided by the mass spectrometer. This instrument, shown schematically in Figure 3.6, is similar to that used by Thomson to measure the e/m ratio of the electron (Section 2.6). In the mass spectrometer, a beam of gaseous atoms or molecules is allowed to flow into a vacuum chamber. This beam is intercepted at one point by a stream of energetic electrons that can cause ionization. The positive ions so formed are accelerated down the tube by a high voltage and pass through two slits, so that there is only a narrow beam of the ions. The system is pumped to a high vacuum to remove most of the gas molecules in the spectrometer. Consequently, the ions do not undergo a significant number of disturbing collisions with other particles in their passage down the tube. When the ions reach the region of the magnet, they are deflected by the magnetic field. The extent to which they are deflected depends on the charge-to-mass ratio, e/m, for the ions. Assuming that only singly charged ions are involved (that is, ions from which only one electron has been removed), the degree to which the ions have been deflected then depends on the mass of the ion. By continuously changing the strength of the magnetic field, or by changing the accelerating voltage, the ions of varying e/m can be caused to enter the slit at the end of the instrument and thereby reach the detector. Thus it is possible to scan over a wide range of e/m. The e/m scale is then easily converted into a mass scale, because the electrical charge of the ion (e) is always the same. A graph of the intensity of signal from the detector versus the mass of the ion is called a mass spectrum.

The mass spectrometer provided the first unambiguous evidence for the existence of isotopes. Suppose we allow a stream of mercury vapor to enter the mass spectrometer. If all the atoms of mercury were in fact identical, all of them upon ionization would possess the same e/m ratio and would therefore appear at the same place in the mass spectrum. But the mass spectrum actually observed for mercury is as shown in Figure 3.7. It is clear that there are several kinds of atoms in mercury, differing slightly in their masses.

Using the mass spectrometer, it is possible to measure relative values of the ratio e/m for ions with great accuracy. It is also possible to measure with high accuracy the relative numbers of the different isotopes of an

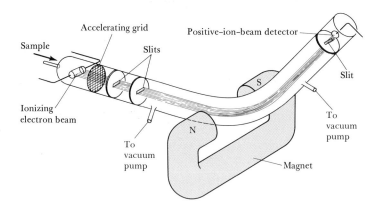

FIGURE 3.6 A schematic diagram of a modern mass spectrometer.

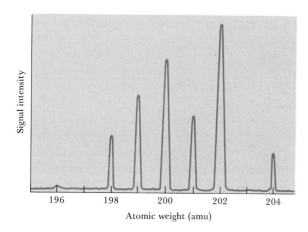

FIGURE 3.7 The mass spectrum of Hg⁺ ions in mercury vapor.

TABLE 3.2 Mass spectrum of mercury

Mass number	Atomic weight (amu)	Fractional abundance
196	195.965	0.0014
198	197.967	0.10039
199	198.967	0.1683
200	199.968	0.2312
201	200.970	0.1323
202	201.970	0.2979
204	203.973	0.0685

element. As we have previously noted, the atomic weight is the weighted average of the masses of all the isotopes of that element. For example, from the mass spectrum of mercury, Figure 3.7, we obtain the data shown in Table 3.2. The atomic weight of mercury, 200.59 amu, is the sum of the masses of the various isotopes, each multiplied by its fractional abundance:

$$AW = (0.0014)(195.965 \text{ amu}) + (0.10039)(197.967 \text{ amu}) + \cdots$$
$$+ (0.0685)(203.973 \text{ amu})$$
$$= 200.59 \text{ amu}$$

Ordinarily, samples of a particular element taken from different sources possess the same isotopic composition. However, for a few elements, for example, lead, that are involved in radioactive decay processes, isotopic composition can vary from sample to sample. Also, samples of matter from other solar systems, or even from another planet in our solar system, might differ in isotopic composition.

3.6 The chemical mole

We indicated earlier that the concept of atomic weights is important because it permits us to count atoms indirectly, by weighing samples. A convenient number of atoms, molecules, or formula units is that number that gives a mass in grams equal to the atomic weight, molecular weight, or formula weight. This quantity is called the mole, abbreviated mol.* Thus, 1 mol of ¹²C atoms is the number of ¹²C atoms in exactly 12 g of ¹²C. A mole of any element is defined as the quantity of that element, as it occurs in nature, that contains the same number of atoms as there are in exactly 12 g of ¹²C. It has been determined experimentally that the number of atoms in this quantity of ¹²C is 6.022×10^{23}. This number is given a special name, Avogadro's number, in honor of Amedeo Avogadro, who proposed that equal volumes of gases at the same temperature and pressure contain equal numbers of molecules (Section 3.4). For most purposes, we will use 6.02×10^{23} for Avogadro's number throughout the text; this number should be committed to memory.

*The term mole comes from the Latin word *moles* meaning "a mass." The term *molecule* is the diminutive form of this word and means "a small mass."

A mole of ions, molecules, or anything else contains Avogadro's number of these objects:

1 mol ^{12}C atoms = 6.02×10^{23} ^{12}C atoms

1 mol H_2O molecules = 6.02×10^{23} H_2O molecules

1 mol NO_3^- ions = 6.02×10^{23} NO_3^- ions

The concept of a mole as being 6.02×10^{23} of something is analogous to the concept of a dozen as 12 of something or a gross as 144 of something. Because a ^{24}Mg atom has twice the mass of a ^{12}C atom, 1 mol of ^{24}Mg has twice the mass of 1 mol of ^{12}C. A ^{12}C atom has a mass of exactly 12 amu, whereas a ^{24}Mg atom has a mass of 24.0 amu. Because 1 mol of ^{12}C atoms weighs 12 g (by definition), 1 mol of ^{24}Mg atoms must weigh 24.0 g. In fact, a mole of atoms of any element has a mass in grams numerically equal to the atomic weight of a single atom:

One ^{12}C atom has a mass of 12 amu; 1 mol ^{12}C weighs 12 g.

One ^{24}Mg atom has a mass of 24.0 amu; 1 mol ^{24}Mg weighs 24.0 g.

One Au atom has a mass of 197 amu; 1 mol Au weighs 197 g.

We can generalize this idea to include molecules and ions: The mass of a mole of formula units of any substance (that is, 6.02×10^{23} of them) is always equal to the formula weight expressed in grams:

One H_2O molecule has a mass of 18.0 amu; 1 mol H_2O weighs 18.0 g.

One NO_3^- ion has a mass of 62.0 amu; 1 mol NO_3^- weighs 62.0 g.

One NaCl unit has a mass of 58.5 amu; 1 mol NaCl weighs 58.5 g.

Further examples of mole relationships are shown in Table 3.3.

The first entries in Table 3.3, those for N and N_2, point out the importance of stating the chemical form of a substance exactly when we use the mole concept. Suppose you read that 1 mol of nitrogen is produced in a particular reaction. You might interpret this statement to mean 1 mol of nitrogen atoms (14.0 g). Unless otherwise stated, what was probably meant is 1 mol of nitrogen molecules, N_2 (28.0 g), because N_2 is the usual chemical form of the element. However, to avoid ambiguity it is always best to state explicitly the chemical form being discussed.

TABLE 3.3 Mole relationships

Name	Formula	Formula weight (amu)	Weight of 1 mol of formula units (g)	Number and kind of particles in 1 mol
Atomic nitrogen	N	14.0	14.0	6.02×10^{23} N atoms
Molecular nitrogen	N_2	28.0	28.0	6.02×10^{23} N_2 molecules $2(6.02 \times 10^{23})$ N atoms
Silver	Ag	108	108	6.02×10^{23} Ag atoms
Silver ions	Ag^+	108[a]	108	6.02×10^{23} Ag^+ ions
Barium chloride	$BaCl_2$	208	208	6.02×10^{23} $BaCl_2$ units 6.02×10^{23} Ba^{2+} ions $2(6.02 \times 10^{23})$ Cl^- ions

[a] Recall that the electron has negligible mass; thus ions and atoms have essentially the same mass.

SAMPLE EXERCISE 3.6

What is the weight of 1 mol of glucose, $C_6H_{12}O_6$?

Solution: By adding the weights of the atoms in glucose, we find it to have a formula weight of 180 amu:

$$6C \text{ atoms} = 6(12.0 \text{ amu}) = 72.0 \text{ amu}$$
$$12H \text{ atoms} = 12(1.0 \text{ amu}) = 12.0 \text{ amu}$$
$$6O \text{ atoms} = 6(16.0 \text{ amu}) = 96.0 \text{ amu}$$
$$180.0 \text{ amu}$$

Hence, 1 mol of $C_6H_{12}O_6$ weighs 180 g. Glucose is sometimes called dextrose; it is also known as blood sugar. It is found widely in nature, occurring, for example, in honey and fruits. Other types of sugars used as food must be converted into glucose in the stomach or liver before they are used by the body as energy sources. Because glucose requires no conversion, it is often given intravenously to patients who need immediate nourishment.

SAMPLE EXERCISE 3.7

How many C atoms are there in 1 mol of $C_6H_{12}O_6$?

Solution: There are 6.02×10^{23} $C_6H_{12}O_6$ molecules in 1 mol. Each molecule contains 6C atoms; hence there are $6(6.02 \times 10^{23})$C atoms:

$$C \text{ atoms} = (1 \text{ mol } C_6H_{12}O_6)$$
$$\times \left(\frac{6.02 \times 10^{23} \text{ molecules}}{1 \text{ mol}}\right)\left(\frac{6C \text{ atoms}}{1 \text{ molecule}}\right)$$
$$= 3.61 \times 10^{24} \text{ C atoms}$$

To illustrate how the mole concept and Avogadro's number allow us to interconvert masses and number of particles, let's calculate the number of copper atoms in a penny. A penny weighs 3 g, and we'll assume that it is 100 percent copper:

$$Cu \text{ atoms} = (3 \text{ g Cu})\left(\frac{1 \text{ mol Cu}}{63.5 \text{ g Cu}}\right)\left(\frac{6.02 \times 10^{23} \text{ Cu atoms}}{1 \text{ mol Cu}}\right)$$
$$= 3 \times 10^{22} \text{ Cu atoms}$$

Notice how we were able to use dimensional analysis (Section 1.5) in a straightforward manner to go from grams to numbers of atoms; the conversion sequence is grams $\longrightarrow$ moles $\longrightarrow$ atoms.

We might reflect momentarily on the number of copper atoms in a penny, 3×10^{22}. This is a tremendously large number. It becomes more impressive when we realize that the entire United States could be covered to a depth of 3 mi with this number of ice cubes, each 1 in. on an edge. We should remember that Avogadro's number is even larger.

SAMPLE EXERCISE 3.8

How many moles of glucose, $C_6H_{12}O_6$, are there in (a) 538 g and (b) 1.00 g of this substance?

Solution: (a) One mol of $C_6H_{12}O_6$ weighs 180 g (Sample Exercise 3.6). Therefore there must be more than 1 mol in 538 g.

$$\text{Moles } C_6H_{12}O_6 =$$
$$(538 \text{ g } C_6H_{12}O_6)\left(\frac{1 \text{ mol } C_6H_{12}O_6}{180 \text{ g } C_6H_{12}O_6}\right)$$
$$= 2.99 \text{ mol}$$

(b) In this case there must be less than 1 mol.

Moles $C_6H_{12}O_6 =$

$$(1.00 \text{ g } C_6H_{12}O_6)\left(\frac{1 \text{ mol } C_6H_{12}O_6}{180 \text{ g } C_6H_{12}O_6}\right)$$

$$= 5.56 \times 10^{-3} \text{ mol}$$

The conversion of mass to moles and of moles to mass is frequently encountered in calculations using the mole concept. Notice that the number of moles is always the mass divided by the mass of one mole (the formula weight expressed in grams).

SAMPLE EXERCISE 3.9

What is the mass, in grams, of 0.433 mol of $C_6H_{12}O_6$?

Solution: Because this is less than 1 mol, the mass will be less than 180 g, the mass of 1 mol.

Grams $C_6H_{12}O_6 =$

$$(0.433 \text{ mol } C_6H_{12}O_6)\left(\frac{180 \text{ g } C_6H_{12}O_6}{1 \text{ mol } C_6H_{12}O_6}\right)$$

$$= 77.9 \text{ g}$$

Notice that the mass of a certain number of moles of a substance is always the number of moles times the mass of one mole.

SAMPLE EXERCISE 3.10

How many glucose molecules are there in 5.23 g of $C_6H_{12}O_6$?

Solution: In this case we need to carry out unit conversions in the order grams $\longrightarrow$ moles $\longrightarrow$ molecules. Because the mass we begin with corresponds to less than a mole, there should be fewer than 6.02×10^{23} molecules.

Molecules $C_6H_{12}O_6 =$

$$(5.23 \text{ g } C_6H_{12}O_6)\left(\frac{1 \text{ mol } C_6H_{12}O_6}{180 \text{ g } C_6H_{12}O_6}\right)$$

$$\times \left(\frac{6.02 \times 10^{23} \text{ } C_6H_{12}O_6 \text{ molecules}}{1 \text{ mol } C_6H_{12}O_6}\right)$$

$$= 1.75 \times 10^{22} \text{ molecules}$$

SAMPLE EXERCISE 3.11

What is the mass, in grams, of 1.00×10^{23} molecules of $C_6H_{12}O_6$?

Solution: In this problem, we need to carry out unit conversions in the order molecules $\longrightarrow$ moles $\longrightarrow$ grams.

Grams $C_6H_{12}O_6 = (1.00 \times 10^{23} \text{ molecules})$

$$\times \left(\frac{1 \text{ mol}}{6.02 \times 10^{23} \text{ molecules}}\right)\left(\frac{180 \text{ g}}{1 \text{ mol}}\right)$$

$$= 29.9 \text{ g}$$

3.7 Empirical formulas from analyses

Before we use the mole concept to determine the masses of substances involved in chemical reactions, let's see how it is used in deducing the formulas of chemical substances. As an example, let us determine the formula of a compound formed between mercury and chlorine. This colorless solid is found on analysis to consist of 73.9 percent mercury and 26.1 percent chlorine by mass. This means that if we had a 100-g sample of the solid, the sample would contain 73.9 g of mercury (Hg) and 26.1 g of chlorine (Cl). We divide each of these weights by the appropriate atomic weight to obtain the number of moles of each element in 100 g:

$$73.9 \text{ g Hg} \left(\frac{1 \text{ mol Hg}}{200.6 \text{ g Hg}} \right) = 0.368 \text{ mol Hg}$$

$$26.1 \text{ g Cl} \left(\frac{1 \text{ mol Cl}}{35.5 \text{ g Cl}} \right) = 0.735 \text{ mol Cl}$$

We then divide the larger number of moles by the smaller to obtain the ratio 1.99 mol of Cl/mole of Hg. Because of experimental errors, the results of an analysis may not lead to exact integers for the ratios of moles. The ratio obtained in this case is very close to 2; the formula for the compound is thus $HgCl_2$. This is called the simplest formula because it uses as subscripts the smallest set of integers that express the correct ratios of atoms present. More commonly, this formula is called the empirical formula.

SAMPLE EXERCISE 3.12

Phosgene, a poison gas used during World War I, contains 12.1 percent C, 16.2 percent O, and 71.7 percent Cl. What is the empirical formula of phosgene?

Solution: For simplicity, we may assume that we have 100 g of material. The number of moles of each element is then

$$\text{Moles C} = (12.1 \text{ g}) \left(\frac{1 \text{ mol C}}{12.0 \text{ g}} \right) = 1.01 \text{ mol C}$$

$$\text{Moles O} = (16.2 \text{ g}) \left(\frac{1 \text{ mol O}}{16.0 \text{ g}} \right) = 1.01 \text{ mol O}$$

$$\text{Moles Cl} = (71.7 \text{ g}) \left(\frac{1 \text{ mol Cl}}{35.5 \text{ g}} \right) = 2.02 \text{ mol Cl}$$

The simplest ratio, found by dividing each number by the smallest, 1.01, is $C:O:Cl = 1:1:2$ and the empirical formula is $COCl_2$. Because other experiments show that the molecular weight of the phosgene molecule is 99 amu, $COCl_2$ is also the molecular formula.

When a compound containing carbon and hydrogen is combusted in an apparatus such as that shown in Figure 3.8, the carbon of the compound is converted to CO_2, and all the hydrogen to H_2O. The amount of CO_2 produced can be measured by determining the mass increase in the CO_2 absorber. Similarly, the amount of H_2O formed is determined from the increase in mass of the water absorption tube. As an example, let's consider an analysis of a sample of ascorbic acid (vitamin C). Combustion of 1.000 g of ascorbic acid produces 1.500 g of CO_2 and 0.405 g of

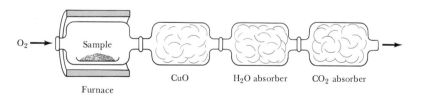

FIGURE 3.8 Apparatus for determining the percentages of carbon and hydrogen in a compound. Copper oxide serves to oxidize traces of carbon and carbon monoxide to carbon dioxide, and hydrogen to water. Magnesium perchlorate, $Mg(ClO_4)_2$, is used to absorb water, whereas sodium hydroxide, NaOH, absorbs carbon dioxide.

H_2O. From these two bits of experimental information we must calculate the quantities of C and H in the 1.000-g sample of ascorbic acid:

$$(1.500 \text{ g CO}_2)\left(\frac{1 \text{ mol CO}_2}{44.00 \text{ g CO}_2}\right)\left(\frac{1 \text{ mol C}}{1 \text{ mol CO}_2}\right)\left(\frac{12.01 \text{ g C}}{1 \text{ mol C}}\right) = 0.409 \text{ g C}$$

$$(0.405 \text{ g H}_2O)\left(\frac{1 \text{ mol H}_2O}{18.0 \text{ g H}_2O}\right)\left(\frac{2 \text{ mol H}}{1 \text{ mol H}_2O}\right)\left(\frac{1.01 \text{ g H}}{1 \text{ mol H}}\right) = 0.045 \text{ g H}$$

From other experiments it can be established that ascorbic acid contains only C, H, and O. Thus the amount of oxygen in the compound must be

$$1.000 \text{ g} - (0.409 \text{ g} + 0.045 \text{ g}) = 0.546 \text{ g}$$

From these data, we can now proceed to calculate the number of moles of each element present in 1 g of ascorbic acid:

$$\text{Moles C} = (0.409 \text{ g C})\left(\frac{1 \text{ mol C}}{12.0 \text{ g C}}\right) = 0.0341 \text{ mol C}$$

$$\text{Moles H} = (0.045 \text{ g H})\left(\frac{1 \text{ mol H}}{1.01 \text{ g H}}\right) = 0.045 \text{ mol H}$$

$$\text{Moles O} = (0.546 \text{ g O})\left(\frac{1 \text{ mol O}}{16.0 \text{ g O}}\right) = 0.0341 \text{ mol O}$$

The relative number of moles of each element can be found by dividing by the smallest number, 0.0341. The ratio of C:H:O is thus 1:1.32:1, which is the same as 3:4:3. The empirical formula is thus $C_3H_4O_3$. In order to determine the molecular formula an experiment must be performed to determine molecular weight. The molecular weight of ascorbic acid is 176 amu. The formula weight for $C_3H_4O_3$ is 3(12.0 amu) + 4(1.0 amu) + 3(16.0 amu) = 88.0 amu. Thus there are two of these formula units in the molecule, and the molecular formula is consequently $C_6H_8O_6$.

3.8 Quantitative information from balanced equations

The mole concept provides a key to placing the quantitative information available in a balanced chemical equation on a practical, macroscopic level. Consider the following balanced equation:

$$2H_2(g) + O_2(g) \longrightarrow 2H_2O(l) \qquad [3.22]$$

The coefficients tell us that two molecules of H_2 react with each molecule of O_2 to form two molecules of H_2O. Therefore $2(6.02 \times 10^{23})$ molecules of H_2 will react with 6.02×10^{23} molecules of O_2 to form $2(6.02 \times 10^{23})$ molecules of H_2O. This is the same as saying that 2 mol of H_2 react with 1 mol of O_2 to form 2 mol of H_2O. The point is that the coefficients in a balanced equation can be interpreted *both* as the *relative numbers of molecules* (or formula units) involved in a reaction *and* as the *relative number of moles*. These interpretations are summarized in Table 3.4. Notice that 4.04 g of H_2 will react with each 32.00 g of O_2 to form 36.04 g of H_2O,

TABLE 3.4 Interpretations of equations

	$2H_2$	$+$	O_2	$\longrightarrow$	$2H_2O$
Molecular ratio:	2 molecules	React with	1 molecule	To form	2 molecules
	$2(6.02 \times 10^{23})$ molecules	React with	6.02×10^{23} molecules	To form	$2(6.02 \times 10^{23})$ molecules
Mole ratio:	2 mol	React with	1 mol	To form	2 mol
Weight ratio:	$2(2.02)$ g $=$ 4.04 g	React with	32.00 g	To form	$2(18.02)$ g $=$ 36.04 g

because these are the masses of 2 mol of H_2, 1 mol of O_2, and 2 mol of H_2O, respectively. Notice also that the sum of the masses of the reactants equals the mass of the product as it must in any chemical reaction according to the law of conservation of mass. The quantities 2 mol of H_2, 1 mol of O_2, and 2 mol of H_2O which are related by Equation [3.22] are called stoichiometrically equivalent quantities. We can represent this as

Mole ratio: 2 mol H_2 $\simeq$ 1 mol O_2 $\simeq$ 2 mol H_2O

Weight ratio: 4.04 g H_2 $\simeq$ 32.00 g O_2 $\simeq$ 36.04 g H_2O

where the symbol $\simeq$ is taken to mean "stoichiometrically equivalent to." These stoichiometric relations can be used to give conversion factors to relate quantities of reactants and products in a chemical reaction. For example, the number of moles of H_2O produced from 1.57 mol of O_2 can be calculated as follows:

$$\text{Moles } H_2O = (1.57 \text{ mol } O_2)\left(\frac{2 \text{ mol } H_2O}{1 \text{ mol } O_2}\right)$$

$$= 3.14 \text{ mol } H_2O$$

As a different example, consider the following reaction:

$$2CuFeS_2(s) + 5O_2(g) \longrightarrow 2Cu(s) + 2FeO(s) + 4SO_2(g) \qquad [3.23]$$

This equation describes a process in the smelting of copper using chalcopyrite ($CuFeS_2$) as the mineral source of the copper. Using the mole concept we can calculate the mass of Cu that can be produced from 1.00 g of chalcopyrite. From Equation [3.23] we can write the following stoichiometric relationships:

Mole ratio: 2 mol $CuFeS_2$ $\simeq$ 2 mol Cu

Weight ratio: $2(183$ g$)$ $CuFeS_2$ $\simeq$ $2(63.5$ g$)$ Cu

Using dimensional analysis we have

$$\text{Grams Cu} = (1.00 \text{ g } CuFeS_2)\left(\frac{127 \text{ g Cu}}{366 \text{ g } CuFeS_2}\right) = 0.347 \text{ g Cu}$$

We can similarly calculate the amount of SO_2 produced in the production of this quantity of copper using the following stoichiometric relationships:

Mole ratio: $2 \text{ mol CuFeS}_2 \simeq 4 \text{ mol SO}_2$

Weight ratio: $2(183 \text{ g}) \text{ CuFeS}_2 \simeq 4(64.1 \text{ g}) \text{ SO}_2$

Using dimensional analysis, we have

$$\text{Grams SO}_2 = (1.00 \text{ g CuFeS}_2)\left(\frac{256 \text{ g SO}_2}{366 \text{ g CuFeS}_2}\right) = 0.700 \text{ g SO}_2$$

It is interesting to note that the mass of SO_2 produced in this reaction is approximately twice the mass of the copper. Consequently, considerable air pollution from sulfur dioxide is often generated in the vicinity of copper smelters (Figure 3.9).

SAMPLE EXERCISE 3.13

How much water is produced in the combustion of 1.00 g of glucose, $C_6H_{12}O_6$:

$$C_6H_{12}O_6(s) + 6O_2(g) \longrightarrow 6CO_2(g) + 6H_2O(l)$$

Solution: For this reaction we have

$1 \text{ mol } C_6H_{12}O_6 \simeq 6 \text{ mol } H_2O$

$180 \text{ g } C_6H_{12}O_6 \simeq 6(18.0 \text{ g}) H_2O$

Therefore,

$$\text{Grams } H_2O = (1.00 \text{ g } C_6H_{12}O_6)\left(\frac{108 \text{ g } H_2O}{180 \text{ g } C_6H_{12}O_6}\right)$$

$$= 0.600 \text{ g } H_2O$$

This type of problem can also be solved in a single step by stepwise conversion of grams of $C_6H_{12}O_6$ to moles of $C_6H_{12}O_6$ to moles of H_2O to grams of H_2O:

$$\text{Grams } H_2O = (1.00 \text{ g } C_6H_{12}O_6)\left(\frac{1 \text{ mol } C_6H_{12}O_6}{180 \text{ g } C_6H_{12}O_6}\right)$$

$$\times \left(\frac{6 \text{ mol } H_2O}{1 \text{ mol } C_6H_{12}O_6}\right)\left(\frac{18.0 \text{ g } H_2O}{1 \text{ mol } H_2O}\right)$$

$$= 0.600 \text{ g } H_2O$$

We may note that an average person ingests 2 L of water daily and eliminates 2.4 L. The difference is produced in metabolism of foodstuffs as above. (Metabolism is a general term used to describe all the processes of a living animal or plant.) The desert rat (kangaroo rat) is able to take great advantage of its metabolic water to help it survive in the dry desert. In fact, it apparently never drinks water.

LIMITING REAGENT

In many situations an excess of one or more substances is available for chemical reaction. Some will therefore be left over when the reaction is complete. For example, consider again the combustion of hydrogen:

$$2H_2(g) + O_2(g) \longrightarrow 2H_2O(l)$$

Suppose that 2 mol of H_2 and 2 mol of O_2 are available for reaction. The balanced equation tells us that only 1 mol of O_2 is required to completely consume 2 mol of H_2, thereby forming 2 mol of H_2O; 1 mol of O_2 will therefore be left over at the end of the reaction. The amount of the substance that is completely consumed in any reaction will determine how much product is formed. This reagent is called the limiting reagent. In this example, H_2 is the limiting reagent.

FIGURE 3.9 Extended exposure to high concentrations of SO_2 and other pollutants can cause extensive damage to plants, animals, and structural materials. This photograph, taken in the Chest Creek Watershed, Clearfield County, Pennsylvania, shows the effects of air and water pollution. Timber in the center background has been killed by fumes from the burning mine-refuse pile at upper center. Stream pollution by mine acid and coal sedimentation is shown at center, whereas soil erosion is evident in the foreground. (*USDA-SCS*)

SAMPLE EXERCISE 3.14

Part of the SO_2 that is introduced into the atmosphere ends up being converted to sulfuric acid, H_2SO_4. The net reaction is

$$2SO_2(g) + O_2(g) + 2H_2O(l) \longrightarrow 2H_2SO_4(l)$$

How much H_2SO_4 can be prepared from 5.0 mol of SO_2, 1.0 mol of O_2, and an unlimited quantity of H_2O?

Solution: The number of moles of O_2 needed for complete consumption of 5.0 mol of SO_2 is

$$\text{Moles } O_2 = (5.0 \text{ mol } SO_2)\left(\frac{1 \text{ mol } O_2}{2 \text{ mol } SO_2}\right)$$
$$= 2.5 \text{ mol } O_2$$

This quantity of O_2 is not available; therefore, all of the SO_2 cannot be consumed; O_2 must be the limiting reagent. We use the quantity of the limiting reagent, O_2, to calculate the quantity of H_2SO_4 prepared:

$$\text{Moles } H_2SO_4 = (1.0 \text{ mol } O_2)\left(\frac{2 \text{ mol } H_2SO_4}{1 \text{ mol } O_2}\right)$$
$$= 2.0 \text{ mol } H_2SO_4$$

We might note that in forming 2.0 mol of H_2SO_4, 2.0 mol of SO_2 are required. Therefore 3.0 mol of SO_2 are left over.

Another approach to the problem of the limiting reagent is to calculate the number of moles of product that could be formed from each of the given amounts of reagents, assuming they were all completely consumed. The reagent that leads to the smallest amount of product is the limiting reagent.

Suppose that in the precipitation reaction described in Sample Exercise 3.3, 3.5 g of $FeBr_3$ were mixed with 6.4 g of Na_2S in solution. How much $Fe_2S_3(s)$ would be formed?

Solution: Our first step, as always, is to write the balanced equation:

$$2FeBr_3(aq) + 3Na_2S(aq) \longrightarrow Fe_2S_3(s) + 6NaBr(aq)$$

Let us now calculate the amount of product that could be formed from each of our given amounts of reactants, assuming that each is the limiting reagent. We can write the following stoichiometric relationships:

Mole ratio:
$$2 \text{ mol } FeBr_3 \simeq 3 \text{ mol } Na_2S \simeq 1 \text{ mol } Fe_2S_3$$

Weight ratio:
$$2(296 \text{ g}) FeBr_3 \simeq 3(78 \text{ g}) Na_2S \simeq 1(208 \text{ g}) Fe_2S_3$$

Using these, we have

$$\text{Grams } Fe_2S_3 = (3.5 \text{ g } FeBr_3)\left(\frac{208 \text{ g } Fe_2S_3}{2 \times 296 \text{ g } FeBr_3}\right)$$
$$= 1.2 \text{ g } Fe_2S_3$$

and

$$\text{Grams } Fe_2S_3 = (6.4 \text{ g } Na_2S)\left(\frac{208 \text{ g } Fe_2S_3}{3 \times 78 \text{ g } Na_2S}\right)$$
$$= 5.7 \text{ g } Fe_2S_3$$

These calculations show that $FeBr_3$ is the limiting reagent, and that mixing of 3.5 g $FeBr_3$ with 6.4 g Na_2S in solution will yield at most 1.2 g Fe_2S_3 as precipitate.

3.9 Molarity and solution stoichiometry

We have seen how it is possible to employ the concept of the mole to determine the quantities of substances that react with one another or to determine the quantity of product that can be obtained from a given mass of reactant. However, it often happens that chemicals are employed not as solid materials but in the form of solutions, particularly aqueous solutions. For example, the reactions described in Sample Exercises 3.15 and 3.16 involve aqueous solutions of the ionic compounds appearing in the balanced equations. It is a great convenience to be able to measure out quantities of dissolved substances by measuring out volumes of the solutions, rather than by weighing out solids or liquids each time and then dissolving these in water.

In discussing solutions it is often convenient to call one component the solvent and the others solutes. The component of a solution whose physical state is preserved when the solution is formed is known as the solvent. For example, when sodium chloride (a solid) is mixed with water, the resultant solution is a liquid. Consequently, water is referred to as the solvent and sodium chloride as the solute. If all components of a solution are in the same state, the one present in greatest amount is called the solvent.

The term concentration is used to designate the amount of solute dissolved in a given quantity of solvent or solution. The method for expressing concentration that is most useful for discussing solution stoichiometry is molarity. The molarity (symbol M) of a solution is defined as the number of moles of solute in a liter of solution (soln):

$$\text{Molarity} = \frac{\text{moles solute}}{\text{volume of soln in liters}} \qquad [3.24]$$

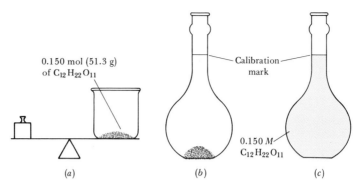

FIGURE 3.10 Procedure for preparation of 1 L of a 0.150 M solution of $C_{12}H_{22}O_{11}$. (a) Weigh out 0.150 mol (51.3 g) of $C_{12}H_{22}O_{11}$ (MW = 342 amu). (b) Add the $C_{12}H_{22}O_{11}$ to a 1-L volumetric flask. (c) Add water to dissolve, then more water until the solution reaches the calibration mark.

A 1.50-molar solution (written $1.50\,M$) contains 1.50 mol of solute in every liter of solution. To make a liter of $0.150\,M$ sucrose, $C_{12}H_{22}O_{11}$, in water requires 0.150 mol of $C_{12}H_{22}O_{11}$. This quantity of solid sucrose (51.3 g) is first dissolved in less than a liter of water. The resulting solution is then diluted to a total volume of 1 L. A volumetric flask, which is a flask calibrated to contain a precise volume of liquid, is used for this purpose. The operation is shown in Figure 3.10.

SAMPLE EXERCISE 3.16

Calculate the molarity of a solution made by dissolving 23.4 g of sodium sulfate (Na_2SO_4) in enough water to form 125 mL of solution.

Solution:

$$\text{Molarity} = \frac{\text{moles } Na_2SO_4}{\text{liters soln}}$$

$$\text{Moles } Na_2SO_4 = (23.4 \text{ g } Na_2SO_4)\left(\frac{1 \text{ mol } Na_2SO_4}{142 \text{ g } Na_2SO_4}\right)$$

$$= 0.165 \text{ mol } Na_2SO_4$$

$$\text{Molarity} = \left(\frac{0.165 \text{ mol } Na_2SO_4}{125 \text{ mL soln}}\right)\left(\frac{1000 \text{ mL soln}}{1 \text{ L soln}}\right)$$

$$= 1.32 \frac{\text{mol } Na_2SO_4}{\text{L soln}}$$

$$= 1.32\,M$$

One advantage to the expression of concentration in molarity is that it allows us to measure out a solution volume of known concentration and readily calculate the number of moles of solute dispensed. Molarity can be used to interconvert volume and moles just as density can be used to interconvert volume and mass (Sample Exercise 1.9). Calculation of the number of moles of HNO_3 in 2.0 L of $0.200\,M$ HNO_3 solution illustrates conversion of volume to moles:

$$\text{Moles } HNO_3 = (2.0 \text{ L soln})\left(0.200 \frac{\text{mol } HNO_3}{\text{L soln}}\right)$$

$$= 0.40 \text{ mol } HNO_3$$

Notice how dimensional analysis can be used in this conversion if we express molarity as moles HNO_3/L soln. Notice also that to obtain moles we multiplied liters and molarity: moles = liters $\times M$. This same expression for moles can be obtained directly by algebraic rearrangement of Equation [3.24]. The use of molarity to convert moles to volume can be illustrated by calculating the volume of $0.30\,M$ HNO_3 solution required to supply 2.0 mol of HNO_3:

$$\text{Liters soln} = (2.0 \text{ mol } HNO_3)\left(\frac{1 \text{ L soln}}{0.30 \text{ mol } HNO_3}\right)$$

$$= 6.7 \text{ L soln}$$

In this case we needed to apply the reciprocal of molarity to convert moles to volume: liters = moles $\times 1/M$.

SAMPLE EXERCISE 3.17

How many grams of Na_2SO_4 are required to make 350 mL of $0.50\,M$ Na_2SO_4?

Solution: Because

$$M \text{ } Na_2SO_4 = \frac{\text{mol } Na_2SO_4}{\text{L soln}}$$

$$\text{Moles } Na_2SO_4 = (0.350 \text{ L soln})\left(0.50\frac{\text{mol } Na_2SO_4}{\text{L soln}}\right)$$

$$= 0.175 \text{ mol } Na_2SO_4$$

Because each mole of Na_2SO_4 weighs 142 g, the required number of grams of Na_2SO_4 is

$$(0.175 \text{ mol } Na_2SO_4)\left(\frac{142 \text{ g } Na_2SO_4}{1 \text{ mol } Na_2SO_4}\right)$$

$$= 24.8 \text{ g } Na_2SO_4$$

It is also possible to work this problem by direct conversion of milliliters to liters to moles to grams. In doing so, we use molarity as a conversion factor between volume and moles:

$$\text{Grams } Na_2SO_4 = (350 \text{ mL soln})\left(\frac{1 \text{ L soln}}{1000 \text{ mL soln}}\right)$$

$$\times \left(0.50\frac{\text{mol } Na_2SO_4}{\text{L soln}}\right)\left(\frac{142 \text{ g } Na_2SO_4}{1 \text{ mol } Na_2SO_4}\right)$$

$$= 24.8 \text{ g } Na_2SO_4$$

DILUTION

It is often convenient to make a solution of a certain concentration from a more concentrated solution. For example, suppose you need to prepare a liter of $0.10\,M$ HNO_3 solution from a solution of $1.0\,M$ HNO_3. The desired solution will contain 0.10 mol of HNO_3. Therefore, you need to remove 0.10 mol of HNO_3 from the $1.0\,M$ solution. There is 0.10 mol of HNO_3 in 100 mL of the $1.0\,M$ solution. Thus, 100 mL is withdrawn from the $1.0\,M$ HNO_3 solution using a pipet and added to a 1-L volumetric flask. It is then diluted to 1 L as shown in Figure 3.11.*

When more solvent is added to a solution, thereby diluting it, the number of moles of solute remains unchanged:

Moles solute before dilution = moles solute after dilution [3.25]

---

*In diluting an acid or base, the acid or base should be added to water, then further diluted by addition of more water. Adding water directly to concentrated acid or base can cause spattering because of the intense heat generated.

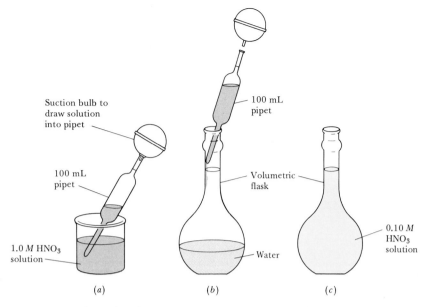

Suction bulb to
draw solution
into pipet

100 mL
pipet

100 mL
pipet

Volumetric
flask

0.10 M
HNO_3
solution

1.0 M HNO_3
solution

Water

(a)   (b)   (c)

FIGURE 3.11 Procedure for preparation of 1 L of 0.10 M HNO_3 by dilution of 1.0 M HNO_3.
(a) Draw 100 mL of the 1.0 M solution into a pipet. (b) Add this amount of 1.0 M HNO_3 to a
small amount of water in a 1-L volumetric flask. (c) Add additional water to dilute the solution
to a total volume of 1 L. The result is a 0.10 M HNO_3 solution.

Because number of moles = $M \times$ liters, we can write:

(Initial molarity)(initial volume) = (final molarity)(final volume)

$$M_{initial}V_{initial} = M_{final}V_{final} \qquad [3.26]$$

SAMPLE EXERCISE 3.18

How much 3.0 M H_2SO_4 would be required to make
500 mL of 0.10 M H_2SO_4?

Solution: Using Equation [3.26], $M_{initial}V_{initial} =$
$M_{final}V_{final}$, we can write

$$V_{initial} = \frac{M_{final}V_{final}}{M_{initial}}$$

$$= \frac{(0.10\ M)(500\ mL)}{(3.0\ M)} = 17\ mL$$

We see that if we start with 17 mL of 3.0 M H_2SO_4
and dilute it to a total volume of 500 mL, the desired
0.10 M solution will be obtained.

TITRATION

The procedure by which a solution of known concentration is added to
another solution until the chemical reaction between the two solutes is
complete is known as a titration. Titrations are widely used in chemistry
to analyze the compositions of mixtures. The solution whose concentration is known is called the standard solution. In titrations, the standard
solution is slowly added from a buret to a solution that contains a known
volume or known mass of solute. The latter solution is commonly referred to as the unknown. The point at which stoichiometrically equivalent quantities of substances have been brought together is known as the
equivalence point of the titration. Some titrations are carried out by
measuring out a known volume of a standard solution and then adding

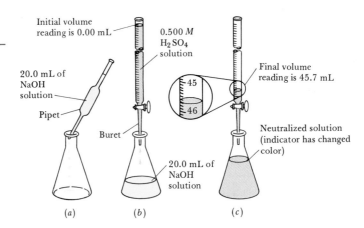

FIGURE 3.12 Procedure for titration of an unknown base against a standardized solution of H_2SO_4. (a) A known quantity of base is added to a flask. (b) An acid-base indicator is added, and standardized H_2SO_4 is added from a buret. (c) Equivalence point is signaled by a color change of the indicator. The concentrations and volumes shown correspond to the example discussed in Sample Exercise 3.20.

the solution of unknown concentration until the equivalence point is reached.

In order to titrate an unknown with a standard solution, there must be some way to determine when the equivalence point of the titration has been reached. In acid-base titrations, organic dyes known as acid-base indicators are used for this purpose. For example, the dye known as phenolphthalein is colorless in acidic solution but is red in basic solution. If phenolphthalein is added to an unknown solution of acid, the solution will be colorless. Standard base can then be added from the buret until the solution barely turns from colorless to red. This indicates that the acid has been neutralized, and the drop of base that caused the solution to become colored has no acid to react with. The solution therefore becomes basic, and the dye turns red. The experimental procedure for the titration of NaOH and H_2SO_4 is summarized in Figure 3.12.

SAMPLE EXERCISE 3.19

What volume of 0.500 M NaCl is required to react completely with 0.200 mol of $Pb(NO_3)_2$? The chemical equation for this reaction is

$$2NaCl(aq) + Pb(NO_3)_2(aq) \longrightarrow$$
$$2NaNO_3(aq) + PbCl_2(s)$$

Solution: According to the reaction equation

$$2 \text{ mol NaCl} \simeq 1 \text{ mol } Pb(NO_3)_2$$

Therefore,

Moles NaCl =

$$(0.200 \text{ mol } Pb(NO_3)_2)\left(\frac{2 \text{ mol NaCl}}{1 \text{ mol } Pb(NO_3)_2}\right)$$

$$= 0.400 \text{ mol NaCl}$$

Because

$$\text{Liters NaCl soln} = \frac{\text{mol NaCl}}{M \text{ NaCl soln}}$$

$$\text{Liters NaCl soln} = \frac{0.400 \text{ mol NaCl}}{0.500 \text{ mol NaCl/L soln}}$$

$$= 0.800 \text{ L}$$

Therefore, 800 mL of the 0.500 M NaCl solution could be measured out and added to the 0.200 mol of $Pb(NO_3)_2$. This problem can also be solved by direct conversion of moles $Pb(NO_3)_2$ to moles NaCl to volume of NaCl solution:

$$\text{Liters NaCl soln} = (0.200 \text{ mol } Pb(NO_3)_2)$$

$$\times \left(\frac{2 \text{ mol NaCl}}{1 \text{ mol } Pb(NO_3)_2}\right)\left(\frac{1 \text{ L NaCl soln}}{0.500 \text{ mol NaCl}}\right)$$

$$= 0.800 \text{ L}$$

SAMPLE EXERCISE 3.20

One method used commercially to peel potatoes is to soak them in a solution of NaOH for a short time, and then spray off the peel after the potatoes are removed from the NaOH. The concentration of NaOH is normally in the range of 3 to 6 M. The NaOH is analyzed periodically to determine its ability to peel potatoes rapidly. In one such analysis, 45.7 mL of 0.500 M H_2SO_4 are required to react completely with a 20.0 mL sample of NaOH solution:

$$H_2SO_4(aq) + 2NaOH(aq) \longrightarrow$$
$$2H_2O(l) + Na_2SO_4(aq)$$

What is the concentration of the NaOH solution?

Solution:

Moles $H_2SO_4 =$

$$(45.7 \text{ mL soln})\left(\frac{1 \text{ L soln}}{1000 \text{ mL soln}}\right)\left(0.500\frac{\text{mol } H_2SO_4}{\text{L soln}}\right)$$
$$= 2.28 \times 10^{-2} \text{ mol } H_2SO_4$$

According to the balanced equation, 1 mol $H_2SO_4 \simeq$ 2 mol NaOH. Therefore,

Moles NaOH =

$$(2.28 \times 10^{-2} \text{ mol } H_2SO_4)\left(\frac{2 \text{ mol NaOH}}{1 \text{ mol } H_2SO_4}\right)$$
$$= 4.56 \times 10^{-2} \text{ mol NaOH}$$

Knowing the number of moles of NaOH present in 20.0 mL of solution allows us to calculate the concentration of this solution:

$$M \text{ NaOH} = \frac{\text{mol NaOH}}{\text{L soln}} =$$
$$\left(\frac{4.56 \times 10^{-2} \text{ mol NaOH}}{20.0 \text{ mL soln}}\right)\left(\frac{1000 \text{ mL soln}}{1 \text{ L soln}}\right)$$
$$= 2.28 \frac{\text{mol NaOH}}{\text{L soln}} = 2.28 \ M$$

SAMPLE EXERCISE 3.21

The quantity of Cl^- in a water supply is determined by titrating the sample against $AgNO_3$:

$$AgNO_3(aq) + Cl^-(aq) \longrightarrow AgCl(s) + NO_3^-(aq)$$

What mass of chloride ion is present in a 10.0-g sample of the water if 20.2 mL of 0.100 M $AgNO_3$ is required to react with all of the chloride in the sample?

Solution: We must first determine the number of moles of $AgNO_3$ required in the titration:

$$(20.2 \text{ mL soln})\left(\frac{1 \text{ L soln}}{1000 \text{ mL soln}}\right)\left(0.100\frac{\text{mol } AgNO_3}{\text{L soln}}\right)$$
$$= 2.02 \times 10^{-3} \text{ mol } AgNO_3$$

From the balanced equation we see that 1 mol of $AgNO_3 \simeq$ 1 mol of Cl^-. Therefore the sample must contain 2.02×10^{-3} mol of Cl^-:

Moles $Cl^- =$

$$(2.02 \times 10^{-3} \text{ mol } AgNO_3)\left(\frac{1 \text{ mol } Cl^-}{1 \text{ mol } AgNO_3}\right)$$
$$= 2.02 \times 10^{-3} \text{ mol } Cl^-$$

The number of moles of Cl^- can then be converted to grams:

Grams $Cl^- =$

$$(2.02 \times 10^{-3} \text{ mol } Cl^-)\left(\frac{35.5 \text{ g } Cl^-}{1 \text{ mol } Cl^-}\right)$$
$$= 7.17 \times 10^{-2} \text{ g } Cl^-$$

We might note that the percentage of Cl^- in the water is

$$\% \ Cl^- = \frac{7.7 \times 10^{-2} \text{ g } Cl^-}{10.0 \text{ g soln}} \times 100 = 0.717\%$$

Chloride ion is one of the major ions in water and sewage. Ocean water contains 1.92 percent Cl^-. Whether or not water containing Cl^- exhibits a salty taste depends on the other ions present. If the accompanying ions are Na^+ ions, a salty taste may be detected with as little as 0.03 percent Cl^-.

Summary

The law of conservation of mass states that there are no detectable changes in mass in any chemical reaction. This indicates that there are the same number of atoms of each type present after a chemical reaction as there were before the reaction. A balanced equation shows equal numbers of each type of atom on each side of the equation and is thereby consistent with the law of conservation of mass. We discussed how equations are balanced by placing coefficients in front of the chemical formulas for the substances involved in the reaction. We have also seen how it is possible to predict the products of simple reactions by analogy to known reactions and by use of the periodic table. Among the reaction types seen in this chapter are the following: (1) Combustion in oxygen, in which an organic compound reacts with oxygen, forming carbon dioxide, water, and possibly other products, depending on the composition of the compound. (2) Neutralization reaction, in which an acid and base react to form water and a salt. (3) Precipitation reaction, in which one of the products of a reaction between two substances in solution is insoluble in the solution.

The coefficients in a balanced equation can be interpreted as either the relative number of formula units involved in the reaction or the relative number of moles. A mole of any substance is Avogadro's number (6.022×10^{23}) of formula units of that substance. The mass of a mole of atoms, molecules, or ions is the formula weight expressed in grams. For example, a single molecule of H_2O weighs 18 amu; a mole of H_2O weighs 18 g. The empirical formula, or simplest formula, of a substance expresses the composition in terms of the smallest possible set of whole-number subscripts denoting the relative numbers of atoms. We have seen how the mole concept can be used to determine the empirical formula of a compound and to calculate the quantities involved in chemical reactions. In dealing with reactions between substances in solution, it is convenient to employ the concept of solution concentration. Molarity is defined as the number of moles of solute per liter of solution. Molarity serves as a conversion factor for interconverting solution volume and number of moles of solute.

Learning goals

Having read and studied this chapter, you should be able to:

1 Balance chemical equations.

2 Predict the products of a chemical reaction, having seen a suitable analogy.

3 Interconvert number of moles, mass in grams, and number of atoms, ions, or molecules.

4 Calculate the empirical formula of a compound, having been given appropriate analytical data such as elemental percentages or the quantity of CO_2 and H_2O produced by combustion.

5 Calculate the molecular formula, having been given the empirical formula and molecular weight.

6 Calculate the mass of a particular substance produced or used in a chemical reaction (mass-mass problems).

7 Determine the limiting reagent in a reaction.

8 Define molarity.

9 Solve problems involving interconversions among molarity, solution volume, and number of moles of solute.

10 Explain what is meant by the term titration.

11 Calculate concentration or mass of solute in a sample from titration data.

Key terms

Among the more important terms and expressions used for the first time in this chapter are the following:

An acid (Section 3.3) is a substance that produces an excess of H^+ ions when it dissolves in water.

The atomic mass unit (amu) (Section 3.4) is based on the value of exactly 12 for the isotope of carbon with six protons and six neutrons in the nucleus.

Atomic weight (Section 3.4) or atomic mass is the average mass of the atoms of an element in atomic mass units. That is, it is the weight of an element that contains the same number of atoms as there are carbon atoms in exactly 12 g of ^{12}C.

Avogadro's number (Section 3.6) is the number of ^{12}C atoms in exactly 12 g of ^{12}C, 6.022×10^{23}.

A balanced chemical equation (Section 3.2) is one that satisfies the law of conservation of mass; it contains equal numbers of atoms of each element on both sides of the equation.

A base (Section 3.3) is a substance that produces an excess of OH^- ions when it dissolves in water.

A combustion reaction (Section 3.3) is one that proceeds with evolution of heat and usually also a flame. Most combustion involves reaction with oxygen, as in the burning of a match.

Solution concentration (Section 3.9) denotes the

number of moles of a solute present in a given quantity of solvent.

The law of conservation of mass (Section 3.1) states that atoms are neither created nor destroyed during a chemical reaction.

The empirical formula (Section 3.7) of any substance expresses the simplest whole-number ratio of the elements present in that substance. For example, the empirical formula of glucose, $C_6H_{12}O_6$, is CH_2O.

The equivalence point (Section 3.9) is the point in a titration at which the added solute just completely reacts with the solute present in solution.

Formula weight (Section 3.4) is the weight of the collection of atoms represented by a chemical formula. For example, the formula weight of NO_2 (46.0 amu) is the sum of the weights of one nitrogen atom and two oxygen atoms. If the formula is the molecular formula of the substance, the formula weight is the molecular weight of the substance.

An indicator (Section 3.9) is a substance added to a solution to indicate by a color change the point at which the added solute has just reacted with all the solute present in solution.

A limiting reagent (Section 3.8) is the reactant present in smallest stoichiometric quantity in a mixture of reactants. The amount of product that can form is limited by the complete consumption of the limiting reagent.

The mass spectrometer (Section 3.5) is an instrument used to measure the precise masses and relative amounts of atomic and molecular ions.

Molarity (Section 3.9) is the concentration of a solution expressed as moles of solute per liter of solution; abbreviated M.

A mole (Section 3.6) is a collection of Avogadro's number (6.022×10^{23}) of objects; for example, a mole of H_2O is 6.022×10^{23} H_2O molecules.

A neutralization reaction (Section 3.3) is one in which an acid and a base react in stoichiometrically equivalent amounts. The product of a neutralization reaction is water and a salt.

A precipitation reaction (Section 3.3) is one occurring between substances in solution in which one of the products is insoluble.

A titration (Section 3.9) is the process of reacting a solution of unknown concentration with one of known concentration (a standardized solution). The procedure is commonly used to determine the concentrations of solutions of unknown concentrations.

EXERCISES

Conservation of mass; chemical equations and chemical reactions

3.1 Cite at least three examples of how processes occurring in our society result in conversion of useful substances into materials that are not readily usable.

3.2 Which of the following equations, as written, is consistent with the law of conservation of mass?

(a) $CF_3H(g) + O(g) \longrightarrow OH(g) + CF_3(g)$
(b) $CCl_4(g) + O_2(g) \longrightarrow CCl_2O(g) + Cl_2(g)$
(c) $H_2SO_4(aq) + 2KOH(aq) \longrightarrow K_2SO_4(aq) + H_2O(l)$
(d) $4C_3H_5ON(g) + 19O_2(g) \longrightarrow$
$\quad 12CO_2(g) + 10H_2O(g) + 4NO_2(g)$

3.3 Balance the following equations:

(a) $Al(s) + Cl_2(g) \longrightarrow AlCl_3(s)$
(b) $P_2O_3(s) + H_2O(l) \longrightarrow H_3PO_3(aq)$
(c) $Ca(OH)_2(aq) + HBr(aq) \longrightarrow CaBr_2(aq) + H_2O(l)$
(d) $Al(s) + S_8(s) \longrightarrow Al_2S_3(s)$
(e) $Mg_2C_3(s) + H_2O(l) \longrightarrow Mg(OH)_2(aq) + C_3H_4(g)$

3.4 Balance the following equations:

(a) $C(s) + F_2(g) \longrightarrow CF_4(g)$
(b) $N_2O_5(s) + H_2O(l) \longrightarrow HNO_3(l)$
(c) $C_2H_4(g) + O_2(g) \longrightarrow CO_2(g) + H_2O(g)$
(d) $La_2O_3(s) + H_2O(l) \longrightarrow La(OH)_3(s)$
(e) $HCl(g) + CaO(s) \longrightarrow CaCl_2(s) + H_2O(g)$
(f) $Pb(NO_3)_2(aq) + NaCl(aq) \longrightarrow$
$\quad PbCl_2(s) + NaNO_3(aq)$

3.5 Write a balanced chemical equation to correspond to each of the following word descriptions. (a) Phosphine, $PH_3(g)$, is combusted in air to form gaseous water and solid diphosphorus pentoxide. (b) Barium metal reacts with methyl alcohol, $CH_3OH(l)$, to form hydrogen gas and dissolved barium methoxide, $Ba(OCH_3)_2$. (c) Boron sulfide, $B_2S_3(s)$, reacts violently with water to form dissolved boric acid, H_3BO_3, and hydrogen sulfide gas, H_2S. (d) Copper metal reacts with hot concentrated sulfuric acid to form aqueous copper(II) sulfate, sulfur dioxide gas, and liquid water. (e) When ammonia gas, NH_3, is passed over hot liquid sodium metal, hydrogen is released and sodium amide, $NaNH_2$, is formed as a solid product.

3.6 Write balanced chemical equations to correspond to each of the following verbal descriptions. (a) Cyanic acid, HCNO, is quite unstable. The gas reacts with water to form ammonia and gaseous carbon dioxide. (b) When solid mercury(II) nitrate is heated, it decomposes to form solid mercury(II) oxide and gaseous nitrogen dioxide and oxygen. (c) When liquid phosphorus trichloride is added to water, it reacts violently to form aqueous phosphorous acid and aqueous hydrogen chloride. (d) When solid potassium nitrate is heated, it decomposes to solid potassium nitrite, and oxygen gas is evolved. (e) When hydrogen sulfide (H_2S) gas is passed over solid hot iron(III) hydroxide, it reacts to form solid iron(III) sulfide and gaseous water.

[3.7] Each of the following equations is balanced except for one missing substance indicated by a blank. In each case, fill in the blank to complete the balancing of the equation. You must deduce the formula and the number of moles of missing substance.

(a) $CaH_2(s) + 2H_2O(l) \longrightarrow Ca(OH)_2(s) + $ _____

(b) $CO(g) + 3H_2(g) \longrightarrow H_2O(g) + $ _____

(c) $Na_2CuCl_4(aq) + $ _____ $\longrightarrow 2NaCuCl_2(aq)$

(d) $10HNO_3(aq) + 4Zn(s) \longrightarrow$
  $NH_4NO_3(aq) + 3H_2O(l) + $ _____

(e) $FeSO_4(s) + H_2S(g) \longrightarrow FeS(s) + $ _____

(f) $2Fe_2S_3(s) + 6H_2O(g) + $ _____ $\longrightarrow$
  $4Fe(OH)_3(s) + 6S(s)$

3.8 Drawing upon the following list of substances as reactants—$BaCl_2$, C_2H_5OH, Na_2CrO_4, O_2, KOH, H_2SO_4, $Ca(OH)_2$, $Pb(NO_3)_2$, HNO_3, and K—write balanced equations for each of the following: (a) a combustion reaction; (b) formation of insoluble $PbCrO_4$ from solution; (c) formation of hydrogen gas; (d) neutralization of sulfuric acid; (e) solid $Ca(OH)_2$, which has a low solubility in water, is dissolved by reaction with another reagent; (f) formation of insoluble $BaSO_4$ from solution.

[3.9] We can sometimes extend our ability to write balanced equations by recognizing the analogies between compounds. From the hints provided, write complete balanced equations for the following. (a) Combustion of nitromethane, $CH_3NO_2(g)$, leads to $NO_2(g)$ as one of the products. (b) Reaction of potassium with liquid ammonia is very much like reaction of this metal with water. (c) Fluorine, like oxygen, can support combustion; for example, methane, $CH_4(g)$, can be made to "burn" in an atmosphere of F_2. (d) Reaction of a metal with an acid solution is like reaction of active metals with water, except that a salt of the metal rather than the hydroxide is the product. For example, Zn reacts with dilute HCl solution.

Atomic and molecular weights; the chemical mole

3.10 Determine the formula weights of each of the following substances: (a) thionyl chloride, $SOCl_2$; (b) thallium(I) oxalate, $Tl_2(C_2O_4)$; (c) coumarin, $C_9H_6O_2$; (d) coniferin, $C_{16}H_{22}O_8 \cdot 2H_2O$; (e) xenon tetrafluoride, XeF_4.

3.11 What is the mass of each of the following: (a) 2 mol of CO_2; (b) 3.58×10^{22} atoms of Kr; (c) 4.83×10^{24} molecules of HCl; (d) 0.0090 mol ethylene, C_2H_4?

3.12 Calculate the total number of carbon atoms present in 33.4 g of isopropyl alcohol, C_3H_7OH. How many moles of C is this? What is the weight percentage of C in this sample?

3.13 Gay-Lussac found that one volume of hydrogen gas reacts with one volume of chlorine gas to form *two* volumes of hydrogen chloride as product. Is this observa-

tion consistent with an assumption that hydrogen and chlorine gas are monatomic? Explain. What does this experiment tell us directly about the formula for hydrogen chloride? Explain.

3.14 We know that chlorine and fluorine are both diatomic gases. One volume of chlorine gas reacts with three volumes of fluorine gas to yield two volumes of product, with all gases measured at the same temperature and pressure. What is the formula for the product?

3.15 Calculate the number of atoms of ^{12}C in exactly 5 g of ^{12}C, using five-place accuracy (Avogadro's number to five places is 6.0220×10^{23} mol^{-1}). Calculate the number of atoms in exactly 5 g of carbon as it is found in nature. Why are these numbers different? Calculate the number of atoms of silver in exactly 5 g of silver.

3.16 What weight of sodium metal has the same number of Na atoms as there are Al atoms in 300 kg of aluminum?

[3.17] Under a certain set of laboratory conditions, 2.1 g of Na react with water to form 1.14 L of H_2 gas. When 3.4 g of another alkali metal is reacted with water under the same conditions, 497 mL of hydrogen gas is evolved. Which of the other alkali metals was reacted?

3.18 In his determination of the atomic weight of the element zinc, Berzelius determined in 1818 that the weight ratio of Zn to O in the oxide of zinc was 4.032. He thought that the formula of the compound was ZnO_2. Assuming an atomic weight of 16.00 for oxygen, what value does this give for the atomic weight of zinc? By comparing this with the presently accepted value, what can you say about Berzelius's assumption? If it was in error, what should it have been?

[3.19] One of the earliest accurate formula weight measurements involved measurement of the weight ratio of $KClO_3$ to KCl, based on decomposition of $KClO_3$: $2KClO_3(s) \longrightarrow 2KCl(s) + 3O_2(g)$. In 1911 Stähler and Meyer measured this ratio and found it to be 1.64382. Using only this ratio and the presently accepted atomic weight of oxygen, calculate the formula weight for KCl. Compare this calculated formula weight with the presently accepted value.

[3.20] Stas reported in 1865 that he had reacted a weighed amount of pure silver with nitric acid and had recovered all the silver as pure silver nitrate, $AgNO_3$. The weight ratio of Ag to $AgNO_3$ was found to be 0.634985. Using only this ratio and the presently accepted values of the atomic weights of silver and oxygen, calculate the atomic weight of nitrogen. Compare this calculated atomic weight with the currently accepted value.

3.21 What is meant by the following expressions: (a) "a mole of argon"; (b) "a mole of potassium bromide"; (c) "a mole of methyl alcohol"?

3.22 Calculate the formula weight of each of the following: (a) C_3H_8; (b) SO_2; (c) SiH_4; (d) B_5H_{11}; (e) H_3PO_2.

3.23 Calculate the formula weight of each of the following: (a) Na_4Pb; (b) $KBrO_3$; (c) Cr_2O_3; (d) $KMnO_4$; (e) $Ca(OH)_2$.

3.24 The element neon consists of three isotopes with masses 19.99, 20.99, and 21.99 amu. These three isotopes are present in nature to the extent of 90.92 percent, 0.25 percent, and 8.83 percent, respectively. From these data, calculate the atomic weight of neon.

[3.25] The element silver consists in nature of two isotopes, ^{107}Ag with atomic mass 106.905 amu, and ^{109}Ag with atomic mass 108.905. The accepted atomic weight of Ag is 107.870. From this calculate the relative amounts of ^{107}Ag and ^{109}Ag in nature.

3.26 The mass spectrum of a sample of lead oxide contains ions of the formula PbO^+. The lead oxide has been prepared from isotopically pure ^{16}O. The ion masses seen and their relative intensities are listed as follows:

PbO^+ ion mass	Fractional intensity
220.002	0.0137
222.056	0.2630
223.050	0.2080
224.055	0.5153

The mass of ^{16}O is 15.9948. Calculate the average atomic weight of lead in this sample.

3.27 The molecule pyridine, C_5H_5N, was found to adsorb on the surfaces of certain metal oxides. A 5.0-g sample of finely divided zinc oxide, ZnO, was found to adsorb 0.068 g pyridine. How many pyridine molecules are adsorbed? What is the ratio of pyridine molecules to formula units of zinc oxide? If the surface area of the oxide is 48 m^2/g, what is the average area of surface per adsorbed pyridine molecule?

3.28 The allowable concentration level of vinyl chloride, C_2H_3Cl, in a chemical plant is 2.05×10^{-6} g per liter. How many molecules of vinyl chloride in each liter does this represent? How many moles per liter?

3.29 It requires about 25 μg minimum of tetrahydrocannabinol, THC, the active ingredient in marijuana, to produce intoxication from smoking substances containing THC. The molecular formula of THC is $C_{12}H_{30}O_2$. How many molecules of THC does this 25 μg represent? How many moles?

3.30 Complete each of the following: (a) 4.2 g of solid iodine contains _____ I_2 molecules. (b) 5.6×10^{17} Fe atoms has a total mass of _____ g. (c) The number of silver atoms in 5.0 g of AgCl is _____. (d) 4.5×10^{-5} moles of glucose, $C_6H_{12}O_6$, has a mass of _____ g. (e) The number of moles of H_2O in 37.5 g is _____.

3.31 Polyethylene, $(C_2H_4)_n$, is a long chainlike molecule made up from ethylene (C_2H_4) monomer units. How many ethylene (C_2H_4) monomer units are in a polyethylene molecule with a mass of 182,000 amu?

Empirical formulas; chemical calculations involving equations

3.32 What are the empirical formulas of compounds with the following compositions: (a) 28.0 percent N, 72.0

percent Ag; (b) 93.22 percent Mn, 6.78 percent C; (c) 10.4 percent C, 27.8 percent S, 61.7 percent Cl; (d) 83.0 percent I, 7.85 percent C, 9.15 percent N.

3.33 Strontium hydroxide is isolated as a hydrate, which means that a certain number of water molecules are included in the solid structure. The formula of the hydrate can be written as $Sr(OH)_2 \cdot xH_2O$, where x indicates the number of moles of water in the solid per mole of $Sr(OH)_2$. When 6.85 g of the hydrate is dried in an oven, 3.13 g of anhydrous $Sr(OH)_2$ is formed. What is the value for x?

[3.34] An oxybromate compound, $KBrO_x$, where x is unknown, is analyzed and found to contain 52.92 percent Br. What is the value for x?

3.35 Which of the following compounds, all used as fertilizers, contains the highest percentage by weight of phosphorus: (a) $Ca_5(PO_4)_3F$; (b) $Ca(H_2PO_4)_2 \cdot H_2O$; (c) $(NH_4)_2HPO_4$.

3.36 In a laboratory, 1.55 g of an organic compound containing carbon, hydrogen, and oxygen is combusted for analysis, as illustrated in Figure 3.8. Combustion resulted in 1.45 g of CO_2 and 0.890 g of H_2O. What is the empirical formula?

3.37 When 0.666 g of a barium compound soluble in water is treated with excess sulfuric acid solution, 0.608 g of $BaSO_4$ is precipitated. What is the percentage of barium in the compound? If the empirical formula contains one barium atom, what is the formula weight?

3.38 The characteristic odor of pineapple is due to ethyl butyrate, a compound containing carbon, hydrogen, and oxygen. Combustion of 2.78 mg of ethyl butyrate leads to formation of 6.32 mg of CO_2 and 2.58 mg of H_2O. What is the empirical formula? The properties of the compound suggest that the molecular weight should be between 100 and 150. What is the likely molecular formula?

3.39 White phosphorus, P_4, burns in excess oxygen to form diphosphorus pentoxide, P_2O_5. Write a balanced chemical equation for this combustion. (a) How many moles of P_2O_5 are formed by combustion of 4.00 g of P_4? (b) When P_2O_5 is added to excess water, it eventually hydrolyzes to H_3PO_4: $P_2O_5(s) + 3H_2O(l) \longrightarrow 2H_3PO_4(aq)$. How many grams of H_3PO_4 are formed from 4.00 g of P_4?

3.40 (a) What weight of NH_3 is formed when 2.55 g of Li_3N reacts with water according to the equation

$$Li_3N(s) + 3H_2O(l) \longrightarrow 3LiOH(s) + NH_3(g)$$

(b) What mass of $CaCO_3(s)$ is required to produce 1.57 g $CO_2(g)$ according to the reaction

$$CaCO_3(s) \longrightarrow CaO(s) + CO_2(g)$$

(c) What mass of CO_2 is formed when 5.53 g of C_2H_5OH is produced in the fermentation of sugar according to the reaction

$$C_6H_{12}O_6(aq) \longrightarrow 2C_2H_5OH(aq) + 2CO_2(g)$$

(d) What mass of Mg is required to react with excess $CuSO_4$ to form 3.46 g Cu_2O according to the reaction

$$2Mg(s) + 2CuSO_4(aq) + H_2O(l) \longrightarrow$$
$$2MgSO_4(aq) + Cu_2O(s) + H_2(g)$$

3.41 (a) In the reaction

$$Mg_2Si(s) + 4H_2O(l) \longrightarrow 2Mg(OH)_2(s) + SiH_4(g)$$

how many moles of SiH_4 are formed by complete reaction of 2.95 g of Mg_2Si? (b) Calculate the weight of water necessary to react with all the Mg_2Si. (c) In a second experiment involving the same reaction, 22.5 mg of Mg_2Si is reacted with 14.0 mg of water. What mass of SiH_4 is formed, assuming that the limiting reagent is completely reacted?

3.42 Magnesium oxide reacts with gaseous phosphorus pentachloride to form solid magnesium chloride and solid diphosphorus pentoxide as products. Write a balanced equation for the reaction. Assuming an excess of $MgO(s)$ and complete reaction, what mass of PCl_5 is required to produce 2.50 kg of diphosphorus pentoxide?

3.43 A sample of aluminum carbide, $Al_4C_3(s)$, is added to a dilute acid solution. The following reaction goes to completion:

$$Al_4C_3(s) + HCl(aq) \longrightarrow AlCl_3(aq) + CH_4(g)$$

Balance this equation. In one particular experiment, the methane produced was collected and found to have a mass of 1.257 g. What mass of $Al_4C_3(s)$ was added to the acid solution? What mass of $AlCl_3$ is formed in solution as the other product?

3.44 A mass of 1.24 g of NaOH is dissolved in water, and 2.05 g $H_2S(g)$ is bubbled in. The following reaction occurs:

$$H_2S(g) + 2NaOH(aq) \longrightarrow Na_2S(aq) + 2H_2O(l)$$

What mass of Na_2S is produced, assuming that the limiting reagent is completely consumed?

3.45 What mass of $AgBr(s)$ is formed when a solution containing 2.45 g of KBr is mixed with a solution containing 5.86 g of $AgNO_3$?

3.46 Several samples of a particular brand of cigarette were analyzed and found to contain an average of 2.25×10^{-5} g of nickel, Ni. The total quantity of nickel metal present in the ashes and cigarette butts of smoked cigarettes was found to be 1.67×10^{-5} g per cigarette. Assuming that the missing nickel was lost from the burning cigarette in the form of gaseous nickel carbonyl, $Ni(CO)_4$, what weight of nickel carbonyl is produced on the average per cigarette?

Molarity; solution stoichiometry

3.47 Calculate the molarity of each of the following solutions: (a) 4.55 g of Na_2CrO_4 in 250 mL of solution; (b) 0.088 mol of H_2SO_4 in 2.5 L of solution; (c) 65.8 kg of NaF in 3.45×10^8 L of solution; (d) 7.6×10^{-3} g of MgF_2 in 100 mL of solution.

3.48 The compound manganese(II) oxalate dihydrate, $MnC_2O_4 \cdot 2H_2O$, is soluble in water at $25°C$ to the extent of 0.0312 g per 100 mL of solution; at $55°C$ the solubility is 0.0643 g per 100 mL of solution. Calculate the molarity of the saturated solution in each case.

3.49 Calculate the mass, in grams, of solute present in each of the following solutions: (a) 240 mL of 0.112 M CdF_2; (b) 125 mL of 0.120 M HNO_3; (c) 6.85 L of 0.0224 M NH_4CN; (d) 5.6×10^{12} m^3 of 2.2×10^{-8} M tetrachlorobiphenyl, $Cl_4C_{12}H_6$.

3.50 What volume of solution is required to obtain: (a) 4.5 g of NH_4Cl from a 0.600 M solution of NH_4Cl; (b) 2.35 mol of H_2SO_4 from a 6.55 M solution of H_2SO_4; (c) 0.088 g of As from a 0.020 M solution of H_3AsO_4; (d) 1.25 g of Cr from a 0.0600 M solution of $K_2Cr_2O_7$.

3.51 A bottle of perchloric acid, $HClO_4$ is analyzed to have a concentration of 2.06 M. (a) Describe how you would use this solution to prepare 1.00 L of 0.100 M perchloric acid solution. (b) What volume of the original solution is required to react completely with 1.65 L of 0.880 M KOH? (c) What volume of the more dilute solution is needed to react completely with 35.0 mL of 0.125 M Na_2CO_3 according to the reaction

$$2HClO_4(aq) + Na_2CO_3(aq) \longrightarrow$$
$$2NaClO_4(aq) + CO_2(g) + H_2O(l)$$

(d) A volume of 54.6 mL of the 0.100 M $HClO_4$ solution is required to neutralize 34.0 mL of an NaOH solution of unknown molarity. What is the concentration of the NaOH solution?

3.52 In the laboratory, 5.55 g of $Sr(NO_3)_2$ is dissolved in enough water to form 0.750 L. A 0.100-L sample is withdrawn from this stock solution and titrated with a 0.0460 M solution of Na_2CrO_4. What volume of Na_2CrO_4 solution is required to precipitate all the $SrCrO_4$?

3.53 Exactly 26.3 mL of a 0.110 M HCl solution was required to neutralize all the base in 46.5 mL of a RbOH solution. What is the molarity of the RbOH solution? What mass of RbOH is required to make up 1.00 L of a solution of this concentration?

3.54 A sample of zinc nitrate is known to contain some zinc chloride as an impurity. When 2.87 g of the material is dissolved in water and titrated with 0.0130 M $AgNO_3$, it requires 2.94 mL to precipitate all the chloride as insoluble AgCl. What is the weight percentage of zinc chloride in the zinc nitrate?

3.55 A sample of solid $Ca(OH)_2$ is allowed to stand in contact with water at $30°C$ for a long time, until the solution contains as much dissolved $Ca(OH)_2$ as it can hold. A 100-mL sample of this solution is withdrawn and titrated with 5.00×10^{-2} M HBr. It requires 48.8 mL of the acid solution for neutralization. What is the concentration of the $Ca(OH)_2$ solution? What is the solubility of $Ca(OH)_2$ in water, at $30°C$, in grams of $Ca(OH)_2$ per 100 mL of solution?

Additional exercises

3.56 Balance the following equations:

(a) $Li_3N(s) + H_2O(l) \longrightarrow NH_3(g) + LiOH(aq)$

(b) $C_3H_7OH(l) + O_2(g) \longrightarrow CO_2(g) + H_2O(g)$

(c) $PBr_3(l) + H_2O(l) \longrightarrow H_3PO_3(aq) + HBr(aq)$

(d) $Mg_3B_2(s) + H_2O(l) \longrightarrow$
$\qquad Mg(OH)_2(aq) + B_2H_6(g)$

(e) $CCl_4(g) + O_2(g) \longrightarrow CCl_2O(g) + Cl_2(g)$

(f) $La(NO_3)_3(aq) + Ba(OH)_2(aq) \longrightarrow$
      $La(OH)_3(s) + Ba(NO_3)_2(aq)$

3.57 Given the following reactions:

$2HCl(aq) + CaCO_3(s) \longrightarrow$
      $CaCl_2(aq) + CO_2(g) + H_2O(l)$
$Zn(s) + 2HCl(aq) \longrightarrow H_2(g) + ZnCl_2(aq)$
$Na_2O(s) + H_2O(l) \longrightarrow 2NaOH(aq)$

predict what will happen in the following cases and write a balanced chemical equation for each reaction. (a) Hydrochloric acid is added to solid $BaCO_3$. (b) Nitric acid is added to solid $CaCO_3$. (c) Solid potassium oxide, K_2O, is added to water. (d) Zinc metal is added to hydrobromic acid, HBr. (e) Barium oxide, $BaO(s)$ is added to water.

3.58 A series of molecules containing the hypothetical element X are analyzed to determine the mass of X per molecule as shown below:

Weight of X per molecule	
Substance 1	22 amu
Substance 2	88 amu
Substance 3	44 amu

What is the most likely value for the atomic weight of element X? Suggest a second possible value.

3.59 The mass spectrum of carbon indicates that it contains 98.89 percent ^{12}C (atomic mass = exactly 12 amu) and 1.11 percent ^{13}C (atomic mass = 13.003 amu). Calculate the average atomic weight of carbon.

3.60 Calculate the weight percentage of hydrogen, sulfur, and oxygen in H_2SO_3.

3.61 Calculate the weight percentage of each element in $K_2Cr_2O_7$.

3.62 Calculate the number of moles in the following: (a) 12 g of Na_2CO_3; (b) 44.2 g of Ar; (c) 65 kg of H_2; (d) 2.3×10^{-7} g of $HgCl_2$.

3.63 What is the empirical formula of a compound between chromium and oxygen that contains 68.4 percent Cr?

3.64 A compound that has a molecular weight of 122 and that contains oxygen is found on analysis to contain 68.8 percent carbon and 4.92 percent hydrogen. Determine the molecular formula.

3.65 Cyclopropane, a substance used with oxygen as a general anesthetic, contains only two elements, carbon and hydrogen. When 1.00 g of this substance is completely combusted, 3.14 g of CO_2 and 1.29 g of H_2O are produced. What is the simplest formula of the cyclopropane?

3.66 Methoxyflurane, an anesthetic developed in the United States during the 1960s, contains fluorine to the extent of 23.0 percent. If there are two fluorine atoms per molecule of methoxyflurane, what is the molecular weight of this substance?

3.67 Halothane, a widely used anesthetic, contains carbon, hydrogen, chlorine, bromine, and fluorine. Its elemental analysis yields the following results: carbon 12.2 percent, hydrogen 0.51 percent, fluorine 28.9 percent, chlorine, 18.0 percent, and bromine 40.4 percent. What is the empirical formula of this substance?

3.68 What mass of ZnO is obtained by heating 80 kg of $ZnSO_3(s)$ according to the following equation:

$ZnSO_3(s) \longrightarrow ZnO(s) + SO_2(g)$

How many grams of $ZnSO_3$ are required to form 2.0 kg of SO_2 in this reaction?

3.69 A mixture of 3.50 g of H_2 and 26.0 g of O_2 is caused to react to form H_2O. How much H_2, O_2, and H_2O remain after reaction is complete?

3.70 A particular coal contains 2.8 percent sulfur by weight. When this coal is burned, the sulfur appears as $SO_2(g)$. This SO_2 is reacted with CaO to form $CaSO_3(s)$. If the coal is burned in a power plant that uses 2000 tons of coal per day, what is the daily production of $CaSO_3$?

3.71 Several samples of a particular brand of cigar were analyzed and found to contain on the average 8.00×10^{-6} g of iron, Fe, per cigar. The quantity of iron remaining in the ashes and butts of smoked cigars was found to average 5.92×10^{-6} g of iron. Assuming that the missing iron was lost from the cigar during smoking as gaseous iron pentacarbonyl, $Fe(CO)_5$, what weight of iron pentacarbonyl is formed on the average during smoking of each cigar?

3.72 In the Solvay process for forming soda ash (sodium carbonate), sodium bicarbonate is formed at one stage in the reaction:

$NH_4HCO_3(aq) + NaCl(aq) \longrightarrow$
      $NaHCO_3(s) + NH_4Cl(aq)$

Assuming complete precipitation, what mass of $NaHCO_3(s)$ could be formed from a solution containing excess NaCl and 4.60 kg of NH_4HCO_3?

3.73 Calculate the empirical formula and molecular formula of each of the following substances: (a) epinephrine (adrenaline), a hormone secreted in the bloodstream in times of danger or stress: 59.0 percent C, 7.1 percent H, 26.2 percent O, and 7.7 percent N; MW about 183 amu; (b) nicotine, a component of tobacco: 74.1 percent C, 8.6 percent H, and 17.3 percent N; MW = 160 ± 5 amu; (c) ethylene glycol, the substance used as the primary component of most antifreeze solutions: 38.7 percent C, 9.7 percent H, and 51.6 percent O; MW = 62.1 amu; (d) caffeine, a stimulant found in coffee: 49.5 percent C, 5.15 percent H, 28.9 percent N, and 16.5 percent O; MW about 195 amu.

[3.74] Chloromycetin is an antibiotic with the formula $C_{11}H_{12}O_5N_2Cl_2$. A 1.03-g sample of an ophthalmic ointment containing chloromycetin was chemically treated so as to convert its chlorine into Cl^- ions. The Cl^- was then precipitated as AgCl. If the AgCl weighed 0.0129 g, calculate the weight percentage of chloromycetin in the sample.

[3.75] The arsenic in a 1.22-g sample of a pesticide was converted to AsO_4^{3-} by suitable chemical treatment. It was then titrated using Ag^+ to form Ag_3AsO_4 as a precipitate. If it took 25.0 mL of 0.102 M Ag^+ to reach the equivalence point in this titration, what is the percentage of arsenic in the pesticide?

[3.76] A tank of impure carbon monoxide contains 85 parts per million by weight of oxygen impurity. The O_2 is to be removed by passing the tank gas over activated cobalt(II) oxide:

$$4CoO(s) + O_2(g) \longrightarrow 2Co_2O_3(s)$$

If the uptake of oxygen is 100 percent efficient, how many grams of the CO can be purified by passage through a column containing 58.8 g of CoO?

3.77 Calculate the molarity of each of the following solutions: (a) 35.0 g of H_2SO_4 in 600 mL of solution; (b) 42.0 mL of 0.550 M HNO_3 solution diluted to 0.500-L volume; (c) a solution formed by dissolving 2.56 g of NaBr in water to form 65.0 mL of solution; (d) a solution formed by mixing 35mL of 0.50 M KBr solution with 65 mL of 0.36 M KBr solution (assume 100 mL total volume).

3.78 A household cleaning solution contains approximately 63 g of Na_3PO_4 per liter of solution. What is the molarity of the solution with respect to this ingredient?

3.79 Describe how you would prepare each of the following solutions: (a) 0.300 L of 0.35 M KOH solution, beginning with solid KOH; (b) 1.00 L of a 0.500 M $AgNO_3$ solution, starting with solid $AgNO_3$; (c) 1 liter of a 0.200 M solution of H_2SO_4, starting with a concentrated H_2SO_4 solution of unknown concentration and a quantity of 0.150 M NaOH solution.

3.80 The objectionable taste of chlorine in highly chlorinated water can be removed by passing the water through a bed of charcoal. The charcoal, which is primarily carbon, reacts with the chlorine as follows:

$$C(s) + Cl_2(g) + H_2O(l) \longrightarrow CO_2(g) + HCl(aq)$$

Balance this equation. Calculate the mass of CO_2 produced by complete reaction of all the chlorine · in 6.0×10^3 m^3 of water containing 3.2×10^{-7} M chlorine.

3.81 Analysis of a sample of seawater taken from the English Channel revealed that there were 3.41×10^{-5} g of phosphorus and 7.98×10^{-4} g of nitrogen per liter of seawater. How many moles each of dissolved phosphorus and nitrogen are there per liter?

3.82 If it requires 64.0 mL of 1.06 M NaOH solution to neutralize 10.0 mL of the H_2SO_4 solution from an auto battery, what is the molarity of the H_2SO_4 solution?

3.83 Aspirin, $C_9H_8O_4$, is produced from salicylic acid, $C_7H_6O_3$, and acetic anhydride, $C_4H_6O_3$:

$$C_7H_6O_3 + C_4H_6O_3 \longrightarrow C_9H_8O_4 + HC_2H_3O_2$$

(a) How much salicylic acid is required to produce 1.5×10^2 kg of aspirin, assuming that all of the salicylic acid is converted to aspirin? (b) How much salicylic acid would be required if only 80 percent of the salicylic acid is converted to aspirin?

3.84 What volume of 0.20 M HBr is required to react with 18.0 g of zinc according to the equation

$$2HBr(aq) + Zn(s) \longrightarrow ZnBr_2(aq) + H_2(g)$$

[3.85] A solid sample of $Zn(OH)_2$ is added to 0.400 L of a 0.550 M solution of HBr. The solution that remains is still acidic. It is then titrated with 0.500 M NaOH solution, and 165 mL of the NaOH solution is required to reach the equivalence point. What was the mass of $Zn(OH)_2$ added to the HBr solution?

4

Energy relationships in chemical systems

In the past two chapters we have considered the classification of matter into elements and compounds, the structure of matter in terms of atoms, ions, and molecules, and the changes of matter in chemical reactions. However, an important feature of chemical reactions is that they also involve changes in energy. Chemistry is therefore concerned with both matter and energy. Our focus in this chapter is on energy and its involvement in chemical processes. Most energy produced in modern society comes from chemical reactions, especially combustion of coal, petroleum products, and natural gas.

For a long period in our history, people relied solely on the energy provided by their own bodies to perform work. Slowly they learned how to utilize animals, wind, water, and fire to enable them to accomplish more work. The largest jump in energy consumption came with the industrial revolution, which was really an energy revolution; the steam engine was the primary invention. Machines were developed that derived their energy largely from coal, then from oil and natural gas. Energy use changed transportation from travel by foot and animal to travel by autos and rockets; it changed communication from face-to-face talking and handwriting to mass printing and telecommunications.

An ample supply of energy at reasonable cost is essential to the working of modern society. This fact lies at the heart of a most critical problem confronting the human race: Where will the energy come from to sustain civilization in the decades to come? In the short century since humankind learned how to utilize fossil fuels as a major energy source—first coal, then oil and gas—we have used up a substantial part of all the reserves that exist. Furthermore, it is becoming increasingly evident that burning a large fraction of all the fossil fuel stores that remain could result in disastrous changes in world climate (see Section 10.5). In the meantime, the demands for energy from an ever-increasing world population continue to grow. In the United States, we are faced with political and economic crises resulting from an excessive dependence on foreign sources of oil and from the lack of a sound long-term energy plan.

The realization that energy may be in short supply, or may become excessively expensive, has in recent years brought about a keener interest in conserving energy. In addition, more attention has been paid to how we in the United States can more effectively utilize coal, which is relatively plentiful. To do this we must convert coal to more useful forms such as synthetic gasoline, synthetic gas, or liquid fuel oils. In addition, more attention will be given to converting plant matter into useful forms of fuel such as alcohols. Measures such as these will be the keystones of energy utilization in the decades to come, until we can arrive at long-range solutions to the problems of nuclear energy use or can develop nuclear fusion as an energy source (see Sections 20.7, 20.8, and 20.9). Thus, chemistry is the key to effective use of the resources on which we will be most dependent for the foreseeable future.

This chapter is really an introduction to thermodynamics, the study of energy, of the forms it can take and the rules that limit interconversion of energy from one form to another. Thermodynamics is based on three laws that summarize our observations of how nature behaves. In this chapter we will be concerned with just the first law of thermodynamics, which states simply that in any change that occurs in nature, the total energy of the universe remains constant. The first law is thus a conservation law; it does for energy what the law of conservation of matter does for the matter involved in a process. Of course, energy may flow from one part of the universe to another, and it may be converted from one form to another. It is precisely these changes we must learn to measure and understand.

4.1 The nature of energy

What is energy? We can't see, touch, or smell it as we can matter; it is a more abstract concept to us. Therefore, it is best that we carefully define energy and consider the concept from a general point of view before we get too deeply into our discussions. Energy is the capacity to do work or to transfer heat. Work (w) is defined and measured by the product of the net force (f) and the distance (d) through which that force moves:

$$w = f \times d \qquad [4.1]$$

That is, work is the movement of an object against some force, and energy is the "something" that is required to perform work. For example, energy is required to perform the work associated with lifting a book against the force of gravity. Work is also required to separate a positively charged ion from a negatively charged one because of the attractive force that exists between them. Matter is the substance of the universe; energy is the mover of the substance.

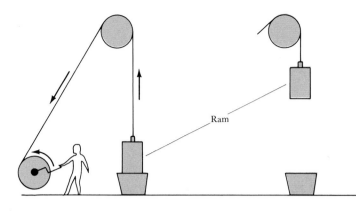

FIGURE 4.1 The ram of the pile driver has potential energy because of its position. This energy is supplied by the person who lifts the ram.

FORMS OF ENERGY

Energy is found in a variety of forms such as heat (thermal energy), light (radiant energy), chemical energy, mechanical energy, and electrical energy. The various forms or types of energy can also be classified as either kinetic or potential energy. Kinetic energy is the energy of motion. The magnitude of the kinetic energy of an object depends on its mass (m) and its velocity (v):

$$E_k = \tfrac{1}{2}mv^2 \qquad\qquad [4.2]$$

This equation tells us what we may have realized intuitively: The larger the mass of a moving object and the greater its velocity or speed, the more work it can do. The SI unit of energy is the joule (J). A joule is 1 kg-m^2/sec^2. This is the kinetic energy possessed by a mass of 2 kg moving at a velocity of 1 m per second:

$$E_k = \tfrac{1}{2}(2 \text{ kg})(1 \text{ m/sec})^2 = 1 \text{ kg-m}^2/\text{sec}^2 = 1 \text{ J}$$

Potential energy is the energy stored in an object by virtue of its position or composition. For example, the ram of a pile driver can be lifted against the force of gravity as shown in Figure 4.1. Once lifted it has the potential to do work; the work is accomplished by allowing the ram to fall. The ram therefore has potential energy. In a similar way two unlike-charged particles that are separated against the force of their electrostatic attraction for each other have potential energy.

Let's return to the pile driver for a few more moments. Work is required to lift the ram of the pile driver. The energy to perform this work could be supplied by the brute force of a person. In our society, machines have replaced people in many such operations, and a simple engine such as that shown in Figure 4.2 could be used to lift the ram. In this case, the energy stored in the coal used to fuel the engine is converted to heat that is transferred to the cylinder. As the gas within the cylinder is heated it expands, thereby moving the piston. This moves the wheel that lifts the ram. Energy is thereby transferred from the coal to the ram. Experience with many such processes tells us that although energy can be transferred from one object to another, and its forms interconverted, no energy is created or destroyed. Observations such as these, based on quantitative

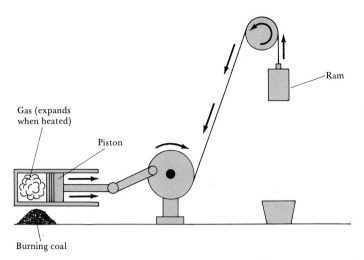

Gas (expands
when heated)

Piston

Burning coal

Ram

FIGURE 4.2 A simple engine that uses the expansion of a gas upon heating to lift the ram of the pile driver. The potential energy of the coal is transferred to the ram.

measurements of heat and work, form the basis of the first law of thermodynamics.

If energy is conserved in any chemical or physical change of matter, you may well wonder why there is any concern over energy resources and why the term "energy crisis" appears in the popular press. The situation with energy is very much like that with matter. In discussing the law of conservation of mass, we spoke of resources being "depleted" in the sense of being converted into forms that are less and less useful to people. The resources become too dilute or can be extracted only through the expenditure of a great deal of energy. Basically the same is true of energy; in any energy transformation some energy is always converted to heat that gets spread throughout the environment and consequently cannot be used to do work. For instance, a common automobile is only about 10 percent efficient in converting the chemical energy of gasoline into mechanical energy used to propel the car; the remainder goes into the environment as heat. Indeed, most of the energy associated with the use of coal and petroleum ends up as heat. When an excessive amount of this heat is concentrated in a small region of the environment, thermal pollution results. For example, the waste heat from an electric power plant may heat the waters of a lake.

4.2 Energy changes in chemical reactions

The combustion of coal mentioned in the introduction of this chapter is only one of many familiar chemical reactions that produce energy. Consider the combustion of a piece of magnesium ribbon:

$$2Mg(s) + O_2(g) \longrightarrow 2MgO(s) + \text{heat and light} \qquad [4.3]$$

Most of us have seen this reaction whether we realize it or not. Flash bulbs are filled with magnesium ribbon and oxygen gas. Passage of an electric current through the magnesium causes it to ignite, producing heat and light.

We can label the magnesium and oxygen atoms in this reaction as our system. Everything around this system, including the container in which the reaction takes place, is called the surroundings. Labeling the portion of the universe that is under examination as the system and the rest of

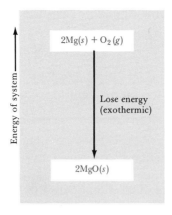

FIGURE 4.3 An energy diagram for an exothermic reaction. Heat is given off by the system to its surroundings, thereby decreasing the energy content of the system.

the universe as the surroundings allows us to clarify what goes on in any energy transfer. Because energy is neither created nor destroyed, any energy lost by the system must be gained by the surroundings, and any energy gained by the system must be supplied by the surroundings. In our example, energy is lost by the system to the surroundings as the atoms of the system rearrange from magnesium metal and oxygen gas into magnesium oxide, MgO. Chemical and physical changes that give off heat to their surroundings are described as being exothermic. Those that absorb heat are labeled as endothermic. We can represent exothermic reactions as shown schematically in Figure 4.3.

An example of an endothermic reaction is the decomposition of water into its elements:

$$\text{Energy} + 2H_2O(l) \longrightarrow 2H_2(g) + O_2(g) \tag{4.4}$$

As we noted in Chapter 2, this reaction can be carried out by supplying electrical energy to the water. This endothermic process is represented schematically in Figure 4.4.

Traditionally, the energy changes accompanying chemical reactions have been measured in calories (cal). A calorie is the amount of energy required to raise the temperature of 1 g of water by 1°C, from 14.5 to 15.5°C.* A kilocalorie, 1000 cal, is the same as the Calorie (capitalized)

*The temperature interval 14.5 to 15.5°C is specified because the energy required to raise the temperature of water by 1°C is slightly different at different temperatures. However, it is constant to three significant figures, 1.00 cal, over the entire liquid range of water, from 0 to 100°C.

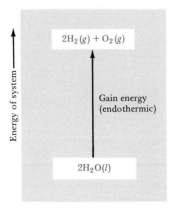

FIGURE 4.4 An energy diagram for an endothermic reaction. Heat is absorbed by the system from the surroundings, thereby increasing the energy content of the system.

used in nutrition. One calorie is the same amount of energy as 4.184 joules:

$$1 \text{ cal} = 4.184 \text{ J} \qquad [4.5]$$

In most places in this text the joule (J) or kilojoule (kJ) is used as the unit of energy.

4.3 Enthalpy

Most chemical reactions, and especially those occurring in living organisms, take place under the essentially constant pressure of the atmosphere. Furthermore, it often happens that the heat given off or absorbed by the reacting system is absorbed or supplied by the surroundings as needed, so as to maintain constant temperature. Thus, for example, the combustion of sugar, an exothermic process, occurs in the body at a constant temperature, about 37°C.

It is useful to discuss chemical changes occurring at constant pressure in terms of a thermodynamic quantity called the heat content, or enthalpy, given by the symbol H. The change in enthalpy of a system during a process that occurs under constant pressure, represented as ΔH (read "delta H"),* is equal to the heat given off or absorbed by the system during that process. To get some feeling for what sort of quantity enthalpy is, let us consider the substance water. A mole of liquid water at 25°C and under normal atmospheric pressure has a certain heat content, or enthalpy. If the amount of water is doubled, to 2 mol, the enthalpy is also doubled. Thus, enthalpy is an extensive property of a system, like volume or mass. We can, however, speak of the enthalpy per mole, just as we speak of the volume per mole or mass per mole (molecular weight). Secondly, we note that the enthalpy depends on the exact state of the water. For example, the enthalpy of liquid water at 50°C, or of gaseous water at 25°C, is different from that of liquid water at 25°C. Once we have specified the relevant conditions, however, the enthalpy is determined. Note that we do not need to recite how the sample came to be in that condition; only its present state matters. Properties of a system that are determined by specifying its state are called state functions. Enthalpy is such a function. So is the temperature, and so also is mechanical potential energy, as illustrated in Figure 4.5.

When a chemical reaction occurs, the enthalpies of the products will, in general, differ from those of the reactants. Thus, there is an overall change in the enthalpy of the system in going from reactants to products. By convention the enthalpy change for a reaction (ΔH_{rxn}) is taken to be the enthalpy or heat content of the products minus that of the reactants:

$$\Delta H_{rxn} = H(\text{products}) - H(\text{reactants}) \qquad [4.6]$$

If the enthalpy of the products is less than that of the reactants, ΔH will be negative in sign; on the other hand, when the products have a larger enthalpy than the reactants, ΔH is positive.

---

*The symbol Δ is commonly used to denote *change*. For example, a change of volume can be represented by ΔV.

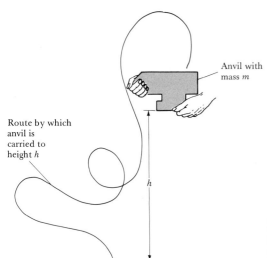

Route by which anvil is carried to height h

Anvil with mass m

h

Floor

FIGURE 4.5 The potential energy of an anvil above a floor is a state function that depends on the height, h. The magnitude of the potential energy of the anvil does not depend on how it proceeded from the floor to its elevated position. The potential energy *change* depends only on its initial and final positions.

The enthalpy change that results from a chemical or physical process might be manifested in any of several ways. For example, heat or light may be evolved, or heat may be absorbed; the energy that the enthalpy change represents may be used in part to produce electrical energy as in a battery; work may be done, as when energy is released in a muscle fiber. You should keep in mind that the enthalpy change in the process is determined simply by specifying the initial and final states of the system. It does not depend on whether the process was carried out to produce only heat, or some mixture of heat and work. For example, combustion of 1 mol of methane, the principal component of natural gas, is represented by the equation

$$CH_4(g) + 2O_2(g) \longrightarrow CO_2(g) + 2H_2O(g) \qquad [4.7]$$

We could allow 1 mol of methane to burn at the nozzle of a burner tip, producing only heat, or we might use it as fuel for a small engine, in which case we obtain both heat and work from the combustion. In either case, the change in enthalpy of the system, consisting of 1 mol of $CH_4(g)$ and 2 mol of $O_2(g)$, is the same, if the final state of the products is the same. In this chapter we will focus attention on processes that are carried out so that all of the enthalpy change appears as heat evolved or absorbed. When the change in enthalpy is negative, there is an overall decrease in enthalpy in proceeding from reactants to products, and heat is evolved. Such processes are exothermic. On the other hand, when ΔH is positive, heat is absorbed by the system, and the process is endothermic.

Let's return to our example of the combustion of methane, Equation [4.7]. It is found experimentally that 802 kJ of heat is produced when 1 mol of CH_4 is burned in a constant-pressure system. We can express this fact as follows:

$$CH_4(g) + 2O_2(g) \longrightarrow CO_2(g) + 2H_2O(g) \qquad \Delta H = -802 \text{ kJ} \qquad [4.8]$$

The negative sign for ΔH tells us that this is an exothermic process. The coefficients in the balanced equation are taken to represent the number of moles of reactants giving the associated enthalpy change. As a consequence of the law of conservation of energy, the amount of heat associated with a reaction is directly proportional to the amount of substances involved. Thus combustion of 1 mol of CH_4 produces 802 kJ of heat, whereas combustion of 2 mol produces 1604 kJ.

SAMPLE EXERCISE 4.1

When barium chlorate, $Ba(ClO_3)_2$, is added to water at $25\,°C$, the salt dissolves and the solution temperature decreases below $25\,°C$. Is the process of dissolving exothermic or endothermic? What is the sign of ΔH?

Solution: The fact that the solution temperature decreases as the salt dissolves indicates that the proc-

ess is one that requires absorption of heat from the surroundings. Thus, it is an endothermic process. To maintain the temperature of the system, consisting of salt plus water, at $25°$, it is necessary to add heat; thus ΔH is positive, as is always the case for an endothermic process.

SAMPLE EXERCISE 4.2

How much heat is produced when 4.50 g of methane gas are burned in a constant-pressure system?

Solution: According to Equation [4.8], 802 kJ are produced when a mole of CH_4 is burned. A mole of

CH_4 has a mass of 16.0 g.

Heat produced $= (4.50 \text{ g } CH_4)$

$$\times \left(\frac{1 \text{ mol } CH_4}{16.0 \text{ g } CH_4}\right)\left(802 \frac{kJ}{\text{mol } CH_4}\right) = 226 \text{ kJ}$$

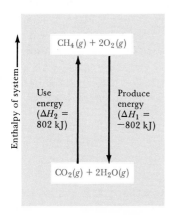

FIGURE 4.6 If a substance is combusted and then reformed from its combustion products, the total change in enthalpy must be zero ($\Delta H_1 + \Delta H_2 = 0$). In general, the net energy change for a series of reactions that eventually regenerate the original substances in their original conditions is zero.

In a related fashion, the enthalpy change for a reaction is equal in magnitude but opposite in sign to ΔH for the reverse reaction. For example:

$$CO_2(g) + 2H_2O(g) \longrightarrow CH_4(g) + 2O_2(g) \qquad \Delta H = 802 \text{ kJ} \qquad [4.9]$$

If more energy were produced by combustion of CH_4 than was required for the reverse reaction, it would be possible to use these processes to create an unlimited supply of energy. Some CH_4 could be combusted, and that portion of the energy necessary to reform CH_4 could be saved. The rest could be used to do work of some type. After CH_4 is reformed, it could again be combusted, and so forth, continually supplying energy. This, of course, is clearly contrary to our experience—such behavior does not obey the law of conservation of energy. The observed situation is shown in Figure 4.6.

The enthalpy change of a reaction also depends on the state of the reactants and products. If the product in the combustion of methane, Equation [4.8], were liquid H_2O instead of gaseous H_2O, ΔH would be -890 kJ instead of -802 kJ. More heat is available for transfer to the surroundings because 88 kJ are released when 2 mol of gaseous water are condensed to the liquid state:

$$2H_2O(g) \longrightarrow 2H_2O(l) \qquad \Delta H = -88 \text{ kJ} \qquad\qquad [4.10]$$

Therefore, the states of the reactants and products must be specified. In addition, we will generally assume that the reactants and products are both at the same temperature, usually 25°C, unless otherwise indicated.

4.4 Hess's law

One specialized statement of the first law of thermodynamics that is especially useful to chemistry is known as Hess's law of constant heat summation. According to Hess's law, if a reaction can be carried out in a series of steps, ΔH for the reaction will be equal to the sum of the enthalpy changes for each step; the enthalpy changes are additive. For example, the enthalpy change for the combustion of methane to form carbon dioxide and liquid water can be calculated from ΔH for the condensation of water vapor and ΔH for the combustion of methane giving gaseous water:

$$CH_4(g) + 2O_2(g) \longrightarrow CO_2(g) + 2H_2O(g) \qquad \Delta H = -802 \text{ kJ}$$

(Add) $\qquad 2H_2O(g) \longrightarrow 2H_2O(l) \qquad\qquad \Delta H = -88 \text{ kJ}$

$$CH_4(g) + 2O_2(g) + 2H_2O(g) \longrightarrow CO_2(g) + 2H_2O(l) + 2H_2O(g)$$
$$\Delta H = -890 \text{ kJ}$$

Net equation:

$$CH_4(g) + 2O_2(g) \longrightarrow CO_2(g) + 2H_2O(l) \qquad \Delta H = -890 \text{ kJ}$$

To obtain the net equation, the sum of the reactants of the two equations are placed on one side of the arrow, and the sum of the products on the other. Because $2H_2O(g)$ occurs on both sides of the arrow, it can be canceled like an algebraic quantity that is on both sides of an equal sign.

Hess's law provides a useful means of calculating energy changes that are difficult to measure directly. For instance, it is not possible to measure directly the heat of combustion of carbon to form carbon monoxide. Combustion of 1 mol of carbon with $\frac{1}{2}$ mol of O_2 produces not only CO, but also CO_2, leaving some carbon unreacted. However, the heat of the reaction forming CO can be calculated as shown in Sample Exercise 4.3.

SAMPLE EXERCISE 4.3

The heat of combustion of C to CO_2 is -393.5 kJ/mol of CO_2, whereas that for combustion of CO to CO_2 is -283.0 kJ/mol of CO_2. Calculate the heat of combustion of C to CO.

Solution: You need first to write out the two combustion reactions and then turn the combustion reaction for CO around, so that CO is the product. The two equations can then be added, as follows:

$$2C(s) + 2O_2(g) \longrightarrow 2CO_2(g)$$
$$\Delta H = -2(393.5) = -787.0 \text{ kJ}$$
$$2CO_2(g) \longrightarrow 2CO(g) + O_2(g)$$
$$\Delta H = 566.0 \text{ kJ}$$

$$2C(s) + O_2(g) \longrightarrow 2CO(g)$$
$$\Delta H = -221.0 \text{ kJ}$$

Note that it is necessary to multiply the first equation through by two to obtain a proper canceling of terms. This requires that ΔH for the process also be multiplied by two, as shown. Remember that when reactions are turned around, the sign of ΔH for the reversed process is the opposite of the original value.

The heat of combustion of $C(s)$ to form $CO(g)$ is $\frac{1}{2}(-221.0 \text{ kJ}) = -110.5 \text{ kJ}$ *per mole of CO formed.*

SAMPLE EXERCISE 4.4

Given the following reactions and their respective enthalpy changes,

$$C_2H_2(g) + \tfrac{5}{2}O_2(g) \longrightarrow 2CO_2(g) + H_2O(l)$$
$$\Delta H = -1299.6 \text{ kJ/mol } C_2H_2$$

$$C(s) + O_2(g) \longrightarrow CO_2(g)$$
$$\Delta H = -393.5 \text{ kJ/mol C}$$

$$H_2(g) + \tfrac{1}{2}O_2(g) \longrightarrow H_2O(l)$$
$$\Delta H = -285.9 \text{ kJ/mol } H_2$$

calculate ΔH for the reaction:

$$2C(s) + H_2(g) \longrightarrow C_2H_2(g)$$

Solution: Because we want to end up with C_2H_2 we turn the first equation around; the sign of ΔH is therefore changed. Because we need to start with $2C(s)$, we multiply the second equation and its ΔH by two. We then add the resultant equations and their enthalpy changes in accordance with Hess's law:

$$2CO_2(g) + H_2O(l) \longrightarrow C_2H_2(g) + \tfrac{5}{2}O_2(g)$$
$$\Delta H = 1299.6 \text{ kJ}$$

$$2C(s) + 2O_2(g) \longrightarrow 2CO_2(g)$$
$$\Delta H = -787.0 \text{ kJ}$$

$$H_2(g) + \tfrac{1}{2}O_2(g) \longrightarrow H_2O(l)$$
$$\Delta H = -285.9 \text{ kJ}$$

$$\overline{2C(s) + H_2(g) \longrightarrow C_2H_2(g)}$$
$$\Delta H = 226.7 \text{ kJ}$$

When the three equations are added, there are $2CO_2$, $\tfrac{5}{2}O_2$, and H_2O on both sides of the arrow. These are canceled in writing the net equation.

The first law of thermodynamics, in the form of Hess's law, teaches us that we can never expect to obtain more (or less) energy from a chemical reaction by changing the method of carrying out the reaction. For example, for the reaction of methane, CH_4, and oxygen, O_2, to form CO_2 and H_2O, we may envision the reaction to occur either directly or with the initial formation of CO, which is then subsequently combusted. This set of choices is illustrated in Figure 4.7. Because ΔH is a state function, either path produces the same change in the enthalpy content of the system. That is, $\Delta H_1 = \Delta H_2 + \Delta H_3$. Again, we may note that if this were not so it would be possible to create energy continuously, in conflict with the first law of thermodynamics.

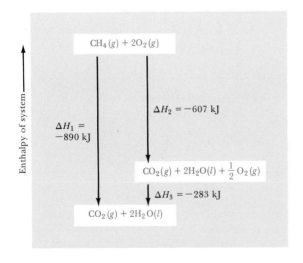

FIGURE 4.7 The quantity of heat generated by combustion of CH_4 is independent of whether the reaction takes place in one or more steps ($\Delta H_1 = \Delta H_2 + \Delta H_3$).

4.5 Heats of formation

The first law of thermodynamics permits calculation of the enthalpy changes for many reactions from a few tabulated values. It is convenient to summarize many of these data in terms of standard heats of formation.* The heat of formation of a compound, ΔH_f, is the enthalpy change involved in forming a mole of that compound from its constituent elements. A standard enthalpy change, $\Delta H°$, is one that takes place with all reactants and products in their standard states. That is, all substances are in the forms most stable at the particular temperature of interest and at standard atmospheric pressure (see Chapter 9). The temperature usually chosen for purposes of tabulating data is 25°C. Thus, for example, the standard heat of formation, $\Delta H_f°$, for ethanol (C_2H_5OH) is the enthalpy change for the following reaction:

$$2C(graphite) + 3H_2(g) + \tfrac{1}{2}O_2(g) \longrightarrow C_2H_5OH(l) \qquad [4.11]$$

The elemental source of oxygen is O_2, not O or O_3, because O_2 is the stable form of oxygen at 25°C and standard atmospheric pressure. Likewise, the elemental source of carbon is graphite and not diamond, because the former is the stable (lowest energy) form at 25°C and standard atmospheric pressure. The conversion of graphite to diamond requires the addition of energy as shown in Equation [4.12]:

$$C(graphite) \longrightarrow C(diamond) \qquad \Delta H° = 1.88 \text{ kJ} \qquad [4.12]$$

A few standard heats of formation are given in Table 4.1; a more complete table is provided in Appendix D. *By convention the standard heat of formation of the stable form of any element is zero.*

The standard enthalpy change for any reaction can be found by summing the heats of formation of all reaction products, taking care to multiply each molar heat of formation by the coefficient of that substance in the balanced equation, then subtracting a similar sum for all the heats of formation of all the reactants. For example, $\Delta H°$ for the combustion of glucose, Equation [4.13], is given by Equation [4.14]:

$$C_6H_{12}O_6(s) + 6O_2(g) \longrightarrow 6CO_2(g) + 6H_2O(l) \qquad [4.13]$$

$$\Delta H_{rxn}° = [6\,\Delta H_f°(CO_2) + 6\,\Delta H_f°(H_2O)] \\ - [\Delta H_f°(C_6H_{12}O_6) + 6\,\Delta H_f°(O_2)] \qquad [4.14]$$

Using the heats of formation recorded in Table 4.1, this process gives:

$$\Delta H_{rxn}° = \left[(6 \text{ mol } CO_2)\left(-393.5\,\frac{kJ}{\text{mol } CO_2}\right) \right.$$
$$\left. + (6 \text{ mol } H_2O)\left(-285.9\,\frac{kJ}{\text{mol } H_2O}\right) \right]$$
$$- \left[(1 \text{ mol } C_6H_{12}O_6)\left(-1260\,\frac{kJ}{\text{mol } C_6H_{12}O_6}\right) \right.$$
$$\left. + (6 \text{ mol } O_2)\left(0\,\frac{kJ}{\text{mol } O_2}\right) \right]$$
$$= -2816 \text{ kJ}$$

---

*It would be more precise to refer to heats of formation as "enthalpy changes of formation." However, the usual practice is to use the more general term "heat of formation."

TABLE 4.1 Standard heats of formation, ΔH_f°, at 25°C

Substance	Formula	ΔH_f° (kJ/mol)
Acetylene	$C_2H_2(g)$	226.7
Ammonia	$NH_3(g)$	−46.19
Benzene	$C_6H_6(l)$	49.04
Calcium carbonate	$CaCO_3(s)$	−1207.1
Calcium oxide	$CaO(s)$	−635.5
Carbon dioxide	$CO_2(g)$	−393.5
Carbon monoxide	$CO(g)$	−110.5
Diamond	$C(s)$	1.88
Ethane	$C_2H_6(g)$	−84.68
Ethanol	$C_2H_5OH(l)$	−277.7
Ethylene	$C_2H_4(g)$	52.30
Hydrogen bromide	$HBr(g)$	−36.23
Hydrogen chloride	$HCl(g)$	−92.30
Hydrogen fluoride	$HF(g)$	−268.6
Hydrogen iodide	$HI(g)$	25.9
Glucose	$C_6H_{12}O_6(s)$	−1260
Methane	$CH_4(g)$	−74.85
Methanol	$CH_3OH(l)$	−238.6
Silver chloride	$AgCl(s)$	−127.0
Sodium bicarbonate	$NaHCO_3(s)$	−947.7
Sodium carbonate	$Na_2CO_3(s)$	−1130.9
Sodium chloride	$NaCl(s)$	−411.0
Sucrose	$C_{12}H_{22}O_{11}(s)$	−2221
Water	$H_2O(l)$	−285.8
Water vapor	$H_2O(g)$	−241.8

The general relationship illustrated by Equation [4.14] follows directly from the fact that ΔH is a state function. This fact allows calculation of the enthalpy change for any reaction from the energies required to convert the initial reactants into elements and then combine the elements into the desired products. This reaction pathway is shown in Figure 4.8 for the combustion of glucose.

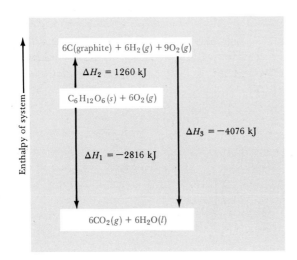

Enthalpy of system

$6C(\text{graphite}) + 6H_2(g) + 9O_2(g)$

$\Delta H_2 = 1260$ kJ

$C_6H_{12}O_6(s) + 6O_2(g)$

$\Delta H_3 = -4076$ kJ

$\Delta H_1 = -2816$ kJ

$6CO_2(g) + 6H_2O(l)$

FIGURE 4.8 Because H is a state function, $\Delta H_1 = \Delta H_2 + \Delta H_3$. Note that ΔH_2 is $-\Delta H_f^\circ(C_6H_{12}O_6)$, while ΔH_3 is $6\,\Delta H_f^\circ(H_2O) + 6\,\Delta H_f^\circ(CO_2)$. This is the same result as given in Equation [4.14].

SAMPLE EXERCISE 4.5

Compare the quantity of heat produced by combustion of 1.00 g of glucose ($C_6H_{12}O_6$) with that produced by 1.00 g of sucrose ($C_{12}H_{22}O_{11}$).

Solution: The example worked in the text gave $\Delta H° = -2816$ kJ for the combustion of a mole of glucose. The molecular weight of glucose is 180 amu. Therefore, the heat produced per gram is

$$\left(-2816 \frac{kJ}{mol}\right)\left(\frac{1 \, mol}{180 \, g}\right) = -15.6 \frac{kJ}{g}$$

For sucrose:

$$C_{12}H_{22}O_{11}(s) + 12O_2(g) \longrightarrow 12CO_2(g) + 11H_2O(l)$$

$$\Delta H°_{rxn} = [12 \, \Delta H°_f(CO_2) + 11 \, \Delta H°_f(H_2O)] - [\Delta H°_f(C_{12}H_{22}O_{11}) + 12 \, \Delta H°_f(O_2)]$$

$$= 12(-393.5 \, kJ) + 11(-285.9 \, kJ) - (-2221 \, kJ)$$

$$= (-4722 - 3145 + 2221) \, kJ$$

$$= -5646 \, kJ/mol \, C_{12}H_{22}O_{11}$$

The molecular weight of sucrose is 342 amu. Therefore, the heat produced per gram of sucrose is

$$\left(-5646 \frac{kJ}{mol}\right)\left(\frac{1 \, mol}{342 \, g}\right) = -16.5 \frac{kJ}{g}$$

Both sucrose and glucose are carbohydrates. As a rule of thumb the energy obtained from the combustion of a gram of carbohydrate is 17 kJ (4 kcal or 4 Cal).

SAMPLE EXERCISE 4.6

The standard enthalpy change for the reaction

$$CaCO_3(s) \longrightarrow CaO(s) + CO_2(g)$$

is 178.1 kJ. From the values for the standard heats of formation given in Table 4.1, calculate the standard heat of formation of $CaCO_3(s)$.

Solution: We have that the standard enthalpy change in the reaction is

$$\Delta H°_{rxn} = [\Delta H°_f(CaO + \Delta H°_f(CO_2)] - \Delta H°_f(CaCO_3)$$

Inserting the known values, we have:

$$178.1 \, kJ = -635.5 \, kJ - 393.5 \, kJ - \Delta H°_f(CaCO_3)$$

$$\Delta H°_f(CaCO_3) = -1207.1 \, kJ/mol$$

4.6 Measurement of energy changes; calorimetry

We have seen that under the most usual conditions for study of chemical reactions, the enthalpy change accompanying a chemical or physical process is equal to the heat evolved or absorbed by the system. We must now consider how these heat changes can be measured. In general, we measure quantities of heat by measuring the amounts of heat that flow from one object to another. The actual temperature change experienced by a body when it absorbs a certain amount of heat is determined by its heat capacity. An object of large heat capacity absorbs more heat when it undergoes a certain temperature change than does an object of small heat capacity. For example, it requires considerably more heat to raise the temperature of the water in an outdoor swimming pool from 15°C to 25°C than it does to produce the same temperature change in a fish tank. The heat capacity can be expressed as $C = Q/\Delta T$, where Q is the total heat flow into or out of an object, and ΔT is the temperature change that heat flow produces. In some cases it is important to specify whether the process occurs under constant pressure or constant volume conditions. Then the heat capacity is labeled C_P or C_V, to denote constant pressure or constant volume conditions, respectively. For liquids and solids, the two values are essentially the same.

SAMPLE EXERCISE 4.7

The heat capacity of iron(III) oxide, Fe_2O_3, is 120 J/K-mol. What quantity of heat is required to increase the temperature of a 2.0-kg brick of Fe_2O_3 from 120°C to 380°C?

Solution: To obtain C, the total heat capacity of the brick, we must multiply the molar heat capacity of Fe_2O_3 by the number of moles of Fe_2O_3 in the 2.0-kg brick:

$$\left(\frac{120 \text{ J}}{\text{K-mol } Fe_2O_3}\right)\left(\frac{1 \text{ mol } Fe_2O_3}{160 \text{ g } Fe_2O_3}\right)\left(\frac{2 \times 10^3 \text{ g } Fe_2O_3}{1 \text{ brick}}\right)$$

$$= \frac{1.50 \text{ kJ}}{\text{K}}$$

The total heat flow into the brick, Q, is the heat capacity of the brick times the temperature change. In this case, $\Delta T = 380°C - 120°C = 260°C = 260 \text{ K}$.

$$Q = \left(\frac{1.50 \text{ kJ}}{\text{K}}\right)260 \text{ K} = 390 \text{ kJ}$$

Substances vary widely in their heat capacities per unit mass. The heat capacities of several substances are listed in Table 4.2. The term **specific heat** refers to the ratio of the heat capacity of a substance to the heat capacity of water at 15°C. Thus, the specific heat is dimensionless.

Notice that water has the highest heat capacity per gram of all the substances listed. We will learn more about the reasons for this when we discuss hydrogen bonding in water (Section 11.5). The fact that liquid water has a relatively high heat capacity is an important factor in determining the earth's climate.

In solar heating systems for homes, the heat collected in solar panels may be stored in the form of hot water or heated rocks. Note that the heat capacities of typical minerals such as MgO or $CaCO_3$ are much lower than those of water. Thus, a much larger mass of rocks is required to provide the same thermal storage capacity.

Measurement of heat effects is known as calorimetry; the techniques and equipment employed in calorimetry depend on the nature of the process being studied. Combustion reactions are generally studied by means of a bomb calorimeter, a device shown schematically in Figure 4.9. The sample to be examined, for instance a fuel, is placed in a cup within a sealed vessel called a bomb. The bomb, which is designed to withstand high pressures, has an inlet valve for adding oxygen under pressure and also has electrical contacts to initiate the combustion reaction. After the sample has been placed in the bomb, the bomb is sealed and filled with oxygen. It is then placed in a large insulated container and covered with an accurately measured quantity of water. The combustion reaction is initiated by passing an electrical current through a fine wire that is in contact with the sample; the sample ignites when the

TABLE 4.2 Heat capacities of some selected substances

Compound	Temperature (°C)	Heat capacity (J/°C-g)	Specific heat
$H_2O(l)$	15	4.184	1.00
$H_2O(s)$	−11	2.03	0.485
$Al(s)$	20	0.89	0.212
$C(s)$	20	0.71	0.17
$Fe(s)$	20	0.45	0.11
$Hg(l)$	20	0.14	0.033
$CaCO_3(s)$	0	0.85	0.20
$MgO(s)$	0	0.87	0.21
$HgS(s)$	0	0.21	0.050

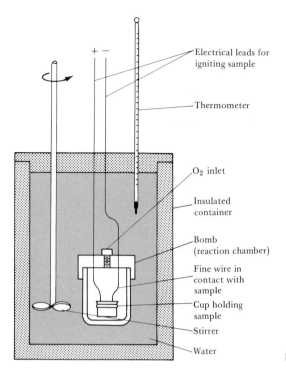

Electrical leads for
igniting sample

Thermometer

O_2 inlet

Insulated
container

Bomb
(reaction chamber)

Fine wire in
contact with
sample

Cup holding
sample

Stirrer

Water

FIGURE 4.9 A bomb calorimeter.

wire gets hot. The temperature of the water is carefully measured before
and after combustion. In keeping with the law of conservation of energy,
the heat lost by the burning sample is gained by its surroundings, namely
the calorimeter (including the water).

The heat absorbed by the calorimeter is determined experimentally
by combusting a sample that gives off a known quantity of heat. For
example, it is known that combustion of exactly 1 g of benzoic acid,
$C_7H_6O_2$, in a bomb calorimeter, produces 26.38 kJ of heat. Suppose that
1 g of benzoic acid is combusted in our calorimeter, and it causes a
temperature increase of 5.022°C. The heat capacity of the calorimeter is
then given by 26.38 kJ/5.022°C = 5.252 kJ/°C. Once we know the
value of the heat capacity of the calorimeter, we can measure temperature changes produced by other reactions, and from these we can calculate the heat, Q, evolved in the reaction:

$$Q_{evolved} = C_{calorimeter} \times \Delta T \qquad [4.15]$$

SAMPLE EXERCISE 4.8

The temperature of a bomb calorimeter is raised
1.17°C when 1.00 g of hydrazine, N_2H_4, is burned in
it. If the calorimeter has a heat capacity of
16.53 kJ/°C, what is the quantity of heat evolved?
What is the heat evolved upon combustion of a mole
of N_2H_4?

Solution: The quantity of heat evolved is just the
product of the temperature change times the heat
capacity, Equation [4.15]:

$$\left(\frac{16.53 \text{ kJ}}{1°C}\right) 1.17°C = 19.3 \text{ kJ heat evolved}$$

Since this is the amount of heat that results from
combustion of 1.00 g of hydrazine, the amount released by combustion of one mole is:

$$\left(\frac{-19.3 \text{ kJ}}{1 \text{ g N}_2\text{H}_4}\right)\left(\frac{32.0 \text{ g N}_2\text{H}_4}{1 \text{ mol N}_2\text{H}_4}\right) = -619 \text{ kJ/mol N}_2\text{H}_4$$

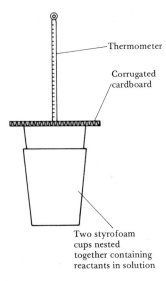

Thermometer

Corrugated cardboard

Two styrofoam cups nested together containing reactants in solution

FIGURE 4.10 A "coffee-cup" calorimeter.

The reactions in a bomb calorimeter occur under constant volume conditions. For this reason, the heats evolved in a bomb calorimeter experiment do not correspond to ΔH. However, it is not difficult to correct the measured heats so that they correspond to ΔH values. We need not concern ourselves here with how this is done. The heat evolved upon combustion of a sample at a constant pressure of 1 atm is termed the heat of combustion. The heat of combustion is normally listed in tables for $CO_2(g)$ and $H_2O(l)$ as products.

For many reactions, such as those occurring in solution, it is a simple matter to control pressure so that ΔH can be measured directly. One constant-pressure calorimeter is shown in Figure 4.10. Such simple, "coffee-cup" calorimeters are often used in freshman chemistry labs to illustrate the principles of calorimetry. Because the calorimeter is not sealed, the reaction occurs under the essentially constant pressure of the atmosphere. The heat change of the reaction is determined from the temperature increase of a known quantity of solution in the calorimeter, as shown in Sample Exercise 4.9.

SAMPLE EXERCISE 4.9

When 50 g of 1.0 M HCl (containing 0.050 mol of HCl) and 50 g of 1.0 M NaOH (containing 0.050 mol of NaOH) are mixed in a "coffee-cup" calorimeter, the temperature of the resultant solution increases from 21.0°C to 27.5°C. Assuming that it takes 4.18 J to increase the temperature of 1.00 g of the solution by 1.00°C, calculate the heat of the reaction

$$HCl(aq) + NaOH(aq) \longrightarrow H_2O(l) + NaCl(aq)$$

Solution: We can ignore the heat capacity of the plastic material from which the cup is made. Thus the total heat capacity of the calorimeter is just the heat capacity of the fluid contents:

$$100 \text{ g} \left(\frac{4.18 \text{ J}}{°C\text{-g}} \right) = 418 \text{ J}/°C.$$

The temperature rise of 6.5°C then represents 6.5°C × 418 J/°C = 2.7 kJ. To put this on a molar basis, we need to know the heat produced per mole of NaOH and HCl:

Heat evolved per mole = −2.7 kJ/0.050 mol
= −54 kJ/mol

4.7 Fuel values of fuels and foods

Most common chemical reactions used to produce heat are combustion reactions. The energy released when a fuel or food is combusted is known as its fuel value. Because all heats of combustion are exothermic, it is common to report fuel values without their associated negative sign. Furthermore, because fuels and foods are usually mixtures, fuel values are reported on a gram rather than a mole basis. For example, the fuel value of octane, C_8H_{18}, a component of gasoline, is the heat produced by combustion of 1 g of this substance as shown in Equation [4.16]:

$$2C_8H_{18}(l) + 25O_2(g) \longrightarrow 16CO_2(g) + 18H_2O(g) \qquad [4.16]$$

Note that in this reaction the product water is considered to be in the gaseous form. This is appropriate because under the conditions that octane would be used as a fuel, water would be vaporized. The enthalpy change for this reaction is $\Delta H = -10{,}920 \text{ kJ}$. Because each mole of C_8H_{18} weighs 114 g, the fuel value of octane is 47.9 kJ/g:

$$\left(\frac{10{,}920 \text{ kJ}}{2 \text{ mol } C_8H_{18}}\right)\left(\frac{1 \text{ mol } C_8H_{18}}{114 \text{ g } C_8H_{18}}\right) = 47.9 \frac{\text{kJ}}{\text{g } C_8H_{18}}$$

In accordance with the first law of thermodynamics, the fuel value of a substance is the same no matter how or where it reacts as long as the products of the reaction are the same. Thus a bomb calorimeter can often be used to measure the fuel values of foods. Unquestionably, this procedure is much easier than having to measure the heat given off in an engine or in our bodies.

FOODS

Most of the energy our bodies need comes from carbohydrates and fats. Carbohydrates are decomposed in the stomach into glucose, $C_6H_{12}O_6$. Glucose is soluble in blood and is known as blood sugar. It is transported by the blood to cells where it reacts with O_2 in a series of steps, eventually producing $CO_2(g)$, $H_2O(l)$, and energy:

$$C_6H_{12}O_6(s) + 6O_2(g) \longrightarrow 6CO_2(g) + 6H_2O(l) \qquad \Delta H^\circ = -2816 \text{ kJ}$$

The breakdown of carbohydrates is rapid, so that their energy is quickly supplied to the body. However, the body stores a very small amount of carbohydrates. The average fuel value of carbohydrates is 17 kJ/g (4 kcal/g).

Like carbohydrates, fats produce CO_2 and H_2O in both their metabolism and their combustion in a bomb calorimeter. The reaction of stearin, $C_{57}H_{110}O_6$, a typical fat, is as follows:

$$2C_{57}H_{110}O_6(s) + 163O_2(g) \longrightarrow$$
$$114CO_2(g) + 110H_2O(l) \qquad \Delta H^\circ = -75{,}520 \text{ kJ}$$

The chemical energy from foods that is not used either to maintain body temperature or for muscular activity or for the organization of the atoms in food into body parts is stored in the form of fats. Fats are well suited to serve as the body's energy reserve for at least two reasons: (1) they are insoluble in water, which permits their storage in the body; (2) they produce more energy per gram than either proteins or carbohydrates, which makes them efficient energy sources on a weight basis. The average fuel value of fats is 38 kJ/g (9 kcal/g).

In the case of proteins, metabolism in the body produces less energy than combustion in a calorimeter because the products are different. Proteins contain nitrogen, which is released in the bomb calorimeter as N_2. In the body this nitrogen ends up mainly as urea, CH_4N_2O. Proteins are used by the body mainly as building materials for organ walls, skin, hair, muscle, and so forth. On the average, the metabolism of proteins produces 17 kJ/g (4 kcal/g).

The food values for a variety of common foods are shown in Table 4.3. The amount of energy the body requires varies considerably depending

TABLE 4.3 Fuel values and compositions of some common foods.

	Approximate composition (%)			Fuel value	
	Protein	Fat	Carbohydrate	kJ/g	kcal/g
Apples (raw)	0.4	0.5	13	2.5	0.59
Beer[a]	0.3	0	1.2	1.8	0.42
Bread (white, enriched)	9	3	52	12	2.8
Cheese (cheddar)	28	37	4	20	4.7
Eggs	13	10	0.7	6	1.4
Fudge	2	11	81	18	4.4
Green beans (frozen)	1.9	—	7.0	1.5	0.38
Hamburger	22	30	—	15	3.6
Milk	3.3	4.0	5.0	3.0	0.74
Peanuts	26	39	22	23	5.5

[a]Beers typically contain 3.5 percent ethanol, which has fuel value.

on such factors as body weight, age, and muscular activity. The average adult requires about 6300 kJ (1500 kcal) a day while at rest in a warm room. When a person is doing average work, his or her energy requirement is increased to about 10,000–13,000 kJ (2500–3000 kcal). This is about the same amount of energy as that consumed by a 100-watt light bulb operating for a 24-hour period.

SAMPLE EXERCISE 4.10

It is estimated that for a person of average weight, running or jogging requires consumption of about 100 Cal/mile. What weight of hamburger provides the fuel value requirements for running 3 miles?

Solution: Recall that the dietary Calorie is equivalent to 1 kcal. The running requires 300 Calories, or 300 kcal.

$$\text{Hamburger required} = 300 \text{ kcal} \times \left(\frac{1 \text{ g hamburger}}{3.6 \text{ kcal}} \right)$$

$$= 83 \text{ g hamburger}$$

Thus, the calories consumed in running 3 miles are replaced by less hamburger than is present in a "quarter-pounder."

FUELS

Several common fuels are compared in Table 4.4. Notice that an increase in the percentage of carbon or hydrogen in the fuel increases the fuel value. For example, the fuel value of bituminous coal is greater than that of wood because of its greater carbon content.

Coal, oil, and natural gas, which are presently our major sources of energy, are known as fossil fuels. All are thought to have formed over millions of years from the decomposition of plants and animals. All are presently being depleted far more rapidly than they are formed. Natural gas consists of gaseous hydrocarbons, compounds of hydrogen and carbon. It varies in composition but contains primarily methane (CH_4), with small amounts of ethane (C_2H_6), propane (C_3H_8), and butane (C_4H_{10}). Oil, which is also known as petroleum, is a liquid composed of hundreds of compounds. Most of these compounds are hydrocarbons,

TABLE 4.4   Fuel values and compositions of some common fuels

	Approximate elemental composition (%)			Fuel value (kJ/g)
	C	H	O	
Wood (pine)	50	6	44	18
Anthracite coal (Pennsylvania)	82	1	2	31
Bituminous coal (Pennsylvania)	77	5	7	32
Charcoal	100	0	0	34
Crude oil (Texas)	85	12	0	45
Gasoline	85	15	0	48
Natural gas	70	23	0	49
Hydrogen	0	100	0	142

with the remainder being mainly organic compounds containing sulfur, nitrogen, or oxygen. Coal, which is solid, contains hydrocarbons of high molecular weight as well as compounds containing sulfur, oxygen, and nitrogen. The sulfur in oil and coal is important from the standpoint of air pollution, as we shall discuss in Chapter 10.

Hydrogen (H_2) is a very attractive fuel because of its high fuel value and because its combustion produces water, a "clean" chemical that produces no negative environmental effects. However, hydrogen cannot be used as a primary energy source because there is so little H_2 in nature. Most hydrogen is produced by the decomposition of water or hydrocarbons. This decomposition requires energy; in fact, because of heat losses, more energy must be used to generate hydrogen than can be reclaimed when the hydrogen is subsequently used as a fuel. However, should large, cheap sources of energy become available because of technical advances, perhaps in areas such as nuclear or solar power generation, a portion of this energy could be used to generate hydrogen. The hydrogen could then serve as a convenient energy carrier. It would be cheaper to transport hydrogen using existing gas pipelines than to transport electrical energy; hydrogen is both portable and storable. Because present industrial technology is based on combustible fuels, hydrogen could replace oil and natural gas as these fuels become scarcer and more expensive.

4.8 Energy usage: trends and prospects

The average energy consumption per person in the United States each day amounts to about 1.3×10^6 kJ. This amount of energy is about 100 times our average food-energy requirement. The use of energy has been increasing each year as shown in Figure 4.11. Presently about 30 percent of the world's annual production of energy is consumed in the United States.

The shifting importance of different sources of energy is shown in Figure 4.12. Until 1850, wood supplied about 90 percent of the energy used in the United States. Coal became increasingly important until it supplied about 75 percent of our energy by 1910. Presently natural gas supplies 31 percent of our energy, whereas 46 percent is supplied by oil, 19 percent by coal, about 2 percent by hydroelectric power, and slightly more than 2 percent by nuclear power.

The increasing demand for energy in the United States is a major factor in the recurrent energy crises, because it has resulted in an increasing dependence on petroleum imports. At least two other factors are involved: (1) concern over the development of "clean" energy sources

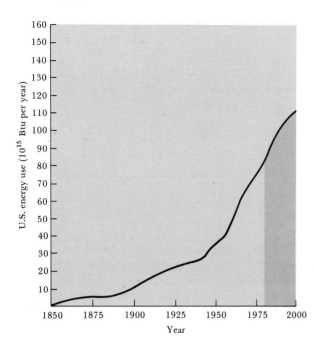

FIGURE 4.11 Energy consumption in the United States by year (1 Btu = 1.05 kJ).

and (2) the rapidly diminishing supplies of domestic natural gas and petroleum. The ultimate problem related to the use of fossil fuels is that we must eventually run out of them. Meanwhile, we will be forced to rely on increasingly expensive sources of these fuels.

FUTURE ENERGY SOURCES

According to some projections, we could run out of oil and natural gas by the end of the twentieth century unless alternate energy sources are found and per capita energy consumption levels off. There is, therefore, considerable interest in developing alternate energy sources. Much research is presently focusing on nuclear and solar energy and on the development of ways to use coal more efficiently. There is also interest in

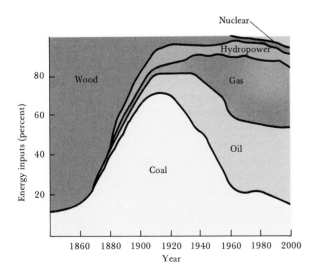

FIGURE 4.12 Pattern of energy consumption in the United States by year. (After S. F. Singer, "Human Energy Production as a Process in the Biosphere." Copyright 1970 by Scientific American Inc. All rights reserved).

increased use of geothermal energy (the heat within the earth) and in use of winds and tides to generate usable power. Experts predict that these three energy sources can each make a small but important contribution to the overall energy picture, but they will have little overall impact for the foreseeable future. We shall not discuss them further. In the following paragraphs we examine coal and solar energy very briefly. We will postpone discussion of nuclear energy until Chapter 20.

Coal is the most abundant fossil fuel; it constitutes 80 percent of the fossil fuel reserves of the United States and 90 percent of those of the world. However, use of coal presents a number of problems. Coal produces more air pollution than other fuels do. It is often expensive and dangerous to mine. Furthermore, most of the remaining rich deposits of coal are in the western United States, whereas energy use is greatest along the East Coast; shipping coal great distances adds significantly to its cost. Some experts feel that coal could be used most effectively if it were converted to a gaseous form, often called "syngas," for synthetic gas. In such a conversion the sulfur is removed, thereby decreasing air pollution when the syngas is burned. Syngas could be easily transported in pipelines and could supplement our diminishing supplies of natural gas. Gasification of coal requires addition of hydrogen to coal. Typically, the coal is pulverized and treated with superheated steam. The product contains a mixture of CO, H_2, and CH_4, all of which can be used as fuels. However, conditions are maintained to maximize production of CH_4. A simplified schematic showing some of the reactions that occur is given in Figure 4.13.

Solar energy is the world's largest energy source. The solar energy incident on only 0.1 percent of the land area of the United States is equivalent to all of the energy that we currently use. The problem with

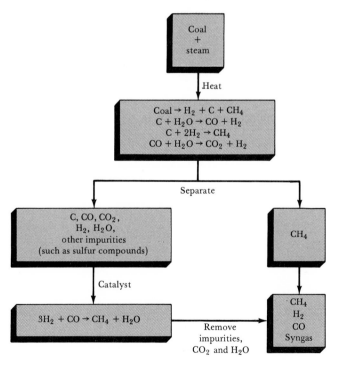

FIGURE 4.13 The basic processes involved in the gasification of coal to form synthetic gas (syngas). A catalyst is a substance that is able to increase the speed of a reaction without being consumed in the reaction.

the use of solar energy is that it is very dilute and fluctuates with time and weather conditions. Devices to convert solar energy to electrical energy are not presently very efficient. One possible way of using solar energy is through "energy plantations." These involve rapid and high-density growth of a mixture of plants. The plants could then be burned to produce energy. Solar energy can presently be used to great advantage to augment traditional means of heating homes; solar water heaters and careful placement of walls and windows of houses can be used to enhance the heat that is captured by a dwelling.

Before we leave this discussion, we should note that we have emphasized meeting the energy crisis by increasing energy supplies. However, we could also attack the problem by cutting back per capita energy use and by reducing energy waste. Such energy conservation is probably our most important option.

FOR REVIEW

Summary

This chapter has been about **energy** and the **first law of thermodynamics**. We have seen that energy can be measured in terms of the ability to accomplish work or transfer heat. An object may possess **potential energy** because of its position relative to another object, or because of its composition. Thus, chemical energy is potential energy, which may be released when the object undergoes a chemical change. An object may possess **kinetic energy** because of its motion relative to another object. The first law of thermodynamics, also referred to as the **law of conservation of energy**, states that in any change that occurs in nature, the total energy of the universe remains constant. It is often convenient to consider one portion of nature, called the **system**, as separate from all the rest, called the **surroundings**. According to the first law of thermodynamics, any energy gained by the system in a change must equal the energy lost from the surroundings. Any process in which heat energy is lost to the surroundings is termed **exothermic**. On the other hand, when heat energy is absorbed by the system from the surroundings, the process is termed **endothermic**.

Heat changes occurring at constant pressure are of special interest. The heat gained or lost by the system in a process occurring at constant pressure is termed the **enthalpy** change, represented by the symbol ΔH. This quantity is negative for an exothermic process, positive for an endothermic process. Enthalpy is a **state function**, which means that the enthalpy of a system is determined by specifying its present condition, and not by the details of how it came to be in that state. If a particular overall change can be described as the sum of several indi-vidual changes, then the enthalpy change for the overall process is equal to the sum of the enthalpy changes associated with the individual steps. This statement is known as **Hess's law of constant heat summation**. In applying Hess's law it is useful to define the **standard heat of formation** of a substance, which is the heat change in the formation of a substance from the elements, all in the states in which they are most stable at the temperature of interest (usually $25\,^{\circ}$C). Using Hess's law, the enthalpy change in any reaction can be described as the sum of the heats of formation of all the products, less the heats of formation of all reactants. In solving problems dealing with enthalpy changes it is important to keep the following points in mind: (1) The enthalpy change in a reaction, ΔH, is directly proportional to the amount of substance that reacts or is produced; (2) ΔH for any reaction is equal in magnitude but opposite in sign to the value of ΔH for the reverse reaction; (3) the heat of formation for any element in its standard state is zero.

Heat changes can be measured in a **calorimeter**, in which the heat evolved or absorbed in a reaction is measured by observing the change in temperature of the calorimeter. The **heat capacity** of an object is the quantity of heat required to produce a unit change in temperature.

The **fuel values** of foods and fuels are defined as the heat energy released when a gram of food or fuel is combusted. In this chapter, we briefly considered the fuel values of some common foods and fuels, including various forms of fossil fuels. Finally, we have considered current and projected rates of fuel consumption, and the prospects for various sources of energy in the future.

Learning goals

Having read and studied this chapter, you should be able to:

1 Distinguish between kinetic energy and potential energy.

2 Distinguish exothermic from endothermic processes.

3 Calculate the enthalpy change in a chemical process, given enthalpy values for other processes that can be combined to yield the process of interest.

4 Calculate the enthalpy change in a chemical reaction, given the enthalpy of formation of each reactant and product.

5 Sketch an energy diagram such as that illustrated in Figure 4.8, given the appropriate enthalpy data.

6 Calculate the heat capacity of a calorimeter, given data regarding temperature changes and heat evolved.

7 Calculate the enthalpy change in a process, given suitable calorimetric data.

8 Describe the fuel values of the major types of food.

9 Perform interconversions between different heat units and between gram and molar bases.

Key terms

Among the more important terms and expressions used for the first time in this chapter are the following:

An **endothermic process** (Section 4.2) is one in which a system absorbs heat from its surroundings.

Energy (Section 4.1) is the ability to do work or to transfer heat.

Enthalpy, H (Section 4.3), is the heat content of a system. Enthalpy change, ΔH, is the heat absorbed or evolved during a reaction that occurs at constant pressure.

An **exothermic process** (Section 4.2) is one in which a system loses heat to its surroundings.

The **first law of thermodynamics** (introduction) states that in any change the total energy of the universe is constant. This law is also known as the **law of conservation of energy**.

Heat capacity (Section 4.5) is defined as the quantity of heat required to cause a 1°C change in temperature. Heat capacity in some cases is expressed on a unit mass basis, for example, as the heat capacity per gram. As an example, the heat capacity of graphite at 20°C is 0.740 J/°C-g.

The **heat of formation** (Section 4.5) of a substance is defined as the heat evolved when a substance is formed from the elements.

Hess's law of constant heat summation (Section 4.4) states that the heat evolved in a given process can be expressed as the sum of the heats of several processes that, when added, yield the process of interest.

The **joule**, J, (Section 4.5) is the SI unit of energy, $1 \text{ kg-m}^2/\text{sec}^2$. A related unit is the **calorie**; $4.184 \text{ J} = 1 \text{ cal}$.

Kinetic energy (Section 4.1) is the energy that an object possesses by virtue of its motion with respect to another object.

Potential energy (Section 4.1) is the energy that an object possesses as a result of its position with respect to another object, or by virtue of its composition. **Chemical energy** is a form of potential energy.

Specific heat (Section 4.6) is the ratio of the heat capacity per gram of a substance to the heat capacity per gram of water.

A **state function** (Section 4.3) is any property of a system that is defined once we have described the state of the system.

The **surroundings** (Section 4.2) are the rest of the universe outside a system we are interested in observing.

A **system** (Section 4.2) is a portion of the universe that we single out for study. We must be careful to state exactly what the system contains, and what transfers of energy it may have with its surroundings.

Thermodynamics is the study of energy, and of the transformations energy may undergo.

EXERCISES

Energy; the first law of thermodynamics

4.1 We have said that energy can be defined as the capacity to do work. Describe some device that uses a dry cell battery as an energy source that serves as an illustration of this definition of energy. Describe another situation in which the dry cell produces mainly heat. What form of energy does a fresh dry cell represent?

4.2 Describe the distinction between kinetic and potential energy. Using a rubber band, a candle, and a brick in your examples, illustrate as many forms of energy as you can think of.

4.3 The moon has a mass of 7.3×10^{22} kg. It moves about the earth with a linear velocity of 1.0×10^5 cm/sec. What is the kinetic energy of the moon relative to earth?

4.4 An energy unit often used in engineering applications is the British Thermal Unit, Btu: $1 \text{ J} = 9.48 \times 10^{-4}$ Btu. What is the kinetic energy in Btu of an object that has a mass of 6.00 kg moving with a speed of 3.20 km/sec?

4.5 A meteorite found near Orgueil, France, in 1864 has a mass of 32 g. Assuming its speed relative to the earth's center of gravity at the time of entry into the earth's atmosphere was 22 km/sec, calculate the kinetic energy of this meteorite. Assuming that when it struck the earth's surface it had a speed of 0.3 km/sec, what change in kinetic energy had occurred? From the law of conservation of energy, what must have happened to the lost kinetic energy? Suggest what forms it might have taken.

4.6 When a highly elastic rubber ball having a mass of 55 g is dropped from a height of 2.0 m onto a thick, hard surface, it bounces to a height of 1.8 m. Describe the net change in potential energy that has occurred at that moment. What form has this energy difference taken? (The potential energy of a mass in the earth's gravitational field is given by $E = mgh$, where m is mass, g is the gravitational-acceleration constant, 9.81 m/sec^2, and h is the height above some reference level.)

4.7 What work is accomplished by a 165-lb person in walking up 16 flights of stairs, with an increase in height of 192 ft? (The gravitational force acting on a mass on earth is given by $f = mg$, where g is the gravitational constant, 9.81 m/sec^2.)

4.8 Starting with a model airplane with a small engine, and a certain amount of fuel, describe all the interconversions of energy involved in the combustion of the fuel, the flight of the model craft, and its return to earth. Using the first law of thermodynamics, account for the "final" form or forms of the energy represented by burning of the fuel.

Energy changes; enthalpy; Hess's law

4.9 Classify the following processes as endothermic or exothermic: (a) discharge of a flashlight battery; (b) melting of ice; (c) evaporation of rubbing alcohol; (d) reaction of sodium metal with water; (e) a lightning discharge.

4.10 Predict the sign of the enthalpy change ΔH for the system in each of the following processes, assuming that heat can flow to or from the surroundings:

(a) $H_2O(l) \longrightarrow H_2O(g)$
(b) $2CH_3OH(g) + 3O_2(g) \longrightarrow 2CO_2(g) + 4H_2O(l)$
(c) $NH_4Cl(s) + H_2O(l) \longrightarrow NH_4Cl(aq)$
     (the temperature decreases)
(d) $2HCl(g) \longrightarrow H_2(g) + Cl_2(g)$

4.11 The heat of combustion of liquid benzene, C_6H_6, to form $CO_2(g)$ and $H_2O(l)$ is -3273 kJ/mol. The heat of combustion of liquid methyl alcohol, CH_3OH, also to form $CO_2(g)$ and $H_2O(l)$ is -728 kJ/mol. Write the balanced equation for both combustion reactions. Calculate the heat of combustion per gram of benzene and of methyl alcohol to form $CO_2(g)$ and $H_2O(g)$. The density of benzene at 25°C is 0.878 g/cm^3; the density of methyl alcohol at 25°C is 0.790 g/cm^3. Assuming the two fuels burn with comparable efficiency in an auto engine, roughly how many times farther would you expect to go on a gallon of benzene as compared with a gallon of methyl alcohol?

4.12 For the reaction

$$2Na(s) + 2H_2O(l) \longrightarrow 2NaOH(aq) + H_2(g)$$
$$\Delta H = -368 \text{ kJ}$$

calculate the heat change that occurs when 3.5 g of $Na(s)$ reacts with excess water.

4.13 The standard enthalpies of formation of gaseous acetylene, C_2H_2, ethylene, C_2H_4, and ethane, C_2H_6, are $+227$ kJ, $+52.3$ kJ, and -84.7 kJ per mole, respectively. Calculate the heat evolved per mole on combustion of each substance; calculate the heat of combustion per kilogram of each substance, to yield $CO_2(g)$ and $H_2O(g)$. Which is the most efficient fuel in terms of heat evolved per unit mass?

4.14 From the following heats of reaction:

$H_2(g) + F_2(g) \longrightarrow 2HF(g)$	$\Delta H = -537$ kJ
$C(s) + 2F_2(g) \longrightarrow CF_4(g)$	$\Delta H = -680$ kJ
$2C(s) + 2H_2(g) \longrightarrow C_2H_4(g)$	$\Delta H = 52.3$ kJ

calculate the heat of reaction of ethylene with F_2:

$$C_2H_4(g) + 6F_2(g) \longrightarrow 2CF_4(g) + 4HF(g)$$

4.15 From the following heats of reaction:

$2SO_2(g) + O_2(g) \longrightarrow 2SO_3(g)$	$\Delta H = -196$ kJ
$2S(s) + 3O_2(g) \longrightarrow 2SO_3(g)$	$\Delta H = -790$ kJ

calculate the heat of the reaction

$$S(s) + O_2(g) \longrightarrow SO_2(g)$$

4.16 Using the heats of formation in Appendix D, calculate the enthalpy changes in each of the following reactions:

(a) $CO(g) + 2NH_3(g) \longrightarrow NH_4CN(s) + H_2O(g)$
(b) $NH_4NO_3(s) \longrightarrow N_2O(g) + 2H_2O(g)$
(c) $P_4O_6(s) + 2O_2(g) \longrightarrow P_4O_{10}(s)$
(d) $KClO_3(s) + 3PCl_3(l) \longrightarrow 3POCl_3(l) + KCl(s)$

4.17 Someone has proposed that the following reaction might occur in the stratosphere:

$$HO(g) + Cl_2(g) \longrightarrow HOCl(g) + Cl(g)$$

Calculate the enthalpy change for this process from the following data:

$Cl_2(g) \longrightarrow 2Cl(g)$	$\Delta H = 242$ kJ
$H_2O_2(g) \longrightarrow 2HO(g)$	$\Delta H = 134$ kJ
$H_2O_2(g) + 2Cl(g) \longrightarrow 2HOCl(g)$	$\Delta H = -209$ kJ

4.18 The heat of formation of anthracene, $C_{14}H_{10}(s)$, is 121.3 kJ/mol. The heat of formation of 9,10-dihydroanthracene, $C_{14}H_{12}(s)$, is 66.4 kJ/mol. Calculate the enthalpy change in the reaction

$$C_{14}H_{10}(s) + H_2(g) \longrightarrow C_{14}H_{12}(s)$$

4.19 From the following data for three prospective fuels, calculate which could provide the most energy per unit volume.

Fuel	Density (g/cm³) at 20°C	Molar heat of combustion
Nitroethane, $C_2H_5NO_2(l)$	1.052	-1348 kJ/mol
Ethanol, $C_2H_5OH(l)$	0.789	-1371 kJ/mol
Diethyl ether, $(C_2H_5)_2O(l)$	0.714	-2727 kJ/mol

4.20 Using the standard heats of formation given in Appendix D, calculate the enthalpy change in each of the following reactions:

(a) $CaO(s) + H_2O(l) \longrightarrow Ca(OH)_2(s)$
(b) $SO_2(g) + 2HCl(g) \longrightarrow H_2O(g) + SOCl_2(l)$
(c) $2NaBr(aq) + Pb(NO_3)_2(aq) \longrightarrow$
$\qquad PbBr_2(s) + 2NaNO_3(aq)$
(d) $2H_2O_2(g) + S(s) \longrightarrow SO_2(g) + 2H_2O(g)$

Calorimetry; fuel values

4.21 A swimming pool contains 240 m³ of water. What quantity of heat input is required to raise the temperature of the water from 16°C to 24°C, assuming that none of the heat supplied is lost to the surroundings?

[4.22] A small "coffee-cup" calorimeter of the type shown in Figure 4.10 contains 150 g of water at 24.6°C. A 110-g block of molybdenum metal is heated to 100°C and then placed in the water in the calorimeter. The contents of the calorimeter come to an average temperature of 28.0°C. What is the heat capacity per gram of the molybdenum metal? (The heat capacity of the water is 4.18 J/°C-g.)

4.23 When a 0.235-g sample of benzoic acid is combusted in a bomb calorimeter, a 3.62°C rise in temperature occurs. When a 0.305-g sample of citric acid, $C_6H_8O_7$, is burned, a 1.83°C temperature rise is seen. Using the value for the heat of combustion of benzoic acid given in Section 4.6, calculate the heat of combustion per mole of citric acid, at constant volume.

4.24 When 50 g of 0.050 M BaI_2 solution is mixed with 50 g of 0.050 M H_2SO_4 solution in a "coffee-cup" calorimeter, the temperature of the mixed solutions increases by 0.55°C. Calculate the heat of the reaction

$$H_2SO_4(aq) + BaI_2(aq) \longrightarrow BaSO_4(s) + 2HI(aq)$$

per mole of $BaSO_4(s)$ formed. Ignore the heat capacity of the plastic cup; assume that the heat capacity of the solution is 4.2 J/°C-g, and that the density of the solution is 1.0 g/cm³.

4.25 When 1.38 g of liquid PCl_3 at 25.55°C is added to 100 g of water at 25.55°C in a "coffee-cup" calorimeter, the temperature of the resulting solution increases to 32.00°C. The heat capacity of the resulting solution may be taken to be 4.18 J/°C-g, and the heat capacity of the plastic cup can be ignored. Calculate the heat of solution of PCl_3 in water.

4.26 A 1-oz serving of a popular "high protein" breakfast cereal contains 6 g of protein, 19 g of carbohydrate,

and 1 g of fat. What is the fuel value in kilojoules and Calories of this quantity of cereal?

4.27 A freshly harvested large Yellow Delicious apple weighs 120 g. Its total fuel value is 62 Cal. Assuming that the fuel value is derived entirely from carbohydrate, calculate the percentage of water in the apple.

4.28 A pound of peanut brittle contains 214 g of carbohydrate, 146 g of fat, and 79 g of protein. Calculate the fuel value per pound. Calculate the fuel value in a 50-g bar. What weight of peanut brittle would an active person with a daily fuel requirement of 2600 kcal need to consume if it were the only source of food?

4.29 The heat of combustion of ethanol, $C_2H_5OH(l)$, is -1371 kJ/mol. A 12-oz (355 mL) bottle of beer contains 3.7 percent ethanol by weight. Assuming the density of the beer to be 1.0 g/cm³, what calorie content does the alcohol in a bottle of beer represent?

4.30 Glycerin, or glycerol, $(CH_2OH)_2CHOH$, has a heat of formation of -666 kJ/mol. This substance is formed in the course of digestion of fats; it is subsequently metabolized to form $CO_2(g)$ and $H_2O(l)$. What is the fuel value in kilojoules and Calories resulting from complete metabolism of 10.0 g of glycerin?

[4.31] Aspirin is produced commercially from salicylic acid, $C_7O_3H_6$. A large shipment of salicylic acid is contaminated with boric oxide, which, like salicylic acid, is a white powder. The heat of combustion of salicyclic acid at constant volume is known to be -3.00×10^3 kJ/mol. Boric oxide, because it is fully oxidized, does not burn. When a 3.556-g sample of the contaminated salicyclic acid is burned in a bomb calorimeter, the temperature increases 2.556°C. From previous measurements, the heat capacity of the calorimeter is known to be 13.62 kJ/°C. What is the amount of boric oxide in the sample, in terms of percent by weight?

4.32 World and U.S. energy requirements are usually expressed in units of quads (1 quad = 10^{15} Btu; 1 Btu = 1.05 kJ). It is estimated that in the year 1985 coal combustion will furnish 19 quads of U.S. energy requirements. Assuming that this is equally divided between anthracite and bituminous coal, what weight of coal in tons, and in kilograms, will be required? (Refer to Table 4.4.)

4.33 In 1977, the "proved" U.S. reserves of natural gas were estimated to be 2.09×10^{11} ft³. Assuming the fuel value of the gas to be 980 kJ/ft³, calculate the total fuel value of this quantity of natural gas. What weight of anthracite coal has an equivalent total fuel value?

Additional exercises

4.34 Distinguish between the following terms: (a) heat of combustion and heat of formation; (b) calorie and Calorie; (c) Hess's law and the first law of thermodynamics; (d) heat capacity and fuel value; (e) work and heat; (f) state function and enthalpy.

4.35 Using the heats of formation given in this chapter and Appendix D, calculate the heats of the three lower reactions in the second box from the top in Figure 4.13. Use your judgment as to which states the substances should be assumed to have, but indicate your choices in each case.

4.36 In 1980, the production of raw sugar from sugarcane required the equivalent of 2×10^8 gal (1 gal = 3.8 L) of fuel oil. The fuel value of fuel oil is estimated to be 46 kJ/g; it has a density of 0.90 g/cm^3. Calculate the total fuel value requirements of the sugar-processing industry.

4.37 It is estimated that between 1980 and 2000, the U.S. energy requirement will grow by 1.5 quad/yr (1 quad = 1.05×10^{15} kJ). What mass of crude oil does this annual increase represent?

4.38 Which possesses more kinetic energy, a 1500-kg car traveling at 85 km/hr or a 2200-kg car traveling at 60 km/hr? How much kinetic energy does each possess in joules; in calories?

4.39 Give two examples of the conversion of chemical energy into some other form of energy.

4.40 Given the following reactions and their associated enthalpy changes:

$$\tfrac{1}{2}H_2(g) + \tfrac{1}{2}Br_2(g) \longrightarrow HBr(g) \qquad \Delta H = -36 \text{ kJ}$$
$$\tfrac{1}{2}H_2(g) \longrightarrow H(g) \qquad \Delta H = 218 \text{ kJ}$$
$$\tfrac{1}{2}Br_2(g) \longrightarrow Br(g) \qquad \Delta H = 112 \text{ kJ}$$

calculate ΔH for the reaction

$$H(g) + Br(g) \longrightarrow HBr(g)$$

4.41 Calculate the enthalpy changes for the following reactions. Indicate which are endothermic and which are exothermic.

(a) $6H_2(g) + P_4(s) \longrightarrow 4PH_3(g)$
(b) $3Fe(s) + 2O_2(g) \longrightarrow Fe_3O_4(s)$
(c) $2HF(g) + Br_2(l) \longrightarrow 2HBr(l) + F_2(g)$
(d) $CaO(s) + CO_2(g) \longrightarrow CaCO_3(s)$

4.42 Given the following reactions and their associated enthalpy changes:

$$CH_4(g) \longrightarrow C(g) + 4H(g) \qquad \Delta H = 1660 \text{ kJ}$$
$$O_2(g) \longrightarrow 2O(g) \qquad \Delta H = 490 \text{ kJ}$$
$$2H(g) + O(g) \longrightarrow H_2O(g) \qquad \Delta H = -930 \text{ kJ}$$
$$C(g) + 2O(g) \longrightarrow CO_2(g) \qquad \Delta H = -1610 \text{ kJ}$$

calculate ΔH for the combustion of CH_4 to form $H_2O(g)$ and $CO_2(g)$

4.43 Diagram the reactions shown in question 4.42 in a fashion similar to Figure 4.8. Show the relationship between reactants ($CH_4 + 2O_2$) and products ($CO_2 + 2H_2O$) and the reaction path through intermediate atoms (C, 4H, 4O).

[4.44] The transformation

$$Na_2SO_4 \cdot 10H_2O(s) \rightleftharpoons Na_2SO_4 \cdot 10H_2O(l)$$

occurs at 32°C. It has been used as a means of storing solar energy. A container with $Na_2SO_4 \cdot 10H_2O(s)$ is heated during the day by solar radiation, and the phase change,

which absorbs heat, occurs as the material is heated above 32°C. At night, when the material cools, heat is released as the reaction is reversed. The material on the right hand side of the equation can be considered as an aqueous solution of 1 mol of Na_2SO_4 in 10 mol of water. Using the heats of formation, $Na_2SO_4 \cdot 10H_2O(s)$, $\Delta H_f^\circ = -4324$ kJ/mol; $Na_2SO_4(aq)$, $\Delta H_f^\circ = -1387$ kJ/mol; $H_2O(l)$, $\Delta H_f^\circ = -286$ kJ/mol, calculate the quantity of heat released per kg of starting salt when the reaction reverses at night. What mass of $Na_2SO_4 \cdot 10H_2O$ would be required to furnish 6×10^4 Btu's (1 Btu = 1.05 kJ), which is about one-third of the average daily heating requirement of a small house near Boston? (*Note:* In practice, much more than the calculated quantity would be required because of system inefficiencies.)

4.45 The heat of formation of $CH_3OH(g)$ is -201.3 kJ/mol. The heat of vaporization of $CH_3OH(l)$ at 25°C is 37.4 kJ/mol. Write the equations for these processes and calculate ΔH for the reaction

$$2C(s) + O_2(g) + 4H_2(g) \longrightarrow 2CH_3OH(l)$$

[4.46] Ammonia, NH_3, boils at -33°C; at this temperature, it has a density of 0.81 g/cm^3. The heat of formation of $NH_3(g)$ is -46.2 kJ/mol, and the heat of vaporization of $NH_3(l)$ is 4.6 kJ/mol. Calculate the heat evolved when 1 L of liquid NH_3 is burned in air to give $N_2(g)$ and $H_2O(g)$. How does this compare with the heat evolved upon complete combustion of a liter of liquid methyl alcohol, CH_3OH, density at 25°C = 0.792 g/cm^3, heat of formation $\Delta H_f^\circ(CH_3OH(l)) = -239$ kJ/mol?

4.47 The heat of combustion at constant volume of glutaric acid, $H_2C_5H_6O_4(s)$, is -2166 kJ/mol. What is the temperature rise in a bomb calorimeter with a heat capacity of 7.74 kJ/°C when 1.80 g of glutaric acid is combusted?

4.48 A low-fat milk contains about 8 g of protein, 12 g of carbohydrate, and 4 g of fat per 8-oz glass. Calculate the approximate fuel value in kilojoules and Calories of a glass of this milk.

4.49 A large coal-burning electric power plant requires 1000 tons per hour of coal for peak operation. The coal has a fuel value of 10,500 Btu/lb. Calculate the total heat generated per hour by coal combustion, in units of kJ (1 Btu = 1.05 kJ). The coal contains 0.92 percent sulfur, which we can assume to be present as elemental sulfur. What heat is evolved per hour as a result of combustion of this sulfur to $SO_2(g)$?

[4.50] A recent U.S. government survey of coal reserves places the quantity of coal recoverable by means of current technology at about 2.5×10^{11} tons. Assuming that this coal has an average fuel value of 31 kJ/g, calculate the total fuel value of the coal. Assuming that the U.S. use of energy stabilizes at about 105 quads per year (1 quad = 1.05×10^{15} kJ), how many years of energy supply does this coal represent? If the average composition of the coal is $C_{135}H_{97}O_9NS$, what mass of CO_2 in kilograms will be produced by this combustion?

5

The electronic
structures of atoms

In earlier chapters, we did not address the question of why atoms form molecules or ions. Why, for example, do hydrogen and oxygen atoms "stick" together to form the familiar water molecule, H_2O? Before we can answer this question and many others like it, we must know more about the atoms themselves. It is especially important to know how electrons are arranged in atoms. We refer to the electron arrangement of an atom as its electronic structure. The goal of this chapter is to give you an understanding of the electronic structure of the simplest atom, the hydrogen atom. In subsequent chapters we will apply these ideas to other atoms and then to the bonding between atoms. A great deal of our understanding of electronic structure has come from studies of the properties of light or radiant energy. We begin our study, then, by considering the characteristics of radiant energy.

5.1 Radiant energy

Different kinds of radiant energy—such as the warmth from a glowing fireplace, the light reflected off snow in the mountains, and the X rays used by a dentist—*seem* very different from one another. Yet they share certain fundamental characteristics. All types of radiant energy, also called electromagnetic radiation, move through a vacuum at a speed of 2.9979250×10^8 m/sec, the "speed of light." This is one of the most accurately known physical constants. For our purposes, however, it will be sufficient to use 3.00×10^8 m/sec.

All radiant energy has wavelike characteristics analogous to those of waves that move through water. A cork bobbing on water as waves pass by is not swept along with the wave. Rather, it moves up and down with the wave motion. Water waves are the result of energy imparted to the water, perhaps by the dropping of a stone, the movement of a boat, or

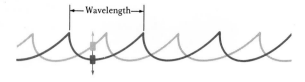

FIGURE 5.1 Characteristics of water waves. The distance between corresponding points on each wave is called the wavelength. The number of times per second that the cork bobs up and down is called the frequency.

the force of wind on the water surface. This energy is expressed as the up and down movement of the water.

If we look at a cross section of a water wave (Figure 5.1), we see that it is periodic in character. That is, the wave form repeats itself at regular intervals. The distance between identical points on successive waves is called the wavelength. As the wave passes, the cork floating on the surface moves up and down. The number of times per second that the cork moves through a complete cycle of upward and then downward motion is called the frequency of the wave.

In a similar way, radiant energy has a characteristic frequency and wavelength associated with it. These are illustrated schematically in Figure 5.2. The frequency of the radiation is the number of cycles that occur in 1 sec as the radiation flows past a given point. Short wavelengths correspond to high frequency, and long wavelengths to low frequency, as shown in Figure 5.2. The wavelength and frequency are thus related.

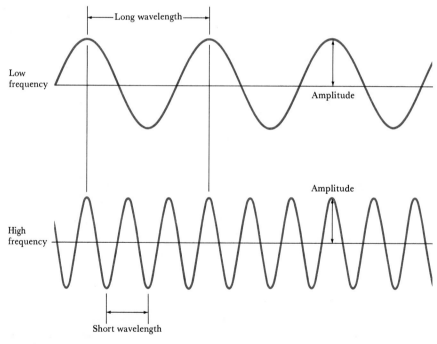

FIGURE 5.2 Wave characteristics of radiant energy. Unlike water waves, radiant energy is not associated with the movement of matter. Instead, it is associated with periodic changes in the electric and magnetic fields that accompany the wave. That is why it is referred to as electromagnetic radiation. Radiant-energy waves move through space with the speed of light. The frequency of the high-frequency radiation in this illustration is three times that of the low-frequency radiation. In the time that the low-frequency, long-wavelength radiation goes through one cycle, the high-frequency, short-wavelength radiation goes through three cycles. The *amplitude* of the wave refers to its intensity. Here both waves have the same amplitude.

TABLE 5.1 Wavelength units for electromagnetic radiation

Unit	Symbol	Length (m)	Type of radiation
Ångstrom	Å	10^{-10}	X ray
Nanometer (or millimicron)[a]	nm (mμ)	10^{-9}	Ultraviolet, visible
Micron	μ	10^{-6}	Infrared
Millimeter	mm	10^{-3}	Infrared
Centimeter	cm	10^{-2}	Microwaves
Meter	m	1	TV, radio

[a]The millimicron is the same size unit as the nanometer; use of the latter is preferred.

The product of frequency, ν (nu), and wavelength, λ (lambda), equals the speed of light, c:

$$\nu\lambda = c \qquad\qquad [5.1]$$

Wavelength is expressed in units of length per cycle; the unit of length chosen depends on the type of radiation as shown in Table 5.1. Figure 5.3 shows the ranges of wavelength that characterize the various types of radiant energy. Frequency is given in units of hertz (Hz), which is cycles per second. For example, the frequency of an AM radio station might be written as 810 kilohertz (kHz), or 810,000 cycles/sec. Because it is understood that cycles are involved, the units are normally given simply as /sec or sec^{-1}.

SAMPLE EXERCISE 5.1

The yellow light given off by a sodium lamp has a wavelength of 589 nm. What is the frequency of this radiation?

Solution: We can rearrange Equation [5.1] to give $\nu = c/\lambda$. We insert the value for c and λ, and then convert nm to m. This gives us

$$\nu = \left(\frac{3.00 \times 10^8 \text{ m/sec}}{589 \text{ nm}}\right)\left(\frac{10^9 \text{ nm}}{1 \text{ m}}\right)$$

$$= 5.09 \times 10^{14}/\text{sec}$$

Note that frequency has units of /sec or sec^{-1}.

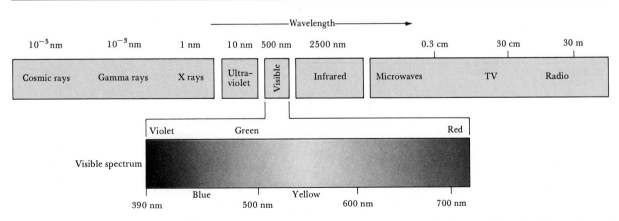

FIGURE 5.3 Wavelengths of electromagnetic radiation characteristic of various regions of the electromagnetic spectrum. Notice that color can be expressed quantitatively by wavelength or frequency.

Radiations of different wavelengths affect matter differently. For example, overexposure of a part of your body to infrared radiation may cause a "heat burn," overexposure to visible and near-ultraviolet light causes sunburn and suntan, and overexposure to X radiation causes tissue damage, possibly even cancer. These diverse effects are due to differences in the energy of the radiation. Radiations of high frequency and short wavelength are more energetic than radiation of lower frequency and longer wavelength. The quantitative relation between frequency and energy was developed at the turn of the century in the far-reaching and revolutionary *quantum theory* of Max Planck (Figure 5.4).

5.2 The quantum theory

The quantum theory is concerned with the rules that govern the gain or loss of energy from an object. Planck's contribution was to see that when we deal with gain or loss of energy from objects in the atomic or subatomic size range, the rules *seem* to be different from those that apply when we are dealing with energy gain or loss from objects of ordinary dimensions. A very crude and fanciful analogy might best illustrate what is involved. Imagine a large dump truck loaded with perhaps 20 tons of fine-grained sand. Let's say that the amount of sand on the truck is measured by driving it onto a supersensitive scale that measures the weight of a 30-ton object to the nearest pound. With this as our measuring device, the gain or loss of sand from the truck might be too small to be measured. For example, a spoonful of sand might be added or deleted with no change in scale reading.

Now imagine, if you will, a tiny little truck, operated by people the size of tiny mites. For this little truck a full load of sand would consist of perhaps a dozen grains of sand of the same size as those carried by the large truck. In this microscopic world, the load on the truck can be added to or decreased only by rolling on or off one or more full grains of sand. On a scale that weighs this tiny truck, even one grain of sand, the smallest piece attainable, represents a substantial fraction of the full load and is easily measurable.

In our analogy the sand represents energy. An object of ordinary, or macroscopic, dimensions, like the dump trucks we have seen on highways, contains energy in so many tiny pieces that the gain or loss of individual pieces is completely unnoticed. On the other hand, an object of atomic dimensions, such as our imaginary little truck, contains such a small amount of energy that the gain or loss of even the smallest possible piece makes a substantial difference. The essence of Planck's quantum theory is that there *is* such a thing as a smallest allowable gain or loss of energy. Even though the amount of energy gained or lost at one time may be very tiny, there is a limit to how small it may be. Planck termed the smallest allowed increment of energy gained or lost a quantum. In our analogy, a single grain of sand represents a quantum of sand.

You should keep in mind that the rules regarding the gain or loss of energy are always the same, whether we are concerned with objects on the size scale of our ordinary experience or with microscopic objects. However, it is only when dealing with matter at the atomic level of size that the impact of the quantum restriction is evident. Humans, being creatures of macroscopic dimensions, had no reason to suppose that the

FIGURE 5.4 Max Planck (1858–1947), physicist. Born in Kiel, Germany, Planck was the son of a law professor at the University of Kiel. When he announced his intention to study physics, Planck was warned that all the major discoveries had already been made in this field. Nevertheless, Planck became a physicist, and in 1892 was named professor of physics at the University of Berlin. In 1900 he presented a paper before the Berlin Physical Society which launched the greatest intellectual revolution in the history of science. (*Library of Congress*)

quantum restriction existed until they devised means of observing the behavior of matter at the atomic level. The major tool for doing this at the time of Planck's work was by observation of the radiant energy absorbed or emitted by matter. An object can gain or lose energy by absorbing or emitting radiant energy. Planck assumed that the amount of energy gained or lost at the atomic level by absorption or emission of radiation had to be a whole number multiple of a constant times the frequency of the radiant energy. Let us call this amount of energy gained or lost ΔE. Then, according to Planck's theory,

$$\Delta E = h\nu, \ 2h\nu, \ 3h\nu, \ \text{etc.} \tag{5.2}$$

The constant h is known as Planck's constant. It has the value of 6.63×10^{-34} joule-sec, or J-sec. The smallest increment of energy at a given frequency, $h\nu$, is called a *quantum* of energy.

SAMPLE EXERCISE 5.2

Calculate the smallest increment of energy (that is, the quantum of energy) that an object can absorb from yellow light whose wavelength is 589 nm.

Solution: We obtain the magnitude of a quantum of energy from Equation [5.2], $\Delta E = h\nu$. The value of Planck's constant is given both in the text above and in the table of physical constants on the back inside cover of the text, $h = 6.63 \times 10^{-34}$ J-sec. The frequency, ν, is calculated from the given wavelength as shown in Sample Exercise 5.1: $\nu = c/\lambda =$

$5.09 \times 10^{14}/\text{sec}$. Thus we have

$$\Delta E = h\nu = (6.63 \times 10^{-34} \text{ J-sec})(5.09 \times 10^{14}/\text{sec}) = 3.37 \times 10^{-19} \text{ J}$$

Planck's theory tells us that an atom or molecule emitting or absorbing radiation whose wavelength is 589 nm cannot lose or gain energy by radiation except in multiples of 3.37×10^{-19} J. It cannot, for example, gain 5.00×10^{-19} J from this radiation because this amount is not a multiple of 3.37×10^{-19} J.

At this point you may be wondering about the practical applications of Planck's quantum theory. A few years after Planck presented his theory, scientists began to see its applicability to a great many experimental observations. It soon became apparent that Planck's theory had within it the seeds of a revolution in the way the physical world is viewed. Let's consider a few applications that are of special importance for chemistry.

THE PHOTOELECTRIC EFFECT

Light shining on a clean metallic surface can cause the surface to emit electrons; this phenomenon, known as the photoelectric effect, can be demonstrated as shown in Figure 5.5. For each metal there is a minimum frequency of light below which no electrons are emitted, regardless of how intense the beam of light. In 1905, Albert Einstein (1879–1955) used the quantum theory to explain the photoelectric effect. He assumed that the radiant energy striking the metal surface is a stream of tiny energy packets. Each energy packet, called a **photon**, is a quantum of energy $h\nu$. Thus radiant energy itself is considered to be quantized. *Photons of high-frequency radiation have high energies, whereas photons of lower-frequency radiation have lower energy:*

$$E_{\text{photon}} = h\nu \tag{5.3}$$

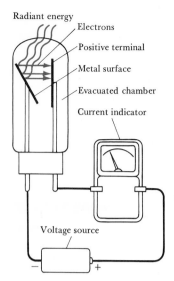

Radiant energy
Electrons
Positive terminal
Metal surface
Evacuated chamber
Current indicator
Voltage source
− +

FIGURE 5.5 The photoelectric effect. When photons of sufficiently high energy strike the metal surface in the tube, electrons are emitted from the metal. The electrons are drawn toward the other electrode, which is a positive terminal. As a result, current flows in the circuit. If the energy of a photon, $h\nu$, is too small, no electrons are freed and no current flows in the circuit. Photocells in automatic doors use the photoelectric effect to generate the electrons that operate the door-opening circuits.

When the photons are absorbed by the metal, their energy is transferred to an electron in the metal. A certain amount of energy is required for the electron to overcome the attractive forces that hold it within the metal. Otherwise it cannot escape from the metal surface, even if the light beam is quite intense. If a photon does have sufficient energy, the electron is emitted. If a photon has more than the minimum energy required to free an electron, the excess appears as the kinetic energy of the emitted electron. Thus the kinetic energy of the emitted electron, E_k, equals the energy supplied by the photon, $h\nu$, minus the binding energy holding the electron in the metal, E_b:

$$E_k = h\nu - E_b \qquad\qquad [5.4]$$

LINE SPECTRA

A particular source of radiant energy may emit a single wavelength, as in the light from a laser (Figure 5.6), or may contain many different wavelengths, as in the radiations from a lightbulb or a star. Radiation composed of a single wavelength is termed monochromatic. When the radiation from a source such as a star is separated into its monochromatic components, a spectrum is produced. This separation of radiations of differing wavelengths can be achieved by dispersing them in a prism, as shown in Figure 5.7. When the white light from a lightbulb is passed through a prism it is dispersed into a continuous range of colors; violet merges into blue, blue into green, and so forth with no blank spots. This rainbow of colors, containing light of all wavelengths, is called a continuous spectrum. The most familiar example of a continuous spectrum is the rainbow, which is produced by the dispersal of sunlight by rain drops or mist.

Not all emitters of light radiate all colors or wavelengths. For example, when hydrogen is placed under reduced pressure in a tube such as that depicted in Figure 5.8 and a high voltage is applied, light is emitted. (If, instead of hydrogen, neon gas were placed in the tube, the familiar red-orange glow of neon lights would be produced; sodium vapor would produce the yellow radiation of many modern streetlights.) When the

FIGURE 5.6 Monochromatic light being emitted from the discharge tube of an argon-gas laser. (*Bell Telephone Laboratories*)

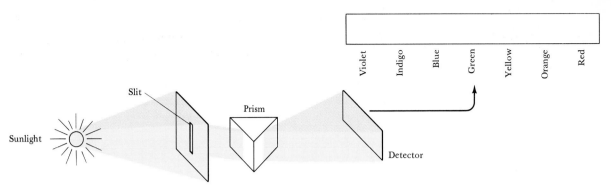

FIGURE 5.7 Production of a continuous visible spectrum by passing a narrow beam of white light through a prism. The white light could be sunlight or light from an incandescent lamp.

light coming from such tubes is separated into its monochromatic components, only certain colors or wavelengths of light are found to be present. A spectrum containing radiation of only specific wavelengths is called a line spectrum (see Figure 5.8).

The spectrum of the radiation emitted by a substance is called an emission spectrum. The line spectrum of hydrogen shown in Figure 5.8 is an emission spectrum. Substances also exhibit absorption spectra. When continuous electromagnetic radiation, such as that from a lightbulb, is passed through a substance, certain wavelengths may be absorbed. The spectrum of the radiation that passes through is called an absorption spectrum. The absorption spectrum of hydrogen consists of what looks like a continuous spectrum interrupted by black lines at 656.3 nm, 486.3 nm, and the other wavelengths found in the emission spectrum. The absorption spectrum of hydrogen is a line spectrum that is complementary to the emission spectrum of this substance.

Each element has a characteristic line spectrum that can be used to identify it. The spectra shown in

Figure 5.9 are from several stars. The lines in the spectra result from absorption of the stars' radiation by the cooler gases at their surfaces. The prominent lines in each are due to hydrogen. It is estimated that 95 percent of the atoms in stars such as our sun are hydrogen atoms. The remaining, fainter lines of the star spectra are due principally to helium, their second most abundant element. It was from a similar spectrum of our sun that helium was first discovered in 1868. Some of the absorption lines of the sun's spectrum could not be matched with those of any then known element; it was concluded that the sun contained an element previously unknown on earth. This element was named helium after *helios,* the Greek word for the sun. Helium was subsequently isolated and characterized in the laboratory in 1895.

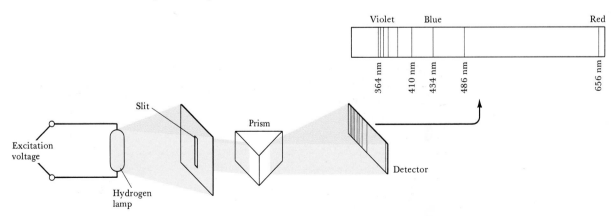

FIGURE 5.8 Production of the emission line spectrum of hydrogen by passing a narrow beam of light emitted by a hydrogen lamp through a prism. The lamp emits light when hydrogen atoms are excited in an electrical discharge.

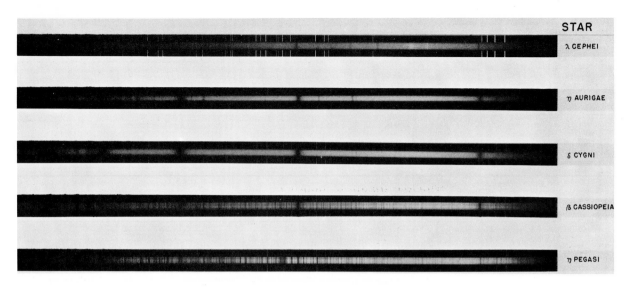

STAR

λ CEPHEI

η AURIGAE

δ CYGNI

β CASSIOPEIA

η PEGASI

FIGURE 5.9 The visible spectra of several stars. The dark bands, which are most prominent in the middle three spectra, correspond to hydrogen-atom absorptions. The Greek letters designate a particular star in the constellation indicated. (*Hale Observatories*)

Although it was not realized at the time by the scientists who gathered the data, line spectra provide beautiful applications of the quantum theory. In 1914 the Danish physicist Niels Bohr (1885–1962) incorporated Planck's theory to explain the line spectrum of hydrogen. The basic idea introduced by Bohr was that the absorptions and emissions of light by hydrogen atoms correspond to energy changes of electrons within the atoms. The fact that only certain frequencies are absorbed or emitted by an atom tells us that only certain energy changes are possible. (We say that the energy changes within an atom are quantized.) Further details of Bohr's model are described in the next section.

5.3 Bohr's model of hydrogen

Bohr's model of the hydrogen atom took into account two important developments that were relatively new at that time. The first was Rutherford's experiments, which established the nuclear nature of the atom (Section 2.6). The other was Einstein's work, which showed that radiant energy could be thought of as a stream of discrete bundles of energy called photons.

Bohr's model consists of a series of postulates that may be summarized as follows:

1 The electron of a hydrogen atom moves about the central proton in a circular orbit. However, an electron in an atom cannot have just any energy; only orbits of certain radii, and having certain energies, are "allowed." This idea is an extension of Planck's concept of the quantization of energy.

2 In the absence of radiant energy, an electron in an atom remains indefinitely in one of the allowed energy states. When radiant energy is present, however, the atom may absorb energy. When this hap-

pens, the electron undergoes a change from one allowed energy state to another. The frequency of the radiant energy absorbed (ν) corresponds exactly to the energy difference (ΔE) between two of the allowed energies: $\Delta E = h\nu$.

We need not concern ourselves with the details of how Bohr used quantum theory to calculate the energies of the electron. The main point is that he was able to calculate a set of allowed energies. Each of these allowed energies corresponds to a circular path of different radius. In Bohr's model, each allowed orbit was assigned an integer n, known as the principal quantum number, that may have values from 1 to infinity. The radius of the electron orbit in these energy states varies as n^2:

$$\text{Radius} = n^2(5.3 \times 10^{-11}\,\text{m}) \tag{5.5}$$

Thus the larger the value of n, the farther the electron from the nucleus. The energy of the electron depends on the orbit it occupies:

$$E_n = -R_H\left(\frac{1}{n^2}\right) \tag{5.6}$$

The constant R_H in Equation [5.6] is called the Rydberg constant; it has the value of 2.18×10^{-18} J. From Equation [5.6] we have that the energy of the electron is -2.18×10^{-18} J when the electron is in the orbit closest to the nucleus, $n = 1$. When it is in the second orbit, $n = 2$, its energy is

$$E_2 = (-2.18 \times 10^{-18}\,\text{J})\left(\frac{1}{2^2}\right) = -5.45 \times 10^{-19}\,\text{J}$$

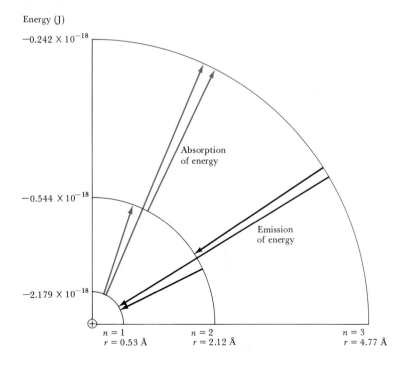

FIGURE 5.10 Radii and energies of the three lowest-energy orbits in the Bohr model of hydrogen. The arrows refer to transitions of the electron from one allowed energy state to another. When the transition takes the electron from a lower- to a higher-energy state, absorption occurs. When the transition is from a higher- to a lower-energy state, emission occurs.

The orbital radii and energies are illustrated for $n = 1, n = 2$, and $n = 3$ in Figure 5.10. Another more common way of representing the allowed energies is shown in Figure 5.11.

The negative sign in Equation [5.6] denotes stability relative to some reference state. In other words, the more negative the value for energy, the more stable the system is. (One way to remember this convention is to think of a ball rolling around on a surface with many hills and valleys. The ball will naturally come to rest in a valley; it is more stable when its potential energy is lowest.) The reference, or zero-energy, state for the electron in hydrogen is chosen to be that in which the electron is completely separated from the nucleus. This, of course, corresponds to an infinitely large value for the principal quantum number n:

$$E_\infty = (-2.18 \times 10^{-18}\,\text{J})\left(\frac{1}{\infty^2}\right) = 0$$

The energy of the electron in any other orbit is then negative relative to this reference state. The lowest energy, or most stable, state, with $n = 1$, is known as the ground state. When the electron is in a higher energy orbit, that is, $n = 2$ or higher, the atom is said to be in an electronically excited state.

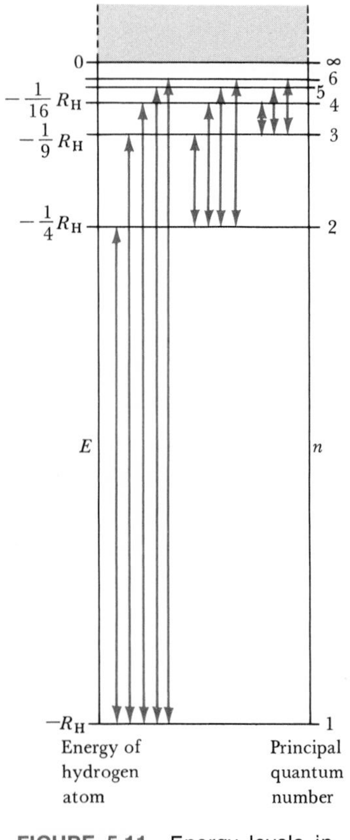

Bohr's model of the hydrogen atom quantitatively explained the observed line spectra of that substance. The absorptions or emissions in the line spectra correspond to transitions of the electron from one orbit to another. Radiant energy is absorbed when the electron moves from one orbit to another having a larger radius; it requires energy to pull the electron away from the nucleus. Conversely, energy is emitted when the electron moves from a larger orbit to another having a smaller radius. The changes in energy, ΔE, are given by the difference between the energy of the final state of the electron, E_f, and the initial state, E_i:

$$\Delta E = E_f - E_i$$

Substituting the expression for the energy of the electron, Equation [5.6], gives

$$\Delta E = \left(\frac{-R_H}{n_f^2}\right) - \left(\frac{-R_H}{n_i^2}\right) = R_H\left(\frac{1}{n_i^2} - \frac{1}{n_f^2}\right)$$

Because $\Delta E = h\nu$, we have

$$\Delta E = h\nu = R_H\left(\frac{1}{n_i^2} - \frac{1}{n_f^2}\right) \tag{5.7}$$

In this expression, n_i and n_f represent the quantum numbers for the initial and final states, respectively. Notice that when the final state quantum number (n_f) is larger than the initial state quantum number (n_i), the term in brackets is positive, and ΔE is positive. This means that the system has absorbed a photon and thus increased in energy (an endothermic process). When n_i is larger than n_f, as happens in emission, energy is given off; ΔE is negative.

FIGURE 5.11 Energy levels in the hydrogen atom from the Bohr model. The arrows refer to transitions of the electron from one allowed energy state to another, as described in Figure 5.10. Only the lowest six energy levels are shown.

Complete removal of the electron from a hydrogen atom, corresponding to a transition from the $n = 1$ (ground) state to the $n = \infty$ state, is known as ionization. This is represented as

$$H(g) \longrightarrow H^+(g) + e^-$$

The energy required for ionization from the ground state is called the ionization energy.

SAMPLE EXERCISE 5.3

Calculate the frequency of the hydrogen line that corresponds to the transition of the electron from the $n = 4$ to the $n = 2$ states.

Solution: We employ Equation [5.7], substituting $n_i = 4$ and $n_f = 2$ because these are the quantum numbers for the initial and final orbits, respectively:

$$\Delta E = h\nu = R_H\left(\frac{1}{n_i{}^2} - \frac{1}{n_f{}^2}\right)$$

$$\nu = \frac{R_H}{h}\left(\frac{1}{n_i{}^2} - \frac{1}{n_f{}^2}\right)$$

$$= \frac{2.18 \times 10^{-18}\,\text{J}}{6.63 \times 10^{-34}\,\text{J-sec}}\left(\frac{1}{4^2} - \frac{1}{2^2}\right)$$

$$= \frac{2.18 \times 10^{-18}\,\text{J}}{6.63 \times 10^{-34}\,\text{J-sec}}\left(\frac{1}{16} - \frac{1}{4}\right)$$

$$= \frac{2.18 \times 10^{-18}\,\text{J}}{6.63 \times 10^{-34}\,\text{J-sec}}\left(-\frac{3}{16}\right)$$

$$= -6.17 \times 10^{14}/\text{sec}$$

The negative sign simply indicates that the light is emitted. This value for ν can be used to calculate the wavelength of the radiation, so we can compare it with the hydrogen spectrum shown in Figure 5.8:

$$\lambda = \frac{c}{\nu} = \frac{3.00 \times 10^8\,\text{m/sec}}{6.17 \times 10^{14}/\text{sec}}$$

$$= 4.86 \times 10^{-7}\,\text{m} = 486\,\text{nm}$$

All of the lines shown in Figure 5.8 correspond to transitions of an electron from higher orbits to the $n = 2$ orbit.

SAMPLE EXERCISE 5.4

Calculate the energy required for ionization of an electron from the ground state of the hydrogen atom.

Solution: The ionization energy may be written as the difference between the final and initial state energies. We have $n_f = \infty$, $n_i = 1$.

$$\Delta E = E_f - E_i = R_H\left(\frac{1}{1^2} - \frac{1}{\infty^2}\right)$$

$$= R_H(1 - 0)$$

This is just equal to R_H, 2.18×10^{-18} J.

It is often useful to express this energy on a molar basis. To do this we simply multiply by Avogadro's number:

$$\left(2.18 \times 10^{-18}\,\frac{\text{J}}{\text{atom}}\right)\left(6.02 \times 10^{23}\,\frac{\text{atoms}}{\text{mol}}\right)$$

$$\times \left(\frac{1\,\text{kJ}}{1000\,\text{J}}\right) = 1.31 \times 10^3\,\text{kJ/mol}$$

Bohr's model was very important because it introduced the idea of quantized energy states for electrons in atoms. This feature is incorporated into our current model of the atom. However, Bohr's model was adequate for explaining only atoms and ions with one electron, such as H, He$^+$, or Li^{2+}. It was not adequate to account for the atomic spectra of other atoms or ions, except in a rather crude way. Consequently, Bohr's model was eventually replaced by a new way of viewing atoms that is called quantum mechanics or wave mechanics. This newer model maintains the concept of quantized energy states, but in addition still further applications of Planck's quantum theory enter the picture.

5.4 Matter waves

In the years following Bohr's development of a model for the hydrogen atom, the dual nature of radiant energy had become a familiar concept. Depending on the experimental circumstances, radiation might appear to have either a wavelike or particlelike (photon) character. Louis de Broglie, a young man working on his Ph.D. thesis in physics at the Sorbonne, in Paris, made a rather daring, intuitive extension of this idea. If radiant energy could under appropriate conditions behave as though it were particulate in nature, could not matter under appropriate conditions possibly show the properties of a wave? Suppose that the electron in orbit around the nucleus of a hydrogen atom could be thought of as a wave, with a characteristic wavelength, as illustrated in Figure 5.2. De Broglie suggested that the electron in its circular path about the nucleus has associated with it a particular wavelength. He went on to propose that the characteristic wavelength of the electron or any other particle depends on its mass, m, and velocity, v:

$$\lambda = \frac{h}{mv} \qquad [5.8]$$

(h is Planck's constant). The quantity mv for any object is called its momentum. De Broglie used the term matter waves to describe the wave characteristics of material particles.

Because De Broglie's hypothesis is perfectly general, any object of mass m and velocity v would give rise to a characteristic matter wave. However, it is easy to see from Equation [5.8] that the wavelength associated with an object of ordinary size, such as a golf ball, is so tiny as to be completely out of the range of any possible observation. This is not so for electrons, because their mass is so small.

SAMPLE EXERCISE 5.5

What is the characteristic wavelength of an electron with a velocity of 5.97×10^6 m/sec? (The mass of the electron is 9.11×10^{-28} g.)

Solution: The value of Planck's constant, h, is 6.63×10^{-34} J-sec (1 J = 1 kg-m²/sec²).

$$\lambda = \frac{h}{mv}$$

$$= \frac{(6.63 \times 10^{-34} \text{ J-sec})}{(9.11 \times 10^{-28} \text{ g})(5.97 \times 10^6 \text{ m/sec})}$$

$$\times \left(\frac{1 \text{ kg-m}^2/\text{sec}^2}{1 \text{ J}}\right)\left(\frac{10^3 \text{ g}}{1 \text{ kg}}\right)$$

$$= 1.22 \times 10^{-10} \text{ m} = 0.122 \text{ nm}$$

By comparing this value with the wavelengths of electromagnetic radiations shown in Figure 5.3, we see that the characteristic wavelength is about the same as that of X rays.

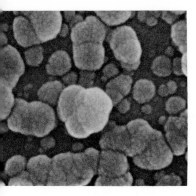

FIGURE 5.12 Gold particles on silicon. The particles are shown magnified over 100,000 times. (*Photograph courtesy of International Scientific Instruments, Inc.*)

Within a few years after De Broglie published his theory, the wave properties of the electron were demonstrated experimentally. Electrons were diffracted by crystals, just as X rays, which are definitely radiant energy, are diffracted. (We shall have more to say about the X-ray diffraction experiment in Chapter 11.)

The technique of electron diffraction has been highly developed. In the electron microscope, the wave characteristics of electrons are used to obtain electron diffraction pictures of tiny objects. The electron microscope is an important technique for studying surface phenomena at the very highest magnifications. An example of an electron microscope picture is shown in Figure 5.12. Pictures such as this are powerful demonstrations that tiny particles of matter can indeed behave as waves.

THE UNCERTAINTY PRINCIPLE

Discovery of the wave properties of matter raised new and interesting questions. If a subatomic particle can exhibit the properties of a wave phenomenon, is it possible to say precisely just where that particle is located? One can hardly speak of the precise location of a wave. The amplitude, or intensity, of a wave can be defined at a certain point, as illustrated in Figure 5.2, but the wave as a whole extends in space. Its location is therefore not defined precisely, at least not in the same sense that one can define the location of a particle. We might logically expect to be able to measure not only a particle's location, but also its direction and speed of motion.

However, the German physicist Werner Heisenberg* concluded that there is a fundamental limitation on just how precisely we can hope to know both the location and the momentum of a particle. Just as in the case of quantum effects, the limitation becomes important only when we deal with matter at the subatomic level, that is, with masses as small as that of an electron. Heisenberg's principle is called the uncertainty principle. When applied to the electrons in an atom, this principle states that it is inherently impossible for us to know both the exact momentum of the electron and its location in space. Thus it is not appropriate to imagine the electrons as moving in well-defined circular orbits about the nucleus, always at the same radius.

De Broglie's hypothesis and Heisenberg's uncertainty principle set the stage for a new and more broadly applicable theory of atomic structure. In this new approach, any attempt to define precisely the instantaneous location and momentum of the electrons is abandoned. The wave nature of the electron is recognized, and its behavior is described in terms appropriate to waves.

5.5 The quantum-mechanical description of the atom

The mathematics employed to determine electron energy levels in even so simple a system as the hydrogen atom by quantum-mechanical methods are quite advanced. There is therefore no point in presenting any mathematical details here. We can, however, understand the idea content of quantum mechanics with the aid of a qualitative description.

*Heisenberg (1901–1976) was one of the leading physicists of the twentieth century. He received the Nobel Prize in physics in 1932.

The approach of quantum mechanics involves a mathematical description of the wave properties of the electron. The equation involved also takes into account the attraction of the electron for the nucleus and the kinetic energy of the electron. Then certain conditions are imposed on the possible solutions to the resultant mathematical equation to make them physically reasonable. The result is a series of solutions that describe the allowed energy states of the electron. These solutions, usually represented by the symbol ψ (the Greek letter *psi*), are called wave functions.*

A wave function provides information about an electron's location in space when it is in a certain allowed energy state. The allowed energy states are the same as those predicted by the Bohr model. However, the Bohr model suggests that the electron is in a circular orbit of some particular radius about the nucleus. In the quantum-mechanical model for hydrogen, it is not so simple to describe the electron's location. The uncertainty principle suggests that if we know the momentum of the electron with high accuracy, our knowledge of its location is very uncertain. Thus, for an individual electron, we cannot hope to specify its location around the nucleus. Rather we must be content with a kind of statistical knowledge. In the quantum-mechanical model we therefore speak of the *probability* that the electron will be in a certain region of space at a given instant. As it turns out, the square of the wave function (ψ^2) at a given point in space represents the probability that the electron will be found at that location. If the value of the wave function is summed over all the points in space around the nucleus, its value must equal 1. This is true because probability of 1 is certainty, and it is certain that the electron must be somewhere in the space around the nucleus.

One way of representing the probability of finding the electron in various regions of an atom is shown in Figure 5.13. In this figure the density of the dots represents the probability of finding the electron; the region of high probability is represented by a high density of dots. These regions correspond to relatively large values for ψ^2. It is common to refer to the electron density in certain regions of space, this being another way of expressing probability; regions where there is a high probability of finding the electron are said to be regions of high electron density. In Section 5.6, we will say more about the ways we can represent electron density.

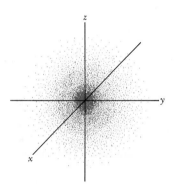

FIGURE 5.13 Electron density distribution in the ground state of the hydrogen atom.

QUANTUM NUMBERS AND ORBITALS

The complete quantum-mechanical solution to the hydrogen atom problem yields a set of wave functions and a corresponding set of energies. Each wave function represents an allowed solution. The term orbital denotes an allowed energy state for the electron. It also refers to the probability function that defines the distribution of electron density in space. Thus an orbital has both a characteristic energy and a characteristic shape. Note that an *orbital* (quantum-mechanical model) is not the same as an *orbit* (Bohr model).

*In mathematics, the term *function* is used to describe a relationship between two variables—for example, time and distance. Thus, the distance (d) covered by an auto moving at a constant speed (s) is a *function* of the amount of time (t) it has been in motion: $d = st$.

The Bohr model introduced a single quantum number (n) to describe an orbit. The quantum-mechanical model introduces three quantum numbers (n, l, and m_l) to describe an orbital. We will consider what information we obtain from each of these and how they are interrelated:

1 The **principal quantum number** (n) can have integral values of 1, 2, 3, and so forth just as in the Bohr model. This quantum number relates to the average distance of the electron from the nucleus and is the quantum-mechanical equivalent of Bohr's principal quantum number. In the hydrogen atom, orbitals possessing the same principal quantum number are of the same energy; as in the Bohr model $E_n = -R_H(1/n^2)$. We shall see in the next chapter that this is not the case for atoms having many electrons.

2 The second quantum number is known as the **azimuthal quantum number,** and is given the symbol l. It defines the shape of the orbital. The possible values for l are limited by the value for n; l can have integral values from 0 to $n - 1$. Thus, for example if $n = 3$, l can have values of 0, 1, or 2, but cannot equal 3 or higher. The value of l is generally designated by the letters s, p, d, f, and g, corresponding to l values of 0, 1, 2, 3, and 4, respectively.*

l	0	1	2	3	4
Designation of orbital	s	p	d	f	g

Table 5.2 shows the possible values of l for each value of n up through $n = 4$.

A collection of orbitals with the same value of n is referred to as an **electron shell.** One or more orbitals with the same set of n and l values is referred to as a **subshell.** For example, the shell with $n = 3$ is composed of three subshells, $l = 0$, 1, and 2 (the allowed values of l for $n = 3$). These subshells are called the $3s$, $3p$, and $3d$ subshells,

*The letters s, p, d, and f came from the words *sharp, principal, diffuse,* and *fundamental.* These words were used to describe certain features of spectra before quantum mechanics was developed.

TABLE 5.2 Relationship among values of n, l, and m_l through $n = 4$

n	l	Orbital designation	m_l	Number of orbitals
1	0	$1s$	0	1
2	0	$2s$	0	1
	1	$2p$	1, 0, −1	3
3	0	$3s$	0	1
	1	$3p$	1, 0, −1	3
	2	$3d$	2, 1, 0, −1, −2	5
4	0	$4s$	0	1
	1	$4p$	1, 0, −1	3
	2	$4d$	2, 1, 0, −1, −2	5
	3	$4f$	3, 1, 2, 0, −1, −2, −3	7

respectively; in each of these designations, the number indicates the value for the principal quantum number (n) and the letter corresponds to the value of l.

3 The third quantum number, labeled m_l, is called the magnetic or orientational quantum number. It describes the orientation of the orbital in space. This quantum number may have integral values ranging from l to $-l$. Thus, when $l = 0$, m_l must be 0. When $l = 1$, m_l can have values of 1, 0, or -1.

The possible values of the three quantum numbers through $n = 4$ are summarized in Table 5.2. Notice that the allowed subshells are $1s$, $2s$, $2p$, $3s$, $3p$, $3d$, and so forth. Also note that each s subshell contains one orbital ($m_l = 0$), and each p subshell contains three orbitals ($m_l = 1$, 0, -1); similarly, each d subshell contains five orbitals, and each f subshell contains seven orbitals. Figure 5.14 shows the number and relative energies of all hydrogen atom orbitals through $n = 3$. When the electron is in the lowest energy orbital (the $1s$ orbital), the hydrogen atom is said to be in its ground state. When it is in any other orbital, the atom is in an excited state. At ordinary temperatures essentially all hydrogen atoms are in their ground states. The electron may be promoted to an excited state orbital by absorption of a photon of appropriate energy.

SAMPLE EXERCISE 5.6

What are the values of n, l, and m_l for orbitals in the $3d$ subshell?

Solution: The number given in the designation of the subshell is the principal quantum number. Thus,

$n = 3$. The letter represents the value for l; s orbitals have $l = 0$, p have $l = 1$, and d have $l = 2$; thus $l = 2$ for the $3d$ subshell. The values of m_l can vary from l to $-l$; thus m_l can equal 2, 1, 0, -1, or -2. Consequently, there are five $3d$ orbitals.

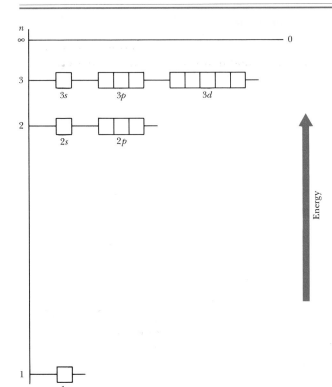

FIGURE 5.14 Orbital energy levels in the hydrogen atom and in hydrogenlike ions (those containing just one electron). Each box represents an orbital; orbitals of the same subshell, such as the $2p$ orbitals, are grouped together. Note that all orbitals with the same value for the principal quantum number, n, have the same energy. This is true only in one-electron systems.

Without referring to Table 5.2, determine the number of orbitals for which $n = 4$; indicate the values of n, l, and m_l for each of these orbitals.

Solution: For $n = 4$, the possible values of l are 0, 1, 2, and 3. These correspond to the 4s, 4p, 4d, and 4f

subshells. There is one 4s orbital ($n = 4$, $l = 0$, $m_l = 0$); there are three 4p orbitals ($n = 4$, $l = 1$, $m_l = 1, 0, -1$); there are five 4d orbitals ($n = 4$, $l = 2, m_l = 2, 1, 0, -1, -2$); there are seven 4f orbitals ($n = 4$, $l = 3$, $m_l = 3, 2, 1, 0, -1, -2, -3$).

5.6 Representations of orbitals

In our discussion of orbitals we have so far emphasized their energies. But the wave function also provides information about the electron's location in space when it is in a particular allowed energy state. We need to examine the ways that we can visualize or picture the orbitals.

THE s ORBITALS

The lowest energy (most stable) orbital, the 1s orbital, is spherically symmetric as shown earlier in Figure 5.13. Figures of this type, showing electron density, are one of the several ways we have to help us visualize orbitals. This figure indicates that the probability of finding the electron around the nucleus decreases as we move away from the nucleus in any direction. When the probability function (ψ^2) for the 1s orbital is graphed as a function of the distance from the nucleus (r), it rapidly approaches zero at large distance, as shown in Figure 5.15. This effect indicates that the electron, which is drawn toward the nucleus by electrostatic attraction, is not likely ever to get very far from the nucleus.

If we similarly consider the 2s and 3s orbitals of hydrogen we find that they are also spherically symmetric. Indeed, *all* s orbitals are spherically symmetric. The manner in which the probability function (ψ^2) varies with r for the 2s and 3s orbitals is shown in Figure 5.15. Notice that for the 2s orbital, ψ^2 goes to zero and then increases again in value before finally approaching zero at a larger value of r. The intermediate regions where ψ^2 goes to zero are called nodal surfaces or simply nodes. The

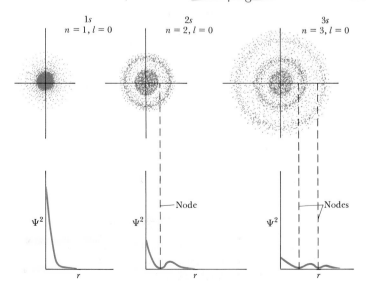

FIGURE 5.15 Electron density distributions in 1s, 2s, and 3s orbitals. The lower part of the figure shows how the electron density, represented by ψ^2, varies as a function of distance from the nucleus. In the 2s and 3s orbitals, the electron-density function drops to zero at certain distances from the nucleus. The spherical surfaces around the nucleus at which ψ^2 is zero are called nodes.

1s

2s

3s

FIGURE 5.16 Contour representations of the 1s, 2s, and 3s orbitals. The spherical surfaces connect points of equal value of ψ^2. The surface encloses 90 percent of the total ψ^2 for each orbital.

number of nodes increases with increasing value for the principal quantum number (n). The 3s orbital possesses two nodes as illustrated in Figure 5.15. Notice also that as n increases the electron is more and more likely to be located further from the nucleus. That is, the size of the orbital increases as n increases.

The most widely used method of representing orbitals is to display a boundary surface that encloses some substantial fraction, say 90 percent, of the total electron density for the orbital. For the s orbitals these contour representations are merely spheres. The contour or boundary surface representations of the 1s, 2s, and 3s orbitals are shown in Figure 5.16. They have the same shape, but they differ in size. The fact that there are nodes within the 2s and 3s surfaces is lost in these representations. This is not a serious disadvantage; it turns out that for most qualitative discussions the most important features of orbitals are their size and shape. These features are adquately represented by the contour diagrams.

THE p ORBITALS

The distribution of electron density for a 2p orbital is shown in Figure 5.17. As we can see from this figure, the electron density is not distributed in a spherically symmetric fashion as in an s orbital. Instead, the electron density is concentrated on two sides of the nucleus separated by a node at the nucleus; we often say that this orbital has two lobes. It is useful to recall that we are making no statement of how the electron is moving within the orbital; Figure 5.17 portrays the *averaged* distribution of the 2p electron in space.

There are three p orbitals in each shell beginning with $n = 2$. For example, there are three 2p orbitals, three 3p orbitals, and so forth. The orbitals of a given principal quantum number have the same size and shape but differ from each other in orientation. The contour surfaces of the three 2p orbitals are shown in Figure 5.18. It is convenient to label these as the $2p_x$, $2p_y$, and $2p_z$ orbitals. The letter subscript indicates the axis along which the orbital is oriented. As it turns out, there is no necessary connection between one of these subscripts and a particular value of m_l. To explain why this is so would require discussion of material beyond the scope of an introductory text.

Just as with the s orbitals, the distance from the nucleus to the center of the electron density moves outward as we go from the 2p to 3p to 4p orbitals. In other words, orbital size increases with increase in the principal quantum number (n). In accurate contour representations of the 3p and higher p orbitals, there are small regions of electron density separating the major lobes. These details of the shapes of the p orbitals are not of major chemical importance. The general overall shape of the p orbitals is more important. We shall therefore always represent the p orbitals as shown in Figure 5.18, regardless of the value for the principal quantum number (n).

THE d AND f ORBITALS

When $l = 2$, the orbital is a d orbital. There are no d orbitals with n lower than 3. This follows from the fact that l can never be larger than $n - 1$. There are five equivalent 3d orbitals corresponding to the five possible

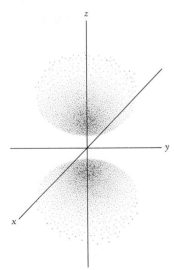

FIGURE 5.17 Electron density distribution in a 2p orbital.

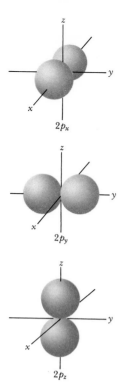

$2p_x$

$2p_y$

$2p_z$

FIGURE 5.18 Contour representations of the three $2p$ orbitals. The three orbitals with differing orientations correspond to different values of the orientational quantum number, m_l. There is no necessary connection between a particular orientation and a particular value for m_l. The only important point to remember is that because there are three possible values for m_l, there are three p orbitals with differing orientations.

values for m_l: 2, 1, 0, -1, and -2. Likewise there are five $4d$ orbitals, and so forth. Just as in the case of the p orbitals, the differing values of m_l correspond to different orientations of orbitals in space. The most useful representations of the $3d$ orbitals are shown in Figure 5.19. Notice that four of the orbitals are of the same shape but have differing orientations. The fifth orbital, labeled the $d_z{}^2$, has a different shape. It is not possible to represent the five d orbitals as having the same shape in an ordinary x, y, z axis system. Although the fifth d orbital looks different, it has the same energy in an atom as any of the other four orbitals.

The representations of higher d orbitals are very much like those for the $3d$. The contour representations shown in Figure 5.19 are commonly employed for all d orbitals, regardless of major quantum number.

There are seven equivalent f orbitals (for which $l = 3$) for each value of n of 4 or greater. The f orbitals are difficult to represent in three-dimensional contour diagrams. We shall have no need to concern ourselves with orbitals having values for l greater than 3.

As we shall see in a later chapter, an understanding of the number and shapes of atomic orbitals is important to a proper understanding of the molecules formed by combining atoms. *You should commit to memory the orbital representations shown in Figures 5.16, 5.18, and 5.19.*

The orbitals we have been describing are those of a hydrogen atom. In the atoms of other elements, there are many electrons moving about a single nucleus. These electrons are attracted by the nucleus and at the same time are repelled by one another. The resulting distribution of electron density and the energies of the allowed energy states of the electrons are thus the product of extremely complex forces. Nevertheless, as we shall see in the following chapter, the electronic structures of atoms having many electrons can be built up by the progressive addition of electrons to orbitals that are very much like those of the hydrogen atom.

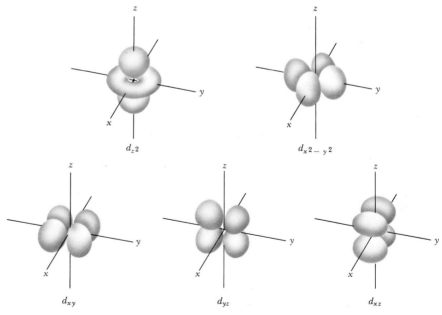

$d_z{}^2$

$d_{x^2-y^2}$

d_{xy}

d_{yz}

d_{xz}

FIGURE 5.19 Contour representations of the five $3d$ orbitals.

Summary

Radiant energy moves through a vacuum at the "speed of light," $c = 3.00 \times 10^8$ m/sec; it has wave-like characteristics that allow it to be described in terms of wavelength (λ) and frequency (ν); these are interrelated: $c = \lambda\nu$. The dispersion of radiation into its component wavelengths produces a spectrum. If all wavelengths are present, the spectrum is said to be continuous; if only certain wavelengths are present, it is called a line spectrum.

The quantum theory describes the minimum amount of radiant energy that an object can gain or lose, $E = h\nu$; this smallest quantity is called a quantum. A quantum of radiant energy is called a photon. The quantum theory was used to explain the photoelectric effect and the line spectrum of the hydrogen atom. The absorptions or emissions of light by an atom, which produce its line spectrum, correspond to energy changes of electrons within the atom; the energy of the electron in an atom is quantized.

Electrons exhibit wave properties and can be described by a wavelength, $\lambda = h/mv$. Discovery of the wave properties of the electron led to the uncertainty principle, which indicates that the position and mometum of a particle as light as an electron can be determined simultaneously with only a limited accuracy.

The ideas described above culminated in our current model of the electronic structures of atoms, in which we speak of the probability of the electron being found at a particular point in space. Although the positions of the electrons are not defined, except in this averaged sense, their energies are precisely known. Each allowed state of an electron in an atom corresponds to a particular set of values for three quantum numbers. Each such allowed energy state is termed an orbital. An orbital is described by a combination of an integer and letters, corresponding to the three values for the quantum numbers. The principal quantum number (n) is indicated by the integers 1, 2, 3, ... This quantum number relates most directly to the size and energy of an orbital. The azimuthal quantum number (l) is indicated by the letters s, p, d, f, and so on, corresponding to values of l of 0, 1, 2, 3, ... The l quantum number defines the shape of the orbital. The magnetic or orientational quantum number (m_l) describes the orientation of the orbital in space. For example, the three $3p$ orbitals are designated $3p_x$, $3p_y$, and $3p_z$, the subscript letters indicating the axis along which the orbital is oriented.

Restrictions in the values of the three quantum numbers gives rise to the following allowed subshells:

$1s$
$2s$, $2p$
$3s$, $3p$, $3d$
$4s$, $4p$, $4d$, $4f$
$\vdots$

There is one orbital in an s subshell, three in a p subshell, five in a d subshell, and seven in an f subshell. The contour representations are the most generally useful way to visualize the spatial characteristics of the orbitals.

Learning goals

Having read and studied this chapter, you should be able to:

1 Describe the wave properties and characteristic speed of propagation of radiant energy (electromagnetic radiation).

2 Use the relationship $\lambda\nu = c$, which relates the wavelength (λ) and frequency (ν) of radiant energy to its speed (c).

3 Describe how the wavelength and frequency differ in the various parts of the electromagnetic spectrum, such as the infrared, visible, and ultraviolet.

4 Explain the essential feature of Planck's quantum theory, namely, that the smallest increment, or quantum, of radiant energy of frequency ν that can be emitted or absorbed is $h\nu$, where h is Planck's constant.

5 Explain how Einstein accounted for the photoelectric effect by considering the radiant energy to be a stream of particlelike photons striking a metal surface. In other words you should be able to explain all the observations about the photoelectric effect using Einstein's model.

6 Explain what is meant by the term *spectrum* and by the expression *line spectrum* in referring to the light emitted or absorbed by an atom.

7 List the assumptions made by Bohr in his model of the hydrogen atom. Most importantly, you should be able to explain how Bohr's model relates to Planck's quantum theory.

8 Explain the concept of an allowed energy state and how this concept is related to the quantum theory.

9 Calculate the energy differences between any two allowed energy states of the electron in hydrogen.

10 Explain the concept of ionization energy.

11 Calculate the characteristic wavelength of a par-

ticle from a knowledge of its mass and velocity.

12 Describe the uncertainty principle and explain the limitations it places on our ability to define simultaneously the location and momentum of a subatomic particle, particularly an electron.

13 Explain the concepts of *orbital, electron density,* and *probability* as used in the quantum-mechanical model of the atom; explain the physical significance of ψ^2.

14 Describe the three quantum numbers used to define an orbital in an atom and list the limitations placed on the values each may have.

15 Describe the correspondence between letter designations and values for the azimuthal quantum number (l).

16 Describe the shapes of s, p, and d orbitals.

Key terms

Among the more important terms and expressions used for the first time in this chapter are the following:

The **azimuthal quantum number,** l (Section 5.5), is one of the quantum numbers that specifies an atomic orbital. The value for l defines the shape of the orbital. The values allowed for l are restricted by the value for the principal quantum number (n); l may take on integral values from 0 to $n - 1$.

The **electron density** (Section 5.5) at a particular point in space in an atom is the probability that the electron will be found in the region immediately around that point (ψ^2).

The **ionization energy** (Section 5.3) for the hydrogen atom is the energy required to lift an electron from its lowest energy, or most stable, orbit to a point infinitely far from the nucleus. In effect, this involves removing the electron from the atom.

A **line spectrum** (Section 5.2) contains radiation only of certain specific wavelengths.

Matter wave (Section 5.4) is the term applied to the wave characteristics of a subatomic particle.

A **node** (Section 5.6), as applied to electron density in atoms, is the locus of points (for example, a plane or a spherical surface) at which the electron density is zero. For example, the node in a 2s orbital, Figure 5.15, is a spherical surface.

An **orbit** (Section 5.3) in the Bohr theory of the hydrogen atom is any one of the allowed energy states of the electron. Each orbit corresponds to a different value for the principal quantum number (n).

An **orbital** (Section 5.5) represents an allowed energy state of an electron in the modern, quantum-mechanical model of the atom. The term orbital is also used to describe the spatial distribution of the electron.

The **orientational quantum number,** m_l (Section 5.5), describes the orientation of an orbital in space. This quantum number may have integral values ranging from l to $-l$. For each value of l there are thus $2l + 1$ values for m_l.

A **photon** (Section 5.2) is a quantum, or smallest increment, of radiant energy, $h\nu$.

The **principal quantum number,** n (Section 5.5), is the quantum number that relates most directly to the size and energy of an atomic orbital. It may take on integer values of 1, 2, 3, ...

A **quantum** (Section 5.2) is the smallest increment of radiant energy that may be absorbed or emitted. The magnitude of the quantum of radiant energy is $h\nu$.

Radiant energy, or **electromagnetic radiation** (Section 5.1), is a form of energy possessing wave character and being propagated through space with a characteristic speed, 3.00×10^8 m/sec.

The term **spectrum** (Section 5.2) refers to the distribution among various wavelengths of the light emitted or absorbed by an object. A continuous spectrum contains radiation distributed over many wavelengths.

A **subshell** (Section 5.5) is designated by a particular set of values for the quantum numbers n and l. For example, we speak of the $2p$ subshell ($n = 2$, $l = 1$), which is composed of three orbitals ($2p_x$, $2p_y$, and $2p_z$).

The **uncertainty principle** (Section 5.4) states that there is an inherent uncertainty in the precision with which we can simultaneously specify the location and momentum of a particle. It is of importance only for the lightest particles such as the electron.

A **wave function** (Section 5.5) is a mathematical description of an allowed energy state, or orbital, for an electron in the quantum-mechanical model for an atom. It is usually symbolized by the Greek letter ψ.

EXERCISES

Radiant energy

5.1 List the following types of electromagnetic radiation in order of increasing wavelength: (a) radiation from a microwave oven; (b) sunlight that has passed through a green filter; (c) radiation from an FM radio station; (d) X rays used by a dentist; (e) the red light from a traffic signal. (Refer to Figure 5.3.)

5.2 (a) What is the wavelength of radiation whose frequency is 5.00×10^{13}/sec? (b) What is the frequency of radiation whose wavelength is 1.50×10^{-17} m? (c) How far can light travel (in meters) in 1.00 minute?

5.3 A neon light strongly emits radiation whose wavelength is 616 nm. What is the frequency of this radiation? Using Figure 5.3, suggest the color associated with this wavelength.

5.4 (a) An FM station operates at a frequency of 96.3×10^6/sec (96.3 megahertz, MHz). Calculate the wavelength of these radiowaves. (b) The antenna length for best reception of radio-frequency transmissions is normally half the wavelength of the signal being received. What length antenna is best suited for reception of a TV station that broadcasts at 210 MHz?

5.5 (a) The distance that light can travel in $365\frac{1}{4}$ days is called a light-year. What is the distance, in meters, of a light-year? What is this distance in miles? (b) A parsec is an astronomical unit of distance equal to 3.26 light-years. What is the distance, in kilometers, of a parsec?

5.6 The mean distance of the moon from earth is 239,000 miles. How long did it take for the TV pictures from the Apollo mission to reach earth from the moon?

Quantum theory

5.7 (a) Calculate the smallest increment of energy (a quantum of energy) that can be emitted or absorbed at a frequency of 5.00×10^{13}/sec. (b) Calculate the energy of a photon if its wavelength is 2.00×10^{-7} m. (c) If the energy of a photon is 1.30×10^{-19} J, what is its frequency?

5.8 Calculate and compare the energy of an X-ray photon ($\lambda = 0.15$ nm) with that of a microwave photon ($\lambda = 1.0$ mm).

5.9 The strong red line of the lithium spectrum occurs at 6708 Å. What is the energy, in kilojoules/photon, of photons of this wavelength? What is the energy in kilojoules/mole?

5.10 If the human eye receives a 1.45×10^{-17} J signal from photons whose wavelength is 550 nm, how many photons have hit the eye?

5.11 Chlorophyll absorbs blue light, $\lambda = 460$ nm, and emits red light, $\lambda = 660$ nm. Calculate the net energy change in chlorophyll when a single photon of 460 nm radiation is absorbed and one of 660 nm is emitted.

5.12 The minimum energy required to remove an electron from the surface of copper metal is 6.69×10^{-19} J. What is the maximum wavelength needed to provide photons of this energy? What happens to the photoemission if the intensity of the radiation of this wavelength is doubled?

5.13 Cesium metal requires radiation with a minimum frequency of 4.60×10^{14}/sec before it can emit an electron from its surface via the photoelectric effect. (a) What is the minimum energy required to produce this photoelectron? (b) What wavelength radiation will provide a photon of this energy? (c) If cesium is irradiated with light whose wavelength is 540 nm, what is the kinetic energy of the emitted electron?

5.14 An Einstein of radiation is 6.02×10^{23} (Avogadro's number) photons. What is the energy of 1.00 Einstein of radiation whose wavelength is 450 nm?

Bohr's model

5.15 Indicate whether energy is emitted or absorbed when the following electron transitions occur in hydrogen: (a) $n = 2$ to $n = 1$; (b) $n = 2$ to $n = 4$; (c) ionization of an electron in the $n = 2$ state.

5.16 Calculate the energy required to excite an electron from the $n = 1$ to $n = 3$ states in a hydrogen atom.

5.17 What wavelength of light is emitted when an electron moves from the $n = 5$ to $n = 2$ states in hydrogen? Compare this with the wavelengths observed in the emission spectrum of hydrogen, Figure 5.8.

5.18 What wavelength of light is emitted when an electron moves from the $n = 5$ to $n = 1$ states in hydrogen? Using Figure 5.3, indicate the region of the spectrum characteristic of the radiation emitted.

5.19 The Li^{2+} ion has only one electron. Would you expect the ionization energy for Li^{2+} to be larger or smaller than that for H? Explain.

5.20 For one-electron ions, the energy of the electron is given by the equation $E_n = -Z^2 R_H (1/n^2)$ where Z is the atomic number of the nucleus. What is the energy required to remove the remaining electron from the He^+ ion? Refer to Sample Exercise 5.4 and compare this value with that for the hydrogen atom. Explain the reason for the difference.

5.21 (a) Calculate the radius of the electron orbit in hydrogen when $n = 3$. (b) The radius of the electron orbit for one-electron ions of nuclear charge Z greater than 1 is given by the equation $r = n^2 (0.53 \times 10^{-8}$ cm$)/Z$. Calculate the radius of the electron orbit for $n = 1$ in Li^{2+}.

Matter waves

5.22 Cite one or more pieces of experimental evidence that show that it is possible for particles of matter to possess wave properties.

5.23 (a) What is the momentum of a neutron moving at a velocity of 3.78×10^3 m/sec? (The mass of a neutron is 1.67×10^{-24} g.) (b) What is the wavelength of a neutron moving at this velocity? (c) If a beam of neutrons has a characteristic wavelength of 2.00×10^{-10} m, what is the velocity of the neutrons?

5.24 What is the momentum of a golf ball of mass 82.5 g moving at a speed of 255 km/hr? What is the characteristic wavelength of this ball?

5.25 What is the characteristic wavelength of an electron moving at one-tenth the speed of light? (The mass of an electron is 9.11×10^{-28} g.)

Wave functions; orbitals; quantum numbers

5.26 Sketch the contour representations of the following orbitals: (a) s; (b) p_x; (c) p_y; (d) d_{xy}; (e) d_{yz}; (f) $d_{x^2-y^2}$.

5.27 Which of the following are incorrect designations for an atomic orbital: $4f$, $2d$, $2s$, $5p$, $1p$, $3f$, $3d$?

5.28 Give the values of n, l, and m_l for each orbital in the $4f$ subshell; for each orbital in the $n = 2$ shell.

5.29 How do the $2p$ and $3p$ orbitals differ?

5.30 Rank the following subshells in a hydrogen atom in order of increasing energy: $4s$, $3d$, $2p$, $1s$, $2s$, $4f$, $3s$.

5.31 What is meant by the term *wave function*? In what sense does the square of the wave function have physical significance?

5.32 Which of the following are permissible sets of quantum numbers for an electron in an atom: (a) $n = 3$, $l = 1$, $m_l = -1$; (b) $n = 3$, $l = 1$, $m_l = 2$; (c) $n = 2$, $l = 2$, $m_l = 0$; (d) $n = 9$, $l = 0$, $m_l = 0$; (e) $n = 4$, $l = -2$, $m_l = 1$? For those combinations that are permissible, write the appropriate designation for the subshell to which the orbital belongs (that is, $1s$, and so on).

5.33 What is the meaning of the word *node*? What is the probability of the electron's being found at a node? Explain.

5.34 What characteristics of orbitals are determined by the value of (a) the principal quantum number, n; (b) the azimuthal quantum number, l; (c) the magnetic quantum number, m_l?

5.35 How many different orbitals are there that can have each of the following designations: (a) $n = 4$; (b) $4p$; (c) $2d$; (d) $3d_{xy}$.

Additional exercises

5.36 (a) What is the characteristic frequency of a photon with energy of 1.45×10^{-19} J? (b) If a photon of wavelength 535 nm was converted entirely to heat, how many times as many calories would result as from conversion of a photon of 0.15 nm wavelength.

5.37 The strong blue line in the mercury spectrum occurs at 4.358×10^{-7} m. (a) What is the energy of a photon of this wavelength? (b) What is the energy, in kJ/mol, of a mole of such photons?

5.38 Most radiation with wavelengths below 400 nm is absorbed by our atmosphere, particularly by ozone, O_3. What is the frequency and energy of a photon whose wavelength is 400 nm?

5.39 (a) How many photons with wavelength of 6.50×10^{-7} m (red light) are required to supply 1 J? (b) How many photons of 4.00×10^{-7} m (blue light) are required to supply 1 J?

5.40 What energy change is associated with the transition of an electron in a hydrogen atom from a $1s$ to a $2s$ orbital?

5.41 The speed of sound in air is 344 m/s at 20°C. Calculate the wavelength of (a) the lowest frequency of a large pipe organ, $\nu = 30$/sec; (b) the highest frequency produced by a piccolo, $\nu = 1.5 \times 10^4$/sec.

5.42 In the photoelectric effect, the minimum energy required to eject an electron from a metal is called the work function of that metal. Lithium metal will emit electrons when the wavelength of radiation is 5200 Å or less. (a) Calculate the work function for lithium; (b) if lithium is irradiated with 3600 Å light, what is the kinetic energy of the emitted electrons?

5.43 Distinguish between the following terms (a) orbit and orbital; (b) wavelength and frequency; (c) s and p orbitals; (d) ground state and excited state; (e) continuous spectrum and line spectrum; (f) principal quantum number and azimuthal quantum number.

[5.44] Calculate the circumference of the circular orbit of the electron with $n = 2$ in the Bohr model for hydrogen. Using the de Broglie relationship, calculate the velocity of the electron in this orbit, assuming that it has a characteristic wavelength that is just one full circumference.

5.45 The equipment designed to search for evidence of biological life on Mars was designed to transmit its findings to Earth from as far away as 4.2×10^8 miles. How long does it take for transmissions to reach Earth from this distance?

5.46 A particular AM radio station broadcasts at a frequency of 1120 kHz. What is the wavelength and energy per photon of this radiation?

5.47 Giant radio telescopes "listening" for radiation from outer space have detected strong radio signals with a wavelength of 21 cm. What is the frequency and energy per photon of this radiation?

5.48 Photocells that are used in "electric-eye" door openers and burglar alarms operate on the principle of the photoelectric effect. If the lamp used as the source of light has an effective wavelength of 540 nm, would it be satisfactory to use chromium as the metal in the photocell? The minimum energy required to eject an electron from the surface of chromium metal is 7.00×10^{-19} J.

5.49 When the light from a neon lamp is dispersed in a prism to form a spectrum, it is found that the spectrum is not continuous; rather, it consists of several sharp lines. Explain in general terms why a line spectrum is produced.

[5.50] (a) What general relationship can you deduce between the principal quantum number n and the number of allowed subshells with that value of n? (b) What relationship can you deduce between n and the number of allowed orbitals with that value of n?

5.51 Draw the contour representations for all the atomic orbitals of principal quantum number 3.

5.52 How many orbitals are there on an atom that have the following designations associated with them: (a) $2p_x$; (b) $4d$; (c) $2s$; (d) $3p$?

5.53 If the de Broglie wavelength of an electron is 3.32×10^{-10} m, at what velocity is the electron moving? (The mass of an electron is 9.11×10^{-28} g.)

5.54 Neutrons moving with kinetic energies in the range 8.0×10^{-21} to 1.6×10^{-20} J are used to study molecular structure. What are the wavelengths of such "radiations"? Compare these wavelengths with the dimensions of molecules (10^{-8} to 10^{-10} m). (The mass of a neutron is 1.67×10^{-24} g.)

5.55 Which of the following transitions in a hydrogen atom would require the lowest-energy photon: $1s \longrightarrow 2p$; $2p \longrightarrow 3s$; $2s \longrightarrow 4s$; $1s \longrightarrow 4d$?

5.56 Predict how the curves for the plots of ψ^2 vs. r would appear for He^+ as compared to those for H shown in Figure 5.16. Explain your answer.

5.57 Write the set of quantum numbers allowed for each of the following orbitals: (a) $1s$; (b) $2p$; (c) $3d$.

6

Periodic relationships among the elements

We might say that the modern era in chemistry had its beginnings with Dalton's atomic theory. It happened that at this same time several other notable discoveries occurred to provide new impetus to experimental studies in chemistry. Many of the elements were isolated for the first time during the 50-year period following Dalton's work (1803–1807). As the quantity of chemical information grew, more attention was given to the possibilities of classifying it in useful ways. The most important product of these attempts at classification is the periodic table, which we introduced in Chapter 2.

In Chapter 5, we saw how the work of physicists led to our present understanding of atomic structure, at least as far as the hydrogen atom is concerned. The application of the concepts outlined there to atoms containing more than one electron proved to be possible in principle, but difficult in practice. While physicists were wrestling with this problem, chemists were making significant progress in understanding atomic structure from another point of view. In brief, they asked what sorts of inferences about the arrangements of electrons in atoms they might be able to make by using the periodic table and the known chemical behavior of the elements. One of the chief contributors to this development was the American chemist Gilbert N. Lewis (1875–1946). Even before Bohr had proposed his theory of the hydrogen atom, Lewis had suggested that electrons in atoms are arranged in shells.

In this chapter we will see why there is a close connection between the periodic table and the ways in which electrons are arranged in atoms. Our first goal will be to use the concepts developed in the last chapter to understand the electronic structures of many-electron atoms. We will then relate these electronic structures to the periodic table and to the chemical behavior of the elements.

The electronic structures of atoms with two or more electrons can be described in terms of orbitals like those we have described for hydrogen. Thus we can continue to use orbital designations such as 1s, $2p_x$, and so forth. Furthermore, each of these orbitals has the same shape as those for hydrogen.

To describe the electronic structure of an atom we need to know how the orbitals are occupied by electrons. In an atom in its ground state, the electrons are always found in the lowest energy orbitals available. We therefore need to consider the relative energies of the orbitals.

The energy of any atom or ion with only one electron depends only on the principal quantum number, n, and on the nuclear charge, Z:

$$E_n = -R_H\left(\frac{Z^2}{n^2}\right)$$

[6.1]

For many-electron atoms, the energy also depends on the azimuthal quantum number, l; that is, the energy differs for the different subshells. It is not possible to express this dependence explicitly in terms of a single equation incorporating n and l. However, we can gain a qualitative understanding of why s, p, d, and f subshells have different energies.

In atoms having many electrons, the electrons are acted upon not only by the attractive force of the nucleus, but also by the repulsive forces operating between electrons. We can simplify the situation by focusing on one electron at a time and considering how it interacts with the average environment created by the nucleus and the other electrons. Electrons that are relatively close to the nucleus screen or shield the nucleus from the electrons further out. That is, there is a decrease in the effective positive charge acting on an electron because of the other electrons between it and the nucleus. We refer to this effect as the screening effect; the resultant positive charge that the electron experiences is called the effective nuclear charge. The effective nuclear charge that a particular electron experiences, Z_{eff}, depends on the number of protons in the nucleus, Z, minus the average number of electrons that are between it and the nucleus, S:

$$Z_{eff} = Z - S$$

[6.2]

This idea is illustrated in Figure 6.1. At some particular instant, the electron shown is at a distance r from the nucleus. Assume that the averaged motion of all the other electrons results in a spherical electron distribution. All the electron density within the sphere of radius r shields the electron from the nucleus; that electron density has the same effect as it would if concentrated at the nucleus. Therefore the electron density between the nucleus and the electron on which we have focused cancels a certain amount of the nuclear charge. Of course each electron is moving about; as it moves closer to or farther from the nucleus, the effective nuclear charge it experiences increases or decreases. However, we are interested in its *average* behavior.

The manner in which the probability function (ψ^2) varies as we move outward from the nucleus differs for orbitals in different subshells. Consider the orbitals for which $n = 3$. The 3s electron distribution extends

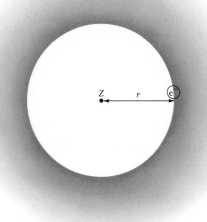

FIGURE 6.1 Shielding of the nuclear charge from an electron by other electrons within an atom. All electronic charge within the sphere of radius r shields the electron located at radius r from the nuclear charge. Electronic charge outside this radius has no shielding effect on this electron. If the atomic number of the nucleus, Z, was 5 and the sphere contained three electrons, the electron on which we are focusing would experience an effective nuclear charge of $5 - 3 = 2$.

closer to the nucleus than does the $3p$; the $3p$ in turn extends closer than does the $3d$. As a result, the other electrons of the atom (that is, the $1s$, $2s$, and $2p$) shield the $3s$ electrons from the nucleus less effectively than they shield the $3p$ or $3d$. Thus the $3s$ electrons experience a larger Z_{eff} than do the $3p$ electrons, and these in turn experience a larger Z_{eff} than do the $3d$ electrons.

The energy of an electron depends on the effective nuclear charge, Z_{eff}, and on the principal quantum number, n, in a way analogous to Equation [6.1]:

$$E_n \propto \frac{Z_{eff}^2}{n^2} \qquad [6.3]$$

where the symbol $\propto$ is read as "proportional to." Since Z_{eff} is larger for the $3s$ electrons, they have a lower energy (that is, they are more stable) than the $3p$, which in turn are lower in energy than the $3d$. As a result, the energy level diagram for the orbitals of many-electron atoms is like that shown in Figure 6.2. Keep in mind that this is a *qualitative* energy level diagram; the exact energies and their spacings differ from one atomic species to another. In all cases, however, the relative energies of the orbitals through $n = 3$ are as shown. Notice that all orbitals of a given subshell, such as the $3d$ orbitals, have the same energy. In scientific jargon, orbitals that have the same energy are said to be degenerate.

SAMPLE EXERCISE 6.1

Based on the energy level diagram of Figure 6.2, would you expect the average distance from the nucleus of a $3d$ electron to be greater or less than that of a $2p$? Explain.

Solution: The energy of the $2p$ orbitals is considerably lower than for a $3d$. This indicates that the average attractive interaction of the $2p$ electron with the nucleus is much greater than for an electron in a $3d$ orbital. The increased attractive interaction is due to a smaller average distance of the $2p$ electron from the nucleus.

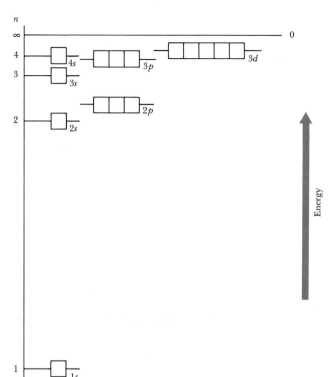

FIGURE 6.2 The ordering of orbital energy levels in many-electron atoms. Note that orbitals with the same value for the principal quantum number, n, but differing values of azimuthal quantum number, l, differ in energy.

With this much background we now have an idea of the ordering of the energies of atomic orbitals in many-electron atoms. We must now consider what rules govern the placement of electrons into these orbitals. However, before the rules can be stated in their most useful form, we must learn about a fourth, and final, quantum number.

6.2 Electron spin and the Pauli exclusion principle

When the atomic spectra of atoms were first studied, certain complicating features of the spectra were noted. Lines that were at first thought to be single lines were found under high resolution to be closely spaced pairs. The only way in which these extra splittings could be accounted for was to introduce a new quantum number in addition to the three quantum numbers we have already encountered. This new quantum number is associated with the electron itself. The electron behaves as though it were spinning on its own axis and thereby acting like a tiny magnet. It was necessary to define an electron-spin quantum number (m_s), which has values of $+\frac{1}{2}$ or $-\frac{1}{2}$, corresponding to the two possible orientations of the electron spin in a magnetic field.

In 1921 Otto Stern and Walter Gerlach succeeded in actually separating a beam of atoms into two groups according to the orientation of the electron spin. Their experiment is diagramed in Figure 6.3. Let us assume that the beam of atoms is hydrogen. As the atoms pass into the region of strong magnetic field, the electron in each atom interacts with the magnetic field, causing the atom to be deflected from its straight-line path. The direction in which the atom is deflected depends on the orientation of the spin of the electron. We expect that there will be equal numbers of electrons with each of the two possible orientations, as is

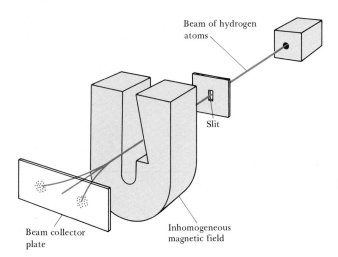

FIGURE 6.3 A diagramatic illustration of the Stern-Gerlach experiment. A beam of hydrogen atoms is allowed to pass through an inhomogeneous magnetic field. Atoms in which the electron-spin quantum number, m_s is $+\frac{1}{2}$ are deflected in one direction, whereas those in which m_s is $-\frac{1}{2}$ are deflected in the other.

Beam of hydrogen atoms

Slit

Beam collector plate

Inhomogeneous magnetic field

found to be the case. The presence of electron spin turns out to be important in determining the electronic structures of atoms. The first to recognize this was a German physicist, Wolfgang Pauli. In 1924 he spelled out what has become known as the Pauli exclusion principle. This principle declares that no two electrons in an atom can have the same set of four quantum numbers, n, l, m_l, and m_s. As we will see in the next section, the Pauli exclusion principle dictates that each orbital can hold a maximum of two electrons. This restriction provides the key to one of the great problems of chemistry—an explanation of the structure of the periodic table of the elements.

6.3 The periodic table and electron configurations

The essential feature of the periodic table is that when elements are arranged in order of increasing atomic number, similarities in physical and chemical properties are repeated at regular intervals (Section 2.7). Elements with similar properties such as the alkali metals, group 1A, are placed together in columns called groups or families.* The horizontal rows of the periodic table are called rows or periods. The first row consists of only two members, H and He. The next two rows, starting with Li and Na, contain eight members each. The following rows, beginning with K, Rb, and Cs, contain 18, 18, and 32 members, respectively. Why do certain elements possess chemical similarities, and why does the periodic table have its particular shape with rows containing 2, 8, 18, 18, and 32 members? The answers to both questions are found in the nature of the arrangements or configurations of electrons in atoms—that is, in how electrons populate the available orbitals.

The atomic number corresponds not only to the number of protons in the nucleus of an atom, but also to the number of electrons in that atom. Thus the periodic table represents an ordering of the elements not only according to nuclear charge, but also according to the number of electrons about the nucleus. Gilbert N. Lewis, who was mentioned in the introduction, reasoned that if the chemical properties of the elements are repeated at intervals, then their electron configurations must be repeat-

*Hydrogen is generally placed in family 1A or 7A. It shows a few remote resemblances to both families but is really not a member of either.

ing in some way. He suggested that electrons in atoms are arranged in shells, and that once a shell of electrons is filled, additional electrons must go into a new shell. He further reasoned that the noble gases, which are chemically very nonreactive, have closed or filled shells of electrons. Chemical behavior might then be understood in terms of the tendency of an atom to achieve a closed shell by gaining, losing, or sharing electrons. For example, we saw in Section 2.9, that sodium, atomic number 11, loses one electron in chemical reactions to form the sodium ion, Na^+. This ion has the stable, closed-shell electron arrangement of the noble gas neon, atomic number 10. These ideas of Lewis generate additional questions: Why do certain electron configurations occur periodically? Why are electrons arranged in shells?

To answer these questions and to account in more detail for the way in which various atomic properties vary with atomic number, we must apply the ideas about orbitals that we have learned about in this and the previous chapter. Our goal is to understand how electrons are arranged in each element of the periodic table. When you have mastered this material, you will be able to describe the arrangement of electrons in any element. This is an important skill; as we have implied, electron arrangements are the key to the chemical behavior of the elements. So let's consider how the electrons in each element populate orbitals, starting with hydrogen and then moving through the periodic table with increasing atomic number.

THE ELECTRON CONFIGURATIONS OF THE ELEMENTS

The most stable, or ground, state of an atom will be that in which all the electrons are in the lowest possible energy states. If there were no restrictions on the possible values for the quantum numbers of the electrons, all the electrons would crowd into the $1s$ orbital, because this is lowest in energy. The Pauli principle, however, tells us that there are limits on the quantum numbers that the electrons may have. The $1s$ orbital corresponds to values $n = 1$, $l = 0$, and $m_l = 0$. The electron that occupies this orbital can have a spin quantum number (m_s) of $+\frac{1}{2}$ or $-\frac{1}{2}$. If one electron has quantum numbers $n = 1$, $l = 0$, $m_l = 0$, and $m_s = +\frac{1}{2}$, then a second electron can have $n = 1$, $l = 0$, $m_l = 0$, and $m_s = -\frac{1}{2}$. Thus, the exclusion principle requires that there can be at most *two electrons* in the $1s$ orbital, or for that matter in any *single atomic orbital*. In hydrogen there is one electron in the $1s$ orbital. In helium, atomic number 2, there are two electrons in this orbital. These electrons must have opposite values for the electron-spin quantum number (m_s). The electrons are said to be paired. Because of this pairing, the magnetic properties of the electrons effectively cancel. If the Stern-Gerlach experiment (Figure 6.3) were performed on a beam of helium atoms, no separation of the beam by the magnetic field would occur as it does with hydrogen.

A particular arrangement of electrons in the orbitals of an atom is referred to as an electron configuration. It is convenient to have a shorthand notation for representing the electron configuration. This is done by writing the symbol for each orbital occupied by an electron, with a superscript to indicate the number of electrons occupying that orbital. For hydrogen the electron configuration is $1s^1$; for helium it is $1s^2$. An-

$1s$

H ⬚ [↑]

He ⬚ [↑↓]

other way of representing the electron configurations of these atoms is shown at left. In this kind of diagram each orbital is represented by a box, and each electron by a half-arrow. A half-arrow pointing upward (↑) represents an electron spinning in one direction ($m_s = +\frac{1}{2}$), whereas a downward half-arrow (↓) represents an electron spinning in the opposite direction ($m_s = -\frac{1}{2}$). We shall refer to representations of this type as orbital diagrams.

The two electrons present in helium complete the filling of orbitals with principal quantum number $n = 1$. Helium therefore possesses a very stable electron configuration, as reflected in its chemical inertness. The electron configurations of lithium and of several elements that follow it in the periodic table are shown in Table 6.1. Recall that a maximum of two electrons can be placed in each orbital. Thus, for lithium, with three electrons, the third electron cannot enter the $1s$ orbital, but must be placed in the next most stable orbital, the $2s$ (refer to Figure 6.2). The change in principal quantum number for the third electron represents a large jump in energy, and a corresponding jump in the average distance of the electron from the nucleus. We may say that it represents the start of a new shell of electrons. As you can see by examining the periodic table, lithium represents the start of a new row of the periodic table. It is the first member of the alkali metals family (group 1A).

The element that follows lithium is beryllium; its electron configuration is $1s^2 2s^2$ (Table 6.1). Boron, atomic number 5, has an electron configuration $1s^2 2s^2 2p^1$. The fifth electron must be placed in a $2p$ orbital, because the $2s$ orbital is filled. Because each of the three $2p$ orbitals are of equal energy, it doesn't matter which $2p$ orbital is occupied. With the next element, carbon, we come to a new situation. We know that the sixth electron must go into a $2p$ orbital, where there is already one electron. However, does this new electron go into the $2p$ orbital that already has one electron, or into one of the others? This question is answered by Hund's rule, which states that electrons occupy degenerate orbitals singly to the maximum extent possible, and with their spins parallel. In the case of carbon, then, the sixth electron goes into one of the other $2p$ orbitals, and with its spin in the same orientation as the other $2p$ elec-

TABLE 6.1 Electron configurations of several lighter elements

Element	Total electrons	Orbital diagram				Electron configuration
Li	3	[↑↓]	[↑]	[ ][ ][ ]	[ ]	$1s^2 2s^1$
Be	4	[↑↓]	[↑↓]	[ ][ ][ ]	[ ]	$1s^2 2s^2$
B	5	[↑↓]	[↑↓]	[↑][ ][ ]	[ ]	$1s^2 2s^2 2p^1$
C	6	[↑↓]	[↑↓]	[↑][↑][ ]	[ ]	$1s^2 2s^2 2p^2$
Ne	10	[↑↓]	[↑↓]	[↑↓][↑↓][↑↓]	[ ]	$1s^2 2s^2 2p^6$
Na	11	[↑↓]	[↑↓]	[↑↓][↑↓][↑↓]	[↑]	$1s^2 2s^2 2p^6 3s^1$
		$1s$	$2s$	$2p$	$3s$	

tron. Hund's rule is based on the fact that electrons repel one another because they have the same electrical charge. By occupying different orbitals, the electrons remain as far as possible from one another in space, thus minimizing electron-electron repulsions. When electrons must occupy the same orbital, the repulsive interaction between the paired electrons is greater than between electrons in different, equivalent orbitals.

Neon, the last member of the second row, has ten electrons. Two electrons fill the $1s$ orbital, two electrons fill the $2s$ orbital, and the remaining six electrons fill the $2p$ orbitals. The electron configuration is thus $1s^2 2s^2 2p^6$ (Table 6.1). In neon, all of the orbitals with $n = 2$ are filled. The filling of the $2s$ and $2p$ orbitals by the eight electrons that they can hold represents a very stable configuration. As a result, neon is chemically quite inert. Lewis, in his model for the electron configurations of elements, noted that the octet of electrons (eight) in the outermost shell of an atom or ion represents an especially stable arrangement.

Sodium, atomic number 11, marks the beginning of a new row of the periodic table. Sodium has a single $3s$ electron beyond the stable configuration of neon. We can abbreviate the electron configuration of sodium as follows:

Na $[Ne]3s^1$

The symbol [Ne] represents the electron configuration of the ten electrons of neon, $1s^2 2s^2 2p^6$. Writing the electron configuration in this manner helps us focus attention on the electron arrangement at the periphery of the atom. The outer electrons are the ones largely responsible for the chemical behavior of an element. For example, we can write the electron configuration of lithium as follows:

Li $[He]2s^1$

By comparing this with the electron configuration for sodium, it is easy to appreciate why lithium and sodium are so similar chemically: They have the same type of outer-shell electron configuration. All the members of the alkali metal family (group 1A) have a single s electron beyond an inner-core noble gas configuration. The outer-shell electrons are often referred to as valence-shell electrons.

SAMPLE EXERCISE 6.2

Draw the orbital diagram representation for the electron configuration of oxygen, atomic number 8.

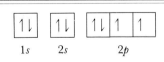

$1s$ $2s$ $2p$

Solution: The ordering of orbitals is as shown in Figure 6.2. Two electrons each go into the $1s$ and $2s$ orbitals. This leaves four electrons for the three $2p$ orbitals. Following Hund's rule, we put one electron into each $2p$ orbital until all three have one each. The fourth electron must then be paired up with one of the three electrons already in a $2p$ orbital, so that the correct representation is

The corresponding electron configuration is written $1s^2 2s^2 2p^4$ or $[He]2s^2 2p^4$. The $1s^2$ or [He] electrons are the inner-shell or core electrons of the oxygen atom. The $2s^2 2p^4$ electrons are the outer-shell or valence electrons.

What is the characteristic outer-shell electron configuration of the group 7A elements, the halogens?

Solution: The first member of the halogen family is fluorine, atomic number 9. The abbreviated form of the electronic configuration for fluorine is

F $[He]2s^22p^5$

Similarly, the abbreviated form of the electron configuration for chlorine, the second halogen, is

Cl $[Ne]3s^23p^5$

From these two examples we see that the characteristic outer-shell electron configuration of a halogen is ns^2np^5, where n ranges from 2 in the case of fluorine to 5 in the case of iodine.

The rare gas element argon marks the end of the row started by sodium. The configuration for argon is $1s^22s^22p^63s^23p^6$. The element following argon in the periodic table is potassium (K), atomic number 19. In all its chemical properties, potassium is very obviously a member of the alkali metal family. The experimental facts about the properties of potassium leave no doubt that the outermost electron of this element occupies an s orbital. But this means that the highest energy electron has *not* gone into a $3d$ orbital, which we might naïvely have expected it to do. In this case the ordering of energy levels is such that the $4s$ orbital is lower in energy than the $3d$ (see Figure 6.2).

Following complete filling of the $4s$ orbital (this occurs in the calcium atom), the next set of equivalent orbitals to become filled is the $3d$. (You'll find it helpful as we go along to refer often to the periodic table on the front inside cover.) Beginning with scandium, and extending through zinc, electrons are added to the five $3d$ orbitals until they are completely filled. Thus the fourth row of the periodic table is ten elements wider than the previous rows because of the insertion of the elements known as the transition metals. Note the position of these ten elements in the periodic table (front inside cover).

In accordance with Hund's rule, electrons are added to the $3d$ orbitals singly until all five orbitals have one electron each. Additional electrons are then placed in the $3d$ orbitals with spin pairing until the shell is completely filled. The orbital diagram representations and electron configurations of two transition elements are as follows:

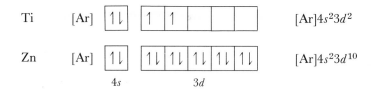

The $3d$ and $4s$ orbital energies are very close together. Occasionally, an electron may be moved from one of these types of orbitals to another. For example, we might expect chromium to have the outer electron configuration $4s^23d^4$, but it is actually $4s^13d^5$. This anomalous behavior is partly due to the special stability associated with precisely half-filled sets of degenerate orbitals. Apparently there is just enough gain in stability in arriving at this arrangement to cause the electron to move from a $4s$ to a $3d$ orbital.

Upon completion of the $3d$ transition series, the $4p$ orbitals begin to be occupied, until the completed octet of outer electrons is again arrived at in krypton (Kr), atomic number 36. Krypton is another of the rare gases. Rubidium (Rb) marks the beginning of the fifth row of the periodic table. This row is in every respect like the previous one, except that the value for n is one greater. The sixth row of the table begins similarly to the previous one: one electron in the $6s$ orbital of cesium (Cs) and two electrons in the $6s$ orbital of barium (Ba). The next element, lanthanum (La), represents the start of the third series of transition elements. But with cerium (Ce), element 58, a new set of orbitals, the $4f$, enter the picture. The energies of the $5d$ and $4f$ orbitals are very close. For lanthanum itself, the $5d$ orbital energy is just a little lower than the $4f$. However, for the elements immediately following lanthanum, the $4f$ orbital energies are a little lower, so that the highest energy electrons go into the $4f$ orbitals.

There are seven equivalent $4f$ orbitals, corresponding to the seven allowed values of m_l ranging from 3 to -3. Thus it requires 14 electrons to completely fill the $4f$ orbitals. The 14 elements corresponding to the filling of the $4f$ orbitals are elements 58–71, known as the rare earth, or lanthanide, elements. In order not to make the periodic table unduly wide, the rare earth elements are set together below the other elements. The properties of the rare earth elements are all quite similar, and they occur together in nature. For many years it was virtually impossible to separate them from one another.

Following completion of the rare earth series, the third transition element series is completed, followed by filling of the $6p$ orbitals. This brings us to radon (Rn), heaviest of the rare gas elements. The final row of the periodic table begins as the one before it. The actinide elements involve completion of the $5f$ electron orbitals. This series consists mainly of elements not found in nature, those that have been synthesized in nuclear reactions.

USING THE PERIODIC TABLE TO WRITE ELECTRON CONFIGURATIONS

Our rather brief survey of electron configurations of the elements has taken us through the entire periodic table. You will find that a familiarity with the general structure of the table will enable you to write down the electronic configuration of any element. There are a few instances in which minor shifts of an electron or two from one orbital to another occur, when orbitals have closely similar energies. We have given as one example the case of chromium, which possesses an outer electron configuration $4s^1 3d^5$, rather than the $4s^2 3d^4$ we might have expected. Another interesting case occurs with copper and its congeners* silver and gold. In copper the configuration is found to be $4s^1 3d^{10}$. Evidently the stability associated with completing the d orbital level causes the electron to move from ns to $(n-1)d$. There are a few other similar instances among the transition elements, lanthanides, and actinides. Although these minor departures from the expected are interesting, they are not of great chemical significance.

---

*A congener is an element in the same family or group of the periodic table as another.

We've seen that the periodic table is structured so that elements with the same outer-shell electron configuration are arranged in columns. The elements can be grouped also in terms of the *type* of orbital into which the electrons are placed. These different groupings are indicated by the shadings on the outline of the periodic table shown in Figure 6.4. The first two groups of elements on the left contain the **active metals,** with outermost *s* electrons. The **transition elements** are those for which the *d* orbitals are incomplete and being filled. The **representative elements** are those for which the outermost *p* orbitals are being filled. The **rare gas,** or noble gas, elements are those for which the octet of outermost electrons has been attained. The two series of elements for which the *f* orbitals are being filled are sometimes called the **inner transition elements.**

SAMPLE EXERCISE 6.4

Write the electron configuration for the element bismuth, atomic number 83.

Solution: We can do this by simply moving across the periodic table one row at a time and writing the occupancies of the orbitals corresponding to each row (refer to Figure 6.4):

First row	$1s^2$
Second row	$2s^2 2p^6$
Third row	$3s^2 3p^6$
Fourth row	$4s^2 3d^{10} 4p^6$
Fifth row	$5s^2 4d^{10} 5p^6$
Sixth row	$6s^2 4f^{14} 5d^{10} 6p^3$

Total: $1s^2 2s^2 2p^6 3s^2 3p^6 3d^{10} 4s^2 4p^6 4d^{10} 4f^{14}$
$$5s^2 5p^6 5d^{10} 6s^2 6p^3$$

Note that 3 is the lowest possible value that n may have for a *d* orbital, and that 4 is the lowest possible value of n for an *f* orbital.

The total of the superscripted numbers should equal the atomic number of bismuth, 83. It does not matter a great deal precisely in which order the orbitals are listed. They may be listed, as shown above, in the order of increasing major quantum number. However, it is also possible to list them in the sequence read from the periodic table: $1s^2 2s^2 2p^6 3s^2 3p^6 4s^2 3d^{10} 4p^6 5s^2 4d^{10} 5p^6 6s^2 4f^{14} 5d^{10} 6p^3$. Writing only the outer electron configuration beyond the nearest rare gas we have $[Xe]6s^2 4f^{14} 5d^{10} 6p^3$.

SAMPLE EXERCISE 6.5

Draw the orbital diagram representation for zirconium, atomic number 40; show only those electrons beyond the krypton inner core.

Solution: Zirconium has four electrons beyond the nearest noble gas, krypton, atomic number 36. Examining the periodic table we see that zirconium is a transition element from the fifth row of the table.

This means that its outermost electrons are in 5*s* and 4*d* orbitals. Two electrons occupy the 5*s* orbital; two must be placed in the five 4*d* orbitals. As indicated by Hund's rule, the 4*d* electrons occupy separate orbitals. Thus we have

A complete list of the electron configurations of the elements is contained in Table 6.2. You can use this table to check your answers as you practice writing electron configurations. We have written these configurations as they would be read off the periodic table. However, they are sometimes written with orbitals of a given principal quantum number

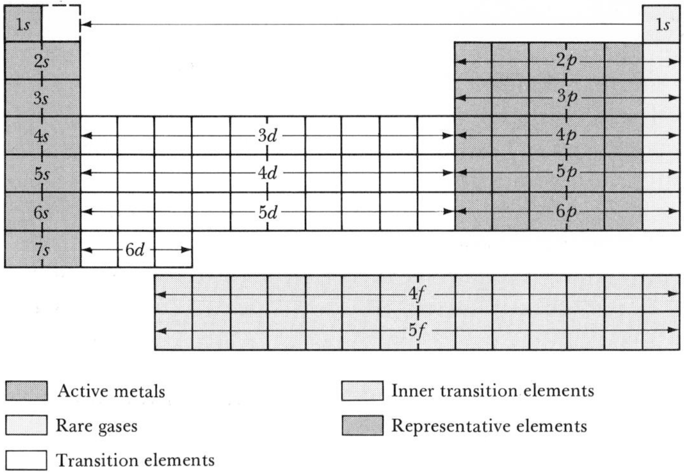

1s

2s

3s

4s — 3d —

5s — 4d —

6s — 5d —

7s — 6d —

2p

3p

4p

5p

6p

1s

4f

5f

■ Active metals
□ Rare gases
□ Transition elements
▨ Inner transition elements
▨ Representative elements

FIGURE 6.4 A block diagram of the periodic table showing the groupings of the elements according to the type of orbital being filled with electrons. The arrangement of elements in this figure is the same as in the periodic table.

TABLE 6.2 The electron configurations of the elements

Atomic number	Symbol	Electron configuration	Atomic number	Symbol	Electron configuration	Atomic number	Symbol	Electron configuration
1	H	$1s^1$	36	Kr	$[Ar]4s^23d^{10}4p^6$	71	Lu	$[Xe]6s^24f^{14}5d^1$
2	He	$1s^2$	37	Rb	$[Kr]5s^1$	72	Hf	$[Xe]6s^24f^{14}5d^2$
3	Li	$[He]2s^1$	38	Sr	$[Kr]5s^2$	73	Ta	$[Xe]6s^24f^{14}5d^3$
4	Be	$[He]2s^2$	39	Y	$[Kr]5s^24d^1$	74	W	$[Xe]6s^24f^{14}5d^4$
5	B	$[He]2s^22p^1$	40	Zr	$[Kr]5s^24d^2$	75	Re	$[Xe]6s^24f^{14}5d^5$
6	C	$[He]2s^22p^2$	41	Nb	$[Kr]5s^14d^4$	76	Os	$[Xe]6s^24f^{14}5d^6$
7	N	$[He]2s^22p^3$	42	Mo	$[Kr]5s^14d^5$	77	Ir	$[Xe]6s^24f^{14}5d^7$
8	O	$[He]2s^22p^4$	43	Tc	$[Kr]5s^24d^5$	78	Pt	$[Xe]6s^14f^{14}5d^9$
9	F	$[He]2s^22p^5$	44	Ru	$[Kr]5s^14d^7$	79	Au	$[Xe]6s^14f^{14}5d^{10}$
10	Ne	$[He]2s^22p^6$	45	Rh	$[Kr]5s^14d^8$	80	Hg	$[Xe]6s^24f^{14}5d^{10}$
11	Na	$[Ne]3s^1$	46	Pd	$[Kr]4d^{10}$	81	Tl	$[Xe]6s^24f^{14}5d^{10}6p^1$
12	Mg	$[Ne]3s^2$	47	Ag	$[Kr]5s^14d^{10}$	82	Pb	$[Xe]6s^24f^{14}5d^{10}6p^2$
13	Al	$[Ne]3s^23p^1$	48	Cd	$[Kr]5s^24d^{10}$	83	Bi	$[Xe]6s^24f^{14}5d^{10}6p^3$
14	Si	$[Ne]3s^23p^2$	49	In	$[Kr]5s^24d^{10}5p^1$	84	Po	$[Xe]6s^24f^{14}5d^{10}6p^4$
15	P	$[Ne]3s^23p^3$	50	Sn	$[Kr]5s^24d^{10}5p^2$	85	At	$[Xe]6s^24f^{14}5d^{10}6p^5$
16	S	$[Ne]3s^23p^4$	51	Sb	$[Kr]5s^24d^{10}5p^3$	86	Rn	$[Xe]6s^24f^{14}5d^{10}6p^6$
17	Cl	$[Ne]3s^23p^5$	52	Te	$[Kr]5s^24d^{10}5p^4$	87	Fr	$[Rn]7s^1$
18	Ar	$[Ne]3s^23p^6$	53	I	$[Kr]5s^24d^{10}5p^5$	88	Ra	$[Rn]7s^2$
19	K	$[Ar]4s^1$	54	Xe	$[Kr]5s^24d^{10}5p^6$	89	Ac	$[Rn]7s^26d^1$
20	Ca	$[Ar]4s^2$	55	Cs	$[Xe]6s^1$	90	Th	$[Rn]7s^26d^2$
21	Sc	$[Ar]4s^23d^1$	56	Ba	$[Xe]6s^2$	91	Pa	$[Rn]7s^25f^26d^1$
22	Ti	$[Ar]4s^23d^2$	57	La	$[Xe]6s^25d^1$	92	U	$[Rn]7s^25f^36d^1$
23	V	$[Ar]4s^23d^3$	58	Ce	$[Xe]6s^24f^15d^1$	93	Np	$[Rn]7s^25f^46d^1$
24	Cr	$[Ar]4s^13d^5$	59	Pr	$[Xe]6s^24f^3$	94	Pu	$[Rn]7s^25f^6$
25	Mn	$[Ar]4s^23d^5$	60	Nd	$[Xe]6s^24f^4$	95	Am	$[Rn]7s^25f^7$
26	Fe	$[Ar]4s^23d^6$	61	Pm	$[Xe]6s^24f^5$	96	Cm	$[Rn]7s^25f^76d^1$
27	Co	$[Ar]4s^23d^7$	62	Sm	$[Xe]6s^24f^6$	97	Bk	$[Rn]7s^25f^9$
28	Ni	$[Ar]4s^23d^8$	63	Eu	$[Xe]6s^24f^7$	98	Cf	$[Rn]7s^25f^{10}$
29	Cu	$[Ar]4s^13d^{10}$	64	Gd	$[Xe]6s^24f^75d^1$	99	Es	$[Rn]7s^25f^{11}$
30	Zn	$[Ar]4s^23d^{10}$	65	Tb	$[Xe]6s^24f^9$	100	Fm	$[Rn]7s^25f^{12}$
31	Ga	$[Ar]4s^23d^{10}4p^1$	66	Dy	$[Xe]6s^24f^{10}$	101	Md	$[Rn]7s^25f^{13}$
32	Ge	$[Ar]4s^23d^{10}4p^2$	67	Ho	$[Xe]6s^24f^{11}$	102	No	$[Rn]7s^25f^{14}$
33	As	$[Ar]4s^23d^{10}4p^3$	68	Er	$[Xe]6s^24f^{12}$	103	Lr	$[Rn]7s^25f^{14}6d^1$
34	Se	$[Ar]4s^23d^{10}4p^4$	69	Tm	$[Xe]6s^24f^{13}$	104	Rf	$[Rn]7s^25f^{14}6d^2$
35	Br	$[Ar]4s^23d^{10}4p^5$	70	Yb	$[Xe]6s^24f^{14}$	105	Ha	$[Rn]7s^25f^{14}6d^3$

gathered together. Thus we might write the electron configuration of arsenic (atomic number 33) as $[Ar]3d^{10}4s^24p^3$ instead of $[Ar]4s^23d^{10}4p^3$ as shown in Table 6.2.

The configurations of many of the heavier elements are not known for certain. The configurations must be deduced by analysis of atomic spectra. These methods are extremely complicated, especially for the heavier elements, in which the energy levels are closely spaced.

6.4 Electron shells in atoms

We have seen how the electron configurations of an atom may be built up by adding electrons to orbitals of successively higher energy, in accordance with the Pauli exclusion principle and Hund's rule. How do our results compare with Lewis' idea of electron shells? Consider the noble gases helium, neon, and argon, whose electron configurations are as follows:

He	$1s^2$
Ne	$1s^22s^22p^6$
Ar	$1s^22s^22p^63s^23p^6$

Very accurate calculations of the total electronic charge distribution in these atoms can be made using large computers. These distributions are shown in Figure 6.5. The quantity plotted on the vertical axis is called the radial electron density. It corresponds to the probability of the electron being located at a particular distance from the nucleus. As Figure 6.5 shows, the radial electron probability does not fall off continuously as we move away from the nucleus. Rather it shows maxima corresponding to distances at which there are higher probabilities of finding electrons. These maxima correspond to the traditional idea of shells of electrons; however, these shells are diffuse and overlap considerably.

Helium shows a single shell, neon two, and argon three. Each of these maxima is due mainly to electrons in the atom that have the same value for the principal quantum number n. Thus, for helium the $1s$ electrons possess a maximum in radial electron density at about 0.3 Å. In argon, the maximum in the $1s$ radial electron density occurs at only 0.05 Å. The

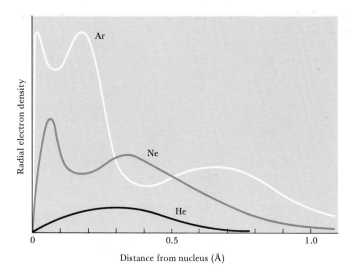

FIGURE 6.5 Radial electron density graphs for the first three rare gas elements, He, Ne, and Ar. The maxima that occur in the radial electron density correspond to electrons with the same value of principal quantum number n.

second maximum, which occurs at larger radial distance, is due to both the $2s$ and $2p$ electrons. The third maximum is due to $3s$ and $3p$ electrons.

The reason for the smaller radial distance of the orbital of the $1s$ electrons in the heavier atom is clear when we recall that the nuclear charge of helium is only 2, whereas that for argon is 18. The $1s$ electrons are the innermost electrons of the atom. The electrons of quantum number $n = 2$ and greater, present in elements beyond helium, therefore do not do much to shield the $1s$ electrons from the increasing nuclear charge. As a result, the size of the $1s$ orbital shrinks steadily as nuclear charge increases. Thus, the calculations show that in many-electron atoms the inner electrons are pulled with ever-increasing force into the region around the nucleus as the nuclear charge increases.

6.5 Ionization energy

Now that we have some understanding of the electronic structures of atoms, we can examine some properties of atoms that are strongly dependent on electron configuration. We will consider three that provide important insights into the chemical behavior of atoms: ionization energy, electron affinity, and atomic size.

Recall from our discussion in the previous chapter that the ionization energy (I) is the energy required to remove an electron from a gaseous atom or ion. The first ionization energy for an element, I_1, is therefore that required for the process shown in Equation [6.4]:

$$M(g) \longrightarrow M(g)^+ + e^- \qquad [6.4]$$

where M is a gaseous, neutral atom. The second ionization energy, I_2, is then the energy for the removal of the second electron, Equation [6.5]:

$$M(g)^+ \longrightarrow M(g)^{2+} + e^- \qquad [6.5]$$

Successive ionization energies are defined in a similar manner. The values of successive ionization energies are known for many elements. Values for the elements sodium through argon are listed in Table 6.3. Remember that a larger value of I corresponds to tighter binding of the electron to the atom or ion.

TABLE 6.3 Successive values of ionization energies (I) for the elements sodium through argon (kJ/mol)[a]

Element	I_1	I_2	I_3	I_4	I_5	I_6	I_7
Na	490	4560	(Inner-shell electrons)				
Mg	735	1445	7730				
Al	580	1815	2740	11,600			
Si	780	1575	3220	4350	16,100		
P	1060	1890	2905	4950	6270	21,200	
S	1005	2260	3375	4565	6950	8490	27,000
Cl	1255	2295	3850	5160	6560	9360	11,000
Ar	1525	2665	3945	5770	7230	8780	12,000

[a]Although the ionization energies are given here in units of kJ/mol, they are also often given in units of electron volts; 1 electron volt is equal to 96.49 kJ/mol.

As we might expect, each successive removal of an electron requires more energy. The reason for this is that the positive nuclear charge that provides the attractive force remains the same, whereas the number of electrons, which produce repulsive interaction, steadily decreases. For example, the electronic configuration for silicon (Si) is $1s^22s^22p^63s^23p^2$. If we look at the successive ionization energies for silicon given in Table 6.3, we see a steady increase from 780 kJ/mol to 4350 kJ/mol for the four values of I that correspond to loss of the four valence-shell electrons with principal quantum number $n = 3$. The fifth electron, however, requires considerably more energy for removal, 16,100 kJ/mol. This sharp increase in ionization energy reflects the fact that the fifth electron is an inner-shell electron. This electron is in a $2p$ orbital and consequently penetrates closer to the nucleus than do the $3s$ and $3p$ electrons. The $2p$ electron in silicon not only has a smaller average distance r from the nucleus, but it also experiences a larger effective nuclear charge because it penetrates the charge distribution of the other electrons.

The effect of change in the principal quantum number can be seen in other comparisons as well. For example, if we compare Mg and Al, we see that the energies required to remove first one and then two electrons from the two metals are not so very different. Yet the energy required to remove the third electron from Mg, 7730 kJ/mol, is much greater than the energy required to remove a third electron from Al, 2740 kJ/mol. The difference in energies must therefore arise predominately from the fact that a third electron removed from Mg is an inner-shell $2p$ electron; in contrast, the third electron removed from Al is an outer-shell $3s$ electron.

These and similar ionization energy data thus support the idea that only the outermost electrons, those beyond the noble gas core, are involved in the sharing and transfer of electrons that give rise to chemical change. The reason for this is that the inner electrons are too tightly bound to the nucleus to be lost from the atom or even shared with another atom.

SAMPLE EXERCISE 6.6

As can be seen in Table 6.3, the energy required to remove an electron from P^{4+} is 6270 kJ/mol, as compared with 16,100 kJ/mol for removal of an electron from Si^{4+}. Account for the large difference.

Solution: The outer electron configuration of phosphorus is $3s^23p^3$. After removal of four of these electrons, the highest-energy electron remaining is a $3s$.

The outer electron configuration of silicon is $3s^23p^2$. After removal of these four electrons the highest-energy electron remaining is a $2p$. It requires considerably less energy to remove the $3s$ electron, which lies largely outside the $1s^22s^22p^6$ core of electrons, than to remove an electron from the $2p$ level, as would be necessary for silicon.

PERIODIC TRENDS IN IONIZATION ENERGIES

It is of interest to observe how the first ionization energies, I_1, vary with atomic number. Figure 6.6 shows a graph of I_1 versus atomic number. It is evident that there is an overall periodicity in this property. Overlooking for the moment the lesser displacements, there is a gradual increase in I_1 with atomic number in any one horizontal row. Thus the alkali

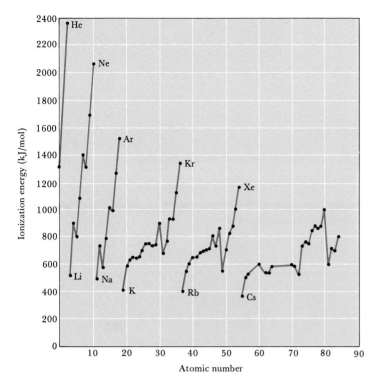

FIGURE 6.6 Ionization energy versus atomic number.

metals show the lowest ionization energy in each row and the rare gas elements the highest. A few simple considerations help to explain this trend. Proceeding along any horizontal row of the table, the electrons that are added to counterbalance the increasing nuclear charge do not completely shield the outermost electrons from the nucleus. Thus the effective nuclear charge increases steadily. For example, the inner $1s^2$ electrons of lithium ($1s^2 2s^1$) shield the outer $2s$ electron from the $3+$ charged nucleus. Consequently, the outer electron experiences an effective nuclear charge of about $1+$. For beryllium ($1s^2 2s^2$), the effective nuclear charge experienced by each outer $2s$ electron is larger; in this case the inner $1s^2$ electrons are shielding a $4+$ nucleus, and each $2s$ electron only partially shields the other from the nucleus. As the effective nuclear charge increases, the electron becomes harder to remove. The irregularities within a given row are primarily associated with the enhanced stability of filled or half-filled subshells.

For elements in any column or family of elements, there is a gradual decrease in ionization energy with increasing atomic number. For example, it requires more energy to remove an electron from a lithium atom than from a potassium atom. When we compare elements of a vertical column, we are comparing elements with the same outer electron configurations but differing values for n, the principal quantum number. As n increases so also does the average distance of the electron from the nucleus. As its average distance from the nucleus increases, the electron becomes easier to remove. Therefore, if all other factors are the same, ionization energy decreases with increasing atomic radius. In terms of the periodic table, the ionization energies vary in the manner shown in Figure 6.7.

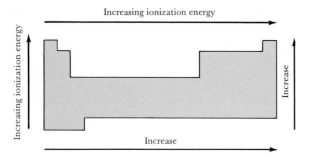

Increasing ionization energy →

Increase →

FIGURE 6.7 Variation of the first ionization energy for the elements in relation to the periodic table.

6.6 Electron affinities

The electron affinity (E) is the energy change that occurs when an electron is added to a gaseous atom or ion.* The process may be represented for a neutral atom as

$$M(g) + e^- \longrightarrow M^-(g) \qquad [6.6]$$

For an ion of $1+$ charge, the equation is

$$M^+(g) + e^- \longrightarrow M(g) \qquad [6.7]$$

For most neutral and for all positively charged species, energy is evolved when the electron is added; E is thus negative in sign. The process shown

*We have defined electron affinity in such a way that a negative E is associated with an exothermic process. Thus the more negative the value for E, the greater the attraction for electrons. However, E is sometimes defined as the energy given off when an electron is added to a gaseous atom or ion. In references that define E in this second way, the more positive the E value, the greater the attraction for electrons.

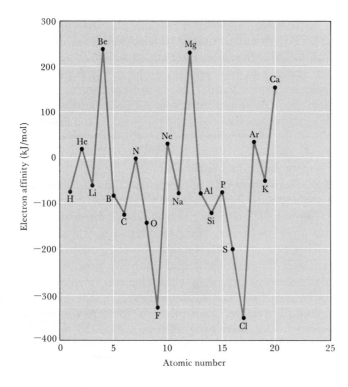

FIGURE 6.8 Electron affinity versus atomic number.

in Equation [6.7] is just the opposite of ionization of the neutral atom. The electron affinity thus bears a close relationship to the ionization energy; the electron affinity of a singly charged positive ion is just the negative of the ionization energy of the corresponding neutral atom. The electron affinities of neutral atoms are quite difficult to measure, and accurate values are now known for only about 50 elements. Figure 6.8 shows the variation of electron affinities for the first 20 elements in the periodic table.

In general, electron affinities become more negative (stronger attraction for an electron) as we move from left to right across any row of the periodic table. The most negative electron affinities are found with the halogens, group 7A. The halogens have an outer electron configuration of the type ns^2np^5. The addition of a single electron gives the stable configuration characteristic of the noble gases. The noble gases, which have filled outer s and p subshells, have no attraction for an additional electron; energy is required to add an electron. Similarly, energy is required to add an electron to the members of the alkaline earth family, group 2A; each member of this family has a filled outer s subshell.

SAMPLE EXERCISE 6.7

Although electron affinities generally become more negative as we move from left to right across any row of the periodic family, the electron affinity of N is more positive than that of C (see Figure 6.8). Rationalize this observation in terms of the electron configurations of N and C.

Solution: The orbital diagram for the outer-shell electrons of C is as follows:

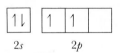

2s   2p

There is an empty $2p$ orbital available to receive the added electron. Nitrogen has a half-filled $2p$ subshell; there is already one electron in each $2p$ orbital. Consequently, when an extra electron is added it must be placed into an orbital that is already occupied by an electron. The resultant electron-electron repulsion causes the nitrogen atom to have less attraction for an extra electron than does carbon. In general, atoms with filled or half-filled subshells have more positive electron affinities (less electron attraction) than elements on either side of them in the periodic table.

TABLE 6.4 Electron affinities of some elements

Element	Ion formed	E(kJ/mol)
H	H⁻	−73
F	F⁻	−332
Cl	Cl⁻	−349
Br	Br⁻	−325
I	I⁻	−295
O	O⁻	−141
O	O²⁻	+710
S	S²⁻	+375

Table 6.4 gives the electron affinities of the halogens. Notice that the values differ very little within the family. As we proceed from fluorine to iodine the added electron is going into a p orbital of increasing major quantum number. The average distance of the electron from the nucleus steadily increases, and electron-nuclear attraction should thus steadily decrease. If this were all that is involved, fluorine would have the highest electron affinity. But the orbitals that hold the outermost electrons of the halogen are increasingly spread out as we proceed from fluorine to iodine. The electron-electron repulsions between these electrons and the added one therefore decrease with increasing atomic weight of the halogen. A lower electron-nuclear attraction is thus counterbalanced by lower electron-electron repulsion. The overall result is that the electron affinities differ very little among the halogens. Other families likewise show little variation in electron affinities among their members.

Note (Table 6.4) that addition of one electron to an oxygen atom results in evolution of energy; that is, it leads to a more stable species than the originally separated atom and electron. However, addition of a second electron to form O^{2-} requires energy, even though it results in a completed octet of electrons about the nucleus; the sign of E is therefore positive. In this case, and for S^{2-} also, the electron-electron repulsions outweigh the electron-nuclear attractions. However, such ions are common in crystalline solids because they are stabilized by electrostatic interactions with cations.

6.7 Atomic sizes

According to the quantum-mechanical model, an atom does not have a sharply defined boundary that determines its size. The electronic charge density in an atom simply drops off with increasing distance from the nucleus, approaching zero at large distance. However, there are techniques that make it possible to measure the distances between atoms in compounds. These can be used to develop a table of atomic radii, which should be fairly good approximate measures of the relative sizes of the atoms. For example, the distance between the centers of the Br atoms in Br_2 is 2.286 Å, making it possible to assign the Br atom a radius of 1.14 Å. Figure 6.9 graphs the atomic radii of the elements versus atomic number.

It is evident that atomic radii show periodic variation. The atomic size generally decreases as we move from the alkali metals (group 1A) toward the halogens (group 7A). Furthermore, we find that among the elements of any one group or family of the periodic table—for example, among the alkali metals—the radius increases regularly with increasing atomic number. These trends are summarized in Figure 6.10.

We can understand these variations using the same line of reasoning employed in discussing the variations in ionization potential. Proceeding from left to right across a horizontal row of the periodic table, the effective nuclear charge experienced by the outermost electrons increases as a result of incomplete shielding. Thus, the orbital containing the electron is contracted. On the other hand, in any vertical row, orbital size increases with increasing value for the principal quantum number.

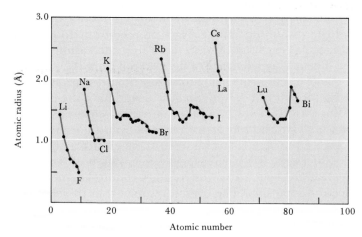

FIGURE 6.9 Atomic radii versus atomic number. The rare gases are not included in this graph because there is no simple way of relating their radii to those of the other elements on the basis of solid-state structure determinations. Gaps in the graph are due to lack of experimental data.

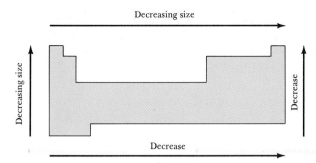

Decreasing size

Decreasing size

Decrease

Decrease

FIGURE 6.10 General periodic trends in atomic size.

6.8 The group 1 and group 2 metals

Our discussion of ionization potentials, electron affinity, and atomic size should give some indication of the way the periodic table can be used to organize and remember facts. Not only do elements in a family possess general similarities, but there are also trends in behavior as we move through a family or from one family to another. In this section we want to use the periodic table and our knowledge of electron configurations to examine the chemistry of the alkali metals, family 1A.

The alkali metal family gets its name from the Arabic word meaning *ashes*. Many compounds of sodium and potassium, the two most abundant alkali metals, were isolated from wood ashes by early chemists. The names soda ash and potash are still sometimes used for Na_2CO_3 and K_2CO_3, respectively.

The alkali metals are all shiny, soft, metallic solids of relatively low density. Some of their physical properties are given in Table 6.5. Notice how their melting points and densities vary in a fairly regular manner with increasing atomic number.

The alkali metals are all very reactive, readily losing one electron to form ions with a 1+ charge:

$$M \longrightarrow M^+ + e^-$$

(The symbol M in this equation and others through this section represents any one of the alkali metals.) This behavior corresponds to their ns^1 outer electron configurations and their low ionization energies. As we have noted in Section 6.5, the ease with which the elements lose electrons increases as we move down a family. Consequently, Cs is especially reactive. Of course, Fr is even more reactive, but it is an extremely rare, radioactive element; Cs is therefore the heaviest alkali metal generally encountered in the laboratory.

Owing to their reactivities, the alkali metals exist in nature only as

TABLE 6.5 Some physical and chemical properties of the alkali metals

Element	Symbol	Atomic number	Atomic weight	Melting point (°C)	Density (g/cm³)	Formula of hydroxide	Formula of chloride
Lithium	Li	3	6.939	181	0.53	LiOH	LiCl
Sodium	Na	11	22.9898	98	0.97	NaOH	NaCl
Potassium	K	19	39.102	63	0.86	KOH	KCl
Rubidium	Rb	37	85.47	39	1.53	RbOH	RbCl
Cesium	Cs	55	132.905	29	1.87	CsOH	CsCl

compounds. The metals can be obtained by passing an electric current through a molten salt, most commonly a chloride. For example, sodium is obtained in this way from NaCl. The electrical energy is used to force an electron onto Na^+ and to remove one from Cl^-:

$$Na^+ + e^- \longrightarrow Na$$

$$2Cl^- \longrightarrow Cl_2 + 2e^-$$

Although the metals have a silvery luster, their surfaces quickly lose this luster and become white when exposed to air. This behavior is due to reaction with oxygen in the air to form oxides:

$$4M(s) + O_2(g) \longrightarrow 2M_2O(s)$$

All except lithium can also form peroxides, M_2O_2; potassium, rubidium, and cesium are able to form superoxides, MO_2. (The superoxide ion has a charge of $1-$; O_2^-.)

The oxides of the alkali metals react readily with water to form hydroxides:

$$M_2O(s) + H_2O(l) \longrightarrow 2M^+(aq) + 2OH^-(aq)$$

These hydroxides are water soluble and completely ionized into M^+ and OH^- ions in aqueous solution. They are the most common laboratory bases.

The alkali metals also combine directly with water, generating hydrogen gas and forming hydroxides:

$$2M(s) + 2H_2O(l) \longrightarrow 2M^+(aq) + 2OH^-(aq) + H_2(g)$$

This reaction is most violent in the case of the heavier members of the family, in keeping with their weaker hold on the single outer-shell electron.

The reactivity of the alkali metals permits them to react directly with most elements. Some further examples follow:

$$2M(s) + H_2(g) \longrightarrow 2MH(s)$$

$$2M(s) + S(s) \longrightarrow M_2S(s)$$

$$2M(s) + Cl_2(g) \longrightarrow 2MCl(s)$$

In all cases the alkali metal exists as $1+$ ions in these compounds.

Almost all compounds of the alkali metals are soluble in water. Consequently, sodium, the most abundant of the alkali metals, is found as Na^+ in seawater. It reaches the sea through the dissolving action of water moving through the ground following rains.

The alkali metal salts are colorless, as are their aqueous solutions. Color is produced when an electron in an atom is excited from one energy level to another by visible radiation. Alkali metal ions, having lost their outermost electrons, have no electrons that can be excited by visible radiation.

The alkaline earth metals, family 2A, are also reactive; however, they lose their electrons less readily than do their alkali metal neighbors. As we have noted in Section 6.5, the ease with which the elements lose electrons decreases as we move across the periodic table from left to right. Berylium and magnesium, the lightest members of the family, are least reactive. Indeed, the chemical reactivity of magnesium is low enough to permit its use as a structural metal. It is nonetheless reasonably reactive; the metal is protected from loss of electrons to substances such as O_2 and H_2O in the environment by a thin and compact surface coat of water-insoluble MgO. Like the alkali metals, the alkaline earth elements are found in nature only as compounds.

In keeping with their outer ns^2 electron configurations, the alkaline earth metals form 2+ ions. These metals have higher melting points and are harder than their alkali metal neighbors; this difference reflects the fact that each alkaline earth atom has two outer-shell electrons available to hold atoms together within the solid.

COMPARISON OF A AND B GROUPS

The label attached to each family in the periodic table has either an A or B designation.* Elements of the same group number but different letter labels may share certain characteristics in common, but in general there is not a close connection. Let's consider the distinction between A and B elements in terms of electron configuration. As an example, locate the group 1A and 1B elements on the periodic table. In both families of group 1 elements the outer electron configuration involves a single s electron. In the case of the alkali metals (group 1A), this s electron is outside a stable noble gas core of electrons. For example, the single $4s$ electron in potassium is outside a set of eight electrons associated with filled $3s$ and $3p$ subshells, $3s^23p^6$. In the group 1B elements, sometimes referred to as the *coinage metals,* the single electron is outside filled $3s$, $3p$, and $3d$ subshells, $3s^23p^63d^{10}$. The presence of the additional ten d electrons in copper, and of course the increase of ten in nuclear charge that goes along with them, has a profound effect on the chemical behavior of the single s electron. Because these d electrons only partially shield the nucleus, the s electron of the 1B group experiences a larger effective nuclear charge than does the s electron of the 1A group. As a result, the 1B elements have higher ionization energies than do the corresponding

---

*The designation of groups as "A" or "B" is purely arbitrary. In fact some periodic tables use a different system from that used in this text; however, the system used in this text is presently the most common one for designating groups.

TABLE 6.6 Comparative values of ionization energies for group 1A, 1B, 2A, and 2B elements (kJ/mol)

1A	I_1	1B	I_1	2A	I_1	I_2	2B	I_1	I_2
K	418	Cu	859	Ca	589	1145	Zn	907	1733
Rb	403	Ag	730	Sr	549	1064	Cd	867	1630
Cs	375	Au	890	Ba	503	965	Hg	994	1805

1A elements. Table 6.6 lists the ionization energies for the group 1 and 2 elements. Compare the values for the A and B groups.

Because the 1B members have a much stronger hold on their outer-shell electron, these metals are much less reactive than are the 1A metals. In fact, the 1B metals can be found in elemental form in nature and are widely used in jewelry.

6.9 Historical development of the periodic table

As we have seen in this chapter, the periodic table arises from the periodic nature of electron configurations. However, the periodic table developed in an entirely empirical manner and was used by generations of chemists before there was any knowledge about electron configurations. Before we bring this chapter to a close, it is instructive to consider briefly the historical development of the periodic table.

During the earliest years of the nineteenth century, many new elements were discovered in a short period of time. By 1830 there were about 56 known elements. The identification of many new elements and the development of their descriptive chemistry naturally led to various attempts at classification. The classification process that occurred in chemistry in those years is common to the development of all science. As the quantity and variety of data increase, some means of orderly classification is sought, simply as a means of managing the large number of facts at hand. When a workable classification scheme is found, it may form the basis for development of a theory that accounts for the regularities observed. In 1869, Dmitri Mendeleev in Russia (Figure 6.11) and Lothar Meyer in Germany, working quite independently of each other, published very similar schemes for classification of the elements. Their tables of the elements were the forerunners of the modern periodic table. Mendeleev arranged the elements in order of increasing atomic weight

FIGURE 6.11 Dmitri Mendeleev. Mendeleev rose from very poor beginnings to a position of great eminence in nineteenth-century science. He was born in Siberia, the youngest child in a family of at least 14. His mother endured great personal sacrifice to make it possible for him to enroll in a university in Saint Petersburg. Mendeleev proved to be a brilliant student in sciences and mathematics and eventually was able to study in France and Germany. He spent most of his career as a professor of chemistry in the University of Saint Petersburg. Despite his eminence as a scientist, he was often in trouble because of his liberal, unorthodox opinions. (*Library of Congress*)

	B 10.81	C 12.01	N 14.01
	Al 26.98	Si 28.09	P 30.97
Zn 65.37	?	?	As 74.92
Cd 112.41	In 114.82	Sn 118.69	Sb 121.75

FIGURE 6.12 A portion of Mendeleev's periodic table showing the symbol for the element and the modern value for atomic mass.

and observed that when this is done, similar chemical and physical properties recur periodically. By arranging the elements so that those with similar characteristics are in vertical groupings, Mendeleev constructed the periodic table.

Although Meyer and Mendeleev came to essentially the same conclusion about the periodicity of properties, Mendeleev must be given credit for more vigorously advancing his ideas and stimulating much new work in chemistry. By sticking to his notion that elements of similar characteristics must be listed in groups, he was forced to leave several spaces in his table blank. For example, arsenic (As) was the element of next highest atomic weight after zinc (Zn). But its placement immediately after Zn in the table would have required that it fall under aluminum (Al). This, however, did not make sense in terms of its properties. Rather, it clearly belonged under phosphorus, as shown in Figure 6.12. This meant that in the table there were two blank spaces that Mendeleev boldly predicted would be filled by as yet undiscovered elements. He gave these elements the names eka-aluminum and eka-silicon and suggested that they might be found in nature with other members of their respective families. For example, eka-aluminum might be found in certain ores containing aluminum, because according to the periodic law it was likely to have properties similar to those of aluminum.

By noting that the properties of elements within a vertical family varied in a regular way with increasing atomic weight, Mendeleev was able to predict the properties of the unknown elements. Thus, the properties of eka-silicon should be intermediate between those of silicon (Si) and tin (Sn). In 1871 Mendeleev predicted the properties for eka-silicon, and not many years later, in 1886, the element germanium (Ge) was discovered. That the element germanium was the eka-silicon predicted by Mendeleev is shown by the data listed in Table 6.7.

SAMPLE EXERCISE 6.8

Note the densities listed below for the elements aluminum and indium. From these values and the data listed in Table 6.7 and Figure 6.12, predict the atomic weight and density for eka-aluminum (gallium). (Note: density of aluminum is 2.70 g/cm³; density of indium is 7.30 g/cm³.)

Solution: We would predict the atomic weight to be intermediate between the two values for the adjacent elements of higher and lower atomic weights, Zn and Ge. The average of the atomic weights of these two elements is 69.0, as compared with observed atomic weight of 69.72. We might estimate the density as intermediate between the values for Al and In, namely, 5.00 g/cm³. The observed value is 5.90 g/cm³. Incidentally, the average of the densities of Zn and Ge, the adjacent elements in the horizontal row, is 6.25 g/cm³, which comes a bit closer to the observed density for Ga.

The development of the periodic classification of the elements did much to systematize the study of chemistry. Nevertheless, many elements did not seem to fit very well into the periodic table. Only much later was it realized that the ordering of the elements should be in terms, not of atomic weight, but rather of atomic number. Today we realize that the periodicity results from the repeating character of electron configurations.

TABLE 6.7 Comparison of the properties for eka-silicon
predicted by Mendeleev with the known properties of
the element germanium

Property	Mendeleev's prediction for eka-silicon	Observed properties of germanium
Appearance	Gray	Grayish-white
Atomic weight	72	72.59
Density (g/cm^3)	5.5	5.35
Specific heat $(J/g$-$K)$	0.31	0.31
Formula of oxide	XO_2	GeO_2
Density of oxide (g/cm^3)	4.7	4.70
Formula of chloride	XCl_4	$GeCl_4$
Density of chloride (g/cm^3)	1.9	1.84

FOR REVIEW

Summary

Our major concern in this chapter has been the relationship between electron configurations and the properties of atoms, especially as organized by the periodic table. We saw that the electron configurations of many-electron atoms can be written by placing electrons into orbitals in the following order:

$1s, 2s, 2p, 3s, 3p, 4s, 3d, 4p, \ldots$

Subshells with a given principal quantum number, such as the $3s$, $3p$, and $3d$ subshells, do not have the same energies. This fact can be understood in terms of the screening effect and the average distance of an electron in each of these subshells from the nucleus.

The Pauli exclusion principle places a limit of two on the number of electrons that may occupy any one atomic orbital. These two electrons differ in their electron-spin quantum number, m_s. As the electrons populate orbitals of equal energy, they do not pair up until each orbital contains one electron; this observation is called Hund's rule. Using the relative energies of the orbitals, the Pauli exclusion principle, and Hund's rule, it is possible to write the electron configuration of any atom. When we do so, we see that the elements in any given family in the periodic table have the same type of electron arrangements in their outermost, incomplete shells. For example, the electron configurations of fluorine and chlorine, which are both members of the halogen family, are $[He]2s^2 2p^5$ and $[Ne]3s^2 3p^5$, respectively. This periodicity in electron configurations, summarized in Figure 6.4, permits us to write the electron configuration of an element from its position in the periodic table.

Many properties of atoms that have chemical significance exhibit periodic character. Among the most important of these are atomic radii, ionization energy, and electron affinity. Electron configurations and the periodic table help us to understand trends in these properties. We also illustrated how electron configurations help us to organize and understand some of the chemistry of the alkali and alkaline earth metals. Finally, we look at the historical origins of the periodic table, reminding ourselves that it predated our present ideas about atomic structure by many years.

Learning goals

Having read and studied this chapter, you should be able to:

1 List the factors that determine the energy of an electron in a many-electron atom. You should be able to explain the fact that electrons with the same value of principal quantum number (n) but differing values of the azimuthal quantum number (l) possess different energies.

2 Explain the concepts of effective nuclear charge and the screening effect as they relate to the energies of electrons in atoms.

3 State the Pauli exclusion principle and Hund's rule and illustrate how they are used in writing the electronic structures for the elements.

4 Explain the basis on which the modern periodic table is constructed.

5 Define the term "group" or "family" in terms of electron configuration.

6 Describe the various blocks of elements in the periodic table in terms of the type of orbital being occupied by electrons in that block (s, p, d, and f blocks).

7 List the names and give the locations in the periodic table for the active metals (*s*-block), representative elements (*p*-block), transition metals (*d*-block), and inner transition metals (*f*-block).

8 Write the electron configuration for any element once you know its place in the periodic table.

9 Write the orbital diagram representation for electron configurations of atoms.

10 Explain the effect of increasing nuclear charge on the radial density function in many-electron atoms.

11 Explain the general variations in first ionization energies among the elements, as depicted in Figure 6.6. You should also be able to explain the observed changes in values of the successive ionization energies for a given atom.

12 Explain the variation in atomic radii with atomic number. You should be able to relate this variation, as shown in Figure 6.9, to the variation in corresponding values of the first ionization energies.

13 Explain the concept of electron affinity and its relationship to ionization energy.

14 Relate the historical development of the periodic table, with emphasis on the logical process employed by Mendeleev and Meyer in placing the elements in the table.

Key terms

Among the more important terms and definitions used for the first time in this chapter are the following:

The **active metals** (Section 6.3), groups 1A and 2A, are those in which electrons occupy only the *s* orbitals of the valence shell.

The **effective nuclear charge** (Section 6.1) is the charge at the nucleus experienced by an electron in a many-electron atom. This charge is not the full nuclear charge, because there is some shielding of the nuclear charge by other electrons in the atom. How much shielding occurs depends on the average distance of the electron from the nucleus, compared with the other electrons in the atom.

The **electron affinity** (Section 6.6) is the energy change that occurs when an electron is added to a gaseous atom or ion.

An **electron configuration** (Section 6.3) is a particular arrangement of electrons in the orbitals of an atom.

Electron spin (Section 6.2) is a property of the electron that makes it behave as though it were a tiny magnet. Associated with the electron spin is a **spin quantum number** (m_s), which may have values of $+\frac{1}{2}$ or $-\frac{1}{2}$.

Hund's rule (Section 6.3) states that electrons must occupy degenerate orbitals one at a time until all orbitals have at least one electron, before pairing of electrons in the orbitals occurs. Note carefully that the rule applies only to orbitals that are **degenerate**, which means that they have the same energy.

The **inner transition elements** (Section 6.3), or **lanthanides** and **actinides**, are those in which the $4f$ or $5f$ orbitals are partially occupied.

The **Pauli exclusion principle** (Section 6.2) states that no two electrons in an atom may have all four quantum numbers, n, l, m_l, and m_s, the same. As a consequence of this principle, there can be no more than two electrons in any one atomic orbital.

The **representative elements** (Section 6.3) are those in which the p orbitals are partially occupied.

The **screening effect** (Section 6.1) is the effect of inner electrons in decreasing the nuclear charge experienced by outer electrons. (Also called the **shielding effect**.)

Transition elements (Section 6.3) are those in which the d orbitals are partially occupied.

Valence-shell electrons (Section 6.3) are the electrons in the outermost shell of an atom.

EXERCISES

Energies of orbitals

6.1 Which orbital in each of the following sets is lower in energy in a many-electron atom: (a) $3p$, $5s$; (b) $2s$, $2p$; (c) $3d$, $3s$; (d) $3d$, $4f$?

6.2 Explain why the effective nuclear charge experienced by a $2s$ electron in beryllium is larger than that experienced by a $2s$ electron in lithium.

6.3 Which of the four quantum numbers n, l, m_l, and m_s is related to the energy of an electron in a hydrogen atom? Which are related to the energy of an electron in a many-electron atom?

6.4 If the ionization energy of hydrogen is 1312 kJ/mol, what is the ionization energy of Li^{2+}?

6.5 Which of the following orbitals are degenerate in a many-electron atom: $2p_x$; $3s$; $3p_x$; $3p_y$?

6.6 In the hydrogen atom the $2s$ and $2p$ orbitals have identical energies, but in the fluorine atom the $2s$ orbital lies at considerably lower energy than the $2p$. Explain.

Electron spin; the Pauli principle

6.7 What is the total number of electrons in an atom that may have the following quantum numbers: (a) $n = 3$,

$l = 1$; (b) $n = 4$, $l = 2$; (c) $n = 3$, $l = 2$, $m_l = 2$; (d) $n = 4$; (e) $n = 3$, $m_s = +\frac{1}{2}$?

6.8 What is the maximum number of electrons that can be put into each of the following subshells: (a) $3d$; (b) $5p$; (c) $4f$; (d) $2s$?

6.9 List the possible values for the four quantum numbers for each of the three electrons in the lithium atom.

Electron configurations of the elements

6.10 Write out the electron configurations for the following atoms using the appropriate noble gas inner core for abbreviation: (a) K; (b) Si; (c) Se; (d) Mn; (e) La.

6.11 Write out the complete electron configurations for the following atoms without using the noble gas inner core abbreviation: (a) Cd; (b) S; (c) Sm; (d) Pb; (e) Kr.

6.12 Draw the orbital diagrams for the electrons beyond the appropriate noble-gas inner core for each of the following elements: (a) P; (b) V; (c) Sr; (d) Lu; (e) I.

6.13 Referring to a periodic table, identify the element that possesses each of the following electron configurations: (a) [He]$2s^2$; (b) [He]$2s^22p^4$; (c) [Ar]$3d^24s^2$; (d) [Ne]$3s^23p^5$; (e) [Kr]$4d^{10}5s^25p^1$; (f) [Kr]$5s^24d^5$.

6.14 Which of the following configurations correspond to ground states and which to excited states: (a) $1s^22s^2$; (b) $1s^23s^1$; (c) [Ne]$3s^23d^1$; (d) [Ar]$4s^23d^2$; (e) $1s^22s^22p^63p^1$?

6.15 Classify each of the following elements as a metal or a nonmetal; also classify each metal as active metal, representative element, transition metal, or inner transition metal: (a) Ca; (b) Fe; (c) Nd; (d) Rb; (e) Se; (f) Xe; (g) Ag; (h) U; (i) Pb.

6.16 What outer electron configuration (beyond the noble-gas core) characterizes each of the following groups of elements (for example, the alkali metals possess an outer-shell configuration of ns^1): (a) the alkaline earths; (b) group 1B; (c) group 5A; (d) the halogens; (e) the noble gases; (f) group 5B?

6.17 When lithium atoms are excited they emit line spectra that are similar in many respects to those produced by excited hydrogen atoms. Which electron or electrons are responsible for the observed line spectra?

Periodicity and atomic properties

6.18 How do each of the following properties vary as we move from left to right across the periodic table: (a) atomic size; (b) ionization energy; (c) electron affinity? How do each of these vary as we move down a family?

6.19 Based on their positions in the periodic table, select the larger atom from each of the following pairs: (a) Li, Na; (b) Li, Be; (c) O, P; (d) N, Si.

6.20 Based on their positions in the periodic table, select the atom with the larger ionization energy from each of the following pairs: (a) N, F; (b) Na, Mg; (c) O, S; (d) Al, C; (e) N, O.

6.21 Indicate which atom in each of the following pairs has the larger size and ionization energy and more negative electron affinity: (a) O, F; (b) N, P; (c) Sc, Ca; (d) Ga, Si.

6.22 Rationalize the low chemical reactivity of the noble gases in terms of their ionization energies and their electron affinities.

6.23 The second ionization energy of Mg is only about twice as great as the first, but the third is ten times as great. Explain why it requires so much more energy to remove the third electron.

6.24 How is the change in radii of the atoms in the series from potassium through krypton related to the change in ionization energies in the same series? Explain the reason for the correlation.

The alkali metals; the periodic table

6.25 Write balanced chemical equations for the reactions that occur (a) when K is added to water; (b) when Li_2O is dissolved in H_2O; (c) when Li reacts with O_2; (d) when H_2 gas is bubbled through molten Na; (e) when Li comes in contact with iodine vapor.

6.26 How does beryllium compare with lithium with respect to each of the following properties: (a) metallic character; (b) atomic size; (c) number of outer-shell electrons; (d) ionization energy; (e) electron affinity; (f) formula of chloride salt.

6.27 Explain why it requires more energy to ionize a $4s$ electron from Zn than from Ca.

6.28 By examining the modern periodic table, find as many examples as you can of violations of Mendeleev's periodic law that the chemical and physical properties of the elements are periodic functions of their atomic *weights*. Why do these violations occur? State the modern version of the periodic law.

6.29 The formula for water is, of course, H_2O. Write the formulas one might expect for the corresponding hydrogen compounds of the other elements in group 6A.

Additional exercises

6.30 How does the average distance from the nucleus of a $2s$ electron in a neon atom compare with that for a $2p$ electron? Explain.

6.31 Which of the electrons in each of the following sets experiences the largest effective nuclear charge in a many-electron atom: (a) 2s, 2p; (b) 2p, 3p; (c) 4f, 3d?

6.32 Write the electron configuration for sodium. Which electrons experience the greatest effective nuclear charge? Which experiences the lowest?

6.33 Calculate and compare the energies of the $1s$ electron in H, He^+, and Li^{2+}. Why do they differ?

6.34 If you could do the Stern-Gerlach type of experiment shown in Figure 6.3 with the first four elements of the periodic table, which would give rise to a separation of beams? How would your results provide experimental evidence for the Pauli exclusion principle?

6.35 The Stern-Gerlach experiment illustrated in Figure 6.3 was first carried out using a beam of silver atoms. Draw the orbital diagram representation for the electron configuration of silver. How many unpaired electrons does this atom have?

6.36 How does the Pauli exclusion principle account

for the existence of inert chemical behavior at atomic numbers 2, 10, 18, 36, and 54?

[6.37] Suppose that the electron could have three values for m_s ($\frac{1}{2}$, 0, and $-\frac{1}{2}$). If, with this exception, the restrictions on quantum numbers were as described in this chapter, how many unpaired electrons would a carbon atom have? What would be its electron configuration?

[6.38] What is the lowest value of the principal quantum number n for which there can be a g subshell? How many electrons are required to fill a g subshell? Are there any elements known whose ground-state electron configurations contain electrons in the g subshell?

6.39 Write the complete electron configuration for each of the following atoms: (a) Cl; (b) Ca; (c) Sc; (d) Sb; (e) Ge; (f) U.

6.40 Draw the orbital diagrams for each of the following elements: (a) As; (b) Cr; (c) Y; (d) K; (e) Al.

6.41 How many unpaired electrons do each of the following atoms possess: (a) C; (b) Ca; (c) Mn; (d) Se?

6.42 Which of the following electron configurations must be incorrect (because they cannot represent either ground-state or excited-state electron configurations): (a) $1s^2 2s^3 2p^1$; (b) $1s^2 2s^1 2p^1$; (c) $1s^2 2s^2 2p^2$; (d) [Ne]$3s^1 3d^1$; (e) [Ne]$3s^2 3d^{12}$; (f) [Ne]$3s^2 3f^4$.

6.43 Identify the specific element or group of elements that can have the following electron configurations: (a) $1s^2 2s^2 2p^6 3s^2 3p^5$; (b) [noble gas]ns^2; (c) [noble gas]$ns^2(n-1)d^7$.

6.44 Name the element that fits each of the following descriptions: (a) alkali metal with a $4s$ valence-shell electron; (b) noble gas with less than eight outer-shell electrons; (c) lightest element with a half-filled p subshell; (d) element with the highest ionization potential; (e) an element with one unpaired $2p$ electron; (f) element with the most negative electron affinity.

6.45 Explain, in terms of electron configurations, why H exhibits properties similar to both Li and F.

6.46 Notice in Figure 6.6 that the first ionization energy of B is slightly less than that of Be. Suggest an explanation for this fact in terms of electron configuration. Similarly, explain why the first ionization energy of O is less than that of N.

6.47 Using a handbook of chemistry or other reference source, make a table of the following properties of the group 2A elements: density, atomic weight, formula of chloride, formula of oxide, first ionization energy, second ionization energy, third ionization energy. Is there a basis in the data for classifying the elements in one group? Explain.

[6.48] The element technetium, atomic number 43, is not observed in nature because it happens to be radioactive. Assuming that it can be synthesized in a nuclear reactor in quantity, predict some of its characteristic properties. These should include electron configuration, density, melting point, and formulas of oxides. Consult a handbook of chemistry for information on related elements.

[6.49] Make a graph of the third ionization energies, I_3, listed in Table 6.3, as a function of atomic number. Discuss the reason for the large value seen for Mg. Discuss also a possible reason for the minimum in I_3 at P.

6.50 Consider these ionization energy data for the elements scandium and gallium:

Element	I_1	I_2	I_3
Sc	646	1235	2375
Ga	576	1971	2950

Offer an explanation for the fact that I_1 is *lower* for Ga than for Sc, despite the fact that I_2 and I_3 are much larger for the group 3A element.

[6.51] There are certain similarities in properties that exist between the first member of any periodic family and the element located one position below it and to the right in the periodic table. For example, in some ways Li resembles Mg, Be resembles Al, and so forth. This observation is known as the diagonal relationship. In terms of what we have learned in this chapter, offer a possible explanation for this relationship.

6.52 Draw the ground-state orbital diagram for carbon. Draw an orbital diagram for an excited state of this atom that possesses four unpaired electrons.

7

Chemical bonding

Deep within the earth below the city of Detroit and beneath the rolling plains of Kansas lie enormous deposits of a white mineral, halite. This substance, also known as sodium chloride, NaCl, was deposited in these and other places millions of years ago, when extensive primordial seas dried upon the changing surface of the earth. Sodium chloride is the most abundant dissolved substance present in seawater and is found in human body tissues in large quantities. However, it is most familiar to us as ordinary table salt. This substance consists of sodium ions and chloride ions, Na^+ and Cl^-.

Water, H_2O, is another very familiar substance. We drink it, swim in it, and use it as a cooling agent. It is found in many places in our environment and is essential to life as we know it. This substance is composed of molecules.

Why is it that some substances are composed of ions while others are composed of molecules? The key to this question is found in the electronic structures of the atoms involved and in the nature of the chemical forces within the compounds. In this chapter and the next we shall examine the relationships between electronic structure, chemical bonding forces, and the properties of substances. As we do this, we shall find it useful to classify chemical forces into three broad groups: (1) ionic bonds, (2) covalent bonds, and (3) metallic bonds.

The term ionic bond refers to the electrostatic forces that exist between particles of opposite charge. As we shall see, ions may be formed from atoms by transfer of one or more electrons from one atom to another. Ionic substances generally result from the interaction of metals from the far left side of the periodic table with the nonmetallic elements from the far right side (excluding the rare gases, group 8A).

The covalent bond results from a sharing of electrons between two atoms. The most familiar examples of covalent bonding are seen in the interactions of nonmetallic elements with one another.

Metallic bonds are found in solid metals such as copper, iron, and aluminum. In the metals, each metal atom is bonded to several neighboring atoms. The bonding electrons are relatively free to move throughout the three-dimensional structure. Metallic bonds give rise to such typical metallic properties as high electrical conductivity and luster. We will postpone further discussion of metallic bonding until Chapter 22.

7.1 Lewis symbols and the octet rule

The term valence is commonly used in discussions of both ionic and covalent bonding. The valence of an element is a measure of its capacity to form chemical bonds. Originally, it was determined by the number of hydrogen atoms with which an element combined. Thus, the valence of oxygen is 2 in H_2O; in CH_4 the valence of carbon is 4.

We speak of valence electrons when referring to the electrons that take part in chemical bonding. These electrons are the ones residing in the outermost electron shell of the atom, the valence shell. Electron-dot symbols (also known as Lewis symbols, after G. N. Lewis) are a simple and convenient way of showing the valence electrons of atoms and keeping track of them in the course of bond formation. The electron-dot symbol for an element consists of the chemical symbol for the element plus a dot for each valence electron. For example, sulfur has an electron configuration $[Ne]3s^23p^4$; its electron-dot symbol therefore shows six valence electrons:

$$\cdot \ddot{\underset{\cdot\cdot}{S}} \cdot$$

Other examples are shown in Table 7.1.

The number of valence electrons of any active metal or representative element is the same as the column number of the element in the periodic table. Thus the electron-dot symbols for both oxygen and sulfur, members of family 6A, show six dots.

The way atoms gain, lose, or share electrons can often be viewed as an attempt on the part of the atoms to achieve the same number of electrons as the noble gas closest to them in the periodic table. The noble gases, you will recall, have very stable electron arrangements, as evidenced by their high ionization energies, low electron affinities, and general lack of chemical reactivity. Because all noble gases (except He) have eight valence electrons, many atoms undergoing reactions also end up with eight valence electrons. This observation has led to what is known as the octet rule. Of course, because He has only two electrons, atoms near it in the periodic table, such as H, will generally tend to obtain two electrons.

TABLE 7.1 Electron-dot symbols

Element	Electron configu- ration	Electron-dot symbols
Li	$[He]2s^1$	Li ·
Be	$[He]2s^2$	· Be ·
B	$[He]2s^22p^1$	· Ḃ ·
C	$[He]2s^22p^2$	· Ċ ·
N	$[He]2s^22p^3$	· N̈ :
O	$[He]2s^22p^4$	: Ö :
F	$[He]2s^22p^5$	· F̈ :
Ne	$[He]2s^22p^6$	: N̈e :

7.2 Ionic bonds

When sodium metal is brought into contact with chlorine gas, Cl_2, a violent reaction ensues. The product of that reaction is sodium chloride, NaCl, a substance composed of Na^+ and Cl^- ions:

$$2Na(s) + Cl_2(g) \longrightarrow 2NaCl(s)$$

These ions are arranged throughout the solid NaCl in a regular three-dimensional array, as shown in Figure 7.1.

The formation of Na^+ from Na and of Cl^- from Cl_2 indicates that an electron has been lost by a sodium atom and gained by a chlorine atom. Such electron transfer to form oppositely charged ions is favored when

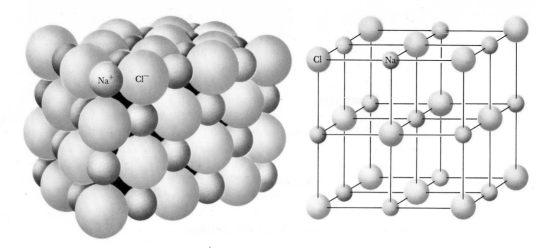

FIGURE 7.1 Two ways of representing the crystal structure of sodium chloride. The structure on the left shows the ions in their correct sizes relative to the distances between them. The larger spheres represent the chloride ions. The structure on the right emphasizes the symmetry of the structure, whereas the one on the left better illustrates the way the ions are packed together throughout the solid.

the atoms involved differ greatly in their attraction for electrons. Our example of NaCl is rather typical for ionic compounds; it involves a metal of low ionization energy and a nonmetal with a high electron affinity. Using electron-dot symbols (and showing a chlorine atom rather than the Cl_2 molecule) we can represent this reaction as follows:

$$Na \cdot + \overset{\cdot\cdot}{\underset{\cdot\cdot}{Cl}}: \longrightarrow Na^+ + :\overset{\cdot\cdot}{\underset{\cdot\cdot}{Cl}}:^-$$

Each ion has an octet of electrons, the octet on Na^+ being the $2s^2 2p^6$ electrons that lie below the single $3s$ valence electron of the Na atom.

The formation of ionic compounds is not merely the result of low ionization energies and high electron affinities, although these factors are very important. The formation of an ionic compound from the elements is always an exothermic process; the compound forms because it is more stable (lower in energy) than its elements. Much of the stability of NaCl results from the packing of the oppositely charged Na^+ and Cl^- ions together as shown in Figure 7.1. A measure of just how much stabilization results from this packing is given by the lattice energy. This quantity is the energy required for 1 mol of the solid ionic substance to be separated completely into ions far removed from one another. We can write the process as

$$NaCl(s) \longrightarrow Na^+(g) + Cl^-(g) \qquad\qquad [7.1]$$

To get a picture of this process, imagine that the lattice shown in Figure 7.1 expands from within, so that the spaces between the ions grow larger and larger, until the ions are very far apart. The energy that would be required for that to occur for a lattice containing 1 mol of Na^+ and 1 mol of Cl^- ions is the lattice energy.

The lattice energy for NaCl(s) amounts to 785 kJ/mol. This is a very large amount of energy, and it accounts for the fact that sodium chloride is a stable, solid substance with a high melting point. We see from the structure of NaCl (Figure 7.1) that each sodium ion is surrounded by six nearest-neighbor chloride ions of opposite charge. Similarly each chloride ion is surrounded by six sodium ions. The attractive force between each ion and its nearest neighbors of opposite charge provides much of the stabilizing lattice energy. Furthermore, each ion also experiences repulsive interactions with ions of like charge in the lattice and is attracted to other ions of opposite charge in addition to its nearest neighbors. The lattice energy is the result of all the electrostatic interactions, taken over the entire lattice. Ionic substances may have arrangements of ions that differ from that shown in Figure 7.1. The lattice arrangements for a few other common ionic substances are shown in Figure 7.2. In each, the ions are arranged to maximize the attractive forces between ions of opposite charge while minimizing the repulsive forces between ions of like charge. The overall structure consequently depends on the charges of the ions and their relative sizes. In all these substances, the forces between ions lead to brittle, crystalline solids with high melting points, properties characteristic of ionic compounds.

The potential energy of two interacting charges is given by Equation [7.2]:

$$E = k\frac{Q_1 Q_2}{d}$$

[7.2]

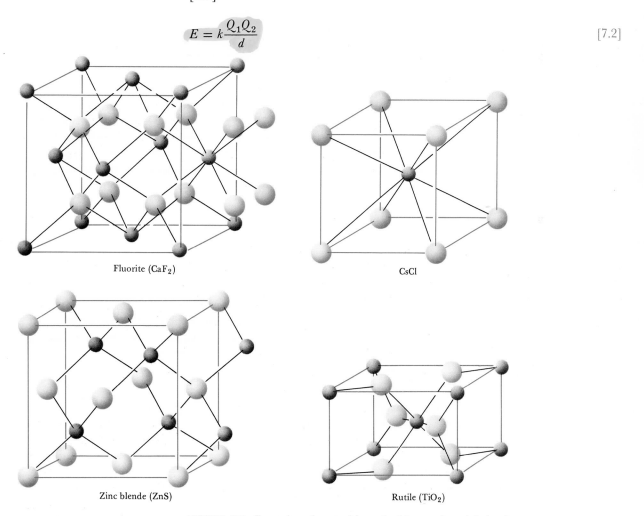

Fluorite (CaF_2)

CsCl

Zinc blende (ZnS)

Rutile (TiO_2)

FIGURE 7.2 Examples of several important types of crystal structure.

Q_1 and Q_2 are the magnitudes of the charges on the particles in coulombs and d is the distance between their centers in meters. The constant k has the value 8.99×10^9 J-m/coul^2. As Equation [7.2] indicates, the attractive interaction between two oppositely charged ions increases as the magnitudes of their charges increase and as the distance between their centers decreases. Thus for a given arrangement of ions, the lattice energy increases as the charges on the ions increase and as their sizes decrease.

Now if lattice energy increases as the charges of the ions increase, why doesn't sodium lose two electrons to form Na^{2+}? The second electron would have to come from the inner shell of the sodium atom. We can see from the large value for the second ionization energy of Na, Table 6.3, that it would require too much energy to form the Na^{2+} ion in chemical compounds. Thus sodium and the other group 1A metals are found in ionic substances only as 1+ ions. Similarly, addition of a second electron to a chloride ion to form a hypothetical Cl^{2-} is never observed. The other group 7A elements (the halogens) are also found only as the 1- ions F^-, Br^-, or I^- in ionic substances.

Magnesium, an element of group 2A, also forms an ionic compound with chlorine, of composition $MgCl_2$. In this instance, the metal achieves a rare-gas configuration by loss of two electrons. It requires much less energy to remove two electrons from Mg (Table 6.3) than from Na because both electrons are in the valence shell of Mg. Of course, it requires more energy to remove two electrons from magnesium than is required for removing just one. This energy is more than recovered, however, in the increased lattice energy of $MgCl_2$, which comes from the higher charge on the metal ion. Thus magnesium and the other group 2A metals do not form ionic compounds in which the metal has a 1+ charge; there is no known MgCl. Similarly, the group 6A elements are found in ionic compounds as O^{2-}, S^{2-}, and so forth, in which the ion possesses a rare-gas configuration. In the formation of the ionic solid MgO, both magnesium and oxygen attain the rare-gas configuration, by transfer of two electrons:

$$Mg + \overset{\cdot\cdot}{\underset{\cdot\cdot}{:}O: \longrightarrow Mg^{2+} + \left[:\overset{\cdot\cdot}{\underset{\cdot\cdot}{O}}: \right]^{2-}$$

Thus there is a balance between ionization energies and lattice energies that determines the magnitude of the positive charge on a metal ion. Similarly, the negative charge on a nonmetal is determined by the balance between electron affinities and lattice energies.

SAMPLE EXERCISE 7.1

Which substance would have the higher lattice energy, NaF or MgO? Explain.

Solution: Magnesium oxide, MgO, would have the higher lattice energy. The electrostatic attraction between oppositely charged ions increases with the charge on the ion. For this reason, it would require a greater amount of energy to separate a mole of Mg^{2+} ions and a mole of O^{2-} ions to infinite separation, as compared with separating a mole of Na^+ and a mole of F^-.

SAMPLE EXERCISE 7.2

Predict the formula of the compound formed between aluminum (Al) and fluorine; between aluminum and oxygen.

Solution: Aluminum, with atomic number 13, has three electrons beyond the inert gas configuration. It might then be expected to lose three electrons to form the Al^{3+} ion. Each fluorine atom attains the noble-gas configuration by accepting one electron to form F^-. As we saw in Chapter 2, the formula of an ionic compound depends on the charges on the ions. The formula may be derived by the principle of electroneutrality: The total charge of the cations must balance that of the anions. Three F^- ions are required to balance the charge of one Al^{3+} ion; the expected formula is thus AlF_3.

In forming a compound with oxygen, each aluminum atom again loses three electrons to form Al^{3+}. Each oxygen atom accepts two electrons to form O^{2-}, thereby achieving the noble-gas configuration (an octet of valence-shell electrons). Two Al^{3+} balance the charge of three O^{2-}; the formula for aluminum oxide is therefore Al_2O_3.

Ionic-bond theory correctly predicts the charges found on many simple ions, based on the notion that attainment of a rare-gas configuration leads to maximum stability. For some elements, however, the rule must be modified, and for others it is not applicable at all. For example, metals of group 1B (Cu, Ag, Au) are observed to occur often as the $1+$ ions (as in CuBr and AgCl). Silver possesses a $4d^{10}5s^1$ outer electron configuration. In forming Ag^+ the $5s$ electron is lost. This leaves a completely filled shell of 18 electrons in the $n = 4$ level. Because it is a completed shell, it is somewhat like a rare-gas arrangement. Similarly, the group 2B elements most commonly are seen as the $2+$ ions (Zn^{2+}, Cd^{2+}, Hg^{2+}) in ionic compounds. The valence-shell s electrons are lost in forming the ions, leaving an electronic arrangement consisting of 18 electrons in the highest occupied level.

For most of the transition metals, the attainment of a rare-gas configuration by loss of electrons is not feasible; that would require the loss of too many electrons. The outer electron configurations of these elements are either $(n - 1)d^x ns^2$ or $(n - 1)d^x ns^1$, where n is 4, 5, or 6, and x may vary from 1 to 10. In forming ions the transition metals lose the valence-shell s electrons first, then as many d electrons as are required to form an ion of particular charge. Most of the transition metals are found in more than one charge state. For example, the element chromium is found in compounds as Cr^{2+} or Cr^{3+}. There are no simple rules to tell which charge state of a transition-metal ion will exist in a particular case.

SAMPLE EXERCISE 7.3

Write the electron configuration for the Co^{2+} ion; for the Co^{3+} ion.

Solution: Cobalt (atomic number 27) has an electron configuration $[Ar]4s^2 3d^7$. To form a $2+$ ion, two electrons must be removed. As discussed in the text above, the $4s$ electrons are removed before the $3d$. Consequently, the Co^{2+} ion has an electron configuration of $[Ar]3d^7$. To form Co^{3+} requires the removal of an additional electron; the electron configuration for this ion is $[Ar]3d^6$.

This is a good point at which to look back to Chapter 2, to review Table 2.5, which lists common ions. You will note when you do so that some common ions are polyatomic. In polyatomic ions, two or more atoms are bound together by predominantly covalent bonds. The group of atoms as a whole then acts as a charged species in forming an ionic compound with an ion of opposite charge. Examples of positively charged polyatomic ions are the vanadyl ion, VO^{2+}, and the familiar ammonium ion, NH_4^+. However, most polyatomic ions are negatively charged. Recall also that positive ions are called cations, whereas negatively charged ions are called anions.

7.3 Sizes of ions

The radii of ions are interesting because they help us understand better how the electrons in ions are attracted to the central nuclear charge. They are also important in many practical ways. For example, the sizes of ions are important in determining the lattice energy in an ionic solid. They also determine how easily the ions can be removed from water in water-softening devices. As another example, many metal ions are important in biological reactions. Biological systems are often very specific; they work well with one metal ion but not with another, even though it has the same charge and seems very similar. It often happens that only a small difference in ionic size is sufficient to cause one metal ion to be biologically active and another not to be so.

When ions are drawn close together, electron-electron repulsions eventually become as large as the attractive forces. The repulsive forces place a limit on the distance of closest approach. We may thus think of ions as having characteristic radii that determine their distances from other ions. From X-ray diffraction studies of ionic solids, which will be described in Chapter 11, it is possible to determine the distances between ions. Using data for a large number of structures, physical chemists have analyzed these distances to obtain a set of ionic radii.

The radii for several ions with rare-gas configurations are listed in Table 7.2. As we might expect, the radii of cations are smaller than the radii of the corresponding neutral atoms from which they are derived, as illustrated in Figure 7.3. For example, the radius of the potassium atom is 2.2 Å, whereas the radius of K^+ is 1.33 Å. Positive ions are formed by removing one or more electrons from the outermost region of the atom. Thus the most spatially extended orbitals are vacated. In addition, removal of an electron decreases the total electron-electron repulsions. On the other hand, the radii of anions are larger than those of the corresponding neutral atoms. For example, the atomic radius of Br^- is 1.96 Å, whereas for bromine atom it is 1.15 Å. Electrons added to atoms to form

TABLE 7.2 Radii (in Å) of ions with rare-gas electron configurations

Group 1A		Group 2A		Group 3		Group 6A		Group 7A	
Li^+	0.68	Be^{2+}	0.30			O^{2-}	1.45	F^-	1.33
Na^+	0.98	Mg^{2+}	0.65	Al^{3+}	0.45	S^{2-}	1.90	Cl^-	1.81
K^+	1.33	Ca^{2+}	0.94	Sc^{3+}	0.68	Se^{2-}	2.02	Br^-	1.96
Rb^+	1.48	Sr^{2+}	1.10	Y^{3+}	0.90	Te^{2-}	2.22	I^-	2.19
Cs^+	1.67	Ba^{2+}	1.31						

negative ions with the rare-gas configuration go into p orbitals that are already partially filled. The increased electron-electron repulsions caused by the additions result in a larger average distance from the nucleus. The electrons spread out in space to minimize their interactions with one another.

SAMPLE EXERCISE 7.4

Based on the data given in Figure 7.3, how would you compare the effective nuclear charge experienced by a $4p$ electron in Br^- with the effective nuclear charge experienced by a $4p$ electron in Br? Explain.

Solution: The effective nuclear charge experienced by a $4p$ electron in Br^- is smaller than in Br. The *actual* nuclear charge is the same in both cases. The only difference is that an extra electron has been added to Br to form Br^-. This extra electron, which goes into the singly occupied $4p$ orbital of Br, acts to some extent to shield the nucleus from the other $4p$ electrons. Thus, they experience a smaller effective nuclear charge. This effect, along with the increased electron-electron repulsions, causes Br^- to have a larger radius than Br.

The effects we have just described are also seen in the variation in radius in an isoelectronic series of ions. The term "isoelectronic" means that the ions possess the same number and arrangement of electrons. For example, in the series O^{2-}, F^-, Na^+, Mg^{2+}, and Al^{3+}, there are ten electrons arranged in the neon electron configuration about each nucleus. The nuclear charge in this series increases steadily in the order listed. With the number of electrons remaining constant, the radius of the ion decreases as the nuclear charge increases, attracting the electrons more strongly toward the nucleus:

———— Increasing nuclear charge (Å) ⟶

O^{2-}	F^-	Na^+	Mg^{2+}	Al^{3+}
1.45	1.33	0.98	0.65	0.45

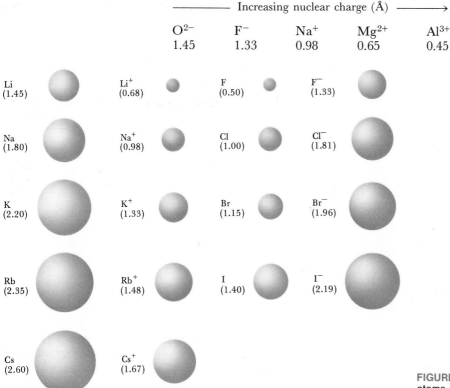

Li (1.45)	Li^+ (0.68)	F (0.50)	F^- (1.33)	
Na (1.80)	Na^+ (0.98)	Cl (1.00)	Cl^- (1.81)	
K (2.20)	K^+ (1.33)	Br (1.15)	Br^- (1.96)	
Rb (2.35)	Rb^+ (1.48)	I (1.40)	I^- (2.19)	
Cs (2.60)	Cs^+ (1.67)			

FIGURE 7.3 The relative sizes of atoms and ions (Å).

We saw in Section 6.7 that as the principal quantum number of an orbital increases, the average distance of the electron from the nucleus increases also. The relative radial extensions of the $1s$, $2s$, and $3s$ orbitals, shown in Figure 5.15, provide a good example. Thus, for ions of the same charge in any one family, the ionic radius increases with increasing period, that is, as the principal quantum number of the outermost occupied orbital increases.

One further comparison worth keeping in mind is the relative sizes of ions from the A and B subgroups. You may recall from the discussion in Section 6.8 that the outermost s electron of the group 1B elements experiences a higher effective nuclear charge. This happens because the ten d electrons added in going from a group 1A element to a group 1B element in the same row (for example, in going from K to Cu) do not completely shield the valence s electron from the nucleus. They also do not completely shield one another from the nucleus. As a result, not only is the atom of the group 1B element smaller, the ion formed by removal of the valence s electron is smaller also. For example, the radius of Cu^+, 0.96 Å, is less than that for the corresponding group 1A ion, K^+, radius 1.33 Å.

SAMPLE EXERCISE 7.5

The radius of Zn^{2+} is 0.74 Å, as compared with 0.99 Å for Ca^{2+}. Account for the smaller radius of Zn^{2+}.

Solution: Zinc occurs in the same horizontal row of the periodic table as Ca, but it has ten additional electrons, and a nuclear charge ten larger than for Ca. As a result of incomplete shielding of the added nuclear charge by the added $3d$ electrons, the effective nuclear charge experienced by all the electrons in the $n = 3$ shell in Zn^{2+} is considerably greater than in Ca^{2+}. As a result, the radius of Zn^{2+} is smaller.

7.4 Covalent bonding

We have seen that ionic substances possess several characteristic properties. They are usually brittle substances with high melting points. They are usually also crystalline, meaning that the solids have flat surfaces that make characteristic angles with one another. Ionic crystals can often be cleaved; that is, they break apart along smooth, flat surfaces. The characteristics of ionic substances result from the ionic forces that maintain the ions in a rigid, well-defined, three-dimensional arrangement such as one of those illustrated in Figures 7.1 and 7.2.

The vast majority of chemical substances do not have the characteristics of ionic materials; we need only think of water, gasoline, banana peelings, hair, antifreeze, and plastic bags as examples. Most of the substances with which we come in daily contact tend to be gases, liquids, or solids with low melting points; many vaporize readily—for example, mothball crystals. Many in their solid forms are plastic rather than rigidly crystalline—for example, paraffin or plastic bags.

For the very large class of substances that do not behave like ionic substances, a different model for the bonding between atoms is required. G. N. Lewis reasoned that an atom might acquire a rare-gas electron configuration by sharing electrons with other atoms. A chemical bond formed by sharing a pair of electrons is called a covalent bond.

The hydrogen molecule, H_2, furnishes the simplest possible example of a covalent bond. Using electron-dot symbols, formation of the H_2 molecule by combination of two hydrogen atoms can be represented as

$$H\cdot \ + \ \cdot H \longrightarrow \boxed{H(\colon)H}$$

The shared pair of electrons provides each hydrogen atom with two electrons in its valence shell (the $1s$) orbital, so that in a sense it has the electron configuration of the rare gas helium (the shared electrons are counted with both atoms). Similarly, when two chlorine atoms combine to form the Cl_2 molecule,

$$:\overset{..}{\underset{..}{Cl}}\cdot \ + \ \cdot\overset{..}{\underset{..}{Cl}}: \longrightarrow \left(:\overset{..}{\underset{..}{Cl}}(\colon)\overset{..}{\underset{..}{Cl}}:\right)$$

each chlorine atom, by sharing in the bonding electron pair, acquires eight electrons (an octet) in its valence shell, and thus achieves the rare-gas electron configuration of argon. The structures shown above for H_2 and Cl_2 are called Lewis structures. In writing Lewis structures, it is the usual practice to show each electron pair shared between atoms as a line, and the unshared electron pairs as pairs of dots. Thus the Lewis structures for H_2 and Cl_2 are shown as follows:

$$H-H \qquad \qquad :\overset{..}{\underset{..}{Cl}}-\overset{..}{\underset{..}{Cl}}:$$

Of course, the shared pairs of electrons are not located in fixed positions between nuclei. Figure 7.4 shows the distribution of electron density in the H_2 molecule. Notice that electron density is concentrated between nuclei. The two atoms are bound into the H_2 molecule principally because of the electrostatic attractions of the two positive nuclei for the concentration of negative charge between them. In the next chapter we shall look more closely at the spatial distribution of electron density within molecules. At that time we shall treat covalent bonds in terms of orbitals. Meanwhile, our discussions shall rely primarily on Lewis structures.

FIGURE 7.4 The electron distribution in the H_2 molecule.

In the Lewis model, the valence of an element is associated with the number of electron pairs shared to complete the octet of electrons. Because the number of valence electrons is the same as the group number for the nonmetals, one might predict that the 7A elements, such as F, would form one covalent bond to achieve an octet; 6A elements such as O would form two covalent bonds; 5A elements such as N would form three covalent bonds; and 4A elements such as C would form four covalent bonds. These predictions are borne out in many compounds. For example, consider the simple hydrides of the nonmetals of the second row (period) of the periodic table:

$$H-\overset{..}{\underset{..}{F}}: \qquad H-\overset{..}{\underset{\underset{\displaystyle H}{|}}{O}}: \qquad H-\overset{..}{\underset{\underset{\displaystyle H}{|}}{N}}-H \qquad H-\overset{\overset{\displaystyle H}{|}}{\underset{\underset{\displaystyle H}{|}}{C}}-H$$

Thus the Lewis model is very successful in accounting for the compositions of compounds of the nonmetals, in which covalent bonding predominates.

Using the Lewis theory and the theory of ionic bonding described earlier, explain the formulas of the following hydrides: NaH; MgH_2; AlH_3; SiH_4; PH_3; H_2S; HCl.

Solution: The hydrides of the metallic elements are ionic compounds consisting of metallic cations and the hydride ion, H^-. These ionic substances are formed by transfer of one or more electrons from the metal to hydrogen atoms. Each hydrogen atom accepts one electron to form H^-. One hydride ion is required for each electron removed from the metal to form the rare-gas configuration. The first three compounds thus correspond to the compositions Na^+H^-; $Mg^{2+}2H^-$; $Al^{3+}3H^-$. The remaining compounds are best formulated as covalent, in which an electron pair is shared between the central atom and each hydrogen. Si, an element of group 4A, requires four electrons to attain the rare-gas configuration of eight valence-shell electrons. Phosphorus requires three, sulfur two, and chlorine one. The formulas of the hydrides are in accord with the number of electrons needed.

The sharing of a pair of electrons constitutes a single covalent bond. In many molecules, atoms attain complete octets by sharing more than one pair of electrons between them. When two electron pairs are shared, two lines are drawn, representing a double bond. A triple bond corresponds to the sharing of three pairs of electrons. Such multiple bonding is found, for example, in the N_2 molecule:

$$:\overset{\cdot}{N}\cdot \; + \; \cdot\overset{\cdot}{N}: \longrightarrow \; :N:::N: \quad (\text{or} \; :N{\equiv}N:)$$

Because each nitrogen atom possesses five electrons in its valence shell, the sharing of three electron pairs is required to achieve the octet configuration. The properties of N_2 are in complete accord with this Lewis structure. Nitrogen gas is a diatomic gas with exceptionally low reactivity. The low chemical reactivity results from the very stable nitrogen-nitrogen bond. Study of the structure of N_2 reveals that the nitrogen atoms are separated by only 1.10 Å. The short N—N bond distance is a result of the triple bond between the atoms. From structure studies of many different compounds in which nitrogen atoms share one or two electron pairs, it has been learned that the average distance between bonded nitrogen atoms varies with the number of shared electron pairs:

N—N	N=N	N≡N
1.47 Å	1.24 Å	1.10 Å

As a general rule, the distance between bonded atoms decreases as the number of shared electron pairs increases.

Carbon dioxide, CO_2, provides a further example of a molecule containing multiple bonds:

$$:\overset{\cdot}{\underset{\cdot}{O}}: \; + \; \cdot\overset{\cdot}{C}\cdot \; + \; :\overset{\cdot}{\underset{\cdot}{O}}: \longrightarrow \overset{\cdot\cdot}{\underset{\cdot\cdot}{O}}::C::\overset{\cdot\cdot}{\underset{\cdot\cdot}{O}} \quad (\text{or} \; \overset{\cdot\cdot}{\underset{\cdot\cdot}{O}}{=}C{=}\overset{\cdot\cdot}{\underset{\cdot\cdot}{O}})$$

7.5 Drawing Lewis structures

We have seen in the foregoing section some simple examples of Lewis structures. It is actually fairly easy to draw the Lewis structures for most compounds and ions formed from nonmetallic elements. Doing so is a key skill that is important in mastering the material in this chapter and the next.

In drawing Lewis structures it is a good idea to follow a regular procedure:

1 Add up the number of valence-shell electrons. (Use the periodic table.) If the species involved is neutral, this amounts to just the sum of the valence electrons for all the atoms present. If the ion is negatively charged, add the number of charges; if it is positively charged, subtract the charge of the ion.

2 Write the symbols for the atoms involved so as to show which atoms are connected to which. This is often very easy; for example, in CH_4, we know that the hydrogens must be connected to carbon. But when the formula is H_3PO_3, or N_2O, or $H_4C_2O_2$, it isn't so immediately obvious which atoms are bonded to which. In these cases you must have more information before you can make an unambiguous choice. You may be asked, however, to work out various possibilities and attempt to judge which of these is the most likely.

3 Draw in a single bond between each pair of atoms to be connected by bonding. Then attempt to put unshared electron pairs on each atom to meet the requirements of the octet rule. If placing all the electrons in this manner uses up all the available electrons, the structure is complete. If, on the other hand, the octet rule cannot be satisfied for each atom in this manner, then double or triple bonds must be formed with a reduction in the number of unshared electron pairs.

These rules are illustrated in the following sample exercise.

SAMPLE EXERCISE 7.7

Draw Lewis structures for the following molecules: (a) PCl_3; (b) HCN; (c) ClO_3^-.

Solution: (a) Phosphorus (group 5A) has five valence electrons and each chlorine (group 7A) has seven. The total number of valence-shell electrons is therefore $5 + (3 \times 7) = 26$. There are various ways the atoms might be arranged. However, in binary (two element) compounds such as PCl_3, the first element listed in the chemical formula is generally surrounded by the remaining atoms. Thus we begin with a skeleton structure that shows single bonds between phosphorus and each chlorine:

Cl—P—Cl
   |
  Cl

(It is not important that we place the atoms in exactly this arrangement; Lewis structures are not drawn to show geometry. However, it is important to show correctly which atoms are bonded to which.) The three single bonds account for six electrons. The remaining 20 electrons are then placed to give an octet of electrons around each atom:

$:\ddot{C}l—\ddot{P}—\ddot{C}l:$
     |
   $:\ddot{C}l:$

(Remember that the bonding electrons are counted for both atoms.)

(b) Hydrogen has one electron, carbon (group 4A) has four, and nitrogen (group 5A) has five. The total number of valence-shell electrons is therefore $1 + 4 + 5 = 10$. Again there are various ways we might choose to arrange the atoms. Because hydrogen can accommodate only one electron pair, it always has only one single bond associated with it in any compound. This fact causes us to reject C—H—N as a possible arrangement. The remaining two possibilities are H—C—N and H—N—C. The first is the arrangement found experimentally. You should have guessed this to be the atomic arrangement because the formula is written with the atoms in this order. Thus we begin with a skeleton structure that shows single bonds between hydrogen, carbon, and nitrogen:

H—C—N

185

These two bonds account for four electrons. If we then place the remaining six electrons around carbon and nitrogen, we cannot achieve an octet:

$$H-\overset{..}{\underset{..}{C}}-\overset{..}{N}$$

We next try a double bond between C and N. Again we are unable to achieve an octet around these atoms. Finally we try a triple bond. The single bond between H and C and the triple bond between C and N together account for eight electrons. The remaining two electrons can be placed on nitrogen to give an octet around both C and N atoms:

$$H-C\equiv N:$$

(c) Chlorine has seven valence electrons, and oxygen (group 6A) has six. An extra electron is added to account for the fact that the ion has a 1− charge. The total number of valence-shell electrons is therefore $7 + (3 \times 6) + 1 = 26$. After putting in the single bonds and distributing the unshared electron pairs we have:

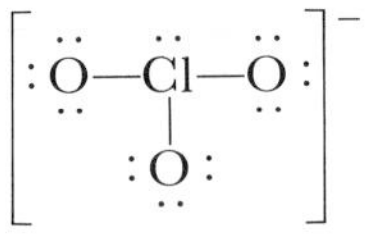

$$\left[:\overset{..}{\underset{..}{O}}-\overset{..}{\underset{\underset{:\overset{..}{\underset{..}{O}}:}{|}}{Cl}}-\overset{..}{\underset{..}{O}}:\right]^{-}$$

(For oxyanions such as ClO_3^-, SO_4^{2-}, NO_3^-, CO_3^{2-}, and so forth, the oxygen atoms surround the central nonmetal atom.)

7.6 Resonance forms

We sometimes encounter substances in which the known arrangement of atoms is not adequately described by a single Lewis structure. The structural chemistry of nonmetallic elements affords several examples. Consider ozone, O_3, about which we shall have much to say in Chapter 10. This fascinating substance consists of bent molecules with both O—O distances the same:

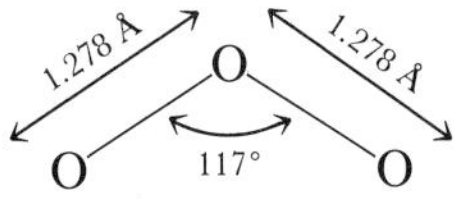

Because each oxygen atom contributes 6 valence-shell electrons, the ozone molecule has 18 valence-shell electrons. In writing the Lewis structure, we find that we must have one double bond to attain an octet of electrons about each atom:

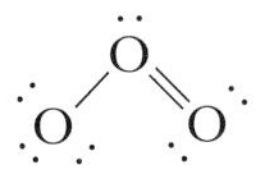

But this structure cannot by itself be correct, because it requires that one O—O bond be different from the other, contrary to the observed structure. However, in drawing the Lewis structure we could just as easily have put the O=O bond on the left:

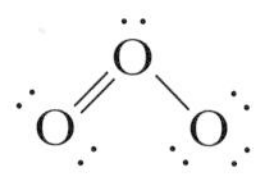

The two alternative Lewis structures for ozone are equivalent except for the placement of electrons. Equivalent Lewis structures of this sort are called resonance forms. To properly describe the structure of ozone, we write both Lewis structures and indicate that the real molecule is described by an average of the structures suggested by the two resonance forms:

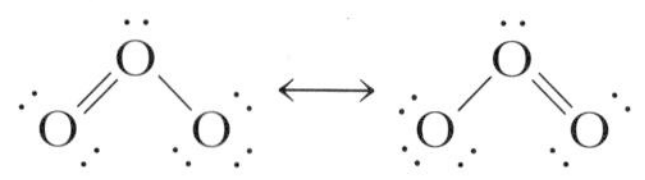

The double-headed arrow is used to indicate that the structures shown are resonance forms.

The fact that we must write more than one Lewis structure to describe a molecule or ion does not imply anything especially different about these species. You must not suppose that the molecule really exists in two or more different forms and oscillates rapidly between them. There is only one form of the molecule, that which is observed experimentally. The fact that we need to write two or more different resonance forms is simply a limitation of the use of Lewis structures in describing the electron distributions in molecules.

One rule that must be followed in writing resonance forms is that the arrangement of the nuclei must be the same in each structure. That is, the same atoms must be bonded to one another in all the structures, so that the only differences are in the arrangements of electrons.

As an additional example of resonance forms, let us draw the Lewis structure for the nitrate ion, NO_3^-, one of the most commonly encountered anions. We find that three equivalent Lewis structures are required in this instance:

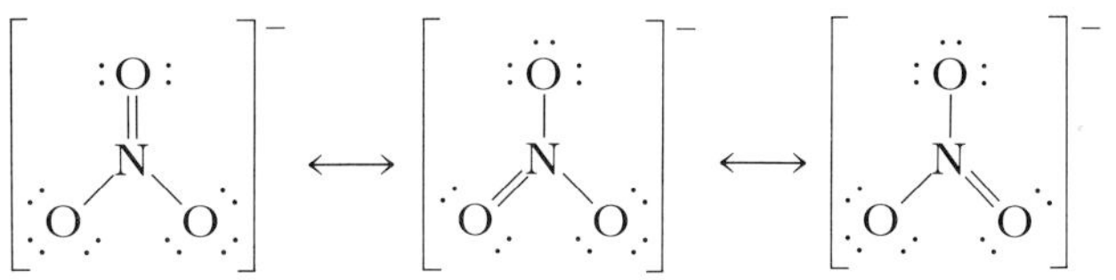

Note that the arrangement of nuclei is the same in each structure; only the placement of electrons differs. All three Lewis structures taken together adequately describe the nitrate ion, which is observed to be planar (all atoms in the same plane), with all three N—O distances equal.

In the examples of resonance structures we have seen so far, all the resonance forms have the same importance in contributing to the overall description of the molecule or ion. In the example in Sample Exercise 7.8, the contributing resonance structures are not equally important.

SAMPLE EXERCISE 7.8

The molecule chlorine dioxide, ClO_2, is bent, with a central chlorine atom bound to two oxygen atoms. The two Cl—O distances are observed to be equal. Describe the ClO_2 molecule in terms of three possible resonance forms.

Solution: The chlorine atom has 7 valence-shell electrons, and oxygen has 6. We therefore have a total of 19 electrons ($2 \times 6 + 7$) to place in this molecule. Note that ClO_2 has an odd number of electrons. This is an unusual situation; it means that one electron must remain unpaired.

We draw the arrangement of atoms in accord with the experimental facts and put one single bond between chlorine and each oxygen:

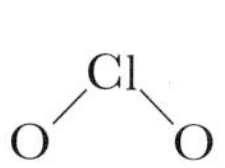

This leaves us with 15 electrons to place. We put electron pairs on the atoms to achieve an octet on each atom, or as close to it as we can get. It turns out that the odd electron must go on one or the other of the atoms, so that three possible structures result:

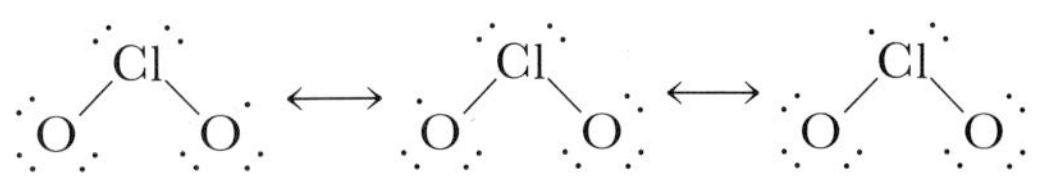

Two of these are equivalent, but the other is different, in that the odd electron is on chlorine rather than oxygen. To find out how much weight to attach to this third resonance structure in comparison with the other two, we would have to have experimental information on how the odd electron is distributed in the real ClO_2 molecule.

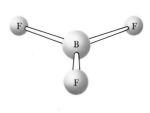

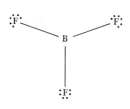

FIGURE 7.5 Geometrical structure and Lewis structure for boron trifluoride, BF_3.

In the Lewis theory, attention is focused on attainment of an octet of electrons about each atom. There are many molecules and ions, however, in which the octet rule is not obeyed. These **exceptions** are of three types.

One fairly small group consists of molecules in which there is an odd number of electrons. In the vast majority of molecules the number of electrons is even, and complete pairing of spins occurs. But in molecules such as ClO_2, NO, and NO_2, the number of electrons is odd. For example, NO contains $7 + 6 = 13$ valence electrons. Obviously, under these conditions complete pairing of all electrons is impossible, and an octet around each atom cannot be achieved.

A second possible failure to obey the octet rule comes when there are fewer than eight electrons about an atom. This is also a relatively rare situation. One example is boron trifluoride, BF_3, a planar molecule with three fluorine atoms around each boron, as shown in Figure 7.5. The properties of BF_3 appear to be best described in terms of single bonds between B and F. The Lewis structure, also shown in the figure, indicates that there are only six electrons about the boron. One of the valence-shell orbitals on the boron must therefore be vacant. The chemical behavior of this molecule reflects this vacancy. It is attracted to other molecules in which there is an atom with an unshared pair of electrons that can be donated to the boron. For example, it reacts with ammonia, NH_3, to form the compound BF_3NH_3:

$$\begin{array}{ccc} H & F & \\ H-N: \ + \ B-F & \longrightarrow & H-N-B-F \\ H & F & \end{array}$$

A bond such as the one between N and B, in which one of the two atoms involved has been the source of the electron pair, is often called a **coordinate covalent**, or **dative, bond**. We should remember that electrons do not "belong" to certain atoms. However, in accounting for the overall placement of electrons we can see that both electrons in the B—N bond can be associated originally with the nitrogen. We shall again consider this type of reaction in discussing acid-base chemistry in Chapter 15.

A third and most numerous class of substances in which the octet rule is not obeyed is that in which there are *more* than eight electrons in the valence shell about an atom. As an example, consider PCl_5. This compound may be made by reaction of PCl_3 with chlorine gas:

$$PCl_3(g) + Cl_2(g) \longrightarrow PCl_5(g)$$

When we draw the Lewis structure for this molecule we are forced to place ten electrons about the central phosphorus atom:

$$\begin{array}{c} Cl \ \ Cl \\ | \diagup \\ Cl-P \\ | \diagdown \\ Cl \ \ Cl \end{array}$$

Among the many other examples of molecules and ions with "expanded" valence shells are PF_5, AsF_6^-, SF_4, and ClF_3. We shall see in Chapter 8 how such structures can be accounted for by the use of additional atomic orbitals. Even without referring to these considerations, however, it is possible to see some logical pattern in their occurrences. We note that:

1 The occurrences of expanded valence shells increase with increasing size of the central atom. There are no instances of such molecules among nonmetals of the second row. For example, the NF_3 molecule is a stable species, but NF_5 is unknown. On the other hand, both PF_3 and PF_5 are well-characterized molecules.

2 Expanded valence shells occur most often when the central atom is bonded to the smallest and most strongly electron-attracting atoms, such as F, Cl, and O.

These observations suggest that expanded valence shells may occur when size considerations allow more atoms to crowd about a central atom than are required by the octet rule. For example, consider the pentahalides of phosphorus. Phosphorus pentafluoride, PF_5, is a very stable species. Phosphorus pentachloride, PCl_5, is reasonably stable, but dissociates in the vapor phase at 300°C into PCl_3 and Cl_2. Phosphorus pentabromide, PBr_5, dissociates in the vapor state even more readily; the compound PI_5 is not known. We see from these observations that the stability of the pentahalide decreases as the size of the halogen atom increases and as its electron attraction decreases.

7.8 Strength of covalent bonds

The stability of a molecule can be related to the strengths of the covalent bonds it contains. The strength of a covalent bond between two atoms is determined by the energy required to break that bond. The bond-dissociation energy is the enthalpy change, ΔH, required to break a particular bond in a mole of gaseous substance. For example, the dissociation energy for the bond between chlorine atoms in the Cl_2 molecule is the energy required to dissociate a mole of Cl_2 into chlorine atoms:

$$:\ddot{C}l\!-\!\ddot{C}l\!: (g) \longrightarrow 2 :\ddot{C}l \cdot (g) \qquad \Delta H = D(\text{Cl}\!-\!\text{Cl}) = 242 \text{ kJ} $$

We have used the designation D(bond type) in this equation and elsewhere to represent bond-dissociation energies.

It is a relatively easy matter to assign bond energies to bonds in diatomic molecules. As we have seen in the example above, the bond energy is just the energy required to break the diatomic molecule into its component atoms. However, for bonds that occur only in polyatomic molecules (such as the C—H bond) we must often utilize average bond energies. For example, the enthalpy change for the process shown below (called "atomization") can be used to define an average bond strength for the C—H bond:

$$ \text{H}\!-\!\overset{\displaystyle \text{H}}{\underset{\displaystyle \text{H}}{\text{C}}}\!-\!\text{H}(g) \longrightarrow \cdot \ddot{\text{C}} \cdot (g) + 4\text{H} \cdot (g) \qquad \Delta H = 1660 \text{ kJ} $$

Since there are four equivalent C—H bonds in methane, the heat of atomization is equal to the total bond energies of the four C—H bonds. Thus the average C—H bond energy, as determined from the reaction above, is $D(\text{C}\!-\!\text{H}) = (1660/4) \text{ kJ/mol} = 415 \text{ kJ/mol}$.

The average C—H bond energy is not the same as the energy required for the process $CH_4(g) \longrightarrow H(g) + CH_3(g)$; that process requires 435 kJ. Use of that bond-dissociation energy in thermochemical calculations will naturally give different results than the use of the average bond energy. The successive bond-dissociation energies for CH_4 are as follows:

$$CH_4 \longrightarrow CH_3 + H \qquad D(CH_3—H) = 435 \text{ kJ}$$
$$CH_3 \longrightarrow CH_2 + H \qquad D(CH_2—H) = 464 \text{ kJ}$$
$$CH_2 \longrightarrow CH + H \qquad D(CH—H) = 422 \text{ kJ}$$
$$CH \longrightarrow C + H \qquad D(C—H) = 339 \text{ kJ}$$

The sum of these energies is 1660 kJ, which yields the average of 415 kJ cited above.

The bond energy for a given set of atoms, say C—H, depends on the remainder of the molecule of which it is a part. However, the variation from one molecule to another is generally small. This supports the idea that the bonding electron pairs are localized between atoms. If we consider C—H bond strengths in many other molecules, we find the average strength is 413 kJ/mol, which compares closely with the 415 kJ/mol value calculated from CH_4.

Table 7.3 lists several average bond energies. Notice that the bond energy is always a positive quantity; energy is always required to break chemical bonds. Conversely, energy is given off when a bond forms between two gaseous atoms or molecular fragments. Of course, the greater the bond energy, the stronger the bond.

Referring to Table 7.3, compare the bond energies for the C—C, C=C, and C≡C bonds. Notice that the bond strength increases as the number of pairs of electrons shared between the atoms increases. We have noted previously, in Section 7.4, that the distance between atoms decreases as we move from a single to a double to a triple bond. For

TABLE 7.3 Average bond energies (kJ/mol)

Single bonds

C—H	413	N—H	391	O—H	463	F—F	155
C—C	348	N—N	163	O—O	146		
C—N	293	N—O	201	O—F	190	Cl—F	253
C—O	358	N—F	272	O—Cl	203	Cl—Cl	242
C—F	485	N—Cl	200	O—I	234		
C—Cl	328	N—Br	243			Br—F	237
C—Br	276			S—H	339	Br—Cl	218
C—I	240	H—H	436	S—F	327	Br—Br	193
C—S	259	H—F	567	S—Cl	253		
		H—Cl	431	S—Br	218	I—Cl	208
Si—H	323	H—Br	366	S—S	266	I—Br	175
Si—Si	226	H—I	299			I—I	151
Si—C	301						
Si—O	368						

Multiple bonds

C=C	614	N=N	418	O_2	495
C≡C	839	N≡N	941		
C=N	615				
C≡N	891			S=O	323
C=O	799			S=S	418
C≡O	1072				

carbon-carbon bonds we have the following average bond lengths and bond strengths:

C—C	C=C	C≡C
1.54 Å	1.34 Å	1.20 Å
348 kJ/mol	614 kJ/mol	839 kJ/mol

In general, longer bonds have lower bond energies. Consequently, stronger bonds are generally associated with smaller atoms.

A molecule with strong chemical bonds generally has less tendency to undergo chemical change than does one with weak bonds. This relationship between strong bonding and chemical stability helps explain the chemical form in which many elements are found in nature. For example, Si—O bonds are among the strongest ones that silicon forms. It is not surprising therefore that SiO_2 and other substances containing Si—O bonds (silicates) are so common; it is estimated that over 90 percent of the earth's crust is composed of SiO_2 and silicates. We will have more to say about these compounds in Chapter 22.

BOND ENERGIES AND CHEMICAL REACTIONS

A knowledge of bond energies is helpful in understanding why some reactions are exothermic (negative ΔH) while others are endothermic (positive ΔH). An exothermic reaction results when the bonds in the product molecules are stronger than those in the reactants. Consider the following reaction:

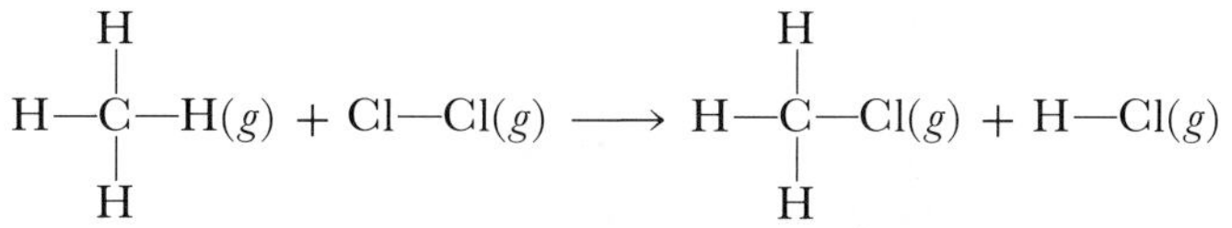

$$\text{H—C—H}(g) + \text{Cl—Cl}(g) \longrightarrow \text{H—C—Cl}(g) + \text{H—Cl}(g)$$

In the course of this reaction, 1 mol of C—H bonds and 1 mol of Cl—Cl bonds must be broken (reactants). Using average bond energies, we can estimate that this requires 655 kJ:

$$\Delta H = D(\text{C—H}) + D(\text{Cl—Cl}) = (413 + 242)\,\text{kJ} = 655\,\text{kJ}$$

However, 1 mol of C—Cl bonds and 1 mol of H—Cl bonds are formed in the products; this produces 759 kJ:

$$\Delta H = -D(\text{C—Cl}) - D(\text{H—Cl}) = (-328 - 431)\,\text{kJ} = -759\,\text{kJ}$$

The overall energy change for the reaction is the difference between the energy required to break the pertinent bonds in the reactants and the energy given up in forming the new bonds in the products:

$$\Delta H = \Sigma D(\text{bonds broken}) - \Sigma D(\text{bonds formed}) \qquad [7.3]$$
$$= 655\,\text{kJ} - 749\,\text{kJ} = -104\,\text{kJ}$$

(where Σ is read "sum of"). The reaction is exothermic because the bonds in the products (especially the H—Cl bond) are stronger than those in the reactants (especially the Cl—Cl bond).

The example above illustrates how bond energies can be used to estimate heats of reactions using Equation [7.3]. In practice, we seldom calculate ΔH in this way, because it can be determined more accurately from heats of formation (Section 4.5). However, we might use bond energies to estimate ΔH for a reaction if the heat of formation of some reactant or product is not known. Whenever one uses bond energies in this fashion, it is important to remember that they refer to gaseous molecules and that they are often just averaged values.

SAMPLE EXERCISE 7.9

Using Table 7.3, estimate ΔH for the following reaction (where we show explicitly the bonds involved in the reactants and products):

$$\text{H}-\overset{\displaystyle \overset{\text{H}}{|}}{\underset{\displaystyle \underset{\text{H}}{|}}{\text{C}}}-\overset{\displaystyle \overset{\text{H}}{|}}{\underset{\displaystyle \underset{\text{H}}{|}}{\text{C}}}-\text{H}(g) + \tfrac{7}{2}\text{O}_2(g) \longrightarrow$$

$$2\text{O}{=}\text{C}{=}\text{O}(g) + 3\text{H}-\text{O}-\text{H}(g)$$

$$\text{C}_2\text{H}_6(g) + \tfrac{7}{2}\text{O}_2(g) \longrightarrow 2\text{CO}_2(g) + 3\text{H}_2\text{O}(g)$$

Solution: Among the reactants, we must break six C—H bonds and a C—C bond in C_2H_6; we also break $\tfrac{7}{2}\text{O}_2$ bonds. Among the products, we form four C=O bonds (two in each CO_2) and six O—H bonds (two in each H_2O). Using Equation [7.3] and data from Table 7.3 we have

$$
\begin{aligned}
\Delta H &= 6D(\text{C}-\text{H}) + D(\text{C}-\text{C}) + \tfrac{7}{2}D(\text{O}_2) \\
&\quad - 4D(\text{C}{=}\text{O}) - 6D(\text{O}-\text{H}) \\
&= 6(413\ \text{kJ}) + 348\ \text{kJ} + \tfrac{7}{2}(495\ \text{kJ}) - 4(804\ \text{kJ}) \\
&\quad - 6(463\ \text{kJ}) \\
&= 4569\ \text{kJ} - 5994\ \text{kJ} = -1415\ \text{kJ}
\end{aligned}
$$

This estimate can be compared with the value of -1428 kJ calculated from more precise thermochemical data: the agreement is excellent.

7.9   Bond polarity; electronegativities

The electron pairs shared between two different atoms are not necessarily shared equally. We can visualize two extreme cases in the degree to which electron pairs are shared. On the one hand, we have bonding between two identical atoms, as in Cl_2 or N_2, where the electron pairs must be equally shared. At the other extreme, illustrated by NaCl, there will be essentially no sharing of electrons. We know that in this case the compound is best described as composed of Na^+ and Cl^- ions. The $3s$ electron of the Na atom is, in effect, transferred completely to chlorine. The bonds occurring in most covalent substances fall somewhere between these extremes. In practice, then, we describe bonds as either ionic or covalent depending on which extreme the bond most closely resembles.

The concept of bond polarity is useful in describing the sharing of electrons between atoms. A nonpolar bond is one in which the electrons are shared equally between two atoms. In a polar covalent bond, one of the atoms exerts a greater attraction for the electrons than the other. If the difference in relative abilities to attract electrons is large enough an ionic bond is formed.

The tendency of an atom to attract electrons to itself in a chemical bond is referred to as electronegativity. We can't readily measure some single property that would give us a value for the electronegativity of an element. Instead, a numerical estimate of electronegativity must be obtained from observations of several different properties.

On theoretical grounds, the best measure of electronegativity for an atom is an average of its ionization energy (Section 6.5) and electron affinity (Section 6.6). Atoms that show high ionization energies also show strong attraction for electrons in bonds. The attraction is even greater when the atom also shows a relatively large electron affinity. The electronegativities of metallic elements are low, because metals have low ionization energies and low electron affinities. In contrast, nonmetals possess high electronegativities as a result of their high ionization energies and relatively high electron affinities.

We noted in Section 6.6 that values of electron affinity are not available for all elements. It is therefore necessary to employ other data to obtain estimates of electronegativities. Figure 7.6 shows electronegativity values for many of the elements. The numbers listed in this table are all based on a particular value chosen for one of the elements. The value of 2.5 for carbon is the most commonly accepted reference; this was the value chosen for carbon by Linus Pauling (1901–), who first developed the concept of electronegativity. The actual value chosen for the reference is not so important; we are interested mainly in the relative values of electronegativity in comparing two elements. Notice that the most electronegative element is fluorine, with an electronegativity of 4.0. The least electronegative element, cesium, has an electronegativity of 0.79. All the other elements have electronegativities that lie between these extremes. The values listed for the transition metals are those for the 2+ state. When the element is in a higher charge state, its electronegativity is higher.

Note that in a horizontal row of the table there is a more or less steady increase in electronegativity in moving from left to right, that is, from the most metallic to the most nonmetallic elements. Notice also that, with a few exceptions, there is an overall decrease in electronegativity with increasing atomic number in any one group of the periodic table. This is what we might expect, because we know that ionization potentials tend to decrease with increasing atomic number in a group, and electron affinities don't change very much. You do not need to memorize numerical values for electronegativity. However, you should know the

1A	2A	3B	4B	5B	6B	7B	8B	8B	8B	1B	2B	3A	4A	5A	6A	7A
H 2.2																
Li 1.0	Be 1.6											B 1.8	C 2.5	N 3.0	O 3.4	F 4.0
Na 0.93	Mg 1.3											Al 1.6	Si 1.9	P 2.2	S 2.6	Cl 3.2
K 0.82	Ca 1.0	Sc 1.4	Ti 1.5	V 1.6	Cr 1.7	Mn 1.6	Fe 1.8	Co 1.9	Ni 1.9	Cu 2.0	Zn 1.6	Ga 1.8	Ge 2.0	As 2.2	Se 2.6	Br 3.0
Rb 0.82	Sr 0.9	Y 1.2	Zr 1.3	Nb 1.6	Mo 2.2	Tc —	Ru 2.2	Rh 2.3	Pd 2.2	Ag 1.9	Cd 1.7	In 1.8	Sn 1.8	Sb 2.0	Te 2.1	I 2.7
Cs 0.79	Ba 0.9								Pt 2.3	Au 2.5	Hg 2.0	Tl 2.0	Pb 2.3	Bi 2.0	Po —	

FIGURE 7.6 Electronegativities of the elements. The values for the transition metals are those for the +2 valence state.

periodic trends so you can predict which of two elements is more electronegative.

Keep in mind that **electronegativities are *approximate* measures of the *relative* tendencies of these elements to attract electrons to themselves in a chemical bond.** The electronegativity varies with the type of chemical environment in which an element is situated. We have noted, for example, that the electronegativities of transition metals vary with their valence. Similarly, the electronegativity of chlorine in its bonding to phosphorus in PCl_3 is likely to be different from its value in the chlorate ion, ClO_3^-, in which it is in a much different bonding situation. Variations of this sort are not so large as to render the concept of electronegativity useless, but we must avoid placing too much reliance on precise values for electronegativities. So long as we remain aware of their limitations, electronegativity values provide a useful guide to polarities in chemical bonds.

The electronegativity difference between two atoms is a measure of the polarity of the bond between them; the greater the difference in electronegativity, the more polar the bond. The shared electron pair is found a greater fraction of the time on the more electronegative of the two atoms. For example, the electronegativity difference between H and F is $4.0 - 2.2 = 1.8$. Consequently, the sharing of electrons is unequal (the bond is polar), with the more electronegative fluorine attracting electron density off the hydrogen. We represent this situation in the following two ways:

$$\overset{\delta+ \quad \delta-}{H\!-\!F} \qquad \text{or} \qquad \overset{\longrightarrow}{H\!-\!F}$$

The $\delta+$ and $\delta-$ are meant to represent partial positive and negative charges, respectively. (The symbol δ is the Greek letter delta.) The arrow represents the pull of electron density off the hydrogen by the fluorine, leaving the hydrogen with a partial positive charge; the head of the arrow points in the direction in which the electrons are attracted, toward the more electronegative atom. We will consider the molecular consequences of bond polarities in the next chapter, after we have discussed molecular shapes.

SAMPLE EXERCISE 7.10

Which of the following bonds is more polar: (a) B—Cl or C—Cl; (b) P—F or P—Cl? Indicate in each case which atom has the partial negative charge.

Solution: (a) The difference in the electronegativities of boron and chlorine is $3.2 - 1.8 = 1.4$; the difference between carbon and chlorine is $3.2 - 2.5 = 0.7$. Consequently, the B—Cl bond is the more polar; the chlorine atom carries the partial negative charge because it has a higher electronegativity. We should be able to reach this same conclusion without using a table of electronegativities; instead, we can rely on periodic trends. Because boron is to the left of carbon in the periodic table, we would predict that it has a lower attraction for electrons. Chlorine, being on the right side of the table, has a strong attraction for electrons. The most polar bond will be the one between the atoms having the lowest attraction for electrons (boron) and the highest attraction (chlorine).

(b) Because fluorine is above chlorine in the periodic table, we would predict it to be more electronegative. Consequently, the P—F bond will be more polar than the P—Cl bond. You should compare the electronegativity differences for the two bonds to verify this prediction.

In light of our discussion of polar covalent bonds, it is useful to examine the similarities between reactions in which electrons are completely transferred and those in which only partial shifts of electron density occur. The reaction between sodium and chlorine atoms to form NaCl involves transfer of an electron from sodium to chlorine. We can imagine the overall reaction as the result of two separate processes:

$$\text{Na} \cdot \longrightarrow \text{Na}^+ + \text{e}^- \qquad\qquad [7.4]$$

$$:\overset{..}{\underset{..}{\text{Cl}}} \cdot + \text{e}^- \longrightarrow [:\overset{..}{\underset{..}{\text{Cl}}}:]^- \qquad\qquad [7.5]$$

Each of these processes is called a half-reaction. Adding these two half-reactions together gives the overall formation of the ionic species from the neutral atoms:

$$\text{Na} + \text{Cl} \longrightarrow \text{Na}^+ + \text{Cl}^-$$

Processes such as the one shown in Equation [7.4], which involve loss of electrons, are called oxidations. Processes such as the one shown in Equation [7.5], which involve gain of electrons, are called reductions. A substance that has lost electrons is said to be oxidized; one that gains electrons is said to be reduced.

The extent of oxidation-reduction, if any, is not so clear in the reaction between hydrogen atoms and chlorine atoms. The product in this case, HCl, is a gaseous substance whose normal boiling point is $-84°C$. Because ionic compounds are solids at room temperature, the bonding in HCl is best described as polar covalent. Because chlorine is more electronegative than hydrogen, the electrons in the H—Cl bond are displaced toward chlorine:

$$\text{H} \cdot + \cdot \overset{..}{\underset{..}{\text{Cl}}}: \longrightarrow \text{H}\overset{\longrightarrow}{:\overset{..}{\underset{..}{\text{Cl}}}:}$$

The H atom in HCl therefore carries a somewhat positive charge and the Cl atom a somewhat negative one. So far as keeping track of electrons is concerned, it would seem a reasonable simplification to assign the shared pair of electrons completely to the chlorine atom:

$$\text{H}\left[:\overset{..}{\underset{..}{\text{Cl}}}:\right.$$

Chlorine would then have eight valence-shell electrons, one more than the neutral chlorine atom. By assigning electrons in this way, we have in effect given a $1-$ charge to the chlorine. Hydrogen, stripped of its electron, is assigned a charge of $1+$.

Charges assigned in this fashion are called oxidation numbers or oxidation states. The oxidation number of an atom may vary from compound to compound. However, in each case it is the charge that results when the electrons in a covalent bond are assigned to the more electronegative atom; it is the charge that an atom would possess if the bonding were ionic. When electrons are shared between like atoms as in H_2, they are divided equally between those atoms. Thus in H_2 each hydrogen atom is assigned one electron; the oxidation number of hydrogen in H_2 is therefore zero. For monoatomic ions, such as Fe^{2+}, the oxidation number is just the charge of the ion ($2+$ in this example).

Although it is possible to determine oxidation numbers using Lewis structures as shown above for HCl and H$_2$, we seldom do this in practice. It is generally easier to determine oxidation numbers using the following set of rules:

1 The oxidation number of a substance in the elemental state is zero. Thus, for Cl$_2$, N$_2$, Na, or P$_4$, the oxidation number of each atom is zero, because electrons involved in bonding must be shared equally between atoms.

2 In a compound, the more electronegative elements are assigned negative oxidation numbers; the less electronegative elements are assigned positive oxidation numbers. The magnitude of the oxidation number corresponds more or less to the valence, or number of shared electron-pair bonds. For example, hydrogen always has an oxidation number of 1. It is -1 when hydrogen is bonded to a less electronegative element, for example, in NaH; it is $+1$ when it is bonded to a more electronegative element. Thus, in HCl, hydrogen is assigned an oxidation number of $+1$, and chlorine an oxidation number of -1.

3 In any molecule or ion, the sum of the positive and negative oxidation numbers must equal the overall charge on the species. For example, in OF$_2$, fluorine is assigned an oxidation number of -1 because it is the more electronegative element. Oxygen must then have an oxidation number of $+2$. In AlO$_2^-$, if oxygen is assigned an oxidation number of -2, then Al must have an oxidation number of $+3$ in order that the overall charge come out -1.

The periodic table provides us with many guidelines to the assignment of oxidation numbers. As shown in Figure 7.7 there is a periodicity in oxidation numbers. All the alkali metals (group 1A) have oxidation numbers of $+1$ in their compounds. There is simply no other commonly observed means of chemical bonding to other elements for these metals other than loss of an electron to form the $+1$ ion. The elements of group 2A are always found in the $+2$ oxidation state. In group 3A, the most commonly encountered element, Al, is always found in the $+3$ oxidation state.

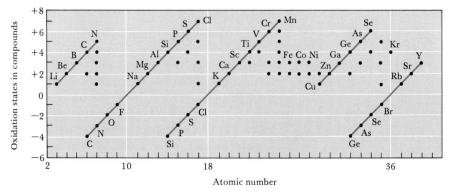

FIGURE 7.7 Common oxidation numbers for the first 39 elements in the periodic table. Notice that the maximum and minimum oxidation states (through which lines have been drawn for emphasis) are a periodic function of atomic number.

The most electronegative element, F, is always found in the -1 oxidation state. Other nonmetals are found in negative oxidation states unless combined with a more electronegative atom. Oxygen is nearly always in the -2 oxidation state. The only common exception to this general rule occurs in peroxides. In the peroxide ion, O_2^{2-}, and in molecular peroxides such as H_2O_2, oxygen has an oxidation state of -1. Oxygen is more electronegative than any element with which it forms oxyanions (NO_3^-, ClO_4^-, and so forth). Consequently, the central atom in these oxyanions always has a positive oxidation state.

SAMPLE EXERCISE 7.11

Predict the maximum and minimum oxidation states for sulfur.

Solution: Sulfur has six valence electrons. Assignment of all of these to a more electronegative atom leads to the maximum oxidation state for sulfur, $+6$.

When sulfur has a share of an octet of electrons and all are assigned to the sulfur, the resultant oxidation state is -2, the lowest that sulfur exhibits. As shown in Figure 7.7, sulfur has other common oxidation states between these extremes.

SAMPLE EXERCISE 7.12

Assign oxidation numbers to each element in the following compounds: (a) $PbCl_2$; (b) NO_3^-; (c) $KMnO_4$; (d) BaO_2.

Solution: (a) The most electronegative element in $PbCl_2$ is Cl; because only one electron can be added to complete the octet around Cl, its oxidation state is -1. Halogens always have an oxidation state of -1 in binary compounds (compounds containing only two different elements) with metals. The sum of the oxidation states of all of the atoms must equal zero (rule 3). Letting x equal the oxidation state of Pb we have $x + 2(-1) = 0$. Consequently, the oxidation state of Pb, x, must be $+2$.

(b) Oxygen has a common oxidation state of -2. In this case the sum of the oxidation numbers is -1,

the net charge on the NO_3^- ion. Letting x equal the oxidation state of N we have $x + 3(-2) = -1$. From this we conclude that the oxidation state of nitrogen is $+5$.

(c) The sum of the oxidation numbers of all the atoms is zero. Potassium, an alkali metal, has an oxidation state of $+1$. The oxidation state of O is -2. If we let x equal the oxidation state of Mn we have $1 + x + 4(-2) = 0$. Thus the oxidation state of Mn is $+7$.

(d) The sum of the oxidation states of all the atoms is zero. Ba, a group 2A metal, must have an oxidation state of $+2$. As a result, oxygen cannot have an oxidation state of -2. We conclude that this compound is a peroxide, oxygen having an oxidation state of -1: $+2 + 2(-1) = 0$.

It is important to keep in mind that the oxidation numbers, or oxidation states, do not correspond to real charges on the atoms, except in the special case of simple ionic substances. Nevertheless, they furnish a useful means of organizing chemical facts, especially in the case of metallic elements. Their most frequent use is in balancing chemical equations for reactions in which changes in oxidation numbers occur.

OXIDATION NUMBERS AND NOMENCLATURE

The rules for naming simple inorganic compounds were discussed in Chapter 2. This is a good point at which to review those rules. Now that the concept of oxidation number has been explained, we can add another widely used method for naming binary compounds. You learned in

Section 2.10 that the name of a binary compound consists of the name of the less electronegative element first, followed by the name of the more electronegative element, modified to have an *-ide* ending. The following examples are illustrative:

BaSe   barium selenide

ZnS   zinc sulfide

MgH_2 magnesium hydride

When one or the other of the elements involved has more than one possible oxidation state, the number of atoms may be included in the name as a Greek prefix (1 = mono-, 2 = di-, 3 = tri-, 4 = tetra-, 5 = penta-, 6 = hexa-, 7 = hepta-), *or* the oxidation state is indicated using a Roman numeral, as in these examples:

MnO_2	manganese dioxide	*or* manganese(IV) oxide
Mn_2O_3	dimanganese trioxide	*or* manganese(III) oxide
P_2O_5	diphosphorus pentoxide	*or* phosphorus(V) oxide
P_2O_3	diphosphorus trioxide	*or* phosphorus(III) oxide
$SnCl_4$	tin tetrachloride	*or* tin(IV) chloride
$SnCl_2$	tin dichloride	*or* tin(II) chloride

OXIDATION-REDUCTION REACTIONS

Many chemical reactions involve atoms that undergo changes in oxidation numbers. Atoms that lose electron density (increase in oxidation number) are said to be oxidized. Atoms that gain electron density (decrease in oxidation number) are said to be reduced. Oxidation and reduction are always discussed together because whenever one substance loses electrons another must gain them. We will take up the topic of balancing oxidation-reduction (or redox) equations in Chapter 19.

SAMPLE EXERCISE 7.13

Is the following reaction an oxidation-reduction reaction? If so, indicate which atom is oxidized and which is reduced:

$$Zn(s) + 2HCl(aq) \longrightarrow ZnCl_2(aq) + H_2(g)$$

Solution: To decide whether this is an oxidation-reduction reaction, we must determine whether any atom undergoes a change in oxidation state. Zinc starts with an oxidation state of 0 and ends up as +2 in $ZnCl_2$. Consequently, the reaction is an oxidation-reduction reaction in which zinc is oxidized. But what is reduced? Hydrogen starts with an oxidation state of +1 in HCl and ends up as zero in H_2. It has gained electrons and has therefore been reduced. The chlorine maintains an oxidation state of −1 throughout the reaction; it does not undergo either oxidation or reduction.

7.11 Binary oxides

With even such simple ideas about bonds as those discussed in this chapter, we can organize and understand many chemical facts. Before taking a closer look at covalent bonding, let's apply some of what we have learned to a discussion of the binary compounds of oxygen.

Oxygen is a very important element that is abundant in air and is found in numerous compounds in the environment. The chemistry of the element is dominated by its high ionization potential and strong attraction for electrons. The term "oxidation," which we have defined as an increase in oxidation number, was originally applied only to reactions of oxygen, O_2. Because of its high electronegativity, oxygen almost invariably gains electron density in its reactions. It thereby oxidizes (removes electron density) from the substance with which it reacts. For example, O_2 reacts with magnesium to produce magnesium oxide, MgO (Mg in a $+2$ oxidation state):

$$2Mg(s) + O_2(g) \longrightarrow 2MgO(s)$$

With the exception of the noble gases He, Ne, Ar, and Kr, oxygen forms compounds with every element. Many elements form more than one oxide. For example, iron forms FeO, Fe_2O_3 and Fe_3O_4, while chlorine forms Cl_2O, Cl_2O_3, ClO_2, Cl_2O_6 and Cl_2O_7. The oxides of the metals are generally ionic substances, and those of the nonmetals are generally covalent.

IONIC OXIDES

Because of their low ionization energies, metals generally combine with oxygen to form ionic solids in which oxygen is present as the oxide ion. Because the oxide ion, O^{2-}, has a high charge and is relatively small, metal oxides tend to have high lattice energies. Consequently, the melting points of these ionic oxides are typically high (around $2000\,°C$), and most are rather insoluble in water.

When a metal oxide does dissolve to any extent in water, the aqueous O^{2-} ion reacts to form hydroxide ions:

$$:\overset{..}{\underset{..}{O}}:^{2-} + H:\overset{..}{\underset{..}{O}}:H \longrightarrow :\overset{..}{\underset{..}{O}}:H^- + :\overset{..}{\underset{..}{O}}:H^-$$

This reaction involves the transfer of H^+ (hydrogen without electrons) from H_2O to the O^{2-} ion.

The solubility of ionic oxides in water is strongly dependent on the size and charge of the associated cation and is greatest for large cations carrying a low charge. The oxides of the alkali metals, such as Na_2O, are quite soluble, as are the oxides of the heavier alkaline earths (especially Ba). They therefore react readily with water to form hydroxides as illustrated by the following examples:

$$Na_2O(s) + H_2O(l) \longrightarrow 2NaOH(aq)$$
$$BaO(s) + H_2O(l) \longrightarrow Ba(OH)_2(aq)$$

As we move from left to right across a period in the periodic table, the solubilities of the metal oxides generally decrease. This trend corresponds to the increasing charge and decreasing size of the metal ions. As we move down a family, the solubilities generally increase; this corresponds to the increasing size of the cation.

Many metal oxides that are water insoluble dissolve in acid solutions owing to the chemical reaction that occurs between O^{2-} ions and H^+

ions to form H_2O. For example, iron(III) oxide, Fe_2O_3, is insoluble in water but dissolves in acids to form soluble salts:

$$Fe_2O_3(s) + 6HCl(aq) \longrightarrow 2FeCl_3(aq) + 3H_2O(l)$$

Metal oxides that dissolve in water to give OH^--containing solutions as well as those that dissolve in acid solutions are called basic oxides.

COVALENT OXIDES

Nonmetals have sufficiently high electronegativities to keep oxygen from removing electrons to form O^{2-} ions. Instead, these elements form molecules containing polar covalent bonds with oxygen. These molecular substances are typically gases, liquids, or solids with low melting points. A notable exception is silicon dioxide, SiO_2, which we shall consider in more detail in Section 11.7. This substance, also known as silica, is relatively hard and has a high melting point (approximately $1600°C$).

Whereas metals tend to form basic oxides, nonmetals tend to form acidic ones. Acidic oxides are ones that dissolve in water to form acidic solutions or dissolve in solutions of bases like NaOH. For example, chlorine(VII) oxide, Cl_2O_7, dissolves in water, reacting to form perchloric acid, $HClO_4$:

$$Cl_2O_7(l) + H_2O(l) \longrightarrow 2HClO_4(aq)$$

Carbon dioxide reacts in a similar way to form an acidic (H^+-containing) solution:

$$CO_2(g) + H_2O(l) \longrightarrow H_2CO_3(aq)$$

This accounts for the acidity of carbonated water. Some nonmetal oxides, CO_2 included, dissolve more readily in aqueous solutions of NaOH than they do in water. Carbon dioxide dissolves in such basic solutions to form hydrogen carbonate ions, HCO_3^-, and carbonate ions, CO_3^{2-}:

$$CO_2(g) + NaOH(aq) \longrightarrow NaHCO_3(aq)$$
$$CO_2(g) + 2NaOH(aq) \longrightarrow Na_2CO_3(aq) + H_2O(l)$$

As the concentration of OH^- increases, less HCO_3^- and more CO_3^{2-} forms. Notice that there is no oxidation-reduction occurring in these reactions. Carbon has the oxidation state of $+4$ in CO_2, HCO_3^-, and CO_3^{2-}.

AMPHOTERIC OXIDES

Certain oxides that are virtually insoluble in water are soluble in both acids and bases. Such oxides are said to be amphoteric. As shown in Figure 7.8, most of the elements that form amphoteric oxides lie near the diagonal line in the periodic table that divides metals from nonmetals. For example, aluminum oxide, Al_2O_3, is amphoteric and reacts with both acidic and basic solutions:

	1A	2A	3A	4A	5A	6A	7A
	Li_2O	BeO	B_2O_3	CO_2	N_2O_5		F_2O
	Na_2O	MgO	Al_2O_3	SiO_2	P_2O_5	SO_3	Cl_2O_7
	K_2O	CaO	Ga_2O_3	GeO_2	As_2O_5	SeO_3	Br_2O_7
	Rb_2O	SrO	In_2O_3	SnO_2	Sb_2O_5	TeO_3	I_2O_7
	Cs_2O	BaO	Tl_2O_3	PbO_2	Bi_2O_5	PoO_3	At_2O_7

(Left axis label: Increasing base character ↓)

FIGURE 7.8 The simplest formulas of the oxides of the representative elements in their maximum oxidation states. Those that are basic are shown without shading, those that are amphoteric with light shading, and those that are acidic with dark shading.

$$Al_2O_3(s) + 6HCl(aq) \longrightarrow 2AlCl_3(aq) + 3H_2O(l)$$

$$Al_2O_3(s) + 6NaOH(aq) \longrightarrow 2Na_3AlO_3(aq) + 3H_2O(l)$$

Chromium(III), which has the same charge as the aluminum ion and has a virtually identical ionic radius, also forms an amphoteric oxide, Cr_2O_3.

SAMPLE EXERCISE 7.14

Write a balanced chemical equation for the reaction that occurs when (a) CaO(s) is added to water; (b) $SO_2(g)$ is bubbled through water; (c) $SO_2(g)$ is bubbled through an aqueous solution of KOH.

Solution: (a) Metal oxides can react with water to form metal hydroxide solutions. In this case we have

$$CaO(s) + H_2O(l) \longrightarrow Ca(OH)_2(aq)$$

(b) Nonmetal oxides are acidic oxides. To the extent to which they dissolve they react with water to form acidic solutions. In this case we have

$$SO_2(g) + H_2O(l) \longrightarrow H_2SO_3(aq)$$

(c) Many acidic oxides, including SO_2, have very low solubilities in water. However, they dissolve in hydroxide solutions to form salts of their oxyanions. In this case,

$$SO_2(g) + KOH(aq) \longrightarrow KHSO_3(aq)$$

or

$$SO_2(g) + 2KOH(aq) \longrightarrow K_2SO_3(aq) + H_2O(l)$$

FOR REVIEW

Summary

In this chapter, we have dealt with the interactions between atoms that lead to formation of chemical bonds. Ionic bonding results from the complete transfer of electrons from one atom to another, with formation of a three-dimensional lattice of charged particles. The stabilities of ionic substances result from the powerful electrostatic attractive forces between an ion and all the surrounding ions of opposite charge. These interactions are measured by the lattice energy.

Covalent bonding results from the sharing of electrons between atoms. The rules that govern this sharing are based on the stability of the rare-gas electron configuration (the octet rule). We can represent shared electron-pair structures of molecules by means of Lewis structures, which show the sharing of electron pairs between atoms. The sharing of one pair of electrons produces a single bond; the sharing of two or three pairs of electrons between atoms produces double and triple bonds, respectively.

It sometimes happens that a single Lewis structure

is inadequate to represent a particular molecule but that an average of two or more Lewis structures does form a satisfactory representation. In these cases, the Lewis structures are referred to as resonance forms. It also sometimes happens that the octet rule is not obeyed; this situation occurs mainly when a large atom is surrounded by small, electronegative atoms like F, O, or Cl. In such instances the large atom often has more than an octet of electrons.

The strengths of covalent bonds increase with the number of electron pairs shared between two atoms. In single bonds, the bond strengths are generally higher between atoms of smaller size.

It is important to recognize that even in covalent bonding, electrons may not be shared equally between two atoms. Electronegativity is a measure of the ability of an atom to compete with other atoms for the electrons shared between them. Highly electronegative elements strongly attract electrons. The electronegativities of the elements, which show a regular periodic relationship, are an important guide to chemical behavior. We shall be using the concept of electronegativity often throughout the text. The difference in electronegativities of bonded atoms is used to determine the polarity of a bond.

Another application of electronegativity is in the assignment of oxidation numbers, formal whole-number charges assigned to atoms in molecules and ions. Although the oxidation numbers do not represent the real charges on atoms except in simple ionic substances, they are of great value in helping us to organize chemical facts and are an aid in the balancing of equations and in the naming of compounds.

Oxidation may be defined as a process in which an atom undergoes an increase in oxidation number. Reduction is a process in which an element undergoes a decrease in oxidation number. In an oxidation-reduction reaction, both oxidation and reduction occur in such a manner as to balance the total increases and decreases in oxidation numbers.

Several ideas presented in this chapter were used in discussing the chemistry of binary oxides. Metal oxides tend to be ionic substances that are either soluble in water to give basic solutions or that can be dissolved by acids; they are called basic oxides. Nonmetal oxides, by contrast, are covalent substances. They tend to be acidic oxides, which dissolve in water to produce acidic solutions or which dissolve in bases. Some oxides are borderline; they are called amphoteric oxides.

Learning goals

Having read and studied this chapter, you should be able to:

1 Determine the number of valence electrons for any atom and write its Lewis symbol.

2 Describe the origin of the energy terms that lead to stabilization of ionic lattices.

3 Describe the more commonly observed arrangements of ions in lattices, as illustrated in Figures 7.1 and 7.2.

4 Predict on the basis of the periodic table the probable formulas of ionic substances formed between common metals and nonmetals.

5 Write the electron configurations of ions.

6 Describe the effects of gain or loss of electrons on atomic radii in producing ionic radii.

7 Explain the concept of an isoelectronic series and the origin of changes in ionic radius within such a series.

8 Describe the basis of the Lewis theory and predict the valence of common nonmetallic elements from their position in the periodic table.

9 Write the Lewis structures for molecules and ions containing covalent bonds, using the periodic table.

10 Write resonance forms for molecules or polyatomic ions that are not adequately described by a single Lewis structure.

11 Explain the significance of electronegativity and in a general way relate the electronegativity of an element to its position in the periodic table.

12 Predict the relative polarities of bonds using either the periodic table or electronegativity values.

13 Relate bond energies to bond strengths and use bond energies to estimate ΔH for reactions.

14 Assign oxidation numbers to atoms in molecules and ions.

15 Give the meaning of the terms oxidation, reduction, and oxidation-reduction reactions.

16 Determine whether oxidation-reduction has occurred in a reaction; if it has, be able to identify the substance that is oxidized and the one that is reduced.

17 Assign acceptable names to simple ionic compounds and ions.

18 Describe the general differences in physical properties between substances with ionic bonds and those with covalent bonds.

19 Describe how the water solubility of a metal oxide is related to cation size and charge.

20 Write balanced chemical equations for the reactions of oxides with water, metallic oxides with acids, and nonmetal oxides with bases.

Among the more important terms and expressions used for the first time in this chapter are the following:

An **amphoteric oxide** is a water-insoluble oxide that dissolves in either an acidic or basic solution.

Bond energy (Section 7.8) is the enthalpy change, ΔH, required to break a chemical bond when a substance is in the gas phase.

Bond polarity (Section 7.9) is a measure of the difference in ability of the two atoms in a chemical bond to attract electrons.

A **cation** (Section 7.2) is a positively charged ion; an **anion** is negatively charged.

A **coordinate covalent bond** (Section 7.7) is a covalent bond in which the electrons are furnished by only one of the bonded atoms.

A **covalent bond** (Section 7.4) is a bond formed between two or more atoms by a sharing of electrons.

A **double bond** (Section 7.4) is a covalent bond involving two electron pairs.

Electronegativity (Section 7.9) is a measure of the ability of an atom that is bonded to another atom to attract electrons to itself.

A **half-reaction** (Section 7.10) is half of an overall oxidation-reduction reaction, corresponding to either the oxidation half or the reduction half.

An **ionic bond** (Section 7.2) is a bond formed on the basis of the electrostatic forces that exist between oppositely charged species in solid lattices made up of ions. The ions are formed from atoms by transfer of one or more electrons.

An **isoelectronic series** (Section 7.3) is a series of atoms or molecules having the same number of electrons.

Lattice energy (Section 7.2) is the energy required to separate completely the ions in an ionic solid.

A **Lewis structure** (Section 7.4) is a representation of covalent bonding in a molecule that is drawn using Lewis symbols. Covalently shared electron pairs are shown as lines, and unshared electron pairs are shown as a pair of dots. Only the valence-shell electrons are shown.

The **octet rule** (Section 7.1) states that bonded atoms tend to possess or share a total of eight valence-shell electrons.

Oxidation (Section 7.10) is the half of an oxidation-reduction process that corresponds to an increase in oxidation number.

Oxidation number (Section 7.10) or oxidation state is a positive or negative whole number assigned to an atom in a molecule or ion on the basis of a set of formal rules; to some degree it reflects the positive or negative character of that atom.

A **polar covalent bond** (Section 7.9) is a covalent bond in which the electrons are not shared equally.

Reduction (Section 7.10) is the half of an oxidation-reduction process that corresponds to a decrease in oxidation number.

Resonance forms (Section 7.6) are individual Lewis structures in cases where two or more Lewis structures are equally good descriptions of a single molecule. The resonance structures in such an instance are "averaged" to give a correct description of the real molecule.

A **triple bond** (Section 7.4) is a covalent bond involving three electron pairs.

Valence (Section 7.1) may be defined as the capacity of an atom for entering into chemical combination with other atoms. Ionic valence is equal to the number of electrons gained or lost in forming the ionic species. Covalence is equal to the number of electrons from an atom that are involved in shared electron-pair bonds with other atoms.

Valence electrons (Section 7.1) are the outer-shell electrons of an atom; these are the ones the atom uses in bonding.

EXERCISES

Valence electrons and Lewis symbols

7.1 Write the electron-dot symbol (Lewis symbol) for each of the following atoms: (a) Ca; (b) Se; (c) Br; (d) B.

7.2 Which of the following atoms or ions has an octet of electrons in the outermost occupied shell: (a) S; (b) O^{2-}; (c) Ne; (d) Ca^{2+}; (e) K?

7.3 Write the electron configuration and the Lewis symbol for each of the following atoms: (a) K; (b) Si; (c) As; (d) Al.

Ionic bonds

7.4 Write the Lewis symbol for each of the following atoms or ions: (a) Ca^{2+}; (b) Ar; (c) S^{2-}; (d) Na; (e) I.

7.5 Predict the chemical formula of the ionic compound formed between each of the following pairs of elements: (a) Sr, Br; (b) Be, O; (c) Al, S; (d) Zn, F; (e) K, Se; (f) Zn, O.

7.6 Write the outer electron configuration for each of the following ions: (a) Mn^{4+}; (b) S^{2-}; (c) Cu^+; (d) Na^+; (e) Cr^{3+}; (f) Pb^{2+}.

7.7 Which of the following ions possess noble-gas electron configurations: (a) Te^{2-}; (b) Ga^+; (c) In^{3+}; (d) Mg^{2+}; (e) Sc^{3+}; (f) Fe^{2+}?

7.8 Explain the following trends in lattice energies: (a) MgO > KCl; (b) MgO > MgS; (c) NaCl > NaBr.

7.9 Draw Lewis (electron-dot) symbols for the ions in the following ionic compounds: (a) BaO; (b) Na_2O; (c) $MgBr_2$; (d) AlF_3.

7.10 Write the formulas for compounds composed of the following pairs of ions: (a) Al^{3+}, NO_3^-; (b) Fe^{3+}, O^{2-}; (c) Cr^{3+}, Cl^-; (d) NH_4^+, SO_4^{2-}; (e) Ca^{2+}, O^{2-}.

7.11 Give the chemical formula for each of the following compounds (refer to Table 2.5 as necessary): (a) calcium bromide; (b) aluminum sulfate; (c) ammonium phosphate; (d) potassium nitride; (e) titanium(IV) fluoride.

Sizes of ions

7.12 Select the larger member of each of the following pairs: (a) Cs^+, Ba^{2+}; (b) Cl, Cl^-; (c) Fe^{2+}, Fe^{3+}; (d) S^{2-}, Cl^-; (e) Zn, Zn^{2+}; (f) Na^+, K^+.

7.13 Select the members of each of the following sets that are isoelectronic with each other: (a) Na^+, K^+, Mg^{2+}; (b) Cl^-, Ar, Br^-, K^+; (c) Se^{2-}, S^{2-}, Rb^+; (d) F, Ne, K, Ca^{2+}.

7.14 Arrange the members of each of the following sets in order of increasing size: (a) Mg^{2+}, Sr^{2+}, Ca^{2+}; (b) K^+, Ca^{2+}, Cl^-, Na^+; (c) O, F, O^{2-}.

7.15 Suggest an explanation for the observation that ionic radii for anions are usually larger than those of cations.

Covalent bonding; Lewis structures

7.16 What are some of the differences in physical properties between ionic substances and covalent substances?

7.17 Label each of the following substances as ionic or covalent: (a) NH_3; (b) K_2S; (c) Br_2; (d) Cl_2O; (e) $BaCl_2$; (f) FeO.

7.18 Draw the Lewis structures for each of the following molecules: (a) SiH_4; (b) H_2S; (c) CO; (d) N_2; (e) H_2O_2; (f) CS_2.

7.19 Draw the Lewis structures for each of the following ions: (a) ClO^-; (b) SO_3^{2-}; (c) PO_4^{3-}; (d) CN^-; (e) NO^+; (f) BF_4^-.

7.20 Draw Lewis structures for each of the following: (a) NOCl; (b) C_2H_2; (c) H_3COH; (d) SO_4^{2-}.

Resonance structures

7.21 Draw the resonance structures for each of the following molecules and ions: (a) SO_3; (b) CO_3^{2-}; (c) N_2O_4 (contains N—N bond); (d) NO_2^-.

7.22 How would you expect the C—O bond distance in CO_3^{2-} to compare with that in CO_2?

7.23 Why is it necessary to use resonance structures to describe adquately the structure of the nitrate ion, NO_3^-, using Lewis structures?

Exceptions to the octet rule

7.24 Which of the following compounds do not obey the octet rule? (a) SF_6; (b) NO_2; (c) CF_4; (d) BCl_3; (e) PCl_5; (f) OF_2; (g) XeF_2. Show the Lewis structures for those that do not.

7.25 Which of the following elements is more likely to show a departure from the octet rule in covalent com-

pounds: (a) C or Si; (b) P or N; (c) B or C. Explain briefly.

7.26 SF_6 is a very stable substance that is used as an electrically insulating gas in high-voltage transformers. Yet SCl_6 is quite unstable and has never been isolated. Explain this difference in stabilities.

Bond energies

7.27 Without looking at Table 7.3, predict which of the bonds in each of the following pairs is stronger: (a) C—F or C—Br; (b) C=O or C—O; (c) O—O or S—S; (d) H—O or H—S. Explain your reasoning for each prediction and then compare bond energies using Table 7.3.

7.28 Using the bond energies tabulated in Table 7.3, estimate ΔH for each of the following reactions:

(a)
$$\begin{array}{c} H \\ | \\ H-C-O-H \\ | \\ H \end{array} + H-I \longrightarrow \begin{array}{c} H \\ | \\ H-C-I \\ | \\ H \end{array} + H-O-H$$

(b)
$$\begin{array}{c} Cl \\ | \\ H-C-Cl \\ | \\ Cl \end{array} + O \longrightarrow \begin{array}{c} O \\ || \\ Cl-C-Cl \end{array} + H-Cl$$

(c) $C\equiv O + H-O-H \longrightarrow H-H + O=C=O$

7.29 Given the following bond-dissociation energies, calculate the average bond strength for the O—H bond:

$$H_2O(g) \longrightarrow OH(g) + H(g) \qquad D(HO-H) = 494 \text{ kJ}$$
$$OH(g) \longrightarrow O(g) + H(g) \qquad D(O-H) = 427 \text{ kJ}$$

Bond polarities; electronegativities

7.30 Without looking at the table of electronegativities (refer to a periodic table instead) arrange the members of each of the following sets in order of increasing electronegativity: (a) O, P, S; (b) Mg, Al, Si; (c) S, Cl, Br; (d) C, Si, N.

7.31 Which of the following bonds is polar? Indicate the more electronegative atom in each polar bond: (a) Cl—I; (b) P—P; (c) C—N; (d) F—F; (e) O—H.

7.32 Arrange the bonds in each of the following sets in order of increasing polarity: (a) HF, HC, HH; (b) PS, SiCl, AlCl.

7.33 Using electronegativities, list the following molecules in order of increasing bond polarity: ICl, BrCl, BrF, ClF.

7.34 Why do the electronegativities of the halogens decrease with increasing atomic number?

7.35 Explain why the electronegativities of the elements increase in a more or less regular way from left to right in each horizontal row of the periodic table.

Oxidation numbers; oxidation-reduction

7.36 Calculate the oxidation numbers of all atoms in each of the following compounds or ions: (a) N_2O_4; (b) CuCl; (c) HSO_3^-; (d) OH^-; (e) Na_2HPO_4; (f) S_8.

7.37 What is the maximum and minimum oxidation state exhibited by each of the following elements: (a) N; (b) Cl; (c) F; (d) Na; (e) Se?

7.38 In each of the following oxidation-reduction equations identify the atom that is oxidized and the one reduced:

(a) $CH_4(g) + 2O_2(g) \longrightarrow CO_2(g) + 2H_2O(l)$
(b) $3H_2S(aq) + 2HNO_3(aq) \longrightarrow$
$\qquad 3S(s) + 2NO(g) + 4H_2O(l)$
(c) $Mg(s) + 2HCl(aq) \longrightarrow MgCl_2(aq) + H_2(g)$
(d) $Cl_2(g) + 2OH^-(aq) \longrightarrow$
$\qquad ClO^-(aq) + Cl^-(aq) + H_2O(l)$

7.39 Identify which of the following are oxidation-reduction reactions:

(a) $2KClO_3(s) \longrightarrow 2KCl(s) + 3O_2(g)$
(b) $BaCl_2(aq) + H_2SO_4(aq) \longrightarrow BaSO_4(s) + 2HCl(aq)$
(c) $2Na(s) + 2H_2O(l) \longrightarrow 2NaOH(aq) + H_2(g)$
(d) $CaCO_3(s) + 2HCl(aq) \longrightarrow$
$\qquad CaCl_2(aq) + CO_2(g) + H_2O(l)$

7.40 Give two different names for each of the following compounds: (a) NF_3; (b) Cr_2O_6; (c) NbF_5; (d) SeO_2; (e) P_2Cl_4; (f) TiF_3.

7.41 Indicate the name of each of the following ions (refer back to Table 2.5 as necessary): (a) Ti^{2+}; (b) Cr^{2+}; (c) Ga^{3+}; (d) Ca^{2+}; (e) IO_2^-; (f) S^{2-}; (g) MnO_4^-; (h) SO_3^{2-}.

7.42 Name each of the following compounds (refer back to Table 2.5 as necessary): (a) OF_2; (b) CdO; (c) PbO_2; (d) $AlPO_4$; (e) $Fe(ClO_2)_2$; (f) $LiBrO_3$.

7.43 Give the chemical formula for each of the following substances: (a) xenon(IV) fluoride; (b) dichromium trioxide; (c) cuprous bromide; (d) nitrogen(I) oxide; (e) gallium(III) sulfide.

Binary oxides

7.44 Write balanced chemical equations for the reactions that occur when: (a) $SrO(s)$ is added to water; (b) $SO_3(g)$ is bubbled through an aqueous solution of KOH; (c) $P_4O_{10}(s)$ is dissolved in water; (d) hydrochloric acid is added to $CoO(s)$; (e) $Ga_2O_3(s)$ is dissolved in aqueous $NaOH$.

7.45 Predict the trend in the acidic character of the XO_2-type oxides of group 4A. Which of these oxides should be the most acidic and which one the least acidic?

7.46 Suggest why NaOH solutions are stored in plastic containers and not in glass? (Glass is made primarily from SiO_2.)

7.47 Which of the following pairs would you expect to be less soluble in water: (a) MgO or SO_2; (b) Na_2O or SiO_2. Explain.

Additional exercises

7.48 The lattice energy of NaF is 916 kJ/mol whereas that for KCl is 715 kJ/mol. Explain why the lattice energy of NaF is larger.

7.49 Explain why NaCl forms a solid ionic compound at room temperature rather than gaseous Na^+Cl^- pairs.

7.50 Which of the following pairs of elements are likely to form ionic compounds: (a) Na, O; (b) P, S; (c) C, O; (d) Ba, I; (f) Sc, F?

7.51 How would you expect the energy of an ionic interaction between positive ion A and negative ion B to be affected by the following changes: (a) the charge on A is doubled; (b) the charge on B is doubled; (c) the charges on both A and B are doubled; (d) the radii of both A and B are simultaneously doubled.

7.52 What is the potential energy of a cation and anion, the first with a charge of $+1.6 \times 10^{-19}$ coul and the second -1.6×10^{-19} coul (this is the charge of an electron), if the distance between their centers is (a) 2.13 Å; (b) 4.62 Å.

7.53 The radius of Al^{3+} is 0.45 Å while that of F^- is 1.33 Å. What is the potential energy of a Al^{3+}—F^- ion pair if the distance between the centers of these two ions is the sum of their ionic radii? The charge on an electron is -1.60×10^{-19} coul.

7.54 How far apart (in meters) are two electrons if their repulsive energy is 1.00×10^{-19} J? The charge on an electron is 1.60×10^{-19} coul.

7.55 If you have an unlabeled bottle containing a white, water-soluble solid that melts at 776°C, which of the following substances might it be: (a) NO_2; (b) KCl; (c) Al? Explain briefly.

7.56 In the 2+ ions of the fourth row of the periodic table, extending from Ca^{2+} to Zn^{2+}, how do the effective nuclear charges seen by the highest electrons vary? How do the ionic radii vary? What relationship is there between these two quantities?

7.57 List three ions that are isoelectronic with Kr.

7.58 What type of bonds (ionic, covalent, or metallic) would you expect to be found in each of the following substances: (a) O_2; (b) Cu; (c) SO_2; (d) $CaCl_2$; (e) NO; (f) NaBr?

[7.59] By comparing the properties of KBr with Br_2 provide some arguments for why it is proper to view KBr as an ionic substance and Br_2 as a covalent (molecular) one. (You should consult a handbook of chemistry for data on the two substances.)

7.60 Write the electron configurations for the following manganese ions: Mn^{2+}, Mn^{4+}, and Mn^{7+}. Write the formulas for oxides in which manganese has each of these oxidation states.

7.61 Write the Lewis structure for each of the following substances: (a) nitrous oxide, N_2O, known as laughing gas and used as an inhalation anesthetic in dental surgery; (b) ethyl alcohol, CH_3CH_2OH, the alcohol of alcoholic beverages (contains a C—C bond); (c) formaldehyde, H_2CO, used to preserve biological samples; (d) urea, H_2NCONH_2, which is excreted in the urine of mammals (contains two N—C bonds).

7.62 Write the electron configuration for each of the following ions: (a) Cl^-; (b) Ba^{2+}; (c) Tl^+; (d) Fe^{3+}; (e) Zn^{2+}.

7.63 Magnesia, MgO, used in making bricks for lining high-temperature furnaces, melts at a very high temperature. By comparison, NaF could not be used for furnace bricks, because it melts at a much lower temperature. Suggest an explanation for this difference in properties.

7.64 Predict the relative S—O bond lengths in SO_3^{2-} and SO_2. Which would you expect to have the greater S—O bond energy?

7.65 Give the chemical formula for each of the following substances: (a) stannous fluoride (in toothpastes); (b) barium sulfate (used in obtaining medical X rays); (c) silver bromide (used in photographic emulsions); (d) potassium iodide (added to "iodized salt").

7.66 Name each of the following substances: (a) $PbBr_2$; (b) $GeCl_4$; (c) SeO_2; (d) $CoCl_2$; (e) KNO_3; (f) BaO_2.

7.67 Give the chemical formula for each of the following substances: (a) nickel(II) chloride; (b) chromium(III) oxide; (c) ferrous bromide; (d) nitrogen dioxide; (e) sodium sulfate; (f) ammonium chloride; (g) aluminum hydroxide; (h) sodium peroxide.

7.68 Assign oxidation numbers to the atoms in each of the following molecules and ions: (a) MnO_2; (b) NO_2^-; (c) N_2H_4.

[7.69] Bacterial decay of organic matter under anaerobic conditions (absence of air) can produce hydrides of the nonmetallic elements. In this manner methane, CH_4, hydrogen sulfide, H_2S, and phosphine, PH_3, are formed. (Phosphine is very reactive and produces a glow when it burns in air. The "will-o'-the-wisp," a glow seen in marshes on a dark night, is probably due to the oxidation of PH_3 released to the air.) Write balanced equations showing the oxidation of each of these hydrides by molecular oxygen. Assume in each case that the highest possible oxidation state is attained by the nonmetal.

7.70 What is the change in oxidation state in the element in colored type during the course of the following chemical transformations? Indicate whether the atom is oxidized or reduced, if either: (a) MnO_4^- is converted to Mn^{2+}; (b) $NaCl$ is treated with an acid to produce HCl; (c) CuO is treated with H_2 to give Cu; (d) H_2O_2 is converted to H_2O.

7.71 Estimate ΔH for the following reactions:
(a) $H-C\equiv C-H(g) + 2H_2(g) \longrightarrow H_3C-CH_3(g)$
(b) $2HCl(g) + F_2(g) \longrightarrow 2HF(g) + Cl_2(g)$
(c) $CO(g) + 3H_2(g) \longrightarrow CH_4(g) + H_2O(g)$

7.72 Atmospheric SO_2, in the presence of moisture, attacks many structural materials including metals and marble ($CaCO_3$). Suggest an explanation.

7.73 Given the following bond dissociation energies, calculate the average bond energy for the Ti—Cl bond.

	ΔH, kJ/mol
$TiCl_4(g) \longrightarrow TiCl_3(g) + Cl(g)$	335
$TiCl_3(g) \longrightarrow TiCl_2(g) + Cl(g)$	423
$TiCl_2(g) \longrightarrow TiCl(g) + Cl(g)$	444
$TiCl(g) \longrightarrow Ti(g) + Cl(g)$	519

[7.74] Using the bond energies in Table 7.3 and the following bond distances, construct a graph of bond energies versus bond distances. What conclusions can you reach from your graph? Bond distances are as follows: C—C (1.54 Å), Si—Si (2.35 Å), N—N (1.45 Å), O—O (1.48 Å), S—S (2.05 Å), F—F (1.42 Å), Cl—Cl (1.99 Å), Br—Br (2.28 Å), I—I (2.67 Å), C—F (1.35 Å), C—O (1.43 Å), and S—Br (2.27 Å).

8

Geometries of molecules; molecular orbitals

Molecules of different substances have diverse shapes; some molecules are flat, others long and thin. Some large molecules such as proteins are found to be globular clusters, while others are coiled. The shape of a molecule, together with the strength and polarity of its bonds, largely determines the physical and chemical properties of that substance. Molecular shape is especially important in biochemical reactions, in which only molecules that are of a certain shape and size may be able to enter into a reaction. For example, small changes in the shapes and sizes of drug molecules may alter activity or may reduce the toxic side effects of a drug that is otherwise beneficial. Those properties of a molecule related to its shape and size are referred to as steric effects.

Diversity in molecular shape arises because atoms attach to one another in various geometric arrangements. The overall molecular shape of a molecule is determined by its *bond angles,* the angles made by the lines joining the nuclei of the atoms in the molecule (the internuclear axes). The distance between the nuclei of two bonded atoms is called the *bond distance.* Figure 8.1 shows the bond angles and bond distances that define the shape and size of the SF_6 molecule. It is possible to determine bond-distance and bond-angle information about a molecule from various experimental techniques. We won't discuss these techniques here, but you should simply accept the fact that by using them the chemist has acquired a great deal of information about molecular geometries. Our concern is with how these geometries can be explained.

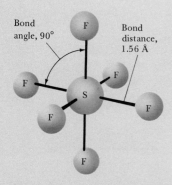

FIGURE 8.1 The geometrical structure of the SF_6 molecule, showing bond-distance and bond-angle information.

8 Geometries of molecules;
molecular orbitals

8.1 The valence-shell
electron-pair repulsion
(VSEPR) model

In Chapter 7, we employed the simple and empirical Lewis model to account for the formulas of covalent compounds. The Lewis structures that we draw are not geometrical figures. They simply describe the bonding connections between atoms in a two-dimensional representation. For example, the Lewis structure for methane, CH_4, does not tell us whether the molecule is planar or tetrahedral. To develop a model that explains three-dimensional structure, we must take into account the electrostatic interactions between electron pairs.

8.1 The valence-shell electron-pair repulsion (VSEPR) model

We have seen that an atom is bonded to other atoms in a molecule by electron pairs that occupy its valence-shell atomic orbitals. These electron pairs are attracted to the central nucleus by the electron-nucleus attractive force. At the same time, though, they are repelled by the other electron pairs about the central atom. As a result, the electron pairs remain as far apart from one another as possible, while maintaining their distance from the central nucleus.

You will remember from Chapter 7 that the most commonly observed electronic arrangement about an atom in covalent compounds is an octet, or four pairs, of electrons. We might then ask how four electron pairs could arrange themselves about a central atom so as to minimize electron-pair repulsions, while maintaining their distance from the central nucleus. Two possibilities are shown in Figure 8.2. As it turns out, the tetrahedral arrangement is the one that has the electron pairs as far apart as possible for a given distance from the central nucleus. (This can be demonstrated using some rather lengthy trigonometric calculations, but we won't go through these.) We therefore consider that an octet of electrons about an atom consists of four electron pairs, with each pair occupying a region of space directed toward the vertex of a tetrahedron. If each of these four electron pairs bonds the central atom to another atom—for example, as in methane—then the observed arrangement of bonded atoms about the central atom is tetrahedral.

We can extend the idea we have just discussed to include cases with differing numbers of electron pairs about the central atom. Table 8.1 lists differing numbers of electron pairs in the valence shell and the

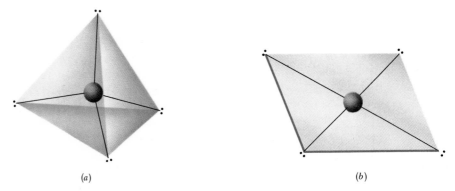

(a)   (b)

FIGURE 8.2 Two possible arrangements of four electron pairs around a central atom: (a) tetrahedral; (b) square planar. For a given distance of the electron pairs from the central atom, the repulsions are lower for the tetrahedral arrangement than for any other, such as the square planar.

TABLE 8.1 Geometries of electron-pair distributions about a central atom as a function of the number of electron pairs

Number of electron pairs	Arrangement of electron pairs		Predicted bond angles	Example
2		Linear	180°	
3		Trigonal planar	120°	
4		Tetrahedral	109.5°	
5		Trigonal bipyramidal	120° 90°	
6		Octahedral	90°	

arrangement that corresponds to maximum separation of the pairs. Our model correctly predicts the structures of molecules with expanded valence shells; PF_5 and PCl_5 are observed, for example, to be trigonal bipyramids, as shown in Table 8.1. Similarly, SF_6 is found to be octahedral, as predicted. Study Table 8.1 closely, paying attention to the geometries associated with each number of electron pairs. Also notice the names (such as tetrahedral and trigonal bipyramidal) associated with each geometry.

The model for molecular geometries we have been discussing is often called the valence-shell electron-pair repulsion (VSEPR) model. It is based on the idea that the arrangement of bonded atoms around a central atom is determined by the repulsions between the electron pairs around that central atom.

It often happens that the electron pairs about a central atom are not all shared with other atoms. For example, remember that in the Lewis structure for water there are two unshared electron pairs about the oxygen:

$$H—\overset{..}{O}:$$
$$\underset{H}{|}$$

These unshared electron pairs are as important as are shared pairs in determining the structure. However, when the structure of a molecule is studied experimentally, only the positions of the atoms are observed, not the positions of electrons. The effects of unshared electron pairs on the geometry must therefore be inferred from what we see of the structures of molecules.

Usually, unshared electron pairs aren't shown explicitly in writing chemical formulas. Before we can use the VSEPR model to predict the geometry of a molecule we must determine whether it has any unshared

TABLE 8.2 Distribution between shared and unshared electron pairs in hydrides of the second-row nonmetals

Lewis structure	Number of shared pairs	Number of unshared pairs	Structure
H—C—H (with H above and below)	4	0	Tetrahedral, $109.5°$
H—N—H (with H below)	3	1	Trigonal pyramidal, $107°$
H—O: (with H below)	2	2	Nonlinear, $104.5°$
H—F:	1	3	Linear

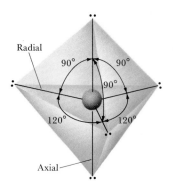

FIGURE 8.3 Trigonal bipyramidal arrangement of five electron pairs about a central atom. There are two geometrically distinct types of electron pairs, the two axial and three radial pairs. The repulsions experienced by the axial pairs are greater. For this reason, unshared electron pairs tend to occupy the radial positions.

valence-shell electrons. The presence of such unshared electrons is revealed when we draw the Lewis structure for the molecule. The Lewis structures and molecular geometries of CH_4, NH_3, H_2O, and HF are shown in Table 8.2. In each of these molecules there is an octet of electrons about the central atom; however, the number of unshared electrons varies. The VSEPR model predicts that the four electron pairs will be arranged in a tetrahedral fashion (see Table 8.1). Consequently, CH_4 can be described as tetrahedral, NH_3 as pyramidal, and H_2O as nonlinear (also referred to as bent or angular). When we describe molecular shape in this way, we are focusing on the positions of the atoms, not the electron pairs.

It is possible to extend the VSEPR model a little further. We can imagine that each electron pair around a central atom has a certain volume requirement, but that the volume requirement of an unshared pair of electrons is larger than for a pair that is shared between two atoms. This is so because the shared pair is held by the attractive forces of two nuclear centers. This should have the effect of contracting the spatial distribution of the electrons. If we use this idea we conclude that the HNH angle in NH_3 should be a little less than the tetrahedral angle of 109.5° found in CH_4, and indeed it is, as shown in Table 8.2. Similarly, the HOH angle in H_2O is reduced from the tetrahedral angle. In the third row hydrides, the corresponding angles in PH_3 and H_2S are reduced even more.

Let's apply the idea that unshared electron pairs have a larger volume requirement to molecules with five electron pairs in the valence shell. In the trigonal bipyramid, shown in Figure 8.3, electron pairs in the axial and radial positions are not equivalent. In the axial position, an electron pair is repelled by three other electron pairs at 90° angles from it. For an electron pair in the radial position, there are only two such 90° interactions and two others much further removed, at 120°. As a result, the repulsions from other electron pairs are larger for a pair located in an axial position than for a pair located in a radial position. Thus we predict that electron pairs with larger volume requirements will go into the radial locations.

The molecule SF_4 represents an example of a molecule with four shared electron pairs and one unshared pair. In the alternative VSEPR structures, shown in Figure 8.4, the unshared electron pair could be in

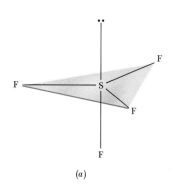

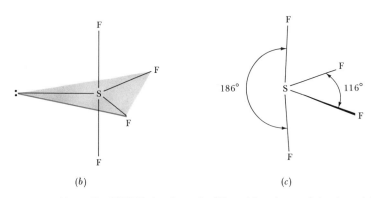

FIGURE 8.4 Alternative VSEPR structures for SF_4 and the observed structure. (a) Unshared pair in axial position. (b) Unshared pair in radial position. (c) Observed structure.

either an axial or radial (equatorial) location. From the observed geometrical structure, also shown in Figure 8.4, we can infer that the unshared electron pair resides in a radial position, as predicted. Note that the "axial" S—F bonds are slightly bent back, suggesting that the unshared electron pair, with its greater volume requirement, is pushing them back.

Using the idea that unshared electron pairs are important in determining the observed geometry, we can predict molecular geometries for various numbers of shared and unshared electron pairs about a central atom. Table 8.2 outlines these possibilities for the case of four electron pairs. In Table 8.3 are listed the geometries to be expected for five or six electron pairs. The guiding rule in arriving at these predictions is that

TABLE 8.3 Relationship between numbers of shared and unshared electron pairs and observed molecular shapes for molecules with more than an octet of electrons

Number of electron pairs about central atom	Geometrical arrangement of electron pairs	Number of bonding pairs	Observed molecular shape		Example	Formula
5	Trigonal bipyramidal	5	Trigonal bipyramidal		PCl_5	AB_5
		4	Seesaw		SF_4	AB_4
		3	T-shaped		ClF_3	AB_3
		2	Linear		XeF_2	AB_2
6	Octahedral	6	Octahedral		SF_6	AB_6
		5	Square pyramidal		BrF_5	AB_5
		4	Square planar		XeF_4	AB_4

the unshared electron pairs have a larger volume requirement than do shared electron pairs and that this determines in some cases where they are placed. Some of the possible arrangements have been omitted from the table because there are no known molecules with those arrangements.

The VSEPR model is very useful in predicting the structures of molecules. The predictions of the model are basically contained in Tables 8.1, 8.2, and 8.3. When the VSEPR model is applied to predicting the structures of molecules or ions containing multiple bonds, one additional rule is required: *Multiple bonds are to be considered the same as single bonds for the purpose of determining the overall geometry.* However, we can imagine that the double bond will occupy more space than a single bond, and will therefore have some effect on bond angles. These ideas are illustrated in Sample Exercise 8.1.

In summary, to use the VSEPR model to predict geometry, you should proceed in the following way:

1. Count the number of valence electrons and write out the Lewis structure for the molecule or ion.
2. From the Lewis structure determine the number of unshared electron pairs and number of bonds around each central atom. (Remember to count multiple bonds as a single bond.)
3. Determine the arrangement of bonds and unshared electron pairs that minimizes electron-pair repulsions (Table 8.1, 8.2, and 8.3).
4. Describe the shape of the molecule in terms of the positions of the atoms (the positions of the bonding electron pairs). The arrangement of the electrons is merely used to predict atomic positions.

SAMPLE EXERCISE 8.1

Using the VSEPR model, predict the geometrical structures of the following: (a) $SnCl_3^-$; (b) CO_2; (c) IF_5; (d) ethylene, C_2H_4.

Solution: (a) The Lewis structure for the $SnCl_3^-$ ion is as follows:

$$\left[:\overset{..}{\underset{..}{Cl}}-\underset{\underset{\overset{|}{\underset{..}{Cl}:}}{}}{Sn}-\overset{..}{\underset{..}{Cl}}: \right]^-$$

The four electron pairs should thus be disposed at the corners of a tetrahedron. One of the corners is occupied by the unshared electron pair. The geometrical arrangement of the atoms is thus pyramidal:

(b) The Lewis structure for carbon dioxide is

$$\overset{..}{\underset{..}{O}}=C=\overset{..}{\underset{..}{O}}$$

The rule that each multiple bond should be treated the same as a single electron pair means that, in effect, we have two electron pairs in the valence shell. The molecule should therefore be linear (Table 8.1). The CO_2 molecule is in fact observed to be linear.

(c) The Lewis structure for IF_5 is

The iodine has six pairs of valence-shell electrons around it, five of them bonding pairs. The electron pairs should point toward the corners of an octahedron. The geometrical arrangement of atoms is therefore square pyramidal (Table 8.3).

(d) The Lewis structure for ethylene is

H H
 \ /
  C=C
 / \
H H

The multiple bond between carbon atoms is treated as a single electron pair, so we have effectively three electron pairs in the valence shell. The three bonds about each carbon are distributed in a plane, with approximately 120° bond angles (Table 8.1). Ethyl-

ene is thus predicted to be planar with approximately 120° bond angles, as observed:

H———————H
   \   /
    C=C
   /   \
H———————H

The experimentally determined H—C—H bond angle is 117°. Thus the spatial demands of the C=C double bond appear to force the C—H bonds together to a slight extent, thereby reducing the H—C—H angle from the "ideal" 120°.

8.2 Dipole moments

The shape of a molecule and the polarity of its bonds together determine the charge distribution in the molecule. A diatomic molecule such as HF, which contains a polar covalent bond (Section 7.9), has a concentration of negative charge on one end and a concentration of positive charge on the other end. A molecule whose centers of positive and negative charges do not coincide is described as being polar. The degree of polarity is measured by the **dipole moment** of the molecule. The dipole moment, μ, is defined as the product of the charge at either end of the dipole, Q, times the distance between the charges, r: $\mu = Qr$. Thus the dipole moment increases as the quantity of charge which is separated increases, or as the distance between the positive and negative centers increases.

Table 8.4 shows the dipole moments of the hydrogen halides. The dipole moments are reported in units of debye, D; a debye is 3.33×10^{-30} coul-m. Notice that the dipole moment decreases as the electronegativity difference decreases.

Experimentally, dipole moments are obtained by observing the effect of the compound on the ability of a capacitor to store electrical charge. The substance is placed between the plates of a capacitor and the plates are charged as shown in Figure 8.5. Molecules with no dipole moments orient themselves randomly between the plates. In contrast, dipolar molecules are preferentially aligned with their positive ends pointing toward the negatively charged plate and their negative ends toward the positive plate. This orientation permits the plates to hold a greater electrical charge.

The dipole moment of a molecule containing more than two atoms depends both on bond polarity and molecular geometry. To have a dipole moment, a polyatomic molecule must have polar bonds. However, even if polar bonds are present, the molecule itself might not have a

TABLE 8.4 Some properties of hydrogen halides

Compound	Electronegativity difference	Dipole moment (D)
HF	1.8	1.91
HCl	1.0	1.03
HBr	0.8	0.79
HI	0.5	0.38

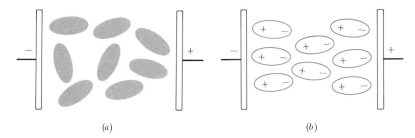

FIGURE 8.5 Molecules with no dipole moments are randomly arranged between the charged plates of a capacitor (a). In contrast, polar molecules tend to become aligned (b).

(a)   (b)

dipole moment if the bonds are arranged so their polarities cancel. This situation is found in the linear CO_2 molecule:

$$\overset{\longleftarrow + \; + \longrightarrow}{O=C=O}$$

Remember that the arrow points toward the more electronegative atom (Section 7.9). Because oxygen has a greater electronegativity than carbon, the bonds are polar, with electron density concentrating on the oxygen atoms. However, the centers of negative and positive charges are found at the same point, on carbon; consequently, the molecule has no net dipole moment. Figure 8.6 shows several further examples of both polar and nonpolar molecules, all of which have polar bonds. Notice that the symmetric arrangement of bonds in the nonpolar molecules leaves them without positive and negative sides; molecules with dipole moments have positive and negative ends, and this allows them to align themselves between the charged plates of a capacitor. Notice also that unshared electrons as well as polar bonds can contribute to molecular polarity.

SAMPLE EXERCISE 8.2

Predict whether the following molecules are polar or nonpolar: (a) ICl; (b) SO_2; (c) SF_6.

Solution: (a) Chlorine is a more electronegative element than iodine. Consequently, ICl will be polar with chlorine as the negative end:

$$\overset{+\longrightarrow}{I-Cl}$$

All diatomic molecules with polar bonds are polar molecules.

(b) Oxygen is more electronegative than sulfur; the molecule therefore has the polar bonds necessary for the molecule to be polar. The molecule has the following resonance forms:

$$:\overset{..}{\underset{..}{O}}-\overset{..}{S}=\overset{..}{O} \longleftrightarrow \overset{..}{O}=\overset{..}{S}-\overset{..}{\underset{..}{O}}:$$

Consequently, the molecule is nonlinear; this geometry follows from the trigonal planar arrangement of electron pairs around sulfur (Table 8.1), there being one unshared pair of electrons on the sulfur atom. Because of the molecular shape, the bond polarities do not cancel and the molecule is polar:

(It is unlikely that the unshared electron pair on sulfur completely cancels the effect of the polar S—O bonds, though it certainly does so to some extent.)

(c) Fluorine is more electronegative than sulfur. The bond dipoles therefore point toward fluorine. The six S—F bonds are arranged in a symmetric octahedral fashion around the central sulfur (there are no unshared valence-shell electron pairs on sulfur):

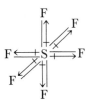

The bond dipoles cancel each other; the molecule does not have a negative and positive side. It is nonpolar with an overall dipole moment of zero.

215

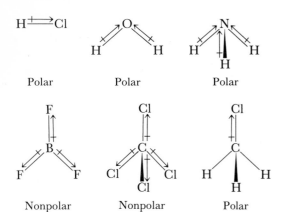

Polar   Polar   Polar

Nonpolar   Nonpolar   Polar

FIGURE 8.6 Examples of molecules with polar bonds. Some of these molecules are nonpolar because their bond polarities cancel each other.

8.3 Hybrid orbitals and molecular shape

The VSEPR theory provides a simple model for predicting the shapes of molecules. However, it does not explain why bonds exist between atoms. The shared and unshared pairs are taken as given; the model is used simply to deduce the shape of the molecule or ion. In developing a theory of covalent bonding, chemists have also approached the problem from another direction. Suppose we take the formula and geometrical structure of the molecule as given. How can we account for the observed geometries in terms of the atomic orbitals used by the atoms in forming bonds to one another?

In the Lewis theory, covalent bonding occurs when atoms share electrons. Such sharing involves a concentration of electron density between nuclei. This buildup of electron density occurs when a valence atomic orbital of one atom merges or overlaps with that of another atom. The orbitals are then said to share a region of space or to *overlap*. We can think of the shared electron pair as occupying an orbital that consists in part of the atomic orbitals contributed by each atom. This idea is illustrated in Figure 8.7, which shows the coming together of two hydrogen atoms to form the H_2 molecule. The overlap of the $1s$ orbital on one H atom with that on the other increases as the atoms draw near. The potential energy decreases because the electrons are simultaneously attracted to both nuclei. At very short internuclear distances the potential energy begins to increase due to increasing repulsions between nuclei and between elec-

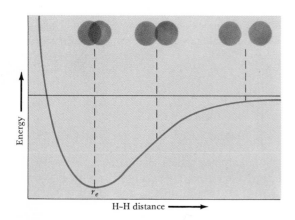

FIGURE 8.7 Formation of the stable H_2 molecule by the overlap of two hydrogen $1s$ orbitals. The minimum in the energy surface, at r_e, represents the equilibrium H—H bond distance in H_2.

trons. The internuclear distance at the minimum in this potential-energy curve corresponds to the observed bond distance. Thus the observed bond distance is a compromise between increased overlap of the atomic orbitals, which draws the atoms together, and the nuclear-nuclear and electron-electron repulsions, which force them apart.

In extending this view of bond formation to polyatomic molecules, we want a model that localizes, as much as possible, the valence electron pairs in the regions between atoms. The problem is to construct such a model, starting with the valence-shell atomic orbitals of the atoms. To see how this might be accomplished, let's consider methane, CH_4. This molecule has a tetrahedral arrangement of hydrogen atoms around a central carbon, as shown in Table 8.2. The valence-shell orbitals on carbon available for bonding are the $2s$ and three $2p$ orbitals. Each hydrogen atom has, of course, a single $1s$ orbital. The three $2p$ orbitals of carbon point along the $x, y,$ and z axes (Figure 5.18), and the $2s$ orbital is spherically symmetrical. It is therefore not possible to employ these four orbitals individually to make the four equivalent bonds between carbon and hydrogen. But if they are not appropriate by themselves for forming the four bonds to hydrogen, perhaps some combination of them is more satisfactory.

We can imagine a stepwise procedure of getting a carbon atom initially in its ground state ready for bonding to four hydrogen atoms. The electronic configuration of the ground state carbon atom, in orbital diagram representation, is as follows:

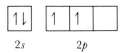

$2s$ $2p$

Notice that a carbon atom in its ground state could at most form two bonds to two other atoms, because there are just two unpaired electrons available. To obtain the capability for forming four bonds, one of the $2s$ electrons must be "promoted" to the vacant $2p$ orbital:

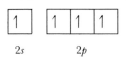

$2s$ $2p$

Because the $2p$ orbital is of higher energy than the $2s$, this promotion costs energy. Next, the mathematical functions that describe the $2s$ and $2p$ orbitals can be combined with one another. The combination yields four equivalent orbitals, called hybrid orbitals, which point toward the corners of a tetrahedron (Figure 8.8). Each one of these hybrid orbitals consists of a certain amount of the $2s$ orbital and a certain amount of the $2p$ orbitals. Because the four hybrid orbitals are made up of one $2s$ and three $2p$ orbitals, they are labeled sp^3 hybrids. (Notice that the superscript 3 relates to the relative proportions of s and p orbitals and not to the number of electrons they contain.) When the $2s$ and $2p$ orbitals are mixed so as to give the four hybrid orbitals we obtain:

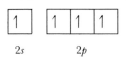

$2sp^3$

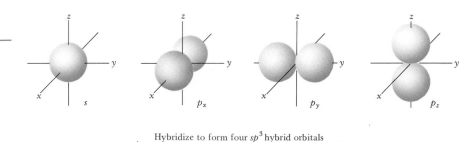

Hybridize to form four sp^3 hybrid orbitals

FIGURE 8.8 Formation of four sp^3 hybrid orbitals from a set of one s and three p orbitals.

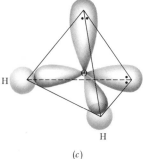

(a)

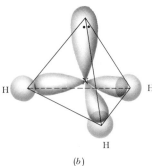

(b)

(c)

FIGURE 8.9 Formation of methane (a), ammonia (b), and water (c) by overlap of hydrogen 1s orbitals with the sp^3 hybrid orbitals of the central atom.

The carbon is now "prepared" to form four equivalent bonds with tetrahedral geometry. The interaction of the four sp^3 hybrid orbitals from carbon with four hydrogen atoms to form methane, CH_4, is shown in Figure 8.9(a).

The first of the steps outlined above, promotion of the $2s$ electron, requires energy. The second does not, because it merely means an averaging to give four equivalent orbitals, each with energy equal to the average of the original set. The overall process does require energy. The reason it occurs is that the system more than recovers this energy in bond formation. As we've seen, a ground-state carbon atom could at most form two bonds to hydrogen, using the two electrons in its $2p$ orbitals. With promotion and hybridization, four bonds are possible. Furthermore, hybridization permits greater orbital overlap, because the hybrid orbitals extend out further from the central nucleus.

In molecules such as NH_3 and H_2O, the central atom has about it four electron pairs in an approximately tetrahedral arrangement. The orbitals used by the central atom can be thought of as sp^3 hybrid orbitals. In NH_3 one of the sp^3 hybrid orbitals contains the unshared electron pair, the other three are employed in the bonds to hydrogen, Figure 8.9(b). In H_2O two of the orbitals contain unshared pairs, two are employed in bonds to hydrogen, Figure 8.9(c). In these examples, the hybrid orbitals employed by the central atom are not pure sp^3 hybrids, because the bond angles about the central atom are not exactly the 109.5° tetrahedral angle, as shown in Table 8.2. In the mathematical formulation of hybrid orbitals, it is possible for the contribution of the s orbital to be a little greater in one hybrid orbital, a little smaller in another, in order to make

TABLE 8.5 Geometrical arrangements characteristic of hybrid orbitals

Atomic orbital set	Hybrid orbital set	Geometrical arrangement	Examples
s,p	sp	Linear (180° angle)	$Be(CH_3)_2$, $HgCl_2$
s,p,p	sp^2	Trigonal planar (120° angles)	BF_3
s,p,p,p	sp^3	Tetrahedral (109.5° angles)	CH_4, $AsCl_4^+$, $TiCl_4$
d,s,p,p	dsp^{2a}	Square planar (90° angles)	$PdBr_4^{2-}$
d,s,p,p,p	dsp^{3a}	Trigonal bipyramidal (120° and 90° angles)	PF_5
d,d,s,p,p,p	d^2sp^{3a}	Octahedral (90° angles)	SF_6, $SbCl_6^-$

[a]Depending on the particular element involved, the d orbital that mixes with the s and p may be of major quantum number one lower or of the same major quantum number. This has no effect on the geometrical characteristics of the resulting hybrid set.

the angles between the hybrid orbitals come as close as possible to the observed bond angles. We shall make no attempt to follow through with these refinements but shall instead employ just the idealized equivalent hybrid set.

Various combinations of atomic orbital sets can be mixed, or hybridized, to obtain different geometries of orbitals about a central atom. Table 8.5 shows several of the more important hybrid orbital combinations and the geometries to which they correspond. The hybrid orbital sets are illustrated in Figure 8.10. Note that expanded valence shells about atoms, that is, those in which there are more than eight electrons in the valence shell, can be accommodated by mixing in d orbitals along with s and p.

The purpose in formulating hybrid orbitals is to provide a convenient model in which we can imagine the electrons to be localized in the region between two atoms. The picture of hybrid orbitals has limited predictive value; that is, we cannot say in advance that in NH_3 the nitrogen uses essentially sp^3 hybrid orbitals. Once given the molecular geometry, however, we can employ the concept of hybridization to describe the atomic orbitals employed by the central atom in bonding.

SAMPLE EXERCISE 8.3

The molecule BeH_2 is known to be linear. Account for the bonding in BeH_2 in terms of the hybrid orbitals employed by Be in bonding to the two hydrogen atoms.

Solution: The orbital diagram for the Be atom is as follows:

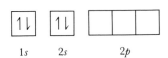

The Be atom in its ground state is incapable of forming bonds with other atoms, because all the electrons are paired. However, suppose one of the electrons is promoted to the $2p$ orbital:

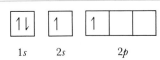

and the $2s$ and one of the $2p$ orbitals are mixed to form two sp hybrid orbitals:

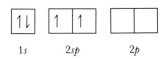

The electrons in the two sp hybrid orbitals can form shared electron pair bonds with two hydrogen atoms. The sp hybrid orbitals are directed at 180° angles from one another, Figure 8.10, so the BeH_2 molecule is linear.

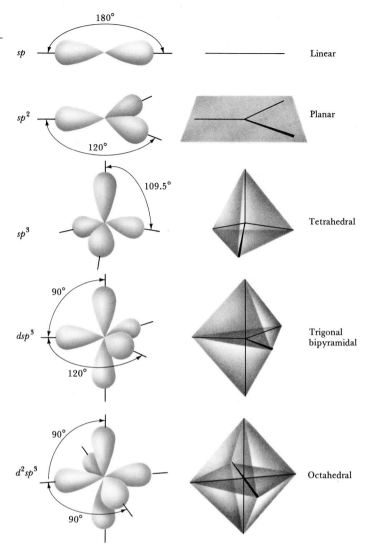

sp 180°   Linear

sp^2 120°   Planar

sp^3 109.5°   Tetrahedral

dsp^3 90°   120°   Trigonal bipyramidal

d^2sp^3 90°   90°   Octahedral

FIGURE 8.10 Geometrical arrangements characteristic of hybrid orbital sets.

SAMPLE EXERCISE 8.4

Indicate the hybridization of orbitals employed by the central atom in each of the following: (a) NH_2^-; (b) SF_4 (see Figure 8.4).

Solution: (a) The Lewis structure for NH_2^- is

$$[H \overset{..}{\underset{..}{N}} H]^-$$

From the VSEPR model we conclude that the four electron pairs around N should be arranged in a tetrahedral fashion. Such a tetrahedral arrangement is characteristic of sp^3 hybridization (Figure 8.10); two of the hybrid orbitals contain unshared electron pairs, the other two contain pairs shared with hydrogen.

(b) As shown in Figure 8.4, there are ten valence-shell electrons around sulfur in SF_4. With an expanded octet of ten electrons, the use of a d orbital on the sulfur is indicated. The trigonal-bipyramidal arrangement of valence-shell electron pairs shown in Figure 8.4 corresponds to dsp^3 hybridization (Figure 8.10). One of the hybrid orbitals contains an unshared electron pair; the other four are involved in bonding to fluorine.

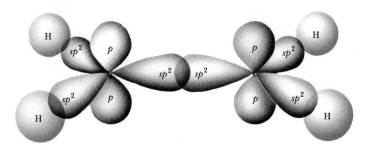

FIGURE 8.11 Hybridization of carbon orbitals in ethylene. The σ bond framework, formed from sp^2 hybrid orbitals on the carbon atoms, determines the observed geometrical structure of the molecule.

8.4 Hybridization in molecules containing multiple bonds

The concept of hybridization may be applied also to molecules containing multiple bonds. For example, we have seen (Sample Exercise 8.1) that ethylene possesses a carbon-carbon double bond and is a planar molecule. The planar arrangement of three bonds about each carbon suggests that the hybrid orbital set it uses to bond to the other carbon and the two hydrogens is sp^2 (Table 8.5). Because the valence orbitals on carbon consist of a $2s$ and *three* $2p$ orbitals, one $2p$ orbital remains unused after forming the sp^2 hybrid set. This is illustrated in Figure 8.11. We notice that the carbon atoms of ethylene are bonded together through overlap of sp^2 hybrid orbitals. The resultant electron density is concentrated symmetrically between the nuclei along the line joining them. Bonds of this kind are called **σ (sigma) bonds.** The bonds between carbon and hydrogen are also σ bonds. Because the hydrogen uses a $1s$ orbital that is spherical, each bond is symmetrical with respect to that particular C—H bond axis.

The formation of the σ bonds gives three electron pairs about carbon in the sp^2 hybrid orbitals, and an electron on each carbon in the $2p$ orbital that is perpendicular to the plane of the molecule. The p orbitals shown on the two carbon atoms in Figure 8.11 can overlap with one another in a sideways fashion. By means of this overlap a second C—C covalent bond is formed, as shown in Figure 8.12. Each carbon has achieved an octet of electrons. The second C—C bond differs from the first, because it results from concentrating electron density above and below the C—C bond axis. This kind of bond is called a **π (pi) bond.**

It is useful to consider the hybridization about carbon in ethylene in terms of an orbital diagram:

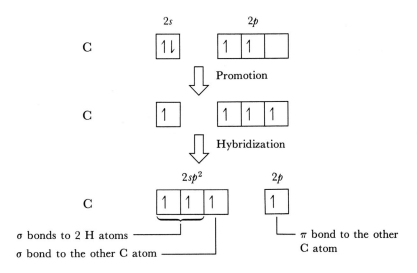

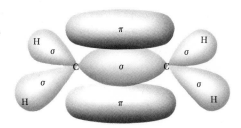

FIGURE 8.12 Formation of the π bond in ethylene by overlap of the $2p$ orbitals on each carbon atom. Note that the centers of charge density in the π bond are above and below the bond axis, whereas in σ bonds the centers of charge density lie on the bond axes.

The first step in our imaginary process of preparing the carbon atom for bonding is again the promotion of one of the $2s$ electrons to the vacant $2p$ orbital. We then form an sp^2 hybrid orbital set from the $2s$ and two of the $2p$ orbitals. The orbitals of the hybrid set are used in bonding as indicated above. By electron pairing in each orbital, carbon gains an octet of electrons in its valence-shell orbitals. Because two pairs of electrons are shared between carbon atoms in ethylene, as compared with only one pair in an ordinary C—C single bond, the C—C bond distance is shorter in ethylene than, for example, in ethane, C_2H_6:

Ethylene
C—C distance = 1.34 Å

Ethane
C—C distance = 1.54 Å

According to the hybrid orbital picture, the double bond between carbons in ethylene consists of one σ and one π bond. This picture of the bonding in ethylene is in good accord with its chemical properties. For example, ethylene reacts with bromine, Br_2, to form dibromoethane:

Ethylene

Dibromoethane

[8.1]

In this reaction the double bond is opened, and a bromine adds to each carbon. The C—C π bond is converted to two C—Br σ bonds.

Acetylene, C_2H_2, is a linear molecule containing a triple bond: H—C≡C—H. Each carbon may be visualized as using sp hybrid orbitals in forming σ bonds with the other carbon and with a hydrogen. Each carbon then has two remaining valence p orbitals at right angles to the

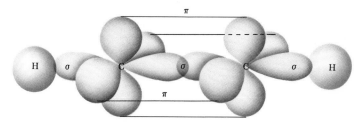

FIGURE 8.13 Formation of two π bonds in acetylene from the overlap of two sets of carbon $2p$ orbitals.

axis of the *sp* hybrid set (Figure 8.13). These overlap to form a pair of π bonds. Thus the triple bond can be thought of as formed from one σ and two π bonds. All double bonds consist of a σ and a π bond, while all triple bonds consist of a σ and two π bonds.

DELOCALIZED ORBITALS

In each of the molecules that we've discussed in this chapter, the bonding electrons are *localized;* by this we mean that the σ and π electrons are associated totally with the two atoms forming the bond. There are some molecules, however, in which the π electrons are free to move over several atoms; we say that such electrons are *delocalized* over those atoms. Benzene, C_6H_6, is an example of such a molecule. This molecule has the following resonance forms:

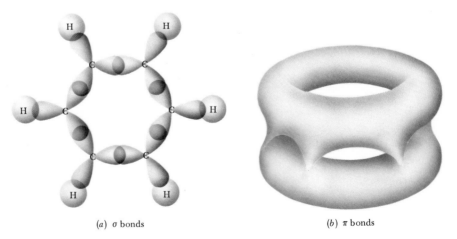

All C—C bonds in benzene are of equal length; the C—C distance is 1.395 Å, intermediate between the values for C—C single bonds (1.54 Å) and C=C double bonds (1.34 Å). The bond angles around each carbon are 120°.

To describe benzene in terms of hybridization of carbon orbitals, we follow the procedure of setting up a hybrid orbital set consistent with the skeletal structure for the molecule. Because each carbon is surrounded by three atoms at 120° angles in a plane, the appropriate hybrid set (Table 8.5) is sp^2, as shown in Figure 8.14(*a*). This leaves a *p* orbital on each carbon perpendicular to the plane of the benzene ring. The situation is

(*a*) σ bonds   (*b*) π bonds

FIGURE 8.14 The σ and π bond networks in benzene, C_6H_6. (*a*) The σ bonds all lie in the molecular plane and are formed from carbon sp^2 hybrid orbitals. (*b*) The π bond network is formed from overlap of a $2p$ orbital on a carbon with the $2p$ orbitals of each of its neighbors. Six π molecular orbitals result.

very much like that in ethylene, except that now we have six carbon $2p$ orbitals, in a cyclic arrangement. The six carbon $2p$ atomic orbitals interact with one another to form π orbitals. Each of the $2p$ orbitals overlaps with two others, one on each adjacent carbon atom, to form a kind of doughnut of electron density above and below the plane of the benzene ring, as illustrated in Figure 8.14(*b*).

Because there is one electron from each $2p$ orbital, there is a total of three electron pairs in the π orbitals formed in this manner. The electrons in the π orbitals of benzene are said to be *delocalized* in the sense that any one electron is free to move around the entire circle of carbon atoms. This delocalization of the π electrons gives benzene a special stability. For example, this substance does not react readily with bromine, as does ethylene.

It is common to represent benzene and related molecules by leaving off the hydrogens attached to the carbon and showing only the carbon-carbon framework. The presence of π electrons may be shown using one of the Lewis structures or by placing a circle in the center of the carbon ring. Thus we can represent benzene as

Benzene and many related molecules are referred to as aromatics. The name aromatics arose because some of the compounds related to benzene possess a spicy fragrance. However, benzene and many other aromatics do not have a pleasant smell. Several aromatic compounds are depicted in Figure 8.15.

Many of the aromatic hydrocarbons are carcinogenic, that is, cancer causing. Benzo[a]pyrene,* shown in Figure 8.15, is among the most potent carcinogens. It is found in significant quantities in urban atmospheres and, most especially, in cigarette smoke. It is regarded as a major cause of lung cancer. The major sources of benzo[a]pyrene in urban atmospheres are coal-burning power plants, engines that burn gasoline and diesel fuel, and incinerators used for refuse disposal. Studies show that for nonsmokers, the incidence of lung cancer among urban dwellers is substantially higher than for rural populations with the same age profile. In both groups the incidence of lung cancer is much higher for cigarette smokers than for nonsmokers.

*The symbol [a] in the name for benzo[a]pyrene is used to designate the manner in which the six-membered rings are connected together.

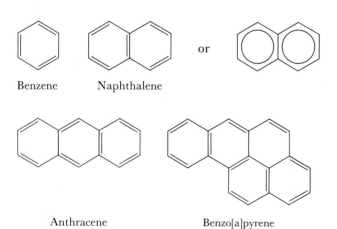

Benzene   Naphthalene

or

Anthracene   Benzo[a]pyrene

FIGURE 8.15 Several aromatic organic structures. Each vertex represents a carbon atom. A hydrogen is attached to each carbon that has only three bonds in the structures as shown.

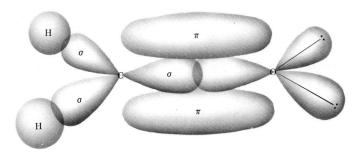

FIGURE 8.16 Formation of σ and π bonds in formaldehyde.

GENERAL CONCLUSIONS

On the basis of all the examples we've seen, we can formulate a few general conclusions that are helpful in using the concept of hybrid orbitals to discuss molecular structures.

1 Every pair of bonded atoms shares one or more pairs of electrons. In every bond at least one pair of electrons is localized in the space between the atoms, in a σ bond. The appropriate set of hybrid orbitals used to form the σ bonds between an atom and its neighbors is determined by the observed geometry of the molecule. The relationship between hybrid orbital set and geometry about an atom is given in Table 8.5.

2 The electrons in σ bonds are localized in the region between two bonded atoms and do not make a significant contribution to the bonding between any other two atoms.

3 When atoms share more than one pair of electrons, the additional pairs are in π bonds. The centers of charge density in a π bond lie above and below the bond axis.

4 π bonds may extend over more than two bonded atoms. Electrons in π bonds that extend over more than two atoms are said to be delocalized.

SAMPLE EXERCISE 8.5

Formaldehyde, which is a planar molecule, has the following Lewis structure:

$$\begin{array}{c} H \\ \diagdown \\ C = \ddot{O} \colon \\ \diagup \\ H \end{array}$$

Describe the bonding in formaldehyde in terms of an appropriate set of hybrid orbitals at the carbon atom.

Solution: Using the VSEPR model, we would predict the bond angles around C to be about 120° (trigonal-planar geometry). The presence of three bonds in a plane about a central atom suggests sp^2 hybrid orbitals for the σ bonds (Table 8.5). There remains a $2p$ orbital on carbon, perpendicular to the plane of the three σ bonds. This orbital overlaps with a similarly oriented orbital on oxygen to form a π bond between carbon and oxygen (see Figure 8.16).

8.5 Molecular orbitals

The models for covalent bonding and molecular geometries discussed in this and the previous chapter are very useful. They provide a nice way of relating the formulas and structures of molecules to the electron configurations of the atoms involved. For example, we can understand why methane has the formula CH_4, and why the arrangement of C—H

bonds about the central carbon is tetrahedral. But in all of this discussion, we have sidestepped a rather important question: Why do atoms combine to form covalent bonds in the first place? The answer to this question has to be expressed in terms of energy.

We have seen that electrons in atoms exist in allowed energy states called atomic orbitals. The quantum theory tell us that, in a similar way, electrons exist in molecules in allowed energy states that are called molecular orbitals. Because molecules are more complex than atoms, it is no surprise that molecular orbitals are more complex than atomic orbitals.

One of the most useful ways of viewing molecular orbitals is to imagine that they are formed by combining the atomic orbitals of the atoms making up the molecule. By examining the formation of molecular orbitals in this fashion, we can understand why some molecules are more stable and others less stable than the separated atoms. There are various rules and restrictions for how atomic orbitals can be combined to form molecular orbitals:

1 When a set of atomic orbitals is combined to form molecular orbitals, the number of molecular orbitals formed is equal to the number of atomic orbitals in the set.

2 The average energy of the molecular orbitals formed by combining a set of atomic orbitals is approximately equal to the average energy of the atomic orbitals. However, some of the molecular orbital energies are lower than the energies of the starting atomic orbitals, while others are higher.

3 The Pauli principle is obeyed for molecular orbitals just as for atomic orbitals; there can be at most two electrons in each molecular orbital, and they must have their spins paired.

4 Atomic orbitals combine most effectively with other atomic orbitals of comparable energy.

5 The effectiveness with which two atomic orbitals combine is proportional to their overlap with one another. Orbitals that have zero overlap thus cannot combine at all.

6 When a molecular orbital is formed by overlap of two nonequivalent atomic orbitals, the bonding molecular orbital contains a greater contribution from the atomic orbital of lower energy. Conversely, the antibonding molecular orbital contains a greater contribution from the higher energy atomic orbital. (Bonding and antibonding orbitals will be explained shortly.)

Let us now illustrate these rules by considering the molecular orbital descriptions of simple diatomic molecules. We begin with two hydrogen atoms coming together to form H_2. The $1s$ orbital on each H atom interacts with the $1s$ orbital on the other atom to form two molecular orbitals (rule 1).

One of the possible interactions or combinations leads to a positive overlap that concentrates electron density between the nuclei. This interaction produces the bonding sigma orbital, labeled σ, shown on the bottom of Figure 8.17. This combination is lower in energy (more stable)

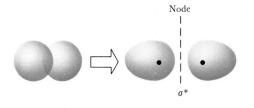

Node

$\sigma*$

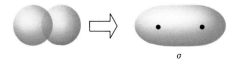

σ

FIGURE 8.17 Contour diagram of the wave functions for the σ and $\sigma*$ molecular orbitals in H_2. Note the presence of a nodal plane midway between the atoms in the $\sigma*$ antibonding orbital.

than the separated atoms. The second molecular orbital that results from the interaction of two $1s$ orbitals is shown on the top of Figure 8.17. This combination leads to negative overlap that concentrates the electron density outside the region between the nuclei. This molecular orbital is the antibonding sigma orbital, labeled $\sigma*$. It is called an antibonding orbital because it is of higher energy (less stable) than the isolated atoms.

The interactions between two $1s$ orbitals to form σ and $\sigma*$ molecular orbitals can be represented by an orbital energy-level diagram as shown in Figure 8.18. Because each isolated hydrogen atom contains one electron, there are two electrons in the H_2 molecule. We use the Pauli exclusion principle (rule 3) in placing these two electrons into the molecular orbitals. Consequently, they occupy the lowest energy orbital available to them, the σ orbital, with their spins paired. The energy of the two electrons in the σ orbital is lower than that of the electrons in the isolated $1s$ orbitals; consequently, the H_2 molecule is more stable than the two separate hydrogen atoms.

Now let's consider the coming together of two helium atoms to form the He_2 molecule. Again, we have two $1s$ orbitals interacting to form a bonding σ and antibonding $\sigma*$ pair of molecular orbitals. In this case, however, each helium atom contributes two electrons, so that the total number to be placed is four. As Figure 8.19 shows, two are placed in the σ orbital, and the other two must be placed in the $\sigma*$ orbital. Using rule 2 we recognize that the $\sigma*$ orbital is destabilized to the same extent that the σ orbital is stabilized. Thus the two electrons that must go into the $\sigma*$ orbital destabilize the He_2 molecule just as much as the two electrons in

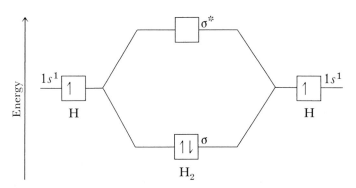

Energy

$1s^1$

H

$1s^1$

H

$\sigma*$

σ

H_2

FIGURE 8.18 Energy-level diagram for the molecular orbitals in the H_2 molecule.

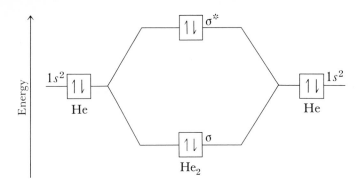

FIGURE 8.19 Energy-level diagram for the molecular orbitals in the He_2 molecule.

the σ orbital stabilize it. We predict from this model that there is no net stabilization in He_2. Laboratory studies have shown that there is no significant tendency for two helium atoms to bond together to form He_2.

From these two examples of the molecular-orbital model, we see that H_2 is stable because electrons can be accommodated in bonding molecular orbitals. The He_2 molecule, on the other hand, is not stable because there are as many electrons in antibonding orbitals as in bonding orbitals.

SAMPLE EXERCISE 8.6

Would you expect the polyatomic ion He_2^+ to be stable relative to a separated He atom and He^+ ion? Explain.

Solution: The energy level diagram for this system is shown in Figure 8.20. Two helium 1s orbitals interact to form bonding and antibonding molecular orbitals. In He_2^+ we have a total of three electrons. Two of these are placed in the bonding orbital, the third in the antibonding orbital, as shown in Figure 8.20.

The stability gained from two electrons in the bonding orbital is greater than the destabilization due to one electron in the antibonding orbital. The He_2^+ molecular ion is therefore predicted to be stable relative to its dissociation products. It has been shown in laboratory studies that He_2^+ is stable relative to separated He and He^+.

In considering the formation of Li_2 from two Li atoms, a new factor enters the picture. The lithium atom has the electron configuration $1s^22s$. In combining two lithium atoms to form the Li_2 molecule we must consider the possible interactions of the 1s orbital on one lithium atom with the 2s orbital on the other. Such an interaction is theoretically possible, but rule 4 tells us that it will not be of importance. The energy difference between the 1s and 2s orbitals is simply too great for a strong

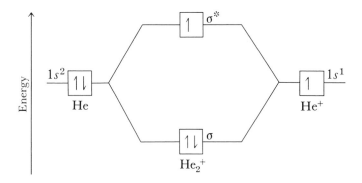

FIGURE 8.20 Energy-level diagram for the molecular orbitals in He_2^+.

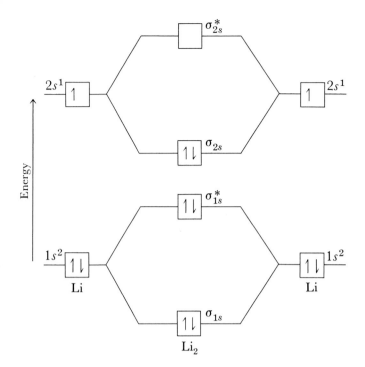

FIGURE 8.21 Energy-level diagram for the Li_2 molecule.

interaction. As a result, the energy-level diagram for the Li_2 molecule is as shown in Figure 8.21. Notice that the molecular orbitals are labeled with subscripts to indicate the set of atomic orbitals from which they are formed. The electrons in the σ_{1s} and σ_{1s}^* orbitals make no net contribution to the bonding. Their average energy is just the energy of the $1s$ electrons in the isolated Li atoms. All of the bonding in the Li_2 molecule is thus due to the $2s$ electrons. This example illustrates the general rule that *filled atomic subshells do not contribute to bonding in molecule formation.* This rule means that whenever an atom has a completed s, p, or d level, the electrons in those atomic orbitals do not contribute to bond formation. It also means that all filled shells do not contribute to bonding. We need only consider the valence-shell electrons. This conclusion is equivalent to the assumption we make when we draw Lewis structures that show only the valence-shell electrons.

Using this rule, we can account for the bonding in the other alkali-metal diatomic molecules. The valence-shell part of the molecular-orbital diagram for Na_2 or K_2 would look just like the valence-shell part for Li_2 (see Figure 8.22). In Figure 8.22, n represents the major quantum

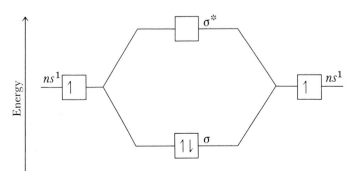

FIGURE 8.22 Energy-level diagram for alkali-metal diatomic molecules.

TABLE 8.6 Bond-dissociation energies for H_2 and the alkali metals in diatomic molecules

Molecule	Major quantum number of s orbital	Bond distance (Å)	Dissociation energy (kJ/mol)
H_2	1	0.75	430
Li_2	2	2.67	100
Na_2	3	3.08	72.4
K_2	4	3.92	49.4
Rb_2	5	4.2	45.2
Cs_2	6	4.7	42.6

number of the valence-shell s orbital. There is no need to show any of the electrons below the valence shell, because they do not contribute to bonding.

Although Li_2 and the other alkali metals are similar to H_2 in terms of the energy-level diagram, the change in major quantum number is of great importance. Table 8.6 shows the bond-dissociation energies of H_2 and the alkali-metal diatomic molecules through Cs_2. (The alkali metals are not normally in the form of diatomic molecules, but they can be studied as diatomic molecules in the gas phase at high temperatures.) The bond-dissociation energy decreases steadily in this series. The major factor responsible for this effect is a decrease in the extent to which the valence-shell s orbitals overlap (rule 5). The inner-shell electrons set a limit on how close the nuclei can draw together, because of repulsions between filled shells. In addition, the s orbitals become more spatially extended with an increase in major quantum number. The overall result is a decrease in the extent of overlap of the s orbitals with increasing major quantum number for the diatomic molecules of Table 8.6.

SAMPLE EXERCISE 8.7

Beryllium, the fourth element of the periodic table, does not form a stable diatomic molecule. Provide a reason for this in terms of molecular orbital formation.

Solution: The electron configuration for Be is $1s^2 2s^2$. The energy-level diagram for Be_2 involves interactions of the $1s$ and $2s$ orbitals, just as for Li_2, as shown above. The four valence-shell electrons in Be_2, however, completely fill both σ_{2s} and σ_{2s}^* orbitals, to leave a net bonding of zero.

OVERLAPS OF p AND s ATOMIC ORBITALS

When we consider the formation of diatomic molecules for the elements beyond beryllium, we need to take account of the ways in which the p orbitals on the atoms may combine with one another and with s orbitals. Figure 8.23 shows the contour diagrams for the molecular orbitals formed by combining s and p atomic orbitals in all the possible ways that lead to the formation of molecular orbitals. These combinations are of two kinds. When the orbitals that make up the combination concentrate electron density about the bond axis, the molecular orbital formed is of the σ type. When the orbitals that make up the combination concentrate electron density above and below the bond axis, the molecular orbital

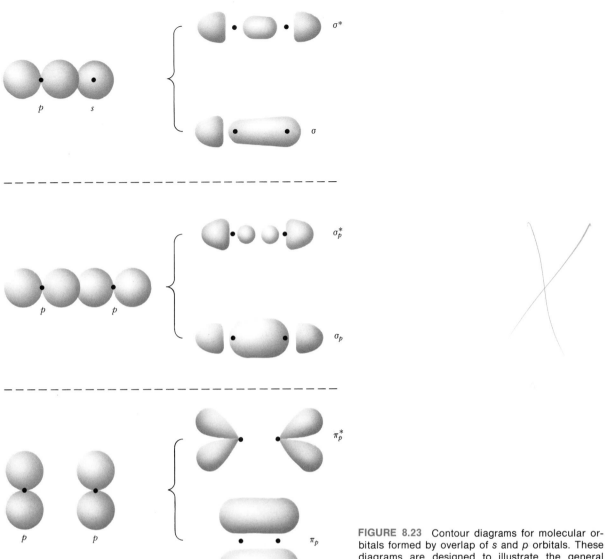

FIGURE 8.23 Contour diagrams for molecular orbitals formed by overlap of s and p orbitals. These diagrams are designed to illustrate the general shapes of bonding and antibonding orbitals and are not accurate representations.

formed from the atomic orbitals is labeled π. (You will remember that we used these same designations in describing bonds formed from overlaps of hybrid atomic orbitals, Section 8.4.)

Both a bonding and an antibonding combination is obtained in each case. Note that the electron density in the bonding orbitals tends to be concentrated in the region between the nuclei, whereas in the antibonding orbital it is concentrated in the regions in back of the nuclei.

8.6 Molecular-orbital diagrams for diatomic molecules

We are now ready to consider the molecular-orbital energies in a diatomic molecule formed from two atoms of an element from the second row of the periodic table. Without specifying which element is involved, let us imagine that two atoms of a second-row element come together to

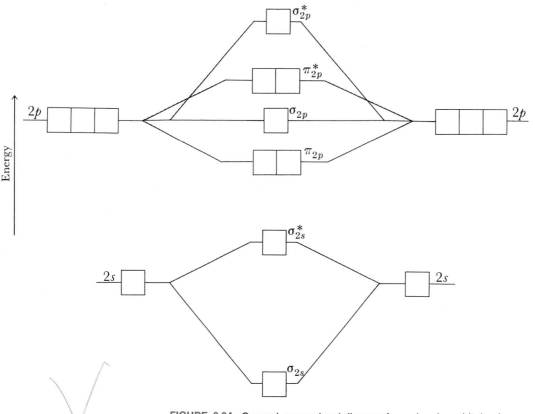

FIGURE 8.24 General energy-level diagram for molecular orbitals of second-row diatomic molecules.

form a molecule. The atomic orbitals on each atom interact with those on the other atom, in accordance with the rules described above. From these interactions, a set of molecular orbitals results, as shown in Figure 8.24. We might expect that the $2s$ orbitals would interact exclusively with one another and the $2p$ orbitals with other $2p$ orbitals. Because there are two possible ways in which the $2p$ orbitals can interact, there are both σ and π molecular orbitals formed in this case. The only complication in this simple picture comes from the fact that the σ_{2s} and σ_{2p} orbitals can interact with one another, as can the σ_{2s}^* and σ_{2p}^*. As a result, the σ_{2p} and σ_{2p}^* orbitals, formed from the $2p$ orbitals, are pushed a little upward in energy; the σ_{2s} and σ_{2s}^* orbitals, formed from the $2s$ orbitals, are pushed downward. These changes in energy result from a certain amount of mixing of the σ_{2s} and σ_{2p} orbitals, and of the σ_{2s}^* and σ_{2p}^* orbitals. Note also that the energy splitting between bonding and antibonding orbitals is larger for the σ_{2p} orbitals than it is for the π_{2p} orbitals. This occurs because the overlap of p orbitals is greater when they are oriented along the axis with respect to one another in σ fashion than when they are oriented in π fashion (rule 5).

Using this basic energy-level diagram we can readily deduce the electronic configuration of any of the second-row diatomic molecules. Some of these molecules are familiar substances, for example, O_2 or N_2. Others are known only in the vapor state at high temperature or under other unusual conditions. In all cases, however, the electronic structures are known from experiments. **Bond order** in these molecules is defined as the excess of bonding electron pairs over antibonding pairs. Thus in Li_2, with two valence electrons, the bond order is one. The molecule Be_2, with four valence-shell electrons, should have a net bond order of zero, because the four electrons are placed in the σ_{2s} and σ_{2s}^* orbitals. However, B_2 should be stable, with a net bonding pair, for a bond order of one. As shown in Table 8.7, the two bonding electrons occupy the pair of π orbitals of equal energy. Hund's rule operates here just as for electrons in atomic orbitals: Electrons occupy degenerate orbitals singly and with spins parallel until all degenerate orbitals have been occupied. Thus B_2 is predicted to have two unpaired electrons, in agreement with experimental results. Pairing of the π electrons comes in C_2, which has a net bond order of two. With nitrogen, the bond order reaches its maximum, three. In view of this high bond order, the exceptional stability of the N_2 molecule and its high bond-dissociation energy are understandable.

The next molecule in our series, O_2, is especially interesting. We have a total of 12 valence electrons to place. As shown in Table 8.7, the last two of these must go into the π_{2p}^* orbitals with spins parallel. Thus the bond order in O_2 is two. Because of the unpaired electrons, molecular oxygen is **paramagnetic**. (A paramagnetic substance is one that is drawn into a magnetic field.) Oxygen is the only commonly available, simple molecule with this property. The paramagnetism of O_2 can be demonstrated by observing the effect of a magnetic field on a tube containing liquid oxygen, as shown in Figure 8.25. When the field is applied, the tube is moved laterally as the sample is drawn into the magnetic field. The prediction of the paramagnetism of O_2 is a most elegant achieve-

TABLE 8.7 Electronic configuration and some experimental data for several second-row diatomic molecules

	B_2	C_2	N_2	O_2	F_2
σ_{2p}^*	[ ]	[ ]	[ ]	[ ]	[ ]
π_{2p}^*	[ ][ ]	[ ][ ]	[ ][ ]	[↑][↑]	[↑↓][↑↓]
σ_{2p}	[ ]	[ ]	[↑↓]	[↑↓]	[↑↓]
π_{2p}	[↑][↑]	[↑↓][↑↓]	[↑↓][↑↓]	[↑↓][↑↓]	[↑↓][↑↓]
σ_{2s}^*	[↑↓]	[↑↓]	[↑↓]	[↑↓]	[↑↓]
σ_{2s}	[↑↓]	[↑↓]	[↑↓]	[↑↓]	[↑↓]
Bond order	One	Two	Three	Two	One
Bond-dissociation energy (kJ/mol)	290	620	941	495	155
Bond distance (Å)	1.59	1.31	1.10	1.21	1.43
Ionization energy (kJ/mol)	—	1150	1495	1205	1700

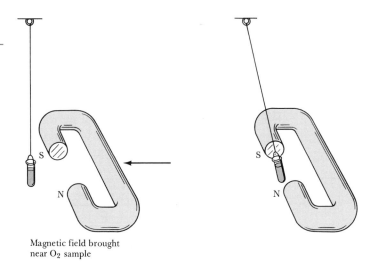

Magnetic field brought near O_2 sample

FIGURE 8.25 Illustration of an experiment that shows the paramagnetic character of O_2. When the magnetic field is applied, the sample of liquid O_2 is drawn into the field.

ment of molecular-orbital theory. The simple Lewis theory (Section 7.4) does not account for the paramagnetism, and there is no other bonding model for O_2 in which this property is explained so naturally.

In the next molecule in our series, F_2, the electrons in the π_{2p}^* orbitals are paired; the net bond order is one. In the hypothetical Ne_2 molecule, all valence molecular orbitals would be filled, for a net bond order of zero. The Ne_2 molecule is nonexistent. It is interesting to compare the electronic structures of the diatomic molecules with their observable properties. One of the most important measures of bond strength is the bond-dissociation energy—that is, the energy required to separate the two atoms of the molecule to a very large distance from one another (Section 7.8). Another important characteristic of a bond is the distance separating the bonded atoms. We have seen in previous discussions (Section 7.4) that bond distances are smaller in the case of multiple bonds. A third property of the diatomic molecules that relates to bonding is the ionization energy—that is, the energy required to remove the highest energy electron from the molecule.

The bond-dissociation energies, bond distances, and ionization energies for the diatomic molecules are listed in Table 8.7. Note that the bond-dissociation energies of molecules with the same bond order are not the same. This should not surprise us, because overlaps differ, and many other factors contribute to the total energy of the molecule. Still, it is roughly true that bond-dissociation energy increases with bond order. Similarly, bond length decreases with increasing bond order.

In addition to the neutral diatomic molecules listed in Table 8.7, there are several known diatomic ions, such as N_2^+ and O_2^+. These species can be produced in the gas phase and their properties studied, even though they are too reactive to permit isolation. We'll see in Chapter 10 that N_2^+ and O_2^+ are important components of the earth's upper atmosphere. It is possible, using the energy-level diagram for molecular orbitals, to predict some of the properties of these ions.

SAMPLE EXERCISE 8.8

Predict the following properties of O_2^+: (a) number of unpaired electrons, (b) bond order, (c) bond-dissociation energy, and (d) bond length.

Solution: O_2^+ has one electron less than O_2. The electron removed from O_2 to form O_2^+ is one of the two unpaired π^* electrons. O_2^+ should therefore have just one unpaired electron left. Because the electron removed has come from an antibonding orbital, the bond order in O_2^+ is larger than for O_2; it is in fact intermediate between O_2 and N_2. Counting a single electron as contributing a bond order of $\frac{1}{2}$, the bond order is $2\frac{1}{2}$. The bond-dissociation energy and bond length should be about midway between that for O_2 and N_2, say, 720 kJ/mol and 1.15 Å, respectively. The observed properties of O_2^+ are: number of unpaired electrons, one; bond length, 1.123 Å; dissociation energy, 625 kJ/mol.

In principle, the energy-level diagram shown in Figure 8.24 is applicable to the diatomic molecules of the third- and higher-row elements. Except for the halogens, however, in which only a single bond is possible, the formation of diatomic molecules among the other elements is not the most stable form of bonding. The molecules P_2 and S_2 have been observed in the high-temperature vapors of these elements, but at lower temperatures other forms of the element are more stable (Section 8.7). Because P—P and S—S single bonds are quite stable, the major reason for the relative instability of the diatomic molecules in these two cases seems to be that their π bonds are not very strong.

POLAR COVALENT MOLECULAR ORBITALS

The concept of molecular orbitals can also be used to understand molecule formation between two different atoms. When the atoms are not very different, the energy-level diagram of Figure 8.24 can be used with some modification. The energy levels of one of the atoms are shifted with respect to the other. Such a diagram should be applicable to such molecules as NO, CO, CN, and so forth.

When the two atoms forming a molecule are appreciably different, extensive changes need to be made in the energy-level diagram. In many instances, it is possible to make simplifying assumptions that work quite well. For example, consider HF, in which the two atoms have quite different electronegativities. This is a simple case because hydrogen has only a single orbital for interaction with the orbitals of fluorine. The energy of the hydrogen 1s orbital is higher than that of the highest valence-shell orbital of fluorine, the 2p, as shown in Figure 8.26. It is *much* higher than the fluorine 2s orbital. Thus, we need consider only the interaction of the hydrogen 1s orbital with the 2p. Of the three 2p orbitals, only one is of correct symmetry to interact with the hydrogen 1s. The interaction leads to a bonding and antibonding pair of orbitals, as shown in Figure 8.26. The other orbitals of fluorine are relatively unperturbed by molecule formation.

The bonding orbital occupied by a pair of electrons in HF is a polar covalent bond. This means that the contributions of the two atomic orbitals to the molecular orbitals are not the same. In the bonding orbital, the fluorine 2p orbital is more important than the hydrogen 1s. As a result, the bonding electron pair is shifted more toward the fluorine

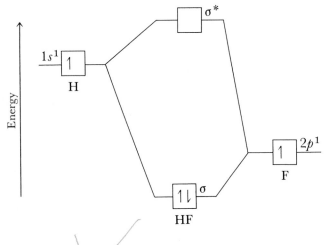

FIGURE 8.26 Energy-level diagram for HF molecular orbitals.

atom than toward hydrogen. In the antibonding orbital, the contribution of the hydrogen $1s$ orbital is greater than that of the fluorine $2p$. Hydrogen fluoride provides an example of rule 6 in the list of rules for formation of molecular orbitals.

8.7 Structures of the nonmetallic elements

Our discussion of molecular shape and bonding has provided the background for us to examine the structures of some common nonmetals. As we do so, we will also take the opportunity to introduce a few properties of these elements, especially ones that reflect molecular structure and bonding.

The halogens, family 7A, all consist of diatomic molecules, with formulas of the type X_2. Fluorine and chlorine are both gases at room temperature. Fluorine has a pale yellow color while chlorine is greenish-yellow. Bromine is a dark red liquid that boils at 58°C. Iodine is a dark violet solid with an almost metallic appearance; it readily forms a violet vapor when gently heated. The diatomic structure of the halogens is maintained not only in the vapor phase, but also in the liquid and solid phases. The diatomic nature of the halogens reflects the fact that each halogen atom has seven valence electrons. Covalent bonding with a single bond between the halogen atoms therefore provides an octet of electrons around each:

$$: \overset{..}{\underset{..}{X}} - \overset{..}{\underset{..}{X}} :$$

Oxygen, the first member of family 6A, also occurs as a diatomic molecule, O_2. This substance is familiar as a common and important component of the atmosphere. The short O—O bond distance (1.21 Å) and the relatively high bond-dissociation energy (495 kJ/mol) of O_2 suggest that it possesses a double bond. The following Lewis structure is consistent with this observation:

$$\overset{..}{\underset{..}{O}} = \overset{..}{\underset{..}{O}}$$

However, this Lewis structure leads to the incorrect prediction that all valence electrons are paired. In fact, the O_2 molecule is paramagnetic; it

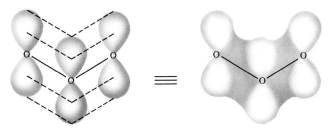

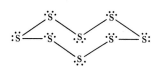

FIGURE 8.27 The delocalized π bond in ozone formed by overlap of $2p$ orbitals on each of the oxygen atoms.

contains two unpaired electrons. Although we can write useful Lewis structures for most molecules, the Lewis structure for O_2 fails to account for the paramagnetism of the molecule. We have seen, in Section 8.6, that molecular-orbital theory correctly predicts both oxygen's bond order of two and its paramagnetism.

Oxygen also occurs as a gaseous triatomic molecule, O_3, known as ozone. This substance is responsible for the pungent odor often detected around electric motors. In contrast to O_2, which is essential for life, O_3 is toxic. Ozone and O_2 are allotropes of oxygen. Allotropes are different forms of the same element in the same state (in this case, both gases). Ozone is less stable and more reactive than O_2. In particular, it is a much stronger oxidizing agent. Although O_2 reacts directly with most elements, high temperatures are often required; O_3 is more likely to react readily at lower temperatures.

The ozone molecule may be represented by the following resonance forms:

$$ \ddot{O} \diagdown \overset{\overset{\displaystyle ..}{O}}{\diagup} \ddot{O} \longleftrightarrow \ddot{O} \diagup \overset{\overset{\displaystyle ..}{O}}{\diagdown} \ddot{O} $$

The molecule is bent, as we would predict using the VSEPR model; the bond angle is found experimentally to be $116.8°$ and the O—O bond distance is 1.278 Å. The σ bonds between the oxygen atoms involve approximately sp^2 hybridization of the valence-shell electrons on the central oxygen atom. A π bond that is delocalized over the three atoms is formed by sideways overlap of p orbitals on each atom (Figure 8.27). The bond-dissociation energy of the molecule, corresponding to the reaction $O_3(g) \longrightarrow O_2(g) + O(g)$, is 107 kJ/mol. This is much lower than the O—O bond energy of O_2. The lower bond energy in O_3 explains why O_3 is more reactive than O_2.

Sulfur, the second member of family 6A, exists in several allotropic forms. The most stable and common allotrope at room temperature is a yellow solid with molecular formula S_8. The S_8 molecule consists of an eight-membered ring of sulfur atoms as shown in Figure 8.28. The S—S—S bond angle is $107.8°$, close to the tetrahedral angle associated with sp^3 hybridization. Each sulfur atom in this structure achieves an octet of electrons by bonding to two other sulfur atoms by single σ-type bonds. By contrast, we have seen that oxygen satisfies its bonding capacity by forming π bonds in the O_2 and O_3 molecules. π bonds are generally more common between smaller atoms like O, N, and C than between larger ones like S, P, and Si.

Nitrogen, the lightest member of family 5A, exists in the earth's at-

FIGURE 8.28 The structure of S_8 molecules as found in the most common allotropic form of sulfur at room temperature.

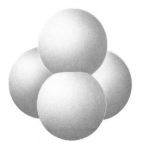

FIGURE 8.29 The structure of P_4 molecules as found in white phosphorus, a common allotropic form of the element.

mosphere as the very stable N_2 molecule. The Lewis structure for N_2 indicates that each nitrogen achieves an octet by forming a triple bond:

$$:N\equiv N:$$

The bond order of three, arising from a σ and two π bonds is also predicted by molecular-orbital theory. The very strong, nonpolar N—N bond results in exceptionally low chemical reactivity. The N—N dissociation energy is 941 kJ/mol and the nitrogen atoms are separated by only 1.10 Å.

In contrast to nitrogen, the next member of family 5A, phosphorus, does not exist as a diatomic molecule, except at high temperatures in the vapor state. Phosphorus forms weak P—P π bonds and consequently seeks to achieve its octet by utilizing only σ bonds; because each atom has five valence electrons, three σ bonds are required. When phosphorus condenses from the vapor state it forms so-called white phosphorus. This reactive allotrope consists of P_4 molecules formed by a tetrahedral arrangement of phosphorus atoms, as shown in Figure 8.29. Each atom in this structure is bonded to three other phosphorus atoms by single bonds. However, this geometry requires that each P—P—P bond angle be only 60°. This angle is much smaller than that in any other molecule that we have discussed thus far. The 60° bond angle requires that the bonding electron pairs in the molecule be crowded rather closely together. The high reactivity of the P_4 molecule is probably due to the strain introduced by electron-electron repulsions.

The heavier elements of group 5A, arsenic, antimony, and bismuth, show increasingly metallic properties with increasing period. The allotropy of arsenic is somewhat similar to that of phosphorus, but for antimony and bismuth the structures of the elements are more complex and possess many of the characteristics of metals. These elements are frequently referred to as metalloids, to indicate that they are on the borderline between metals and nonmetals in their properties. The group 5A elements exemplify the rule that *in any one group, the metallic characteristics of the elements increase with increasing atomic number*. In the periodic table shown on the inside front cover, the heavy line that runs diagonally from just below B to Po represents the approximate dividing line between metallic and nonmetallic elements.

Carbon, the lightest of the group 4A elements, exists in two major allotropic forms, graphite and diamond. The structure of diamond is shown in Figure 8.30. You can see that each carbon atom is surrounded by four other carbon atoms arranged at the corners of a tetrahedron. Each carbon can therefore be described as using sp^3 hybrid orbitals in forming σ bonds with each of its four neighboring carbon atoms. The stability of the diamond lattice comes from the fact that the carbon atoms are interconnected in a three-dimensional array of strong carbon-carbon single bonds. Diamond is a very hard and brittle material. In fact, industrial-grade diamonds are employed in the blades of saws for the most demanding cutting jobs. The melting point of diamond, above 3500°C, is higher than that for any other element. It is exceptionally inert chemically. However, when heated to about 1800°C in the absence of air, diamond converts to graphite, the other allotropic form of the

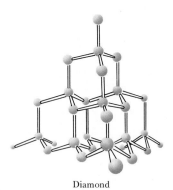

Diamond

FIGURE 8.30 Structure of diamond, a major allotropic form of carbon.

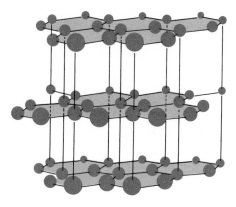

FIGURE 8.31 The structure of graphite.

element. Diamonds also burn at about 900°C when heated in air or in oxygen. (Naturally, there is not much interest in carrying out experiments of this kind.)

The structure of graphite is shown in Figure 8.31. The carbon atoms are arranged in layers; each carbon atom is surrounded by three other carbon atoms, all at the same distance of separation. The distance between adjacent carbon atoms in the plane is 1.42 Å, very close to the C—C distance in benzene (1.395 Å). The distance between adjacent layers is 3.41 Å, too great a distance for a covalent bond to exist. Graph-

FIGURE 8.32 A piece of graphite photographed with an electron microscope (magnification about 15 million times). The bright bands are layers of carbon atoms that are only 3.41 Å apart. (*P. A. Marsch and A. Voet, J. M. Huber Corp.*)

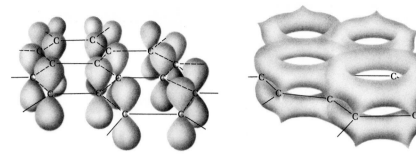

FIGURE 8.33 Formation of the π bonds in graphite.

ite occurs in grey, shiny plates and is usually found in masses of thin, easily separated sheets (see Figure 8.32). The layers easily slide past one another when rubbed, giving the substance a greasy feel. Graphite has been used as a lubricant and in making the "lead" in pencils.

The properties of graphite are due to the nature of the bonding between carbon atoms. Because the individual layers in graphite are not directly bonded together, they slide past one another easily and are readily separated. We can understand the bonding within each layer by supposing that each carbon is bonded to its three neighbors by sp^2 hybrid σ bonds. The remaining $2p$ orbital is employed in π bonding to the same three atoms, as illustrated in Figure 8.33. But because those three neighboring atoms are also involved in π bonding to two other carbons, the network of π bonds extends over essentially the entire plane. The electrons that occupy the π bonds are free to move from one bond to the next. Thus graphite represents an extension to an entire plane of the kind of delocalization we saw in benzene. In benzene the π electrons are free to move about the circumference of the ring. In graphite they are free to move over the entire plane. Because of this freedom of motion, graphite is a good conductor of heat and electricity in directions along the planes of carbon atoms. (If you have ever taken apart a flashlight battery, you know that the central electrode in the battery is made of graphite.) Although graphite readily conducts heat and electricity along the planes, it is an insulator in the direction normal to (perpendicular to) the planes. This is so because there is no means by which electrons can move easily from one plane to the other.

The structures of elemental silicon and germanium are the same as for diamond. No graphitelike allotrope of these elements is known. This observation parallels our expectation that π bonds between these larger atoms will be of low strength. Tin exists in two allotropic forms. White tin is the more metallic in character; this is the more stable form at room temperature and above. However, at lower temperatures, the element converts to a gray form that has the diamond structure. The conversion of tin from the metallic form to a gray powder was a great nuisance when tin was used for construction in cold climates. This phenomenon was referred to as "tin disease." Tin is another example of an element on the borderline between metallic and nonmetallic, with perhaps more metallic character in its properties. Lead, the heaviest element of group 4A, is quite markedly metallic in character.

Summary

In this chapter, we've applied the basic principles of chemical bonding to several important areas of chemical structure and behavior. The three-dimensional structures of molecules are determined by the distances between bonded atoms and by the directions of chemical bonds with respect to one another around a particular atom. The **valence-shell electron-pair repulsion (VSEPR) model** explains these relative directions in terms of the repulsions that exist between electron pairs. According to this model, electron pairs around an atom orient themselves so as to minimize electrostatic repulsions; that is, they remain as far apart as possible. By recognizing that unshared electron pairs take up more space (exert greater repulsive forces) than shared electron pairs, it is possible to account for the departures of bond angles from the idealized values and to explain many other aspects of molecular structure. The shape of a molecule and the bond polarities determine whether or not a molecule will be polar. The degree of polarity of a molecule is measured by its **dipole moment.**

The Lewis model for covalent bonding introduced in Chapter 7 can be extended to account very nicely for the geometrical properties of molecules. We can imagine that the atoms in a molecule are bonded to one another by electron pairs that occupy pairs of overlapping atomic orbitals. The extent to which the atomic orbitals share the same region of space, called overlap, is important in determining the amount of stability that results from bond formation. The bonds directed along the internuclear axes are called σ **bonds.** It is possible to formulate orbitals on an atom that are directed toward each of the other atoms surrounding it by forming **hybrid orbitals.** These orbitals are made up of mixtures of the familiar s, p, and d atomic orbitals. Depending on the particular number of other atoms bonded to an atom and their arrangement in space, a particular set of hybrid orbitals can be formulated that has the necessary directional characteristics. For example, sp^3 hybrid orbitals are directed toward the corners of a tetrahedron.

In addition to the σ bonds, which determine the geometry of the bonding around a particular atom, there may be also π **bonds** constructed from remaining, unhybridized atomic orbitals. Thus double bonds, consisting of a σ and a π bond, or a triple bond, consisting of a σ and two π bonds, may be formed. In some molecules the π bonds may extend, or be delocalized, over several atoms. Delocalization of the π electrons in a cyclic structure, such as in benzene, or throughout a plane, as in graphite, leads to a special stability.

The coming together of atoms to form molecules may be viewed also as the coming together of atomic orbitals to form **molecular orbitals.** Atomic orbitals may combine with one another in various ways. The rules for combining atomic orbitals on atoms to form molecular orbitals allow us to account very well for the observed properties of the diatomic molecules formed by the first several elements of the periodic table. The molecular-orbital model is particularly impressive in explaining the fact that the O_2 molecule contains two unpaired electrons. Many of the ideas presented in this chapter are important in understanding the structures of the nonmetallic elements.

Learning goals

Having read and studied this chapter, you should be able to:

1 Relate the number of electron pairs in the valence shell of an atom in a molecule to the geometrical arrangement around that atom.

2 Explain why unshared electron pairs exert a greater repulsive interaction on other pairs than do shared electron pairs.

3 Predict the geometrical structure of a molecule or ion from its Lewis structure.

4 Determine whether or not a molecule has a dipole moment.

5 Explain the concept of hybridization and its relationship to geometrical structure.

6 Assign a hybridization to the valence orbitals of an atom in a molecule, knowing the number and geometrical arrangement of the atoms to which it is bonded.

7 Formulate the bonding in a molecule in terms of σ bonds and π bonds, from its Lewis structure.

8 Explain the concept of delocalization in π bonds.

9 Explain the concept of orbital overlap.

10 Describe how molecular orbitals are formed by overlap of atomic orbitals.

11 Explain the relationship between bonding and antibonding molecular orbitals.

12 Construct the molecular-orbital energy-level diagram for a diatomic molecule or ion built from elements of the first or second row and predict the bond order and number of unpaired electrons.

13 Describe the structures of the common nonmetallic elements.

14 Relate the structures and bonding of nonmetallic elements to their properties.

Key terms

Among the more important terms and expressions used for the first time in this chapter are the following:

An **allotrope** (Section 8.7) is one of the forms of an element, when that element is capable of existing in more than one form. For example, O_2 and O_3 are allotropes of oxygen.

An **antibonding molecular orbital** (Section 8.5) is a molecular orbital in which electron density is concentrated in regions at the ends of atoms rather than between them. Such orbitals, designed as σ^* or π^*, are less stable (of higher energy) than bonding molecular orbitals.

A **bonding molecular orbital** (Section 8.5) is one in which the electron density is concentrated in the internuclear region. The energy of a bonding molecular orbital is lower than the energy of the separate atomic orbitals from which it forms.

Bond order (Section 8.6) is expressed as the number of bonding electron pairs shared between two atoms, less the number of antibonding electron pairs.

The **dipole moment** (Section 8.2) is a measure of the separation between centers of positive and negative charges in polar molecules.

Hybridization (Section 8.3) refers to the mixing of different types of atomic orbitals to produce a set of equivalent hybrid orbitals.

A **molecular orbital** (Section 8.5) is an allowed state for an electron in a molecule. A molecular orbital is entirely analogous to an atomic orbital, which is an allowed state for an electron in an atom. A molecular orbital may be classified as σ or π, depending on the disposition of electron density with respect to the internuclear axis.

The term **overlap** (Section 8.3) refers to the extent to which atomic orbitals on different atoms share the same region of space to form a molecular orbital. When overlap is large, a strong bond may be formed.

Paramagnetism (Section 8.6) is a property that a substance may possess if it contains one or more unpaired electrons. A paramagnetic substance is drawn into a magnetic field.

A **pi (π) bond** (Section 8.4) is a covalent bond in which electron density is concentrated above and below the line joining the bonding atoms.

A **sigma (σ) bond** (Section 8.3) is a covalent bond in which electron density is concentrated along the internuclear axis.

The **valence-shell electron-pair repulsion (VSEPR) model** (Section 8.1) accounts for the geometrical arrangments of shared and unshared electron pairs around a central atom in terms of the repulsions between electron pairs.

EXERCISES

The VSEPR model

8.1 Without referring to Table 8.1, indicate the angles characteristic of the maximum separation of (a) two electron pairs; (b) three electron pairs; (c) four electron pairs; (d) six electron pairs.

8.2 Using the VSEPR model, predict the geometrical structures of the following: (a) CF_4; (b) SO_3^{2-}; (c) H_2CO; (d) SeF_4; (e) CS_2; (f) SO_3.

8.3 Predict the molecular shape of each of the following: (a) CH_2Cl_2; (b) $POCl_3$ (has three PCl bonds); (c) ClO_2^-; (d) H_3O^+; (e) N_2O (NNO); (f) $TiCl_4$.

8.4 Urea, the end-product of protein metabolism in animals, has the following Lewis structure:

$$H-\overset{\displaystyle ..}{\underset{\displaystyle H}{N}}-\overset{\displaystyle ..}{\underset{\displaystyle :O:}{C}}-\overset{\displaystyle ..}{\underset{\displaystyle H}{N}}-H$$

Describe the geometry of this molecule, indicating the approximate bond angles around the N and C atoms.

8.5 Explain the variation in H—M—H bond angles among the following compounds: CH_4 ($109.5°$); NH_3 ($107°$); H_2O ($104.5°$).

8.6 Which member of each of the following pairs would you predict to have the larger bond angle: (a) NF_3 or BF_3; (b) H_2O or NH_3; (c) C_2H_2 or C_2H_4; (d) OF_2 or H_2O? Explain.

[8.7] The three species NO_2^+, NO_2, and NO_2^- all have a central nitrogen atom. The ONO bond angles in the three species are $180°$, $134°$, and $115°$, respectively. Explain this variation in bond angle.

Dipole moments

8.8 How does the dipole moment of a diatomic molecule, AB, change as the electronegativity difference between A and B decreases?

8.9 Predict whether the following molecules are polar or nonpolar: (a) $CHCl_3$; (b) CF_4; (c) BF_3; (d) NH_3; (e) CO; (f) SF_4.

8.10 Would you expect CS_2 to have a higher dipole moment than SCO or a lower one? Explain.

8.11 Explain how the dipole moment of PF_3Cl_2 could be used to deduce whether the Cl atoms occupy axial or equatorial positions.

8.12 (a) The bond length in the HBr molecule is 1.41 Å. What is the dipole moment, in units of debyes, of a pair of opposite charges 1.60×10^{-19} coul in magnitude separated by this distance (-1.60×10^{-19} coul is the charge of an electron). (b) The observed dipole moment of HBr is 0.79 D. Explain why the dipole moment that you have calculated in (a) is larger than the dipole moment observed for HBr.

Hybrid orbitals

8.13 Without referring to Table 8.5, indicate the angles between the following orbitals: (a) sp hybrids; (b) sp^2 hybrids; (c) sp^3 hybrids; (d) dsp^2 hybrids; (e) dsp^3 hybrids; (f) d^2sp^3 hybrids; (g) unhybridized p orbitals.

8.14 Indicate the hybrid orbitals employed by carbon in each of the following compounds: (a) CH_4; (b) C_2H_4; (c) C_2H_2; (d) H_2CO. Indicate the number of π bonds in each of these molecules.

8.15 Indicate the type of hybrid orbitals employed by the central atom in each of the following species: (a) SiH_4; (b) PCl_3; (c) AsF_5; (d) ICl_4^+.

Consider the acetic acid molecule which gives vinegar its characteristic odor:

H :O:
| ||
H—C—C—Ö—H
| ..
H

(a) What are the approximate bond angles around each carbon atom? (b) What is the hybridization around each carbon atom? (c) What is the total number of σ bonds in the entire molecule? (d) What is the total number of π bonds?

8.17 Explain what is meant by *overlap* of atomic orbitals.

8.18 What relationship exists between the concepts of delocalized bonding and resonance? Use both concepts to explain the bonding in the CO_3^{2-} ion.

Molecular orbitals

8.19 Show, using drawings, how two p orbitals can overlap to form either a σ or π molecular orbital. Sketch the σ, σ^*, π, and π^* orbitals that result from interaction of p orbitals.

8.20 The acetylide ion, C_2^{2-}, occurs in calcium carbide. Draw the molecular-orbital energy-level diagrams for both C_2^{2-} and C_2 and compare them with regard to their expected relative bond energies, bond lengths, and magnetism.

8.21 Using the energy-level diagram for homonuclear diatomic molecules, predict the bond order of SO and indicate whether the molecule should be paramagnetic. Why is SO less stable than O_2?

8.22 Why is the bonding orbital formed from two atomic orbitals lower in energy than the starting atomic orbitals?

8.23 Explain why removal of an electron from O_2 strengthens the bond whereas removal of an electron from N_2 weakens it.

Structures of the nonmetallic elements

8.24 What experimental evidence can you cite to support the formulation of N_2 as having a triple bond?

8.25 What are the two allotropic forms of oxygen? How do they differ from one another in terms of structure and properties?

8.26 Why doesn't phosphorus normally occur as P_2 molecules in a manner analogous to nitrogen, which forms the stable N_2 molecule?

8.27 Explain why silicon does not possess a graphite-like allotrope.

8.28 Explain why the bond angle in ozone is less than the ideal $120°$ angle associated with three pairs of electrons.

8.29 Discuss the bonding in O_3 both in terms of resonance and in terms of delocalized molecular orbitals.

8.30 Account for (a) the electrical conductivity of graphite; (b) the hardness of diamond; (c) the reactivity of white phosphorus.

Additional exercises

8.31 Using VSEPR theory, predict the geometry of each of the following: (a) SO_4^{2-}; (b) $AlCl_6^{3-}$; (c) $SnCl_2$; (d) PH_4^+; (e) ICl_3; (f) OCN^-.

8.32 The reaction $BF_3 + NH_3 \longrightarrow F_3BNH_3$, causes the BF bond angles to change. Explain.

8.33 Explain why a dipole may orient itself by rotating in an electric field but not undergo a net migration from one charged plate to the other.

8.34 (a) Can a molecule have a dipole moment if it has no polar covalent bonds? (b) Can a molecule have polar covalent bonds and still have no dipole moment?

8.35 What is the dipole moment, in units of coul-m, of a pair of opposite charges 1.60×10^{-19} coul in magnitude separated by a distance of 1.00 Å (-1.60×10^{-19} coul is the charge of an electron)? What is the dipole moment in units of Debyes?

8.36 Two different structures for dichloroethylene are shown below:

Cl   H   Cl   Cl
  \   /   \   /
   C=C   C=C
  /   \   /   \
H   Cl   H   H

(Such substances, which have the same molecular formulas but different atomic arrangements, are called isomers.) One of these isomers has a dipole moment of 1.74 D, the other 0 D. Match each structure with its associated dipole moment.

8.37 Indicate which of the following molecules will have a dipole moment: (a) Br_2; (b) BrCl; (c) CO_2; (d) H_2S; (e) $HCCl_3$.

8.38 Indicate the type of hybrid orbital set that can be formed from each of the following sets of atomic orbitals and describe the geometry of the set: (a) $2s$, $2p_x$; (b) $4s$, $4p_x$, $4p_y$, $4p_z$; (c) $5s$, $5p_x$, $5p_y$, $5p_z$, $4d_{x^2-y^2}$, $4d_{z^2}$.

[8.39] Nicotine, the familiar stimulant found in tobacco, has the structure shown below:

CH₃ structure of nicotine

Indicate the number of carbon atoms possessing approximate sp, sp^2, and sp^3 hybridizations. Do the same for the nitrogen atoms. (Note that, in keeping with usual practice, unshared pairs on N are not shown.)

8.40 Indicate the number of p orbitals available for π bond formation if an atom has each of the following types of hybridization: (a) sp^3; (b) sp^2; (c) sp.

8.41 What change in hybridization of orbitals occurs at the central atom in each of the following reactions?

(a) $AsCl_3 + Cl^- \longrightarrow AsCl_4^-$

(b) $SF_6 \longrightarrow SF_5^+ + F^-$

8.42 Distinguish between the following terms: (a) localized and delocalized bonds; (b) hybridized and unhybridized orbital; (c) σ and π molecular orbital; (d) σ and σ^* molecular orbital.

8.43 Benzene, C_6H_6, is sometimes abbreviated as

What does the hexagon represent, and what is the purpose of the circle?

[8.44] Explain why double and triple bonds are counted as a single pair of electrons when applying the VSEPR model.

8.45 Why does the potential energy experienced by two H atoms go through a minimum and then begin to increase as the atoms come closer together (Figure 8.8)?

8.46 Describe the molecular shape and bonding in hydrazine, H_2NNH_2: (a) draw the Lewis structure for this molecule; (b) use the VSEPR model to predict the molecular geometry; (c) indicate the type of hybrid orbitals employed by nitrogen; (d) label the covalent bonds as σ or π types.

[8.47] In each of the following pairs of molecules, one is stable and well known, the other is unstable or unknown; identify the unstable member of each pair and explain why it is the less stable compound: (a) C_2H_4, Si_2H_4; (b) PF_5, NF_5; (c) GeF_6, SF_6; (d) Mg_2O, Na_2O.

8.48 When aluminum reacts with chlorine a white solid with composition 20.2 percent Al and 79.8 percent Cl results. The compound is nonpolar and has a molecular weight of 267. What is the molecular formula for this substance? On the basis of what you have learned about bonding in this chapter, suggest the structure of the compound and indicate the nature of the bonding. (Hints: Chlorine atoms bonded to Al have unshared electron pairs; the chlorines in this compound are in two distinct kinds of bonding environments.)

8.49 How many σ and how many π bonds constitute a triple bond?

8.50 Draw the σ, σ^*, π, and π^* molecular orbitals, showing the nodal planes, if any, that each possesses.

8.51 Using the molecular-orbital scheme for homonuclear diatomic molecules, determine the bond order in the CN^- ion. How does this compare with its Lewis structure?

8.52 The ions O_2^-, O_2^{2-}, and O_2^+ occur in several compounds. Compare these three ions with O_2 by listing the four in order of increasing bond length.

8.53 Both H_2 and Li_2 have bond orders of one. Why is the bond energy of Li_2 less than that of H_2?

8.54 Using a homonuclear molecular-orbital scheme, determine the bond orders of NO^-, NO, and NO^+. Which would you predict to have the shortest N—O bond?

8.55 How and why do the orbital energy diagrams of heteronuclear diatomic molecules differ from those for homonuclear diatomic molecules?

8.56 The carbon-oxygen bond lengths in the series of molecules, CH_3OH, H_2CO, CO are 1.43 Å, 1.23 Å, and 1.13 Å, respectively. Write the Lewis structure for each molecule. How does the C—O bond order vary in this series? Is this variation consistent with the observed variation in carbon-oxygen bond length? Explain.

9

The properties of gases

In the past several chapters we have learned about the electronic structures of atoms and about how atoms come together to form molecules or ionic substances. We commonly observe matter, however, not on the atomic and molecular level, but as a solid, liquid, or gas. In the next few chapters we will be considering some of the important characteristics of these states of matter. We will be interested in learning why substances are found in one or the other state, what forces operate between atoms, ions, or molecules in these states, what transitions may occur between states, and about some of the characteristic properties of matter in each state.

9.1 Characteristics of gases

Under appropriate conditions, most substances can exist in any of the three states of matter. The substance H_2O, for example, is familiar to us as liquid water, ice, or water vapor. Frequently a substance exists in all three separate states of matter, or phases, at the same time. A Thermos flask containing a mixture of ice and water at 0°C has a certain pressure of water vapor in the gas phase over the liquid and solid phases.

Normally the three states of matter differ very obviously from one another. Gases differ dramatically from solids and liquids in several respects. A gas expands to fill its container, whereas the volumes of solids and liquids are not determined by the container. The corollary of this is that gases are highly compressible. When pressure is applied to a gas, its volume readily contracts. Liquids and solids, on the other hand, are not very compressible at all. Great pressures must be applied to cause the volume of a liquid or solid to diminish by even as little as 5 percent.

Two or more gases form homogeneous mixtures in all proportions, regardless of how different the gases may be. Liquids, on the other hand, often do not form homogeneous mixtures. For example, when water and gasoline are poured into a bottle, the water vapor and gasoline vapors above the liquids form a homogeneous gas mixture. The two liquids, by contrast, remain largely separate; each dissolves in the other to only a slight extent.

The characteristic properties of gases arise because the individual molecules of a gas are relatively far apart. In a liquid, the individual molecules are close together and take up perhaps 70 percent of the total space. By comparison, in the air we breathe, the molecules take up only about 0.1 percent of the volume. In liquids, the molecules are constantly in contact with neighbors. They experience attractive forces for one another; this is what keeps the liquid together. However, when a pair of molecules come close together, repulsive forces prevent any closer approach. These attractive and repulsive forces differ from one substance to another. The result is that different liquids behave differently. By contrast, the molecules of a gas are well separated and are not much influenced by one another. As we shall see in more detail later, gas molecules are in constant motion, and they frequently collide. On the average, though, they remain fairly far apart. For example, in air, the average distance between molecules is about ten times as great as the sizes of the molecules themselves. Each molecule thus tends to behave as though the others weren't there. The relative degree of isolation of the molecules causes different gases to behave similarly, even though they are made up of different molecules.

To define properly the state, or condition, of a gas, it is necessary to assign values to certain variable quantities—namely, temperature, volume, quantity of gas, and pressure. We have already discussed temperature scales (Section 1.3). The absolute temperature scale, K, is the appropriate one to employ in working problems involving gases. Volume is measured in an appropriate volume unit, usually liters. The quantity of gas is measured in terms of the number of moles. Now let's consider pressure, what it is, and how it is measured.

9.2 Pressure

In general terms, pressure carries with it the idea of a force, something that tends to move something else in a given direction. Pressure is in fact the force that acts on a given area ($P = F/A$). A gas exerts force on the walls of any container; for example, the gas in an inflated balloon exerts pressure on the wall of the balloon.

To understand better the concept of pressure and the units in which it is expressed, consider the aluminum cylinder illustrated in Figure 9.1. Because of the gravitational force, this cylinder exerts a downward pressure upon the surface on which it rests. The mass of this cylinder is 170 g, and the cross-sectional area of the cylinder base is $\pi r^2 = (3.14)(1.00 \text{ cm})^2 = 3.14 \text{ cm}^2 = 3.14 \times 10^{-4} \text{ m}^2$.

According to Newton's second law of motion, the force exerted by an object is the product of its mass, m, times its acceleration, a: $F = ma$. The acceleration due to the gravitational force of earth is 981 cm/sec^2. Thus the force with which earth attracts the aluminum cylinder is

$$(170 \text{ g})(981 \text{ cm/sec}^2) = 1.67 \times 10^5 \text{ g-cm/sec}^2$$

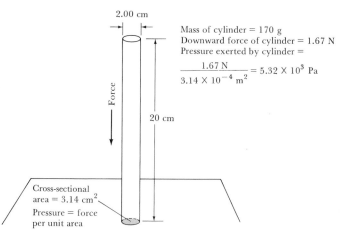

Mass of cylinder = 170 g
Downward force of cylinder = 1.67 N
Pressure exerted by cylinder =

$$\frac{1.67\ \text{N}}{3.14 \times 10^{-4}\ \text{m}^2} = 5.32 \times 10^3\ \text{Pa}$$

2.00 cm

Force

20 cm

Cross-sectional
area = 3.14 cm^2
Pressure = force
per unit area

FIGURE 9.1 Pressure exerted by an aluminum cylinder on the surface below.

To express this in SI units we must convert to kg and m:

$$\left(\frac{1.67 \times 10^5\ \text{g-cm}}{\text{sec}^2}\right)\left(\frac{1\ \text{kg}}{1000\ \text{g}}\right)\left(\frac{1\ \text{m}}{100\ \text{cm}}\right) = 1.67\ \text{kg-m/sec}^2$$

A kg-m/sec^2 is the SI unit for force; it is called the newton, abbreviated N: 1 N = 1 kg-m/sec^2. Now let us suppose that the aluminum cylinder is standing upright as illustrated in Figure 9.1. We then ask, what pressure does it exert on the surface beneath it? We have just calculated the force with which it is attracted toward earth; we know that this force is exerted over an area of 3.14 cm^2 (3.14 × 10^{-4} m^2) because that is the cross-sectional area of the cylinder base. Thus the pressure exerted by the cylinder is

$$\text{Pressure} = P = \text{force/area} = \frac{1.67\ \text{N}}{3.14 \times 10^{-4}\ \text{m}^2}$$
$$= 5.32 \times 10^3\ \text{N/m}^2$$

The newton per square meter is the standard unit of pressure in SI units. It is given the name pascal (abbreviated Pa) after Blaise Pascal (1623–1662), a French mathematician and scientist: 1 Pa = 1 N/m^2.

Like the aluminum cylinder in our example above, earth's atmosphere is also pulled toward the earth by gravitational attraction. As a result it exerts a pressure on the earth's surface that we call atmospheric pressure. The actual amount of air over a given place on the earth varies slightly from time to time. However, a standard atmosphere of pressure (abbreviated atm) has been defined as 1.01325 × 10^5 Pa = 101.325 kPa. That is, a column of earth's atmosphere reaching all the way to outer space exerts a downward pressure at the bottom of that column of 101.325 × 10^3 kg/m-sec^2.

Gas pressures are most commonly expressed by chemists in units of atmospheres (atm) or of millimeters of mercury (mm Hg). To see how the latter unit is arrived at we must understand the principle of the mercury barometer, illustrated in Figure 9.2. A barometer is formed by filling a glass tube more than 76 cm long, which is closed at one end, with mercury and inverting it in a dish of mercury. Care must be taken

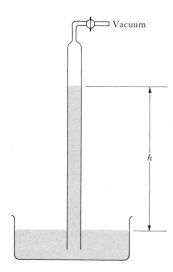

Vacuum

h

FIGURE 9.2 A mercury barometer.

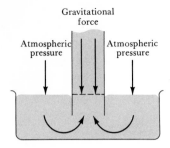

Gravitational force

Atmospheric pressure — Atmospheric pressure

FIGURE 9.3 Balancing of atmospheric pressure and gravitational force on mercury in a mercury barometer.

that no air gets into the tube. When the tube is inverted in this manner, some of the mercury runs out, but a column remains.

The mercury surface outside the tube experiences the full force of the earth's atmosphere over each unit area, but the surface of the mercury within the tube does not. The height of the column of mercury in the tube depends on the pressure of the atmosphere pushing down on the mercury surface outside the tube. Suppose that we look more closely at the base of the tube (Figure 9.3) and imagine a surface inside the tube that is level with the surface of the mercury outside (the dashed line in Figure 9.3). The force which the atmosphere exerts on the outside surface is transferred through the liquid to the surface within the tube. There is then "1 atmosphere" of pressure per unit area pushing mercury upward across this imaginary boundary. But at the same time, the mercury above the boundary in the tube is being forced downward by the same gravitational force that is pulling the atmosphere toward earth. When the downward force of the liquid mercury equals the upward force of mercury being pushed by the atmosphere there will be no net flow of mercury. Standard atmospheric pressure corresponds to a pressure sufficient to support a column of mercury 760 mm in height. Thus we can say that one atmosphere (1 atm) corresponds to 760 mm Hg: 1 atm = 760 mm Hg. One mm Hg pressure is also referred to as a torr, after the Italian scientist Evangelista Torricelli (1608–1647), who invented the barometer: 1 mm Hg = 1 torr.

SAMPLE EXERCISE 9.1

(a) Convert 0.605 atm to millimeters of mercury (mm Hg). (b) Convert 3.5×10^{-4} mm Hg to atmospheres.

Solution: (a) Because 1 atm = 760 mm Hg, conversion of atm to mm Hg is made by multiplying the number of atm by the factor 760 mm Hg/1 atm:

$$(0.605 \text{ atm}) \left(\frac{760 \text{ mm Hg}}{1 \text{ atm}} \right) = 460 \text{ mm Hg}$$

Notice that the units cancel in the required manner.
(b) To convert from mm Hg to atm, we must multiply by the conversion factor 1 atm/760 mm Hg:

$$(3.5 \times 10^{-4} \text{ mm Hg}) \left(\frac{1 \text{ atm}}{760 \text{ mm Hg}} \right)$$
$$= 4.6 \times 10^{-7} \text{ atm}$$

SAMPLE EXERCISE 9.2

Convert a pressure of 735 mm Hg to kPa.

Solution: From the material discussed above we know that 1 atm = 101.3 kPa = 760 mm Hg. Thus, the conversion factor we want is of the form 101.3 kPa/760 mm Hg. We use this conversion factor to convert the pressure given:

$$(735 \text{ mm Hg}) \left(\frac{101.3 \text{ kPa}}{760 \text{ mm Hg}} \right) = 98.0 \text{ kPa}$$

In countries that use the metric system, for example, Canada, atmospheric pressure is expressed in weather reports in units of kPa.

In this text we will ordinarily express gas pressure in units of atm or mm Hg. However, you should be able to convert gas pressures from one set of units to another.

Joseph Louis Gay-Lussac (1778–1850) is one of those extraordinary figures in the early history of modern science who can truly be called an adventurer. He was interested in lighter-than-air balloons and in 1804 made an ascent to 23,000 ft. This exploit set the altitude record for several decades, but Gay-Lussac had other reasons for making the flight: He tested the variation of the earth's magnetic field and sampled the composition of the atmosphere as a function of elevation.

To control lighter-than-air balloons properly, Gay-Lussac needed to know more about the properties of gases. He therefore was led to carry out several experiments. The most important of these led to his discovery in 1808 of the law of combining volumes (Section 3.4). Recall that this law states that the volumes of gases that react with one another at the same pressure and temperature are in the ratios of small whole numbers.

It was most especially Gay-Lussac's work that led Amedeo Avogadro to propose in 1811 that *equal volumes of gases at the same temperature and pressure contain equal numbers of molecules.* We saw in Chapter 3 the importance of both Gay-Lussac's and Avogadro's work in setting the stage for a correct appreciation of atomic weights. Let us now consider how their results can help us understand the nature of the gaseous state. Suppose that we have three 1-L bulbs containing H_2, N_2, and Ar, respectively (Figure 9.4), and that each gas is at the same pressure and temperature. According to Avogadro's hypothesis, those bulbs contain *equal numbers of gaseous particles.* Note that one of the gases is made up of atoms, and the other two of diatomic molecules; note also that the masses of the substances in the bulbs differ greatly. As far as pressure, temperature, and volume relationships are concerned, however, what counts is that there are equal numbers of gaseous particles in the bulbs, at the same temperature.

The similarities in the behaviors of these three gases go further. When heated at constant pressure they expand the same amount for each degree rise in temperature. When compressed at constant temperature, they contract to the same degree for a given increase in pressure. In other words, the variables that determine the physical properties of the gas are not sensitive to the particular chemical composition of the individual gas particles.

Experiments with a large number of gases reveal that the variables temperature (T), pressure (P), volume (V), and quantity of gas (n) are sufficient to define the state (Section 4.3) of many gaseous substances.

	Ar	N₂	H₂
Volume	1 L	1 L	1 L
Pressure	1 atm	1 atm	1 atm
Temperature	0°C	0°C	0°C
Mass of gas	1.783 g	1.250 g	0.0899 g
Number of gas molecules	2.688×10^{22}	2.688×10^{22}	2.688×10^{22}

FIGURE 9.4 Demonstration of Avogadro's hypothesis. Note that argon gas consists of argon atoms; we can regard these as one-atom molecules.

The relationship of the variables to one another is expressed in the ideal-gas equation:

$$PV = nRT \qquad [9.1]$$

An equation of this type is referred to as an equation of state, because when we have specified all the variables we have defined the state of the system. Although real gases don't exactly obey the ideal-gas equation, the difference between ideal and real behavior is often so small that it can be ignored. Thus, we may use the ideal-gas equation to solve problems of practical importance involving gases.

In Equation [9.1] temperature must be expressed in an absolute temperature scale, normally K. The quantity of gas, n, is ordinarily expressed as the number of moles; R is called the gas constant. The numerical value of R is determined by the units chosen for the other variables. Numerical values for R in a few of the more important units are listed in Table 9.1. The value based on volume expressed in liters and pressure in atmospheres is the one we will use for calculations in this chapter.

As a simple application of the ideal-gas equation let us calculate the volume of exactly 1 mol of gas at 0°C (273.15 K) and 1 atm pressure. (These are the so-called standard temperature and pressure, STP.) We rearrange the terms in Equation [9.1] to solve for V and insert all the required numerical values:

$$V = \frac{nRT}{P} = \frac{(1 \text{ mol})(0.08206 \text{ L-atm/mol-K})(273.15 \text{ K})}{1 \text{ atm}}$$

$$= 22.41 \text{ L}$$

According to the ideal-gas equation and Avogadro's hypothesis, 1 mol of any gas will occupy this volume at STP.

The ideal-gas equation relates four distinct and independent properties of a gas to one another. We can best appreciate the properties of gases as expressed in the ideal-gas equation by considering the relationship between any two variables when the other two are held fixed. We will examine first a family of experiments in which n is held constant and in which one other of the remaining three variables is also fixed.

TABLE 9.1 Numerical values of the gas constant, R, in various units

Units	Numerical value
Liter-atm/K-mol	0.08206
Calories/K-mol	1.987
Joules/K-mol[a]	8.314

[a] SI units

PRESSURE-VOLUME RELATIONSHIP AT CONSTANT TEMPERATURE; BOYLE'S LAW

If both n and T are fixed, then the product PV is a constant:

$$PV = nRT = \text{constant} = c \qquad [9.2]$$

We can rearrange this as follows:

$$V = c/P \qquad [9.3]$$

Equation [9.3] tells us that, *at constant temperature, the volume of a gas is inversely proportional to pressure*. This relationship between pressure and volume at constant temperature was first made clear in experiments performed by Robert Boyle (1627–1691) and is known as Boyle's law.

Equation [9.3] expresses the very important fact that a gas is compressible; the more it is pressed upon, the denser it gets. As explained in

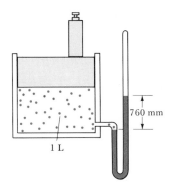
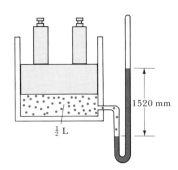

FIGURE 9.5 Gas pressure versus volume.

Section 1.3, density is a measure of the mass contained within a unit volume. Consider the experiment shown in Figure 9.5. We have a certain quantity of gas, let us say dry air, in the cylinder. The gas, consisting primarily of O_2 and N_2 molecules, has a certain mass. Let us say that the gas is at about room temperature, 25°C, and at a pressure of 1 atm, and that the volume of the cylinder is exactly 1 L. One liter of dry air under these conditions has a mass of 1.185 g. Its density is therefore 1.185 g/L. Now, let us compress the gas to half the volume, thereby doubling its pressure to 2 atm. There is still 1.185 g of air in the cylinder, because the act of compressing the gas does not change the number of gas molecules. But the density is now 1.185 g/0.5 L, or 2.370 g/L. Thus, the density has doubled in the course of doubling the pressure of the gas. If we squeezed the gas still further, the density would continue to increase. Thus, at constant temperature, *gas density is directly proportional to pressure*.

SAMPLE EXERCISE 9.3

The pressure of nitrogen gas in a 12.0-L tank at 27°C is 2300 lb/in.². What volume would the gas in this tank have at 1 atm pressure (14.7 lb/in.²)?

Solution: Let us begin by making up a table that lists the initial and final values for the pressure, temperature, and volume of the gas. (It is good practice always to convert temperature to the Kelvin scale.)

	Temperature (K)	Pressure (lb/in.²)	Volume (L)
Initial	300	2300	12.0
Final	300	14.7	?

We can reason from Equation [9.2] that if the quantity of gas (n) and temperature (T) do not change, the constant in the equation must remain constant, even though P and V individually change. Thus if we have two different sets of conditions for the same quantity of gas at constant temperature, we can write

$$P_1V_1 = P_2V_2$$

In our example, P_1 is 2300 lb/in.², P_2 is 14.7 lb/in.², V_1 is 12 L, and V_2 is unknown. Inserting all the known quantities and solving for V_2 we obtain

$$(2300 \text{ lb/in.}^2)(12.0 \text{ L}) = (14.7 \text{ lb/in.}^2)(V_2)$$

$$V_2 = \left(\frac{2300 \text{ lb}}{1 \text{ in.}^2}\right)(12.0 \text{ L})$$

$$\times \left(\frac{1 \text{ in.}^2}{14.7 \text{ lb}}\right)$$

$$= 1880 \text{ L}$$

We can also approach this problem from a slightly different viewpoint. We recognize that if the pressure decreases, volume must increase. Assuming that there is an inverse proportionality between pressure and volume, we look for a pressure factor that will increase the volume when the pressure decrease occurs. This pressure factor should be the ratio of the higher to the lower pressure:

$$V = (12.0 \text{ L})(\text{pressure factor})$$

$$= (12.0 \text{ L})\left(\frac{2300 \text{ lb/in.}^2}{14.7 \text{ lb/in.}^2}\right)$$

$$= 1880 \text{ L}$$

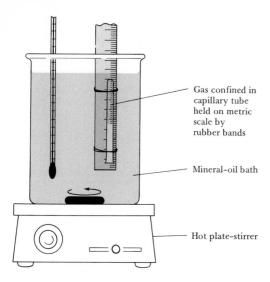

FIGURE 9.6 A simple apparatus for observing the effect of varying temperature on a gas confined at constant pressure. To maintain strictly constant pressure, the gas-liquid interface inside the tube should be kept a constant distance below the surface as temperature is varied. This can be done by raising or lowering the metric scale.

Gas confined in capillary tube held on metric scale by rubber bands

Mineral-oil bath

Hot plate-stirrer

VOLUME-TEMPERATURE RELATIONSHIP; CHARLES'S LAW

TABLE 9.2 Student data for the volume of gas as a function of temperature (see Figure 9.7)

Temperature (°C)	Length of gas column (mm)
26	28
52	30
82	33
114	36
158	40
172	42

When P and n are maintained constant, so that volume and temperature are the variables, the ideal-gas equation can be written as

$$V = \left(\frac{nR}{P}\right) T = \text{constant} \times T \qquad [9.4]$$

This relationship indicates that the volume of a gas under constant pressure conditions should be directly proportional to the absolute temperature. An experiment often performed in the general chemistry laboratory to test this equation is illustrated in Figure 9.6. A quantity of gas is trapped in a capillary tube, which is inverted in a mineral-oil bath. The volume of the gas is proportional to the length of gas in the tube, which is easily measured. Table 9.2 shows some student data; these are graphed in Figure 9.7, after converting temperatures to K.

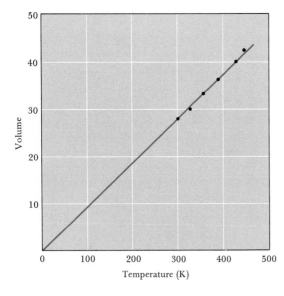

FIGURE 9.7 Volume (length of gas in capillary tube in millimeters) as a function of absolute temperature.

Because the length of the gas space in the capillary is proportional to volume, we can label the vertical axis in the figure as volume. Note that the volume is linearly related to absolute temperature, as we expect from Equation [9.4]. Note also that the extrapolated (extended) relationship passes through the origin. That is, the gas is predicted to have zero volume at 0 K. Of course, this condition is never fulfilled, because all gases liquefy or solidify before reaching this temperature. We will have more to say about departures from the ideal-gas equation in a later section. For now, the important point to note is that *at constant pressure, the volume of a given quantity of gas is proportional to absolute temperature*. This relationship is often called Charles's law, after J. A. C. Charles (1746–1823), a French scientist who preceded Gay-Lussac in experiments with gases.

SAMPLE EXERCISE 9.4

A large natural-gas storage tank is arranged so that the pressure is maintained at 2.2 atm. On a cold day in December, when the temperature is $-15\,°C\,(4\,°F)$, the volume of gas in the tank is 28,500 ft³. What is the volume of the same quantity of gas on a warm July day when the temperature is $31\,°C\,(88\,°F)$?

	Temperature (K)	Pressure (atm)	Volume (ft³)
Initial	258	2.2	28,500
Final	304	2.2	?

Solution: We recognize intuitively that when the gas is heated at constant pressure it will expand. Thus we will need to multiply the gas volume at the lower temperature by a factor greater than one. In gas-law problems, temperatures must always be expressed in terms of an absolute temperature scale, that is, in degrees K. After converting temperatures to degrees K, we have the following values for initial and final conditions:

Because pressure and quantity of gas are constant, we know that the volume is proportional to absolute temperature, as stated in Equation [9.4]. Thus, the volume at the higher temperature, call it V_2, must be larger than the volume at the lower temperature, V_1:

$$V_2 = V_1\left(\frac{T_2}{T_1}\right) = (28,500 \text{ ft}^3)\left(\frac{304 \text{ K}}{258 \text{ K}}\right)$$

$$= 33,600 \text{ ft}^3$$

BEHAVIOR OF GASES AT CONSTANT VOLUME

We know that when a confined gas is heated at constant volume the pressure increases. For example, a popcorn kernel bursts open under the pressure of steam that forms within the kernel when it is heated in oil. We could make quantitative measurements of the change in pressure of a confined gas by placing the gas in a steel container fitted with a pressure gauge, and then varying the temperature. We would find that the pressure increases linearly with absolute temperature, perhaps as shown by the sample data labeled A in Figure 9.8. If the experiment were repeated with a different-sized sample of the same gas we might obtain the results labeled B in the figure. Note that in both cases the extrapolated pressure at 0 K is zero.

If both n and V in Equation [9.1] are fixed, the pressure varies with temperature as expressed in Equation [9.5]:

$$P = \left(\frac{nR}{V}\right)T = \text{constant} \times T \qquad [9.5]$$

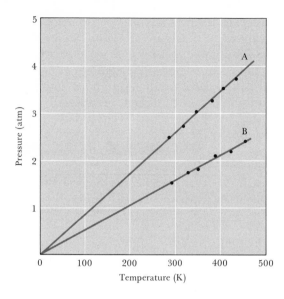

FIGURE 9.8 Variation of gas pressure with temperature under constant volume conditions.

Thus, the ideal-gas equation predicts a linear relationship between pressure and absolute temperature, extrapolating to zero pressure at 0 K. Again, we must remind ourselves that real gases lose their gaseous properties before absolute zero is reached.

SAMPLE EXERCISE 9.5

Why do the two samples of gas for which data are shown in Figure 9.8 show two different linear relationships?

Solution: Inspection of Equation [9.5] shows us that the linear relationship between P and T passes through the origin, and has slope nR/V. (You may wish to review linear equations, Appendix A.4.) In our example, both R and V are the same for the two samples, but the number of moles of gas, n, is different. The slope of the pressure versus temperature relationship for a given volume is proportional to the amount of gas that is confined. There is more gas in sample A.

SAMPLE EXERCISE 9.6

The gas pressure in an aerosol can is 1.5 atm at 25 °C. Assuming that the gas inside obeys the ideal-gas equation, what would the pressure be if the can were heated to 450 °C?

Solution: Let us proceed, as in Sample Exercise 9.4, by writing down the initial and final conditions of temperature, pressure, and volume that the problem gives us. (Remember that we must convert temperatures to degrees K.)

	Volume	Pressure (atm)	Temperature (K)
Initial	V_1	1.5	298
Final	V_1	P_2	723

Because the pressure increases in proportion to the absolute temperature, the pressure at the higher temperature, P_2, will equal P_1 times a ratio of absolute temperatures that is greater than 1:

$$P_2 = (1.5 \text{ atm})\left(\frac{723 \text{ K}}{298 \text{ K}}\right) = 3.6 \text{ atm}$$

It is evident from this example why aerosol cans carry the warning not to incinerate.

EFFECTS OF CHANGING THE QUANTITY OF GAS; DALTON'S LAW OF PARTIAL PRESSURES

An increase in the quantity of gas, n, while maintaining constant temperature, causes an increase in the product PV; this follows directly from Equation [9.1]. If the increase occurs at constant volume, then

$$P = \left(\frac{RT}{V}\right)n = \text{constant} \times n \qquad [9.6]$$

We see that the pressure of a gas under conditions of constant volume and temperature is directly proportional to the number of moles of gas. Suppose the gas with which we are concerned is not a single kind of gas particle but is rather a mixture of two or more different substances. We might expect that the total pressure exerted by the gas mixture is the sum of pressures due to the individual components. Each of the individual components, if present alone under the same temperature and volume conditions as the mixture, would exert a pressure that we term the *partial pressure*. John Dalton was the first to observe that the *total pressure of a mixture of gases is just the sum of the pressures that each gas would exert if it were present alone*:

$$P_t = P_1 + P_2 + P_3 + \cdots \qquad [9.7]$$

Each of the gases obeys the ideal-gas equation. Thus we can write

$$P_1 = n_1\left(\frac{RT}{V}\right), \qquad P_2 = n_2\left(\frac{RT}{V}\right), \qquad P_3 = n_3\left(\frac{RT}{V}\right), \text{ etc.}$$

All of the gases experience the same temperature and volume. Therefore, by substituting into Equation [9.7], we obtain

$$P_t = \frac{RT}{V}(n_1 + n_2 + n_3 + \cdots) \qquad [9.8]$$

That is, the total pressure at constant temperature and volume is determined by the total number of moles of gas present, whether that total represents just one substance or a mixture.

SAMPLE EXERCISE 9.7

If a 0.20-L sample of O_2 at 0°C and 1.0 atm pressure and a 0.10-L sample of N_2 at 0°C and 2.0 atm pressure are both placed in a 0.40-L container at 0°C, what is the total pressure in the container?

Solution: We can solve this problem by calculating the pressure that each gas would exert if it were alone in the 0.40-L container. Because the quantity of each gas and the temperature are constant, the only quantities that vary are pressure and volume. Using the same type of approach employed in Sample Exercise 9.3, we have

$$P_1 V_1 = P_2 V_2$$

$$P_2 = P_1\left(\frac{V_1}{V_2}\right)$$

$$P_{O_2} = (1.0 \text{ atm})\left(\frac{0.20 \text{ L}}{0.40 \text{ L}}\right) = 0.5 \text{ atm}$$

$$P_{N_2} = (2.0 \text{ atm})\left(\frac{0.10 \text{ L}}{0.40 \text{ L}}\right) = 0.5 \text{ atm}$$

According to Dalton's law the total pressure is the sum of each of the partial pressures exerted by the gases individually:

$$P_t = P_{O_2} + P_{N_2} = 0.5 \text{ atm} + 0.5 \text{ atm} = 1.0 \text{ atm}$$

9.4 Typical problems involving the ideal-gas equation

We have two reasons for being interested in gases. In the first place, the properties of the gaseous state are important for our understanding of the nature of matter on the atomic and molecular level. We will deal with this aspect of gases in later sections. A second, very practical reason for studying gases is that they may be reactants and products in chemical reactions. Just as we must learn to deal with pure solid or liquid reagents, or with solutions of reagents, so we must also learn to express quantities of gases under various conditions in terms of volume and pressure. In this section we will review several typical problems involving gases that might be encountered in the chemical laboratory or related situations.

SAMPLE EXERCISE 9.8

A quantity of helium gas occupies a volume of 16.5 L at 78°C and 45.6 atm. What is its volume at STP?

Solution: It is best to begin problems of this sort by writing down all we know of the initial and final values of temperature, pressure, and volume. (Remember that you must always convert temperatures to the absolute temperature scale, K.)

	Pressure (atm)	Volume (L)	Temperature (K)
Initial	45.6	16.5	351
Final	1	V_2	273

We can then proceed to calculate the unknown final volume by multiplying the initial volume by the appropriate ratios of the final and initial temperature and pressure. The temperature of the gas has decreased; this would lead to a lower volume. Thus we want to multiply by the ratio 273 K/351 K. On the other hand, the decrease in pressure from 45.6 to 1 atm will cause an increase in volume; thus, we want the ratio 45.6 atm/1 atm. Our final volume is thus

$$V_2 = \left(\frac{273 \text{ K}}{351 \text{ K}}\right)\left(\frac{45.6 \text{ atm}}{1 \text{ atm}}\right)(16.5 \text{ L}) = 585 \text{ L}$$

We can approach this problem a little differently by rearranging the ideal gas equation:

$$\frac{PV}{T} = nR$$

As long as the total quantity of gas, n, is constant, PV/T is a constant. If we represent the initial and final conditions of pressure, temperature, and volume by subscripts 1 and 2, we can write:

$$\frac{P_1V_1}{T_1} = nR = \text{constant} = \frac{P_2V_2}{T_2}$$

$$\frac{P_1V_1}{T_1} = \frac{P_2V_2}{T_2}$$

Putting in all the quantities we know, we obtain:

$$\frac{(45.6 \text{ atm})(16.5 \text{ L})}{351 \text{ K}} = \frac{(1 \text{ atm})(V_2)}{273 \text{ K}}$$

$$V_2 = \left(\frac{45.6 \text{ atm}}{1 \text{ atm}}\right)\left(\frac{273 \text{ K}}{351 \text{ K}}\right)(16.5 \text{ L}) = 585 \text{ L}$$

Note that previous problems involved the effects of changing conditions upon a variable such as volume. In such problems R is not needed; it cancels from the calculations. In the remaining examples we will need to employ R. For example, we can use the ideal-gas equation to calculate the magnitude of one of the variables, given values for the other three. This sort of calculation is shown in the next exercise.

SAMPLE EXERCISE 9.9

A flashbulb of volume 2.6 cm^3 contains O_2 gas at a pressure of 2.3 atm at a temperature of 26°C. How many moles of O_2 does the flashbulb contain?

Solution: Because we know volume, temperature,

and pressure, the only unknown quantity in the ideal-gas equation, Equation [9.1], is the number of moles, n. We rearrange the equation and, *after converting all known quantities to appropriate units,* insert those quantities into the expression for n:

$$PV = nRT$$

$$n = \frac{PV}{RT}$$

The values we can assign to all quantities involved are as follows:

$P = 2.3$ atm

$V = 2.6 \, \text{cm}^3 = 2.6 \times 10^{-3} \, \text{L}$

$n = ?$

$R = 0.0821$ L-atm/mol-K

$T = 26°C = 299$ K

Thus we have

$$n = \frac{(2.3 \, \text{atm})(2.6 \times 10^{-3} \, \text{L})}{(0.0821 \, \text{L-atm/mol-K})(299 \, \text{K})}$$

$$= 2.4 \times 10^{-4} \, \text{mol O}_2$$

Some of the most useful calculations using the ideal-gas equation involve measurements and calculation of the gas density. Density has the units of mass per unit volume. We can arrange the gas equation to obtain

$$\frac{n}{V} = \frac{P}{RT}$$

Now n/V has the units of moles per liter. Suppose we multiply the numerator in this equation by molecular weight (MW), which is the number of grams in 1 mol of a substance:

$$\frac{n(\text{MW})}{V} = \frac{P(\text{MW})}{RT} \tag{9.9}$$

But the product of the two quantities on the left is density, because the units multiply as follows:

$$\frac{\text{Moles}}{\text{Liter}} \times \frac{\text{grams}}{\text{mole}} = \frac{\text{grams}}{\text{liter}}$$

Thus the density of the gas is given by the expression on the right in Equation [9.9]:

$$d = \frac{P(\text{MW})}{RT} \tag{9.10}$$

or, rearranging

$$\text{MW} = \frac{dRT}{P} \tag{9.11}$$

Sample Exercises 9.10 and 9.11 illustrate the uses of these relations.

SAMPLE EXERCISE 9.10

What is the density of carbon dioxide gas at 745 mm Hg and 65°C?

Solution: The molecular weight of carbon dioxide, CO_2, is $12.0 + (2)(16.0) = 44.0$ g/mol. If we are to use 0.0821 L-atm/K-mol for R we must convert pressure to atmospheres. We have then

$$d = \frac{\left(\frac{745}{760} \, \text{atm}\right)(44.0 \, \text{g/mol})}{(0.0821 \, \text{L-atm/K-mol})(338 \, \text{K})}$$

$$= 1.56 \, \text{g/L}$$

The problem can be turned around a bit to determine the molecular weight of a gas from its density as shown in Sample Exercise 9.11.

A large flask fitted with a stopcock is evacuated and weighed; its mass is found to be 134.567 g. It is then filled to a pressure of 735 mm Hg at 31 °C with a gas of unknown molecular weight, and then reweighed; its mass is 137.456 g. The flask is then filled with water and again weighed; its mass is now 1067.9 g. Assuming that the ideal-gas equation applies, what is the molecular weight of the unknown gas? (The density of water at 31 °C is 0.997 g/cm³.)

Solution: First we must determine the volume of the flask. This is given by the difference in weights of the empty flask and the flask filled with water, divided by the density of water at 31 °C, 0.997 g/cm³:

$$V = \frac{1067.9 \text{ g} - 134.6 \text{ g}}{0.997 \text{ g/cm}^3} = 936 \text{ cm}^3$$

The next task is to calculate the number of moles of gas that will occupy this volume at the temperature and pressure indicated.

$$n = \frac{PV}{RT}$$

We must be careful to insert all quantities in the appropriate units. When this is done we obtain:

$$n = \frac{\left(\frac{735}{760} \text{ atm}\right)(0.936 \text{ L})}{(0.0821 \text{ L-atm/K-mol})(304 \text{ K})}$$
$$= 0.0363 \text{ mol}$$

We now have the number of moles of gas. We know also that this number of moles weighs 137.456 g − 134.567 g = 2.889 g.

The molecular weight is then simply the mass divided by the number of moles which that mass represents:

$$\frac{2.889 \text{ g}}{0.0363 \text{ mol}} = 79.6 \text{ g/mol}$$

An experiment that often comes up in the course of laboratory work is the determination of the number of moles of gas collected from a chemical reaction. Sometimes this gas is collected over water. For example, potassium chlorate ($KClO_3$) may be decomposed by heating solid $KClO_3$ in a test tube with an arrangement as shown in Figure 9.9. The balanced equation for the reaction is

$$2KClO_3(s) \longrightarrow 2KCl(s) + 3O_2(g)$$

The oxygen gas is collected in a bottle that is initially filled with water and inverted in a water pan.

The volume of gas collected is measured by raising or lowering the bottle as necessary until the water levels inside and outside the bottle are the same. When this condition is met, the pressure inside the bottle is equal to the atmospheric pressure outside. But the total pressure inside is

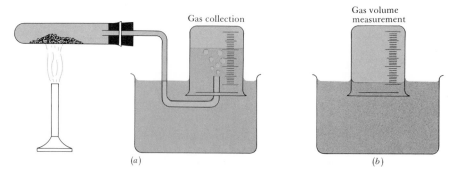

FIGURE 9.9 (a) Collection of a gas over water. (b) When the gas has been collected, the bottle is raised or lowered to equalize pressures inside and outside before measuring volume of gas collected.

the sum of the pressure of gas collected and the water vapor in equilibrium with liquid water:

$$P_{total} = P_{gas} + P_{H_2O}$$

The pressure exerted by water vapor, P_{H_2O}, at various temperatures is shown in Appendix C.

SAMPLE EXERCISE 9.12

Suppose that 0.200 L of oxygen gas is collected over water as in Figure 9.9 at a temperature of 26°C and a pressure of 750 mm Hg. What volume would the O_2 gas collected occupy when dry, at the same temperature and pressure?

Solution: The pressure of O_2 gas in the vessel is the difference between the total pressure, 750 mm Hg, and the vapor pressure of water at 26°C, 25 mm (Appendix C):

$$P_{O_2} = 750 - 25 = 725 \text{ mm Hg}$$

If the water were dried from the gas sample, while at the same time maintaining the same total pressure, the O_2 pressure after drying would be 750 mm Hg. The corrected O_2 volume would thus be less for the dry gas, because its partial pressure is greater than for the wet gas:

$$V_{O_2} = (0.200 \text{ L})\left(\frac{725 \text{ mm Hg}}{750 \text{ mm Hg}}\right) = 0.193 \text{ L}$$

The collection of a gas over water is often done in an experiment to determine the number of moles of gaseous product. To compute the number of moles of gas collected, a correction must be applied for the partial pressure of water vapor in the collection bottle. We shall go through a complete analysis of such an experiment in Sample Exercise 9.13 to show how the stoichiometric relationships come together.

SAMPLE EXERCISE 9.13

A 2.55-g sample of ammonium nitrite (NH_4NO_2) is heated in a test tube connected as shown in Figure 9.9. The ammonium nitrite is expected to decompose according to the equation

$$NH_4NO_2(s) \longrightarrow N_2(g) + 2H_2O(g)$$

If it does decompose in this way, what volume of N_2 will be collected in the flask? The water and gas temperature are 26°C, and the barometric pressure is 745 mm Hg.

Solution: We begin by calculating the number of moles of N_2 gas formed:

$$2.55 \text{ g NH}_4\text{NO}_2\left(\frac{1 \text{ mol NH}_4\text{NO}_2}{64.0 \text{ g NH}_4\text{NO}_2}\right)$$

$$\times \left(\frac{1 \text{ mol N}_2}{1 \text{ mol NH}_4\text{NO}_2}\right) = 0.0398 \text{ mol N}_2$$

To predict the volume of N_2 gas that is to be collected, we might be tempted to just calculate the volume that would be occupied by 0.0398 mol of N_2 at 745 mm Hg. But we must also take into account the

partial pressure of water vapor at 26°C, because the gas within the bottle is saturated with water vapor. From a table of water-vapor pressure versus temperature in the laboratory manual or in a chemistry handbook (see also Appendix C), we can determine that the vapor pressure of water at 26°C is 25 mm Hg. The pressure of nitrogen gas in the flask when the water levels inside and out have been equalized is thus $745 - 25 = 720$ mm Hg. We must calculate the predicted volume of gas using this pressure ($720/760 = 0.947$ atm). (Pressure must be expressed in atm when we employ the value for the gas constant, R, expressed in L-atm/mol-K). Rearranging Equation [9.1], we obtain

$$V = \frac{nRT}{P}$$

Inserting all known quantities in correct units, we obtain

$$V = \frac{(0.0398 \text{ mol})(0.0821 \text{ L-atm/K-mol})(299 \text{ K})}{0.947 \text{ atm}}$$

$$= 1.03 \text{ L}$$

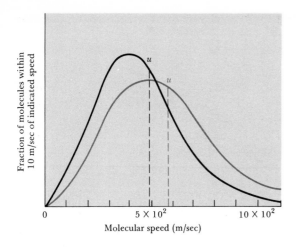

FIGURE 9.10 Distribution of molecular speeds for nitrogen at 0°C (black line) and 100°C (color line).

Fraction of molecules within 10 m/sec of indicated speed

Molecular speed (m/sec)

5×10^2 10×10^2

9.5 Kinetic-molecular theory of gases

The work of Boyle, Charles, Avogadro, Dalton, and others led to empirical relationships that are embodied in the ideal-gas equation. But an empirical equation is not the same as a model. To *understand* gases we must have a theoretical model, capable of being expressed in mathematical language, that yields a relationship between the variables that define the state of the gas. That is, we need a model that eventually yields the ideal-gas equation.

It had been suggested in 1738 by the Swiss physicist Daniel Bernoulli (1700–1782) and later by others that the properties of gases as expressed in Boyle's law and Charles's law could be accounted for if it were assumed that the particles of a gas are in ceaseless motion but have no forces whatever between them. In this model, the temperature of the gas was related to the average energy of motion of the gas particles. These ideas culminated in a new theory, called the kinetic-molecular theory of gases, which was published in its most complete and satisfactory form by Rudolf Clausius in 1857.

The scientists who developed the kinetic-molecular theory needed to take account of the fact that the molecules of a gas move with varying speeds. Gas molecules move about at random in their container, colliding with one another and with the walls of the enclosure. At any one instant some of them are moving rapidly, others slowly. It is possible to make measurements of the distribution of molecular speeds within a gas. For nitrogen at 0°C the distribution curve looks like the black line in Figure 9.10. For nitrogen at 100°C the distribution is like that shown by the color line in the figure. Note that at higher temperatures the distribution curve is shifted toward higher speeds.

Figure 9.10 also shows the value of the root mean square (rms) speed, u, of the molecules at each temperature. This quantity represents the square root of the average squared speed of the molecules.* The rms speed is important because the average kinetic energy of the gas mole-

---

*The quantity u is *not* the same as the average speed. To illustrate, suppose that we have four molecules, with speeds 4, 6, 10, and 12. The average speed is $\frac{1}{4}(4 + 6 + 10 + 12) = 8.00$. However, the square root of the average of the squares of the speeds is $[\frac{1}{4}(16 + 36 + 100 + 144)]^{\frac{1}{2}} = \sqrt{74} = 8.60$. This latter quantity is the rms speed, u.

cules, ϵ, is given by one-half the product of mass, m, times the average squared speed, u^2:

$$\epsilon = \tfrac{1}{2}mu^2 \qquad\qquad [9.12]$$

When the temperature of a gas increases, u^2 increases, and so also does the average kinetic energy. We will see in a moment that the manner in which the average kinetic energy of gas molecules varies with temperature is one of the important assumptions of the kinetic-molecular theory.

In the kinetic theory, several assumptions are made about the detailed nature of gases:

1 Gases consist of large numbers of molecules in ceaseless motion. A molecule will be taken to mean a particle consisting of one or more atoms that move together as a unit. When we say that the particles are in ceaseless motion, we mean that the average energy of all the particles does not change with time, so long as the temperature of the gas remains constant.
2 The volume of all the molecules of the gas is very small in comparison to the total volume in which the gas is contained. In other words, most of the space occupied by the gas is empty space.
3 The time during which a collision between two molecules occurs is negligibly short in comparison with the time between collisions. This means that at any one instant, if we could "freeze" the action in a gas, a very small fraction of the molecules would be undergoing collision with another molecule.
4 No attractive or repulsive forces operate between the molecules up to the point at which they collide. In other words, the gas molecules behave like tiny billiard balls.
5 The average kinetic energy of the molecules is proportional to absolute temperature.

Beginning with these assumptions it is possible to derive the ideal-gas equation. Rather then proceed through a derivation, let's consider in more qualitative terms how the ideal-gas equation might follow. In the kinetic-molecular theory the pressure that a gas exerts results from collisions of the gas molecules with the walls of the container. Pressure is defined as force per unit area (Section 9.2). The force per unit area increases with the number of collisions per unit time and with an increase in the momentum, mu, of the colliding molecule. The number of collisions per unit time is proportional to the speed of the molecules, u. As the speed increases, the time required for the molecule to traverse the distance from one wall to another is smaller, and so collisions occur more frequently. The collision frequency also increases with the number of molecules per unit volume, n/V. When we put these considerations together we find that the pressure is proportional to the terms shown in Equation 9.13:

$$P \propto (n/V)(u)(mu) \qquad\qquad [9.13]$$

$$P \propto \frac{nmu^2}{V} \qquad\qquad [9.14]$$

Now let us consider assumption 5, that the average kinetic energy is proportional to temperature:

$$\epsilon = \tfrac{1}{2}mu^2 \propto T \qquad\qquad [9.15]$$

Using Equation [9.15], we can substitute T for mu^2 in Equation [9.14]. This means that pressure is proportional to absolute temperature:

$$P \propto \frac{nT}{V} \qquad\qquad [9.16]$$

Let us now convert the proportionality sign to an equal sign by expressing n as the number of moles of gas, V as the volume in liters, and then inserting a proportionality constant, R, called the molar gas constant:

$$P = \frac{nRT}{V} \qquad\qquad [9.17]$$

This, of course, is the familiar ideal-gas equation, Equation [9.1].

The kinetic-molecular theory provides simple explanations for the empirical observations of gas properties as expressed in the various gas laws. Consider, for example, what happens when the volume of a gas is increased at constant temperature. The fact that temperature remains constant means that the average kinetic energy of the gas molecules remains unchanged. This in turn means that the rms speed of the molecules, u, is unchanged. However, if the volume is increased, the molecules must move a longer distance between collisions; there will be fewer collisions per unit time with the container walls, and pressure decreases. Thus, the model accounts in a simple way for Boyle's law.

Now consider how the pressure of a gas might change with an increase in temperature. An increase in temperature means an increase in the average kinetic energy of the molecules and an increase in u. If there is no change in volume this means that there will be more collisions with the walls per unit time. Furthermore, there is a larger average change in momentum on each collision. These considerations would lead to a pressure increase. However, if the volume is increased, the number of collisions per unit time decreases. A sufficiently large increase in volume causes a reduction in the number of collisions per unit time to balance the effect of the increased momentum change on each collision. Thus the kinetic-molecular theory accounts for the effects of changing temperature on pressure and volume, as expressed in Charles's law.

Assumption 5 of the kinetic-molecular theory is quite important. It is actually not so much an assumption as a definition of what we mean by absolute temperature. The average kinetic energies of the atoms, molecules, or ions from which *any* substance is formed are proportional to absolute temperature. Thus the absolute temperature is a measure of the heat energy per mole of the substance. We can imagine that there is a temperature at which the average energy of motion of the gas molecules would be zero. This corresponds to absolute zero. Let's consider what happens when a cold gas comes in contact with a hot container wall. We expect, of course, that the gas will be heated. By this we mean that the average kinetic energy of the gas molecules is increased. This increase in average kinetic energy occurs when the gas molecules collide with the container walls. Some of the higher average kinetic energy of the wall molecules is transferred on collision to the gas molecules. This process continues until the gas molecules and container molecules possess the same average kinetic, or thermal, energy. The overall result is that additional kinetic energy is imparted to the gas molecules, with a corresponding decrease in kinetic energy of the container wall molecules.

SAMPLE EXERCISE 9.14

A sample of O_2 gas initially at STP is transferred from a 2-L container to a 1-L container at constant temperature. What effect does this change have on (a) the average speed of the O_2 molecules, (b) the average kinetic energy of O_2 molecules, (c) the total number of collisions of O_2 molecules with the container walls in a unit time, and (d) the number of collisions of O_2 molecules with a unit area of container wall in a unit time?

Solution: (a) The average speed of molecules is determined only by the absolute temperature. Therefore the average speed is not changed by the com-

pression of O_2 from 2 to 1 L at constant temperature. (b) The average kinetic energy of O_2 molecules does not change either, because this is also dependent only on temperature. (c) The total number of collisions with the container walls in a unit time must increase, because the molecules are moving within a smaller volume but with the same average speed as before. Under these conditions they must encounter a wall more frequently. (d) The number of collisions with a unit area of wall increases, because the total number of collisions with the walls is higher and the area of wall is smaller than before.

9.6 Molecular effusion and diffusion; Graham's law

We have already made reference to the fact that the molecules of a gas do not all move at the same speed. Instead, the molecules are distributed over a range of speeds, as shown for nitrogen at two different temperatures in Figure 9.10. The distribution of molecular speeds depends on the mass of the gas molecules. Because the average kinetic energy of the molecules in any gas is determined only by temperature, it follows that the quantity $\frac{1}{2}mu^2$, which is kinetic energy, must have the same value for two gases at the same temperature, though their masses may differ. This in turn means that molecules of larger mass must have smaller average speeds. From the kinetic-molecular theory it can be shown that the rms speed, u, is given by Equation [9.18]. Note that u is proportional to the square root of the absolute temperature, and *inversely* proportional to the square root of the molecular weight:

$$u = \sqrt{\frac{3RT}{MW}}$$

[9.18]

This relationship tells us that lighter molecules have higher speeds. In fact, the entire distribution of molecular speeds is skewed to higher values for gases of lower molecular weights, as shown for several gases in Figure 9.11.

The dependence of molecular speeds on mass has several interesting consequences. Thomas Graham discovered in about 1830 that the effusion rates of gases are inversely related to the square roots of their molec-

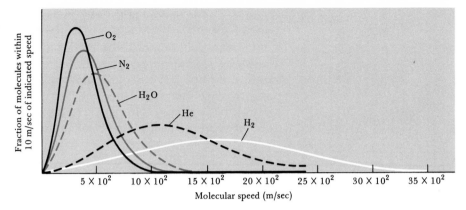

FIGURE 9.11 The distribution of molecular speeds for different gases at 25°C.

ular weights. (Effusion refers to the escape of a gas through a tiny hole.*) Assume that we have two gases at the same initial pressure contained in identical containers, each with an identical pinhole in one wall. Let the rate of effusion be called r. Graham's law states that

$$\frac{r_1}{r_2} = \sqrt{\frac{MW_2}{MW_1}} \qquad [9.19]$$

This equation follows from our previous discussion if we assume that the rate of effusion is proportional to the average speed of the molecules. Because R and T are constant, we have from Equations [9.18] and [9.19]

$$\frac{r_1}{r_2} = \frac{u_1}{u_2} = \sqrt{\frac{MW_2}{MW_1}} \qquad [9.20]$$

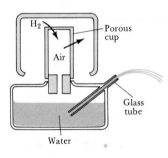

FIGURE 9.12 A hydrogen-fountain demonstration of the greater rate of diffusion of hydrogen as compared with air. A large container filled with H_2 gas is placed over the porous cup containing air. Hydrogen diffuses into the cup more rapidly than the molecules of air diffuse outward. As a result the pressure inside the vessel increases, and water is pushed out the glass tube, which is open to the outside.

We see many applications of these ideas in the properties of gases. One popular lecture demonstration, the hydrogen fountain, is illustrated in Figure 9.12. It makes use of the fact that the diffusion of hydrogen through the walls of the porous cup is faster than diffusion of atmospheric gases out of the cup through the wall. Thus an excess pressure builds up in the enclosure. Another example is seen in the behavior of toy balloons that have been filled with helium. Helium diffuses outward through the balloon surface more rapidly than the atmospheric gases diffuse in, and so the balloon collapses much more rapidly than it would if filled with air.

In the course of the effort during World War II to develop the atomic bomb, it was necessary to separate the relatively low-abundance uranium isotope ^{235}U (0.7 percent) from the much more abundant ^{238}U (99.3 percent). This was done by converting the uranium into a volatile compound, UF_6, which boils at 56°C. The gaseous UF_6 was allowed to diffuse from one chamber into a second through a porous barrier. Because of the slight difference in molecular weights, the relative rates of

---

*Effusion is related to, but is not quite the same as, diffusion. The latter term refers to the spread of one substance throughout a space, or throughout a second substance. For example, the molecules of a perfume diffuse through a room.

passage through the barrier for $^{235}UF_6$ and $^{238}UF_6$ are not exactly the same. The ratio of diffusion rates is given by the square root of the ratio of molecular weights, Equation [9.19]:

$$\frac{r_{235}}{r_{238}} = \sqrt{\frac{352}{349}} = 1.0043$$

Thus the gas initially appearing on the opposite side of the barrier would be very slightly enriched in the lighter molecule. The diffusion process was repeated thousands of times, leading to a nearly complete separation of the two nuclides of uranium.

9.7 Molecular collisions; mean free paths

We can see from the horizontal scale of Figure 9.11 that the speeds of molecules are really quite high. Translated into more familiar units, the average speed of N_2 at room temperature, 515 m/sec, corresponds to 1850 km/hr, or 1150 mi/hr. Yet we know that molecules don't travel through the atmosphere from one place to another at these speeds. For example, if a vial of a substance possessing a strong odor is opened at one end of a room, it is some time, perhaps a few minutes, before the odor is noted by a person at the other end. Therefore, the movement of gas molecules in the atmosphere is much slower than the molecular speeds would suggest. We must ascribe these slower diffusion rates to collisions between molecules.

According to the assumptions made in deriving the ideal-gas laws, the actual volumes of the gas molecules are negligible in comparison with the volume of the space enclosing the gas. At the same time, however, we know that gas molecules can't have *zero* volume. If they did, there could be no collisions between molecules, and there would be nothing to impede the flow of gas molecules from one place to another at rates commensurate with their speeds. But because molecules do have finite volumes, they collide with one another and thus suffer interruptions in their paths. The collisions are in fact very frequent for a gas at atmospheric pressure—about 10^{10} times per second for each molecule. The paths of the gas molecules are therefore interrupted very often. Diffusion of gas molecules, depicted in Figure 9.13, thus consists of a random motion, first in this direction, then that, at one instant at high speed, the next at low speed, with no overall direction to the motion of the gas as a whole except when there is a pressure difference. The average distance traveled by a molecule between collisions is referred to as the mean free path. This distance depends on the effective radius of the molecules, because larger molecules are more likely to undergo a collision. It depends also on the number of molecules in a unit volume—the larger the number of molecules per unit volume, the more likely is collision. Table 9.3 lists some values of experimentally determined mean free paths for several gases at STP. As you can see from these values, the molecules of a gas at STP do not travel very far before undergoing collision. By contrast, at an elevation of about 100 km in the earth's atmosphere, the mean free paths of nitrogen and oxygen molecules are on the order of 10 cm, more than a million times longer than at the surface.

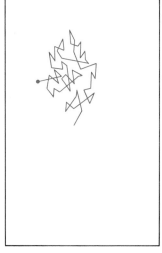

FIGURE 9.13 Schematic illustration of the diffusion of a gas molecule. For the sake of clarity all the other gas molecules in the container are not shown. The path of the molecule of interest begins at the dot. Each short segment of line represents travel between collisions. The path traveled by the molecule is often described as a "random walk."

TABLE 9.3 Mean free paths for several gases at 0°C, 1 atm

Gas	Mean free path (nm)
Carbon dioxide	39.7
Carbon monoxide	58.4
Argon	63.5
Nitrogen	60.0
Oxygen	64.7
Hydrogen	112.3
Helium	179.8

Thermal conductivity is an important property of gases. It is a measure of the rate at which heat energy can be transferred through a gas. We might expect that thermal conductivity would be greatest for the gases whose molecules have the highest average speeds, and this is indeed found to be the case. Thermal conductivity is also proportional to the mean free path. Among the gases in Table 9.2, carbon dioxide has the highest molecular weight and thus has the lowest average speed per molecule for a given temperature. It also has the shortest mean free path of the gases listed. We can conclude that carbon dioxide has the lowest thermal conductivity of any of the gases in the table. On the other hand, helium should have the highest thermal conductivity.

9.8 Departures from the ideal-gas equation

Although the ideal-gas equation is a very useful description of gases, all real gases fail to obey the relationship to greater or lesser degree. The extent to which a real gas departs from ideal behavior may be seen by rearranging Equation [9.1] slightly:

$$\frac{PV}{RT} = n \qquad [9.21]$$

For a mole of gas, $n = 1$; the quantity PV/RT should therefore equal 1 if the gas is ideal. Figure 9.14 shows the quantity PV/RT plotted as a function of pressure for a few gaseous substances, as compared with the expected behavior of an ideal gas. It is clear from the figure that real gases are simply not ideal. However, the pressures shown are very high; at more ordinary pressures, in the range from 1 to 10 atm, the deviations from ideal behavior are not so large, and the ideal-gas equation can be used without serious error.

We can explain the data for real gases illustrated in Figure 9.14 by taking account of two properties of real gases that are ignored in the kinetic molecular theory: (1) the molecules of a gas possess finite volumes and (2) at short distances of approach they exert attractive forces upon one another.

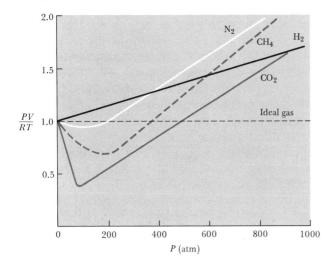

FIGURE 9.14 PV/RT versus pressure for several gases at 300 K. The data for CO_2 pertain to a temperature of 313 K, because CO_2 liquifies under high pressure at 300 K.

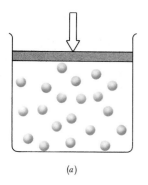 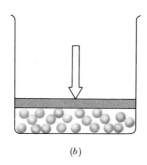

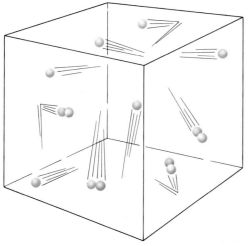

FIGURE 9.15 Illustration of the effect of finite volume of gas molecules on the properties of a real gas at high pressure. In (a), at low pressure, the volume of the gas molecules is small compared with the container volume. In (b), at high pressure, the volume of the gas molecules themselves is a large fraction of the total space available.

When gases are contained at relatively low pressure, say 1 atm, the volume of the space they occupy is very large in comparison with the volumes of the gas molecules themselves. At increasingly high pressures, however, the volume taken up by the molecules becomes a larger fraction of the total. This effect is illustrated in Figure 9.15. Thus, at pressures of several hundred atmospheres, the free volume in which the gas molecules can move is considerably smaller than the volume of the container. As a result the value of V that *should* be used in the product PV is smaller than the volume of the container. Hence, when we use the container volume in the ideal-gas equation we obtain a product PV that is larger than it should be. The data for H_2 in Figure 9.14 illustrate this situation. Notice that PV/RT increases steadily with increasing pressure. This is due entirely to the finite volume of the H_2 molecules.

The attractive forces between molecules come into play at short distances, when the molecules undergo collisions with one another. Because of these attractive forces, the molecules tend to stick together upon collision. If we could stop the action at any one instant in a real gas we might see something like that illustrated in Figure 9.16. Some of the molecules, having come together in collisions, are "paired up" for short times, because of the attractive forces. This means that the total number of particles present in the gas is decreased. Thus the product PV is smaller than it should be for an ideal gas.

From the data illustrated in Figure 9.14 we can guess that the attractive forces between molecules are greatest for CO_2. The product PV

FIGURE 9.16 Illustration of the effects of intermolecular attractive forces. Upon collision, molecules may form pairs that are held together for short periods by the attractive forces between them. Thus, the total number of independent particles in the gas and the quantity PV/RT are reduced.

shows a substantial negative departure from the ideal-gas relationship over a wide range of pressure. The attractive forces are less important for CH_4, and less important still for N_2. Because there is very little attractive interaction between H_2 molecules the quantity PV/RT for this gas is continuously larger than that expected for an ideal gas.

We have seen that the effects of finite molecular volumes, on the one hand, and of attractive intermolecular forces, on the other, work in opposite directions in influencing the value of PV/RT. At the highest pressures the effects of finite molecular volumes become dominant, because the free volume becomes small in relationship to the total volume. Thus for all gases the ratio PV/RT eventually rises above the value expected for an ideal gas.

For an ideal gas the ratio PV/RT is exactly 1 for a mole of gas at all temperatures. However, the departure of a real gas from the ideal relationship varies with temperature. The positive deviation due to the finite volume of the gas molecules is not much affected by temperature, because molecular volumes do not change significantly with temperature. At higher temperatures, however, the molecules possess higher average kinetic energies. As the motional energies of the molecules grow larger as compared with the intermolecular attractions, the molecules tend to stick together less upon collision. For this reason, we find that negative departures of real gases from the predictions of the ideal-gas laws are smaller at higher temperatures.

THE VAN DER WAALS EQUATION

Engineers and scientists who work with gases at high pressures often cannot use the ideal-gas equation to predict the pressure-volume properties of gases, because departures from the ideal-gas behavior are too large. Various equations of state have been developed to predict more realistically the pressure-volume behavior of real gases. These equations, while more realistic, are also considerably more complicated than the simple ideal-gas equation, Equation [9.1]. Equation [9.22] shows the van der Waals equation, named after Johannes van der Waals who presented it in 1873:

TABLE 9.4 Van der Waals constants for gas molecules

Substance	a (L²·atm/mol²)	b (L/mol)
He	0.0341	0.02370
Ne	0.211	0.0171
Ar	1.34	0.0322
Kr	2.32	0.0398
Xe	4.19	0.0510
H_2	0.244	0.0266
N_2	1.39	0.0391
O_2	1.36	0.0318
Cl_2	6.49	0.0562
CO_2	3.59	0.0427
CH_4	2.25	0.0428
CCl_4	20.4	0.1383

$$\left(P + \frac{an^2}{V^2}\right)(V - nb) = nRT \qquad [9.22]$$

This equation differs from the ideal-gas equation by the presence of two correction terms; one corrects the volume, another modifies the pressure. The term nb in the expression $(V - nb)$ is a correction for the finite volume of the gas molecules; the van der Waals constant b, different for each gas, has units of liters/mole. It is a measure of the actual volume occupied by the gas molecules. Values of b for several gases are listed in Table 9.4. Note that b increases with an increase in mass of the molecule or in the complexity of its structure.

The correction to the pressure takes account of the intermolecular attractions between molecules. Notice that it consists of the constant a, different for each gas, times the quantity $(n/V)^2$. The units of n/V are moles/liter. This quantity is squared because the number of molecular pairings, as illustrated in Figure 9.16, is proportional to the square of the number of molecules per unit volume. Values of the van der Waals constant a are listed in Table 9.4 for several gases. Notice that a increases with an increase in molecular weight and with an increase in complexity of molecular structure.

SAMPLE EXERCISE 9.15

Compare the van der Waals constants a and b for N_2 and CO_2 and account for the differences.

Solution: We note that the constant b, which is a measure of molecular volume, is larger for CO_2; 0.0427 L/mol as compared with 0.0391 L/mol for N_2. Because CO_2 is a larger molecule, consisting of three atoms rather than two, the larger value of b for CO_2 is expected. Furthermore, because CO_2 is both larger and more complex than N_2, we expect that the attractive forces between CO_2 molecules upon colli-

sion will be greater than the analogous attractive forces for N_2. Thus, the value for a should be larger for CO_2 than for N_2, as observed. In effect, under the same conditions of temperature and pressure, there are more molecular pairs per unit volume of CO_2 than for N_2.

Note also that the relative values of a and b are consistent with the relative magnitudes of the departures from the ideal-gas equation, as illustrated in Figure 9.14.

To get some feeling for the magnitudes of the departures from ideal behavior, let's calculate these departures for CO_2 at STP.

SAMPLE EXERCISE 9.16

Calculate the correction terms to pressure and volume for CO_2 at STP, using the data in Table 9.4, and compare with the ideal-gas values for P and V.

Solution: From Equation [9.21] we see that the volume correction term is given by nb. Since $n = 1$, nb equals 0.0427 L, which is to be compared with 22.4 L, the molar volume of an ideal gas at STP. The correction to volume is thus

$$\frac{0.0427}{22.4} \times 100 = 0.191\% \simeq 0.2\%$$

The correction to pressure is given by an^2/V^2. Inserting the value of a from Table 9.4, $n = 1$ and $V = 22.4$ L, we obtain

$$\frac{an^2}{V^2} = \frac{\left(\dfrac{3.59 \text{ L}^2\text{-atm}}{\text{mol}^2}\right)1 \text{ mol}^2}{(22.4 \text{ L})^2} = 0.007 \text{ atm}$$

The required correction to pressure is thus $0.007 \times 100 = 0.7\%$. We conclude that the ideal-gas law is obeyed by CO_2 at STP conditions to within 1 percent.

Summary

Many substances are capable of existing in any one of the three states of matter—solid, liquid, or gas. This chapter has been concerned with the gaseous state. To describe the state or condition of a gas, it is necessary to specify four variables: pressure, temperature, volume, and quantity of gas. Volume is usually measured in liters (L), and temperature in the Kelvin scale. Pressure is defined as the force per unit area. It is expressed in SI units as pascals, Pa (1 Pa = 1 N/m^2 = 1 kg/m-sec^2), or more commonly in millimeters of mercury (mm Hg). One standard atmosphere pressure equals 101.325 kPa, or 760 mm Hg. A barometer is often used to measure the atmospheric pressure.

The ideal-gas equation, $PV = nRT$, is the equation of state for an ideal gas. Most gases at pressures of about 1 atm and temperatures of 300 K and above obey the ideal-gas equation reasonably well. We can use the ideal-gas equation to calculate variations in one variable when one or more of the others are changed. For example, for a constant quantity of gas at constant temperature, the pressure of the gas is inversely proportional to the volume (Boyle's law). Similarly, for a constant quantity of gas at constant pressure, the volume of a gas is directly proportional to temperature (Charles's law). In gas mixtures, the total pressure is the sum of the partial pressures that each gas would exert if it were present alone under the same conditions. (Dalton's law of partial pressures). In all applications of the ideal-gas equation we must remember to convert temperatures to the absolute temperature scale, Kelvin.

It is important to be able to use the ideal-gas equation to solve problems involving gases as reactants or products in chemical reactions. From the gas density, d, under given conditions of pressure and temperature, it is possible to calculate the molecular weight of the gas: MW = dRT/P. In calculating the quantity of gas collected over water, correction must be made for the partial pressure of water vapor in the container.

The kinetic-molecular theory accounts for the properties of an ideal gas in terms of a set of assumptions about the nature of gases. Briefly these assumptions are that molecules are in ceaseless, chaotic motion; that the volume of gas molecules is negligible in relation to the volume of their container; that the gas molecules have no attractive forces for one another; and finally, that the average kinetic energy of the gas molecules is proportional to absolute temperature.

The molecules of a gas do not all have the same kinetic energy at a given instant. Their speeds are distributed over a wide range; the distribution varies with the molecular weight of the gas and with temperature. The root mean square (rms) speed, u, varies in proportion to the square root of absolute temperature, and inversely with the square root of molecular weight: $u = (3RT/\text{MW})^{\frac{1}{2}}$. It follows that the rate at which a gas escapes (effuses) through a tiny hole is inversely proportional to the square root of its molecular weight (Graham's law). Molecules in a real gas possess finite volume and thus undergo frequent collisions with one another. These frequent collisions limit the rate at which a gas molecule can diffuse through a space occupied by other gas molecules and determine the thermal conductivity of a gas.

The extent of nonideality of a real gas can be seen by examining the quantity PV/RT for 1 mol of the gas, as a function of pressure; this quantity is exactly 1 for an ideal gas at all pressures. Real gases depart from the ideal behavior because the molecules possess finite volume (leads to $PV/RT > 1$), or because the molecules experience attractive forces for one another upon collision (leads to $PV/RT < 1$). The van der Waals equation is an equation of state for gases that attempts to correct the ideal-gas equation to take account of two properties of real gases.

Learning goals

Having read and studied this chapter, you should be able to:

1 Describe the general characteristics of gases as compared with other states of matter and list the ways in which gases are distinctly different.

2 List the variables that are required to define the state of a gas.

3 Define 1 atmosphere, millimeters of mercury, and kilopascals, the most important units in which pressure is expressed. You should also understand the principle of operation of a barometer.

4 Explain the way in which pressure, volume, and temperature are related in the ideal-gas equation. That is, you should remember Equation [9.1].

5 Solve problems involving changes in the condition or state of a gas. You should be able to explain how one variable is affected by a change in another, when the other variables are maintained constant.

6 Calculate the quantity of a gas under a given set of conditions that is required as a reactant or formed as product in a chemical reaction.

7 Correct for the effects of water vapor pressure in calculating the quantity of a gas collected over water.

8 Explain the concept of gas density and describe how it is related to temperature, pressure, and molecular weight.

9 Calculate molecular weight, given gas density under defined conditions of temperature and pressure. You should also be able to calculate gas density under stated conditions, knowing molecular weight.

10 List and explain the assumptions on which the kinetic theory of gases is based.

11 Describe graphically how gas molecules are distributed over a range of speeds and how that distribution changes with temperature.

12 Describe how the relative rates of diffusion or effusion of two gases depend on their relative molecular weights (Graham's law).

13 Explain the concept of mean free path and how it relates to the rates of diffusion of molecules in the gas state and to thermal conductivity.

14 Explain the origin of deviations shown by real gases from the relationship $PV/RT = 1$ for a mole of ideal gas.

15 List the two major factors responsible for deviations of gases from ideal behavior.

16 Explain the origins of the correction terms to P and V that appear in the van der Waals equation of state for a gas.

Key terms

Among the more important terms and expressions used for the first time in this chapter are the following:

According to Avogadro's hypothesis (Section 9.3), equal volumes of gases at the same temperature and pressure contain equal numbers of molecules.

A barometer (Section 9.2) is a device for measuring atmospheric pressure in terms of the height of a liquid column sustained by that pressure.

According to Boyle's law (Section 9.3), at constant temperature, the product of the volume and pressure of a given amount of gas is a constant.

According to Charles's law (Section 9.3), (1) at constant pressure, the volume of a given quantity of gas is proportional to absolute temperature, and (2) at constant volume, the pressure of a given quantity of gas is proportional to absolute temperature.

Dalton's law of partial pressures (Section 9.3) states that the total pressure of a mixture of gases is just the sum of the pressures that each gas would exert if it were present alone.

Diffusion (Section 9.6) refers to the rate at which a substance spreads into and throughout a space. Thus, a gas might diffuse throughout a room, or atmospheric oxygen might diffuse through the waters of a lake. Effusion refers to the rate at which a gas escapes through an orifice or hole.

The gas constant, R (Section 9.3), is the constant of proportionality in the ideal-gas equation.

Graham's law (Section 9.6) states that the relative rate of effusion of two gases is inversely proportional to the square root of the ratio of their molecular weights.

The ideal-gas equation (Section 9.3) is an equation of state for gases that embodies Boyle's law, Charles's law, and Avogadro's hypothesis in the form $PV = nRT$.

The kinetic-molecular theory of gases (Section 9.5) consists of a set of assumptions about the nature of gases. These assumptions, when translated into mathematical form, yield the ideal-gas equation.

The mean free path (Section 9.7) in a gas sample is the average distance traveled by a gas molecule between collisions.

Pressure (Section 9.2) is a measure of the force exerted on a unit area. In work with gases, pressure is most commonly expressed in units of atmospheres (atm) or of millimeters of mercury (mm Hg)—760 mm Hg = 1 atm; in SI units, pressure is expressed in pascals (Pa).

The root mean square (rms) speed, u, of a gas is given by the square root of the average of the squared speeds of the gas molecules.

The standard atmosphere (Section 9.2) is defined as 760 mm Hg, or in SI units, 101.325 kPa.

Standard temperature and pressure (STP) (Section 9.4), 0°C and 1 atm pressure, are frequently used reference conditions for a gas.

The thermal conductivity of a gas (Section 9.7) is a measure of the rate at which heat energy can be transferred through it.

The torr (Section 9.2) is a unit of pressure (1 torr = 1 mm Hg).

The van der Waals equation (Section 9.8) is an equation of state for real gases containing terms that correct for the existence of attractive forces between molecules and for their finite volumes.

EXERCISES

Introduction; pressure

9.1 Give three examples of substances other than water that can be observed in more than one of the three states of matter.

9.2 The compressibility, which is the change in volume of a substance in response to a change in pressure, is much greater for gases than for liquids or solids. Explain why.

9.3 What variables must be specified to define properly the state or condition of a gas? Indicate a set of units in which each of these variables can be properly expressed.

9.4 Mercury has a density of 13.59 g/cm³. Using the approach described in the text for the aluminum cylinder, calculate the force per unit area of a column of mercury 76.0 cm in height at the earth's surface, in units of kilopascals.

9.5 Suppose that the mercury in the barometer of Figure 9.2 were replaced by a liquid metal alloy with a density of 5.73 g/cm³. The density of mercury is 13.6 g/cm³. What height of column of liquid alloy would be supported by 1 atm pressure?

[9.6] Water-well pumps, such as the familiar pitcher pump, which depend on creating a vacuum to lift water to the surface, cannot lift water from a depth greater than about 32 feet. Why is this so?

The ideal-gas equation; gas law problems

9.7 The ideal-gas equation is called an "equation of state." What does this expression mean?

9.8 An airtight cylinder contains 3.67 g of a gas in a volume of 2.52 L. What is the density of this gas after the pressure is doubled at constant temperature?

9.9 Which of the following statements, if any, are false? Correct any false statements. (a) At constant temperature and volume, pressure is inversely proportional to the number of moles of gas. (b) At constant volume the pressure of a given amount of gas increases in proportion to the absolute temperature. (c) An increase in volume of a given quantity of gas at constant pressure arises from an increase in absolute temperature. (d) At constant volume the pressure of a gas is inversely proportional to temperature.

9.10 Suppose you were responsible for filling a 12.0-L weather balloon with 1.00 atm pressure of helium each day and releasing it. Assuming no waste, how many 25-L tanks of helium at a pressure of 2250 lb/in.² would you require in a year (all temperatures about 27°C)? (1 atm = 14.7 lb/in.²)

9.11 The compression ratio for an internal combustion engine is the ratio of the pressure of gas in the cylinder when the piston is all the way in as compared with when the gas is introduced. If the compression ratio is 8.0, and the volume of the cylinder at the time the gas is admitted is 0.25 L, what is the volume at the time of maximum compression (assume temperature constant)?

9.12 The density of a particular gas in a cylinder with a volume of 6.50 L is found to be 1.45 g/L. The gas is compressed at constant temperature until the volume is 3.20 L. What is the density of the gas under the new conditions?

9.13 A sample of CO_2 gas occupies a volume of 64.0 cm³ at 130°C, 0.372 atm pressure. What is the pressure of the gas if the volume is reduced to 12.0 cm³ at constant temperature?

9.14 A helium gas bubble is trapped in a cavity of 0.172 cm³ volume inside a mineral sample at a pressure of 127 atm. What pressure does the helium exert after release from the mineral and storage in a bulb of 125-cm³ volume? (Assume constant temperature.)

9.15 A student obtains the following data using the apparatus illustrated in Figure 9.6:

Temperature (°C)	16	55	85	103	126	163
Length of gas column (mm)		31	35	38	40	43	47

Make a graph of the data. Is Charles's law obeyed? Extrapolate the data carefully to zero volume; at what temperature does this intercept occur? What special significance does this temperature have?

9.16 A gas exerts a pressure of 137 kPa at 27°C. The temperature of the gas is increased to 148°C with no significant volume change. What is the gas pressure at the higher temperature?

9.17 It requires 0.182 mol of O_2 to exert a pressure of 1.50 atm in a particular tank at 25°C. What quantity of O_2 would be required to exert a pressure of 17.2 atm in the same tank at 100°C?

9.18 Complete each of the following statements: (a) When a gas is heated at constant volume from 28°C to 274°C, the pressure increases from _____ atm to 12.3 atm. (b) A gas that occupies a volume of 6.75 L at 89.0 atm will occupy a volume of _____ L at 6.85 mm Hg pressure, if temperature remains constant. (c) A gas that exerts a pressure of 1.65 atm when confined to a volume of 88.0 cm³ at 25°C will exert a pressure of _____ atm when the volume is increased to 6.85 L at 25°C.

9.19 Consider the arrangement of bulbs shown in Figure 9.17. Each of the bulbs contains a gas at the pressure shown. What is the pressure in the system when all the stopcocks are opened, assuming that the temperature remains constant? (We can neglect the volume of the capillary tubing connecting the bulbs.)

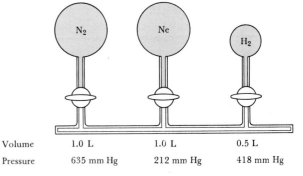

Volume	1.0 L	1.0 L	0.5 L
Pressure	635 mm Hg	212 mm Hg	418 mm Hg

FIGURE 9.17

9.20 Complete each of the following: (a) A 2.5-L flask contains N_2 at a pressure of 635 mm Hg, at 29°C. The volume of this N_2 sample at STP would be _____ L. (b) The density of a sample of Ar gas at STP is _____ g/L. (c) A large underground storage cavern for natural gas, of volume 3.0×10^7 m³, could contain _____ mol of methane, CH_4, at 1.15 atm pressure and 12°C.

9.21 What is the partial pressure of H_2 in a gas mixture containing 1.5 g of H_2, 85.2 g of O_2, and 17.0 g of Ar confined to a volume of 3.20 m^3 at 88°C? What is the total pressure in the vessel?

9.22 The density of dry air at 30.0°C, 720 mm Hg, is 1.104 g/L. Calculate the average molecular weight of the air.

9.23 Suppose that the volume of an engine cylinder at the time gas is admitted is 0.46 L and that at the time of firing the volume is 0.082 L. Suppose also that the temperature of the incoming gas is 35°C and that at the time of firing the gas is at a temperature of 420°C. What is the ratio of the pressure of the gas at time of firing to that at time of admission?

[9.24] A large meteorological balloon weighing 46 g has a volume of 265 L. It is filled with helium at 1.00 atm pressure and 25°C. The lifting capacity of the balloon is equal to the difference in mass between the helium in the balloon and an equivalent volume of outside air, less the mass of the balloon itself. Calculate the lifting capacity of the balloon. (The average molecular weight of air is 29.0 g/mol.) Assuming that a mass less than the lifting capacity is attached to the balloon, it will rise when released. Would the balloon be expected to rise to a higher or lower maximum altitude if it is made of rubber, and is therefore capable of expansion or contraction, as compared with its behavior if it has a fixed volume?

9.25 A typical ranch house has a volume of 13,000 ft^3. In winter the house has a water-vapor pressure of only 4 mm Hg. A humidifier is installed to increase the water-vapor pressure to 9 mm Hg, which represents about 40 percent humidity. What mass of water vapor must the humidifier produce to increase the humidity to the desired level? Assume the house temperature averages 22°C.

9.26 Magnesium can be used as a "getter" in evacuated enclosures, to react with the last traces of oxygen. (The magnesium is usually heated by passing an electric current through a wire or ribbon of the metal). If an enclosure of 0.283 L has a partial pressure of O_2 of 6.5×10^{-6} mm Hg at 27°C, what mass of magnesium will react according to the equation

$$2Mg(s) + O_2(g) \longrightarrow 2MgO(s)$$

9.27 When 6.52 g of $Mg_3N_2(s)$ is reacted with 0.185 g of H_2O, what volume of $NH_3(g)$ at 715 mm Hg and 30°C is formed by the reaction

$$Mg_3N_2(s) + 3H_2O(l) \longrightarrow 3MgO(s) + 2NH_3(g)$$

9.28 A 3.25-g sample of $KClO_3$ is decomposed according to the equation

$$2KClO_3(s) \longrightarrow 2KCl(s) + 3O_2(g)$$

Assuming 100 percent decomposition, what volume of O_2 should be collected over water at 22°C, when atmospheric pressure is 740 mm Hg?

9.29 Calculate the volume of O_2 at 710 mm Hg pressure and 36°C required to react with 6.50 g of CuS according to the equation

$$CuS(s) + 2O_2(g) \longrightarrow CuSO_4(s)$$

Kinetic molecular theory; Graham's law; molecular speeds

9.30 What change or changes in the state of a gas bring about each of the following effects? (a) The number of impacts per unit time on a given area of container wall increases. (b) The average energy of impact of molecules with the wall of the container decreases. (c) The average distance between gas molecules increases. (d) The average speed of molecules in the gas mixture is increased.

9.31 What experimental observations justify the assumption that the volumes of gas molecules are negligible in comparison with the volume of the gas container?

9.32 What is the ratio of average kinetic energies of H_2 to N_2 molecules at 25°C? What is the ratio of the rms speeds of the molecules of the two gases?

9.33 A sample of N_2 gas is initially at 1 atm pressure and 300 K. It is heated to 600 K and its volume is simultaneously doubled. What effect do these changes have on (a) the rms speed of the N_2 molecule; (b) the average kinetic energy of N_2 molecules; and (c) the number of collisions per unit time *with a unit area* of container wall?

9.34 For H_2 the root mean square speed at 0°C is 1693 m/sec. What is the root mean square speed of HCl molecules at 0°C?

9.35 Suppose we have two 1-L flasks, one containing N_2 at STP, the other containing SF_6 at STP. How do these systems differ with respect to: (a) the average kinetic energies of the molecules; (b) the total number of collisions occurring per unit time with the container walls; (c) the shapes of the distribution curves of molecular speeds; and (d) the relative rates of effusion through a pinhole leak?

9.36 Suppose a nozzle system is being designed for a rocket in which the thrust will come from reaction of H_2 with F_2, a highly exothermic reaction. The two reacting substances will be effused into the reaction zone through a series of tiny holes, all the same size. What should be the ratio of the number of holes for F_2 to those for H_2 to maintain a 1 : 1 stoichiometry in the reaction zone?

9.37 Which gas in each of the following pairs should have the higher thermal conductivity: (a) CO or He; (b) N_2 or CO_2; (c) Ar or O_2?

Departures from ideal-gas behavior

9.38 Which of the following gases would you expect to show the largest negative departure from the PV/RT relationship expected for an ideal gas: C_2H_6, N_2, He, or CH_4? For which of these gases should the correction for finite volume of the gas molecules be largest?

9.39 For each of the following pairs of gases, indicate which you would expect to deviate more from the PV/RT relationship expected for an ideal gas: (a) O_2 or UF_6; (b) BF_3 or $SiCl_4$; (c) CO_2 or SO_2.

9.40 It turns out that the van der Waals constant b is equal to four times the total volume actually occupied by the molecules of a mole of a gas. Using this figure, calculate the fraction of the volume in a container actually occupied by Ar atoms at STP; at 100 atm pressure and 0°C. (Assume for simplicity that the ideal-gas equation still holds.)

[9.41] Figure 9.18 shows the PV/RT behavior for CH_4 as a function of temperature. Explain why the negative departures from the ideal-gas relationship decrease with increasing temperature. (A rather more subtle question: Why is the positive departure at high pressures larger when the gas is at lower temperatures?)

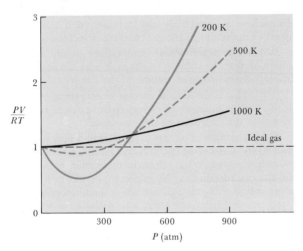

PV/RT

Ideal gas

200 K
500 K
1000 K

300   600   900

P (atm)

FIGURE 9.18

Additional exercises

9.42 (a) A plastic bucket has a flat bottom of area 3.2×10^3 cm². When filled with water it weighs 9.60 kg. What pressure, in kilopascals, does the bucket exert on a flat surface on which it rests? (b) A phonograph stylus resting on a record has an effective weight of 2.5 g. The area of the stylus making contact with the record averages 2.7×10^{-4} mm². What pressure, in kilopascals, does the stylus exert on the record?

9.43 A mercury manometer, illustrated in Figure 9.19, is a device for measuring pressure in B by determining the difference in heights of mercury columns in the two arms, and knowing the pressure, A, that operates on the left arm. What is the pressure in B when: (a) $h = 0, A = 760$ mm Hg; (b) $h = 759$ mm, $A = 1$ atm; (c) $h = 350$ mm, $A =$

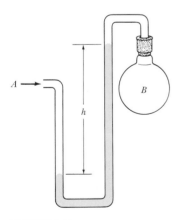

FIGURE 9.19

409 mm Hg;   (d) $h = 3.6$ mm, $A = 650.5$ mm Hg, (e) $h = 0.65$ m, $A = 207.5$ kPa.

[9.44] Suppose that the manometer illustrated in Figure 9.19 has mineral oil (density 0.752 g/cm³) as fluid, in place of mercury, which has a density of 13.6 g/cm³. The oil has no significant partial pressure of its own. (a) When h is 16.5 cm and $A = 1.00$ atm pressure, what is the pressure in B? (b) A pure liquid is placed in B and all gases other than the vapor of the liquid are removed from B. At equilibrium, the value of h for the mineral oil manometer is 0.65 m when A is 135 mm Hg. What is the pressure exerted by the vapor of the pure liquid in B?

9.45 Complete each of the following: (a) A pressure of 237 kPa corresponds to _____ atm. (b) A pressure of 6.25×10^{-7} torr corresponds to _____ atm. (c) A pressure of 273 mm Hg corresponds to _____ atm. (d) A pressure of 838 mm Hg corresponds to _____ kPa.

9.46 A sample of hydrogen gas at 26°C has a volume of 8.73 L at a pressure of 735 mm Hg. What is the volume of the gas sample if the pressure is increased to 1675 mm Hg?

9.47 Complete each of the following: (a) 3.45 mol of N_2 would occupy a volume of _____ L at a pressure of 835 kPa, 320 K. (b) A gas sample weighing 1.32 g occupies a volume of 1.15 L at 27°C and 92.4 kPa pressure. The molecular weight of the gas is _____ g/mol. (c) 6.4 mol of N_2 is confined to a 25.0-L cylinder at 37°C. The pressure of the gas is _____ g/mol.

9.48 A McLeod gauge is a device used to measure the pressure of gases at low pressures in a so-called vacuum line. A large volume of gas from the vacuum line system is compressed to a much smaller volume at higher pressures that can be directly measured. In this manner a 0.68-L sample of gas at unknown pressure is compressed to a volume of 0.047 cm³, at which time it is under a pressure of 600 mm Hg. What is the pressure in the vacuum line system?

9.49 A volume of 1.86 L of H_2 is trapped in a nuclear reactor at 870°C and 125 atm. What is the partial pressure of the H_2 if it is released into a containment vessel of 2.8×10^4 m³ at 25°C?

9.50 Suppose the experiment illustrated in Figure 9.6 is carried out on a gas sample that has a length of gas column in the capillary of 42 mm at 25°C. At a higher temperature the length of the gas column is 63 mm. What is the temperature of the oil bath at this point.

9.51 Calculate the density of Ar gas at 1.65 atm and 13°C.

9.52 In the Dumas bulb technique for determining the molecular weight of an unknown liquid, one vaporizes a sample of a liquid that boils below 100°C in a boiling water bath and determines the mass of vapor required to just fill the bulb (see Figure 9.20). From the following data, calculate the molecular weight of the unknown liquid: mass of unknown vapor, 0.912 g; volume of bulb, 342 cm³; pressure, 742 mm Hg; temperature, 99°C.

9.53 A piece of solid magnesium is reacted with dilute hydrochloric acid to form hydrogen gas:

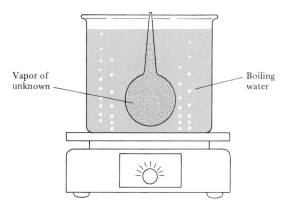

FIGURE 9.20

$$Mg(s) + 2HCl(aq) \longrightarrow MgCl_2(aq) + H_2(g)$$

What volume of H_2 is collected over water at $28\,°C$ (see Appendix C) by reaction of 1.25 g of Mg with excess acid solution? The barometer records an atmospheric pressure of 748 mm Hg.

[9.54] A glass vessel fitted with a stopcock has a mass of 337.428 g when evacuated. When filled with Ar it has a mass of 339.712 g. When evacuated and refilled with a mixture of Ne and Ar, under the same conditions of temperature and pressure, it weighs 339.146 g. What is the mole percentage of Ne in the gas mixture?

9.55 Complete each of the following: (a) A gas that has a density of 1.84 g/L at 52°C and 600 mm Hg pressure has a molecular weight of _____ g/mol. (b) 1280 cm³ of CO is collected over water at 18°C, total pressure 745 mm Hg. The partial pressure of CO is _____ mm Hg (see Appendix C). The volume of CO corrected to STP is _____ cm³. (c) The density of an equimolar mixture of He and Ar at 82°C and 600 mm Hg is _____ g/L.

9.56 A 1.86-g sample of lead nitrate, $Pb(NO_3)_2$, is heated in an evacuated cylinder with a volume of 1.62 L. The salt decomposes when heated according to the equation

$$2Pb(NO_3)_2(s) \longrightarrow 2PbO(s) + 4NO_2(g) + O_2(g)$$

Assuming complete decomposition, what is the pressure in the cylinder after decomposition and cooling to a final temperature of 30°C?

[9.57] Consider the experiment illustrated in Figure 9.21. A gas is confined in the leftmost cylinder under 1 atm pressure, and the other cylinder is evacuated. When the stopcock is opened, the gas expands to fill both cylinders. Only a very small temperature change is noted when this expansion occurs. Explain how this observation re-

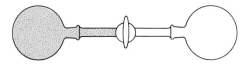

FIGURE 9.21

lates to assumption 4 of the kinetic molecular theory, Section 9.5.

9.58 Calculate the ratio of rates of effusion of: (a) He and Ar; (b) CO and CO_2.

9.59 A gas of unknown molecular weight is allowed to effuse through a small opening under constant pressure conditions. It required 88 sec for 1 L of the gas to effuse. Under identical experimental conditions it required 28 sec for 1 L of O_2 gas to effuse. Calculate the molecular weight of the unknown gas.

9.60 Suppose it has been suggested that a relatively rare isotope of carbon, ^{13}C, could be separated from the more abundant ^{12}C by using a diffusion process similar to that described in the text for UF_6, but using either CO or CO_2. Calculate the relative rates of diffusion for ^{12}CO and ^{13}CO and similarly for $^{12}CO_2$ and $^{13}CO_2$ compounds. Which substance would give the greater degree of separation?

9.61 The PV/RT behaviors of N_2 and ethylene, C_2H_4, are shown in Figure 9.22. What is the significance of the greater negative departure of the curve for ethylene as compared with N_2? Which of the two gases has the larger volume per molecule? How is that deduced from the curves?

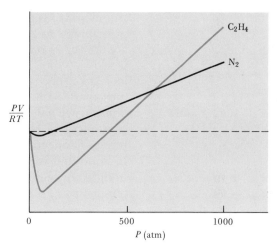

FIGURE 9.22

[9.62] Using the van der Waals equation, with the correction constants given in Table 9.4, calculate the volume occupied by 1 mol of Ar at 200 atm pressure, 273 K. (Hint: You can solve this problem by using a method of successive approximations. Use the ideal-gas equation value for V in the pressure correction term, then solve for V in the $(V - nb)$ term. Use this value for V in a new calculation of the correction to the pressure, recalculate V, and so forth, until reasonable consistency is obtained.) Compare the calculated volume with that predicted by the ideal-gas equation. If the van der Waals constant b is taken to be four times the actual volume occupied by 1 mol of Ar atoms, what fraction of the volume is occupied by the gas itself?

10

Chemistry of the atmosphere

In the previous chapters of this text we have dealt for the most part with principles that govern the chemical and physical behavior of matter. We are now in a position to apply these principles to an understanding of the world in which we live. We start in this chapter with a fairly detailed look at the chemistry of earth's atmosphere. We will be concerned for the most part with quite simple molecules whose structures and mode of bonding we have treated in some detail. The chemical properties of these substances in the atmosphere can therefore be readily understood in terms of concepts we have already discussed. However, you should keep in mind that there is great variation in the chemical processes occurring in different regions of the atmosphere, because the conditions of pressure, temperature, solar radiation, and so forth, vary greatly.

10.1 Earth's atmosphere

The atmosphere of our planet is in many ways like the liquid sea of water that covers three-fourths of the earth's surface. Creatures that live in the depths of the oceans experience an environment that is vastly different from that experienced by those that live near the surface. In a similar way, the environment that we experience at the bottom of the atmospheric sea is very different from that which we would experience if we were a few hundred kilometers up, above most of the earth's atmosphere. Because most of us have never been very far from the earth's surface, we tend to take for granted the many ways in which the atmosphere determines the environment in which we live. In this section, we will examine some of the important physical characteristics of our planet's atmosphere in light of what we know of the properties of gases.

The temperature of the atmosphere varies in a rather complex manner as a function of altitude, as shown in Figure 10.1. Just above the surface the temperature normally decreases with increasing altitude and

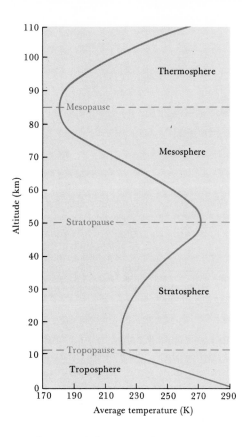

FIGURE 10.1 Temperature variations in the atmosphere at altitudes below 110 km. (*Adapted from "U.S. Standard Atmosphere, 1962." Washington, D.C.: Government Printing Office*)

reaches a minimum value at an elevation of about 12 km. In this region, called the tropopause, the temperature is about 215 K ($-60\,°$C). Above this elevation the temperature increases to about 275 K in the region of 50 km and then begins again to decrease. The altitude at which the temperature reaches a maximum is called the stratopause. Above the stratopause the temperature drops to an even lower value than at the tropopause. The region of this second temperature minimum is called the mesopause. Above the mesopause, the temperature rises rapidly in the region called the thermosphere. Note that the regions of temperature minima and maxima are denoted by the suffix -*pause*. The regions between these are denoted by the suffix -*sphere*. The boundaries between different regions are important because mixing of the atmosphere across the boundaries is relatively slow. Thus, for example, pollutant gases generated in the troposphere find their way into the stratosphere only very slowly.

The troposphere is the region of the atmosphere in which nearly all of us live out our entire lives. When Shakespeare has Hamlet speak of "this most excellent canopy, the air, look you, this brave o'erhanging firmament, this majestical roof fretted with golden fire," he is speaking of the troposphere. Howling winds and soft breezes, rain, sunny skies, all that we normally think of as weather occurs in this region. Even when we fly in a modern jet aircraft between distant cities, we are still in the troposphere, though we may be near the tropopause.

In contrast to the temperature changes that occur in the atmosphere, the pressure of the atmosphere decreases in a quite regular way with

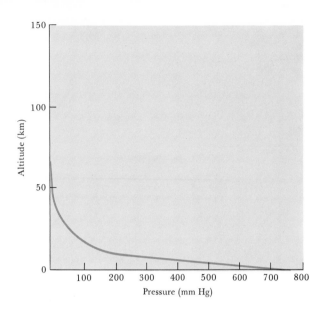

increasing elevation, as shown in Figure 10.2. We see from this illustration that atmospheric pressure drops off much more rapidly at lower elevations than at higher. The explanation for this characteristic of the atmosphere lies in its compressibility. Gases are very different from liquids in this regard. As a result of the atmosphere's compressibility, the pressure decreases from an average value of 760 mm Hg at sea level to 2.3×10^{-3} mm Hg at 100 km, to only 1.0×10^{-6} mm Hg at 200 km.

COMPOSITION OF THE ATMOSPHERE

The atmosphere is an extremely complex system. Its temperature and pressure change over a wide range with altitude, as we have already seen. The atmosphere is subjected to bombardment by radiation and energetic particles from the sun and by cosmic radiation from outer space. This barrage of energy has profound chemical effects, especially on the outer reaches of the atmosphere. In addition, because of the earth's gravitational field, lighter atoms and molecules tend to rise to the top. As a result of all these factors, the composition of the atmosphere is not constant. However, it is useful to know the composition of the atmosphere in the region near the earth's surface. Table 10.1 shows the composition of dry air near sea level. We note that although traces of a great many substances are present, only a few dominate. The two diatomic molecules N_2 and O_2 make up about 99 percent of the entire atmosphere. Essentially all the remainder, with the exception of carbon dioxide, is made up of the monatomic rare gases.

Note that the contribution of each component of the atmosphere listed in Table 10.1 is given in terms of its mole fraction. This is simply the total number of moles of a particular component in a given sample of air, divided by the total number of moles of all the components in that sample. The partial pressure of a given component in the atmosphere is given by the total atmospheric pressure times the mole fraction of that component. This statement comes from Dalton's law of partial pressures

TABLE 10.1   Composition of dry air near sea level

Component[a]	Content (mole fraction)	Molecular weight
Nitrogen	0.78084	28.013
Oxygen	0.20948	31.998
Argon	0.00934	29.948
Carbon dioxide	0.000330	44.0099
Neon	0.00001818	20.183
Helium	0.00000524	4.003
Methane	0.000002	16.043
Krypton	0.00000114	83.80
Hydrogen	0.0000005	2.0159
Nitrous oxide	0.0000005	44.0128
Xenon	0.000000087	131.30

[a]Ozone, sulfur dioxide, nitrogen dioxide, ammonia, and carbon monoxide are present as trace gases in variable amounts.

and Avogadro's hypothesis. By using it we can write the total pressure P_t for a gas mixture consisting of gases with mole fractions X_1, X_2, and so on, as:

$$P_t = X_1 P_t + X_2 P_t + X_3 P_t + \cdots$$
$$= P_t(X_1 + X_2 + X_3 + \cdots)$$

[10.1]

Because the X's must all add up to 1, this equation is clearly correct.

SAMPLE EXERCISE 10.1

What is the partial pressure of CO_2 in dry air when the total dry air pressure (P_t) is 735 mm Hg?

Solution: Referring to Table 10.1, we see that the mole fraction of CO_2 is 3.30×10^{-4}. This means that the fractional contribution of CO_2 to a unit total pressure of 1 would be 3.30×10^{-4}:

Pressure $CO_2 = (735$ mm Hg $P_t)$
$$\times \left(\frac{3.30 \times 10^{-4} \text{ mm Hg } CO_2}{1 \text{ mm Hg } P_t} \right)$$
$$= 0.243 \text{ mm Hg } CO_2$$

10.2   The outer regions

Although the upper reaches of the atmosphere contain only a small fraction of the atmospheric mass, the upper atmosphere plays an important role in determining the conditions of life at the earth's surface. These upper layers form the outer bastion of defense against the hail of radiation and high-energy particles with which the planet is continually bombarded. In absorbing these assaults, the molecules and atoms of the atmosphere undergo chemical change.

Up to an altitude of about 90 km, the average molecular weight of the atmosphere remains about the same as that at sea level, 28.96 amu. Above about 90 km the average molecular weight decreases as shown in Figure 10.3. This sharp decline means, of course, that the composition of the atmosphere is not constant. The changing composition has two causes. The first is called diffusive separation; the lightest molecules and atoms experience the least gravitational attraction to earth. Over a long

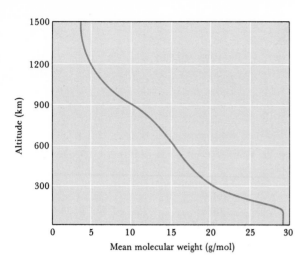

Altitude (km)

Mean molecular weight (g/mol)

FIGURE 10.3 Mean molecular weight of the atmosphere as a function of altitude.

time period, they drift to the top of the atmosphere. The element helium, which is a *very* minor constituent of the atmosphere at sea level (Table 10.1), is a major constituent in the region between 500 and 1000 km. (Remember, however, that there are very few atoms or molecules of *any* kind at these elevations.)

The second cause of the changing composition of the atmosphere is chemical in origin. Electromagnetic radiation and high-energy particles from the sun bombard the atmosphere and are absorbed. The energy absorptions that occur cause dissociation and ionization of atoms and molecules. For example, oxygen molecules are caused to dissociate into oxygen atoms:

$$O_2(g) + h\nu \longrightarrow 2O(g) \tag{10.2}$$

(The symbol $h\nu$ as used here denotes that the chemical process shown results from absorption of a photon.) As processes of this type become sufficiently extensive, the average molecular weight of the atmosphere decreases. For example, a gas consisting of O_2 molecules has a molecular weight of 32 g/mol. A gas that consisted of just O atoms would have a molecular weight of 16 g/mol. A gas consisting of a mixture of O_2 and O atoms has an average molecular weight of between 32 and 16 g/mol. We must now examine in more detail the nature of the chemical reactions that occur as a result of the absorption of radiation.

PHOTODISSOCIATION

The sun emits radiant energy over a wide range of wavelengths. The shorter-wavelength, higher-energy radiations in the ultraviolet range of the spectrum are sufficiently energetic to cause chemical changes. We have already seen, in Section 5.2, that electromagnetic radiation can be pictured as a stream of photons. The energy of each photon is given by the relationship $E = h\nu$, where h is Planck's constant and ν is the frequency of the radiation. For a chemical change to occur when radiation falls on the earth's atmosphere, two conditions must be met. First, there must be photons with energy at least as large as that required to break a chemical bond, remove an electron, or otherwise accomplish whatever chemical process is being considered. Second, molecules must absorb these photons. This requirement means that the energy of the photons is converted into some other form of energy within the molecule.

One of the most important processes occurring in the upper atmosphere is dissociation of the oxygen molecule as a result of the absorption of a photon (photodissociation), as shown in Equation [10.2]. The minimum energy required to cause this change is determined by the dissociation energy of O_2, 495 kJ/mol. We calculate the longest-wavelength photon having sufficient energy to dissociate the O_2 molecule in Sample Exercise 10.2.

SAMPLE EXERCISE 10.2

What is the wavelength of a photon corresponding to a molar bond-dissociation energy of 495 kJ/mol?

Solution: We must first calculate the energy required on a per molecule basis and then determine the wavelength of a photon with that energy:

$$\left(495 \times 10^3 \frac{J}{mol}\right)\left(\frac{1 \text{ mol}}{6.022 \times 10^{23} \text{ molecules}}\right)$$

$$= 8.22 \times 10^{-19} \frac{J}{molecule}$$

The energy of the photon is given by $E = h\nu$. Rearranging we have:

$$\nu = \frac{E}{h} = \left(\frac{8.22 \times 10^{-19} \text{ J}}{6.625 \times 10^{-34} \text{ J-sec}}\right)$$

$$= 1.24 \times 10^{15}/\text{sec}$$

Recall from Section 5.1 that the product of frequency and wavelength of radiation equals the velocity of light:

$$\nu\lambda = c = 3.00 \times 10^8 \text{ m/sec}$$

Therefore, rearranging this equation we have

$$\lambda = \frac{c}{\nu} = \left(\frac{3.00 \times 10^8 \text{ m/sec}}{1.24 \times 10^{15}/\text{sec}}\right)\left(\frac{10^9 \text{ nm}}{1 \text{ m}}\right)$$

$$= 242 \text{ nm}$$

The calculations in Sample Exercise 10.2 tell us that any photon of wavelength *shorter* than 242 nm will have sufficient energy to dissociate the O_2 molecule. (Remember that shorter wavelength means higher energy!)

The second condition that must be met before dissociation actually occurs is that the photon must be absorbed by O_2. Fortunately for us, O_2 absorbs much of the high-energy, short-wavelength radiation from the solar spectrum before it reaches the lower atmosphere. As it does so, atomic oxygen (O) is formed. The oxygen composition of the atmosphere as a function of altitude is illustrated in Figure 10.4. At higher elevations the dissociation of O_2 is very extensive; at 400 km only 1 percent of the oxygen is in the form of O_2; the other 99 percent is in the form of atomic oxygen. At 130 km, O_2 and O are just about equally abundant. Below this elevation O_2 is more abundant than O.

Recall from our discussion of the electronic structures of diatomic molecules that the bond-dissociation energy of N_2 is very high (Table 8.7). Thus, only photons of very short wavelength possess sufficient energy to cause dissociation of this molecule. Furthermore, N_2 does not readily absorb photons, even when they do possess sufficient energy. The overall result is that very little atomic nitrogen is formed in the upper atmosphere by dissociation of N_2.

One of the most interesting photochemical processes in the atmosphere is photodissociation of water. The partial pressure of water in the

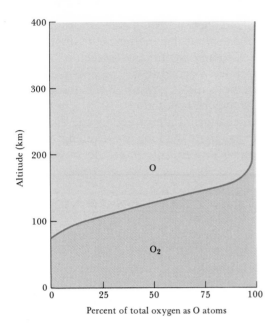

FIGURE 10.4 Oxygen composition of the atmosphere as a function of altitude.

atmosphere is quite appreciable near the surface, but decreases very rapidly with increasing altitude. The water-vapor level in the stratosphere at 30 km is only about three parts per million (3 ppm).* Although it has not been possible to measure the water-vapor level at still higher altitudes, it seems quite certain that it is never more than a few parts per million. The amount of water that finds its way to the top of the atmosphere from the surface is thus very small. Once in the upper atmosphere, however, water undergoes photodissociation:

$$H_2O(g) + h\nu \longrightarrow H(g) + OH(g)$$
$$OH(g) + h\nu \longrightarrow H(g) + O(g)$$

[10.3]

Some scientists think that in the early stages of the earth's history, when it had no oxygen atmosphere, photodissociation of water was the means by which the planet acquired an oxygen atmosphere.

10.3 Ionization processes

In 1901 Guglielmo Marconi carried out a sensational experiment, by receiving in St. John's, Newfoundland, a radio signal transmitted from Land's End, England, some 2900 km away. Because radio waves were thought to travel in straight lines, it had been assumed that radio communication over large distances on earth would be impossible. The fact that Marconi's experiment was successful suggested that the earth's atmosphere in some way substantially affected radio-wave propagation. His discovery led to intensive study of the upper atmosphere. In about 1924 the existence of electrons in the upper atmosphere was established by experimental studies.

*Parts per million is a very commonly used unit for expressing the relative concentrations of trace constituents. A concentration of one part per million means that in a million parts of total atmosphere, one part by volume is due to the trace constituent. From Dalton's law of partial pressures and Avogadro's hypothesis (Section 9.3), this also means that of a million total molecules, one molecule on the average is due to the trace constituent.

TABLE 10.2 Ionization processes, ionization energies, and maximum wavelength of a photon capable of causing ionization

Process	Ionization energy (kJ/mol)	λ_{max} (nm)
$N_2 + h\nu \longrightarrow N_2^+ + e^-$	1495	80.1
$O_2 + h\nu \longrightarrow O_2^+ + e^-$	1205	99.3
$O + h\nu \longrightarrow O^+ + e^-$	1313	91.2
$NO + h\nu \longrightarrow NO^+ + e^-$	890	134.5

For each electron present in the upper atmosphere, there is a corresponding positively charged ion. The electrons in the upper atmosphere are a result of ionization of molecules, which is caused mainly by solar radiation (photoionization), and to a lesser extent by high-energy electrons and protons that stream into the earth's atmosphere from the sun. We shall concern ourselves here only with ionization caused by radiation. For photoionization to occur, a photon must be absorbed by the molecule, and this photon must have enough energy to cause removal of the highest-energy electron. Some of the more important ionization processes occurring in the upper atmosphere, that is, above about 90 km, are shown in Table 10.2, along with the ionization energies and λ_{max}, the maximum wavelength of a photon capable of causing ionization. Photons with energies sufficient to cause ionization have wavelengths in the short, or high-energy, region of the ultraviolet. These wavelengths are completely filtered out of the radiation reaching earth as a result of absorption by the upper atmosphere.

REACTIONS OF ATMOSPHERIC IONS

The molecular ions formed by photoionization are very reactive species. Without any additional input of energy from solar radiation, they react very rapidly on contact with a variety of charged species and neutral molecules. As we discuss these reactions, bear in mind that only processes that are exothermic, that is, those in which heat is evolved, need to be considered. Endothermic processes, those that require an input of energy, take place so slowly as to be of no interest in atmospheric chemistry.

Dissociative recombination   One of the obvious possibilities for reaction of a molecular ion is recombination with an electron to yield the neutral molecule. A great deal of energy is released when the electron recombines with the molecular ion; in fact, this amount of energy is equal to the ionization potential of the neutral molecule, because recombination is just the reverse of ionization. Unless there is a way to transfer this excess energy—for example, by collision with another molecule—the excess energy causes dissociation of the molecule. The likelihood of getting rid of the excess energy by collision with another molecule is very small in the upper atmosphere. The density of molecules is relatively low, and collisions do not occur frequently. As a result, essentially all the recombinations of electrons with molecular ions result in dissociation:

$$N_2^+(g) + e^- \longrightarrow N(g) + N(g) \qquad [10.4]$$

$$O_2^+(g) + e^- \longrightarrow O(g) + O(g) \qquad [10.5]$$

$$NO^+(g) + e^- \longrightarrow N(g) + O(g) \qquad [10.6]$$

A reaction of this type is referred to as dissociative recombination. It is the principal source of the atomic nitrogen present in the upper atmosphere.

Charge transfer   When a molecular ion undergoes collision with a neutral species, there may be transfer of an electron. For example, when an N_2^+ ion encounters an O_2 molecule, an electron is transferred from O_2 to N_2^+:

$$N_2^+(g) + O_2(g) \longrightarrow N_2(g) + O_2^+(g) \qquad [10.7]$$

For such a charge-transfer reaction to occur, the ionization energy of the molecule losing the electron must be less than the ionization energy of the molecule formed after transfer occurs. For example, the ionization energy of O_2 is less than that of N_2. When this condition is met, the reaction is exothermic, as illustrated in Figure 10.5. The excess energy of reaction is carried off in the kinetic energy of the molecules. On the basis of the data on ionization energies in Table 10.2, the reaction shown in Equation [10.7] and the following reactions should be exothermic:

$$O^+(g) + O_2(g) \longrightarrow O(g) + O_2^+(g) \qquad [10.8]$$

$$O_2^+(g) + NO(g) \longrightarrow O_2(g) + NO^+(g) \qquad [10.9]$$

$$N_2^+(g) + NO(g) \longrightarrow N_2(g) + NO^+(g) \qquad [10.10]$$

SAMPLE EXERCISE 10.3

Using Hess's law (Section 4.4) and the data in Table 10.2, calculate the energy change in the reaction shown in Equation [10.8].

Solution: We can write Equation [10.8] as the sum of two processes, for each of which the heat is given in Table 10.2 (recall that when we reverse the direction of a process, we reverse the sign of the energy change):

$$O^+(g) + e^- \longrightarrow O(g) \qquad E = -1313 \text{ kJ/mol}$$
$$O_2(g) \longrightarrow O_2^+(g) + e^-$$
$$ E = +1205 \text{ kJ/mol}$$

$$\overline{O^+(g) + O_2(g) \longrightarrow O(g) + O_2^+(g)}$$
$$ E = -108 \text{ kJ/mol}$$

The negative sign for E denotes an exothermic process.

Because the N_2 molecule has the highest ionization energy of any of the species present in the upper atmosphere, N_2^+ is capable of undergoing charge transfer with every molecule with which it collides. Charge-transfer reactions are extremely rapid. As a result, although N_2^+ is produced extensively by photoionization, there is very little N_2^+ present in the upper atmosphere. These gas-phase charge-transfer reactions represent the simplest examples of a very important type of oxidation-reduction reaction known as electron transfer. Reactions in which an electron is transferred from one chemical species to another are of great importance in all areas of chemistry, including biochemistry. We shall encounter other examples of this type of reaction in later chapters.

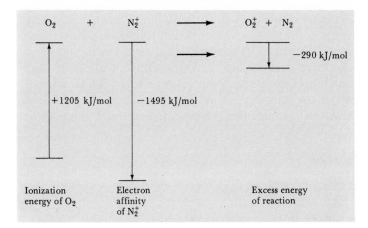

$$O_2 \quad + \quad N_2^+ \quad \longrightarrow \quad O_2^+ \ + \ N_2$$

$$\longrightarrow \qquad -290 \text{ kJ/mol}$$

$+1205$ kJ/mol -1495 kJ/mol

Ionization
energy of O_2

Electron
affinity
of N_2^+

Excess energy
of reaction

FIGURE 10.5 Illustration of the energy relations in a charge-transfer reaction. Recall (Section 6.6) that the electron affinity of a positive ion (N_2^+) is just the ionization potential, with reversed sign, of the corresponding neutral molecule (N_2).

Atom transfer In the simple charge-transfer reaction, all chemical bonds remain intact. An electron is simply transferred from one species to another. There is a class of reactions, however, in which an atom is transferred. The two important reactions of this type occurring in the upper atmosphere are

$$O^+(g) + N_2(g) \longrightarrow NO^+(g) + N(g) \qquad [10.11]$$

$$N_2{}^+(g) + O(g) \longrightarrow NO^+(g) + N(g) \qquad [10.12]$$

As before, these reactions are of significance because they are exothermic and proceed very readily. Notice that the product of both reactions is NO^+. Because the ionization energy of NO is lower than that of the other abundant species present in the upper atmosphere (Table 10.2), once the NO^+ ion is formed it does not become neutralized to any extent by a charge-transfer reaction. The only means for removal of the ion is the dissociative-recombination reaction, Equation [10.6]. As a result, even though the neutral NO molecule is only a minor constituent of the upper atmosphere, present to the extent of about one part per million in total concentration, NO^+ is the most abundant ion in the upper atmosphere.

At this point, let's pause to see where we have been. All of the chemical processes we have discussed so far are important in the atmosphere above about 100 km. Note again that this is far above the troposphere, the only part of the atmosphere with which we have any direct contact. Yet if it were not for the absorptions of short-wavelength solar radiation in the upper atmosphere, life on earth as we know it would be impossible. However, not all the short-wavelength radiation is filtered out by N_2, O_2, and NO. To learn how some of the remaining short-wavelength radiation is filtered out, we must focus our attention on a lower region of the atmosphere.

10.4 Ozone in the upper atmosphere

At an elevation of about 90 km, most of the short-wavelength solar radiation capable of causing ionization has been absorbed. As a result, the concentration of ions and electrons drops off very rapidly at about this elevation. Radiation capable of causing dissociation of the O_2 molecule remains sufficiently intense, however, so that photodissociation of O_2,

Equation [10.2], remains important down to 30 km. The chemical processes that occur in the region below about 90 km following photodissociation of O_2 are very different than processes that occur at higher elevations. In the mesosphere and stratosphere the concentration of O_2 is much greater than that of atomic oxygen (Figure 10.4). Thus, when O atoms are formed in the mesosphere and stratosphere, they undergo frequent collisions with O_2 molecules. These collisions lead to formation of ozone, O_3:

$$O(g) + O_2(g) \longrightarrow O_3^*(g) \qquad \text{[10.13]}$$

The asterisk over the O_3 denotes that the ozone molecule contains an excess of energy. Reaction of O with O_2 to form O_3 results in release of 105 kJ/mol. This energy must be gotten rid of by the O_3 molecule in a very short time, or else it will simply fly apart again into O_2 and O. This decomposition is shown in Equation [10.14] as the reverse of the process by which O_3 is formed. The double arrows, $\rightleftharpoons$, indicate that the reaction is reversible; that is, that it may occur in either direction. The energy-rich O_3 molecule can get rid of the excess energy by colliding with another atom or molecule and transferring some of the excess energy to it. Let us represent the atom or molecule undergoing the collision as M. (Nearly always M is O_2 or N_2, because these are the most abundant molecules.) The transfer of energy can then be represented as in the second reaction, Equation [10.15]:

$$O(g) + O_2(g) \rightleftharpoons O_3^* \qquad \text{[10.14]}$$
$$\underline{O_3^*(g) + M(g) \longrightarrow O_3(g) + M^*(g)} \qquad \text{[10.15]}$$
$$O(g) + O_2(g) + M(g) \longrightarrow O_3(g) + M^*(g) \qquad \text{[10.16]}$$

The reaction in Equation [10.15] competes with the reverse version of Equation [10.14]. When collisions are not very frequent, the reverse reaction in Equation [10.14] wins out; most of the O_3^* molecules formed fall apart into O and O_2 before they undergo a stabilizing collision. However, when the number of collisions per unit time is high, formation of O_3 via Equation [10.15] is favored. Because the number of molecules per unit volume increases rapidly with decreasing elevation, the frequency of stabilizing collisions is greater at the lower elevations. However, at much lower altitudes, the solar radiation energetic enough to produce dissociation of O_2 becomes largely absorbed. The overall result of the opposing factors is a maximum in the rate of production of ozone at about 50 km altitude. The ozone molecule, once formed, does not stay around very long; ozone itself is capable of absorbing solar radiation, with the result that it is decomposed into O_2 and O. Because the energy required for this process is only 105 kJ/mol, photons of wavelength shorter than 1140 nm are sufficiently energetic. The strongest and most important absorptions are of photons with wavelengths from about 200 to 310 nm. Radiation in this wavelength range is not strongly absorbed by any species other than ozone. If it were not for the layer of ozone in the stratosphere, therefore, these short-wavelength, high-energy photons would penetrate to the earth's surface. Plant and animal life as we know it could not survive in the presence of this high-energy radiation. The "ozone shield" is thus essential for our continued well-being.

The photodecomposition of ozone reverses the reaction leading to its formation. We thus have a cyclic process of ozone formation and decomposition, summarized as follows:

$$O_2(g) + h\nu \longrightarrow O(g) + O(g)$$

$$O(g) + O_2(g) + M(g) \longrightarrow O_3(g) + M^*(g) \qquad \text{(heat released)}$$

$$O_3(g) + h\nu \longrightarrow O_2(g) + O(g)$$

$$O(g) + O(g) + M(g) \longrightarrow O_2(g) + M^*(g) \qquad \text{(heat released)}$$

The overall result of this cycle is that ultraviolet radiation from the sun is converted into heat energy. The ozone cycle in the stratosphere is responsible for the temperature rise that reaches its maximum at the stratopause, as illustrated in Figure 10.1.

The scheme described above for the life and death of ozone molecules accounts for some but not all of the known facts about the ozone layer. Many chemical reactions involving substances other than just oxygen are involved. In addition, the effects of turbulence and winds in mixing up the stratosphere must be considered. A very complicated picture results. It is quite certain, however, that the oxides of nitrogen are important in the ozone cycle.

Nitric oxide, NO, and its close relative nitrogen dioxide, NO_2, are present in the stratosphere in low concentrations. Ozone reacts with NO to form NO_2 and O_2; then NO_2 reacts with atomic oxygen to regenerate NO and form O_2. The NO is then ready again to react with O_3. The overall reaction involving NO is simply:

$$\begin{aligned} O_3(g) + NO(g) &\longrightarrow NO_2(g) + O_2(g) \\ NO_2(g) + O(g) &\longrightarrow NO(g) + O_2(g) \\ \hline O_3(g) + O(g) &\longrightarrow 2O_2(g) \end{aligned} \qquad [10.17]$$

We see from this sequence of reactions that NO serves the function of increasing the rate of decomposition of O_3. There is no net change in the chemical state of the NO. We have here a very simple example of a catalyst, a substance that has the effect of increasing the rate of a chemical reaction, without itself undergoing a net chemical change.

The overall result of ozone formation and removal reactions, coupled with atmospheric turbulence and other factors, is to produce an ozone profile in the upper atmosphere as shown in Figure 10.6.

Supersonic transport (SST) planes have been the subject of intense controversy during the past several years because of their possible effect on the ozone layer. In any internal-combustion engine, the temperatures attained may be sufficiently high so that there is some reaction of the nitrogen and oxygen of the atmosphere to form NO:

$$N_2(g) + O_2(g) \longrightarrow 2NO(g) \qquad [10.18]$$

This reaction is quite endothermic at ordinary temperatures, but it does proceed at a measurable rate at high temperatures such as those encountered in intensely hot flames. Some NO is therefore produced in the engines of an SST. The SST planes are designed to fly at stratospheric

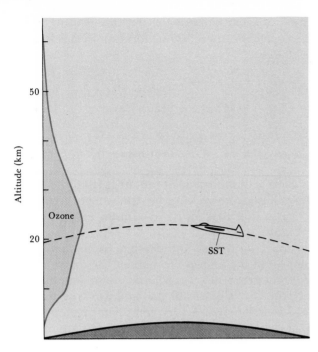

FIGURE 10.6 The ozone profile of the atmosphere.

altitudes (Figure 10.6). Thus, the NO produced in the engines of the SST's would be released to the stratosphere at an altitude where there is a fairly high concentration of ozone. The conclusion of many experts is that a worldwide fleet of SST planes would produce enough NO to decrease significantly the total ozone content of the stratosphere, especially over those regions of the earth directly under the flight paths of the planes. Thus, the SST flights would result in an increased amount of high-energy ultraviolet radiation at sea level. Among the detrimental effects of a higher level of ultraviolet radiation would be an increased incidence of skin cancer among those exposed to the sun's rays.

A few years ago it was recognized that the chlorofluoromethanes, principally CF_2Cl_2 and $CFCl_3$, may have an adverse effect on the ozone layer. These substances have been widely used as propellent gases in spray cans, and as refrigerant gases. They are quite inert chemically; there seems to be no relatively rapid chemical process that removes chlorofluoromethanes from the lower atmosphere. The lifetimes of these molecules in the atmosphere are thus controlled by the rate at which they diffuse into the stratosphere and become subject to the action of ultraviolet light. The action of high-energy light with wavelengths in the range of 190 to 225 nm results in photolysis, or light-induced rupture, of a carbon-chlorine bond:

$$CF_xCl_{4-x}(g) + h\nu \longrightarrow CF_xCl_{3-x}(g) + Cl(g) \qquad [10.19]$$

There may be further photochemical breakdown of the CF_xCl_{3-x} fragment. Calculations suggest that the rate of chlorine-atom formation will be maximized at an altitude of about 30 km. The atomic chlorine produced by photolysis is capable of rapid reaction with ozone to form chlorine oxide and molecular oxygen. Chlorine oxide is capable of reaction with atomic oxygen to reform atomic chlorine:

$$Cl(g) + O_3(g) \longrightarrow ClO(g) + O_2(g) \qquad [10.20]$$
$$ClO(g) + O(g) \longrightarrow Cl(g) + O_2(g) \qquad [10.21]$$
$$\overline{O_3(g) + O(g) \longrightarrow 2O_2(g)} \qquad [10.17]$$

This pair of reactions is analogous to those involving nitric oxide to produce the net reaction shown in Equation [10.17]. In both cases the original species is regenerated. The overall result is reaction of ozone with atomic oxygen to form molecular oxygen. Many uncertainties are involved in any quantitative estimate of how much ozone destruction might be due to chlorofluoromethanes. Because rates of diffusion of molecules into the stratosphere from the earth's surface are likely to be very slow, it may be several decades before the full impact of the chlorofluoromethanes is felt. Additional research on this problem is in progress in many laboratories. In the meantime, the use of the chlorofluoromethanes as aerosol propellent gases has been substantially reduced.

10.5 Chemistry of the troposphere

In the previous sections, we have described the photodissociation of oxygen and ozone. These two processes, and, to a lesser extent, other photodissociations and photoionizations, result in essentially complete absorption of all solar radiation of less than about 300 nm wavelength at the altitude of the tropopause. Because the major constituents of the atmosphere do not interact with radiation of wavelength longer than 300 nm, the photochemical reactions that occur in the troposphere are entirely those of minor atmospheric constituents. Many of the minor constituents occur to only a slight extent in the natural environment but exhibit much higher local concentrations in certain areas as a result of human activities. Table 10.3 is a summary of information on several minor atmospheric constituents. Some of these are of interest for their role as air pollutants. We shall discuss here the most important characteristics of a few of these minor constituents and their chemical role as air pollutants.

TABLE 10.3 Sources and typical concentrations of some minor atmospheric constituents

Minor constituent	Sources	Typical concentrations
Carbon dioxide (CO_2)	Decomposition of organic matter; release from the oceans; fossil-fuel combustion	320 ppm throughout troposphere
Carbon monoxide (CO)	Decomposition of organic matter; industrial processes; fuel combustion	0.05 ppm in nonpolluted air; 1 to 50 ppm in urban traffic areas
Methane (CH_4)	Decomposition of organic matter; natural-gas seepage	1 to 2 ppm throughout troposphere
Nitric oxide (NO)	Electrical discharges; internal-combustion engines; combustion of organic matter	0.01 ppm in nonpolluted air; 0.2 ppm in smog atmospheres
Ozone (O_3)	Electrical discharges; diffusion from stratosphere; photochemical smog	0 to 0.01 ppm in nonpolluted air; 0.5 ppm in photochemical smog
Sulfur dioxide (SO_2)	Volcanic gases; forest fires; bacterial action; fossil-fuel combustion; industrial processes (roasting of ores, and so on)	0 to 0.01 ppm in nonpolluted air; 0.1 to 2 ppm in polluted urban environment

TABLE 10.4 Concentrations of atmospheric pollutants likely to be exceeded about 50 percent of the time in a typical urban atmosphere

Pollutant	Concentration (ppm)
Carbon monoxide	10
Hydrocarbons	3
Sulfur dioxide	0.08
Nitrogen oxides	0.05
Total oxidants (ozone and others)	0.02

SULFUR COMPOUNDS

Sulfur-containing compounds are present to some extent in the natural, unpolluted atmosphere. They originate in the bacterial decay of organic matter, in volcanic gases, and from other sources listed in Table 10.3. Some scientists think that a certain amount of sulfur dioxide may also originate in the oceans. The concentration of sulfur-containing compounds in the atmosphere as a result of distribution of material from natural sources is very small in comparison with the concentrations that may build up in urban and industrial environments as a result of human activities. Sulfur compounds, chiefly sulfur dioxide, SO_2, are among the most unpleasant and harmful of the common pollutant gases. Table 10.4 shows the concentrations of several pollutant gases in a *typical* urban environment (not one that is particularly affected by smog). According to these data, the level of sulfur dioxide would be 0.08 ppm or higher about half the time. This concentration is considerably lower than that of other pollutants, notably carbon monoxide. Nevertheless, sulfur dioxide is regarded as the most serious health hazard among the pollutants shown, especially for persons with respiratory difficulties. Studies of the medical case histories of large population segments in urban environments have shown clearly that those living in the most heavily polluted parts of cities have higher levels of respiratory diseases and shorter life expectancies. One industrial process that may produce very high local levels of SO_2 is the roasting, or smelting, of ores. By this process a metal sulfide is oxidized, driving off SO_2, as in the following example:

$$2ZnS(s) + 3O_2(g) \longrightarrow 2ZnO(s) + 2SO_2(g) \qquad [10.22]$$

Smelting operations account for about 8 percent of the total SO_2 released in the United States. About 80 percent of the SO_2 generated comes from the combustion of coal and oil. The extent to which SO_2 emissions are a problem in the burning of fossil fuels depends on the level of sulfur concentration in the coal or oil. Oil that is burned in the power plants of electrical generating stations is the nonvolatile residue that remains after the low boiling fractions have been distilled off. Some oil, such as that from the Middle East, is relatively low in sulfur, whereas Venezuelan oil is relatively high. Because of concern about SO_2 pollution, low-sulfur oil is in greater demand and is consequently more expensive.

Coals vary considerably in their sulfur content. Much of the coal lying in beds east of the Mississippi is relatively high in sulfur content, up to 6 percent by weight. Much of the coal lying in the western states has a lower sulfur content. (This coal, however, also has a lower heat content per unit weight of coal, so that the difference in sulfur content on the basis of a unit amount of heat produced is not as large as is often assumed.)

Altogether, more than 30 million tons of SO_2 are released into the atmosphere in the United States each year. This material does a great deal of damage to both property and human health. Not all the damage, however, is caused by SO_2 itself; it is likely, in fact, that SO_3, formed by oxidation of SO_2, is the major culprit.

Sulfur dioxide is not readily oxidized in clean dry air. It is, however, very rapidly oxidized to sulfur trioxide, SO_3, by O_2 in the presence of metal oxide dust particles. Studies show that the reaction between SO_2 and O_2 occurs on the surface of a dust particle. In some way, the surface of the metal oxide promotes the reaction. The surface of the metal oxide thus serves as a heterogeneous catalyst. The term *heterogeneous* refers to the fact that the reaction occurs on a surface, whereas ordinarily it would have occurred in the gas phase. The term *catalyst* indicates that the surface of the metal oxide increases the rate of reaction over the rate of homogeneous reaction. (There is more on heterogeneous catalysis in Chapter 13.) Very finely divided solids suspended in air are referred to as particulate matter. The levels of particulate matter are generally quite high in polluted air. In the stack gases issuing from power plants, for example, there is a considerable quantity of metal-containing fly ash. It is reasonable to suppose from the studies with metal oxides that these particles serve as catalytic surfaces for oxidation of SO_2.

A second pathway for SO_2 oxidation is via fog or cloud droplets; it is known that oxidation of SO_2 dissolved in water is quite rapid. Finally, there is a possibility for photochemical oxidation of SO_2. We have already observed that solar radiation reaching the troposphere contains no wavelengths shorter than about 300 nm. This radiation is not sufficiently energetic to produce either ionization or dissociation of most molecules. However, it may be possible for absorption of a photon to cause a transition of an electron within the molecule from one allowed orbital to another. The molecule is then considered to be in an *excited state*. While in this state it is more reactive than a molecule in the normal ground state. The photochemical oxidation of SO_2 may thus be written as a two-step process:

$$SO_2(g) + h\nu \longrightarrow SO_2^{\#}(g) \qquad\qquad [10.23]$$

$$SO_2^{\#}(g) + O_2(g) \longrightarrow SO_3(g) + O(g) \qquad\qquad [10.24]$$

In this scheme $SO_2^{\#}$ denotes electronically excited SO_2. The details of the process by which SO_3 is formed are still not clear; the reactions shown are simply an indication of the overall effect, which is to produce SO_3.

It is evident from the above discussion that SO_2 may be oxidized to SO_3 by any of several pathways, depending on the particular nature of the atmosphere. Once SO_3 is formed it dissolves in water droplets, forming sulfuric acid, H_2SO_4:

$$SO_3(g) + H_2O(l) \longrightarrow H_2SO_4(aq) \qquad\qquad [10.25]$$

The phenomenon of "acid rain" has been known for a long time in the Scandinavian countries and other parts of northern Europe. The dominant contributor to the acidity in the rain is sulfuric acid. The acidity has caused fish populations in many freshwater lakes to decline, and it has measurably affected other parts of the network of interdependent living things within the lakes. There is evidence that acid rain is now affecting many northern lakes in the United States. Acid rain is strongly corrosive when it comes in contact with metals, paints, and similar substances. An example of this corrosive effect is shown in Figure 10.7.

FIGURE 10.7 Statue of the River Nile, Rome. The extensive decay of the stonework is due to the presence of SO_2 in the atmosphere. The SO_2 reacts with limestone, $CaCO_3$, and calcium sulfate, $CaSO_4$, is eventually formed, leading to a powdering or blistering of the surface. Wind and weather certainly contribute to the decay of buildings and statuary, but the major culprit is the chemical process that destroys the limestone.

In some areas, where the atmosphere also contains ammonia, NH_3, an acid-base reaction may occur, producing ammonium hydrogen sulfate, $NH_4(HSO_4)$, or ammonium sulfate, $(NH_4)_2SO_4$:

$$NH_3(g) + H_2SO_4(aq) \longrightarrow NH_4(HSO_4)(aq \text{ or } s) \qquad [10.26]$$

$$NH_4(HSO_4)(aq) + NH_3(g) \longrightarrow (NH_4)_2SO_4(aq \text{ or } s) \qquad [10.27]$$

The thick haze that overlays many heavily industrial areas consists largely of a dispersal of ammonium sulfate formed in this manner.

Aside from the damage to human health, billions of dollars each year are lost as a result of corrosion resulting from SO_2 pollution. Obviously we all want this noxious gas removed from the environment. But removal of sulfur from coal or oil is difficult and, therefore, expensive. Rather than attempt to remove sulfur from fuel before it is burned, the sulfur dioxide formed when the fuel is combusted may be removed. There are many possible ways of doing this. One way involves blowing powdered limestone, $CaCO_3$, into the combustion chamber. The carbonate (limestone) is decomposed into lime (CaO) and carbon dioxide:

$$CaCO_3(s) \longrightarrow CaO(s) + CO_2(g) \qquad [10.28]$$

The lime then reacts with SO_2 to form calcium sulfite:

$$CaO(s) + SO_2(g) \longrightarrow CaSO_3(s) \qquad [10.29]$$

Only about half the SO_2 is removed by contact with the dry solid. It is necessary to "scrub" the furnace gas with an aqueous suspension of lime to remove the $CaSO_3$ formed, and to remove any unreacted SO_2. This process, which is illustrated in Figure 10.8, is difficult to engineer, reduces the heat effectiveness of the fuel, and leaves an enormous solid waste disposal problem. An electric power plant that would serve the

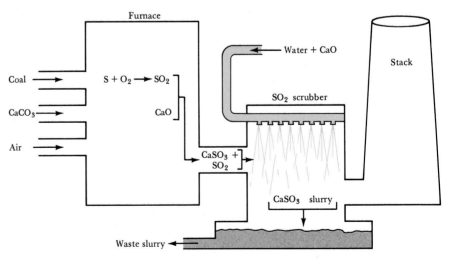

FIGURE 10.8 A common method for removing SO_2 from combusted fuel. Powdered limestone decomposes into CaO, which reacts with SO_2 to form $CaSO_3$. The $CaSO_3$ and any unreacted SO_2 enter a purification chamber where a shower of CaO and water precipitates the $CaSO_3$ into a watery residue called slurry and converts the remaining SO_2 to $CaSO_3$.

needs of a population of about 150,000 people would produce about 160,000 tons per year of solid waste if it were equipped with the purification system just described. This is three times the normal fly-ash waste from a plant of this size. Various schemes may be employed to recover elemental sulfur or some other industrially useful chemical from the SO_2, but as yet no process has been found sufficiently attractive from an economic point of view to warrant large-scale development. Pollution by sulfur dioxide remains a major problem and will probably continue to remain so for some time.

NITROGEN OXIDES; PHOTOCHEMICAL SMOG

The atmospheric chemistry of the nitrogen oxides is interesting because these substances are components of smog, a phenomenon with which city dwellers are all too familiar. The term smog refers to a particularly unpleasant condition of pollution in certain urban environments, which occurs when weather conditions produce a relatively stagnant air mass. The smog made famous by Los Angeles, but now common in many other urban areas as well, is more accurately described as a photochemical smog, because photochemical processes play an essential role in its formation.

Nitric oxide, NO, is formed in small quantities in the cylinders of internal-combustion engines via the direct combination of nitrogen with oxygen, as described by Equation [10.18]. Prior to installation of control measures, typical emission levels of NO_x were 4 g per mile. (The x is either 1 or 2; both NO and NO_2 are formed, though NO predominates.) The most recent auto-emission standards call for NO_x emission levels of less than 1 g per mile.

Nitric oxide is oxidized very slowly in air. A certain amount of NO_2 is present, however; this is formed either directly in the automobile engine or by slow oxidation of NO. The dissociation of NO_2 into NO and O requires 304 kJ/mol. This requirement corresponds to a photon wavelength of 393 nm. In sunlight, NO_2 undergoes dissociation to NO and O:

$$NO_2(g) + h\nu \longrightarrow NO(g) + O(g) \qquad [10.30]$$

The atomic oxygen formed undergoes several possible reactions. One of these is with the abundant molecular oxygen to form ozone, as described earlier:

$$O(g) + O_2(g) + M(g) \longrightarrow O_3(g) + M^*(g) \qquad [10.16]$$

Ozone is capable of rapidly oxidizing NO to NO_2, a reaction we have also seen earlier in connection with Equation [10.17]:

$$O_3(g) + NO(g) \longrightarrow NO_2(g) + O_2(g) \qquad [10.31]$$

To see the significance of these reactions for smog formation, we must look at Figure 10.9, which shows the time dependence of the concentrations of various smog components. For the moment look at just the curves for the nitrogen oxides. In the early morning hours, the NO_2 concentration is low. As auto traffic builds up, and NO is formed, oxida-

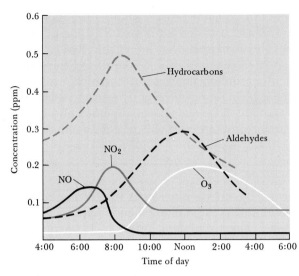

FIGURE 10.9 Concentration of smog components as a function of time of day. (*After P. A. Leighton, "Photochemistry of Air Pollution." Academic Press, 1961*)

tion by ozone, Equation [10.31], takes over. Note that the ozone level does not noticeably increase during this period.

In addition to nitrogen oxides and carbon monoxide, an automobile engine also emits as pollutants unburned and partially burned hydrocarbons, compounds made up entirely of carbon and hydrogen. A typical engine without emission controls emits about 10–15 g of such organic compounds per mile. The newest standards require that hydrocarbon emissions be less than 0.4 g per mile. Table 10.5 shows typical concentrations of trace constituents in photochemical smog. The most important organic compounds in this list for smog formation are olefins and aldehydes. An olefin is an organic compound, a type of hydrocarbon, containing a double bond between carbon atoms. Ethylene, C_2H_4, is the simplest member of the series. Aldehydes are compounds containing a carbon-oxygen double bond on a carbon atom at the end of a hydrocarbon chain. Formaldehyde, acetaldehyde, and propionaldehyde are examples:

$$
\begin{array}{ccc}
\overset{\displaystyle O}{\underset{\displaystyle \parallel}{}} & \overset{\displaystyle O}{\underset{\displaystyle \parallel}{}} & \overset{\displaystyle O}{\underset{\displaystyle \parallel}{}} \\
H\!-\!C\!-\!H & CH_3\!-\!C\!-\!H & CH_3\!-\!CH_2\!-\!C\!-\!H \\
\text{Formaldehyde} & \text{Acetaldehyde} & \text{Propionaldehyde}
\end{array}
$$

Note that in Figure 10.9 the hydrocarbon levels decrease as aldehyde levels increase. This occurs because some of the atomic oxygen produced in photodissociation of NO_2 reacts with organic compounds, eventually producing aldehydes through a complex series of reactions. Ozone is also capable of reaction with olefins to eventually yield aldehydes. Many of the compounds formed by reaction of atomic oxygen and ozone with organic compounds are free radicals, molecular fragments that contain an unpaired electron. They are very reactive and lead to a complex chemistry in the polluted atmosphere. One group of molecules formed in all this is the peroxyacylnitrates (PAN), especially unpleasant substances that cause eye irritation and breathing difficulties:

$$
\overset{\displaystyle O}{\underset{\displaystyle \parallel}{}}
$$
$$
R\!-\!C\!-\!O\!-\!O\!-\!NO_2
$$

TABLE 10.5 Typical concentrations of trace pollutants during a photochemical smog (levels are subject to wide variation from one situation to another)

Constituent	Concentration (ppm)
NO_x	0.2
NH_3	0.02
CO	40
O_3	0.5
CH_4	2
C_2H_4	0.5
Higher olefins[a]	0.25
C_2H_2 (acetylene)	0.25
Aldehydes	0.6
SO_2	0.2

[a] Higher olefins are compounds that contain a hydrocarbon chain attached to one of the carbons of the C=C bond.

In this diagram, R represents an organic group such as CH_3, C_6H_5, and so on.

During the afternoon hours, the smog atmosphere contains relatively high concentrations of ozone, as shown in Figure 10.9. Much of the nitrogen oxides have been converted into PAN and several other related compounds. In addition, nitric acid is formed by dissolution of NO_2 in water droplets if these are present. Besides the gas phase components, an aerosol (a dispersion of tiny droplets in the air) usually forms and reduces visibility markedly. This aerosol consists mainly of organic compounds produced in the smog atmosphere. The entire mixture is thoroughly unpleasant and harmful to health.

Reduction or elimination of smog requires that the essential ingredients for its formation be removed from automobile exhaust. The stricter emission standards in effect since 1975 are designed to reduce drastically the levels of two of the major ingredients of smog—NO_x and hydrocarbons. Whether the means for meeting these standards now in use will be successful remains to be seen. Emission-control systems are not notably successful in poorly maintained autos.

CARBON MONOXIDE

In terms of total mass, carbon monoxide is the most important of all the pollutant gases. The level of CO present in fresh, nonpolluted air is small, probably on the order of 0.05 to 0.1 ppm. The estimated total amount of CO in the earth's atmosphere is about 5.2×10^{14} g. In the United States alone, however, about 1×10^{14} g of CO are produced each year. The CO is formed mostly in the incomplete combustion of fossil fuels. The major sources of CO in the United States are automobile and power-plant emissions.

The total amount of CO generated by humans each year is estimated to be about 30 percent of that present in the atmosphere. However, the atmospheric level as a whole has not increased to the extent that these figures would imply. This suggests that there is a *sink* for CO, that is, a process that consumes it. Although CO is susceptible to oxidation to CO_2 by atmospheric oxygen, the reaction is extremely slow. The major pathway for removal of CO from the air over landmasses seems to be consumption by microorganisms in the soil. In addition, some of the gaseous CO may dissolve in the oceans. Very little is known about this process; in fact, it is not clear whether the oceans serve as a source or sink of CO. Finally, CO may be removed from the lower atmosphere by diffusion into the stratosphere, where it is removed by reactions with more reactive molecules and atoms present there. Experts estimate that the average residence time of CO in the atmosphere is about 6 months. Thus, the averaged concentration of CO in the total atmosphere is not large. However, this compound is a serious pollution hazard in two special situations: (1) in urban centers where traffic density is high and (2) in the cigarettes smoked by large numbers of Americans.

Carbon monoxide is a relatively unreactive molecule, and it might be supposed that it would not be a health hazard. It does have the unusual ability, however, of binding very strongly to hemoglobin, the iron-containing protein that is responsible for oxygen transport in the blood. Hemo-

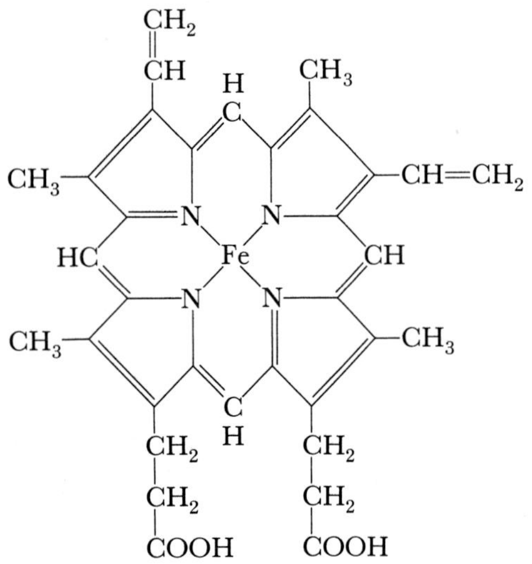

FIGURE 10.10 The structure of the heme molecule.

globin consists of four protein chains loosely held together in a cluster. Each chain has within its folds a heme molecule. The structure of heme is shown in Figure 10.10. The important characteristics of heme are that the iron is situated in the center of a plane of coordinating nitrogen atoms. An oxygen molecule reacts with the iron atom to form a species called oxyhemoglobin. Under appropriate conditions the oxygen is released from the iron. The equilibrium between hemoglobin and oxyhemoglobin may be illustrated graphically as shown in Figure 10.11. Oxygen is picked up by hemoglobin in the lungs, and released in the tissues where it is needed for cell metabolism, that is, for the chemical processes occurring in the cell.

Carbon monoxide also happens to bind very strongly to the iron in hemoglobin. The complex is called carboxyhemoglobin and is represented as COHb. The affinity of human hemoglobin for CO is about 210 times greater than for O_2. As a result, a relatively small quantity of CO can inactivate a substantial fraction of the hemoglobin in the blood for oxygen transport. For example, a person breathing air that contains only 0.1 percent of CO takes in enough CO after a few hours of breathing to convert up to 60 percent of the hemoglobin into COHb, thus reducing the blood's normal oxygen-carrying capacity by 60 percent.

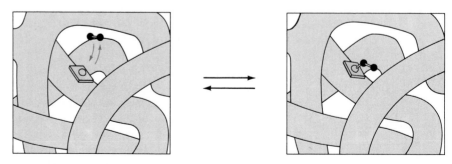

FIGURE 10.11 The equilibrium between hemoglobin and oxyhemoglobin. The grey regions represent the protein chain. The heme molecule is attached to the protein at a particular point.

TABLE 10.6 Carboxyhemoglobin (COHb) percentages in the blood of persons under various conditions

	COHb (%)
Continuous exposure, 10 ppm CO	2.0
Continuous exposure, 30 ppm CO	5.0
Nonsmokers, Chicago (1970)	2.0
Smokers, Chicago (1970)	5.8
Nonsmokers, Milwaukee (1969–1971)	1.1
Smokers, Milwaukee (1969–1971)	5.0

Considerable evidence has accumulated that chronic exposure to CO in polluted air for a long period of time may have adverse health effects. Under normal conditions, a nonsmoker breathing unpolluted air has about 0.3–0.5 percent carboxyhemoglobin, COHb, in the bloodstream. This small amount arises mainly from the production of small amounts of CO in the course of normal body chemistry, and from the small amount of CO present in clean air. Exposure to higher concentrations of CO causes the COHb level to increase. This doesn't happen instantly, but requires several hours. Similarly, when the CO level is suddenly decreased, it requires several hours for the COHb concentration to level off at a lower value. Table 10.6 shows the percentages of COHb in blood that are typical of various groups of people. It is interesting to note that the federal Clean Air Act of 1971 was designed to control CO emissions so that the COHb percentage would remain below 1.5 percent in active nonsmokers.

The CO concentration in city traffic often reaches 50 ppm and may go as high as 140 ppm in traffic jams. Persons who work in areas with high traffic density, such as policemen, guards in traffic tunnels, and cab drivers, show abnormally high percentages of COHb as compared with the population as a whole. The most serious souce of carbon monoxide poisoning, however, comes from cigarette smoking. The inhaled smoke from cigarettes contains about 400 ppm of CO. The effect of smoking on COHb percentage is evident from the data in Table 10.6. A study of a group of San Francisco dock workers presented further proof of the dramatic relationship between smoking and COHb percentage. Nonsmokers in the group averaged 1.3 percent COHb; light smokers (less than half a pack per day) averaged 3.0 percent; moderate smokers averaged 4.7 percent; and heavy smokers (two packs or more per day) averaged 6.2 percent COHb.

There is widespread evidence that chronic exposure to CO impairs performance on standardized tests. Thus, it is most definitely not a good idea to smoke heavily before and during a test. In addition, motor performance is also impaired by high COHb percentages. For example, there is evidence that drivers responsible for traffic accidents have, on the average, higher than normal percentages of COHb in their blood. As has been mentioned, a chronically high level of COHb means that a certain fraction of the hemoglobin in the blood is not available for oxygen transport. This in turn means that the heart must work that much harder to ensure an adequate supply of oxygen. It is not surprising, therefore, that many medical researchers believe that chronic exposure to CO is a contributing factor in heart disease and in heart attacks.

WATER VAPOR, CARBON DIOXIDE, AND CLIMATE

We have seen how the atmosphere makes life as we know it possible on earth by screening out harmful short-wavelength radiation. In addition, the atmosphere is essential in maintaining a reasonably uniform and moderate temperature on the surface of the planet. The two atmospheric components of major importance in maintenance of the earth's surface temperature are carbon dioxide and water.

About 71 percent of all the solar radiation that strikes the outer atmosphere is eventually absorbed by the planet. The remainder is reflected back into space. The temperature of the planet as seen from outer space is determined by the amount of absorbed energy. The maximum in intensity of solar radiation occurs in the visible portion of the spectrum. Because the earth's atmosphere is reasonably transparent in this region of the spectrum, most absorption of solar radiation takes place at the earth's surface. Thus, although the absorption of short-wavelength photons in the upper atmosphere is essential for the continued existence of living things, the absorption does not represent a large fraction of all the energy absorbed by the planet.

The earth is in overall thermal balance with its surroundings. This means that the planet radiates energy into space at a rate equal to the rate at which it absorbs energy from the sun. The sun is a very hot body with a temperature of about 6000 K. As seen from outer space the earth is relatively cold, with a temperature of about 254 K. The distribution of wavelengths in the radiation emitted from an object is determined by its temperature. The radiation emitted by relatively cold objects is in the low-energy, or long-wavelength, region of the spectrum. This means that the maximum in the wavelength of radiation from the earth is in the far infrared region, around 12,000 nm (Figure 5.3). But although the troposphere is transparent to visible light, it is not at all transparent to infrared radiation. Figure 10.12 shows the distribution of radiation from the earth's surface and, on the same scale, the wavelengths absorbed by water vapor and carbon dioxide. Clearly, these atmospheric gases absorb much of the outgoing radiation from the earth's surface. It is indeed fortunate for us that they do so; they serve to maintain a livably uniform temperature at the surface by holding in, as it were, the infrared radiation from the surface.

The partial pressure of water vapor in the atmosphere varies greatly from place to place and time to time, but, in general, it is highest near the surface and drops off very sharply with increased elevation. Carbon dioxide, by contrast, is uniformly distributed throughout the atmosphere, at a concentration of about 330 ppm. Because water vapor ab-

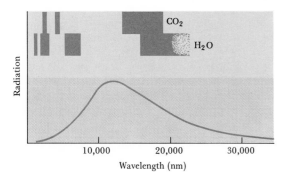

FIGURE 10.12 Long-wavelength radiation from earth, as compared with the absorption of infrared radiation by carbon dioxide and water.

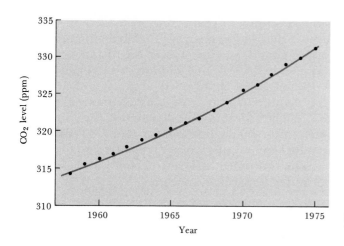

FIGURE 10.13 Annual mean CO_2 level in the atmosphere, 1958–1975.

sorbs infrared radiation so strongly, it plays the major role in maintaining the atmospheric temperature at night, when the surface is emitting radiation into space and not receiving energy from the sun. In very dry desert climates, where the water vapor concentration is unusually low, it may be extremely hot during the day but very cold at night. In the absence of an extensive layer of water vapor to absorb and then radiate back part of the infrared radiation, the surface loses this radiation into space and cools off very rapidly.

Carbon dioxide plays a secondary, but very important, role in maintaining the surface temperature. John Tyndall, in 1861, was the first to suggest that it might be possible to account for the earth's past climatic history in terms of variations in the concentration of carbon dioxide in the atmosphere. Periods of unusually high temperatures, the so-called interglacial periods, could be accounted for in terms of unusually high CO_2 levels. The glacial periods, on the other hand, could have been caused by unusually low levels of carbon dioxide. It is now generally recognized that climate is the result of extraordinarily complex, interacting factors, and it seems unlikely that any one factor could have been responsible for drastic climatic change. Nevertheless, interest in the possible climatic effects of changing carbon dioxide levels has increased of late, as a result of new information on the impact of human activities on the atmosphere.

The worldwide combustion of fossil fuels, principally coal and oil, on a prodigious scale in the modern era has materially increased the carbon dioxide level of the atmosphere. From measurements such as those graphed in Figure 10.13 it is clear that the CO_2 concentration in the atmosphere is steadily increasing. From a knowledge of the infrared-absorbing characteristics of CO_2 and water, and using a theoretical model for the atmosphere, it has been estimated that if the CO_2 level were to double from its present level, the average surface temperature of the planet would increase 3°C. On the basis of present and expected future rates of fossil-fuel use, about 70 percent of the present known reserves of coal and essentially all the oil will have been consumed by about 2050. This amount of fuel consumption should just about double the atmospheric CO_2 level. If the calculated effect of a doubling of CO_2 level on surface temperature is correct, this means that the earth's temperature will be 3°C higher within 75 years. Such a small change may seem insignificant, but it is not. Major changes in global climate could result from a temperature change of this or even smaller magnitude. Because so many factors go into determining climate, it is not possible to predict with certainty precisely what changes will occur. It is clear, however, that humanity has acquired the potential, by changing the CO_2 concentration in the atmosphere, for substantially altering the climate of the planet. Unfortunately, if it should turn out for the worst, as seems altogether likely, there is little or nothing we can presently visualize that could be done about it. The continued high rate of combustion of fossil fuels is therefore a matter for long-range concern.

FOR REVIEW

Summary

In this chapter we've examined the physical and chemical properties of earth's atmosphere. The complex temperature variations in the atmosphere give rise to several regions, each with characteristic properties. The lowest of these regions, the troposphere, extends from the surface to about 11 km. Above the troposphere, in order of increasing altitude, are the stratosphere, mesosphere, and thermosphere. In the upper reaches of the atmosphere, only the simplest chemical species can survive the bombardment of highly energetic particles and radiation from the sun. The average molecular weight of the atmosphere at high elevations is lower than at the earth's surface, because the lightest atoms and molecules diffuse upward and because of photodissociation.

Absorption of radiation may also lead to ionization. Molecular ions formed by photoionization may recombine with electrons and thus form again the neutral molecule. However, the energy of recombination is so high that the molecule exists only momentarily before dissociating. Reaction of an electron with a molecular ion is thus referred to as dissociative recombination. Molecular ions may also react by removing an electron from another neutral molecule in a charge-transfer reaction. Another possible reaction pathway is atom transfer.

Ozone is produced in the mesosphere and stratosphere as a result of reaction of atomic oxygen with O_2. Ozone is itself decomposed by absorption of a photon or by reaction with an active species such as NO. Human activities could result in the addition to the stratosphere of atomic chlorine, which is capable of reacting with ozone in a catalytic cycle to convert ozone to O_2. A marked reduction in the ozone level in the upper atmosphere would have serious adverse consequences, because the ozone layer filters out certain wavelengths of ultraviolet light that are not taken out by any other atmospheric component.

In the troposphere, the lower atmosphere in which we live, the chemistry of trace atmospheric components is of major importance. Many of these minor components are pollutants; sulfur dioxide is one of the more noxious and prevalent. It is oxidized in air to form sulfur trioxide, which upon dissolving in water forms sulfuric acid. To control this source of pollution, it is necessary to prevent SO_2 from escaping from industrial operations in which it is formed. One method for doing this involves reacting the SO_2 with CaO to form calcium sulfite, $CaSO_3$.

Photochemical smog is a complex mixture of components in which both nitrogen oxides and ozone play important roles. The smog components are generated mainly in automobile engines, and smog control consists largely of controlling emissions from automobiles.

Carbon monoxide is found in high concentrations in the exhaust of automobile engines and in cigarette smoke. This compound is a health hazard because of its ability to form a strong bond with hemoglobin and thus reduce the capacity of blood for oxygen transfer from the lungs.

Carbon dioxide and water vapor are the only components of the atmosphere that strongly absorb infrared radiation. The level of carbon dioxide in the atmosphere is thus of importance in determining worldwide climate. As a result of the extensive combustion of fossil fuels (coal, oil, and natural gas) the carbon dioxide level of the atmosphere is steadily increasing. Burning of a large fraction of the known reserves of fossil fuels could increase the CO_2 concentration in the atmosphere sufficiently to produce marked, and almost certainly undesirable, changes in the global climate.

Learning goals

Having read and studied this chapter, you should be able to:

1 Sketch the manner in which the atmospheric temperature varies with altitude and list the names of the various regions of the atmosphere and the boundaries between them.

2 Sketch the manner in which atmospheric pressure decreases with elevation and explain in general terms the reason for the decrease.

3 Describe the composition of the atmosphere with respect to the four most abundant components.

4 Explain why the mean molecular weight of the atmosphere decreases above about 120 km.

5 Explain what is meant by the term photodissociation and calculate the maximum wavelength of a photon that is energetically capable of producing photodissociation, given the dissociation energy of the bond to be broken in the process.

6 Describe the way in which the element oxygen is distributed between the atomic and molecular form as a function of altitude and explain the reasons for this distribution.

7 Explain what is meant by photoionization and relate the energy requirement for photodissociation to the ionization energy of the species undergoing ionization.

8 Explain the process known as dissociative recombination and explain why it occurs.

9 Explain the conditions under which reaction between a molecular ion and a neutral molecule leads to charge transfer.

10 Write atom-transfer reactions that account for the presence of NO^+ in the upper atmosphere.

11 Explain the presence of ozone in the mesosphere and stratosphere in terms of appropriate chemical reactions.

12 Describe how nitrogen oxides or atomic chlorine function in the stratosphere as ozone-removal agents.

13 Explain how atomic chlorine might appear in the stratosphere as a product of the chlorofluoromethanes.

14 List the names and chemical formulas of the more important pollutant substances present in the troposphere and in urban atmospheres.

15 List the major sources of sulfur dioxide as an atmospheric pollutant and the various means by which SO_2 may be oxidized to SO_3.

16 List the more important reactions of nitrogen oxides and ozone that occur in smog formation.

17 Explain why carbon monoxide constitutes a health hazard.

18 Explain why the concentration of carbon dioxide in the troposphere has an effect on the average temperature at the earth's surface.

Key terms

Among the more important terms and expressions used for the first time in this chapter are the following:

Acid rain (Section 10.5) refers to rainwater that has become excessively acidic because of absorption of pollutant oxides, notably SO_3, produced by human activities.

In an atom-transfer reaction (Section 10.3), an atom is transferred from one species to another.

Carboxyhemoglobin (Section 10.5) is a complex formed between carbon monoxide and hemoglobin, in which CO is bound to the iron atom.

A catalyst (Section 10.4) is a substance that affects the rate of a chemical reaction but does not itself undergo a net, overall chemical change.

A charge-transfer reaction (Section 10.3) is one in which an electron is transferred from a neutral atom or molecule to a positively charged atom or molecule.

Chlorofluoromethanes (Section 10.4) are compounds of the general formula CF_xCl_{4-x}, $x = 1, 2$, or 3, used as propellant gases in aerosol spray cans and in refrigeration units.

Dissociative recombination (Section 10.3) is the process by which an electron recombines with an ionized molecule. The energy of recombination is released by rupture of a chemical bond in the neutral species.

Hemoglobin (Section 10.5) is an iron-containing protein responsible for oxygen transport in the blood.

A heterogeneous catalyst (Section 10.5) is a substance that catalyzes a reaction occurring in another phase. For example, a solid substance might act as a heterogeneous catalyst for a gas-phase reaction.

Photochemical smog (Section 10.5) is a complex mixture of undesirable substances produced by the action of sunlight on an urban atmosphere polluted with automobile emissions. The major starting ingredients are nitrogen oxides and organic substances, notably olefins and aldehydes.

Photodissociation (Section 10.2) refers to the breaking of a molecule into two or more neutral fragments as a result of absorption of light.

Photoionization (Section 10.3) refers to the removal of an electron from an atom or molecule by absorption of light.

The troposphere (Section 10.1) is the region of earth's atmosphere extending from the surface to about 11 km altitude. The regions of the atmosphere extending above the troposphere are, in order of increasing altitude, the stratosphere, mesosphere, and thermosphere.

Earth's atmosphere, processes in the outer regions

10.1 Name the regions of the atmosphere, indicate the altitude interval for each, and describe the variation in temperature in that region.

10.2 Name the boundaries between the regions of the earth's atmosphere and indicate the temperature in the boundary region.

10.3 The troposphere, which extends to only 11 km, contains about 80 percent of the total atmospheric mass. Explain this in terms of the compressibility of gases.

10.4 From the data in Table 10.1 calculate the partial pressures in mm Hg of helium and methane in the atmosphere when total pressure is 1.00 atm.

10.5 The dissociation energy of a carbon-bromine bond is typically about 210 kJ/mol. What is the maximum wavelength of photon that can cause C—Br bond dissociation?

10.6 In CF_3Cl, the C—Cl bond dissociation energy is 339 kJ/mol. In CCl_4 the C—Cl bond dissociation energy is 293 kJ/mol. What is the range of wavelengths of photons that can cause C—Cl bond rupture in one molecule but not in the other?

10.7 At 100 km altitude the atmospheric pressure is 2.3×10^{-3} mm Hg. Estimate the temperature from Figure 10.1 and calculate the number of molecules per cubic centimeter at this elevation.

[10.8] Suppose that on another planet with a sun like ours the atmosphere consisted of 55 percent Ar and 45 percent O_2. What would be the average molecular weight at the surface? What would be the average molecular weight at 150 km elevation, assuming that all the O_2 were dissociated into O atoms?

10.9 What conditions must be met in order for radiant energy to produce the change described in Equation [10.2]?

10.10 The dissociation energy of N_2 is 941 kJ/mol. Calculate the wavelength of the lowest-energy photon that can cause this photodissociation.

10.11 In the region around 90–120 km the daytime and nighttime concentrations of electrons, ions, and many high energy species differ greatly. Explain.

10.12 Write the reaction which is the most important source of each of the following species in the upper atmosphere: (a) O; (b) N; (c) NO^+.

[10.13] Suggest one or more reactions by which NO could be formed in the upper atmosphere. Would you expect these to be rapid reactions? Would total pressure affect the rate of formation of NO by each?

[10.14] Calculate the overall heat absorbed or evolved in each of the following changes (the bond dissociation energy of NO is 682 kJ/mol):

(a) $NO^+(g) + O_2(g) \longrightarrow O_2^+(g) + NO(g)$
(b) $N_2(g) + O^+(g) \longrightarrow N(g) + NO^+(g)$
(c) $O_2^+(g) + e^- \longrightarrow O(g) + O(g)$
(d) $O_2(g) + N(g) \longrightarrow NO(g) + O(g)$

10.15 Why is there very little N_2^+ present in the upper atmosphere as compared with NO^+, even though N_2 is about 1 million times more abundant than NO?

Chemistry of the stratosphere

10.16 Explain why oxygen atoms exist in the atomic state for much longer average times at 120 km elevation than at 50 km elevation.

10.17 Describe the means by which ozone, O_3, is formed in the stratosphere. What is the biological significance at the earth's surface of the ozone layer in the stratosphere?

10.18 In the reaction $O_3^* + M \longrightarrow O_3 + M^*$, what is the significance of the asterisk? What is the function of M? What chemical species does M represent?

10.19 Using the thermodynamic data in Appendix D, calculate the overall enthalpy change in each step in the catalytic cycle that converts O_3 to O_2 (see Equation [10.17]).

$NO(g) + O_3(g) \longrightarrow NO_2(g) + O_2(g)$
$NO_2(g) + O(g) \longrightarrow O_2(g) + NO(g)$

10.20 Using the thermodynamic data in Appendix D, calculate the overall enthalpy change in each step in the following catalytic cycle:

$SO_2(g) + O_3(g) \longrightarrow SO_3(g) + O_2(g)$
$SO_3(g) + O(g) \longrightarrow SO_2(g) + O_2(g)$

On the basis of your results, indicate whether $SO_2(g)$ is at least a possible catalyst for the decomposition of ozone in the stratosphere.

10.21 In general, carbon-bromine bonds are somewhat weaker than carbon-chlorine bonds (see Table 7.3). Would you expect that the long-range environmental problems associated with use of CF_2Cl_2 might be averted by using a compound such as CF_3Br? Explain.

10.22 What is the expected geometry of the $CFCl_3$ molecule? In one extension of the VSEPR model it is said that electron pairs in bonds to more electronegative elements have a smaller volume requirement than do those in bonds to less electronegative elements. Using this idea predict the way in which the bond angles about carbon in $CFCl_3$ will deviate from idealized values.

10.23 What is the expected geometry for the CF_2Cl_2 molecule? Using the VSEPR model, predict the way in which the bond angles about carbon will deviate from the idealized values (see question 10.22).

10.24 Beginning with the intact chlorofluoromethane CF_2Cl_2, write equations showing how a catalytic effect for destruction of ozone may be established in the stratosphere.

10.25 It has recently been pointed out that today there may be increased amounts of NO in the troposphere as compared with the past because of massive use of nitrogen-containing compounds in fertilizers. Assuming that NO can eventually diffuse into the stratosphere, what role might it play in affecting the conditions of life on earth? Explain.

Chemistry of the troposphere

10.26 In a particular urban environment, the ozone concentration is 0.26 ppm. Assuming a temperature of 26°C and an atmospheric pressure of 740 mm Hg at the time, calculate the partial pressure of ozone, and the number of O_3 molecules per cubic meter.

10.27 In a particular urban environment the NO concentration is 0.92 ppm. If the atmospheric pressure at the time is 710 mm Hg and the temperature is 30°C, calculate the partial pressure of NO, and the number of NO molecules per cubic meter.

10.28 Compare typical concentrations of CO, SO_2, and NO in nonpolluted air (Table 10.3) and urban air (Table 10.4) and indicate in each case at least one possible source of the higher values in Table 10.4.

10.29 U.S. production of zinc has recently been averaging about 600,000 tons per year. Assuming that all of this is formed by roasting of ZnS, calculate the total mass in grams of SO_2 formed.

10.30 In the metallurgy of copper, Cu_2S is roasted to form copper metal, with release of SO_2. Assuming that this process is involved in producing all of the 1.6 million tons of Cu produced annually in the U.S., calculate the mass in grams of SO_2 released through copper smelting.

10.31 Describe all of the pathways by which SO_2 in the atmosphere may be oxidized to SO_3. What is the environmental significance of this oxidation?

10.32 For each of the following gases, make a list of known or possible naturally occurring sources: (a) CO; (b) SO_2; (c) CH_4; (d) NO.

10.33 Assuming an overall efficiency of about 22 percent, how much calcium carbonate would be required to remove the SO_2 formed in burning a ton of oil containing 1.7 percent sulfur?

10.34 Assuming an SO_2 concentration of 0.087 ppm over an urban area of 580 km², and assuming the SO_2 to be evenly distributed to an elevation of 1200 m, calculate the total mass of SO_2 present. Atmospheric pressure is 740 mm Hg, and the temperature is 24°C.

10.35 From Figure 10.9 we see that the concentration of NO_2 in an urban atmosphere increases to a maximum rather early in the day, then decreases to a lower, steady value. What is the reason for this behavior?

10.36 Indicate the major source or sources of high carbon monoxide concentrations in air and discuss what might be done to lower carbon monoxide concentrations.

10.37 Why is continuous exposure to low concentrations of carbon monoxide in the air a health hazard?

[10.38] We have noted that the affinity of carbon monoxide for hemoglobin is about 210 times that of O_2. Assume that a person is inhaling air that contains 120 ppm of CO. If all the hemoglobin leaving the lungs carries either oxygen or carbon monoxide, calculate the fraction in the form of carboxyhemoglobin.

10.39 Even if all the problems associated with particulates and sulfur dioxide and carbon monoxide emissions could be solved, the extensive worldwide combustion of the earth's fossil fuel reserves may produce a serious environmental risk. What is this risk, and how does it arise?

Additional exercises

[10.40] In terms of the chemical processes described in this chapter, explain why the temperature of the atmosphere increases dramatically above about 90 km.

10.41 What factors are responsible for the fact that the mean molecular weight of the atmosphere decreases above about 150 km?

[10.42] Suppose that it were possible to provide the moon with several "instant oceans" by filling several of the large maria (lunar "seas") with water. What do you think would happen over a long period of time to the water-vapor atmosphere established on the moon? Describe several stages of the changes that would lead to whatever you think would be the final result.

10.43 Atomic nitrogen can be produced in the upper atmosphere by (a) photoionization of N_2; (b) atom transfer between N_2 and O^+; (c) dissociative recombination of NO^+; (d) atom transfer between N_2^+ and O. Write balanced chemical equations for each process.

10.44 The temperature profile in the atmosphere shows a maximum at about 50 km (Figure 10.1). Explain the origin of this maximum.

10.45 It is said that atomic chlorine serves as a catalyst for the decomposition of ozone according to the reaction $O_3 + O \longrightarrow 2O_2$. What is a catalyst? How does Cl serve as a catalyst in this reaction? What overall chemical change occurs for Cl?

[10.46] In terms of the energy requirements, explain why photodissociation of oxygen is more important than photoionization of oxygen at altitudes below about 90 km.

[10.47] Experiments have been performed in which metals such as sodium or barium have been released into the atmosphere at altitudes of about 120 km. Assuming that the metals are present in the atomic form, what reactions would you expect to occur with the ionic species present? Explain.

10.48 Describe a process that is a major source of each of the following species, writing an equation where appropriate: (a) atomic oxygen at 120 km elevation; (b) the OH radical at 150 km; (c) O at 500 m altitude over a city in daytime; (d) NO^+ at 100 km altitude; (e) O_3 at 50 km elevation.

10.49 In a recent study carried out in Sweden, it was found that in a particular location far from industrial activity the content of sulfate in rainfall amounted to 150 mg/m² per year. Calculate the total mass of sulfate falling in a square mile per year.

10.50 Except for two substances, the components of the earth's atmosphere are transparent to long-wavelength, infrared radiation. What are these two substances? In what way does the absorption of infrared radiation affect the earth's climate? Explain how increased levels of infrared-absorbing substances in the atmosphere could lead to a higher average surface temperature.

10.51 On an annual basis, rain falling over large regions of the world is from 5 to 30 times more acidic than the lowest value expected for rain in an unpolluted atmosphere. What is the origin of this increased acidity?

11

Liquids, solids, and intermolecular forces

The physical properties of any substance depend on its physical state. We have seen (Section 2.1) that matter exists in three states, gaseous, liquid, or solid. Some properties characteristic of these states are listed in Table 11.1. In Chapter 9 we discussed the gaseous state in some detail. We now turn our attention to the properties of liquids and solids. You will remember that we could explain the most important properties of gases in terms of the kinetic-molecular theory. This same theory, with some modifications, can also help us to understand the characteristics of liquids and solids. In this chapter we will also examine those properties of the molecules themselves that are important in determining whether a substance is a gas, liquid, or solid under a given set of conditions. A key concept in comparing the properties of substances is that of intermolecular forces, the attractive forces responsible for keeping molecules together in liquids and solids. In this chapter we will be exploring the relationships among structure, intermolecular forces, and physical properties.

11.1 The kinetic-molecular theory for liquids and solids

In the kinetic-molecular theory a gas is viewed as a collection of widely separated molecules in chaotic, constant motion. The translational, or kinetic, energies of the molecules are large in comparison with the attractive forces between them. Indeed, in the derivation of the ideal-gas equation attractive forces are ignored altogether.

There is a great deal of free volume in a gas, that is, volume not occupied by the molecules themselves. Because this is so, gases can be readily compressed upon application of an external force. By contrast, the molecules in a liquid are in quite close contact and are kept near to one another by the attractive forces between them. The liquid state is often referred to as a condensed state. The molecules of a liquid do move about with respect to one another, as in a gas. The major difference between the liquid and gaseous states of matter has to do with the

TABLE 11.1 Some characteristic properties of the states of matter

State	Property
Gas	1. Assumes both the volume and shape of container. 2. Is compressible. 3. Diffusion occurs rapidly. 4. Flows readily.
Liquid	1. Assumes the shape of the portion of the container it occupies. 2. Does not expand to fill container. 3. Is virtually incompressible. 4. Diffusion occurs slowly. 5. Flows readily.
Solid	1. Retains its own shape and volume. 2. Is virtually incompressible. 3. Diffusion occurs extremely slowly. 4. Does not flow.

amount of space available for translational motion. In liquids the free volume is very small; typically it might be about 3 percent of the total volume of a liquid. In a gas, even at 10 atm pressure, the free volume is about 99 percent of the total.

Because the amount of free space is so small, and the molecules of a liquid are already in quite close contact, it requires considerable force to compress a liquid still further. Thus, liquids are much less compressible than gases; they have definite volumes, independent of the shape or size of their container. However, the molecules of a liquid move freely with respect to one another. Liquids thus do not have a definite shape; they can be poured, and they flow to assume the shape of their container.

The same sorts of attractive forces that cause molecules to form a liquid also operate in the formation of solids. The difference between liquids and solids is largely one of order. In a liquid the molecules, though close together, move about in more or less chaotic fashion. Each molecule is somewhat influenced in its motions by its nearest neighbors, but there is no long-range order. In a solid, on the other hand, each molecule takes up a certain position relative to its neighbors, and this regular arrangement extends throughout the solid. The transition from a liquid to a solid is thus rather like the change that occurs on a military parade ground when the troops are called to formation.

Solids, like liquids, are not very compressible, because there is little free volume. When pressure is applied to a solid, its volume can be reduced only by forcing the molecules closer together. The intermolecular forces that oppose compression can be very large.

Because the particles of a solid are not free to undergo long-range movement, solids are rigid. The molecules within a solid have little or no translational energy, because they are constrained to remain in their locations. However, they may have vibrational energy; that is, they may move back and forward periodically about the position they occupy in the lattice. The amount of vibrational energy in a solid is proportional to absolute temperature, just as the amount of translational energy in a gas is proportional to absolute temperature. Figure 11.1 illustrates schematically the comparisons among the three states of matter based on the kinetic-molecular model. This figure depicts most of the comparative features of the three states of matter that we have just discussed.

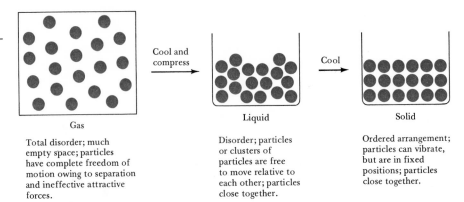

Total disorder; much empty space; particles have complete freedom of motion owing to separation and ineffective attractive forces.

Liquid

Disorder; particles or clusters of particles are free to move relative to each other; particles close together.

Solid

Ordered arrangement; particles can vibrate, but are in fixed positions; particles close together.

FIGURE 11.1 A molecular-level comparison of gases, liquids, and solids. The density of particles in the gas phase is exaggerated as compared with the usual situation.

SAMPLE EXERCISE 11.1

At about 6°C the gas, liquid, and solid phases of benzene can all coexist in a stable mixture. Under these conditions, the molar volume of $C_6H_6(g)$ is 470 L, of $C_6H_6(l)$ is 88.7 mL, and of $C_6H_6(s)$ is 86.6 mL. Account for the relative volumes of 1 mol of benzene in the three states.

Solution: In gaseous benzene, the molecules are well separated from one another and experience few close encounters. The average attractive force between molecules is thus very low. In the liquid state the molecules are in close contact and experience strong intermolecular forces that keep them together. However, there is no long-range order in the liquid, and molecules can move with respect to one another in the small amount of free volume that the liquid possesses. In solid benzene the molecules are packed in a regular lattice array, held in position by the intermolecular forces between them. The free volume in the solid is even less than in the liquid form; the molar volume is therefore slightly smaller.

11.2 Equilibria between phases

We know that when a substance is capable of existing in more than one of the three states of matter, it readily undergoes conversion from one state, or phase, to another. For example, water exists on our planet as a liquid in oceans, rivers, and lakes, as a gas in the earth's atmosphere, and as a solid in snow cover and glaciers. Interconversions between these three forms of water go on continuously. We must now look in more detail at how the interconversions between two different phases of matter occur.

Let us carry out an experiment in which we begin with a quantity of ethyl alcohol, also called ethanol, in a closed container, with a large space above the liquid. Suppose that we could somehow begin the experiment with all the ethanol in liquid form. The pressure in the container under these conditions would be zero, as illustrated in Figure 11.2(a). We would find that the pressure in the container quickly rises as a function of time. After a short time the pressure would attain a constant value, termed the vapor pressure of liquid ethanol, Figure 11.2(b).

We can account for these observations by using the kinetic-molecular theory. We know that the molecules of liquid ethanol are in constant motion. At any given temperature some of the molecules possess more than the average thermal energy, some less. That is, there is a distribution of molecular energies in the liquid similar to the distribution in a gas, illustrated in Figure 9.10.

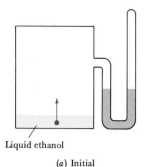
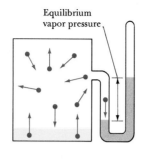

(a) Initial

Liquid ethanol

(b) At equilibrium

Equilibrium vapor pressure

FIGURE 11.2 Schematic representation of the vaporization of ethanol in a closed container. At equilibrium the rate of condensation and the rate of vaporization are equal. This produces a stable vapor pressure, one that does not change with time so long as the temperature remains constant.

At any instant some of the molecules at the surface of the liquid possess sufficient energy to escape from the attractive forces of their neighbors. Transfer of molecules into the gas phase is called vaporization or evaporation. The movement of molecules from the liquid to the gas phase goes on continuously. However, as the number of gas-phase molecules increases, the probability increases that a molecule from the gas phase will strike the liquid surface and stick there. We call this process condensation. Eventually, the two opposing processes occur at an equal rate, as illustrated in Figure 11.3. The pressure in the gas phase at this point becomes constant.

The system we have just discussed provides a simple example of dynamic equilibrium, in which two opposing processes take place at equal rates. To the observer it may appear that nothing is taking place; in our example of ethyl alcohol, the pressure remains constant at some value determined by the temperature of the liquid. In fact, a great deal is happening; molecules continuously pass from the liquid state to the gas state, and from the gas state to liquid state. All equilibria between matter in different phases possess this dynamic character.

Although equilibria can exist simultaneously between all three states of matter, we usually encounter equilibria occurring between just two states, as in our example of the equilibrium between liquid and gaseous ethanol. Figure 11.4 summarizes the possible two-phase equilibria and the terms associated with them. Some of these are more familiar than others, but a little thought will bring to mind several examples of each possible kind of phase equilibrium.

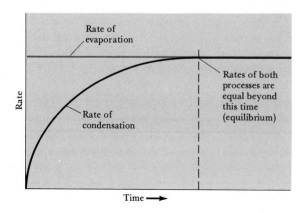

Rate of evaporation

Rate

Rate of condensation

Rates of both processes are equal beyond this time (equilibrium)

Time →

FIGURE 11.3 A comparison of the rates of evaporation and condensation of a liquid as a function of time as the system approaches equilibrium.

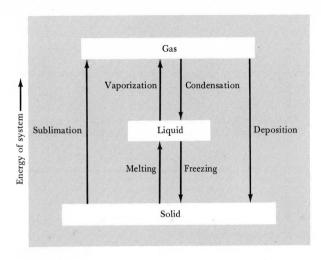

FIGURE 11.4 The possible phase changes and the terms associated with them.

SAMPLE EXERCISE 11.2

Using water as the substance, give examples of all the possible equilibria between two different phases.

Solution: *Vaporization* of liquid water occurs from lakes or other water bodies in the summer months. *Condensation* of water vapor occurs on a cold window surface or on the lawn in the early morning hours. Ice *melts* in the spring, and lakes *freeze* in winter. When air moves over snow or ice at temperatures below freezing, the solid *sublimes*. Formation of snow crystals in clouds in the winter provide an example of *deposition*.

The change of matter from one state to another is called a phase change. Each of the phase changes diagramed in Figure 11.4, when it occurs at constant temperature, has associated with it an enthalpy change. As we have already seen, energy is required to overcome the attractive forces between molecules in vaporization, the phase change from the liquid to the gaseous state. Similarly, we expect that sublimation will require energy; the attractive forces that bind the molecules to one another in the solid must be overcome to produce a gas of widely separated particles. Melting also requires energy; the regularity of the solid state, in which molecules are packed to attain the most effective intermolecular interactions, is disrupted when the solid melts to form the liquid state. Thus, vaporization, sublimation, and melting are all endothermic processes; an input of heat is required for them to proceed. For example, an input of 2.26 kJ is required to cause the vaporization of 1 g of water at 100°C. The enthalpies of vaporization, sublimation, and melting are thus positive quantities.

Condensation is just the reverse process of vaporization. Similarly, deposition is the reverse of sublimation, and freezing is the reverse of melting. Hess's law tells us that in any cyclic process the overall enthalpy change must be zero (Section 4.4). Thus it follows that the heats of condensation, deposition, and freezing are the same in magnitude but opposite in sign from the heats of vaporization, sublimation, and melting, respectively.

In terms of the kinetic molecular theory, explain the origin of the heat of deposition.

Solution: Deposition is the process in which molecules from the gas state come together to form a solid. As the gas-phase molecules encounter the surface and come under the attractive influences of the other molecules of the solid, they attain a lower potential energy state. That is, the kinetic energies of the gas-phase molecules are, in part, converted to potential energy. In the process heat is evolved. If the temperature of the entire system of gas phase and solid is to remain constant, heat must be removed from the system; thus, deposition is exothermic. The heat associated with deposition is opposite in sign to, but of equal magnitude with, the heat associated with the reverse process, sublimation.

11.3 Properties of liquids

We have seen that molecules may escape from the surface of a liquid into the gas phase by vaporization or evaporation. The weaker the attractive forces between molecules in the liquid phase, the more readily vaporization occurs. When two liquids are compared, the one that evaporates more readily is described as being the more volatile. For example, alcohol (ethanol) is more volatile than water. The rate of vaporization increases with increasing temperature. Figure 11.5 provides a graphic explanation for this behavior. In that figure the distribution of kinetic energies of the particles at the surface of the liquid is compared at two temperatures with the attractive forces at the surface. Notice that the distribution curves are like those shown earlier for gases (Figure 9.10); average kinetic energy increases with increasing temperature. Therefore, as the temperature of a liquid is increased, there is an increase in the number of particles having sufficient kinetic energy to escape from the surface. Loss of particles with high kinetic energy causes the average kinetic energy of the particles remaining in the liquid to decrease. Anyone getting out of a swimming pool, particularly on a windy day, has experienced the cooling that accompanies evaporation. The molar heats of vaporization (ΔH_v) of some common liquids at their boiling points are summarized in Table 11.2.

The rate of metabolic heat generation is such that an adult human's body temperature would rise 1 to 30°F per hour (depending on the level of activity) if the heat were not dissipated. This rise would result in a heat stroke at 106°F and death at 110 to 112°F. Because heat does not flow from colder to hotter objects, cooling by simple conductive heat loss from the body is not possible when the outside temperature is greater than body temperature. Our bodies rely for cooling on the heat absorbed by evaporation of water; we are cooled by sweating. Thus, one of the significant roles played by water in our bodies is that of a cooling agent. We may note that water is particularly well suited for this role; it takes more energy to vaporize a gram of water than to vaporize an equal mass of any known liquid substance. Comparing water and mercury (Table 11.2) we have, for water,

$$\left(40.7 \, \frac{kJ}{mol}\right)\left(\frac{1 \, mol}{18.0 \, g}\right) = 2.26 \, kJ/g$$

For mercury we have:

$$\left(59.3 \, \frac{kJ}{mol}\right)\left(\frac{1 \, mol}{200.5 \, g}\right) = 0.296 \, kJ/g$$

Note also that when the surrounding air is humid and thus contains many water molecules, the net rate of evaporation of water is slower, and we feel more uncomfortable.

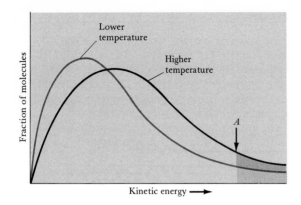

FIGURE 11.5 Distribution of kinetic energies of surface molecules of a hypothetical liquid compared with the average attractive force experienced by each molecule at the liquid surface, A. The shaded areas represent the fraction of molecules with sufficient kinetic energy to escape from the surface; that is, those with an energy greater than A.

TABLE 11.2 Heats of vaporization of some common liquids at their boiling points

Substance	Formula	ΔH_v (kJ/mol)	Boiling point (°C)
Benzene	C_6H_6	30.8	80.2
Ethanol	C_2H_5OH	39.2	78.3
Ether	$C_2H_5OC_2H_5$	26.0	34.6
Mercury	Hg	59.3	356.9
Methane	CH_4	10.4	−164
Water	H_2O	40.7	100

VAPOR PRESSURE

Liquids vary greatly in their vapor pressures at any given temperature. Gasoline or water evaporates fairly quickly when exposed to moving air, whereas engine oil exposed to the air at room temperature seems not to be volatile at all. The volatility of a liquid is determined by the magnitude of the intermolecular forces that constrain the molecules to remain together in the liquid. (We will consider the origins of intermolecular forces in Section 11.5.) Volatility increases with increasing temperature, as the average kinetic energy of the molecules increases in relation to the intermolecular forces. Figure 11.6 depicts the variation in vapor pressure with temperature for four common substances that differ greatly in volatility. Note that in all cases the vapor pressure increases quite nonlinearly with increasing temperature.

As we shall see in a few of the end-of-chapter exercises, it is usually possible to represent the dependence of the vapor pressure, P, on temperature as a log function of the form

$$\log P = \frac{-\Delta H_v}{2.303\ RT} + C$$

In this equation T is absolute temperature, R is the gas constant (Table 9.1), ΔH_v is the enthalpy of vaporization, and C is a constant. This equation, called the Clausius-Clapeyron equation, tells us that a graph of log P versus 1/T should be linear. The slope of the line can be used to calculate the enthalpy of vaporization.

BOILING POINTS

A liquid is said to boil when vapor bubbles form in the interior of the liquid; this condition occurs when the vapor pressure equals the external

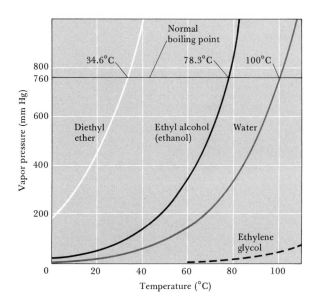

FIGURE 11.6 Equilibrium vapor pressures of four common liquids shown as a function of temperature. The temperature at which the vapor pressure is 760 mm Hg is the normal boiling point of the liquid.

pressure acting on the liquid's surface. Consequently, the boiling point of a liquid depends on pressure. From Figure 11.6 it can be seen that the boiling point of water at 1 atm pressure (760 mm Hg) is 100°C. The boiling point of a liquid at 1 atm pressure is called its **normal boiling point.** At 650 mm Hg water boils at 96°C.

SAMPLE EXERCISE 11.4

What is the boiling point of ethanol at 400 mm Hg?   Solution: From Figure 11.6 it can be seen that the boiling point must be about 64°C, as compared to the normal boiling point of 78.3°C.

The fact that boiling points vary with pressure is utilized in the pressure cooker. The time required to cook food depends on the temperature. As long as water is present, the maximum temperature of the food is the boiling point of water. Pressure cookers are sealed and allow steam to escape only when it exceeds a predetermined pressure; the pressure above the water can therefore increase above atmospheric pressure. The higher pressure causes water to boil at a higher temperature; thereby allowing the food to get hotter. It therefore cooks more rapidly. The effect of pressure on boiling point also explains why it takes longer to cook food at higher elevations than at sea level; at higher altitudes the atmospheric pressure is lower, and water boils at a lower temperature.

VISCOSITY

Some liquids literally flow like molasses, whereas others flow quite easily. The resistance of liquids to flow is referred to as their viscosity. The larger the viscosity, the more slowly the liquid flows. Liquids such as molasses or motor oils are relatively viscous; water and organic liquids such as carbon tetrachloride are not. Viscosity can be measured by measuring how long it takes a certain amount of a liquid to flow through a

TABLE 11.3 Viscosities and surface tensions of some common liquids at 20°C, and of water at several temperatures

Substance	Formula	Viscosity $(N\text{-}sec/m^2)^a$	Surface tension, γ (J/m^2)
Benzene	C_6H_6	0.65×10^{-3}	2.89×10^{-2}
Ethanol	C_2H_5OH	1.20×10^{-3}	2.23×10^{-2}
Ether	$C_2H_5OC_2H_5$	0.23×10^{-3}	1.70×10^{-2}
Glycerin	$C_3H_8O_3$	1490×10^{-3}	6.34×10^{-2}
Mercury	Hg	1.55×10^{-3}	46×10^{-2}
Water at:	H_2O		
20°C		1.00×10^{-3}	7.29×10^{-2}
40°C		0.652×10^{-3}	6.99×10^{-2}
60°C		0.466×10^{-3}	6.70×10^{-2}
80°C		0.356×10^{-3}	6.40×10^{-2}

[a]This is the SI unit: $1 \text{ N-sec}/m^2 = 1 \text{ kg/m-sec}$.

thin tube, under gravitational force. In another method, the rate of fall of steel spheres through the liquid is measured. The spheres fall more slowly through the more viscous liquids.

Viscosity is related to the ease with which individual molecules of the liquid can move with respect to one another. It thus depends on the attractive forces between molecules, and on whether there are structural features that cause the molecules to become entangled. Viscosity decreases with increasing temperature, because at higher temperature the greater average kinetic energy of the molecules more easily overrides the attractive forces between molecules. The viscosities of some common liquids are listed in Table 11.3.

SURFACE TENSION

Liquids have a tendency to assume a minimum surface area. This minimum is achieved when the liquid has a spherical shape. For example, when water is placed on a waxy surface, it "beads up," forming distorted spheres. The state of minimum energy is that state of minimum surface area; energy must therefore be supplied to increase the surface area. The energy required to increase the surface area of a liquid by a unit amount is known as its surface tension. The origin of surface tension is an imbalance of forces at the surface of the liquid, as shown in Figure 11.7. There is a net inward pull on the surface that contracts the surface and makes the liquid behave almost as if it had a skin. This effect permits a carefully placed needle to float on the surface of water or some insects to "walk" on water even though their densities are greater than that of water.

The forces between like molecules that are reflected in a substance's vapor pressure, boiling point, heat of vaporization, viscosity, and surface tension are called cohesive forces. The forces between unlike substances, such as water and glass, are called adhesive forces. In a glass tube, the adhesive forces between water and glass are sufficiently strong relative to the cohesive forces for water to form a concave-upward surface (Figure 11.7). Such a curved surface on a liquid is known as a meniscus. For mercury, cohesive forces are greater than the adhesive forces with glass;

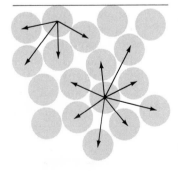

FIGURE 11.7 A molecular-level view of the unbalanced intermolecular forces on the surface of a liquid that result in surface tension.

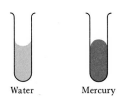

Water   Mercury

FIGURE 11.8 A comparison between the shape of the meniscus of water in a glass tube and the shape of the mercury meniscus.

mercury does not adhere to glass and a concave-downward meniscus is observed (Figure 11.8).

When a liquid adheres to or wets the walls of a tube as water adheres to glass, the liquid is drawn up the tube. This phenomenon is known as capillary rise, or capillary action. Wetting of the walls of a tube tends to increase the surface area of the liquid. Surface tension tends to reduce this area, and consequently the liquid is drawn up the tube. One method of determining surface tension involves measuring the height to which a liquid rises in a tube of known radius. The surface tensions of several common liquids are given in Table 11.3. The surface tension of 7.26×10^{-2} J/m^2 for water indicates that an energy of 7.26×10^{-2} J must be supplied to increase the surface area of a given amount of water by 1 m^2. This is a very small amount of energy.

Surface tension generally decreases with increasing temperature, as illustrated by the data for water in Table 11.3. For the majority of compounds the dependence of surface tension (γ) on temperature can be expressed as

$$\gamma = a - bT \tag{11.1}$$

where a and b are constants and T is the temperature in degrees Celsius.

CRITICAL TEMPERATURE AND PRESSURE

We have seen that the transition from the gaseous to liquid state is promoted by the intermolecular attractive forces that cause condensation. On the other hand, condensation is opposed by the kinetic energies of the molecules, which keep them in independent motion.

As the temperature is raised, the kinetic energies of molecules increase in relation to intermolecular attractions, and a gas becomes more difficult to liquefy. Consequently, the pressure required to condense a gas to a liquid increases. Finally, a temperature is reached at which no amount of pressure, however great, causes the gas to pass from gas to liquid state. The highest temperature at which a gas liquefies is called its critical temperature. The critical pressure is the pressure required to bring about liquefaction at this critical temperature. The critical temperatures and pressures of gases are of considerable practical importance to engineers and others working with gases. The relative values of these quantities among different gases also provide some indication of the relative importance of intermolecular forces. Table 11.4 shows the critical temperatures and pressures for the noble gas elements and a few substances composed of nonpolar diatomic molecules. The critical temperature and pressure increase in the same way as the boiling point.

11.4 Crystalline solids

Solids are rigid; they cannot be poured like liquids or compressed like gases. Solids such as quartz or diamond possess highly regular crystalline shapes or cleavage planes. These facts suggest a regular atomic arrangement within the solid. Indeed, a statement of this regular arrangement is sometimes included as part of the definition of solids. For our purposes, we shall divide materials that we would ordinarily call solids because of their rigidity into two groups: crystalline solids and amorphous solids.

TABLE 11.4 Molecular masses, normal boiling points, enthalpies of vaporization, and critical temperatures and pressures of the noble gas elements and a few substances composed of nonpolar diatomic molecules

Substance	Molecular mass (amu)	Normal boiling point (K)	ΔH_v (kJ/mol)[a]	Critical temperature (K)	Critical pressure (atm)
He	4	4.2	0.081	5.2	2.26
Ne	20	27	1.76	44.4	25.9
Ar	40	87	6.52	151	48
Kr	84	121	9.03	210	54
Xe	131	164	12.64	290	58
Rn	222	211	16.78	377	62
H_2	2	20	0.903	33.2	12.8
N_2	28	77	5.58	126	33.5
O_2	32	90	6.82	154	49.7
Cl_2	71	239	20.40	417	76.1

[a] Measured at the temperature of the normal boiling point.

Crystalline solids are characterized by a regular three-dimensional arrangement of atoms. Amorphous solids lack this regular atomic-level organization. Familiar materials of the second type include substances such as rubber and glass, which are composed of large or complicated molecules. In some texts, amorphous solids are referred to as supercooled liquids because they have the molecular disorder of liquids. In fact, glass is capable of flowing, as revealed by a careful examination of the window panes of very old houses. The panes are thicker at the bottom than at the top because the glass has flowed under the continued influence of the force of gravity. In this section our focus is on crystalline solids.

X-RAY STUDY OF CRYSTAL STRUCTURE

Much of what we know about the internal molecular-level regularity of crystalline solids is revealed by their interaction with X rays. X rays, you will recall, are electromagnetic waves of short wavelength and high energy (Section 5.1). They were first discovered in 1895 by the German physicist Wilhelm Roentgen. However, it was not until 1913 that they were used to determine the location of atoms within a crystalline solid. In that year, the Englishman W. L. Bragg and his father, W. H. Bragg, determined the location of the zinc and sulfur atoms in a ZnS crystal from a mathematical analysis of the X-ray diffraction pattern of this substance.

Diffraction is a phenomenon that results from scattering of light waves by a regular arrangement of points or lines. It is observed when the wavelength of the light rays is comparable to the distances that separate the points or lines. A crystalline solid is a regular array of particles in three dimensions. The particles may be ions, molecules, or atoms; that is not important for the moment. We will call the sites at which these particles are located the *lattice sites*. The three-dimensional array of these sites is called the space lattice, or simply the *lattice*. When light rays with a wavelength of the magnitude of the distances between the lattice points impinge on the crystal, scattering of the rays by the individual

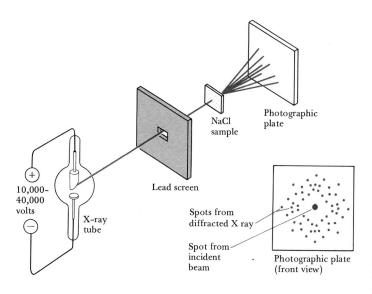

NaCl sample

Photographic plate

Lead screen

10,000–
40,000
volts

+

−

X-ray
tube

Spots from
diffracted X ray

Spot from
incident
beam

Photographic plate
(front view)

FIGURE 11.9 The X-ray diffraction pattern for NaCl
and the experimental method by which it is obtained.

particles of the lattice may occur. The distances between lattice sites in a solid is on the order of 1 to 3 or 4 Å, that is, 0.10 to 0.30 or 0.40 nm. A glance at Figure 5.3 reminds us that X rays possess these short wavelengths.

In the X-ray diffraction experiment, X rays of a single wavelength are allowed to fall on a single crystal, as illustrated in Figure 11.9. A photographic plate or other detection device is used to determine the angles at which the X rays are scattered* from the crystal. The photographic plate shows a series of dots in a regular array, indicating that X rays are scattered from the crystals at only certain angles. All of the various angles at which scattering, or diffraction, occur can be determined from measurements made on the photographic plate.

To understand the significance of these measurements we must understand the idea of constructive and destructive interference. When two rays of light of the same wavelength come together, they may be in phase, as shown in Figure 11.10. When this occurs, the amplitudes of the rays add to give a more intense ray; this is termed **constructive interference.** On the other hand, when the two rays are out of phase, as illustrated in Figure 11.10, **destructive interference** occurs, and the resultant ray has zero amplitude.

Rays scattered from the lattice points of the solid come together at the surface of the film. If they are in phase, constructive interference occurs, and a spot is seen on the film. If they are out of phase, destructive interference results. To see how the angle of scattering from the crystal affects the different rays we need to look in detail at the paths taken by two rays reflected from adjacent lattice sites. The geometrical situation is illustrated in Figure 11.11.

The incoming rays are in phase at AB. Wave ACA' is scattered or reflected by an atom in the first layer of the solid, whereas wave BEB' is reflected by an atom in the second layer. If these two waves are to be in phase at $A'B'$, the extra distance covered by BEB' must be a whole-

*Each atom absorbs some of the energy of the X rays and then reemits it in all directions, thereby scattering the X rays.

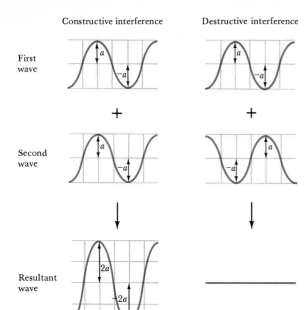

Constructive interference   Destructive interference

First wave

Second wave

Resultant wave

FIGURE 11.10 Constructive and destructive interference of waves.

number multiple of the wavelength, λ. In Figure 11.11(b), the extra distance, DEF, is 2λ. Now notice that the triangle CDE is a right triangle. Using trigonometry, it can be shown that the distance DE is $d \sin \theta$, where d is the distance between the planes and θ is the angle between the incoming wave and the plane. Because the distance DEF is twice DE, we have

$$DEF = 2\lambda = 2d \sin \theta$$

It can be shown that the general equation for constructive interference is

$$n\lambda = 2d \sin \theta, \text{ where } n = 1, 2, 3, 4, \ldots . \qquad [11.2]$$

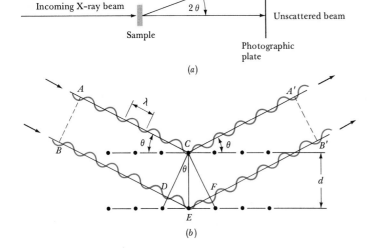

(a)

(b)

FIGURE 11.11 Scattering of X rays by atoms in parallel planes. (a) The experimental arrangement with the X-ray beam scattered at an angle of 2θ from the unscattered beam. (b) The atomic-level view of the atoms and waves. The incoming X rays of wavelength λ are diffracted by atoms, which are represented by dots. The atoms are arranged in planes that are separated by a distance d. The incoming X rays make an angle of θ with the planes and are scattered at an angle of 2θ from the unscattered beam, which is not shown in (b).

This relationship, known as the Bragg law, allows determination of the spacing between planes from the known wavelength of the light and experimentally determined values of θ at which constructive interference occurs.

SAMPLE EXERCISE 11.5

When a crystal of cesium metal is irradiated with X rays of wavelength 1.591 Å, a series of reflections due to constructive interference occurs at angles θ of 8.63°, 17.46°, and 26.75°. What is the separation between the planes of atoms in the solid giving rise to these reflections?

Solution: We wish to solve for the quantity d in Equation [11.2]. Upon rearranging the equation, we have

$$d = \frac{n\lambda}{2 \sin \theta}$$

(Notice that we have values for all quantities except d and n.) Let us assume that $n = 1$ for the reflection giving rise to the smallest value of θ, 2 for the next, and 3 for the largest value. Inserting the appropriate values of $\sin \theta$ and n, we obtain the same value of d, 5.30 Å, in all three cases. This represents the distance between the plane of cesium atoms giving rise to the reflections.

CRYSTAL LATTICES AND UNIT CELLS

The order characteristic of crystalline solids allows us to convey a picture of an entire crystal by looking at only a small part of it. That portion of the three-dimensional arrangement of particles, the space lattice, that is consistent with the compound formula and generates the entire crystal by simple displacement in three dimensions is called the unit cell. A simple two-dimensional example is afforded by a sheet of wallpaper. The smallest area that contains the repeating pattern of the sheet is the "unit cell" for the pattern. The unit cell may be considered the "fundamental building block" of the crystal lattice because the lattice can be constructed by stacking unit cells. The NaCl lattice, which we have discussed on previous occasions, provides a convenient example. A two-dimensional portion, representing a slice through the lattice and showing the Na^+ and Cl^- ions simply as point charges, is shown in Figure 11.12.

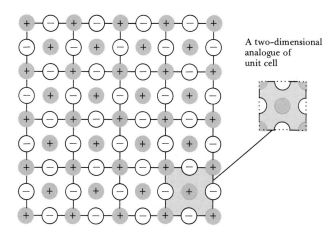

A two-dimensional analogue of unit cell

FIGURE 11.12 A cross-section of the space lattice of NaCl, showing the repeating pattern of its unit cell.

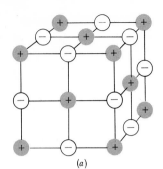

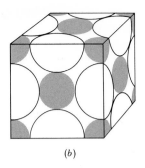

FIGURE 11.13 Two representations of the unit cell of sodium chloride. In (a) the ions are shown as point charges. In (b) we get a better feel for the relative sizes of the Na^+ and Cl^- ions and how ions are shared between unit cells.

(a)

(b)

Two representations of the unit cell of NaCl are given in Figure 11.13. As shown in Figure 11.13(b), the particles at the corners, edges, and faces of the unit cell do not lie wholly within it. Sample Exercise 11.6 amplifies on this point.

SAMPLE EXERCISE 11.6

Determine the net number of Na^+ and Cl^- ions in the NaCl unit cell (Figure 11.13).

Solution: There is $\frac{1}{8}$ of a Na^+ on each corner (each Na^+ on a corner is shared by eight cubes which intersect at that point). There is $\frac{1}{2}$ of a Na^+ on each face; $\frac{1}{4}$ of a Cl^- on each edge; and a whole Cl^- in the center of the cube. Because this may be hard for you to visualize, the sharing of particles on the corners and faces of a unit cell is shown in Figure 11.14. For NaCl we therefore have the following:

Na^+: $(\frac{1}{8} Na^+$ per corner)(8 corners) $= 1\, Na^+$
$(\frac{1}{2} Na^+$ per face)(6 faces) $= 3\, Na^+$

Cl^-: $(\frac{1}{4} Cl^-$ per edge)(12 edges) $= 3\, Cl^-$
$(1\, Cl^-$ per center)(1 center) $= 1\, Cl^-$

Thus the unit cell contains $4\, Na^+$ and $4\, Cl^-$. This result agrees with the compound's stoichiometry: one Na^+ for each Cl^-.

SAMPLE EXERCISE 11.7

If the unit cell of NaCl is 5.64 Å on an edge, calculate the density of NaCl.

Solution: The volume of the unit cell is $(5.64\text{ Å})^3$. Because each unit cell contains $4\, Na^+$ and $4\, Cl^-$ (Sample Exercise 11.6), its mass is

$4(23.0\text{ amu}) + 4(35.5\text{ amu}) = 234.0\text{ amu}$

The density is mass/volume:

$$\text{Density} = \frac{234.0\text{ amu}}{(5.64\text{ Å})^3}\left(\frac{1\text{ g}}{6.02 \times 10^{23}\text{ amu}}\right)$$
$$\times \left(\frac{1\text{ Å}}{10^{-8}\text{ cm}}\right)^3$$
$$= 2.17\text{ g/cm}^3$$

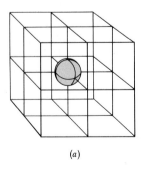

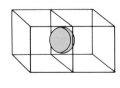

(a)

(b)

FIGURE 11.14 Particles at the corners and faces of unit cells are shared by other unit cells. Here we see the sharing of a corner atom by eight unit cells (a) and the sharing of a face-centered atom by two unit cells (b).

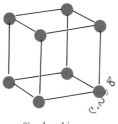

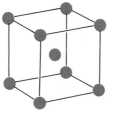

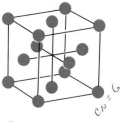

Simple cubic   Body-centered cubic   Face-centered cubic

FIGURE 11.15 Three types of cubic cells.

Any crystalline solid can be described in terms of a three-dimensional array of unit cells in a lattice arrangement. As examples, the cubic unit cells illustrated in Figure 11.15 are characterized by three sides of equal length, and 90° angles between the sides. When particles are located only at the corners of this cell, the unit cell is described as simple cubic. When particles are located at the corners and at the center of the unit cell, the cell is known as body-centered cubic. A third type of cubic cell has particles located at each corner, as well as at the center of each face; this is known as a face-centered cubic cell.

CLOSE PACKING

We frequently visualize the unit cells of crystals in terms of points connected by imaginary lines. The points represent atoms, ions, or molecules, and the lines help us visualize the symmetry of the crystal. It is important to remember, however, that the particles of a solid actually take up much more of the space of the crystal lattice than these representations show. The unit cell of NaCl shown in Figure 11.13(b) and the unit cell of CH_4 shown in Figure 11.16 provide a more realistic picture of how the particles actually pack within the solid. Notice in Figure 11.16 that the CH_4 molecules can be approximated as spheres. Many crystals, particularly those of metals, can be visualized as consisting of equal-sized spheres packed together in space.

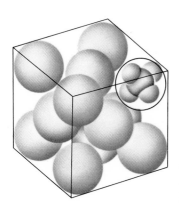

FIGURE 11.16 The unit cell of solid methane. Each large sphere represents a CH_4 molecule as shown at the upper right.

"There are several ways," Dr. Breed said to me, "in which certain liquids can crystallize—can freeze—several ways in which their atoms can stack and lock in an orderly, rigid way."

That old man with spotted hands invited me to think of the several ways in which cannonballs might be stacked on a courthouse lawn, of the several ways in which oranges might be packed into a crate.

"So it is with atoms in crystals, too; and two different crystals of the same substance can have quite different physical properties." *

It is instructive to consider how equal-sized spheres can pack most efficiently (that is, with the minimum amount of empty space). Such arrangements, referred to as closest-packed structures, are quite common. The most efficient arrangement of a layer of equal-sized spheres is

*Excerpt from *Cat's Cradle*, by Kurt Vonnegut, Jr. Copyright © 1963 by Kurt Vonnegut, Jr. Reprinted by permission of Delacorte Press/Seymour Lawrence and Donald C. Farber for Kurt Vonnegut, Jr.

shown in Figure 11.17(*a*). Each sphere is surrounded by six others in the layer. The most efficient arrangement of the spheres in a second layer is in the depressions of the first layer, labeled *A* in Figure 11.17(*a*). This is shown in Figure 11.17(*b*). The spheres of the third layer sit in depressions in the second layer. However, there are two types of depressions, and they lead to two different structures. If the spheres of the third layer are placed immediately above those of the first layer, in the positions labeled *B* in Figure 11.17(*b*), the structure known as the hexagonal close-packed structure results—Figure 11.17(*c*). If we consider more than three layers of spheres, the arrangement of the layers in this structure can be represented as ABABAB. . . . The second type of close-packed structure results if the third-layer spheres are placed in the positions labeled *C* in Figure 11.17(*b*). The resultant structure is known as the cubic close-packed structure—Figure 11.17(*d*). In this case the stacking sequence can be represented as ABCABC. . . . Although it is not obvious from Figure 11.17, the cubic-close-packed arrangement of spheres has a face-centered cubic unit cell. In Figure 11.18, the close-packed structures are viewed from a perspective that permits the unit cells to be seen more clearly. In both of the close-packed structures, each sphere has twelve nearest neighbors (that is, a coordination number of twelve).

When unequal-sized spheres are packed in a lattice, the large particles sometimes assume one of the close-packed arrangements with small particles occupying the holes between the large spheres. For example, in Li_2O the oxide ions assume a cubic close-packed structure, whereas the Li^+ ions occupy small cavities that exist between oxide ions. We shall consider this particular view of crystals further in Chapter 22 when we discuss the structures of silicates, the primary structural materials of our planet.

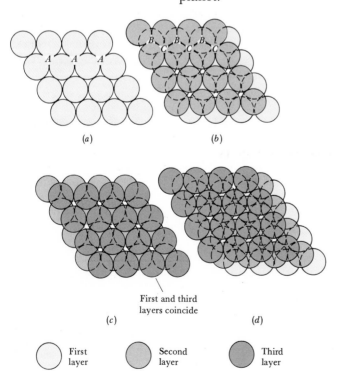

(*a*)　　　　(*b*)

First and third layers coincide

(*c*)　　　　(*d*)

○ First layer　　◔ Second layer　　● Third layer

FIGURE 11.17 The closest packing of equal-sized spheres. (*a*) One layer. (*b*) Two superimposed layers. (*c*) Three superimposed layers in hexagonal close-packed arrangement. (*d*) Three superimposed layers in cubic close-packed arrangement.

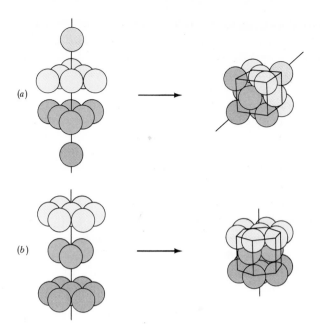

FIGURE 11.18 The stacking of spheres by means of (a) cubic close-packed structure and (b) hexagonal close-packed structure, drawn so as to show the unit cell of each lattice.

CRYSTAL DEFECTS

Although we have generally pictured crystalline solids as being composed of perfectly ordered arrays of particles, this is merely a useful abstraction like that of an ideal gas. Real crystals contain imperfections whose number and type can play an important role in determining the properties of the solid. For example, a crystal lattice that has many sites where particles are missing (vacancies) can be more readily deformed than can a perfect crystal lattice of the same substance. It is not hard to appreciate that structural imperfections can occur readily. In forming a solid, the structural units can be thought of as a mob that has been required to assume a military drill formation. The faster crystal formation occurs, the greater the chance for defects.

11.5 Intermolecular forces

The fact that matter exists in the liquid or solid forms at all is evidence that attractive forces exist between atoms, molecules, and ions. We will refer to these as intermolecular forces, even though in some instances it is actually ions that are involved. Intermolecular forces are all electrostatic in nature. They are based upon Coulomb's law of attraction between unlike charges and repulsion between like charges (Section 2.6). The magnitudes of intermolecular forces vary over a wide range. The forces between neutral molecules are generally much smaller than those that cause atoms to be bound together in the same molecule. For example, the heat required to vaporize a mole of liquid HCl is 16 kJ/mol, whereas the heat required to dissociate a mole of HCl into H and Cl atoms is 431 kJ/mol. Similarly, for HI the enthalpy of vaporization is 20 kJ/mol, the HI bond dissociation energy is 294 kJ/mol.

We can distinguish several varieties of intermolecular force; each has a characteristic dependence on the distance between the interacting particles. The distance dependence is important, because it determines how close two particles must approach before the interaction between them becomes significant. An intermolecular force that varies as $1/d^2$, say, where d is the distance between centers, will operate over a much larger range of distance than an intermolecular attraction that varies as $1/d^6$. As an example, let us suppose that $d = 1$. Then $1/d^2$ and $1/d^6$ both equal 1. Now let the distance d double, to 2. Then $1/d^2$ equals 0.25, whereas $1/d^6$ equals only 0.016. It is clear that as the distance d increases the intermolecular force that varies as $1/d^6$ decreases much more rapidly than the intermolecular force that varies as $1/d^2$. The classes of intermolecular force we are about to discuss are illustrated in Figure 11.19.

ION-ION FORCES

Ion-ion forces exist in a solid ionic lattice, as discussed in Section 7.2. The ions are arranged in the solid lattice to maximize the attractive interactions between unlike charges, while minimizing the repulsive interactions between like charges. Ion-ion forces also operate in molten salts. As in any liquid, the ions are free to move with respect to one another, and there is no long-range order. On the average, however, the particles surrounding each cation will be mainly anions, and those surrounding each anion will be mostly cations. As indicated in Section 7.2, the attractive energy of ions varies as Q_1Q_2/d, where Q_1 and Q_2 are the charges on the ions, and d is their distance of separation. To obtain the electrostatic attractive energy of a solid or liquid ionic substance, all the possible attractive and repulsive interactions must be summed over the entire volume. This can be done for a solid of known geometry;

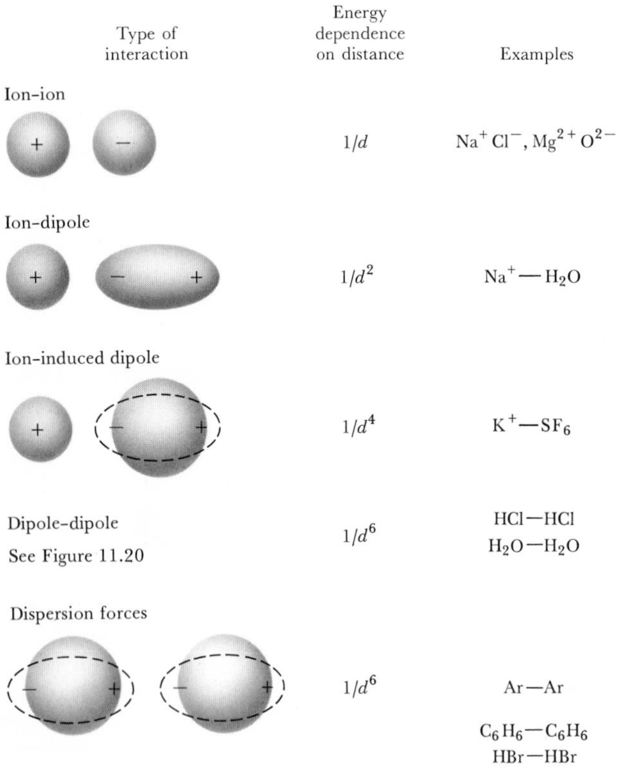

Type of interaction	Energy dependence on distance	Examples
Ion–ion	$1/d$	$Na^+Cl^-, Mg^{2+}O^{2-}$
Ion–dipole	$1/d^2$	$Na^+ — H_2O$
Ion–induced dipole	$1/d^4$	$K^+ — SF_6$
Dipole–dipole See Figure 11.20	$1/d^6$	$HCl — HCl$ $H_2O — H_2O$
Dispersion forces	$1/d^6$	$Ar — Ar$ $C_6H_6 — C_6H_6$ $HBr — HBr$

FIGURE 11.19 Illustration of the various types of intermolecular forces. Note that the ion-induced dipole and dispersion forces depend upon a distortion (dashed lines) of the electron distribution in a nonpolar atom or molecule.

TABLE 11.5 Boiling points of some ionic liquids and the sums of ionic radii

Substance	$r_+ + r_-$ (Å)	Boiling point (°C)
LiF	2.01	1717
NaCl	2.79	1465
KBr	3.29	1398
RbI	3.67	1304
Li_2O	2.13	2563
MgO	2.10	3260
MgF_2	1.98	2226
$MgCl_2$	2.46	1437
$MgBr_2$	2.61	1158

the resulting quantity is known as the lattice energy (Section 7.2). For an ionic liquid such a treatment is not possible because of the disorder that prevails. Nevertheless, we can make a few predictions; for example, we can guess that the cohesive energy of the ionic liquid (that is, the energy that keeps the particles together in the liquid state) will increase with the charges on the ions and will be larger for ions of smaller radius. The normal boiling point of a liquid is a crude measure of the extent of the attractive energies that hold the particles of a liquid together. Table 11.5 shows the boiling points for several ionic liquids.

Notice that in the series formed by the first four compounds listed, the boiling point decreases steadily with increasing sum of the ionic radii. This is what we expect from Coulomb's law; the centers of large ions cannot approach as closely, and thus their electrostatic attractive energies are smaller.

SAMPLE EXERCISE 11.8

Account for the variation in boiling points in the series of ionic liquids, LiF, Li_2O, MgO, listed in Table 11.5.

Solution: In this series the sum of ionic radii is nearly constant. However, the ionic charges increase from $(+1:-1)$ to $(+1:-2)$ to $(+2:-2)$ in the series. This means that the average Coulomb attractive energy increases in the series; thus, higher temperatures are required for boiling to occur.

ION-DIPOLE FORCES

An ion-dipole force exists between an ion and a neutral polar molecule that possesses a permanent dipole moment. Recall that polar molecules are those to which a positive and a negative end can be assigned owing to molecular shape and to charge separation caused by unequal sharing of electrons (Section 8.2). A simple example is HCl ($\mu = 1.03$ D). The end of the dipole possessing an opposite charge from that of the ion is attracted to the ion, as illustrated in Figure 11.19. Ion-dipole forces vary as $1/d^2$, where d is the distance from the center of the ion to the midpoint of the dipole. Ion-dipole forces are especially important in solutions of ionic substances in polar liquids, as, for example, in a solution of NaCl in water. We will have more to say about such solutions later (Section 12.2).

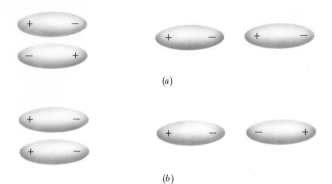

(a)

(b)

FIGURE 11.20 Variation in the dipole-dipole interaction with orientation. In (a) the dipoles are aligned so as to produce an attractive interaction. In (b) the interactions are repulsive.

An ion may exert a strong polarizing effect on a nearby nonpolar molecule, thereby inducing a dipole moment. This occurs because the electrons of the nonpolar molecule are affected by the nearby ionic charge. For example, if the ion is positively charged, the electrons of the adjacent molecule are drawn toward it. In the process, the electron cloud of the nonpolar molecule is distorted, so that the molecule acquires an induced dipole moment, as illustrated in Figure 11.19. The attractive energy of interaction between an ion and an induced dipole varies as $1/d^4$.

DIPOLE-DIPOLE FORCES

Dipole-dipole forces exist between polar molecules. The sign and magnitude of the interaction varies with the relative orientations of the two dipoles, as illustrated in Figure 11.20. In a crystalline solid the dipolar molecules tend to pack so as to attain the most stable, lowest energy orientation. Other factors such as the shapes of the molecules and location of the dipole within the molecule also play a role.

Table 11.6 lists the dipole moments and boiling points for a few organic liquids of comparable molecular weights. These data suggest that the boiling point increases with increasing magnitude of molecular dipole moment. Assuming that two dipolar molecules in the liquid state move freely with respect to one another, they will sometimes be in an orientation that is attractive, sometimes in an orientation that is repulsive. The net effect, averaged over time, is an attractive energy that varies as μ^4/d^6, where d is the distance between the centers of the dipoles, and μ is the dipole moment. Thus, dipole-dipole forces in a liquid are significant only between polar molecules that are very close together.

TABLE 11.6 Molecular masses, dipole moments, and boiling points of several simple organic substances

Substance	Molecular mass (amu)	Dipole moment, μ (D)	Boiling point (K)
Propane, $CH_3CH_2CH_3$	44	0.0	231
Dimethyl ether, CH_3OCH_3	46	1.3	249
Methyl chloride, CH_3Cl	50	2.0	249
Acetaldehyde, CH_3CHO	44	2.7	293
Acetonitrile, CH_3CN	41	3.9	355

Table 11.4 contains the masses, normal boiling points, and enthalpies of vaporization of several substances composed either of atoms or simple, nonpolar molecules. The fact that these substances can be liquefied, some of them at quite high absolute temperatures, tells us that there must be attractive interactions between the molecules. Yet none of the various types of attractive forces we have discussed can apply here. The solution to this longstanding puzzle was provided in 1930 by Fritz London, who applied the new theory of quantum mechanics to the problem. London's contribution was essentially to distinguish what we see on the average from what we would see if we could freeze the charge distribution in a collection of molecules at any particular instant. In a collection of helium atoms, the average distribution of electronic charge about each nucleus is spherically symmetrical. But the electrons are in constant motion. Because each electron constantly experiences repulsive interactions with other electrons on the same atom, *and on other adjacent atoms,* the motion of any one electron is at least partly determined by the motions of all its near neighbors. Suppose it happens that at some instant the electrons in a given helium atom are slightly displaced, so that the atom possesses an instantaneous dipole moment. This instantaneous dipole moment would induce a similar dipole moment on an adjacent atom, because of the somewhat synchronized motions of the electrons, as illustrated in Figure 11.21. The result is an attractive interaction between the two atoms, called the London dispersion force.

London's analysis showed that dispersion forces between two molecules should vary as $1/d^6$, where, as usual, d is the distance between the molecular centers (Figure 11.19). Thus, this force is significant only when the molecules are close together.

The strength of the attractive dispersion forces depends on how easily the electron cloud is distorted or polarized. In general, the larger the molecule, the farther its electrons are from the nuclei, and consequently the greater its polarizability. Therefore, the magnitude of the London dispersion forces increases with increase in molecular size. Because molecular size and mass generally parallel each other, it is often suggested that dispersion forces become more important with increasing molecular

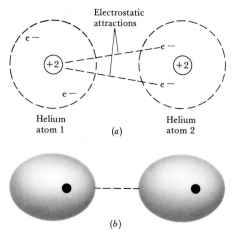

Electrostatic
attractions

Helium
atom 1

(a)

Helium
atom 2

(b)

FIGURE 11.21 Two schematic representations of an instantaneous dipole on two adjacent helium atoms, showing the electrostatic attraction between them.

mass. The truth of this generalization is evident in Table 11.4; the boiling points and enthalpies of vaporization of the noble gas elements increase steadily with increase in atomic mass. A similar trend is evident among the diatomic molecules. However, we must expect that the shapes of the molecules involved will also play a role. Note, for example, that although O_2 has a slightly lower mass than Ar, it has a slightly higher boiling point and enthalpy of vaporization. As another example, n-pentane and neopentane, illustrated in Figure 11.22, have the same molecular formula, C_5H_{12}, yet the boiling point of n-pentane* is 27°C higher than that of neopentane. The difference can be traced to the different shapes of the two molecules. The overall attraction between molecules is greater in the case of n-pentane because there are more sites of interaction; the molecules are able to come in contact over the entire length of the molecule. In the case of neopentane, less contact is possible between molecules.

GENERAL COMPARISONS

Figure 11.19 summarizes the various intermolecular forces we have discussed. The interactions between neutral particles (consisting mainly of the dipole-dipole and dispersion forces) are often referred to in general as van der Waals forces. We have not yet said very much about the relative magnitudes of these different attractive interactions. It is difficult to make generalizations, because the relative magnitudes of the various contributions to the total attractive forces vary greatly from one substance to another. In the noble gases, only the dispersion forces contribute. In HCl, on the other hand, both dipole-dipole and dispersion forces can play a role. In this particular case it turns out that the dispersion-force contribution is about five times more important than the dipole-dipole contribution. On the other hand, in H_2O, where the molecular dipole moment is larger (1.87 D for H_2O versus 1.03 D for HCl), and the molecule is less polarizable, the dipole-dipole contribution is four times larger than the dispersion-force contribution.

*The n in n-pentane is an abbreviation for the word "normal." A normal hydrocarbon is one whose carbon atoms are arranged in a straight chain.

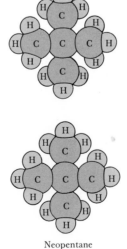

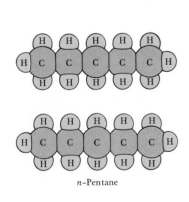

n-Pentane

Neopentane

FIGURE 11.22 Illustration of the effect of molecular shape on intermolecular attraction. The boiling point of n-pentane is 36.2°C, whereas that of neopentane is 9.5°C.

The dipole moment of HCl is 1.03 D, that of HCN is 2.98 D. Predict the relative importance of dipole-dipole and dispersion-force contributions to the attractive intermolecular forces in HCN.

Solution: In our discussion of dipole-dipole forces we saw that the dipole-dipole interaction varies as μ^4/d^6. We can assume that HCl and HCN are about the same size, so that d should be nearly constant. Because the dipole moment for HCN is about 2.9 times larger that that for HCl, we expect that the dipole-dipole interaction should be about $(2.9)^4$ or about 70 times larger. On the other hand, the dispersion-force interaction should be about the same for the two substances. (HCl is more massive, but the $C\equiv N$ triple bond in HCN is more polarizable than single bonds.) We learned above that the dispersion-force contribution in HCl is about five times larger than the dipole-dipole. Because the dipole-dipole contribution is predicted to be about 70 times larger in HCN than in HCl, we expect that the dipole-dipole contribution will be about 10 to 15 times larger than the dispersion-force contribution to the intermolecular attractive energy in HCN.

Which of the following substances is most likely to exist as a gas at room temperature and normal atmospheric pressures: P_4O_{10}, Cl_2, AgCl, or I_2?

Solution: In essence, the question asks which substance has the weakest intermolecular attractive forces, because the weaker these forces, the more likely the substance is to exist as a gas at any given temperature and pressure. We should therefore select Cl_2, because this is both a nonpolar molecule and also has the lowest molecular weight. In fact Cl_2 does exist as a gas at room temperature and normal atmospheric pressure, whereas the others are solids. Of the other substances, AgCl is least likely to be a gas because it exists as Ag^+ and Cl^- ions with very strong ionic bonds holding the ions within the solid.

11.6 Hydrogen bonding Our analysis of intermolecular attractive forces would lead us to predict that in a series of molecules of similar shape and polarity, melting and boiling points should increase with increasing molecular mass. Figure 11.23 is a graph of the boiling points of the simple hydrides of the group 4A and 6A elements as a function of molecular mass. For the group 4A hydrides there is a general increase in boiling point with increasing mass as expected. The same behavior is followed by the group 6A hydrides, with the notable exception of H_2O. Clearly, the boiling point of this substance is much higher than would be predicted on the basis of its molecular mass. A similar abnormally high boiling point is observed when the boiling point of NH_3 is compared with those of the other group 5A element hydrides, or when HF is compared with the hydrides of the other group 7A elements.

Water has many other unusual characteristics that distinguish it from other substances of comparable molecular mass and polarity. Among these are its high melting point, high heat capacity, high heat of vaporization, and exceptional ability to dissolve ionic substances. The origin of these unusual properties is a special type of intermolecular interaction called hydrogen bonding. This type of intermolecular attractive force is most important in substances in which a hydrogen atom is bonded to nitrogen, oxygen, or fluorine. The electronegativity (EN) of hydrogen is 2.2, much less than the electronegativity of nitrogen (EN = 3.0), oxygen (EN = 3.4), or fluorine (EN = 4.0). As a result, the bond between hydrogen and any of these three elements is quite polar, with hydrogen at

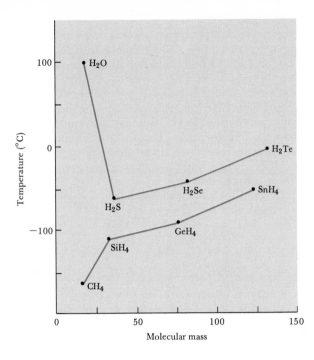

FIGURE 11.23 Boiling points of the group 4A and 6A hydrides as a function of molecular mass.

the positive end. We can think of each bond as having a bond dipole moment. The bond dipole moments have been estimated for these three bonds as follows:

Bond	N—H	O—H	F—H
	←—+	←—+	←—+
Dipole moment (D)	1.0	1.6	2.3

Each of these bond dipoles is capable of interacting with an unshared electron pair on the nitrogen, oxygen, or fluorine atom of an adjacent molecule. It is this electrostatic interaction between the X—H bond dipole of one molecule and the unshared electron pair of another molecule that we call hydrogen bonding.

It is possible to have hydrogen bonding arrangements of the kind X—H···Y for all the cases where X and Y are N, O, or F. (In this manner of depicting the hydrogen bond, the dotted line segment refers to the hydrogen bond between the X—H dipole and the unshared electron pair of Y.) Thus, it is possible to have hydrogen bonding arrangements of the following kinds:

F—H···F   F—H···O   F—H···N   O—H···F   O—H···O

O—H···N   N—H···F   N—H···O   N—H···N

As an example of the first of these arrangements, Figure 11.24 shows the structure of crystalline HF, in which the arrangement of the H—F molecules with respect to one another is determined by hydrogen bonding. Hydrogen bonding is responsible for the rather open structure for ice (Figure 11.25), which causes it to have a *lower* density than liquid water. By contrast, for most substances the solid phase is *more* dense than the liquid. This unusual property of water has some important conse-

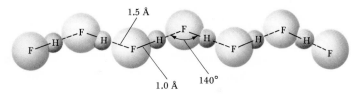

1.5 Å

1.0 Å 140°

FIGURE 11.24 The structure of crystalline HF showing the hydrogen bonding between HF molecules. The 140° H—F—H angle is probably determined in part by the directionality of the lone-pair electrons in HF and in part by crystal packing forces.

quences. First of all, it permits ice to float on water. When ice forms in cold weather, it covers the top of the water, thereby insulating the water below. If ice were more dense than water, ice forming at the top of a lake would fall to the bottom, and the lake could freeze solid. Most aquatic life could not survive under these circumstances. The expansion of water upon freezing is also what causes water pipes to break in freezing weather. The low density of ice compared to water can be understood in terms of hydrogen-bonding interactions between water molecules. The interactions in the liquid are random. However, when water freezes, the molecules assume the ordered arrangement shown in Figure 11.25. This structure, which extends in all directions in space, permits the maximum number of hydrogen-bonding interactions between the H_2O molecules. Because the structure has large hexagonal holes, ice is more open and less dense than the liquid. Application of pressure depresses the melting point of ice because the pressure causes the solid to revert to the more dense liquid.

As an illustration of the case in which X and Y may be different, we might have an F—H···O hydrogen bond in a mixture of HF and H_2O. In many organic substances, an O—H or N—H bond is hydrogen-bonded to an atom on an adjacent molecule that is quite different from itself. For example, we find that in proteins, the structure of the protein chain is determined by hydrogen bonding interactions of the form:

Hydrogen bond

$$\text{N—H} \text{------} : \overset{..}{\text{O}} = \text{C}$$

The energies of hydrogen bonds vary from about 4 or 5 kJ/mol to 25 kJ/mol or so. Thus, they are not more than a few percent of the energies of ordinary chemical bonds (see Table 7.3). Nevertheless, hydro-

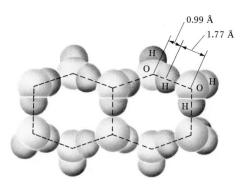

0.99 Å

1.77 Å

FIGURE 11.25 The arrangement of water molecules in ice. Each hydrogen atom on one water molecule is oriented toward a nonbonding pair of electrons on an adjacent water molecule. Distances between the centers of the bonded atoms are shown. In the full structure each oxygen atom is hydrogen bonded to two O—H groups.

gen bonding has very important consequences for the properties of many substances, particularly in biological systems. If the hydrogen bond is indeed the result of an electrostatic interaction between the X—H bond dipole and the unshared electron pair on Y, we would expect that the strength of hydrogen bonding should increase as the X—H bond dipole increases. Thus, if Y is the same, we would expect hydrogen bonding strengths to increase in the series

$$N—H\cdots Y < O—H\cdots Y < F—H\cdots Y$$

This is indeed found to be the case. But what property of Y is important? The atom Y must possess an unshared electron pair that attracts the positive end of the X—H dipole. The electron pair must not be too diffuse in space; if the electrons occupy too large a volume the X—H dipole does not experience a strong, directed attraction. For this reason, we find that hydrogen bonding is not very strong unless Y is one of the smaller atoms N, O, or F. Among these three elements, we find that hydrogen bonding is stronger when the electron pair is not attracted too strongly to its own nuclear center. The ionization energy of the electrons on Y is a good measure of this aspect. Thus, for example, the ionization energy of an unshared-pair electron on nitrogen in a covalent molecule is less than the corresponding value for oxygen. Nitrogen is thus a better donor of the electron pair to the X—H bond. For a given X—H, hydrogen-bond strength increases in the order

$$X—H\cdots F < X—H\cdots O < X—H\cdots N$$

When X and Y are the same, the energy of hydrogen bonding increases in the order

$$N—H\cdots N < O—H\cdots O < F—H\cdots F$$

When the Y atom carries a negative charge, the electron pair is especially able to form strong hydrogen bonds. The hydrogen bond in the $F—H\cdots F^-$ ion is among the strongest known; the reaction

$$F^-(g) + HF(g) \longrightarrow FHF^-(g)$$

has a ΔH of about $-155\ kJ/mol$.

SAMPLE EXERCISE 11.11

Arrange the following hydrogen bonds in the order of increasing strength: $O—H\cdots Cl$; $O—H\cdots N$; $N—H\cdots O$; $F—H\cdots O$.

Solution: The weakest of these hydrogen bonding interactions should occur in the first case, $O—H\cdots Cl$, because chlorine is a larger atom from the third row of the periodic table and should thus not be a good electron-pair donor in the hydrogen

bond. $O—H\cdots N$ and $F—H\cdots O$ should be about comparable in hydrogen-bond strength, because the larger F—H bond dipole is offset by the better donor character of nitrogen as compared with oxygen. Both of these should be stronger than $N—H\cdots O$ because of the smaller bond dipole of the N—H bond. Thus the order is

$$O—H\cdots Cl < N—H\cdots O < O—H\cdots N \sim F—H\cdots O$$

List the substances $BaCl_2$, H_2, CO, HF, and Ne in order of increasing boiling points.

Solution: The boiling point reflects the attractive forces in the liquid. These are stronger for ionic substances than for molecular ones, and so $BaCl_2$ has the highest boiling point. The intermolecular forces of the remaining substances depend on molecular weight, polarity, and hydrogen bonding. The other molecular weights are H_2 (2), CO (28), HF (20), and Ne (20). The boiling point of H_2 should be the lowest, because it is nonpolar and has the lowest molecular weight. The molecular weights of CO, HF, and Ne are roughly the same. HF has hydrogen bonding, and so it has the highest boiling point of the three. CO, which is slightly polar and has the highest molecular weight, is next. Ne, which is nonpolar, comes last of these three. The predicted boiling points are therefore

$$H_2 < Ne < CO < HF < BaCl_2$$

The actual normal boiling points are H_2 (20 K), Ne (27 K), CO (83 K), HF (293 K), and $BaCl_2$ (1813 K).

11.7 Bonding in solids

The properties of solids are determined not only by the arrangement of the particles in the lattice but also by the types of forces that exist between the particles. In Table 11.7, solids are organized according to the types of particles within the lattice and the types of forces that hold these particles together. It is important to study this table carefully because it summarizes a great deal of information about solids, relating atomic-level rearrangements and forces to the bulk properties of the crystal.

Solid argon and methane are examples of atomic solids and molecular solids, respectively. Because the interparticle forces are of the weak van der Waals type, such solids normally have relatively low melting points and exhibit trends in melting points that are similar to those discussed for the boiling points of molecular substances. Most substances that are gases or liquids at room temperature form molecular solids at low temperature.

TABLE 11.7 Crystal classifications

Crystal classification	Form of unit particles	Forces between particles	Properties	Examples
Atomic	Atoms	London dispersion forces	Soft, very low melting point, poor thermal and electrical conductors	Rare gases— Ar, Kr
Molecular	Polar or non-polar molecules	Van der Waals forces (London dispersion, dipole-dipole forces, hydrogen bonds)	Fairly soft, low to moderately high melting point, poor thermal and electrical conductors	Methane, CH_4; sugar, $C_{12}H_{22}O_{11}$; dry ice, CO_2
Ionic	Positive and negative ions	Electrostatic attraction	Hard and brittle, high melting point, poor thermal and electrical conductors	Typical salts—for example, NaCl, $Ca(NO_3)_2$
Covalent (network)	Atoms that are connected in covalent-bond network	Covalent bond	Very hard, very high melting point, poor thermal and electrical conductors	Diamond, C; quartz, SiO_2.
Metallic	Atoms	Metallic bond	Soft to very hard, low to very high melting point, excellent thermal and electrical conductors, malleable and ductile	All metallic elements—for example, Cu, Fe, Al, W

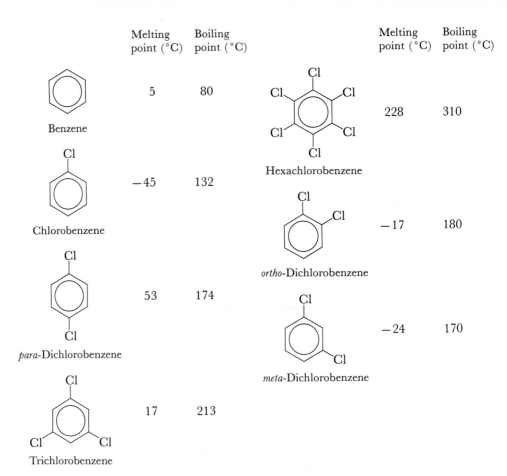

	Melting point (°C)	Boiling point (°C)		Melting point (°C)	Boiling point (°C)
Benzene	5	80	Hexachlorobenzene	228	310
Chlorobenzene	−45	132	ortho-Dichlorobenzene	−17	180
para-Dichlorobenzene	53	174	meta-Dichlorobenzene	−24	170
Trichlorobenzene	17	213			

FIGURE 11.26 Melting and boiling points for a series of chlorine-substituted benzenes.

The properties of molecular solids depend not only on the magnitudes of the forces that operate between molecules but also on the abilities of the molecular units to pack efficiently in three dimensions. For example, consider the melting and boiling points of several chlorine-substituted benzene molecules, as given in Figure 11.26. These are all planar molecules; the chlorine atoms have a larger radius than the carbon or hydrogen atoms that make up the benzene molecule. The attractive forces between molecules increase with an increase in molecular mass, as evidenced by the increase in boiling points with increasing number of chlorine atoms. The melting points, however, do not vary in the same smooth manner. For example, chlorobenzene melts at a much lower temperature than benzene itself, because it is a less symmetrical molecule and cannot pack as efficiently in the solid state. When all other factors are about equal, more symmetrical molecules melt at a higher temperature than do unsymmetrical ones.

Ionic solids have higher melting points than do atomic and molecular solids, because electrostatic bonds are stronger than van der Waals forces. Ionic solids are also harder and fracture when struck rather than simply deforming.

In covalent or network solids the lattice units are joined by covalent bonds. Such materials are consequently much stronger than molecular solids. Diamond, whose structure and bonding was discussed in Section

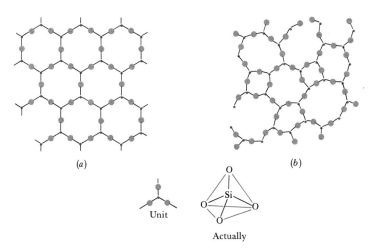

(a)

(b)

Unit

O
|
Si
O   O
|
O

Actually

FIGURE 11.27 A schematic comparison of (a) crystalline SiO_2 (quartz) and (b) amorphous SiO_2 (quartz glass). The small dots represent silicon atoms; the large ones represent oxygen atoms. The structure is actually three-dimensional and not planar as drawn. The unit shown as the basic building block (silicon and three oxygens) actually has four oxygens, the fourth coming out of the plane of the paper and capable of bonding to other silicon atoms.

8.7, is an example of this type of lattice (see Figure 8.30). Quartz (see Figure 11.27) is another example.

Bonding in **metals** differs from the bonding in other solids. Each atom in a metallic lattice typically has eight to twelve atoms adjacent to it, reflecting a close-packing arrangement. The bonding is too strong to be of the van der Waals type, and yet there are not enough valence electrons for ordinary covalent bonds between the atoms. We can visualize the valence electrons as delocalized, somewhat as outlined for the π electrons in graphite (Section 8.7). This model allows us to picture metals as composed of metal atoms held together by electrons that are distributed throughout all the spaces between the atoms. The electrons are free to move through the orbitals that extend over the entire metal. However, they maintain a uniform average distribution. Metals vary greatly in the strength of metallic bonding, as evidenced by such physical properties as melting and boiling points. For example, in platinum, with a melting point of 1770°C and a boiling point of 3824°C, the metallic bonding is very strong; in cesium, melting point 29°C and boiling point 678°C, the metallic bonding is comparatively weak. The properties and structures of metals will be examined more closely in Chapter 22.

AMORPHOUS SOLIDS

As noted in Section 11.4, not all solids have a regular arrangement of particles. For example, the material obtained when quartz, SiO_2, is melted and then rapidly cooled is amorphous. Quartz has a three-dimensional structure like that of diamond (Section 8.7). When quartz is melted (at approximately 1600°C) it becomes a viscous, tacky liquid. Although the silicon-oxygen network remains largely intact, many Si—O bonds are broken, and the orderliness of the quartz is lost. If the melt is rapidly cooled, the atoms are unable to return to their orderly arrangement, and an amorphous solid known as quartz glass or silica glass results. The lack of molecular-level regularity in this glass is shown in Figure 11.27, where its structure is compared with that of quartz.

Even when a solid lacks any long-range order, small regions of regularity known as crystallites may exist. Their existence permits description of these solids in terms of the degree of crystallinity. Such crystallites are often found, for example, in synthetic polymers, large molecules composed of many molecular parts fused together.

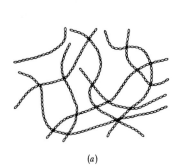

(a)

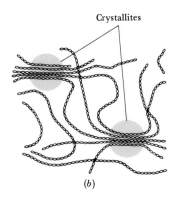

Crystallites

(b)

FIGURE 11.28 (a) An amorphous polymer compared with (b) a polymer containing regions of crystallinity, or crystallites.

The simplest synthetic polymer is polyethylene, a waxy-feeling substance used in packaging films, wire insulation, and molded articles. Polyethylene is formed by causing ethylene, C_2H_4, molecules to fuse together to form chains. Typically 700–2000 C_2H_4 molecules combine to form a chain containing 1400–4000 carbon atoms:

Ethylene   Polyethylene

The regular layering of these chains produces crystallites as shown in Figure 11.28.

The properties of a plastic* like polyethylene are determined by at least three factors: (1) the length of the polymer chain; (2) the degree of crystallinity; and (3) the extent of bonding between chains. As the length of a polymer chain increases, intermolecular attractive forces increase, thus making the polymer mechanically stronger and harder. Mechanical strength and hardness also increase as the degree of crystallinity increases. The regular arrangement of chains in crystallites permits closer approach of the molecules, thereby increasing intermolecular attractions.

*Although the term "plastic" has come to mean a certain type of synthetic material, the term is most precisely used to describe any material that changes shape when a force is exerted and maintains its distortion upon removal of the force. Materials like rubber that distort but return to their original shape when the force is removed are called elastomers.

TABLE 11.8   Properties of polyethylene as a function of crystallinity

	Degree of crystallinity (%)				
	55	62	70	77	85
Melting point (°C)	109	116	125	130	133
Density (g/cm³)	0.92	0.93	0.94	0.95	0.96
Stiffness[a]	25	47	75	120	165
Yield stress[a]	1700	2500	3300	4200	5100

[a]These tests reflect the increased mechanical strength of the polymer with increased crystallinity. The physical units for the stiffness test are psi $\times 10^{-3}$ (psi = pounds per square inch), while those for the yield stress test are psi. Discussion of the exact meaning and significance of these tests is beyond the scope of this text.

The effect of the degree of crystallinity on the properties of polyethylene can be seen in Table 11.8. The degree of crystallinity depends on the conditions of the polymerization.

To soften a plastic or make it more pliable a substance with low molecular weight known as a **plasticizer** may be added during the course of fabrication. The plasticizer molecules occupy positions between polymer strands, thereby interfering with intermolecular forces between chains and lowering the degree of crystallinity. The oily film that develops on the insides of the windows of new cars left standing in the sun is due in part to loss of plasticizers from the plastics in the cars' interiors. Continual loss of the plasti-

cizer leaves the plastics brittle, causing them eventually to crack.

In contrast to the weakening of bonds between chains by use of a plasticizer, the bonds can be strengthened by replacing the intermolecular bonds with covalent bonds between the chains. These bonds are known as cross-links. The process of vulcanizing rubber, which increases its rigidity, involves cross-linking of polymer chains.

11.8 Phase diagrams

We began this chapter with a consideration of phase changes. Let us conclude it by returning to that subject, to see how we might represent the equilibria that exist between the various phases as a function of temperature and pressure. First we should consider an experimental question: How do we measure the temperatures at which phase changes occur?

When a substance is heated, its temperature increases unless it is undergoing a phase change such as from solid to liquid. During the time a pure substance is melting or boiling its temperature remains constant. In Figure 11.29 the temperature of a sample of water (starting as ice) is plotted as a function of time as heat is added at a uniform rate. The resultant diagram is known as a heating curve.* We understand the increasing temperature as indicating that added energy is used to increase the average kinetic energy of the particles. At the melting and boiling points, temperature remains constant because the added energy is used to overcome attractive forces between particles; it increases the potential energy of the system.

The existence of a temperature plateau in the heating curve indicates that a phase change is occurring. If we wanted to learn how pressure

*The experiment might also be carried out by first heating a substance to the highest temperature of interest and then letting it lose heat to its surroundings at a constant rate. In this case the variation in temperature as a function of time is called a cooling curve.

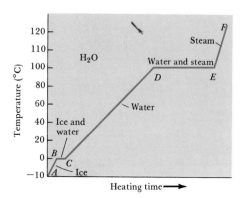

FIGURE 11.29 The heating curve for water with heat added at a constant rate. The rate of climb (slope) of the portions *AB, CD,* and *EF* depends on the heat capacities of ice (2.092 J/g-°C), water (4.184 J/g-°C), and water vapor (1.841 J/g-°C), respectively. The lengths of the level segments *BC* and *DE* are related to the heat of fusion (6.02 kJ/mol) and heat of vaporization (40.67 kJ/mol), respectively.

affected the phase changes for water, we would need to carry out the heating-curve experiment at various pressures and note how the temperatures of the plateaus changed with pressure. From such data we could then construct a **phase diagram,** in which the conditions for equilibria between the various phases of a substance are represented simultaneously on a single graph.

The phase diagram of water is shown in Figure 11.30. Those of other substances are qualitatively similar. The conditions of equilibrium exist only along the lines that separate the areas labeled solid, liquid, and gas. For example, the liquid and vapor phases of H_2O are in equilibrium at a water vapor pressure of 0.57 atm (433 mm Hg) and 85°C, point A. The line through points C and D represents the vapor pressure of liquid water, while the line through points D and E represents the vapor pressure of ice. The line through B and D represents the conditions of temperature and water-vapor pressure at which liquid water is in equilibrium with ice. As we noted in Section 11.5, the melting point of ice decreases with increasing pressure. Therefore, the line through points B and D slopes to the left as pressure increases. For most substances, the melting point increases with increasing temperature, and the line representing the solid-liquid equilibrium would slope to the right. Point D, which represents the single point on the diagram where all three phases are in equilibrium with each other, is known as the **triple point.** For water this point is at 0.0098°C and 0.006 atm (4.58 mm Hg). The areas between the equilibrium lines—the gas, liquid, and solid states—represent nonequilibrium conditions under which a single phase is present. For example, at 0.5 atm and 60°C, water exists entirely as a liquid. Point B represents the temperature at which ice melts when the external pressure is 1 atm; this is known as the normal melting point of the ice. Point C represents the normal boiling point of water.

To understand the utility of the diagram, consider what happens to water if its temperature is held at 85°C while the vapor pressure is increased, beginning at 0.5 atm. Figure 11.30 shows that water exists entirely as a gas at 0.5 atm and 85°C. When the pressure reaches 0.57 atm, point A, the water vapor condenses and an equilibrium between liquid and vapor is established. The water-vapor pressure cannot increase beyond this value. As we continue to exert pressure on the equilibrium mixture of vapor and liquid, the vapor is continuously converted

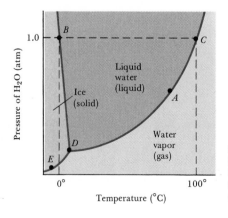

FIGURE 11.30 The phase diagram of H_2O showing the normal melting point, the normal boiling point, and the triple point. The axes are not drawn exactly to scale; the triple point should actually be almost on the horizontal axis, a fraction above 0°C.

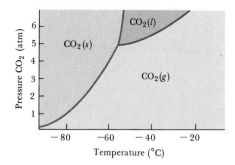

FIGURE 11.31 The phase diagram of CO_2.

to liquid until no vapor remains. Still higher pressures are then exerted on the pure liquid phase.

Consider now what happens to H_2O when the pressure is maintained at 0.57 atm and the temperature increases from $-10°C$ to $70°C$. At $-10°$ the H_2O exists as a solid. At just above $0°C$ ($0.005°C$) it melts, and a solid-liquid equilibrium exists. Above this temperature, the water exists entirely as a liquid until the temperature reaches $85°C$. At that temperature a liquid-gas equilibrium is established. Above $85°C$ the H_2O exists entirely as a gas.

SAMPLE EXERCISE 11.13

Referring to the phase diagram for CO_2 (Figure 11.31), determine the phase in which CO_2 exists at a CO_2 pressure of 3 atm and a temperature of $-80°C$.

Solution: The CO_2 exists as a solid under these conditions because the point falls in the area marked s. Because this point does not fall on a line, it does not represent an equilibrium condition.

SAMPLE EXERCISE 11.14

What is the lowest pressure at which solid CO_2 is able to melt? (Refer to Figure 11.31; the triple point of CO_2 is $-57°C$ and 5.2 atm.)

Solution: The lowest pressure is 5.2 atm; below this the CO_2 sublimes. It is for this reason that solid CO_2 (dry ice) is such a convenient coolant. It sublimes rather than melts as it absorbs sufficient energy at ordinary pressures.

In order for ice to sublime, its temperature must be below $0.0098°C$ (the triple-point temperature), and the water-vapor pressure must be below its equilibrium vapor pressure. These conditions are employed in the preparation of freeze-dried foods. The food is frozen and then introduced into a vacuum chamber. Water is thereby removed from the food through sublimation.

FOR REVIEW

Summary

In this chapter we have concerned ourselves with the properties of liquids and solids. We can understand the physical properties of matter in these states in terms of the kinetic-molecular theory, which we used earlier to explain the properties of gases. In a liquid the intermolecular forces keep molecules in close proximity, though they retain freedom to move with respect to one another. The free volume in a liquid is thus small, and liquids are not very compressible. The particles that make up a solid lattice are even more restrained than in a liquid; they occupy specific locations in a three-dimensional arrangement. Thus, solids possess long-range order and retain both their shape and volume. Any substance may exist in more than one state of matter, or phase. The equilibria between phases are dynamic; that is, there is continuous transfer of particles from one phase to the other. Equilibrium in such a dynamic system occurs when the rates of transfer between the phases are equal. The change of matter from one state of matter to another is termed a phase change. Conversions of a solid to a liquid (melting), solid to a gas (sublimation), or liquid to a gas (vaporization) are all endothermic processes; that is, the enthalpies of melting, sublimation, or vaporization are all positive. The reverse processes, conversion of a liquid to a solid (freezing), gas to a solid (deposition), or gas to a liquid (condensation) are all exothermic; thus, the enthalpy changes for these processes are all negative.

The vapor pressures of liquids increase nonlinearly with temperature. Boiling occurs when the vapor pressure equals the externally applied pressure. The normal boiling point is the temperature at which the vapor pressure of the liquid equals 1 atm.

Physical properties of liquids, such as viscosity and surface tension, are related to the intermolecular forces between molecules and to their sizes and shapes.

The structures of crystalline solids can be determined by observations of X-ray diffraction patterns. The Bragg law, $n\lambda = 2d \sin \theta$, expresses the angles at which constructive interference of the diffracted rays occurs, in terms of the distance d between planes in the crystal. The unit cell is the smallest portion of the space lattice that can, by simple displacement, reproduce the three-dimensional structure. In many solid structures the particles have a close-packing arrangement, in which spherical particles are arranged so as to leave the minimal amount of free volume. Two closely related forms of close packing, cubic and hexagonal, are possible. In both, each close-packed unit is in contact with 12 equivalent nearest neighbors.

The intermolecular forces that keep the particles of a liquid or solid together are essentially electrostatic in nature. They include ion-ion, ion-dipole, ion-induced dipole, dipole-dipole, and London disperson forces. The relative importance of each of these contributions to the intermolecular attractions depends on the character of each substance. The London dispersion forces increase with increasing molecular mass and complexity. Intermolecular forces are important in determining the critical temperature and critical pressure of a substance; the critical temperature increases with increasing intermolecular attractive forces. Hydrogen bonding is an important source of intermolecular attractions in compounds containing O—H, N—H, and F—H bonds, and in mixtures of substances containing these. The unusual properties of water, for example, its high boiling point and high heat of sublimation,

are due to extensive O—H⋯O hydrogen bonding in both the liquid and solid forms.

Solids may be classified according to the type of bonding between the units of the solid as atomic or molecular, ionic, covalent network, or metallic. The magnitude of the forces operating between the units of the lattice ranges from very small, as with the rare gases, to very large, as in diamond (a covalent network structure), MgO (ionic), or W (metallic). Amorphous solids are characterized by a lack of long-range order. Nevertheless, they are classed as solids rather than liquids because the particles are not free to move extensively with respect to one another. Polymers form a large and important class of amorphous solids.

The equilibria between the solid, liquid, and gas phases of a substance as a function of temperature and pressure are displayed on a phase diagram. Equilibrium between any two phases on such a diagram is indicated by a line. The point on such a diagram at which all three phases coexist in equilibrium is called the triple point.

Learning goals

Having read and studied this chapter, you should be able to:

1 Distinguish between gases, liquids, and solids on a molecular level.

2 Employ the kinetic-molecular theory and the concept of intermolecular attractions to explain the properties of each phase, such as surface tension, viscosity, vapor pressure, and boiling and melting points.

3 Explain the nature of the equilibria that may exist between phases. Account for the enthalpy changes that accompany phase changes.

4 Explain the relation between pressure, vapor pressure, temperature, and boiling point.

5 Explain the meaning of the terms critical temperature and critical pressure and account for the variation in critical temperatures of different substances in terms of intermolecular forces.

6 Explain the origin of the diffraction patterns obtained when X rays impinge on a crystal.

7 Use Bragg's law to calculate spacings in a crystal given the appropriate experimental data.

8 Determine the number of particles in a unit cell and use this to calculate the density of the substance.

9 Describe the various types of intermolecular attractive forces, indicate how each arises, and indicate the manner in which each varies with distance.

10 Predict, for any particular substance of known structure, which types of intermolecular forces may be operative and which particular type is of major importance.

11 Describe the nature of the hydrogen bond and distinguish those molecular systems in which hydrogen bonding is likely to be important.

12 Predict the type of solid (atomic, molecular, ionic, covalent network, or metallic) formed by a substance and predict its general properties.

13 Distinguish between crystalline and amorphous solids.

14 Draw a phase diagram of a substance, given appropriate data.

15 Calculate the heat absorbed or emitted when a given quantity of substance changes from one condition to another, given the needed heat capacities and enthalpy changes associated with phase changes.

16 Use a phase diagram to predict what phases are present at any given temperature and pressure.

Key terms

Among the more important terms and expressions used for the first time in this chapter are the following:

An amorphous solid (Section 11.4, 11.7) is a solid whose molecular arrangement lacks a regular and long-range pattern.

The boiling point (Section 11.3) of a liquid is the temperature at which its vapor pressure equals the external pressure. The normal boiling point is the temperature at which the liquid boils when the external pressure is 1 atm (that is, the temperature at which the vapor pressure of the liquid is 1 atm).

The Bragg law (Section 11.4) relates the angles at which X rays are scattered from a crystal to the spacing between the layers of particles.

Capillary action (Section 11.3) is the term used to describe the process by which a liquid rises in a tube because of a combination of adhesion with the walls of the tube and cohesion between liquid particles.

Close packing (Section 11.4) refers to the most efficient packing of spheres in a three-dimensional array. There are two closely similar forms, cubic close packing and hexagonal close packing.

Critical pressure (Section 11.3) is the pressure at which a gas at its critical temperature is converted to the liquid state.

Critical temperature (Section 11.3) is the highest temperature at which it is possible to convert the gaseous form of a substance to a liquid. The critical temperature increases with an increase in the magnitude of intermolecular forces.

A crystalline solid (Section 11.4) (or simply a crystal) is a solid whose internal arrangement of atoms, molecules, or ions shows a regular repetition in any direction through the solid.

A dynamic equilibrium (Section 11.2) is a state of balance in which opposing processes occur at the same rate.

Hydrogen bonds (Section 11.6) are intermolecular attractions between molecules containing hydrogen bonded to oxygen, nitrogen, or fluorine.

Intermolecular forces (Section 11.1, 11.5) are the short-range attractive forces operating between the particles that make up the units of a liquid or solid substance. These same forces also cause gases to liquefy or solidify at low temperatures.

London dispersion forces (Section 11.5) are intermolecular forces resulting from attractions between induced dipoles.

The melting point (Section 11.7) of a solid (or the freezing point of a liquid) is the temperature at which solid and liquid phases coexist in equilibrium. The normal melting point is the melting point at 1 atm pressure.

A meniscus (Section 11.3) is the curved upper surface of a liquid column.

A phase change (Section 11.2) represents the conversion of a substance from one state of matter to another. The phase changes we consider are melting and freezing (solid ↔ liquid), sublimation and deposition (solid ↔ gas), and vaporization and condensation (liquid ↔ gas).

A phase diagram (Section 11.8) is a graphic representation of the equilibria between the solid, liquid, and gaseous phases of a substance as a function of temperature and pressure.

Surface tension (Section 11.3) is the intermolecular, cohesive attraction that causes a liquid surface to become as small as possible.

The triple point (Section 11.8) of a substance is the temperature at which solid, liquid, and gas phases coexist in equilibrium.

A unit cell (Section 11.4) is the smallest portion of a crystal that reproduces the structure of the entire crystal when repeated in different directions in space. It is the repeating unit or "building block" of the crystal lattice.

The van der Waals forces are the attractive interactions that operate between neutral particles. They consist mainly of dipole-dipole and London dispersion forces.

Vaporization (Section 11.2), or evaporation, is a phase change in which a liquid is converted to a gas.

Vapor pressure (Section 11.2) is the pressure exerted by a vapor in equilibrium with its liquid or solid phase.

Viscosity (Section 11.3) is a measure of the resistance of fluids to flow.

X ray diffraction (Section 11.4) refers to the scattering of X rays by the units of a regular crystalline solid. The scattering patterns obtained can be used to deduce the arrangements of particles in the solid lattice.

EXERCISES

Kinetic-molecular theory of liquids and solids; phase changes

11.1 Suppose that a drop of liquid bromine is added to a 1-L vessel containing air and to another 1-L vessel containing water (bromine is soluble in water). Compare the relative rates at which the bromine would diffuse through the 1-L volume in each case. Explain in terms of the kinetic-molecular theory.

11.2 Pure acetic acid, $HC_2H_3O_2$, sometimes called glacial acetic acid, melts at 16.6°C. The liquid has a density of 1.05 g/cm^3, and the solid acid has a density of 1.09 g/cm^3. Account for the difference in densities.

11.3 As a general rule, liquids are slightly more compressible than the corresponding solid. Why is this so?

11.4 Explain why the heat of vaporization is positive.

11.5 Crystals containing a radioactive form of solid iodine are placed on one side of a closed container, as illustrated in Figure 11.32, and ordinary solid iodine is placed on the other side. After a few days the radioactive form is found in the solid iodine on both sides of the container. Explain how this experiment provides evidence for the dynamic character of equilibria between phases.

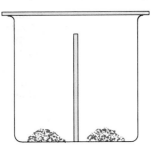

FIGURE 11.32

11.6 Ethyl chloride, C_2H_5Cl, has a normal boiling point of 12°C. When liquid ethyl chloride under pressure is sprayed onto a surface at atmospheric pressure, the surface is cooled considerably. Explain.

11.7 The enthalpy of melting of ice at 0°C is 6.01 kJ/mol, and the enthalpy of vaporization of water at 0°C is 44.86 kJ/mol. Calculate the enthalpy of sublimation of ice at 0°C.

11.8 The molar enthalpy of sublimation of a solid is always larger than the molar heat of vaporization of the corresponding liquid. Explain.

11.9 It requires 3.38 kJ/mol to heat a mole of solid CuO from 0°C to 80°C. There are no phase changes in this temperature range. In what form is the added heat stored in the CuO?

Properties of liquids

11.10 Explain how each of the following affects the vapor pressure of a liquid: (a) surface area; (b) temperature; (c) intermolecular attractive forces; (d) volume of liquid.

11.11 Explain why the boiling point of a substance is more pressure dependent than is its melting point.

11.12 Suppose the atmospheric pressure at a high mountain camp is 500 mm Hg. Determine from Figure 11.6 the temperature at which diethyl ether, ethanol, and water will boil.

11.13 Describe the molecular origin of viscosity in liquids. Indicate the factors that lead to high viscosity.

11.14 Would you expect a liquid to evaporate more rapidly into a vacuum or into air at the same temperature? Explain.

11.15 Account for the observed variation in viscosities of the series of straight chain hydrocarbons, as listed below:

Compound	Formula	Viscosity (N-sec/m^2)
Pentane	C_5H_{12}	0.225×10^{-3}
Hexane	C_6H_{14}	0.313×10^{-3}
Heptane	C_7H_{16}	0.397×10^{-3}
Octane	C_8H_{18}	0.546×10^{-3}

11.16 Describe some of the characteristics of a substance on the molecular level that might lead to high viscosity.

[11.17] Test the relationship expressed in Equation [11.1] using the data for water listed in Table 11.3. Is the equation applicable?

11.18 Indicate whether the following properties increase in magnitude, decrease in magnitude, or remain unaffected by an increase in the strength of intermolecular forces: (a) vapor pressure; (b) normal boiling point; (c) normal melting point; (d) surface tension; (e) viscosity; (f) heat of fusion; (g) heat of vaporization; (h) molecular weight.

11.19 The critical temperature and pressure of CO_2 are 31°C and 73 atm, respectively. For CS_2 the corresponding quantities are 279°C and 78 atm, respectively. Explain the differences in critical temperatures and pressures.

11.20 In a boiler used to generate steam for an electric power plant, water is heated to boiling as it passes through tubes in the fire chamber. Graph the following data for the vapor pressure versus temperature of water and estimate the temperature of the resulting steam if the pressure in the system is 172 atm.

Temperature (°C)	Vapor pressure (atm)
330	127
340	144
350	163
360	184
370	207

[11.21] Test the applicability of the Clausius-Clapeyron equation using the following vapor pressure versus temperature data for mercury:

Temperature (°C)	Vapor pressure of Hg (mm Hg)
50.0	0.01267
60.0	0.02524
70.0	0.04825
80.0	0.0880
90.0	0.1582

If the Clausius-Clapeyron equation is obeyed, use the slope of the line to calculate ΔH_v for mercury in this temperature range.

[11.22] We might guess that the Clausius-Clapeyron equation would be applicable also to the vapor pressure data for a solid. Use this equation to estimate the heat of sublimation of ice from the following data:

Temperature (°C)	Vapor pressure (mm Hg)
−20.0	0.640
−16.0	1.132
−12.0	1.632
−8.0	2.326
−4.0	3.280
0.0	4.579

Crystalline solids

11.23 Would you classify a "nonbreakable" plastic pocket comb as an amorphous or crystalline solid? Explain.

11.24 From which of the following materials would you expect to obtain well-defined X-ray diffraction patterns such as the one illustrated in Figure 11.9: (a) a sugar crystal; (b) KBr; (c) liquid water; (d) pure iron; (e) ice; (f) a section of a rubber stopper?

11.25 A crystal with spacings of 2.68 Å between planes is subjected to X-ray analysis, using X rays of wavelength 1.65 Å. Calculate the first three angles θ at which constructive interference occurs.

11.26 The first-order ($n = 1$) reflection of X rays of wavelength 1.68 Å is found at an angle of 8.7°. What is the distance between planes of particles in a solid that causes this diffraction?

11.27 CsBr has a body-centered cubic type of crystal structure. The density of CsBr at 20°C is 4.428 g/cm³. Calculate the length of the unit cell side and the distances between the centers of the Cs^+ and Br^- ions.

11.28 Krypton crystallizes in the face-centered cubic lattice in which the edge length of the cell is 5.59 Å. Calculate the density of solid krypton.

11.29 Potassium fluoride has the NaCl-type of lattice. The density of KF at 25° is 2.468 g/cm³. Calculate the dimensions of the KF unit cell, and the nearest-neighbor distance.

11.30 The circles in Figure 11.33 show a portion of a two-dimensional structure. What is the significance of the area bounded by the colored lines?

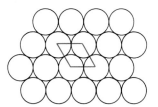

FIGURE 11.33

[11.31] If a substance is composed of atoms arranged in a simple cubic unit cell, determine the fraction of the volume of the unit cell that is occupied by atoms, assuming the atoms on each corner touch each other.

11.32 How many cannonballs are required to make a three-sided pyramidal stack on the courthouse lawn, if the pyramid is to have a base of four on each edge? What form of close packing does the pyramid represent?

11.33 Argon crystallizes in a cubic close-packed structure. The unit cell length is 5.25 Å. Calculate the effective radius of the argon atom in the structure. (Hint: Recall that the cubic close-packed arrangement corresponds to face-centered cubic.)

Intermolecular forces

11.34 Indicate some of the ways in which intermolecular forces differ from the *intra*molecular forces that hold atoms together in chemical combination.

11.35 Using the thermodynamic data listed in Appendix D, calculate the enthalpy changes in the following processes at 25°C:

$$Br_2(l) \longrightarrow Br_2(g)$$
$$Br_2(g) \longrightarrow 2Br(g)$$

Discuss the relative values of these enthalpies in terms of the forces involved in each case.

11.36 Indicate which of the following substances is most likely to exist as a crystalline solid at room temperature and which would be least readily liquefied under pressure: HF; PCl_3; Fe_2; $FeCl_2$; SO_2.

11.37 From the following list of substances indicate which is expected to have the highest boiling point and which the lowest: CO_2; Ar; CF_4; RbCl; SiF_4. Explain your answers.

11.38 What is the nature of the major attractive intermolecular force in each of the following: (a) $I_2(s)$; (b) $MgO(l)$; (c) $CH_3CN(l)$ (dipole moment 3.9 D); (d) $HF(l)$ (dipole moment 1.82 D).

11.39 What are the required structural features for a substantial hydrogen-bonding contribution to the intermolecular attractive forces?

11.40 List the following compounds in the expected order of increasing energy of the hydrogen bonding interaction between molecules: H_2S; CH_3NH_2; C_6H_5OH (phenol).

11.41 The boiling points of the halides of lithium are as follows: LiF, 1717°C; LiCl, 1383°C; LiBr, 1289°C; LiI, 1178°C. Is this the expected order? Explain in detail.

11.42 Make a graph of the boiling points listed in problem 11.41 versus the sum of the anion and cation radii (Table 7.2). The normal boiling point for LiOH is 1626°C. Using the graph, estimate an "effective" ionic radius for OH^- in molten LiOH.

11.43 The boiling points of the alkali metal chlorides vary as follows: LiCl, 1383°C; NaCl, 1465°C; KCl, 1437°C; RbCl, 1381°C; CsCl, 1324°C. What is the nature of the most important intermolecular force operating in these liquids? Are the boiling points in the expected order?

11.44 Indicate all the various types of intermolecular attractive forces that may operate in each of the following pure substances or mixtures: (a) $CH_3OH—H_2O(l)$; (b) $Xe(l)$; (c) $C_6H_6(s)$; (d) $ClF(l)$; (e) $Ca(NO_3)_2(l)$. In each case indicate if you can the type of attractive force that makes the major contribution.

11.45 Predict the order of increasing viscosity among the following liquids, all at a common temperature: (a) propanol, $CH_3CH_2CH_2OH$; (b) propane, $CH_3CH_2CH_3$; (c) propane 1,3-diol, $HOCH_2CH_2CH_2OH$. Explain your reasoning.

11.46 The boiling points of the fluorides of second-row elements of the periodic table are: LiF, 1717°C; BeF_2, 1175°C; BF_3, −101°C; CF_4, −128°C; NF_3, −120°C; OF_2, −145°C; F_2, −188°C. Account for this variation in terms of the nature and strengths of the intermolecular forces that operate.

11.47 In dichloromethane, CH_2Cl_2, dipole moment 1.60 D, the dispersion-force contribution to the intermolecular attractive forces is about five times larger than the dipole-dipole contribution. How would you expect the relative importance of the two kinds of intermolecular attractive force to vary in dibromomethane, dipole moment 1.43 D? In difluoromethane, dipole moment 1.93 D?

11.48 The critical temperatures for the hydrogen halides vary as follows: HF, 188°C; HCl, 51°C; HBr, 90°C; HI, 151°C. Explain this observed variation in terms of the intermolecular attractive forces operative in each case.

Bonding in solids

11.49 Compare the following group of substances with respect to hardness electrical conductivity, and melting point: $BaCl_2$, Ni, SCl_2, C (diamond).

11.50 For each of the following pairs of substances, predict which will have the higher melting point and indicate why: (a) $CuBr_2$, Br_2; (b) CO_2, SiO_2; (c) S, Cr; (d) CsBr, CaF_2.

11.51 What is the simplest formula of the following polymer?

Would you expect this substance to form a crystalline or amorphous solid? Briefly justify your answer.

[11.52] The structure of zinc blende, ZnS, is shown in Figure 7.2 as an example of an ionic structure. This same compound is, however, often listed as an example of a covalent network structure. What is the distinction between these two types of structure? What properties of the substance itself, or of the elements involved in the structure, would you look up to try to make a decision as to which description is most appropriate in a given case?

11.53 Why does hexachlorobenzene, Figure 11.26, have higher melting and boiling points than benzene?

11.54 On the basis of the data presented in Figure 11.26, predict the melting and boiling points of pentachlorobenzene (five chlorines) and the tetrachlorobenzene (four chlorines) compound with the most symmetrical structure (show which structure you choose). About which of the two predicted quantities for each compound do you feel the more confident? Why?

Phase diagrams

11.55 Two pans containing water are on two adjacent burners on a stove. One is boiling gently, the other contains a mixture of water and ice. When both are rapidly stirred, the temperature in each is constant as a function of time. If the pressure is 1 atm, what is the temperature in each pan? Why is it constant as a function of time?

11.56 Plateaus in heating curves represent temperatures at which phase changes occur. When heat is added to liquid benzene at 1 atm pressure, 80°C, the temperature remains constant so long as liquid benzene is present. What is happening to the heat energy added at this temperature?

11.57 Roughly sketch a heating curve for propanol, $CH_3CH_2CH_2OH$, given the following data for this substance: normal melting point, −127°C; normal boiling point, 97°C; heat of fusion, 5.18 kJ/mol; heat of vaporization, 41.7 kJ/mol; heat capacity of $CH_3CH_2CH_2OH(s)$,

142 J/mol-°C; heat capacity of the liquid, 170 J/mol-°C; heat capacity of the gas, 108 J/mol-°C.

11.58 How much energy is required to convert 26.0 g of ice at −20.0°C to steam at 120°C (see caption to Figure 11.29)?

11.59 The normal melting and boiling points of xenon are −112 and −108°C, respectively, and its triple point is at −121°C, at a pressure of 282 mm Hg. Sketch the phase diagram for xenon, showing the three points given above and indicating the areas in which each phase is stable.

Additional exercises

11.60 The molar heats of sublimation of benzene, C_6H_6, and the monohalobenzenes, C_6H_5X, are as follows:

Compound	Formula	ΔH_s (kJ/mol)
Benzene	C_6H_6	33.85
Fluorobenzene	C_6H_5F	34.61
Chlorobenzene	C_6H_5Cl	41.04
Bromobenzene	C_6H_5Br	44.43
Iodobenzene	C_6H_5I	49.58

Account for this variation in terms of the nature of the intermolecular forces that are likely to be most important.

11.61 Calculate the net number of particles in the face-centered cubic and body-centered cubic unit cells.

11.62 What X-ray wavelength would be defracted at an angle of 7.60° if the interplanar spacing is 2.10 Å?

11.63 The enthalpy of vaporization of water at 0°C is 44.86 kJ/mol, whereas at 100°C the enthalpy of vaporization of water is 40.65 kJ/mol. Account for the fact that water has a lower enthalpy of vaporization at the higher temperature.

[11.64] From the following data for the vapor pressure of $NH_3(l)$ versus temperature, use the Clausius-Clapeyron equation to estimate the molar heat of vaporization of $NH_3(l)$.

Temperature (°C)	Vapor pressure of $NH_3(l)$ (atm)
−60.0	0.2161
−54.0	0.3167
−50.0	0.4034
−44.0	0.5693
−40.0	0.7083
−34.0	0.9676

Note that in working this problem you do *not* need to change the units of pressure. Why is this so?

[11.65] The critical temperature and pressure of $CClF_3$ (Freon 13) are 29°C and 39 atm, respectively. For $CClH_3$ (methyl chloride) the corresponding values are 143°C and 66 atm. What do these values tell us about the relative intermolecular attractive forces between molecules in the two substances? Are the results surprising? If so, why?

11.66 For ethanol, the values of a and b appropriate for Equation [11.1] are a, 24.05×10^{-3} J/m^2, and b, 0.0832×10^{-3} J/m^2-°C. Calculate the surface tension of ethanol at 28°C; at 60°C. Account for the variation in surface tension with temperature in terms of the kinetic-molecular theory.

11.67 Explain the following observed temperature variation in viscosity of glycerol, $HOCH_2CH_2(OH)CH_2OH$: 20°C, 1.49 N-sec/m^2; 25°C, 0.942 N-sec/m^2; 30°C, 0.622 N-sec/m^2.

11.68 Draw a sketch of the unit cell for the two-dimensional array shown in Figure 11.34, assuming that it is extended in both directions.

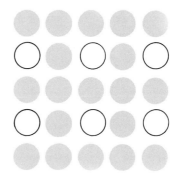

FIGURE 11.34

11.69 Determine the number of formula units present in each of the lattice cells shown in Figure 7.2.

11.70 A crystal has a simple cubic lattice with a unit cell length of 4.06 Å. Using X rays of 1.78 Å, calculate the first three diffraction angles at which the X rays are diffracted by the planes that parallel the unit cell faces.

11.71 A clean, very dry silica or glass surface has many Si—OH groups on it. Describe the nature of the adhesive forces that might operate between such a surface and CH_3OH.

11.72 Hydrogen peroxide, H_2O_2, melts at −0.4°C and boils at 151°C. In comparison with water, do these data suggest that hydrogen bonding might be important? Draw a diagram that shows the nature of the hydrogen-bonding interactions that could occur between molecules in this substance.

11.73 Explain the phenomenon of critical temperature in terms of the kinetic-molecular theory.

[11.74] The normal boiling points and molar heats of vaporization at the boiling points vary among the hydrogen halides as follows:

Compound	Boiling point (K)	ΔH_v(kJ/mol)
HF	292	7.5
HCl	188	16.1
HBr	207	17.6
HI	238	19.7

Account for the variation in boiling point among the compounds. In light of your explanation for the variation in boiling points, how can you account for the abnormally low heat of vaporization for HF? (Hint: The density of HF vapor just above the boiling point indicates a molecular weight much greater than the formula weight of 20.)

11.75 The critical temperatures of the boron trihalides vary as follows: BF_3, $-12°C$; BCl_3, $179°C$; BBr_3, $300°C$. Explain this variation in terms of the nature of the intermolecular attractive forces in these systems.

11.76 Irradiation of polyethylene with X rays introduces cross-linking into the material. What effect would this have on the properties of the polyethylene?

11.77 Consider the phase diagram in Figure 11.35. State which phase or phases are present at each point represented by a letter.

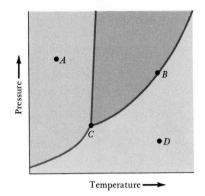

FIGURE 11.35

Solutions

Very few of the materials that we encounter in everyday life are pure substances; most are mixtures. Many of these mixtures are homogeneous; that is, their components are uniformly intermingled on a molecular level. We have seen that homogeneous mixtures are called solutions (Sections 2.2 and 3.9). Examples of solutions abound in the world around us. The air we breathe is a homogeneous mixture of several gaseous substances. The familiar metal brass is a solution of zinc in copper. The oceans are a solution of many dissolved substances in water. The fluids that run through our bodies are solutions, carrying a great variety of essential nutrients, salts, and so forth.

Solutions may be gaseous, liquid, or solid; examples of each kind are given in Table 12.1. Recall that the solvent is the component whose phase is retained when the solution forms (Section 3.11); if all components are in the same phase, the one in greatest amount is called the solvent. Other components are called solutes. Liquid solutions are the most common, and it is on this type of solution that we focus our attention in this chapter.

In Chapter 11 we considered the various types of intermolecular forces that exist between molecular and ionic particles. In this chapter we will see that these forces are also involved in the interactions between solutes and solvents. We will examine the solution process, the factors that determine the amount of solute that can dissolve in a given quantity of sol-

TABLE 12.1 Examples of solutions

State of solution	State of solvent	State of solute	Example
Gas	Gas	Gas	Air
Liquid	Liquid	Gas	Oxygen in water
Liquid	Liquid	Liquid	Alcohol in water
Liquid	Liquid	Solid	Salt in water
Solid	Solid	Gas	Hydrogen in platinum
Solid	Solid	Liquid	Mercury in silver
Solid	Solid	Solid	Silver in gold (certain alloys)

vent, and some properties of the solutions that result. We will be particularly concerned with aqueous solutions of ionic substances, because of their central importance in chemistry and in our daily lives. Near the end of the chapter we will consider a type of mixture, known as a colloid, that is on the borderline between heterogeneous mixtures and solutions. Before we examine these topics, however, it is useful to discuss ways of describing the concentrations of solutions, that is, the amount of solute dissolved in a given quantity of solvent or solution.

12.1 Ways of expressing concentration

The concentration of a solution can be expressed either qualitatively or quantitatively. The terms dilute and concentrated are used to describe a solution qualitatively. A solution with a relatively small concentration of solute is said to be dilute; one with a large concentration is said to be concentrated.

Several quantitative expressions of concentration are employed in chemistry. We discussed one of these, molarity (M), in Section 3.9. We defined molarity as the number of moles of solute in a liter of solution, Equation [12.1]:

$$\text{Molarity} = \frac{\text{moles solute}}{\text{volume of soln in liters}} \qquad [12.1]$$

One of the simplest ways of expressing concentration is in terms of the weight percentage. The weight percentage of a component of a solution is given by Equation [12.2]. If a solution of hydrochloric acid contains 36 percent HCl by weight, then it has 36 g of HCl for each 100 g of solution:

$$\text{Wt \% of component} = \frac{\text{mass of component in soln}}{\text{total mass of soln}} \times 100 \qquad [12.2]$$

SAMPLE EXERCISE 12.1

A solution is made containing 6.9 g of $NaHCO_3$ per 100 g of water. What is the weight percentage of solute in this solution?

Solution:

$$\text{Wt \% of solute} = \frac{\text{mass solute}}{\text{mass soln}} \times 100$$

$$= \frac{6.9\ \text{g}}{6.9\ \text{g} + 100\ \text{g}} \times 100 = 6.5\%$$

Notice that the mass of solution is the sum of the mass of solvent and the mass of solute. The weight percentage of solvent in this solution is $(100 - 6.5)\% = 93.5\%$.

The mole fraction of a component of a solution is given by Equation [12.3].

$$\text{Mole fraction of component} = \frac{\text{moles component}}{\text{total moles all components}} \qquad [12.3]$$

This concept was used in Chapter 9 when we discussed Dalton's law of partial pressures. The symbol X is commonly used for mole fraction, with a subscript to indicate the component on which attention is being focused. For example, the mole fraction of HCl in a hydrochloric acid solution can be represented as X_{HCl}. The mole fractions of all components of a solution will total 1.

SAMPLE EXERCISE 12.2

Calculate the mole fraction of HCl in a solution of hydrochloric acid containing 36 percent HCl by weight.

Solution: Assume that there are 100 g of solution (you can verify for yourself that assuming any other quantity will not change the result, though it can make the arithmetic more difficult). The solution therefore contains 36 g of HCl and 64 g of H_2O.

$$\text{Moles HCl} = (36 \text{ g HCl})\left(\frac{1 \text{ mol HCl}}{36.5 \text{ g HCl}}\right)$$
$$= 0.99 \text{ mol HCl}$$

$$\text{Moles } H_2O = (64 \text{ g } H_2O)\left(\frac{1 \text{ mol } H_2O}{18 \text{ g } H_2O}\right)$$
$$= 3.6 \text{ mol } H_2O$$

$$X_{HCl} = \frac{\text{moles HCl}}{\text{moles } H_2O + \text{moles HCl}}$$

$$= \frac{0.99}{3.6 + 0.99} = \frac{0.99}{4.6} = 0.22$$

The molality (m) of a solution is defined as the number of moles of solute in a kilogram of solvent:

$$\text{Molality} = \frac{\text{moles solute}}{\text{mass of solvent in kilograms}} \qquad [12.4]$$

Notice the difference between molarity and molality. These two ways of expressing concentration are similar enough to be easily confused. Molality is defined in terms of the mass of solvent, whereas molarity is defined in terms of the volume of solution. A 1.50 molal (written 1.50 m) solution contains 1.50 moles of solute for every kilogram of solvent. (When water is the solvent, the molality and molarity of a dilute solution are numerically about the same, because 1 kg of solvent is nearly the same as 1 kg of solution, and 1 kg of the solution has a volume of about 1 L.)

SAMPLE EXERCISE 12.3

What is the molality of a solution made by dissolving 5.0 g of toluene (C_7H_8) in 225 g of benzene (C_6H_6)?

Solution:

$$\text{Molality} = \frac{\text{moles } C_7H_8}{\text{kilograms } C_6H_6} =$$

$$\left(\frac{5.0 \text{ g } C_7H_8}{225 \text{ g } C_6H_6}\right)\left(\frac{1 \text{ mol } C_7H_8}{92.0 \text{ g } C_7H_8}\right)\left(\frac{1000 \text{ g } C_6H_6}{1 \text{ kg } C_6H_6}\right)$$

$\uparrow$ Convert grams C_7H_8 to moles

$\uparrow$ Convert grams C_6H_6 to kilograms

$$= 0.24 \text{ } m$$

For most purposes in ordinary chemical laboratory work molarity is the most useful expression of concentration. However, the other ways we have just considered find use in special situations, and you should be familiar with them. Sample Exercise 12.4 shows by way of example how the various expressions of concentration are related.

SAMPLE EXERCISE 12.4

Given that the density of a solution of 5.0 g of toluene and 225 g of benzene (Sample Exercise 12.3) is 0.876 g/mL, calculate the concentration of the solution in (a) molarity, (b) mole fraction of solute, and (c) weight percentage of solute.

Solution: (a) The total mass of the solution is equal to the mass of the solvent plus the mass of the solute:

Mass soln = 5.0 g + 225 g = 230 g

The density of the solution is used to convert the mass of the solution to its volume:

$$\text{Milliliters soln} = (230 \text{ g})\left(\frac{1 \text{ mL}}{0.876 \text{ g}}\right) = 263 \text{ mL}$$

Density must be known in order to interconvert molarity and molality, because one is on a mass basis whereas the other is on a volume basis. The number of moles of solute must be known to calculate either molarity or molality:

$$\text{Moles C}_7\text{H}_8 = (5.0 \text{ g C}_7\text{H}_8)\left(\frac{1 \text{ mol C}_7\text{H}_8}{92.0 \text{ g C}_7\text{H}_8}\right)$$

$$= 0.054 \text{ mol}$$

Molarity is moles of solute per liter of solution:

$$\text{Molarity} = \frac{\text{moles C}_7\text{H}_8}{\text{liter soln}}$$

$$= \left(\frac{0.054 \text{ mol C}_7\text{H}_8}{263 \text{ mL soln}}\right)\left(\frac{1000 \text{ mL soln}}{1 \text{ L soln}}\right)$$

$$= 0.21 \ M$$

Compare this value with the molality of the same solution calculated in Sample Exercise 12.3.

(b) The mole fraction of solute is expressed as:

$$X_{\text{C}_7\text{H}_8} = \frac{\text{moles C}_7\text{H}_8}{\text{moles C}_7\text{H}_8 + \text{moles C}_6\text{H}_6}$$

We have already calculated the number of moles of C_7H_8.

$$\text{Moles C}_6\text{H}_6 = (225 \text{ g C}_6\text{H}_6)\left(\frac{1 \text{ mol C}_6\text{H}_6}{78.0 \text{ g C}_6\text{H}_6}\right)$$

$$= 2.88 \text{ mol}$$

$$X_{\text{C}_7\text{H}_8} = \frac{0.054 \text{ mol}}{0.054 \text{ mol} + 2.88 \text{ mol}} = \frac{0.054}{2.93}$$

$$= 0.018$$

(c) The weight percentage of solute is calculated as follows:

$$\text{Wt \% C}_7\text{H}_8 = \frac{5.0 \text{ g C}_7\text{H}_8}{5.0 \text{ g C}_7\text{H}_8 + 225 \text{ g C}_6\text{H}_6} \times 100$$

$$= \frac{5.0 \text{ g}}{230 \text{ g}} \times 100 = 2.2\%$$

The concentration units that we have just discussed are the ones that we will use through the rest of this chapter and elsewhere in this text. However, one further concentration expression, called normality, may be encountered in other places. Normality (abbreviated N) is defined as the number of equivalents of solute per liter of solution. An equivalent is defined according to the type of reaction being examined. For acid-base reactions, an equivalent of an acid is the quantity that supplies 1 mol of H^+; an equivalent of a base is the quantity reacting with 1 mol of H^+. In an oxidation-reduction reaction, an equivalent is the quantity of substance that gains or loses 1 mol of electrons. The masses of 1 equivalent of several substances are given in Table 12.2. An equivalent is always defined in such a way that 1 equivalent of reagent A will react with 1

TABLE 12.2 Equivalent-mass relationships

Reactant	Product	Reaction type	Mass of 1 mol of reactant (g)	Mass of 1 equivalent of reactant (g)
$KMnO_4$	Mn^{2+}	Reduction (5 e⁻)	158.0	$158.0/5 = 31.6$
$KMnO_4$	MnO_2	Reduction (3 e⁻)	158.0	$158.0/3 = 52.7$
$Na_2C_2O_4$	CO_2	Oxidation (2 e⁻)	134.0	$134.0/2 = 67.0$
H_2SO_4	SO_4^{2-}	Acid (2 H⁺)	98.0	$98.0/2 = 49.0$
$Al(OH)_3$	Al^{3+}	Base (3 OH⁻)	78.0	$78.0/3 = 26.0$

equivalent of reagent B. For example, in an oxidation-reduction reaction, 31.6 g of $KMnO_4$ is stoichiometrically equivalent to 67.0 g of $Na_2C_2O_4$ (refer to Table 12.2). Likewise, in an acid-base reaction, 49.0 g of H_2SO_4 is stoichiometrically equivalent to 26.0 g of $Al(OH)_3$.

Where $KMnO_4$ is reduced to Mn^{2+}, thereby gaining five electrons, we have

1 mol $KMnO_4$ = 5 equivalents of $KMnO_4$

Therefore, if 1 mol of $KMnO_4$ is dissolved in sufficient water to form 1 L of solution, the concentration of the solution can be expressed as either 1 M or 5 N. Normality is always a whole-number multiple of molarity. In oxidation-reduction reactions, the whole number is the number of electrons gained or lost by one formula unit of the substance. In acid-base reactions the whole number is the number of H⁺ or OH⁻ available in a formula unit of the substance.

SAMPLE EXERCISE 12.5

What is the molarity and normality of a solution of H_2SO_4 made by dissolving 5.00 g of H_2SO_4 in enough water to make 200 mL of solution? The H_2SO_4 is used as an acid, forming SO_4^{2-}.

Solution: The molecular weight of H_2SO_4 is 98.0 amu. The molarity is

Molarity =
$$\left(\frac{5.00 \text{ g } H_2SO_4}{200 \text{ mL soln}}\right)\left(\frac{1000 \text{ mL soln}}{1 \text{ L soln}}\right)\left(\frac{1 \text{ mol } H_2SO_4}{98.0 \text{ g } H_2SO_4}\right)$$

$$= 0.255 \frac{\text{mol } H_2SO_4}{\text{L soln}} = 0.255 \text{ } M$$

Because H_2SO_4 has two available hydrogen ions, there are 2 chemical equivalents (equiv) in a mole and the normality is twice the molarity, 0.510 N:

$$\text{Normality} = \left(\frac{5.00 \text{ g } H_2SO_4}{200 \text{ mL soln}}\right)\left(\frac{1000 \text{ mL soln}}{1 \text{ L soln}}\right)$$
$$\times \left(\frac{2 \text{ equiv } H_2SO_4}{1 \text{ mol } H_2SO_4}\right)\left(\frac{1 \text{ mol } H_2SO_4}{98.0 \text{ g } H_2SO_4}\right)$$
$$= 0.510 \frac{\text{equiv } H_2SO_4}{\text{L soln}} = 0.510 \text{ } N$$

12.2 The solution process

A solution is formed when one substance disperses uniformly throughout another. With the exception of gas mixtures, all solutions involve substances in a condensed phase. We learned in Chapter 11 that substances in the liquid and solid state experience intermolecular attractive forces that hold the individual particles together. Intermolecular forces also operate between a solute particle and the solvent that surrounds it. Thus,

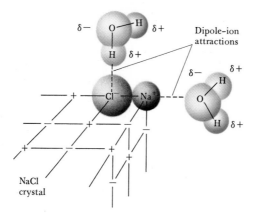

FIGURE 12.1 Interactions between H_2O molecules and the Na^+ and Cl^- ions of a NaCl crystal.

to understand solutions we must understand the kinds of attractive forces that exist between solute and solvent.

Any of the various kinds of intermolecular forces that we discussed in Chapter 11 can operate between solute and solvent particles in a solution. As a general rule, we expect solutions to form when the attractive forces between solute and solvent are comparable in magnitude with those that exist between the solute particles themselves or between solvent particles themselves. For example, the ionic substance NaCl dissolves readily in water. When NaCl is added to water, the water molecules orient themselves on the surface of the NaCl crystals as shown in Figure 12.1. The positive end of the water dipole is oriented toward the Cl^- ions, while the negative end of the water dipole is oriented toward the Na^+ ions. The ion-dipole attractions between Na^+ and Cl^- ions and water molecules are sufficiently strong to pull these ions from their positions in the crystal. Notice that the corner Na^+ ion is held in the crystal by only three adjacent Cl^- ions. In contrast, an Na^+ ion on the edge of the crystal has four nearby Cl^- ions, and a Na^+ ion in the interior of the crystal has six surrounding Cl^- ions. The corner Na^+ ion is therefore particularly vulnerable to removal from the crystal. Once this Na^+ ion

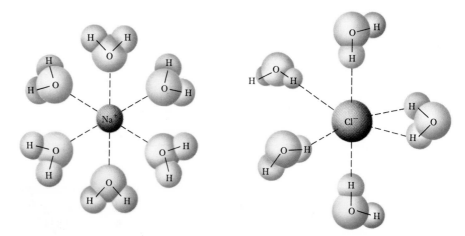

FIGURE 12.2 Hydrated Na^+ and Cl^- ions. The negative ends of the water dipole point toward the positive ion. The positive ends of the water dipole point toward the negative ion. We do not know whether one or both positive hydrogens touch the negative ion.

has been removed, adjacent Cl^- ions are similarly exposed and are therefore removed more easily than before.

Once removed from the crystal, the Na^+ and Cl^- ions are surrounded by water molecules as shown in Figure 12.2. Such interactions between solute and solvent molecules are known as solvation. When the solvent is water it is known as hydration.

Sodium chloride dissolves in water because the water molecules interact with the Na^+ and Cl^- ions sufficiently strongly to overcome the attraction between Na^+ and Cl^- ions in the crystal. To form a solution, the water molecules must also separate to make room for the ions. Therefore, we can visualize three types of interaction that take place in the solution process: (1) solute-solute interactions; (2) solute-solvent interactions; and (3) solvent-solvent interactions. The enthalpy changes accompanying each of these interactions when a salt dissolves in water are shown in Figure 12.3. Energy is required in processes (a) and (b) to overcome solute-solute and solvent-solvent attractions. Energy is released in process (c), when solute and solvent interact with one another. As shown in Figure 12.3, the net solution process can be either exothermic or endothermic. For example, ammonium nitrate, NH_4NO_3, dissolves by an endothermic reaction ($\Delta H = 26.4\ kJ/mol$). It has consequently been used to make instant ice packs, which are used to treat athletic injuries. The solid NH_4NO_3 is placed inside a thin-walled plastic bag. This, in turn, is sealed inside a thicker-walled bag together with some water. The small bag can be broken by kneading the larger bag. The resultant solution gets quite cold.

Figure 12.3 shows us that the heat of solution is the net result of enthalpy terms that are rather large when ionic solutions are involved. The overall heat may be exothermic or slightly endothermic. In both examples shown in Figure 12.3, however, there is a large exothermic heat term due to the attractive interaction of the ions with the solvent molecules—the solvation term, (c). We can understand from a diagram of this kind why NaCl does not dissolve in a nonpolar liquid such as gasoline.

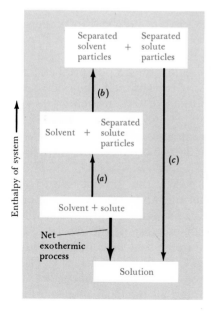

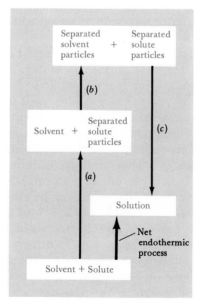

FIGURE 12.3 Analysis of the enthalpy changes accompanying the solution process: process (a) represents the enthalpy required to separate solute particles; process (b) represents the enthalpy required to separate solvent particles; process (c) represents the enthalpy released when solute and solvent particles interact with each other. The figure on the left shows a net exothermic heat of solution, whereas the one on the right shows a net endothermic heat of solution.

The hydrocarbon molecules of the gasoline could experience only weak ion-induced dipole attractions (see Figure 11.19), and these would not go very far toward compensating for the energies required to separate the ionic particles from one another.

By reasoning in a similar way we can understand why a polar liquid such as water does not dissolve to a great extent in a nonpolar liquid such as carbon tetrachloride, CCl_4. The water molecules experience strong attractive hydrogen-bonding interactions with one another (Section 11.6); these attractive forces would need to be overcome to disperse the water molecules throughout the nonpolar liquid. But there are no compensating attractive forces of comparable magnitude between water and carbon tetrachloride molecules. Consider the other possible case, the dissolving of carbon tetrachloride as a solute in water as solvent. The attractive forces between carbon tetrachloride molecules are mainly of the London-dispersion-force type, since the CCl_4 molecules are non-polar. There would thus not be much loss in attractive energy by dispersing the CCl_4 molecules in water as solvent. In this case, however, water molecules must separate from one another to some degree, to make a space for the solute CCl_4 molecules. This separation costs energy that is not recovered in the form of the attractive interactions between water and the CCl_4 molecules. Thus solution formation does not occur readily in this case, either.

Finally let us consider yet another example that raises a new idea. When two nonpolar substances such as CCl_4 and hexane, C_6H_{14}, are mixed, they readily dissolve in one another in all proportions. The attractive forces between molecules in both these cases are of the London-dispersion-force type. We note that CCl_4 boils at 77°C, whereas C_6H_{14} boils at 69°C. Thus, it is reasonable to suppose that the magnitudes of the attractive forces between molecules are comparable in the two substances. This means that when they are mixed there is little or no energy change upon solution formation. Yet this process occurs spontaneously; that is, it occurs to an appreciable extent without any extra input of energy from outside the system.

Experience tells us that two distinct factors are involved in processes that occur spontaneously. The most obvious has to do with energy. If you let go of a book you have in hand, it falls to the floor. The book is acted upon by the force of gravity. At its initial height it has a potential energy higher than when it is on the floor. Unless it is restrained, the book falls and thereby loses energy. This leads us to the first basic principle identifying spontaneous processes and influencing their direction: *Processes in which the energy content of the system decreases tend to occur spontaneously.* Spontaneous processes tend to be exothermic. Change occurs in the direction that leads to a lower energy content.

On the other hand, we can think of processes that do not result in a lower energy, or that may even be endothermic, and yet still occur spontaneously. The mixing of CCl_4 and C_6H_{14} provides a simple example. All such processes are characterized by an increase in the disorder, or randomness, of the system. To illustrate, suppose that we could suddenly remove a barrier that separated 500 mL of CCl_4 from 500 mL of C_6H_{14}, as in Figure 12.4(a). Before the barrier is removed each liquid occupies a volume of 500 mL. We know that we can find all the CCl_4 molecules in

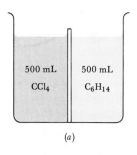

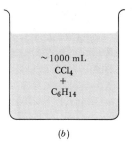

(a)

(b)

FIGURE 12.4 Formation of a homogeneous solution between CCl_4 and C_6H_{14} upon removal of a barrier separating the two liquids. The solution in (b) is more disordered, or random in character, than the separate liquids before solution formation (a).

the 500 mL to the left, all the C_6H_{14} molecules in the 500 mL to the right of the barrier. When equilibrium has been established after removal of the barrier, the two liquids together occupy a volume of about 1000 mL.* Formation of a homogeneous solution has resulted in an increased disorder, or randomness, in that the molecules of each substance are now distributed in a volume twice as large as that which they occupied before mixing. This example illustrates our second basic principle: *Processes in which the disorder of the system increases tend to occur spontaneously.*

When molecules of different types are mixed an increase in disorder occurs spontaneously unless the molecules are restrained by sufficiently strong intermolecular forces or by physical barriers. Thus, because of the strong bonds holding the sodium and chloride ions together, sodium chloride does not spontaneously dissolve in gasoline. On the other hand, gases spontaneously expand unless restrained by their containers; in this case intermolecular forces are too weak to restrain the molecules. We shall discuss spontaneous processes again in Chapter 18; at that time we shall consider the balance between the tendency toward lower energy and toward increased disorder in greater detail. For the moment we need to be aware that formation of a solution is always favored by the increase in disorder that accompanies mixing. Consequently, a solution will form unless solute-solute or solvent-solvent interactions are too strong relative to the solute-solvent interactions.

In all of our discussions of solutions we must be careful to distinguish the *physical* process of solution formation from chemical processes that lead to a solution. For example, zinc metal is dissolved on contact with hydrochloric acid solution. The following chemical reaction occurs:

$$Zn(s) + 2HCl(aq) \longrightarrow H_2(g) + ZnCl_2(aq) \qquad [12.5]$$

In this instance the chemical form of the substance being dissolved is changed. If the solution is evaporated to dryness, $Zn(s)$ is not recovered as such; instead $ZnCl_2(s)$ is recovered. In contrast, when $NaCl(s)$ is dissolved in water, it can be recovered by evaporation of its solution to dryness. Our focus throughout this chapter is on solutions from which the solute can be recovered unchanged from the solution.

SOLUBILITY

As the solution process involving a solid solute proceeds, and the concentration of solute particles in solution increases, the chances of their colliding with the surface of the solid increases (Figure 12.5). Such a collision may result in the solute particle becoming attached to the solid. This process, which is the opposite of the solution process, is called crystallization. Thus, two opposing processes occur in a solution in contact with undissolved solute. This situation is represented in Equation [12.6] by use of a double arrow:

$$\text{Solute + solvent} \xrightleftharpoons[\text{crystallize}]{\text{dissolve}} \text{solution} \qquad [12.6]$$

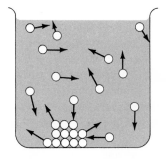

FIGURE 12.5 Movement of solute particles in solvent containing excess solute. Both dissolution and crystallization occur.

*There may be a slight change in the total volume upon mixing, but that is unimportant for our example.

When the rates of these opposing processes become equal, there will be no further net increase in the amount of solute in solution. This is an example of dynamic equilibrium, similar to that discussed in Section 11.2, where the processes of evaporation and condensation were considered. A solution that is in equilibrium with undissolved solute is said to be saturated. Additional solute will not dissolve if added to a saturated solution. The amount of solute needed to form a saturated solution in a given quantity of solvent is known as the solubility of that solute. For example, the solubility of NaCl in water at 0°C is 35.7 g per 100 mL of water. This is the maximum amount of NaCl that can be dissolved in water to give a stable, equilibrium solution at that temperature. If less solute is added than the equilibrium amount, the solution is said to be unsaturated. In some cases it is possible to prepare solutions that contain more solute than the equilibrium amount. Such solutions, which are said to be supersaturated, are unstable, and under the proper conditions solute will crystallize from them to give saturated solutions. Sodium acetate, $NaC_2H_3O_2$, forms supersaturated solutions very easily. The solubility of this substance at 0°C is 119 g per 100 mL of H_2O. It is more soluble at higher temperatures. If a hot solution containing more than 119 g of $NaC_2H_3O_2$ per 100 mL of H_2O is slowly cooled to 0°C, the excess solute remains dissolved; the resultant solution is supersaturated. Addition of a small crystal of $NaC_2H_3O_2$ gives a surface on which crystallization can start, and crystallization occurs until the concentration of the solution drops to the saturation level.

12.3 Solubilities and molecular structure

Our discussion of the factors that lead to formation of solutions enables us to understand many observations regarding solubilities. As a simple example, consider the data in Table 12.3 for the solubilities of various simple gases in water. Note that solubility increases with increasing molecular mass. The attractive forces between the gas and solvent molecules are mainly of the London-dispersion-force type, which increase with increasing size and molecular mass of the gas molecules. When chemical reaction occurs between the gas and solvent, much higher gas solubilities result. We will encounter instances of this in later chapters, but as a simple example, the solubility of Cl_2 in water under the same conditions given in Table 12.3 is 0.102 M. This is much higher than would be predicted from the trends in the table, based just on molecular mass. We can infer from this that the dissolving of Cl_2 in water is accompanied by some kind of chemical process. The use of chlorine as a bactericide in municipal water supplies and swimming pools is based on its chemical reaction with water (Section 17.5).

Polar liquids tend to dissolve readily in polar solvents. Thus, for example, acetone, a polar molecule whose structure is shown at left, mixes in all proportions with water. Pairs of liquids that mix in all proportions are said to be miscible, while liquids that do not mix are termed immiscible. Water and hexane, C_6H_{14}, for example, are immiscible. Diethyl ketone (see structure at left), which is similar to acetone but has a higher molecular mass, dissolves in water to the extent of about 47 g per liter of water at 20°C, but is not completely miscible.

$$\underset{\text{Acetone}}{CH_3\overset{\overset{\displaystyle O}{\|}}{C}CH_3}$$

$$\underset{\text{Diethyl ketone}}{CH_3CH_2\overset{\overset{\displaystyle O}{\|}}{C}CH_2CH_3}$$

TABLE 12.3 Solubilities of several gases in water at 20° C, with 1 atm gas pressure

Gas	Solubility (M)
N_2	6.9×10^{-4}
CO	1.04×10^{-3}
O_2	1.38×10^{-3}
Ar	1.50×10^{-3}
Kr	2.79×10^{-3}

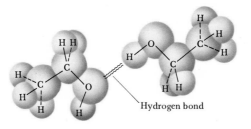

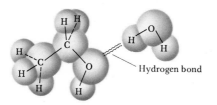

FIGURE 12.6 Hydrogen-bonding interactions between ethanol molecules and between water and ethanol molecules.

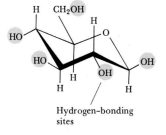

FIGURE 12.7 Structure of glucose. Color spheres indicate sites capable of hydrogen bonding with water.

Hydrogen-bonding interactions between solute and solvent may lead to high solubility. For example, water is completely miscible with ethanol, CH_3CH_2OH. The CH_3CH_2OH molecules are able to form hydrogen bonds with water molecules as well as with each other. This is shown in Figure 12.6. Because of this hydrogen-bonding ability, the solute-solute, solvent-solvent, and solute-solvent forces are not appreciably different within a mixture of CH_3CH_2OH and H_2O. There is no significant change in the environment of the molecules as they are mixed. The increase in disorder accompanying mixing therefore plays a significant role in formation of the solution.

The number of carbon atoms in an alcohol affects its solubility in water as shown in Table 12.4. As the length of the carbon chain increases, the OH group becomes an ever smaller part of the molecule and the molecule becomes more like a hydrocarbon. The solubility of the alcohol decreases correspondingly. If the number of OH groups along the carbon chain increases, more solute-water hydrogen bonding is possible and solubility generally increases. In the case of glucose, $C_6H_{12}O_6$, there are five OH groups on a six-carbon framework that make the molecule very soluble in water (83 g dissolve in 100 mL of water at 17.5°C). The glucose molecule is shown in Figure 12.7.

TABLE 12.4 Solubilities of some alcohols in water

Alcohol	Solubility in H_2O (moles/100 g H_2O at 20°C)
CH_3OH (methanol)	∞ [a]
CH_3CH_2OH (ethanol)	∞
$CH_3CH_2CH_2OH$ (propanol)	∞
$CH_3CH_2CH_2CH_2OH$ (butanol)	0.11
$CH_3CH_2CH_2CH_2CH_2OH$ (pentanol)	0.030
$CH_3CH_2CH_2CH_2CH_2CH_2OH$ (hexanol)	0.0058
$CH_3CH_2CH_2CH_2CH_2CH_2CH_2OH$ (heptanol)	0.0008

[a] Infinity symbol indicates that there is no real limit to the solubility of this alcohol in water.

355

Examination of pairs of substances such as those listed in the preceding paragraphs has led to an important generalization: *Substances with similar intermolecular attractive forces tend to be soluble in one another*. This generalization is often simply stated as *"likes dissolve likes."* Nonpolar substances are soluble in nonpolar solvents, whereas ionic and polar solutes are soluble in polar solvents. Network solids like diamond and quartz are not soluble in either polar or nonpolar solvents because of the strong bonding forces within the solid.

SAMPLE EXERCISE 12.6

Predict whether each of the following substances is more likely to dissolve in carbon tetrachloride, CCl_4, which is a nonpolar solvent, or in water: C_7H_{16}, $NaHCO_3$, HCl, I_2.

Solution: Both C_7H_{16} and I_2 are nonpolar. We would therefore predict that they would be more soluble in CCl_4 than in H_2O. On the other hand, $NaHCO_3$ is ionic and HCl is polar covalent. Water would be a better solvent than CCl_4 for these two substances.

Vitamins B and C are water soluble; the A, D, E, and K vitamins are soluble in nonpolar solvents and in the fatty tissue of the body (which is nonpolar). Because of their water solubility, vitamins B and C are not stored to any appreciable extent in the body, and so foods containing these vitamins should be included in the daily diet. In contrast, the fat-soluble vitamins are stored in sufficient quantities to keep vitamin-deficiency diseases from appearing even after a person has subsisted for a long period on a vitamin-deficient diet. With the ready availability of vitamin supplements, cases of hypervitaminosis, illness caused by an excessive amount of vitamins, are now being seen by physicians in this country. Because only the fat-soluble vitamins are stored, they are the

ones for which true hypervitaminosis has been observed. The different solubility patterns of the water-soluble vitamins and the fat-soluble ones can be rationalized in terms of the structures of the molecules. The chemical structures of vitamin A (retinol) and of vitamin C (ascorbic acid) are shown below.

Note that the vitamin A molecule is an alcohol with a very long carbon chain. It is nearly nonpolar, and because the OH group is such a small part of the molecule, the molecule mainly resembles a hydrocarbon. The situation is like that of the long-chain alcohols listed in Table 12.4. In contrast, the vitamin C molecule is smaller and has more OH groups that can form hydrogen bonds with water. It is somewhat like the glucose example discussed above.

Vitamin A

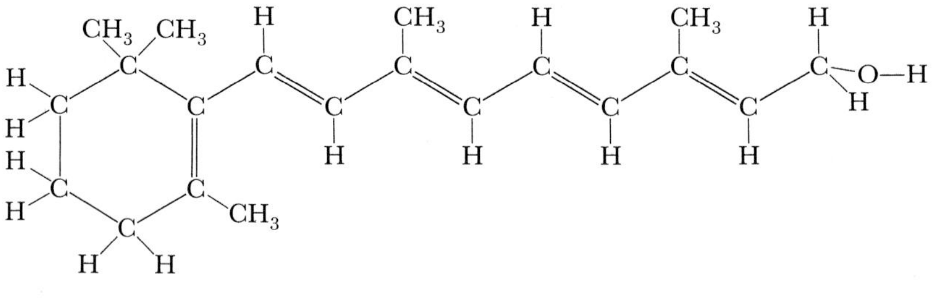

Vitamin C

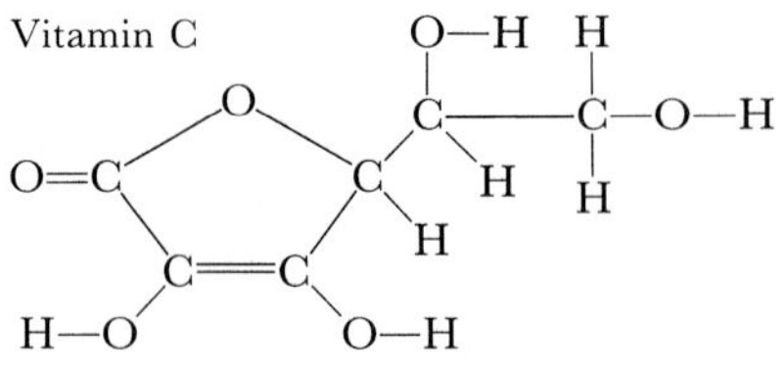

The solubilities of substances in water are of special concern to us, because water is of such central importance in our lives. In this section we will consider the effects of temperature and pressure on solubilities, with special attention to solutions in water.

The solubility of a gas in any solvent is increased as the pressure of the gas over the solvent increases. By contrast, the solubilities of solids and liquids are not appreciably affected by pressure. We can understand the effect of pressure on the solubility of a gas by considering the equilibrium that operates, illustrated in Figure 12.8. Suppose we have a gaseous substance distributed between the gas and solution phases. When equilibrium is established, the rate at which gas molecules enter the solution equals the rate at which solute molecules escape from the solution to enter the gas phase. The small arrows in Figure 12.8(a) represent the rates of these opposing processes. This is another example of dynamic equilibrium, which we have encountered before (Section 11.2). Now suppose that we exert added pressure on the piston and compress the gas above the solution, as shown in Figure 12.8(b). If we reduce the volume to half its original value, the pressure of the gas would increase to about twice its original value. But this would mean that the rate at which gas molecules strike the surface to enter the solution phase would increase. Thus the solubility of the gas in the solution would increase until there is again an equilibrium; that is, until the rates at which gas molecules enter the solution equals the rate at which solute molecules escape from the solvent, as indicated by the arrows in Figure 12.8(b). This means that the solubility of the gas should increase in direct proportion to the pressure. The relationship between pressure and solubility is expressed in terms of a simple equation known as **Henry's law:**

$$C_g = kP_g \qquad [12.7]$$

where C_g is the solubility of the gas in the solution phase, P_g is the pressure of the gas over the solution, and k is a proportionality constant, known as the Henry's-law constant. As an example, the solubility of pure nitrogen gas in water at 25°C and 0.78 atm pressure is $5.3 \times 10^{-4}\ M$. If the partial pressure of N_2 is doubled, Henry's law predicts that the solubility in water is also doubled to $1.06 \times 10^{-3}\ M$.

The effect of pressure on solubility is utilized in the production of carbonated beverages like champagne, beer, and soda pop. These are bottled under a carbon dioxide pressure slightly greater than 1 atm. When the bottles are opened to the air, the partial pressure of CO_2 above the solution is decreased, and CO_2 bubbles out of the solution.

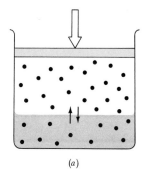

(a)

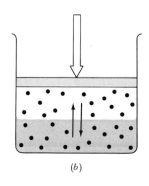

(b)

FIGURE 12.8 Effect of pressure on the solubility of a gas. When the pressure is increased, as in (b), the rate at which gas molecules enter the solution increases; the result is that the concentration of solute molecules at equilibrium increases in proportion to the pressure.

Deep-sea divers rely on compressed air for their oxygen supply. According to Henry's law, the solubilities of gases increase with pressure. If a diver is suddenly exposed to atmospheric pressure, where the solubility of gases is less, bubbles form in the blood stream and in other fluids of the body. These bubbles affect nerve impulses and give rise to the disease known as "the bends," or decompression sickness. Nitrogen is the main problem because it has the highest partial pressure in air and because it can be removed only through the respiratory system. Oxygen is consumed in metabolism. Substitution of helium for nitrogen minimizes this effect, because helium has a much lower solubility in biological fluids than does N_2. Cousteau's divers on *Conshelf III* used a mixture of 98 percent helium and 2 percent oxygen. At the high pressures (10 atm) experienced by the divers, this percentage of oxygen gives an oxygen partial pressure of about 0.2 atm, which is the partial pressure in normal air at 1 atm. If the oxygen partial pressures become too great, the urge to breathe is reduced, CO_2 is not removed from the body, and this leads to CO_2 poisoning.

The solubilities of several common gases in water as a function of temperature are graphed in Figure 12.9. These solubilities correspond to a constant pressure of 1 atm of gas over the solution. Note that in general, solubility decreases with increasing temperature. If a glass of cold tap water is warmed, bubbles of air are seen on the side of the glass. Similarly, a carbonated beverage like soda pop goes "flat" as it is allowed to warm; as the temperature of the solution increases, CO_2 escapes from the solution. The decreased solubility of O_2 in water as temperature increases is one of the effects of the thermal pollution of lakes and streams. The effect is particularly serious in deep lakes, because warm water is less dense than cold water. It therefore tends to remain on top of the cold water, at the surface. This situation impedes the dissolving of oxygen into the deeper layers, thus stifling the respiration of all aquatic life needing oxygen. Fish may suffocate and die in these circumstances.

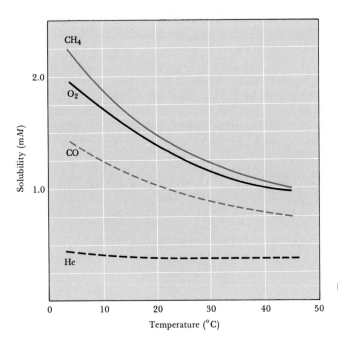

FIGURE 12.9 Solubilities of several gases in water as a function of temperature. Note that solubilities are in units of millimoles/L, for a constant pressure of 1 atm in the gas phase.

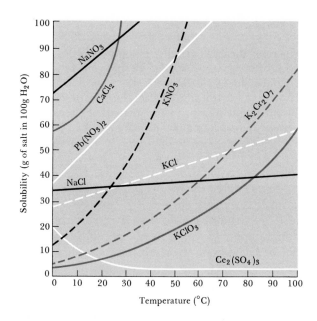

FIGURE 12.10 The solubilities of several common ionic solids shown as a function of temperature.

Figure 12.10 shows the effect of temperature on the solubilities of several ionic substances in water. Note that for most solids the solubility increases with increasing temperature. The effect of temperature on solubility depends on the enthalpy change for the solution process. For solutes that dissolve with endothermic heats of solution, solubility increases as temperature increases. We can understand these effects in terms of a principle first put forward by Henri Louis LeChatelier (1850–1931), a French industrial chemist. LeChatelier's principle can be stated as follows: If a system at equilibrium is disturbed by a change in temperature, pressure, or the concentration of one of the components, the system will tend to shift its equilibrium position so as to counteract the effect of the disturbance. Let us consider a solution in equilibrium with undissolved solute. Assume that the process of solution is endothermic; that is, that heat is absorbed from the surroundings as the solution is formed. Then the equilibrium can be represented as follows:

$$\text{Solute} + \text{solvent} + \text{heat} \rightleftharpoons \text{solution} \qquad [12.8]$$

Now according to LeChatelier's principle, if we add heat to this system, the equilibrium will shift in such a direction as to undo the effect of the added heat. That is, it will shift in the direction in which heat is absorbed—to the right. Thus, an increase in temperature, which represents an addition of heat to the system, results in increased solubility. On the other hand, if the heat of solution were negative—that is, if the solution process were exothermic—solubility would decrease with increasing temperature. This case is illustrated by the data for $Ce_2(SO_4)_3$ in Figure 12.10.

LeChatelier's principle is very useful in making predictions about the effects of changes on equilibrium systems. We will examine it more closely in Section 14.3.

Use LeChatelier's principle to predict the effect of increased gas pressure on the solubility of a gas.

Solution: We can answer this by reference to Figure 12.8. The equilibrium system is disturbed by an increase in the gas pressure. According to LeChatelier's principle, the system will respond by shifting the equilibrium so as to return the pressure toward its original value. This happens when more of the gas dissolves in the solution phase. Thus, we predict that solubility will increase with increasing pressure, in agreement with our previous conclusion.

12.5 Electrolyte solutions

Electrolytes are substances that dissolve in water to give a solution that conducts electricity. An example is sodium chloride; it dissolves in water to give hydrated Na^+ and Cl^- ions that are able to move freely throughout the solution. A solution labeled as $1.0 M$ NaCl actually contains 1.0 mol of Na^+ and 1.0 mol of Cl^- per liter of solution. Because NaCl ionizes completely in solution it is referred to as a strong electrolyte. On the other hand, when glucose, $C_6H_{12}O_6$, or ethanol, C_2H_5OH, dissolve, they are not fragmented into ions but maintain their molecular structure in solution. These substances are referred to as nonelectrolytes. Some substances that contain highly polar, but not completely ionic, bonds ionize incompletely on dissolving in water. An example is $HgCl_2$, which forms an equilibrium of the form

$$HgCl_2(aq) \rightleftharpoons HgCl^+(aq) + Cl^-(aq) \qquad [12.9]$$

Only a small fraction of the $HgCl_2$ is present in solution as ions. This compound is an example of a weak electrolyte. The qualitative abilities of solutions to conduct an electric current can be determined using a device such as that shown in Figure 12.11. Ions in the solution complete the electric circuit and thereby permit the light bulb to glow.

It is very important to distinguish solubility from whether a substance is a strong or weak electrolyte. For example, silver chloride, AgCl, is only very slightly soluble in water. Therefore, if we shake some solid AgCl with water for a long time until equilibrium is established, most of the AgCl will remain as an insoluble solid. All of the AgCl that does dissolve, however, is present in solution as Ag^+ and Cl^- ions. We say, therefore, that AgCl is a strong electrolyte.

In Section 3.3 we introduced acids, bases, and salts. You should review that section at this time. We can distinguish acids and bases according to the degree to which they ionize in solution. Acids and bases that are incompletely ionized in solution are referred to as weak acids and weak bases. The terms weak acid and weak base have no relation to how reactive the acid or base is. Hydrofluoric acid, HF, is a weak electrolyte and is therefore called a weak acid. A $0.1 M$ solution of HF is 8 percent ionized. However, HF is a very reactive acid that vigorously attacks many substances, including glass.

The following generalizations are useful in recognizing which substances are strong electrolytes and which weak:

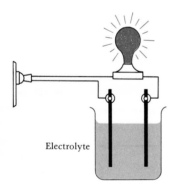

Electrolyte

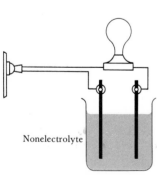

Nonelectrolyte

FIGURE 12.11 A simple device that can be used to distinguish between electrolyte and nonelectrolyte solutions.

1 *Acids:* The common strong acids are HCl, HBr, HI, HNO_3, H_2SO_4, and $HClO_4$. Nearly all other acids are weak electrolytes.

2 *Bases:* The common strong bases are the hydroxides of the alkali metals and the alkaline earths (except Be). NH_3 is a weak electrolyte.

3 *Salts:* Most common salts (that is, ionic compounds) are strong electrolytes. The exceptions are found mainly among the heavy metals, as in the example of $HgCl_2$ cited above.

SAMPLE EXERCISE 12.8

Classify each of the following substances as nonelectrolyte, weak electrolyte, or strong electrolyte: $CaCl_2$, HNO_3, CH_3OH (methanol), $HC_2H_3O_2$ (acetic acid), KOH, H_2O_2.

Solution: Only one of the substances, $CaCl_2$, is a salt. It is a strong electrolyte. Two of the substances, HNO_3 and $HC_2H_3O_2$, are acids; HNO_3 is a common strong acid (strong electrolyte); because $HC_2H_3O_2$ is not a common strong acid, our best guess would be that it is a weak acid (weak electrolyte). This is correct. There is one base, KOH. It is one of the common strong bases (a strong electrolyte) because it is a hydroxide of an alkali metal. The remaining compounds, CH_3OH and H_2O_2, are neither acids, bases, nor salts. They are nonelectrolytes.

SAMPLE EXERCISE 12.9

What is the concentration of all species in a 0.1 M solution of $Ba(NO_3)_2$?

Solution: $Ba(NO_3)_2$ is a salt and as such we expect it to be a strong electrolyte. It ionizes into Ba^{2+} and the polyatomic nitrate ion, NO_3^-. A 0.1 M solution of $Ba(NO_3)_2$ is 0.1 M in Ba^{2+} and 0.2 M in NO_3^- (since 1 mol of $Ba(NO_3)_2$ supplies 1 mol of Ba^{2+} and 2 mol of NO_3^-).

12.6 Reactions in aqueous solution

In Section 3.3 we discussed two important reaction types, neutralization and precipitation. We should now examine these and related reactions in more detail, keeping in mind the ionic character of the solution species. Consider first the neutralization reaction between a strong acid such as HCl and a strong base such as NaOH. The form in which we previously wrote the equation for this reaction, Equation [3.18], is known as a molecular equation. Because we know that both HCl and NaOH as well as the reaction product NaCl are completely ionized in solution, we can more realistically write the equation as:

$$Na^+(aq) + OH^-(aq) + H^+(aq) + Cl^-(aq) \longrightarrow$$
$$Na^+(aq) + Cl^-(aq) + H_2O(l) \qquad [12.10]$$

This form is known as the ionic equation. Notice that in Equation [12.10] Na^+ and Cl^- ions appear on both sides of the arrow. Such ions, which go through the reaction unchanged, are known as spectator ions. We can omit them from both sides to obtain the net ionic equation:

$$H^+(aq) + OH^-(aq) \longrightarrow H_2O(l) \qquad [12.11]$$

This equation expresses the essential feature of the neutralization reaction, the combination of the H^+ and OH^- ions to form water. It is the feature common to all neutralization reactions between strong acids and strong bases. In writing net ionic equations, keep in mind that only strong electrolytes are written in the ionic form. Solids, gases, nonelectrolytes, and weak electrolytes are written in the "molecular" form.

By considering the ionic nature of solutions of salts that are strong electrolytes we can gain a better understanding also of precipitation reactions. Consider, for example, what happens when we mix aqueous solutions of $AgNO_3$ and $NaCl$. Each of these substances is present in solution in the form of separated ions. The ions might combine with one another in four possible precipitation reactions:

$$Na^+(aq) + NO_3^-(aq) \xrightarrow{\quad\times\quad} NaNO_3(s)$$

$$Ag^+(aq) + NO_3^-(aq) \xrightarrow{\quad\times\quad} AgNO_3(s)$$

$$Na^+(aq) + Cl^-(aq) \xrightarrow{\quad\times\quad} NaCl(s)$$

$$Ag^+(aq) + Cl^-(aq) \longrightarrow AgCl(s)$$

The first three of these reactions do not occur because the salt formed on the right is soluble in water. However, the last reaction does occur, because $AgCl$ is only very slightly soluble in water. The ionic equation for the precipitation reaction is thus

$$Na^+(aq) + Cl^-(aq) + Ag^+(aq) + NO_3^-(aq) \longrightarrow$$
$$Na^+(aq) + NO_3^-(aq) + AgCl(s)$$

Again we note that Na^+ and NO_3^- are spectator ions. When they are omitted we obtain the net ionic equation for the precipitation reaction:

$$Ag^+(aq) + Cl^-(aq) \longrightarrow AgCl(s) \qquad [12.12]$$

The same net ionic equation applies to the reaction of the soluble salts $AgClO_3$ and KCl (do you see why this is so?).

In order to predict whether precipitation will occur when two solutions are mixed, we must have more knowledge of solubilities than presented so far in this text. The rule "likes dissolve likes" allows us to predict correctly that a substance like $AgCl$ is more likely to be soluble in water than in a nonpolar solvent. However, it doesn't permit us to predict the extent to which $AgCl$ will dissolve in water. In fact, only 1.3×10^{-5} mol of $AgCl$ dissolves in a liter of water at $25\,°C$. For all practical purposes, the compound is insoluble. In our discussions, any substance whose solubility is less than 0.01 mol per liter at $25\,°C$ will be referred to as being insoluble. Those substances whose solubilities are 0.01 to 0.1 M will be considered as moderately soluble, and those whose solubilities are greater than 0.1 M will be labeled as being soluble. We shall consider solubility from a quantitative point of view in Chapter 17, but the following are some qualitative generalizations concerning the solubilities of common ionic compounds in water:

1 Salts of the alkali metals are *soluble*.
2 Ammonium (NH_4^+) salts are *soluble*.

3 Salts containing nitrate (NO_3^-), chlorate (ClO_3^-), perchlorate (ClO_4^-), and acetate ($C_2H_3O_2^-$) are *soluble*.

4 All chlorides (Cl^-), bromides (Br^-), and iodides (I^-) are *soluble* except for those of Pb^{2+}, Hg_2^{2+}, and Ag^+, which are *insoluble*.

5 All sulfates (SO_4^{2-}) are *soluble* except for those of Sr^{2+}, Ba^{2+}, Hg_2^{2+}, Hg^{2+}, and Pb^{2+}, which are *insoluble*. The sulfate salts of Ca^{2+} and Ag^+ are *moderately soluble*.

6 All hydroxides (OH^-) are *insoluble* except for those of the alkali metals, which are *soluble*, and the hydroxides of Ca^{2+}, Ba^{2+}, and Sr^{2+}, which are *moderately soluble*.

7 All sulfites (SO_3^{2-}), carbonates (CO_3^{2-}), chromates (CrO_4^{2-}), and phosphates (PO_4^{3-}) are *insoluble* except for those of NH_4^+ and the alkali metals, which are *soluble*.

8 All sulfides (S^{2-}) are *insoluble* except for those of NH_4^+, the alkali metals, and the alkaline earths, which are *soluble*.

SAMPLE EXERCISE 12.10

Write balanced molecular, ionic, and net ionic equations for reactions (if any) that occur when solutions of the following compounds are mixed: (a) $BaCl_2$ and Na_2SO_4; (b) KCl and Na_2SO_4.

Solution: (a) Both $BaCl_2$ and Na_2SO_4 are soluble and ionize to give Ba^{2+}, Cl^-, Na^+, and SO_4^{2-} ions. The possible precipitation products are $BaSO_4$ and $NaCl$. The $BaSO_4$ is insoluble according to rule 5 in the list just given. Therefore, the molecular equation is

$$BaCl_2(aq) + Na_2SO_4(aq) \longrightarrow$$
$$BaSO_4(s) + 2NaCl(aq)$$

The ionic equation is

$$Ba^{2+}(aq) + 2Cl^-(aq) + 2Na^+(aq) + SO_4^{2-}(aq) \longrightarrow$$
$$BaSO_4(s) + 2Na^+(aq) + 2Cl^-(aq)$$

Both $Na^+(aq)$ and $Cl^-(aq)$ are spectator ions. The net ionic equation is

$$Ba^{2+}(aq) + SO_4^{2-}(aq) \longrightarrow BaSO_4(s)$$

(b) Both reactants are soluble and ionize in solution. There are no possible insoluble salts resulting from reaction. Both $NaCl$ and K_2SO_4 are soluble. Therefore, there is no reaction.

The neutralization and precipitation reactions we have just considered proceed essentially to completion. The "driving force," that is, the factor that causes the reaction to occur, is formation of water in the first instance and formation of an insoluble salt in the second.

One additional source of "driving force" for completion of reactions in aqueous solution is formation of a volatile nonelectrolyte. The complete ionic and net ionic equations for three common examples of this reaction type are as follows:

1 A carbonate reacting with an acid:

$$2Na^+(aq) + CO_3^{2-}(aq) + 2H^+(aq) + 2Cl^-(aq) \longrightarrow$$
$$2Na^+(aq) + 2Cl^-(aq) + CO_2(g) + H_2O(l)$$

$$CO_3^{2-}(aq) + 2H^+(aq) \longrightarrow CO_2(g) + H_2O(l) \qquad [12.13]$$

2 **A sulfite reacting with an acid:**

$$Na^{2+}(aq) + SO_3^{2-}(aq) + 2H^+(aq) + 2Br^-(aq) \longrightarrow$$
$$Na^{2+}(aq) + 2Br^-(aq) + SO_2(g) + H_2O(l)$$

$$SO_3^{2-}(aq) + 2H^+(aq) \longrightarrow SO_2(g) + H_2O(l) \qquad [12.14]$$

3 **A sulfide reacting with an acid:**

$$MnS(s) + 2H^+(aq) + SO_4^{2-}(aq) \longrightarrow$$
$$Mn^{2+}(aq) + SO_4^{2-}(aq) + H_2S(g)$$

$$MnS(s) + 2H^+(aq) \longrightarrow Mn^{2+}(aq) + H_2S(g) \qquad [12.15]$$

SAMPLE EXERCISE 12.11

Write balanced complete ionic and net ionic equations for any reactions that occur when the following compounds are mixed: (a) $FeCO_3(s)$ and $HCl(aq)$; (b) $NiS(s)$ and $HCl(aq)$; (c) $NaOH(aq)$ and $H_3PO_4(aq)$.

Solution: (a) This reaction is similar to Equation [12.13], but the reactant carbonate is a solid, insoluble in water.

Complete ionic:

$$FeCO_3(s) + 2H^+(aq) + 2Cl^-(aq) \longrightarrow$$
$$Fe^{2+}(aq) + 2Cl^-(aq) + H_2O(l) + CO_2(g)$$

Net ionic:

$$FeCO_3(s) + 2H^+(aq) \longrightarrow$$
$$Fe^{2+}(aq) + H_2O(l) + CO_2(g)$$

(b) This reaction is of the type of Equation [12.15]. Most sulfides will react with acids even though they are insoluble. Similarly, solid carbonates and hydroxides will generally react with acids, as exemplified in (a).

Complete ionic:

$$NiS(s) + 2H^+(aq) + 2Cl^-(aq) \longrightarrow$$
$$Ni^{2+}(aq) + 2Cl^-(aq) + H_2S(g)$$

Net ionic:

$$NiS(s) + 2H^+(aq) \longrightarrow Ni^{2+}(aq) + H_2S(g)$$

Notice that in writing the net ionic equation, in both (a) and (b), both solids and gases are included. The only species that goes through the reaction unchanged is the Cl^- ion.

(c) This is the familiar acid-base reaction summarized by equation (12.11).

Complete ionic:

$$3Na^+(aq) + 3OH^-(aq) + 3H^+(aq) + PO_4^{3-}(aq) \longrightarrow$$
$$3H_2O(l) + 3Na^+(aq) + PO_4^{3-}(aq)$$

Net ionic:

$$H^+(aq) + OH^-(aq) \longrightarrow H_2O(l)$$

CHEMICAL REACTIONS: A RECONSIDERATION

This is a good point at which to summarize the important things we have learned so far about various types of chemical change that might occur upon mixing substances in solution. It is of course possible that no reaction at all will occur when solutions of ionic substances are mixed. For example, mixing aqueous solutions of calcium chloride and potassium nitrate leads to no chemical change. Both salts are strong electrolytes, neither is notably acidic or basic, and no insoluble salt is formed by any combination of the ions present. We do observe a net chemical change when one or more of the following occur:

1 *A precipitate is formed.* Several examples of this type are given above.

2 *A gas is formed.* Equations [12.13], [12.14], and [12.15] provide examples of this reaction type.

3 *A nonelectrolyte is formed.* The neutralization reaction is an example of this type of reaction, since $H_2O(l)$ is a nonelectrolyte.

4 *A weak electrolyte is formed.* As an example of this reaction type, consider the reaction of sodium acetate, $NaC_2H_3O_2$, with hydrochloric acid. The net ionic equation describes formation of acetic acid, a weak electrolyte:

$$H^+(aq) + C_2H_3O_2^-(aq) \rightleftharpoons HC_2H_3O_2(aq) \qquad [12.16]$$

We will encounter many more examples of this reaction type in Chapters 15 and 16, when we discuss aqueous equilibria involving acids, bases, and other substances in aqueous solution.

5 *One or more electrons are transferred.* You will recall that we saw examples of oxidation-reduction equations in Section 7.10. Clearly, when substances undergo changes in their oxidation numbers, a chemical reaction has occurred. As we proceed through the text from this point, oxidation-reduction reactions occurring in an aqueous medium will be written in the form of net ionic equations.*

6 *One or more covalent bonds are made or broken.* Many of the reactions in this category—for example, the combustion reactions discussed in Section 3.3—do not occur in an aqueous medium. However, many such reactions do occur in water. For example, an alcohol such as ethanol is oxidized by potassium dichromate to form acetaldehyde:

$$2H^+(aq) + 3C_2H_5OH(aq) + Cr_2O_7^{2-}(aq) \longrightarrow$$
$$3C_2H_4O(aq) + Cr_2O_3(s) + 4H_2O(l) \qquad [12.17]$$

Notice that in this reaction we not only have new covalent bonds formed, we also have changes in oxidation numbers and formation of a nonelectrolyte. This example points up the fact that any particular reaction might contain elements of more than one of the categories we have just defined. The main point you should keep in mind is that when one or more of the "driving forces" for reaction that we have just outlined is present, a reaction occurs, as opposed to simply a mixing with no net chemical change.

12.7 Colligative properties

Many properties of a solution depend not only on the concentration of solute but also on its particular nature. For example, the density of a solution depends on the identity of both solute and solvent as well as on the concentration of solute. However, for certain kinds of solutes, a number of physical properties of the solution depend on the number of solute particles present in solution but not on the identity of the solute.

*The techniques used to balance oxidation-reduction equations are described in Section 19.1.

Such properties are called colligative properties. They include vapor-pressure lowering, boiling-point elevation, freezing-point depression, and osmotic pressure.

VAPOR-PRESSURE LOWERING

If a nonvolatile solute is added to a solvent, the vapor pressure of the solvent is decreased. For example, solutions of sugar or salt both have lower vapor pressures than does pure water. Experiments have shown that the vapor-pressure lowering depends on the concentration of solute particles. For example, 1.0 mol of a nonelectrolyte such as glucose and 0.5 mol of NaCl will produce essentially the same vapor-pressure lowering in a given quantity of water. There are 1.0 mol of particles in both solutions, because 0.5 mol of NaCl dissolves giving 0.5 mol of Na^+ and 0.5 mol of Cl^-. Quantitatively, the vapor pressure of solutions containing nonvolatile solutes is given by Raoult's law, Equation [12.18], where P_A is the vapor pressure of the solution, X_A is the mole fraction of the solvent, and P_A° is the vapor pressure of the pure solvent:

$$P_A = X_A P_A^\circ \qquad [12.18]$$

For example, the vapor pressure of water is 17.5 mm Hg at 20°C. Imagine holding the temperature constant while adding glucose, $C_6H_{12}O_6$, to the water so that the resulting solution has $X_{H_2O} = 0.80$ and $X_{C_6H_{12}O_6} = 0.20$. According to Equation [12.18], the vapor pressure of water over the solution will be 14 mm Hg:

$$P_{H_2O} = (0.80)(17.5 \text{ mm Hg}) = 14 \text{ mm Hg}$$

The vapor-pressure lowering, ΔP_A, is $P_A^\circ - P_A$. Using Equation [12.18] we can write this as

$$\begin{aligned} \Delta P_A &= P_A^\circ - X_A P_A^\circ \\ &= P_A^\circ(1 - X_A) \end{aligned} \qquad [12.19]$$

But since $X_A + X_B = 1$, where X_B is the mole fraction of solute particles we can write

$$\begin{aligned} X_B &= 1 - X_A \\ \Delta P_A &= X_B P_A^\circ \end{aligned} \qquad [12.20]$$

In our example, the vapor pressure will be lowered by 3.5 mm Hg:

$$\Delta P_A = (0.20)(17.5 \text{ mm Hg}) = 3.5 \text{ mm Hg}$$

It is easy to explain this effect qualitatively. At a given temperature, the average kinetic energy of all particles in the solution is the same. Essentially the same number of particles are at the surface in both the solution and pure solvent. However, in the case of a nonvolatile solute, only a fraction of these particles are volatile and have sufficient kinetic energy to escape from the solution.

Calculate the vapor-pressure lowering caused by the addition of 100 g of sucrose, $C_{12}H_{22}O_{11}$, to 1000 g of water if the vapor pressure of the pure water at 25°C is 23.8 mm Hg.

Solution:

$$X_{C_{12}H_{22}O_{11}} = \frac{\text{moles } C_{12}H_{22}O_{11}}{\text{moles } H_2O + \text{moles } C_{12}H_{22}O_{11}}$$

$$\text{Moles } C_{12}H_{22}O_{11} = (100 \text{ g})\left(\frac{1 \text{ mol } C_{12}H_{22}O_{11}}{342 \text{ g } C_{12}H_{22}O_{11}}\right)$$

$$= 0.292 \text{ mol}$$

$$\text{Moles } H_2O = (1000 \text{ g})\left(\frac{1 \text{ mol } H_2O}{18.0 \text{ g } H_2O}\right)$$

$$= 55.5 \text{ mol}$$

$$X_{C_{12}H_{22}O_{11}} = \frac{0.292}{55.5 + 0.292} = \frac{0.292}{55.8}$$

$$\Delta P_A = \left(\frac{0.292}{55.8}\right)(23.8 \text{ mm Hg}) = 0.125 \text{ mm Hg}$$

BOILING-POINT ELEVATION

We saw in Section 11.2 that the vapor pressure of a liquid increases with increasing temperature; it boils when its vapor pressure is the same as the external pressure over the liquid. Because nonvolatile solutes lower the vapor pressure of the solution, a higher temperature is required to cause the solution to boil. Figure 12.12 shows the variation in vapor pressure of a pure liquid with temperature (the black line), as compared with a solution (the color line). * Because the vapor pressure of the solution is lower than that of the solvent at all temperatures, in accordance with Raoult's law, a higher temperature is required to attain a vapor pressure of 1 atm. The increase in boiling point, ΔT_b (relative to the boiling point of the pure solvent), is directly proportional to the number of solute particles per mole of solvent molecules. We know that molality expresses the number of moles of solute per 1000 g of solvent, which represents a fixed number of moles of solvent. Thus, ΔT_b is proportional to molality, as shown in Equation [12.21]:

*Both lines refer to the usual laboratory situation, in which we have a constant total gas pressure of 1 atm over the system.

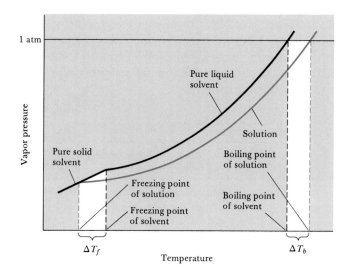

FIGURE 12.12   Vapor-pressure-versus-temperature curves of a pure solvent and a solution of a nonvolatile solute, under a constant total pressure of 1 atm. The vapor pressure of the solid solvent is unaffected by the presence of solute if the solid solvent freezes out without containing a significant concentration of solute, as is usually the case.

$$\Delta T_b = K_b m \qquad [12.21]$$

The magnitude of K_b, which is called the **molal boiling-point-elevation constant,** depends on the solvent. Some typical values for several common solvents are given in Table 12.5.

For water, K_b is $0.52\,°C/m$; therefore, a $1\,m$ aqueous solution of sucrose or any other aqueous solution that is $1\,m$ in nonvolatile solute particles will boil at a temperature $0.52\,°C$ higher than pure water. It is important to notice that the boiling-point elevation is proportional to the number of solute particles present in a given quantity of solution. When NaCl dissolves in water 2 mol of solute particles, 1 mol of Na^+ and 1 mol of Cl^-, are formed for each mole of NaCl that dissolves. Thus, a $1\,m$ solution of NaCl in water causes a boiling-point elevation twice as large as a $1\,m$ solution of a nonelectrolyte such as sucrose.

FREEZING-POINT DEPRESSION

The freezing point corresponds to the temperature at which the vapor pressures of the solid and liquid phases are the same. The freezing point of a solution is lowered because the solute is not normally soluble in the solid phase of the solvent. For example, when aqueous solutions freeze, the solid that separates out is almost always pure ice. Thus, the vapor pressure of the solid is unaffected by the presence of solute. On the other hand, if the solute is nonvolatile, the vapor pressure of the solution is reduced in proportion to the mole fraction of solute, Equation [12.20]. This means that the temperature at which the solution and solid phases will have the same vapor pressure is reduced, as illustrated in Figure 12.12. Note that the point that represents the equilibrium between solid and liquid lies to the left of the corresponding point for pure solvent, at a lower temperature. Like the boiling-point elevation, the decrease in freezing point, ΔT_f, is directly proportional to the molality of the solute:

$$\Delta T_f = K_f m \qquad [12.22]$$

The values of K_f, **the molal freezing-point-depression constant,** for several common solvents are given in Table 12.5. For water K_f is $1.86\,°C/m$; therefore a $0.5\,m$ aqueous solution of NaCl or any aqueous solution that is $1\,m$ in nonvolatile solute particles will freeze $1.86\,°C$ lower than pure water. The freezing-poing lowering caused by solutes explains the use of antifreeze (Sample Exercise 12.13) and the use of calcium chloride, $CaCl_2$, to melt ice on roads during winter.

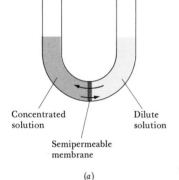

Concentrated solution

Dilute solution

Semipermeable membrane

(a)

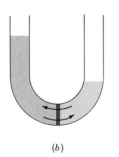

(b)

FIGURE 12.13 Osmosis: (a) net movement of solvent from the solution with low solute concentration into the solution with high solute concentration; (b) osmosis stops when the column of solution on the left becomes high enough to exert sufficient pressure to stop the osmosis.

TABLE 12.5 Molal boiling-point-elevation and freezing-point-depression constants

Solvent	Normal boiling point (°C)	$K_b(°C/m)$	Normal freezing point (°C)	$K_f(°C/m)$
Water, H_2O	100.0	0.52	0.0	1.86
Benzene, C_6H_6	80.1	2.53	5.5	5.12
Carbon tetrachloride, CCl_4	76.8	5.02	−22.3	29.8
Ethanol, C_2H_5OH	78.4	1.22	−114.6	1.99
Chloroform, $HCCl_3$	61.2	3.63	−63.5	4.68

SAMPLE EXERCISE 12.13

Calculate the freezing point and the boiling point of a solution of 100 g of ethylene glycol ($C_2H_6O_2$) in 900 g of H_2O.

Solution:

$$\text{Molality} = \frac{\text{moles } C_2H_6O_2}{\text{kilograms } H_2O} =$$

$$\left(\frac{100 \text{ g } C_2H_6O_2}{900 \text{ g } H_2O}\right)\left(\frac{1 \text{ mol } C_2H_6O_2}{62.0 \text{ g } C_2H_6O_2}\right)\left(\frac{1000 \text{ g } H_2O}{1 \text{ kg } H_2O}\right)$$

$$= 1.79 \text{ } m$$

$$\Delta T_f = K_f m = \left(1.86 \frac{°C}{m}\right)(1.79 \text{ } m) = 3.33°C$$

Freezing point = (normal f.p. of solvent) − ΔT_f

$$= 0.00°C - 3.33°C = -3.33°C$$

$$\Delta T_b = K_b m = \left(0.52 \frac{°C}{n}\right)(1.79 \text{ } m) = 0.93°C$$

Boiling point = (normal b.p. of solvent) + ΔT_b

$$= 100.00°C + 0.93°C$$
$$= 100.93°C$$

Ethylene glycol is the main component of antifreeze. It keeps the fluid in the cooling system of a car from freezing by lowering the freezing point of water. Because it also raises the boiling point of water, it is also useful in the summer to prevent what antifreeze advertisements call "boil-over."

SAMPLE EXERCISE 12.14

List the following solutions in order of their expected freezing points: 0.05 m $CaCl_2$; 0.15 m NaCl; 0.10 m HCl; 0.05 m $HC_2H_3O_2$; 0.10 m $C_{12}H_{22}O_{11}$.

Solution: First notice that $CaCl_2$, NaCl, and HCl are strong electrolytes, $HC_2H_3O_2$ is a weak electrolyte, and $C_{12}H_{22}O_{11}$ is a nonelectrolyte. The molality of each solution in total particles is as follows:

0.05 m $CaCl_2$ (0.15 m in particles)

0.15 m NaCl   (0.30 m in particles)

0.10 m HCl   (0.20 m in particles)

0.05 m $HC_2H_3O_2$ (between 0.05 and 0.10 m in particles)

0.10 m $C_{12}H_{22}O_{11}$ (0.10 m in particles)

The freezing points are expected to run from 0.15 m NaCl (lowest freezing point), to 0.10 m HCl to 0.05 m $CaCl_2$ to 0.10 m $C_{12}H_{22}O_{11}$ to 0.05 $HC_2H_3O_2$ (highest freezing point).

Because NaCl is a strong electrolyte, the expected freezing-point depression of a 0.100 m aqueous solution of NaCl is (0.200 m) (0.186°C/m) = 0.372°C. However, the measured value is slightly less than this, 0.348°C. Although strong electrolytes like NaCl give nearly the effect expected from the number of ions present in solution, the effect is always slightly less than that of a corresponding number of nonelectrolyte particles. This is due to the residual electrostatic attraction between ions, even when separated by solvent molecules. This attraction makes solutions of electrolytes behave as though their ion concentrations were less than they actually are.

OSMOSIS

Certain materials, including many membranes in biological systems, are semipermeable, meaning that they allow some particles to pass through, but not others. Osmosis is the net movement of solvent molecules but not solute particles through a semipermeable membrane from a more dilute solution into a more concentrated one. Consider two solutions separated by a semipermeable membrane as shown in Figure 12.13. The net migration of solvent occurs from the right arm to the left arm of the apparatus

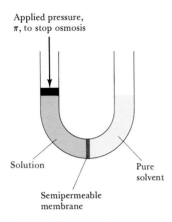

Applied pressure, π, to stop osmosis

Solution

Pure solvent

Semipermeable membrane

FIGURE 12.14 Applied pressure on the left of apparatus stops net movement of solvent from the right side of the semipermeable membrane. This applied pressure is known as the osmotic pressure of the solution.

as if the solutions were trying to equalize their concentrations. Pressure can be applied to the left arm of the apparatus, as shown in Figure 12.14, to prevent the osmosis. The pressure required to stop osmosis from a pure solvent into a solution is known as the osmotic pressure, π, of the solution. The osmotic pressure is related to concentration as shown in Equation [12.23]:

$$\pi = MRT \qquad [12.23]$$

where M is molarity, R is the ideal gas constant, and T is the temperature on the Kelvin scale.

If two solutions of identical osmotic pressure are separated by a semipermeable membrane, no osmosis will occur. The two solutions are said to be isotonic. If one solution is of lower osmotic pressure, it is described as being hypotonic with respect to the more concentrated solution. The more concentrated solution is said to be hypertonic with respect to the dilute solution.

Osmosis plays a very important role in living systems. For example, the membranes of red blood cells are semipermeable. Placement of red blood cells in a solution that is hypertonic relative to the intracellular solution (the solution within the cells) causes water to move out of the cell as shown in Figure 12.15. This causes the cell to shrivel, a process known as crenation. Placement of the cell in a solution that is hypotonic relative to the intracellular fluid causes water to move into the cell. This causes rupturing of the cell, a process known as hemolysis. Persons needing replacement of body fluids or nutrients who cannot be fed orally are administered solutions by intravenous (or IV) infusion, meaning slow addition to the veins. To prevent crenation or hemolysis of red blood cells, the IV solutions must be isotonic with the intracellular fluids of the cells.

SAMPLE EXERCISE 12.15

The average osmotic pressure of blood is 7.7 atm at 25 °C. What concentration of glucose, $C_6H_{12}O_6$, will be isotonic with blood?

Solution:

$$\pi = MRT$$

$$M = \frac{\pi}{RT} = \frac{7.7 \text{ atm}}{\left(0.082 \dfrac{\text{L-atm}}{\text{K-mol}}\right)(298 \text{ K})}$$

$$= 0.31 \, M$$

In clinical situations, the concentrations of solutions are generally expressed in terms of weight percentages. The weight percentage of a 0.31 M solution of glucose is 5.3 percent. The concentration of NaCl that is isotonic with blood is 0.16 M since NaCl ionizes to form two particles, Na^+ and Cl^- (a 0.155 M solution of NaCl is 0.310 M in particles). A 0.16 M solution of NaCl is 0.9 percent in NaCl. Such a solution is known as a physiological saline solution.

There are many interesting examples of osmosis. A cucumber placed in concentrated brine loses water via osmosis and shrivels into a pickle. A carrot that has become limp because of water loss to the atmosphere can be placed in water. Water moves into the carrot through osmosis, making it once again firm. People eating a lot of salty food experience water retention in tissue cells and intercellular spaces because of osmosis. The

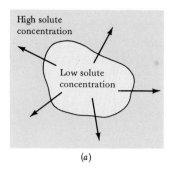
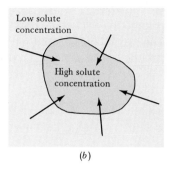

resultant swelling or puffiness is called edema. Movement of water from soil into plant roots and subsequently into the upper portions of the plant is due at least in part to osmosis. The preservation of meat by salting and of fruit by adding sugar protects against bacterial action. Through the process of osmosis, a bacterium on salted meat or candied fruit loses water, shrivels, and dies.

In osmosis, water moves from an area of high water concentration (low solute concentration) into an area of low water concentration (high solute concentration). Such movement of a substance from an area where its concentration is high to an area where it is low is spontaneous. Biological cells transport not only water, but other select materials through their membrane walls. This permits entry of nutrients and allows for disposal of waste materials. In some cases, substances must be moved from an area of low concentration to one of high concentration. This movement is called active transport. It is not spontaneous but requires expenditure of energy by the cell.

DETERMINATION OF MOLECULAR WEIGHT

The colligative properties of solutions provide a useful means of experimentally determining molecular weights. Any of the four colligative properties could be used to determine molecular weight. The procedures are illustrated in Sample Exercises 12.16 and 12.17.

SAMPLE EXERCISE 12.16

A solution of an unknown nonvolatile nonelectrolyte was prepared by dissolving 0.250 g in 40.0 g of CCl_4. The normal boiling point of the resultant solution was increased by 0.357°C. Calculate the molecular weight of the solute.

Solution: Using Equation [12.21] we have

$$\text{Molality} = \frac{\Delta T_b}{K_b} = \frac{0.357°C}{5.02°C/m} = 0.0711\ m$$

Thus the solution contains 0.0711 mol of solute per kilogram of solvent. The solution was prepared from 0.250 g solute and 40.0 g of solvent. The number of grams of solute in a kilogram of solvent is therefore

$$\frac{\text{Grams solute}}{\text{Kilograms }CCl_4} = \left(\frac{0.250\ \text{g solute}}{40.0\ \text{g }CCl_4}\right)\left(\frac{1000\ \text{g }CCl_4}{1\ \text{kg }CCl_4}\right)$$

$$= \frac{6.25\ \text{g solute}}{1\ \text{kg }CCl_4}$$

Notice that a kilogram of solvent contains 6.25 g, which from the ΔT_b measurement must be 0.0711 m. Therefore

$$0.0711\ \text{mol} = 6.25\ \text{g}$$

$$1\ \text{mol} = \frac{6.25\ \text{g}}{0.0711} = 87.9\ \text{g}$$

Therefore

$$MW = 87.9\ \text{amu}$$

SAMPLE EXERCISE 12.17

The osmotic pressure of 0.200 g of hemoglobin in 20.0 mL of solution is 2.88 mm Hg at 25°C. Calculate the molecular weight of hemoglobin.

Solution:

$$\pi = MRT$$

$$M = \frac{\pi}{RT} = \frac{2.88 \text{ mm Hg}}{\left(0.0821 \dfrac{\text{L-atm}}{\text{K-mol}}\right)(298 \text{ K})}\left(\frac{1 \text{ atm}}{760 \text{ mm Hg}}\right)$$

$$= 1.55 \times 10^{-4} \frac{\text{mol}}{\text{L}}$$

A liter of solution contains

$$\left(\frac{0.200 \text{ g solute}}{20.0 \text{ mL soln}}\right)\left(\frac{1000 \text{ mL soln}}{1 \text{ L soln}}\right) = 10.0 \text{ g}$$

Therefore

$$1.55 \times 10^{-4} \text{ mol} = 10.0 \text{ g}$$

$$1 \text{ mol} = \frac{10.0 \text{ g}}{1.55 \times 10^{-4}} = 6.45 \times 10^4 \text{ g}$$

Therefore

$$MW = 64{,}500 \text{ amu}$$

Osmotic pressure is particularly useful for measuring the molecular weights of high-molecular-weight substances.

12.8 Colloids

When finely divided clay particles are dispersed through water, they do not remain suspended but eventually settle out of the water. The dispersed clay particles are much larger than molecules and consist of many thousands or even millions of atoms. In contrast, the dispersed particles of a solution are of molecular size. Between these extremes is the situation in which dispersed particles are larger than molecules, but not so large that the components of the mixture separate under the influence of gravity. These intermediate types of dispersions or suspensions are called colloidal dispersions or simply colloids. Thus colloids are on the dividing line between solutions and heterogeneous mixtures. Like solutions, colloids can be either gases, liquids, or solids. Examples of each are listed in Table 12.6.

The size of the dispersed particle is the property used to classify a mixture as a colloid. Colloid particles range in diameter from approximately 10 Å to 2000 Å, whereas solute particles are smaller. The colloid particle may consist of many atoms, ions, or molecules or may even be a single giant molecule. For example, the hemoglobin molecule, which

TABLE 12.6 Types of colloids

Phase of colloid	Dispersing (solventlike) substance	Dispersed (solutelike) substance	Colloid type	Example
Gas	Gas	Gas	—	None (all are solutions)
Gas	Gas	Liquid	Aerosol	Fog
Gas	Gas	Solid	Aerosol	Smoke
Liquid	Liquid	Gas	Foam	Whipped cream
Liquid	Liquid	Liquid	Emulsion	Milk
Liquid	Liquid	Solid	Sol	Paint
Solid	Solid	Gas	Solid foam	Marshmallow
Solid	Solid	Liquid	Solid emulsion	Butter
Solid	Solid	Solid	Solid sol	Ruby glass

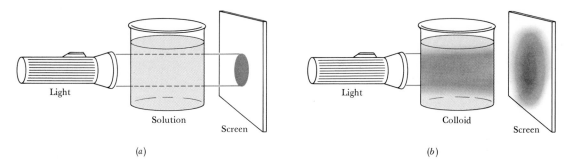

FIGURE 12.16 Comparison of the action of light on a true solution and a colloid. The light beam is shown in (a) as passing through the solution without scattering. However, the particles of a colloid scatter the light, so that the outline of the beam is visible, as illustrated by the colored section in (b). The beam is diffused, and thus may not appear distinctly on the screen.

carries oxygen in blood, has molecular dimensions of $65 \times 55 \times 50$ Å and a molecular weight of 64,500 amu. Colloids can be made either by breaking up large pieces of matter or by clustering small particles into colloidal size.

Even though colloid particles are so small that the dispersion appears uniform even under a microscope, they are large enough to scatter light very effectively. Consequently, colloids usually appear cloudy or opaque. Furthermore, because they scatter light, a light beam can be seen as it passes through a colloidal suspension as shown in Figure 12.16. This scattering of light by colloidal particles is known as the Tyndall effect. Such light scattering makes it possible to see the light beam coming from the projection housing in a smoke-filled theatre or the light beam from an automobile on a dusty dirt road.

HYDROPHILIC AND HYDROPHOBIC COLLOIDS

The most important colloids are those in which the dispersing medium is water. Such colloids are frequently referred to as hydrophilic (water loving) or hydrophobic (water hating). Hydrophilic colloids are most like the solutions that we have previously examined. In the human body, the extremely large molecules that make up such important substances as enzymes and antibodies are kept in suspension by interaction with surrounding water molecules. The molecules fold so that polar or charged groups can interact with water molecules at the periphery of the molecules. These hydrophilic groups generally contain oxygen or nitrogen. Some examples are shown in Figure 12.17.

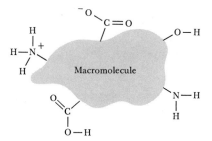

FIGURE 12.17 Examples of hydrophilic groups at the surface of a giant molecule (macromolecule) that help keep the macromolecule suspended in water.

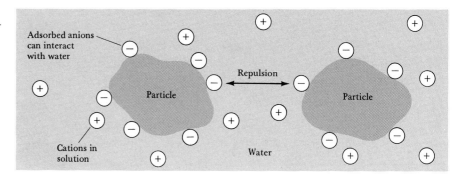

FIGURE 12.18 Schematic representation of the stabilization of a hydrophobic colloid by adsorbed ions.

Hydrophobic colloids can be prepared in water only if they are stabilized in some way. Otherwise, their natural lack of affinity for water causes them to separate from the water. Hydrophobic colloids can be stabilized by adsorption of ions on their surface as shown in Figure 12.18. These adsorbed ions can interact with water, thereby stabilizing the colloid. At the same time the mutual repulsion between colloid particles with adsorbed ions of the same charge keeps the particles from colliding and thereby getting larger. Hydrophobic colloids can also be stabilized by the presence of other hydrophilic groups on their surfaces. For example, small droplets of oil are hydrophobic. They do not remain suspended in water; instead they separate, forming an oil slick on the surface of the water. Addition of sodium stearate, whose structure is shown at left, or any similar substance having one end that is hydrophilic (polar or charged) and one that is hydrophobic (nonpolar), will stabilize a suspension of oil in water as shown in Figure 12.19. The hydrophobic ends of the stearate ions interact with the oil droplet, whereas the hydrophilic ends point out toward the water with which they interact.

The cleansing action of a soap or detergent is due to its ability to stabilize hydrophobic colloids. These cleansing agents contain substances such as sodium stearate. The action of such molecules in removing dirt from a piece of cloth is illustrated in Figure 12.20.

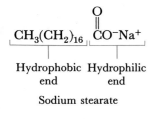

Sodium stearate

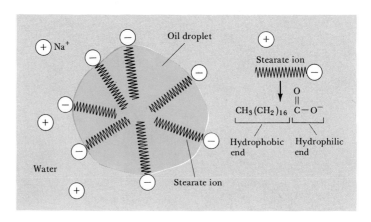

FIGURE 12.19 Stabilization of an emulsion of oil in water by stearate ions.

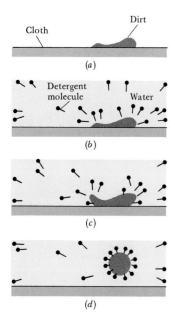

FIGURE 12.20 Schematic representation of the action of detergent ions on dirt. (a) Dirt on a piece of cloth. (b) Detergent molecules in water attach their hydrophobic ends to the dirt and (c) begin to lift it. (d) Detergent surrounds the dirt and holds it in suspension so that it can be washed away.

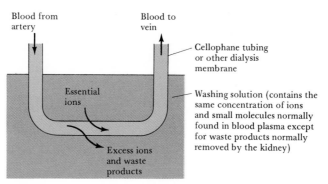

FIGURE 12.21 An illustration of dialysis as applied in the operation of an artificial-kidney machine. A dialysis membrane permits small particles to pass from an area of high concentration to one of low concentration. Colloid-sized particles, such as hemoglobin, do not pass through the membrane.

REMOVAL OF COLLOID PARTICLES

It is often desirable to remove colloidal particles from a dispersing medium, as in the removal of smoke from stacks or a sol from a liquid. This removal, which cannot be accomplished by simple filtration, is sometimes referred to as coagulation. It is normally accomplished in one of two ways: (1) by heating or (2) by adding an electrolyte to the mixture. When a suspension is heated, the increased molecular motion of the particles increases collisions between particles. Whenever they stick together following collision, the colloid particle increases in size. As the particle increases in size, it eventually settles out of solution. The effect of added electrolytes is to neutralize the charge of ions adsorbed on the surface of hydrophobic colloids. The effect of electrolytes is seen in the depositing of suspended clay in a river as it mixes with salt water. This results in the formation of river deltas wherever rivers empty into oceans or other salty bodies of water.

Semipermeable membranes can be used to separate ions from colloidal particles; the ions can pass through the membrane, whereas the colloid particles cannot. This type of separation is known as dialysis. This process is used in the purification of blood in artificial kidney machines (see Figure 12.21).

In the genetic disease known as sickle-cell anemia, hemoglobin molecules are abnormal, in that they have lower solubility, especially in the unoxygenated form. Consequently as much as 85 percent of the hemoglobin in the red blood cells crystallizes from solution. This distorts the cells into a sickle shape as shown in Figure 12.22. These clog the capillaries, thus causing gradual deterioration of the vital organs. The disease is hereditary, and if both parents carry the defective genes, it is likely that the child will possess only abnormal hemoglobin. Such children seldom survive more than a few years after birth. The reason for the insolubility of hemoglobin in sickle-cell anemia can be traced to a change in one part of the molecule. The normal atom group shown at right is replaced at one place in the hemoglobin molecule by the abnormal combination. The normal group of atoms has a polar group that can interact with water, thereby enhancing the solubility of the hemoglobin. The abnormal group of atoms is nonpolar (hydrophobic), and its presence leads to the aggregation of this defective form of hemoglobin into particles too large to remain suspended in biological fluids.

$$-CH_2-CH_2-\overset{\overset{\displaystyle O}{\|}}{C}-OH \qquad -\underset{\underset{\displaystyle CH_3}{|}}{CH}-CH_3$$

Normal   Abnormal

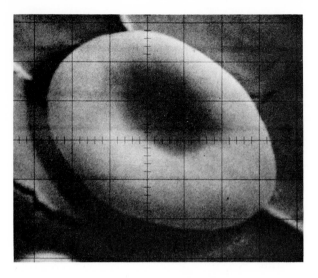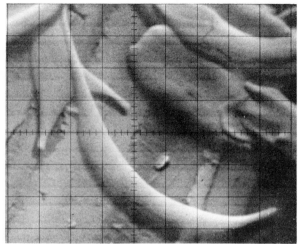

FIGURE 12.22 Electron micrograph showing a normal red blood cell (left; magnification 10,000 times) and sickled red blood cells (right; magnification 5000 times). (*Courtesy Philips Electronic Instruments, Inc.*)

FOR REVIEW

Summary

Solutions are homogeneous mixtures of atoms, ions, or molecules. The relative amounts of solute and solvent in a solution can be described qualitatively (dilute or concentrated solutions) or quantitatively (weight percentage, molarity, molality, normality, and mole fraction were discussed).

The extent to which a solute will dissolve in a particular solvent depends on the relative magnitudes of solute-solute, solute-solvent, and solvent-solvent attractive forces as well as on the changes in disorder accompanying mixing. The rule "likes dissolve likes" was found to be useful in rationalizing solubilities. It is possible to change the solubility of a solute by changing temperature and pressure. If the solution process is endothermic, an increase in temperature promotes solubility. With a gas, an increase in pressure promotes solubility. These effects can be understood in terms of LeChatelier's principle.

Substances that exist in solution as ions are called electrolytes. Those substances that are completely ionized in solution are called strong electrolytes. Reactions occur between electrolytes if an insoluble substance, a gas, or a nonelectrolyte can form. Net ionic equations focus attention on the particular species that actually undergo some change during the reaction.

The presence of a solute in a solvent lowers the vapor pressure and the freezing point and increases the boiling point of the solvent. These changes are termed colligative properties. The magnitude of the change depends on the total concentration of solute particles in solution and not on their characteristics. Osmotic pressure is the pressure that must be applied to a solution to prevent transfer of solvent molecules from a pure solvent through a semipermeable membrane. It is a colligative property because it is proportional to the total concentration of solute particles. Colligative properties can be used to calculate the molecular weights of nonvolatile nonelectrolytes.

True solutions can be differentiated from colloids on the basis of particle size. Colloids play an important role in many chemical and biological systems.

Learning goals

Having read and studied this chapter, you should be able to:

1 Define molarity, molality, mole fraction, normality, and weight percentage and calculate concentrations in any of these concentration units.

2 Convert concentration in one concentration unit into any other (given the density of the solution where necessary).

3 Give definitions of the qualitative terms used to describe solutions: dilute, concentrated, saturated, unsaturated, and supersaturated.

4 Describe the solution process, including the bonds made and broken when a substance dissolves.

5 Describe the energy changes that occur in the solution process in terms of the attractive forces

that operate in the solvent and solute and the attractive forces between solvent and solute.

6 Describe the role of disorder in the solution process.

7 Rationalize the solubilities of substances in various solvents in terms of their molecular structures and intermolecular forces.

8 Discuss the effects of pressure and temperature on solubilities of gases.

9 Relate the effect of temperature on the solubilities of salts in water to the enthalpy change in the solution process.

10 State LeChatelier's principle and apply it to rationalize the effects of temperature and pressure on solubilities.

11 Predict which substances are electrolytes and which are nonelectrolytes; predict which electrolytes are strong electrolytes and which are weak.

12 Predict the course of reactions involving formation of a precipitate, gas, or nonelectrolyte, and write net ionic equations for these reactions.

13 Describe the effect of solute concentration on solvent vapor pressure; state Raoult's law.

14 Explain the origin of the colligative properties—boiling-point elevation, freezing-point lowering, and osmotic pressure—of solutions.

15 Determine the molecular weight of a solute from the magnitude of the effect of a known concentration of solute on one of the colligative properties of a solvent.

16 Explain the difference between the magnitude of changes in colligative properties caused by electrolytes compared to those caused by nonelectrolytes.

17 Describe how a colloid differs from a true solution.

Key terms

Among the more important terms and expressions used for the first time in this chapter are the following:

Colligative properties (Section 12.7) are those properties of a solvent (vapor-pressure lowering, freezing-point lowering, boiling-point elevation, osmotic pressure) that depend on the total concentration of solute particles present.

Colloids (Section 12.8) are particles that are larger than normal molecules, but that are nevertheless small enough to remain suspended in a dispersing medium indefinitely.

Dialysis (Section 12.8) is the separation of small solute particles from colloid particles by means of a semipermeable membrane.

An electrolyte (Section 12.5) is a solute that gives a solution containing ions; such a solution will conduct an electrical current. Strong electrolytes (strong acids, strong bases, and most common salts) are completely ionized in solution. Weak electrolytes are partially ionized in solution.

Henry's law (Section 12.4) states that the solubility of a gas in a liquid, C_g, is proportional to the pressure of gas over the solution: $C_g = kP_g$.

LeChatelier's principle (Section 12.4) states that any attempt to change the conditions of a system at equilibrium results in a shift in the position of the equilibrium in the direction that tends to offset the change.

Molality (Section 12.1) is the concentration of a solution expressed as moles of solute per kilogram of solvent; abbreviated m.

Mole fraction (Section 12.1) is the ratio of the number of moles of one component of a solution to the total number of moles of all substances present in the solution; abbreviated X, with a subscript identifying the component.

A net ionic equation (Section 12.6) is an equation for a reaction involving ions that is obtained by eliminating spectator ions (those ions that go through the reaction unchanged and appear on both sides of the overall ionic equation).

Normality (Section 12.1) is the concentration of a solution expressed as equivalents of solute per liter of solution; abbreviated N. An equivalent is defined according to the type of reaction being considered; 1 equivalent of a given reactant will react with 1 equivalent of a second reactant.

Osmosis (Section 12.7) is the net movement of solvent through a semipermeable membrane toward the solution with greatest solute concentration. The osmotic pressure of a solution is the pressure that must be applied to a solution to stop osmosis from pure solvent into the solution.

Raoult's law (Section 12.7) states that the partial pressure of a solvent over a solution, P_A, is given by the vapor pressure of the pure solvent, P_A°, times the mole fraction of solvent in the solution, X_A: $P_A = X_A P_A^\circ$.

A saturated solution (Section 12.2) is a solution in which undissolved solute and dissolved solute are in equilibrium. The solubility of a substance is the amount of solute that dissolves to form a saturated solution. Solutions containing less solute than this are said to be unsaturated, whereas those containing more are said to be supersaturated.

Solvation (Section 12.2) is the clustering of solvent molecules around a solute particle. When the solvent is water, this clustering is known as hydration.

The **Tyndall effect** (Section 12.8) describes the visible path of a light beam passing through a colloidal dispersion caused by scattering of light by the colloid particles.

The **weight percentage** (Section 12.1) of a solution is the number of grams of solute it contains in each 100 g of solution.

EXERCISES

Concentrations of solutions

12.1 Calculate the number of moles of solute present in each of the following solutions: (a) 256 mL of 0.358 M $Ca(NO_3)_2$; (b) 4.00×10^4 L of 0.0567 M HBr; (c) 450 g of an aqueous NaCl solution that is 0.565 percent NaCl by weight.

12.2 Calculate the molarity in each of the following solutions: (a) one containing 16.0 g $CaCl_2$ in 4.00×10^2 mL of solution; (b) 16.0 mL of a 6.0 M H_3PO_4 solution diluted to 462 mL; (c) an aqueous HCl solution in which $X_{HCl} = 0.160$, density 1.18 kg/L.

12.3 For each of the following solutions—(i) 3.6 g of NaCl in 340 g of water; (ii) 4.45 g of KBr in 564 g of CH_3CH_2OH; (iii) 35.2 g of ethylene glycol $(HOCH_2CH_2OH)$ in 65.0 g water; (iv) 2.4 g NaCl, 3.6 g of KBr in 82 g water—(a) calculate the weight fraction of each solute; (b) calculate the mole fractions of solute and solvent; (c) calculate the molality of each solution.

12.4 For each of the following solutions—(i) 12.0 g of acetone, C_3H_6O, in 75.0 g of water; (ii) 3.22 g of $La(NO_3)_3$ in 1.46 kg of water; (iii) 1.26 g of KBr in 560 g of liquid ammonia, NH_3; (iv) 16.8 g of acetone (C_3H_6O), 1.65 g of $Al(NO_3)_3$, 142 g of H_2O—(a) calculate the weight fraction of each solute; (b) calculate the mole fraction of each component present; (c) calculate the molality of each solution.

12.5 Describe how you would prepare each of the following aqueous solutions, starting with solid KBr: (a) 2.40 L of 2.00×10^{-2} M solution of KBr; (b) 150 g of a 0.420 m solution of KBr; (c) 2.5 L of a 14 percent by weight solution of KBr; (d) a 0.200 M solution of KBr that would contain just enough KBr to precipitate 26.0 g of AgBr from a solution containing 0.44 mol of $AgNO_3$.

12.6 What mass of $La(NO_3)_3$ is dissolved in each 1.00×10^2 g of water in each of the following: (a) a 0.050 m solution; (b) an aqueous solution in which $X_{La(NO_3)_3} = 0.0105$; (c) a solution that is 1.26 percent $La(NO_3)_3$ by weight.

12.7 A solution containing 66.0 g of acetone, C_3H_6O, and 46.0 g of H_2O has a density of 0.926 g/mL. Calculate (a) the weight and mole percentages of solute and solvent; (b) the molarity of acetone in the water.

12.8 A solution is made up by dissolving 1.68 g of benzoic acid, $C_7H_6O_2$, in 206 mL of CCl_4, density 1.59 g/cm³; another is prepared by dissolving the same quantity of benzoic acid in 206 mL of ethanol (C_2H_5OH), density 0.782 g/mL. (a) Calculate the mole fractions and molalities of solute in each case. (b) Assuming the density of the solution is the same as that of pure solvent, calculate the molarity in each case. How are molarity and molality related in each of these solvents?

12.9 (a) In a certain reaction Sn^{4+} is reduced to Sn^{2+}. What is the normality of a 0.36 M solution of Sn^{4+}? What is the molarity of a 0.42 N solution of Sn^{4+}? (b) What is the molarity of a 0.105 N solution of H_3PO_4 reacted with NaOH solution, in which all three hydrogens of the H_3PO_4 react?

12.10 In each of the following reactions indicate the normality of a 0.1 M solution of the reagent underlined:

(a) $\underline{H_3PO_4}(aq) + \mathbf{3}NaOH(aq) \longrightarrow$
$Na_3PO_4(aq) + 3H_2O(l)$

(b) $BaCl_2(aq) + \underline{H_2SO_4}(aq) \longrightarrow BaSO_4(s) + 2HCl(aq)$

(c) $MnS(s) + \underline{2HCl}(aq) \longrightarrow MnCl_2(aq) + H_2S(g)$

12.11 Methylcyclopentadienyl manganese tricarbonyl (MMT for short), $C_9H_7O_3Mn$, has been used as an antiknock agent in lead-free gasolines in place of tetraethyl lead. It is employed at a concentration level of 0.031 g per gallon of gasoline. If the gasoline has a density of 0.78 kg/L, calculate the molarity and molality of MMT in the gasoline (assume the density of the solution is the same as that of the gasoline itself).

The solution process; solubility and structure

12.12 Indicate the type of solute-solvent intermolecular attractive force (Section 11.5) that should be most important in each of the following solutions: (a) KCl in water; (b) benzene, C_6H_6, in CCl_4; (c) HF in water; (d) acetonitrile (CH_3CN) in acetone, C_2H_6O (see Table 11.6).

12.13 When CsCl dissolves in water, 19.87 kJ of heat is absorbed from the surroundings per mole of salt. We know that the cation and anion both interact strongly with water molecules, in the manner illustrated for NaCl in Figures 12.1 and 12.2. Why is the overall process endothermic in spite of this interaction?

12.14 Explain in terms of the enthalpy changes in the various steps involved (Figure 12.3) why KBr is not soluble in CCl_4.

12.15 Offer an explanation, in terms of the intermolecular forces involved, for each of the following observations: (a) $CoCl_2$ is soluble to the extent of 540 g per 1 kg of ethyl alcohol. (b) Water is completely miscible with dioxane, for which the structural formula is

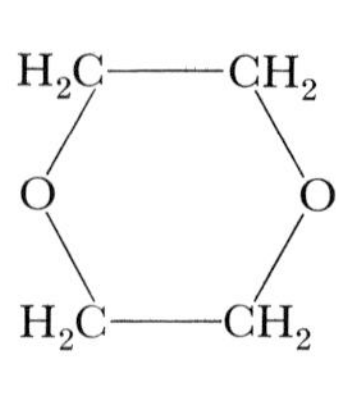

H_2C————CH_2
⟋   ⟍
O O
⟍   ⟋
H_2C————CH_2

However, water is not soluble in cyclohexane, for which the structural formula is

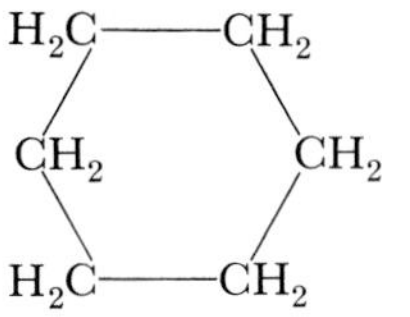

$$H_2C-CH_2$$
$$CH_2 \quad CH_2$$
$$H_2C-CH_2$$

(c) Chloroform $CHCl_3$, is soluble in water to the extent of only 1 g per 100 g of water, but is miscible with ethyl alcohol.

12.16 Which of the following observed processes proceeds spontaneously because of an increase in the randomness, or disorder, of the process: (a) evaporation of solid I_2 crystals; (b) mixing of $Br_2(l)$ and $CCl_4(l)$; (c) precipitation of $BaSO_4(s)$ from aqueous solution; (d) alignment of iron filings in a magnetic field.

12.17 Give an example of a process that occurs spontaneously because of a decrease in the energy content of the system, although it results in a *decrease* in randomness or disorder.

12.18 Indicate a reasonable explanation for the following relative solubilities in water, in terms of solute-solvent interactions: (a) ethyl cyanide (CH_3CH_2CN), very soluble, but not miscible; (b) ethyl alcohol (CH_3CH_2OH), miscible; (c) ethyl chloride (CH_3CH_2Cl), 0.5 g/100 g H_2O; (d) ethyl sulfide (CH_3CH_2SH), 1.5 g/100 g H_2O

12.19 In terms of the concepts developed in this chapter, indicate why each of the following commercial products is formulated as it is: (a) gas-line antifreeze additive to the gas tank consists almost entirely of methyl alcohol; (b) the solvent in a water-repellent spray for treating shoes is methylene chloride, CH_2Cl_2; (c) A wax remover for cross-country skis consists of benzene and similar aromatic hydrocarbons (Section 8.4).

12.20 Which of the following pairs of liquids would you expect to be miscible and which nonmiscible? Explain in each case: (a) H_2O and $HOCH_2CH(OH)CH_2OH$; (b) $C(CH_2OH)_4$ and C_7H_{16}; (c) $Hg(l)$ and $NaI(l)$; (d) CH_2Cl_2 and $CH_3CH_2OCH_2CH_3$.

Effect of temperature and pressure on solubility

12.21 State Henry's law. From the data listed in Table 12.3, which has the larger Henry's-law constant in water at $20°C$, N_2 or O_2?

12.22 Using Henry's law and the partial pressure of O_2 in the air (Table 10.1), calculate the molar solubility of O_2 in the surface waters of a lake at $20°C$ from the data in Table 12.3.

12.23 The solubility of CO_2 in water at $18°C$, with 1 atm CO_2 pressure over the solution, is 0.0414 M. What is the molar solubility of the CO_2 present in the atmosphere at this temperature (see Table 10.1).

12.24 The Henry's-law constant for helium gas in water at $30°C$ is $3.7 \times 10^{-4} M/atm$; that for N_2 at $30°C$ is $6.0 \times 10^{-4} M/atm$. If the two gases are each present at 0.5 atm pressure over water solution, calculate the solubility of each gas.

12.25 Using Figure 12.9, calculate the Henry's-law constant for CH_4 and CO at $20°C$ and $40°C$.

[12.26] In terms of the kinetic-molecular theory, explain why the solubility of gases in liquids generally decreases with an increase in temperature.

12.27 The dissolving of $CuCl_2$ in water is accompanied by the evolution of 46.4 kJ/mol $CuCl_2$. Using LeChatelier's principle, predict the effect of temperature on the solubility of $CuCl_2$.

Electrolyte solutions

12.28 Classify each of the following substances as a nonelectrolyte, weak electrolyte, or strong electrolyte in water: (a) benzene; (b) ethyl alcohol; (c) HF; (d) $Ba(NO_3)_2$; (e) H_2SO_4; (f) CH_3NH_2: (g) $HClO$; (h) O_3; (i) $LiOH$. Identify those that are acids, bases, or salts.

12.29 A 0.01 M solution of $HgCl_2$, when placed in the beaker as illustrated in Figure 12.11, causes the bulb to glow much more weakly than does a 0.01 M solution of KCl. What conclusion can you draw from this?

12.30 For each of the following solutions, indicate the predominant form in which the solute is present: (a) acetone in water; (b) $CaCl_2$ in molten NaCl; (c) $CaCl_2$ in water; (d) NaOH in water; (e) $HClO_4$ in water.

12.31 Indicate the total concentration of all solute species present in each of the following aqueous solutions: (a) 0.1 M NaOH; (b) 0.35 M $CaBr_2$; (c) 0.14 M CH_3CH_2OH; (d) a mixture of 50 mL of 0.10 M $KClO_3$ and 50 mL of 0.20 M Na_2SO_4.

Reactions in aqueous solution

12.32 Summarize the "driving forces" that can cause aqueous reactions to proceed essentially to completion and give one example of each.

12.33 Using the following reagents—KOH, $MgCl_2$, $HClO_4$, $BaCO_3$—write balanced complete ionic and net ionic equations to describe (a) a neutralization reaction; (b) a precipitation reaction; (c) a reaction in which a gas is evolved.

12.34 Write balanced ionic and net ionic equations for the reaction that occurs when each of the following pairs is mixed: (a) $H_2SO_4(aq)$ with $BaCl_2(aq)$; (b) $AgNO_3(aq)$ with $Na_2CrO_4(aq)$; (c) $Ag_2S(s)$ with $HCl(aq)$; (d) $K_2SO_3(aq)$ with $H_2SO_4(aq)$; (e) $KOH(aq)$ plus $HNO_3(aq)$. (You may need to refer to the solubility generalizations given in the text.)

12.35 Write balanced molecular, complete ionic, and net ionic equations for each of the following reactions in an aqueous solution: (a) $BaSO_3(s) + HBr(aq)$; (b) $Cd(NH_2)_2(s) + HCl(aq)$; (c) $Mn(OH)_2(s) + HClO_4$.

[12.36] Suppose you have a solution that might contain any or all of the following cations: Ni^{2+}, Ag^+, Sr^{2+}, and Mn^{2+}. Addition of HCl solution causes a precipitate to form. After filtering off the precipitate, H_2SO_4 solution is added to the resultant solution, and another precipitate forms. This is filtered off, and a solution of Na_2CrO_4 is added to the resulting solution. No precipitate is observed.

Which ions are present in the original solution in moderately high concentrations?

12.37 Which of the following salts, insoluble in pure water, would you expect to dissolve upon addition of moderately concentrated HBr solution: (a) $SrCO_3$; (b) $MnSO_3$; (c) $BaSO_4$; (d) CaF_2. Explain.

Colligative properties

12.38 Which of the following properties is a colligative property: (a) heat of vaporization of a liquid; (b) boiling point of a liquid; (c) osmotic pressure of a solution; (d) density of a solution; (e) freezing-point lowering of a solution; (f) variation in solubility of a solute with temperature.

12.39 For nonvolatile solutes, what is the relationship between the vapor pressure over a solution and the mole fraction of solvent in the solution? Why is the vapor-pressure lowering referred to as a colligative property?

12.40 What mass of each of the following ionic substances in 1.00 kg of water would lead to the same vapor-pressure lowering of the solvent as 120 g of sucrose, $C_{12}H_{22}O_{11}$: (a) KBr; (b) $SrCl_2$; (c) HI?

12.41 Using Raoult's law, calculate the vapor pressure of water at 60°C over a solution containing 465 g of glycerin, $C_3H_9O_3$ and 148 g of water (see Appendix C).

[12.42] A constant humidity in a closed container can be achieved by placing in the chamber an aqueous solution that has the desired vapor pressure of water. Provide a recipe for preparing 1 kg of a solution of glycerin $(C_3H_9O_3)$ and water that has a vapor pressure of 14.5 mm Hg at 24°C (see Appendix C).

12.43 A "canned heat" product used to warm chafing dishes consists of a homogeneous mixture of methyl alcohol, CH_3OH, and paraffin that has an average formula $C_{24}H_{50}$. What mass of CH_3OH should be added to 48.0 kg of the paraffin in formulating the mixture if the vapor pressure of the methanol over the mixture at 24°C is to be 15 mm Hg? The vapor pressure of methanol at 24°C is 177 mm Hg.

12.44 List the following aqueous solutions in order of their expected boiling points: 0.03 m glycerin; 0.020 m KBr; 0.018 m $MgCl_2$; 0.030 m benzoic acid $(C_6H_5CO_2H)$, a weak acid.

12.45 Calculate the freezing points and boiling points of each of the following solutions: (a) 1.2 m glucose in ethanol; (b) 3.5 g of CCl_4 in 128 g of benzene; (c) 1.4 g of KNO_3 in 33.6 g of water; (d) 1.86 g of Li_2CrO_4 in 60.0 g of water.

[12.46] Calculate the vapor pressure over a 1.50 m solution of glucose at 60°C (see Appendix C). (Note that you will need to convert from units of molality to units of mole fraction.)

12.47 A dilute sugar solution prepared by mixing 8.0 g of an unknown sugar with 200 g of water is found to have an osmotic pressure of 2.86 atm at 25°C. What is the molecular weight of the sugar?

[12.48] A lithium salt used in a lubricating grease has the formula $LiC_nH_{2n-1}O_2$. The salt is soluble in water to the extent of 0.036 g per 100 g of water at 25°C. The os-

motic pressure of this solution is found to be 57.1 mm Hg. Assuming that molality and molarity in such a dilute solution are the same and that the lithium salt is completely ionized, determine an appropriate value of n in the formula for the salt.

Colloids

12.49 Indicate whether each of the following is a hydrophilic or hydrophobic colloid: (a) the butterfat in homogenized milk; (b) Jello; (c) colloidal gold in water; (d) eggnog.

12.50 Glucose, which has the structure shown in Figure 12.7, is not soluble in benzene. Explain how it might be possible to stabilize a colloidal suspension of glucose in benzene using sodium stearate. Draw a sketch to show how the stabilization would occur.

12.51 Greases used in lubrication consist of mixtures of soaps such as sodium stearate or barium stearate with lubricating oils. The soap acts as a thickener for the oil. Suggest a structure for the colloid; that is, how might the soap molecules be oriented in the oil?

12.52 It is possible to precipitate gold from aqueous solution in such a manner that the gold particles are extremely small. Although the density of gold is 19.3 g/cm^3, such a sol (Table 12.6) is stable indefinitely. In terms of the kinetic-molecular theory, how is it possible for the gold particles to remain dispersed?

12.53 Hair shampoos usually contain a detergent such as sodium lauryl sulfate, $CH_3(CH_2)_{11}OSO_3^-Na^+$. Explain how such a substance might act in removing dirt from hair. Explain how it might be effective in removing naturally occurring oils from the hair.

Additional exercises

12.54 The concentration of H_2SO_4 in a bottle labeled "concentrated sulfuric acid" is 18 M. The solution has a density of 1.84 g/mL. What is the mole fraction and weight percentage of H_2SO_4 in this solution?

12.55 Calculate the molarity of each of the following solutions: (a) one that contains 12.0 g of H_2SO_4 in 600 mL of solution; (b) a 2.0 N solution of H_2SO_4 (used to form SO_4^{2-}); (c) one made by diluting 8.0 mL of 3.0 M H_2SO_4 to a total volume of 35.0 mL.

12.56 Toluene can be added to gasolines to improve octane ratings. Addition of 0.060 gal of toluene (C_7H_8) per gallon of gasoline increases the octane rating of the fuel by one. The density of gasoline is about 0.78 kg/L, whereas that of toluene is 0.87 kg/L. In 1980 U.S. drivers burned about 1.10×10^9 gal of gasoline. If toluene were added to increase the octane number of this quantity of gasoline by one, what mass of toluene would be needed?

12.57 Caffeine has the following molecular structure:

Caffeine

It is soluble in water to the extent of 1.35 g/100 g water at 16°C, 46 g/100 g water at 65°C. (a) What structural features of the caffeine molecule account for its solubility in water? (b) Is the solution of caffeine an endothermic or exothermic process?

12.58 Describe the procedure you would use to determine the solubility of potassium nitrate, KNO_3, in ethyl alcohol at 25°C. What additional experiments would you perform to determine whether the enthalpy change on dissolving is exothermic or endothermic?

12.59 Which of the following ions is likely to be most strongly hydrated in aqueous solution: K^+, Cs^+, Ca^{2+}?

12.60 Indicate whether each of the following proceeds with an increase or decrease in randomness or disorder: (a) crystallization of KBr from alcohol solution; (b) evaporation of liquid CCl_4; (c) melting of solid, crystalline sulfur.

12.61 The following data give the solubility of radon (Rn) in water with 1 atm pressure of the gas over the solution at various temperatures: 20°C, $9.91 \times 10^{-3} M$; 30°C, $7.27 \times 10^{-3} M$; 40°C, $5.62 \times 10^{-3} M$. Is the solution of radon in water an exothermic or endothermic process? Explain.

12.62 A gas sample consisting of a mixture of the noble-gas elements contains 3.5×10^{-6} mole fraction of radon. This gas at a total pressure of 80 atm is shaken in contact with water at 30°C. Using the data of problem 12.61, calculate the molar concentration of radon in the water.

[12.63] Butylated hydroxytoluene (BHT) has the following molecular structure:

BHT

It is widely used as a preservative in a variety of foods, including dried cereals, cooking oils, and canned goods. The average person in the United States consumes about 2 mg of BHT daily. In terms of its structure, would you expect it to be readily excreted from the body or found stored in body fats? (Incidentally, BHT is not known to have any harmful properties; it is, in fact, a known antiviral agent.)

12.64 Using LeChatelier's principle, present an argument for why the solubility of an ionic substance in water should not vary greatly as a function of pressure.

12.65 Distinguish between: (a) molarity and molality; (b) solution and emulsion; (c) hydrophilic and hydrophobic; (d) saturated and supersaturated; (e) electrolyte and nonelectrolyte; (f) strong acid and weak acid; (g) osmosis and dialysis.

[12.66] $Cu(CN)_2$ is insoluble in water, but $CuCl_2$ is soluble. HCN is a very weak acid, but very soluble, in water. Write the net ionic equation for reaction of $Cu(CN)_2(s)$ with aqueous HCl.

12.67 Account for each of the following observations, and write a balanced net ionic chemical equation to represent what occurs: (a) a solid forms upon addition of sodium hydroxide solution to an aqueous solution of manganese chloride; (b) limestone formations, composed mainly of $CaCO_3$, dissolve when in contact with acidic ground water; (c) chloric acid, $HClO_3$, was first prepared by Gay-Lussac, who reacted a barium chlorate solution with dilute H_2SO_4.

12.68 What is the osmotic pressure of 0.64 g of aspirin $(C_9H_8O_4)$ in 200 mL of aqueous solution at 25°C?

12.69 Explain how NaCl or $CaCl_2$ works to melt ice on highways.

12.70 Calculate the freezing and boiling points of a solution formed from 163 g of glycerin, $C_3H_9O_3$, in 3.46 kg of water.

12.71 A sample of seawater taken from the Arctic Ocean freezes at -1.98°C; a sample taken from the middle of the Atlantic Ocean freezes at -2.08°C. What is the total molality of ionic solutes present in each of these solutions?

12.72 Camphor, an organic substance melting at 180°C, was at one time widely used to obtain estimates of molecular weights of organic compounds, because it has a large freezing-point-lowering constant, 37.7°C/m. When 0.125 g of sulfur is finely ground with 3.62 g of camphor, the melting point of the camphor is lowered by 5.08°C. What is the molecular weight of the sulfur in camphor? To what molecular formula does this correspond?

[12.73] Cetyl alcohol, $CH_3(CH_2)_{14}CH_2OH$, has only a low solubility in water, but a tiny amount does dissolve when the alcohol is placed in contact with water. How would you expect this substance to be distributed in the water? Explain.

12.74 Adrenaline, the hormone that triggers release of extra glucose molecules at times of stress or emergency, contains 59.0 percent C, 26.2 percent O, 7.1 percent H, and 7.6 percent N. A solution of 0.64 g of adrenaline in 36.0 g of CCl_4 causes an elevation of 0.49°C in the boiling point. What are the molecular weight and molecular formula of adrenaline?

Rates of chemical reactions

Chemistry is by its very nature concerned with change. Substances with well-defined properties are converted by chemical reactions into other materials with different properties. Chemists want to know which new substances are formed from a given set of starting reactants. However, it is equally important to know how rapidly chemical reactions occur and to understand the factors that control their speeds. For example, what factors are important in determining how rapidly foods spoil? What determines the rate at which steel rusts? How does one design a rapidly setting material for dental fillings? What factors control the rate at which fuel burns in an auto engine, and how does burning rate determine the pollutant content of the engine exhaust?

The area of chemistry concerned with the speeds, or rates, at which chemical reactions occur is called kinetics. In this chapter we learn how to express and determine the rates at which reactions occur. We shall also learn how reaction rates are affected by variables such as concentration, temperature, and the presence of catalysts.

13.1 Reaction rate

The speed of any event is measured by the change occurring in a given interval of time. For example, the speed of an automobile expresses its change in position in a certain time; the associated units are usually miles per hour (mi/hr). Similarly, the speed or rate of a reaction is expressed as the change in the concentration of a reactant or product in a certain time; the units with which we express this speed are usually molarity per second (M/sec). As an example, consider the reaction that occurs when butyl chloride, C_4H_9Cl, is placed in water. The resulting reaction produces butyl alcohol, C_4H_9OH, and hydrochloric acid:

$$C_4H_9Cl(l) + H_2O(l) \longrightarrow C_4H_9OH(aq) + HCl(aq) \qquad [13.1]$$

Suppose that we begin with a 0.1000 M solution of C_4H_9Cl in water and then begin to measure the C_4H_9Cl concentration at different times after the solution is mixed. We could in this way collect the data shown in the first two columns of Table 13.1. The average rate of the reaction is given by the change in concentration of C_4H_9Cl within a corresponding interval of time:

$$\text{Average rate} = \frac{\text{change in concentration of } C_4H_9Cl}{\text{corresponding time interval}} = -\frac{\Delta[C_4H_9Cl]}{\Delta t} \quad [13.2]$$

The brackets around C_4H_9Cl in this equation indicate the concentration of that substance. The Greek letter delta, Δ, is read "change in"; $\Delta[C_4H_9Cl]$ is the change in the concentration of C_4H_9Cl:

$$\Delta[C_4H_9Cl] = [C_4H_9Cl]_{final} - [C_4H_9Cl]_{initial} \quad [13.3]$$

Likewise, Δt, the corresponding time interval, is the amount of time between the beginning and the end of the interval. The negative sign in Equation [13.2] indicates that the concentration of C_4H_9Cl is decreasing with time.

Notice in Table 13.1 that during the first interval of 50 sec, the concentration of C_4H_9Cl decreases from 0.1000 M to 0.0905 M. Thus the average rate over this 50-sec interval is

$$\text{Average rate} = -\frac{(0.0905 - 0.1000)\, M}{(50 - 0)\ \text{sec}} = 1.90 \times 10^{-4}\, M/\text{sec}$$

We can calculate the average rates over other time intervals in the same fashion. Average rates over other intervals are shown in the third column of Table 13.1. Notice that the average rate steadily decreases as the reaction proceeds. At some point the reaction stops; that is, there is no longer any change in concentration with time.

We can also see this decrease in rate if we display the data graphically, as in Figure 13.1. The dots represent the experimental data from the first

TABLE 13.1 Rate data for reaction of C_4H_9Cl with water

Time (sec)	$[C_4H_9Cl]$ (M)	Average rate (M/sec)
0	0.1000	
		1.90×10^{-4}
50	0.0905	
		1.70×10^{-4}
100	0.0820	
		1.58×10^{-4}
150	0.0741	
		1.40×10^{-4}
200	0.0671	
		1.22×10^{-4}
300	0.0549	
		1.01×10^{-4}
400	0.0448	
		0.80×10^{-4}
500	0.0368	
		0.56×10^{-4}
800	0.0200	
10,000	0	

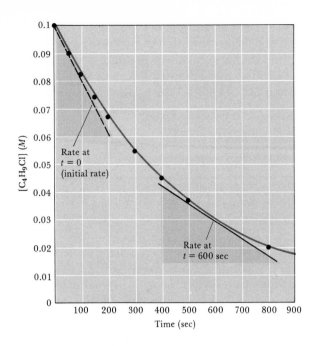

FIGURE 13.1 Concentration of butyl chloride as a function of time. The reaction rate at any time is given by the slope of the tangent to the curve at that time. Tangents have been drawn that touch the curve at $t = 0$ and at $t = 600$ sec.

two columns of Table 13.1; the color line is the smooth curve drawn to connect the data points. Using such a curve we can determine the instantaneous rate; this is the rate at a particular time as opposed to the average rate over an interval of time. The instantaneous rate is obtained from the straight-line tangent that touches the curve at the point of interest. We have drawn two such tangents on Figure 13.1, one at $t = 0$ and the other at $t = 600$ sec. The slopes of these tangents give the instantaneous rates at these times.* For example, at 600 sec we have

$$\text{Instantaneous rate} = -\frac{(0.017 - 0.042)\,M}{(800 - 400)\,\text{sec}} = 6.2 \times 10^{-5}\,M/\text{sec}$$

We will usually refer to the instantaneous rate merely as the rate.

Whenever we discuss either the rate or average rate of a reaction we need to define which substance we are using as a reference. In the reaction

$$2HI(g) \longrightarrow H_2(g) + I_2(g)$$

we can measure the rate of disappearance of HI or the appearance of either H_2 or I_2. Because 2 mol of HI disappear for each mole of H_2 or I_2 that forms, the rate of disappearance of HI is twice the rate of appearance of H_2 or I_2:

$$\left. -\frac{\Delta[HI]}{\Delta t} = 2\frac{\Delta[H_2]}{\Delta t} = 2\frac{\Delta[I_2]}{\Delta t} \right\} \qquad [13.4]$$

*You may wish to briefly review the idea of graphical determination of slopes by referring to Appendix A.

SAMPLE EXERCISE 13.1

(a) Using the data in Table 13.1, calculate the average rate of disappearance of C_4H_9Cl over the time interval from 50 to 150 sec. (b) Using Figure 13.1, estimate the instantaneous rate of disappearance of C_4H_9Cl at $t = 0$ (the initial rate). (c) How is the rate of disappearance of C_4H_9Cl related to the rate of appearance of C_4H_9OH?

Solution: (a) From Table 13.1 we have

$$\text{Average rate} = -\frac{\Delta[C_4H_9Cl]}{\Delta t}$$

$$= -\frac{(0.0741 - 0.0905)\,M}{(150 - 50)\,\text{sec}}$$

$$= 1.64 \times 10^{-4}\,M/\text{sec}$$

(b) The initial rate is given by the slope of the dashed line in Figure 13.1. The slope of a straight line is given by the change in the vertical axis divided by the corresponding change in the horizontal axis. The straight line falls from $[C_4H_9Cl] = 0.100$ to $0.060\,M$ in the time change from 0 to 200 sec. Thus the instantaneous rate is

$$\text{Rate} = -\frac{(0.060 - 0.100)\,M}{(200 - 0)\,\text{sec}}$$

$$= 2.0 \times 10^{-4}\,M/\text{sec}$$

(c) Because 1 mol of C_4H_9OH forms for each mole of C_4H_9Cl that disappears,

$$-\frac{\Delta[C_4H_9Cl]}{\Delta t} = \frac{\Delta[C_4H_9OH]}{\Delta t}$$

13.2 Dependence of reaction rate on concentrations

The decreasing rate of reaction with passing time that is evident in Figure 13.1 is quite typical of reactions. Reaction rates diminish as the concentrations of reactants diminish. Conversely, rates generally increase when reactant concentrations are increased.

One way of studying the effect of concentration on reaction rate is to determine the way in which the rate at the beginning of a reaction depends on the starting concentrations. To illustrate this approach, consider the following reaction:

$$NH_4^+(aq) + NO_2^-(aq) \longrightarrow N_2(g) + 2H_2O(l) \qquad [13.5]$$

We might study the rate of this reaction by measuring the concentration of NH_4^+ or NO_2^- as a function of time or by measuring the volume of N_2 collected. Because of the 1:1 stoichiometry of the reaction, all of these rates will be equal. If we determine the initial reaction rate (the instantaneous rate at $t = 0$) for various starting concentrations of NH_4^+ and NO_2^- we could collect the data shown in Table 13.2. These data indicate

TABLE 13.2 Rate data for the reaction of ammonium and nitrite ions in water at 25°C

Experiment number	Initial NO_2^- concentration (M)	Initial NH_4^+ concentration (M)	Observed initial rate (M/sec)
1	0.0100	0.200	5.4×10^{-7}
2	0.0200	0.200	10.8×10^{-7}
3	0.0400	0.200	21.5×10^{-7}
4	0.0600	0.200	32.3×10^{-7}
5	0.200	0.0202	10.8×10^{-7}
6	0.200	0.0404	21.6×10^{-7}
7	0.200	0.0606	32.4×10^{-7}
8	0.200	0.0808	43.3×10^{-7}

that changing either $[NH_4^+]$ or $[NO_2^-]$ changes the reaction rate. Notice that if we double $[NO_2^-]$ while holding $[NH_4^+]$ constant, the rate doubles (compare experiments 1 and 2). If $[NO_2^-]$ is increased by a factor of four (compare experiments 1 and 3) the rate changes by a factor of four, and so forth. These results indicate that the rate is directly proportional to $[NO_2^-]$. When $[NH_4^+]$ is similarly varied while $[NO_2^-]$ is held constant, the rate is affected in the same manner. We conclude that the rate is also directly proportional to the concentration of NH_4^+. We can express the overall concentration dependence in the following way:

$$\text{Rate} = k[NH_4^+][NO_2^-] \qquad [13.6]$$

The proportionality constant, k, in Equation [13.6] is called the rate constant. We can evaluate the magnitude of k using the data in Table 13.2. Using the results of experiment 1, and substituting into Equation [13.6], we have

$$5.4 \times 10^{-7}\, M/\text{sec} = k(0.0100\, M)(0.200\, M)$$

Solving for k gives

$$k = \frac{5.4 \times 10^{-7}\, M/\text{sec}}{(0.0100\, M)(0.200\, M)} = 2.7 \times 10^{-4}/M\text{-sec}$$

You may wish to satisfy yourself that this same value of k is obtained using any of the other experimental results given in Table 13.2. You might also note that given $k = 2.7 \times 10^{-4}/M$-sec and using Equation [13.6], we can calculate the rate for any concentration of NH_4^+ and NO_2^-. Suppose $[NH_4^+] = 0.100\, M$ and $[NO_2^-] = 0.100\, M$, then

$$\text{Rate} = (2.7 \times 10^{-4}/M\text{-sec})(0.100\, M)(0.100\, M) = 2.7 \times 10^{-6}\, M/\text{sec}$$

An equation like [13.6] that relates the rate of a reaction to concentration is called a rate law. *The rate law for any chemical reaction must be determined experimentally; it cannot be predicted by merely looking at the chemical equation.* The following are some additional examples of rate laws:

$$2N_2O_5(g) \longrightarrow 4NO_2(g) + O_2(g) \qquad \text{Rate} = k[N_2O_5] \qquad [13.7]$$

$$CHCl_3(g) + Cl_2(g) \longrightarrow CCl_4(g) + HCl(g) \quad \text{Rate} = k[CHCl_3][Cl_2]^{\frac{1}{2}} \quad [13.8]$$

$$H_2(g) + I_2(g) \longrightarrow 2HI(g) \qquad \text{Rate} = k[H_2][I_2] \qquad [13.9]$$

The rate laws for a great many reactions have the general form

$$\text{Rate} = k[\text{reactant 1}]^n[\text{reactant 2}]^m \cdots \qquad [13.10]$$

As noted above, the proportionality constant, k, in a rate law is called the rate constant. Its value does not change with changes in concentration; however, as we will see in Section 13.4, k does change with temperature. For a given set of reactant concentrations, reaction rate increases as k increases. The exponents n and m in Equation [13.10] are called the reaction orders and their sum is the overall reaction order. For the reaction of NH_4^+ with NO_2^-, the rate law, Equation [13.6], contains the concentration of NH_4^+ raised to the first power. Thus the reaction is said to be first order in NH_4^+. Similarly, it is also first order in NO_2^-. The overall reaction order is two.

In a great many rate laws the reaction orders are zero, one, or two. However, reaction orders can be fractional or even negative. If a reaction is zero order in a particular reactant, changing its concentration will have no influence on rate as long as some of that reactant is present. On the other hand, if the reaction is first order in a reactant, changes in the concentration of that substance will produce proportional changes in the rate; doubling the concentration will double the rate, and so forth. When the rate law is second order in a particular reactant, doubling its concentration changes the rate by a factor of $2^2 = 4$; tripling its concentration causes the rate to increase by a factor of $3^2 = 9$.

SAMPLE EXERCISE 13.2

The initial rate of a hypothetical reaction $A + B \longrightarrow C$ was measured for several different starting concentrations of A and B, with the results given below:

Experiment number	[A] (M)	[B] (M)	Initial rate (M/sec)
1	0.100	0.100	4.0×10^{-5}
2	0.100	0.200	4.0×10^{-5}
3	0.200	0.100	16.0×10^{-5}

Using these data, determine (a) the rate law for the reaction, (b) the magnitude of the rate constant, and (c) the rate of the reaction when [A] = 0.050 M and [B] = 0.100 M

Solution: (a) Experiments 1 and 2 indicate that the concentration of B has no influence on the reaction rate. The reaction is therefore zero order in B. Experiments 1 and 3 indicate that doubling A increases the rate fourfold. This result indicates that rate is proportional to $[A]^2$; the reaction is second order in A. The rate law is

$$\text{Rate} = k[A]^2[B]^0 = k[A]^2$$

(b) Using the rate law and the data from experiment 1 we have

$$k = \frac{\text{rate}}{[A]^2} = \frac{4.0 \times 10^{-5}\,M/\text{sec}}{(0.100\,M)^2}$$
$$= 4.0 \times 10^{-3}/M\text{-sec}$$

(c) Using the rate law from part (a) and the rate constant from part (b) we have

$$\text{Rate} = k[A]^2$$
$$= (4.0 \times 10^{-3}/M\text{-sec})(0.050\,M)^2$$
$$= 1.0 \times 10^{-5}\,M/\text{sec}$$

Because [B] is not part of the rate law, its concentration is immaterial to the rate, provided there is at least some B present to react with A.

13.3 Rate equations relating time and concentration

The rate law, Equation [13.10], tells us how the rate of a reaction changes as we change reactant concentrations. Such rate laws can be converted into equations that tell us what the concentrations of the reactants or products are at any time during the course of a reaction. The mathematics required involve calculus. We don't expect you to be able to perform the calculus operations; however, you should be able to use the resulting equations. We will apply this conversion to two of the simplest rate laws—those that are first order overall and those that are second order overall.

FIRST-ORDER REACTIONS

The conversion of methyl isonitrile to acetonitrile provides a simple example of a first-order reaction:

$$H_3C-N\equiv C: \longrightarrow H_3C-C\equiv N: \qquad [13.11]$$

Methyl isonitrile   Acetonitrile

In this reaction the molecule undergoes an intramolecular rearrangement. The reaction is first order with respect to isonitrile (we use R to represent the methyl group, CH_3):

$$\text{Rate} = -\frac{\Delta[RNC]}{\Delta t} = k[RNC] \qquad [13.12]$$

Notice the units in this equation; [RNC] has units of molarity, and t is in units of seconds. Thus,

$$\text{Rate} = \frac{(M)}{(\text{sec})} = k(M)$$

Rearranging,

$$k = \frac{(M)}{(\text{sec})}\frac{1}{(M)} = \frac{1}{\text{sec}} \ (\text{or } \sec^{-1})$$

Equation [13.12] can be rearranged and calculus used to obtain an equation that relates the concentration of isonitrile at the start of the reaction $[RNC]_0$, to its concentration at any other time t, $[RNC]_t$:

$$\log\left(\frac{[RNC]_0}{[RNC]_t}\right) = \frac{kt}{2.30} \qquad [13.13]$$

In general, for a reactant species A,

$$\log\left(\frac{[A]_0}{[A]_t}\right) = \frac{kt}{2.30} \qquad [13.14]$$

Using the properties of logarithms (Appendix A.2), Equation [13.14] can be rearranged to give

$$\log[A]_t = \left(-\frac{k}{2.30}\right)t + \log[A]_0 \qquad [13.15]$$

One important fact about Equation [13.15] is that it is in the form of an equation for a straight line. Straight-line equations are of the type

$$y = ax + b \qquad [13.16]$$

(See Appendix A.4.) Equation [13.15] has this form with $y = \log[A]_t$, $a = -k/2.30$ (the slope), $x = t$, and $b = \log[A]_0$ (the intercept).

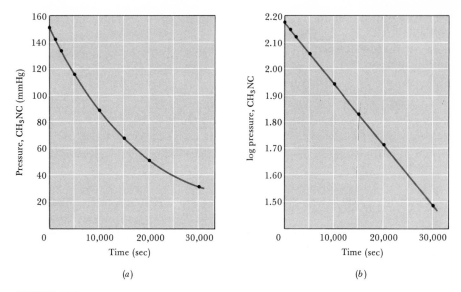

FIGURE 13.2 (a) Variation in the pressure of methyl isonitrile, CH_3NC, with time during the reaction $CH_3NC \rightarrow CH_3CN$. (b) The data from (a) plotted in a linear form as log of CH_3NC pressure as a function of time.

Figure 13.2 shows how the concentration of isonitrile varies with time as it rearranges in the gas phase at 198.9°C. In Figure 13.2(a) we have plotted the pressure of CH_3NC as a function of time. Figure 13.2(b) shows the same data in straight-line form, as the logarithm of pressure as a function of time. Pressure is a legitimate unit of concentration for a gas because the number of molecules per unit volume is directly proportional to pressure. The slope of the linear plot is $-2.22 \times 10^{-5}/\text{sec}$. (You should verify this for yourself, remembering that your result may vary slightly from ours because of the inaccuracies associated with reading the graph.) Because the slope equals $-k/2.30$,

$$k = -2.30(-2.22 \times 10^{-5}/\text{sec}) = 5.11 \times 10^{-5}/\text{sec}$$

For a first-order reaction, Equations [13.14] and [13.15] can be used to determine (1) the concentration of a reactant remaining at any time after the reaction has started, (2) the time required for a given fraction of sample to react, or (3) the time required for a reactant concentration to reach a certain level.

SAMPLE EXERCISE 13.3

The first-order rate constant for hydrolysis of a certain insecticide in water at 12°C is 1.45/yr. A quantity of this insecticide is washed into a lake in June, leading to an overall concentration of 5.0×10^{-7} g/cm^3 water. Assuming that the effective temperature of the lake is 12°C, (a) what is the concentration of the insecticide in June of the following year; (b) how long will it take for the concentration of the insecticide to drop to 3.0×10^{-7} g/cm^3?

Solution: (a) Substituting $k = 1.45/\text{yr}$, $t = 1$ yr, and $[\text{insecticide}]_0 = 5.0 \times 10^{-7}$ g/cm^3 into Equation [13.15] gives

$$\log [\text{insecticide}]_t = -\frac{1.45/\text{yr}}{2.30}(1.00 \text{ yr})$$
$$+ \log (5.0 \times 10^{-7})$$
$$= -0.630 - 6.30 = -6.93$$
$$[\text{insecticide}]_t = 10^{-6.93} = 1.2 \times 10^{-7} \text{ g/cm}^3$$

(The concentration units for $[A]_0$ and $[A]_t$ must be the same.)

(b) Again substituting into Equation [13.15], with $[insecticide]_t = 3.0 \times 10^{-7}$ g/cm³, gives

$$\log (3.0 \times 10^{-7}) = -\frac{1.45/\text{yr}}{2.30}t$$
$$+ \log (5.0 \times 10^{-7})$$

Solving for t gives

$$t = -\frac{2.30}{1.45/\text{yr}}[\log (3.0 \times 10^{-7})$$
$$- \log (5.0 \times 10^{-7})]$$
$$= -\frac{2.30}{1.45/\text{yr}}(-6.52 + 6.30) = 0.35 \text{ yr}$$

HALF-LIFE

The half-life of a reaction, $t_{\frac{1}{2}}$, is the time required for the concentration of the reactant to decrease to halfway between its initial and final values. In the cases we'll be considering, the final concentration is zero. To obtain an expression for $t_{\frac{1}{2}}$ for a first-order reaction, we begin with Equation [13.14]. The half-life corresponds to the time when $[A]_t = \frac{1}{2}[A]_0$. Inserting these quantities into the equation, we have

$$\log \frac{[A]_0}{\frac{1}{2}[A]_0} = \left(\frac{k}{2.30}\right)t_{\frac{1}{2}}$$

$$\log 2 = \left(\frac{k}{2.30}\right)t_{\frac{1}{2}}$$

$$t_{\frac{1}{2}} = \frac{(2.30)\log 2}{k} = \frac{0.693}{k} \qquad [13.17]$$

Notice that $t_{\frac{1}{2}}$ is independent of the initial concentration of reactant. This result tells us that if we measure reactant concentration at *any* time in the course of a first-order reaction, the concentration of reactant will be half of that measured value at a time $0.693/k$ later. The concept of half-life is widely used in describing radioactive decay. This application is discussed in detail in Section 20.4.

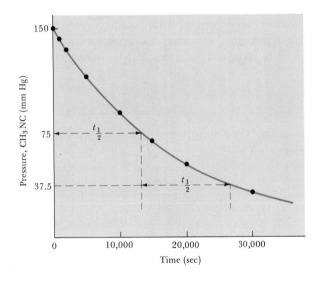

FIGURE 13.3 Pressure of methyl isonitrile as a function of time. Two successive half-lives of the rearrangement reaction, Equation [13.11], are shown.

The data for the first-order rearrangement of methyl isonitrile at 198.9°C are graphed in Figure 13.3. The first half-life is shown at 13,320 sec. At a time 13,320 sec later, the isonitrile concentration has decreased to $\frac{1}{2}$ of $\frac{1}{2}$, or $\frac{1}{4}$ the original concentration. *It is a characteristic of a first-order reaction that the concentration of the reactant decreases by factors of $\frac{1}{2}$ in a series of regularly spaced time intervals.*

SECOND-ORDER REACTIONS

For a reaction that is second order in just one reactant, A, the rate is given by

$$\text{Rate} = k[A]^2$$

Relying on calculus, this rate law can be used to derive the following equation:

$$\frac{1}{[A]_t} = \frac{1}{[A]_0} + kt \qquad [13.18]$$

In this case, a linear form of the data is obtained by plotting $1/[A]_t$ versus t. The resultant line has a slope of k and an intercept of $1/[A]_0$. One way to distinguish between first- and second-order rate laws is to graph both $\log [A]_t$ and $1/[A]_t$ against t. If the $\log [A]_t$ plot is linear, the reaction is first order; if the $1/[A]_t$ plot is linear, the reaction is second order.

Using Equation [13.18], it can be shown that the half-life of a second-order reaction is given by the expression $t_{\frac{1}{2}} = 1/k[A]_0$. It is *not* independent of the initial concentration of reactant as is $t_{\frac{1}{2}}$ for a first-order reaction. Thus a constant half-life is indicative of a first-order reaction, but not a second-order one.

SAMPLE EXERCISE 13.4

The following data were obtained for the gas-phase decomposition of nitrogen dioxide at 300°C, $2NO_2(g) \longrightarrow 2NO(g) + O_2(g)$:

Time (sec)	$[NO_2](M)$
0	0.0100
50	0.0079
100	0.0065
200	0.0048
300	0.0038

Is the reaction first or second order in NO_2?

Solution: To test whether the reaction is first or second order, we can construct plots of $\log [NO_2]$ and $1/[NO_2]$ against time. In doing so, we will find it useful to prepare the following table from the data given:

Time	$[NO_2]$	$\log [NO_2]$	$1/[NO_2]$
0	0.0100	−2.00	100
50	0.0079	−2.10	127
100	0.0065	−2.19	154
200	0.0048	−2.32	208
300	0.0038	−2.42	263

As Figure 13.4 shows, only the plot of $1/[NO_2]$ versus time is linear. Thus the reaction obeys a second-order rate law: Rate $= k[NO_2]^2$. From the slope of this straight-line graph we have that $k = 0.543/M$-sec.

A reaction may also be second order by having a first-order dependence of the rate on each of two reagents, that is, rate $= k[A][B]$. It is possible to derive an expression for the variation in concentrations of A and B with time. However, we will not consider this and other more complicated rate laws in this text.

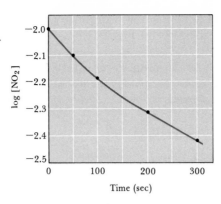
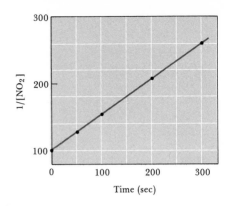

FIGURE 13.4 Kinetic data from Sample Exercise 13.4 for the reaction $2NO_2(g) \rightarrow 2NO(g) + O_2(g)$ at 300°C. The plot of log $[NO_2]$ versus time is not linear; consequently, the reaction is not first order in NO_2. The plot of $1/[NO_2]$ versus time is linear; the reaction is second order in NO_2.

13.4 The temperature dependence of reaction rates

The rates of most chemical reactions increase as the temperature rises. We see examples of this generalization in many biological processes around us. The rate at which grass grows and the metabolic activity of the common housefly are both greater in warm weather than in the cold of winter. As another example, food cooks more rapidly in boiling water than in merely hot water. However, it is dangerous to place too much emphasis on the apparent rates at which biological systems operate as a function of temperature, because they are very complex and are adapted to operate optimally in a narrow temperature range. To obtain a clear understanding of how temperature affects reaction rates, we must examine simple reaction systems. As an example, let us consider the reaction about which we have already learned quite a bit, the first-order rearrangement of methyl isonitrile, Equation [13.11]. Figure 13.5 shows the

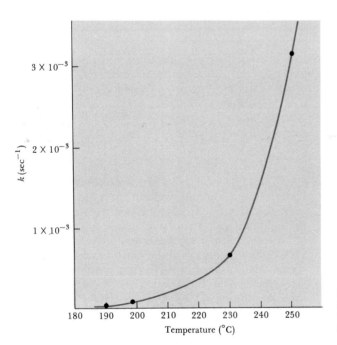

FIGURE 13.5 Variation in the first-order rate constant for rearrangement of methyl isonitrile as a function of temperature. (The four indicated points are used in connection with Sample Exercise 13.5.)

experimentally determined rate constant for this reaction as a function of temperature. It is evident that the rate of the reaction increases rapidly with temperature. Furthermore, the increase is nonlinear.

As we seek an explanation for this behavior, perhaps the first question to ask is, why do *any* reactions go slowly? What keeps reactions from simply occurring immediately? If the methyl isonitrile molecules are going eventually to rearrange into acetonitrile, why don't they all do it at once?

ACTIVATION ENERGY

We know from the kinetic-molecular theory of gases that with increasing temperature the average energy of the gas molecules increases. The fact that the rate of the methyl isonitrile reaction increases with increasing temperature suggests that perhaps the rearrangement is related to the kinetic energies of the molecules. Svante Arrhenius suggested in 1888 that before reaction can occur, a certain minimum amount of energy must be available to "propel" the molecules from one chemical state into another. The situation is rather like that shown in Figure 13.6. The boulder will be in a lower (or more stable) potential-energy state in valley B than in valley A. Before it can come to rest in B, however, it must acquire the energy to overcome the barrier blocking its passage from the one state into the other. In the same way, molecules may require a certain minimum energy to overcome the forces that tend to keep them as they are, if they are to form the new chemical bonds that will result in a different arrangement. In our methyl isonitrile example, we might imagine that for rearrangement to occur, the $N \equiv C$ portion of the molecule must turn over:

$$H_3C-N\equiv C: \longrightarrow \left[H_3C \cdots \overset{\cdot\cdot}{\underset{\cdot\cdot}{\overset{C}{\underset{N}{|||}}}} \right] \longrightarrow H_3C-C\equiv N: \qquad [13.19]$$

Even though the bonding may be more stable in the product acetonitrile than in the starting compound, energy is required to force the molecule through the relatively unstable intermediate state to the final result. The energy of the molecule as it proceeds along this reaction pathway is shown in Figure 13.7. Arrhenius called the energy barrier between the starting molecule and the highest energy along the reaction pathway the

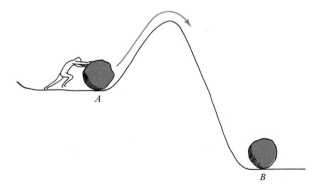

FIGURE 13.6 Illustration of the potential-energy profile for a boulder. The boulder must be moved over the energy barrier before it can come to rest in the lower energy location, *B*.

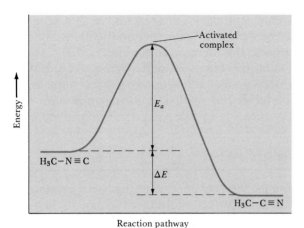

FIGURE 13.7 Energy profile for the rearrangement of methyl iso-nitrile. The molecule must surmount the activation-energy barrier before it can form the product, acetonitrile.

activation energy, E_a. The particular arrangement of atoms that has the maximum energy is often called the **activated complex**.

The conversion of $H_3CN\equiv C$ to $H_3CC\equiv N$ is exothermic; Figure 13.7 therefore shows the product as having a lower energy than the reactant. Notice that the reverse reaction is then endothermic; for that reaction the activation barrier is equal to the sum of ΔE and E_a for the forward reaction.

Energy is transferred between molecules through collisions. Thus, within a certain period of time, any particular isonitrile molecule might acquire enough energy to overcome the energy barrier and be converted into acetonitrile. At any given temperature only a small fraction of collisions will occur with sufficient energy to overcome the barrier to reaction. However, as shown in Figure 9.10, the distribution of molecular speeds is more spread out toward higher values when the gas is at a higher temperature. The distribution of kinetic energies of molecules changes in a similar way, as shown in Figure 13.8. This graph shows that at the higher temperature a larger fraction of molecules possesses the minimum energy needed for reaction.

In addition to the requirement that the reactant species collide with sufficient energy to begin to rearrange bonds, there is also an orientational requirement. The relative orientations of the molecules during their collisions may determine whether the energy gets to the right place for reaction or whether atoms are suitably oriented to form new bonds.

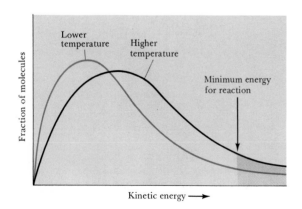

FIGURE 13.8 The distribution of kinetic energies in a sample of gas molecules at two different temperatures. At the higher temperature a larger number of molecules have higher energies. Thus, a larger fraction in any one instant will have more than the minimum energy required for reaction.

Thus only a fraction of the collisions possessing enough energy for reaction actually produce products.

In a mixture of H_2 and I_2 at ordinary temperatures and pressures, there are about 10^{10} collisions per second. If every collision between H_2 and I_2 resulted in formation of HI, the reaction would be over in much less than a second. Instead, at room temperature the reaction proceeds very slowly. Obviously, every collision does not lead to reaction. In fact, only about one in every 10^{13} collisions is effective. Only a small fraction of the collisions occurs with suitable orientation and with sufficient energy to carry the molecule over the energy barrier to products. As the temperature increases, the number of collisions increases as does the fraction that are sufficiently energetic for reaction. With each 10°C rise in temperature, the rate of this reaction triples.

THE ARRHENIUS EQUATION

Arrhenius noted that the increase in rate with increasing temperature for most reactions is nonlinear, as in the example shown in Figure 13.5. He found that most reaction-rate data obeyed the equation

$$\log k = \log A - \frac{E_a}{2.30\,RT} \qquad [13.20]$$

This equation is called the Arrhenius equation. The term E_a is the activation energy, which we have already defined; R is the gas constant (8.314 J/K-mol), and T is absolute temperature; A is constant, or nearly so, as temperature is varied. It is called the frequency factor; it is related to the frequency of collisions and the probability that the collisions are favorably oriented for reaction. Notice that as the magnitude of E_a increases, k becomes smaller. Thus reaction rates decrease as the energy barrier increases.

Equation [13.20] has the form of a straight line, in which one variable is $\log k$, and the other is $1/T$. The slope of the line is given by $-E_a/2.30\,R$; the intercept, at $(1/T) = 0$, is $\log k = \log A$. Thus equation [13.20] can be used to determine E_a from a graph of $\log k$ versus $1/T$.

It is sometimes convenient to further manipulate Equation [13.20] to give the relationship between the rate constants at two different temperatures, T_1 and T_2. At T_1 we have

$$\log k_1 = \log A - \frac{E_a}{2.30\,RT_1}$$

At T_2,

$$\log k_2 = \log A - \frac{E_a}{2.30\,RT_2}$$

Subtracting $\log k_2$ from $\log k_1$ gives

$$\log k_1 - \log k_2 = \left(\log A - \frac{E_a}{2.30\,RT_1}\right) - \left(\log A - \frac{E_a}{2.30\,RT_2}\right)$$

Simplifying this equation and rearranging it gives

$$\log \frac{k_1}{k_2} = \frac{E_a}{2.30\,R}\left(\frac{1}{T_2} - \frac{1}{T_1}\right)$$ [13.21]

Equation [13.21] provides a convenient means of calculating the rate constant at some temperature, T_1, when we know the activation energy and the rate constant, k_2, at some other temperature, T_2.

SAMPLE EXERCISE 13.5

The following table shows the rate constant for rearrangement of methyl isonitrile at various temperatures (these are the data that are graphed in Figure 13.5).

Temperature (°C)	$k(\text{sec}^{-1})$
189.7	2.52×10^{-5}
198.9	5.25×10^{-5}
230.3	6.30×10^{-4}
251.2	3.16×10^{-3}

(a) From these data calculate the activation energy for the reaction. (b) What is the magnitude of the rate constant at 430.0 K?

Solution: (a) We must first convert temperatures to the absolute temperature scale, K. We then take the inverse of these temperatures, and obtain the corresponding log values for k. This gives us the following table:

$T(\text{K})$	$1/T(\text{K})$	$\log k$
462.7	2.160×10^{-3}	-4.60
471.9	2.118×10^{-3}	-4.29
503.3	1.986×10^{-3}	-3.20
524.2	1.907×10^{-3}	-2.50

A graph of these data results in a straight line, as shown in Figure 13.9. The data points shown in the graph lie very close to the best straight line through all four points. The slope of the line is obtained by choosing two well-separated points, as shown, and reading off the coordinates of each:

$$\text{Slope} = \frac{-2.45 - (-4.45)}{0.00190 - 0.00214} = -8330$$

The numerator in this equation has no units, because logs have no units. The denominator has the units of $1/T$, that is, $1/\text{K}$. Thus, the overall units for the slope are K. The slope is equal to $-E_a/2.30\,R$. We want the value for the molar gas constant R in J/K-mol (Table 9.1), which is 8.314. Thus we obtain

$$\text{Slope} = \frac{-E_a}{2.30\,R}$$

$$E_a = -(\text{slope})(2.30\,R)$$

$$= -(-8330\text{ K})(2.30)\left(8.31\,\frac{\text{J}}{\text{K-mol}}\right)$$

$$\times \left(\frac{1\text{ kJ}}{1000\text{ J}}\right)$$

$$= 159\text{ kJ/mol}$$

(b) To determine the rate constant, k_1, at 430 K, we apply Equation [13.21] with $E_a = 159$ kJ/mol, $k_2 = 2.52 \times 10^{-5}/\text{sec}$, $T_2 = 462.7$ K, and $T_1 = 430.0$ K:

$$\log \frac{k_1}{2.52 \times 10^{-5}/\text{sec}} = \frac{159\text{ kJ/mol}}{(2.30)(8.31\text{ J/K-mol})}$$

$$\times \left(\frac{1}{462.7\text{ K}} - \frac{1}{430.0\text{ K}}\right)\left(\frac{10^3\text{ J}}{1\text{ kJ}}\right)$$

$$= -1.366$$

$$\frac{k_1}{2.52 \times 10^{-5}/\text{sec}} = 10^{-1.366} = 4.31 \times 10^{-2}$$

$$k_1 = (2.52 \times 10^{-5}/\text{sec})(4.31 \times 10^{-2})$$

$$= 1.09 \times 10^{-6}/\text{sec}$$

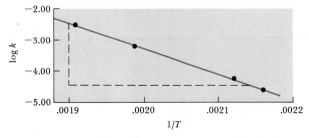

FIGURE 13.9 Log of the rate constant for rearrangement of methyl isonitrile as a function $1/T$. The linear relationship is predicted from the Arrhenius equation.

13.5 Reaction mechanisms

A knowledge of how the rate of a chemical reaction depends on concentrations, temperature, and pressure has many practical applications. For example, it is necessary to have such information to design a chemical plant to produce a certain desired substance. To understand how a certain pollutant chemical such as a herbicide behaves when released into the environment, it is necessary to know how it reacts under various conditions in nature, and at what speeds. Aside from such practical reasons for studying the kinetics of chemical reactions, there is also a more fundamental goal. A knowledge of the rate law and the activation energy for a chemical reaction can be of help in arriving at a reaction mechanism, a detailed picture of how the reaction occurs. The mechanism describes the path or sequence of steps by which the reaction occurs; it describes the order in which bonds are broken and formed and the changes in relative positions of atoms in the course of the reaction. The formulation of detailed mechanisms for chemical reactions presents one of the great challenges of chemistry. Once a reaction mechanism has been devised, the chemist can use it to predict new reaction possibilities and to test them by additional experiments.

The simplest sort of a mechanism is one that occurs in a single step. For example, a molecule may gain sufficient energy for some of its bonds to break or rearrange. Such is the case in the conversion of methyl isonitrile to acetonitrile:

$$H_3C-N\equiv C: \longrightarrow \left[H_3C\cdots \underset{N}{\overset{C}{|\!|\!|}} \right] \longrightarrow H_3C-C\equiv N:$$

A single-step process in which only one reacting molecule participates, as in this example, is said to be unimolecular.

Single-step reactions can also involve two (bimolecular) or three reactant molecules (termolecular). The reaction between nitric oxide, NO, and ozone, O_3, to form nitrogen dioxide, NO_2, and molecular oxygen, O_2, is a bimolecular process:

$$NO(g) + O_3(g) \longrightarrow NO_2(g) + O_2(g) \qquad [13.22]$$

The reaction requires the collision of two molecules, NO and O_3. Each collision that occurs with sufficient energy and suitable orientation results in the transfer of an oxygen atom from O_3 to NO.

Single-step reactions involving simultaneous collision of three molecules (termolecular processes) are generally less probable than unimolecular or bimolecular ones. The chance that four or more molecules will collide simultaneously with any regularity is even more remote; consequently, such collisions are never proposed as part of a reaction mechanism.

Although the rate law for a reaction cannot, in general, be predicted from the coefficients of the balanced chemical equation, the rate law of any single-step reaction is predictable from its chemical equation. For example, consider the general unimolecular process

$$A \longrightarrow products \qquad [13.23]$$

It seems clear that as the number of A molecules increases, the number that decompose in a given interval of time will increase. Thus the rate of a unimolecular process will be first order:

$$\text{Rate} = k[A] \qquad [13.24]$$

In the case of bimolecular reactions, the rate law will be second order as in the following examples:

$$A + B \longrightarrow \text{products} \qquad \text{Rate} = k[A][B] \qquad [13.25]$$
$$A + A \longrightarrow \text{products} \qquad \text{Rate} = k[A]^2 \qquad [13.26]$$

The second-order rate law follows from the fact that the rate of collision between A and B molecules is proportional to the concentrations of A and of B.

In general, the order for each reactant in a single-step reaction is equal to the coefficient for that reactant in the chemical equation. Of course, we cannot tell by merely looking at a balanced chemical equation whether that reaction occurs in a single step or in a number of sequential steps.

MULTISTEP MECHANISMS

Many chemical reactions occur in a number of unimolecular and bimolecular steps; indeed, this is generally the rule rather than the exception. For example, consider the reaction between nitrogen dioxide, NO_2, and carbon monoxide, CO, to form nitric oxide, NO, and carbon dioxide, CO_2:

$$NO_2(g) + CO(g) \longrightarrow NO(g) + CO_2(g) \qquad [13.27]$$

At temperatures below $225\,°C$ the experimentally determined rate law for this process is second order in NO_2 and zero order in CO:

$$\text{Rate} = k[NO_2]^2 \qquad [13.28]$$

Because the order of the reaction does not correspond to the coefficients in the chemical equation, we know this reaction cannot proceed in a single step involving the coming together of NO_2 and CO molecules. It has been proposed that the reaction occurs in two bimolecular steps:

$$NO_2(g) + NO_2(g) \longrightarrow NO_3(g) + NO(g) \qquad [13.29]$$
$$NO_3(g) + CO(g) \longrightarrow NO_2(g) + CO_2(g) \qquad [13.30]$$

Each of these steps is called an elementary reaction. In the first step, two NO_2 molecules collide, and an oxygen atom is transferred from one to the other. The resultant NO_3 then rapidly transfers an oxygen atom to CO to produce CO_2 and reform NO_2.

The reaction steps in a multistep mechanism must always add to give the chemical equation for the overall reaction. In the present example, the sum of the two steps is

$$2NO_2(g) + NO_3(g) + CO(g) \longrightarrow NO_3(g) + NO(g) + NO_2(g) + CO_2(g)$$

Simplifying by eliminating substances that appear on both sides of the equation gives the net overall chemical equation:

$$NO_2(g) + CO(g) \longrightarrow NO(g) + CO_2(g)$$

Because NO_3 is neither a reactant nor a product of the overall reaction but is formed in one elementary reaction and consumed in the next, it is called an **intermediate.** Multistep mechanisms involve one or more intermediates.

We now have to find how the experimentally observed rate law for an overall reaction is related to the rate laws for the elementary processes that comprise its mechanism. The rate law for the first step of the mechanism we are considering is given by*

$$Rate = k_1[NO_2]^2 \tag{13.31}$$

For the second step the rate is given by

$$Rate = k_2[NO_3][CO] \tag{13.32}$$

The overall reaction can be no faster than its slowest step. If one of the steps is much slower than others, it determines the overall reaction rate; this slow step is called the **rate-determining step.** The situation is analogous to the progression of auto traffic through a tunnel. The rate of traffic flow is largely determined by the speeds of the slowest cars.

When a mechanism includes a rate-determining step, the rate law for the overall reaction will depend on the rate law for that slow elementary step. For example, in the reaction between NO_2 and CO, the first step is much slower than the second step. Consequently, the rate law for the overall reaction is the same as the rate law for the first step:

$$Rate = k[NO_2]^2$$

We should now see why the rate law for a reaction cannot ordinarily be determined from the coefficients in the chemical equation. The rate law depends not on the overall reaction, but on the relative rates of the elementary steps.

The basic procedure in establishing the mechanism of a chemical reaction is first to determine the rate law and then to postulate one or more elementary reactions that account for the rate law and for other experimental facts. When an intermediate is involved in the rate-determining step, the relationship between the rate law for that step and the observed rate law for the overall process is not a direct one. *The experimental rate law will always include only substances present in measurable concentrations.* Intermediates are normally encountered in low, unknown concentrations and are not included in the experimental rate law. As an

---

*The subscript 1 on k indicates that k is the rate constant for the first step in the set of reactions that describes the overall reaction.

example, consider the gas-phase reaction of chlorine, Cl_2, with chloro-form, $CHCl_3$:

$$Cl_2(g) + CHCl_3(g) \longrightarrow HCl(g) + CCl_4(g) \qquad [13.33]$$

The experimentally observed rate law is

$$\text{Rate} = k[Cl_2]^{\frac{1}{2}}[CHCl_3] \qquad [13.34]$$

The following mechanism is proposed for this reaction:

$$Cl_2(g) \rightleftharpoons 2Cl(g) \qquad \text{(fast)} \qquad [13.35]$$
$$Cl(g) + CHCl_3(g) \longrightarrow HCl(g) + CCl_3(g) \qquad \text{(slow)} \qquad [13.36]$$
$$Cl(g) + CCl_3(g) \longrightarrow CCl_4(g) \qquad \text{(fast)} \qquad [13.37]$$

The second step is rate determining. Thus the overall rate is determined by the rate law for that step:

$$\text{Rate} = k_2[Cl][CHCl_3] \qquad [13.38]$$

To express this rate law in terms of Cl_2 molecules instead of intermediate Cl atoms, we must relate the concentrations of Cl_2 and Cl. Because each Cl_2 produces 2Cl, $[Cl_2]$ is proportional to $[Cl]^2$. Setting the proportionality constant equal to K, we have

$$[Cl]^2 = K[Cl_2] \qquad [13.39]$$

(We have to anticipate a bit here from Chapter 14, Section 14.2. We need only note that a relationship between $[Cl_2]$ and $[Cl]$ exists that is based on the stoichiometry of Equation [13.35]. In general, the concentration of an intermediate can be related to the concentrations of the molecule or molecules from which it forms.) From Equation [13.39] it follows that

$$[Cl] = K^{\frac{1}{2}}[Cl_2]^{\frac{1}{2}} \qquad [13.40]$$

Substituting Equation [13.40] into the rate expression for the rate-determining step, Equation [13.38], we have

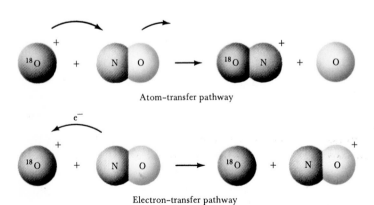

Atom–transfer pathway

Electron–transfer pathway

FIGURE 13.10 Alternative pathways for the reaction $O^+ + NO \rightarrow O + NO^+$. The results of experiments using ^{18}O labeling show that the reaction proceeds according to the electron-transfer pathway.

$$\text{Rate} = k_2 K^{\frac{1}{2}} [Cl_2]^{\frac{1}{2}} [CHCl_3] \qquad\qquad [13.41]$$

The two constants in this equation can be lumped together as one constant, $k = k_2 K^{\frac{1}{2}}$, giving the observed rate law:

$$\text{Rate} = k[Cl_2]^{\frac{1}{2}}[CHCl_3]$$

SAMPLE EXERCISE 13.6

It has been proposed that the conversion of ozone into O_2 proceeds in two steps:

$$O_3(g) \rightleftharpoons O_2(g) + O(g)$$
$$O_3(g) + O(g) \longrightarrow 2O_2(g)$$

(a) Write the equation for the overall reaction; (b) identify the intermediate, if any; (c) write the rate law for each elementary step in the proposed mechanism.

Solution: (a) Adding the two elementary steps gives

$$2O_3(g) + O(g) \longrightarrow 3O_2(g) + O(g)$$

Since $O(g)$ appears in equal amounts on both sides of the equation, it can be eliminated to give the net overall chemical equation:

$$2O_3(g) \longrightarrow 3O_2(g)$$

(b) The intermediate is $O(g)$. It is neither an original reactant nor a final product, but is formed in one step and used in the other.

(c) Step 1: Rate $= k_1[O_3]$
    Step 2: Rate $= k_2[O_3][O]$

The observed rate law is Rate $= k[O_3]^2[O_2]^{-1}$. (You should be able to conclude from this rate law that the first step cannot be the rate-determining one.) It turns out that the second step is rate determining. Because $O(g)$ is not an original reactant, it does not appear in the experimental rate law. However, it is found that the concentration of O depends on the concentrations of O_3 and O_2: $[O] = K[O_3][O_2]^{-1}$. Thus the rate law for the rate-determining step is consistent with the experimental rate law:

$$\text{Rate} = k_2[O_3][O] = k_2[O_3]K[O_3][O_2]^{-1}$$
$$= k_2K[O_3]^2[O_2]^{-1}$$

The observed rate constant, k, equals k_2K.

The fact that a mechanism gives a rate law that agrees with the observed one does not prove that that mechanism is correct. Often several mechanisms can be envisioned that give rise to the same rate law. To distinguish between such mechanisms, chemists might search for proposed intermediates or carry out other types of studies. As an example, consider the following reaction:

$$O^+(g) + NO(g) \longrightarrow NO^+(g) + O(g) \qquad [13.42]$$

This reaction occurs in the upper atmosphere. The rate law for this reaction is Rate $= k[O^+][NO]$, and the reaction is believed to occur in a single elementary step. However, we may ask whether the reaction involves the breaking of the NO bond and the forma-

tion of a new bond between N and O^+, or whether it merely involves the transfer of an electron from NO to O^+. These two possibilities are shown in Figure 13.10. The question of which of these mechanisms is operative is answered by performing the reaction using O^+ that has been highly enriched in the rare isotope ^{18}O. Because this isotope is present to the extent of only 0.2 percent in nature, the NO contains only 0.2 percent $N^{18}O$. By measuring the location of ^{18}O in the reaction products, the two mechanisms can be distinguished (see Figure 13.10). The experiment indicates that no ^{18}O is incorporated into NO^+; thus the reaction proceeds by the electron-transfer pathway rather than by the atom-transfer pathway.

13.6 Catalysis

A catalyst is a substance that acts to change the speed of a chemical reaction without itself undergoing a permanent chemical change in the process. Nearly all catalysts increase reaction rates. Catalysts are very common; most reactions occurring in the human body, the atmosphere, the oceans, or in industrial chemical processes are affected by catalysts.

If you have been exposed to chemical laboratory work, it is likely that you have carried out the reaction in which oxygen is produced by heating of potassium chlorate, $KClO_3$:

$$2KClO_3(s) \xrightarrow{\triangle} 2KCl(s) + 3O_2(g) \qquad [13.43]$$

In the absence of a catalyst, $KClO_3$ does not readily decompose in this manner, even on strong heating. However, mixing black manganese dioxide, MnO_2, with the $KClO_3$ before heating causes the reaction to occur much more readily. The MnO_2 can be recovered largely unchanged from this reaction, so it is clear that the overall chemical process is still the same. Thus, MnO_2 acts as a catalyst for decomposition of $KClO_3$. As another example, we know that a cube of sugar, when dissolved in water at $37°C$, does not undergo oxidation at a significant rate. The sugar could be recovered essentially unchanged from the solution after several days. Yet when sugar is ingested into the human body it is rapidly oxidized and soon ends up mostly as carbon dioxide and water:

$$C_{12}H_{22}O_{11}(aq) + 12O_2(aq) \longrightarrow 12CO_2(aq) + 11H_2O(l) \qquad [13.44]$$

The oxidation of sugar in the biochemical system has been greatly speeded up by the presence of one or more catalysts. These catalysts are enzymes, protein molecules that act to catalyze specific biochemical reactions. (Enzymes are discussed at length in Chapter 25.)

Much industrial chemical research is devoted to the search for new and more effective catalysts for reactions of commercial importance. Extensive research efforts also are devoted to finding means of inhibiting or removing certain catalysts that promote undesirable reactions, such as those involved in corrosion of metals, aging, and tooth decay.

HOMOGENEOUS CATALYSIS

A catalyst that is present in the same phase as the components of a chemical reaction is known as a **homogeneous catalyst.** For example, a homogeneous catalyst for a reaction occurring in solution would itself be dissolved in the solution.

In Chapter 10 we considered a simple example of homogeneous catalysis, the action of NO in promoting the decomposition of ozone, O_3. The NO acts as a catalyst by reaction with O_3 to form NO_2 and O_2. The NO_2 thus formed then reacts with atomic oxygen present in the stratosphere to reform NO and yield O_2 as the other product. The sequence of reactions and the overall result are as follows:

$$NO(g) + O_3(g) \longrightarrow NO_2(g) + O_2(g)$$
$$NO_2(g) + O(g) \longrightarrow NO(g) + O_2(g)$$
$$\overline{\quad O_3(g) + O(g) \longrightarrow 2O_2(g) \quad}$$

In this example, NO acts as a catalyst for O_3 decomposition because it speeds up the rate of the overall reaction without itself undergoing any net, or overall, chemical change; it is used in one step and reformed in the next.

As another example, hydrogen peroxide, H_2O_2, when dissolved in water undergoes slow decomposition, leading to oxygen and water:

$$2H_2O_2(aq) \longrightarrow 2H_2O(l) + O_2(g) \qquad [13.45]$$

This reaction has a very strong tendency to occur in the direction shown. By this we mean that eventually essentially all the hydrogen peroxide present in solution will have decomposed to oxygen and water. In the absence of a catalyst, however, the reaction occurs at an extremely slow rate. Many different substances are capable of catalyzing the reaction; among these is bromine, Br_2. The bromine reacts with hydrogen peroxide in acidic solution, forming bromide ion and liberating oxygen:

$$Br_2(aq) + H_2O_2(aq) \longrightarrow 2Br^-(aq) + 2H^+(aq) + O_2(g) \qquad [13.46]$$

We recognize this as a simple oxidation-reduction reaction (Chapter 7) in which bromine is reduced to bromide ion, and the oxygen of hydrogen peroxide is oxidized from the -1 oxidation state in H_2O_2 to the zero oxidation state of O_2. If this reaction were all that were involved, bromine would not be a catalyst, because it undergoes chemical change in this reaction. It happens that hydrogen peroxide reacts with bromide ion in acidic solution to form bromine:

$$2Br^-(aq) + H_2O_2(aq) + 2H^+(aq) \longrightarrow Br_2(l) + 2H_2O(l) \qquad [13.47]$$

In this oxidation-reduction reaction, bromide ion is oxidized by hydrogen peroxide to bromine; the oxygen of hydrogen peroxide is reduced, forming water. The overall sum of reaction Equations [13.46] and [13.47] is just Equation [13.45]. (Add these two reactions together yourself to make sure you see that reaction Equation [13.45] results.) We see that bromine is indeed a catalyst in the reaction, because it speeds the overall reaction without itself undergoing any net, or overall, change.

On the basis of the Arrhenius expression for a chemical reaction, Equation [13.20], the rate constant k is determined by the activation energy E_a and the frequency factor A. A catalyst may affect the rate of reaction by altering the value for either E_a or A. The most dramatic

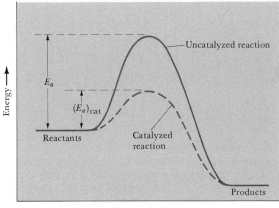

Reaction pathway

FIGURE 13.11 Energy profile for a catalyzed and uncatalyzed reaction. The catalyst functions in this example to lower the activation energy for reaction. Notice that the energies of reactants and products are unchanged by the catalyst.

catalytic effects come from lowering of E_a. As a general rule, *a catalyst lowers the overall activation energy for chemical reaction.* The lowering of E_a by a catalyst is shown schematically in Figure 13.11.

A catalyst usually lowers the overall activation energy for reaction by providing a completely different pathway for reaction. The two examples given above involve a reversible, cyclic reaction of the catalyst with the reactants. For example, in the decomposition of hydrogen peroxide, two successive reactions of H_2O_2, with bromine and then with bromide, are involved. Because these two reactions together serve as a catalytic pathway for hydrogen peroxide decomposition, *both* of these reactions must have significantly lower activation energies than the uncatalyzed decomposition, as shown schematically in Figure 13.12.

The catalysts we have so far discussed (and this includes enzymes, which are discussed in more detail in Chapter 25) are homogeneous catalysts. However, a great many reactions are catalyzed by substances that exist in a different phase from the reactants.

HETEROGENEOUS CATALYSIS

A heterogeneous catalyst exists in a different phase from the reactant molecules. For example, a reaction between molecules in the gas phase might be catalyzed by a finely divided metal oxide. In the absence of a catalyst, the reaction would occur slowly in the gas phase. However, when the catalyst is present, the reaction occurs more rapidly on the surface of the solid catalyst.

We saw in Section 10.5 an example of heterogeneous catalysis, in the oxidation of SO_2 to SO_3. Many industrially important reactions occurring in the gas phase are catalyzed by solid surfaces. Reactions occurring in solution may also be catalyzed by solids. Heterogeneous catalysts are often composed of finely divided metal or metal oxides. Because the catalyzed reaction occurs on the surface, special methods are often used to prepare catalysts so that they have very large surface areas.

The initial step in heterogeneous catalysis is usually adsorption of reactants. As noted in Chapter 2, the term *adsorption* should be distinguished from *absorption*. Adsorption refers to binding of molecules to a surface, whereas absorption refers to uptake of molecules into the interior

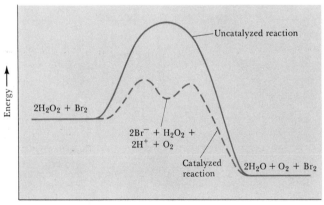

FIGURE 13.12 Energy profile for the uncatalyzed decomposition of hydrogen peroxide, and for the reaction as catalyzed by Br_2. The catalyzed reaction involves two successive steps, each of which has a lower activation energy than the uncatalyzed reaction.

of another substance. Adsorption occurs because the atoms or ions at the surface of a solid are extremely reactive. Unlike their counterparts in the interior of the substance, they have unfulfilled valence requirements. The unused bonding capability of surface atoms or ions may be utilized to bond molecules from the gas or solution phase to the surface of the solid. In practice, not all the atoms or ions of the surface are reactive; various impurities may be adsorbed at the surface, and these may occupy many potential reaction sites and block further reaction. The places where reacting molecules may become adsorbed are called active sites. The number of active sites per unit amount of catalyst depends on the nature of the catalyst, on its method of preparation, and on its treatment before use.

As an example of heterogeneous catalysis, consider the hydrogenation of ethylene to form ethane:

$$\underset{\text{Ethylene}}{\begin{array}{c} H \\ \diagdown \\ H \diagup \end{array} C = C \begin{array}{c} H \\ \diagup \\ \diagdown H \end{array}} + H_2 \longrightarrow \underset{\text{Ethane}}{\begin{array}{c} H \\ | \\ H - C - \\ | \\ H \end{array} \begin{array}{c} H \\ | \\ C - H \\ | \\ H \end{array}} \qquad [13.48]$$

In the absence of a catalyst, this reaction does not occur at all readily. However, in the presence of a very finely divided metal such as nickel, palladium, or platinum, the reaction occurs rather easily at room temperature, under a few hundred atmospheres of hydrogen pressure. The mechanism by which reaction occurs is shown diagrammatically in Figure 13.13. Both ethylene and hydrogen are adsorbed at the metal surface, Figure 13.13(a). The adsorption of hydrogen results in breaking of the H—H bond and formation of two M—H bonds, where M represents the metal surface, Figure 13.13(b). The hydrogen atoms are relatively free to move about the surface. When they encounter an adsorbed ethylene, the hydrogen may become bound to the carbon, Figure 13.13(c). The carbon thus acquires four σ bonds about it, which reduces its tendency to remain adsorbed at the metal. When the other carbon also acquires a hydrogen, the ethane molecule is released from the surface, Figure 13.13(d). The active site is ready to adsorb another ethylene molecule, and thus begin the cycle again.

CATALYSIS AND AIR POLLUTION

Heterogeneous catalysis plays a major role in the fight against urban air pollution. Two components of automobile exhausts that are involved in the formation of photochemical smog are nitrogen oxides and unburned hydrocarbons of various types (Section 10.5). In addition, automobile exhausts may contain considerable quantities of carbon monoxide. Even with the most careful attention to engine design and fuel characteristics, it is not possible under normal driving conditions to reduce the contents of these pollutants to an acceptable level in the exhaust gases coming from the engine. It is therefore necessary somehow to remove them from the exhaust gases before they are vented to the air. This removal is accomplished in the catalytic converter.

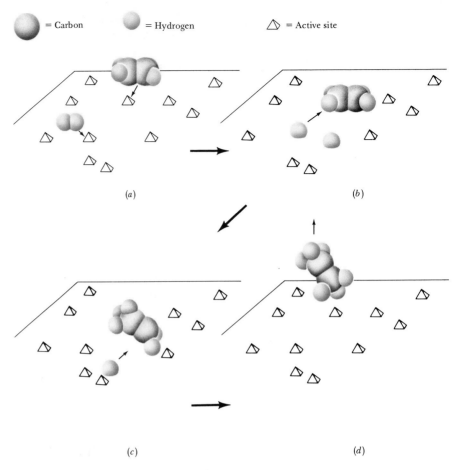

FIGURE 13.13 Mechanism for reaction of ethylene with hydrogen on a catalytic surface. (*a*) The hydrogen and ethylene are adsorbed at the metal surface. (*b*) The H—H bond is broken to give adsorbed hydrogen atoms. (*c*) These migrate to the adsorbed ethylene and bond to the carbon atoms. (*d*) As C—H bonds are formed, the adsorption of the molecule to the metal surface is decreased, and ethane is released.

The catalytic converter, illustrated in Figure 13.14, must perform two distinct functions: (1) oxidation of CO and unburned hydrocarbons to carbon dioxide and water and (2) reduction of nitrogen oxides to nitrogen gas:

$$CO, \text{ hydrocarbons } (C_xH_y) \xrightarrow{O_2} CO_2 + H_2O$$
$$NO, NO_2 \longrightarrow N_2$$

These two functions require two distinctly different catalysts. The development of a successful catalyst system represents a very difficult challenge. The catalysts must be effective over a wide range of operating temperatures; they must continue to be active in spite of the poisoning action of various gasoline additives emitted along with the exhaust; they must be physically rugged enough to withstand gas turbulence and the mechanical shocks of driving under various conditions for thousands of miles.

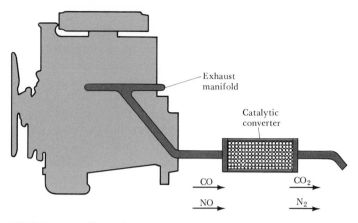

FIGURE 13.14 Illustration of the arrangement and functions of a catalytic converter.

Catalysts that promote the combustion of CO and hydrocarbons are, in general, the transition-metal oxides and noble metals such as platinum. As an example, a mixture of two different metal oxides such as CuO and Cr_2O_3 might be used. These materials are supported on a structure, Figure 13.15, which allows the best possible contact between the flowing exhaust gas and the catalyst surface. Either bead or honeycomb structures made from alumina, Al_2O_3, and impregnated with the catalyst may be employed. Such catalysts operate by first adsorbing oxygen gas, also present in the exhaust gas. This adsorption weakens the O—O bond in O_2, so that oxygen atoms are in effect available for reaction with adsorbed CO to form CO_2. Hydrocarbon oxidation probably proceeds somewhat similarly, with the hydrocarbons first being adsorbed by rupture of a C—H bond.

Reduction of nitrogen oxides is favored thermodynamically. That is, the decomposition of NO, for example, to yield N_2 and O_2 is favored, but

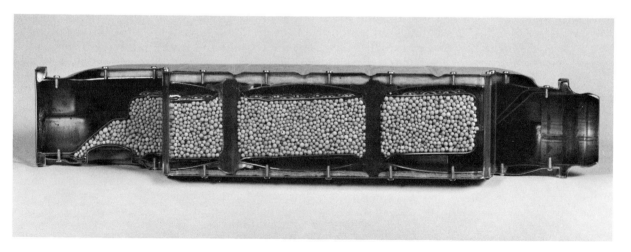

FIGURE 13.15 Cross-section view of a catalytic converter. The beads within the converter are impregnated with a catalyst that promotes the combustion of CO and hydrocarbons. (*General Motors Corp.*)

the reaction is extremely slow. A catalyst is therefore necessary. The most effective catalysts are transition-metal oxides and noble metals, the same kinds of materials that catalyze the oxidation of CO and hydrocarbons. The catalysts that are most effective in one reaction, however, are usually much less effective in the other. It is therefore necessary to have two different catalytic components.

As an illustration of the difficulties faced in the design of a catalytic converter, one of the frequent products of nitrogen oxide reduction in exhausts is ammonia. This comes about because water is present in the exhaust gases, and it undergoes reaction with CO in the presence of a catalyst to yield H_2:

$$H_2O(g) + CO(g) \longrightarrow H_2(g) + CO_2(g) \tag{13.49}$$

The hydrogen in turn is adsorbed on the catalyst and reacts with NO to form either N_2 or NH_3:

$$2NO(g) + 2H_2(g) \longrightarrow N_2(g) + 2H_2O(g) \tag{13.50}$$

$$2NO(g) + 5H_2(g) \longrightarrow 2NH_3(g) + 2H_2O(g) \tag{13.51}$$

Ammonia itself would, of course, be an undesirable pollutant. Furthermore, it is capable of reacting with oxygen on the catalyst surface to form NO and H_2O, putting us right back where we started. Obviously, therefore, the catalyst system for reduction of nitrogen oxides must not lead to much ammonia formation.

An additional problem that has recently come to light is the possibility of oxidation of SO_2 to SO_3 in the catalytic converter. Gasolines contain small quantities of sulfur that ordinarily appear in the exhaust gases as SO_2. In the catalytic converter, the oxidation of SO_2 to SO_3 is catalyzed. Because SO_3 dissolves in water to form highly corrosive sulfuric acid (Section 10.5), the appearance of SO_3 in the exhaust gases could cause corrosion and health problems in the vicinity of heavy traffic areas.

The activity of catalytic converters decreases with use, because of loss of active catalyst, cracking and fractures due to repeated heating and cooling, and poisoning of the catalysts. One of the most active poisons is the lead that comes from the tetramethyl lead, $Pb(CH_3)_4$, or tetraethyl lead, $Pb(C_2H_5)_4$, added to gasoline. Because of the severe catalyst poisoning that results from use of leaded fuels, cars built since 1975 have been engineered to discourage the use of leaded gas.

FOR REVIEW

Summary

In this chapter we've learned how to express in a quantitative way the relationship between the rate of a chemical reaction and the concentrations of reactants. The rate constant for a reaction is a proportionality constant between the reaction rate and the appropriate function of reactant concentrations. That appropriate function must be determined by experiments in which the concentrations of reactants are varied, and the effect on reaction rate noted. The equation that relates the reaction rate to the rate constant times some function of reactant concentrations is called the rate law. In a first-order reaction, the reaction rate is proportional to the concentration of a single reactant, raised to the first power. In such a reaction, a graph of log $[X]_t$, where $[X]_t$ is the concentration of reactant X at time t, versus time yields a straight line of slope $-k/2.3$, where k is the first-order rate constant. For a first-order reaction the half-life, $t_{\frac{1}{2}}$, is given by $0.693/k$. Reactions more complex than first-order ones yield different expressions for rate constant and half-life.

In coming together to form products, reactants require a certain minimum amount of energy, called the activation energy. When the activation energy is appreciable, only a very small fraction of all collisions between reactants provides the energy required to surmount the activation-energy barrier and form the products of the reaction. From the manner in which the rate constant for a reaction varies with temperature, it is possible to determine the magnitude of the activation energy for a reaction using the Arrhenius equation: $\log k = \log A - E_a/2.30\,RT$.

From a knowledge of the rate law for a reaction, and with other information such as the activation energy, results of isotope labeling experiments, and so on, a detailed picture of how the reaction proceeds may be formulated. The overall reaction may proceed in a series of elementary steps. One of these may be much slower than the others and thereby be the rate-determining step. The sequence of elementary steps by which the reaction proceeds is known as its mechanism.

The activation-energy barrier to formation of products from reactants can be lowered by use of a catalyst, a substance that increases the rate of reaction without itself undergoing a net chemical change. Catalysts may be either homogeneous, that is, in the same phase with the reactants, or heterogeneous, in a separate phase. Heterogeneous catalysts are particularly important in large-scale industrial chemical processes, and in applications such as the catalytic converter of an automobile.

Learning goals

Having read and studied this chapter, you should be able to:

1. Express the rate of a given reaction in terms of the variation in concentration of a reactant or product substance with time.

2. Calculate the average rate over an interval of time, given the concentrations of a reactant or product at the beginning and end of that interval.

3. Calculate instantaneous rates from a graph of reactant or product concentrations as a function of time.

4. Explain the meaning of the term rate constant and state the units associated with rate constants for first- and second-order reactions.

5. Calculate rate, rate constants, or reactant concentration, given two of these together with the rate law.

6. Use Equation [13.14] or [13.15] (for first-order reactions) or [13.18] (for second-order reactions) to determine (a) the concentration of a reactant or product at any time after a reaction has started, (b) the time required for a given fraction of sample to react, or (c) the time required for a reactant concentration to reach a certain level.

7. Use Equations [13.15] and [13.18] to determine graphically whether the rate law for a reaction is first or second order.

8. Explain the concept of reaction half-life and describe the relationship between half-life and rate constant for a first-order reaction.

9. Explain the concept of activation energy and how it relates to the variation of reaction rate with temperature.

10. Determine the activation energy for a reaction from a knowledge of how the rate constant varies with temperature (the Arrhenius equation).

11. Explain what is meant by the mechanism of a reaction using the terms elementary steps, rate-determining step, and intermediate.

12. Derive the rate law for a reaction that has a rate-determining step, given the elementary steps and their relative speeds; or, conversely, choose a plausible mechanism for a reaction given the rate law.

13. Describe the effect of a catalyst on the energy requirements for a reaction.

14. Relate the factors that are important in determining the activity of a heterogeneous catalyst.

15. Explain the functions of a catalytic converter in an automobile exhaust system.

Key terms

Among the more important terms and expressions used for the first time in this chapter are the following:

The activated complex (Section 13.4) is the particular arrangement of reactant and product molecules at the point of maximum energy in the rate-determining step of a reaction.

Activation energy (Section 13.4), E_a, is the minimum energy that must be supplied by the reactants in a chemical reaction to overcome the barrier to formation of products.

The Arrhenius equation (Section 13.4) relates the rate constant for a reaction to the frequency factor, A, the activation energy, E_a, and the temperature, T: $\log k = \log A - E_a/2.30\,RT$.

A catalyst (Section 13.6) is a substance that acts to change the speed of a chemical reaction without itself undergoing a permanent change in the process.

A catalytic converter (Section 13.6), or catalytic muffler, is a catalyst system built into the exhaust system of an automobile for the purpose of removing CO, NO, and unburned hydrocarbons from the exhaust.

An elementary reaction (Section 13.5) is a single-step unimolecular, bimolecular, or termolecular reaction. A multistep mechanism involves two or more such reactions.

An enzyme (Section 13.6) is a protein molecule that acts to catalyze specific biochemical reactions.

A first-order reaction (Section 13.3) is one in which the reaction rate is proportional to the concentration of a single reactant, raised to the first power.

The half-life (Section 13.3) of a rection is the time required for the concentration of a reactant substance to decrease to halfway between its initial and final values.

Heterogeneous catalyst (Section 13.6) is a catalyst that is in a different phase from that of the reactant substances.

A homogeneous catalyst (Section 13.6) is a catalyst that is in the same phase as the reactant substances.

An intermediate (Section 13.5) is a substance formed in one elementary step of a multistep mechanism and consumed in another; it is neither a reactant nor an ultimate product of the overall reaction.

A mechanism (Section 13.5) for a chemical reaction is a detailed picture, or model, of how the reaction occurs; that is, the order in which bonds are broken and formed, and the changes in relative positions of the atoms as reaction proceeds.

The rate constant (Section 13.2) is a constant of proportionality between the reaction rate and the concentrations of reactants that appear in the rate law.

The rate-determining step (Section 13.5) in a chemical reaction is the slowest step in a reaction that proceeds via a series of elementary steps from reactants to products.

A rate law (Section 13.2) is an equation in which the reaction rate is set equal to a mathematical expression involving the concentrations of reactants (and sometimes of products also).

The reaction order (Section 13.2) is the sum of the powers to which all the reactants appearing in the rate expression are raised.

Reaction rate (Section 13.1) is defined in terms of the decrease in concentration of a reactant molecule or the increase in concentration of a product molecule with time. It can be expressed as either the average rate over a period of time or the instantaneous rate at a particular time.

EXERCISES

Reaction rates, rate laws, and rate constants

13.1 The rate of disappearance of H^+ was measured for the following reaction:

$$CH_3OH(aq) + HCl(aq) \longrightarrow CH_3Cl(aq) + H_2O(l)$$

The following data were collected:

Time (min)	$[H^+](M)$
0	1.85
79	1.67
158	1.52
316	1.30
632	1.00

Calculate the average rate of reaction for the time interval between each measurement.

13.2 The rearrangement of methyl isonitrile, CH_3NC, was studied in the gas phase at 215°C, and the following data were obtained:

Time (sec)	Pressure CH_3NC (mm Hg)
0	502
2000	335
5000	180
8000	95.5
12,000	41.7
15,000	22.4

(a) Using these data make a graph of pressure CH_3NC versus time. Draw tangents to the curve at $t = 0, 5000$, and 10,000 sec. Determine the rates at these times. (b) Given that the rate law is first order in CH_3NC, determine the value of the rate constant at $t = 0$, 5000, and 10,000 sec.

13.3 Using the data provided in question 13.1, make a graph of $[H^+]$ versus time. Draw tangents to the curve at $t = 100$ and 500 min. Determine the rates at these times.

13.4 In each of the following reactions, how is the rate of disappearance of each reactant related to the rate of appearance of each product?

(a) $3H_2(g) + N_2(g) \longrightarrow 2NH_3(g)$
(b) $2NOCl(g) \longrightarrow 2NO(g) + Cl_2(g)$
(c) $HI(g) + CH_3I(g) \longrightarrow CH_4(g) + I_2(g)$

13.5 The average rate of disappearance of ozone in the reaction $2O_3(g) \longrightarrow 3O_2(g)$ is found to be 9.0×10^{-3} atm/sec over a measured interval of time. What is the rate of appearance of O_2 during this interval?

13.6 (a) In the reaction

$$2NO(g) + 2H_2(g) \longrightarrow N_2(g) + 2H_2O(g)$$

how is the rate of disappearance of NO related to the rate of appearance of N_2? (b) This reaction is first order in H_2 and second order in NO. Write the rate law. (c) If the concentrations of reactants are expressed in moles/liter, what are the units for k?

13.7 Consider the hypothetical reaction A + B $\longrightarrow$ products. For each of the possible rate laws listed below, indicate the reaction order with respect to A, with respect to B, and the overall reaction order: (a) Rate = $k[A][B]$; (b) Rate = $k[A]^2$; (c) Rate = $k[A][B]^2$.

13.8 What units would each of the rate constants in question 13.7 have if concentration is expressed as M and rate as M/sec?

13.9 For each rate law in question 13.7, (a) indicate how the rate changes if the concentration of A is doubled while the concentration of B remains constant; (b) if the concentration of B is doubled while the concentration of A is held constant.

13.10 When the concentration of a substance is doubled, what effect does it have on the rate if the order with respect to that reactant is (a) 0; (b) 1; (c) 2; (d) 3; (e) $\frac{1}{2}$.

13.11 Consider a hypothetical reaction A + 2B $\longrightarrow$ C for which the following kinetic data were obtained:

Experiment	[A](M)	[B](M)	$\Delta[C]/\Delta t$(M/sec)
1	0.20	0.10	7.0×10^{-5}
2	0.40	0.10	7.0×10^{-5}
3	0.20	0.20	2.8×10^{-4}

(a) Write the rate law for this reaction. (b) What is the order of the reaction with respect to A; with respect to B; what is the overall reaction order? (c) What is the numerical value for the rate constant?

13.12 The following data were collected for the gas-phase reaction between nitric oxide and bromine at 273°C:

$$2NO(g) + Br_2(g) \longrightarrow 2NOBr(g)$$

Experiment	[NO](M)	[Br$_2$](M)	Initial rate of appearance of NOBr (M/sec)
1	0.10	0.10	12
2	0.10	0.20	24
3	0.20	0.10	48
4	0.30	0.10	108

(a) Determine the rate law. (b) Calculate the value of the rate constant. (c) How is the rate of appearance of NOBr related to the rate of disappearance of Br_2?

13.13 The decomposition of a substance is found to be first order. If it takes 2.5×10^3 sec for the concentration of that substance to fall to half its original value, what is the magnitude of the rate constant?

13.14 The decomposition of a substance is found to be first order. If $k = 8.6 \times 10^{-3}$/sec, what is the half-life for the reaction?

13.15 The reaction

$$SO_2Cl_2(g) \longrightarrow SO_2(g) + Cl_2(g)$$

is first order in SO_2Cl_2. Using the following kinetic data, determine the magnitude of the first-order rate constant:

Time (sec)	Pressure, SO$_2$Cl$_2$ (atm)
0	1.000
2500	0.947
5000	0.895
7500	0.848
10000	0.803

13.16 Sucrose, $C_{12}H_{22}O_{11}$, which is more commonly known as table sugar, reacts in dilute acid solutions to form two simpler sugars, glucose and fructose. Both of these sugars have the molecular formula $C_6H_{12}O_6$, though they differ in molecular structure. The reaction is

$$C_{12}H_{22}O_{11}(aq) + H_2O(l) \longrightarrow 2C_6H_{12}O_6(aq)$$

The rate of this reaction was studied at 23°C and in 0.5 M HCl; the following data were obtained:

Time (min)	[C$_{12}$H$_{22}$O$_{11}$](M)
0	0.316
39	0.274
80	0.238
140	0.190
210	0.146

Is the reaction first order or second order with respect to the concentration of sucrose? What is the numerical value for the rate constant?

13.17 A reaction that is second order in A is 50 percent complete after 350 min. If $[A]_0 = 1.35\ M$, what is the rate constant?

13.18 The first-order rate constant for the gas-phase decomposition of N_2O_5 to NO_2 and O_2 at 65°C is 4.87×10^{-3}/sec. (a) If we start with 0.500 mol of N_2O_5 in a 500 mL container, how many moles of N_2O_5 will remain after 5.00 min? (b) How long will it take for the quantity of N_2O_5 to drop to 0.100 mol?

13.19 The first-order rate constant for the decomposition of a certain antibiotic in water at 20°C is 1.29/yr. (a) If a $5.0 \times 10^{-3}\ M$ solution of this antibiotic is stored at

20°C for 1 month, what is the concentration of the antibiotic? After 1 year? (b) How long will it take for the concentration of the antibiotic to reach $1.0 \times 10^{-3} M$?

Effects of temperature; activation energy

13.20 Draw energy profiles (like those in Figures 13.7, 13.11, and 13.12) for single-step reactions with the following values of E_a and ΔE: (a) $E_a = 100 \text{ kJ/mol}$; $\Delta E = -50 \text{ kJ/mol}$; (b) $E_a = 50 \text{ kJ/mol}$; $\Delta E = -120 \text{ kJ/mol}$; (c) $E_a = 120 \text{ kJ/mol}$; $\Delta E = 50 \text{ kJ/mol}$.

13.21 (a) Which of the reactions in question 13.20 would be the fastest? Which would be the slowest? (b) In which case would the *reverse* reaction be the fastest? Which reverse reaction would be the slowest?

13.22 Sketch the energy profile for each of the following reactions: (a) The uncatalyzed decomposition of $H_2O_2(aq)$ into $O_2(g)$ and $H_2O(l)$ for which $E_a = 75.3 \text{ kJ/mol}$ and $\Delta E = -98.1 \text{ kJ/mol}$. (b) The reaction

$$O_3(g) + NO(g) \longrightarrow NO_2(g) + O_2(g)$$

for which $E_a = 10.3 \text{ kJ/mol}$ and $\Delta E = -199.6 \text{ kJ/mol}$.

13.23 The rate of the reaction

$$CH_3COOC_2H_5(aq) + OH^-(aq) \longrightarrow$$
$$CH_3COO^-(aq) + C_2H_5OH(aq)$$

was measured at several temperatures and the following data collected:

Temperature (°C)	k(1/M-sec)
15	0.0521
25	0.101
35	0.184
45	0.332

Using these data, construct a graph of $\log k$ versus $1/T$. Using your graph, determine the value of E_a.

13.24 The temperature dependence of the rate constant for the reaction

$$CO(g) + NO_2(g) \longrightarrow CO_2(g) + NO(g)$$

is tabulated below:

Temperature (K)	k(1/M-sec)
600	0.028
650	0.22
700	1.3
750	6.0
800	23

Calculate E_a and A.

13.25 The activation energy, E_a, for a reaction is 70.0 kJ/mol. What is the effect on rate if the temperature increases (a) from 25.0°C to 35.0°C; (b) from 100°C to 110°C? (Assume E_a and A do not change with temperature.)

13.26 The gas-phase decomposition of HI into H_2 and I_2 is found to have $E_a = 182 \text{ kJ/mol}$. The rate constant at 700°C is $1.57 \times 10^{-3} M/\text{sec}$. What is the value of k at (a) 600°C; (b) 800°C?

13.27 The rate constant, k, for a reaction is $1.0 \times 10^{-3}/\text{sec}$ at 25°C. Calculate the rate constant for the same reaction at 100°C if (a) $E_a = 67.0 \text{ kJ/mol}$; (b) $E_a = 134 \text{ kJ/mol}$.

13.28 For the reaction

$$HI(g) + CH_3I(g) \longrightarrow CH_4(g) + I_2(g)$$

$k = 1.7 \times 10^{-2}/\text{sec}$ at 430 K and $9.5 \times 10^{-2}/\text{sec}$ at 450 K. What is the numerical value of the activation energy, E_a?

Reaction mechanisms; catalysis

13.29 The following mechanism has been proposed for the reaction of NO with Br_2 to form NOBr:

$$NO(g) + Br_2(g) \longrightarrow NOBr_2(g)$$
$$NOBr_2(g) + NO(g) \longrightarrow 2NOBr(g)$$

(a) Identify the intermediate. (b) Write the rate law for each step of the mechanism. (c) What would the rate law for the overall reaction be if the first step were very slow and the second step very fast? (d) The observed rate law is Rate $= k[NO]^2[Br_2]$. If the proposed mechanism is correct, what can we conclude about the relative speeds of steps one and two?

13.30 The rate law for the reaction

$$2NO(g) + O_2(g) \longrightarrow 2NO_2(g)$$

is $k[NO]^2[O_2]$. The following mechanism has been proposed:

$$NO(g) + O_2(g) \rightleftharpoons NO_3(g) \quad \text{(fast)}$$
$$NO_3(g) + NO(g) \longrightarrow 2NO_2(g) \quad \text{(slow)}$$

Explain how the proposed mechanism leads to the observed rate law.

13.31 The decomposition of hydrogen peroxide is catalyzed by iodide ion. The catalyzed reaction is thought to proceed by a two-step mechanism:

$$H_2O_2(aq) + I^-(aq) \longrightarrow H_2O(l) + IO^-(aq) \quad \text{(slow)}$$
$$IO^-(aq) + H_2O_2(aq) \longrightarrow$$
$$H_2O(l) + O_2(g) + I^-(aq) \quad \text{(fast)}$$

Make a sketch of the energy profile for the uncatalyzed decomposition of H_2O_2 and for the catalyzed reaction that is consistent with the mechanism described above.

13.32 The activity of a heterogeneous catalyst is highly dependent on its method of preparation and prior treatment. Explain why this is so.

13.33 When D_2 is reacted with ethylene, C_2H_4, in the presence of a finely divided catalyst, ethane with two deuteriums, CH_2D—CH_2D, is formed. (Deuterium, D, is an isotope of hydrogen of mass 2.) There is very little ethane formed in which two deuteriums are bound to one carbon, for example, $CH_3—CHD_2$. Explain why this is so in terms of the sequence of steps involved in hydrogenation.

Additional exercises

13.34 (a) List three factors that can be varied to change the speed of a particular reaction. (b) Explain, on a molecular level, how each exerts its influence.

13.35 What factors determine whether a collision will lead to chemical reaction?

13.36 Calculate the average rate of reaction over the first 100 sec if the concentration of a reactant is $0.200\ M$ at the beginning of the reaction and $0.187\ M$ 100 sec later.

13.37 (a) How does the rate of disappearance of N_2O_5 relate to the rate of appearance of NO_2 and the rate of appearance of O_2 in the reaction

$$2N_2O_5(g) \longrightarrow 4NO_2(g) + O_2(g)$$

(b) If the rate law for this reaction is $k[N_2O_5]$, how does k differ for the disappearance of N_2O_5 and the appearance of NO_2?

13.38 A particular reaction is found to have the following rate law: Rate $= k[A]^2[B]$. Which terms in this rate law are changed by each of the following changes: (a) the concentration of A is doubled; (b) a catalyst is added; (c) the concentration of A is increased by a factor of 2, whereas the concentration of B is decreased by a factor of 4; (d) temperature is increased.

13.39 The oxidation of NO by O_3, a reaction of importance in formation of photochemical smog (Section 10.5), is first order in each of the reactants. The second-order rate constant at 28°C is $1.5 \times 10^7/M$-sec. If the concentrations of NO and O_3 are each $2 \times 10^{-8}\ M$, what is the rate of oxidation of NO?

13.40 Explain why the rate law for a reaction cannot in general be written using the coefficients of the balanced chemical equation as the reaction orders.

13.41 The anaerobic (in absence of air) breakdown of glucose to form lactic acid proceeds as follows:

$$C_6H_{12}O_6(aq) \longrightarrow 2HC_3H_5O_3(aq)$$

This reaction occurs in muscles during strenuous activity when the energy demands of the cells exceed the O_2 available for normal oxidative breakdown of glucose into CO_2 and H_2O. The lactic acid accumulates, contributing to muscle fatigue and triggering heavy breathing to supply O_2 necessary to decompose the lactic acid. Suppose that in a laboratory experiment you are assigned the task of determining the rate of the conversion of glucose to lactic acid at 25°C. Suggest some procedure that could be used in obtaining data.

13.42 The reaction of $(CH_3)_3CBr$ with hydroxide ion proceeds with the formation of $(CH_3)_3COH$:

$$(CH_3)_3CBr(aq) + OH^-(aq) \longrightarrow$$
$$(CH_3)_3COH(aq) + Br^-(aq)$$

This reaction was studied at 55°C, and the following data obtained:

Experiment	$[(CH_3)_3CBr](M)$	$[OH^-](M)$	Initial rate (M/sec)
1	0.10	0.10	1.0×10^{-3}
2	0.20	0.10	2.0×10^{-3}
3	0.10	0.20	1.0×10^{-3}

Write the rate law for this reaction and compute the rate constant.

13.43 Urea, NH_2CONH_2, is the end product in protein metabolism in animals. The decomposition of urea in $0.1\ M$ HCl occurs according to the reaction

$$NH_2CONH_2(aq) + H^+(aq) + 2H_2O(l) \longrightarrow$$
$$2NH_4^+(aq) + HCO_3^-(aq)$$

The reaction is first-order in urea. When [urea] $= 0.200\ M$, the rate at 61.05°C is $8.56 \times 10^{-5}\ M/sec$. (a) What is the numerical value for the rate constant, k? (b) What is the concentration of urea in this solution after 5.00×10^3 sec if the starting concentration is $0.500\ M$?

13.44 The half-life for a first-order reaction is 75 min. What is the value of the rate constant in sec^{-1}?

13.45 The gas-phase decomposition of HI into H_2 and I_2 yielded the following data:

Time (hr)	$[HI](M)$
0	1.00
2.0	0.50
4.0	0.33
6.0	0.25

Is the reaction first or second order in HI? What is the numerical value for the rate constant?

13.46 Hydrogen sulfide, H_2S, is a common and troublesome pollutant in industrial waste waters. One pathway for removal is oxidation by dissolved oxygen:

$$2H_2S + O_2 \longrightarrow 2S + 2H_2O$$

Assuming that this is a second-order reaction, first order in each reactant, how would the rate of removal of H_2S depend on the concentration of dissolved oxygen? Suppose that the concentration of dissolved H_2S in polluted water is $5 \times 10^{-6}\ M$. The concentration of dissolved oxygen in water in equilibrium with air is about $2.6 \times 10^{-4}\ M$. If the second-order rate constant for the oxidation of H_2S is $4 \times 10^{-5}/M$-sec, what is the initial rate at which the H_2S is oxidized?

13.47 The reaction of Cl_2 with H_2S in water

$$Cl_2 + H_2S \longrightarrow S + 2HCl$$

is first order in each reactant. The second-order rate constant at 28°C has the value $3.5 \times 10^{-2}/M$-sec. If at a given time the concentration of H_2S is $5 \times 10^{-5}\ M$, and that of Cl_2 is $0.05\ M$, what is the rate of formation of HCl?

13.48 The reaction

$$F + H_2 \longrightarrow HF + H$$

has been studied in the gas phase at various temperatures and has been found to have an activation energy of 22 kJ/mol. The overall change in energy in the reaction, ΔE, is -130 kJ/mol. Draw a diagram of the energy profile of the system as a function of reaction coordinate.

13.49 The first-order rate constant for hydrolysis of a particular organic chloride compound in water varies with temperature as follows:

Temperature (K)	Rate constant (sec^{-1})
300	1.0×10^{-5}
320	5.0×10^{-5}
340	2.0×10^{-4}
355	5.0×10^{-4}

From these data calculate the activation energy in units of kJ/mol.

[13.50] The rate at which fireflies flash varies with the temperature of the insect, because a certain rate-controlling reaction proceeds at speeds that are temperature dependent. The period of the firefly flash was found to be 16.3 sec at 21.0°C, and 13.0 sec at 27.8°C. What is the activation energy for the reaction that controls the rate of firefly flash?

13.51 The activation energy for the reaction

$$2NO_2(g) \longrightarrow 2NO(g) + O_2(g)$$

is 113.8 kJ/mol. If $k = 0.75/M$-sec at 600°C, what is the value of k at 500°C?

13.52 The gas-phase decomposition of NO_2Cl is believed to occur in two steps:

$$NO_2Cl(g) \longrightarrow NO_2(g) + Cl(g)$$
$$NO_2Cl(g) + Cl(g) \longrightarrow NO_2(g) + Cl_2(g)$$

(a) Indicate whether each elementary reaction is unimolecular, bimolecular, or termolecular. (b) Identify the intermediate in this mechanism. (c) What is the rate law for the overall reaction if the first step is rate determining?

13.53 The gas-phase reaction of chlorine with carbon monoxide to form phosgene

$$Cl_2 + CO \longrightarrow COCl_2$$

obeys the rate law

$$\text{Rate} = \frac{\Delta[COCl_2]}{\Delta t} = k[Cl_2]^{\frac{3}{2}}[CO]$$

A mechanism involving the following series of steps is consistent with the rate law:

$$Cl_2 \rightleftharpoons 2Cl$$
$$Cl + CO \rightleftharpoons COCl$$
$$COCl + Cl_2 \rightleftharpoons COCl_2 + Cl$$

Assuming that this mechanism is correct, which of the above steps is the slow, or rate-determining, step? Explain.

13.54 The rate law for the reaction

$$NO + O_3 \longrightarrow NO_2 + O_2$$

is of the form

$$\text{Rate} = k[NO][O_3]$$

Which of the following mechanisms is consistent with this rate law? Explain.

(a) $NO \longrightarrow N + O$ (slow)
$N + O_3 \longrightarrow NO_2 + O$ (fast)
$O + O \longrightarrow O_2$ (fast)
(b) $NO + O_3 \longrightarrow NO_2 + O_2$
(c) $O_3 \longrightarrow O_2 + O$ (slow)
$O + NO \longrightarrow NO_2$ (fast)

14

Chemical equilibrium

It often happens that chemical reactions do not proceed to completion. By this we mean that a mixture of reactants is not entirely converted into products. Instead, after a time, reactant concentrations no longer decrease. The reaction system at this point consists of a mixture of reactant and product substances. A chemical system in such a condition is said to be in a state of chemical equilibrium. We have already encountered instances of simple equilibrium conditions. For example, in a closed container, the vapor above a liquid achieves an equilibrium with the liquid phase (Section 11.2). The rate at which molecules escape from the liquid to the gas phase equals the rate at which molecules of the gas phase strike the surface and become part of the liquid. As another example, solid sodium chloride may be in equilibrium with the ions dissolved in water (Section 12.2). The rate at which ions leave the solid surface equals the rate at which other ions are removed from the liquid to become part of the solid. We know from these examples that equilibrium is not a static condition in which nothing is happening. Rather, it is dynamic; opposing processes are occurring at equal rates. In this chapter we shall consider chemical equilibrium and examine the principles on which the idea of equilibrium is based. To demonstrate the importance of equilibrium considerations and to illustrate the most basic concepts involved, we begin with a discussion of the Haber process for synthesizing ammonia.

14.1 The Haber process

Of all the chemical reactions that humans have learned to carry out and control for their own purposes, the synthesis of ammonia from hydrogen and atmospheric nitrogen is perhaps the most important. This is especially the case in the present world situation, with food shortages becoming more critical each year. The growth of plant matter requires a substantial store of nitrogen in the soil, in a form usable by plants. The quantity of food required to feed the ever-increasing human population far exceeds that which could be produced if we relied solely on naturally available nitrogen in the soil. Huge quantities of fertilizer rich in nitrogen are required to sustain the high-yield agriculture practiced in the United States today. Great amounts of fertilizer must be used throughout the world if the gap between food supplies and human needs is to be narrowed. The only widely available source of nitrogen is the N_2 present in the atmosphere. The problem thus becomes one of "fixing" atmospheric N_2, that is, converting it to a form that plants can use. This process is called *nitrogen fixation.*

The N_2 molecule is exceptionally unreactive. Its lack of reactivity is due in large measure to the strong triple bond between the nitrogen atoms (Section 7.4). Because this bond is so strong, there is very little tendency for the molecule to engage in chemical reactions that will result in loss of the $N\equiv N$ triple bond and formation of other bonds to nitrogen. For this reason, the process of fixation is not easy to achieve. In nature the fixation of N_2 is carried out by a special group of nitrogen-fixing bacteria that grow on the roots of certain plants, for example, clover or alfalfa. Here we are interested in only one particular fixation reaction, the Haber process.

Fritz Haber, a German chemist, investigated the energy relations in the reaction between nitrogen and hydrogen and convinced himself that it should be possible to form ammonia in a reasonable yield from these two starting substances. The chemical reaction involved is

$$N_2(g) + 3H_2(g) \rightleftharpoons 2NH_3(g) \qquad [14.1]$$

The double arrow indicates the reversible character of the reaction. The NH_3 can form from N_2 and H_2, but it can also decompose into these elements.

Haber's research was of great interest to the German chemical industry. Germany was preparing for World War I, and nitrogen compounds figured heavily in the manufacture of explosives. Without a synthetic source of these nitrogen compounds Germany would be greatly handicapped. By 1913, Haber had designed a process that worked, and for the first time ammonia was produced on a large scale from atmospheric nitrogen. In the following year, World War I began.

From these unhappy beginnings as a major factor in international warfare the Haber process has become the world's principal source of fixed nitrogen. It is estimated that 18 million tons of ammonia were formed via the Haber process in the United States in 1978.* The ammonia produced in the Haber process can be applied directly to the soil. It may also be converted into ammonium salts, for example, ammonium sulfate, $(NH_4)_2SO_4$, or ammonium hydrogen phosphate, $(NH_4)_2HPO_4$, which are then used as fertilizers.

---

*The industrial fixation of nitrogen, chiefly by the Haber process, now accounts for more than 30 percent of all the nitrogen fixed on the planet.

In designing the process that bears his name, Haber had to face two separate questions. First, is there a catalyst that will allow the reaction to occur at a reasonable speed under practically attainable conditions? After much long and difficult searching, Haber found a suitable catalyst; we shall return to this aspect of his work in a later section. Second, assuming that a catalyst can be found, to what extent will nitrogen be converted into ammonia? Let us now consider this latter question, which relates to chemical equilibrium.

14.2 The equilibrium constant

The Haber process consists of putting together N_2 and H_2 in a high-pressure tank at a total pressure of several hundred atmospheres, in the presence of a catalyst, and at a temperature of a few hundred degrees centigrade. Under these conditions, the two gases react to form ammonia. But the reaction does not lead to complete consumption of the N_2 and H_2. Rather, at some point the reaction appears to stop, with all three components of the reaction mixture present at the same time. The manner in which the concentration of H_2, N_2, and NH_3 vary with time in this situation is shown in Figure 14.1(a). The condition of the system in which all concentrations have achieved steady values is referred to as chemical equilibrium. The relative amounts of N_2, H_2, and NH_3 present at equilibrium do not depend on the amounts of catalyst present. However, they do depend on the relative amounts of H_2 and N_2 with which the reaction was begun. Furthermore, if only ammonia is placed into the tank under the usual reaction conditions, at equilibrium there is again a mixture of N_2, H_2, and NH_3. The variations in concentrations as a function of time for this situation are shown in Figure 14.1(b). By comparing the two parts of Figure 14.1, we can see that at equilibrium the relative concentrations of H_2, N_2, and NH_3 are the same, regardless of whether the starting mixture was a 3:1 mixture of H_2 and N_2 or pure NH_3. Equilibrium is thus a condition of the system that can be approached from both directions. This suggests to us that equilibrium is *not* a static condition. On the contrary, both the forward reaction in Equation [14.1], in which ammonia is formed, and the reverse reaction in which H_2 and N_2 are formed from ammonia, are going on at precisely the same rates. Thus the net rate of change in the system is zero.

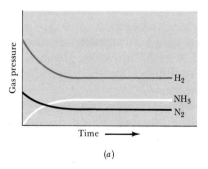

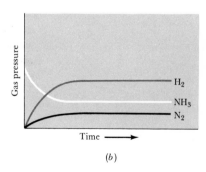

(a)   (b)

FIGURE 14.1 Variation in gas pressures in formation of the equilibrium $N_2 + 3H_2 \rightleftharpoons 2NH_3$. (a) The equilibrium is approached beginning with H_2 and N_2 in the ratio 3:1. (b) The equilibrium is approached beginning with NH_3.

The fact that it is possible to arrive at the same equilibrium condition from two different starting points tells us also that we have a true equilibrium. It is very important to distinguish this from another condition that is often called *metastable equilibrium*. Suppose we place a mixture of N_2, H_2, and NH_3 together in a reaction vessel in which there is no catalyst. After observing that the concentrations of the reactants were not changing with time, we might be tempted to conclude that the definition of equilibrium has been met; the rates of the forward and reverse reactions are equal. But they seem to be equal only because they are practically zero. The true equilibrium condition is reached only by increasing the temperature to speed up reaction or by introducing a catalyst.

If we were systematically to change the relative amounts of N_2, H_2, and NH_3 in the starting mixture of gases, and then analyze the gas mixtures at equilibrium, it would be possible to determine what sort of "law" governs the equilibrium state. Studies of this kind were carried out on other chemical systems by chemists in the nineteenth century, long before Haber's work. In 1864, Cato Maximilian Guldberg and Peter Waage proposed their **law of mass action**. This law expresses the relative concentrations of reactants and products at equilibrium in terms of a quantity called the equilibrium constant. Suppose we have the general reaction

$$j\mathrm{A} + k\mathrm{B} \rightleftharpoons p\mathrm{R} + q\mathrm{S} \qquad [14.2]$$

where A, B, R, and S are the chemical species involved, and j, k, p, and q are their coefficients in the balanced chemical equation. According to the law of mass action, the equilibrium condition is expressed by the equation

$$K = \frac{[\mathrm{R}]^p[\mathrm{S}]^q}{[\mathrm{A}]^j[\mathrm{B}]^k} \qquad [14.3]$$

where K is a constant, called the **equilibrium constant**, and the square brackets signify the *concentration* of the species within the brackets. The law of mass action applies only to a system that has attained equilibrium. In general, the equilibrium constant is given by the concentrations of all reaction products multiplied together, each raised to the power of its coefficient in the balanced equation, divided by the concentrations of all reactants multiplied together, each raised to the power of its coefficient in the balanced equation. (Remember the convention is to write the concentration terms for the *products* in the *numerator* and those for the *reactants* in the *denominator*.)

The equilibrium constant is a true constant. Its value at any given temperature does not depend on the initial concentrations of reactants and products. It also does not matter whether there are other substances present, so long as they do not consume a reactant or product through chemical reaction. The value of the equilibrium constant does, however, vary with temperature.

TABLE 14.1 Initial and equilibrium concentrations (molarities) of NO_2 and N_2O_4 in the gas phase at 100°C

Experiment	Initial N_2O_4 concentration (M)	Initial NO_2 concentration (M)	Equilibrium N_2O_4 concentration (M)	Equilibrium NO_2 concentration (M)	K_c
1	0.0	0.0200	0.00140	0.0172	0.211
2	0.0	0.0300	0.00280	0.0243	0.211
3	0.0	0.0400	0.00452	0.0310	0.213
4	0.0200	0.0	0.00452	0.0310	0.213

As an illustration of the law of mass action, consider the gas phase equilibrium between dinitrogen tetroxide and nitrogen dioxide:

$$N_2O_4(g) \rightleftharpoons 2NO_2(g) \qquad [14.4]$$

Because NO_2 is a dark brown gas, and N_2O_4 is colorless, the amount of NO_2 in the mixture can be measured by measuring the intensity of the brown color of the gas mixture.

Following the rule given above, the equilibrium-constant expression for reaction Equation [14.4] is

$$K = \frac{[NO_2]^2}{[N_2O_4]} \qquad [14.5]$$

Suppose that to determine the numerical value for K, and to verify that it is indeed constant as the concentrations of NO_2 and N_2O_4 change, three samples of NO_2 were placed in sealed glass vessels. In addition a sample of N_2O_4 was placed in a fourth vessel. The vessels were allowed to remain at 100°C until no further change in the color of the gas was noted. The mixture of gases was then analyzed to determine the concentrations of both NO_2 and N_2O_4. The results are given in Table 14.1.

To evaluate the equilibrium constant, K, the equilibrium concentrations are inserted into the equilibrium-constant expression, Equation [14.5]. When the concentration unit is molarity, as in the present case, we will label the equilibrium constant as K_c.

For example, using the first set of data,

$$[NO_2] = 0.0172\,M \quad [N_2O_4] = 0.00140\,M$$
$$K_c = \frac{[NO_2]^2}{[N_2O_4]} = \frac{(0.0172)^2}{(0.00140)} = 0.211$$

Proceeding in the same way, the values of K_c for the other samples were calculated, as listed in Table 14.1. Note that the value for K_c is essentially constant, even though the initial concentrations vary. Furthermore, the results of experiment 4 show that equilibrium can be attained beginning with N_2O_4 as well as with NO_2. That is, equilibrium can be approached from either direction. Figure 14.2 shows how both experiments 3 and 4 result in the same equilibrium mixture even though one begins with 0.0400 M NO_2 and the other with 0.0200 M N_2O_4.

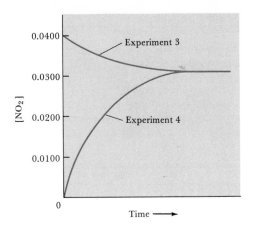

FIGURE 14.2 The same equilibrium mixture is produced starting with either 0.400 M NO_2 (Experiment 3) or 0.0200 M N_2O_4 (Experiment 4).

SAMPLE EXERCISE 14.1

Write the equilibrium-constant expression for each of the following reactions:

(a) $N_2(g) + 3H_2(g) \rightleftharpoons 2NH_3(g)$
(b) $2NH_3(g) \rightleftharpoons N_2(g) + 3H_2(g)$
(c) $2HI(g) \rightleftharpoons H_2(g) + I_2(g)$

Solution: (a) As indicated by Equation [14.3], the equilibrium-constant expression has the form of a quotient. The numerator of this quotient is obtained by multiplying the equilibrium concentrations of products, each raised to a power equal to its coefficient in the balanced equation. The denominator is obtained similarly using the equilibrium concentrations of reactants:

$$K = \frac{[NH_3]^2}{[N_2][H_2]^3}$$

(b) This reaction is just the reverse of that given in (a). Placing products over reactants we have

$$K = \frac{[N_2][H_2]^3}{[NH_3]^2}$$

Notice that this expression is just the reciprocal of that given in (a). It is a general rule that the equilibrium-constant expression for a reaction written in one direction is the reciprocal of the one for the reverse reaction. This also means that the numerical value of the equilibrium constant for a reaction written in one direction is the reciprocal of the value of the equilibrium constant for the reaction written in the reverse direction.

(c) $K = \dfrac{[H_2][I_2]}{[HI]^2}$

One of the first tasks confronting Haber and his coworkers when they set to work on the problem of ammonia synthesis was the determination of the numerical value of the equilibrium constant for the synthesis of NH_3 at various temperatures. If the value of K for this reaction is very small, this means that the amount of NH_3 formed would be small relative to the initial amounts of N_2 and H_2 used. This situation can be described by saying that the equilibrium for

$$N_2(g) + 3H_2(g) \rightleftharpoons 2NH_3(g)$$

lies to the left, that is, toward the reactant side. Clearly, if the equilibrium were too far to the left, it would not be possible to develop a satisfactory synthesis for ammonia.

The value of K can be calculated if we know the equilibrium concentrations of all reactants and products involved in the reaction. These might be obtained by a direct experimental measurement. This was the method described in constructing Table 14.1. Alternatively, the equilibrium concentrations can be calculated if we know the concentrations at the beginning of the reaction and the equilibrium concentration of at least one species. These approaches are described in Sample Exercises 14.3 and 14.4.

SAMPLE EXERCISE 14.2

The reaction of N_2 with O_2 to form NO might be considered as a means of "fixing" nitrogen:

$$N_2(g) + O_2(g) \rightleftharpoons 2NO(g)$$

The value for the equilibrium constant for this reaction at 25°C is $K_c = 1 \times 10^{-30}$. Describe the feasibility of this reaction for nitrogen fixation.

Solution: Because K_c is so small, very little NO will form at 25°C. The equilibrium is said to lie to the left, favoring the reactants. Consequently this reaction is an extremely poor choice for nitrogen fixation, at least at 25°C.

SAMPLE EXERCISE 14.3

In one of their experiments, Haber and coworkers introduced a mixture of hydrogen and nitrogen into a reaction vessel and allowed the system to attain chemical equilibrium at 472°C. The equilibrium mixture of gases was analyzed and found to contain 0.1207 M H_2, 0.0402 M N_2, and 0.00272 M NH_3. From these data calculate the equilibrium constant, K_c, for

$$N_2(g) + 3H_2(g) \rightleftharpoons 2NH_3(g)$$

Solution:

$$K_c = \frac{[NH_3]^2}{[N_2][H_2]^3} = \frac{(0.00272)^2}{(0.0402)(0.1207)^3} = 0.105$$

SAMPLE EXERCISE 14.4

A mixture of 5.00×10^{-3} mol H_2 and 1.00×10^{-2} mol I_2 are placed in a 5.00-L container at 448°C and allowed to come to equilibrium. Analysis of the equilibrium mixture shows that the concentration of HI is 1.87×10^{-3} M. Calculate K_c at 448°C for the reaction

$$H_2(g) + I_2(g) \rightleftharpoons 2HI(g)$$

Solution: The initial concentrations of H_2 and I_2 must be calculated:

$$[H_2]_i = \frac{5.00 \times 10^{-3} \text{ mol}}{5.00 \text{ L}} = 1.00 \times 10^{-3} M$$

$$[I_2]_i = \frac{1.00 \times 10^{-2} \text{ mol}}{5.00 \text{ L}} = 2.00 \times 10^{-3} M$$

The equilibrium concentrations of H_2 and I_2 can be calculated from these initial concentrations and the equilibrium concentration of HI. During the course of the reaction the concentration of HI changes from 0 to 1.87×10^{-3} M. The balanced equation indicates that 2 mol of HI form from each mole H_2.

Thus the amount of H_2 consumed is

$$\left(1.87 \times 10^{-3} \frac{\text{mol HI}}{\text{L}}\right)\left(\frac{1 \text{ mol } H_2}{2 \text{ mol HI}}\right)$$

$$= 0.935 \times 10^{-3} \text{ mol } H_2/\text{L}$$

The equilibrium concentration of H_2 is the initial concentration minus that consumed:

$$[H_2] = 1.00 \times 10^{-3} M - 0.935 \times 10^{-3} M$$
$$= 0.065 \times 10^{-3} M$$

The same line of argument gives the equilibrium concentration of I_2:

$$[I_2] = 2.00 \times 10^{-3} M - 0.935 \times 10^{-3} M$$
$$= 1.065 \times 10^{-3} M$$

In calculating the equilibrium concentrations in problems of this type, it is often convenient to set up a table showing the initial concentration, the change, and the equilibrium concentration:

$$
\begin{array}{lccc}
& H_2(g) & + \quad I_2(g) & \rightleftharpoons \qquad 2HI(g) \\
\text{Initial:} & 1.00 \times 10^{-3}\,M & 2.00 \times 10^{-3}\,M & 0\,M \\
\text{Change:} & -0.935 \times 10^{-3}\,M & -0.935 \times 10^{-3}\,M & +1.87 \times 10^{-3}\,M \\
\text{Equilibrium:} & 0.065 \times 10^{-3}\,M & 1.065 \times 10^{-3}\,M & 1.87 \times 10^{-3}\,M
\end{array}
$$

The concentrations we started with are printed in color. The concentration changes and finally the equilibrium concentrations of H_2 and I_2 were then determined as described above.

Once we have the equilibrium concentrations of each reactant and product we can calculate the equilibrium constant:

$$
K_c = \frac{[HI]^2}{[H_2][I_2]} = \frac{(1.87 \times 10^{-3})^2}{(0.065 \times 10^{-3})(1.065 \times 10^{-3})}
$$
$$
= 50.5
$$

CONCENTRATION UNITS AND EQUILIBRIUM CONSTANTS

As we have seen, the square brackets around a chemical symbol, as in $[NH_3]$, represent the concentration of that substance. Molarity is the common concentration unit for reactions occurring in solution. For gas-phase reactions, the concentration units used are either molarity or atmospheres of pressure. When the concentration is expressed in molarity, we denote the equilibrium constant as K_c. When the units are atmospheres, we write K_p. Since the numerical values of K_c and K_p will generally be different, we must take care to indicate which we are using by means of these subscripts.

The ideal-gas equation permits us to convert between atmospheres and molarity and therefore convert between K_p and K_c:

$$
PV = nRT
$$
$$
P = \left(\frac{n}{V}\right)RT = MRT \tag{14.6}
$$

where n/V (the number of moles per liter) is concentration in molarity, M. As a result of the relationship between pressure and molarity expressed in Equation [14.6], a general expression relating K_p and K_c can be written:

$$
K_p = K_c(RT)^{\Delta n} \tag{14.7}
$$

The quantity Δn in this equation is the change in the number of moles of gas upon going from reactants to products; Δn is equal to the number of moles of gaseous products minus the number of moles of gaseous reactants. For example, in the reaction

$$
H_2(g) + I_2(g) \rightleftharpoons 2HI(g)
$$

there are 2 mol of HI (the coefficient in the balanced equation); there are also 2 mol of gaseous reactants ($1H_2 + 1I_2$). Therefore $\Delta n = 2 - 2 = 0$, and $K_p = K_c$ for this reaction.

SAMPLE EXERCISE 14.5

Using the value of K_c obtained in Sample Exercise 14.3, calculate K_p for

$$N_2(g) + 3H_2(g) \rightleftharpoons 2NH_3(g)$$

at 472°C.

Solution: There are 2 mol of gaseous products ($2NH_3$); and there are 4 mol of gaseous reactants ($1N_2 + 3H_2$). Therefore $\Delta n = 2 - 4 = -2$. (Remember that Δ functions are always based on products minus reactants.) The temperature, T, is $273 + 472 = 745$ K. The value for the ideal-gas constant, R, is 0.0821 L-atm/K-mol. The value of K_c from Sample Exercise 14.3 is 0.105. We therefore have

$$K_p = \frac{P_{NH_3}^2}{P_{N_2}P_{H_2}^3} = K_c(RT)^{\Delta n}$$
$$= (0.105)(0.0821 \times 745)^{-2}$$
$$= 2.81 \times 10^{-5}$$

Concentration units can be carried through the calculation of the equilibrium constant to give units for K. For example, for the reaction $N_2O_4(g) \rightleftharpoons 2NO_2(g)$ we have $K = [NO_2]^2/[N_2O_4]$. When concentration is in molarity, the units of the equilibrium constant are $M^2/M = M$; when concentration is in atmospheres, the units are $atm^2/atm = atm$. Attaching units to the equilibrium constant has the advantage of clearly indicating the units in which concentration is expressed. Nevertheless, the more common practice is to write equilibrium constants as dimensionless quantities. We have adopted this practice in this text.

14.3 Heterogeneous equilibria

Many equilibria of importance, such as in the hydrogen-nitrogen-ammonia system, involve substances all in the same phase. Such equilibria are said to be homogeneous. On the other hand, it is possible for the substances in equilibrium to be in different phases, giving rise to heterogeneous equilibria. As an example, consider the decomposition of calcium carbonate:

$$CaCO_3(s) \longrightarrow CaO(s) + CO_2(g) \qquad [14.8]$$

This system involves a gas in equilibrium with two solids. If we write the equilibrium constant expression for this process in the usual way we obtain:

$$K = \frac{[CaO][CO_2]}{[CaCO_3]} \qquad [14.9]$$

This example presents us with a problem we have not encountered previously: How do we express the concentration of a solid substance? Because the calcium carbonate or calcium oxide are present as pure solids, their "concentration" is a constant. The number of moles per liter of either of these solids is not changed, whether we have a large amount of solid present or just a little. The concentration of a pure substance, liquid or solid, can be expressed as the density divided by the molecular weight:

$$\frac{\text{Density}}{\text{MW}} = \frac{g/cm^3}{g/mol} = \frac{mol}{cm^3}$$

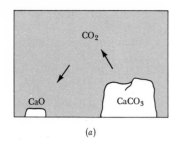

(a)

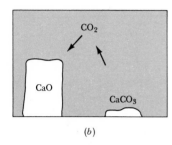

(b)

FIGURE 14.3 Heterogeneous equilibrium in $CaCO_3$ decomposition. Assuming the temperature in each case to be the same, the equilibrium pressure of CO_2 is the same in (a) as in (b).

The density of a pure liquid or solid is a constant at any given temperature and in fact changes very little with temperature. Thus, the effective concentration of a pure liquid or solid is a constant. The equilibrium-constant expression for Equation [14.8] then simplifies to

$$K = \frac{[CO_2](\text{constant 1})}{(\text{constant 2})}$$

where constant 1 is the concentration of CaO, and constant 2 is the concentration of $CaCO_3$. Moving the constants to the left-hand side of the equation, we have

$$K' = K\frac{(\text{constant 2})}{(\text{constant 1})} = [CO_2] \qquad [14.10]$$

As a practical matter, the overall effect is the same as if we set the concentrations of both solids equal to one in the equilibrium-constant expression.

Equation [14.10] tells us that it doesn't matter at all how much $CaCO_3$ or CaO are present, so long as there is some of each in the system. As shown in Figure 14.3, we would have the same pressure of CO_2 in the system when we have a large excess of CaO as when we have an excess of $CaCO_3$. On the other hand, if one of the three ingredients is missing, we cannot have an equilibrium.

Thus we see that any pure solids or liquids that might be involved in an equilibrium have the same effect on the equilibrium no matter how much solid or liquid is present. The "concentrations" of pure solids and liquids are incorporated into the equilibrium constant. What this means when we write equilibrium-constant expressions is that the concentrations of pure solids and liquids are absent from the expression.

SAMPLE EXERCISE 14.6

Write the equilibrium-constant expression for each of the following reactions:

(a) $CO_2(g) + H_2(g) \rightleftharpoons CO(g) + H_2O(l)$
(b) $SnO_2(s) + 2CO(g) \rightleftharpoons Sn(s) + 2CO_2(g)$

Solution: (a) The equilibrium-constant expression is

$$K = \frac{[CO]}{[CO_2][H_2]}$$

(Because H_2O is a pure liquid its concentration does not appear in the equilibrium-constant expression.)

(b) The equilibrium-constant expression is

$$K = \frac{[CO_2]^2}{[CO]^2}$$

(Because SnO_2 and Sn are both pure solids, they do not appear in the equilibrium-constant expression.)

14.4 Uses of equilibrium constants

We have seen that the magnitude of K indicates the extent to which a reaction will proceed. If K is very large, the reaction will tend to proceed far to the right; if K is very small, very little reaction will occur and the equilibrium mixture will contain mainly reactants. The equilibrium constant also allows us (1) to predict the direction in which a reaction mixture will proceed to achieve equilibrium and (2) to calculate the concentrations of reactants and products once equilibrium has been reached.

Suppose we place a mixture of 2.00 mol of H_2, 1.00 mol of N_2, and 2.00 mol of NH_3 in a 1-L container at 472 K. Will there be a reaction between N_2 and H_2 to form more NH_3? If we insert the starting concentrations of N_2, H_2, and NH_3 into the equilibrium-constant expression, we have

$$\frac{[NH_3]^2}{[N_2][H_2]^3} = \frac{(2.00)^2}{(1.00)(2.00)^3} = 0.500$$

We have seen, in Sample Exercise [14.3], that at this temperature $K_c = 0.105$. Therefore, the quotient $[NH_3]^2/[N_2][H_2]^3$ will need to change from 0.500 to 0.105 to move the system toward equilibrium. This change can happen only if $[NH_3]$ decreases and $[N_2]$ and $[H_2]$ increase. Thus the reaction proceeds toward equilibrium with the formation of N_2 and H_2 from the NH_3; the reaction proceeds from right to left.

When we substitute reactant and product concentrations into the equilibrium-constant expression as we did above, the result is known as the reaction quotient and is represented by the letter Q. The reaction quotient will equal the equilibrium constant, K, only if the concentrations are ones for the system at equilibrium: $Q = K$ only at equilibrium. We have seen that when the reaction quotient is larger than K, substances on the right side of the chemical equation will react to form substances on the left; the reaction moves from right to left in approaching equilibrium: if $Q > K$ the reaction moves right to left. Conversely, if $Q < K$, the reaction will move toward equilibrium with the formation of more products (from left to right).

SAMPLE EXERCISE 14.7

At 448°C the equilibrium constant, K_c, for the reaction

$$H_2(g) + I_2(g) \rightleftharpoons 2HI(g)$$

is 50.5. Predict the direction in which the reaction will proceed to reach equilibrium at 448°C if we start with 2.0×10^{-2} mol of HI, 1.0×10^{-2} mol of H_2, and 3.0×10^{-2} mol of I_2 in a 2.0-L container.

Solution: The starting concentrations are

$[HI] = 2.0 \times 10^{-2} \text{ mol}/2.0 \text{ L} = 1.0 \times 10^{-2} M$
$[H_2] = 1.0 \times 10^{-2} \text{ mol}/2.0 \text{ L} = 5.0 \times 10^{-3} M$
$[I_2] = 3.0 \times 10^{-2} \text{ mol}/2.0 \text{ L} = 1.5 \times 10^{-2} M$

The reaction quotient is

$$Q = \frac{[HI]^2}{[H_2][I_2]} = \frac{(1.0 \times 10^{-2})^2}{(5.0 \times 10^{-3})(1.5 \times 10^{-2})} = 1.3$$

Because $Q < K_c$, $[HI]$ will need to increase and $[H_2]$ and $[I_2]$ decrease to reach equilibrium; the reaction will proceed from left to right.

CALCULATION OF EQUILIBRIUM CONCENTRATIONS

If we know the value of the equilibrium constant for a particular reaction, we can calculate the concentrations of substances present in the reaction mixture at equilibrium. The complexity of the calculation will depend on several factors including the complexity of the balanced equation for the reaction and what concentrations we are given. The following examples illustrate the general types of problems.

SAMPLE EXERCISE 14.8

What is the partial pressure of NH_3 that is in equilibrium with N_2 and H_2 at $500°C$ if the equilibrium partial pressure of H_2 is 0.733 atm and that of N_2 is 0.527 atm $(K_p = 1.45 \times 10^{-5}$ at $500°C$ for $N_2(g) + 3H_2(g) \rightleftharpoons 2NH_3(g))$.

Solution:

$$K_p = \frac{P^2_{NH_3}}{P_{N_2}P^3_{H_2}} = 1.45 \times 10^{-5}$$

Because K_p, P_{N_2}, and P_{H_2} are given, we can rearrange this equation and solve for P_{NH_3}:

$$P^2_{NH_3} = K_p P_{N_2} P^3_{H_2} = (1.45 \times 10^{-5})(0.527)(0.733)^3$$
$$= 3.01 \times 10^{-6}$$
$$P_{NH_3} = \sqrt{3.01 \times 10^{-6}} = 1.73 \times 10^{-3} \text{ atm}$$

SAMPLE EXERCISE 14.9

A 1-L container is filled with 0.50 mol of HI at $448°C$. The value of the equilibrium constant, K_c, for the reaction

$$H_2(g) + I_2(g) \rightleftharpoons 2HI(g)$$

at this temperature is 50.5. What are the concentrations of H_2, I_2, and HI in the vessel at equilibrium?

Solution: In this case we are not given any of the equilibrium concentrations, only the starting concentrations: $[H_2] = [I_2] = 0$, and $[HI] = 0.50 M$. The problem is related to the one we worked in Sample Exercise 14.4: We must use the balanced chemical equation to write an expression for equilibrium concentrations in terms of initial concentrations. It is useful to set up a table as in Sample Exercise 14.4. We can define x as the amount of HI that reacts forming H_2 and I_2. For each x HI that decomposes, $(x/2)$ H_2 and $(x/2)$ I_2 form

	$H_2(g)$	+	$I_2(g)$	$\rightleftharpoons$	$2HI(g)$
Initial:	$0\,M$		$0\,M$		$0.50\,M$
Change:	$(x/2)\,M$		$(x/2)\,M$		$-x\,M$
Equilibrium:	$(x/2\,)M$		$(x/2)\,M$		$(0.50-x)\,M$

We can substitute the equilibrium concentrations into the equilibrium expression and solve for the single unknown, x:

$$K_c = \frac{[HI]^2}{[H_2][I_2]} = \frac{(0.50-x)^2}{(x/2)^2} = 50.5$$

This equation is second order in x (it contains x^2 as the highest power of x); such equations can always be solved by use of the quadratic formula (Appendix A.3). However, a quicker solution can be obtained in this particular case by taking the square root of both sides of the equation:

$$\frac{0.50 - x}{x/2} = \sqrt{50.5} = 7.11$$

Solving for x:

$$0.50 - x = \frac{x}{2}(7.11) = 3.56x$$
$$0.50 = x + 3.56x = 4.56x$$
$$x = \frac{0.50}{4.56} = 0.11\,M$$

Thus the equilibrium concentrations are as follows:

$$[H_2] = \frac{x}{2} = \frac{0.11\,M}{2} = 0.055\,M$$
$$[I_2] = \frac{x}{2} = \frac{0.11\,M}{2} = 0.055\,M$$
$$[HI] = 0.50\,M - 0.11\,M = 0.39\,M$$

TABLE 14.2 Effect of temperature and total pressure on the percentage of ammonia present at equilibrium, beginning with a 3:1 H_2/N_2 mixture

Temperature (°C)	Total pressure (atm)			
	200	300	400	500
400	38.7	47.8	54.9	60.6
450	27.4	35.9	42.9	48.8
500	18.9	26.0	32.2	37.8
600	8.8	12.9	16.9	20.8

14.5 Factors affecting equilibrium: LeChatelier's principle

In developing his process for making ammonia from N_2 and H_2 Haber sought to know the factors that might be varied to increase the yield of NH_3. Using the values of the equilibrium constant at various temperatures he calculated the equilibrium amounts of NH_3 formed under a variety of conditions. The results of some of his calculations are shown in Table 14.2. Notice that the yield of NH_3 decreases with increasing temperature and increases with increasing pressure. These results can be qualitatively understood in terms of LeChatelier's principle (Section 12.4). In this section we will use LeChatelier's principle to make qualitative predictions about the response of a system at equilibrium to various changes in external conditions. We will consider three ways that a chemical equilibrium can be shifted: (1) adding or removing a reactant or product; (2) changing the pressure; and (3) changing the temperature.

CHANGE IN REACTANT OR PRODUCT CONCENTRATIONS

A system at equilibrium is in a dynamic state; the forward and reverse processes are occurring at equal rates, and the system is in a state of balance. An alteration in the conditions of the system may cause the state of balance to be disturbed. If this occurs, the equilibrium is shifted until a new state of balance is attained. LeChatelier's principle states that the shift will be in the direction that minimizes or reduces the effect of the change. Therefore, *if a chemical system is at equilibrium, and we add a substance (either a reactant or a product), the reaction will shift so as to reestablish equilibrium by consuming part of that added substance. Conversely, removal of a substance will result in the reaction moving in the direction that forms more of that substance.*

For example, addition of hydrogen to an equilibrium mixture of H_2, N_2, and NH_3 would cause the system to shift in such a way as to reduce the hydrogen pressure toward its original value. This can occur only if the equilibrium is shifted in the direction of forming more NH_3. At the same time, the quantity of N_2 would also be reduced slightly. Addition of more N_2 to an equilibrium system would similarly cause a shift in the direction of forming more ammonia. On the other hand, LeChatelier's principle tells us that if we add NH_3 to the system at equilibrium, the shift will be in such a direction as to reduce the NH_3 concentration toward its original value; that is, some of the added ammonia will decompose to form N_2 and H_2.

We can reach the same conclusions by considering the effect that adding or removing a substance has on the reaction quotient (Section 14.4). For example, removal of NH_3 from an equilibrium mixture gives

$$\frac{[NH_3]^2}{[N_2][N_2]^3} = Q < K$$

Because $Q < K$, the reaction shifts from left to right, forming more NH_3 and decreasing $[N_2]$ and $[H_2]$, to restore a new equilibrium that is still governed by K.

If the products of a reaction can be removed continuously, then the reacting system can be continuously shifted to form more products. The yield of NH_3 in the Haber process can be increased dramatically by liquefying the NH_3; the liquid NH_3 is removed and the N_2 and H_2 are recycled to form more NH_3. The general way this is accomplished is shown in Figure 14.4. If a reaction is operated so that equilibrium cannot be achieved because of escape of products, or if the equilibrium constant is very large, the reaction will proceed essentially to completion. In such instances the chemical equation for the reaction is usually given with a single arrow: reactants $\longrightarrow$ products.

EFFECT OF PRESSURE AND VOLUME CHANGES

If a system is at equilibrium and the total pressure is increased by reducing the volume, the system responds by shifting in the direction that occupies the smaller

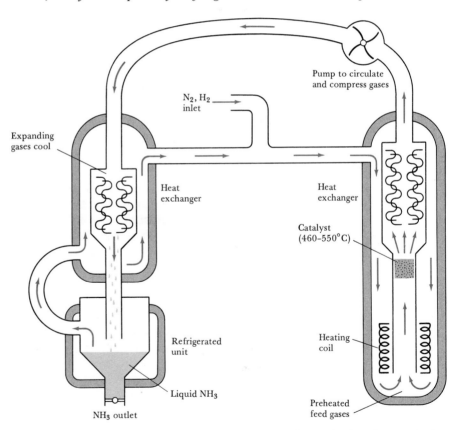

FIGURE 14.4 Schematic diagram summarizing the industrial production of ammonia. Incoming N_2 and H_2 gases are heated to approximately 500°C and passed over a catalyst. The resultant gas mixture is allowed to expand and cool, causing NH_3 to liquefy. Unreacted N_2 and H_2 gases are recycled.

volume. In practice, this means that the shift is in the direction that decreases the number of moles of gas. Conversely, decreasing the pressure by increasing the volume causes a shift in the direction that produces more gas molecules. For the reaction

$$N_2(g) + 3H_2(g) \rightleftharpoons 2NH_3(g)$$

there are 2 mol of gas on the right side of the chemical equation ($2NH_3$) and 4 mol of gas on the left ($1N_2 + 3H_2$). Consequently, an increase in the total pressure leads to the formation of NH_3. Increasing the pressure shifts the reaction toward the side with fewer gas molecules. In the case of the reaction

$$H_2(g) + I_2(g) \rightleftharpoons 2HI(g)$$

changing the total pressure will not influence the position of the equilibrium. In this case there are the same number of gaseous product molecules as there are gaseous reactants.

It is important to keep in mind that a change in the total pressure does not change the value of K, so long as the temperature remains constant. Rather, the pressure changes the concentrations of the gaseous substances. In Sample Exercise 14.3, we calculated K_c for an equilibrium mixture at $472°C$ that contained $[H_2] = 0.1207\ M$, $[N_2] = 0.0402\ M$, and $[NH_3] = 0.0027\ M$. The value of K_c is 0.105. Consider what happens if we double the pressure of this equilibrium mixture by reducing the volume by half. If there were no shift in equilibrium, this volume change would cause the concentrations of all substances to double, giving new concentrations of $[H_2] = 0.2414\ M$, $[N_2] = 0.0804\ M$, and $[NH_3] = 0.00544\ M$. The reaction quotient would then no longer equal the equilibrium constant:

$$Q = \frac{[NH_3]^2}{[N_2][H_2]^3} = \frac{(0.00544)^2}{(0.0804)(0.2414)^3} = 2.62 \times 10^{-2}$$

Because $Q < K_c$, the system is no longer at equilibrium. Equilibrium will be reestablished by increasing $[NH_3]$ and decreasing $[N_2]$ and $[H_2]$ until $Q = K_c = 0.105$. Therefore the equilibrium shifts to the right as LeChatelier's principle predicted.

EFFECT OF TEMPERATURE CHANGES

Changes in concentrations or total pressure can cause shifts in equilibrium without changing the equilibrium constant. In contrast, almost every equilibrium constant changes in value with change in temperature. Consider the decomposition of carbon dioxide into carbon monoxide and oxygen:

$$2CO_2(g) \rightleftharpoons 2CO(g) + O_2(g) \qquad [14.11]$$

From a knowledge of the heats of formation of $CO_2(g)$ and $CO(g)$, we can conclude that the forward reaction is highly endothermic; using the values of $\Delta H_f°$ from Appendix D, $\Delta H°$ for the overall reaction is calcu-

TABLE 14.3 Percent dissociation of CO_2 into CO and O_2 as a function of temperature

Temperature (K)	Percent dissociation
1500	0.048
2000	2.05
2500	17.6
3000	54.8

TABLE 14.4 Variation in K_p for the equilibrium $N_2 + 3H_2 \rightleftharpoons 2NH_3$ as a function of temperature

Temperature (°C)	K_p
300	4.34×10^{-3}
400	1.64×10^{-4}
450	4.51×10^{-5}
500	1.45×10^{-5}
550	5.38×10^{-6}
600	2.25×10^{-6}

lated to be 566 kJ. At room temperature and thereabouts, CO_2 has no observable tendency to dissociate into CO and O_2. However, at high temperatures the equilibrium shifts to the right as shown in Table 14.3. Clearly, the equilibrium constant for reaction Equation [14.11] is very dependent on temperature and increases with increasing temperature.

The equilibrium constants for exothermic reactions, that is, those in which heat is evolved, decrease with increase in temperature. By contrast, equilibrium constants for endothermic reactions increase with increase in temperature.

Following the line of reasoning employed in Section 12.4, we can deduce the rules for the temperature dependence of the equilibrium constant by applying LeChatelier's principle. When heat is added to a system by increasing the temperature, the equilibrium should shift in such a direction as to undo partially the effect of the added heat. It shifts, therefore, in the direction in which heat is absorbed. If a reaction is exothermic in the forward direction, it must be endothermic in the reverse direction. Thus, when heat is added to an equilibrium system that is exothermic in the forward direction, the equilibrium shifts in the reverse direction, in the direction of reactants. In summary, the rule is that *when heat is added at constant pressure to an equilibrium system, the equilibrium shifts in the direction that absorbs heat*. Conversely, if heat is removed from an equilibrium system, the equilibrium shifts in the direction that evolves heat.

SAMPLE EXERCISE 14.10

Consider the following reaction

$$N_2O_4(g) \rightleftharpoons 2NO_2(g) \qquad \Delta H° = 58.0 \text{ kJ}$$

In what direction will the equilibrium shift when each of the following changes is made to a system at equilibrium: (a) add N_2O_4; (b) remove NO_2; (c) increase pressure; (d) increase volume; (e) decrease temperature.

Solution: LeChatelier's principle can be applied to determine the effects of each of these changes: (a) The system will adjust so as to decrease the concentration of the added N_2O_4; the reaction consequently shifts toward the formation of more products (shifts toward the right side of the equation).

(b) The system will adjust to this change by forming more NO_2; the equilibrium shifts toward the product side of the equation, to the right. (c) The system will establish a new equilibrium that has a smaller volume (fewer gas molecules); consequently, the reaction shifts to the left. (d) The system will shift in the direction that occupies a larger volume (more gas molecules); it moves to the right. (e) The system will adjust to a new equilibrium position by shifting in the direction that produces heat. The reaction is endothermic in the forward direction (left to right). It therefore shifts to the left, with the formation of more N_2O_4 in order to produce heat. Note that only this last change effects the numerical value of the equilibrium constant, K.

SAMPLE EXERCISE 14.11

Using the standard heat of formation data in Appendix D, determine the enthalpy change for the reaction

$$N_2(g) + 3H_2(g) \rightleftharpoons 2NH_3(g)$$

From this determine how the equilibrium constant for the reaction should change with temperature.

Solution: Recall that the standard enthalpy change for a reaction is given by the standard molar enthalpies of the reactants, each multiplied by its coefficient in the balanced chemical equation, less the same quantities for the reactants. $\Delta H_f°$ for $NH_3(g)$ at 25°C is -46.19 kJ/mol. The $\Delta H_f°$ values for $H_2(g)$ and $N_2(g)$ are zero by definition, because the enthalpies of formation of the elements in their normal states at 25°C are defined as zero (Section 4.5). Because 2 mol of NH_3 are formed, the total enthalpy change is

$$2 \text{ mol}(-46.19 \text{ kJ/mol}) - 0 = -92.38 \text{ kJ}$$

The reaction in the forward direction is exothermic. An increase in temperature should therefore cause the reaction to shift in the *reverse* direction, that is, in the direction of less NH_3 and more N_2 and H_2. This is what occurs, as reflected in the values for K_p pre-

sented in Table 14.4. Notice that K_p changes very markedly with change in temperature, and that it is larger at lower temperatures. This is a matter of great practical importance. To form ammonia at a reasonable rate, higher temperatures are required. Yet at higher temperatures, the equilibrium constant is smaller, so the percentage conversion to ammonia is smaller. To compensate for this, higher pressures are needed, because high pressure favors ammonia formation.

14.6 The relationship between chemical equilibrium and chemical kinetics

It is very important to be clear on the distinction between a system at equilibrium and the rate at which the system approaches equilibrium in the first place. Imagine that we have a system initially containing only reactant molecules and no products. As the system begins to change, the only reaction occurring is formation of products. As products accumulate, however, the reverse reaction also begins to occur. In many systems, this reverse reaction is so slow that it is completely unimportant. The system then just keeps on changing until essentially all the reactants are converted to products. Reactions of this sort are said to proceed to completion. Even though the reverse reaction, conversion of products to reactants, is possible in principle, it is not observed.

In many other reactions, on the other hand, the reverse reaction occurs more rapidly. After a time, reactant molecules are being formed just as rapidly as they are themselves reacting to form products. When the rates of the two opposing processes are equal, the overall rate of change in the system is zero, and we have chemical equilibrium. A true chemical equilibrium always involves a balancing of equal and opposite rate processes. This is true regardless of the particular mechanism or pathway by which the reaction proceeds. Figure 14.5 shows an energy profile for the single-step reaction between reactants A and B and products C and D:

$$A + B \rightleftharpoons C + D \qquad [14.12]$$

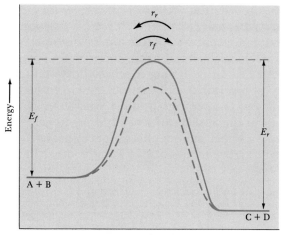

Progress of reaction

FIGURE 14.5 Schematic illustration of chemical equilibrium in the reaction $A + B \rightleftharpoons C + D$. When equilibrium is attained, the rate of the forward reaction, r_f, equals the rate of the reverse reaction, r_r. The dashed line refers to the energy profile for a catalyzed reaction in which the activation energy is lowered. The rates of forward and reverse reactions in the catalyzed reaction are increased to the same degree.

At equilibrium, the rate at which reactants cross the energy barrier to form products equals the rate at which products cross the energy barrier in the reverse direction to form reactants. Because we have chosen the situation where the reaction occurs in a single, bimolecular step in each direction, the rate law in each direction is second order. Thus

r_f = rate of forward reaction = $k_f[A][B]$

r_r = rate of reverse reaction = $k_r[C][D]$

At equilibrium these two rate processes must be equal:

$$k_f[A][B] = k_r[C][D]$$

If we rearrange this equation we obtain

$$\frac{k_f}{k_r} = \frac{[C][D]}{[A][B]} = K \qquad [14.13]$$

Thus the equilibrium constant is just the ratio of rate constants for the forward and reverse reactions.* Recall that the rate of a reaction decreases as the height of the energy barrier increases (Section 13.4). The barrier for the forward reaction (E_f) in Figure 14.5 is lower than for the reverse reaction (E_r). Thus, k_f must be larger than k_r, so that K should be a large number. This is in keeping with the fact that the energies of the products are lower than the energies of the reactants.

THE EFFECTS OF CATALYSTS

Suppose that we add a catalyst to the reaction system described in Equation [14.12] so that the energy barrier to reaction is lowered, as shown by the dashed line in Figure 14.5. The rates of *both* the forward and reverse reactions are increased by the presence of a catalyst. In fact, the two rate constants are affected to precisely the same degree. In other words, it is impossible for a catalyst to lower the barrier for just the forward reaction and not the reverse reaction. Because the forward and reverse reaction rate constants are affected to exactly the same degree, their ratio is unchanged. We therefore have the rule that *a catalyst may change the rate of approach to equilibrium, but it does not change the value of the equilibrium constant.*

The rate at which a reaction approaches equilibrium is an important practical consideration. As an example, let us again consider the synthesis of ammonia from N_2 and H_2. In designing a process for ammonia synthesis, Haber had to deal with a rather serious problem. He wished to synthesize ammonia at the lowest temperature possible, consistent with a reasonable reaction rate. But in the absence of a catalyst, hydrogen and nitrogen do not react with one another at a significant rate either at room temperature or even at much higher temperatures. On the other hand,

---

*The relationship between rate constants and the equilibrium constant is somewhat more complicated if the reaction occurs by a multistep mechanism. However, even in that case the equilibrium constant can be related to a ratio of rate constants. Regardless of the mechanism, the equilibrium-constant expression can be written directly from the balanced equation for the overall reaction. By contrast, rate laws must be determined by experiment.

Haber had to cope with a rapid decrease in equilibrium constant with increasing temperature as shown in Table 14.4. At temperatures sufficiently high to give a satisfactory reaction rate, the amount of ammonia formed was too small. The solution to this dilemma was to develop a catalyst that would produce a reasonably rapid approach to equilibrium, at a sufficiently low temperature so that the equilibrium constant was still reasonably large. The development of a catalyst thus became the focus of Haber's research efforts.

After trying different substances to see which would be most effective, Haber finally settled on iron mixed with metal oxides. Variants of the original catalyst formulations are still employed. With these catalysts it is possible to obtain a reasonably rapid approach to equilibrium at temperatures around 400–500°C, and with gas pressures of 200–600 atm. The high pressures are needed to obtain a satisfactory degree of conversion at equilibrium. You can see from Table 14.2 that if an improved catalyst could be found, one that would lead to sufficiently rapid reaction at temperatures lower than 400–500°C, it would be possible to obtain the same degree of equilibrium conversion at much lower pressures. This would result in a great savings in the cost of equipment for ammonia synthesis. In view of the growing need for nitrogen as fertilizer, the fixation of nitrogen is a process of ever-increasing importance, worthy of additional research effort.

The formation of NO from N_2 and O_2 provides another interesting example of the practical importance of changes in equilibrium constant and reaction rate with temperature. The balanced equation and the standard enthalpy change for the reaction are

$$\tfrac{1}{2}N_2(g) + \tfrac{1}{2}O_2(g) \rightleftharpoons NO(g) \qquad \Delta H° = 90.4 \text{ kJ} \qquad [14.14]$$

The reaction is endothermic, that is, heat is absorbed when NO is formed from the elements. By applying LeChatelier's principle, we deduce that an increase in temperature will shift the equilibrium in the direction of more NO. The equilibrium constant, K_c, for formation of 1 mol of NO from the elements at 300 K is only about 10^{-15}. On the other hand, at a much higher temperature, about 2400 K, the equilibrium constant is much larger, about 0.05. The manner in which K_c for reaction Equation [14.14] varies with temperature is shown in Figure 14.6.*

This graph helps to explain why NO is a pollution problem. In the cylinder of a modern high-compression auto engine, the temperatures during the fuel-burning part of the cycle may be on the order of 2400 K. Because there is a fairly large excess of air in the cylinder, these conditions provide an opportunity for the formation of some NO. After the combustion, however, the gases are quickly cooled. As the temperature drops the equilibrium of Equation [14.14] shifts strongly to the left, that is, in the direction of N_2 and O_2. But the lower temperatures also mean that the rate of the reaction also is decreased. The NO formed at high temperatures is essentially "frozen" in that form as the gas cools.

The gases exhausting from the cylinder are still quite hot, perhaps

*It is necessary to use a log scale for K_c in Figure 14.6 because the values of K_c vary over such a huge range.

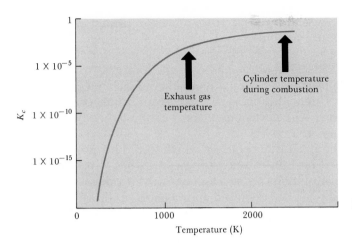

FIGURE 14.6 Variation of the equilibrium constant for the reaction $\frac{1}{2}N_2(g) + \frac{1}{2}O_2(g) \rightleftharpoons NO(g)$ as a function of temperature.

1200 K. At this temperature, as shown in Figure 14.6, the equilibrium constant for formation of NO is much smaller. However, the rate of conversion of NO to N_2 and O_2 is too slow to permit much loss of NO before the gases are cooled still further. Getting the NO out of the exhaust gases, described in Chapter 10, depends on finding a catalyst that will work at the temperatures of the exhaust gases, and that will cause conversion of NO into something harmless. If a catalyst could be found which would convert the NO back into N_2 and O_2 the equilibrium would be sufficiently favorable. It has not proved possible to find a catalyst capable of withstanding the grueling conditions found in automobile exhaust systems that can catalyze the conversion of NO into N_2 and O_2. Instead, the catalysts used are designed to catalyze reaction of NO with H_2 or CO, as described in Section 13.6.

FOR REVIEW

Summary

If the reactants and products of a reaction are kept in contact, a chemical reaction can achieve a state of dynamic balance in which forward and reverse reactions are occurring at equal rates. This condition is known as chemical equilibrium. A system at equilibrium does not change with time. For such a system, the ratio of products to reactants, each concentration raised to the power corresponding to the coefficient in the balanced equation, is called the equilibrium constant, K: $K = $ [products]/[reactants]. The equilibrium constant changes with temperature but is not affected by changes in relative concentrations of any reacting substances or by pressure or the presence of catalysts. In heterogeneous equilibria, the concentrations of pure solids or liquids are absent from the equilibrium-constant expression.

If concentrations are expressed in molarity, we label the equilibrium constant K_c; if the concentration units are atmospheres, we use K_p. K_c and K_p are related by the following equation: $K_p = K_c(RT)^{\Delta n}$.

A large value for K_c or K_p indicates that the equilibrium mixture contains more products than reactants. A small value for the equilibrium constant means the equilibrium lies toward the reactant side.

The reaction quotient, Q, is found by substituting reactant and product concentrations into the equilibrium-constant expression. If the system is at equilibrium, $Q = K$. However, if $Q \neq K$, nonequilibrium conditions apply; if $Q < K$, the reaction will move toward equilibrium by forming more products (the reaction moves from left to right); if $Q > K$, the reaction will proceed from right to left. Knowledge of the value of K_c or K_p permits calculation of the equilibrium concentrations of reactants and products.

LeChatelier's principle indicates that if we disturb a system that is at equilibrium, the equilibrium will shift to minimize the disturbing influence. The effects of adding (or removing) reactants or products, and of changing pressure, volume, or temperature can be predicted using this principle. The equilibrium constant changes value with changes in temperature. Catalysts affect the speed at which equilibrium is reached but do not affect K (do not affect the position of equilibrium).

Learning goals

Having read and studied this chapter, you should be able to:

1. Write the equilibrium-constant expression for a balanced chemical equation, whether heterogeneous or homogeneous.

2. Numerically evaluate K_c (or K_p) from a knowledge of the equilibrium concentrations (or pressures) of reactants or products or from the initial concentrations and the equilibrium concentration of at least one substance.

3. Interconvert K_c and K_p.

4. Calculate the reaction quotient, Q, and by comparison with the value of K_c or K_p, determine whether a reaction is at equilibrium. If it is not at equilibrium, you should be able to predict in which direction it will shift to reach equilibrium.

5. Use the equilibrium constant to calculate equilibrium concentrations. (You will also need to know either the equilibrium concentrations of all but one substance or the initial concentrations together with the equilibrium concentration of one substance.)

6. Explain how the relative equilibrium quantities of reactants and products are shifted by changes in temperature, pressure, or the concentrations of substances involved in the equilibrium reaction.

7. Explain how the change in equilibrium constant with change in temperature is related to the heat change in the reaction.

8. Describe the effect of a catalyst on a system as it approaches equilibrium.

Key terms

Among the more important terms and expressions used for the first time in this chapter are the following:

Chemical equilibrium (Section 14.2) is a state of a chemical system in which the rate of formation of products equals the rate of formation of reactants from products.

The Haber process (Section 14.1) refers to the catalyst system and conditions of temperature and pressure developed by Fritz Haber and co-workers for the formation of NH_3 from H_2 and N_2.

Heterogeneous equilibrium (Section 14.3) refers to the equilibrium state established between substances in two or more different phases, for example, between a gas and solid or between a solid and liquid.

Homogeneous equilibrium (Section 14.3) is the state of equilibrium established between reactant and product substances all in the same phase, for example, all gases, or all dissolved in solution.

The law of mass action (Section 14.2) provides the rules according to which the equilibrium constant is expressed in terms of the concentrations of reactants and products, in accordance with the balanced chemical equation for the reaction.

LeChatelier's principle (Section 14.5) tells us that when we bring to bear some disturbing influence on a system at chemical equilibrium, the relative concentrations of reactants and products will shift so as to undo partially the effects of the disturbance.

The reaction quotient, Q, (Section 14.4) is the value that is obtained when concentrations of reactants and products are inserted into the equilibrium-constant expression. If the concentrations are equilibrium concentrations, $Q = K$; otherwise, $Q \neq K$.

EXERCISES

Equilibrium-constant expressions

14.1 Write the equilibrium-constant expression for each of the following reactions:

(a) $2SO_2(g) + O_2(g) \rightleftharpoons 2SO_3(g)$
(b) $ZnO(s) + H_2(g) \rightleftharpoons Zn(s) + H_2O(g)$
(c) $H_2(g) + CO_2(g) \rightleftharpoons CO(g) + H_2O(g)$
(d) $4NH_3(g) + 3O_2(g) \rightleftharpoons 2N_2(g) + 6H_2O(g)$
(e) $NH_4NO_2(s) \rightleftharpoons N_2(g) + 2H_2O(g)$

14.2 Write the equilibrium-constant expression for each of the following reactions. In each case indicate whether the reaction is homogeneous or heterogeneous.

(a) $2H_2S(g) + 3O_2(g) \rightleftharpoons 2H_2O(g) + 2SO_2(g)$
(b) $2NO_2(g) \rightleftharpoons 2NO(g) + O_2(g)$
(c) $PbCO_3(s) \rightleftharpoons PbO(s) + CO_2(g)$
(d) $S(s) + O_2(g) \rightleftharpoons SO_2(g)$
(e) $H_2(g) + Cl_2(g) \rightleftharpoons 2HCl(g)$

Calculation of K_c and K_p

14.3 Gaseous hydrogen iodide is introduced into a container at 425°C where it partly decomposes to hydrogen and iodine:

$$2HI(g) \rightleftharpoons H_2(g) + I_2(g)$$

At equilibrium the resultant mixture is analyzed, and it is found that $[H_2] = 4.79 \times 10^{-4}\,M$, $[I_2] = 4.79 \times 10^{-4}\,M$, and $[HI] = 3.53 \times 10^{-3}\,M$. What is the value of K_c at this temperature?

14.4 A sample of chlorine gas is placed in a vessel and heated to 1400 K. The chlorine dissociates:

$$Cl_2(g) \rightleftharpoons 2Cl(g)$$

When equilibrium is reached at this temperature it is found that $P_{Cl_2} = 1.00$ atm and $P_{Cl} = 2.97 \times 10^{-2}$ atm. What is the value of K_p at 1400 K?

14.5 The equilibrium constant for the reaction

$$SO_2(g) + \tfrac{1}{2}O_2(g) \rightleftharpoons SO_3(g)$$

is $K_c = 20.4$ at 700°C. (a) What is the value of K_c for

$$SO_3(g) \rightleftharpoons SO_2(g) + \tfrac{1}{2}O_2(g)$$

(b) What is the value of K_c for

$$2SO_2(g) + O_2(g) \rightleftharpoons 2SO_3(g)$$

(c) What is the value of K_p for

$$2SO_2(g) + O_2(g) \rightleftharpoons 2SO_3(g)$$

14.6 At 700 K, carbon dioxide and hydrogen react to form carbon monoxide and water:

$$CO_2(g) + H_2(g) \rightleftharpoons CO(g) + H_2O(g)$$

At this temperature $K_c = 0.11$. What is the value of K_p?

14.7 At 1000 K, $K_c = 278$ for the reaction

$$2SO_2(g) + O_2(g) \rightleftharpoons 2SO_3(g)$$

What is the value of K_p at this temperature?

14.8 For the equilibrium

$$C(s) + CO_2(g) \rightleftharpoons 2CO(g)$$

$K_p = 167.5$ at 1000°C. What is the value for K_c at this temperature?

14.9 At temperatures near 800°C, steam passed over hot coke (a form of carbon obtained from coal) reacts to form CO and H_2:

$$C(s) + H_2O(g) \rightleftharpoons CO(g) + H_2(g)$$

The mixture of gases that results is an important industrial fuel called water gas. When equilibrium is achieved at 800°C, $[H_2] = 4.0 \times 10^{-2}\,M$, $[CO] = 4.0 \times 10^{-2}\,M$, and $[H_2O] = 1.0 \times 10^{-2}\,M$. Calculate K_c and K_p at this temperature.

14.10 A mixture of 0.100 mol of NO, 0.050 mol of H_2, and 0.100 mol of H_2O are placed in a 1.00-L vessel. The following equilibrium is established:

$$2NO(g) + 2H_2(g) \rightleftharpoons N_2(g) + 2H_2O(g)$$

At equilibrium, $[NO] = 0.070\,M$. (a) Calculate the equilibrium concentrations of H_2, N_2, and H_2O. (b) Calculate K_c.

14.11 A mixture of 1.000 mol of SO_2 and 1.000 mol of O_2 are placed in a 1.000-L vessel and kept at 1000 K until equilibrium is reached. At equilibrium the vessel is found

to contain 0.919 mol of SO_3. Calculate K_c for the reaction

$$2SO_2(g) + O_2(g) \rightleftharpoons 2SO_3(g)$$

14.12 A sample of nitrosyl bromide, NOBr, decomposes according to the following equation:

$$2NOBr(g) \rightleftharpoons 2NO(g) + Br_2(g)$$

An equilibrium mixture in a 5.00-L vessel at 100°C contains 3.22 g of NOBr, 3.08 g of NO, and 4.19 g of Br_2. (a) Calculate K_c; (b) calculate K_p; (c) what is the total pressure exerted by the mixture of gases?

14.13 A mixture of 0.34 mol of H_2 and 0.22 mol of Br_2 are heated together in a 1.00-L vessel at 700 K. These substances react as follows:

$$H_2(g) + Br_2(g) \rightleftharpoons 2HBr(g)$$

At equilibrium, the vessel is found to contain 0.14 mol of H_2. What is the numerical value for K_c at 700 K?

Reaction quotient

14.14 As shown in Table 14.4, K_p for the equilibrium

$$N_2(g) + 3H_2(g) \rightleftharpoons 2NH_3(g)$$

is 4.51×10^{-5} at 450°C. Each of the mixtures listed below may or may not be at equilibrium at 450°C. Indicate in each case whether the mixture is at equilibrium; if it is not at equilibrium, indicate the direction (toward product or toward reactants) in which the mixture must shift to achieve equilibrium: (a) 100 atm NH_3, 30 atm N_2, 500 atm H_2; (b) 30 atm NH_3, 600 atm H_2, no N_2; (c) 26 atm NH_3, 42 atm H_2, 202 atm N_2; (d) 100 atm NH_3, 60 atm H_2, 5 atm N_2.

14.15 At 100°C the equilibrium

$$COCl_2(g) \rightleftharpoons CO(g) + Cl_2(g)$$

has an equilibrium constant, K_c, equal to 2.19×10^{-10}. Are the following mixtures of CO, Cl_2, and $COCl_2$ at equilibrium? If they are not at equilibrium, indicate the direction the reaction must proceed to reach equilibrium. (a) $[CO] = 1.0 \times 10^{-3}\,M$, $[Cl_2] = 1.0 \times 10^{-3}\,M$, $[COCl_2] = 2.19 \times 10^{-1}\,M$; (b) $[CO] = 3.31 \times 10^{-6}\,M$, $[Cl_2] = 3.31 \times 10^{-6}\,M$, $[COCl_2] = 5.00 \times 10^{-2}\,M$; (c) $[CO] = 4.50 \times 10^{-7}\,M$, $[Cl_2] = 5.73 \times 10^{-6}\,M$, $[COCl_2] = 8.57 \times 10^{-2}\,M$.

Equilibrium concentrations

14.16 At 2000°C, K_c for the equilibrium

$$N_2(g) + O_2(g) \rightleftharpoons 2NO(g)$$

is 4.1×10^{-4}. If the equilibrium concentrations of N_2 and NO are $0.10\,M$ and $0.050\,M$, respectively, what is the equilibrium concentration of O_2?

14.17 The equilibrium constant, K_p, is 3.5×10^4 at 1495 K for the reaction

$$H_2(g) + Br_2(g) \rightleftharpoons 2HBr(g)$$

If an equilibrium mixture at this temperature contains H_2 at a partial pressure of 0.10 atm and Br_2 at 0.40 atm, what is the partial pressure of HBr?

14.18 At 1558 K the equilibrium constant, K_c, for the reaction

$$Br_2(g) \rightleftharpoons 2Br(g)$$

is 1.04×10^{-3}. If a 0.200-L vessel contains 4.53×10^{-2} mol of Br_2 at equilibrium, how many moles of Br are present?

14.19 For the equilibrium

$$C(s) + CO_2(g) \rightleftharpoons 2CO(g)$$

$K_p = 167.5$ at $1000\,°C$. What is the partial pressure of CO_2 that is in equilibrium with CO whose partial pressure is 0.500 atm?

14.20 The equilibrium constant for the reaction

$$I_2(g) + Br_2(g) \rightleftharpoons 2IBr(g)$$

has a value of 280 at $150\,°C$. Suppose that a quantity of IBr is placed in a closed reaction vessel and the system allowed to come to equilibrium. When equilibrium is attained, the pressure of IBr is 0.20 atm. What are the pressures of $I_2(g)$ and $Br_2(g)$ at this point?

14.21 The following reaction can occur when N_2 and O_2 come in contact at high temperatures:

$$N_2(g) + O_2(g) \rightleftharpoons 2NO(g)$$

At 2400 K, the equilibrium constant, K_c, for this reaction is 2.5×10^{-3}. What are the equilibrium concentrations of N_2, O_2, and NO (in moles per liter) if equilibrium is obtained at 2400 K starting with a reaction mixture of 0.20 mol of N_2 and 0.20 mol of O_2 in a 5.00-L vessel?

[**14.22**] At 1500 K, the equilibrium constant for the reaction

$$N_2(g) + O_2(g) \rightleftharpoons 2NO(g)$$

is $K_c = 1.0 \times 10^{-5}$. Calculate the equilibrium concentrations of N_2, O_2, and NO if, before any reaction, 0.500 mol of NO is placed in a 2.00-L container.

14.23 At $21.8\,°C$, the equilibrium constant, K_c, is 1.2×10^{-4} for the following reaction:

$$NH_4HS(s) \rightleftharpoons NH_3(g) + H_2S(g)$$

Calculate the equilibrium concentrations of NH_3 and H_2S if a sample of solid NH_4HS is placed in a closed vessel and allowed to decompose until equilibrium is reached at $21.8\,°C$.

14.24 The equilibrium constant for the reaction

$$H_2(g) + I_2(g) \rightleftharpoons 2HI(g)$$

has a value of 54.6 at 699 K. What are the equilibrium concentrations of H_2, I_2, and HI obtained at 699 K beginning with a reaction mixture of $0.0700\,M$ H_2 and $0.0500\,M$ I_2?

LeChatelier's principle

14.25 Consider the following equilibrium system:

$$C(s) + CO_2(g) \rightleftharpoons 2CO(g) \qquad \Delta H° = 119.8\,kJ$$

If the reaction is at equilibrium, what would be the effect of (a) adding $CO_2(g)$; (b) adding $C(s)$; (c) adding heat;

(d) increasing the pressure on the system by decreasing the volume; (e) adding a catalyst; (f) removing $CO(g)$?

14.26 In the reaction

$$6CO_2(g) + 6H_2O(l) \rightleftharpoons C_6H_{12}O_6(s) + 6O_2(g)$$
$$\Delta H° = 2816\,kJ$$

how is the equilibrium yield of $C_6H_{12}O_6$ affected by (a) increasing P_{CO_2}; (b) increasing temperature; (c) removing CO_2; (d) increasing the total pressure; (e) removing part of the $C_6H_{12}O_6$?

14.27 How do each of the following changes affect the numerical value of the equilibrium constant for an exothermic reaction: (a) remove a reactant or product; (b) increase the total pressure; (c) decrease the temperature.

14.28 The solubility of $CuSO_4$ in water, in grams $CuSO_4$ per 100 g of H_2O, varies with temperature as follows:

Temperature (°C)	Solubility
15	19.3
25	22.3
30	25.5
50	33.6
60	39.0
80	53.5

Is the dissolving of $CuSO_4$ an exothermic or endothermic process? Explain.

14.29 As we all know, ice is less dense at $0\,°C$ than liquid water. Apply LeChatelier's principle to determine how the melting temperature of ice will vary with an increase in pressure.

Chemical equilibrium and kinetics

14.30 Consider the following reaction which occurs in a single step and is reversible:

$$CO(g) + Cl_2(g) \rightleftharpoons COCl(g) + Cl(g)$$

Kinetic studies indicate that the rate constant for the forward reaction, k_f, is $1.38 \times 10^{-28}/M$-sec; for the reverse reaction, $k_r = 9.3 \times 10^{10}/M$-sec (both rate constants at $25\,°C$). What is the value for the equilibrium constant for this reaction at $25\,°C$?

14.31 For the reaction

$$2SO_2(g) + O_2(g) \longrightarrow 2SO_3(g)$$

the standard enthalpy change is

$$\Delta H° = -196.6\,kJ$$

The activation energy for the uncatalyzed reaction is about $160\,kJ/mol$. Sketch the energy profile for this reaction, as in Figure 14.5. Sulfur dioxide is produced in the cylinder of an auto engine by oxidation of the small quantity of sulfur present in gas. The catalytic mufflers installed in cars since 1975 result in conversion of a large portion of this SO_2 to SO_3 (Section 13.6). Using a dotted line, sketch on your figure an energy profile which might apply for SO_2 oxidation in a catalytic muffler.

14.32 In the reaction

$$NO(g) + O_3(g) \rightleftharpoons NO_2(g) + O_2(g)$$

The rate law for the forward reaction is

$$r_f = k_f[NO][O_3]$$

and for the reverse reaction

$$r_r = k_r[NO_2][O_2]$$

Using these expressions, write the equilibrium condition in terms of opposing rates, and show how K_c relates to k_f and k_r.

14.33 Are any of the following statements false? For those that are, discuss the sense in which they are incorrect. (a) If a catalyst increases the rate of a forward reaction by a factor of 1000 over the uncatalyzed rate, it increases the rate of the reverse reaction by a factor of 1000 also. (b) A catalyst can promote the formation of product in some reactions by inhibiting the reverse reaction in an equilibrium. (c) Although heterogeneous catalysts must affect the rates of both forward and reverse reactions to an equal extent, homogeneous catalysts can be made to affect the rate of just the forward or just the reverse step. (d) All fast reactions have large equilibrium constants.

Additional exercises

14.34 Write the equilibrium-constant expression for each of the following reactions:

(a) $PCl_5(g) \rightleftharpoons PCl_3(g) + Cl_2(g)$
(b) $ZnSO_3(s) \rightleftharpoons ZnO(s) + SO_2(g)$
(c) $CO(g) + 2H_2(g) \rightleftharpoons CH_3OH(l)$
(d) $2H_2(g) + O_2(g) \rightleftharpoons 2H_2O(g)$
(e) $3O_2(g) \rightleftharpoons 2O_3(g)$

14.35 Indicate which reactions in question 14.34 are homogeneous and which are heterogeneous.

14.36 Throughout this chapter we have used gas-phase reactions to illustrate the concept of equilibrium. However, solution reactions could also be used. Write the equilibrium-constant expression for each of the following solution reactions:

(a) $PbCl_2(s) \rightleftharpoons Pb^{2+}(aq) + 2Cl^-(aq)$
(b) $HCN(aq) \rightleftharpoons H^+(aq) + CN^-(aq)$
(c) $NH_3(aq) + H_2O(l) \rightleftharpoons NH_4^+(aq) + OH^-(aq)$
(d) $Ag^+(aq) + 2NH_3(aq) \rightleftharpoons Ag(NH_3)_2^+(aq)$
(e) $Zn(s) + Cu^{2+}(aq) \rightleftharpoons Zn^{2+}(aq) + Cu(s)$
(f) $BaSO_4(s) \rightleftharpoons Ba^{2+}(aq) + SO_4^{2-}(aq)$

14.37 Assuming that we choose to associate units with equilibrium constants, what units would we associate with K_c and with K_p for each of the following reactions?

(a) $N_2O_4(g) \rightleftharpoons 2NO_2(g)$
(b) $N_2(g) + 3H_2(g) \rightleftharpoons 2NH_3(g)$
(c) $CO_2(g) + H_2(g) \rightleftharpoons CO(g) + H_2O(g)$
(d) $NH_4Cl(s) \rightleftharpoons NH_3(g) + HCl(g)$

14.38 A mixture of CH_4 and H_2O is passed over a nickel catalyst at 1000 K. The emerging gas has the composition $[CO] = 0.0616\,M$, $[H_2] = 0.260\,M$, and $[CH_4] = [H_2O] = 0.538\,M$. Assuming that equilibrium has been reached, calculate K_c for the reaction

$$CH_4(g) + H_2O(g) \rightleftharpoons CO(g) + 3H_2(g)$$

14.39 A mixture of 3.0 mol of SO_2, 4.0 mol of NO_2, 1.0 mol of SO_3, and 4.0 mol of NO is placed in a 2.0-L vessel. The following reaction takes place:

$$SO_2(g) + NO_2(g) \rightleftharpoons SO_3(g) + NO(g)$$

When equilibrium is reached at 700°C, the vessel is found to contain 1.0 mol of SO_2. (a) Calculate the equilibrium concentrations of SO_2, NO_2, SO_3, and NO. (b) Calculate the value of K_c for this reaction at 700°C.

[14.40] A 0.831-g sample of SO_3 is placed in a 1.00-L container and heated to 1100 K. The SO_3 undergoes decomposition to SO_2 and O_2:

$$2SO_3(g) \rightleftharpoons 2SO_2(g) + O_2(g)$$

At equilibrium, the total pressure in the container is 1.300 atm. Find the values of K_p and K_c for this reaction at 1100 K.

[14.41] PCl_5 is placed in a 2.00-L flask at 250°C. The following reaction occurs:

$$PCl_5(g) \rightleftharpoons PCl_3(g) + Cl_2(g)$$

At equilibrium, the total pressure of the mixture is 2.00 atm. The partial pressure of PCl_5 at equilibrium is 0.37 atm. Calculate K_p at 250°C.

14.42 A mixture of 1.000 mol of N_2 and 3.000 mol of H_2 are placed in a 1.00-L vessel at 600°C. At equilibrium it is found that the mixture contains 0.371 mol of NH_3. Calculate K_c at 600°C for the reaction

$$N_2(g) + 3H_2(g) \rightleftharpoons 2NH_3(g)$$

14.43 Nitric oxide, NO, rapidly oxidizes to nitrogen dioxide, NO_2, even at room temperature. At 1000 K, 0.0400 mol of NO and 0.0600 mol of O_2 are placed in a 2.00-L vessel. At equilibrium the concentration of NO_2 was $2.2 \times 10^{-3}\,M$. (a) Calculate the equilibrium concentrations of NO and O_2; (b) Calculate the equilibrium constant, K_c, for the reaction

$$2NO(g) + O_2(g) \rightleftharpoons 2NO_2(g)$$

14.44 Calculate K_p for the reaction

$$2SO_2(g) + O_2(g) \rightleftharpoons 2SO_3(g)$$

if at a particular temperature and a total pressure of 88.0 atm the equilibrium mixture consists of 56.6 mole percent SO_2, 10.6 mole percent O_2, and 32.8 mole percent SO_3.

[14.45] Write the equilibrium-constant expression for the equilibrium

$$C(s) + CO_2(g) \rightleftharpoons 2CO(g)$$

The table below shows the relative mole percentages of $CO_2(g)$ and $CO(g)$ at a total pressure of 1 atm for several temperatures. Calculate the value of K_c at each temperature. Is the reaction exothermic or endothermic? Explain?

Temperature (°C)	CO$_2$ (%)	CO (%)
850	6.23	93.77
950	1.32	98.68
1050	0.37	99.63
1200	0.06	99.94

14.46 Nitric oxide, NO, reacts readily with chlorine gas as follows:

$$2NO(g) + Cl_2(g) \rightleftharpoons 2NOCl(g)$$

At 700 K, the equilibrium constant, K_p, for this reaction is 0.26. Predict the behavior of each of the following mixtures at this temperature: (a) $P_{NO} = 0.15$ atm, $P_{Cl_2} = 0.31$ atm, and $P_{NOCl} = 0.11$ atm; (b) $P_{NO} = 0.12$ atm, $P_{Cl_2} = 0.10$ atm, and $P_{NOCl} = 0.050$ atm; (c) $P_{NO} = 0.15$ atm, $P_{Cl_2} = 0.20$ atm, and $P_{NOCl} = 5.10 \times 10^{-3}$ atm.

14.47 Consider the reaction

$$2CO(g) + O_2(g) \rightleftharpoons 2CO_2(g) \qquad \Delta H° = -514.2 \text{ kJ}$$

In which direction will the reaction move if (a) CO$_2$ is added; (b) CO$_2$ is removed; (c) the volume is increased; (d) the pressure is increased; (e) the temperature is increased?

14.48 For the reaction shown in question 14.47, what effect does increasing temperature have on the magnitude of the equilibrium constant?

14.49 The evaporation of any liquid requires energy added as heat. (a) How is the liquid-vapor equilibrium—for instance, $H_2O(l) \rightleftharpoons H_2O(g)$—affected by increasing temperature? (b) How does the equilibrium constant vary as the quantity of the liquid is increased?

14.50 NiO is to be reduced to nickel metal in an industrial process by use of the reaction

$$NiO(s) + CO(g) \rightleftharpoons Ni(s) + CO_2(g)$$

At 1600 K the equilibrium constant for the reaction is 600. If a CO pressure of 150 mm Hg is to be employed in the furnace, and total pressure never exceeds 760 mm Hg, will reduction occur?

14.51 Consider the reaction

$$CO(g) + 2H_2(g) \rightleftharpoons CH_3OH(l)$$

Using the thermochemical data in Appendix D, determine whether the equilibrium constant for this reaction increases or decreases with increasing temperature. Assuming equal pressures of CO and H$_2$, how would the extent of conversion of the gas mixture to methanol (CH$_3$OH) vary with total pressure?

[14.52] Suppose there is a region in outer space where initially the hydrogen molecule concentration is 10^2 molecules/cm^3, the N$_2$ concentration is 1 molecule/cm^3, and the temperature is 100 K. At this temperature, K_p for the reaction

$$N_2(g) + 3H_2(g) \rightleftharpoons 2NH_3(g)$$

is approximately 6×10^{37}. Assuming that equilibrium is attained, is a significant fraction of the N$_2$ converted to NH$_3$?

14.53 A solution that contains $1.0 \times 10^{-3} M$ Ag$^+$ is added to an equal volume of a solution containing $2.00 \times 10^{-4} M$ Cl$^-$. (a) Calculate the reaction quotient for the reaction

$$Ag^+(aq) + Cl^-(aq) \rightleftharpoons AgCl(s)$$

(b) If $K = 1.4 \times 10^{10}$ for this reaction, will solid AgCl form in the solution?

[14.54] At 1558 K the equilibrium constant, K_c, for the reaction

$$Br_2(g) \rightleftharpoons 2Br(g)$$

is 1.04×10^{-3}. (a) Calculate the equilibrium concentration of Br atoms if the initial concentration of Br$_2$ is $1.00 M$; (b) calculate the fraction of the initial concentration of Br$_2$ that is dissociated into atoms.

14.55 At 400 K, the equilibrium constant, K_p, for the following reaction is 6.0×10^{-9}:

$$NH_4Cl(s) \rightleftharpoons NH_3(g) + HCl(g)$$

What are the equilibrium vapor pressures of NH$_3$ and HCl that are produced by the decomposition of solid NH$_4$Cl at 400 K?

[14.56] Consider the equilibrium described in Sample Exercise 14.9 ([HI] = 0.39 M, [I$_2$] = [H$_2$] = 0.055 M, K_c = 50.5). If 6.0×10^{-2} mol of HI is added to this mixture, what will the new equilibrium concentrations be?

14.57 Suppose you worked at the U.S. Patent Office and a patent application came across your desk in which it was claimed that a newly developed catalyst was much superior to the Haber catalyst for ammonia synthesis, because the catalyst led to much greater equilibrium conversion of N$_2$ and H$_2$ into NH$_3$ than the Haber catalyst under the same conditions. What would be your response?

14.58 At 1200 K, the approximate temperature of automobile exhaust gases (Figure 14.6), the equilibrium constant for the reaction

$$2CO_2(g) \rightleftharpoons 2CO(g) + O_2(g)$$

is about 1×10^{-13} atm. Assuming that the exhaust gas (total pressure 1 atm) contains 0.2 percent CO by volume, 12 percent CO$_2$, and 3 percent O$_2$, is the system at equilibrium with respect to the above reaction? Based on your conclusion, would the CO concentration in the exhaust be lowered or increased by a catalyst that speeded up the above reaction?

14.59 For the single-step reaction

$$NO(g) + O_3(g) \underset{k_r}{\overset{k_f}{\rightleftharpoons}} NO_2(g) + O_2(g)$$

$K_p = 1.32 \times 10^{10}$ at 1000 K. If $k_f = 6.26 \times 10^8/M$-sec at this temperature, calculate k_r.

15

Acids and bases

One of the most important classifications of substances is in terms of acid and base properties. From the earliest days of experimental chemistry it was recognized that certain substances, called acids, possess a sour taste and are capable of dissolving active metals such as zinc. Acids also cause vegetable dyes to turn a characteristic color; for example, litmus, which is obtained from certain lichens (a composite plant made up of an alga and a fungus), turns red on contact with acids. Like acids, bases possess a set of characteristic properties that can be used to identify them. Whereas acids have a sour taste (the sour taste of lemons is due to the presence of citric acid in lemon juice), bases have a characteristic bitter taste. Bases also feel slippery to the touch. Like acids, bases also cause litmus to change its color, but, whereas acids cause litmus to turn red, bases cause it to turn blue. Bases also react with many dissolved metal salts to form precipitates.

The fact that all acids and all bases show certain characteristic chemical properties suggests that there must be an essential feature common to all the members of each class. Lavoisier proposed that acids were oxygen-containing substances. In fact, Lavoisier derived the name *oxygen* from the Greek word for "acid former." However, careful studies by a number of other scientists showed that hydrochloric acid contains no oxygen. By 1830 it became evident that hydrogen was the one element present in all acids. It was subsequently shown that aqueous solutions of both acids and bases conduct electrical currents. In the 1880s, the Swedish chemist Svante Arrhenius (1859–1927) suggested the existence of ions to explain this electrical conduction. Shortly after, he proposed that acids are substances that form H^+ ions in water solutions and that bases produce OH^- ions. These definitions of acids and bases were the ones that were presented in Section 3.3 and that we have used in subsequent discussions.

In this chapter we shall take a closer look at acids and bases. We shall see how the properties of these substances can be understood in terms of their structures and bonding. The concept of equilibrium introduced in the last chapter will figure prominently in these discussions. We shall see also that the properties of acids and bases, as we generally encounter them, are very much dependent on the fact that water is the solvent in which those properties are observed. To appreciate how really remarkable an aqueous (water) acid solution is, let's begin by considering a very common chemical reagent, hydrochloric acid.

15.1 Water and acidic solutions

Figure 15.1 is a reproduction of the label from a bottle of reagent grade* concentrated hydrochloric acid. Although this label might at first seem to be rather dull subject matter, it contains fascinating and even amazing data. Consider first some properties of HCl. It is a gaseous substance, formed commercially by the controlled reaction of hydrogen and chlorine:

$$H_2(g) + Cl_2(g) \longrightarrow 2HCl(g)$$

Hydrogen chloride liquefies under 1 atm pressure at $-84\,°C$, and freezes at $-112°$. Liquid HCl is a difficult and most unpleasant substance to work with, but some enterprising and persistent persons have learned that the liquid is a very poor conductor of electricity. This suggests that there are very few ions present in the liquid. Dry HCl gas is soluble to only a limited extent in a dry organic solvent such as benzene. The solutions do not conduct electric current. Furthermore, these solutions lack other properties that we associate with hydrochloric acid. For example, addition of an active metal such as zinc to them produces no evident chemical reaction. In contrast, zinc reacts readily with aqueous solutions of HCl to produce hydrogen gas:

$$Zn(s) + 2HCl(aq) \longrightarrow H_2(g) + ZnCl_2(aq) \qquad [15.1]$$

*The term "reagent grade" denotes that a substance is of high purity.

KEEP FROM CHILDREN

DANGER!
CAUSES BURNS.

BEFORE USING, READ MANUFACTURING CHEMISTS' ASSOCIATION, INC., SAFETY DATA SHEET SD-39

Do not get in eyes, on skin, on clothing. Avoid breathing vapor. Keep container closed. Use with adequate ventilation. Wash thoroughly after handling.

FIRST AID: In case of contact, immediately flush eyes or skin with plenty of water for at least 15 minutes while removing contaminated clothing and shoes. Call a physician. Wash clothing before re-use.

SPILL OR LEAK: Flush away by flooding with water applied quickly to entire spill. Neutralize washings with lime or soda ash.

POISON
ANTIDOTE

If taken internally: Call Physician Immediately. Give at once large draughts of water containing Milk of Magnesia, or milk, soap and water, magnesium oxide, white of eggs beaten up with water, or olive oil. Avoid Carbonates. Give No Emetics.

DU PONT
REG. U.S. PAT OFF

6 POUNDS
(2.721 kg)

HYDROCHLORIC ACID
REAGENT

KEEP THIS BOTTLE IN A COOL PLACE AND REMOVE CAP CAREFULLY TO AVOID SPURTING

E. I. DU PONT DE NEMOURS & COMPANY (INC.)
INDUSTRIAL CHEMICALS DEPARTMENT
Wilmington, Del., U. S. A.

MEETS A. C. S. SPECIFICATIONS

HCl . . . . {Min. 37.0% / Max. 38.0%
Sp. Gr. 60°/60°F, Min. . 1.19
Maximum of Impurities
Free Cl 0.00005%
Sulfites (SO$_3$) . . 0.00008%
Sulfates (SO$_4$) . 0.00008%
Hy. Met. (as Pb) 0.00002%
Residue after
    Ignition 0.0004%
Fe 0.00001%
As 0.0000005%
NH$_4$ 0.0003%
Br 0.005%
Cu 0.00005%
Ni 0.00005%
Color, APHA . 10
Extractable Organic
    Substances Passes A.C.S.
    Test (Approx. 0.0005%)

IC-20902 1-72   PRINTED IN U.S.A.

FIGURE 15.1 Reproduction of a label from a bottle of reagent grade, concentrated hydrochloric acid. (*Courtesy of E. I. du Pont de Nemours & Co.*)

With these facts in mind let us now examine the label shown in Figure 15.1. We note that concentrated aqueous hydrochloric acid consists of 37 percent by weight of hydrogen chloride; HCl is obviously very soluble in water. At 15°C, 1 L of water dissolves up to 450 L of dry HCl gas at 1 atm pressure! This high solubility is nicely shown by the hydrogen chloride fountain, a popular lecture demonstration, illustrated in Figure 15.2. The molarity of concentrated hydrochloric acid solution is about 12 M, as shown in Sample Exercise 15.1. To have some appreciation for just how remarkable this solubility is, you should know that argon, which has essentially the same molecular weight as HCl, is soluble in water at 15°C and 1 atm pressure to the extent of only 0.002 M.

SAMPLE EXERCISE 15.1

From the data shown in Figure 15.1, calculate the molarity of a concentrated hydrochloric acid solution.

Solution: We see that the specific gravity, or density, of the solution is 1.19 g/cm³. Using this and the stated weight percentage, we have

$$\left(\frac{1.19 \text{ g soln}}{1 \text{ cm}^3}\right)\left(\frac{1000 \text{ cm}^3}{1 \text{ L}}\right)\left(\frac{0.37 \text{ g HCl}}{1 \text{ g soln}}\right)$$
$$\times \left(\frac{1 \text{ mol HCl}}{36.5 \text{ g HCl}}\right) = \frac{12 \text{ mol HCl}}{1 \text{ L soln}}$$
$$= 12 \ M$$

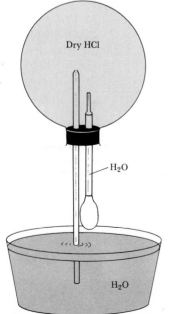

FIGURE 15.2 The hydrogen chloride fountain. The large flask is filled with dry hydrogen chloride gas at 1 atm pressure. When a small amount of water is introduced into the flask by squeezing the medicine dropper, the HCl dissolves in it very rapidly. This causes the pressure in the flask to decrease, and water is forced up the glass tube. The spray of water continues until the flask is almost completely filled.

Dry HCl

H_2O

H_2O

In contrast with the solutions of HCl in dry benzene, aqueous solutions of HCl, that is, hydrochloric acid solutions, strongly conduct an electrical current. This behavior indicates that hydrochloric acid contains ions. As we have noted, this observation led Arrhenius to define an acid as any substance capable of producing an excess of H^+ ions in water. Arrhenius's definition of an acid is quite reasonable, but it doesn't tell us much about what actually happens when an acid dissolves in water. To appreciate the nature of acidic solutions, and thus understand more clearly why acids behave as they do, we must look more closely into the question of how and why a substance such as HCl reacts with water to form ions.

NATURE OF THE HYDRATED PROTON

What is it about water that causes a molecule such as HCl to come apart, or dissociate, into H^+ and Cl^- ions? The first and most obvious point to make is that water is a polar liquid (Section 11.5). Consequently, the oxygen atom, which bears a partial negative charge, is attracted to the partially positive hydrogen end of the polar HCl molecule. Secondly, the water molecule has unshared electron pairs on the oxygen atom; these are capable of forming a covalent bond to the hydrogen ion in the following manner:

$$\text{Cl}-\text{H}---\overset{..}{\text{O}}-\text{H} \longrightarrow \text{Cl}^- + \left[\text{H}-\overset{..}{\underset{\text{H}}{\text{O}}}-\text{H}\right]^+ \qquad [15.2]$$

The reaction between HCl and H_2O shown in Equation [15.2] involves the transfer of a proton (an H^+ ion) from HCl to the water mole-

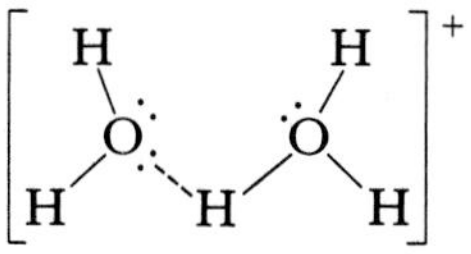

$H_5O_2^+$

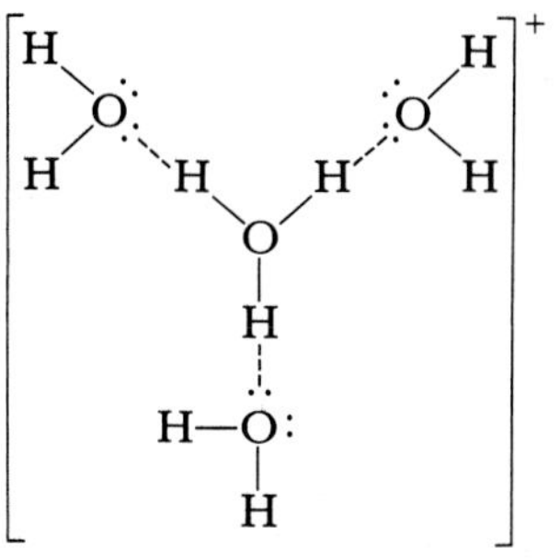

$H_9O_4^+$

FIGURE 15.3 Two possible forms for the proton in water, in addition to H_3O^+. Experimental evidence indicates the existence of both these species.

cule. This reaction produces the H_3O^+ ion, called the *hydronium ion*.

The situation in water is actually much more complex than Equation [15.2] suggests. We've learned (Section 11.3) that hydrogen bonds exist throughout liquid water. The existence of this hydrogen-bond network is responsible for many of the special properties of water, for example, its high polarity and high melting and boiling points. Much research has been devoted to learning how H^+ ions fit into the complex structure of liquid water. Experimental studies show that, in part, the H^+ ions must exist as hydronium ions. In fact, it is possible to isolate salts of the form $H_3O^+Cl^-$, $H_3O^+ClO_4^-$, and others, in which there is clearly an H_3O^+ ion in the solid lattice. But just as water molecules are strongly hydrogen bonded to one another, the H_3O^+ ion in solution is hydrogen bonded to other water molecules. Thus, ions such as the two shown in Figure 15.3 are possible and have been shown to form.

We must conclude from these observations that no single species can adequately represent the proton in solution. Both $H^+(aq)$ and $H_3O^+(aq)$ are used to represent the hydrated, or aquated, hydrogen ion, that is, the hydrogen ion surrounded by solvent water. Thus we may write the reaction of HCl with water to form hydrochloric acid in either of the following ways:

$$HCl(aq) + H_2O(l) \longrightarrow H_3O^+(aq) + Cl^-(aq) \qquad [15.3]$$

$$HCl(aq) \longrightarrow H^+(aq) + Cl^-(aq) \qquad [15.4]$$

The hydrated proton will be represented through the remainder of this chapter as $H^+(aq)$. However, regardless of whether $H^+(aq)$ or $H_3O^+(aq)$ is used to represent the hydrated proton, you should keep in mind that *acidic solutions are formed by a chemical reaction in which an acid transfers a proton to water.*

15.2 Brønsted-Lowry theory of acids and bases

The nature of the reaction between an acid and water as we have just described it was first appreciated by the Danish chemist Johannes Brønsted (1879–1947) and the English chemist Thomas M. Lowry (1874–1936). Brønsted and Lowry recognized that acid-base behavior could be described in terms of the ability of substances to transfer protons. In 1923 Brønsted and Lowry independently proposed that *acids be defined as substances that are capable of donating a proton, and bases as substances capable of accepting a proton.* In these terms, when HCl dissolves in water, it acts as an acid in donating a proton to the solvent. The solvent, H_2O, then acts as a base in accepting the proton (see Equation [15.2]).

In earlier discussions, we have applied the term base to substances that produce an excess of OH^- ions in aqueous solution. Notice, however, that the OH^- ion is an acceptor of protons; it reacts readily with the hydrated proton to form water:

$$H^+(aq) + OH^-(aq) \rightleftharpoons H_2O(l) \qquad [15.5]$$

Likewise, we have seen (Section 3.3) that aqueous solutions of ammonia are basic because NH_3 reacts with H_2O to form NH_4^+ and OH^-:

$$H_2O(l) + NH_3(aq) \rightleftharpoons NH_4^+(aq) + OH^-(aq) \qquad [15.6]$$

In this reaction the H_2O gives a proton to NH_3; H_2O is the acid, NH_3 is the base.

The reactions that we have cited above as examples of proton-transfer reactions are reversible. For example, when NH_4^+ (as from NH_4Cl) and OH^- (as from NaOH) are mixed they form H_2O and NH_3:

$$NH_4^+(aq) + OH^-(aq) \rightleftharpoons NH_3(aq) + H_2O(l) \qquad [15.7]$$

The net ionic equation is just the reverse of the reaction between NH_3 and H_2O, Equation [15.6]. In this reverse reaction, Equation [15.7], NH_4^+ acts as the proton donor and OH^- as the proton acceptor. Thus we see that as the reaction proceeds in one direction, H_2O is the acid, and NH_3 is the base. In the other direction, NH_4^+ is the acid, and OH^- is the base. This example illustrates that every acid has associated with it a conjugate base* formed from the acid by the loss of a proton. For example, the conjugate base of NH_4^+ is NH_3; the conjugate base of H_2O is OH^-. Similarly, every base has a conjugate acid formed from the base by addition of a proton. H_2O is the conjugate acid of OH^-. An acid and a base, such as H_2O and OH^-, that differ only by the presence or absence of a proton are called a conjugate acid-base pair.

*The word "conjugate" means joined together as a pair, or coupled.

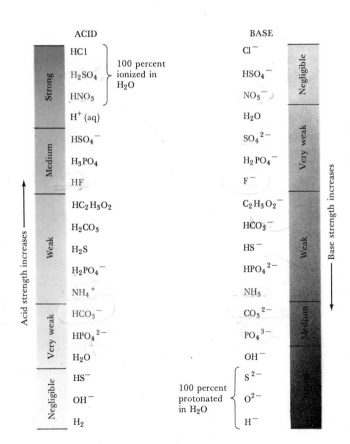

FIGURE 15.4 Relative strengths of some common conjugate acid-base pairs, which are listed opposite one another in the two columns. The conjugate bases of strong acids are weak bases; the conjugate bases of weak acids are strong bases. The strongest acid that can exist in water is the aquated proton, H+(aq). Any substance that gives up a proton more readily than the solvent simply loses that proton to the solvent, forming H+ (aq). Such strong acids, for example, HNO_3, are essentially completely ionized in water. Similarly, any base that is a stronger proton acceptor than OH− removes protons from the water to become essentially completely protonated.

The more readily a substance gives up a proton, the less readily its conjugate base will accept a proton. In other words, *the stronger an acid, the weaker its conjugate base; the weaker an acid, the stronger its conjugate base.* For example, HCl is a good proton donor because its conjugate base, Cl⁻, has less attraction for protons than water does. The proton is therefore transferred to H_2O to form $H^+(aq)$.

Figure 15.4 displays some common acids and their conjugate bases. Notice that strong acids have weak conjugate bases and weak acids have strong conjugate bases. $H^+(aq)$ is the strongest proton donor that can exist at equilibrium in aqueous solution. Thus acids listed above $H^+(aq)$ in Figure 15.4 completely transfer protons to water to form $H^+(aq)$. Likewise, $OH^-(aq)$ is the strongest base that can exist at equilibrium in aqueous solution. Any stronger proton acceptor will completely react with water, removing a proton to form OH^- ions.

SAMPLE EXERCISE 15.2

Hydrocyanic acid, HCN, is ionized in water by the reaction

$$HCN(aq) \rightleftharpoons H^+(aq) + CN^-(aq)$$

to a lesser extent than is an HF solution of the same concentration. What is the conjugate base of HCN? Is it a stronger or weaker base than F^-?

Solution: The conjugate base of HCN is CN^-, the ion that remains after a proton has been lost to the solvent. Because HCN dissociates to a lesser extent than does HF, this means that the tendency of the reverse reaction to occur is greater. In the reverse reaction the conjugate base, CN^-, accepts a proton from the solvent. In other words, CN^- is a stronger conjugate base than is F^-.

15.3 The dissociation of water and the pH scale

In the presence of an acid such as HCl, water acts as a proton acceptor; in the presence of a base such as NH_3 it acts as a proton donor. An important feature of water is that it is also capable of acting as a proton donor and proton acceptor toward itself. The process by which this occurs is called autoionization:

$$H-\overset{..}{O}: + H-\overset{..}{O}: \rightleftharpoons \left[H-\overset{..}{O}-H \right]^+ + :\overset{..}{O}-H^- \qquad [15.8]$$
$$\quad | \qquad\quad | \qquad\qquad\quad | $$
$$\quad H \qquad\quad H \qquad\qquad\quad H$$

This reaction amounts to a spontaneous ionization of the solvent. It occurs only to a very small extent. At room temperature, only about one out of every 10^8 molecules is in the ionic form at any one instant. We know that water is a strongly hydrogen-bonded liquid. Thus, the hydrogen of one molecule may be attracted to an unshared pair of electrons on the oxygen of an adjacent molecule. Occasionally the hydrogen will transfer to the other molecule. Perhaps simultaneously there will be a transfer of H^+ from some other molecule to the oxygen that is losing a hydrogen ion. As a result of these ready transfers of H^+ from one molecule to another, there is, on the average, a certain very small fraction of molecules in the ionized form. No one molecule remains in that condition for long; the equilibria are extremely rapid. It has been found that,

on the average, a proton transfers from one molecule to another in water
at the rate of about 1000 times per second.

By expressing the hydrated proton as $H^+(aq)$ rather than $H_3O^+(aq)$,
we can rewrite Equation [15.8] as

$$H_2O(l) \rightleftharpoons H^+(aq) + OH^-(aq) \qquad [15.9]$$

The equilibrium expression for this autoionization reaction can be
written as

$$K = \frac{[H^+][OH^-]}{[H_2O]}$$

The concentration of water in aqueous solutions is typically very large,
about 55 M, and remains essentially constant for dilute solutions. It is
therefore customary to exclude the concentration of water from equilibrium-constant expressions for aqueous solutions, just as we exclude the
concentrations of pure solids and liquids from the equilibrium-constant
expressions for heterogeneous reactions (Section 14.3). Thus we can write
the equilibrium-constant expression for the autoionization of water as

$$K[H_2O] = K_w = [H^+][OH^-]$$

The product of two constants, K and $[H_2O]$, defines a new constant, K_w.
This important equilibrium constant is called the **ion-product constant**
for water. K_w has the value of 1.0×10^{-14} at 25°C. This is an important
equilibrium constant. You should memorize this expression:

$$K_w = [H^+][OH^-] = 1.0 \times 10^{-14} \qquad [15.10]$$

Equation [15.10] is valid for aqueous solutions as well as for pure
water. A solution for which $[H^+] = [OH^-]$ is said to be *neutral*. In most
solutions H^+ and OH^- concentrations are not equal. As the concentration of one of these ions increases, the concentration of the other must
decrease so that the ion-product equals 1.0×10^{-14}. In acidic solutions,
$[H^+]$ exceeds $[OH^-]$; in basic solutions, the reverse is true, $[OH^-]$
exceeds $[H^+]$.

SAMPLE EXERCISE 15.3

Calculate the values of $[H^+]$ and $[OH^-]$ in a neutral
solution at 25°C.

Solution: By definition, in a neutral solution, $[H^+]$
equals $[OH^-]$. Let us call the concentration of each
of these species in neutral solution x. Using Equation
[15.10], we have

$$[H^+][OH^-] = (x)(x) = 1.0 \times 10^{-14}$$
$$x^2 = 1.0 \times 10^{-14}$$
$$x = 1.0 \times 10^{-7} = [H^+] = [OH^-]$$

In an acid solution, $[H^+]$ is greater than $1 \times 10^{-7} M$;
in a basic solution it is less than $1 \times 10^{-7} M$.

SAMPLE EXERCISE 15.4

Calculate the concentration of $H^+(aq)$ in (a) a solution in which $[OH^-]$ is $0.010\,M$; (b) a solution in which $[OH^-]$ is $2.0 \times 10^{-9}\,M$.

Solution:

(a) Using Equation [15.10] we have

$$[H^+][OH^-] = 1.0 \times 10^{-14}$$

$$[H^+] = \frac{1.0 \times 10^{-14}}{[OH^-]} = \frac{1.0 \times 10^{-14}}{0.010}$$

$$= 1.0 \times 10^{-12}\,M$$

This solution is basic because $[H^+] < [OH^-]$.

(b) In this instance

$$[H^+] = \frac{1.0 \times 10^{-14}}{[OH^-]} = \frac{1.0 \times 10^{-14}}{2.0 \times 10^{-9}}$$

$$= 5.0 \times 10^{-6}\,M$$

This solution is acidic because $[H^+] > [OH^-]$.

pH

In almost every area of pure and applied chemistry the acid-base properties of water are of importance. As examples, the fate of pollutant chemicals in a water body, the rapidity with which a metal object immersed in water corrodes, and the suitability of an aquatic environment for support of fish and plant life are all critically dependent on the acidity or basicity of the water. The concentration of $H^+(aq)$ in such solutions is often expressed in terms of pH. The pH is defined as the negative log in base 10, of the hydrogen-ion concentration.*

$$pH = -\log [H^+] = \log \left(\frac{1}{[H^+]} \right) \qquad [15.11]$$

Notice that a change in $[H^+]$ by a factor of 10 results in a unit change in pH. (If you need a review of exponential notation and of the use of logs, see Appendix A.) As an example of the use of Equation [15.11], let us calculate the pH of a neutral solution, that is, one in which $[H^+] = [OH^-] = 1 \times 10^{-7}$ (Sample Exercise 15.5). The pH is given by

$$pH = -\log [H^+] = -\log (1 \times 10^{-7}) = -0 - (-7) = 7$$

Thus, the pH of a neutral solution is 7.

Because pH is simply another means of expressing $[H^+]$, acidic and basic solutions can be distinguished on the basis of their pH values:

pH < 7 in acidic solutions

pH > 7 in basic solutions

pH = 7 in neutral solutions

You should keep in mind that the pH is a measure only of the equilibrium concentration of dissociated hydrogen ion present as $H^+(aq)$. The pH values characteristic of several familiar solutions are shown in Figure 15.5.

*Usually you will see pH defined as $-\log [H^+]$, occasionally as $-\log [H_3O^+]$. As discussed in Section 15.1, the same species is involved in all cases.

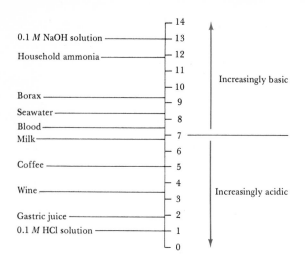

0.1 M NaOH solution — 13
Household ammonia — 12

Increasingly basic

Borax — 9
Seawater — 8
Blood — 7
Milk — 7

Coffee — 5

Wine — 3

Increasingly acidic

Gastric juice — 2
0.1 M HCl solution — 1

FIGURE 15.5 Values of pH for some more commonly encountered solutions. The pH scale in this figure is shown to extend from 0 to 14, because nearly all solutions commonly encountered have pH values in that range. In principle, however, the pH values for strongly acidic solutions can be less than 0, and for strongly basic solutions can be greater than 14.

With a log table (Appendix B) or log function on a calculator, it is a simple matter to convert from concentration of $H^+(aq)$ to pH, and vice versa, as outlined in Sample Exercises 15.5 and 15.6.

SAMPLE EXERCISE 15.5

Calculate the pH values for the two solutions described in Sample Exercise 15.4.

Solution: In the first instance we found $[H^+]$ to be 1.0×10^{-12}. The pH of this solution is given by:

$$pH = -\log (1.0 \times 10^{-12}) = -(-12.00) = 12.00$$

The pH of the second solution is given by

$$pH = -\log (5.0 \times 10^{-6})$$
$$= -(\log 5 + \log 10^{-6})$$
$$= -(0.699 - 6.00) = 5.30$$

SAMPLE EXERCISE 15.6

A sample of freshly pressed apple juice has a pH of 3.76. Calculate $[H^+]$.

Solution: From the equation defining pH, Equation [15.11], we have $-\log [H^+] = 3.76$. Thus, $\log [H^+] = -3.76$. To find $[H^+]$ we need to find the antilog of -3.76. Some calculators have an antilog or $\log^{-1}$ function which makes the calculation quite simple. Other calculators rely on 10^x or y^x functions to find antilogs: antilog $(-3.76) = 10^{-3.76}$. If you are relying on a log table, the simplest way to take the antilog is to write the log as a sum of an integer and a positive decimal fraction: $-3.76 = -4.00 + 0.24$. The antilog of -4 is 1×10^{-4}. From a log table we find that the antilog of 0.24 is approximately 1.7. Thus $[H^+] = 1.7 \times 10^{-4} M$.

INDICATORS

Various means are available for quantitatively estimating pH. The simplest is the use of an indicator. An indicator is a colored substance, usually derived from plant material, that can exist in either an acid or base form. The two forms are differently colored. By adding a small amount of an indicator to a solution and noting its color, it is possible to

TABLE 15.1 Some of the more common acid-base indicators

Name	pH interval for color change	Acid color	Base color
Methyl violet	0–2	Yellow	Violet
Methyl yellow	1.2–2.3	Red	Yellow
Methyl orange	2.9–4.0	Red	Yellow
Methyl red	4.2–6.3	Red	Yellow
Bromthymol blue	6.0–7.6	Yellow	Blue
Thymol blue	8.0–9.6	Yellow	Blue
Phenolphthalein	8.3–10	Colorless	Pink
Alizarin yellow G	10.1–12.0	Yellow	Red

determine whether it is in the acid or base form. If one knows the pH at which the indicator turns from one form to the other, one can then determine from the observed color whether the solution has a higher or lower pH than this value. For example, litmus, one of the most common indicators, changes color in the vicinity of pH 7. However, the color change is not very sharp. Red litmus indicates a pH of about 5 or lower, and blue litmus, a pH of about 8.2 or higher. Many other indicators change color at various pH values between 1 and 14. Some of the more commonly used are listed in Table 15.1. We see from this table that methyl orange, for example, changes color over the pH interval from 2.9 to 4.0. Below pH 2.9 it is in the acid form, which is red. In the interval from pH 2.9 to 4.0 it is gradually converted to its basic form, which has a yellow color. By pH 4.0 the conversion is complete, and the solution is yellow. Paper tape that is impregnated with various indicators and that comes complete with a comparator color scale is widely used for approximate determinations of pH.

The pH meter is a widely used, simple instrument for rapid and accurate determination of pH. It is so common that if you go on to further study in chemistry or in an applied science you are almost certain to encounter one. A complete understanding of how a pH meter works requires a knowledge of electrochemistry, a subject we take up in Chapter 19. However, we can say at this point that a pH meter consists of a pair of electrodes that are placed in the solution to be measured and a sensitive meter for measuring small voltages, on the order of millivolts. A typical pH meter with a pair of electrodes is shown in Figure 15.6. When the electrodes are placed in the solution, they form an electrochemical cell (something like a battery) that has a voltage. The voltage of the cell is dependent on [H$^+$]; thus, by measuring the voltage we obtain a measure of [H$^+$]. Electrodes that can be used with pH meters come in all shapes and sizes, depending on their intended use, but fundamentally

they are nearly all the same. One of the electrodes is a reference electrode. The one that is actually sensitive to H$^+$(aq) is almost always a so-called glass electrode. The wire in the inner compartment of the electrode is in contact with a solution of known and fixed H$^+$(aq) concentration. The wall of the compartment is formed of a special thin glass that is permeable to H$^+$(aq). As a result, the voltage that this electrode, together with a reference electrode, generates when placed in a solution depends on [H$^+$] in the solution.

To extend the range of possible pH measurements, much research has gone into the development of electrodes that can be used with very small quantities of solution. It is now possible to insert electrodes into single living cells in order to monitor pH of the cell medium. The pH meter is also widely used outside the laboratory. Pocket-sized models are available for use in environmental studies, monitoring of industrial effluents, and in agricultural work.

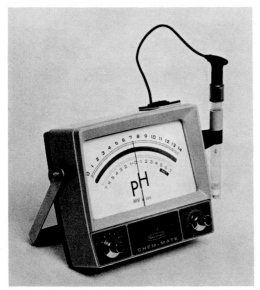

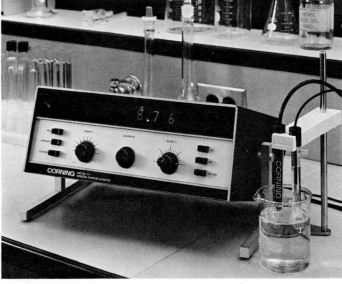

FIGURE 15.6 (a) A pH meter of the type normally used for student work. (b) A research-type pH meter with digital display. (a, courtesy of Beckman Instruments, Inc.; b, courtesy of Corning Glass Works)

OTHER pX SCALES

The negative log is a convenient way of expressing the magnitudes of numbers that are generally very small. We use the convention that the negative log of a quantity is labeled p(quantity). For example, one can express the concentration of OH⁻ as pOH:

$$pOH = -\log[OH^-]$$

By taking the log of both sides of Equation [15.10] and multiplying through by -1 we can obtain

$$pH + pOH = -\log K_w = 14 \qquad [15.12]$$

This expression is often convenient to use. We will see later (Section 15.7) that the pX notation is also useful in dealing with equilibrium constants.

15.4 Strong acids and strong bases

The pH of an aqueous solution depends on the ability of the solute to give protons to, or remove them from, water. In this section and the following ones we will examine the extent to which these proton-transfer reactions occur.

In terms of the Brønsted-Lowry concept, a strong aqueous acid is any substance that reacts completely with water to form $H^+(aq)$; a weak acid is a substance that only partly reacts in this fashion. The number of strong acids is not very large; the six most important ones are listed in Table 15.2. We recommend that you commit them to memory if you have not already done so.

TABLE 15.2 Common strong acids and bases

Acids	Bases
HCl (hydrochloric acid)	Hydroxides and oxides of 1A metals
HBr (hydrobromic acid)	Hydroxides and oxides of 2A metals (except Be)
HI (hydroiodic acid)	
HNO_3 (nitric acid)	
$HClO_4$ (perchloric acid)	
H_2SO_4 (sulfuric acid)	

We can consider aqueous solutions of all strong acids to consist entirely of ions, with no significant concentration of neutral solute molecules remaining; such acids are said to be completely ionized or dissociated. For example, a $0.10\,M$ aqueous solution of HNO_3, nitric acid, contains $[H^+] = 0.10\,M$, and $[NO_3^-] = 0.10\,M$; the concentration of HNO_3 is virtually zero.

Strong bases are strong proton acceptors. The most common strong bases are NaOH and KOH, which are ionic compounds containing OH^- ions in the solid lattice. They are strong electrolytes, dissolving in water as would any other ionic substance (Section 12.5). For example, a $0.10\,M$ aqueous solution of NaOH contains $0.10\,M\ Na^+(aq)$ and $0.10\,M$ $OH^-(aq)$ with no undissociated NaOH:

$$NaOH(s) \xrightarrow{\text{H}_2\text{O}} Na^+(aq) + OH^-(aq) \qquad [15.13]$$

All of the hydroxides of the alkali metals (family 1A) are strong electrolytes, but the compounds of Li, Rb, and Cs are too expensive to be encountered commonly in the laboratory. The hydroxides of all of the alkaline earths (family 2A) except Be are also strong electrolytes. However, they have limited solubilities and are consequently used only when high solubility is not critical. $Mg(OH)_2$ has an especially low solubility (9×10^{-3} g/L of water at $25\,^\circ C$). The least expensive and most common of the alkaline earth hydroxides is $Ca(OH)_2$, whose solubility at $25\,^\circ C$ is 0.97 g/L.

Basic solutions are also created when substances react with water to form $OH^-(aq)$. The most common of these is the oxide ion. Ionic metal oxides, especially Na_2O and CaO, are often used in industry when a strong base is needed. Each mole of O^{2-} reacts with water to form 2 mol of OH^-, leaving virtually no O^{2-} remaining in the solution:

$$O^{2-}(aq) + H_2O(l) \longrightarrow 2OH^-(aq) \qquad [15.14]$$

SAMPLE EXERCISE 15.7

What is the pH of a solution of (a) $0.010\,M$ HCl; (b) $0.010\,M\ Ca(OH)_2$?

Solution: (a) HCl is a strong acid. Consequently, $[H^+] = 0.010\,M$ and $pH = -\log(0.010) = 2.00$. (b) $Ca(OH)_2$ is a strong base; each $Ca(OH)_2$ forms $2OH^-$ ions. Consequently, $[OH^-] = 0.020\,M$ and

$$[H^+] = \frac{1.00 \times 10^{-14}}{[OH^-]} = \frac{1.00 \times 10^{-14}}{0.020}$$

$$= 5.0 \times 10^{-13}$$

$$pH = -\log(5.0 \times 10^{-13}) = 12.30$$

15.5 Weak acids

Most substances that are acidic in water are actually weak acids. The extent to which an acid ionizes in an aqueous medium can be expressed by the equilibrium constant for the ionization reaction. In general we can represent any acid by the symbol HX (or XH), where X⁻ is the formula for the conjugate base that remains when the proton ionizes. The ionization equilibrium is then given by Equation [15.15]:

$$HX(aq) \rightleftharpoons H^+(aq) + X^-(aq) \tag{15.15}$$

The corresponding equilibrium-constant expression is

$$K_a = \frac{[H^+][X^-]}{[HX]} \tag{15.16}$$

The equilibrium constant is often given the symbol K_a and is called the acid-dissociation constant.

Table 15.3 shows the names, structures, and values of K_a for several weak acids. A more complete table is given in Appendix E. Note that many weak acids are compounds composed largely of carbon and hydrogen. Generally speaking, hydrogen atoms bound to carbon are not ionized in an aqueous medium. The ionizable hydrogens are in most instances bound to oxygen. The smaller the value for K_a, the weaker the acid. For example, phenol is the weakest acid listed in Table 15.3.

From the value for K_a it is possible to calculate the concentration of $H^+(aq)$ in a solution of a weak acid. For example, consider acetic acid, $HC_2H_3O_2$, the substance that gives the characteristic odor and acidic properties to vinegar. Let us calculate the concentration of $H^+(aq)$ in a $0.10\,M$ solution of acetic acid.

TABLE 15.3 Some weak acids in water at 25°C[a]

Acid	Formula	Lewis structure	Conjugate base	K_a
Hydrofluoric	HF	H—F	F⁻	6.8×10^{-4}
Hydrocyanic	HCN	H—C≡N	C≡N⁻	4.9×10^{-10}
Acetic	$HC_2H_3O_2$		$C_2H_3O_2^-$	1.8×10^{-5}
Benzoic	$HC_7H_5O_5$		$C_7H_5O_2^-$	6.5×10^{-5}
Nitrous	HNO_2	H—O—N=O	NO_2^-	4.5×10^{-4}
Phenol	HOC_6H_5		$C_6H_5O^-$	1.3×10^{-10}
Ascorbic (vitamin C)	$HC_6H_7O_6$		$C_6H_7O_6^-$	8.0×10^{-5}

[a] In cases where hydrogen is present in more than one chemical environment in the molecule, the one that ionizes is shown in color.

The first step in solving any equilibrium problem is to write the equation for the equilibrium reaction. The ionization equilibrium for acetic acid can be written as

$$HC_2H_3O_2(aq) \rightleftharpoons H^+(aq) + C_2H_3O_2^-(aq) \qquad [15.17]$$

Note from the Lewis structure for acetic acid shown in Table 15.3 that the hydrogen that ionizes is attached to the oxygen atom. We write this hydrogen separate from the others in the formula to emphasize that this one hydrogen is readily ionized.

The second step is to write the equilibrium-constant expression and the value for the equilibrium constant, if that is known. From Table 15.3 we have $K_a = 1.8 \times 10^{-5}$. Thus, we can write the following:

$$K_a = \frac{[H^+][C_2H_3O_2^-]}{[HC_2H_3O_2]} = 1.8 \times 10^{-5} \qquad [15.18]$$

As the third step, we need to express the concentrations that make up the equilibrium-constant expression. This can be done with a little accounting, which can be done below the expression for the equilibrium:

$$HC_2H_3O_2(aq) \rightleftharpoons H^+(aq) + C_2H_3O_2^-(aq)$$

Initial:	$0.10\ M$	$0\ M$	$0\ M$
Equilibrium:	$(0.10 - x)\ M$	$x\ M$	$x\ M$

Because we seek to find a value for $[H^+]$, let us call this quantity x. The concentration of acetic acid before any of it dissociates is $0.10\ M$. The equation for the equilibrium tells us that for each molecule of $HC_2H_3O_2$ that dissociates, one $H^+(aq)$ and one $C_2H_3O_2^-(aq)$ are formed. Thus, if x moles per liter of $H^+(aq)$ are formed at equilibrium, x moles per liter of $C_2H_3O_2^-(aq)$ must also have formed, and x moles per liter of $HC_2H_3O_2$ must have been dissociated. This gives rise to the equilibrium concentrations shown above.

As the fourth step of the problem, we need to substitute the equilibrium concentrations into the equilibrium-constant expression. The substitution gives the following equation:

$$K_a = \frac{[H^+][C_2H_3O_2^-]}{[HC_2H_3O_2]} = \frac{(x)(x)}{(0.10 - x)} = 1.8 \times 10^{-5} \qquad [15.19]$$

Because this equation has only one unknown it can be solved using algebra. However, the solution is a little tedious, because it requires use of the quadratic formula (Appendix A.3). By taking account of what is actually occurring in the solution, we can make things simpler for ourselves. Because the value of K_a is small, we might guess that x will be quite small. (In other words, perhaps only a small fraction of the $HC_2H_3O_2$ is actually ionized.) Indeed, if we solve the problem using the quadratic formula we find that $x = 1.3 \times 10^{-3}\ M$. Now you know that if a small number is subtracted from a much larger one, the result is approximately equal to the larger number. In our example we have

$$(0.10 - x) = (0.10 - 0.0013) \simeq 0.10$$

We can therefore make the approximation of ignoring x relative to 0.10 in the denominator of Equation [16.8]. This leads us to the following simplified expression:

$$K_a = \frac{(x)(x)}{(0.10)} = 1.8 \times 10^{-5}$$

Solving for x we have

$$x^2 = (0.10)(1.8 \times 10^{-5}) = 1.8 \times 10^{-6}$$
$$x = \sqrt{1.8 \times 10^{-6}} = 1.3 \times 10^{-3}\,M = [H^+]$$

From the value calculated for x we see that our simplifying approximation is quite reasonable. This type of approximation can be used whenever conditions in solution are such that only a small fraction of acid ionizes. As a general rule, if the quantity x, which is subtracted from the initial concentration of the acid, is more than about 5 percent of the initial value, it is best to use the quadratic formula. In cases of doubt, assume that the approximation is valid and solve for x in the simplified equation. Compare this approximate value of x with the initial concentration of acid. If it is more than about 5 percent as large, the problem should be reworked using the quadratic formula. For example, if the initial concentration of acid were 0.050 M, and x turned out in a given case to be 0.0016 M, then

$$\left(\frac{0.0016\,M}{0.050\,M}\right)(100) = 3.2\%$$

SAMPLE EXERCISE 15.8

Calculate the pH of a 0.20 M solution of HCN (refer to Table 15.3 or Appendix E for value of K_a).

Solution: Proceeding as in the example worked out above, we write:

$$HCN(aq) \rightleftharpoons H^+(aq) + CN^-(aq)$$

$$K_a = \frac{[H^+][CN^-]}{[HCN]} = 4.9 \times 10^{-10}$$

Let $x = [H^+]$ at equilibrium. Then we have the following concentrations:

	HCN(aq) $\rightleftharpoons$	H$^+$(aq) +	CN$^-$(aq)
Initial:	0.20 M	0 M	0 M
Equilibrium:	(0.20 − x) M	x M	x M

Substituting into the equilibrium constant expression:

$$K_a = \frac{(x)(x)}{(0.20 - x)} = 4.9 \times 10^{-10}$$

We next make the simplifying approximation that x, the amount of acid that dissociates, is small in comparison with the initial concentration of acid; that is,

$$(0.20 - x) \simeq 0.20$$

Thus

$$\frac{x^2}{0.20} = 4.9 \times 10^{-10}$$

Solving for x we have

$$x^2 = (0.20)(4.9 \times 10^{-10})$$
$$= 0.98 \times 10^{-10}$$
$$x = \sqrt{0.98 \times 10^{-10}}$$
$$= 0.99 \times 10^{-5} = 9.9 \times 10^{-6} = [H^+]$$
$$pH = -\log [H^+] = -\log (9.9 \times 10^{-6})$$
$$= 5.00$$

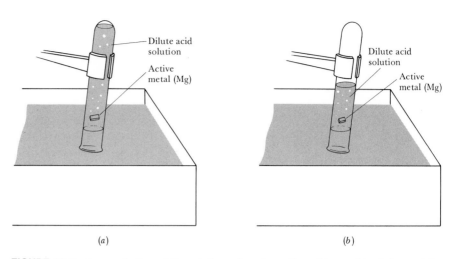

(a)   (b)

FIGURE 15.7 Demonstration of the relative rates of reaction of two acid solutions of the same concentration with an active metal. The solution in (a) is that of a weak acid, in (b) that of a strong acid. Reaction produces $H_2(g)$, which collects in the tube. From the relative amounts of gas collected in the two tubes after a period of time, it is evident that reaction is faster in (b). This indicates that even though the concentrations of acid are the same in the two tubes, the concentration of H^+ (aq) is much greater in (b).

The result obtained in Sample Exercise 15.8 is typical of the behavior of weak acids; the concentration of $H^+(aq)$ is only a small fraction of the concentration of the acid in solution. Thus, those properties of the acid solution that relate directly to the concentration of $H^+(aq)$, such as electrical conductivity or rate of reaction with an active metal, are much less in evidence for a solution of a weak acid than for a solution of a strong acid. Figure 15.7 illustrates an experiment often carried out in the chemistry laboratory to demonstrate the difference in concentration of $H^+(aq)$ in weak and strong acid solutions of the same concentration. The rate of reaction with the active metal is much faster for the solution of a strong acid. Reactions in which the rate depends on $H^+(aq)$ are common.

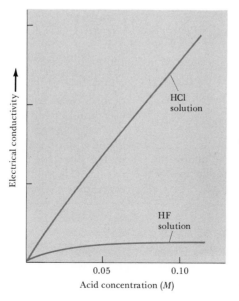

FIGURE 15.8 Electrical conductivity versus concentration for solutions of HCl, a strong acid, and HF, weak acid. The conductivity of the HCl solution is not completely linear with concentration because of attractive forces between the ions at higher concentrations (Section 12.7). The conductivity for the HF solution is quite nonlinear with concentration and much lower than for HCl because only a fraction of the HF molecules ionize. It is nonlinear with concentration because the fraction of molecules ionizing changes with concentration.

455

Figure 15.8 illustrates an experiment in which the electrical conductivity of an HCl solution is compared with the conductivity of an HF solution. The conductivity of the solution of the strong acid increases approximately in proportion to the concentration. This is what one would expect; because all the acid molecules ionize, the concentration of ions in solution is directly proportional to the concentration of acid. The conductivity of the solution of the weak acid is very much less than that for a strong acid and does not vary linearly with the acid concentration. The nonlinearity of the graph arises from the fact that the percentage of acid ionized varies with the acid concentration. This is illustrated in Sample Exercise 15.9.

SAMPLE EXERCISE 15.9

Calculate the percentage of HF molecules ionized in a 0.10 M HF solution; in a 0.010 M HF solution.

Solution: The equilibrium reaction and equilibrium concentrations can be written as follows:

$$HF(aq) \rightleftharpoons H^+(aq) + F^-(aq)$$

Initial: 0.10 M 0 M 0 M

Equilibrium: (0.10 − x) M x M x M

The equilibrium-constant expression is as follows:

$$K_a = \frac{[H^+][F^-]}{[HF]} = \frac{(x)(x)}{(0.10 - x)} = 6.8 \times 10^{-4}$$

We might be tempted to try solving this equation using the same approximation used in earlier examples, that is, by neglecting the concentration of acid that ionizes in comparison with the initial concentration (by neglecting x in comparison with 0.10). However K_a is large enough in this case to make that a poor approximation. We must therefore rearrange our equation, and write it in standard quadratic form:

$$x^2 = (0.10 - x)(6.8 \times 10^{-4})$$
$$= 6.8 \times 10^{-5} - (6.8 \times 10^{-4})x$$
$$x^2 + (6.8 \times 10^{-4})x - 6.8 \times 10^{-5} = 0$$

Solving this equation using the quadratic formula gives

$$x = [H^+] = [F^-] = 7.9 \times 10^{-3} M$$

(You should solve this problem using the quadratic formula in order to satisfy yourself about the answer and to be sure that you know how to do a problem of this sort. You might also see what answer you would get by making the simplifying approximation of neglecting x with respect to 0.10.)

From our result we can calculate the percent of molecules ionized:

$$Percent\ ionized = \left(\frac{concentration\ ionized}{original\ concentration}\right)(100)$$

$$= \left(\frac{7.9 \times 10^{-3} M}{0.10 M}\right)(100) = 7.9\%$$

Proceeding similarly for the 0.010 M solution we have

$$\frac{x^2}{(0.010 - x)} = 6.8 \times 10^{-4}$$

Solving the resultant quadratic expression we obtain

$$x = [H^+] = [F^-] = 2.3 \times 10^{-3} M$$

The percentage of molecules ionized is

$$\left(\frac{0.0023}{0.010}\right)(100) = 23\%$$

Notice that in diluting the solution by a factor of 10, the percentage of molecules ionized increases by a factor of 3. We could have arrived at this conclusion qualitatively by applying LeChatelier's principle (Section 14.5) to the equilibrium. There are more "particles" or reaction components on the right side of the equation. Dilution causes the reaction to shift in the direction of the larger number of particles, because this counters the effect of the decreasing concentration of particles.

POLYPROTIC ACIDS

Many substances are capable of furnishing more than one proton to water. Substances of this type are called **polyprotic acids**. As an example, sulfurous acid, H_2SO_3, may react with water in two sucessive steps:

TABLE 15.4 Acid-dissociation constants of some common polyprotic acids at 25°C

Name	Formula	K_{a1}	K_{a2}	K_{a3}
Carbonic	H_2CO_3	4.3×10^{-7}	5.6×10^{-11}	
Oxalic	$H_2C_2O_4$	5.9×10^{-2}	6.4×10^{-5}	
Phosphoric	H_3PO_4	7.5×10^{-3}	6.2×10^{-8}	4.2×10^{-13}
Sulfurous	H_2SO_3	1.7×10^{-2}	6.4×10^{-8}	

$$H_2SO_3(aq) \rightleftharpoons H^+(aq) + HSO_3^-(aq) \qquad K_a = 1.7 \times 10^{-2} \qquad [15.20]$$

$$HSO_3^-(aq) \rightleftharpoons H^+(aq) + SO_3^{2-}(aq) \qquad K_a = 6.4 \times 10^{-8} \qquad [15.21]$$

The values for K_a in each case show that the reactions are incomplete. Notice that loss of the second proton occurs much less readily than the first, as shown by the smaller value for K_a in the second reaction. This trend is intuitively reasonable; on the basis of electrostatic attractions we would expect the positively charged proton to be lost more readily from the neutral H_2SO_3 molecule than from the negatively charged HSO_3^- ion.

The successive acid-dissociation constants of polyprotic acids are sometimes labeled K_{a1}, K_{a2}, and so forth. This notation is often simplified to K_1, K_2, and so forth. For example, the equilibrium constant for the loss of a proton from HSO_3^-, Equation [15.21], can be labeled as K_{a2} or K_2, because this proton is the second one removed from the neutral acid, H_2SO_3. The acid-dissociation constants for a few common polyprotic acids are given in Table 15.4; a more complete list is found in Appendix E. Notice that the K_a values for successive losses of protons from these acids differ by at least a factor of 10^3.

Because K_{a1} is so much larger than subsequent dissociation constants for these polyprotic acids, almost all the $H^+(aq)$ in the solution comes from the first ionization reaction. As long as successive K_a values differ by a factor of 10^3 or more, it is possible to obtain a satisfactory estimate of the pH of polyprotic acid solutions by considering only K_{a1}.

SAMPLE EXERCISE 15.10

The solubility of CO_2 in pure water at 25°C and 1 atm pressure is 0.0037 M. The common practice is to assume that all of the dissolved CO_2 is in the form of H_2CO_3, which is produced by reaction between the CO_2 and H_2O:

$$CO_2(aq) + H_2O(l) \rightleftharpoons H_2CO_3(aq)$$

What is the pH of a 0.0037 M solution of H_2CO_3?

Solution: H_2CO_3 is a polyprotic acid; the two acid dissociation constants, K_{a1} and K_{a2} (Table 15.4), differ by more than a factor of 10^3. Consequently, the pH can be determined by considering only K_{a1}, thereby treating the acid as if it were a monoprotic acid. Proceeding as in Sample Exercises 15.8 and 15.9, we can write the equilibrium reaction and equilibrium concentrations as follows:

$$H_2CO_3(aq) \rightleftharpoons H^+(aq) + HCO_3^-(aq)$$

Initial:	0.0037 M	0 M	0 M
Equilibrium:	$(0.0037 - x)\,M$	$x\,M$	$x\,M$

The equilibrium-constant expression is as follows:

$$K_{a1} = \frac{[H^+][HCO_3^-]}{[H_2CO_3]} = \frac{(x)(x)}{(0.0037 - x)} = 4.3 \times 10^{-7}$$

Because K_{a1} is small, we make the simplifying approximation that x is small so that $0.0037 - x \simeq 0.0037$. Thus

457

$$\frac{(x)(x)}{0.0037} = 4.3 \times 10^{-7}$$

Solving for x we have

$$x^2 = (0.0037)(4.3 \times 10^{-7}) = 1.6 \times 10^{-9}$$
$$x = \sqrt{1.6 \times 10^{-9}} = 4.0 \times 10^{-5}\,M = [H^+]$$
$$= [HCO_3^-]$$

The small value of x indicates that our simplifying assumption was justified. The pH is therefore

$$pH = -\log[H^+] = -\log(4.0 \times 10^{-5}) = 4.40$$

If we had been asked to solve for $[CO_3^{2-}]$, we would need to use K_{a2}. Let's illustrate that calculation. Using the values of $[HCO_3^-]$ and $[H^+]$ calculated above, and setting $[CO_3^{2-}] = y$, we have the following initial and equilibrium concentration values:

	$HCO_3^-(aq)$	$\rightleftharpoons$	$H^+(aq)$	$+$	$CO_3^{2-}(aq)$
Initial:	$4.0 \times 10^{-5}\,M$		$4.0 \times 10^{-5}\,M$		$0\,M$
Equilibrium:	$(4.0 \times 10^{-5} - y)M$		$(4.0 \times 10^{-5} + y)M$		$y\,M$

Assuming that y is small compared to 4.0×10^{-5}, we have:

$$K_{a2} = \frac{[H^+][CO_3^{2-}]}{[HCO_3^-]} = \frac{(4.0 \times 10^{-5})(y)}{(4.0 \times 10^{-5})}$$
$$= 5.6 \times 10^{-11}$$
$$y = 5.6 \times 10^{-11}\,M = [CO_3^{2-}]$$

The value calculated for y is indeed very small in comparison with 4.0×10^{-5}, showing that our as-sumption was justified. It also shows that the ioniza-tion of the HCO_3^- is negligible in comparison with that of H_2CO_3 as far as production of H^+ is con-cerned. However, it is the *only* source of CO_3^{2-}, which has a very low concentration in the solution.

Our calculations thus tell us that in a solution of carbon dioxide in water most of the CO_2 is in the form of CO_2 or H_2CO_3, a small fraction ionizes to form H^+ and HCO_3^-, and an even smaller fraction ionizes to give CO_3^{2-}.

TABLE 15.5 Some weak bases in water at 25°C

Base	Formula	Lewis structure	Conjugate acid	K_b
Ammonia	NH_3	H—N—H, H	NH_4^+	1.8×10^{-5}
Pyridine	C_5H_5N	N:	$C_5H_5NH^+$	1.7×10^{-9}
Hydroxylamine	H_2NOH	H—N—OH, H	H_3NOH^+	1.1×10^{-8}
Methylamine	NH_2CH_3	H—N—CH$_3$, H	$NH_3CH_3^+$	4.4×10^{-4}
Nicotine	$C_{10}H_{14}N_2$	H_2C——CH_2, HC CH_2, N, CH_3, N	$HC_{10}H_{14}N_2^+$	7×10^{-7} 1.4×10^{-11}
Hydrosulfide ion	HS^-	$[H—S:]^-$	H_2S	1.8×10^{-7}
Carbonate ion	CO_3^{2-}	$\left[\begin{array}{c} :O: \\ C \\ :O\quad O: \end{array}\right]^{2-}$	HCO_3^-	1.8×10^{-4}
Hypochlorite	ClO^-	$[:Cl—O:]^-$	$HClO$	3.3×10^{-7}

Many substances behave as weak bases in water. They are able to remove protons from water thereby generating OH^- ions. However, the reaction reaches equilibrium before it goes to completion. The most commonly encountered weak base is ammonia:

$$NH_3(aq) + H_2O(l) \rightleftharpoons NH_4^+(aq) + OH^-(aq) \qquad [15.22]$$

The equilibrium-constant expression for this reaction can be written as

$$K = \frac{[NH_4^+][OH^-]}{[NH_3][H_2O]} \qquad [15.23]$$

Because the concentration of water is essentially constant even when moderate concentrations of other substances are present, the $[H_2O]$ term is incorporated into the equilibrium constant, giving

$$K[H_2O] = K_b = \frac{[NH_4^+][OH^-]}{[NH_3]} \qquad [15.24]$$

K_b is called the base-dissociation constant, by analogy with the acid dissociation constant, K_a, which describes weak acids. Table 15.5 lists the names, structures, and K_b values for several weak bases in water. Appendix E includes a more extensive list. Notice that the bases listed in Table 15.5 are of two general types: nitrogen-containing compounds and anions. Most of the weak bases that we will encounter are of these types. Also notice that these bases contain unshared pairs of electrons. These are necessary to form a bond with H^+.

SAMPLE EXERCISE 15.11

Calculate the concentration of OH^- in a 0.15 M solution of NH_3.

Solution: We use essentially the same procedure here as used in solving problems involving the dissociation of acids. The first step is to write the equilibrium expression and the corresponding equilibrium-constant expression:

$$NH_3(aq) + H_2O(l) \rightleftharpoons NH_4^+(aq) + OH^-(aq)$$
$$K_b = \frac{[NH_4^+][OH^-]}{[NH_3]} = 1.8 \times 10^{-5}$$

We then tabulate the equilibrium concentrations involved in the equilibrium:

	$NH_3(aq) + H_2O(l) \rightleftharpoons$	$NH_4^+(aq) +$	$OH^-(aq)$
Initial:	0.15 M	0 M	0 M
Equilibrium:	$(0.15 - x)M$	x M	x M

(Notice that we ignore the concentration of H_2O, because this is not involved in the equilibrium-constant expression.) Inserting these quantities into the equilibrium-constant expression gives the following:

$$K_b = \frac{[NH_4^+][OH^-]}{[NH_3]} = \frac{(x)(x)}{(0.15 - x)} = 1.8 \times 10^{-5}$$

Because K_b is small we can neglect the small amount of NH_3 that reacts with water, as compared with the total NH_3 concentration; that is, we can neglect x in comparison with 0.15 M. Then we have

$$\frac{x^2}{0.15} = 1.8 \times 10^{-5}$$
$$x^2 = (0.15)(1.8 \times 10^{-5}) = 0.27 \times 10^{-5}$$
$$x = \sqrt{2.7 \times 10^{-6}} = 1.6 \times 10^{-3} M = [OH^-]$$

Notice that the value obtained for x is only about 1 percent of the NH_3 concentration, 0.15 M. Therefore our neglect of x in comparison with 0.15 is justified.

AMINES

The weak nitrogen bases listed in Table 15.5 belong to a family known as amines. These compounds can be thought of as being formed by replacing one or more of the N—H bonds in NH_3 with N—C bonds. (In hydroxylamine, H_2NOH, one of the N—H bonds of NH_3 has been replaced by a N—OH bond.) Like ammonia, such amines are able to extract a proton from the water molecule by forming a N—H bond. The following equation illustrates this behavior:

$$H-\overset{\cdot\cdot}{\underset{\underset{H}{|}}{N}}-CH_3(aq) + H_2O(l) \rightleftharpoons \left[H-\overset{\overset{H}{|}}{\underset{\underset{H}{|}}{N}}-CH_3 \right]^+ (aq) + OH^-(aq) \qquad [15.25]$$

Many amines with low molecular weights have unpleasant, often "fishy" odors. Amines as well as NH_3 are produced by anaerobic (absence of O_2) decomposition of dead animal or plant matter, thereby contributing to their odor. One such amine, $H_2N(CH_2)_5NH_2$, is known as cadaverine.

Many drugs, including quinine, codeine, caffeine, and amphetamine (Benzedrine) are amines. Like other amines, these substances are weak bases; the amine nitrogen is readily protonated by treatment with an acid. The resulting products are called acid salts. If we use A as abbreviation for an amine, the acid salt formed by reaction with hydrochloric acid would be written as AH^+Cl^-. (It is sometimes written as $A \cdot HCl$ and referred to as a hydrochloride.) For example, amphetamine hydrochloride is the acid salt formed by treating amphetamine with HCl:

Amphetamine

Amphetamine hydrochloride

Such acid salts are less volatile, more stable, and generally more water soluble than the corresponding neutral amines. Many drugs that are amines are sold and administered as acid salts.

ANIONS OF WEAK ACIDS

A second common class of weak bases is composed of the anions of weak acids. Consider, for example, an aqueous solution of sodium acetate, $NaC_2H_3O_2$. This salt dissolves in water to give Na^+ and $C_2H_3O_2^-$ ions. The $C_2H_3O_2^-$ ion is the conjugate base of a weak acid, acetic acid. Consequently, the $C_2H_3O_2^-$ ion is basic and reacts to a slight extent with water ($K_b = 5.6 \times 10^{-10}$):

$$C_2H_3O_2^-(aq) + H_2O(l) \rightleftharpoons HC_2H_3O_2(aq) + OH^-(aq) \qquad [15.26]$$

SAMPLE EXERCISE 15.12

Calculate the pH of a 0.010 M solution of sodium hypochlorite, NaClO.

Solution: NaClO is an ionic compound consisting of Na^+ and ClO^- ions. As such it is a strong electrolyte. The hypochlorite ion, ClO^-, supplied by this salt is a weak base. The base dissociation constant for ClO^- is given in Table 15.5, $K_b = 3.3 \times 10^{-7}$. We can write the reaction between ClO^- and water, and the equilibrium concentrations present in this solution as follows:

$$\text{ClO}^-(aq) + \text{H}_2\text{O}(l) \rightleftharpoons \text{HOCl}(aq) + \text{OH}^-(aq)$$

| Initial: | 0.010 M | 0 M | 0 M |
| Equilibrium: | $(0.010 - x)M$ | $x\,M$ | $x\,M$ |

Because K_b is small, we anticipate that x will be small, so that $(0.010 - x) \simeq 0.010$. Using this approximation, we have

$$K_b = \frac{[\text{HOCl}][\text{OH}^-]}{[\text{ClO}^-]} = \frac{(x)(x)}{0.010} = 3.3 \times 10^{-7}$$

Solving for x:

$$x^2 = (0.010)(3.3 \times 10^{-7}) = 3.3 \times 10^{-9}$$

$$x = [\text{OH}^-] = \sqrt{3.3 \times 10^{-9}} = 5.7 \times 10^{-5}\,M$$

$[\text{H}^+]$ can be obtained using the ion-product constant for water:

$$[\text{H}^+] = \frac{1.0 \times 10^{-14}}{[\text{OH}^-]} = \frac{1.0 \times 10^{-14}}{5.7 \times 10^{-5}}$$

$$= 1.8 \times 10^{-10}\,M$$

$$\text{pH} = -\log[\text{H}^+] = -\log(1.8 \times 10^{-10}) = 9.75$$

Thus we see that this solution of NaClO is slightly basic.

15.7 Relation between K_a and K_b

We've seen in a qualitative way that strong acids have weak conjugate bases, while weak acids have strong conjugate bases. The fact that this qualitative relationship exists suggests that we might be able to find a quantitative relationship. Let's explore this matter by considering the NH_4^+ and NH_3 conjugate acid-base pair. Each of these species reacts with water as follows:

$$\text{NH}_4^+(aq) \rightleftharpoons \text{NH}_3(aq) + \text{H}^+(aq) \qquad [15.27]$$

$$\text{NH}_3(aq) + \text{H}_2\text{O}(l) \rightleftharpoons \text{NH}_4^+(aq) + \text{OH}^-(aq) \qquad [15.28]$$

Each of these equilibria is expressed by a characteristic dissociation constant:

$$K_a = \frac{[\text{NH}_3][\text{H}^+]}{[\text{NH}_4^+]} \qquad K_b = \frac{[\text{NH}_4^+][\text{OH}^-]}{[\text{NH}_3]}$$

Now notice something very interesting and important. When Equations [15.27] and [15.28] are added together, the NH_4^+ and NH_3 species cancel, and we are left with just the autoionization of water:

$$\text{NH}_4^+(aq) \rightleftharpoons \text{NH}_3(aq) + \text{H}^+(aq)$$
$$\underline{\text{NH}_3(aq) + \text{H}_2\text{O}(l) \rightleftharpoons \text{NH}_4^+(aq) + \text{OH}^-(aq)}$$
$$\text{H}_2\text{O}(l) \rightleftharpoons \text{H}^+(aq) + \text{OH}^-(aq)$$

To determine what we should do about the equilibrium constants for the added reactions, we make use of a rule that can be derived from the general principles governing chemical equilibria: *When two reactions are added to give a third reaction, the equilibrium constant for the third reaction is given by the product of the equilibrium constants for the two added reactions.* Thus in general,

If reaction 1 + reaction 2 = reaction 3
then $K_1 \times K_2 = K_3$

Applying this to our present example, if we multiply K_a and K_b, we obtain the following result:

$$K_a \times K_b = \left(\frac{[NH_3][H^+]}{[NH_4^+]} \right)\left(\frac{[NH_4^+][OH^-]}{[NH_3]} \right)$$
$$= [H^+][OH^-] = K_w$$

Thus, the result of multiplying K_a times K_b is just the ion-product constant, K_w, Equation [15.10]. This is, of course, just what we would expect, because addition of Equations [15.27] and [15.28] gave us just the autoionization equilibrium for water, for which the equilibrium constant is K_w.

The relationship we have just found is so important that it should be emphasized and called to special attention: *The product of the acid-dissociation constant for an acid and the base-dissociation constant for its conjugate base is the ion-product constant for water:*

$$K_a \times K_b = K_w \qquad\qquad [15.29]$$

As the strength of an acid increases (larger K_a), the strength of its conjugate base must decrease (smaller K_b), so that the product $K_a \times K_b$ remains equal to 1.0×10^{-14}.

Because of Equation [15.29], we can calculate K_a for any weak acid if we know K_b for its conjugate base. Likewise, we can calculate K_b for a weak base if we know K_a for its conjugate acid. As a practical consequence, ionization constants are often listed for only one member of a conjugate acid-base pair. For example, Appendix E does not contain K_b values for the anions of weak acids because these can be readily calculated from the tabulated K_a values for their conjugate acids.

SAMPLE EXERCISE 15.13

Calculate (a) the base-dissociation constant, K_b, for the fluoride ion, F^-, and (b) the acid-dissociation constant, K_a, for the ammonium ion, NH_4^+.

Solution: (a) K_b for F^- is not included in Table 15.5 or in Appendix E. However, K_a for its conjugate acid, HF, is given in Table 15.3 and Appendix E as $K_a = 6.8 \times 10^{-4}$. We can therefore use Equation [15.29] to calculate K_b:

$$K_b = \frac{K_w}{K_a} = \frac{1.0 \times 10^{-14}}{6.8 \times 10^{-4}} = 1.5 \times 10^{-11}$$

(b) K_b for NH_3 is listed in Table 15.5 and in Appendix E as $K_b = 1.8 \times 10^{-5}$. Using Equation [15.21], we can calculate K_a for the conjugate acid, NH_4^+:

$$K_a = \frac{K_w}{K_b} = \frac{1.0 \times 10^{-14}}{1.8 \times 10^{-5}} = 5.6 \times 10^{-10}$$

If you have the occasion to look up the values for acid or base dissociation constants in a chemistry handbook, you may find them expressed as pK_a or pK_b, that is, as $-\log K_a$ or $-\log K_b$ (Section 15.3). Equation [15.29] can be put in terms of pK_a and pK_b by taking the negative log of both sides:

$$pK_a + pK_b = pK_w = 14.00 \qquad\qquad [15.30]$$

This form is particularly useful when the tabulated pK value is that for the conjugate acid or base of the substance of interest. It is not uncommon to find the dissociation constants for bases tabulated as pK_a values for the corresponding conjugate acids. As an example, morphine, a nitrogen-containing base, is listed as the protonated cation, with $pK_a = 7.87$. This means that for the reaction

$$C_{17}H_{19}O_3NH^+(aq) \rightleftharpoons C_{17}H_{19}O_3N(aq) + H^+(aq)$$

the equilibrium constant, K_a, has the value K_a = antilog $(-7.87) = 10^{-7.87} = 1.3 \times 10^{-8}$. The reaction

$$C_{17}H_{19}O_3N(aq) + H_2O(l) \rightleftharpoons$$
$$C_{17}H_{19}O_3NH^+(aq) + OH^-(aq)$$

is described by equilibrium constant K_b. Using Equation [15.30] and $pK_a = 7.87$, we have

$$pK_b = 14.00 - pK_a = 14.00 - 7.87 = 6.13$$

Thus K_b = antilog $(-6.13) = 10^{-6.13} = 7.4 \times 10^{-7}$.

15.8 Acid-base properties of salt solutions

Even before you began this chapter you were undoubtably aware of many substances that are acidic such as HNO_3, HCl, and H_2SO_4 and others that are basic such as $NaOH$ and NH_3. However, our recent discussions have indicated that ions can also exhibit acidic or basic properties. For example, we calculated K_a for NH_4^+ and K_b for F^- in Sample Exercise 15.13. Such behavior implies that salt solutions can be acidic or basic. Before proceeding with further discussions of acids and bases, let's summarize some features of salts that should bring their acid and base properties into sharper focus.

We can assume that when salts dissolve in water they are completely ionized; nearly all salts are strong electrolytes. Consequently, the acid-base properties of salt solutions are due to the behavior of the cations and anions. Many ions are able to react with water to generate $H^+(aq)$ or $OH^-(aq)$. This type of reaction is often called hydrolysis.

The anions of weak acids, HX, are basic, and consequently they react with water to produce OH^- ions:

$$X^-(aq) + H_2O(l) \rightleftharpoons HX(aq) + OH^-(aq) \qquad [15.31]$$

In contrast, the anions of strong acids, such as the NO_3^- ion, exhibit no significant basicity; these ions do not hydrolyze, and consequently do not influence pH.

Anions of polyprotic acids such as HCO_3^- that still have ionizable protons are capable of acting as either proton donors or proton acceptors (that is, either acids or bases). Their behavior toward water will be determined by the relative magnitudes of K_a and K_b for the ion, as shown in Sample Exercise 15.14.

SAMPLE EXERCISE 15.14

Predict whether the salt Na_2HPO_4 will form an acidic or basic solution on dissolving in water.

Solution: The two possible reactions that HPO_4^{2-} may undergo on addition to water are

$$HPO_4^{2-}(aq) \rightleftharpoons H^+(aq) + PO_4^{3-}(aq) \qquad [15.32]$$
$$HPO_4^{2-}(aq) + H_2O \rightleftharpoons$$
$$H_2PO_4^-(aq) + OH^-(aq) \qquad [15.33]$$

Depending on which of these has the larger equilibrium constant, the ion will cause the solution to be acidic or basic. The value of K_a for reaction Equation [15.32], as shown in Table 15.4, is 4.2×10^{-13}. We must calculate the value of K_b for reaction Equa-

tion [15.33] from the value of K_a for the conjugate acid formed, $H_2PO_4^-$. We make use of the relationship shown in Equation [15.29]:

$$K_a \times K_b = K_w$$

We want to know K_b for the base HPO_4^{2-}, knowing the value of K_a for the conjugate acid $H_2PO_4^-$:

$$K_b(HPO_4^{2-}) \times K_a(H_2PO_4^-) = K_w = 1.0 \times 10^{-14}$$

Because K_a for $H_2PO_4^-$ is 6.2×10^{-8} (Table 15.4), we calculate K_b for HPO_4^{2-} to be 1.6×10^{-7}. This is considerably larger than K_a for HPO_4^{2-}; thus, the reaction shown in Equation [15.33] would predominate over that in Equation [15.32], and the solution would be basic.

All cations, except those of the alkali metals and the heavier alkaline earths (Mg^{2+}, Ca^{2+}, Sr^{2+}, and Ba^{2+}), act as weak acids in water solution. The alkali metals and heavier alkaline earths, which are the cations that form strong bases (Table 5.3, Section 15.4), do not undergo hydrolysis. Consequently, these ions do not influence pH. That the NH_4^+ ion should be acidic is not surprising; it is the conjugate acid of the weak base NH_3. The behavior of NH_4^+ in water is as follows:

$$NH_4^+(aq) \rightleftharpoons H^+(aq) + NH_3(aq) \qquad \text{[15.34]}$$

It might be more surprising that most metal ions, including Al^{3+} and the transition metal ions, form weakly acidic solutions. We present this observation now as a point of fact. The reasons for this behavior will become evident in Section 15.10.

The pH of a solution of a salt can be qualitatively predicted by considering the cation and anion of which the salt is composed. A convenient way to do this is to consider the relative strengths of the acids and bases from which the salt is derived:*

1 *Salt derived from a strong base and a strong acid.* Examples are NaCl and $Ca(NO_3)_2$. Neither cation nor anion hydrolyzes. The solution has a pH of 7.

2 *Salt derived from a strong base and a weak acid.* In this case the anion is a relatively strong conjugate base. Examples are NaClO and $Ba(C_2H_3O_2)_2$. The anion hydrolyzes to produce $OH^-(aq)$ ions. The solution has a pH above 7.

3 *Salt derived from a weak base and a strong acid.* In this case the cation is a relatively strong conjugate acid. Examples are NH_4Cl and $Al(NO_3)_3$. The cation hydrolyzes to produce $H^+(aq)$ ions. The solution has a pH below 7.

4 *Salt derived from a weak base and a weak acid.* Examples are $NH_4C_2H_3O_2$, NH_4CN, and $FeCO_3$. Both cation and anion hydrolyze. The pH of the solution depends upon the extent to which each ion hydrolyzes. The pH of a solution of NH_4CN is greater than 7 because CN^- ($K_b = 2.0 \times 10^{-5}$) is a stronger base than NH_4^+ ($K_a = 5.6 \times 10^{-10}$) is an acid. Consequently, CN^- hydrolyzes to a greater extent than NH_4^+ does.

15.9 Acid-base character and chemical structure

From our discussion to this point, we have seen that when any substance is dissolved in water one of three things can happen. The $H^+(aq)$ concentration might increase, in which case the substance behaves as an acid; it might decrease (with corresponding increase in OH^-), in which case the substance is acting as a base; or, there might be no change in $[H^+]$, an indication that the substance possesses neither acid nor base character. It would be very helpful to have further guidelines as to how acid or base

*These rules apply to what can be called normal salts. These salts are ones that contain no ionizable protons on the anion. The pH of an acid salt (such as $NaHCO_3$ and NaH_2PO_4) is affected not only by the hydrolysis of the anion but also by its acid-dissociation as well, as shown in Sample Exercise 15.14.

characteristics relate to chemical structure, so that we might be better able to predict how a compound will behave on dissolving in water. However, we must expect that any simple rules we might formulate won't always work. Many different factors contribute to ionization in a polar solvent such as water. The best we can hope for are a few rules that are *almost always* obeyed.

EFFECTS OF BOND POLARITY AND BOND STRENGTH

When a substance HX transfers a proton to the solvent, an ionic rupture of the H—X bond occurs. Such a reaction will occur most readily when the H—X bond is already polarized in the following sense:

$$\overset{\longrightarrow}{\text{H—X}}$$

For example, compare NH_4^+ and CH_4. These two species have the same electronic structure, that is, they are isoelectronic. Both consist of a central atom with an octet of electrons bonding four hydrogens. The difference is in the nuclear charge of the central atom. Because the nuclear charge of N is one greater than for C, the electron pairs shared with the hydrogens are more closely attracted to N in NH_4^+ than to C in CH_4. That is, the N—H bonds are more polarized than the C—H bonds. Correspondingly, ammonium ion is an acid in water, whereas methane is not:

$$NH_4^+(aq) \rightleftharpoons H^+(aq) + NH_3(aq) \qquad K_a = 5.6 \times 10^{-10} \qquad [15.35]$$

$$CH_4(aq) \rightleftharpoons H^+(aq) + CH_3^-(aq) \qquad \text{no reaction} \qquad [15.36]$$

One other factor of major importance in determining whether a substance acts as an acid is the strength of the H—X bond. Very strong bonds are less easily ionized than weaker ones. This factor is of importance in the case of the hydrogen halides. The H—F bond is the most polar of any H—X bond. One might therefore expect that HF would be a very strong acid, if the first rule were all that mattered. However, the energy required to dissociate HF into H and F atoms is much higher than for the other hydrogen halides, as shown in Table 7.3. As a result, HF is a weak acid, whereas all the other hydrogen halides are strong acids in water.

HYDROXIDES AND OXYACIDS

Many of the acids commonly encountered involve one or more O—H bonds. For example, H_2SO_4 contains two such bonds:

$$\text{H—O—S—O—H}$$

Such substances in which OH groups and possibly additional oxygen atoms are bound to a central atom are referred to as **oxyacids**. Let's consider, then, an OH group bound to some other atom Y, which might in turn have other groups attached to it:

$$\overset{\diagdown}{\underset{\diagup}{\text{Y}}}\!\!-\!\!\text{O}\!-\!\text{H}$$

At one extreme, Y might be a metal such as Na, K, or Mg. The pair of electrons shared between Y and O is then completely transferred to oxygen, and an ionic compound involving OH^- is formed. Because of the charge that surrounds it, the oxygen of the OH^- ion does not strongly attract to itself the electron pair it shares with hydrogen. That is, the O—H bond in OH^- is not strongly polarized. There is therefore no tendency for the hydrogen of OH^- to be transferred to the solvent as $H^+(aq)$. Such compounds therefore behave as bases.

When Y is an element of intermediate electronegativity, around 2.0, the bond to O is more covalent in character, and the substance does not readily lose OH^-. Elements with electronegativities in this range include B, C, P, As, and I (Figure 7.6). Examples of acids of such elements include orthoboric acid, hypoiodous acid, and methanol, the structures of which are shown at left.

Such substances might behave as acids in water, depending on the ease with which the proton is lost from oxygen. As a general rule, the more strongly the group Y attracts the electron pair it shares with the oxygen, the more polar the OH bond will be, and the more acidic the substance. In the three examples just given, the central atom does not strongly attract the electron pair it shares with oxygen. The acid-dissociation constant for orthoboric acid is 6.5×10^{-10}; for hypoiodous acid, 2.3×10^{-11}; no acidic or basic character is observed for methyl alcohol in water.

$$\overset{\displaystyle \text{O}-\text{H}}{\underset{\displaystyle}{\text{H}-\text{O}-\text{B}-\text{O}-\text{H}}}$$

Orthoboric acid

$$\text{I}-\text{O}-\text{H}$$

Hypoiodous acid

$$\text{CH}_3-\text{O}-\text{H}$$

Methanol

SAMPLE EXERCISE 15.15

Draw the Lewis structure for orthosilicic acid, $Si(OH)_4$. What do you predict for the acid-base properties of this substance?

Solution: The Lewis structure for $Si(OH)_4$ is as follows:

$$\overset{\displaystyle \text{O}-\text{H}}{\underset{\displaystyle \text{O}-\text{H}}{\text{H}-\text{O}-\text{Si}-\text{O}-\text{H}}}$$

Because silicon is an element of intermediate electronegativity (Figure 7.6), we would expect that $Si(OH)_4$ would not be strongly acidic or basic. By analogy with orthoboric acid, we might guess that it would be weakly acidic, as in fact it is:

$$K_a = 2 \times 10^{-10}$$

As the electronegativity of Y increases, or as groups with greater electron-attracting ability are placed on Y, the acidic properties of the substance increase. It is possible to relate the acid strengths of oxyacids both to the electronegativity of Y and to the number of groups attached to it. *For acids that have the same structure, but differ in the electronegativity of the central atom, Y, acid strength increases with increasing electronegativity.* Examples are shown in Table 15.6.

TABLE 15.6 Acid-dissociation constants (K_a) of oxyacids in comparison with electronegativity values (EN) of central atom Y

H—O—Y	K_a	EN of Y	$\overset{\overset{\text{O}}{\|}}{\text{H—O—Y—OH}}$	K_{a1}	EN of Y
HOCl	3×10^{-8}	3.2	H_2SO_3	1.7×10^{-2}	2.6
HOBr	2×10^{-9}	3.0	H_2SeO_3	3.5×10^{-3}	2.6
HOI	2×10^{-11}	2.7	H_2CO_3	4.3×10^{-7}	2.5
HOCH$_3$	~ 0	2.5[a]			

[a] This value is the electronegativity for carbon.

In a series of acids that have the same central atom, Y, but differing numbers of attached groups, the acid strength increases with increasing oxidation number of the central atom. For example, in the series of oxyacids of chlorine extending from hypochlorous to perchloric acid, acid strength steadily increases:

Acid	H—O—Cl:	H—O—Cl—O:	H—O—Cl—O:	H—O—Cl—O:
	Hypochlorous	Chlorous	Chloric	Perchloric
Chlorine oxidation numbers	+1	+3	+5	+7

Increasing acid strength →

In this series the ability of chlorine to withdraw electrons from the OH group, and thus make the O—H bond even more polar, increases as electron-withdrawing oxygen atoms are added to the chlorine.

15.10 The Lewis theory of acids and bases

For a substance to be a proton acceptor (that is, a base in the Brønsted-Lowry sense), that substance must possess an unshared pair of electrons for binding the proton. For example, we have seen that NH_3 acts as a proton acceptor. Using Lewis structures we can write the reaction between H^+ and NH_3 as follows:

$$H^+ + :\overset{\overset{\text{H}}{\|}}{\underset{\underset{\text{H}}{\|}}{\text{N}}}\text{—H} \longrightarrow \left[\overset{\overset{\text{H}}{\|}}{\underset{\underset{\text{H}}{\|}}{\text{H—N—H}}} \right]^+ \qquad [15.37]$$

G. N. Lewis was the first to notice this aspect of acid-base reactions. He proposed a definition of acid and base that emphasizes the shared electron pair: *An acid is defined as an electron-pair acceptor and a base as an electron-pair donor.*

Every base that we have discussed thus far, whether it be OH^-, H_2O, an amine, or an anion, is an electron-pair donor. Everything that is a base in the Brønsted-Lowry sense (a proton acceptor) is also a base in the Lewis sense (an electron-pair donor). However, in the Lewis theory, a base can donate its electron pair to something other than H^+. The Lewis definition therefore greatly increases the number of species that can be considered as acids; H^+ is a Lewis acid, but not the only one. For example, consider the reaction between NH_3 and BF_3. This reaction occurs because BF_3 has a vacant orbital in its valence shell (Section 7.7). It therefore acts as an electron-pair acceptor (a Lewis acid) towards NH_3 which donates the electron pair:

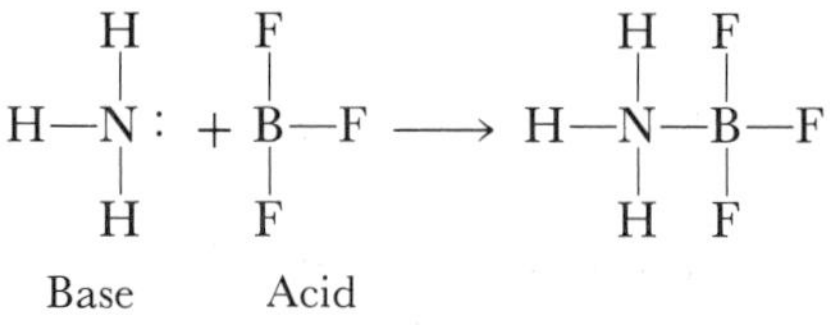

[15.38]

Our emphasis throughout this chapter has been on water as the solvent and on the proton as the source of acidic properties. In such cases we find the Brønsted-Lowry definition of acids and bases to be the most useful one to use. In fact, when we speak of a substance as being acidic or basic, we are usually thinking of aqueous solutions and using these terms in the Arrhenius or Brønsted-Lowry sense. The advantage of the Lewis theory is that it allows us to treat a wider variety of reactions, including ones that do not involve proton transfer, as acid-base reactions. In order to avoid confusion, a substance like BF_3 is rarely called an acid unless it is clear from the context that we are using the term in the sense of the Lewis definition. Instead, such substances which function as electron-pair acceptors are referred to explicitly as "Lewis acids."

Lewis acids include molecules like BF_3 that have an incomplete octet of electrons. In addition, many simple cations can function as Lewis acids. For example, Fe^{3+} interacts strongly with cyanide ions to form the ferricyanide ion, $Fe(CN)_6^{3-}$:

$$Fe^{3+} + 6:C\equiv N:^- \longrightarrow [Fe(:C\equiv N)_6]^{3-}$$

Compounds with multiple bonds can behave as Lewis acids. For example, the reaction of carbon dioxide with water to form carbonic acid, H_2CO_3, can be pictured as an attack by a water molecule on CO_2, in which the water acts as an electron-pair donor, and the CO_2 as an electron pair acceptor:

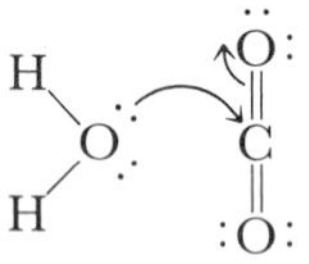

The electron pair of one of the carbon-oxygen π bonds is moved onto the oxygen to leave a vacant orbital on the carbon, which can act as electron-pair acceptor. We have shown the shift of these electrons with arrows. After forming the initial "adduct, " a proton moves from one oxygen to another, thereby forming carbonic acid:

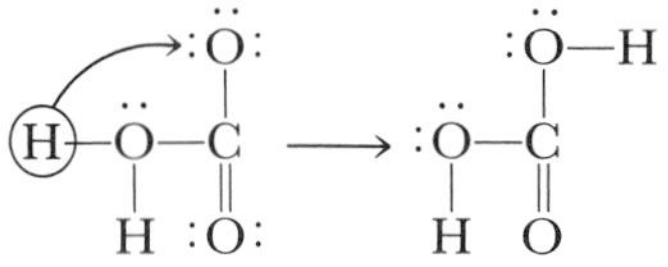

TABLE 15.7	Ionic-charge/ ionic-radius ratio for metal ions of various charges	
Metal ion	Charge/ionic radius
Li^+	1.5
Na^+	1.0
Cu^{2+}	2.8
Mg^{2+}	3.1
Ca^{2+}	2.1
Zn^{2+}	2.7
Al^{3+}	6.7
Cr^{3+}	4.8
Fe^{3+}	4.7

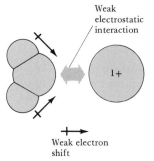

Charge/radius small

Weak electrostatic interaction

1+

Weak electron shift

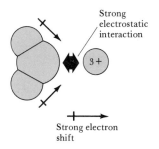

Charge/radius large

Strong electrostatic interaction

3+

Strong electron shift

FIGURE 15.9 Interaction of a water molecule with a cation of 1+ charge or 3+ charge. The interaction is much stronger with the smaller ion of higher charge.

A similar kind of Lewis acid-base reaction takes place when any oxide of a nonmetal dissolves in water to form an acidic solution.

HYDROLYSIS OF METAL IONS

The Lewis theory is also helpful in explaining why solutions of many metal ions show acidic properties (Section 15.8). For example, a solution of a salt such as $Cr(NO_3)_3$ is quite acidic. A solution of $ZnCl_2$ is likewise acidic, though to a lesser extent. To understand why this is so, we must examine the interaction between a metal ion and water molecules.

Because metal ions are positively charged, they attract the unshared electron pairs of water molecules. It is primarily this interaction, referred to as hydration, that causes salts to dissolve in water, as explained in Section 12.2. The strength of attraction increases with the charge of the ion and is strongest for the smallest ions. The ratio of ionic charge to ionic radius provides a good measure of the extent of hydration. This ratio is listed in Table 15.7 for a selection of metal ions. The process of hydration is a Lewis acid-base interaction, in which the metal ion acts as a Lewis acid, and the water molecules as Lewis bases. When the water molecule interacts with the positively charged metal ion, electron density is drawn from the oxygen, as illustrated in Figure 15.9. This flow of electron density causes the O—H bond to become more polarized; as a result, water molecules bound to the metal ion, M, are more acidic than those in the bulk solvent. The hydrated metal ion thus acts as a source of protons:

$$M(H_2O)_n{}^{z+} \rightleftharpoons M(H_2O)_{(n-1)}(OH)^{(z-1)+} + H^+(aq) \qquad [15.39]$$

In this equation z is the charge on the metal ion, and n is the number of hydrating water molecules. In the case of the 3+ metal ions listed in Table 15.7, n is 6; for the other ions it is probably closer to 4, although the exact number is difficult to determine. The hydrolysis reaction shown in Equation [15.39] represents the behavior of an acid in just the same way as Equation [15.15], which applies to HX. The "acid" in Equation [15.39] is not just a single molecule, but a collection of molecules. For example, it might be $Fe(H_2O)_6{}^{3+}$, which we usually represent merely as $Fe(aq)^{3+}$; this ion has a pK_a of 2.2. The main point is that such an ion is able to transfer a proton to water and thus act as an acidic species. The effect is greatest for the smallest and most highly charged ions, as shown in Figure 15.9.

FOR REVIEW

Summary

In this chapter we have considered the general properties of acidic and basic solutions, with emphasis on water as the solvent. We have seen that an acid solution is created when a substance reacts with water in such a way as to increase the concentration of solvated hydrogen ions, which are represented as $H^+(aq)$ or $H_3O^+(aq)$. The concentration of $H^+(aq)$ is often expressed on the pH scale: $pH = -\log [H^+]$. Solutions of pH less than 7 are acidic; those with pH greater than 7 are basic.

Water spontaneously ionizes to a slight degree (autoionization), forming $H^+(aq)$ and $OH^-(aq)$. The extent of ionization is expressed by the ion-product constant for water: $K_w = [H^+][OH^-] = 1.0 \times 10^{-14}$. This relationship describes not only pure water, but aqueous solutions as well. Because the concentration of water is effectively constant in dilute solutions, $[H_2O]$ is omitted from this equilibrium-constant expression as well as from others associated with reactions in aqueous solution.

Through most of this chapter we have relied on the Brønsted-Lowry theory of acids and bases. According to this theory, an acid is a proton (H^+) donor, a base is a proton acceptor. Reaction of an acid with water results in the formation of $H^+(aq)$ and the conjugate base of the acid. Strong acids have conjugate bases that are weaker than H_2O. Such acids are strong electrolytes, ionizing completely in solution. The common strong acids are HCl, HBr, HI, HNO_3, $HClO_4$ and H_2SO_4. Weak acids are substances for which the reaction with water is incomplete, and an equilibrium is established. The extent to which the reaction proceeds is expressed by the acid-dissociation constant, K_a. Polyprotic acids are acids such as H_2SO_3 that have more than one ionizable proton. These acids have more than one acid-dissociation constant: K_{a1}, K_{a2}, and so forth, which decrease in magnitude in the order $K_{a1} > K_{a2} > K_{a3}$.

Aside from the ionic hydroxides such as NaOH, bases produce an increase of OH^- by reaction with water. Strong bases have conjugate acids that are no stronger than H_2O. The common strong bases are the hydroxides and oxides of the alkali metals and alkaline earths. Weak bases include H_2O, NH_3, amines, and the anions of weak acids. The extent to which a weak base reacts with water to generate OH^- and the conjugate acid of the base is measured by the base-dissociation constant, K_b.

The stronger an acid, the weaker its conjugate base; the weaker an acid, the stronger its conjugate base. This qualitative observation is expressed quantitatively by the expression $K_a \times K_b = K_w$ (where K_a and K_b are dissociation constants for conjugate acid-base pairs).

The acid-base properties of salts can be ascribed to the behavior of their respective cations and anions. The reaction of ions with water with a resultant change in pH is called hydrolysis. The cations of strong bases (the alkali metal ions and alkaline earth metal ions) and the anions of strong acids do not undergo hydrolysis.

The tendency of a substance to show acidic or basic characteristics in water can be correlated reasonably well with chemical structure. Acid character requires the presence of a highly polar H—X bond, promoting loss of hydrogen as H^+ on reaction with water. Basic character, on the other hand, requires

the presence of an available pair of electrons. By considering the effects of changes in structure, it is possible to predict how a given structural change is likely to alter the acidity or basicity.

In the Lewis theory of acids and bases, the emphasis is on the shared electron pair rather than on the proton. An acid is defined as an electron-pair acceptor, a base as an electron-pair donor. The Lewis theory is more general than the Brønsted-Lowry model, because it applies to cases in which the proton is the acid, and to others as well.

Learning goals

Having read and studied this chapter, you should be able to:

1 Explain the process that occurs when an acid dissolves in water.

2 Describe the forms in which the proton exists in water.

3 Define an acid, base, conjugate acid, and conjugate base in terms of the Brønsted-Lowry theory of acids and bases.

4 Explain what is meant by the autoionization of water and write the ion-product-constant expression.

5 Explain what is meant by pH and calculate pH from a knowledge of $[H^+]$ or $[OH^-]$; also be able to perform the reverse operations.

6 Calculate $[OH^-]$ from pOH and K from pK, and be able to perform the reverse operations.

7 Identify the common strong acids and bases.

8 Write the acid-dissociation-constant expression for any weak acid in water.

9 Calculate $[H^+]$ for a weak acid solution in water, knowing acid concentration and K_a.

10 Write the base-dissociation-constant expression for a weak base in water.

11 Calculate $[H^+]$ for any weak base solution in water, knowing base concentration and K_b.

12 Explain the relationship between an acid and its conjugate base or between a base and its conjugate acid and calculate K_b from a knowledge of K_a, or vice versa.

13 Predict whether a particular salt solution will be acidic, basic, or neutral.

14 Explain how acid strength relates in a general way to the nature of the H—X bond.

15 Predict the relative acid strengths of oxyacids and oxyanions.

16 Define an acid or base in terms of the Lewis acid-base theory.

17 Predict the relative acidities of solutions of metal salts from a knowledge of metal-ion charges and ionic radii.

Key terms

Among the more important terms and expressions used for the first time in this chapter are the following:

An acid-base indicator (Section 15.3) is a substance whose color changes in passing from an acidic to a basic form, or vice versa.

The acid-dissociation constant, K_a (Section 15.5), is an equilibrium constant that expresses the extent to which an acid transfers a proton to solvent water.

Autoionization of water (Section 15.3) is the process whereby water spontaneously forms low concentrations of $H^+(aq)$ and $OH^-(aq)$ ions by proton transfer from one water molecule to another.

The base-dissociation constant, K_b (Section 15.6) is an equilibrium constant that expresses the extent to which a base reacts with solvent water, accepting a proton and forming $OH^-(aq)$.

A Brønsted acid (Section 15.2) is any substance capable of acting as a source of protons.

A Brønsted base (Section 15.2) is any substance capable of acting as a proton acceptor.

A conjugate acid (Section 15.2) is a substance formed by addition of a proton to a Brønsted base.

A conjugate base (Section 15.2) is a substance formed by loss of a proton from a Brønsted acid.

Hydrolysis (Section 15.8) is a process in which a cation or anion reacts with water so as to change the pH.

The ion-product constant (Section 15.3) for water, K_w, is the product of the aquated hydrogen ion and hydroxide ion concentrations: $[H^+][OH^-] = K_w = 1 \times 10^{-14}$.

A Lewis acid (Section 15.10) is defined as an electron-pair acceptor.

A Lewis base (Section 15.10) is defined as an electron-pair donor.

An oxyacid (Section 15.9) is a compound in which one or more OH groups, and possibly additional oxygen atoms, are bonded to a central atom.

The term pH (Section 15.3) is defined as the negative log in base 10 of the aquated hydrogen-ion concentration: $pH = -\log [H^+]$.

A polyprotic acid (Section 15.5) is a substance capable of dissociating more than one proton in water; H_2SO_4 is an example.

EXERCISES

The hydrated proton

15.1 Hydrogen chloride gas is extremely soluble in water but not very soluble in benzene. Explain.

15.2 Describe the state of the proton in a dilute aqueous solution of hydrogen iodide. Is it different from the state of a proton in a dilute solution of hydrogen chloride? Explain.

15.3 When very concentrated perchloric acid solutions are cooled to low temperature, a solid hydrate whose composition conforms to the formula $HClO_4 \cdot 2H_2O$ is formed. Assuming that the compound is actually ionic, what are the likely structures of the anion and cation? Draw the Lewis structures of each.

15.4 Gaseous hydrogen iodide is very soluble in water. A solution containing 52.4 percent HI by weight has a density of 1.60 g/cm^3. What is the molarity of this solution?

Brønsted acids and bases

15.5 Give the conjugate base of each of the following proton sources: (a) HCN; (b) H_2SO_4; (c) $HC_2H_3O_2$; (d) NH_4^+; (e) NH_3; (f) HSO_3^-; (g) OH^-.

15.6 Give the conjugate acid of each of the following bases: (a) OH^-; (b) HPO_4^{2-}; (c) H_2O; (d) F^-; (e) NH_3.

15.7 Identify the acid and the base in each of the following reactions:

(a) $NH_4^+(aq) + OH^-(aq) \rightleftharpoons NH_3(aq) + H_2O(l)$
(b) $HCN(g) + H_2O(l) \rightleftharpoons H_3O^+(aq) + CN^-(aq)$
(c) $O^{2-}(aq) + H_2O(l) \rightleftharpoons 2OH^-(aq)$
(d) $H^-(aq) + H_2O(l) \rightleftharpoons H_2(g) + OH^-(aq)$
(e) $HC_2O_4^-(aq) + CO_3^{2-}(aq) \rightleftharpoons$
    $C_2O_4^{2-}(aq) + HCO_3^-(aq)$

15.8 In each of the following pairs, indicate which is the stronger Brønsted base in water: (a) Cl^- or F^-; (b) HS^- or S^{2-}; (c) NO_3^- or NO_2^-; (d) OH^- or NH_3.

15.9 Using Figure 15.4, predict whether the following reactions proceed to the right to any appreciable extent ($K > 1$):

(a) $HCO_3^-(aq) + F^-(aq) \rightleftharpoons CO_3^{2-}(aq) + HF(aq)$
(b) $HSO_4^-(aq) + NH_3(aq) \rightleftharpoons$
    $SO_4^{2-}(aq) + NH_4^+(aq)$
(c) $HF(aq) + NO_3^-(aq) \rightleftharpoons HNO_3(aq) + F^-(aq)$
(d) $NH_4^+(aq) + PO_4^{3-}(aq) \rightleftharpoons$
    $NH_3(aq) + HPO_4^{2-}(aq)$

Autoionization of water; pH

15.10 Indicate whether the following solutions are acidic, basic, or neutral at $25°C$: (a) $1.00 \times 10^{-9} M \ H^+$; (b) $1.0 \times 10^{-3} M \ H^+$; (c) $1.0 \times 10^{-7} M \ H^+$; (d) $5.88 \times 10^{-5} M \ H^+$; (e) $5.2 \times 10^{-7} M \ OH^-$; (f) $9.4 \times 10^{-9} M \ OH^-$; (g) $3.0 \times 10^{-3} M \ OH^-$.

15.11 If $K_w = 1.14 \times 10^{-15}$ at $0°C$, what is the value of $[H^+]$ for a neutral solution at this temperature?

15.12 Calculate the pH that corresponds to each of the following concentrations of $H^+(aq)$ or $OH^-(aq)$: (a) $5.0 \times 10^{-2} M$ H^+; (b) $6.92 \times 10^{-5} M$ H^+; (c) $2.50 \times 10^{-3} M$ OH^-; (d) $8.8 \times 10^{-8} M$ OH^-. Indicate in each case whether the solution is acidic, basic, or neutral.

15.13 Calculate $[H^+]$ for solutions with the following pH values: (a) 2.44; (b) 3.0; (c) 12.87; (d) 6.55; (e) 8.05; (f) 6.97. Indicate in each case whether the solution is acidic, basic, or neutral.

15.14 Calculate $[H^+]$ for the following solutions: (a) urine, pH = 6.1; (b) saliva, pH = 6.6; (c) lemon juice, pH = 2.3; (d) a solution of baking soda, pH = 8.5; (e) a household ammonia solution, pH = 11.9. Indicate in each case whether the solution is acidic, basic, or neutral.

15.15 (a) The $OH^-(aq)$ concentration in a certain solution is $6.8 \times 10^{-4} M$. What is the pOH? (b) The pOH of a solution is 4.30. What is $[OH^-]$? (c) The pH of a solution is 5.8. What is pOH? (d) If the pOH of a solution is 2.62, what is the pH? (e) If $K_a = 5.82 \times 10^{-4}$ for a certain acid, what is the value of pK_a for this substance? (f) If pK_a for an acid has a value of 5.97, what is the value of the acid-dissociation constant, K_a?

15.16 A sample of lake water is filtered and tested with indicators to determine its pH. It turns methyl yellow to a yellow color and turns methyl red to a red color. Indicate the highest and lowest pH values the solution can have.

Strong acids and bases

15.17 What is the pH of each of the following solutions: (a) $0.010 M$ NaOH; (b) $5.5 \times 10^{-2} M$ HNO_3; (c) $0.020 M$ $Ca(OH)_2$; (d) $3.7 \times 10^{-5} M$ HCl?

15.18 Calculate the mass of solid NaOH required to prepare 0.500 L of NaOH solution with pH = 12.00.

15.19 Calculate the pH of a solution that contains (a) 0.640 g of KOH in 0.800 L of solution; (b) 0.807 g of HI in 0.200 L of solution.

15.20 Without performing the calculation, predict whether the pH of a solution of HF will be greater or smaller than that of an equimolar solution of HCl. Explain briefly.

15.21 In terms of the Brønsted theory, state the difference between a strong acid and a weak acid in water.

15.22 Without looking at Table 15.2, list the six common strong acids; list the common strong bases.

[15.23] The pH of a $1.0 \times 10^{-10} M$ solution of HCl is not 10.00. Why not?

Weak acids

15.24 On the basis of the data in Table 15.3, which of the following acids is the strongest acid in water and which is the weakest: (a) hydrocyanic acid; (b) benzoic acid; (c) phenol; (d) ascorbic acid?

15.25 Write the acid-dissociation-constant expressions for the following weak acids: (a) nitrous acid, HNO_2; (b) hydrogen oxalate anion, $HC_2O_4^-$.

15.26 Calculate the concentration of $H^+(aq)$ in each of the following solutions (K_a values are given in Appendix E): (a) $0.010 M$ hydrazoic acid, HN_3; (b) $0.10 M$ benzoic acid, $HC_7H_5O_2$; (c) $2.5 M$ cyanic acid, HCNO.

15.27 Benzoic acid, $HC_7H_5O_2$, is used as a preservative in certain foods to retard spoilage caused by molds and bacteria. How many grams of benzoic acid are there in a 3.00-L solution of this acid if the pH of the solution is 2.80? (K_a is given in Appendix E.)

[15.28] In all of the calculations done in the text, we have ignored H_2O as a source of $H^+(aq)$ whenever we calculated the pH of solutions of weak or strong acids. Under what conditions might you need to consider $H^+(aq)$ from water?

15.29 Ascorbic acid, better known as vitamin C, has the formula $HC_6H_7O_6$ (see Table 15.3). What is the pH of a solution formed by dissolving a 500-mg tablet of this substance in enough water to form 0.200 L of solution?

15.30 What is the percent ionization of propionic acid, $HC_3H_5O_2$, in solutions of the following concentrations: (a) $1.00 M$; (b) $0.100 M$; (c) $0.0100 M$. (K_a is given in Appendix E.)

15.31 A $0.0100 M$ solution of a weak acid, HX, is 18.5 percent ionized. Using this information, calculate $[H^+]$, $[X^-]$, $[HX]$, and K_a for HX.

15.32 Citric acid, present in citrus fruits, is a triprotic acid (one that can supply three protons to water) with the formula $H_3C_6H_5O_7$. Write the chemical equations for the three dissociation equilibria and the appropriate equilibrium-constant expression for each step. Which dissociation constant will have the largest value and which the smallest?

15.33 Which of the following acids is strongest and which is weakest (see Table 15.4): (a) HCO_3^-; (b) HSO_3^-; (c) $H_2PO_4^-$; (d) HPO_4^{2-}?

[15.34] Rhubarb owes its sour taste to the presence of oxalic acid, $H_2C_2O_4$, for which $K_{a1} = 5.9 \times 10^{-2}$ and $K_{a2} = 6.4 \times 10^{-5}$. If there are 2.4 mg of oxalic acid per gram of rhubarb and 180 g of rhubarb are cooked with sufficient water to make 1.0 L of liquid sauce, what pH should the sauce have?

Weak bases; K_a–K_b relationship

15.35 Write balanced net ionic equations for the reactions of each of the following substances with water. Write the base-dissociation-constant expression for each substance: (a) methylamine, CH_3NH_2; (b) cyanide ion, CN^-.

15.36 Calculate $[OH^-]$ and pH for each of the following solutions (K_b values are listed in Appendix E): (a) $0.010 M$ pyridine; (b) $0.020 M$ methylamine; (c) $0.10 M$ ammonia.

15.37 Dimethylamine, $(CH_3)_2NH$, is used in the insecticide Sevin. What is the pH of a $0.010 M$ aqueous solution of dimethylamine if this substance has a pK_b of 3.267?

15.38 Using the necessary values of K_a from Appendix E, calculate K_b for each of the following species: (a) formate ion, CHO_2^-; (b) phosphate ion, PO_4^{3-}; (c) cyanide ion, CN^-; (d) lactate ion, $C_3H_5O_2^-$.

15.39 Using the necessary values of K_b from Appendix E, calculate K_a for each of the following species: (a) methylammonium ion, $CH_3NH_3^+$; (b) pyridinium ion, $C_5H_5NH^+$.

15.40 Novocain is the brand name for procaine hydrochloride, an acid salt of an amine whose formula we shall represent as $PrnH^+Cl^-$. This substance is widely used as a local anesthetic in surgery. The neutral amine, procaine, has a pK_b of 5.05. What is the pH of a $0.010\,M$ aqueous solution of procaine hydrochloride?

15.41 Calculate K_{b1} and K_{b2} for SO_3^{2-}. K_{b1} is the equilibrium constant for the addition of the first proton to SO_3^{2-} upon reaction with water:

$$SO_3^{2-}(aq) + H_2O(l) \rightleftharpoons HSO_3^-(aq) + OH^-(aq)$$

K_{b2} pertains to the addition of the second proton:

$$HSO_3^-(aq) + H_2O(l) \rightleftharpoons H_2SO_3(aq) + OH^-(aq)$$

Why can we ignore the K_{b2} process in calculating the pH of a Na_2SO_3 solution?

15.42 Strong bases degrade protein structure; thus warning labels are required on any household products whose pH exceeds 11 (or whose contents will form a solution exceeding this pH). Should a product containing $0.020\,M$ Na_2CO_3 contain such a warning?

15.43 Calculate the percentage of pyridine that forms pyridinium ion, $C_5H_5NH^+$, in a $0.10\,M$ aqueous solution of pyridine. (See Appendix E for K_b.)

15.44 Calculate the pH of each of the following solutions: (a) $0.10\,M$ sodium cyanide, NaCN; (b) $0.15\,M$ calcium acetate, $Ca(C_2H_3O_2)_2$; (c) $0.10\,M$ methylammonium chloride, $CH_3NH_3^+Cl^-$; (d) $0.020\,M$ sodium phosphate, Na_3PO_4.

Salt solutions

15.45 Classify the solutions of the following salts as acidic, basic, or neutral: (a) $NaNO_3$; (b) NH_4NO_3; (c) $AlBr_3$; (d) $Cu(NO_3)_2$; (e) Na_2SO_3; (f) $KClO_4$.

15.46 Arrange the following ions in the order of increasing base strength: (a) CN^-; (b) $C_2H_3O_2^-$; (c) Br^-; (d) NO_2^-.

15.47 Predict whether each of the following salts will form an acidic or basic solution in water: (a) $NaHCO_3$; (b) NH_4I; (c) $MgHPO_4$; (d) $NaHC_2O_4$.

15.48 Calculate the pH of each of the following solutions: (a) $0.10\,M$ Na_2SO_3; (b) $0.30\,M$ NH_4Cl; (c) $0.10\,M$ NaBrO.

15.49 Which member of each of the following pairs of ions would you expect to form the more acidic solutions: (a) Fe^{2+} or Fe^{3+}; (b) Co^{2+} or Cs^+; (c) Al^{3+} or Ga^{3+}? Explain the reason for your choice in each case.

15.50 Predict whether the following salts will form an acidic or basic solution on dissolving in water: (a) NaH_2PO_4; (b) $NaClO_4$; (c) NaHS; (d) $NaHSO_4$.

Acid-base character and chemical structure

15.51 All other factors being equal, what relationship is there between the acidity of a substance and (a) the polarity of the H—X bond; (b) the strength of the H—X bond.

15.52 Which member of each of the following pairs is the stronger acid: (a) NH_3 or H_3O^+; (b) NH_4^+ or CH_4; (c) HCl or HF? Explain in terms of the structure of each substance.

15.53 Draw the Lewis structure for nitrous acid, HNO_2, and nitric acid, HNO_3. Which is the stronger acid in water? Why?

15.54 Which member of each of the following pairs is the stronger acid: (a) H_2SO_4 or H_2SO_3; (b) H_2SO_4 or H_2SeO_4; (c) HBrO or HClO; (d) H_3PO_4 or $H_2PO_4^-$; (e) H_2SO_3 or H_2CO_3? Explain.

Lewis acids and bases

15.55 Prepare a table in which you compare the definitions of acids and bases according to the Lewis, Brønsted-Lowry, and Arrhenius theories. Which is more general; that is, which includes the others within its scope? Explain.

15.56 With the aid of Lewis structures, if necessary, identify the Lewis acid and the Lewis base in each of the following reactions:

(a) $BF_3(g) + F^-(aq) \longrightarrow BF_4^-(aq)$
(b) $Ag^+(aq) + 2NH_3(aq) \rightleftharpoons Ag(NH_3)_2^+(aq)$
(c) $HF(aq) + H_2O(l) \rightleftharpoons H_3O^+(aq) + F^-(aq)$
(d) $HF(aq) + OH^-(aq) \longrightarrow H_2O(l) + F^-(aq)$
(e) $SO_3(aq) + H_2O(l) \rightleftharpoons H_2SO_3(aq)$

15.57 Describe the process that causes a solution of $Al(NO_3)_3$ to be acidic.

15.58 Which member of each of the following pairs of salts would you expect to produce the more acidic solution: (a) LiI or ZnI_2; (b) $CaCl_2$ or $FeCl_3$; (c) NH_4Cl or NaCl; (d) $FeCl_2$ or $FeCl_3$?

Additional exercises

15.59 When pH changes by two units, say from 2 to 4, by what factor does $[H^+]$ change?

15.60 Classify each of the following compounds as strong acid, strong base, weak acid, or weak base: HBr, KOH, H_3PO_4, CH_3NH_2, NH_4^+, CN^-, Rb_2O.

15.61 Arrange the following $0.10\,M$ solutions in order of decreasing pH: KOH, $HClO_4$, H_2SO_4, NH_4Br, KCN, KBr.

15.62 Solutions of $0.1\,M$ concentration of formic acid, hydrocyanic acid, and phosphoric acid are allowed to react with magnesium as indicated in Figure 15.7. Using Appendix E, predict which will react most rapidly.

15.63 What is the pH of each of the following solutions: (a) $0.015\,M$ NaOH; (b) $1.05 \times 10^{-3}\,M$ HBr; (c) $0.0100\,M$ HClO; (d) $0.150\,M$ NaCN; (e) $0.200\,M$ CH_3NH_2; (f) $0.150\,M$ NH_4Br; (g) 4.85 g of KOH in 58.0 mL of solution; (h) 3.0 mL of $6.0\,M$ HNO_3 that is diluted to a total volume of 20.0 mL.

15.64 Calculate the number of moles of each of the following substances that must be present in 500 mL of solution to form a solution with pH = 3.0: (a) HCl; (b) $HC_2H_3O_2$; (c) HF.

15.65 Calculate the percent ionization of each of the following substances in the solutions indicated: (a) 0.010 M HCN; (b) 0.010 M NH_3.

15.66 Soaps are Na^+ and K^+ salts of a family of weak acids isolated from animal fats and known as fatty acids. On the basis of this information, do you expect a soap solution to be acidic, basic, or neutral? Do soaps have any of the characteristic properties of acids and bases listed in the introductory paragraph of this chapter? Which ones do you recognize?

15.67 Liquid ammonia undergoes autoionization analogous to that of water:

$$2NH_3(aq) \rightleftharpoons NH_4^+(aq) + NH_2^-(aq)$$

At $-50°C$, the ion product for this reaction, $[NH_4^+][NH_2^-]$, is 1.0×10^{-33}. What is the concentration of NH_4^+ in a liquid ammonia solution at $-50°C$ if 1.0×10^{-4} mol of $NaNH_2$ is dissolved, forming 0.500 L of solution?

15.68 Hemoglobin is involved in a series of equilibria involving protonation-deprotonation and oxygenation-deoxygenation. The overall reaction is approximately as follows:

$$HbH^+(aq) + O_2(aq) \rightleftharpoons HbO_2(aq) + H^+(aq)$$

(where Hb stands for hemoglobin and HbO_2 for oxyhemoglobin). (a) $[O_2]$ is higher in the lungs and lower in the tissues. What effect does high $[O_2]$ have on the position of this equilibrium? (b) The normal pH of blood is 7.4. What is $[H^+]$ in normal blood? Is the blood acidic, basic, or neutral? (c) If the blood pH is lowered by the presence of large amounts of acidic metabolism products, a condition called acidosis results. What effect does lowering the pH of blood have on the ability of hemoglobin to transport O_2?

15.69 The dye bromthymol blue is a weak acid whose ionization can be represented as

$$HBb(aq) \rightleftharpoons H^+(aq) + Bb^-(aq)$$

Which way will this equilibrium shift when NaOH is added? The acid form of the dye is yellow whereas its conjugate base is blue. What color is the NaOH solution containing this dye?

15.70 Predict the effect that each of the following substances has on the pH of an aqueous solution of HF: (a) NaF; (b) NaOH; (c) NaCl; (d) HCl; (e) NaCN.

[15.71] Although we think of NH_3 as a base, it can donate a proton if a strong enough base is present. (a) Could such a reaction occur in aqueous solution? Explain. (b) The conjugate base of NH_3 is NH_2^-. Do you think that NH_2^- is a weaker or stronger base than OH^-? Explain. (c) Calculate the pH of a solution obtained by adding 0.30 g of $NaNH_2$ to sufficient water to form 0.500 L of solution.

15.72 Saccharin, a sugar substitute, is a weak acid with $pK_a = 11.68$ at $25°C$. It ionizes in aqueous solution as follows:

$$HNC_7H_4SO_3(aq) \rightleftharpoons H^+(aq) + NC_7H_4SO_3^-(aq)$$

What is the pH of a 0.10 M solution of this substance?

[15.73] What are the concentrations of H^+, HSO_4^-, and SO_4^{2-} in a 0.100 M solution of H_2SO_4?

15.74 Formic acid, $HCHO_2$, is the principal substance responsible for the stings that result from an ant bite. Much of the sting is due to $H^+(aq)$. Calculate the pH in the vicinity of an ant bite if the ant injects 1 μmol (that is 10^{-6} mol) of formic acid into such a volume that 1 μL of solution is formed. (See Appendix E for K_a value.)

15.75 (a) Calculate the pH of a 0.500 M solution of malonic acid, $H_2C_3H_2O_4$. (See Appendix E for K_a values.) What are the concentrations of $HC_3H_2O_4^-$ and of $C_3H_2O_4^{2-}$ in this solution?

[15.76] What are the concentrations of H^+, $H_2PO_4^-$, HPO_4^{2-}, and PO_4^{3-} in a 0.10 M solution of H_3PO_4?

15.77 Morphine is a weak base containing a basic nitrogen atom. Its pK_b is 6.1. (a) What is the pH of a $1.0 \times 10^{-3} M$ aqueous solution of morphine? (b) What is the pH of a $1.0 \times 10^{-3} M$ solution of the acid-salt of morphine?

[15.78] Many moderately large organic molecules containing basic nitrogen atoms are not very soluble in water as the neutral molecule but are frequently much more soluble as the acid salt. Assuming that the pH in the stomach is 2.5, indicate whether each of the following compounds would be present in the stomach as the neutral compound or in the protonated form: nicotine, $K_b = 7 \times 10^{-7}$; caffeine, $K_b = 4 \times 10^{-14}$; strychnine, $K_b = 1 \times 10^{-6}$; quinine, $K_b = 1.1 \times 10^{-6}$.

[15.79] Amino acids contain an amino group, $-NH_2$, located on the carbon atom that also contains a carboxylic acid group, $-COOH$. Glycine, the simplest amino acid could exist in water in either form I or II below:

$$\underset{I}{H_2N-CH_2-\overset{\overset{O}{\|}}{C}-OH} \qquad \underset{II}{{}^+H_3N-CH_2-\overset{\overset{O}{\|}}{C}-O^-}$$

K_a for the carboxylic acid group of glycine is 4.3×10^{-3}, and K_b for the amino group is 6.0×10^{-5}. (a) What is the pH of a 0.10 M aqueous solution of glycine? (b) In what forms, other than I and II, can glycine exist in solution? (c) Which form of glycine would you expect to exist in a solution with pH = 10; with pH = 2?

[15.80] A 1.00 m solution of HF freezes at $-1.90°C$. Based on the extent of freezing point lowering (Section 12.7), calculate the fraction of HF dissociated at this temperature. What is the value of K_a for HF at $-1.90°C$?

15.81 Cocaine is a weak amine base. If a $5.0 \times 10^{-3} M$ solution of cocaine has a pH of 10.04, what is the value of K_b for this substance?

Aqueous equilibria

Water is the solvent of preeminent importance on our planet. It is also, in a sense, the solvent of life. It is difficult to imagine how living matter in all its complexity could exist with any liquid other than water as solvent. Water occupies its position of importance not only because of its abundance, but also because of its exceptional ability to dissolve a wide variety of substances. Aqueous solutions encountered in nature, such as biological fluids or seawater, contain many solutes. Consequently, many equilibria can take place in these solutions. In Chapter 15 we discussed weak acid and base equilibria. However, we restricted our discussions to solutions containing a single solute. In this chapter we shall consider acid-base equilibria in aqueous solutions that contain two or more solutes. Furthermore, we shall broaden our treatment of aqueous equilibria to include other kinds of reactions, especially those involving slightly soluble salts.

16.1 The common-ion effect

When $NaC_2H_3O_2$ is added to a solution of $HC_2H_3O_2$, the pH of the solution increases ($[H^+]$ is reduced). This result isn't surprising because $C_2H_3O_2^-$ is a weak base; like any other base it should increase the pH. However, it is instructive to view this effect from the perspective of LeChatelier's principle. Like most salts, $NaC_2H_3O_2$ is a strong electrolyte. Consequently, it completely ionizes in aqueous solution to form Na^+ ions and $C_2H_3O_2^-$ ions. $HC_2H_3O_2$ is a weak electrolyte that dissociates as follows:

$$HC_2H_3O_2(aq) \rightleftharpoons H^+(aq) + C_2H_3O_2^-(aq) \qquad [16.1]$$

The addition of $C_2H_3O_2^-$, from $NaC_2H_3O_2$, causes this equilibrium, Equation [16.1], to shift to the left, thereby decreasing the equilibrium concentration of $H^+(aq)$. The $C_2H_3O_2^-$ is said to decrease or repress the dissociation of $HC_2H_3O_2$. In general, the dissociation of a weak electrolyte is decreased by adding to the solution a strong electrolyte that has an ion in common with the weak electrolyte. This shift in equilibrium position, which occurs when we add an ion that is common to an equilibrium reaction, is called the common-ion effect.

SAMPLE EXERCISE 16.1

Suppose we add 8.20 g, or 0.100 mol, of sodium acetate, $NaC_2H_3O_2$, to 1 L of a 0.100 M solution of acetic acid, $HC_2H_3O_2$. What is the pH of the resultant solution?

Solution: Because $NaC_2H_3O_2$ is a strong electrolyte, the only equilibrium that we need to consider is that for dissociation of acetic acid. This equilibrium reaction and the concentrations of the species involved are summarized below:

$$HC_2H_3O_2(aq) \rightleftharpoons H^+(aq) + C_2H_3O_2^-(aq)$$

Initial:	0.100 M	0 M	0.100 M
Equilibrium:	(0.100 − x)M	x M	(0.100 + x)M

Notice that the equilibrium concentration of $C_2H_3O_2^-$ is the initial amount coming from the added $NaC_2H_3O_2$ (0.100 M), plus the amount (x) formed by dissociation of acetic acid.

The equilibrium constant expression is

$$K_a = 1.8 \times 10^{-5} = \frac{[H^+][C_2H_3O_2^-]}{[HC_2H_3O_2]}$$

(The value of the equilibrium constant is taken from Appendix E.) Addition of the acetate salt does not change the value of the equilibrium constant. Substituting the equilibrium concentrations in this equilibrium-constant equation, we obtain:

$$\frac{(x)(0.100 + x)}{(0.100 - x)} = 1.8 \times 10^{-5}$$

We can simplify this equation by ignoring x relative to 0.100. This simplification gives us

$$\frac{x(0.100)}{(0.100)} = 1.8 \times 10^{-5}$$

$$x = 1.8 \times 10^{-5} M = [H^+]$$

$$pH = -\log(1.8 \times 10^{-5}) = 4.74$$

Earlier (Section 15.5) we calculated that in a 0.10 M solution of $HC_2H_3O_2$, $[H^+]$ is $1.3 \times 10^{-3} M$, corresponding to pH of 2.89.

SAMPLE EXERCISE 16.2

Calculate the fluoride concentration and pH of a solution containing 0.10 mol of HCl and 0.20 mol of HF in a liter of solution.

Solution: This solution consists of a strong electrolyte (HCl) and a weak electrolyte (HF) which share a common ion (H^+). Consequently the exercise is of the same general sort as Sample Exercise 16.1, except that the common ion is a cation rather than an anion.

Let's define the equilibrium concentration of F^- as x. The equilibrium concentration of $H^+(aq)$ is then the amount supplied by the strong acid, HCl (0.10 M), plus the amount formed by the dissociation of HF (x). The equilibrium process involving the weak acid and the concentrations of the species involved in that equilibrium are summarized as follows:

$$HF(aq) \rightleftharpoons H^+(aq) + F^-(aq)$$

Initial:	0.20 M	0.10 M	0
Equilibrium:	(0.20 − x)M	(0.10 + x)M	x M

The equilibrium constant, from Appendix E, is 6.8×10^{-4}. Thus we have

$$K_a = 6.8 \times 10^{-4} = \frac{(0.10 + x)(x)}{(0.20 - x)}$$

If we assume x is small relative to 0.10 or 0.20 M, this expression simplifies to give

$$\frac{(0.10)x}{(0.20)} = 6.8 \times 10^{-4}$$

$$x = \frac{0.20}{0.10}(6.8 \times 10^{-4})$$

$$= 1.4 \times 10^{-3} M = [F^-]$$

The concentration of $H^+(aq)$ comes almost entirely from the strong acid, HCl:

$$[H^+] = (0.10 + x)M = 0.10 M$$

Thus pH = 1.00.

Both Sample Exercise 16.1 and Sample Exercise 16.2 involve weak acids. However, weak bases could have been used just as easily to illustrate the common-ion effect. For example, a mixture of aqueous NH_3 and NH_4Cl have the NH_4^+ ion in common; a mixture of aqueous NH_3 and NaOH have the OH^- ion in common. In each case the pertinent equilibrium is dissociation of the weak base, NH_3:

$$NH_3(aq) + H_2O(l) \rightleftharpoons NH_4^+(aq) + OH^-(aq) \qquad [16.2]$$

COMMON IONS GENERATED BY ACID-BASE REACTIONS

Consider a solution formed by mixing a weak acid such as $HC_2H_3O_2$ and a strong base such as NaOH. The net ionic equation for the resultant proton transfer is

$$HC_2H_3O_2(aq) + OH^-(aq) \rightleftharpoons C_2H_3O_2^-(aq) + H_2O(l) \qquad [16.3]$$

The Na^+ ion is a spectator ion in this process (Section 12.5). Because this reaction is just the reverse of the base-dissociation reaction for $C_2H_3O_2^-$, the equilibrium constant is the reciprocal of K_b for $C_2H_3O_2^-$:

$$K = 1/(K_b \text{ for } C_2H_3O_2^-) = (K_a \text{ for } HC_2H_3O_2)/K_w$$
$$= \frac{1.8 \times 10^{-5}}{1.0 \times 10^{-14}} = 1.8 \times 10^9$$

The large equilibrium constant indicates that the reaction proceeds to a large extent toward products. Reactions between strong acids and strong bases, strong acids and weak bases, and weak acids and strong bases (as in the present example) proceed virtually to completion.

If the number of moles of $HC_2H_3O_2$ exceed the number of moles of NaOH, the reaction between them produces $C_2H_3O_2^-$ ions leaving the excess $HC_2H_3O_2$ unreacted. For example, consider a one-liter solution that initially contains 0.20 mol $HC_2H_3O_2$ and 0.10 mol NaOH. The reaction, Equation [16.3], consumes 0.10 mol $HC_2H_3O_2$ producing 0.10 mol $C_2H_3O_2^-$ and leaving 0.10 mol $HC_2H_3O_2$ unreacted. The resultant solution, which is 0.10 M in $C_2H_3O_2^-$ and 0.10 M in $HC_2H_3O_2$, is identical to a solution produced by mixing 0.10 mol $NaC_2H_3O_2$ and 0.10 mol $HC_2H_3O_2$ to form a liter of solution. As shown in Sample Exercise 16.1, such a solution has a pH of 4.74.

SAMPLE EXERCISE 16.3

Calculate the pH of a solution produced by mixing 0.60 L of 0.10 M NH_4Cl with 0.40 L of 0.10 M NaOH.

Solution: NH_4Cl is a strong electrolyte supplying NH_4^+ and Cl^- ions. NaOH is a strong electrolyte supplying Na^+ and OH^- ions. The molar quantities of these ions present in the solution before any reaction occurs are

Moles NH_4^+ = moles Cl^- = $M_{NH_4Cl} \times V_{NH_4Cl}$
$$= \left(0.10 \frac{mol}{L}\right)(0.60 \text{ L}) = 0.060 \text{ mol}$$

Moles Na^+ = moles OH^- = $M_{NaOH} \times V_{NaOH}$
$$= \left(0.10 \frac{mol}{L}\right)(0.40 \text{ L}) = 0.040 \text{ mol}$$

The weakly acidic NH_4^+ and the basic OH^- react as follows:

$$NH_4^+(aq) + OH^-(aq) \rightleftharpoons$$
$$NH_3(aq) + H_2O(l)$$
[16.4]

The Na^+ ion and Cl^- ion are not acidic or basic and are merely spectator ions. The reaction shown in Equation [16.4], involving a weak acid and a strong base, is virtually complete. Thus the 0.040 mol of OH^- consumes 0.040 mol of NH_4^+ producing 0.040 mol of NH_3. The remaining NH_4^+ is the original amount minus the amount consumed: 0.060 mol − 0.040 mol = 0.020 mol. The total volume of the solution is the sum of the two original solutions: 0.60 L + 0.40 L = 1.00 L. Thus the concentrations of NH_3 and NH_4^+ are

$$[NH_3] = \frac{0.040 \text{ mol}}{1.00 \text{ L}} = 0.040 \, M$$

$$[NH_4^+] = \frac{0.020 \text{ mol}}{1.00 \text{ L}} = 0.020 \, M$$

The pH of a solution containing NH_4^+ and NH_3 can be calculated by considering the common-ion effect of NH_4^+ on the dissociation equilibrium of the weakly basic NH_3. It can also be calculated by con-

sidering the effect of NH_3 on the dissociation equilibrium of the weakly acidic NH_4^+. The same result will be obtained in either case. Using the reverse reaction of that shown in Equation [16.4], the pertinent species are summarized below with x defined as the equilibrium concentration of OH^-:

$$NH_3(aq) + H_2O(l) \rightleftharpoons NH_4^+(aq) + OH^-(aq)$$

Initial:	0.040 M	0.020 M	0
Equilib-rium:	(0.040 − x)M	(0.020 + x)M	x M

Substituting the equilibrium concentrations in the equilibrium-constant expression (K_b for NH_3 is taken from Appendix E) and assuming x to be small compared with 0.020 and 0.040 gives:

$$K_b = 1.8 \times 10^{-5} = \frac{(0.020)(x)}{0.040}$$

$$x = \frac{0.040}{0.020}(1.8 \times 10^{-5}) = 3.6 \times 10^{-5} = [OH^-]$$

$$pOH = -\log(3.6 \times 10^{-5}) = 4.44$$

$$pH = 14.00 - pOH = 14.00 - 4.44 = 9.56$$

16.2 Buffer solutions

Many aqueous solutions resist a change in pH upon addition of small amounts of acid or base. Such solutions are called buffer solutions, and are said to be buffered. Human blood, for example, is a complex aqueous medium with a pH buffered at about 7.4. Any significant variation of the pH from this value results in a severe pathological response and, eventually, death. As another example, the chemical behavior of seawater is determined in very important respects by its pH, buffered at about 8.1 to 8.3 near the surface. Addition of a small amount of an acid or base to either blood or seawater does not result in a large change in pH. Compare, for example, the behavior of a liter of seawater and a liter of pure water upon addition of 0.1 mL of 1 M HCl solution (1×10^{-4} mol of HCl), Figure 16.1. The pH of pure water changes by 3 units, from 7 to 4. The pH of seawater changes by only 0.6 pH units. Substances already dissolved in seawater limit the change in $[H^+]$ on addition of HCl.

Buffers ordinarily require two species, an acidic one to react with added OH^- and a basic one to react with added H^+. It is, of course, necessary that these acidic and basic species not consume each other through a neutralization reaction. These requirements are fulfilled by an acid-base conjugate pair such as $HC_2H_3O_2$–$C_2H_3O_2^-$ or NH_4^+–NH_3. The $HC_2H_3O_2$–$C_2H_3O_2^-$ buffer mixture can be prepared by adding sodium acetate, $NaC_2H_3O_2$, to a solution of acetic acid, $HC_2H_3O_2$. The NH_4^+–NH_3 buffer mixture can be prepared by adding ammonium chloride, NH_4Cl, to a solution of ammonia, NH_3. In general, a buffer mixture consists of an aqueous solution of an acid-base conjugate pair prepared by mixing a weak acid or base with a salt of that acid or base.

To understand how a buffer works, let's consider a solution of

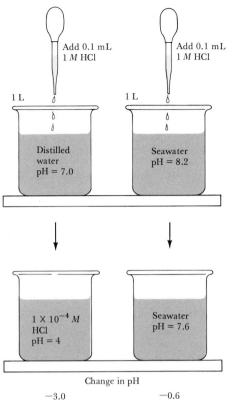

FIGURE 16.1 A comparison of the effect of added acid on the pH of distilled water as compared with the same quantity of seawater.

$HC_2H_3O_2$ and $NaC_2H_3O_2$. Ionization of $HC_2H_3O_2$ is governed by the following equilibrium reaction:

$$HC_2H_3O_2(aq) \rightleftharpoons H^+(aq) + C_2H_3O_2^-(aq) \qquad [16.5]$$

The $C_2H_3O_2^-$ in this equilibrium comes from both $HC_2H_3O_2$ and $NaC_2H_3O_2$. This mixture can either react with surplus H^+ ions or release them, according to the circumstances. For example, if a small quantity of acid is added to the solution, the equilibrium shifts to the left; acetate ion reacts with the added H^+. The solution thereby limits pH change due to added acid. On the other hand, if a small quantity of a base is added, it reacts with H^+. This reaction causes the equilibrium of Equation [16.5] to shift to the right; $HC_2H_3O_2$ dissociates to form more H^+. The solution thereby also resists change in pH due to added base.

Two important characteristics of a buffer are buffering capacity and pH. Buffering capacity is the amount of acid or base the buffer can neutralize before the pH begins to change to an appreciable degree. This capacity depends on the amount of acid and base from which the buffer is made. The pH of the buffer depends on K_a for the acid and on the relative concentrations of the acid and base that comprise the buffer.

To better understand these characteristics, let's consider a buffer mixture consisting of a weak acid HX and a corresponding salt MX, where M could be Na^+, K^+, and so forth. The acid dissociation equilibrium is

$$HX(aq) \rightleftharpoons H^+(aq) + X^-(aq) \qquad [16.6]$$

and the corresponding acid dissociation constant expression is

$$K_a = \frac{[H^+][X^-]}{[HX]}$$ [16.7]

Let us now take the logarithm in base 10 of both sides of the equation:

$$\log K_a = \log [H^+] + \log \frac{[X^-]}{[HX]}$$ [16.8]

Multiplying through on both sides by -1, we have

$$-\log K_a = -\log [H^+] - \log \frac{[X^-]}{[HX]}$$ [16.9]

Because $-\log K_a = pK_a$ and $-\log [H^+] = pH$, we have

$$pK_a = pH - \log \frac{[X^-]}{[HX]}$$ [16.10]

$$pH = pK_a + \log \frac{[X^-]}{[HX]}$$ [16.11]

In general,

$$pH = pK_a + \log \frac{[base]}{[acid]}$$ [16.12]

This relationship, which is called the Henderson-Hasselbach equation, is very useful in dealing with buffers. Notice that $pH = pK_a$ when the concentration of the acid and its conjugate base are equal. The pH range of most buffers is limited to the vicinity of the pK_a of the acid. For this reason, one usually tries to select a buffer whose acid has a pK_a close to the desired pH.

Although the pH of a particular buffer mixture depends on the relative concentrations of the acid and its conjugate base, its buffering capacity depends on the amounts of the acid and base. A 1-L solution that is $1 M$ in $HC_2H_3O_2$ and $1 M$ in $NaC_2H_3O_2$ will have the same pH as a 1-L solution that is $0.1 M$ in $HC_2H_3O_2$ and $0.1 M$ in $NaC_2H_3O_2$. However, the first solution has a greater buffering capacity because it contains more $HC_2H_3O_2$ and $C_2H_3O_2^-$.

SAMPLE EXERCISE 16.4

What is the pH of a buffer mixture composed of equal concentrations of NH_4Cl and NH_3?

Solution: Using Equation [16.12], we have

$$pH = pK_a + \log \frac{[NH_3]}{[NH_4^+]}$$

Because $[NH_3] = [NH_4^+]$,

$$\frac{[NH_3]}{[NH_4^+]} = 1 \quad \text{and} \quad \log \frac{[NH_3]}{[NH_4^+]} = 0$$

Thus

$$pH = pK_a + 0$$

The pK_a for NH_4^+ is related to K_b for NH_3 through the relationship $K_a \times K_b = K_w$ (Section 15.7). From Appendix E we have $K_b = 1.8 \times 10^{-5}$.

$$K_a = \frac{K_w}{K_b} = \frac{1.0 \times 10^{-14}}{1.8 \times 10^{-5}} = 5.6 \times 10^{-10}$$

Thus

$$pH = pK_a = -\log(5.6 \times 10^{-10}) = 9.25$$

SAMPLE EXERCISE 16.5

(a) How is the pH of an aqueous solution of NH_3 affected by the addition of NH_4Cl? (b) What concentration of NH_4Cl must be added to a 0.10 M solution of NH_3 to adjust the pH to 9.00?

Solution: (a) NH_4Cl is a strong electrolyte supplying NH_4^+ and Cl^- ions. NH_3 is a weak base that reacts with water as follows:

$$NH_3(aq) + H_2O(l) \rightleftharpoons NH_4^+(aq) + OH^-(aq)$$

The presence of NH_4^+, which is a weak acid, shifts the ionization equilibrium to the left, lowering $[OH^-]$ and thereby lowering pH. We have here simply another example of the common-ion effect; aqueous solutions of NH_3 and NH_4Cl have the NH_4^+ ion in common.

(b) A pH of 9.00 corresponds to $[H^+] = 1.0 \times 10^{-9}$. As shown in Sample Exercise 16.4, the pK_a for NH_4^+ is 9.25. The NH_4^+ and NH_3 are a conjugate acid-base pair. Thus we have

$$pH = pK_a + \log \frac{[base]}{[acid]}$$

$$9.00 = 9.25 + \log \frac{[NH_3]}{[NH_4^+]}$$

$$\log \frac{[NH_3]}{[NH_4^+]} = 9.00 - 9.25 = -0.25$$

$$\frac{[NH_3]}{[NH_4^+]} = 10^{-0.25} = 0.56$$

To calculate NH_4^+ from this ratio, we can assume that the concentration of NH_3 is 0.10 M. From the examples worked out in Chapter 15 (see, for example, Sample Exercise 15.11), we know that only a small fraction of the NH_3 reacts with water; by Le Chatelier's principle, addition of NH_4^+ to the solution will reduce the extent of reaction even more. Thus, all the NH_4^+ present can be considered to be that which is added. Therefore, we have

$$[NH_4^+] = \frac{[NH_3]}{0.56} = \frac{0.10\ M}{0.56} = 0.18\ M$$

ADDITION OF ACIDS OR BASES TO BUFFERS

To further illustrate the buffer action of a weak acid and its conjugate base, we should compare its behavior with the action of a solution that is not a buffer. We saw in Sample Exercise 16.1 that the pH of a solution that is 0.100 M in acetic acid and 0.100 M in sodium acetate is 4.74. A solution of this same pH is obtained by addition of 1.8×10^{-5} mol of HCl to a liter of water. Because HCl is a strong acid, $1.8 \times 10^{-5}\ M$ HCl has a $H^+(aq)$ concentration of $1.8 \times 10^{-5}\ M$ and thus a pH of 4.74. Now suppose that 1.0 mL of 10 M HCl solution, that is 0.010 mol of HCl, is added to a liter of each of these two solutions of pH 4.74, as shown in Figure 16.2.

The HCl added to the $HC_2H_3O_2$–$C_2H_3O_2^-$ buffer solution reacts with the $C_2H_3O_2^-$ ion:

$$H^+(aq) + C_2H_3O_2^-(aq) \rightleftharpoons HC_2H_3O_2(aq)$$

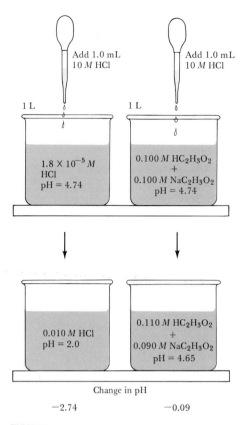

FIGURE 16.2 Comparison of the effect of added acid on a buffer solution of pH 4.74 as compared with an HCl solution of pH 4.74.

We have noted (Section 16.1) that the equilibrium constants for reactions between strong acids and weak bases are large. Consequently, such reactions proceed virtually to completion. Thus we can assume that all the added 0.010 mol of HCl reacts with 0.010 mol of $C_2H_3O_2^-$ to form 0.010 mol of $HC_2H_3O_2$. Of course, 0.010 mol of NaCl also forms, but it is neither acidic nor basic. As shown in Figure 16.2 addition of the HCl simply changes the relative amounts of acetic acid and acetate ion we have to deal with in setting up the problem: Before the addition of HCl the solution was 0.100 M each in $HC_2H_3O_2$ and $C_2H_3O_2^-$. After addition, the solution is 0.110 M in $HC_2H_3O_2$ and 0.090 M in $C_2H_3O_2^-$. Using the method outlined in Sample Exercise 16.3, or using Equation [16.12], the pH is calculated to be 4.65, about 0.09 units lower than before addition (Figure 16.2). Clearly, this is a much smaller change than occurs on addition of acid to the dilute HCl solution. In that case, the pH drops to 2.0.

We might have chosen to add 0.01 mol of hydroxide to the solutions illustrated in Figure 16.2. This addition would have caused the pH of the dilute HCl solution to go from 4.74 to essentially 12 (you should be able to explain why this is so), whereas the pH of the buffer solution would have increased by only about 0.09 pH units. Thus, a buffer solution responds to approximately the same degree, but in opposite direction, to acid or base addition.

SAMPLE EXERCISE 16.6

A liter of solution containing 0.100 mol of $HC_2H_3O_2$ and 0.100 mol of $NaC_2H_3O_2$ was prepared to provide a buffer of pH 4.74. Calculate the pH of this solution after 0.020 mol of NaOH are added. (Assume the total volume is still 1 L.)

Solution: The OH^- provided by the strong base NaOH reacts nearly completely with the weak acid $HC_2H_3O_2$:

$$HC_2H_3O_2(aq) + OH^-(aq) \rightleftharpoons$$
$$H_2O(l) + C_2H_3O_2^-(aq)$$

The 0.020 mol of OH^- therefore react with 0.020 mol of $HC_2H_3O_2$, producing 0.020 mol of $C_2H_3O_2^-$. Thus the final solution contains 0.080 mol of $HC_2H_3O_2$ (the original amount, 0.100 mol, minus that which reacts, 0.020 mol), and 0.120 mol of $C_2H_3O_2^-$ (the original 0.100 mol plus that formed in the reaction, 0.020 mol). Using Equation [16.12], we have

$$pH = pK_a + \log \frac{[C_2H_3O_2^-]}{[HC_2H_3O_2]}$$
$$= 4.74 + \log \frac{0.120}{0.080} = 4.74 + 0.18 = 4.92$$

Blood is an important example of a buffered solution. Human blood is slightly basic with a pH of about 7.39 to 7.45. In a healthy person the pH never departs more than perhaps 0.2 pH units from the average value. Whenever pH falls below about 7.4 the condition is called *acidosis;* when pH rises above 7.4, the condition is called *alkalosis.* Death normally results if the pH falls below 7.0 or rises above 7.8. Acidosis is the more common tendency, because ordinary metabolism produces several acids.

The body uses three primary methods to control blood pH: (1) The blood contains several buffers, including H_2CO_3–HCO_3^- and $H_2PO_4^-$–HPO_4^{2-} pairs, and hemoglobin-containing conjugate acid-base pairs. (2) The kidneys serve to absorb or release $H^+(aq)$. The pH of urine is normally about 5.0 to 7.0. Acidosis is accompanied by increased loss of body fluids as the kidneys work to reduce $H^+(aq)$. (3) The concentration of $H^+(aq)$ is also altered by the rate at which CO_2 is removed from the lungs. The pertinent equilibria are

$$H^+(aq) + HCO_3^-(aq) \rightleftharpoons H_2CO_3(aq) \rightleftharpoons$$
$$H_2O(l) + CO_2(g)$$

Removal of CO_2 shifts these equilibria to the right, thereby reducing $H^+(aq)$.

Acidosis or alkalosis disrupts the mechanism by which hemoglobin transports oxygen in blood. Hemoglobin (Hb) is involved in a series of equilibria whose overall result is approximately

$$HbH^+(aq) + O_2(aq) \rightleftharpoons HbO_2(aq) + H^+(aq)$$

In acidosis, this equilibrium is shifted to the left and the ability of hemoglobin to form oxyhemoglobin, HbO_2, is decreased. The lesser amount of O_2 thereby available to cells in the body causes fatigue and headaches; if great enough, it also triggers air hunger (the feeling of being "out of breath" that causes deep breathing).

Temporary acidosis occurs during strenuous exercise, when energy demands exceed the oxygen available for complete oxidation of glucose to CO_2. In this case the glucose is converted to an acidic metabolism product, lactic acid, $CH_3CHOHCOOH$. Acidosis also occurs when glucose is unavailable to the cells. This situation can arise, for example, during starvation or as a result of diabetes. In the case of diabetes, glucose is unable to enter the cells because of inadequate insulin, the substance responsible for passage of glucose from the bloodstream to the interior of cells. When glucose is unavailable, the body relies for energy on stored fats, which produce acidic metabolism products.

16.3 Titration curves

Many acid-base reactions are used in chemical analyses. For example, the carbonate content in a sample can be determined by a procedure that involves titration with a strong acid such as HCl. Titrations were described briefly in Chapter 3 (Section 3.11). In that earlier discussion we noted that acid-base indicators can be used to signal the equivalence point (that is, the point at which stoichiometrically equivalent quantities of acid and base have been brought together). But given the variety of indicators, changing colors at different pH's, which indicator is best for a particular titration? This question can be answered by examining a graph of pH changes during a titration. A graph of pH as a function of the volume of added titrant is called a titration curve.

STRONG ACID-STRONG BASE

The titration curve produced when a strong base is added to a strong acid has the general shape shown in Figure 16.3. This curve depicts the pH change that occurs as 0.100 M NaOH is added to 50.0 mL of 0.100 M HCl. The pH can be calculated at various stages in the course of the titration. The pH starts out low; pH = 1.00 for 0.100 M HCl. As NaOH is added, the pH increases slowly at first and then rapidly in the vicinity of the equivalence point. The pH at the equivalence point in any acid-base titration is the pH of the resultant salt solution. In this example, NaCl is formed, which does not hydrolyze (Section 15.8); the equilibrium point therefore occurs at pH 7. Because the pH change is very large near the equivalence point, the indicator for the titration need not change color precisely at 7.0. Most strong acid-strong base titrations are carried out using phenolphthalein as an indicator because its color change is dramatic. From Table 15.1 we see that this indicator changes color in the pH range 8.3 to 10. Thus a slight excess of NaOH must be present to cause the observed color change. However, it requires such a tiny excess of base to make the color change occur that no serious error is introduced. Similarly, methyl red, which changes color in the slightly acid range, could also be used. The pH intervals of color change for these two indicators are shown in Figure 16.3.

Titration of a solution of a strong base with a solution of a strong acid would yield an entirely analogous curve of pH versus added acid. In this case, however, the pH would be high at the outset of the titration, and low at its completion.

SAMPLE EXERCISE 16.7

Calculate the pH of the titration solution when 49.9 mL of 0.100 M NaOH solution have been added.

Solution: The number of moles of OH⁻ in 49.9 mL 0.100 M NaOH is

$$0.0499 \text{ L soln} \left(\frac{0.100 \text{ mol OH}^-}{1 \text{ L soln}} \right)$$
$$= 4.99 \times 10^{-3} \text{ mol OH}^-$$

Thus there remains

$$(5.00 \times 10^{-3}) - (4.99 \times 10^{-3})$$
$$= 1.0 \times 10^{-5} \text{ mol H}^+(aq)$$

in 0.0999 L solution. The concentration of $H^+(aq)$ is thus

$$\frac{1.0 \times 10^{-5} \text{ mol}}{0.0999 \text{ L}} = 1.0 \times 10^{-4} M$$

The corresponding pH is 4.0.

TITRATIONS INVOLVING A WEAK ACID OR BASE

Titration of a weak acid by a strong base results in pH curves that look similar to those for strong acid-strong base titrations. Figure 16.4 shows the pH curve produced when 0.100 M NaOH is added to 50.0 mL of 0.100 M acetic acid. There are three noteworthy differences between this curve and that for a strong acid-strong base titration: (1) The weak acid

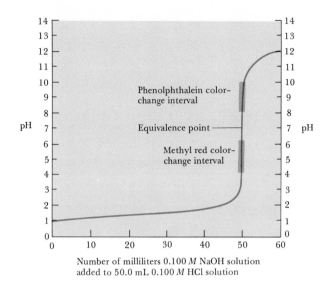

FIGURE 16.3 The pH curve for titration of a solution of a strong acid with a solution of a strong base, in this case HCl and NaOH.

Number of milliliters 0.100 M NaOH solution added to 50.0 mL 0.100 M HCl solution

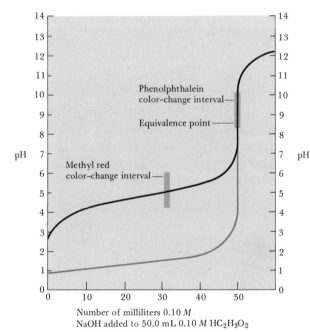

Number of milliliters 0.10 M NaOH added to 50.0 mL 0.10 M HC₂H₃O₂

FIGURE 16.4 The black line shows the graph of pH versus added 0.10 M NaOH solution in the titration of 0.10 M acetic acid solution. The color line segment shows the graph of pH versus added base for the titration of 0.10 M HCl.

will have a higher initial pH. The pH of $0.100 \ M \ HC_2H_3O_2$ is 2.89 (see Section 15.5 where [H⁺] is calculated, $[H^+] = 1.3 \times 10^{-3} \ M$). (2) The pH will rise more rapidly in the early part of the titration, but more slowly near the equivalence point. The weaker the acid, the less marked the pH change near the equivalence point (see Figure 16.5). (3) The pH at the equivalence point is *not* 7. In the present example the solution contains $0.050 \ M \ NaC_2H_3O_2$ at the equivalence point (remember that the total volume of solution is doubled by addition of the NaOH solution to the $HC_2H_3O_2$ solution). The pH of this solution is 8.71 (this might be a good time to review how the pH of a $NaC_2H_3O_2$ solution can be calculated; an example of a similar calculation is given in Section 15.6, Sample Exercise 15.12).

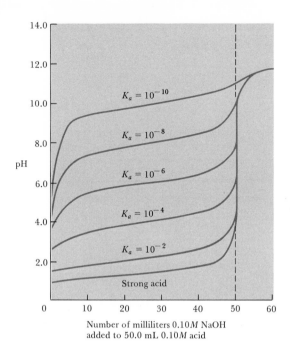

FIGURE 16.5 Influence of acid strength upon titration curves. Each curve represents titration of 50.0 mL of 0.10 M acid with 0.10 M NaOH.

SAMPLE EXERCISE 16.8

Calculate the pH in the titration of acetic acid by sodium hydroxide after 30.0 mL of 0.100 M NaOH solution have been added to 50.0 mL of 0.100 M acetic acid solution.

Solution: The total number of moles of $HC_2H_3O_2$ originally in solution is

$$\left(\frac{0.100 \text{ mol } HC_2H_3O_2}{1 \text{ L soln}}\right)(0.0500 \text{ L soln})$$

$$= 5.00 \times 10^{-3} \text{ mol } HC_2H_3O_2$$

Similarly, 30.0 mL of 0.100 M NaOH solution contains 3.00×10^{-3} mol of OH^-. During the titration, this OH^- reacts with the acetic acid, forming 3.00×10^{-3} mol of $C_2H_3O_2^-$ and leaving 2.00×10^{-3} mol of $HC_2H_3O_2$, in 80 mL solution:

$$OH^-(aq) + HC_2H_3O_2(aq) \longrightarrow$$
$$H_2O(l) + C_2H_3O_2^-(aq)$$

The resulting molarities are thus

$$[HC_2H_3O_2] = \frac{2.00 \times 10^{-3} \text{ mol } HC_2H_3O_2}{0.0800 \text{ L}}$$

$$= 0.0250 \ M$$

$$[C_2H_3O_2^-] = \frac{3.00 \times 10^{-3} \text{ mol } C_2H_3O_2^-}{0.0800 \text{ L}}$$

$$= 0.0375 \ M$$

The value for pK_a of acetic acid is $-\log(1.8 \times 10^{-5}) = 4.74$. Inserting these quantities into Equation [16.12], we find

$$pH = 4.74 + \log\frac{(0.0375)}{(0.0250)}$$

$$= 4.91$$

By proceeding in similar fashion other points on the titration curve shown in Figure 16.4 could be calculated. Incidentally, you might note that at the halfway point in the titration, when $[C_2H_3O_2^-]$ equals $[HC_2H_3O_2]$, the pH equals pK_a, which in this case is 4.74.

The fact that the pH at the equivalence point in the titration of the weak acid is considerably higher than it is in the titration of the strong acid is important. We saw earlier that in titrating 0.10 M HCl with 0.10 M NaOH, either phenolphthalein or methyl red could be used as indicator. Although the pH of color change does not correspond precisely to the equivalence point for either indicator, both were close

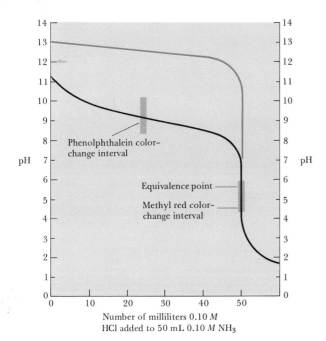

FIGURE 16.6 The black line shows the graph of pH versus added 0.10 M HCl solution in the titration of 0.10 M ammonia. The color line segment shows the graph of pH versus added acid for the titration of 0.10 M NaOH.

Phenolphthalein color–change interval

Equivalence point

Methyl red color–change interval

pH

Number of milliliters 0.10 M
HCl added to 50 mL 0.10 M NH$_3$

enough that no significant error would be introduced by using either of them (Figure 16.3). In a titration of acetic acid with NaOH, phenolphthalein is an ideal indicator, because it changes color just at the pH of the equivalence point. However, methyl red is not a good choice. The pH at the midrange of its color change is 5.2. At this pH we are still far short of the equivalence point as shown in Figure 16.4.

If the acid to be titrated were weaker than acetic acid, the titration curve would look similar to that shown in Figure 16.4. However, the pH at the beginning of the reaction would be higher, and the pH at the equivalence point would also be higher. This means that the pH change at the equivalence point would not be as sharp, and the end point would be more difficult to detect (see Figure 16.5). With very weak acids, for which K_a is less than about 1×10^{-8}, an acid-base titration is not really a good quantitative procedure when using indicators.

Titration of a weak base, for example 0.10 M NH$_3$, with a strong acid such as 0.10 M HCl solution, leads to the titration curve shown in Figure 16.6. In this particular example, the equivalence point occurs at pH 5.3. In this case methyl red would be an ideal indicator, but phenolphthalein would be a poor choice.

In the case of weak acids containing more than one ionizable proton, reaction with OH$^-$ will occur in a series of steps. An important example that we will encounter again in the next chapter is carbonic acid, H$_2$CO$_3$. The neutralization of H$_2$CO$_3$ proceeds in two stages:

$$H_2CO_3(aq) + OH^-(aq) \rightleftharpoons H_2O(l) + HCO_3^-(aq)$$
$$HCO_3^-(aq) + OH^-(aq) \rightleftharpoons H_2O(l) + CO_3^{2-}(aq)$$

Figure 16.7 shows how the relative abundances of H$_2$CO$_3$, HCO$_3^-$, and CO$_3^{2-}$ vary as a function of the solution pH. As OH$^-$ is added to H$_2$CO$_3$ (as pH increases), [H$_2$CO$_3$] decreases while [HCO$_3^-$] increases. By pH 8.5 H$_2$CO$_3$ has been nearly completely converted to HCO$_3^-$. At this stage Figure 16.7 shows [H$_2$CO$_3$] = 0, and [CO$_3^{2-}$] = 0; the concentrations of these species are too small even to register on the graph. As pH increases beyond 8.5, [HCO$_3^-$] de-

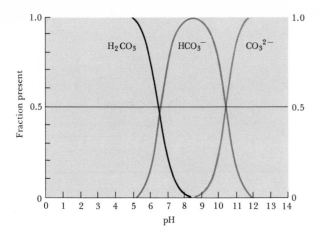

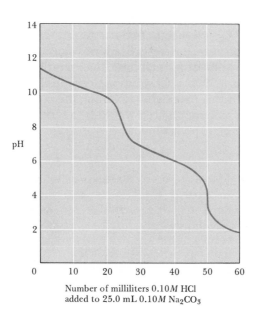

FIGURE 16.7 Relative abundances of the various species derived from dissolved CO_2 as a function of pH.

creases while $[CO_3^{2-}]$ increases. Thus it is only after the first reaction goes nearly to completion that the second reaction starts.

When the neutralization steps of a polyprotic acid or a polybasic base are sufficiently separated, the substance will exhibit a pH titration curve with multiple equivalence points. Figure 16.8 shows the titration curve for CO_3^{2-}. This ion is neutralized in two steps, each exhibiting a distinct equivalence point on the titration curve.

16.4 Solubility equilibria

The equilibria that we have considered thus far in this chapter have involved acids and bases. Furthermore, they have been homogeneous; that is, all the species involved in an equilibrium have been in the same phase. In this section we will consider the equilibria involved in another important type of solution reaction, the dissolution or precipitation of slightly soluble salts. Such reactions are heterogeneous.

For an equilibrium to exist between a solid substance and its solution, the solution must be saturated and in contact with undissolved solid. As an example, consider a saturated solution of $BaSO_4$ that is in contact

Number of milliliters 0.10M HCl
added to 25.0 mL 0.10M Na$_2$CO$_3$

FIGURE 16.8 Titration curve for the titration of 25.0 mL of 0.10 M Na$_2$CO$_3$ with 0.10 M HCl.

with solid $BaSO_4$. The chemical equation for the relevant equilibrium can be written as

$$BaSO_4(s) \rightleftharpoons Ba^{2+}(aq) + SO_4{}^{2-}(aq) \qquad [16.13]$$

The solid is an ionic compound (as are others that we will discuss in this section). Such compounds are almost invariably strong electrolytes; to the extent that they dissolve, these compounds are present in solutions as ions. We observed in Section 14.3 that for heterogeneous reactions the concentration of the solid is constant. It is thus incorporated into the equilibrium constant. Thus the equilibrium-constant expression for the dissolution of $BaSO_4$ can be written as

$$K = [Ba^{2+}][SO_4{}^{2-}]$$

When concentration is expressed in molarity, the equilibrium constant is called the solubility-product constant and designated as K_{sp}:

$$K_{sp} = [Ba^{2+}][SO_4{}^{2-}] \qquad [16.14]$$

The rules for writing the solubility-product expression are the same as those for the writing of any equilibrium-constant expression: *The solubility product is equal to the product of the concentrations of the ions involved in the equilibrium, each raised to the power of its coefficient in the equilibrium equation.*

SAMPLE EXERCISE 16.9

Write the expression for the solubility-product constant for $Ca_3(PO_4)_2$.

Solution: We first write the equation for the solubility equilibrium:

$$Ca_3(PO_4)_2(s) \rightleftharpoons 3Ca^{2+}(aq) + 2PO_4{}^{3-}(aq)$$

Following the rule stated above, the power to which the Ca^{2+} concentration is raised is 3, the power to which the $PO_4{}^{3-}$ concentration is raised is 2. The resulting expression for K_{sp} is

$$K_{sp} = [Ca^{2+}]^3[PO_4{}^{3-}]^2$$

It is important to distinguish carefully between solubility and solubility product. The solubility is the quantity of substance that dissolves in a given quantity of water. It is often expressed as grams of solute per 100 g of water, or in terms of molarity. The solubility of a substance can be changed by adding some other substance to the solution. For example, the addition of Na_2SO_4 lowers the solubility of $BaSO_4$; the $SO_4{}^{2-}$ supplied by the Na_2SO_4 shifts the solubility equilibrium, Equation [16.13], to the left. This behavior is another example of the common-ion effect. However, the presence of Na_2SO_4 does not affect K_{sp} for $BaSO_4$.* K_{sp} is a true equilibrium constant; its value at a particular temperature is the same, regardless of whether other substances are present in solution. Nevertheless, K_{sp} and solubility are related and one can be calculated from the other.

*This is strictly true only for very dilute solutions. As we will see in the next chapter, the values of equilibrium constants are somewhat altered when the total concentration of ionic substances in water is increased. However, we shall ignore these effects, which are taken into consideration only for very accurate work.

SAMPLE EXERCISE 16.10

Solid barium sulfate is shaken in contact with pure water at 25°C for several days. Each day a sample is withdrawn and analyzed for its barium concentration. After several days, the value of $[Ba^{2+}]$ is constant, indicating that equilibrium has been reached. The concentration of Ba^{2+} is $1.04 \times 10^{-5} M$. What is K_{sp} for $BaSO_4$?

Solution: The analysis provides values for both $[Ba^{2+}]$ and $[SO_4^{2-}]$. Because all of the Ba^{2+} and SO_4^{2-} ions come from $BaSO_4$, there must be one Ba^{2+} for each SO_4^{2-}. Consequently, $[Ba^{2+}] = [SO_4^{2-}] = 1.04 \times 10^{-5} M$. Thus we have

$$K_{sp} = [Ba^{2+}][SO_4^{2-}]$$
$$= (1.04 \times 10^{-5})(1.04 \times 10^{-5})$$
$$= 1.08 \times 10^{-10}$$

SAMPLE EXERCISE 16.11

The K_{sp} for CaF_2 is 3.9×10^{-11}. What is the solubility of CaF_2 in water in grams per liter?

Solution: The solubility equilibrium involved is

$$CaF_2(s) \rightleftharpoons Ca^{2+}(aq) + 2F^-(aq)$$

For each mole of CaF_2 that dissolves, 1 mol of Ca^{2+} and 2 mol of F^- enter the solution. Thus if we let x equal the solubility of CaF_2 in moles per liter, the molar concentrations of Ca^{2+} and F^- will be $[Ca^{2+}] = x$ and $[F^-] = 2x$.

The K_{sp} expression for this reaction is

$$K_{sp} = [Ca^{2+}][F^-]^2 = 3.9 \times 10^{-11}$$

Substituting $[Ca^{2+}] = x$ and $[F^-] = 2x$ and solving for x we have

$$(x)(2x)^2 = 3.9 \times 10^{-11}$$
$$4x^3 = 3.9 \times 10^{-11}$$
$$x = 2.1 \times 10^{-4} M$$

Thus the molar solubility of CaF_2 is 2.1×10^{-4} mol/L. The mass of CaF_2 that dissolves in a liter of solution is

$$\left(\frac{2.1 \times 10^{-4} \text{ mol } CaF_2}{1 \text{ L soln}}\right)\left(\frac{78.1 \text{ g } CaF_2}{1 \text{ mol } CaF_2}\right)$$
$$= 1.6 \times 10^{-2} \text{ g } CaF_2/\text{L soln}$$

Appendix E contains K_{sp} values for a variety of salts at 25°C. From these equilibrium constants we can calculate solubilities under a variety of conditions. If a single solute is dissolved in water, the relative concentrations of the ions are determined by the formula of the solute. In the case of CaF_2, the concentration of F^- must be twice the concentration of Ca^{2+}.

$$CaF_2(s) \rightleftharpoons Ca^{2+}(aq) + 2F^-(aq) \qquad [16.15]$$

However, it is possible to alter the relative concentrations of Ca^{2+} and F^- by adding a soluble salt containing one of these ions. For example, we might add either $Ca(NO_3)_2$ or NaF. Either will lower the solubility of the CaF_2. If $Ca(NO_3)_2$ is added, the equilibrium concentration of Ca^{2+} is the sum of the Ca^{2+} concentrations from the two sources.

SAMPLE EXERCISE 16.12

What is the molar solubility of CaF_2 in a solution containing $0.010 M$ NaF?

Solution: As in Sample Exercise 16.11, we have

$$K_{sp} = [Ca^{2+}][F^-]^2 = 3.9 \times 10^{-11}$$

The value of K_{sp} is unchanged by the fact that the solution initially contains $0.010 M$ NaF. In Sample

Exercise 16.11, the relative concentrations of Ca^{2+} and F^- were determined entirely by the solubility of CaF_2. In the present exercise we must take account of the fact that there is a second source of F^-. Let's again let the molar solubility of CaF_2 be x. Then the concentration of Ca^{2+} is x and the concentration of F^- derived from CaF_2 is $2x$. But there is in addition a $0.010 M$ contribution to the F^- concentration from the dissolved NaF. We thus have

$[Ca^{2+}] = x$ $[F^-] = 0.010 + 2x$

$K_{sp} = [Ca^{2+}][F^-]^2 = [x][0.010 + 2x]^2$

$\qquad = 3.9 \times 10^{-11}$

This would be a messy problem to solve exactly, but fortunately it is possible to greatly simplify matters. Even without the common-ion effect of 0.010 M F$^-$, the solubility of CaF$_2$ in water is not very great, as illustrated by the small value of x obtained in Sample Exercise 16.11. We know from application of LeChatelier's principle that the solubility of CaF$_2$ will be even smaller in the presence of 0.010 M NaF.

We can therefore safely assume that the 0.010 M F$^-$ concentration from the NaF is much greater than the small additional contribution resulting from the solubility of CaF$_2$. That is, we can neglect $2x$ in comparison with 0.010. We then have

$3.9 \times 10^{-11} = (x)(0.010)^2$

$\qquad x = 3.9 \times 10^{-7} M = [Ca^{2+}]$

This value for x represents the solubility of CaF$_2$ in a solution that is 0.010 M in NaF. Note that it is much smaller than 2.1×10^{-4} M, the solubility of CaF$_2$ in pure water.

16.5 Criteria for precipitation or dissolution

Equilibrium can be achieved starting with the substances on either side of the chemical equation. The equilibrium between BaSO$_4(s)$, Ba$^{2+}(aq)$, and SO$_4{}^{2-}(aq)$, Equation [16.13], can be achieved starting with solid BaSO$_4$. It can also be achieved starting with soluble salts containing Ba^{2+} and SO$_4{}^{2-}$, say BaCl$_2$ and Na$_2$SO$_4$. When these two solutions are mixed, a precipitate of BaSO$_4$ will form if the product of ion concentrations $Q = [Ba^{2+}][SO_4{}^{2-}]$ is greater than K_{sp}.*

The possible relationships between Q and K_{sp} are summarized as follows:

if $Q > K_{sp}$ precipitation occurs until $Q = K_{sp}$

if $Q = K_{sp}$ equilibrium exists (saturated solution)

if $Q < K_{sp}$ solid dissolves until $Q = K_{sp}$

*The use of the reaction quotient, Q, to determine the direction a reaction must proceed to reach equilibrium was discussed earlier, in Section 14.4. In the present case the equilibrium expression contains no denominator and therefore is really not a quotient. Thus Q is often referred to simply as the ion product.

SAMPLE EXERCISE 16.13

Will a precipitate form when 0.100 L of $3.0 \times 10^{-3} M$ Pb(NO$_3$)$_2$ is added to 0.400 L of $5.0 \times 10^{-3} M$ Na$_2$SO$_4$?

Solution: The possible reaction products are PbSO$_4$ and NaNO$_3$. Sodium salts are quite soluble; however, PbSO$_4$ has a K_{sp} of 1.6×10^{-8} (Appendix E). To determine whether the PbSO$_4$ precipitates we must calculate $Q = [Pb^{2+}][SO_4{}^{2-}]$, and compare it with K_{sp}.

When the two solutions are mixed, the total volume becomes 0.100 L + 0.400 L = 0.500 L. The number of moles of Pb^{2+} in 0.100 L of $3.0 \times 10^{-3} M$ Pb(NO$_3$)$_2$ is

$(0.100 \text{ L})\left(3.0 \times 10^{-3} \dfrac{\text{mol}}{\text{L}}\right) = 3.0 \times 10^{-4} \text{ mol}$

The concentration of Pb^{2+} in the 0.500-L mixture is therefore

$[Pb^{2+}] = \dfrac{3.0 \times 10^{-4} \text{ mol}}{0.500 \text{ L}} = 6.0 \times 10^{-4} M$

The number of moles of SO$_4{}^{2-}$ is

$(0.400 \text{ L})\left(5.0 \times 10^{-3} \dfrac{\text{mol}}{\text{L}}\right) = 2.0 \times 10^{-3} \text{ mol}$

Therefore $[SO_4{}^{2-}]$ in the 0.500-L mixture is

$[SO_4{}^{2-}] = \dfrac{2.0 \times 10^{-3} \text{ mol}}{0.500 \text{ L}} = 4.0 \times 10^{-3} M$

We then have

$Q = [Pb^{2+}][SO_4{}^{2-}] = (6.0 \times 10^{-4})(4.0 \times 10^{-3})$
$\qquad = 2.4 \times 10^{-6}$

Because $Q > K_{sp}$, precipitation of PbSO$_4$ will occur.

SAMPLE EXERCISE 16.14

What concentration of OH^- must be exceeded in a 0.010 M solution of $Ni(NO_3)_2$ in order to precipitate $Ni(OH)_2$? (Assume that the added OH^- does not change the concentration of Ni^{2+}.)

Solution: From Appendix E we have $K_{sp} = 1.6 \times 10^{-14}$ for $Ni(OH)_2$. Any OH^- in excess of that in a saturated solution of $Ni(OH)_2$ will cause some $Ni(OH)_2$ to precipitate. For a saturated solution we have

$$K_{sp} = [Ni^{2+}][OH^-]^2 = 1.6 \times 10^{-14}$$

Thus if $Q = [Ni^{2+}][OH^-]^2 > 1.6 \times 10^{-14}$ precipitation will occur. Letting $[OH^-] = x$ and using $[Ni^{2+}] = 0.010 \, M$ we have:

$$(0.010)x^2 > 1.6 \times 10^{-14}$$

$$x^2 > \frac{1.6 \times 10^{-14}}{0.010} = 1.6 \times 10^{-12}$$

$$x > \sqrt{1.6 \times 10^{-12}} = 1.3 \times 10^{-6} \, M$$

Thus any OH^- above $1.3 \times 10^{-6} \, M$ will cause precipitation of $Ni(OH)_2$. This concentration of OH^- corresponds to a solution pH of 8.11 (pH = 14 − pOH = 14 − 5.89 = 8.11).

SOLUBILITY AND pH

The solubility of any substance whose anion is basic will be affected to some extent by the pH of the solution. For example, consider $Mg(OH)_2$, for which the solubility equilibrium is

$$Mg(OH)_2(s) \rightleftharpoons Mg^{2+}(aq) + 2OH^-(aq) \qquad [16.16]$$

The value of K_{sp} for $Mg(OH)_2$ is 1.8×10^{-11}. Suppose that solid $Mg(OH)_2$ is equilibrated with a solution buffered at a pH of 9.0. Then pOH is 5.0, that is, $[OH^-] = 1.0 \times 10^{-5}$. Inserting this value for $[OH^-]$ into the solubility-product expression, we have

$$K_{sp} = [Mg^{2+}][OH^-]^2 = 1.8 \times 10^{-11}$$
$$[Mg^{2+}][1.0 \times 10^{-5}]^2 = 1.8 \times 10^{-11}$$
$$[Mg^{2+}] = 0.18 \, M$$

Thus, $Mg(OH)_2$ is quite soluble in a buffered, slightly basic medium. If the solution were made more acidic, the solubility of $Mg(OH)_2$ would increase, because the OH^- concentration decreases with increasing acidity. The Mg^{2+} concentration would thus increase to maintain the equilibrium condition.

The solubility of almost any salt is affected if the solution is made sufficiently acidic or basic. The effects are very noticeable, however, only when one or both ions involved is moderately acidic or basic. The metal hydroxides we've just discussed are good examples of compounds involving a strong base, the hydroxide ion. As an additional example, the fluoride ion of CaF_2 is a weak base; it is the conjugate base of the weak acid HF. As a result, CaF_2 is more soluble in acidic solutions than in neutral or basic ones, because of the reaction of F^- with H^+ to form HF. The solution process can be considered as two consecutive reactions:

$$CaF_2(s) \rightleftharpoons Ca^{2+}(aq) + 2F^-(aq) \qquad [16.17]$$

$$F^-(aq) + H^+(aq) \rightleftharpoons HF(aq) \qquad [16.18]$$

The equation for the overall process is

$$CaF_2(s) + 2H^+(aq) \rightleftharpoons Ca^{2+}(aq) + 2HF(aq) \qquad \text{[16.19]}$$

Qualitatively we can understand what occurs in terms of LeChatelier's principle: The solubility equilibrium is driven to the right because the free F^- concentration is reduced by reaction with H^+. The reduction of $[F^-]$ causes Q to be reduced so that it becomes smaller than K_{sp}. Thus more CaF_2 dissolves.

These examples illustrate a general rule: *The solubility of slightly soluble salts containing basic anions increases as $[H^+]$ increases (as pH is lowered).* Salts with anions that do not hydrolyze (the anions of strong acids) are largely unaffected by pH.

SAMPLE EXERCISE 16.15

Which of the following substances will be more soluble in acidic solution than in basic solution? (a) $Ni(OH)_2(s)$; (b) $CaCO_3(s)$; (c) $BaSO_4(s)$; (d) $AgCl(s)$.

Solution: (a) $Ni(OH)_2(s)$ will be more soluble in acidic solution, because of reaction of H^+ with the OH^- ion, forming water:

$$Ni(OH)_2(s) \rightleftharpoons Ni^{2+}(aq) + 2OH^-(aq)$$
$$2OH^-(aq) + 2H^+(aq) \rightleftharpoons 2H_2O(l)$$

Overall: $$\overline{Ni(OH)_2(s) + 2H^+(aq) \rightleftharpoons} $$
$$Ni^{2+}(aq) + 2H_2O(l)$$

(b) Similarly, $CaCO_3(s)$ reacts with acid, liberating gaseous CO_2:

$$CaCO_3(s) \rightleftharpoons Ca^{2+}(aq) + CO_3^{2-}(aq)$$
$$CO_3^{2-}(aq) + 2H^+(aq) \rightleftharpoons H_2CO_3(aq)$$
$$\longrightarrow CO_2(g) + H_2O(l)$$

Overall: $$\overline{CaCO_3(s) + 2H^+(aq) \longrightarrow}$$
$$Ca^{2+}(aq) + CO_2(g) + H_2O(l)$$

(c) The solubility of $BaSO_4$ is largely unaffected by changes in solution pH, because SO_4^{2-} is a rather weak base and thus has little tendency to combine with a proton. However, $BaSO_4$ is slightly more soluble in strongly acidic solutions.

(d) The solubility of $AgCl$ is unaffected by changes in pH, because Cl^- is the anion of a strong acid.

Ions can be separated from each other on the basis of their solubilities. One widely used procedure for separating metal ions is based on the relative solubilities of their sulfides. Metals that form insoluble metal salts will not precipitate unless $[S^{2-}]$ is sufficiently high for Q to exceed K_{sp}. The sulfide concentration can be adjusted by regulating the pH of the solution. For example, CuS can be precipitated from a mixture of Cu^{2+} and Zn^{2+} by bubbling H_2S gas into a properly acidified solution of these ions. CuS ($K_{sp} = 6.3 \times 10^{-36}$) is less soluble than ZnS ($K_{sp} = 1.1 \times 10^{-21}$). Consequently, $[S^{2-}]$ can be regulated to cause CuS to precipitate while not exceeding K_{sp} for ZnS.

When H_2S is bubbled through an aqueous solution at 25°C and 1 atm, a saturated solution of H_2S forms that is approximately 0.1 M in H_2S. H_2S is a weak diprotic acid:

$$H_2S(aq) \rightleftharpoons H^+(aq) + HS^-(aq)$$
$$K_{a1} = 5.7 \times 10^{-8} \qquad \text{[16.20]}$$

$$HS^-(aq) \rightleftharpoons H^+(aq) + S^{2-}(aq)$$
$$K_{a2} = 1.3 \times 10^{-13} \qquad \text{[16.21]}$$

These equations can be combined to give an equation for the overall dissociation of H_2S into S^{2-}:

$$H_2S(aq) \rightleftharpoons 2H^+(aq) + S^{2-}(aq) \qquad \text{[16.22]}$$

The equilibrium-constant expression for this overall dissociation process is

$$K = \frac{[H^+]^2[S^{2-}]}{[H_2S]} = K_{a1} \times K_{a2}$$
$$= 7.4 \times 10^{-21} \qquad \text{[16.23]}$$

Substituting the solubility of H_2S, 0.1 M, into this expression gives

$$\frac{[H^+]^2[S^{2-}]}{(0.1)} = 7.4 \times 10^{-21}$$

$$[H^+]^2[S^{2-}] = (0.1)(7.4 \times 10^{-21})$$
$$= 7 \times 10^{-22} \qquad \text{[16.24]}$$

Equation [16.24] can be used to calculate the concentration of S^{2-} in saturated solutions of H_2S at various pH's.

Now consider how we can calculate the pH that will allow us to prevent precipitation of ZnS while precipitating CuS from a solution that is 0.10 M in Zn^{2+}, 0.10 M in Cu^{2+}, and saturated with H_2S. The maximum concentration of S^{2-} that can be present in the solution before precipitation of ZnS occurs can be calculated from the solubility-product expression for this substance:

$$[Zn^{2+}][S^{2-}] = K_{sp} = 1.1 \times 10^{-21}$$

$$[S^{2-}] = \frac{1.1 \times 10^{-21}}{[Zn^{2+}]} = \frac{1.1 \times 10^{-21}}{0.10}$$

$$= 1.1 \times 10^{-20}\ M$$

The concentration of hydrogen ions necessary to give $[S^{2-}] = 1.1 \times 10^{-20}\ M$ can be calculated using Equation [16.24]

$$[H^+]^2[S^{2-}] = 7 \times 10^{-22}$$

$$[H^+]^2 = \frac{7 \times 10^{-22}}{[S^{2-}]} = \frac{7 \times 10^{-22}}{1.1 \times 10^{-20}} = 6 \times 10^{-2}$$

$$[H^+] = \sqrt{6 \times 10^{-2}} = 0.24\ M$$

$$pH = -\log(2.4 \times 10^{-1}) = 0.6$$

Thus ZnS will not precipitate if pH is 0.6 or lower. However, at pH 0.6, where $[S^{2-}] = 1.1 \times 10^{-20}$, the ion product for a 0.10 M Cu^{2+} solution would be

$$Q = [Cu^{2+}][S^{2-}] = (0.10)(1.1 \times 10^{-20}) = 1.1 \times 10^{-21}$$

Because Q exceeds the K_{sp} of CuS (that is, 6.3×10^{-36}), CuS will precipitate under these conditions. Indeed, $[S^{2-}]$ must be less than $6.3 \times 10^{-35}\ M$ to prevent the precipitation of CuS from this solution. Even under strongly acidic conditions $[S^{2-}]$ will be high enough to cause precipitation of CuS. Thus a fairly broad range of sulfide concentration will allow for effective separation of Cu^{2+} (as CuS) from Zn^{2+}.

EFFECT OF COMPLEX FORMATION ON SOLUBILITY

It is a characteristic property of metal ions that they are able to act as Lewis acids, or electron-pair acceptors, toward water molecules, which act as Lewis bases, or electron-pair donors (Section 15.10). Lewis bases other than water can also interact with metal ions, particularly with transition-metal ions. Such interactions can have a dramatic effect on the solubility of a metal salt. For example, AgCl, whose $K_{sp} = 1.82 \times 10^{-10}$, will dissolve in the presence of aqueous ammonia because of the interaction between Ag^+ and the Lewis base NH_3. This process can be viewed as the sum of two reactions, the solubility equilibrium of AgCl and the Lewis acid-base interaction between Ag^+ and NH_3:

$$AgCl(s) \rightleftharpoons Ag^+(aq) + Cl^-(aq) \qquad [16.25]$$

$$Ag^+(aq) + 2NH_3(aq) \rightleftharpoons Ag(NH_3)_2^+(aq) \qquad [16.26]$$

Overall: $\quad AgCl(s) + 2NH_3(aq) \rightleftharpoons Ag(NH_3)_2^+(aq) + Cl^-(aq) \qquad [16.27]$

The presence of NH_3 drives the top reaction, the solubility equilibrium of AgCl, to the right as $Ag^+(aq)$ is removed to form $Ag(NH_3)_2^+$.

For a Lewis base such as NH_3 to increase the solubility of a metal salt, it must be able to interact more strongly with the metal ion than water does. The NH_3 must displace solvating H_2O molecules (Sections 12.2 and 15.10) in order to form $Ag(NH_3)_2^+$:

$$Ag^+(aq) + 2NH_3(aq) \rightleftharpoons Ag(NH_3)_2^+(aq) \qquad [16.28]$$

An assembly of a metal ion and the Lewis bases bonded to it, such as $Ag(NH_3)_2^+$, is called a complex ion. The stability of a complex ion in aqueous solution can be judged by the magnitude of the equilibrium constant for the formation of the complex ion from the hydrated metal ion. For example, the equilibrium constant for formation of $Ag(NH_3)_2^+$, Equation [16.28], is 1.7×10^7:

$$K_f = \frac{[\mathrm{Ag(NH_3)_2}^+]}{[\mathrm{Ag}^+][\mathrm{NH_3}]^2} = 1.7 \times 10^7 \qquad [16.29]$$

Such equilibrium constants are called **formation constants,** K_f. The formation constants for a few complex ions are shown in Table 16.1.

The general rule is that metal salts will dissolve in the presence of a suitable Lewis base such as NH_3 if the metal forms a sufficiently stable complex with the base. However, we will not consider such equilibria from a quantitative point of view. The ability of metal ions to form complexes is an extremely important aspect of their chemistry. In Chapter 23 we will take a much closer look at complex ions. In that chapter and others we shall see applications of complex ions to areas such as biochemistry, metallurgy, and photography.

AMPHOTERISM

Many metal hydroxides and oxides that are relatively insoluble in neutral water dissolve in *either* a strongly acidic or a strongly basic medium. Such substances are said to be **amphoteric.** Examples include the hydroxides and oxides of Al^{3+}, Cr^{3+}, Zn^{2+}, and Sn^{2+}.

The dissolution of these species in acidic solutions should be anticipated based on the earlier discussions in this section. We have seen that acids promote the dissolving of compounds with basic anions. What makes amphoteric oxides and hydroxides special is that they dissolve in strongly basic solutions. This behavior results from the formation of complex anions containing several (typically four) hydroxides bound to the metal ion:

$$\mathrm{Al(OH)_3}(s) + \mathrm{OH}^-(aq) \rightleftharpoons \mathrm{Al(OH)_4}^-(aq) \qquad [16.30]$$

Amphoterism is often interpreted in terms of the behavior of the water molecules that surround the metal ion and that are bonded to it by Lewis acid-base interactions (Section 15.10). For example, $Al^{3+}(aq)$ is more accurately represented as $Al(H_2O)_6{}^{3+}(aq)$; six water molecules are bonded to the Al^{3+} in aqueous solution. As discussed in Section 15.10, this hydrated ion is a weak acid. As a strong base is added, the $Al(H_2O)_6{}^{3+}$ loses protons in a stepwise fashion, eventually forming the neutral and water-insoluble $Al(H_2O)_3(OH)_3$. This substance then dissolves upon removal of an additional proton to form the anion $Al(H_2O)_2(OH)_4{}^-$. The reactions that occur are as follows:

TABLE 16.1 Formation constants for some metal complex ions in water at 25°C

Complex ion	K_f	Equilibrium equation
$\mathrm{Ag(NH_3)_2}^+$	1.7×10^7	$\mathrm{Ag}^+(aq) + 2\mathrm{NH_3}(aq) \rightleftharpoons \mathrm{Ag(NH_3)_2}^+(aq)$
$\mathrm{Ag(CN)_2}^-$	1×10^{21}	$\mathrm{Ag}^+(aq) + 2\mathrm{CN}^-(aq) \rightleftharpoons \mathrm{Ag(CN)_2}^-(aq)$
$\mathrm{Ag(S_2O_3)_2}^{3-}$	2.9×10^{13}	$\mathrm{Ag}^+(aq) + 2\mathrm{S_2O_3}^{2-}(aq) \rightleftharpoons \mathrm{Ag(S_2O_3)_2}^{3-}(aq)$
$\mathrm{Cu(NH_3)_4}^{2+}$	5×10^{12}	$\mathrm{Cu}^{2+}(aq) + 4\mathrm{NH_3}(aq) \rightleftharpoons \mathrm{Cu(NH_3)_4}^{2+}(aq)$
$\mathrm{Cu(CN)_4}^{2-}$	1×10^{25}	$\mathrm{Cu}^{2+}(aq) + 4\mathrm{CN}^-(aq) \rightleftharpoons \mathrm{Cu(CN)_4}^{2-}(aq)$
$\mathrm{Ni(NH_3)_6}^{2+}$	5.5×10^8	$\mathrm{Ni}^{2+}(aq) + 6\mathrm{NH_3}(aq) \rightleftharpoons \mathrm{Ni(NH_3)_6}^{2+}(aq)$
$\mathrm{Fe(CN)_6}^{4-}$	1×10^{35}	$\mathrm{Fe}^{2+}(aq) + 6\mathrm{CN}^-(aq) \rightleftharpoons \mathrm{Fe(CN)_6}^{4-}(aq)$
$\mathrm{Fe(CN)_6}^{3-}$	1×10^{42}	$\mathrm{Fe}^{3+}(aq) + 6\mathrm{CN}^-(aq) \rightleftharpoons \mathrm{Fe(CN)_6}^{3-}(aq)$

$$Al(H_2O)_6^{3+}(aq) + OH^-(aq) \rightleftharpoons Al(H_2O)_5(OH)^{2+}(aq) + H_2O(l)$$

$$Al(H_2O)_5(OH)^{2+}(aq) + OH^-(aq) \rightleftharpoons Al(H_2O)_4(OH)_2^{+}(aq) + H_2O(l)$$

$$Al(H_2O)_4(OH)_2^{+}(aq) + OH^-(aq) \rightleftharpoons Al(H_2O)_3(OH)_3(s) + H_2O(l)$$

$$Al(H_2O)_3(OH)_3(s) + OH^-(aq) \rightleftharpoons Al(H_2O)_2(OH)_4^{-}(aq) + H_2O(l)$$

Further proton removals are possible, but each successive reaction occurs less readily than the one before. As the charge on the ion becomes more negative, it becomes increasingly difficult to remove a positively charged proton. Addition of an acid reverses these reactions. The proton adds in a stepwise fashion to convert the OH^- groups to H_2O, eventually reforming $Al(H_2O)_6^{3+}$. The common practice is to simplify the equations for these reactions by excluding the bound H_2O molecules. Thus we usually write Al^{3+} instead of $Al(H_2O)_6^{3+}$, $Al(OH)_3$ instead of $Al(H_2O)_3(OH)_3$, $Al(OH)_4^{-}$ instead of $Al(H_2O)_2(OH)_4^{-}$, and so forth.

The extent to which an insoluble metal hydroxide reacts with either acid or base varies with the particular metal ion involved. Many metal hydroxides, for example, $Ca(OH)_2$, $Fe(OH)_2$, and $Fe(OH)_3$, are capable of dissolving in acidic solution but do not react with excess base. These hydroxides are not amphoteric.

The purification of aluminum ore in the manufacture of aluminum metal provides an interesting application of the property of amphoterism. As we have seen, $Al(OH)_3$ is amphoteric, whereas $Fe(OH)_3$ is not. Aluminum occurs in large quantities as the ore bauxite, which is essentially Al_2O_3 with additional water molecules. The ore is contaminated with Fe_2O_3 as an impurity. When bauxite is added to a strongly basic solution, the Al_2O_3 dissolves, because the aluminum forms complex ions such as $Al(OH)_4^{-}$. The Fe_2O_3 impurity, however, is not amphoteric and remains as a solid. The solution is filtered, getting rid of the iron impurity. Aluminum hydroxide is then precipitated by addition of acid. The purified hydroxide receives further treatment and eventually yields aluminum metal.

When writing reactions of amphoteric substances, we often show them as the oxides rather than the hydrated hydroxides. Although the various ways of showing the reactions are confusing at first, the only real differences are in the number of water molecules shown. Sample Exercise 16.16 shows how we can view the reaction of a metal oxide with base.

SAMPLE EXERCISE 16.16

Write the balanced net-ionic equation for the reaction between $Al_2O_3(s)$ and aqueous solutions of NaOH that causes the $Al_2O_3(s)$ to dissolve. (H_2O is also involved as a reactant.)

Solution: NaOH is a strong electrolyte that provides the necessary OH^- ions. The reaction product can be taken to be $Al(OH)_4^{-}$, an assembly of Al^{3+} and four OH^- ions. The unbalanced equation is

$$Al_2O_3(s) + OH^-(aq) \rightleftharpoons Al(OH)_4^{-}(aq)$$

Two Al are needed among the products:

$$Al_2O_3(s) + OH^-(aq) \rightleftharpoons 2Al(OH)_4^{-}(aq)$$

To balance the charge on both sides of the equation requires $2OH^-$. Sufficient H_2O is then added to balance the O and H counts. The balanced equation is

$$Al_2O_3(s) + 2OH^-(aq) + 3H_2O(l) \rightleftharpoons 2Al(OH)_4^{-}(aq)$$

16.6 Qualitative analyses for metallic elements

In this chapter we have seen several examples of equilibria involving metal ions in aqueous solution. In this final section we look very briefly at how solubility equilibria and complex formation can be used to detect the presence of particular metal ions in solution. Before the development of modern analytical instrumentation, it was necessary to analyze mixtures of metal in a sample by so-called "wet" chemical methods. For example, a metallic sample that might contain several metallic elements was dissolved in a concentrated acid solution. This solution was then tested in a systematic way for the presence of various metallic ions.

Qualitative analysis involves simply determining the presence or absence of a particular metal ion. It should be distinguished from quantitative analysis, which involves determining how much of a given substance is present. "Wet" methods of qualitative analysis are no longer so important as a means of analysis. However, they are frequently used in freshman chemistry laboratory programs as a means of illustrating equilibria, teaching the properties of common metal ions in solution, and developing laboratory skills. Typically, such analyses proceed in three stages: (1) The ions are separated into broad groups on the basis of solubility properties. (2) The individual ions within each group are then separated by selectively dissolving members in the group. (3) The ions are then identified by means of specific tests.

A scheme in common use divides the common cations into five groups as shown in Figure 16.9. The order of addition of reagents is important. The most selective separations—that is, those that involve the smallest number of ions—are carried out first. The reactions that are used must proceed so far toward completion that any concentration of cations remaining in the solution is too small to interfere with subsequent tests.

Let's take a closer look at each of these five groups of cations, examining briefly the logic used in this qualitative analysis scheme:

1 *Insoluble chlorides:* Of the common metal ions only Ag^+, Hg_2^{2+}, and Pb^{2+} form insoluble chlorides. Thus when dilute HCl is added to a mixture of cations, only AgCl, Hg_2Cl_2, and $PbCl_2$ will precipitate, leaving the other cations in solution. The absence of a precipitate indicates that the starting solution contains no Ag^+, Hg_2^{2+}, or Pb^{2+}.

2 *Acid-insoluble sulfides:* After any insoluble chlorides have been removed, the remaining, now acidic solution is treated with H_2S. As we saw in Section 16.5, the dissociation of H_2S is repressed in acidic solutions so that the concentration of free S^{2-} is very low. Consequently, only the most insoluble metal sulfides, CuS, Bi_2S_3, CdS, HgS, As_2S_3, Sb_2S_3, and SnS_2, can precipitate. (Note the very small values of K_{sp} for some of these sulfides in Appendix E.) Those metal ions whose sulfides are somewhat more soluble, for example ZnS or NiS, remain in solution.

3 *Base-insoluble sulfides:* After the solution is filtered to remove any acid-insoluble sulfides, the remaining solution is made slightly basic, and more H_2S is added. In basic solutions the concentration of S^{2-} is higher than in acidic solutions. Thus the ion products for many of the more soluble sulfides are caused to exceed their K_{sp} values and precipitation occurs. The metal ions precipitated at this stage are Al^{3+},

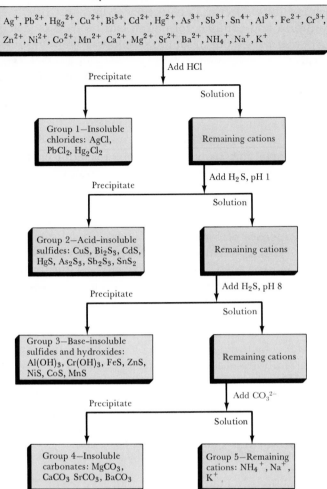

Aqueous solution of common cations

$Ag^+, Pb^{2+}, Hg_2^{2+}, Cu^{2+}, Bi^{3+}, Cd^{2+}, Hg^{2+}, As^{3+}, Sb^{3+}, Sn^{4+}, Al^{3+}, Fe^{2+}, Cr^{3+},$ $Zn^{2+}, Ni^{2+}, Co^{2+}, Mn^{2+}, Ca^{2+}, Mg^{2+}, Sr^{2+}, Ba^{2+}, NH_4^+, Na^+, K^+$

Add HCl

Precipitate   Solution

Group 1—Insoluble chlorides: AgCl, $PbCl_2$, Hg_2Cl_2

Remaining cations

Add H_2S, pH 1

Precipitate   Solution

Group 2—Acid–insoluble sulfides: CuS, Bi_2S_3, CdS, HgS, As_2S_3, Sb_2S_3, SnS_2

Remaining cations

Add H_2S, pH 8

Precipitate   Solution

Group 3—Base–insoluble sulfides and hydroxides: $Al(OH)_3$, $Cr(OH)_3$, FeS, ZnS, NiS, CoS, MnS

Remaining cations

Add CO_3^{2-}

Precipitate   Solution

Group 4—Insoluble carbonates: $MgCO_3$, $CaCO_3$ $SrCO_3$ $BaCO_3$

Group 5—Remaining cations: NH_4^+, Na^+, K^+

FIGURE 16.9 Qualitative-analysis scheme for separating common cations into groups.

$Cr^{3+}, Fe^{3+}, Zn^{2+}, Ni^{2+}, Co^{2+},$ and Mn^{2+}. (Actually the Al^{3+} and Cr^{3+} ions do not form insoluble sulfides, but instead are precipitated as the insoluble hydroxides at the same time.)

4   *Insoluble carbonates:* At this point the solution contains only metal ions from periodic table groups 1A and 2A. Addition of carbonate ion to a slightly basic solution causes precipitation of the group 2A elements Mg^{2+}, Ca^{2+}, Sr^{2+}, and Ba^{2+}, because these metals form insoluble carbonates.

5   *The alkali metal ions and NH_4^+:* The ions that remain after removal of the insoluble carbonates form a small group in which each ion can be tested for by an individual test. For example, the flame test is useful to show the presence of K^+, because the flame turns a characteristic violet color if K^+ is present.

Additional separation and testing is necessary to determine which ions are present within each of the groups. As an example, consider the ions of the insoluble chloride group. The precipitate containing the metal chlorides is boiled in water. It happens that $PbCl_2$ is relatively soluble in hot water, whereas $AgCl$ and Hg_2Cl_2 are not. The hot solution is filtered and a solution of Na_2CrO_4 added to the filtrate. If Pb^{2+} is present, a yellow precipitate of $PbCrO_4$ forms. The test for Ag^+ consists of treating the metal chloride precipitate with dilute ammonia. Only Ag^+ forms an ammonia complex. If $AgCl$ is present in the precipitate it will dissolve in the ammonia solution:

$$AgCl(s) + 2NH_3(aq) \rightleftharpoons Ag(NH_3)_2{}^+(aq) + Cl^-(aq) \qquad [16.31]$$

After treatment with ammonia, the solution is filtered, and the filtrate made acidic by adding nitric acid. The nitric acid removes ammonia from solution by forming $NH_4{}^+$, thus releasing Ag^+, which should reform the $AgCl$ precipitate:

$$Ag(NH_3)_2{}^+(aq) + Cl^-(aq) + 2H^+(aq) \rightleftharpoons AgCl(s) + 2NH_4{}^+(aq) \qquad [16.32]$$

The analyses for individual ions in the acid-insoluble and base-insoluble sulfides are a bit more complex, but the same general principles are involved. The detailed procedures for carrying out such analyses are given in laboratory manuals.

FOR REVIEW

Summary

In this chapter we've considered several types of important equilibria occurring in aqueous solution. Our primary emphasis has been on acid-base equilibria in solutions containing two or more solutes and on solubility equilibria. We observed that the dissociation of a weak acid or base is repressed by the presence of a strong electrolyte that provides an ion common to the equilibrium. This phenomenon is an example of the **common-ion effect.**

A particularly important type of acid-base mixture is one involving a weak conjugate acid-base pair. Such mixtures function as **buffers.** Addition of small amounts of additional acid or base to a buffered solution causes only small changes in pH, because the buffer reacts with the added acid or base. (Recall that strong acid-strong base, strong acid-weak base, and weak acid-strong base reactions proceed essentially to completion.) Buffer solutions are usually prepared from a weak acid and a salt of that acid or from a weak base and a salt of that base. Two important characteristics of a buffer solution are its buffering capacity and its pH.

The plot of the pH of an acid (or base) as a function of the volume of added base (or acid) is called a **titration curve.** Titration curves aid in selecting a proper pH indicator for an acid-base titration. The titration curve of a strong acid-strong base titration exhibits a large change in pH in the immediate vicinity of the equivalence point; at the equivalence point for this titration pH = 7. For strong acid-weak base or weak acid-strong base titrations, the pH change in the vicinity of the equivalence point is not as large. Furthermore, the pH at the equivalence point is not 7 in either of these cases.

The equilibrium between a solid salt and its ions in solution provides an example of heterogeneous equilibrium. The **solubility-product constant,** K_{sp}, is an equilibrium constant that expresses quantitatively the extent to which the salt dissolves. Addition to the solution of an ion common to a solubility equilibrium causes the solubility of the salt to decrease. This phenomenon is another example of the common-ion effect.

Comparison of the ion product, Q, with the value of K_{sp} can be used to judge whether a precipitate will form when solutions are mixed or whether a slightly soluble salt will dissolve under various conditions. Solubility is affected by the common-ion effect, by pH, and by the presence of certain Lewis bases that react with metal ions to form stable **complex ions.** Solubility is affected by pH when one or more of the ions involved in the solubility equilibrium is an acid or base. For example, the solubility of MnS is increased on addition of acid, because S^{2-} is basic. Amphoteric metal hydroxides are those slightly soluble metal hydroxides that dissolve on addition of either acid or base. The reactions that give rise to the amphoterism are acid-base reactions involving the OH^- or H_2O groups bound to the metal ion. Complex-ion formation in aqueous solution involves the displacement by Lewis bases such as NH_3 or CN^- of water molecules attached to the metal ion. The extent to which such complex formation occurs is expressed quantitatively by the formation constant for the complex ion.

The fact that the ions of different metallic elements vary a great deal in the solubilities of their salts, in their acid-base behavior, and in their tendencies to form complexes can be used to separate and detect the presence of metal-ions in mixtures. Qualitative analysis refers to determination of the presence or absence of a metal ion in a mixture of metal ions in solution. The analysis usually proceeds by separation of the ions into groups on the basis of precipitation reactions, and then by analyses for individual metal ions within each group.

Learning goals

Having read and studied this chapter, you should be able to:

1 Predict qualitatively and calculate quantitatively the effect of an added common ion on the pH of an aqueous solution of a weak acid or base.

2 Calculate the concentrations of each species present in a solution formed by mixing an acid and a base.

3 Describe how a buffer solution of a particular pH is made and how it operates to control pH.

4 Calculate the change in pH of a simple buffer solution of known composition caused by adding a small amount of strong acid or base.

5 Describe the form of the titration curves for titration of a strong acid by a strong base, a weak acid by a strong base, or a weak base by a strong acid.

6 Calculate the pH at any point, including the equivalence point, in an acid-base titration.

7 Set up the expression for the solubility-product constant for a salt.

8 Calculate K_{sp} from solubility data and solubility from the value for K_{sp}.

9 Calculate the effect of an added common ion on the solubility of a slightly soluble salt.

10 Predict whether a precipitate will form when two solutions are mixed, given appropriate K_{sp} values.

11 Explain the effect of pH on a solubility equilibrium involving a basic or acidic ion.

12 Formulate the equilibrium between a metal ion and a Lewis base to form a complex ion of a metal.

13 Describe how complex formation can affect the solubility of a slightly soluble salt.

14 Explain the origin of amphoteric behavior and write equations describing the dissolution of an amphoteric metal hydroxide in either an acidic or basic medium.

15 Explain the general principles that apply to the groupings of metal ions in the qualitative analysis of an aqueous mixture.

Key terms

Among the more important terms and expressions used for the first time in this chapter are the following:

Amphoterism (Section 16.5) is a term used to describe the ability of certain slightly soluble metal hydroxides to dissolve in either an acidic or basic medium. The solubility results from formation of a complex ion via an acid-base reaction.

A buffer solution (Section 16.2) is one that undergoes a limited change in pH upon addition of a small amount of acid or base.

The common-ion effect (Section 16.1) refers to the effect of an ion common to an equilibrium in shifting the equilibrium. For example, added Na_2SO_4 decreases the solubility of the slightly soluble salt $BaSO_4$, or added $NaC_2H_3O_2$ decreases the percent ionization of $HC_2H_3O_2$.

The formation constant (Section 16.5) for a metal ion complex is the equilibrium constant for formation of the complex from the metal ion and base species present in solution. It is a measure of the tendency of the complex to form.

A metal ion complex (Section 16.5) consists of a metal ion and a well-defined group of ions or neutral molecules bound to the ion via a Lewis acid-base interaction.

Qualitative analysis (Section 16.6) refers to determining the presence or absence of a particular substance that might be present in a mixture.

The solubility-product constant (Section 16.4) is an equilibrium constant related to the equilibrium between a solid salt and its ions in solution. It provides a quantitative measure of the solubility of a slightly soluble salt.

Common-ion effect; buffers

16.1 What is the pH of a 1.0-L solution containing 0.10 mol of benzoic acid, $HC_7H_5O_2$, and 0.15 mol of sodium benzoate, $NaC_7H_5O_2$?

16.2 (a) Calculate the pH of a 0.035 M solution of propionic acid, $HC_3H_5O_2$. (b) Calculate the pH of this solution after sufficient sodium propionate, $NaC_3H_5O_2$, has been added to make the solution 0.050 M in $C_3H_5O_2^-$. (c) If a solution of $HC_3H_5O_2$ and $NaC_3H_5O_2$ is 0.035 M in $HC_3H_5O_2$ and has a pH of 4.80, what is the concentration of $NaC_3H_5O_2$?

16.3 Calculate the pH of each of the following solutions: (a) 0.20 M nitrous acid, HNO_2, and 0.20 M sodium nitrate, $NaNO_3$; (b) 0.10 M benzoic acid, $HC_7H_5O_2$, and 0.10 M nitric acid, HNO_3; (c) 0.010 M pyridine, C_5H_5N, and 0.010 M pyridinium chloride, $C_5H_5NH^+Cl^-$; (d) 0.10 M sodium hydroxide, $NaOH$, and 0.20 M ammonia, NH_3.

16.4 Consider the solution produced by mixing 0.50 L of 0.20 M $HClO_4$ with 0.50 L of 0.10 M $NaC_2H_3O_2$. (a) Write the net ionic equation for the reaction that occurs when these solutions are mixed. (b) Calculate the equilibrium constant for this reaction. (c) Calculate the concentrations of Na^+, H^+, ClO_4^-, $C_2H_3O_2^-$, and $HC_2H_3O_2$ in the solution at equilibrium.

16.5 Calculate the concentrations of H^+, F^-, Cl^-, and HF in a solution formed by mixing 0.050 L of 0.10 M NaF with 0.020 L of 0.20 M HCl.

16.6 Explain how a buffer solution acts to limit changes in pH on addition of acid or base.

16.7 Suppose you have two vessels, one containing a buffer solution of pH 7, the other a KBr solution of pH 7. Explain how you would test a small sample from each vessel to determine which is the buffer solution.

16.8 Explain with the aid of equations how a solution of 0.1 M $HC_2H_3O_2$ and 0.1 M $NaC_2H_3O_2$ acts as a buffer.

16.9 Explain what is meant in speaking of the *capacity* of a buffer?

16.10 Which of the following solutions has the greatest buffering capacity and which has the least: (a) 0.01 M $HC_2H_3O_2$ and 0.01 M $NaC_2H_3O_2$, pH = 4.74; (b) 1.8×10^{-5} M HBr, pH = 4.74; (c) 0.10 M $HC_2H_3O_2$ and 0.10 M $NaC_2H_3O_2$, pH = 4.74. Explain your answers.

16.11 The main buffer in blood is hemoglobin. However, the H_2CO_3–HCO_3^- system is also important. Calculate the $[HCO_3^-]$ to $[H_2CO_3]$ ratio in blood (pH 7.4). Is this carbonate buffer mixture better able to buffer against acid or against base?

16.12 The optimal buffering capacity of a buffer solution made up from an acid and its conjugate base (or from a base and its conjugate acid) occurs when the conjugate acid-base pair are present in equal concentrations. Explain why this is so.

16.13 Consider 1 L of a buffer mixture that is 0.100 M

in formic acid, $HCHO_2$, and 0.100 M in sodium formate, $NaCHO_2$. Calculate the change in pH that occurs when (a) 1×10^{-3} mol of HCl is added; (b) 1.0×10^{-2} mol of HCl is added; (c) 1×10^{-3} mol of NaOH is added; (d) 1.0×10^{-2} mol of NaOH is added.

16.14 One liter of a buffer solution contains 0.15 mol of acetic acid and 0.10 mol of sodium acetate. (a) What is the pH of this buffer; (b) what is the pH after addition of 0.01 mol of HNO_3; (c) what is the pH after addition of 0.01 mol of KOH?

Titration curves

16.15 How does titration of a weak acid with a strong base differ from titration of a strong acid with a strong base, with respect to each of the following points: (a) quantity of base required to reach the equivalence point; (b) pH at the beginning of the titration; (c) pH at the equivalence point; (d) pH after addition of a slight excess of base; (e) choice of indicator for determining the equivalence point?

16.16 How many milliliters of 0.025 M NaOH are required to reach the equivalence point in titrating each of the following solutions: (a) 25 mL of 0.040 M HBr; (b) 41 mL of 0.022 M $HC_2H_3O_2$.

16.17 (a) How many moles of HCl are required to completely neutralize 35.0 mL of 0.120 M KOH? (b) How many milliliters of 0.100 M HNO_3 are needed to completely neutralize 40.0 mL of 0.120 M NaOH? (c) How many milliliters of 0.100 M HNO_3 are required to neutralize 0.510 g of NaOH? (d) If it requires 20.5 mL of HCl to neutralize 30.0 mL of 0.100 M NaOH, what is the concentration of HCl?

16.18 Fifty milliliters of 0.100 M HCl is titrated with 0.100 M NaOH. Calculate the pH of the solution after the following volume of NaOH has been added: (a) 0.0 mL; (b) 25.0 mL; (c) 49.9 mL; (d) 50.0 mL; (e) 50.1 mL; (f) 75.0 mL.

16.19 Calculate the pH of a solution containing 50.0 mL of 0.100 M benzoic acid after the following volume of 0.100 M NaOH has been added: (a) 0.0 mL; (b) 25.0 mL; (c) 49.9 mL; (d) 50.0 mL; (e) 50.1 mL; (f) 75.0 mL.

16.20 Calculate the pH at the equivalence point for titration of 0.10 M solutions of each of the following bases by 0.10 M perchloric acid, $HClO_4$: (a) potassium hydroxide, KOH; (b) pyridine, C_5H_5N; (c) ethylamine; $C_2H_5NH_2$.

16.21 Which of the indicators listed in Table 15.1 might be used in each of the titrations described in exercise 16.20?

Solubility equilibria

16.22 Write the expression for the solubility product constant for the solubility equilibrium of each of the following compounds: (a) AgBr; (b) MgC_2O_4; (c) $Cd(IO_3)_2$; (d) Ag_2SO_4.

16.23 If the molar solubility of $Mg(OH)_2$ in water at a particular temperature is 1.8×10^{-4} mol/L, what is its K_{sp} at this temperature?

16.24 For each salt listed below, we have given the solubility in grams per liter of solution. From these data calculate the value for K_{sp} in each case. (a) $AgIO_3$, 0.0283 g/L; (b) MgF_2, 0.0729 g/L; (c) Ag_2CO_3, 3.51×10^{-2} g/L.

16.25 The K_{sp} for $NiCO_3$ is 6.6×10^{-9}. What is the molar solubility (that is, moles per liter) of this substance in water?

16.26 Using the K_{sp} values given in Appendix E, calculate the solubility in grams per liter of solution for each of the following substances: (a) $SrCO_3$; (b) $Cd(OH)_2$; (c) Cu_2S.

16.27 Milk of magnesia is a water suspension of magnesium hydroxide, $Mg(OH)_2$. What is the pH of a saturated solution of $Mg(OH)_2$?

16.28 The cheapest strong base available is $Ca(OH)_2$, known commonly as slaked lime. One of the main factors that limits its applications is its rather low solubility. Using K_{sp} from Appendix E, calculate (a) the number of grams of $Ca(OH)_2$ that will dissolve in 100 mL of water and (b) the pH of a saturated solution of $Ca(OH)_2$.

16.29 Calculate the molar solubility of $PbSO_4$ in (a) pure water; (b) $0.010\,M$ Na_2SO_4; (c) $0.010\,M$ $Pb(NO_3)_2$.

16.30 Calculate the molar solubility of Ag_2CrO_4 in (a) pure water; (b) $0.010\,M$ Na_2CrO_4; (c) $0.010\,M$ $AgNO_3$.

16.31 Should a precipitate of $PbCl_2$ form when 0.10 L of $0.10\,M$ $Pb(NO_3)_2$ is added to 0.10 L of $0.05\,M$ NaCl?

16.32 Calculate the pH at which $Ca(OH)_2$ just begins to precipitate when NaOH is added to a $0.10\,M$ solution of $CaCl_2$.

16.33 Will $Mg(OH)_2$ precipitate from solution if the pH of a $0.010\,M$ solution of $MgCl_2$ is adjusted to 10.0?

16.34 Will a precipitate form when 30.0 mL of $0.010\,M$ $CaCl_2$ is mixed with 70.0 mL of 1.0×10^{-3} M Na_2CO_3?

Controlled precipitation; dissolution

16.35 Which of the following salts will be more soluble in acid solution than in pure water: (a) CdS; (b) AgCl; (c) CuI; (d) $Co(OH)_2$; (e) $PbSO_4$; (f) $MnCO_3$?

16.36 What pH is necessary to produce a saturated solution of H_2S ($[H_2S] = 0.10\,M$) in which the concentration of sulfide ion, S^{2-}, is (a) 7×10^{-20} M; (b) 7×10^{-10} M; (c) 7×10^{-4} M?

16.37 A solution that is $0.10\,M$ in Zn^{2+} and $0.010\,M$ in H^+ is saturated with H_2S ($[H_2S] = 0.10\,M$). Will ZnS precipitate?

16.38 Calculate the solubility of CoS (in moles per liter) in a $0.10\,M$ solution of H_2S if $[H^+] = 0.10\,M$.

16.39 Which of the following metal hydroxides are amphoteric: (a) CsOH; (b) $Cr(OH)_3$; (c) $Fe(OH)_3$; (d) $Ba(OH)_2$; (e) $Al(OH)_3$?

16.40 Complete the following equations:

(a) $Zn(OH)_2(s) + 2OH^-(aq) \longrightarrow$
(b) $Zn(OH)_2(s) + 2H^+(aq) \longrightarrow$
(c) $Zn(OH)_4{}^{2-}(aq) + 2H^+(aq) \longrightarrow$
(d) $Zn(OH)_4{}^{2-}(aq) + 4H^+(aq) \longrightarrow$
(e) $Zn^{2+}(aq) + 2OH^-(aq) \longrightarrow$
(f) $Zn^{2+}(aq) + 4OH^-(aq) \longrightarrow$

16.41 Explain the formation of a metal ion complex such as $Ag(NH_3)_2{}^+$ in terms of Lewis acid-base theory.

16.42 Write the formation-constant expression for each of the following complex ions: (a) $Co(NH_3)_6{}^{3+}$; (b) $Ag(CN)_2{}^-$.

Qualitative analysis

16.43 A solution containing an unknown number of metal ions is treated with dilute HCl; no precipitate forms. The pH is adjusted to about 1, and H_2S is bubbled through. Again no precipitate forms. The pH of the solution is then adjusted to about 8. Again H_2S is bubbled through. This time a precipitate forms. The filtrate from this solution is treated with sodium carbonate. No precipitate forms. Which metal ions discussed in Section 16.6 are possibly present? Which are definitely absent within the limits of these tests?

16.44 A student who is in a great hurry to finish his laboratory work decides that his qualitative analysis unknown contains a metal ion from the insoluble carbonate group, group 4. He therefore tests his sample directly with Na_2CO_3, skipping earlier tests for the metal ions in groups 1–3. He observes a precipitate and concludes that a metal ion from group 4 is indeed present. Why is this possibly an erroneous conclusion? (Hint: Consider the K_{sp} values in Appendix E.)

[16.45] What concentration of Pb^{2+} remains in a solution after $PbCl_2$ has been precipitated from a solution that is $0.1\,M$ in Cl^- at $25\,°C$? Will PbS form if the filtrate from the $PbCl_2$ precipitate is saturated with H_2S at pH = 1?

16.46 In the course of various qualitative-analysis procedures the following mixtures are encountered: (a) $Al(OH)_3$ and $Fe(OH)_3$; (b) AgCl and $PbCl_2$; (c) Hg^{2+} and Zn^{2+}; (d) Na^+ and Ca^{2+}. Suggest how each mixture might be separated.

16.47 What is the minimum concentration of Ag^+ that can be detected as Ag_2SO_4 if sufficient H_2SO_4 is added to adjust the concentration of $SO_4{}^{2-}$ to $0.1\,M$?

Additional exercises

16.48 Consider the solution produced by mixing 10.0 mL of $0.100\,M$ HCl with 20.0 mL of $0.200\,M$ sodium acetate. (a) Write the net ionic equation for the reaction. (b) Calculate the pH of the solution.

16.49 Write balanced net ionic equations for the reaction between (a) hydrochloric acid and sodium formate; (b) nitric acid and calcium hydroxide; (c) hydrofluoric acid and potassium hydroxide; (d) nitrous acid and ammonia.

[16.50] Calculate the equilibrium constant, K, for the following reaction:

$$Mg(OH)_2(s) + 2H^+(aq) \rightleftharpoons Mg^{2+}(aq) + 2H_2O(l)$$

16.51 Calculate the pH of the solution formed by mixing (a) 50.0 mL of 0.100 M HNO$_3$ and 50.0 mL of 0.100 M HC$_2$H$_3$O$_2$; (b) 50.0 mL of 0.100 M NaC$_2$H$_3$O$_2$ and 50.0 mL of 0.100 M HC$_2$H$_3$O$_2$; (c) 50.0 mL of 0.100 M NaOH and 25.0 mL of 0.100 M HC$_2$H$_3$O$_2$; (d) 50.0 mL of 0.100 M NaOH and 50.0 mL of 0.100 M HC$_2$H$_3$O$_2$; (e) 25.0 mL of 0.100 M NaOH and 50.0 mL of 0.100 M HC$_2$H$_3$O$_2$; (f) 25.0 mL of 0.100 M HCl and 50.0 mL of 0.100 M NaC$_2$H$_3$O$_2$.

16.52 A hypothetical weak acid, HA, was combined with NaOH in the following proportions: 0.25 mol of HA, 0.05 mol of NaOH. The mixture was diluted to a total volume of 1.00 L and the pH measured. If pH = 6.0, what is the pK_a of the acid?

16.53 A solution containing 0.030 mol of a weak acid, HX, and 0.40 mol of a salt NaX is diluted to a total volume of 0.200 L. The pH of the solution is 5.32. What is the acid-dissociation constant, K_a?

16.54 A 0.25 M solution of dimethylamine, (CH$_3$)$_2$NH, containing an unknown concentration of dimethylammonium chloride, (CH$_3$)$_2$NH$_2^+$Cl$^-$, has a pH of 10.38. What is the concentration of dimethylammonium ion in the solution?

16.55 0.500-L of a 0.10 M solution of a weak acid, HX, has a pH of 4.35. What is the pH of the solution after 0.010 mol of NaX is added?

16.56 How many moles of sodium hypochlorite, NaClO, should be added to 0.500 L of 0.100 M hypochlorous acid, HClO, to produce a buffer with pH = 6.77? Assume that no volume change occurs when the NaClO is added.

16.57 Which of the following conjugate acid-base pairs is the best choice to prepare a buffer with pH = 4.19: (a) HCO$_3^-$ and CO$_3^{2-}$; (b) NH$_4^+$ and NH$_3$; (c) HC$_7$H$_5$O$_2$ (benzoic acid) and C$_7$H$_5$O$_2^-$; (d) HCN and CN$^-$?

16.58 What is the pH of a buffer mixture composed of equal concentrations of the following substances: (a) HCO$_3^-$ and CO$_3^{2-}$; (b) HC$_3$H$_5$O$_3$ (lactic acid) and C$_3$H$_5$O$_3^-$; (c) CH$_3$NH$_2$ (methylamine) and CH$_3$NH$_3^+$; (d) HPO$_4^{2-}$ and PO$_4^{3-}$?

16.59 One liter of a buffer solution is prepared so that it contains 0.10 mol of HC$_2$H$_3$O$_2$ and 0.10 mol of NaC$_2$H$_3$O$_2$. When 1 mol of HCl is added to this solution, the pH changes drastically. Why did the buffer solution fail to control the pH?

16.60 Consider a 0.20-L buffer mixture that is 0.050 M in lactic acid, HC$_3$H$_5$O$_3$, and 0.040 M in sodium lactate, NaC$_3$H$_5$O$_3$. (a) What is the pH of this buffer? (b) What is the pH of the buffer mixture after 10 mL of 0.10 M HCl is added? (c) What is the pH of the buffer after 10 mL of 0.10 M NaOH is added?

16.61 Phosphate buffer, consisting of H$_2$PO$_4^-$ and HPO$_4^{2-}$, helps control the pH of blood. Many carbonated beverages also use this phosphate buffer system, although at a much higher concentration than found in blood. What is the pH of a buffer made by dissolving 12 g each of sodium dihydrogen phosphate, NaH$_2$PO$_4$, and sodium hydrogen phosphate, Na$_2$HPO$_4$, in sufficient water to form 0.500 L of solution?

[16.62] A biochemist needs 0.400 L of an acetic acid-sodium acetate buffer with pH = 4.44. Solid sodium acetate, NaC$_2$H$_3$O$_2$, and glacial acetic acid are available. Glacial acetic acid is 99 percent HC$_2$H$_3$O$_2$ by weight and has a density of 1.05 g/mL. If the buffer is to be 0.30 M in HC$_2$H$_3$O$_2$, how many grams of NaC$_2$H$_3$O$_2$ and how many milliliters of glacial acetic acid must be used?

16.63 What is the pH at the equivalence point in the following titrations: (a) 0.100 M HCl with 0.100 M methylamine; (b) 0.100 M NaOH with 0.100 M ascorbic acid; (c) 0.100 M HI with 0.100 M KOH.

16.64 Calculate the concentration of each ion present in the solution formed by adding 25.0 mL of 0.100 M NaOH to 50.0 mL of 0.100 M formic acid.

16.65 If 50.0 mL of 0.10 M Na$_2$CO$_3$ is titrated with 0.10 M HCl, calculate (a) the pH at the start of the titration; (b) the volume of HCl required to reach the first equivalence point; (c) the volume of HCl required to reach the second equivalence point.

[16.66] Equivalent quantities of 0.1 M solutions of an acid, HA, and a base, B, are mixed. The pH of the resulting solution is 8.8. (a) Write the equilibrium equation and equilibrium-constant expression for the reaction between HA and B. (b) If K_a for HA is 5.0×10^{-6}, what is the value of the equilibrium constant for the reaction between HA and B?

[16.67] Sodium nitrite, NaNO$_2$, is added to a 0.10 M solution of phenol, C$_6$H$_5$OH. From the known K_a values for phenol and nitrous acid (Appendix E), calculate the concentration of added NaNO$_2$ at which the pH of the solution is equal to 7.0.

16.68 Determine the number of grams of magnesium oxalate, MgC$_2$O$_4$, that will dissolve in (a) 0.500 L of H$_2$O; (b) 0.500 L of 0.10 M MgCl$_2$ solution.

16.69 Explain the following observations: (a) PbCO$_3$ is less soluble in Na$_2$CO$_3$ solution than in H$_2$O; (b) PbCO$_3$ is more soluble in HNO$_3$ solution than in H$_2$O.

16.70 Write the net ionic equation for the reaction that occurs when (a) lead nitrate is mixed with potassium sulfate; (b) hydrogen sulfide gas is bubbled through a solution of silver nitrate; (c) nitric acid is added to calcium carbonate; (d) aqueous ammonia is added to silver chloride.

16.71 Ground water that comes in contact with the mineral fluorite, CaF$_2$, can become saturated with CaF$_2$. (a) What is the concentration of F$^-$ in such water solution? (b) Compare this fluoride concentration with that found in a fluoridated water supply containing 1 g of F$^-$ per 10^6 g of water (that is, one part per million of F$^-$).

[16.72] Cadmium sulfide, CdS, is used commercially as a yellow paint pigment known as cadmium yellow. It is prepared by saturating a slightly acidic solution of Cd^{2+} with H$_2$S gas. The presence of Fe^{2+}, a common impurity, can result in the precipitate being contaminated with black FeS, thereby ruining the color of the pigment. (a) If a solution containing 0.10 M Cd^{2+} and 1×10^{-4} M Fe^{2+} is saturated with H_2S ($[H_2S] = 0.10$ M), what pH is needed to keep the Fe^{2+} from precipitating? (b) If a $HC_2H_3O_2-C_2H_3O_2$ buffer is used to control pH, what ratio of $HC_2H_3O_2$ to $C_2H_3O_2^-$ is required?

16.73 The concentration of Cl^- in a certain water supply is 2.0×10^{-6} M. Will a precipitate of AgCl form when 10.0 mL of this solution is added to 40.0 mL of 1.0 M $AgNO_3$?

16.74 A solution is prepared by mixing 10.0 mL of 0.24 M $CaCl_2$ with 10.0 mL of a solution that is 0.20 M in NH_3 and 0.050 M in NH_4^+. Will $Ca(OH)_2$ precipitate?

16.75 What is the concentration of S^{2-} in a saturated solution of H_2S if (a) pH = 2.0; (b) pH = 7.0; (c) pH = 11.0?

16.76 How many liters of water would be required to dissolve 1 g of HgS?

16.77 The pH of a solution that is 0.020 M in Hg^{2+}, 0.020 M in Ni^{2+}, and 0.020 M in Pb^{2+} is adjusted to 2.0 and then saturated with H_2S gas. Which metal ions, if any, will precipitate?

[16.78] The solubility of magnesium oxalate dihydrate, $MgC_2O_4 \cdot 2H_2O$, is 0.042 g per 100 mL of solution. The acid dissociation constants of oxalic acid are listed in Appendix E. Calculate the pH of a saturated solution of $MgC_2O_4 \cdot 2H_2O$.

[16.79] Calculate the molar solubility of $Fe(OH)_3$ in water. (This problem is not as simple as it might first appear; consider carefully the concentration of OH^-.)

[16.80] What pH range will permit PbS to precipitate while leaving Mn^{2+} in solution if the solution is 0.010 M in Pb^{2+} and 0.010 M in Mn^{2+} prior to saturation with H_2S?

[16.81] Using the K_f value for $Ag(NH_3)_2^+$ that is listed in Table 16.1, calculate the concentration of Ag^+ present in solution at equilibrium when concentrated NH_3 is added to a 0.010 M solution of $AgNO_3$ until the ammonia concentration is 0.20 M. (Neglect the small concentration change in Ag^+ that accompanies the change in the volume of the solution upon addition of NH_3.)

17

Chemistry of natural waters

In the previous two chapters we've focused on equilibria occurring in solutions, mainly in water. The emphasis on water as a solvent is highly appropriate; we live on a planet in which the aqueous environment, called the hydrosphere, dominates.

In this chapter, we shall see how the water on the surface of the planet is divided between saline, or salty, water and fresh water. We'll learn about the composition of ocean water, including something of its physical and chemical properties. The equilibria between the oceans and the gases of the atmosphere are important for the existence of life in the oceans and even on the landmasses. The oceans represent a vast, largely untapped reservoir of chemical and mineral resources.

In the course of being used, the limited fresh water at our disposal becomes contaminated. We'll see how water taken from the environment must be treated so that it is fit for human consumption and other uses. The water we return to the environment should be (but is not always) treated to remove dissolved organic matter and other substances that may be harmful to human health or to other units of the environment. Many substances released in waste waters are toxic, or may become toxic, as a result of chemical changes.

Water-pollution problems often involve complex chemical considerations. The community in which you live may have to cope with problems very different from those encountered somewhere else. As a voting citizen you may have to form an opinion regarding a bond issue to build a new water-treatment plant, or you may have to decide whether a particular industrial operation in your community should be forced to add an expensive water-treatment facility. The answers to questions of this sort are usually not clear-cut, but the better your understanding of the chemical principles involved, the better your chances of forming a sound judgment.

The physical properties of water and the relationships between its solid, liquid, and gaseous phases are important in determining the roles that water plays in our natural environment. It is well to begin, then, by recalling some of the special characteristics of water described in preceding chapters. For a substance of such low molecular weight, water possesses unusually high melting and boiling points (Section 11.5). Methane, CH_4, which has about the same molecular weight as water, boils at 89 K, as compared with 373 K for water. Water also has an unusually high specific heat, 4.184 J/g-°C. The specific heat of most simple organic liquids is only about half this large. This means that a body of water can absorb a certain amount of heat with a smaller temperature increase than is the case for other liquids. The heat of vaporization of water is also exceptionally high; more heat is required to cause the evaporation of a gram of water than is needed to evaporate a gram of any other liquid. About one-third of all the energy that earth receives from the sun is used up in the evaporation of water from the surface of the oceans and other water bodies. The high heat of vaporization of water is thus obviously important in determining the conditions of life on earth. As we shall see in Section 17.3, the heat of vaporization of water is also important in determining the feasibility of recovering fresh water from salt-laden ocean water.

Ice, the solid form of water, has a very interesting and unusual structure, shown in Figure 11.25. Because this structure has so much open space, ice is less dense than liquid water. Water is one of the very few substances for which the solid phase is less dense than the liquid.

All of the unusual physical properties of water in both the liquid and solid forms are related to the formation of hydrogen bonds, as described in Section 11.5.

17.1 Seawater

All of the vast layer of salty water that covers so much of the earth is connected and is of more or less constant composition. For this reason, oceanographers (scientists whose major interest is the sea) speak in terms of a world ocean, rather than of the separate oceans we learn about in geography books. The world ocean is indeed huge. Its volume is 1.35 billion cubic kilometers. It covers about 72 percent of the earth's surface. Almost all the water on earth, 97.2 percent, is in the world ocean. About 2.1 percent is in the form of ice caps and glaciers. All of the fresh water, in lakes, rivers, and ground water, amounts to only 0.6 percent. The remaining 0.1 percent is in the form of brine wells and brackish (salty) waters.

CHEMICAL COMPOSITION OF SEAWATER

Seawater is often referred to as saline water. The salinity of seawater is defined as the mass in grams of dry salts present in 1 kg of seawater. In the world ocean, the salinity varies from 33 to 37, with an average of about 35. To put it another way, this means that seawater contains about 3.5 percent dissolved salts. The list of elements present in seawater is very long. However, most are present only in very low concentrations. Table 17.1 lists the 11 ionic species present in seawater at concentrations greater than 0.001 g/kg, or 1 part per million (ppm) by weight. (To

TABLE 17.1 Ionic constituents of seawater present in concentrations greater than 0.001 g/kg (1 ppm) by weight

Ionic constituent	g/kg seawater	Concentration (M)
Chloride, Cl^-	19.35	0.55
Sodium, Na^+	10.76	0.47
Sulfate, SO_4^{2-}	2.71	0.028
Magnesium, Mg^{2+}	1.29	0.054
Calcium, Ca^{2+}	0.412	0.010
Potassium, K^+	0.40	0.010
Carbon dioxide[a]	0.106	2.3×10^{-3}
Bromide, Br^-	0.067	8.3×10^{-4}
Boric acid, H_3BO_3	0.027	4.3×10^{-4}
Strontium, Sr^{2+}	0.0079	9.1×10^{-5}
Fluoride, F^-	0.001	7×10^{-5}

[a] CO_2 is present in seawater as HCO_3^- and CO_3^{2-}.

convert from g/kg to ppm, multiply by 1000.) In a lower range of concentration, the elements nitrogen, lithium, rubidium, phosphorus, iodine, iron, zinc and molybdenum are present in amounts ranging from 1 ppm to 0.01 ppm. At least 50 other elements have been identified at still lower concentrations.

So far as can be determined, the composition of the oceans is constant with time. Of course, chemical data on the composition of seawater has been available for less than 100 years. If the composition were changing, it would probably be doing so on a time scale much longer than 100 years. However, there are other lines of evidence that also suggest that the chemical makeup of the oceans has not changed materially over a long period.

The constancy of composition of the oceans represents a balance of input and output processes. Water continually flows into the oceans from rivers, bringing in water with very different mineral composition from that already present. For example, the weathering of rocks leads to incorporation of aluminum, silicon, iron, or calcium in the river water. In the sea, these elements eventually become part of a biological cycle or are removed by precipitation. The average abundance of many elements in the ocean thus represents a balance between the input and output rates. The result is a more or less constant composition of the oceans as a function of time.

CARBON DIOXIDE IN SEAWATER

The carbon dioxide present in the atmosphere is in equilibrium with carbon dioxide dissolved in the ocean. The equilibrium between atmospheric CO_2 and the layer of ocean water extending to a depth of about 100 m is established in a period of about 2 years.* On the other hand, the surface waters are brought into equilibrium with deeper waters more slowly, over a period of perhaps several thousand years.

*This means that if we could somehow keep track of all the CO_2 molecules released into the atmosphere during a short time period, the fraction of them dissolved in the upper part of the ocean would be constant after a period of about two years.

Dissolved CO_2 is an important component of the oceanic buffer system. The solubility of CO_2 in seawater is much higher than it is in pure water. Carbon dioxide is about 15 times more soluble in seawater than is O_2, and 30 times more soluble than is N_2, assuming the same pressure of gas over the solution in each case. To understand why CO_2 is so soluble in seawater, we must examine all the equilibria involved:

$$CO_2(g) \rightleftharpoons CO_2(aq) \tag{17.1}$$

$$CO_2(aq) + H_2O(l) \rightleftharpoons H_2CO_3(aq) \tag{17.2}$$

$$H_2CO_3(aq) \rightleftharpoons H^+(aq) + HCO_3^-(aq) \tag{17.3}$$

$$HCO_3^-(aq) \rightleftharpoons H^+(aq) + CO_3^{2-}(aq) \tag{17.4}$$

Both of the ionization equilibria, Equations [17.3] and [17.4], are shifted to the right as pH increases, that is, as H^+ decreases. Because seawater is slightly alkaline, most of the CO_2 dissolved is in the form of HCO_3^- and CO_3^{2-}. As a result, the solubility equilibrium for gaseous CO_2 is thus also shifted to the right, in accordance with LeChatelier's principle.

We saw in Figure 16.7, that the form in which the dissolved CO_2 exists at equilibrium depends very much on the pH of the medium. Figure 16.7 is applicable to CO_2 dissolved in an aqueous solution that is relatively dilute with respect to ionic substances. However, when the total concentration of ionic substances in water is increased, the values for equilibrium constants are altered.* For example, the ionization equilibria of carbonic acid in seawater are shifted from the values for dilute aqueous solution, as shown in Figure 17.1. The solid lines correspond to those shown in Figure 16.7. The dashed lines apply to seawater. We see that at the pH of seawater, about 8, there is a considerably higher concentration of CO_3^{2-} than would be present in a dilute aqueous solution at the same pH. If oceanographers are to understand properly the equilibria involving carbonate and bicarbonate in seawater, they must employ the corrected values.

*When the concentration of ions in solution is high, there is a high probability that ions of opposite charge will be found close together in the solution. Thus the assumption that each ion moves more or less independently of the others no longer holds true. As a result, the equilibrium constants are shifted in value. In accurate work, ionic equilibria in aqueous solutions of fairly high concentration must be corrected for the so-called interionic attractions. There are standard methods for doing this, but we need not concern ourselves with them.

SAMPLE EXERCISE 17.1

The total concentration of CO_2 in a sample of seawater brought up from a depth of 150 m is 0.0043 M. The pH of the sample is 8.3. Using Figure 17.1, determine the concentration of H_2CO_3, HCO_3^-, and CO_3^{2-} in this sample.

Solution: We employ Figure 17.1 to determine the fractions of each of the three species in seawater of pH 8.3. These are H_2CO_3, about 0; HCO_3^-, 0.86; CO_3^{2-}, 0.14. These fractions are then taken times the

total CO_2 concentration to obtain the following approximate molar concentrations:

$$[H_2CO_3] = 0$$
$$[HCO_3^-] = 0.86(0.0043 \ M) = 0.0037 \ M$$
$$[CO_3^{2-}] = 0.14(0.0043 \ M) = 0.0006 \ M$$

We know, of course, that the concentration of H_2CO_3 is not precisely zero; it is merely quite small relative to the concentrations of the other two forms.

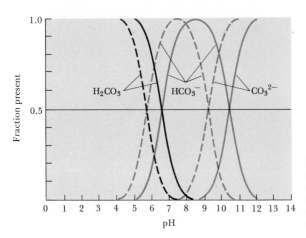

FIGURE 17.1 The dashed lines show the distribution of dissolved CO_2 in seawater between various ionization species, as a function of pH. The solid lines represent the same equilibria in water containing a very low concentration of dissolved ionic substances, as shown in Figure 16.7. Note that CO_2 is, in effect, more acidic in seawater than it is in pure water.

One of the most important equilibria involving dissolved CO_2 in seawater is that for formation of solid $CaCO_3$. The equilibrium between solid $CaCO_3$ and the Ca^{2+} and CO_3^{2-} ions present in the oceans is important in the development of many marine biological forms, for example, corals and shellfish. The solubility product for $CaCO_3$ in seawater at 20°C has the value 6.0×10^{-7}, whereas for $CaCO_3$ in equilibrium with pure water at this temperature it is 2.8×10^{-9}. The solubility equilibrium is shifted toward increased solubility in seawater because of the effects of all the other ions present in the medium. The increase of more than 100-fold in solubility in seawater is another example of the effects of interionic attraction in aqueous media of high concentration.

The ocean appears to be supersaturated with respect to $CaCO_3$ at depths of about 1 km or less. This means that the solubility product for $CaCO_3$ is exceeded. However, the rate at which $CaCO_3$ is removed by precipitation or incorporation into animal shells or skeletal formations is quite small. At lower depths, where the concentration of Ca^{2+} is lower, the ocean is undersaturated. Carbonate shells formed near the surface sink to lower depths when the organism dies and dissolve there. At water depths greater than about 3 to 4 km, the degree of undersaturation with respect to $CaCO_3$ seems to be quite large. As a result, there is very little $CaCO_3$ in sediments taken from the ocean bottom at these depths and below.

BIOCHEMICAL PROCESSES IN THE SEA

Plants and animals present in the ocean exert an important influence on the composition of seawater. The simplest elements of the food chain are the phytoplankton, minute plants in which CO_2, water, and other nutrients are converted by photosynthesis into plant organic matter. Analysis of the composition of phytoplankton shows that carbon, nitrogen, and phosphorus are present in atomic ratios of 108:16:1, as illustrated in Figure 17.2. Thus, about 108 molecules of CO_2 are required for each 16 atoms of nitrogen (usually in the form of nitrate ion) and for each single atom of phosphorus (usually present as hydrogen phosphate ion, HPO_4^{2-}). Because of its high solubility in seawater, CO_2 is always present in excess. The concentration of nitrogen or phosphorus is therefore

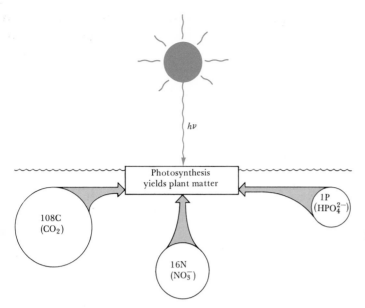

FIGURE 17.2 Schematic diagram of photosynthesis in phytoplankton at or near the surface of the sea. The relative amounts of required nutrients are indicated in the circles. The chemical forms shown are not the only ones usable, but are simply the more abundant ones.

the limiting factor in the rate of formation of organic matter via photosynthesis.

The concentrations of nitrate and phosphate are complicated functions of ocean currents and many other factors. Vertical mixing of water is most important. This is so because the deep ocean water is higher in phosphate and nitrate than is the water near the surface. A typical profile of phosphate concentration as a function of depth is shown in Figure 17.3. The curve for nitrate concentration is very similar.

Photosynthesis occurs in the photosynthetic zone near the surface, where the sun's rays are strongest. Thus, in the water extending to a depth of about 150 m, phosphate and nitrate concentrations are depleted because of the photosynthesis that has already occurred. At lower depths, dead plant and animal matter decomposes, restoring the phosphate and nitrate levels.

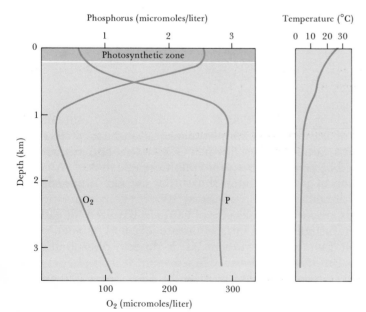

FIGURE 17.3 Concentrations of dissolved phosphorus (mostly as phosphate) and O_2 as a function of depth in the sea. The scale of phosphorus concentration is along the top of the figure; the oxygen concentration scale is along the bottom.

In the photosynthetic zone, the oxygen concentration is high because oxygen is released during photosynthesis. At lower depths, the oxygen level drops sharply because the oxygen is used up in oxidizing dead plant and animal matter. As shown in Figure 17.3, the oxygen concentration is at a minimum at about 1 km depth, the same region in which the phosphate level is restored to its highest values.

All the higher forms of life in the ocean are ultimately related to the phytoplankton at the base of the food chain. Thus, the abundance of fish is coupled very closely to the conditions that determine the rate of photosynthesis. In certain regions of the ocean, where seasonal upwellings of nutrient-rich lower water occur, the rate of photosynthesis may become very high. For example, the Grand Banks in the northern Atlantic, off the coast of Newfoundland, are one of the richest fishing regions in the world ocean. The abundance of fish there is due to an upwelling of nutrient-rich deep water.

17.2 Raw materials from seawater

The sea is a vast storehouse of chemicals. Each cubic mile of seawater contains 1.5×10^{11} kg of dissolved solids. The sea is so vast that if a substance is present in seawater to only the extent of 1 part per billion by weight, there are still 5×10^9 kg of it in the world ocean. Nevertheless, the ocean is not used very much as a source of raw materials, because the costs of extracting the desired substances from the water are too high. Only three substances are recovered from seawater in commercially important amounts: sodium chloride, bromine, and magnesium.

SODIUM CHLORIDE

Because sodium chloride is the most abundant dissolved substance in seawater, it is perhaps not too surprising that a considerable amount (about 4×10^{10} kg each year) of pure sodium chloride is recovered from the sea. Seawater is filtered and then allowed to evaporate partially until the solubility of NaCl is exceeded. The solid NaCl that crystallizes is quite pure, but it may be recrystallized from fresh water for still higher purity, depending on the intended use.

BROMINE

Seawater is the major source of the over 2.3×10^8 kg of bromine produced worldwide each year. From Table 17.1 we see that bromide ion is present in seawater at a concentration of only 8.3×10^{-4} M. Recovery of pure bromine thus represents a considerable concentration. The first step in removal of bromine is addition of sulfuric acid to the seawater to adjust the pH to 3.5. Then chlorine gas is added in slight excess as compared with the amount of bromine present. An oxidation-reduction reaction occurs between the dissolved chlorine gas and bromide ion:

$$\text{Cl}_2(aq) + 2\text{Br}^-(aq) \rightleftharpoons \text{Br}_2(aq) + 2\text{Cl}^-(aq) \qquad [17.5]$$

We might expect that this reaction would occur, because chlorine is a more electronegative element than bromine.

Bromine is freed by passing the seawater through a tower packed with wood strips, while air is swept through in the reverse direction. To recover the bromine from this airstream, it is treated with SO_2 and steam. A solution of hydrobromic acid and sulfuric acid is formed:

$$SO_2(g) + Br_2(g) + 2H_2O(l) \rightleftharpoons 2HBr(aq) + H_2SO_4(aq) \qquad [17.6]$$

Bromine can be recovered from this solution by again treating with just the right amount of chlorine, as in Equation [17.5], and again sweeping the gaseous bromine out with air. The bromine is recovered from the airstream by allowing it to pass over a cold surface. The boiling point of liquid bromine is 59°C, so it can be separated from water by distillation. The dilute sulfuric acid solution left after bromine is removed, Equation [17.6], is used to acidify the incoming fresh batch of seawater.

MAGNESIUM

Magnesium is the second most abundant metallic element in seawater. The largest plant for production of magnesium from seawater in the United States is operated by the Dow Chemical Company at Freeport, Texas. In the process used, Mg^{2+} is precipitated from seawater in large settling ponds (Figure 17.4) as $Mg(OH)_2$, $K_{sp} = 1.8 \times 10^{-11}$, by addition of lime, CaO. The calcium oxide employed in the process is obtained from oyster shells in nearby Galveston Bay. Oyster shells are composed of calcium carbonate. They are washed, then heated in a kiln to form lime:

$$CaCO_3(s) \longrightarrow CaO(s) + CO_2(g) \qquad [17.7]$$

FIGURE 17.4 An aerial view of settling ponds (*right of center*) used in precipitating $Mg(OH)_2$ from seawater. (*Courtesy of Dow Chemical U.S.A.*)

FIGURE 17.5 A view of a row of cells in which molten $MgCl_2$ is electrolyzed to form Mg. The photo shows only the tops of the cells. The round vertical rods are carbon anodes. The rectangular bars are copper lines that carry up to 100,000 amps of current to the cells. (*Courtesy of Dow Chemical U.S.A.*)

Magnesium hydroxide is formed in the following reaction:

$$Mg^{2+}(aq) + CaO(s) + H_2O(l) \longrightarrow Mg(OH)_2(s) + Ca^{2+}(aq) \qquad [17.8]$$

The magnesium hydroxide that precipitates out is contaminated with Ca^{2+}, Na^+, and HCO_3^- ions. The solid is filtered and then treated with a mixed HCl and H_2SO_4 solution. The $Mg(OH)_2$ dissolves as the solution is made acidic (see Section 16.5):

$$Mg(OH)_2(s) + 2H^+(aq) \longrightarrow Mg^{2+}(aq) + 2H_2O(l) \qquad [17.9]$$

Most of the sodium ion impurity crystallizes as NaCl, and calcium ion is precipitated as $CaSO_4$. The solution of Mg^{2+} is filtered and then concentrated in an evaporator. Solid $MgCl_2$ is eventually recovered. It is dissolved in a mixture of molten metal chlorides at 700°C in electrolysis cells (see Figure 17.5). In these cells electrical energy is supplied to cause the formation of magnesium metal and chlorine gas from the molten metal chloride (electrolysis is discussed in more detail in Chapter 19):

$$MgCl_2(l) \xrightarrow{\text{electrical energy}} Mg(l) + Cl_2(g) \qquad [17.10]$$

The molten metal is cast into ingots of 99.9 percent purity.

17.3 Desalination

Although it may seem strange at first, the most valuable component of seawater may prove in the long run to be the water itself. Shortages of fresh water have developed even in countries such as the United States that are relatively well supplied with rainfall. In many regions of the United States, the demand for fresh water for home, agricultural, and industrial uses exceeds the available supply. In countries such as Israel or

Kuwait, where rainfall is low, the supply is totally inadequate to meet the demands brought on by modernization and increasing population. Inevitably, humankind must look to the oceans as the source of the needed water.

Because of its high salt content, seawater is unfit for human consumption and indeed for most of the uses to which we put water. In the United States, the salt content of municipal water supplies is restricted by health codes to no more than about 0.05 percent. This is much lower than the 3.5 percent dissolved salts present in seawater, or the 0.5 percent or so present in brackish water found underground in some regions. The removal of salts from seawater or brackish water to the extent that the water becomes usable is called desalination.

There are many different ways to desalinate water, and any one of them can be made the basis of a large production plant. The challenge is to carry out the desalination with the absolute minimum energy requirement and with the least possible investment in equipment and facilities. This is important because a nation or region that must rely upon desalinated water to a large extent must compete economically with other nations that may have more abundant and cheaper sources of fresh water. A small nation such as Kuwait, situated on the Persian Gulf, which has almost no fresh-water resources, can afford to depend to a large extent on desalinated water only because it derives a very high per capita income from its oil production.

DESALINATION BY DISTILLATION

Water can be separated from dissolved salts by distillation, as described in Section 2.3. This process takes advantage of the fact that water is a volatile substance, whereas the salts are nonvolatile. The principle of distillation is simple enough, but there are many problems associated with carrying out the process on a large scale. For example, as water is distilled from a vessel containing seawater, the salts become more and more concentrated and eventually precipitate out. This causes formation of scale, which in turn causes poor heat transfer through the vessel, plugging of pipes, and so forth. The obvious solution is to discard the seawater after a certain amount of water has been distilled from it and begin again with a new batch. But unless this is done very carefully, all the heat values stored in the hot seawater will be lost, and additional heat must be then supplied to heat up the cold, incoming seawater. The heat lost represents wasted energy and higher costs. In addition, if the distillation is carried out at atmospheric pressure, the water must be heated to 100°C; at lower pressures, the boiling point of water would be lowered, and less heat would be required.

One rather successful attempt to get around some of these difficulties is called the multistage flash-distillation process, which is illustrated in Figure 17.6. To see how this method works, let's begin at A with warm seawater, which we'll call brine. The brine is pumped under pressure through the coils of a condenser in B, then into C, and then into D, growing hotter in each chamber. The heat comes from condensing steam that forms on the coils in each chamber. The condensed steam, which is fresh water, is collected and pumped off. In region E the heated brine is

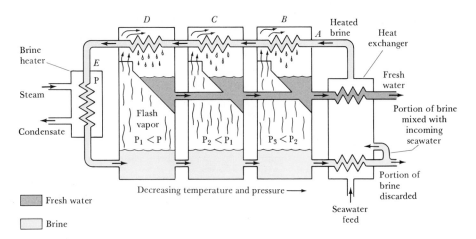

FIGURE 17.6 Schematic diagram of a multistage flash-distillation unit for desalination.

heated still further by steam that passes around the coils. The steam used at this point represents a large fraction of the total energy input to the system. From E the heated brine passes into chamber D, which is held at a lower pressure. Because the pressure is lower in this chamber, some of the brine is "flashed off," or distilled to form water vapor. We know that energy is required to evaporate water. As water evaporates from our bodies the surface that remains is cooled. In the same way, the brine that remains after some water has evaporated, or flashed off, is cooler. It then passes into chamber C, in which the pressure is a little lower still. A bit more of the water evaporates, and the brine is cooled still further. In each succeeding stage, the brine becomes more concentrated in salts and lower in temperature. At the end, a portion of the brine, which is now about 7 percent salt by weight, is mixed with incoming seawater. The other portion is discarded to prevent the salt concentration from getting too high.

A large multistage flash-distillation unit is shown in Figure 17.7. This particular unit is designed to produce about 9 million liters of fresh water per day. The efficiency of a multistage flash-distillation system is limited more than anything else by the formation of scale in the hot-brine circulating system. The major culprits are calcium carbonate and magnesium hydroxide. Various additives have been employed to inhibit formation of these precipitates and thus permit use of higher temperatures. At higher temperatures, however, calcium sulfate precipitation becomes a problem. The chemist thus has an important role to play in improving this particular approach to desalination.

The large heat requirements of any distillation process are a major factor in the overall cost. In a typical multistage flash-distillation unit, the cost of the steam represents about 40 percent of the cost of the water. Many other schemes for desalinating water that bypass the need for vaporization have been advanced. In one method the water is removed from seawater by freezing. As ice freezes out of seawater, the dissolved salts are left behind. Of course the freezing process also requires energy, as anyone who has made ice cubes in a home refrigerator knows. Large-scale processes using the freezing technique are currently being tested.

FIGURE 17.7 A multistage flash-distillation unit. This unit is capable of producing about 9 million liters of fresh water each day. (*Courtesy of Aqua-Chem, Inc., Milwaukee, Wisconsin*)

DESALINATION BY REVERSE OSMOSIS

In the reverse-osmosis method of desalination, water is separated from dissolved salts by means of a membrane that is permeable to the passage of water, but not of dissolved salts. As we learned in Section 12.6, the phenomenon of osmosis depends on having a selective membrane that permits the flow of water through it, but not substances dissolved in the water. If such a membrane is placed between a brine solution and pure water, the tendency to equalize concentrations on each side would cause water to flow through the membrane into the brine solution. The flow may be countered by applying a pressure on the side of the brine solution. If the pressure is sufficiently high, the flow may be stopped altogether. The pressure required to balance the tendency of water to flow across the membrane is called the osmotic pressure. For seawater at standard conditions, the osmotic pressure is about 25 atm.

SAMPLE EXERCISE 17.2

Using the ion concentrations given in Table 17.1, calculate the osmotic pressure of seawater at 0°C. (Refer back to Section 12.6, if necessary.)

Solution: From Section 12.6 we have that osmotic pressure, π, is given by $\pi = MRT$, where M is molarity, R is the ideal gas constant, 0.0821 L-atm/K-mol, and T is the absolute temperature. Recall that osmotic pressure is a colligative property and will therefore depend on the concentration of all ions in solution. If we add the concentrations of all ions in Table 17.1, we have $M = 1.12$. Thus

$$\pi = MRT$$
$$= \left(1.12 \ \frac{\text{mol}}{\text{L}}\right)\left(0.0821 \ \frac{\text{L-atm}}{\text{K-mol}}\right)(273 \text{ K})$$
$$= 25.1 \text{ atm}$$

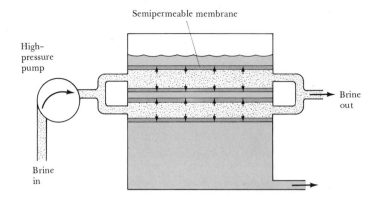

Semipermeable membrane

High–
pressure
pump

Brine
out

Brine
in

FIGURE 17.8 Schematic diagram of a reverse-osmosis process for desalination of brackish water or seawater. The pressure applied by the high-pressure pump is sufficient to overcome the osmotic pressure of the brine solution compared with fresh water. As a result, fresh water flows across the semipermeable membrane. To prevent the buildup of salt near the membrane surface, the pump must continually circulate brine through the tubes. In practice the tubes may be very small in diameter, and the apparatus may consist of many thousands of tubes.

If the pressure on the brine solution is increased beyond the osmotic pressure, the direction in which water tends to flow across the membrane is reversed, and fresh water passes from the brine into the fresh water side. This process, called reverse osmosis, is illustrated in Figure 17.8. Seawater or brackish water is pumped at high pressure into chambers lined with semipermeable membranes. As the water passes across the membrane the local concentration of salt at the wall of the membrane increases. This causes the osmotic pressure to increase, and decreases the flow of fresh water. To prevent this, seawater must be continually pumped through. The flow of water through the membrane is proportional to the applied pressure. The maximum pressure that can be applied is limited by the characteristics of the membranes. Under excessive pressure the membranes may rupture, become fouled with impurities present in the water, or leak through an excessive amount of dissolved salts. In a typical reverse-osmosis unit, the tubes illustrated in Figure 17.8 are constructed of a porous material lined on the inside with an extremely thin film of cellulose acetate. The cellulose acetate (from which cellophane and photographic film are made) acts as the semipermeable membrane. The unit consists of many such tubes arranged in parallel array. The rate of water flow through the membrane is rather low. For example, using brackish water of about 0.5 percent dissolved salts, and a pressure of 50 atm, about 700 L of fresh water per square meter of membrane can be obtained per day. Because it requires a great many small tubes to produce a large surface area, the reverse-osmosis process has not yet been used to produce large quantities of usable water. This process does appear to have promise, with development of improved membranes, for purification of brackish waters, because brackish water requires lower pressures than does seawater.

17.4 Fresh water

We've seen that the total amount of fresh water on earth is not a large fraction of the total water present. What there is of it can be traced to evaporation of water from the oceans and the land, and to transpiration through the leaves of plants. The water vapor that accumulates in the atmosphere is transported via global atmospheric circulation to other latitudes, where it falls as rain or snow. The water that falls on land runs off in rivers or collects in lakes or underground caverns. Eventually it is evaporated or carried via streams and rivers back to the oceans.

BOD AND WATER QUALITY

Fresh water, of course, isn't all water. It contains dissolved gases (principally O_2, N_2, and CO_2), a variety of cations (mainly Na^+, K^+, Mg^{2+}, Ca^{2+}, and Fe^{2+}) and a variety of anions (mainly Cl^-, SO_4^{2-}, and HCO_3^-). Suspended solids such as tiny clay particles are also likely to be present.

The amount of dissolved oxygen present is an important indicator of the quality of the water. Oxygen is necessary for fish and much other aquatic life. Cold-water fish require about 5 ppm of dissolved oxygen for survival. Water fully saturated with air at 1 atm and 20°C contains about 9 ppm of O_2. Aerobic bacteria consume dissolved oxygen as they utilize it to oxidize organic materials as food to meet their energy requirements. The organic material that the bacteria are able to oxidize is said to be **biodegradable**. This oxidation occurs by a complex set of chemical reactions, and the organic material disappears gradually. The carbon, hydrogen, oxygen, nitrogen, sulfur, and phosphorus in the biodegradable material ends up mainly as CO_2, H_2O, NO_3^-, SO_4^{2-}, and phosphates. These oxidation reactions sometimes reduce the amount of dissolved oxygen to the point where the aerobic bacteria can no longer survive. Anaerobic bacteria then take over the decomposition process, forming products such as CH_4, NH_3, H_2S, and PH_3 that contribute to the offensive odors of some polluted waters.

The amount of oxygen required to decompose all of the biodegradable organic wastes in water is called the **biological oxygen demand (BOD)**. The BOD indicates the organic pollution load of the water. The standard test for such organic material is the 5-day BOD test. In this test the contaminated water is diluted in air-saturated distilled water to ensure an excess of oxygen. The amount of dissolved oxygen in the resultant solution is measured. The solution is then stored for 5 days at 20°C, after which time the amount of dissolved oxygen is again measured. The 5-day BOD, BOD_5, is calculated from the amount of dissolved oxygen consumed. The 5-day BOD is usually about three-quarters of the total BOD of the water. Drinkable water normally has a BOD_5 of 1.5 ppm O_2 or less. Raw, untreated sewage usually has a BOD_5 range of from 100 to 400 ppm O_2.

The major elements required by plants are carbon, hydrogen, oxygen, nitrogen, and phosphorus. The carbon, hydrogen, and oxygen are readily supplied by CO_2 and H_2O. The nitrogen and phosphorus are much scarcer, and a lack of these elements often limits the growth of plants (Section 17.1). Consequently, they are the principal elements contained in fertilizers.

Lakes are born in a nutrient-poor, or **oligotrophic,** condition and consequently support little plant life. As a result, the water is clear and saturated with dissolved oxygen. As the nutrient supply, especially of nitrates and phosphates, increases, the lake becomes nutrient-rich, or **eutrophic.** The nutrients promote growth of water weeds and algae. As these plants die their decay consumes dissolved oxygen. This oxygen depletion causes the generation of smelly chemicals from anaerobic decay. Furthermore, the dead plants and algae fill the bottom of the lake, causing it to become shallow and to eventually fill and die. Eutrophication is a natural aging process by which lakes are converted slowly (over thousands of years) into marshes and eventually into meadows or forests. Human activity has greatly accelerated eutrophication in many lakes owing especially to sewage and fertilizers entering the water.

An estimated 8000 L of water per person are used daily in the United States for personal use, for agriculture, and for industry. This represents about one-third of the available water. About 10 percent of this water is used for domestic purposes, the remainder being used in agriculture or industry. It requires about 200 L of water to process a pound of sugar, 700 L to grow a pound of corn (including rainfall), and 1200 L to manufacture a pound of synthetic rubber.

Domestic water use is increasing faster than population growth, because of the growing use of devices such as automatic dishwashers, clothes washers, and garbage disposals. A dishwasher uses about 40 L of water per load, twice that needed to wash dishes by hand. A typical American uses about 200 L of water each day, not counting lawn and garden watering.

17.5 Water treatment

The water needed for domestic uses, for agriculture, or for industrial processes is taken from naturally occurring lakes, rivers, and underground sources or from reservoirs. Much of the water that finds its way into municipal water systems is "used" water; it has already passed through one or more sewage systems or industrial plants. Consequently, it is usually necessary to treat the water before it is distributed to our faucets.

After we have used the water, it is once again necessary to treat the water so that it does not pollute the lakes and rivers to which it is returned. The importance of treating sewage and waste water from industrial uses becomes all the more apparent when one considers that many waters receive repeated use in the course of their journey to the sea. For example, the water taken by the city of New Orleans from the Mississippi River contains the effluents of many cities that lie along the banks of the Mississippi, Ohio, and Missouri rivers.

TREATMENT OF MUNICIPAL WATER SUPPLIES

Municipal water treatment usually involves five steps: coarse filtration, sedimentation, sand filtration, aeration, and sterilization. Figure 17.9 shows a typical treatment process.

After coarse filtration through a screen, the water is allowed to stand in large settling tanks in which finely divided sand and other minute particles can settle out. To aid removal of very small particles, the water may first be made slightly basic by adding CaO, and then $Al_2(SO_4)_3$ is added. The aluminum sulfate reacts with OH^- ions to form a spongy, gelatinous precipitate of $Al(OH)_3$. This precipitate settles slowly, carrying suspended particles down with it, thereby removing nearly all finely divided matter and most bacteria. The water is then filtered through a sand bed. Following filtration the water may be sprayed into the air to hasten the oxidation of dissolved organic substances.

The final stage of the operation normally involves treating the water with a chemical agent to ensure the destruction of bacteria. Ozone, O_3, is most effective, but it must be generated at the place where it is used.

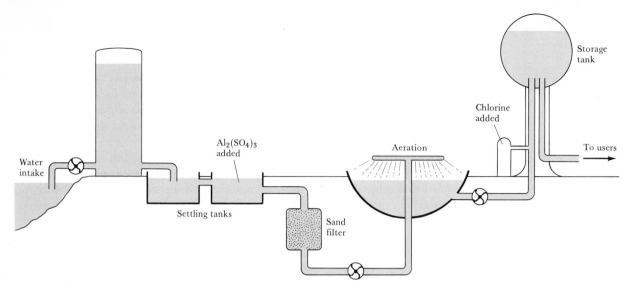

FIGURE 17.9 The common steps in treating public drinking water.

Chlorine, Cl_2, is therefore more convenient to use. Chlorine can be shipped in tanks as the liquefied gas and dispensed from the tanks through a metering device directly into the water supply. The amount used depends on the presence of other substances with which the chlorine might react and on the concentrations of bacteria and viruses to be removed. The sterilizing action of chlorine is probably due not to Cl_2 itself, but to hypochlorous acid, which forms when chlorine reacts with water:

$$Cl_2(aq) + H_2O(l) \longrightarrow HOCl(aq) + H^+(aq) + Cl^-(aq) \qquad [17.11]$$

Although chlorine has been used to sterilize water for many years with no obvious harmful effects on those who use the water, a potential hazard has recently been discovered. Studies of the water supplies in several American cities have shown the presence of minute amounts of chloroform, $CHCl_3$, and carbon tetrachloride, CCl_4. These substances are known to be toxic. Although the levels at which they are present in the water supply are very low, the possibility exists that long-term consumption of water containing them may damage the liver and kidneys. They are believed to have been formed by reaction of organic pollutant molecules present in the water with the chlorine used for sterilization.

WATER SOFTENING

The water treatment described thus far should remove all the substances potentially harmful to health. Sometimes additional treatment is used to reduce the concentrations of Ca^{2+} and Mg^{2+}, which are responsible for water hardness. These ions react with soaps to form an insoluble material. Although they do not form precipitates with detergents, they adversely affect the performance of such cleaning agents. In addition, mineral deposits may form when water containing these ions is heated. When water containing Ca^{2+} and bicarbonate ions is heated, some carbon dioxide is driven off. The result is a shift in pH to higher values, and formation of insoluble calcium carbonate:

$$Ca^{2+}(aq) + 2HCO_3^-(aq) \xrightarrow{heat} CaCO_3(s) + CO_2(g) + H_2O(l) \qquad [17.12]$$

The solid $CaCO_3$ coats the surfaces of hot-water systems and the insides of teakettles, thereby reducing heating efficiency. Deposits of scale can be especially serious in boilers in which water is heated under pressure in pipes running through a furnace. Formation of scale reduces the efficiency of heat transfer and may result in melting of the pipes.

Not all municipal water supplies require water softening. In those that do, the water is generally taken from underground sources in which the water has had considerable contact with limestone ($CaCO_3$) and other minerals containing Ca^{2+}, Mg^{2+}, and Fe^{2+}. The lime-soda process is used for large-scale municipal water-softening operations. The water is treated with "lime," CaO (or "quicklime," $Ca(OH)_2$), and "soda ash," Na_2CO_3. These chemicals cause precipitation of calcium as $CaCO_3$ and of magnesium as $Mg(OH)_2$. The role of Na_2CO_3 is to increase the pH and to provide a source of CO_3^{2-}, if needed. If the water already contains a high concentration of bicarbonate ion, calcium can be removed as $CaCO_3$ simply by increasing the pH by addition of $Ca(OH)_2$:

$$Ca^{2+}(aq) + 2HCO_3^-(aq) + [Ca^{2+}(aq) + 2OH^-(aq)] \longrightarrow$$
$$2CaCO_3(s) + 2H_2O(l) \qquad [17.13]$$

Lime is used only to the extent that bicarbonate is present: 1 mol of $Ca(OH)_2$ for each 2 mol of HCO_3^-. When bicarbonate is not present, addition of Na_2CO_3 causes removal of Ca^{2+} as $CaCO_3$. The strongly basic carbonate ion (Sections 15.2 and 15.6) also serves to raise the pH enough to cause precipitation of $Mg(OH)_2$:

$$Mg^{2+}(aq) + 2CO_3^{2-}(aq) + 2H_2O(l) \longrightarrow$$
$$2HCO_3^-(aq) + Mg(OH)_2(s) \qquad [17.14]$$

There are two problems with the lime-soda process. First, formation of $CaCO_3$ and $Mg(OH)_2$ precipitates may take a long time, and they may not settle out very well. Second, the pH of the resulting water is too high. Usually $Al_2(SO_4)_3$ (alum) is added to remove the precipitates. In basic solution, Al^{3+} forms $Al(OH)_3$, a gelatinous precipitate that carries finely divided solids with it as it settles out of solution. To prevent the remaining $Mg(OH)_2$ and $CaCO_3$ from later precipitating out and causing trouble, CO_2 is bubbled through the water. This has the effect of lowering the pH to around 8, thus preventing further precipitation.

SEWAGE TREATMENT

Municipal sewage treatment is generally divided into three stages referred to as primary, secondary, and tertiary treatments. About 10 percent of sewage handled by public sewers receives no treatment, about 30 percent receives only primary treatment, while about 60 percent is subject to secondary treatment as well. Tertiary treatment is presently rare,

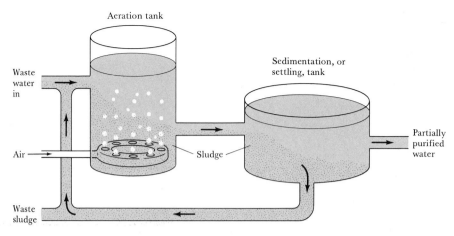

FIGURE 17.10 The activated-sludge process.

but it is expected to become more common as communities upgrade their treatment facilities to meet federal water-pollution standards.

Primary treatment consists first of screening the incoming sewage to filter out debris and larger suspended solids. Then the sewage is passed into settling or sedimentation tanks where suspended solids, called sludge, settle out. If the water receives no secondary treatment it is then often treated with chlorine before being dumped back into the natural water system. Primary treatment removes about 60 percent of the suspended solids and 35 percent of the BOD.

Secondary treatment is based on aerobic decomposition of organic material. The most common type of secondary treatment is known as the activated-sludge method. In this method the waste from primary treatment is passed into an aeration tank where air is blown through it as shown in Figure 17.10. This aeration results in rapid growth of aerobic bacteria that feed on the organic wastes in the water. The bacteria form a mass called activated sludge. This sludge settles out in sedimentation tanks and the liquid effluent is discharged, often after chlorination. Most of the activated sludge is returned to the aeration tank, where it aids in decomposing the organic wastes in the incoming water. After secondary treatment, about 90 percent of suspended solids and 90 percent of BOD has been removed.

Water that has received only primary and secondary treatment may contain relatively large quantities of phosphorus and nitrogen. These can cause damage to natural waters by promoting excessive growth of algae. In addition, many chemicals present in sewage are not affected by secondary treatment and simply pass through and are released to the environment. The cost of removing the many metals and organic substances that might be present in waste water is high. As a result, very little waste water receives a general tertiary treatment, in which such contaminants are removed.

The substances that might contaminate waste water when it is returned to the environment are derived from all the many substances flushed down toilets, ground up in garbage disposal units, and rinsed down the drains in hospitals, stores, factories, and laboratories. In addition to these effluents, natural waters receive the water from storm drains and the runoff from cattle feedlots and from farmland dosed with fertilizers, insecticides, and weed-killing chemicals. We cannot consider here the long list of known contaminants. By way of example, we'll look at the characteristics of some common metallic elements likely to be present.

17.6 Pollutant metallic elements

Table 17.2 is a list of some of the more common (and in some cases more troublesome) metallic elements likely to be present in contaminated natural waters. For good measure we've added two nonmetallic elements, arsenic and selenium. The table lists the typical chemical form or oxidation state for each element. In any particular situation, the element might be present in some other form, depending on the source.

You may be surprised to note from Table 17.2 that many of the metals that can act as pollutants are actually essential in human nutrition. Copper affords a good example; the toxicity of this element is relatively low. An absence of copper(II) in the diet produces an anemic, or iron-deficient, condition, because copper is used in the body along with iron in some metabolic processes. The minimum dietary requirement seems to be on the order of 2 mg of copper per day. But a much higher intake, say about 50 mg per day or more, causes diarrhea, vomiting, and other miserable symptoms.

TABLE 17.2 Pollutant elements in water supplies

Element	Common chemical state	Essential in nutrition	Toxicity	Toxic effects	Sources	U.S. Public Health Service limits per liter
Arsenic	AsO_2^-	No	High	Kidney failure, mental disturbance	Fossil-fuel combustion, detergents, smelting, pesticides	0.05 mg
Cadmium	Cd^{2+}	No	High	High blood pressure, kidney damage, red blood cell loss	Metal plating, mining, cigarette smoke	0.01 mg
Chromium	CrO_4^{2-}	Yes	Medium	Suspected carcinogen	Electroplating	0.05 mg
Copper	Cu^{2+}	Yes	Low	Liver damage	Mining, metal plating, copper pipes	1 mg
Iron	Fe^{2+}, Fe^{3+}	Yes	Low	Excessive intake may increase susceptibility to infection	Mineral sources, corroded metal	0.3 mg[a]
Lead	Pb^{2+}	No	High	Anemia, kidney failure, mental retardation (children), convulsions	Lead piping, lead paints, auto emissions from leaded gasoline	0.05 mg
Manganese	Mn^{2+}	Yes	Low	Not well characterized	Industrial waste, acid mine drainage	0.05 mg[a]
Mercury	Hg^{2+}, Hg_2^{2+}, CH_3Hg^+	No	High	Neurological damage, paralysis, insanity, blindness, birth defects	Chemical plant wastes, discarded mercury batteries	0.002 mg
Silver	Ag^+	No	Medium	Discoloration of skin and eyes	Electroplating, manufacturing	0.05 mg
Selenium	SeO_3^{2-}, SeO_4^{2-}	Yes	High	Liver damage, mental disturbance	Minerals, smelting operations	0.01 mg
Zinc	Zn^{2+}	Yes	Low	Not characterized	Metal-plating wastes, acid mine drainage	5 mg

[a]The limits on manganese and iron are not determined by their toxicity, but because they stain clothing and ceramic plumbing fixtures.

The metals vary a great deal in their toxicities and in the variety of toxic effects they bring about. This is so because they differ in the kinds of chemical reactions they undergo with biochemical systems. Although not all of the biochemical processes involved are understood, we have quite good evidence about what is happening in some cases. Cadmium, for instance, owes its high toxicity to the fact that it is chemically similar to zinc, a metallic element that is essential in many biochemical reactions. Cadmium is apparently enough like zinc to take its place in biochemical systems, but once there it fails to perform precisely as zinc would.

It is worthwhile to consider mercury as a toxic substance in some detail, because several important points that could apply to any pollutant are involved. In the first place, *the toxicity of a substance may depend greatly on its chemical state*. Metallic mercury has a small but finite vapor pressure. If mercury metal is allowed to stand open in a poorly ventilated room for a long time, persons regularly occupying that room may inhale enough mercury over a period of time to show toxic symptoms of poisoning. However, ingestion of a small amount of mercury, as from a bit of silver amalgam a dentist uses to fill a cavity, is not considered a serious hazard; the metal passes through the system without undergoing chemical change. Compounds of mercury(I), such as calomel, Hg_2Cl_2, are not especially toxic, because they have very low solubility in water. The insoluble salts pass through the digestive system without any significant transfer to the bloodstream. Mercuric ion, Hg^{2+} is a very dangerous form of the element. When taken into the body as Hg^{2+}, the element affects the central nervous system, producing symptoms of insanity. Years ago mercuric nitrate, a water-soluble salt of mercury, was used to soften the fur used in the making of felt hats. The phrase "mad as a hatter" originated from the symptoms displayed by hat workers suffering from mercury poisoning.

The most toxic of all forms of mercury are those compounds in which the element is combined with an organic group. Among these are the methyl mercury ion, CH_3Hg^+, and dimethyl mercury, $(CH_3)_2Hg$. The latter compound is a volatile, strong-smelling substance that boils at 96°C. It is readily absorbed through the skin or may be inhaled as the vapor. It is regarded as one of the most poisonous substances known.

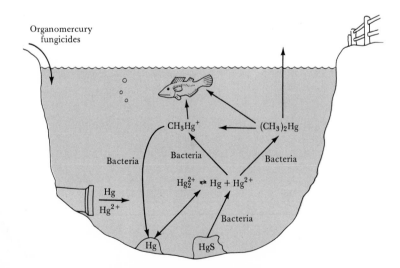

FIGURE 17.11 Chemical interconversions of mercury in natural waters polluted by mercury, sewage wastes, and other contaminants. Mercury in the water may originate from sewage-disposal outlets or in the runoff from farmlands on which organomercury fungicides have been applied. The mercury may be present in the water as the element or as Hg_2^{2+} or Hg^{2+} ion. It may also react with other substances to form an insoluble precipitate. However, bacteria act upon the mercury in these forms to convert it into dimethyl mercury, $(CH_3)_2Hg$, or methyl mercury ion, CH_3Hg^+. These species are taken up by aquatic life and are eventually concentrated in the fatty tissues of fish.

These observations show clearly that the toxicity of mercury varies a great deal with its chemical state. The second thing we must realize is that in nature *a substance may undergo chemical transformation that converts it from a relatively harmless form to a deadly one.* Figure 17.11 shows a diagram that illustrates how this happens in the case of mercury. For many years mercury metal was used in electrolysis cells for production of chlorine and sodium hydroxide. The metal escaped into the environment as the free element or as Hg^{2+}. The tiny amount of metal that escaped found its way to the bottoms of lakes and waterways. There it could react, possibly with some form of sulfur to form insoluble HgS, or possibly to form some other insoluble salt. However, the bottoms of rivers, lakes, and waterways contain abundant bacterial life, and after a time the sulfide is oxidized to sulfate, and Hg^{2+} is released to the water. In addition, if Hg_2^{2+} ion should be formed, it is capable of undergoing disproportionation* to mercury metal and Hg^{2+}:

$$Hg_2^{2+}(aq) \rightleftharpoons Hg(l) + Hg^{2+}(aq) \qquad [17.15]$$

This interconversion reaction can be catalyzed by microorganisms and thus serve as an additional source of Hg^{2+}. Other sources of mercury might also find their way into waters. For example, certain mercury-containing organic compounds have long been used to prevent fungus growth on seeds, in paper and paint products, and in other applications. These mercury compounds are toxic in their own right.† In addition, they may also be converted into other even more toxic forms.

The same waters that contain the mercury impurities also contain sewage wastes, and thus large numbers of bacteria that act upon the organic compounds may be present in the waters. These bacteria are able to react with mercury(II), adding one or two methyl groups to the metal and forming CH_3Hg^+ and $(CH_3)_2Hg$. Thus the mercury that might have entered the environment in any one of a number of forms becomes converted to highly toxic forms.

The third factor that is important in considering the effects of a pollutant is often referred to as biological concentration. Both CH_3Hg^+ and $(CH_3)_2Hg$ have a tendency to accumulate in organisms; the presence of the one or two methyl groups increases solubility in organic substances. The result is that the two species accumulate in the plants and tiny organisms on which fish feed and eventually concentrate in the fish themselves. The concentration of mercury in fish may be as much as 1000 times that in the waters from which the fish are taken. This means that, when waters contain mercury compounds at concentrations in the range of a few parts per billion, the fish that live in those waters may contain one or more parts per million of the element. U.S. and Canadian health authorities have set an upper limit of 0.5 parts per million on the mercury content of fish which may be sold or given away. In 1970 it was

*A disproportionation reaction is one in which a substance is converted into two different chemical forms, one of higher and the other of lower oxidation state than the reactant form. For example, in Equation [17.15], mercury in the +1 oxidation state is converted into the zero and +2 oxidation states.

†In February 1976, the Environmental Protection Agency moved to ban production of nearly all mercury-containing pesticides on the grounds that their continued use poses an unreasonable hazard to humans and to the environment in general.

necessary to suspend commercial fishing in several waterways in the United States and Canada because this limit was exceeded.

Both CH_3Hg^+ and $(CH_3)_2Hg$ have a tendency to accumulate in humans who eat contaminated fish, just as they accumulated in the fish in the first place. Because they are at least partly organic in character, these compounds pass very readily through biological defenses and attack the central nervous system. They have very long retention times in the body as compared with many other poisons, and thus their effects are cumulative. Persons eating a diet high in contaminated fish acquire over a period of time ever-increasing levels of mercury.

From this example of mercury as a pollutant, you can see that the question of whether a particular substance will be a serious pollutant in the environment is really an entire set of complicated questions. Unfortunately, too many substances have been released to the environment before the proper questions were even asked, let alone answered. For example, the water in Minimata Bay in Japan was for many years polluted with mercury wastes from a chemical factory. Over a period of more than 10 years, more than 50 people in Minimata died of mercury poisoning as a result of eating fish caught in Minimata Bay. In addition, many children born in Minimata during that time were seriously deformed and brain-damaged.

FOR REVIEW

Summary

In this chapter we've learned something of the nature and distribution of the water that covers our planet. Most of this water is in the oceans. Seawater contains about 3.5 percent by weight of dissolved salts. These salts, along with dissolved carbon dioxide, establish a buffer system that maintains the pH of seawater in the vicinity of 8.0.

The varieties of biological life that live in the sea depend on the growth of photosynthetic phytoplankton as the base of the food chain. The photosynthetic zone is near the water's surface, where the sun's rays can penetrate. The major ingredients needed for photosynthesis are CO_2 and suitable forms of nitrogen and phosphorus. Usually the availability of one or the other of these latter elements limits the rate of photosynthesis.

The sea is a vast storehouse of chemicals that remains largely untapped. Sodium chloride, the most abundant salt present in seawater, is recovered by partial evaporation. The elements bromine and magnesium are produced mainly from seawater.

Because most of the world's water is in the oceans, it is perhaps inevitable that humankind must eventually look to the seas for fresh water. Desalination refers to the removal of dissolved salts from seawater, brine, or brackish water, so as to render it fit for human consumption. Among the many means by which desalination may be accomplished are distillation, freezing of ice from the water, and reverse osmosis. In reverse osmosis, pressure is applied to salt water to induce the flow of water, but not the dissolved salts, through a semipermeable membrane.

Fresh water that is available from rivers, lakes, and underground sources may require treatment to render it fit for use. The several steps which may be used in water treatment are coarse filtration, sedimentation, sand filtration, aeration, sterilization, and softening.

Waste-water treatment is applied to sewage waters or to water that has been used in an industrial operation. Municipal waste waters are given a primary treatment to remove insoluble scum, grease, and other materials. Secondary treatment consists of aeration of sewage sludge to promote the growth of microorganisms that feed on the organic compounds present in sewage. Eventually, clear water is separated from the mass of microorganisms. The water is lower in biological oxygen demand (BOD) than before treatment. However, it may still contain many substances that are toxic to aquatic life or to humans or that cause excessive growth of algae in natural waters. The many substances that remain in waters after secondary treatment can be removed only by extensive additional processing, referred to as tertiary treatment.

As examples of pollutant substances present in waters, we have considered some of the metallic elements. The toxicity of a particular element varies with its chemical state. For this reason, it is important to understand all the chemical transformations that an element may undergo once it has been released to the environment. An element may be released in a relatively nontoxic chemical state and then be converted in nature to a highly toxic form. In evaluating the toxicity of an element, it is important to consider also the possibility that biological concentration may occur. When a toxic substance is stored cumulatively or preferentially in various species in the food chain, its concentration in the highest species in the food chain may be a thousand times what it is in the environment as a whole.

Leaning goals

Having read and studied this chapter, you should be able to:

1 List the more abundant ionic species present in seawater.

2 Explain the nature of the equilibrium between atmospheric CO_2 and the CO_2 dissolved in the oceans; in particular, you should be able to explain how the ionization equilibria involving H_2CO_3 are shifted by the slightly alkaline nature of seawater.

3 Describe how the growth of phytoplankton is related to the availability of nutrient nitrogen and phosphorus.

4 Describe how the concentrations of oxygen, phosphate, and nitrate as a function of ocean depth are linked to photosynthesis and decay of plant matter.

5 Write and explain the chemical reactions involved in the extraction of bromine and magnesium from seawater.

6 Explain the principles involved in the multistage flash-distillation and reverse-osmosis processes for desalination of seawater.

7 Describe the BOD test for water and explain how it is related to water purity.

8 List and explain the various stages of treatment that may be applied to a fresh-water supply.

9 Describe the chemical principles involved in the lime-soda process for reducing water hardness.

10 List and explain the stages in treatment of waste water.

11 List some of the more common metallic elements that might be present in a contaminated source of fresh water.

12 List and explain the three considerations described in the text that are important in determining the impact of a pollutant substance on the environment.

Key terms

Among the more important terms and expressions used for the first time in this chapter are the following:

Biological concentration (Section 17.6) refers to the tendency of a pollutant substance to accumulate in plant and animal tissues in concentrations much higher than in the immediate environment.

The biological oxygen demand (BOD) (Section 17.4) is the total capacity of a water sample for consumption of oxygen for biological oxidations. It is measured by the amount of oxygen gas taken up by a given quantity of water during a 5-day period at $20°C$.

Desalination (Section 17.3) refers to the removal of salts from seawater, brine, or brackish water, so as to render it fit for human consumption.

The lime-soda process (Section 17.5) is a method for removal of Mg^{2+} and Ca^{2+} ions from water to reduce water hardness. The substances added to the water are "lime," CaO (or "quicklime," $Ca(OH)_2$), and "soda ash," Na_2CO_3, in amounts determined by the concentrations of the offending ions.

Multistage flash distillation (Section 17.3) is a method for desalination that involves distillation of saline water in several stages, for maximum possible conservation of heat stored in the water.

The photosynthetic zone (Section 17.1) is the region of the oceans extending from the surface to a depth of about 150 m, in which the photosynthetic growth of phytoplankton occurs.

Phytoplankton (Section 17.1) are microscopic plants that abound in the water near the ocean surface. They utilize photosynthesis to consume CO_2 and suitable forms of nitrogen and phosphorus in forming plant matter. They form the base of the food chain for biological life in the oceans.

Reverse osmosis (Section 17.3) is a method for desalination of saline water that involves application of pressure to the saline water to force passage of water, but not dissolved salts, through a semipermeable membrane.

Salinity (Section 17.1) is a measure of the salt content of seawater, brine, or brackish water. It is equal to the weight in grams of dissolved salts present in 1 kg of seawater, brine, or brackish water.

Water hardness (Section 17.5) refers to the presence in a water supply of Ca^{2+}, Mg^{2+} (and sometimes Fe^{2+}) ions. These ions form insoluble precipitates with soaps and are responsible for scale formation when the water is heated.

Seawater

17.1 It is estimated that the rivers of the world bring a total of 4×10^{15} g of dissolved salts to the oceans each year. What fraction of the total salts dissolved in the oceans is this annual influx?

17.2 What is the molarity of Na^+ in a solution of NaCl whose salinity is 5, if the solution has a density of 1.0 g/mL?

17.3 Phosphorus is present in seawater to the extent of 0.07 ppm by weight (that is, 0.07 g P per 10^6 g H_2O). If this phosphorus is present as phosphate, PO_4^{3-}, calculate the corresponding molar concentration of phosphate.

17.4 Seawater is slightly basic (pH = 8.0). (a) Assuming that interionic interactions in the seawater can be ignored, and using $K_{sp} = 4 \times 10^{-38}$, calculate the molar solubility of $Fe(OH)_3$ in seawater. (b) The average content of Fe^{3+} in river water is about 1 ppm (that is, 1 gram Fe^{3+} per 10^6 g of water). Again ignoring interionic interactions in seawater, will $Fe(OH)_3$ precipitate when the river water reaches the sea?

17.5 Explain the following observations: (a) seawater has a lower vapor pressure than fresh water; (b) HCO_3^- is more abundant in river water than in seawater; (c) the concentration of nitrate ion in the open sea varies as shown in Figure 17.12.

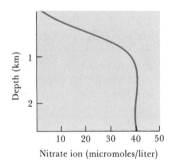

FIGURE 17.12 Concentration of nitrate ion as a function of depth.

Raw materials from seawater

17.6 Assuming a 10 percent efficiency of recovery, how many liters of seawater must be processed to obtain 10^8 kg of bromine in a commercial production process, assuming the bromine concentration listed in Table 17.1?

17.7 In the production of magnesium from seawater, at which stage does an oxidation-reduction reaction occur? Write this reaction. What is oxidized? What is reduced?

17.8 Write balanced chemical reactions for each of the following processes: (a) oxidation of Br^- by Cl_2; (b) the decomposition of $CaCO_3$ to form lime; (c) the dissolution of $Mg(OH)_2$ in acid solutions.

17.9 What mass of CaO is needed to precipitate 5.0×10^6 g of $Mg(OH)_2$ according to Equation [17.8]?

Desalination

17.10 What is the approximate osmotic pressure at 20°C of a solution of brackish water which has a salinity of 7 if we assume the salt is entirely NaCl?

[17.11] Suppose that in the large multistage flash-distillation unit shown in Figure 17.7 the efficiency of use of heat values in the steam applied to the brine heater is 20 percent and that the steam is formed with a 20 percent efficiency from burning oil. The heat of combustion of petroleum oil is -46 kJ/g. If the unit is to produce 4 million liters of fresh water per day, and if the heat of vaporization of water in the units is 2.5 kJ/g of H_2O, how much oil must be burned each day?

17.12 List the problems encountered in desalination by (a) distillation; (b) reverse osmosis.

17.13 What is osmosis? How does it differ from reverse osmosis?

Water quality and treatment

17.14 What naturally occurring impurities are commonly present in fresh, unpolluted water?

17.15 (a) What ions are commonly responsible for the hardness of water? (b) What makes these ions objectionable?

17.16 Explain what is meant by the following terms and abbreviations: (a) biodegradable; (b) aerobic decay; (c) anaerobic decomposition; (d) BOD; (e) BOD_5.

17.17 The most abundant elements in organic compounds are C, H, O, N, S, and P. (a) What are the products formed by these elements during aerobic decomposition? (b) What products are formed in anaerobic decomposition?

17.18 The following organic anion is found in most detergents:

$$H_3C-(CH_2)_9-\overset{\overset{\displaystyle H}{|}}{\underset{\underset{\displaystyle CH_3}{|}}{C}}-\bigcirc-SO_3^-$$

Assume that this anion undergoes aerobic decomposition in the following manner:

$$2C_{18}H_{29}O_3S^-(aq) + 51O_2(aq) \longrightarrow$$
$$36CO_2(aq) + 28H_2O(l) + 2H^+(aq) + 2SO_4^{2-}(aq)$$

What is the total BOD of a water sample containing 1.0 g of this substance per 100 L of water?

17.19 Two 10-mL samples of waste water were diluted to 300 mL with aerated distilled water containing sufficient inorganic nutrients to ensure complete biochemical oxidation of any biodegradable organic substances present. The samples were seeded with bacteria. One sample was then analyzed immediately for dissolved oxygen and a second analyzed after it had been incubated for 5 days at 20°C. The initial analysis revealed the sample to contain 7.90 ppm of dissolved oxygen. The second analysis, performed 5 days later, indicated that 1.40 ppm of dissolved oxygen was present. Taking sample dilution into effect, calculate the 5-day BOD of the waste water. (Ignore any BOD associated with the dilution and preliminary treatment.)

17.20 Explain how $Al_2(SO_4)_3$ and CaO are used in the purification of water.

17.21 Distinguish between primary, secondary, and tertiary sewage treatment.

[17.22] In a particular water supply, the concentration of Ca^{2+} is $2.2 \times 10^{-3} M$, and the concentration of bicarbonate ion, HCO_3^-, is $1.3 \times 10^{-3} M$. What weights of $Ca(OH)_2$ and Na_2CO_3 are needed to reduce the level of Ca^{2+} to one-fourth its original level if 1.0×10^7 L of water must be treated?

17.23 The manufacture of 1 gal of gasoline requires about 25 gal of water. How much water is used to produce the gasoline required to drive 3000 mi if your automobile uses 1 gal each 22 mi?

Pollutant elements

17.24 What is the maximum total mass of each of the following metals that can pass each day through a municipal water-treatment system furnishing 1.0×10^7 L of water per day, if the U.S. Public Health Service limits in Table 17.2 are not to be exceeded: (a) zinc; (b) cadmium; (c) manganese?

17.25 As discussed in the text, zinc is not a toxic element under normal conditions. Yet it happens that objects made of zinc metal can under certain circumstances be the cause of a toxic metal pollution, for example, galvanized pipe, when used with water that is a little on the acidic side. By thinking of this problem in terms of periodic relationships, and with the aid of Table 17.2, suggest the source of this toxicity.

17.26 Describe three considerations of major importance in evaluating the possible toxic effects of a substance released to the environment.

Additional exercises

17.27 Explain why the concentration of Na^+ in seawater is higher than that of Ca^{2+} although calcium is more abundant than sodium in the earth's crust.

17.28 In discussing water quality, how is the phrase parts per million defined?

17.29 Why is the presence of blue-green algae on the surface of a lake an undesirable sign? What will happen when the algae die?

17.30 If Hg^{2+} ions are present in water, what are some of the chemical products that may result? Of these, which are potentially the most harmful to humans?

17.31 Explain briefly how a substance can be concentrated as it moves along a food chain (for example, as it moves from water to plant to lower animal to human).

17.32 List five elements essential for the growth of aquatic organisms. Which two usually limit growth?

17.33 The average daily mass of O_2 taken up by sewage discharged in the United States is 59 g per person. How many liters of water at 9 ppm O_2 are totally depleted of oxygen in 1 day by a population of 50,000 people?

17.34 Urea, $CH_4N_2O_2$, is the end product of protein metabolism in animals. Assume that aerobic bacteria can decompose it as follows:

$$CH_4N_2O(aq) + 4O_2(aq) \longrightarrow$$
$$H_2O(l) + CO_2(aq) + 2H^+(aq) + 2NO_3^-(aq)$$

What is the BOD of a body of water of total volume 3.0×10^6 L if 30 kg of urea enter?

17.35 How many moles of $Ca(OH)_2$ and of Na_2CO_3 should be added to soften 10^3 L of water in which $[Ca^{2+}] = 5.0 \times 10^{-4} M$ and $[HCO_3^-] = 7.0 \times 10^{-4} M$?

17.36 What is the osmotic pressure at 27°C of a solution of 0.100 M NaCl? What pressure must be applied to purify this solution by reverse osmosis?

[17.37] One method of removing phosphate, PO_4^{3-}, in tertiary water treatment is to treat the water with CaO. Using chemical equations and appropriate K_{sp} values, suggest how this treatment removes phosphate.

17.38 Using the solubility-product constant for $CaCO_3$ in seawater (6.0×10^{-7}), determine the concentration of Ca^{2+} in the sample of seawater described in Sample Exercise 17.1, assuming that the sample was in equilibrium with solid $CaCO_3$.

17.39 Why does the K_{sp} value for $CaCO_3$ in seawater at 20°C (6.0×10^{-7}) differ from that in pure water at the same temperature (2.8×10^{-9})?

17.40 Complete and balance a chemical equation corresponding to each of the following verbal descriptions: (a) when chlorine is added to water, it forms hypochlorous acid; (b) hypochlorous acid reacts with ammonia dissolved in water to form chloramine, NH_2Cl; (c) when alum, $Al_2(SO_4)_3$, is added to slightly alkaline water, a gelatinous precipitate forms; (d) when lime, CaO, is added to water, the water becomes more basic; (e) mercury metal reacts in water with mercury(II); (f) a sample of water containing Ca^{2+} and bicarbonate ion forms a precipitate when heated.

17.41 The overall elemental composition of phytoplankton can be represented approximately by the formula $C_{108}H_{266}N_{16}O_{109}P$. Suppose that in a particular stretch of water the concentration of nitrogen is limiting for growth of phytoplankton; also assume that the nitrogen concentration in the zone from the surface to 100 m averages 1.0×10^{-6} mol of nitrogen atoms per liter. If half of this nitrogen is converted to phytoplankton, what is the total mass of plant matter in a square kilometer of ocean in a region from the surface to a depth of 100 m?

17.42 At what concentration, in ppm Hg^{2+}, would $HgO \cdot H_2O$ begin to precipitate from a natural lake having a pH of 7.8? K_{sp} for the following reaction is 1.6×10^{-23}:

$$HgO \cdot H_2O(s) \rightleftharpoons Hg^{2+}(aq) + 2OH^-(aq)$$

18

Free energy, entropy, and equilibrium

Two of the major questions concerning any chemical reaction are: **How far toward completion does the reaction proceed? How rapidly does the reaction approach equilibrium?** We get the answer to the first question from a knowledge of the equilibrium constant. We learn about the second from a study of the reaction rate.

If we want to carry out a reaction that has a favorable equilibrium constant, that is, a reaction that proceeds with conversion of a sizable fraction of reactants into products, then we need only worry about getting the reaction rate into a convenient range. This may not be easy, but hard work and ingenious research might eventually lead to a catalyst that can speed up a slow reaction or to some means of controlling a reaction that is too rapid. Haber's discovery of a suitable catalyst for ammonia synthesis (Chapter 14) provides an excellent example of successful research of this type. However, if Haber had attempted to fix nitrogen by reacting N_2 and O_2 at 400–500°C, instead of reacting N_2 and H_2, he would never have developed a successful process. The equilibrium constant for the reaction of N_2 with O_2 to form NO is so small in this temperature range that the amount of NO present at equilibrium would have been too small for practical purposes. Even if one had a catalyst that would enable rapid reaction of N_2 with O_2 to form NO, the reaction would still have no practical consequences. There is therefore a need for some way to predict in advance whether a reaction can proceed to any significant extent before coming to equilibrium.

When we first introduced chemical equilibrium, in Chapter 14, we defined it from a kinetic point of view: Equilibrium occurs when opposing reactions occur at equal rates. However, equilibrium also has a basis in thermodynamics, the area of science dealing with energy relationships. In this chapter we shall see that it is possible to predict the position of an equilibrium using certain principles and concepts from thermodynamics.

Thermodynamics is based on several fundamental laws that summarize our experience with energy changes. The first law of thermodynamics (Chapter 4) states that energy is conserved. By this we mean that energy is neither created nor destroyed in processes such as the falling of a brick, the melting of an ice cube, or a chemical reaction. Energy flows from one part of nature to another, or is converted from one form or another, but the total remains constant. The first law is often stated in the form $\Delta E = q - w$, where ΔE represents the change in energy of the system in a process, q is the heat absorbed by the system from its surroundings in the process, and w is the work done by the system on its surroundings during the process. In applying this formula, q is positive for an endothermic process and is negative for an exothermic process. Work, w, is a positive quantity when the system does work on its surroundings (for example, a flashlight battery runs a toy) and is negative when work is done by the surroundings on the system (for example, a gas is compressed).

Once we specify a particular process or change, the first law helps us to balance the books, so to speak, on the heat released, work done, and so forth. However, it says nothing about whether the process or change we specify can in fact occur. That question is encompassed in the second law of thermodynamics.

The second law of thermodynamics expresses the notion that there is an inherent direction in which any system not at equilibrium moves. To reverse that direction requires addition of energy to the system. For example, if you hold a brick in your hand and let it go, the brick falls to the floor. Water placed in a freezer compartment converts to ice. A shiny nail left outdoors turns to rust. Every one of these processes proceeds without needing to be driven by an outside source of energy; such processes are said to be spontaneous. For every spontaneous process, we can imagine a reverse process occurring. For example, we can imagine a brick moving from the floor into your hand, ice cubes melting at $-10°C$, or a rusty iron nail being transformed into a shiny one. It is inconceivable that any of these occurrences is spontaneous. If we saw a film in which these things happened, we would conclude that the film was being run backward. Our years of observing nature at work have impressed us with a simple rule: *Processes that are spontaneous in one direction are not spontaneous in the reverse direction.*

Consider a reaction about which we had much to say in Chapter 14:

$$N_2(g) + 3H_2(g) \rightleftharpoons 2NH_3(g) \qquad [18.1]$$

When we mix N_2 and H_2 at any temperature, say 472 K, the reaction proceeds in the forward direction; this process is spontaneous. However, if we place a mixture of 1.00 mol of N_2, 3.00 mol of H_2, and 1.00 mol of NH_3 in a 1-L container at 472 K, it is not immediately obvious whether or not the formation of more NH_3 should be spontaneous. Nevertheless, if we have the equilibrium constant, which at this temperature happens to be $K_c = 0.105$, then we can predict the direction in which the reaction proceeds to reach equilibrium. In this case

$$Q = \frac{[NH_3]^2}{[N_2][H_3]^3} = \frac{(1.00)^2}{(1.00)(3.00)^3} = 0.0370$$

Because the reaction quotient is smaller than K_c, the system moves spontaneously toward equilibrium by formation of NH_3 (Section 14.4). The opposite process, conversion of NH_3 into N_2 and H_2, is not spontaneous for this particular reaction mixture at 472 K.

The process by which a system reaches equilibrium is a spontaneous change. The process may be fast, or it may be slow; thermodynamics has nothing to say about how rapidly such a process occurs. For example, a mixture of $H_2(g)$ and $Cl_2(g)$ will react at 25°C to form $HCl(g)$. The equilibrium constant for this reaction, K_p, is very large, 5.2×10^{16}. When $H_2(g)$ and $Cl_2(g)$ are mixed, the reaction between them is spontaneous, yet this mixture may show no apparent reaction even over a period of centuries. The reaction is extremely slow unless initiated by a suitable catalyst, a spark, or a ray of light. Once initiated in one of these ways, this reaction proceeds with explosive speed nearly to completion.

What factors make a process spontaneous? We considered that question briefly in Section 12.2. In that earlier discussion we noted that two basic principles are involved: Processes in which the energy content of the system decreases (exothermic processes) tend to occur spontaneously. Furthermore, spontaneity characterizes processes in which the randomness or disorder of the system increases. In the next section we consider these factors, especially the matter of disorder, in greater detail.

18.2 Spontaneity, enthalpy, and entropy

The spontaneous motion of a brick released from your hand is toward the ground. As the brick falls, it loses potential energy. This potential energy is first converted into kinetic energy, the energy of motion of the brick. When the brick hits the floor, its kinetic energy is converted into heat. The overall result of the brick's fall is thus a conversion of the potential energy of the brick into heat in its surroundings. Our experience with other simple mechanical systems is similar: objects fall, clocks run down, stretched rubber bands contract. All of these phenomena can be summarized by saying that such systems seek a resting place of minimum energy.

It is clear that the tendency for a system to achieve the lowest possible energy is one of the driving forces that determines the behavior of molecular systems. For example, just as a brick possesses potential energy because of its position relative to the floor, so also a chemical substance possesses potential energy relative to other substances because of the arrangements of nuclei and electrons. When these arrangements change, energy may be released. For example, the combustion of propane (bottled gas), which is clearly a spontaneous process, is strongly exothermic:

$$C_3H_8(g) + 5O_2(g) \longrightarrow 3CO_2(g) + 4H_2O(l) \qquad \Delta H° = -2202 \text{ kJ} \qquad [18.2]$$

The rearrangements in space of nuclei and electrons in going from propane and oxygen to carbon dioxide and water lead to a lower chemical potential energy, and so heat is evolved. Reactions that are exothermic are generally spontaneous. However, it is clear that the tendency toward minimum energy cannot be the only factor that determines spontaneity in molecular processes. It is instructive to consider some spontaneous processes that are not exothermic.

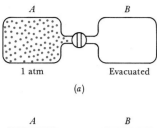

A B

1 atm   Evacuated

(a)

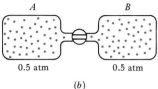

A B

0.5 atm   0.5 atm

(b)

FIGURE 18.1 Expansion of an ideal gas into an evacuated space. In (a), flask A holds an ideal gas at 1 atm pressure, whereas flask B is evacuated. In (b), the stopcock connecting the flasks has been opened. The ideal gas expands to occupy both flasks A and B at a pressure of 0.5 atm.

A bit of thinking brings to mind several processes that are spontaneous even though they are not exothermic. For example, consider an ideal gas confined at 1 atm pressure to a 1-L flask, as shown in Figure 18.1. The flask is connected via a closed stopcock to another 1-L flask that is evacuated. Now suppose the stopcock is opened. Is there any doubt about what would happen? We intuitively recognize that the gas would expand into the second flask until the pressure were equally distributed in both flasks, at 0.5 atm. In the course of expanding from the 1-L flask into the larger volume, the ideal gas neither absorbs nor emits heat. Nevertheless, the process is spontaneous. The reverse process, in which the gas that is evenly distributed between the two flasks suddenly moves entirely into one of the flasks, leaving the other vacant, is inconceivable. Yet this is also a process that would involve no emission or absorption of heat. It is evident that some other factor than heat emitted or absorbed is important in making the process of gas expansion spontaneous.

As another example, consider the melting of ice cubes at room temperature. The process

$$H_2O(s) \longrightarrow H_2O(l) \hspace{3cm} [18.3]$$

at $27\,°C$ is highly spontaneous, as we all know. Yet this is an endothermic change. The melting of ice above $0\,°C$ thus represents an example of a spontaneous, endothermic process.

A similar type of process, discussed in Chapter 12, is the endothermic dissolving of many salts in water. If we add solid potassium chloride, KCl, to a glass of water at room temperature and stir, we can feel the solution growing colder as the salt dissolves. Thus the process that we can write as

$$KCl(s) \xrightarrow{\text{H}_2\text{O}} KCl(aq) \hspace{3cm} [18.4]$$

is endothermic, and yet spontaneous.

The three processes just described have something in common that accounts for the fact that they are spontaneous. In each instance, the products of the process are in a more random or disordered state than the reactants. Let's consider each case in turn.

When we have a gas confined to a 1-L volume as in Figure 18.1(a), we can specify the location of each and every gas molecule as being in that liter of space. After the gas has expanded, we can't be sure which gas molecules are at any one instant in the original volume, and which are on the other side. We must therefore say that the location of each and every gas molecule is specified as being in the entire 2-L space. In other words, the gas molecules, because they can be anywhere within a 2-L space, are more randomized than when they were confined to a 1-L space.

The molecules of water that make up an ice crystal are held rigidly in place in the ice crystal lattice (Figure 18.2). When the ice melts, the water molecules are free to move about with respect to one another and to turn over. Thus, in liquid water the individual water molecules are

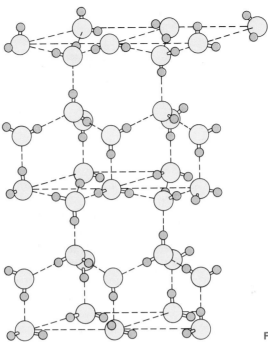

FIGURE 18.2 The structure of ice.

more randomly distributed than in the solid. The highly ordered solid
structure is replaced by the highly disordered liquid structure.

A similar situation applies when KCl dissolves in water, although here
we must be a little careful not to take too much for granted. In solid KCl,
the K^+ and Cl^- are in a highly ordered, crystalline state. When the solid
dissolves, the ions are free to move about in the water. They are obviously in a much more random and disordered state than before. At the
same time, though, water molecules are held around the ions, as water of
hydration (Section 12.3), as illustrated in Figure 18.3. These water molecules, are in a *more* ordered state than before, because they are confined to
the immediate environment of the ions. Thus the dissolving of a salt
involves both ordering and disordering processes. It happens that the
disordering processes are dominant, so the overall effect is an increase in
disorder upon dissolving a salt in water.

SAMPLE EXERCISE 18.1

Indicate in each of the following cases whether the
process shown results in an increase or decrease in
randomness or disorder.

(a) $4Fe(s) + 3O_2(g) \longrightarrow 2Fe_2O_3(s)$

(b) $Ag^+(aq) + Cl^-(aq) \longrightarrow AgCl(s)$

(c) $H_2O(l) \longrightarrow H_2O(g)$

Solution: Process (a) results in a decrease in randomness, because a gas is converted into part of a
solid lattice. The units of the solid oxide lattice are
much more highly ordered and confined to specific
locations than are the molecules of a gas. (Note that

this reaction is spontaneous even though there is an
overall decrease in randomness. This is so because
the reaction is highly exothermic. The combined effects of enthalpy change and change in randomness
are discussed later in Section 18.5.)

Process (b) also represents a decrease in randomness, because the ions that are free to move about the
volume of the solution form a solid lattice in which
they are confined to highly regular locations.

Process (c) occurs with an increase in randomness
or disorder, because the gaseous water molecules are
distributed throughout a much larger volume than
in the liquid state.

As these examples illustrate, spontaneity is associated with an increase in randomness or disorder of a system. The randomness is expressed by a thermodynamic quantity called entropy, given the symbol S. The more random a system, the larger its entropy. Like enthalpy, entropy is a state function (Section 4.5). The change in entropy for a process, $\Delta S = S_{final} - S_{initial}$, depends only on the initial and final states of the system and not on the particular pathway by which it changes from one state to another.

THE SECOND LAW OF THERMODYNAMICS

Our introduction of the concept of entropy allows us to reexamine the second law of thermodynamics and its implications. In the discussion of spontaneity in Section 18.1, it was noted that the second law has to do with the direction in which processes move; it is associated with the idea that processes that are spontaneous in one direction are not spontaneous in the opposite direction. This idea applies not only to chemical changes but to other processes as well.

We all know that heat flows spontaneously from a hot object to a cold one. We also know that to cause heat to flow in the reverse direction, from a cold object to a hotter one or from a system at some temperature to surroundings at a higher temperature, requires an input of energy. For example, it requires electrical energy to maintain a refrigerator at a lower temperature than the surrounding kitchen.

A related but less obvious point is that heat cannot be completely converted into work. There is always some heat that is transferred to the surroundings. For example, in a steam turbine, the heat energy con-

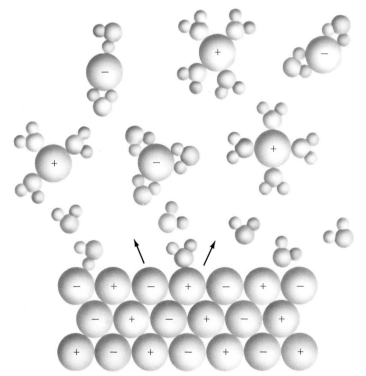

FIGURE 18.3 Changes in degree of order in the ions and solvent molecules on dissolving of an ionic solid in water. The ions become more randomized and the water molecules that hydrate the ions become less randomized.

tained in high-temperature steam is converted into electrical energy; the kinetic energy of the steam molecules is converted into the kinetic energy of the moving turbine blades, and eventually into electrical energy. But not all the kinetic energy of the steam molecules can be converted into the kinetic energy of the turbine. Some energy is lost to the surroundings as heat. Every electrical power plant is a source of waste heat; the laws of thermodynamics tell us that this must be so. Indeed, the fact that heat cannot be completely converted into work is one of the earliest statements of the second law.

There are many ways to state the second law. In chemical contexts it's usually expressed in terms of entropy. To develop such a statement, let's think in terms of an isolated system, one that doesn't exchange energy or matter with its surroundings. When a process occurs spontaneously in an isolated system, the system always ends up in a more random state. For example, when a gas expands, as in Figure 18.1, there is no exchange of heat, work, or matter with the surroundings; this is an isolated system. The spontaneous expansion corresponds to an increase in entropy.

In the real world we rarely deal with isolated systems. We are usually concerned with systems that exchange energy with their surroundings in the form of heat or work. When such a system changes spontaneously, it may undergo either an increase or decrease in its entropy. However, the second law tells us that *the universe as a whole must increase in entropy in any spontaneous change.* As an example, consider the oxidation of iron to $Fe_2O_3(s)$:

$$4Fe(s) + 3O_2(g) \longrightarrow 2Fe_2O_3(s) \qquad [18.5]$$

As discussed in Sample Exercise 18.1, this chemical process results in a decrease in the degree of randomness; that is, ΔS for the process is negative. But when the process occurs, some change also occurs in the surroundings. For example, the reaction is exothermic; heat is therefore evolved and absorbed by the surroundings. In fact, the change that occurs in the surroundings causes an increase in the entropy of the surroundings that is larger than the decrease that occurs in the system itself. For any spontaneous process the sum of the entropy change in the system and that in the surroundings (which is the entropy change in the universe caused by that process) must be positive:

$$\Delta S_{universe} = \Delta S_{system} + \Delta S_{surroundings} > 0 \qquad [18.6]$$

No process that produces order (lower entropy) in a system can proceed without producing an even larger disorder (higher entropy) in its surroundings. Furthermore, the disorder thereby introduced into the surroundings will always exceed the order achieved in the system. Thus while energy is conserved (the first law), entropy continues to increase (the second law).

The consequences of the statement of the second law that is given above are quite profound. We humans, for example, are very complex, highly organized, and well-ordered systems. We have a very low entropy content as compared with the same amount of carbon dioxide, water, and several other simple chemicals into which our bodies might be decomposed. But all of the thousands of chemical reactions necessary to produce one adult human have caused a very large increase in entropy of the rest of the universe. Thus the overall entropy change necessary to form and maintain a human, or for that matter any other living system, is positive.

In a similar way, the human activities that produce such an impressive ordering of the world around us—formation of copper metal from a widely dispersed copper ore; production from sand of silicon used in transistors; production of the paper on which this book is printed from trees—have along the way used up a great deal of energy that has been converted, in a sense, to disorder—coal and oil burned to form CO_2 and H_2O; a sulfide ore roasted to form SO_2 that pollutes the atmosphere; radioactive wastes scattered in the environment. Modern human society is, in effect, using up its limited storehouse of energy-rich materials in its headlong rush to exploit technology.

In recent years a few social scientists have begun to appreciate the importance of thermodynamic considerations in the economic laws that must eventually rule human activities. A leading economic theorist, Nicholas Georgescu-Roegen, published in 1971 a book entitled *The Entropy Law and the Economic Process*. He argues that the human race must eventually learn to live within the bounds of the energy supply that reaches earth daily from the sun, because it will soon have exhausted the supply of readily available energy of other sorts.

18.3 A molecular interpretation of entropy

It is useful to develop a qualitative sense of how entropy changes in a system depend on changes in structure, physical state, and so forth. In Section 18.2 some examples of spontaneous, endothermic processes were considered. We saw, for example, that the increase in volume that occurs when a gas expands results in an increase in the randomness of the system (positive ΔS). In a similar fashion, the distribution of a liquid or solid solute in a solution is accompanied by an increase in entropy. For example, ΔS is positive when KCl dissolves in water. However, the dissolving of a gas, such as CO_2 in H_2O, causes the gas molecules to move in a much smaller volume; consequently, for this process the entropy of the system decreases (negative ΔS). Similarly, a decrease in the number of gaseous particles as the result of a reaction causes a decrease in entropy (negative ΔS). For example, ΔS for the following reaction is negative:

$$2NO(g) + O_2(g) \longrightarrow 2NO_2(g) \tag{18.7}$$

Entropy changes can also be associated with molecular motions within a substance. A molecule consisting of more than one atom can engage in several types of motion. The entire molecule can move in one direction or another as in the movements of gas molecules. We call such movement translational motion. The atoms within a molecule may also undergo vibrational motion in which they move periodically toward and away from each other, much as a tuning fork vibrates about its equilibrium shape. Figure 18.4 shows the vibrational motions possible for the water molecule. In addition, molecules may possess rotational motion, as though they were spinning like a top. The rotational motion of the water molecule is also illustrated in Figure 18.4. These forms of motion are ways that the molecule has of storing energy. As the temperature of a system increases, the amounts of energy stored in these forms of motion increase.

To see what this has to do with entropy, let's imagine that we begin with a pure substance that forms a perfect crystalline lattice at the lowest temperature possible, absolute zero. At this stage, none of the kinds of motion that we have been talking about are present. The individual atoms and molecules are as well defined in position and in terms of energy as they can ever be. Let us say that the entropy of our substance at

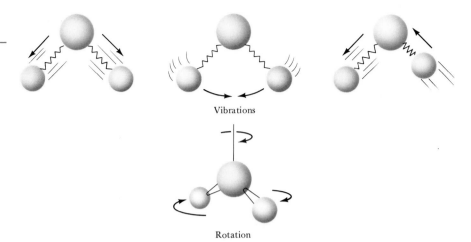

Vibrations

Rotation

FIGURE 18.4 Examples of vibrational and rotational motion, as illustrated for the water molecule. Vibrational motions involve periodic displacements of the atoms with respect to one another, a phenomenon similar to the vibrations of the arms of a tuning fork. Rotational motions involve the spinning of a molecule about an axis.

this point is zero.* As the temperature is raised, the units of the solid lattice begin to acquire energy. In a crystalline solid, the molecules or atoms that occupy the lattice places are constrained to remain more or less in place. Nevertheless they may store energy in the form of vibrational motion about their lattice positions. Instead of all the molecules necessarily being in the lowest possible energy state, there is a kind of expansion in the number of possible energies that the lattice atoms or molecules may have. This increase in possible energy states is not unlike the expansion of the gas illustrated in Figure 18.1. The entropy of the gas increases on expansion because the volume through which the gas molecules move is larger. The entropy of the lattice increases with temperature because the number of possible energy states in which the molecules or atoms are distributed is larger.

*The third law of thermodynamics states that the entropy of a perfect crystal at 0 K is zero.

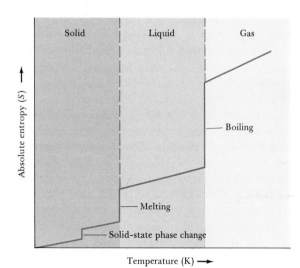

FIGURE 18.5 Entropy changes that occur as the temperature of a substance rises from absolute zero.

It is instructive to follow what happens to the entropy of our substance as we continue to heat it. Let's suppose that at some temperature a phase change occurs, converting the substance from one solid form to another. This means that the arrangement of the lattice units changes in some way, possibly so that the lattice is less regular.* This type of phase change occurs sharply at one temperature, just as do other types of phase changes, for example, from a solid to a liquid. When the change occurs, there is a change in entropy, because the two lattice arrangements do not have precisely the same degrees of randomness.

Figure 18.5 shows the variation in entropy with temperature for our sample. Note that the change in S with temperature is gradual up to the solid-state phase change and that there is then a sharp increase in S at that temperature. At temperatures above the phase change, the entropy increases with increasing temperature, up to the melting point of the solid.

When the solid melts, the units of the lattice are no longer confined to specific locations relative to other units, but are free to move about the entire volume of the unit. This added freedom of motion for the individual molecules adds greatly to the entropy content of the substance. At the temperature of melting, we therefore see a large increase in entropy content. After all the solid has melted, the temperature again increases, and with it the entropy.

SAMPLE EXERCISE 18.2

Figure 18.5 shows that the entropy of a liquid increases as its temperature increases. What factors are responsible for the increase in entropy?

Solution: The average kinetic energy of the molecules in a liquid increases with temperature. As temperature increases, more molecules at any given instant possess higher energies. This "expansion" in the energies that the molecules possess is measured by the increase in entropy. The increase in S with increasing temperature results from increased energy of motion of all kinds within the liquid.

At the boiling point of the liquid there is again a big increase in entropy. The increase in this case results largely from the increased volume in which the molecules may be found. This is intuitively in line with our earlier ideas about entropy, because an increase in volume means an increase in randomness. It is less likely that a given molecule will be found in a given volume element when there is a great expansion from the liquid to the gaseous state.

As the gas is heated, the entropy increases steadily, because more and more energy is being stored in the gas molecules. The distribution of molecular speeds is spread out toward higher values, as illustrated in Figure 9.10. Again, the idea of an expansion in the range of energies in which molecules may be found helps us remember that increased average energy means increased entropy.

*As an example of a solid-state phase change, gray tin converts at 13°C to another solid form called white tin. White tin is stable above the transition temperature, gray tin below it. White tin has a higher entropy than gray tin.

SAMPLE EXERCISE 18.3

For each of the following pairs of substances, indicate which has the higher entropy and suggest a possible reason: (a) 1 mol of NaCl(s) and 1 mol of HCl(g) at 25°C; (b) 2 mol of HCl(g) and 1 mol of HCl(g) at 25°C; (c) 1 mol of HCl(g) and 1 mol of Ar(g) at 25°C; (d) 1 mol of N_2(s) at 24 K and 1 mol of N_2(g) at 298 K.

Solution: (a) Gaseous HCl has the higher entropy per mole, because it has acquired a high degree of randomness as a result of being in the gaseous state.

(b) The sample containing 2 mol of HCl has twice the entropy of the sample containing 1 mol. (c) The HCl sample has the higher entropy, because the HCl molecule is capable of storing energy in more ways than is Ar. It may rotate, or the H—Cl distance may change periodically in a vibrational motion. (d) The gaseous N_2 sample has the higher entropy, because the entropy increases resulting from melting and then boiling of N_2 are included in its total entropy content.

SAMPLE EXERCISE 18.4

Predict whether the entropy change of the system in each of the following reactions is positive or negative.

(a) $H_2O(l) \longrightarrow H_2O(g)$ (at 25°C)

(b) $CaCO_3(s) \longrightarrow CaO(s) + CO_2(g)$

(c) $N_2(g) + 3H_2(g) \longrightarrow 2NH_3(g)$

(d) $N_2(g) + O_2(g) \longrightarrow 2NO(g)$

(e) $Ag^+(aq) + Cl^-(aq) \longrightarrow AgCl(s)$

Solution: (a) The entropy change in this process is positive, because the single substance involved is going from the liquid to the gaseous state. We have already seen that the phase change from liquid to gas results in an increase in entropy (Figure 18.5).

(b) The entropy change here is positive, because a solid is converted into a solid and a gas. Gaseous substances generally possess more entropy than solids, and so whenever the products contain more moles of gas than the reactants, the entropy change is probably positive.

(c) The entropy change in formation of ammonia from nitrogen and hydrogen is negative, because there are fewer moles of gas in the product than in the reactants.

(d) This represents a case in which the entropy change will be small, because the same number of moles of gas is involved in the reactants and the product. The sign of ΔS is impossible to predict based on our discussions thus far, but we can predict that ΔS will be small.

(e) The entropy change in the precipitation of a salt from solution is negative. The ions in solution are free to move about the entire volume of the liquid, whereas in the solid lattice they are more confined.

18.4 Calculation of entropy changes

The enthalpy change in a chemical reaction is often easily measured in a calorimeter, as described in Section 4.7. There is no comparable easy means for measuring the change in entropy. Nevertheless, various types of measurements have made it possible to determine the absolute entropy at any temperature for a great number of substances. These entropies are based on the fact that the entropies of pure solids are zero at absolute zero. A table of absolute entropy values (usually written $S°$) is included in Appendix D. These entropies are expressed in units of joules per degree Kelvin per mole: J/K-mol.

The entropy change in a chemical reaction is given by the sum of the entropies of the products less the sum of entropies of reactants. Thus, in the overall reaction

$$aA + bB + \cdots \rightleftharpoons pP + qQ + \cdots \qquad [18.8]$$

The total entropy change is given by

$$\Delta S^\circ = [pS^\circ(\text{P}) + qS^\circ(\text{Q}) + \cdots] - [aS^\circ(\text{A}) + bS^\circ(\text{B}) + \cdots]. \qquad [18.9]$$

In other words we sum the absolute entropies of all the products, multiplying each by the coefficient of the product in the balanced equation, and then subtract the same sort of sum of entropies of the reactants.

SAMPLE EXERCISE 18.5

Calculate ΔS° for the synthesis of ammonia from $N_2(g)$ and $H_2(g)$:

$$N_2(g) + 3H_2(g) \longrightarrow 2NH_3(g)$$

Solution: Using Equation [18.9] we have

$$\Delta S^\circ = 2S^\circ(NH_3) - [S^\circ(N_2) + 3S^\circ(H_2)]$$

Substituting the appropriate S° values from Appendix D:

$$\Delta S^\circ = (2 \text{ mol})\left(192.5\ \frac{\text{J}}{\text{K-mol}}\right)$$
$$-\left[(1 \text{ mol})\left(191.5\ \frac{\text{J}}{\text{K-mol}}\right)\right.$$
$$\left. + (3 \text{ mol})\left(130.58\ \frac{\text{J}}{\text{K-mol}}\right)\right]$$
$$= -198.2 \text{ J/K}$$

The value for ΔS° is negative, as we predicted in Sample Exercise 18.4(c).

18.5 The free-energy function

We still haven't attempted to use thermodynamics to predict whether or not a given reaction will be spontaneous. We have seen that spontaneity involves two thermodynamic concepts, entropy and enthalpy. Before we can make the predictions we would like, we must introduce a third function that interrelates entropy and enthalpy. That function is called free energy, or Gibbs free energy after the American mathematician J. Willard Gibbs (1839–1903) who first proposed it (Figure 18.6). The free energy, G, is related to enthalpy and entropy by the expression

$$G = H - TS \qquad [18.10]$$

where T is the absolute temperature. Free energy, like the enthalpy and entropy functions to which it is related, is a state function.

For a process occurring at constant temperature and pressure, the change in free energy is given by the expression

$$\Delta G = \Delta H - T\Delta S \qquad [18.11]$$

A process that is driven spontaneously toward equilibrium both by decreasing energy (negative ΔH) and increasing randomness (positive ΔS) will have a negative ΔG. Indeed, there is a simple relationship between the sign of ΔG for a reaction and the spontaneity of that reaction operated at constant temperature and pressure:

1 If ΔG is negative, the reaction is spontaneous in the forward direction.
2 If ΔG is zero, the reaction is at equilibrium; there is no driving force tending to make the reaction go in either direction.

If ΔG is positive, the reaction in the forward direction is nonspontaneous; work must be supplied from the surroundings to make it occur. However, the reverse reaction will be spontaneous.

An analogy is often drawn between the free-energy change in a spontaneous reaction and the potential-energy change in a boulder rolling down a hill. Potential energy in a gravitational field "drives" the boulder until it reaches a state of minimum potential energy in the valley, Figure 18.7(a). Similarly, the free energy of a chemical system decreases (negative ΔG) until it reaches a minimum value, Figure 18.7(b). When this minimum is reached, a state of equilibrium exists. In any spontaneous process at constant temperature and pressure, the free energy always decreases. As shown in Figure 18.7(b), the equilibrium condition can be approached by a spontaneous change from either direction, from the product side or the reactant side.

As an example of these ideas, let's return to the synthesis of ammonia from nitrogen and hydrogen. Imagine that we have a certain number of moles of nitrogen and three times that number of moles of hydrogen in a reaction vessel that permits us to maintain a constant temperature and pressure. We know from our earlier discussions that the formation of ammonia will not be complete; an equilibrium will be reached in which the reaction vessel contains some mixture of N_2, H_2, and NH_3. The free energy of the system decreases (negative ΔG) until this equilibrium is attained. Once the equilibrium has been reached, there is no further spontaneous formation of NH_3. The equilibrium condition is the minimum free energy available to the system at this temperature and pressure. To form more NH_3 from N_2 and H_2 once equilibrium has been reached requires an increase in free energy (positive ΔG).

It is not necessary that we reach equilibrium beginning only with N_2 and H_2. We could reach the same equilibrium by beginning with an

FIGURE 18.6 Josiah Willard Gibbs (1839–1903) was the first person to be awarded a Ph.D. in science from an American university (Yale, 1863). He went on to become one of the foremost mathematical scientists of his day. From 1871 until his death he held the chair of mathematical physics at Yale University. He made significant contributions to thermodynamics and is credited with laying much of the theoretical foundation that led to the development of chemical thermodynamics. (*Culver Pictures*)

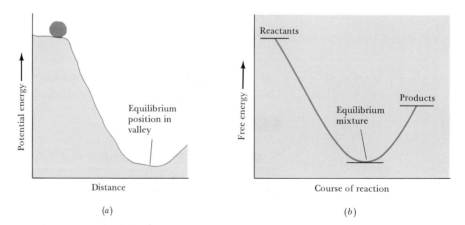

(a)   (b)

FIGURE 18.7 Analogy between the potential energy change of a boulder rolling down a hill (a) and the free-energy change in a spontaneous reaction (b). The equilibrium position in (a) is given by the minimum potential energy available to the system. The equilibrium position in (b) is given by the minimum free energy available to the system.

appropriate amount of NH_3. Ammonia held at a constant temperature and pressure will decompose to form N_2 and H_2 until an equilibrium is attained. This process is also spontaneous, involving a decrease in free energy as the system approaches equilibrium. The equilibrium can be approached from either the reactant side or product side, as indicated in a general fashion in Figure 18.7(b).

CALCULATION OF $\Delta G°$

We have noted that free energy is a state function. This means that it is possible to tabulate standard free energies of formation for substances, just as it is possible to tabulate standard enthalpies of formation. It is important to remember that standard values for these functions imply a particular set of conditions, or standard states (Section 4.5). The standard state for gaseous substances is 1 atm pressure. For solid substances the standard state is the pure solid; for liquids, the pure liquid. For substances in solution, the standard state is normally a concentration about 1 M; in accurate work it may be necessary to make certain corrections, but we need not worry about these. The temperature usually chosen for purposes of tabulating data is $25°C$. Just as for the standard heats of formation, the free energies of elements in their standard states are arbitrarily set to zero. This arbitrary choice of reference point has no effect on the quantity in which we are really interested, namely the *difference* in free energy between reactants and products. The rules about standard states are summarized in Table 18.1. A table of standard free energies of formation is to be found in Appendix D.

The standard free energies of formation for substances are useful in calculating the standard free-energy change in a chemical process. For the general reaction

$$aA + bB + \cdots \longrightarrow pP + qQ + \cdots \qquad [18.12]$$

the standard free-energy change is

$$\Delta G° = (p\Delta G_f°(P) + q\Delta G_f°(Q) + \cdots) - (a\Delta G_f°(A) + b\Delta G_f°(B) + \cdots) \qquad [18.13]$$

In this expression $\Delta G_f°(P)$ represents the standard free energy of formation of product P, and all the other $\Delta G°$ have similar meanings. Stated verbally, the standard free-energy change for a reaction equals the sum of the standard free-energy values per mole of each product, each multiplied by the corresponding coefficient in the balanced equation, less the same sort of sum for the reactants.

What use can be made of this standard free-energy change for a chemical reaction? The quantity $\Delta G°$ tells us whether a mixture of reactants and products, each present under standard conditions, would spontaneously react in the forward direction to produce more products ($\Delta G°$ negative) or in the reverse direction to form more reactants ($\Delta G°$ positive). Because standard free-energy values are readily available for a large number of substances, the standard free-energy change is easy to calculate for any reaction system of interest.

TABLE 18.1 Conventions used in establishing standard free-energy values

State of matter	Standard state
Solid	Pure solid
Liquid	Pure liquid
Gas	1 atm pressure[a]
Solution	Usually 1 M[b]
Elements	Standard free energy of formation of the element in its normal state is defined as zero

[a]Neglecting nonideal gas behavior.
[b]Neglecting nonideality of solutions.

SAMPLE EXERCISE 18.6

Determine the standard free-energy change for the following reaction at 298 K:

$$N_2(g) + 3H_2(g) \rightleftharpoons 2NH_3(g)$$

Solution: Using Appendix D we find that the standard free energies for the three substances of interest are as follows: $N_2(g)$, $\Delta G_f^\circ = 0.0$; $H_2(g)$, $\Delta G_f^\circ = 0.0$; $NH_3(g)$, $\Delta G_f^\circ = -16.66 \text{ kJ/mol}$.

The standard free-energy change for the reaction of interest is

$$\Delta G^\circ = 2\Delta G_f^\circ(NH_3) - [3\Delta G_f^\circ(H_2) + \Delta G_f^\circ(N_2)]$$

Inserting numerical quantities we obtain:

$$\Delta G^\circ = -33.32 \text{ kJ}$$

The fact that ΔG° is negative tells us that a mixture of H_2, N_2, and NH_3 at 25°C, each present at a pressure of 1 atm, would react spontaneously to form more ammonia. (Remember, however, that this says nothing about the rate at which the reaction occurs.)

FREE ENERGY AND TEMPERATURE

It is worthwhile to examine Equation [18.11] closely to see how the free-energy function depends on both the enthalpy and entropy changes for a given process. If it were not for entropy effects, all exothermic reactions, those in which ΔH is negative, would be spontaneous. The entropy contribution, represented by the quantity $-T\Delta S$, may increase or decrease the tendency of the reaction to proceed spontaneously. When ΔS is positive, meaning that the final state is more random or disordered than the initial state, the term $-T\Delta S$ makes a negative contribution to ΔG; that is, it increases the tendency of the reaction to occur spontaneously. When ΔS is negative, however, the term $-T\Delta S$ decreases the tendency of the reaction to occur spontaneously.

When ΔH and $-T\Delta S$ are of opposite sign, the relative importance of the two terms determines whether ΔG is negative or positive. In these instances, temperature is an important consideration. Both ΔH and ΔS are in principle capable of changing with temperature. In practice, however, the changes that occur are not very large unless very large temperature changes are involved. The only quantity in the equation

$$\Delta G = \Delta H - T\Delta S$$

that changes markedly with temperature is therefore $-T\Delta S$. At high temperatures the entropy term becomes relatively more important.

TABLE 18.2   Effect of temperature on reaction spontaneity

ΔH	ΔS	ΔG	Reaction characteristics	Examples
−	+	Always negative	Reaction is spontaneous at all temperatures: reverse reaction is always non-spontaneous	$2O_3(g) \longrightarrow 3O_2(g)$
+	−	Always positive	Reaction is nonspontaneous at all temperatures; reverse reaction occurs	$3O_2(g) \longrightarrow 2O_3(g)$
−	−	Negative at low temperatures; positive at high temperatures	Reaction is spontaneous at low temperatures; reverse reaction becomes spontaneous at high temperatures	$CaO(s) + CO_2(g) \longrightarrow CaCO_3(s)$
+	+	Positive at low temperatures; negative at high temperatures	Reaction is nonspontaneous at low temperatures but becomes spontaneous as temperature is raised	$CaCO_3(s) \longrightarrow CaO(s) + CO_2(g)$

Various possible situations for the relative signs of ΔH and ΔS are shown in Table 18.2, along with examples of each. By applying the concepts we have developed for predicting entropy changes it is often possible to predict how ΔG will change with change in temperature.

SAMPLE EXERCISE 18.7

(a) Predict the direction in which $\Delta G°$ for the equilibrium

$$N_2(g) + 3H_2(g) \rightleftharpoons 2NH_3(g)$$

will change with increase in temperature. (b) Calculate $\Delta G°$ at 500°C assuming that $\Delta H°$ and $\Delta S°$ do not change with temperature.

Solution: (a) In Sample Exercise 18.5 we saw that the change in $\Delta S°$ for the equilibrium of interest is negative. This means that the term $-T\Delta S°$ is positive and grows larger with increasing temperature. The standard free-energy change, $\Delta G°$, is the sum of the negative quantity $\Delta H°$ and the positive quantity, $-T\Delta S°$. Because only the latter grows larger with increasing temperature, $\Delta G°$ grows less negative.

(b) The $\Delta H_f°$ and $S°$ values necessary to calculate $\Delta H°$ and $\Delta S°$ for this reaction can be taken from Appendix D. We have previously performed these calculations in Sample Exercise 14.9 (Section 14.5) and in Sample Exercise 18.5, giving $\Delta H° = -92.38$ kJ and $\Delta S° = -198.2$ J/K. To calculate $\Delta G°$ given $\Delta H°$ and $\Delta S°$ we use the following relationship:

$$\Delta G° = \Delta H° - T\Delta S°$$

(Recall that the superscript ° indicates that the process is operated under standard-state conditions.) Assuming that $\Delta H°$ and $\Delta S°$ do not change with temperature, and using $T = 500 + 273 = 773$ K, we have

$$\Delta G° = -92.38 \text{ kJ} + (773 \text{ K})\left(-198.2 \frac{J}{K}\right)\left(\frac{1 \text{ kJ}}{10^3 \text{ J}}\right)$$
$$= -92.38 \text{ kJ} + 153.21 \text{ kJ}$$
$$= 60.83 \text{ kJ}$$

Notice that we changed $T\Delta S°$ from units of joules to kilojoules so it could be added to $\Delta H°$, which is in units of kilojoules.

In Sample Exercise 18.6, we calculated $\Delta G°$ for this reaction at 298 K: $\Delta G_{298}° = -33.32$ kJ. Thus we see that increasing the temperature from 298 K to 773 K changes $\Delta G°$ from -33.32 kJ to $+60.83$ kJ. Of course, the result at 773 K is not as accurate as that at 298 K, because $\Delta H°$ and $\Delta S°$ do change slightly with temperature. Nevertheless, the result should be a very reasonable approximation. The positive increase in $\Delta G°$ with increasing temperature is in agreement with the qualitative prediction made in part (a) of this exercise. Our result indicates that a mixture of $N_2(g)$, $H_2(g)$, and $NH_3(g)$, each at 1 atm pressure (standard-state conditions), will react spontaneously at 298 K to form more $NH_3(g)$; however, this same reaction is not spontaneous at 773 K. In fact, at 773 K, the reaction will proceed in the opposite direction, to form more $N_2(g)$ and $H_2(g)$.

18.6 Free energy and the equilibrium constant

Although it is very valuable to have a ready means of determining $\Delta G°$ for a reaction from tabulated values, we usually want to know about the direction of spontaneous change for systems that are not at standard conditions. For any chemical process the general relationship between the free-energy change under standard conditions, $\Delta G°$, and the free-energy change under any other conditions is given by the following expression:

$$\Delta G = \Delta G° + 2.303 \, RT \log Q \qquad [18.14]$$

R in this expression is the ideal-gas-equation constant, 8.314 J/K-mol; T is the absolute temperature; and Q is the reaction quotient (Section 14.4) that corresponds to the chemical reaction and particular reaction mixture of interest.

SAMPLE EXERCISE 18.8

Calculate ΔG at 298 K for the following reaction if the reaction mixture consists of 1.0 atm N_2, 3.0 atm H_2, and 1.0 atm NH_3:

$$N_2(g) + 3H_2(g) \longrightarrow 2NH_3(g)$$

Solution: For the balanced equation and set of concentrations given, the reaction quotient Q is

$$Q = \frac{P_{NH_3}^2}{P_{N_2}P_{H_2}^3} = \frac{(1.0)^2}{(1.0)(3.0)^3} = 3.7 \times 10^{-2}$$

ΔG_{298}° was calculated in Sample Exercise 18.6:
$\Delta G_{298}^\circ = -33.32$ kJ. Thus we have

$$\Delta G = \Delta G^\circ + 2.303\, RT \log Q$$

$$= (-33.32 \text{ kJ}) + 2.303\left(8.314\frac{J}{K}\right)(298 \text{ K})$$

$$\times \left(\frac{1 \text{ kJ}}{10^3 \text{ J}}\right) \log(3.7 \times 10^{-2})$$

$$= -33.32 \text{ kJ} + (-8.17 \text{ kJ})$$
$$= -41.49 \text{ kJ}$$

The free-energy change becomes more negative, changing from -33.32 kJ to -41.49 kJ, as the pressures of N_2, H_2, and NH_3 are changed from 1.0 atm each (standard-state conditions, ΔG°) to 1.0 atm, 3.0 atm, and 1.0 atm, respectively. The larger negative value for ΔG when the pressure of H_2 is increased from 1.0 atm to 3.0 atm indicates a larger "driving force" to produce NH_3. This result bears out the prediction of LeChatelier's principle, which indicates that increasing P_{H_2} should shift the reaction more to the product side, thereby forming more NH_3.

When a system is at equilibrium, ΔG must be zero, and the reaction quotient Q must by definition equal K (Section 14.4). Thus for a system at equilibrium (when $\Delta G = 0$ and $Q = K$), Equation [18.14] transforms as follows:

$$\Delta G = \Delta G^\circ + 2.303\, RT \log Q \qquad [18.14]$$

$$0 = \Delta G^\circ + 2.303\, RT \log K$$

$$\Delta G^\circ = -2.303\, RT \log K \qquad [18.15]$$

From Equation [18.15] we can readily see that if ΔG° is negative, $\log K$ must be positive. A positive value for $\log K$ means that $K > 1$. On the other hand if ΔG° is positive, $\log K$ is negative, which means that $K < 1$. To summarize:

ΔG° negative	$K > 1$
ΔG° zero	$K = 1$
ΔG° positive	$K < 1$

It is possible from a knowledge of ΔG° for a reaction to calculate the value for the equilibrium constant, using Equation [18.15]. Some care is necessary, however, in the matter of units. When dealing with gases, the concentrations of reactants should be expressed in units of atmospheres. The concentrations of solids are 1 if they are pure solids; the concentrations of pure liquids are 1 if they are pure liquids; if a mixture of liquids is involved, the concentration of each is expressed as mole fraction. For substances in solution, concentrations in moles per liter are appropriate.

From standard free energies of formation, calculate the equilibrium constant for the reaction

$$N_2(g) + 3H_2(g) \rightleftharpoons 2NH_3(g)$$

at 25°C.

Solution: The equilibrium constant for this reaction is written as

$$K = \frac{P_{NH_3}^2}{P_{N_2}P_{H_2}^3}$$

where the gas concentrations are expressed in atmospheres pressure. The standard free-energy change for the reaction was determined in Sample Exercise 18.6 to be -33.32 kJ. Inserting this into Equation [18.15] we obtain

$$-33,320 \text{ J} = -2.30(8.314 \text{ J/K-mol})(298 \text{ K}) \log K_p$$
$$\log K_p = 5.85$$

Taking the antilog,

$$K_p = 7.0 \times 10^5$$

This is a large equilibrium constant. Compare its magnitude with the equilibrium constants at higher temperature, as listed in Table 14.4. If a catalyst could be found that would permit reasonably rapid reaction of N_2 with H_2 at room temperature, high pressures would not be required to force the equilibrium toward NH_3.

ΔG AND WORK

The free energy has one further interesting property that comes from the fact that it is related to the degree of spontaneity of a process. Any process that occurs spontaneously can be utilized for the performance of useful work, at least in principle. For example, the falling of water in a waterfall is certainly a spontaneous process; it is also one from which it is possible to extract work, by causing the water to turn the blades of a turbine as it falls. Similarly, the burning of gasoline in the cylinders of a car leads to production of useful work in the motion of the car. How much work is extracted from a particular process depends on how it is carried out. For example, we might burn a liter of gasoline in an open container and extract no useful work at all. In an automobile engine, the overall efficiency for production of work is low, perhaps 20 percent. If the gasoline were caused to react with oxygen under other, more favorable conditions, the amount of work we could extract would be higher. In practice, we never achieve the maximum amount of work possible from a theoretical point of view. However, it is useful to know what the maximum possible work derivable from a process is, so that we might have a measure of our success in extracting work from processes of practical importance. Thermodynamics tells us that *the maximum possible work that can be derived from a spontaneous process occurring at constant temperature and pressure is equal to the free-energy change.*

For processes that are not spontaneous, the free-energy change is a measure of the minimum amount of work that must be done to cause the process to occur. In actual cases, more than the theoretical minimum amount of work must be done, because of inefficiencies in the way in which the change is caused to occur.

Free-energy considerations are very important in thinking about many nonspontaneous reactions we might wish to carry out for our own purposes, or that occur in nature. For example, we might wish to extract a metal from an ore. If we look at a reaction such as

$$Cu_2S(s) \longrightarrow 2Cu(s) + S(s)$$
$$\Delta G° = +86.2 \text{ kJ}; \Delta H° = +79.5 \text{ kJ} \qquad [18.16]$$

we find that it is endothermic and highly nonspontaneous. Clearly, then, we cannot hope to obtain copper metal from Cu_2S merely by trying to catalyze the reaction shown in Equation [18.16]. Instead, we must "do work" on the reaction in some way, in order to force it to occur as we wish. We might do this by coupling the reaction we've written with another reaction, so that we arrive at an overall reaction that *is* spontaneous. Consider, for example, the reaction

$$S(s) + O_2(g) \longrightarrow SO_2(g)$$
$$\Delta G° = -300.1 \text{ kJ}; \Delta H° = -296.9 \text{ kJ} \qquad [18.17]$$

This is an exothermic, spontaneous reaction. Adding reaction Equations [18.16] and [18.17], we obtain:

$$Cu_2S(s) \longrightarrow 2Cu(s) + S(s) \qquad \Delta G° = +86.2 \text{ kJ}; \Delta H° = +79.5 \text{ kJ}$$
$$\underline{S(s) + O_2(g) \longrightarrow SO_2(g) \qquad \Delta G° = -300.1 \text{ kJ}; \Delta H° = -296.9 \text{ kJ}}$$
$$Cu_2S(s) + O_2(g) \longrightarrow 2Cu(s) + SO_2(g)$$
$$\Delta G° = -213.9 \text{ kJ}; \Delta H° = -217.4 \text{ kJ}$$

The free-energy change for the overall reaction is the sum of the free-energy changes of the two reactions, and similarly for the overall enthalpy change. Because the negative free-energy change for the second reaction is larger than the positive free-energy change for the first, the overall reaction has a large and negative standard free-energy change.

The coupling of two or more reactions together to cause a nonspontaneous chemical process to occur is very important in biochemical systems. Many of the reactions that are absolutely essential to the maintenance of life do not occur within the human body spontaneously. These necessary reactions are made to occur, however, by coupling them with reactions that are spontaneous and energy releasing. The energy releases that accompany the metabolism of foodstuffs provide the primary source of necessary free energy. For example, a compound such as glucose, $C_6H_{12}O_6$, is oxidized in the body, and a substantial amount of energy is released.

$$C_6H_{12}O_6(s) + 6O_2(g) \longrightarrow 6CO_2(g) + 6H_2O(l)$$
$$\Delta G° = -2880 \text{ kJ}; \Delta H° = -2800 \text{ kJ} \qquad [18.18]$$

This energy is employed to "do work" in the body. However, some means is necessary to couple, or connect, the energy released by glucose oxidation to the reactions that require energy. One means of accomplishing this is shown graphically in Figure 18.8. Adenosine triphosphate (ATP) is a high-energy molecule. When ATP is converted to a lower-

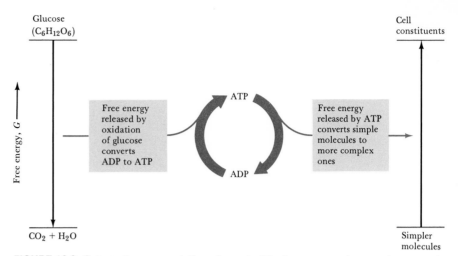

FIGURE 18.8 Schematic representation of a part of the free-energy changes that occur in cell metabolism. The oxidation of glucose to CO_2 and H_2O produces free energy. This released free energy is used to convert ADP into the more energetic ATP. The ATP is then used, as needed, as an energy source to convert simple molecules into more complex cell constituents. When ATP releases its free energy, it is converted to ADP.

energy molecule, adenosine diphosphate (ADP), energy to drive other chemical reactions becomes available. The energy released in glucose oxidation is used in part to reconvert ADP back to ATP. Thus, the ATP-ADP interconversions act as a means of storing energy and releasing it to drive needed reactions. The coupling of reactions so that the free energy released in one may be utilized by another requires particular enzymes as catalysts. In the chapter dealing with the biosphere, we shall have occasion to examine the energy relationships in living systems in more detail.

FOR REVIEW

Summary

In this chapter we examined the concept of equilibrium from a thermodynamic viewpoint. We have seen that any system will spontaneously approach equilibrium by a process that reduces the free energy of the system. The free-energy function, G, combines the two state functions enthalpy, H, and entropy, S. For processes that occur at constant temperature and pressure $\Delta G = \Delta H - T\Delta S$.

The enthalpy change of a system, ΔH, is a measure of the potential energy changes occurring. Exothermic processes (negative ΔH) tend to occur spontaneously. The entropy change of a system, ΔS, is a measure of the changes in the randomness or disorder of the system. Processes that produce an increase in the randomness of the system (positive ΔS) tend to occur spontaneously.

Entropy increases in a system are associated with an increase in the number of ways the particles of the system can be distributed among possible energy states or spatial arrangements. For example, an increase in volume, in translational energy, or in number of particles all lead to an increase in entropy. The standard entropy change in a system, $\Delta S°$, can be calculated from tabulated absolute entropy values, $S°$. Any spontaneous process causes an increase in the entropy of the universe. If the entropy of a system decreases, there must be an even larger entropy increase in its surroundings:

$$\Delta S_{system} + \Delta S_{surroundings} > 0$$

The free-energy change for a process occurring at constant temperature and pressure relates directly to reaction spontaneity. ΔG is negative for all spontaneous processes, whereas a positive ΔG indicates a nonspontaneous process. ΔG is zero at equilibrium. The free-energy change is the maximum useful work that can be performed by a spontaneous process. If a process is nonspontaneous, ΔG is the minimum amount of work that must be done to cause the process to occur.

The standard free-energy change, $\Delta G°$, for any process can be calculated from tabulated standard free energies of formation, $\Delta G_f°$; it can also be calculated from standard enthalpy and entropy changes: $\Delta G° = \Delta H° - T\Delta S°$. Temperature changes will change the value of ΔG and can also change its sign.

The free-energy change under nonstandard conditions is related to the standard free-energy change: $\Delta G = \Delta G° + 2.303\, RT \log Q$. At equilibrium ($\Delta G = 0$, $Q = K$), $\Delta G° = -2.303\, RT \log K$. Thus the standard free-energy change is related to the equilibrium constant. Consequently, we can understand the position of a chemical equilibrium in terms of the $\Delta H°$ and $T\Delta S°$ functions of which $\Delta G°$ is composed.

Learning goals

Having read and studied this chapter, you should be able to:

1 Define the term spontaneity and apply it in identifying spontaneous processes.

2 Describe how entropy is related to randomness or disorder.

3 State the second law of thermodynamics.

4 Predict whether the entropy change in a given process is positive, negative, or near zero.

5 Describe how and why the entropy of a substance changes with increasing temperature or when a phase change occurs, starting with the substance as a pure solid at $0\,K$.

6 Calculate $\Delta S°$ for any reaction from tabulated absolute entropy values, $S°$.

7 Define free energy in terms of enthalpy and entropy.

8 Explain how the sign of the free-energy change, ΔG, determines whether or not a process is spontaneous in the forward direction.

9 Calculate the standard free-energy change at constant temperature and pressure, $\Delta G°$, for any process from tabulated values for the standard free energies of reactants and products.

10 List the usual conventions regarding standard states in setting the values for standard free energies.

11 Predict how ΔG will change with temperature, given the signs for ΔH and ΔS.

12 Estimate $\Delta G°$ at any temperature given $\Delta S_{298}°$ and $\Delta H_{298}°$.

13 Calculate the free-energy change under nonstandard conditions, ΔG, given $\Delta G°$, temperature, and the data needed to calculate the reaction quotient.

14 Calculate $\Delta G°$ from K and perform the reverse operation.

15 Describe the relationship between ΔG and the work that can be derived from a spontaneous process.

Key terms

Among the more important terms and expressions used for the first time in this chapter are the following:

Absolute entropy, $S°$ (Section 18.4), refers to the entropy of a particular substance at some temperature and in some particular state, referenced to a zero entropy for the pure solid substance at a temperature of absolute zero.

Entropy (Section 18.2) is a thermodynamic function associated with the number of different, equivalent energy states or spatial arrangements in which a system may be found. It is a thermodynamic state function, which means that once we specify the conditions for a system, that is, the temperature, pressure, and so on, the entropy is defined.

Free energy (Section 18.5) is a thermodynamic state function that combines enthalpy and entropy, in the form $G = H - TS$. For a change occurring at constant temperature and pressure, the change in free energy is $\Delta G = \Delta H - T\Delta S$.

The **second law of thermodynamics** (Section 18.1) is a statement of our experience that there is a direction to the way events occur in nature: When a process occurs spontaneously in one direction, it is nonspontaneous in the reverse direction. It is possible to state the second law in many different forms, but they all relate back to the same idea about spontaneity. One of the most common statements found in chemical contexts is that in any spontaneous process the entropy of the universe increases.

A **spontaneous process** (Section 18.1) is one that is capable of proceeding in a given direction, as written or described, without needing to be driven by an outside source of energy. A process may be spontaneous even though it is very slow. ΔG is negative for all spontaneous processes.

Vibrational and rotational energies (Section 18.3) refer to the storing of energies in molecules in the form of vibrational motions between the atoms of the molecules or rotational motions of the molecule as a whole.

EXERCISES

Spontaneity and entropy

18.1 Which of the following processes are spontaneous: (a) a rock rolling uphill; (b) rusting of a nail; (c) putting Humpty-Dumpty together again; (d) assembling letters to make a book; (e) osmosis of water into NaCl solution; (f) development of an adult human from a newborn baby?

18.2 For each of the following pairs, indicate which substance you would expect to possess the larger absolute entropy (consider each substance at 1 atm pressure): (a) 1 mol of $He(g)$ at 25 K or 1 mol of $He(g)$ at 25°C; (b) 1 mol of $H_2(g)$ at 25°C or 2 mol of $H(g)$ at 25°C; (c) 1 mol of $H_2O(l)$ at 100°C or 1 mol of $H_2O(g)$ at 100°C; (d) 1 mol of $HCl(g)$ at 25°C or 1 mol of $HCl(aq)$ at 25°C; (e) 1 mol of $CH_4(g)$ at 25°C or 1 mol of $C_2H_6(g)$ at 25°C. Give a reason for your choice in each case.

18.3 How would you expect the absolute entropy (per mole) of each of the following substances to compare with that of $F_2(g)$ if each of these substances is at the same temperature and pressure as $F_2(g)$: (a) $F_2(s)$; (b) $OF_2(g)$; (c) $F(g)$; (d) $LiF(s)$?

18.4 Using Appendix D, compare the absolute entropy values at 298 K for the substances in each of the following pairs: (a) $O_2(g)$ and $O_3(g)$; (b) C(diamond) and C(graphite); (c) $Br_2(g)$ and $Br_2(l)$; (d) $NaCl(s)$ and $MgCl_2(s)$. Explain the origin of the difference in $S°$ in each case.

Entropy changes

18.5 Predict the sign of ΔS for the system in each of the following processes: (a) evaporation of 1 mol of $CCl_4(l)$; (b) mixing 10 g of alcohol with 25 g of water; (c) precipitation of $AgCl(s)$ upon mixing $NaCl(aq)$ and $AgNO_3(aq)$.

18.6 Predict whether the entropy change of the system is positive or negative for each of the following reactions:

(a) $CaH_2(s) + H_2O(l) \longrightarrow Ca(OH)_2(s) + H_2(g)$
(b) $Ag_2O(s) \longrightarrow 2Ag(s) + \frac{1}{2}O_2(g)$
(c) $2Fe(s) + \frac{3}{2}O_2(g) \longrightarrow Fe_2O_3(s)$
(d) $Cl_2(g) \longrightarrow 2Cl(g)$
(e) $NaCl(s) \longrightarrow NaCl(aq)$
(f) $HCl(aq, 1\,M) \longrightarrow HCl(aq, 0.1\,M)$

18.7 Using tabulated $S°$ values from Appendix D, calculate $\Delta S°$ for each of the following reactions:

(a) $CO(g) + 2H_2(g) \longrightarrow CH_3OH(g)$
(b) $2HCl(g) + Br_2(l) \longrightarrow 2HBr(g) + Cl_2(g)$
(c) $2SO_2(g) + O_2(g) \longrightarrow 2SO_3(g)$
(d) $2NO_2(g) \longrightarrow N_2O_4(g)$

In each case, does the sign of $\Delta S°$ agree with your expectations? Explain.

18.8 When the insecticide DDT is applied it is sprayed over large areas. It later moves in the environment through soil into plants and water supplies. In lakes it concentrates in the fatty tissue of fish. If we think of DDT

as the "system," describe the entropy change associated with each of the above processes. In terms of cleaning up DDT in the environment, what entropy change to the DDT system is required? How can such an entropy change be brought about? Is the process feasible?

18.9 According to the second law of thermodynamics, the entropy of the universe increases during any spontaneous process. Yet in the freezing of water the entropy of the water decreases. Explain why these statements are not contradictory.

Free energy

18.10 The reaction

$$H_2(g) + \tfrac{1}{2}O_2(g) \longrightarrow H_2O(l)$$

proceeds spontaneously at 25°C even though there is a decrease in disorder in the system. Explain how this can occur.

18.11 How does ΔG relate to spontaneity? To reaction speed? To extent of reaction? To the direction from which the reaction approaches equilibrium?

18.12 Calculate $\Delta G°_{298}$ for

$$H_2O(l) \longrightarrow H_2O(g)$$

using data from Appendix D. Why is $\Delta G°_{298}$ positive?

18.13 Using data from Appendix D, calculate the standard free-energy change for each of the following processes. In each case indicate whether the reaction is spontaneous under standard conditions.

(a) $ZnO(s) + CO(g) \longrightarrow Zn(s) + CO_2(g)$
(b) $2CO_2(g) + H_2O(g) \longrightarrow C_2H_2(g) + \frac{5}{2}O_2(g)$
(c) $H_2(g) + I_2(g) \longrightarrow 2HI(g)$

18.14 Calculate $\Delta G°_{298}$ for

$$H_2O_2(g) \longrightarrow H_2O(g) + \tfrac{1}{2}O_2(g)$$

given that $\Delta H°_{298} = -106$ kJ and $\Delta S°_{298} = +58$ J/K for this process. Would you expect $H_2O_2(g)$ to be very stable at 298 K? Explain briefly.

18.15 Using the data in Appendix D calculate $\Delta H°$, $\Delta S°$, and $\Delta G°$ for each of the following reactions. In each case show that $\Delta G° = \Delta H° - T\Delta S°$.

(a) $2SO_2(g) + O_2(g) \longrightarrow 2SO_3(g)$
(b) $CH_4(g) + 2O_2(g) \longrightarrow CO_2(g) + 2H_2O(l)$
(c) $CaO(s) + CO_2(g) \longrightarrow CaCO_3(s)$
(d) $Fe_2O_3(s) + 3CO(g) \longrightarrow 3CO_2(g) + 2Fe(s)$

18.16 Calculate $\Delta G°$ for the process diamond $\longrightarrow$ graphite at 25°C. If the process is spontaneous, why don't diamonds in rings turn into graphite?

Free energy and temperature

18.17 Classify each of the following reactions as belonging to one of the four possible types summarized in Table 18.2:

(a) $N_2(g) + 3F_2(g) \longrightarrow 2NF_3(g)$
$$\Delta H^\circ = -249 \text{ kJ}; \Delta S^\circ = -278 \text{ J/K}$$

(b) $N_2(g) + 3Cl_2(g) \longrightarrow 2NCl_3(g)$
$$\Delta H^\circ = +460 \text{ kJ}; \Delta S^\circ = -275 \text{ J/K}$$

(c) $N_2F_4(g) \longrightarrow 2NF_2(g)$
$$\Delta H^\circ = 85 \text{ kJ}; \Delta S^\circ = 198 \text{ J/K}$$

(d) $2H_2O(l) \longrightarrow 2H_2(g) + O_2(g)$
$$\Delta H^\circ = 572 \text{ kJ}; \Delta S^\circ = 329 \text{ J/K}$$

18.18 (a) Using Appendix D, predict how ΔG° for the following process will change with increasing temperature:

$$CO(g) + 2H_2(g) \longrightarrow CH_3OH(g)$$

(b) Calculate ΔG° at 500°C assuming that ΔH° and ΔS° do not change with temperature.

18.19 The following reaction is employed in the metallurgy of iron to obtain iron metal from its oxide:

$$Fe_2O_3(s) + 3CO(g) \longrightarrow 3CO_2(g) + 2Fe(s)$$

At what temperature does this reaction become spontaneous under standard conditions?

Free energy and concentration

18.20 (a) Explain qualitatively how ΔG changes for the following reaction as the partial pressure of N_2 is increased:

$$N_2(g) + 3H_2(g) \longrightarrow 2NH_3(g)$$

(b) Calculate ΔG for this reaction if $P_{N_2} = 5.0$ atm, $P_{NH_3} = 0.10$ atm, and $P_{H_2} = 2.0$ atm.

18.21 The standard free-energy change for the following reaction at 25°C is -118.4 kJ/mol:

$$KClO_3(s) \longrightarrow KCl(s) + \tfrac{3}{2}O_2(g)$$

At what pressure of O_2 gas does $\Delta G = 0$ at 25°C?

18.22 Calculate ΔG at 298 K for the following reaction if the reaction mixture consists of 2.0 atm of $CO(g)$, 2.0 atm of $O_2(g)$, and 10.0 atm of $CO_2(l)$:

$$2CO(g) + O_2(g) \longrightarrow 2CO_2(g)$$

Free energy and equilibrium

18.23 Calculate the equilibrium constant at 298 K for each of the following reactions, using data from Appendix D:

(a) $2HI(g) + Cl_2(g) \rightleftharpoons 2HCl(g) + I_2(g)$
(b) $H_2(g) + I_2(g) \rightleftharpoons 2HI(g)$

18.24 Calculate ΔG° at 298 K for the ionization of H_2O:

$$H_2O(l) \rightleftharpoons H^+(aq) + OH^-(aq)$$

How does the sign and magnitude of ΔG° relate to the position of the equilibrium?

18.25 In the reaction

$$PCl_3(g) + Cl_2(g) \rightleftharpoons PCl_5(g)$$

a certain equilibrium mixture at 460 K contains $P_{PCl_3} = 0.25$ atm, $P_{Cl_2} = 0.25$ atm, and $P_{PCl_5} = 0.50$ atm. From this information calculate ΔG°.

18.26 Consider a weak acid HX. If a 0.10 M solution of this acid has a pH of 5.83 at 25°C, what is ΔG° for the acid dissociation?

18.27 Based on your general chemical knowledge, predict which of the following reactions will have $K > 1$.

(a) $C(s) + O_2(g) \rightleftharpoons CO_2(g)$
(b) $2Na(s) + Cl_2(g) \rightleftharpoons 2NaCl(s)$
(c) $Al_2O_3(s) \rightleftharpoons 2Al(s) + \tfrac{3}{2}O_2(g)$
(d) $2N(g) \rightleftharpoons N_2(g)$
(e) $2H_2O(g) + CO_2(g) \rightleftharpoons CH_4(g) + 2O_2(g)$

In each case, predict the sign for ΔH° and ΔS° and discuss briefly how these functions determine the magnitude of K.

[18.28] The reaction

$$SO_2(g) + 2H_2S(g) \rightleftharpoons 3S(s) + 2H_2O(g)$$

is the basis of a suggested method for removal of SO_2 from power-plant stack gases. The standard free energy for each substance is given in Appendix D. (a) What is the equilibrium constant at 298 K for the reaction as written? (b) Is this reaction at least in principle a feasible method of removing SO_2? (c) Assuming that the H_2O vapor pressure is 25 mm Hg and adjusting the conditions so that P_{SO_2} equals P_{H_2S}, calculate the equilibrium SO_2 pressure in this system. (d) The reaction as written is exothermic. Would you expect the process to be more or less effective at higher temperatures?

Additional exercises

18.29 What is a state function?

18.30 Without looking in the summary or within the text of the present chapter, give an equation that relates (a) ΔG to ΔH and ΔS; (b) ΔG° to ΔG; (c) ΔG° to K.

18.31 A friend approaches you with a new idea for a catalyst that she says will make the reaction between N_2 and O_2 to form NO_2 spontaneous at 298 K under standard conditions. You realize that such a process would be a very useful means to fix nitrogen for making fertilizers. What's wrong with your friend's idea?

18.32 Proteins are long-chain biomolecules made from amino acids. The protein chain is held in specific arrangements by hydrogen bonds between N—H and $>$C=O groups along the chain (Section 11.5). When a protein is denatured, as when an egg is boiled, the increased thermal energy causes rupture of hydrogen bonds, and the regular arrangement along the protein chain is lost. What are the signs of ΔH and ΔS in the protein system associated with denaturation?

18.33 When a rubber band is stretched, the long-chain molecules of the rubber become increasingly aligned along the direction of the stretch. What are the signs of ΔH and ΔS in the rubber-band system associated with stretching of a rubber band?

18.34 Explain briefly why all tabulated S° values for all pure substances are positive.

18.35 In each of the following pairs, which item has the greater entropy: (a) a shuffled deck of cards or a deck arranged by suit; (b) toys spread randomly through a toy box or the same toys spread randomly throughout the house; (c) an assembled television set or its unassembled parts?

[18.36] Entropy has been called "time's arrow," in that it indicates the direction in which time is flowing. Discuss this description in light of the second law of thermodynamics.

18.37 The solubility of $MgCl_2$ at 20°C is 6 M. Which of the following processes are spontaneous and which non-spontaneous?

(a) $MgCl_2(s) \longrightarrow MgCl_2(aq, 6\ M)$
(b) $MgCl_2(s) \longrightarrow MgCl_2(aq, 8\ M)$
(c) $MgCl_2(s) \longrightarrow MgCl_2(aq, 4\ M)$

18.38 Indicate whether each of the following statements is true or false. If it is false, describe in what respect it is false. (a) All exothermic reactions are spontaneous. (b) In most spontaneous reactions the entropy of the universe increases. (c) If an endothermic reaction has a positive ΔS, then the free-energy change associated with the process will grow increasingly negative as temperature increases.

18.39 If energy is conserved in any process, why is there an energy crisis?

18.40 Predict the signs of ΔH, ΔS, and ΔG for the following processes: (a) ice melts at 0°C; (b) ice melts at 25°C; (c) ice melts at −25°C.

18.41 Calculate $\Delta S°$ and $\Delta G°$ for each of the following reactions using data tabulated in Appendix D:

(a) $2NO(g) + O_2(g) \longrightarrow 2NO_2(g)$
(b) $C_2H_4(g) + H_2(g) \longrightarrow C_2H_6(g)$
(c) $2NaHCO_3(s) \longrightarrow Na_2CO_3(s) + CO_2(g) + H_2O(g)$

18.42 Assume that the evaporation of water at 100°C and 1 atm is an equilibrium process throughout ($\Delta G = 0$). If $\Delta H_v = 40.7$ kJ/mol at 100°C, what is ΔS for this process at this temperature?

[18.43] Suppose that we have the following change taking place in a biological polymer: native state $\longrightarrow$ denatured stated. (a) If we find that the equilibrium shifts to the right as the temperature is increased, according to LeChatelier's principle, what is the sign for ΔH? (b) If, at a temperature of 60°C, ΔG is negative and ΔH is positive, what must be true of the sign and magnitude of ΔS? What does this suggest in terms of the structure of the polymer?

18.44 On the basis of the $\Delta H°$ and $\Delta S°$ values given for each reaction below, indicate (a) which are spontaneous under standard conditions at 25°C and (b) which can be expected to be spontaneous at high temperatures.

(a) $KClO_3(s) \longrightarrow KCl(s) + \frac{3}{2}O_2(g)$
    $\Delta H° = -44.7$ kJ; $\Delta S° = +59.1$ J/K
(b) $2Al(s) + 3Cl_2(g) \longrightarrow 2AlCl_3(s)$
    $\Delta H° = -332$ kJ; $\Delta S° = -93.4$ J/K
(c) $NOCl(g) \longrightarrow NO(g) + \frac{1}{2}Cl_2(g)$
    $\Delta H° = +37.6$ kJ; $\Delta S° = +58.5$ J/K

18.45 Would you judge the following reaction to be potentially significant in removal of NO from auto exhaust in a catalytic muffler at 1000°C?

$2NO(g) + 3H_2O(g) \longrightarrow 2NH_3(g) + \frac{5}{2}O_2(g)$

(There is quite a lot of $H_2O(g)$ in the exhaust.)

18.46 (a) Given K_a for acetic acid at 298 K (Appendix E), calculate $\Delta G°$ for the dissociation of this substance. (b) What is the value of ΔG at equilibrium? (c) What is the value of ΔG when $[H^+] = 1.0 \times 10^{-7}\ M$, $[C_2H_3O_2^-] = 1.0 \times 10^{-7}\ M$, and $[HC_2H_3O_2] = 2.0\ M$?

18.47 $\Delta G_f°$ for urea, $CO(NH_2)_2(s)$, is −197.2 kJ/mol. What is the equilibrium constant, K_p, for the following reaction at 25°C?

$CO(NH_2)_2(s) + H_2O(g) \rightleftharpoons CO_2(g) + 2NH_3(g)$

18.48 At what temperature will K_p equal unity for the reaction

$O_2(g) \rightleftharpoons 2O(g)$

Assume that $\Delta H°$ and $\Delta S°$ do not vary with temperature.

18.49 Evaporating NH_3 and HCl from aqueous ammonia ("ammonium hydroxide") and hydrochloric acid, respectively, produce the white haze often seen on laboratory windows and glassware:

$NH_3(g) + HCl(g) \longrightarrow NH_4Cl(s)$

Calculate the equilibrium constant, K_p, for this reaction at 25°C.

18.50 The oxidation of glucose, $C_6H_{12}O_6$, in body tissue produces CO_2 and H_2O. In contrast, anaerobic decomposition, which occurs during fermentation, produces ethyl alcohol, C_2H_5OH. (a) Compare the equilibrium constants for the following reactions:

$C_6H_{12}O_6(s) + 6O_2(g) \rightleftharpoons 6CO_2(g) + 6H_2O(l)$
$C_6H_{12}O_6(s) \rightleftharpoons 2C_2H_5OH(l) + 2CO_2(g)$

The standard free energy of formation of $C_6H_{12}O_6(s)$ is −912 kJ/mol. (b) Compare the maximum amounts of work that can be obtained from these two processes under standard conditions.

[18.51] The relationship between the temperature of a reaction, its standard enthalpy change, and the equilibrium constant at that temperature can be expressed in terms of the following linear equation:

$$\log K = \frac{-\Delta H°}{2.30\ RT} + \text{constant}$$

(a) Explain how this equation can be used to determine $\Delta H°$ experimentally from the equilibrium constants at several different temperatures. (b) Derive the above equation using relationships given in this chapter. What is the constant equal to?

[18.52] The potassium-ion concentration in blood plasma is about $5.0 \times 10^{-3}\ M$ while the concentration in muscle-cell fluid is much greater, 0.15 M. The plasma and intracellular fluid are separated by the cell membrane, which we assume is permeable only to K^+. (a) What is ΔG for the transfer of 1 mol of K^+ from blood plasma to the cellular fluid at body temperature, 37°C? (b) What is the minimum amount of work that must be used to transfer this K^+?

19

Electrochemistry

Many important chemical processes utilize electricity, whereas others can be used to produce it. Because of the importance of electricity in modern society, it is useful for us to examine the subject area of electrochemistry, which deals with the relationships that exist between electricity and chemical reactions. As we shall see, our discussions of electrochemistry will provide insight into such diverse topics as the construction and operation of batteries, spontaneity of chemical reactions, electroplating, and the corrosion of metals. Because electricity involves the flow of electrons, electrochemistry focuses on reactions in which electrons are transferred from one substance to another. Such reactions are known as oxidation-reduction or "redox" reactions.

19.1 Oxidation-reduction reactions

Oxidation-reduction reactions were briefly discussed in Section 7.10. At that time we defined oxidation as an increase in oxidation number (loss of electrons) and reduction as a decrease in oxidation number (gain of electrons). If one substance gains electrons and is thereby reduced, another substance must lose electrons and be thereby oxidized. Oxidation and reduction must occur simultaneously; there cannot be one without the other. For example, consider the reaction between iron and hydrogen chloride, Equation [19.1]:

$$\overset{0}{Fe}(s) + 2\overset{+1\,-1}{HCl}(g) \longrightarrow \overset{+2\,-1}{FeCl_2}(s) + \overset{0}{H_2}(g) \qquad [19.1]$$

The oxidation number of each element is given above the symbol for the element. If you have forgotten the rules for finding oxidation numbers, it would be useful to refer back to Section 7.10 and review them. By looking at the oxidation states in Equation [19.1], we see that iron is oxidized while HCl is simultaneously reduced.

In discussing oxidation-reduction reactions, it is often useful to refer to the substance causing oxidation as the oxidizing agent or oxidant. Oxidizing agents possess an affinity for electrons and cause other substances to be oxidized by abstracting electrons from them. Because the oxidizing agent gains electrons, it is reduced. Similarly, a substance that causes reduction is called a reducing agent or reductant. In Equation [19.1], HCl is the oxidizing agent, whereas Fe is the reducing agent. The substance reduced in a reaction is always the oxidizing agent, whereas the substance oxidized is always the reducing agent.

HALF-REACTIONS

Although oxidation and reduction must take place simultaneously, it is often convenient to consider them as separate processes. For example, the oxidation of Sn^{2+} by Fe^{3+}, Equation [19.2],

$$Sn^{2+}(aq) + 2Fe^{3+}(aq) \longrightarrow Sn^{4+}(aq) + 2Fe^{2+}(aq) \qquad [19.2]$$

can be considered to consist of two processes: (1) the oxidation of Sn^{2+}, Equation [19.3]; and (2) the reduction of Fe^{3+}, Equation [19.4].

Oxidation: $\qquad\qquad\qquad Sn^{2+}(aq) \longrightarrow Sn^{4+}(aq) + 2e^- \qquad [19.3]$

Reduction: $\qquad 2Fe^{3+}(aq) + 2e^- \longrightarrow 2Fe^{2+}(aq) \qquad\qquad [19.4]$

Such equations, which show either oxidation or reduction alone, are known as half-reactions. As shown in Equations [19.3] and [19.4], the number of electrons lost in an oxidation half-reaction must equal the number gained in a reduction half-reaction. When this condition is met and the half-reactions are balanced, they can be added to give the balanced total oxidation-reduction equation.

BALANCING EQUATIONS BY THE METHOD OF HALF-REACTIONS

As we have suggested above, oxidation-reduction equations can be balanced using half-reactions. As an example, let's consider the reaction that occurs between permanganate ion, MnO_4^-, and oxalate ion, $C_2O_4^{2-}$, in acidic water solutions. When MnO_4^- is added to an acidified solution of $C_2O_4^{2-}$, the deep purple color of the MnO_4^- ion fades. Bubbles of CO_2 form, and the solution takes on the pale pink color of Mn^{2+}. We can therefore write the rough, unbalanced equation as follows:

$$MnO_4^-(aq) + C_2O_4^{2-}(aq) \longrightarrow Mn^{2+}(aq) + CO_2(g) \qquad [19.5]$$

Experiments would also show that H^+ is consumed and H_2O produced in the reaction. We shall see that this fact can be deduced in the course of balancing the equation.

To complete and balance Equation [19.5] by the method of half-reactions, we proceed through the following three steps. *Step 1* is to write two incomplete half-reactions, one involving the oxidant and the other involving the reductant.

$MnO_4^-(aq) \longrightarrow Mn^{2+}(aq)$

$C_2O_4^{2-}(aq) \longrightarrow CO_2(g)$

In *step 2*, the half-reactions are completed and balanced separately. First the atoms undergoing oxidation or reduction are balanced, then the remaining elements, and finally the charge. If the reaction occurs in acidic water solution, H^+ and H_2O can be added to either reactants or products to balance hydrogen and oxygen. Similarly, in basic solution, the equation can be completed using OH^- and H_2O. These species are in large supply in the respective solutions, and their formation as products or their utilization as reactants can easily go undetected experimentally. In the permanganate half-reaction, we already have one manganese atom on each side of the equation. However, we have four oxygens on the left and none on the right side; four H_2O molecules are needed among the products to balance the four oxygen atoms in MnO_4^-:

$$MnO_4^-(aq) \longrightarrow Mn^{2+}(aq) + 4H_2O(l)$$

The eight hydrogen atoms that this introduces among the products can then be balanced by adding $8H^+$ to the reactants:

$$8H^+(aq) + MnO_4^-(aq) \longrightarrow Mn^{2+}(aq) + 4H_2O(l)$$

At this stage there are equal numbers of each type of atom on both sides of the equation, but the charge still needs to be balanced. The total charge of the reactants is $+8 - 1 = +7$, while that of the products is $+2 + 4(0) = +2$. To balance the charge, five electrons are added to the reactant side.*

$$5e^- + 8H^+(aq) + MnO_4^-(aq) \longrightarrow Mn^{2+}(aq) + 4H_2O(l) \qquad [19.6]$$

Proceeding similarly with the oxalate half-reaction we have

$$C_2O_4^{2-}(aq) \longrightarrow 2CO_2(g)$$

Charge is balanced by adding two electrons among the products:

$$C_2O_4^{2-}(aq) \longrightarrow 2CO_2(g) + 2e^- \qquad [19.7]$$

In *step 3*, we multiply each equation by an appropriate factor so that the number of electrons gained in one half-reaction equals the number of electrons lost in the other. The half-reactions are then added to give the overall balanced equation. In our example, the MnO_4^- half-reaction must be multiplied by 2 while the $C_2O_4^{2-}$ half-reaction must be multiplied by 5:

$$10e^- + 16H^+(aq) + 2MnO_4^-(aq) \longrightarrow 2Mn^{2+}(aq) + 8H_2O(l)$$
$$\underline{5C_2O_4^{2-}(aq) \longrightarrow 10CO_2(g) + 10e^-}$$
$$16H^+(aq) + 2MnO_4^-(aq) + 5C_2O_4^{2-}(aq) \longrightarrow$$
$$2Mn^{2+}(aq) + 8H_2O(l) + 10CO_2(g) \qquad [19.8]$$

The balanced equation is the sum of the balanced half-reactions.

*Although the oxidation numbers of the elements need not be used in balancing a half-reaction by this method, oxidation numbers can be used as a check. In this example MnO_4^- contains manganese in a $+7$ oxidation state. Because manganese changes from a $+7$ to a $+2$ oxidation state, it must gain five electrons just as we have already concluded.

The equations for reactions that occur in basic solution can be balanced initially as if they occurred in acidic solution. The H^+ ions can then be "neutralized" by adding an equal number of OH^- ions to both sides of the equation. This procedure is shown in Sample Exercise 19.1.

SAMPLE EXERCISE 19.1

Complete and balance the following equations:

(a) $Cr_2O_7^{2-}(aq) + Cl^-(aq) \longrightarrow$
$\quad\quad Cr^{3+}(aq) + Cl_2(g)$ (acidic solution)

(b) $CN^-(aq) + MnO_4^-(aq) \longrightarrow$
$\quad\quad CNO^-(aq) + MnO_2(s)$ (basic solution)

Solution: (a) The incomplete and unbalanced half-reactions are

$$Cr_2O_7^{2-}(aq) \longrightarrow Cr^{3+}(aq)$$
$$Cl^-(aq) \longrightarrow Cl_2(g)$$

The half-reactions are first balanced with respect to the elements. In the first half-reaction the presence of $Cr_2O_7^{2-}$ among the reactants requires two Cr^{3+} among the products. The 7 oxygen atoms in $Cr_2O_7^{2-}$ are balanced by adding $7H_2O$ to the products. The 14 hydrogen atoms in $7H_2O$ are then balanced by adding $14H^+$ among the reactants:

$$14H^+(aq) + Cr_2O_7^{2-}(aq) \longrightarrow 2Cr^{3+}(aq) + 7H_2O(l)$$

Charge is balanced by adding electrons to the left side of the equation so that the total charge is the same on both sides:

$$6e^- + 14H^+(aq) + Cr_2O_7^{2-}(aq) \longrightarrow$$
$$2Cr^{3+}(aq) + 7H_2O(l)$$

In the second half-reaction, two Cl^- are required to balance one Cl_2:

$$2Cl^-(aq) \longrightarrow Cl_2(g)$$

We add two electrons to the right side to attain charge balance:

$$2Cl^-(aq) \longrightarrow Cl_2(g) + 2e^-$$

This second half-reaction must be multiplied by 3 to equalize electron loss and gain in the two half-reactions. The two half-reactions are then added to give the balanced equation:

$$14H^+(aq) + Cr_2O_7^{2-}(aq) + 6Cl^-(aq) \longrightarrow$$
$$2Cr^{3+}(aq) + 7H_2O(l) + 3Cl_2(g)$$

(b) The incomplete and unbalanced half-reactions are

$$CN^-(aq) \longrightarrow CNO^-(aq)$$
$$MnO_4^-(aq) \longrightarrow MnO_2(s)$$

The equations may be balanced initially as if they took place in acidic solution. The resultant balanced half-reactions are

$$CN^-(aq) + H_2O(l) \longrightarrow$$
$$CNO^-(aq) + 2H^+(aq) + 2e^-$$
$$3e^- + 4H^+(aq) + MnO_4^-(aq) \longrightarrow$$
$$MnO_2(s) + 2H_2O(l)$$

Because H^+ cannot exist in any appreciable concentration in basic solution, it is removed from the equations by the addition of an appropriate amount of $OH^-(aq)$. In the CN^- half-reaction, $2OH^-(aq)$ is added to both sides of the equation to "neutralize" the $2H^+(aq)$. The $2OH^-(aq)$ and $2H^+(aq)$ form $2H_2O(l)$:

$$2OH^-(aq) + H_2O + CN^-(aq) \longrightarrow$$
$$CNO^-(aq) + 2H_2O(l) + 2e^-$$

The half-reaction can be simplified because H_2O occurs on both sides of the equation. The simplified equation is

$$2OH^-(aq) + CN^-(aq) \longrightarrow CNO^-(aq) + H_2O(l) + 2e^-$$

For the MnO_4^- half-reaction, $4OH^-(aq)$ is added to both sides of the equation:

$$3e^- + 4H_2O(l) + MnO_4^-(aq) \longrightarrow$$
$$MnO_2(s) + 2H_2O(l) + 4OH^-(aq)$$

Simplifying gives

$$3e^- + 2H_2O(l) + MnO_4^-(aq) \longrightarrow$$
$$MnO_2(s) + 4OH^-(aq)$$

The top equation is multiplied by 3 and the bottom one by 2 to equalize electron loss and gain in the two half-reactions. The half-reactions are then added:

$$6OH^-(aq) + 3CN^-(aq) \longrightarrow$$
$$3CNO^-(aq) + 3H_2O(l) + 6e^-$$
$$6e^- + 4H_2O(l) + 2MnO_4^-(aq) \longrightarrow$$
$$2MnO_2(s) + 8OH^-(aq)$$

$$6OH^-(aq) + 3CN^-(aq) +$$
$$4H_2O(l) + 2MnO_4^-(aq) \longrightarrow$$
$$3CNO^-(aq) + 3H_2O(l) + 2MnO_2(s) + 8OH^-(aq)$$

The overall equation can be simplified, because H_2O and OH^- occur on both sides. The simplified equation is

$$3CN^-(aq) + H_2O(l) + 2MnO_4^-(aq) \longrightarrow$$
$$3CNO^-(aq) + 2MnO_2(s) + 2OH^-(aq)$$

The method of half-reactions is not the only method of balancing oxidation-reduction reactions. We have stressed this method because it is useful in studying other material in this chapter. Another popular method, the oxidation-number method, is considered below. To illustrate this approach, consider the following equation which was balanced earlier using half-reactions:

$$MnO_4^-(aq) + H^+(aq) + C_2O_4^{2-}(aq) \longrightarrow Mn^{2+}(aq) + H_2O(l) + CO_2(g)$$

To balance this equation by the oxidation-number method, we proceed as follows:

1. Determine the oxidation number for each element on both sides of the equation, thereby determining the elements that have undergone oxidation and reduction. In the present example, hydrogen is +1 in both reactants and products while oxygen is −2; neither is oxidized or reduced. However, Mn changes from +7 in MnO_4^- to +2 in Mn^{2+}; C changes from +3 in $C_2O_4^{2-}$ to +4 in CO_2.

2. Determine the change in oxidation number for each element undergoing oxidation and reduction. These changes can be conveniently shown above arrows connecting the pertinent elements in the reactants and products:

In this example the oxidation number of Mn changes by five units while that for each C changes by one unit. However, there are two carbons in $C_2O_4^{2-}$, so that the total change for each $C_2O_4^{2-}$ is two, one for each carbon.

3. Using the oxidation-number changes found in step 2, determine the simplest ratio of moles of oxidant and reductant that will lead to equal gains and losses in oxidation numbers. The procedure is equivalent to requiring that the number of electrons lost by $C_2O_4^{2-}$ equals the number gained by MnO_4^-. We must have $5C_2O_4^{2-}$ for each $2MnO_4^-$ so that each substance is involved in a total change in oxidation numbers of 10: 2×5 for Mn and $5 \times 2 \times 1$ for C.

4. Having determined coefficients for species being oxidized and reduced, balance the number of atoms of the remaining elements in the usual manner. In this example, the $5C_2O_4^{2-}$ contain 10 carbon atoms, requiring $10CO_2$ among the products. There are 28 oxygen atoms among the reactants (8 in $2MnO_4^-$ and 20 in $5C_2O_4^{2-}$), which require there to be $8H_2O$ among the products; there are then 28 oxygen atoms in the products (8 in $8H_2O$ and 20 in $10CO_2$). Finally, $8H_2O$ require the presence of $16H^+$ among the reactants:

$$2MnO_4^-(aq) + 16H^+(aq) + 5C_2O_4^{2-}(aq) \longrightarrow 2Mn^{2+}(aq) + 8H_2O(l) + 10CO_2(g)$$

In cases where an incomplete oxidation-reduction equation is given, H^+ and H_2O (in acidic solutions) or OH^- and H_2O (in basic solutions) can be added to complete the balancing.

19.2 Voltaic cells

In principle, the energy released in any spontaneous redox reaction can be directly harnessed to perform electrical work. This task is accomplished through a voltaic (or galvanic) cell, which is merely a device in which electron transfer is forced to take place through an external pathway rather than directly between reactants.

One such spontaneous reaction occurs when a piece of zinc is placed in contact with a solution containing Cu^{2+}. As the reaction proceeds, the blue color that is characteristic of $Cu^{2+}(aq)$ ions fades, and copper metal begins to deposit on the zinc. At the same time, the zinc begins to dissolve. These transformations are shown in Figure 19.1 and are summarized by Equation [19.9]:

$$Zn(s) + Cu^{2+}(aq) \longrightarrow Zn^{2+}(aq) + Cu(s) \qquad [19.9]$$

Figure 19.2 shows a voltaic cell that utilizes the oxidation-reduction reaction between Zn and Cu^{2+}, Equation [19.9]. Although the experimental design shown in Figure 19.2 is more complex than that in Figure 19.1, it is important to recognize that the chemical reaction involved is

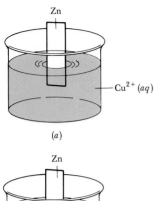

(a)

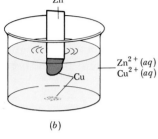

(b)

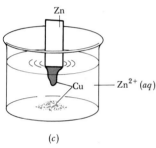

(c)

FIGURE 19.1 (a) A strip of zinc is placed in a colored solution containing Cu^{2+} ions. (b) Electrons are transferred from zinc to Cu^{2+}, and the zinc metal is eaten away and copper metal deposits. During this process, the color characteristic of Cu^{2+} ions in the solution fades. (c) After reaction has gone to completion, zinc is in excess, and the color characteristic of Cu^{2+} has disappeared.

the same in both cases. The major difference between the two arrangements is that in Figure 19.2 the zinc metal and $Cu^{2+}(aq)$ are no longer in direct contact. Consequently, reduction of the Cu^{2+} can occur only by a flow of electrons through the wire that connects Zn and Cu (the external circuit).

The two solid metals that are connected by the external circuit are called electrodes. By definition, the electrode at which oxidation occurs is called the anode; the electrode at which reduction occurs is called the cathode.* In our present example, Zn is the anode and Cu is the cathode:

Oxidation; anode $\qquad\qquad\qquad Zn(s) \longrightarrow Zn^{2+}(aq) + 2e^-$

Reduction; cathode $\qquad Cu^{2+}(aq) + 2e^- \longrightarrow Cu(s)$

The voltaic cell may be regarded as composed of two half-cells, one corresponding to the oxidation process and one corresponding to the reduction process. Electrons become available as zinc metal is oxidized at the anode. They flow through the external circuit to the cathode, where they are consumed as $Cu^{2+}(aq)$ is reduced. The sign convention for the electrodes in a voltaic cell is chosen with reference to the external circuit. Electrons move spontaneously from the negative to the positive electrode; thus the anode is negative, and the cathode is positive.

As the cell pictured in Figure 19.2 operates, oxidation of Zn introduces additional Zn^{2+} ions into the anode compartment. Unless a means is provided to neutralize this positive charge, no further oxidation can take

---

*To help remember these definitions, it is useful to note that anode and oxidation both begin with a vowel, whereas cathode and reduction both begin with a consonant.

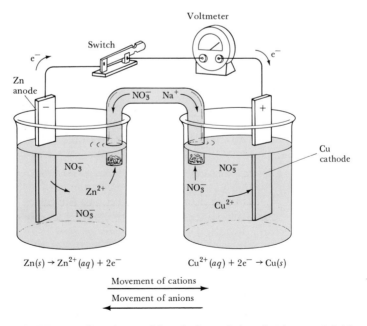

$$Zn(s) \rightarrow Zn^{2+}(aq) + 2e^-$$

$$Cu^{2+}(aq) + 2e^- \rightarrow Cu(s)$$

Movement of cations

Movement of anions

FIGURE 19.2 Complete and functioning voltaic cell using a salt bridge to complete the electrical circuit.

place. Similarly, the reduction of Cu^{2+} at the cathode leaves an excess of negative charge in solution in that compartment. Electrical neutrality is maintained by a migration of ions through a "salt bridge" as illustrated in Figure 19.2. A salt bridge consists of a U-shaped tube that contains an electrolyte solution such as $NaNO_3(aq)$ whose ions will not react with other ions in the cell or with the electrode materials. The ends of the U-tube may be loosely plugged with glass wool or the electrolyte incorporated into a gel so that the electrolyte solution does not run out when the U-tube is inverted. As oxidation and reduction proceed at the electrodes, ions from the salt bridge migrate to neutralize charge in the anode and cathode compartments. Anions migrate toward the anode and cations toward the cathode. In fact, no measurable electron flow will occur through the external circuit unless a means is provided for ions to migrate through the solution from one electrode compartment to another, thereby completing the circuit.

SAMPLE EXERCISE 19.2

The following oxidation-reduction reaction is spontaneous in the direction indicated:

$$Cr_2O_7^{2-}(aq) + 14H^+(aq) + 6I^-(aq) \longrightarrow$$
$$2Cr^{3+}(aq) + 3I_2(s) + 7H_2O(l)$$

A solution containing $K_2Cr_2O_7$ and H_2SO_4 is poured into one beaker and a solution of KI is poured into another. A salt bridge is used to join the beakers. A metallic conductor that will not react with either solution (such as platinum foil) is suspended in each solution, and the two conductors are connected with wires through a voltmeter or some other device to detect an electric current. The resultant voltaic cell generates an electric current. Indicate the reaction occurring at the anode, the reaction at the cathode, the direction of electron and ion migrations, and the signs of the electrodes.

Solution: The half-reactions for the given chemical equation are

$$Cr_2O_7^{2-}(aq) + 14H^+(aq) + 6e^- \longrightarrow$$
$$2Cr^{3+}(aq) + 7H_2O(l)$$

$$6I^-(aq) \longrightarrow 3I_2(s) + 6e^-$$

The first half-reaction is the reduction process, which by definition occurs at the cathode. The second half-reaction is an oxidation, which occurs at the anode. The I^- ions are the source of electrons and the $Cr_2O_7^{2-}$ ions are the receptors. Consequently, the electrons flow through the external circuit from the electrode immersed in the KI solution (the anode) to the electrode immersed in the $K_2Cr_2O_7$–H_2SO_4 solution (the cathode). The electrodes themselves do not react in any way; they merely provide a means of transferring electrons from or to the solutions. The cations move through the solutions toward the cathode, while the anions move toward the anode. The anode is the negative electrode and the cathode the positive one.

19.3 Cell emf

A voltaic cell may be thought to possess a "driving force" or "electrical pressure" that pushes electrons through the external circuit. This driving force is called the **electromotive force** (abbreviated **emf**); emf is measured in units of volts and is also referred to as the voltage or cell potential. One volt (V) is the emf required to impart 1 J of energy to a charge of 1 coul:

$$1 V = 1\frac{J}{coul} \tag{19.10}$$

The accurate determination of cell emf requires the use of special apparatus. The measurement must be made in such a manner that only a negligible current flows. If a significant current is allowed to flow, the apparent voltage of the cell is lowered as a result of the internal resistance of the cell, and because of changes in concentrations around the electrodes.

The voltaic cell pictured in Figure 19.2 generates an emf of 1.10 V when operated under standard-state conditions. You will recall from Section 18.5 (see Table 18.1) that standard-state conditions include 1 M concentrations for reactants and products that are in solution and 1 atm pressure for those that are gases. In the present example, the cell would be operated with $[Cu^{2+}]$ and $[Zn^{2+}]$ both at 1 M.* The emf generated by a cell is given the symbol E. When operated under standard-state conditions it is called the standard emf, $E°$:

$$Zn(s) + Cu^{2+}(aq) \longrightarrow Zn^{2+}(aq) + Cu(s) \qquad E° = 1.10 \text{ V} \qquad [19.11]$$

The emf of any voltaic cell depends on the nature of the chemical reactions taking place in the cell, the concentrations of reactants and products, and the temperature of the cell, which we shall take to be 25°C unless otherwise noted.

STANDARD ELECTRODE POTENTIALS

Just as an overall cell reaction can be thought of as the sum of two half-reactions, the emf of a cell can be thought of as the sum of two half-cell potentials: that due to the loss of electrons at the anode (the oxidation potential, E_{ox}) and that due to electron gain at the cathode (the reduction potential, E_{red}):

$$E_{cell} = E_{ox} + E_{red} \qquad\qquad [19.12]$$

It is impossible directly to measure an isolated oxidation or reduction potential. However, if one half-reaction is arbitrarily assigned a standard half-cell potential, standard potentials of other half-reactions can be determined relative to that reference. The half-reaction that corresponds to reduction of H^+ to form H_2 has been chosen as the reference and assigned a standard reduction potential of exactly 0 volts:

$$2H^+(1\,M) + 2e^- \longrightarrow H_2(1 \text{ atm}) \qquad E°_{red} = 0 \text{ V} \qquad [19.13]$$

Figure 19.3 illustrates a voltaic cell constructed to use the redox reaction between Zn and H^+, Equation [19.14]:

$$Zn(s) + 2H^+(aq) \longrightarrow Zn^{2+}(aq) + H_2(g) \qquad\qquad [19.14]$$

*Actually, solutions of 1 M represent the standard state only if they behave ideally. Salt dissolved in aqueous medium at concentrations on the order of 1 M do not in general behave ideally; substantial corrections need to be made to allow for nonideal behavior. However, in this introduction to electrochemistry we will not concern ourselves with these corrections.

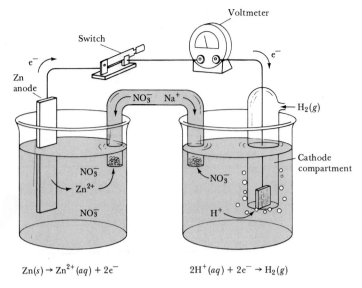

$$Zn(s) \rightarrow Zn^{2+}(aq) + 2e^-$$ $$2H^+(aq) + 2e^- \rightarrow H_2(g)$$

FIGURE 19.3 Voltaic cell using a hydrogen electrode.

Oxidation of zinc occurs in the anode compartment while reduction of H^+ occurs in the cathode compartment. The standard hydrogen electrode, which is designed to operate under standard-state conditions ($[H^+] = 1\ M$ and $P_{H_2} = 1$ atm), consists of a platinum wire and a piece of platinum foil covered with finely divided platinum that serves as an inert surface for the cathode reaction. This electrode is encased in a glass tube so that hydrogen gas can bubble over the platinum. The voltaic cell generates a standard emf (E°_{cell}) of 0.76 V. By using the defined standard reduction potential of H^+ ($E^\circ_{red} = 0$), it is possible to calculate the standard oxidation potential of Zn:

$$E^\circ_{cell} = E^\circ_{ox} + E^\circ_{red}$$
$$0.76\ V = E^\circ_{ox} + 0$$

Thus a standard oxidation potential of 0.76 V can be assigned to Zn:

$$Zn(s) \longrightarrow Zn^{2+}(aq) + 2e^- \qquad E^\circ_{ox} = 0.76\ V \qquad [19.15]$$

The standard potentials for other half-reactions can be established from other cell emfs in a similar fashion.

The half-cell potential for any oxidation is equal in magnitude but opposite in sign to that of the reverse reduction. Thus, for example,

$$Zn^{2+}(aq) + 2e^- \longrightarrow Zn(s) \qquad E^\circ_{red} = -0.76\ V \qquad [19.16]$$

By convention, half-cell potentials are tabulated as standard reduction potentials, also referred to merely as electrode potentials. Table 19.1 contains a selection of electrode potentials; a more complete list is found in Appendix F. These electrode potentials may be combined to calculate the standard emfs of a large variety of voltaic cells.

TABLE 19.1 Standard electrode potentials in water at 25°C

Standard potential (V)	Reduction half-reaction
2.87	$F_2(g) + 2e^- \longrightarrow 2F^-(aq)$
1.51	$MnO_4^-(aq) + 8H^+(aq) + 5e^- \longrightarrow Mn^{2+}(aq) + 4H_2O(l)$
1.36	$Cl_2(g) + 2e^- \longrightarrow 2Cl^-(aq)$
1.33	$Cr_2O_7^{2-}(aq) + 14H^+(aq) + 6e^- \longrightarrow 2Cr^{3+}(aq) + 7H_2O(l)$
1.23	$O_2(g) + 4H^+(aq) + 4e^- \longrightarrow 2H_2O(l)$
1.09	$Br_2(l) + 2e^- \longrightarrow 2Br^-(aq)$
0.96	$NO_3^-(aq) + 4H^+(aq) + 3e^- \longrightarrow NO(g) + 2H_2O(l)$
0.80	$Ag^+(aq) + e^- \longrightarrow Ag(s)$
0.77	$Fe^{3+}(aq) + e^- \longrightarrow Fe^{2+}(aq)$
0.68	$O_2(g) + 2H^+(aq) + 2e^- \longrightarrow H_2O_2(aq)$
0.59	$MnO_4^-(aq) + 2H_2O(l) + 3e^- \longrightarrow MnO_2(s) + 4OH^-(aq)$
0.54	$I_2(s) + 2e^- \longrightarrow 2I^-(aq)$
0.40	$O_2(g) + 2H_2O(l) + 4e^- \longrightarrow 4OH^-(aq)$
0.34	$Cu^{2+}(aq) + 2e^- \longrightarrow Cu(s)$
0	$2H^+(aq) + 2e^- \longrightarrow H_2(g)$
−0.28	$Ni^{2+}(aq) + 2e^- \longrightarrow Ni(s)$
−0.44	$Fe^{2+}(aq) + 2e^- \longrightarrow Fe(s)$
−0.76	$Zn^{2+}(aq) + 2e^- \longrightarrow Zn(s)$
−0.83	$2H_2O(l) + 2e^- \longrightarrow H_2(g) + 2OH^-(aq)$
−1.66	$Al^{3+}(aq) + 3e^- \longrightarrow Al(s)$
−2.71	$Na^+(aq) + e^- \longrightarrow Na(s)$
−3.05	$Li^+(aq) + e^- \longrightarrow Li(s)$

SAMPLE EXERCISE 19.3

Given that E°_{cell} is 1.10 V for the Zn–Cu^{2+} cell shown in Figure 19.2, and that E°_{ox} is 0.76 V for the oxidation of zinc, Equation [19.15], calculate E°_{red} for the reduction of copper

$$Cu^{2+}(aq) + 2e^- \longrightarrow Cu(s)$$

Solution:

$$E^\circ_{cell} = E^\circ_{ox} + E^\circ_{red}$$
$$1.10 \text{ V} = 0.76 \text{ V} + E^\circ_{red}$$
$$E^\circ_{red} = 1.10 \text{ V} - 0.76 \text{ V} = 0.34 \text{ V}$$

SAMPLE EXERCISE 19.4

Using standard electrode potentials tabulated in Table 19.1, calculate the standard emf for the cell described in Sample Exercise 19.2:

$$Cr_2O_7^{2-}(aq) + 14H^+(aq) + 6I^-(aq) \longrightarrow$$
$$2Cr^{3+}(aq) + 3I_2(s) + 7H_2O(l)$$

Solution: The standard emf of the cell, E°_{cell}, is the sum of the standard oxidation and reduction potentials for the appropriate half-reactions:

$$E^\circ_{cell} = E^\circ_{ox} + E^\circ_{red}$$

The half reactions and their potentials are as follows:

$$Cr_2O_7^{2-}(aq) + 14H^+(aq) + 6e^- \longrightarrow$$
$$2Cr^{3+}(aq) + 7H_2O(l) \quad E^\circ_{red} = 1.33 \text{ V}$$
$$6I^-(aq) \longrightarrow 3I_2(s) + 6e^-$$
$$E^\circ_{ox} = -0.54 \text{ V}$$

$$Cr_2O_7^{2-}(aq) + 14H^+(aq) + 6I^-(aq) \longrightarrow$$
$$2Cr^{3+}(aq) + 7H_2O(l) + 3I_2(s) \quad E^\circ_{cell} = 0.79 \text{ V}$$

Notice that E°_{ox} for I$^-$ has the opposite sign from the reduction potential of I$_2$ listed in Table 19.1. Also notice that even though the iodide half-reaction must be multiplied by 3 in order to obtain the balanced equation for the reaction, the half-cell potential is *not* multiplied by 3. The standard potential is an intensive property; it does not depend on the quantity of reactants and products, but only on their concentration. Thus it does not matter whether there are 6 mol of I^- or 1 mol as long as the concentration of iodide is 1 M.

Just as the sum of the half-reactions gives the chemical equation for the overall cell reaction, the sum of the corresponding oxidation and reduction potentials gives the cell emf. In this case $E^\circ_{cell} = 0.79$ V; a positive emf is characteristic of all voltaic cells.

OXIDIZING AND REDUCING AGENTS

Half-cell potentials indicate the ease with which a species is oxidized or reduced. *The more positive the E° value for a half-reaction, the greater the tendency for that reaction to occur as written.* A negative reduction potential indicates that the species is more difficult to reduce than $H^+(aq)$, whereas a negative oxidation potential indicates that the species is more difficult to oxidize than H_2. Examination of the half-reactions in Table 19.1 shows that F_2 is the most easily reduced species and consequently the strongest oxidizing agent listed:

$$F_2(g) + 2e^- \longrightarrow 2F^-(aq) \qquad E^\circ_{red} = 2.87 \text{ V} \qquad [19.17]$$

Lithium ion, Li^+, is the most difficult to reduce and therefore the poorest oxidizing agent:

$$Li^+(aq) + e^- \longrightarrow Li(s) \qquad E^\circ_{red} = -3.05 \text{ V} \qquad [19.18]$$

The most frequently encountered oxidizing agents are the halogens, oxygen, and oxyanions such as MnO_4^-, $Cr_2O_7^{2-}$, and NO_3^- whose central atoms have high positive oxidation states. Metal ions in high positive oxidation states such as Ce^{4+}, which is readily reduced to Ce^{3+}, are also employed as oxidizing agents.

Among the substances listed in Table 19.1, lithium is the most easily oxidized and consequently the strongest reducing agent:

$$Li(s) \longrightarrow Li^+(aq) + e^- \qquad E^\circ_{ox} = 3.05 \text{ V} \qquad [19.19]$$

Fluoride ion, F^-, is the most difficult to oxidize and therefore the poorest reducing agent:

$$2F^-(aq) \longrightarrow F_2(g) + 2e^- \qquad E^\circ_{ox} = -2.87 \text{ V} \qquad [19.20]$$

H_2 and a variety of metals having positive oxidation potentials are employed as reducing agents. Some metal ions in low oxidation states such as Sn^{2+}, which is oxidized to Sn^{4+}, also function as reducing agents. Solutions of reducing agents are difficult to store for extended periods because of the ubiquitous presence of O_2, a good oxidizing agent. For example, developer solutions used in photography are mild reducing agents; consequently they have only a limited shelf life because they are readily oxidized by O_2 from the air.

SAMPLE EXERCISE 19.5

Using Table 19.1, determine which of the following species is the strongest oxidizing agent: MnO_4^- (in acid solution), $I_2(s)$, $Zn^{2+}(aq)$.

Solution: The strongest oxidizing agent will be the species that is most readily reduced. Therefore we should compare reduction potentials. From Table 19.1 we have:

$$MnO_4^-(aq) + 8H^+(aq) + 5e^- \longrightarrow$$
$$Mn^{2+}(aq) + 4H_2O(l) \qquad E_{red}^\circ = \quad 1.51 \text{ V}$$
$$I_2(s) + 2e^- \longrightarrow 2I^-(aq) \qquad E_{red}^\circ = \quad 0.54 \text{ V}$$
$$Zn^{2+}(aq) + 2e^- \longrightarrow Zn(s) \qquad E_{red}^\circ = -0.76 \text{ V}$$

Because the reduction of MnO_4^- has the highest positive potential, MnO_4^- is the strongest oxidizing agent of the three.

19.4 Spontaneity and extent of redox reactions

We have observed that voltaic cells utilize redox reactions that proceed spontaneously. Conversely, any reaction that can occur in a voltaic cell to produce a positive emf must be spontaneous. Consequently, it is possible to decide whether a redox reaction will be spontaneous by using half-cell potentials to calculate the emf associated with it: *A positive emf indicates a spontaneous process whereas a negative emf indicates a nonspontaneous one.*

SAMPLE EXERCISE 9.6

Using the standard electrode potentials listed in Table 19.1, determine whether the following reactions are spontaneous under standard conditions:

(a) $Cu(s) + 2H^+(aq) \longrightarrow Cu^{2+}(aq) + H_2(g)$

(b) $Cl_2(g) + 2I^-(aq) \longrightarrow 2Cl^-(aq) + I_2(s)$

Solution: (a) We utilize Table 19.1 to obtain the necessary half-reaction potentials. Because the overall reaction converts Cu to Cu^{2+}, we use the standard oxidation potential for Cu. Because the overall reaction converts H^+ to H_2 we use the reduction potential for H^+. Adding these we obtain the standard potential for the overall reaction:

$$Cu(s) \longrightarrow Cu^{2+}(aq) + 2e^-$$
$$E_{ox}^\circ = -0.34 \text{ V}$$
$$2e^- + 2H^+(aq) \longrightarrow H_2(g) \qquad E_{red}^\circ = \quad 0 \text{ V}$$
$$\overline{Cu(s) + 2H^+(aq) \longrightarrow Cu^{2+}(aq) + H_2(g)}$$
$$E^\circ = -0.34 \text{ V}$$

Because the standard potential is negative, the reaction is not spontaneous in the direction written. Copper does not react with acids in this fashion. However, the reverse reaction is spontaneous: Cu^{2+} can be reduced by H_2.

(b) We write equations to obtain the standard potential for the overall reaction:

$$2e^- + Cl_2(g) \longrightarrow 2Cl^-(aq) \qquad E_{red}^\circ = \quad 1.36 \text{ V}$$
$$2I^-(aq) \longrightarrow I_2(s) + 2e^-$$
$$E_{ox}^\circ = -0.54 \text{ V}$$
$$\overline{Cl_2(g) + 2I^-(aq) \longrightarrow 2Cl^-(aq) + I_2(s)}$$
$$E^\circ = \quad 0.82 \text{ V}$$

This reaction is spontaneous and could be used to build a voltaic cell. It is often used as a qualitative test for the presence of I^- in aqueous solution. The solution is treated with a solution of Cl_2. If I^- is present, I_2 forms. If CCl_4 is added, the I_2 dissolves in the CCl_4, imparting a characteristic purple color to the solution.

EMF AND FREE-ENERGY CHANGE

We've seen that the free-energy change, ΔG, accompanying a chemical process is a measure of its spontaneity (Chapter 18). Because the cell emf indicates whether a redox reaction is spontaneous, we might expect some relationship to exist between the cell emf and the free-energy change.

FIGURE 19.4 Michael Faraday (1791–1867) was born in England, one of ten children of a poor blacksmith. At the age of 14 he was apprenticed to a bookbinder who, with uncommon leniency, gave the young man time to read and even to attend lectures. In 1812 he became an assistant in Humphrey Davy's laboratory in the Royal Institution. He eventually succeeded Davy as the most famous and influential scientist in England. During his scientific career he made an amazing number of important discoveries in chemistry and physics. He developed methods for liquefying gases, discovered benzene, and formulated the quantitative relationships between electrical current and the extent of chemical reaction in electrochemical cells that either produce or use electricity. In addition, he worked out the design of the first electric generator and laid the theoretical foundations for the development of our modern theory of electricity. (Culver Pictures)

Indeed, this is the case; the emf, E, and the free-energy change, ΔG, are related by Equation [19.21]:

$$\Delta G = -n\mathcal{F}E \qquad [19.21]$$

In this equation n is the number of moles of electrons transferred in the reaction, and $\mathcal{F}$ is Faraday's constant, named after Michael Faraday (Figure 19.4). Farady's constant is the electrical charge on 1 mol of electrons; consequently, this quantity of charge is referred to as a faraday:

$$1\mathcal{F} = 96{,}500 \; \frac{\text{coul}}{\text{mol e}^-} = 96{,}500 \; \frac{\text{J}}{\text{V-mol e}^-} \qquad [19.22]$$

For the situation in which reactants and products are in their standard states, Equation [19.21] is modified to relate ΔG° and E°:

$$\Delta G^\circ = -n\mathcal{F}E^\circ \qquad [19.23]$$

SAMPLE EXERCISE 19.7

Use the standard electrode potentials given in Table 19.1 to calculate the standard free-energy change, ΔG°, for the following reaction:

$$2\text{Br}^-(aq) + \text{F}_2(g) \longrightarrow \text{Br}_2(l) + 2\text{F}^-(aq)$$

Solution:

$$2\text{Br}^-(aq) \longrightarrow \text{Br}_2(l) + 2e^-$$
$$E^\circ_{ox} = -1.09 \text{ V}$$
$$\underline{\text{F}_2(g) + 2e^- \longrightarrow 2\text{F}^-(aq) \qquad E^\circ_{red} = \quad 2.87 \text{ V}}$$
$$2\text{Br}^-(aq) + \text{F}_2(g) \longrightarrow \text{Br}_2(l) + 2\text{F}^-(aq)$$
$$E^\circ = \quad 1.78 \text{ V}$$

Notice that two electrons are transferred in the reaction so $n = 2$.

$$\Delta G^\circ = -n\mathcal{F}E^\circ$$
$$= -(2 \text{ mol e}^-)\left(96{,}500 \; \frac{\text{J}}{\text{V-mol e}^-}\right)(1.78 \text{ V})$$
$$= -3.44 \times 10^5 \text{ J} = -344 \text{ kJ}$$

Notice also that whereas a negative sign for ΔG° indicates spontaneity, a positive sign for E° indicates the same.

In Chapter 18, we discussed the significance of ΔG for a chemical system. Among other things, we saw that the standard free-energy change is related to the equilibrium constant, K, through Equation [19.24]:

$$\Delta G^\circ = -2.30 \, RT \log K \qquad\qquad [19.24]$$

This relationship indicates that E° should also be related to the equilibrium constant:

$$-n\mathfrak{F}E^\circ = -2.30 \, RT \log K$$

$$E^\circ = \frac{+2.30 \, RT}{n\mathfrak{F}} \log K \qquad\qquad [19.25]$$

When $T = 298$ K, this equation can be simplified by substituting the numerical values for R and $\mathfrak{F}$:

$$E^\circ = \frac{+2.30(8.314 \, \text{J/K-mol})(298 \, \text{K})}{n(96,500 \, \text{J/V-mol})} \log K$$

$$= \frac{0.0591 \, \text{V}}{n} \log K \qquad\qquad [19.26]$$

Thus the standard emf generated by a cell increases as the equilibrium constant for the cell reaction increases.

SAMPLE EXERCISE 19.8

Using standard electrode potentials, calculate the equilibrium constant at 25°C for the reaction

$$O_2(g) + 4H^+(aq) + 4Fe^{2+}(aq) \longrightarrow$$
$$4Fe^{3+}(aq) + 2H_2O(l)$$

Solution:

$$O_2(g) + 4H^+(aq) + 4e^- \longrightarrow 2H_2O(l)$$
$$E^\circ_{\text{red}} = \quad 1.23 \text{ V}$$

$$4Fe^{2+}(aq) \longrightarrow 4Fe^{3+}(aq) + 4e^-$$
$$E^\circ_{\text{ox}} = -0.77 \text{ V}$$
<hr>
$$O_2(g) + 4H^+(aq) + 4Fe^{2+}(aq) \longrightarrow$$
$$4Fe^{3+}(aq) + 2H_2O(l) \qquad E^\circ = \quad 0.46 \text{ V}$$

From the half-reactions given above we see that $n = 4$. Using Equation [19.26] we have

$$\log K = \frac{nE^\circ}{0.0591 \text{ V}}$$

$$= \frac{4(0.46 \text{ V})}{0.0591 \text{ V}}$$

$$= 31.2$$

$$K = 2 \times 10^{31}$$

Thus Fe^{2+} ions are stable in acidic solutions only in the absence of O_2 (unless a suitable reducing agent is present).

EMF AND CONCENTRATION

In practice, voltaic cells are unlikely to be operated under standard-state conditions. However, the emf generated under nonstandard conditions can be calculated from E°, temperature, and the concentrations of reactants and products in the cell. The equation that permits this calculation can be derived from the relationship between ΔG and ΔG°, given earlier in Section 18.6:

$$\Delta G = \Delta G^\circ + 2.30\,RT \log Q \qquad [19.27]$$

Because $\Delta G = -n\mathcal{F}E$, Equation [19.21], we can write:

$$-n\mathcal{F}E = -n\mathcal{F}E^\circ + 2.30\,RT \log Q$$

Solving this equation for E gives

$$E = E^\circ - \frac{2.30\,RT}{n\mathcal{F}} \log Q \qquad [19.28]$$

This relationship is known as the **Nernst equation** after Hermann Walther Nernst (1864–1941), a German chemist who established a good part of the theoretical foundations of electrochemistry. At 298 K, the quantity $2.30\,RT/\mathcal{F}$ is equal to 0.0591 V-mol, so that the Nernst equation can be written in the following simplified form:

$$E = E^\circ - \frac{0.0591}{n} \log Q \qquad [19.29]$$

As an example of how Equation [19.29] might be used, consider the following reaction:

$$Zn(s) + Cu^{2+}(aq) \longrightarrow Zn^{2+}(aq) + Cu(s) \qquad E^\circ = 1.10 \text{ V}$$

In this case $n = 2$ and the Nernst equation gives

$$E = 1.10 \text{ V} - \frac{0.0591 \text{ V}}{2} \log \frac{[Zn^{2+}]}{[Cu^{2+}]} \qquad [19.30]$$

Recall that Q includes expressions for the species in solution, but not for solids. Experimentally it is found that the emf generated by a cell is independent of the size or shape of the solid electrodes used. From Equation [19.30] it is evident that the emf of a cell based on this chemical reaction increases as $[Cu^{2+}]$ increases and as $[Zn^{2+}]$ decreases. For example, when $[Cu^{2+}]$ is $5.0\,M$ and $[Zn^{2+}]$ is $0.050\,M$, we have

$$E = 1.10 \text{ V} - \frac{0.0591 \text{ V}}{2} \log \frac{(0.050)}{(5.0)}$$

$$= 1.10 \text{ V} - \frac{0.0591 \text{ V}}{2}(-2.00) = 1.16 \text{ V}$$

In general, if the concentrations of reactants increase relative to the concentrations of products, the cell reaction becomes more highly spontaneous and the emf increases. Conversely, if the concentrations of products increase relative to reactants, emf decreases. As a cell operates, reactants are consumed and products form. Resultant decreases in reactant concentrations and increases in product concentrations cause the emf to decrease.

SAMPLE EXERCISE 19.9

Calculate the emf generated by the cell described in Sample Exercise 19.2 when $[Cr_2O_7{}^{2-}] = 2.0\,M$, $[H^+] = 1.0\,M$, $[I^-] = 1.0\,M$, and $[Cr^{3+}] = 1.0 \times 10^{-5}\,M$:

$$Cr_2O_7{}^{2-}(aq) + 14H^+(aq) + 6I^-(aq) \longrightarrow$$
$$2Cr^{3+}(aq) + 3I_2(s) + 7H_2O(l)$$

Solution: The standard emf for this reaction was calculated in Sample Exercise 19.4: $E° = 0.79$ V. As you will see if you refer back to that exercise, n is 6. The reaction quotient, Q, is

$$Q = \frac{[Cr^{3+}]^2}{[Cr_2O_7{}^{2-}][H^+]^{14}[I^-]^6} = \frac{(1.0 \times 10^{-5})^2}{(2.0)(1.0)^{14}(1.0)^6}$$
$$= 5.0 \times 10^{-11}$$

Using Equation [19.29] we have

$$E = 0.79\text{ V} - \frac{0.0591\text{ V}}{6} \log(5.0 \times 10^{-11})$$

$$= 0.79\text{ V} - \frac{0.0591\text{ V}}{6}(-10.30)$$

$$= 0.79\text{ V} + 0.10\text{ V}$$

$$= 0.89\text{ V}$$

This result is qualitatively what we expect: Because the concentration of $Cr_2O_7{}^{2-}$ (a reactant) is above $1\,M$, whereas the concentration of Cr^{3+} (a product) is below $1\,M$, the emf is greater than $E°$.

SAMPLE EXERCISE 19.10

If the measured voltage in a Zn–H$^+$ cell such as that shown in Figure 19.3 is 0.45 V at 25°C when $[Zn^{2+}]$ is $1\,M$ and P_{H_2} is 1 atm, what is the concentration of H$^+$?

Solution: The cell reaction is

$$Zn(s) + 2H^+(aq) \longrightarrow Zn^{2+}(aq) + H_2(g)$$

The standard emf is $E° = 0.76$ V. Applying Equation [19.29] with $n = 2$ gives

$$0.45 = 0.76 - \frac{0.0591}{2} \log \frac{[Zn^{2+}]P_{H_2}}{[H^+]^2}$$

$$= 0.76 - \frac{0.0591}{2} \log \frac{1}{[H^+]^2}$$

Since

$$\log \frac{1}{x^2} = -\log x^2 = -2 \log x$$

we can write

$$0.45 = 0.76 - \frac{0.0591}{2}(-2 \log[H^+])$$

Solving for $\log[H^+]$:

$$\log[H^+] = \frac{0.45 - 0.76}{0.0591} = -5.25$$

$$[H^+] = 10^{-5.25} = 5.7 \times 10^{-6}\,M$$

This example shows how a voltaic cell whose cell reaction involves H$^+$ can be used to measure $[H^+]$ or pH. A pH-meter (Section 15.3) is merely a specially designed voltaic cell with a voltmeter calibrated to read pH directly.

19.5 Some commercial voltaic cells

Voltaic cells that receive wide use are convenient energy sources whose primary virtue is portability. Although any spontaneous redox reaction can be used as the basis of a voltaic cell, making a commercial cell that utilizes a particular redox reaction can require considerable ingenuity. Salt-bridge cells such as those that we have been discussing provide considerable insight into the operation of voltaic cells. However, they are generally unsuitable for commercial use, because they have high internal resistances. As a result, if we attempt to draw a large current from them, their voltage drops sharply. Furthermore, the cells that we have pictured so far lack the compactness and ruggedness required for portability.

Voltaic cells cannot yet compete with other common energy sources on the basis of cost alone. The cost of electricity from a common flashlight battery is on the order of $80 per kilowatt-hour. By comparison, electrical energy from power plants normally costs the consumer less than 10¢ per kilowatt-hour.

In this section we shall consider some common batteries. A battery merely consists of one or more voltaic cells. When the cells are connected in series (that is, with the positive terminal of one attached to the negative terminal of another), the battery produces an emf that is the sum of the emfs of the individual cells.

LEAD-STORAGE BATTERY

One of the most common batteries is the lead-storage battery used in automobiles. A 12-V lead-storage battery consists of six cells, each producing 2 V. The anode of each cell is composed of lead; the cathode is composed of lead dioxide, PbO_2, packed on a metal grid. Both electrodes are immersed in sulfuric acid. The electrode reactions that occur during discharge are as follows:

Anode:
$$Pb(s) + HSO_4^-(aq) \longrightarrow PbSO_4(s) + H^+(aq) + 2e^-$$
Cathode:
$$\frac{PbO_2(s) + HSO_4^-(aq) + 3H^+(aq) + 2e^- \longrightarrow PbSO_4(s) + 2H_2O(l)}{Pb(s) + PbO_2(s) + 2H^+(aq) + 2HSO_4^-(aq) \longrightarrow 2PbSO_4(s) + 2H_2O(l)}$$

$$[19.31]$$

The reactants Pb and PbO_2 between which electron transfer occurs, serve as the electrodes. Because the reactants are solids, there is no need to separate the cell into anode and cathode compartments; there is no

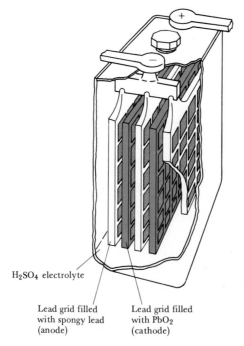

H$_2$SO$_4$ electrolyte

Lead grid filled
with spongy lead
(anode)

Lead grid filled
with PbO$_2$
(cathode)

FIGURE 19.5 A lead-storage cell.

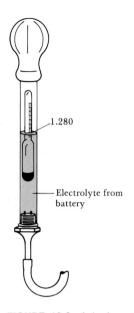

FIGURE 19.6 A hydrometer of the type used to measure the density of the electrolyte in a lead-storage battery.

way the Pb and PbO_2 can come into direct physical contact unless one electrode plate touches another. To keep the electrodes from touching, wood or glass-fiber spacers are placed between them. To increase the current output, each cell contains a number of anode and cathode plates as shown in Figure 19.5.

As Equation [19.31] indicates, H_2SO_4 is used up during the discharge of a lead storage battery. Because sulfuric acid has a high density, the density of the electrolyte in the cell decreases during discharge. When fully charged the electrolyte has a density of 1.25–1.30 g/cm^3. If the density falls below 1.20 g/cm^3, the battery needs recharging. A service-station attendant can test the density using a hydrometer. This device, shown in Figure 19.6, has a float that sinks to a depth that is a function of the density of the liquid in which it is immersed. In some so-called lifetime batteries, the H_2SO_4 is incorporated into a gel and each cell is permanently sealed.

One advantage of the lead-storage battery is that it can be recharged. During recharging, an external source of energy is used to reverse the direction of the spontaneous redox reaction, Equation [19.31]. Thus the overall process during charging is as follows:

$$2PbSO_4(s) + 2H_2O(l) \longrightarrow Pb(s) + PbO_2(s) + 2H^+(aq) + 2HSO_4^-(aq)$$
$$[19.32]$$

The energy necessary for recharging the battery is provided in an automobile by a generator that is driven by the engine. The recharging is made possible by the fact that $PbSO_4$ formed during discharge adheres to the electrodes. Thus as the external source forces electrons from one electrode to another, the $PbSO_4$ is converted to Pb at one electrode and to PbO_2 at the other; these, of course, are the materials of a fully charged cell. If the battery is charged too rapidly, water may be decomposed to form H_2 and O_2. Besides the explosive potential of H_2–O_2 mixtures, this secondary reaction can shorten the lifetime of the battery. The evolution of these gases can dislodge Pb, PbO_2, or $PbSO_4$ from the plates. Solids accumulate as a sludge at the bottom of the battery. In time they may form a short circuit that renders the cell useless.

DRY CELL

The common **dry cell** is familiar because of its wide use in flashlights and portable radios. Because of this use, it is often referred to as a flashlight battery. It is also known as the Leclanché cell after its inventor who patented it in 1866. In the acid version, the anode consists of a zinc can that is in contact with a paste of MnO_2, NH_4Cl, and carbon. An inert cathode, consisting of a graphite rod, is immersed in the center of the paste as shown in Figure 19.7. The cell has an exterior layer of cardboard or metal to seal the cell against the atmosphere. The electrode reactions are complex, and the cathode reaction appears to vary with the rate of discharge. The reactions at the electrodes are generally represented as shown below:

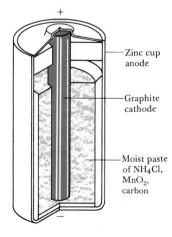

FIGURE 19.7 A cutaway view of a dry cell.

Anode: $\quad Zn(s) \longrightarrow Zn^{2+}(aq) + 2e^-$ $\qquad$ [19.33]

Cathode:

$$2NH_4^+(aq) + 2MnO_2(s) + 2e^- \longrightarrow$$
$$Mn_2O_3(s) + 2NH_3(aq) + H_2O(l) \qquad [19.34]$$

Only a fraction of the cathode material, that near the electrode, is electrochemically active because of the limited mobility of the chemicals in the cell.

In the alkaline version of the dry cell, NH_4Cl is replaced by KOH. The anode reaction still involves oxidation of Zn, and the cathode reaction involves the reduction of MnO_2. The alkaline version lasts longer than the acid version because it eliminates the problem of corrosion of the zinc anode by acidic NH_4Cl. However, alkaline dry cells are more costly than acid dry cells. Both the acid and alkaline dry cells produce approximately 1.5 V.

NICKEL-CADMIUM BATTERIES

Because dry cells are not rechargeable, they have to be replaced frequently. Thus a rechargeable cell, the nickel-cadmium battery, has become increasingly popular, especially for use in battery-operated tools and calculators. The electrode reactions occurring within this cell during discharge are as follows:

Anode: $\qquad Cd(s) + 2OH^-(aq) \longrightarrow Cd(OH)_2(s) + 2e^- \qquad [19.35]$

Cathode: $\quad NiO_2(s) + 2H_2O(l) + 2e^- \longrightarrow Ni(OH)_2(s) + 2OH^-(aq) \quad [19.36]$

As in the lead storage cell, the reaction products adhere to the electrodes. This permits the reactions to be readily reversed during charging. Because no gases are produced during either charging or discharging, the battery can be sealed.

FUEL CELLS

Many substances, for example, H_2 or CH_4, are used as fuels. The heat released in their reaction with oxygen is the source of energy. The thermal energy released by combustion is often subsequently converted to electrical energy. Because combustion reactions are oxidation-reduction reactions, it would be possible to use such reactions to generate an electrical current directly if suitable voltaic cells could be designed. It is, in fact, advantageous to perform such a direct conversion to electrical energy. In producing electricity via combustion, heat is used to convert water to steam. The steam is used to drive a turbine that drives a generator. Some energy is lost as waste heat whenever energy is converted from one form to another or from one material to another. Typically, a maximum of only 40 percent of the energy from combustion is converted to electricity; the remainder is lost as heat. Direct production of electricity from fuels via a voltaic cell should yield a higher rate of conversion of the chemical energy of the fuels into electricity. Voltaic cells that utilize conventional fuels are known as fuel cells.

A great deal of research has gone into attempts to develop practical fuel cells. One of the problems encountered is the high operating tempera-

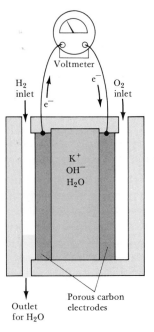

FIGURE 19.8 A cross-section of a H_2-O_2 fuel cell.

tures of most cells, which not only siphons off energy but also accelerates corrosion of cell parts. A low-temperature cell has been developed that uses H_2, but the present cost of the cell makes it too expensive for large-scale use. However, it has been used in special situations, such as in space vehicles. For example, a H_2–O_2 fuel cell was used as the primary source of electrical energy on the *Apollo* moon flights. The weight of the fuel cell sufficient for 11 days in space was approximately 500 lb. This may be compared to the several tons that would have been required for an engine-generator set.

The electrode reactions in the H_2–O_2 fuel cell are as follows:

Anode: $\quad\quad\quad 2H_2(g) + 4OH^-(aq) \longrightarrow 4H_2O(l) + 4e^-$

Cathode: $\quad\quad 4e^- + O_2(g) + 2H_2O(l) \longrightarrow 4OH^-(aq)$

$$\overline{\quad\quad\quad\quad\quad 2H_2(g) + O_2(g) \longrightarrow 2H_2O(l) \quad\quad\quad} \quad\quad \text{[19.37]}$$

The cell is illustrated in Figure 19.8. The electrodes are composed of hollow tubes of porous, compressed carbon impregnated with catalyst; the electrolyte is KOH. Because the reactants are supplied continuously, a fuel cell does not "go dead."

19.6 Electrolysis and electrolytic cells

We have seen that spontaneous oxidation-reduction reactions are used to make voltaic cells, electrochemical devices that generate electricity. Conversely, it is possible to use electrical energy to cause nonspontaneous oxidation-reduction reactions to occur. For example, electricity can be used to decompose molten sodium chloride into its component elements:

$$2NaCl(l) \longrightarrow 2Na(l) + Cl_2(g)$$

Such processes, which are driven by an outside source of electrical energy, are referred to as electrolysis reactions and are performed in electrolytic cells.

An electrolytic cell consists of two electrodes in a molten salt or aqueous solution, as shown in Figure 19.9. The cell is driven by a battery or some other source of direct electrical current. The battery acts as an electron pump, pushing electrons into one electrode and pulling them from the other. Withdrawing electrons from an electrode gives it a positive charge, whereas adding electrons to an electrode makes it negative. In the electrolysis of molten NaCl, shown in Figure 19.9, Na^+ ions pick up electrons at the negative electrode and are thereby reduced. As the Na^+ ions in the vicinity of this electrode are depleted, additional Na^+ ions migrate in. In a related fashion, there is a net movement of Cl^- ions to the positive electrode, where they give up electrons and are thereby oxidized. Just as in voltaic cells, the electrode at which reduction occurs is called the cathode, whereas the electrode at which oxidation occurs is called the anode.

Anode: $\quad\quad\quad\quad\quad\quad\quad 2Cl^-(l) \longrightarrow Cl_2(g) + 2e^-$

Cathode: $\quad\quad\quad\quad 2Na^+(l) + 2e^- \longrightarrow 2Na(l)$

$$\overline{\quad\quad\quad 2Na^+(l) + 2Cl^-(l) \longrightarrow 2Na(l) + Cl_2(g) \quad\quad} \quad\quad \text{[19.38]}$$

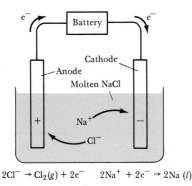

$$2Cl^- \rightarrow Cl_2(g) + 2e^- \qquad 2Na^+ + 2e^- \rightarrow 2Na\,(l)$$

FIGURE 19.9 The electrolysis of molten sodium chloride.

Electrolysis is used in the commercial preparation of sodium. Molten NaCl is electrolyzed in a specially designed cell called the Downs cell. This cell, shown in Figure 19.10, is designed to keep Na and Cl_2 from coming in contact and re-forming NaCl. The Na is also prevented from coming in contact with air and forming an oxide. Calcium chloride, $CaCl_2$, is added to the NaCl to lower the melting point of the molten liquid from its normal melting point of 804°C to around 600°C.

ELECTROLYSIS OF AQUEOUS SOLUTIONS

Sodium cannot be prepared by electrolysis of aqueous solutions of NaCl, because water is more easily reduced than $Na^+(aq)$:

$$2H_2O(l) + 2e^- \longrightarrow H_2(g) + 2OH^-(aq) \qquad E^\circ_{red} = -0.83\text{ V}$$
$$Na^+(aq) + e^- \longrightarrow Na(s) \qquad E^\circ_{red} = -2.71\text{ V}$$

Consequently, H_2 is produced at the cathode.

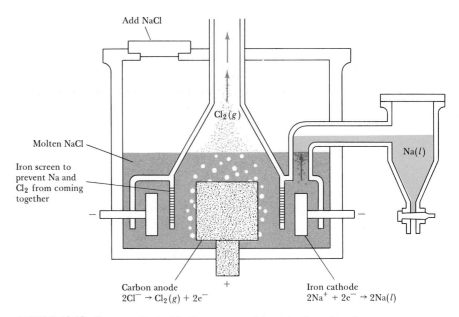

FIGURE 19.10 Downs cell used in the commercial production of sodium.

The possible anode reactions are the oxidation of Cl^- and of H_2O:

$$2Cl^-(aq) \longrightarrow Cl_2(g) + 2e^- \qquad E^\circ_{ox} = -1.36 \text{ V}$$
$$2H_2O(l) \longrightarrow 4H^+(aq) + O_2(g) + 4e^- \qquad E^\circ_{ox} = -1.23 \text{ V}$$

These standard oxidation potentials are not greatly different, but they do suggest that H_2O should be oxidized more readily than Cl^-. However, the voltage required for a reaction is sometimes much greater than that indicated by the electrode potentials. The additional voltage required to cause electrolysis is called the overvoltage. The overvoltage is believed to be caused by slow reaction rates at the electrodes. Overvoltages for the deposition of metals are low, but those required for the liberation of hydrogen gas or oxygen gas are usually high. In the present case the overvoltage for H_2 formation is sufficiently high to permit oxidation of Cl^- rather than H_2O. Consequently, electrolysis of aqueous solutions of NaCl, known as brines, produces H_2 and Cl_2 unless the concentration of Cl^- is quite low:

Anode: $\qquad\qquad\qquad 2Cl^-(aq) \longrightarrow Cl_2(g) + 2e^-$

Cathode: $\qquad\qquad 2H_2O(l) + 2e^- \longrightarrow H_2(g) + 2OH^-(aq)$

$$2Cl^-(aq) + 2H_2O(l) \longrightarrow$$
$$Cl_2(g) + H_2(g) + 2OH^-(aq) \qquad\qquad [19.39]$$

The Na^+ ion is merely a spectator ion (Section 12.6) in the electrolysis. This process is used commercially because all of the products (H_2, Cl_2, and NaOH) are commercially important chemicals.

Electrode potentials can be used to determine the minimum potential required for an electrolysis. In the case of the formation of H_2 and Cl_2 from a brine solution under standard conditions, a minimum voltage of 2.06 V is required.

$$E^\circ = E^\circ_{ox}(Cl^-) + E^\circ_{red}(H_2O)$$
$$= -1.23 \text{ V} + (-0.83 \text{ V}) = -2.06 \text{ V}$$

The emf calculated above is negative, reminding us that the process is not spontaneous but must be driven by an outside source of energy. Higher voltages than those calculated are invariably needed. One reason is the resistance of the cell; another is the overvoltage phenomenon discussed above.

SAMPLE EXERCISE 19.11

Explain why the electrolysis of an aqueous solution of $CuCl_2$ produces $Cu(s)$ and $Cl_2(g)$. What is the minimum emf required for this process under standard conditions?

Solution: At the cathode we can envision either reduction of Cu^{2+} or of H_2O:

$$Cu^{2+}(aq) + 2e^- \longrightarrow Cu(s) \qquad E^\circ_{red} = 0.34 \text{ V}$$
$$2H_2O(l) + 2e^- \longrightarrow H_2(g) + 2OH^-(aq)$$
$$E^\circ_{red} = -0.83 \text{ V}$$

The electrode potentials indicate that reduction of Cu^{2+} occurs more readily. Reduction of H_2O is made even more difficult because of the overvoltage for H_2 formation.

At the anode, we can envision either the oxidation of Cl^- or H_2O. As in the case of NaCl solutions, Cl_2 is usually produced because of the overvoltage for O_2 formation.

The minimum emf required for this electrolysis under standard conditions is 0.49 V:

$$E° = E°_{red}(Cu^{2+}) + E°_{ox}(Cl^-)$$
$$= 0.34\ V + (-0.83\ V) = -0.49\ V$$

REDUCTION OF ALUMINUM

The electrolysis of aqueous solutions of the active metals such as Na, Ca, Mg, and Al, which have quite negative standard reduction potentials, leads to the formation of H_2 rather than the metal. Consequently, these active metals are obtained by electrolysis of their molten salts. We have already briefly examined the electrolytic production of sodium. Let's examine one further example, the formation of aluminum. The electrolytic process used commercially to produce aluminum is known as the Hall process, named after its inventor Charles M. Hall (Figure 19.11). Aluminum oxide, Al_2O_3, is mined as the mineral bauxite. After purification, the Al_2O_3 is dissolved in molten cryolite, Na_3AlF_6, producing a solution that will conduct an electric current. The molten mixture is

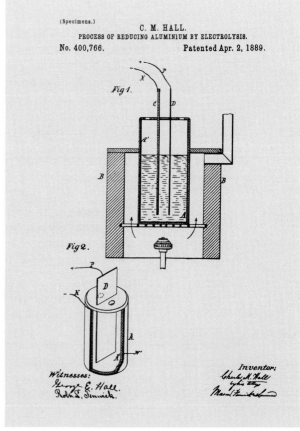

FIGURE 19.11 A photograph of Charles M. Hall (1863–1914) as a young man and the patent diagram of Hall's device for reducing aluminum. (*Courtesy of ALCOA*)

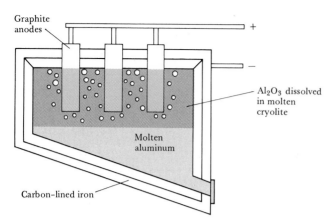

FIGURE 19.12 A typical Hall-process electrolysis cell used to reduce aluminum. Because molten aluminum is more dense than the molten mixture of Na_3AlF_6 and Al_2O_3, the metal collects at the bottom of the cell.

Graphite anodes

Al$_2$O$_3$ dissolved in molten cryolite

Molten aluminum

Carbon-lined iron

electrolyzed using carbon electrodes as shown in Figure 19.12. The electrode reactions are:

Anode: $\quad C(s) + 2O^{2-} \longrightarrow CO_2(g) + 4e^-$ [19.40]

Cathode: $\quad 3e^- + Al^{3+} \longrightarrow Al(l)$ [19.41]

Charles M. Hall began work on the problem of reducing aluminum in about 1885, after he had learned from a professor of the difficulty of reducing ores. Prior to the development of his electrolytic process, aluminum was obtained by chemical reduction using sodium or potassium as a reducing agent. This procedure caused the cost of aluminum to be very high. As late as 1852, the cost of aluminum was $545 a pound. During the Paris Exposition in 1855, aluminum was exhibited as a rare metal in spite of the fact that it is the third most abundant element on the earth. The cost of aluminum by then had fallen to $90 a pound, which still made it more expensive than gold or silver. It is reported that the very rich of that era could flaunt their wealth by using aluminum eating utensils. Hall, who was 21 years old when he started his research, utilized handmade and borrowed equipment in his studies, and his laboratory was a woodshed near his home. In about a year's time he was able to solve the problem. Strangely, Paul Heroult, who was the same age as Hall, made the same discovery in France at approximately the same time. The basic discovery was that Al_2O_3 dissolves in cryolite, a rare mineral found in a small region in Greenland, to produce a conducting solution. As a result of the discovery of Hall and Heroult, large-scale reduction of aluminum became commercially feasible, and aluminum became a common and familiar metal. Its price subsequently fell as low as 15¢ a pound. Even today aluminum sells for less than 80¢ a pound.

ELECTROLYSIS WITH ACTIVE ELECTRODES

In our discussion of the electrolysis of molten NaCl and of NaCl solutions we considered the electrodes to be inert. Consequently, the electrodes did not themselves undergo reaction but merely served as the surfaces at which oxidation and reduction occurred. However, in the Hall electrolysis used to produce aluminum, the anode does undergo reaction, Equation [19.40]. Thus, possible electrode reactions include not only oxidation and reduction of solute and solvent but also of the electrodes themselves. When aqueous solutions are electrolyzed using metal electrodes, the electrode will be oxidized if its oxidation potential is greater than that for water. For example, copper is oxidized more readily than water:

$$Cu(s) \longrightarrow Cu^{2+}(aq) + 2e^- \qquad E^\circ_{ox} = -0.34\text{ V}$$
$$2H_2O(l) \longrightarrow 4H^+(aq) + O_2(g) + 4e^- \qquad E^\circ_{ox} = -1.23\text{ V}$$

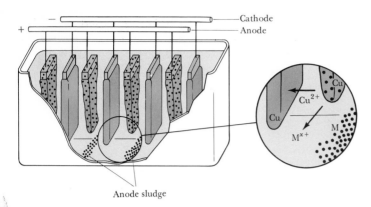

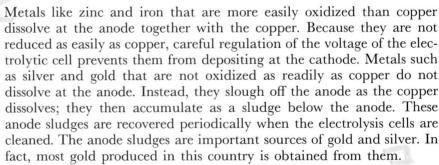

FIGURE 19.13 Electrolysis cell for the refining of copper. Notice that as the anodes dissolve away, the cathodes grow in size.

One of the many interesting applications of such an electrolysis process is the refining or purification of copper metal. Commercially, copper compounds are reduced using chemical reducing agents. For example, cuprous sulfide, Cu_2S, can be reduced by blowing air through the molten material as shown in Equation [19.42]:

$$Cu_2S(l) + O_2(g) \longrightarrow 2Cu(l) + SO_2(g) \qquad [19.42]$$

The copper metal obtained in this way is known as blister copper; it is about 99 percent pure, containing iron, zinc, gold, and silver among the impurities. Certain impurities greatly lower the electrical conductivity of the metal. Therefore, if it is to be used to make electrical wiring, it must be further purified. This purification is done electrolytically. The blister copper is made the anode in an electrolytic cell as shown in Figure 19.13. Thin sheets of pure copper serve as the cathode, and an aqueous mixture of H_2SO_4 and $CuSO_4$ is added to the cell to permit conduction of electric current through the solution. As the current flows, copper dissolves from the anode and deposits on the cathode:

Anode: $\qquad\qquad Cu(s) \longrightarrow Cu^{2+}(aq) + 2e^- \qquad$ [19.43]

Cathode: $\qquad Cu^{2+}(aq) + 2e^- \longrightarrow Cu(s) \qquad$ [19.44]

Metals like zinc and iron that are more easily oxidized than copper dissolve at the anode together with the copper. Because they are not reduced as easily as copper, careful regulation of the voltage of the electrolytic cell prevents them from depositing at the cathode. Metals such as silver and gold that are not oxidized as readily as copper do not dissolve at the anode. Instead, they slough off the anode as the copper dissolves; they then accumulate as a sludge below the anode. These anode sludges are recovered periodically when the electrolysis cells are cleaned. The anode sludges are important sources of gold and silver. In fact, most gold produced in this country is obtained from them.

Another interesting application of electrolysis is in the plating of metals. For example, if some metal other than copper were made the cathode in the electrolytic cell that we have just described, it would become coated with copper. The "plating out" of one metal on another in an electrolytic cell is known as electroplating. The object to be electroplated is made the cathode of the electrolytic cell. The plating metal is made

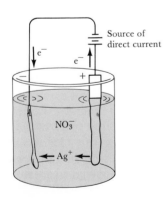

FIGURE 19.14 Electroplating of an object with silver. The object to be plated is made the cathode, whereas the plating metal is made the anode.

the anode as illustrated in Figure 19.14. Electroplating is used to protect objects against corrosion and to improve their appearance. For example, many exposed parts on automobiles such as bumpers and door handles are electroplated with chromium.

19.7 Quantitative aspects of electrolysis

The quantity of chemical reaction occurring in an electrolytic cell is directly proportional to the quantity of electricity passed into the cell. For example, 1 mol of electrons will plate out 1 mol of Na metal, whereas 2 mol of electrons will plate out 2 mol of Na metal:

$$Na^+ + e^- \longrightarrow Na$$

Similarly, it requires 2 mol of electrons to produce 1 mol of copper from Cu^{2+} and 3 mol of electrons to produce 1 mol of aluminum from Al^{3+}:

$$Cu^{2+} + 2e^- \longrightarrow Cu$$
$$Al^{3+} + 3e^- \longrightarrow Al$$

The quantity of charge passing through an electrical circuit, such as that in an electrolytic cell, is generally measured in coulombs. As noted in Section 19.4, there are 96,500 coulombs (coul) in a faraday:

$$1\mathcal{F} = 96,500 \text{ coul} = \text{charge of 1 mol of electrons} \qquad [19.45]$$

In terms of other, perhaps more familiar electrical units, a coulomb is the quantity of electrical charge passing a point in a circuit in 1 sec when the current is 1 ampere (amp).* Therefore, the number of coulombs passing through a cell can be obtained by multiplying the amperage and the elapsed time in seconds:

$$\text{Coulombs} = \text{amperes} \times \text{seconds} \qquad [19.46]$$

These ideas are applied in Sample Exercises 19.12 and 19.13. Although these exercises involve electrolytic cells, the same relationships can be applied to voltaic cells.

*Conversely, current is the rate of flow of electricity. An ampere is the current associated with the flow of 1 coul past a point each second.

SAMPLE EXERCISE 19.12

Calculate the amount of aluminum produced in 1.00 hr by the electrolysis of molten $AlCl_3$ if the current is 10.0 amp.

Solution: Using Equation [19.46], we can write:

Coulombs = (10.0 amp)(1.00 hr)

$$\times \left(\frac{3600 \text{ sec}}{1 \text{ hr}}\right)\left(\frac{1 \text{ coul}}{1 \text{ amp-sec}}\right)$$

$$= 3.60 \times 10^4 \text{ coul}$$

The half-reaction for the reduction of Al^{3+} is

$$Al^{3+} + 3e^- \longrightarrow Al$$

The amount of aluminum produced depends on the number of available electrons: 1 mol Al ≏ 3 $\mathcal{F}$. We can therefore write

Grams Al = $(3.60 \times 10^4 \text{ coul})\left(\dfrac{1 \mathcal{F}}{96,500 \text{ coul}}\right)$

$$\times \left(\frac{1 \text{ mol Al}}{3 \mathcal{F}}\right)\left(\frac{27.0 \text{ g Al}}{1 \text{ mol Al}}\right)$$

$$= 3.36 \text{ g}$$

A constant current was passed through a solution of Cu^{2+} for a period of 5.00 min. During this time the cathode increased in mass by 1.24 g. How many amperes of current was used?

Solution: In this case the half-reaction that we are focusing on is

$$Cu^{2+} + 2e^- \longrightarrow Cu$$

Therefore, 1 mol Cu $\simeq$ 2 $\mathfrak{F}$. Using this information, we can calculate the amperes as follows:

$$Coulombs = (1.24 \text{ g Cu})\left(\frac{1 \text{ mol Cu}}{63.5 \text{ g Cu}}\right)$$
$$\times \left(\frac{2 \text{ } \mathfrak{F}}{1 \text{ mol Cu}}\right)\left(\frac{96,500 \text{ coul}}{1 \text{ } \mathfrak{F}}\right)$$
$$= 3.77 \times 10^3 \text{ coul}$$

$$Amperes = \frac{coul}{sec}$$
$$= \left(\frac{3.77 \times 10^3 \text{ coul}}{5.00 \text{ min}}\right)$$
$$\times \left(\frac{1 \text{ min}}{60 \text{ sec}}\right)\left(\frac{1 \text{ amp-sec}}{1 \text{ coul}}\right)$$
$$= 12.6 \text{ amp}$$

ELECTRICAL WORK

The maximum electrical work that can be obtained from a voltaic cell is given by the product of the emf of the cell (E) and the electrical charge $(n\mathfrak{F})$ that it delivers:*

$$W_{max} = n\mathfrak{F}E \qquad [19.47]$$

A similar equation can be written for the minimum work required to cause a chemical reaction in an electrolytic cell:

$$W_{min} = n\mathfrak{F}E$$

The movement of 1 coul of electrical charge using an emf of 1 V corresponds to 1 J of energy:

$$1 \text{ J} = 1 \text{ coul-V} \qquad [19.48]$$

Electrical energy is sometimes expressed in units of kilowatt-hours (kwh). The watt is a unit of electrical power, that is, the rate of energy expenditure:

$$1 \text{ watt} = \frac{1 \text{ J}}{sec} \qquad [19.49]$$

Thus a watt-sec is a joule. A kilowatt-hour, which is also a unit of energy, is 3.6×10^6 J:

$$1 \text{ kwh} = (1000 \text{ watts})(1 \text{ hr})\left(\frac{3600 \text{ sec}}{1 \text{ hr}}\right)\left(\frac{1 \text{ J/sec}}{1 \text{ watt}}\right) = 3.6 \times 10^6 \text{ J}$$

---

*Equation [19.47] follows from $\Delta G = -n\mathfrak{F}E$, Equation [19.21], and the fact that ΔG is the maximum work a system can accomplish at constant temperature and pressure (Section 18.6). The negative sign is absent from Equation [19.47] if we define positive W as work done by the system on the surroundings, which requires a spontaneous process (positive E) to accomplish.

SAMPLE EXERCISE 19.14

Calculate the minimum number of kilowatt-hours of electricity required to produce 1000 kg of aluminum by electrolysis of Al^{3+} if the required emf is 4.5 volts.

Solution:

$$\text{Coulombs} = (1000 \text{ kg Al})\left(\frac{1000 \text{ g Al}}{1 \text{ kg Al}}\right)$$

$$\times \left(\frac{1 \text{ mol Al}}{27.0 \text{ g Al}}\right)\left(\frac{3 \text{ }\mathcal{F}}{1 \text{ mol Al}}\right)$$

$$\times \left(\frac{96,500 \text{ coul}}{1 \text{ }\mathcal{F}}\right)$$

$$= 1.07 \times 10^{10} \text{ coul}$$

A kilowatt-hour is a measure of energy. Electrical energy is obtained by multiplying coulombs and voltage, Equation [19.48]. When we do this and then apply a unit conversion factor to obtain kilowatt-hours we have:

$$\text{Kilowatt-hours} = (1.07 \times 10^{10} \text{ coul})(4.5 \text{ V})$$

$$\times \left(\frac{1 \text{ J}}{1 \text{ coul-V}}\right)\left(\frac{1 \text{ kwh}}{3.6 \times 10^{6} \text{ J}}\right)$$

$$= 1.33 \times 10^{4} \text{ kwh}$$

This quantity of energy does not include the energy used to mine, transport, and process the aluminum ore, and to keep the electrolysis bath molten during electrolysis. A typical electrolytic cell used to reduce aluminum is only 40 percent efficient, 60 percent of the electrical energy being dissipated as heat. It therefore requires on the order of 33 kwh of electricity to produce 1 kg of aluminum. The aluminum industry consumes about 2 percent of the electrical energy generated in the United States. Because this is used mainly for reduction of aluminum, recycling this metal would save large quantities of energy. Interestingly, the United States uses 5 percent of the world's production of aluminum to make food and beverage cans.

SAMPLE EXERCISE 19.15

A 12-V lead-storage battery contains 410 g of lead in its anode plates and a stoichiometrically equivalent amount of PbO_2 in the cathodes. (a) What is the maximum number of coulombs of electrical charge it can deliver without being recharged? (b) For how many hours could the battery deliver a steady current of 1.0 amp assuming the current does not fall during discharge? (c) What is the maximum electrical work that the battery can accomplish in kilowatt-hours?

Solution: (a) The lead anode undergoes a two-electron oxidation:

$$Pb \longrightarrow Pb^{2+} + 2e^{-}$$

Consequently, $2 \text{ }\mathcal{F} \simeq 1 \text{ mol Pb}$. Using this relationship we have

$$\text{Coulombs} = (410 \text{ g Pb})\left(\frac{1 \text{ mol Pb}}{207 \text{ g Pb}}\right)$$

$$\times \left(\frac{2 \text{ }\mathcal{F}}{1 \text{ mol Pb}}\right)\left(\frac{96,000 \text{ coul}}{1 \text{ }\mathcal{F}}\right)$$

$$= 3.8 \times 10^{5} \text{ coul}$$

Although the emf of a cell is independent of the masses of the solid reactants involved in the cell, the total electrical charge the cell can deliver does depend on these quantities. The size and surface area further affects the rate at which electrical charge can be delivered.

(b) We calculate hours as follows:

$$\text{Hours} = (3.8 \times 10^{5} \text{ coul})\left(\frac{1 \text{ amp-sec}}{1 \text{ coul}}\right)$$

$$\times \left(\frac{1 \text{ hr}}{3600 \text{ sec}}\right)\left(\frac{1}{1.0 \text{ amp}}\right)$$

$$= 1.1 \times 10^{2} \text{ hr}$$

This battery might be described as a 110 amp-hr battery.

(c) The maximum work is given by the product $n\mathcal{F}E$, Equation [19.47]:

$$\text{Kilowatt-hours} = (12 \text{ V})(3.8 \times 10^{5} \text{ coul})$$

$$\times \left(\frac{1 \text{ J}}{1 \text{ coul-V}}\right)\left(\frac{1 \text{ kwh}}{3.6 \times 10^{6} \text{ J}}\right)$$

$$= 1.3 \text{ kwh}$$

Before we close our discussion of electrochemistry, let's apply some of what we have learned to a very important problem, the corrosion of metals. Corrosion reactions are redox reactions in which a metal is attacked by some substance in its environment and converted to an unwanted compound.

One of the most familiar corrosion processes is the rusting of iron. From an economic standpoint this is a significant process. It is estimated that up to 20 percent of the iron produced annually in this country is used to replace iron objects that have been discarded because of rust damage.

The rusting of iron is known to involve oxygen; iron does not rust in water unless O_2 is present. Rusting also involves water; iron does not rust in oil, even if it contains O_2, unless H_2O is also present. Other factors such as the pH of the solution, the presence of salts, contact with metals more difficult to oxidize than iron, and stress on the iron can accelerate rusting.

The corrosion of iron is generally believed to be electrochemical in nature. A region on the surface of the iron serves as an anode at which the iron undergoes oxidation:

$$Fe(s) \longrightarrow Fe^{2+}(aq) + 2e^- \qquad E^{\circ}_{ox} = 0.44 \text{ V} \qquad [19.50]$$

The electrons so produced migrate through the metal to another portion of the surface that serves as the cathode. Here oxygen can be reduced:

$$O_2(g) + 4H^+(aq) + 4e^- \longrightarrow 2H_2O(l) \qquad E^{\circ}_{red} = 1.23 \text{ V} \qquad [19.51]$$

Notice that H^+ is involved in the reduction of O_2. As the concentration of H^+ is lowered (that is, as pH is increased), the reduction of O_2 becomes less favorable. It is observed that iron in contact with a solution whose pH is above 9–10 does not corrode. In the course of the corrosion, the Fe^{2+} formed at the anode is further oxidized to Fe^{3+}. The Fe^{3+} forms the hydrated iron(III) oxide known as rust:*

$$4Fe^{2+}(aq) + O_2(g) + 4H_2O(l) + 2xH_2O(l) \longrightarrow$$
$$2Fe_2O_3 \cdot xH_2O(s) + 8H^+(aq) \qquad [19.52]$$

Because the cathode is generally the area having the largest supply of O_2, the rust often deposits there. If you look closely at a shovel after it has stood outside in the moist air with wet dirt adhered to its blade, you may notice that pitting has occurred under the dirt but that rust has formed elsewhere, where O_2 is more readily available. The corrosion process is summarized by way of illustration in Figure 19.15.

The enhanced corrosion caused by the presence of salts is usually quite evident on autos in areas where there is heavy salting of roads during winter. The effect of salts is readily explained by the voltaic mechanism:

---

*Frequently, metal compounds that are obtained from aqueous solution have water associated with them. For example, copper(II) sulfate crystallizes from water with 5 mol of water per mole of $CuSO_4$. We represent this formula as $CuSO_4 \cdot 5H_2O$. Such compounds are called hydrates. Rust is a hydrate of iron(III) oxide with a variable amount of water of hydration. We represent the variable water content by writing the formula as $Fe_2O_3 \cdot xH_2O$.

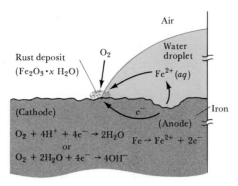

FIGURE 19.15 Corrosion of iron in contact with water.

The ions of a salt provide the electrolyte necessary for completion of the electrical circuit.

The presence of anodic and cathodic sites on the iron requires two different chemical environments on the surface. These can occur through the presence of impurities or lattice defects (perhaps introduced by strain on the metal). At the sites of such impurities or defects, the atomic-level environment around the iron atom may permit the metal to be either more or less easily oxidized than at normal lattice sites. Thus these sites may serve as either anodes or cathodes. Ultrapure iron, prepared in such a way as to minimize lattice defects, is far less susceptible to corrosion than is ordinary iron.

Iron is often covered with a coat of paint or another metal such as tin, zinc, or chromium to protect its surface against corrosion. "Tin cans" are produced by applying a thin layer of tin over steel. The tin protects the iron only as long as the protective layer remains intact. Once it is broken and the iron exposed to air and water, tin actually promotes the corrosion of the iron. It does so by serving as the cathode in the electrochemical corrosion. As shown by the following half-cell potentials, iron is more readily oxidized than tin:

$$Fe(s) \longrightarrow Fe^{2+}(aq) + 2e^- \qquad E^\circ_{ox} = 0.44 \text{ V} \qquad [19.53]$$
$$Sn(s) \longrightarrow Sn^{2+}(aq) + 2e^- \qquad E^\circ_{ox} = 0.14 \text{ V} \qquad [19.54]$$

The iron therefore serves as the anode and is oxidized as shown in Figure 19.16.

"Galvanized iron" is produced by coating iron with a thin layer of zinc. The zinc protects the iron against corrosion even after the surface

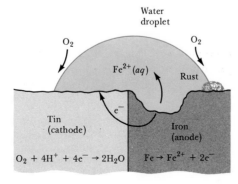

FIGURE 19.16 Corrosion of iron in contact with tin.

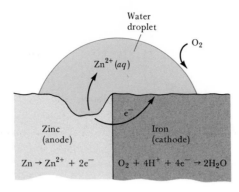

FIGURE 19.17 Cathodic protection of iron in contact with zinc.

coat is broken. In this case the iron serves as the cathode in the electrochemical corrosion because zinc is oxidized more easily than iron:

$$Zn(s) \longrightarrow Zn^{2+}(aq) + 2e^- \qquad E°_{ox} = 0.76 \text{ V} \qquad [19.55]$$

The zinc therefore serves as the anode and is corroded instead of the iron, as shown in Figure 19.17. Such protection of a metal by making it the cathode in an electrochemical cell is known as cathodic protection. Underground pipelines are often protected against corrosion by making the pipeline the cathode of a voltaic cell. Pieces of an active metal such as magnesium are buried along the pipeline and connected to it by wire as shown in Figure 19.18. In moist soil, where corrosion can occur, the active metal serves as the anode and the pipe experiences cathodic protection.

Although our discussions have centered on iron, this is not the only metal subject to corrosion. One thing that may be surprising in light of our discussions is the fact that an aluminum can, disposed of carelessly in the environment, corrodes so much more slowly than a steel can. On the

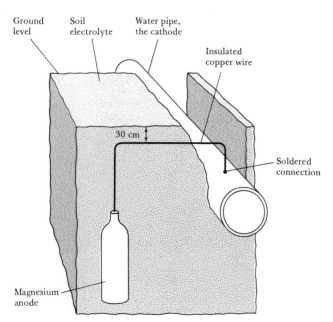

FIGURE 19.18 Cathodic protection of an iron water pipe. The magnesium anode is surrounded by a mixture of gypsum, sodium sulfate, and clay to promote conductivity of ions. The pipe, in effect, is the cathode of a voltaic cell.

basis of the standard oxidation potentials of aluminum ($E^{\circ}_{ox} = 1.66$ V) and iron ($E^{\circ}_{ox} = 0.44$ V), we would expect the aluminum to be much more readily corroded. The slow corrosion of aluminum is explained by the formation of a thin, compact oxide coating that forms on its surface. This protects the underlying metal from further corrosion. Magnesium, which also has a large oxidation potential, is similarly protected. The oxide coat on iron is too porous to offer similar protection. However, when iron is alloyed with chromium, a protective oxide coating does form. Such alloys are called stainless steels.

SAMPLE EXERCISE 19.16

Predict the nature of the corrosion that would take place if an iron gutter were nailed to a house using aluminum nails.

Solution: A voltaic cell can be formed at the point of contact of the two metals. The metal that is most easily oxidized will serve as the anode, whereas the other metal serves as the cathode. By comparing standard oxidation potentials of Al and Fe we see that Al will be the anode:

$$Al(s) \longrightarrow Al^{3+}(aq) + 3e^{-} \qquad E^{\circ}_{ox} = 1.66 \text{ V}$$
$$Fe(s) \longrightarrow Fe^{2+}(aq) + 2e^{-} \qquad E^{\circ}_{ox} = 0.44 \text{ V}$$

The gutter will thus be protected against corrosion in the vicinity of the nail, because the iron serves as the cathode. However, the nail would quickly corrode, leaving the gutter on the ground.

What do you think would happen if aluminum siding were nailed to a house using iron nails?

FOR REVIEW

Summary

We have seen that oxidation-reduction (or redox) reactions can be considered to consist of two half-reactions, one due to oxidation (loss of electrons) and the other due to reduction (gain of electrons). The substance oxidized is called the reducing agent, or reductant, whereas the substance reduced is called the oxidizing agent, or oxidant. If the half-reactions are balanced, they can be added to obtain the balanced redox equation. The number of electrons lost by the reducing agent must equal the number gained by the oxidizing agent. Consequently, different types of atoms and total charges must each balance.

Spontaneous redox reactions can be used to generate electricity in voltaic cells. Conversely, electricity can be used to bring about nonspontaneous reactions in electrolytic cells. In either type of cell, the electrode at which oxidation occurs is called the anode, whereas the electrode at which reduction occurs is called the cathode.

A voltaic cell may be thought to possess a "driving force" that moves the electrons through the external circuit, from anode to cathode. This driving force is called the electromotive force (emf), and is measured in volts. The emf of a cell can be regarded as being composed of two parts: that due to oxidation at the anode and that due to reduction at the cathode: $E_{cell} = E_{ox} + E_{red}$.

Oxidation potentials (E_{ox}) and reduction potentials (E_{red}) can be assigned to half-reactions by defining the standard hydrogen electrode as a reference:

$$2H^{+}(1 \ M) + 2e^{-} \longrightarrow H_2(1 \text{ atm}) \qquad E^{\circ} = 0$$

Standard reduction potentials are referred to merely as standard electrode potentials and are tabulated for a great variety of reduction half-reactions. The oxidation potential for an oxidation half-reaction will be of the same magnitude as, but opposite in sign to, the electrode potential for the reverse reduction process. The more positive the potential associated with a half-reaction, the greater the tendency for that reaction to occur as written. Electrode potentials can be used to determine the maximum voltages generated by voltaic cells or the minimum voltages required in electrolytic cells. They can also be used to predict whether certain redox reactions are spontaneous (positive E). The emf is related to free-energy changes: $\Delta G = -n\mathfrak{F}E$, where $\mathfrak{F}$ is Faraday's constant, 96,500 J/V-mol.

The emf of a cell varies in magnitude with temperature and with the concentrations of reactants and products. The Nernst equation relates emf under nonstandard conditions to the standard emf:

$$E = E° - \frac{2.30\, RT}{n\mathcal{F}} \log Q$$

At equilibrium

$$E = 0 \text{ and } E° = \frac{2.30\, RT}{n\mathcal{F}} \log K$$

Thus standard emfs are related to equilibrium constants. The maximum electrical work that can be obtained from a voltaic cell is the product of the total charge it delivers, $n\mathcal{F}$, and its emf, E: $W_{max} = n\mathcal{F}E$.

To illustrate the principles involved in voltaic cells, simple cells utilizing salt bridges were discussed. Commercial cells need to be more rugged. Three common batteries were discussed: the lead-storage battery, the nickel-cadmium battery, and the common dry cell. The first two are rechargeable whereas the dry cell is not.

In an electrolytic cell an external source of electricity is used to "pump" electrons from the anode to the cathode. The products produced during electrolysis can generally be predicted by comparing electrode potentials associated with possible oxidation and reduction processes. However, because of the overvoltage phenomenon, some reactions (such as those that generate H_2 and O_2) occur less readily than electrode potentials would suggest.

The extent of chemical reaction in either an electrolytic or voltaic cell can be related to the quantity of electricity that passes through the external circuit. The amount of electrical charge possessed by a mole of electrons is known as a faraday (abbreviated $\mathcal{F}$), 96,500 coul; 1 coul = 1 amp-sec.

Our knowledge of electrochemistry allows us to design batteries and to bring about desirable redox reactions such as those involved in electroplating and in the reduction and refining of metals. Electrochemical principles also help us to understand and combat corrosion. Corrosion of a metal such as iron can be shown to be electrochemical in origin. A metal can be protected against corrosion by putting it in contact with another metal that more readily undergoes oxidation. This process is known as cathodic protection.

Learning goals

Having read and studied this chapter, you should be able to:

1 Recognize redox reactions and identify the reductant and oxidant.

2 Complete and balance redox equations using the method of half-reactions.

3 Diagram simple voltaic and electrolytic cells, labeling anode, cathode, the directions of ion and electron movements, and the signs of the electrodes.

4 Calculate the emf generated by a voltaic cell or the minimum emf required to cause an electrolytic cell reaction to proceed, having been given appropriate electrode potentials.

5 Use electrode potentials to predict whether a reaction will be spontaneous.

6 Interconvert $E°$, $\Delta G°$, and K for redox reactions.

7 Use the Nernst equation to calculate emfs under nonstandard conditions.

8 Use the Nernst equation to calculate the concentration of an ion, given E, $E°$, and the concentrations of the remaining ions.

9 Describe the lead-storage battery, the dry cell, and the nickel-cadmium cell.

10 Calculate the third quantity, having been given any two of the following: time, current, amount of substance produced or consumed in an electrolysis reaction.

11 Calculate the maximum electrical work performed by a voltaic cell (or the minimum electrical work to cause a reaction in an electrolytic cell), having been given emf and electrical charge or information from which they can be determined.

12 Describe the phenomenon of corrosion and explain the principle that underlies cathodic protection.

Key terms

Among the more important terms and expressions used for the first time in this chapter are the following:

An anode (Section 19.2) is an electrode at which oxidation occurs.

A cathode (Section 19.2) is an electrode at which reduction occurs.

Cathodic protection (Section 19.8) is a means of protecting a metal against corrosion by making it the cathode in a voltaic cell. This can be done by attaching a more active metal.

Corrosion (Section 19.8) is the process by which a metal is oxidized by substances in its environment.

A faraday (Section 19.4) is the total charge of a mole of electrons, 96,500 coul.

A half-reaction (Section 19.1) is an equation for either oxidation or reduction that explicitly shows the electrons involved (for example, $2H^+(aq) + 2e^- \longrightarrow H_2(g)$).

An oxidant, or oxidizing agent (Section 19.1), is a substance that is reduced and thereby causes the oxidation of some other substance.

A reductant, or reducing agent (Section 19.1), is a substance that is oxidized and thereby causes the reduction of some other substance.

A standard electrode potential, $E°$ (Section 19.3) is the reduction potential of a half-reaction with all solution species at $1\ M$ concentration and all gaseous species at 1 atm, measured relative to the hydrogen electrode, for which $E°$ is exactly 0 V.

EXERCISES

Oxidation-reduction reactions

19.1 Complete and balance the following half-reactions, indicating in each case whether oxidation or reduction is occurring:

(a) $SO_4^{2-}(aq) \longrightarrow SO_2(g)$ (acidic solution)
(b) $Fe(s) \longrightarrow Fe^{2+}(aq)$ (acidic solution)
(c) $NO_2^-(aq) \longrightarrow NO_3^-(aq)$ (acidic solution)
(d) $O_2(g) \longrightarrow OH^-(aq)$ (basic solution)
(e) $Cr(OH)_3(s) \longrightarrow CrO_4^{2-}(aq)$ (basic solution)

19.2 Indicate whether the following reactions involve oxidation-reduction. If they do, identify the substance oxidized and the substance reduced.

(a) $H_2SO_4(aq) + 2KOH(aq) \longrightarrow K_2SO_4(aq) + 2H_2O(l)$
(b) $MnO_2(s) + 4HCl(aq) \longrightarrow$
    $MnCl_2(aq) + Cl_2(g) + 2H_2O(l)$
(c) $Na_2SO_4(aq) + BaCl_2(aq) \longrightarrow 2NaCl(aq) + BaSO_4(s)$
(d) $6KOH(aq) + 3Br_2(aq) \longrightarrow$
    $5KBr(aq) + KBrO_3(aq) + 3H_2O(l)$

19.3 Balance the following equations, identifying in each case the substance oxidized and the substance reduced:

(a) $Fe(s) + H^+(aq) + SO_4^{2-}(aq) \longrightarrow$
    $Fe^{3+}(aq) + SO_2(g) + H_2O(l)$
(b) $Cl_2(g) + OH^-(aq) \longrightarrow$
    $Cl^-(aq) + ClO^-(aq) + H_2O(l)$
(c) $Zn(s) + NO_3^-(aq) + H^+(aq) \longrightarrow$
    $Zn^{2+}(aq) + NH_4^+(aq) + H_2O(l)$
(d) $MnO_4^-(aq) + S^{2-}(aq) + H_2O(l) \longrightarrow$
    $MnO_2(s) + S(s) + OH^-(aq)$
(e) $H_2O_2(aq) + MnO_4^-(aq) + H^+(aq) \longrightarrow$
    $O_2(g) + Mn^{2+}(aq) + H_2O(l)$

19.4 Complete and balance the following equations:

(a) $Cr_2O_7^{2-}(aq) + CH_3OH(aq) \longrightarrow$
    $HCO_2H(aq) + Cr^{3+}(aq)$ (acidic solution)
(b) $MnO_4^-(aq) + I^-(aq) \longrightarrow$
    $MnO_2(s) + I_2(s)$ (basic solution)
(c) $Cr(OH)_4^-(aq) + H_2O_2(aq) \longrightarrow$
    $CrO_4^{2-}(aq)$ (basic solution)
(d) $NO_3^-(aq) + I_2(aq) \longrightarrow$
    $IO_3^-(aq) + NO_2(g)$ (acidic solution)
(e) $As(s) + NO_3^-(aq) \longrightarrow$
    $H_3AsO_3(aq) + NO(g)$ (acidic solution)
(f) $H_2O_2(aq) + ClO_2(aq) \longrightarrow$
    $ClO_2^-(aq) + O_2(g)$ (basic solution)

19.5 Hydrazine (N_2H_4) and dinitrogen tetraoxide (N_2O_4) forms a self-igniting mixture that has been employed as a rocket propellant. The reaction products are N_2, H_2, and H_2O. (a) Write a balanced chemical equation for this reaction. (b) Which substance serves as reducing agent and which as oxidizing agent?

[19.6] Oxalic acid, $H_2C_2O_4$, occurs in certain vegetables, for example, rhubarb and spinach. This acid is toxic but occurs below toxic limits in foods. However, it may be concentrated during certain types of food processing. The concentration of $H_2C_2O_4$ in a sample may be determined by titration with permanganate ion (MnO_4^-) in an acid solution forming $CO_2(g)$ and $Mn^{2+}(aq)$. (a) Write a balanced net ionic equation for this reaction. (b) If a 50.0-g sample requires 22.0 mL of $0.0500\ M$ MnO_4^- solution to reach an end point in the titration, what is the weight percentage of $H_2C_2O_4$ in the sample?

Voltaic cells; emf; spontaneity

19.7 Sketch a voltaic cell based on the following reaction:

$$Fe(s) + Cu^{2+}(aq) \longrightarrow Fe^{2+}(aq) + Cu(s)$$

Label the anode and cathode and identify the positive and the negative terminal. Also indicate the directions of ion and electron movements and calculate the emf generated by the cell under standard conditions.

19.8 A solution of $1\ M$ $Ni(NO_3)_2$ is placed in a beaker with a strip of Ni metal. A solution of $1\ M$ $AgNO_3$ is placed in a second beaker together with a piece of Ag. The two beakers are connected by a salt bridge, and the two electrodes are linked by a wire leading to a voltmeter. (a) Which electrode serves as the anode and which as the cathode? (b) Which electrode gains mass and which loses mass as the voltaic cell operates? (c) Which electrode is positive and which negative? (d) What is the emf generated by the cell under standard conditions?

19.9 The following oxidation-reduction reaction is spontaneous in the direction indicated:

$$5Fe^{2+}(aq) + MnO_4^-(aq) + 8H^+(aq) \longrightarrow$$
$$5Fe^{3+}(aq) + Mn^{2+}(aq) + 4H_2O(l)$$

A solution containing $KMnO_4$ and H_2SO_4 is poured into one beaker while a solution of $FeSO_4$ is poured into another. A salt bridge is used to join the beakers. A platinum foil is placed in each solution, and the two solutions connected by a wire that passes through a voltmeter. (a) Indicate the reactions occurring at the anode and at the cathode, the direction of electron movement through the external circuit, the direction of ion migrations through the solutions, and the signs of the electrodes. (b) Calculate the emf of the cell under standard conditions.

19.10 Consider the following half-reactions:

$$Cr^{3+}(aq) + 3e^- \longrightarrow Cr(s)$$
$$MnO_2(s) + 4H^+(aq) + 2e^- \longrightarrow Mn^{2+}(aq) + 2H_2O(l)$$

(a) What spontaneous reaction can take place that involves these two half-reactions? (b) What is the standard emf of a voltaic cell based on these reactions?

19.11 Which of the following will not act as reducing agents (that is, cannot be readily oxidized): (a) Na^+; (b) Cl^-; (c) SO_4^{2-}; (d) Cl_2. Explain briefly.

19.12 Which of the following does not ordinarily function as oxidizing agents (that is, cannot be readily reduced): (a) F^-; (b) ClO_3^-; (c) Na; (d) Cl_2?

19.13 (a) Arrange the following species in order of increasing strength as oxidizing agents: $Cr_2O_7^{2-}$, H_2O_2, Cu^{2+}, Cl_2, O_2. (b) Arrange the following species in order of increasing strength as reducing agents: Zn, I^-, Sn^{2+}, H_2O_2, Al.

19.14 Which of the following substances can be oxidized by MnO_4^- in acidic solution: (a) Cl^-; (b) Cl_2; (c) Cr^{3+}; (d) Fe^{2+}; (e) Cu?

19.15 Determine whether the following processes are spontaneous under standard conditions:

(a) $O_2(g) + 4Cl^-(aq) + 4H^+(aq) \longrightarrow$
$\qquad 2H_2O(l) + 2Cl_2(g)$
(b) $Cr(s) + Al^{3+}(aq) \longrightarrow Al(s) + Cr^{3+}(aq)$
(c) $2Br^-(aq) + Cl_2(g) \longrightarrow Br_2(l) + 2Cl^-(aq)$
(d) $Sn^{2+}(aq) + 2Fe^{3+}(aq) \longrightarrow Sn^{4+}(aq) + 2Fe^{2+}(aq)$
(e) $Cu(s) + Cl_2(g) \longrightarrow Cu^{2+}(aq) + 2Cl^-(aq)$

19.16 Write the balanced chemical equation for each of the following processes and indicate whether the reaction will be spontaneous under standard conditions (in each case the reaction occurs in acidic solutions): (a) oxidation of Fe^{2+} to Fe^{3+} by Br_2 (which forms Br^-); (b) oxidation of Cr^{3+} to $Cr_2O_7^{2-}$ by MnO_4^- (reduced to Mn^{2+}); (c) reduction of Cd^{2+} by Fe.

Relationships between $E°$, $\Delta G°$, and K

19.17 Using electrode potentials, calculate $\Delta G°$ for the following reactions:

(a) $FeCl_2(aq) + Zn(s) \longrightarrow Fe(s) + ZnCl_2(aq)$
(b) $Pb(s) + PbO_2(s) + 2H^+(aq) + 2HSO_4^-(aq) \longrightarrow$
$\qquad 2PbSO_4(s) + 2H_2O(l)$
(c) $Br_2(l) + 2Cl^-(aq) \longrightarrow Cl_2(g) + 2Br^-(aq)$
(d) $Ca(s) + 2H_2O(l) \longrightarrow$
$\qquad Ca^{2+}(aq) + 2OH^-(aq) + H_2(g)$

19.18 What magnitude of equilibrium constant is associated with a standard emf of (a) 1.00 V; (b) 0.10 V, assuming $n = 1$?

19.19 Using electrode potentials, calculate the equilibrium constant for each of the following reactions:

(a) $Mn(s) + Zn^{2+}(aq) \longrightarrow Zn(s) + Mn^{2+}(aq)$
(b) $Ni(s) + 2H^+(aq) \longrightarrow Ni^{2+}(aq) + H_2(g)$

(c) $10Cl^-(aq) + 2MnO_4^-(aq) + 16H^+(aq) \longrightarrow$
$\qquad 2Mn^{2+}(aq) + 8H_2O(l) + 5Cl_2(g)$
(d) $Cu^{2+}(aq) + 2I^-(aq) \longrightarrow I_2(s) + Cu(s)$

19.20 Cytochrome, a complicated molecule that we shall represent as CyFe^{2+}, reacts with the air we breathe to supply energy required to synthesize adenosine triphosphate, ATP. The ATP is employed by the body as an energy source to drive other reactions (Section 18.6). At pH 7 the following electrode potentials pertain to this oxidation of CyFe^{2+}:

$$O_2(g) + 4H^+(aq) + 4e^- \longrightarrow 2H_2O(l) \qquad E = 0.82 \text{ V}$$
$$CyFe^{3+}(aq) + e^- \longrightarrow CyFe^{2+}(aq) \qquad E = 0.22 \text{ V}$$

(a) What is ΔG for the oxidation of CyFe^{2+} by air? (b) If the synthesis of 1 mol of ATP from adenosine diphosphate, ADP, requires a ΔG of 37.7 kJ, how many moles of ATP are synthesized per mole of O_2?

Nernst equation

19.21 Predict the effect of an increasing concentration of iodide ion on the emf generated by a voltaic cell utilizing the following reaction:

$$Cl_2(g) + 2I^-(aq) \longrightarrow 2Cl^-(aq) + I_2(s)$$

What is the effect of increasing the concentration of chloride ion?

19.22 Calculate the emf of a Zn–Ag$^+$ cell under the following conditions:

$$Zn(s) + 2Ag^+(0.5 \text{ M}) \longrightarrow Zn^{2+}(0.01 \text{ M}) + 2Ag(s)$$

19.23 A voltaic cell is constructed that is based on the following reaction:

$$Sn^{2+}(aq) + Pb(s) \longrightarrow Pb^{2+}(aq) + Sn(s)$$

If the concentration of Sn^{2+} in the cathode compartment is 1.00 M and the cell generates an emf of 0.22 V, what is the concentration of Pb^{2+} in the anode compartment? If the anode compartment contains $[SO_4^{2-}] = 1.00$ M, what is the K_{sp} of the PbSO$_4$?

19.24 Consider the Zn–Cu cell, Figure 19.2. Indicate whether the following changes increase, decrease, or have no effect on the emf of this cell: (a) $[Cu^{2+}]$ is increased to 3 M; (b) NaOH is added to the cathode compartment to precipitate $Cu(OH)_2$; (c) the size of the zinc electrode is doubled; (d) $[Zn^{2+}]$ is decreased to 0.1 M; (e) the temperature of the cell is increased.

19.25 A primitive pH meter is constructed using zinc and hydrogen electrodes as shown in Figure 19.3. The anode compartment contains 1.0 M Zn^{2+} while $P_{H_2} = 1.0$ atm in the cathode compartment. The cell generates an emf of 0.702 V. What is the pH of the solution in the cathode compartment?

Commercial voltaic cells

19.26 In both the lead-storage cell and the nickel-cadmium cell, the reactants are solids. (a) How does this simplify the design of the cell? (b) What advantage does such a cell have in terms of voltage drop during use? (Consider the Nernst equation.)

19.27 Why will the liquid in a discharged lead-storage battery freeze more readily in very cold weather than will a fully charged one?

19.28 Bumping and shaking knocks $PbSO_4$ from the electrodes of a lead-storage battery. How does this shorten the battery's life?

19.29 What would happen if HNO_3 were substituted for H_2SO_4 in the lead-storage battery? What if Na_2SO_4 were substituted for H_2SO_4?

Electrolysis

19.30 Why are different products obtained when molten $AlCl_3$ and aqueous $AlCl_3$ are electrolyzed with inert electrodes? Predict the products in each case.

19.31 Predict the products formed when the following are electrolyzed with inert electrodes: (a) molten $MgCl_2$; (b) $1.0\,M$ $CuSO_4$ (aq); (c) $1.0\,M$ $AgNO_3(aq)$; (d) $1.0\,M$ $HBr(aq)$.

19.32 Sketch a cell for the electrolysis of $NiCl_2$ solution using inert electrodes. Indicate the directions in which ions and electrons move. Give the electrode reactions and label the anode and cathode, indicating which is positive and which is negative.

19.33 Sketch a cell for the electrolysis of HBr using nickel electrodes. Give the electrode reactions, labeling the anode and cathode. Calculate the minimum emf needed to cause electrolysis under standard conditions.

19.34 Using Faraday's constant, $\mathcal{F}$, calculate the charge on a single electron.

19.35 Determine the value of the faraday from the mass of iodine, 4.8285 g, released by the passage of 3671.3 coul of electricity through an aqueous solution of HI. Compare this result with the accepted value.

19.36 How many faradays would be required to reduce 1 mol of each of the following to the indicated product: (a) MnO_4^- to Mn^{2+}; (b) $Cr_2O_7^{2-}$ to Cr^{3+}; (c) Fe^{3+} to Fe^{2+}; (d) Zn^{2+} to Zn?

19.37 A Cd^{2+} solution is electrolyzed using a current of 0.500 amp. How many grams of cadmium metal plate out in 20.0 min?

19.38 How many minutes are required to plate 10.0 g of nickel metal from a solution of $NiSO_4$ using a current of 1.50 amp?

19.39 How many grams of aluminum can be obtained in the Hall process in the time it takes the anode to lose 1.50 kg of carbon?

19.40 In the electrolyses of aqueous NaCl solutions, how many liters of Cl_2 gas (measured at STP) are generated by a current of 2.00 amp for a period of 1.00 hr?

[19.41] If 0.250 L of 0.500 M $NiSO_4$ solution is electrolyzed using 3.00 amp for 40.0 min, what are the final concentrations of each ion remaining in the solution? (Assume the volume of the solution does not change.)

[19.42] Making stamping masters for phonograph records involves placing a fine coating of silver on a plastic record to give it a surface that conducts an electrical current. The record is then electroplated with nickel.

(a) Should the record be made the anode or cathode in the electrolytic cell? (b) How much time is required to provide a 0.0010 cm coating of nickel on both faces of the record using a current of 0.20 amp if the record has a diameter of 30.0 cm? (The density of nickel is 8.90 g/cm³.) Ignore the nickel deposited on the edges.

19.43 Compare the current required to produce 1.00 kg of metal per hour by reduction of each of the following: (a) Al^{3+}; (b) Cu^{2+}; (c) Ag^+.

Electrical work

19.44 What is the maximum electrical work, in joules, that a Zn-Cu cell (Figure 19.2) can accomplish under standard conditions if 0.100 mol of Zn is consumed?

[19.45] How long will a 25-watt light bulb burn if powered by a lead-storage cell that produces 2.0 V if during the duration of its discharge 15 g of Pb is converted to $PbSO_4$?

Corrosion

19.46 The zinc on galvanized iron has been called a "sacrificial anode." What does this mean? Does chromium act similarly on iron objects that are plated with chromium?

19.47 Chrome plating actually relies on an undercoating of nickel for the protection of iron. The chromium layer keeps the nickel from tarnishing and gives a hard, bright surface. Does the nickel protect iron by cathodic protection?

19.48 Amines are compounds related to ammonia. One of their characteristics is their ability to function as Brønsted bases (Section 15.6). Suggest how they protect against corrosion when added to antifreeze as corrosion inhibitors.

19.49 Describe the nature of the corrosion that can occur if a copper pipe is fitted under the soil to a galvanized steel pipe.

Additional exercises

19.50 Complete and balance the following equations:

(a) $I_2(s) + H_3AsO_3(aq) \longrightarrow$
$\quad I^-(aq) + H_3AsO_4(aq)$ (acidic solution)
(b) $Br_2(l) \longrightarrow BrO_3^-(aq) + Br^-(aq)$ (basic solution)
(c) $Fe^{3+}(aq) + I^-(aq) \longrightarrow$
$\quad Fe^{2+}(aq) + I_2(s)$ (acidic solution)
(d) $ClO_3^-(aq) + I_2(s) \longrightarrow$
$\quad IO_3^-(aq) + Cl^-(aq)$ (acidic solution)
(e) $Fe^{2+}(aq) + Cr_2O_7^{2-}(aq) \longrightarrow$
$\quad Fe^{3+}(aq) + Cr^{3+}(aq)$ (acidic solution)
(f) $CrO_4^{2-}(aq) + HSnO_2^-(aq) \longrightarrow$
$\quad HSnO_3^-(aq) + CrO_2^-(aq)$ (basic solution)

19.51 (a) The process by which nitrogen locked in the soil is lost to the atmosphere is called denitrification. The following reaction is one means by which denitrification may occur in acidic soils rich in plant matter:

$$C_6H_{12}O_6(aq) + NO_3^-(aq) \longrightarrow CO_2(g) + N_2(g)$$

Complete and balance this equation. (b) Aqua regia, a mixture of nitric and hydrochloric acid, is one of the few reagents capable of dissolving gold. Complete and balance the equation for this process:

$$Au(s) + NO_3^-(aq) + Cl^-(aq) \longrightarrow$$
$$AuCl_4^-(aq) + NO_2(g)$$

(c) Aluminum can be etched with hydrochloric acid. Areas to be protected from the acid are painted. When the acid is applied it attacks the exposed aluminum. Balance the equation for this process:

$$Al(s) + H^+(aq) \longrightarrow Al^{3+}(aq) + H_2(g)$$

(d) Copper sulfate is used in some antifungicide sprays and is added to swimming pools in order to kill algae. It can be prepared by the action of hot sulfuric acid on copper. Complete and balance the equation for this process:

$$Cu(s) + HSO_4^-(aq) \longrightarrow Cu^{2+}(aq) + SO_2(g)$$

(e) Using H_2O_2 as an oxidizing agent has the advantage of forming only H_2O as the by-product. In World War II, the German V2 rockets used H_2O_2 as an oxidizing agent for a mixture of methanol (CH_3OH) and hydrazine (N_2H_4). The methanol is oxidized to CO_2 and H_2O while hydrazine is oxidized to N_2 and H_2O. Write balanced equations for these two reactions.

19.52 Copper will not dissolve in hydrochloric acid, but it does dissolve in nitric acid. For the reaction in concentrated nitric acid the unbalanced chemical equation is

$$Cu(s) + NO_3^-(aq) + H^+(aq) \longrightarrow$$
$$Cu^{2+}(aq) + NO_2(g) + H_2O(l)$$

(a) Balance this chemical equation. (b) Identify the oxidizing and reducing agents in this reaction. (c) Explain why Cu is able to dissolve in HNO_3, but not in HCl.

19.53 Iodic acid, HIO_3, can be prepared by oxidizing iodine with concentrated nitric acid. The HIO_3 settles out of the reaction mixture as a white solid. The unbalanced equation is

$$I_2(s) + H^+(aq) + NO_3^-(aq) \longrightarrow HIO_3(s) + NO_2(g)$$

Calculate the quantity of NO_2 produced in the formation of 10.0 g of HIO_3.

19.54 Cyanogen, $(CN)_2$, may be made by the oxidation of aqueous sodium cyanide by aqueous copper(II) sulfate. Insoluble copper(I) cyanide is also formed. (a) Write a balanced net ionic equation for this reaction. (b) What quantity of cyanogen can be prepared using 5.0 g of copper(II) sulfate if the reaction goes to completion?

19.55 A cell is constructed that consists of Sn in contact with 1.0 M Sn^{2+} in one compartment and Fe in contact with 1.0 M Fe^{2+} in the other. (a) Which electrode serves as the anode? (b) What is the standard emf of the cell?

19.56 A common shorthand way of representing cells is to list reactants and products from left to right in the following form:

anode | anode solution || cathode solution | cathode

A double vertical line represents a salt bridge or porous barrier. A single vertical line represents a change of phase such as from solid to solution. (a) Write the half-reactions and overall cell reaction represented by $Fe \mid Fe^{2+} \parallel Ag^+ \mid Ag$. (b) Write the half-reactions and overall cell reaction represented by $Mg \mid Mg^{2+} \parallel H^+ \mid H_2 \mid Pt$. (c) Use the line notation to represent a cell based on the following reaction:

$$Cu^{2+}(aq) + Zn(s) \longrightarrow Cu(s) + Zn^{2+}(aq)$$

19.57 The zinc-silver oxide cell used in hearing aids and electric watches is based on the following half-reactions:

$$Zn^{2+}(aq) + 2e^- \longrightarrow Zn(s) \qquad E° = -0.763 \text{ V}$$
$$Ag_2O(s) + H_2O(l) + 2e^- \longrightarrow 2Ag(s) + 2OH^-(aq)$$
$$E° = 0.344 \text{ V}$$

(a) What substance is oxidized and what is reduced in the cell during discharge? (b) What is the positive electrode and what is the negative electrode? (c) What emf does this cell generate under standard conditions?

19.58 (a) Which member of the following pairs is the stronger oxidizing agent: H_2O or Al; MnO_4^- or Cl_2? (b) Which member of the following pairs is the stronger reducing agent: Sn or Fe; H_2 or H_2O?

19.59 Using Table 19.1, (a) suggest one or more reducing agents capable of reducing Ag^+ to Ag but not I_2 to I^- under standard conditions; (b) suggest an oxidizing agent capable of oxidizing Cl^- to Cl_2 but not F^- to F_2 under standard conditions.

19.60 Predict whether the following reactions will be spontaneous in acid solution under standard conditions: (a) oxidation of Fe^{2+} to Fe^{3+} by Br_2 (to form Br^-); (b) oxidation of Cr^{3+} to $Cr_2O_7^{2-}$ by MnO_4^- (reduced to Mn^{2+}).

19.61 Calculate the emf of voltaic cells that are based on the following reactions:

(a) $Sn^{2+}(1.0\,M) + Ni(s) \longrightarrow Sn(s) + Ni^{2+}(0.0010\,M)$
(b) $2H^+(0.10\,M) + Fe(s) \longrightarrow$
$$H_2(0.10 \text{ atm}) + Fe^{2+}(0.020\,M)$$

19.62 Calculate the emf of a cell containing $5.0 \times 10^{-3}\,M$ Cr^{3+} in one compartment and 2.0 M Cr^{3+} in the other if Cr electrodes are used in both. Which is the anode compartment?

19.63 Calculate $\Delta G°$ and the equilibrium constant for each reaction described in question 19.61.

[19.64] In a fully charged lead-storage cell, the electrolyte solution consists of an aqueous solution that is 38 percent H_2SO_4 by mass and has a density of 1.286 g/mL. (a) Calculate $E°$ and E for this cell. (b) What is the value of E when the H_2SO_4 solution is 5 percent by weight and has a density of 1.025 g/mL?

[19.65] The standard potential for the reduction of AgSCN is 0.0895 V:

$$AgSCN(s) + e^- \longrightarrow Ag(s) + SCN^-(aq)$$

Using this value together with another electrode potential, calculate K_{sp} for AgSCN.

19.66 Primary cells are ones like the ordinary dry cell

that cannot be recharged. Storage cells are ones like the nickel-cadmium cell or the lead-storage battery that can be recharged. Is the Zn–Cu^{2+} cell, Figure 19.2, a primary cell or a storage cell? Explain briefly.

19.67 Calculate the electrode potential for reduction of H_2O at pH 7. (Assume $P_{H_2} = 1$ atm.)

19.68 What is the sign of the PbO_2 electrode of the lead-storage cell during discharge? During charging?

[19.69] (a) How many grams of silver are required in order to plate a coin whose diameter is 2.4 cm and whose thickness is 0.10 cm with a silver plate whose thickness is 1.0×10^{-3} cm? The density of silver is 10.5 g/cm^3. (b) How long will it take to plate 150 such coins if the electrolytic cell is 85 percent efficient and the current is 0.200 amp?

19.70 Explain why electrolysis of an aqueous solution of Na_2SO_4 containing litmus develops a blue color at the cathode and a red color at the anode.

19.71 What current is being drawn from a lead-storage cell if 75 mg of lead is oxidized per minute?

19.72 Peroxyborate bleaches, such as found in Bora-teem, have replaced older "chlorine" bleaches in many bleaching agents. Sodium peroxyborate, $NaBO_3$, can be prepared by electrolytic oxidation of borax ($Na_2B_4O_7$) solutions:

$$Na_2B_4O_7(aq) + 10NaOH(aq) \longrightarrow$$
$$4NaBO_3(aq) + 5H_2O(l) + 8Na^+(aq) + 8e^-$$

How many grams of $NaBO_3$ can be prepared in 24.0 hr if the current is 20.0 amp?

19.73 How long could a nickel-cadmium cell be used in a child's mechanical toy if a current of 2.0 amp is drawn from a cell containing 2.5 g of NiO_2?

19.74 If charging efficiency is 85 percent, how much time is required to convert 50.0 g of $PbSO_4$ into PbO_2 using a current of 10.0 amp?

19.75 What is the maximum amount of work, in joules, that a 12-V lead-storage battery can accomplish if it is rated at 100 amp-hr?

19.76 A copper coin can be cleaned by suspending it on a copper wire attached to the negative pole of a battery, placing it in a 2.5 percent NaOH solution, and inserting a graphite electrode connected to the positive terminal. Explain how the coin gets clean.

[19.77] A family owns an antique set of silverware that has a fine, dark coating of Ag_2S in the crevices of the pattern that adds to its beauty. The set is placed in a galvanized container together with soap and water in order to be cleaned. The Ag_2S disappears as the set sits in the container, leaving the silverware with the appearance of a new rather than an antique set. Explain the electrochemical processes that have occurred.

[19.78] A voltaic cell is constructed with one half-cell containing Ag as an electrode in contact with $AgNO_3$ solution. The other half-cell contains Cd as an electrode in contact with $Cd(NO_3)_2$. (a) Diagram the cell, showing the direction of electron movement and ion migrations and labeling the anode and cathode. (b) Which electrode will gain weight and which will lose? (c) What is the standard potential generated by the cell? (d) Given the following quantities, calculate the total electrical charge that the cell can deliver: mass of silver electrode, 1.00 g; mass of cadmium electrode, 1.00 g; mass of $AgNO_3$ in solution, 1.00 g; mass of $Cd(NO_3)_2$ in solution, 1.00 g.

[19.79] Several years ago, a unique proposal was made to raise the *Titanic*. The plan involved placing pontoons within the ship using a surface-controlled submarine-type vessel. The pontoons would contain cathodes and would be filled with hydrogen gas formed by the electrolysis of water. It has recently been estimated that it would require about 7×10^8 mol of H_2 to provide the bouyancy to lift the ship (*Journal of Chemical Education*, vol. 50, p. 61, 1973). (a) How many coulombs of electrical charge would be required? (b) What is the minimum voltage required to generate H_2 and O_2 if the pressure on the gases at the depth of the wreckage (2 mi) is 300 atm? (c) What is the minimum electrical energy required to raise the *Titanic* by electrolysis? (d) What is the minimum cost of the electrical energy required to generate the necessary H_2 if the electricity costs 23¢ per kilowatt-hour?

[19.80] Edison's invention of the light bulb and its public demonstration in December 1879 generated considerable demand for the distribution of electricity to homes. One problem was how to measure the amount of electricity consumed by each household. Edison invented a coulometer (described in the *Journal of Chemical Education*, vol. 49, p. 627, 1972) that could be used with an AC current. Zinc plated out at the cathode of the coulometer. Every month the cathode was removed and weighed to determine the quantity of electricity used. If the cathode increased in mass by 1.62 g and the coulometer drew 0.35 percent of the current entering the home, how many coulombs of electricity were used in that month?

20

Nuclear chemistry

As we have progressed through this book, our focus has been on chemical reactions, specifically reactions in which electrons play a dominant role. In this chapter, we shall consider nuclear reactions, changes in matter whose origin is the nucleus of the atom. Some experts predict that we shall have to depend more and more on nuclear energy to replace our dwindling supplies of fossil fuels and to meet our rising energy demands. Thus our consideration of nuclear chemistry continues a minor theme of energy generation started in the last chapter. Even before we begin, you should have some awareness of the controversy surrounding nuclear energy—how do you feel about having a nuclear power plant in your town? Because the topic of nuclear energy evokes such emotional reaction, it is difficult to sift fact from opinion and begin to weigh pros and cons rationally. It is therefore important for any educated person of our time to have some understanding of nuclear reactions and the use of radioactive substances.

However, before we get too deeply involved in our discussions, it is useful to review and extend slightly some ideas introduced in Section 2.6. First, we should recall that there are two subatomic particles that reside in the nucleus, the proton and the neutron. We shall refer to these particles as nucleons. Recall also that all atoms of a given element have the same number of protons; this number is known as the element's atomic number. However, the atoms of a given element can have different numbers of neutrons and therefore different mass numbers; the mass number is the total number of nucleons in the nucleus. Atoms with the same atomic number but different mass numbers are known as isotopes. The different isotopes of an element are distinguished by citing their mass numbers. For example, the three naturally occurring isotopes of uranium are identified as uranium-233, uranium-235, and uranium-238, where the numbers given are the mass numbers. These isotopes are also labeled, using chemical symbols, as $^{233}_{92}U$, $^{235}_{92}U$, and $^{238}_{92}U$. The superscript is the mass number, the subscript is the atomic number. Different isotopes have different natural abundances. For example, 99.3 percent of the naturally occurring uranium is uranium-238, whereas 0.7 percent is uranium-235, and only a trace is uranium-233. One reason that it now becomes important to distinguish between different isotopes is that the nuclear properties of an atom depend on the number of both protons and neutrons in its nucleus. In contrast, we have found that an atom's chemical properties are unaffected by the number of neutrons in the nucleus. Now let's begin to discuss the reactions that a nucleus can undergo.

20.1 Nuclear reactions: an overview

There are several ways in which a nucleus can undergo a reaction and thereby change its identity. Some nuclei are unstable and spontaneously emit particles and electromagnetic radiation. Such spontaneous emission from the nucleus of the atom is known as radioactivity. The discovery of this phenomenon by Henri Becquerel in 1896 was described in Section 2.6. Those isotopes that are radioactive are known as radioisotopes. An example is uranium-238, which spontaneously emits alpha rays; these rays consist of a stream of helium-4 nuclei known as alpha particles. When a uranium-238 nucleus loses an alpha particle, the remaining fragment has an atomic number of 90 and a mass number of 234. It is therefore a thorium-234 nucleus. We can represent this reaction by the following nuclear equation:

$$\require{mhchem} {}^{238}_{92}\text{U} \longrightarrow {}^{234}_{90}\text{Th} + {}^{4}_{2}\text{He} \qquad [20.1]$$

When a nucleus spontaneously decomposes in this way, it is said to have decayed, or undergone radioactive decay.

Notice in Equation [20.1] that the sum of the mass numbers is the same on both sides of the equation ($238 = 234 + 4$). Likewise, the sum of the atomic numbers or nuclear charges on both sides of the equation is equal ($92 = 90 + 2$). Mass numbers and atomic numbers are similarly balanced in writing other nuclear equations. In writing nuclear equations, we are not concerned with the chemical form of the atom in which the nucleus resides. The radioactive properties of the nucleus are essentially independent of the state of chemical combination of the atom. It makes no difference whether we are dealing with the atom in the form of an element or of one of its compounds.

Another way a nucleus can change identity is to be struck by a neutron or by another nucleus. Nuclear reactions that are induced in this way are known as nuclear transmutations. Such a transmutation occurs when the chlorine-35 nucleus is struck by a neutron (${}^{1}_{0}\text{n}$); this collision produces a sulfur-35 nucleus and a proton (${}^{1}_{1}\text{p}$ or ${}^{1}_{1}\text{H}$). The nuclear equation for this reaction is shown in Equation [20.2]:

$${}^{35}_{17}\text{Cl} + {}^{1}_{0}\text{n} \longrightarrow {}^{35}_{16}\text{S} + {}^{1}_{1}\text{H} \qquad [20.2]$$

Notice again that the sum of mass numbers and of atomic numbers is the same on both sides of the equation. By bombarding nuclei with various particles, it is possible to prepare nuclei not found in nature. The sulfur-35 produced in Equation [20.2] is such an isotope.

Write a balanced nuclear equation for the nuclear transmutation in which an aluminum-27 nucleus is struck by a helium-4 nucleus producing a phosphorus-30 nucleus and a neutron.

Solution: By referring to a periodic table or a list of elements, we find that aluminum has an atomic number of 13; its chemical symbol is therefore $^{27}_{13}\text{Al}$. The atomic number of phosphorus is 15; its chemical symbol is therefore $^{30}_{15}\text{P}$. The balanced equation is

$$^{27}_{13}\text{Al} + ^{4}_{2}\text{He} \longrightarrow ^{30}_{15}\text{P} + ^{1}_{0}\text{n}$$

We shall take a closer look at radioactive decay in sections that follow. We shall also discuss the preparation of nuclei via nuclear transmutation. In Sections 20.8 and 20.9, we shall discuss two other types of nuclear reactions. One is known as nuclear fission and the other as nuclear fusion. Fission involves the fragmentation of a large nucleus into two roughly equal-sized nuclei. Fusion involves the combination of two small nuclei to form a larger nucleus.

20.2 Radioactivity

As we were discussing radioactivity in the last section, two general questions might have occurred to you: Which nuclei are radioactive and what types of radiation do they emit? These are important questions, and we shall now examine them. Let us first consider the types of radiation involved in the phenomenon of radioactivity.

TYPES OF RADIOACTIVE DECAY

Emission of radiation is one of the ways by which an unstable nucleus is transformed into a stable one with less energy. The emitted radiation is the carrier of the excess energy. In Section 2.6, we discussed the three most common types of radiation emitted by radioactive substances: alpha (α), beta (β), and gamma (γ) rays.

As we noted in Section 20.1, alpha rays consist of streams of helium-4 nuclei known as alpha particles. Equation [20.3] gives another example of this type of radioactive decay:

$$^{222}_{86}\text{Rn} \longrightarrow ^{218}_{84}\text{Po} + ^{4}_{2}\text{He} \qquad [20.3]$$

Beta rays consist of streams of electrons. Because the beta particles are electrons, they are represented as $^{0}_{-1}\text{e}$. The superscript zero reflects the exceedingly small mass of the electron by comparison to the mass of a nucleon. The subscript -1 indicates the negative charge of the particle, which is opposite that of the proton. Iodine-131 is an example of an isotope that undergoes decay by beta emission. This reaction is summarized by Equation [20.4]:

$$^{131}_{53}\text{I} \longrightarrow ^{131}_{54}\text{Xe} + ^{0}_{-1}\text{e} \qquad [20.4]$$

Emission of a beta particle has the effect of converting a neutron within the nucleus into a proton, thereby increasing the atomic number of the nucleus by one:

$$\ce{^1_0n -> ^1_1p + ^0_{-1}e}$$

[20.5]

However, just because electrons are ejected from the nucleus, we need not think that the nucleus is composed of these particles, any more than we consider a match to be composed of sparks simply because it gives them off when struck. The electrons come into being when the nucleus is disrupted.

Gamma rays consist of electromagnetic radiation of very short wavelength (that is, high-energy photons). The position of gamma rays in the electromagnetic spectrum is shown in Figure 5.3. Gamma rays can be represented as $\ce{^0_0\gamma}$. Such radiation changes neither the atomic number nor the mass number of a nucleus. It almost always accompanies other radioactive emission, because it represents the energy lost when the remaining nucleons reorganize into more stable arrangements. Generally we shall not show the gamma rays when writing nuclear equations.

Two other types of radioactive decay that occur are positron emission and electron capture. A positron is a particle that has the same mass as an electron but an opposite charge.* The positron is represented as $\ce{^0_1e}$. Carbon-11 is an example of an isotope that decays by positron emission:

$$\ce{^{11}_6C -> ^{11}_5B + ^0_1e}$$

[20.6]

Emission of a positron can be thought of as converting a proton into a neutron as shown in Equation [20.7]. The atomic number of the nucleus is thereby decreased by one:

$$\ce{^1_1p -> ^1_0n + ^0_1e}$$

[20.7]

Electron capture involves capture of an electron from the electron cloud surrounding the nucleus. Rubidium-81 undergoes decay in this fashion as shown in Equation 20.8:

$$\ce{^{81}_{37}Rb + ^0_{-1}e} \text{ (orbital electron) } \ce{-> ^{81}_{36}Kr}$$

[20.8]

Electron capture has the effect of converting a proton within the nucleus into a neutron as shown in Equation [20.9]:

$$\ce{^1_1P + ^0_{-1}e -> ^1_0n}$$

[20.9]

*The positron has a very short life, because it is annihilated when it collides with an electron: $\ce{^0_1e + ^0_{-1}e -> 2^0_0\gamma}$.

SAMPLE EXERCISE 20.2

Write balanced nuclear equations for the following reactions: (a) thorium-230 undergoes alpha decay; (b) thorium-231 undergoes decay to form protactinium-231.

Solution: (a) The information given in the problem can be summarized as

$$\ce{^{230}_{90}Th -> ^4_2He + X}$$

The remaining product, X, must be deduced. Because mass numbers must have the same sum on both sides of the equation, we deduce that X has a mass number of 226. Similarly, the atomic number of X must be 88. Element number 88 is radium (refer to periodic table or list of elements). The equation is therefore as follows:

$$\ce{^{230}_{90}Th -> ^4_2He + ^{226}_{88}Ra}$$

(b) In this case we must determine what type of particle is emitted in the course of the radioactive decay. We can write the following equation:

$$^{231}_{90}\text{Th} \longrightarrow {}^{231}_{91}\text{Pa} + \text{X}$$

The atomic numbers are obtained from a list of elements such as that given on the front inside cover. In order for the mass numbers to balance, X must have a mass number of 0. Its atomic number must be -1. The particle with these characteristics is the beta particle (electron). We therefore write the following:

$$^{231}_{90}\text{Th} \longrightarrow {}^{231}_{91}\text{Pa} + {}^{0}_{-1}\text{e}$$

NUCLEAR STABILITY

There is no single rule that will allow us to predict whether a particular nucleus is radioactive and how it might decay. However, we can list some empirical observations that are helpful in making predictions:

1 All nuclei with 84 or more protons are unstable. For example, all isotopes of uranium, atomic number 92, are radioactive.

2 Nuclei with a total of 2, 8, 20, 50, 82, or 126 protons or neutrons are generally more stable than nuclei found near them in the periodic table. For example, there are three stable nuclei with an atomic number of 18, two with 19, five with 20, and one with 21; there are three stable nuclei with 18 neutrons, none with 19, four with 20, and none with 21. Thus there are more stable nuclei with 20 protons or 20 neutrons than with 18, 19, or 21. The numbers 2, 8, 20, 50, 82, and 126 are called **magic numbers.** Just as enhanced chemical stability is associated with the presence of 2, 10, 18, 36, 54, or 86 electrons, the noble-gas configurations, enhanced nuclear stability is associated with the magic number of nucleons.

3 Nuclei with even numbers of both protons and neutrons are generally more stable than those with odd numbers of nucleons, as shown in Table 20.1.

4 The stability of a nucleus can be correlated to a certain degree with its neutron-to-proton ratio. All nuclei with two or more protons contain neutrons. Neutrons are apparently involved in some way in holding protons together within the nucleus. As shown in Figure 20.1, the number of neutrons necessary to create a stable nucleus increases rapidly as the number of protons increases; the neutron-to-proton ratios of stable nuclei increase with increasing atomic number. The area within which all stable nuclei are found is known as the **belt of stability.**

TABLE 20.1 The number of stable isotopes with even and odd numbers of protons and neutrons

Number of stable isotopes	Protons	Neutrons
157	Even	Even
52	Even	Odd
50	Odd	Even
5	Odd	Odd

SAMPLE EXERCISE 20.3

Would you expect the following nuclei to be radioactive: $^{4}_{2}\text{He}$, $^{39}_{20}\text{Ca}$, $^{210}_{85}\text{At}$?

Solution: Helium-4 has a magic number of both protons and neutrons (two each). We would therefore expect $^{4}_{2}\text{He}$ to be stable.

Calcium-39 has an even number of protons (20) and an odd number of neutrons (19); 20 is one of the magic numbers. Nevertheless, we should suspect that this nuclide is radioactive because the neutron-to-proton ratio is less than 1. This would place it below the belt of stability.

Astatine-210 is radioactive. Recall that there are no stable nuclei beyond atomic number 83.

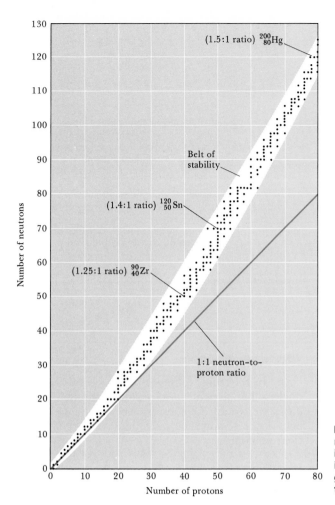

FIGURE 20.1 A plot of the number of neutrons versus the number of protons in stable nuclei. As the atomic number increases, the neutron-to-proton ratio of the stable nuclei increases. The stable nuclei are located in an area of the graph known as the belt of stability. The majority of radioactive nuclei occur outside this belt.

The type of radioactive decay that a particular radioisotope will undergo depends to a large extent on its neutron-to-proton ratio compared to those of nearby nuclei that are within the belt of stability. Consider a nucleus whose high neutron-to-proton ratio places it above the belt of stability. This nucleus can lower its ratio and move toward the belt of stability by emitting a beta particle. Beta emission decreases the number of neutrons and increases the number of protons in a nucleus as shown in Equation [20.5].

Nuclei that have low neutron-to-proton ratios and that therefore lie below the belt of stability either emit positrons or undergo electron capture. Both modes of decay decrease the number of protons and increase the number of neutrons in the nucleus as shown in Equations [20.7] and [20.9]. Positron emission is more common than electron capture among the lighter nuclei; however, electron capture becomes increasingly common as nuclear charge increases.

Alpha emission is found primarily among nuclei whose atomic numbers are greater than 83. These nuclei would lie beyond the upper right edge of Figure 20.1, outside the belt of stability. Emission of an alpha particle moves the nucleus diagonally toward the belt of stability by decreasing both the number of protons and the number of neutrons by two.

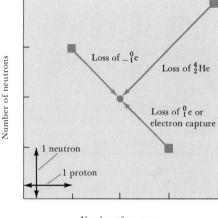

FIGURE 20.2 The result of alpha emission ($^4_2$He), beta emission ($^0_{-1}$e), positron emission ($^0_1$e), and electron capture on the number of protons and neutrons in a nucleus. The squares represent unstable nuclei, and the circle represents a stable one. Moving from right to left or from bottom to top, each tick mark represents an additional proton or neutron, respectively. Moving in the reverse direction indicates the loss of a proton or neutron.

The result of each type of radioactive decay relative to a stable nucleus is shown in Figure 20.2.

SAMPLE EXERCISE 20.4

By referring to Figure 20.1, predict the mode of radioactive decay of the following nuclei: (a) $^{20}_{11}$Na; (b) $^{97}_{40}$Zr; (c) $^{235}_{92}$U.

Solution: To answer this question we use the guidelines given above in the text.

(a) This nucleus has a neutron-to-proton ratio below 1. It therefore lies below the belt of stability. It can gain stability either by positron emission or electron capture. Because the atomic number is small we might predict that the nucleus undergoes positron emission. If we refer to a standard reference like the *Handbook of Chemistry and Physics,* we find that this prediction is correct. The nuclear reaction is

$$^{20}_{11}\text{Na} \longrightarrow {}^0_1\text{e} + {}^{20}_{10}\text{Ne}$$

(b) In referring to Figure 20.1, we find that this nucleus has a neutron-to-proton ratio that is too high. We would therefore predict that it undergoes beta decay. Again the prediction is correct. The nuclear reaction is

$$^{97}_{40}\text{Zr} \longrightarrow {}^0_{-1}\text{e} + {}^{97}_{41}\text{Nb}$$

(c) This nucleus lies outside the belt of stability to the upper right. We might therefore predict that it would undergo alpha emission. Again the prediction is correct. The nuclear equation is

$$^{235}_{92}\text{U} \longrightarrow {}^4_2\text{He} + {}^{231}_{90}\text{Th}$$

At this point we should note that our guidelines don't always work. For example thorium-233, $^{233}_{90}$Th, which we might expect to undergo alpha decay, undergoes beta decay instead. Furthermore, a few radioactive nuclei actually lie within the belt of stability. For example, both $^{146}_{60}$Nd and $^{148}_{60}$Nd are stable and lie in the belt of stability; however $^{147}_{60}$Nd, which lies between them, is radioactive.

RADIOACTIVE SERIES

Some nuclei, like uranium-238, cannot gain stability by a single emission. Consequently, a series of successive emissions occur. As shown in Figure 20.3, uranium-238 decays to thorium-234, which is radioactive and decays to protactinium-234. This nucleus is also unstable and subsequently decays. Such successive reactions continue until a stable nucleus, lead-206, is formed. A series of nuclear reactions that begins with an

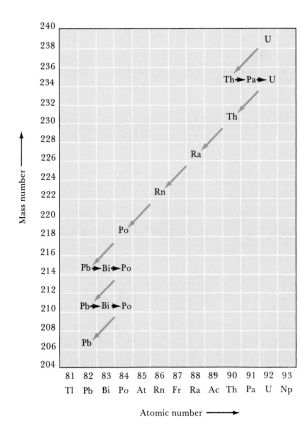

FIGURE 20.3 The nuclear disintegration series for uranium-238. The $^{238}_{92}U$ nucleus decays to $^{234}_{90}$ Th. Subsequent decay processes eventually form the stable $^{206}_{82}Pb$ nucleus. Each of the color arrows corresponds to the loss of an alpha particle. Each black arrow corresponds to the loss of a beta particle.

unstable nucleus and terminates with a stable one is known as a radioactive series or a nuclear disintegration series. Altogether there are three such series found in nature. In addition to the series that begins with uranium-238 and terminates with lead-206, there is one that begins with uranium-235 and ends with lead-207. The third series begins with thorium-232 and ends with lead-208.

20.3 Preparation of new nuclei

In 1919 Rutherford performed the first artificial conversion of one nucleus into another. He succeeded in converting nitrogen-14 into oxygen-17 using the high velocity alpha (α) particles emitted by radium. The reaction is shown in Equation [20.10]:

$$^{14}_{7}N + {}^{4}_{2}He \longrightarrow {}^{17}_{8}O + {}^{1}_{1}H \qquad [20.10]$$

This reaction demonstrated that nuclear reactions can be induced by striking nuclei with particles such as alpha particles. Such reactions have permitted synthesis of hundreds of radioisotopes in the laboratory. As noted in Section 20.1, these conversions of one nucleus into another are called nuclear transmutations. It is common to represent such conversions by listing, in order, the target nucleus, the bombarding particle, the ejected particle, and the product nucleus. Written in this fashion, Equation [20.10] is $^{14}_{7}N(\alpha, p)^{17}_{8}O$. The alpha particle, proton, and neutron are abbreviated as α, p, and n, respectively.

Write the balanced nuclear equation for the process summarized as $^{27}_{13}\text{Al}(n, \alpha)^{24}_{11}\text{Na}$.

Solution: The n is the abbreviation for a neutron, whereas α represents an alpha particle. The neutron is the bombarding particle, and the alpha particle is a product. Therefore the nuclear equation is

$$^{27}_{13}\text{Al} + {}^{1}_{0}\text{n} \longrightarrow {}^{4}_{2}\text{He} + {}^{24}_{11}\text{Na}$$

Charged particles such as alpha particles must be moving very fast in order to overcome the electrostatic repulsion between them and the target nucleus. The higher the nuclear charge on either the projectile or the target, the faster the projectile must be moving to bring about a nuclear reaction. Therefore, many methods have been devised to accelerate charged particles using strong magnetic and electrostatic fields. These particle accelerators bear such names as the cyclotron and synchrotron. The cyclotron is illustrated in Figure 20.4. The hollow D-shaped electrodes are called "dees." The projectile particles are introduced into a vacuum chamber within the cyclotron. The particles are then accelerated by making the dees alternately positively and negatively charged. Magnets placed above and below the dees keep the particles moving in a spiral path until they are finally deflected out of the cyclotron and emerge to strike a target substance. Particle accelerators have been used mainly to probe the secrets of nuclear structure and to synthesize heavy elements.

Most synthetic isotopes used in quantity in medicine and scientific research are made using neutrons as projectiles. Because neutrons are neutral, they are not repelled by the nucleus; consequently, they do not need to be accelerated, as do the charged particles, in order to cause nuclear reactions (indeed they cannot be so accelerated). The necessary neutrons are produced by the reactions that occur in nuclear reactors (Section 20.8). Cobalt-60, used in radiation therapy for cancer, is produced by neutron capture. Iron-58 is placed in a nuclear reactor where it is bombarded by neutrons. The sequence of reactions shown in Equations [20.11] through [20.13] takes place:

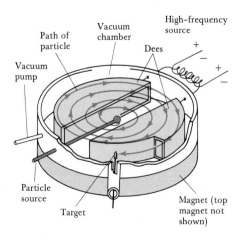

FIGURE 20.4 Schematic representation of a cyclotron.

$$^{58}_{26}\text{Fe} + ^{1}_{0}\text{n} \longrightarrow ^{59}_{26}\text{Fe} \qquad \text{[20.11]}$$
$$^{59}_{26}\text{Fe} \longrightarrow ^{59}_{27}\text{Co} + ^{0}_{-1}\text{e} \qquad \text{[20.12]}$$
$$^{59}_{27}\text{Co} + ^{1}_{0}\text{n} \longrightarrow ^{60}_{27}\text{Co} \qquad \text{[20.13]}$$

TRANSURANIUM ELEMENTS

Artificial transmutations have been used to produce the elements from atomic number 93 to 105. These are known as the transuranium elements, because they occur immediately following uranium in the periodic table. Elements 93 (neptunium) and 94 (plutonium) were first discovered in 1940. They were produced by bombarding uranium-238 with neutrons as shown in Equations [20.14] and [20.15]:

$$^{238}_{92}\text{U} + ^{1}_{0}\text{n} \longrightarrow ^{239}_{92}\text{U} \longrightarrow ^{239}_{93}\text{Np} + ^{0}_{-1}\text{e} \qquad \text{[20.14]}$$
$$^{239}_{93}\text{Np} \longrightarrow ^{239}_{94}\text{Pu} + ^{0}_{-1}\text{e} \qquad \text{[20.15]}$$

Elements with larger atomic numbers are normally formed in small quantities in particle accelerators. For example, curium-242 is formed when a plutonium-239 target is struck with accelerated alpha particles:

$$^{239}_{94}\text{Pu} + ^{4}_{2}\text{He} \longrightarrow ^{242}_{96}\text{Cm} + ^{1}_{0}\text{n} \qquad \text{[20.16]}$$

20.4 Half-life

Again our discussions may have raised some questions in your mind. For example, why is it that some radioisotopes, like uranium-238, are found in nature, whereas others are not and must be synthesized? The key to this question is the fact that different nuclei undergo decay at different speeds. Uranium-238 undergoes decay very slowly, whereas many other nuclei such as sulfur-35 decay rapidly. To better understand the phenomenon of radioactivity, it is important to consider the rates of radioactive decay.

Radioactive decay is a first-order process. As shown in Section 13.3, first-order processes have characteristic half-lives. The half-life is the time required for half of any given quantity of a substance to react. The rates of decay of nuclei are commonly discussed in terms of their half-lives.

Each isotope has its own characteristic half-life. For example, the half-life of strontium-90 is 29 yr. If we started with 10.0 g of strontium-90, only 5.0 g of that isotope would remain after 29 yr. The other half of the strontium-90 would have been converted to yttrium-90 as shown in Equation [20.17]:

$$^{90}_{38}\text{Sr} \longrightarrow ^{90}_{39}\text{Y} + ^{0}_{-1}\text{e} \qquad \text{[20.17]}$$

After another 29-yr period half of the remaining 5.0 g of strontium-90 would likewise decay. The loss of strontium-90 as a function of time is shown in Figure 20.5.

Half-lives as short as millionths of a second and as long as billions of years have been observed. The half-lives of some important radioisotopes are listed in Table 20.2. One important feature of half-lives for nuclear decay is that they are unaffected by external conditions such as tempera-

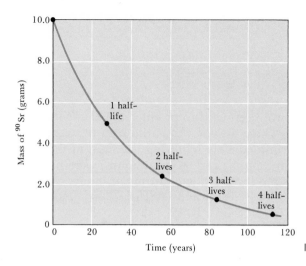

FIGURE 20.5 The decay of a 10.0-g sample of $^{90}_{38}Sr$ ($t_{\frac{1}{2}}$ = 28.8 yr).

ture, pressure, or state of chemical combination. Therefore, unlike chemical toxins, radioactive atoms cannot be rendered harmless by chemical reaction or by any other practical treatment. As far as we know now our only choice is merely to allow these nuclei to lose activity at their own characteristic rates. In the meantime, of course, we must take precautions to isolate the radioisotopes as much as possible because of the damage radiation can cause (see Section 20.7).

SAMPLE EXERCISE 20.6

The half-life of cobalt-60 is 5.3 yr. How much of a 1.000-mg sample of cobalt-60 is left after a 15.9-yr period?

Solution: A period of 15.9 yr is three half-lives for cobalt-60. At the end of one half-life, 0.500 mg of cobalt-60 remains; 0.250 mg remains at the end of two half-lives, and 0.125 mg at the end of three half-lives.

DATING

Because the half-life of any particular nuclide is so constant, it can serve as a molecular clock and can be used to determine the ages of different objects. For example, carbon-14 has been used to determine the age of

TABLE 20.2 Some important radioactive isotopes and their half-lives and type of decay

	Isotope	Half-life	Type of decay
Natural radioisotopes	$^{238}_{92}U$	4.5×10^9 yr	Alpha
	$^{235}_{92}U$	7.1×10^8 yr	Alpha
	$^{232}_{90}Th$	1.4×10^{10} yr	Alpha
	$^{40}_{19}K$	1.3×10^9 yr	Alpha
	$^{14}_{6}C$	5700 yr	Beta
Synthetic radioisotopes	$^{239}_{94}Pu$	24,000 yr	Alpha
	$^{137}_{55}Cs$	30 yr	Beta
	$^{90}_{38}Sr$	28.8 yr	Beta
	$^{131}_{53}I$	0.022 yr	Beta

organic materials (see Figure 20.6). The procedure is based on the formation of carbon-14 by neutron capture in the upper atmosphere:

$$^{14}_{7}\text{N} + ^{1}_{0}\text{n} \longrightarrow ^{14}_{6}\text{C} + ^{1}_{1}\text{H}$$ [20.18]

This reaction provides a small but reasonably constant source of carbon-14. The carbon-14 is radioactive, undergoing beta decay with a half-life of 5700 yr:

$$^{14}_{6}\text{C} \longrightarrow ^{14}_{7}\text{N} + ^{0}_{-1}\text{e}$$ [20.19]

In using radiocarbon dating, it is generally assumed that the ratio of carbon-14 to carbon-12 in the atmosphere has been constant for at least 50,000 yr. The carbon-14 is incorporated into carbon dioxide, which is in turn incorporated, through photosynthesis, into more complex carbon-containing molecules within plants. The plants are eaten by animals, and the carbon-14 thereby becomes incorporated within them as well. Because a living plant or animal has a constant intake of carbon compounds, it is able to maintain a ratio of carbon-14 to carbon-12 that is identical with that of the atmosphere. However, once the organism dies, it no longer ingests carbon compounds to replenish the carbon-14 that is lost through radioactive decay. The ratio of carbon-14 to carbon-12 therefore decreases. If the ratio diminishes to half that of the atmosphere, we can conclude that the object is one half-life, or 5700 yr, old. The method cannot be used to date objects that are over about 20,000–50,000 yr old. After this length of time, the radioactivity is too low to be measured accurately. The radiocarbon-dating technique has been checked by comparing the ages of trees determined by counting their

FIGURE 20.6 Ancient manuscripts and other cultural artifacts made of organic materials are often dated by means of carbon-14 analysis.

rings and by radiocarbon analysis. As the tree grows, it adds a ring each year. The old growth no longer replenishes the supply of carbon. Thus the carbon-14 decays while the concentration of carbon-12 remains constant. Most of the wood used in these tests was from California bristlecone pines, which reach ages up to 2000 yr. By using trees that died at a known time thousands of years ago, it is possible to make comparisons back to about 5000 B.C. The two dating methods agree to within about 10 percent.

Other isotopes can be similarly used to date other types of objects. For example, it takes 4.5×10^9 yr for half of a sample of uranium-238 to decay to a stable product, lead-206. The age of rocks containing uranium can be determined by measuring the ratio of lead-206 to uranium-238. If the lead-206 had somehow become incorporated into the rock by normal chemical processes instead of by radioactive decay, the rock would contain large amounts of the more abundant isotope lead-208. In the absence of large amounts of this "geonormal" isotope of lead, it is assumed that all of the lead-206 was at one time uranium-238.

The oldest rocks found on the earth are approximately 3×10^9 yr old. This age indicates that the crust of the earth has been solid for at least this length of time. Prior to the crystallization of rock, lead-206 and uranium-238 could separate. It is estimated that it required $1-1.5 \times 10^9$ yr for the earth to cool and its surface to become solid. This places the age of earth at $4.0-4.5 \times 10^9$ yr.

CALCULATIONS BASED ON HALF-LIFE

So far our discussion has been mainly qualitative. We now consider the topic of half-lives from a more quantitative point of view. This permits us to answer questions of the following types: How do we determine the half-life of uranium-238? We certainly don't sit around for 4.5×10^9 yr waiting for half of it to decay! Similarly, how do we quantitatively determine the age of an object that can be dated by radiometric means?

The rate of radioactive decay of any radioisotope is first order. It can therefore be described by Equation [20.20],

$$\text{Rate} = kN \qquad\qquad [20.20]$$

where N is the number of nuclei of a particular radioisotope, and k is the first-order rate constant. This equation can be transformed into Equation [20.21]. (Note that the equations that follow are the same as those we encountered in Section 13.3, in dealing with first-order chemical processes.)

$$k = \frac{2.30}{t} \log \frac{N_0}{N_t} \qquad\qquad [20.21]$$

where t is the time interval during which decay is measured, k is the rate constant, N_0 is the initial number of nuclei (at zero time), and N_t is the number remaining after the time interval. The relationship between the rate constant, k, and half-life, $t_{\frac{1}{2}}$, is given by Equation [20.22]:

$$k = \frac{0.693}{t_{\frac{1}{2}}}$$

[20.22]

If Equation [20.22] is substituted into Equation [20.21], the following relationship results:

$$\frac{0.693}{t_{\frac{1}{2}}} = \frac{2.30}{t} \log \frac{N_0}{N_t}$$

[20.23]

Rearranging this expression gives Equation [20.24]:

$$t = 3.32 t_{\frac{1}{2}} \log \frac{N_0}{N_t}$$

[20.24]

This equation is used in performing calculations such as those shown in Sample Exercises 20.7 and 20.8.

SAMPLE EXERCISE 20.7

A rock contains 0.257 mg of lead-206 for every milligram of uranium-238. How old is the rock?

Solution: The amount of uranium-238 in the rock when it was first formed is assumed to be the 1.000 mg present at the time of the analysis plus the quantity that decayed to lead-206.

Original $^{238}_{92}U = 1.000$ mg $+ \frac{238}{206}(0.257$ mg$)$

$= 1.297$ mg

Using Equation [20.24] and the half-life of $^{238}_{92}U$, 4.5×10^9 yr, gives the following result:

$t = (3.32)(4.5 \times 10^9 \text{ yr}) \log \frac{1.297}{1.000}$

$= (3.32)(4.5 \times 10^9 \text{ yr})(0.113)$

$= 1.7 \times 10^9$ yr

SAMPLE EXERCISE 20.8

If we start with 1.000 g of strontium-90, 0.953 g will still remain after 2.00 yr. (a) What is the half-life of strontium-90? (b) How much strontium-90 would remain after 5.00 yr?

Solution: (a) Equation [20.24] can be solved for $t_{\frac{1}{2}}$ using $N_0 = 1.000$ g, $N_t = 0.953$ g, and $t = 2.00$ yr.

$t_{\frac{1}{2}} = \dfrac{t}{3.32 \log \dfrac{N_0}{N_t}}$

$= \dfrac{2.00 \text{ yr}}{3.32 \log \dfrac{1.000}{0.953}} = \dfrac{2.00 \text{ yr}}{3.32 \log 1.049}$

$= \dfrac{2.00 \text{ yr}}{3.32(0.0209)} = 28.8$ yr

(b) Again using Equation [20.24] we have

$\log \dfrac{N_0}{N_t} = \dfrac{t}{3.32 t_{\frac{1}{2}}}$

$= \dfrac{5.00 \text{ yr}}{(3.32)(28.8 \text{ yr})} = 0.0522$

$\dfrac{N_0}{N_t} = 1.13$

$N_t = \dfrac{N_0}{1.13} = \dfrac{1.000 \text{ g}}{1.13} = 0.885$ g

A variety of methods have been devised to detect emissions from radioactive substances. Becquerel discovered radioactivity because of the effect of radiation on photographic plates. Photographic plates and film have long been used to detect radioactivity. The radiation affects photographic film in the same way as ordinary light. With care, film can be used to give a quantitative measure of activity. The greater the extent of exposure to radiation, the darker the area of the developed negative. People who work with radioactive substances carry film badges to record the extent of their exposure to radiation.

Radioactivity can also be detected and measured using a device known as a Geiger counter. The operation of the Geiger counter is based on the ionization of matter caused by radiation (Section 20.7). The ions and electrons produced by the ionizing radiation permit conduction of an electrical current. The basic design of a Geiger counter is shown in Figure 20.7. It consists of a metal tube filled with gas. The cylinder has a "window" made of material that can be penetrated by alpha, beta, or gamma rays. In the center of the tube is a wire. The wire is connected to one terminal of a source of direct current and the metal cylinder is attached to the other terminal. Current flows between the wire and metal cylinder whenever ions are produced by entering radiation. The current pulse created when radiation enters the tube is amplified; each pulse is counted as a measure of the amount of radiation.

Certain substances that are electronically excited by radiation can also be used as means for detecting and measuring radiation. Some substances excited by radiation give off light (fluoresce) as electrons return to their lower energy states. For example, dials of luminous watches are painted with a mixture of ZnS and a tiny quantity of $RaSO_4$. The zinc sulfide fluoresces when struck by the radioactive emissions from the radium. An instrument known as a scintillation counter can be used to detect and measure fluorescence and thereby the radiation that causes it.

RADIOTRACERS

Because radioisotopes can be detected so readily, they can be used to follow an element through its chemical reactions. For example, the incorporation of carbon atoms from CO_2 into glucose in photosynthesis, Equation [20.25], has been studied using CO_2 containing carbon-14:

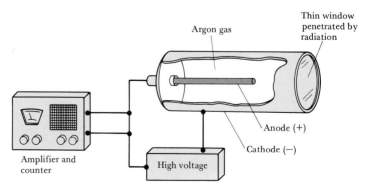

FIGURE 20.7 Schematic representation of a Geiger counter.

FIGURE 20.8 Rosalyn S. Yalow was born in New York City in 1921. She is a graduate of Hunter College in New York and received her Ph.D. in physics from the University of Illinois (Urbana) in 1945. For the past several years she has carried out research in medical physics and is chief of the Nuclear Medicine Service at the Veterans Administration Hospital in Bronx, New York. Among her many awards and recognitions is the Nobel Prize in medicine and physiology, which she was awarded in 1977.

$$6CO_2 + 6H_2O \xrightarrow[\text{chlorophyll}]{\text{sunlight}} C_6H_{12}O_6 + 6O_2 \qquad\qquad [20.25]$$

The CO_2 is said to be labeled with the carbon-14. Detection devices such as Geiger counters can then be used to follow the carbon-14 as it moves from the CO_2 through the various intermediate compounds to glucose.

Such use of radioisotopes is made possible by the fact that all isotopes of an element have essentially identical chemical properties. If a small quantity of a radioisotope is mixed with the naturally occurring stable isotopes of the same element, all of the isotopes will go through the same reactions together. Where the element goes is then revealed by the radioactivity of the radioisotope. Because the radioisotope can be used to trace the path of the element, it is called a radiotracer.

Radiotracers have found wide use as a diagnostic tool in medicine. For example, iodine-131 has been used to test the activity of the thyroid gland. This gland is the only important user of iodine in the body. A solution of NaI containing a small quantity of iodine-131 is drunk by the patient. The ability of the thyroid to take up the iodine is determined using a Geiger tube placed close to the thyroid, in the neck region. A normal thyroid will absorb about 12 percent of the iodine within a few hours. When radioisotopes are used in this manner, only a very small amount is used so that the patient does not receive a harmful dose of radioactivity.

In 1977 Rosalyn S. Yalow (Figure 20.8) received the Nobel Prize in medicine for her pioneering work in developing the technique of radioimmunoassay (RIA). This technique is an extraordinarily sensitive method involving radiotracers for detecting the presence of minute amounts of drugs, hormones, peptides, antibiotics, and a host of other substances. RIA methods can be used, for example, to provide an early indication of pregnancy or to detect substances that signal the early stages of a disease. The tests are carried out on a small sample of tissue, blood, or other fluid taken from the subject.

20.6 Mass-energy conversions

So far we have said little about the energies associated with nuclear reactions. These energies can be considered with aid of Einstein's famous equation relating mass and energy:

$$E = mc^2 \qquad [20.26]$$

In this equation E stands for energy, m for mass, and c for the speed of light, 3.00×10^8 m/sec. This equation states that the mass and energy of an object are proportional. The greater an object's mass, the greater its energy. Because the proportionality constant in the equation, c^2, is such a large number, small changes in mass are accompanied by large changes in energy.

The mass changes accompanying chemical reactions are too small to detect. For this reason, it is possible to speak of the conservation of mass in chemical reactions. The calculation of the loss of mass that accompanies the combustion of a mole of CH_4 illustrates this point. This calculation is shown in Sample Exercise 20.9.

SAMPLE EXERCISE 20.9

Calculate the mass that disappears from the system when a mole of CH_4 is combusted:

$$CH_4(g) + 2O_2(g) \longrightarrow CO_2(g) + 2H_2O(l)$$

The system loses 890 kJ of energy in this process ($\Delta E = -890$ kJ).

Solution: From Equation [20.26] we can write that the change in mass, Δm, in the system is proportional to its change in energy, ΔE:

$$\Delta E = c^2 \, \Delta m$$

Rearranging to solve for Δm we have

$$\Delta m = \frac{\Delta E}{c^2}$$

Substituting the value given for ΔE and remembering that a negative sign is associated with an exothermic process, we have:

$$\Delta m = \frac{-890 \text{ kJ}}{(3.00 \times 10^8 \text{ m/sec})^2}\left(\frac{1000 \text{ J}}{1 \text{ kJ}}\right)$$
$$\times \left(\frac{1 \text{ kg-m}^2/\text{sec}^2}{1 \text{ J}}\right)$$
$$= -9.89 \times 10^{-12} \text{ kg}$$

The negative sign indicates mass loss. This mass change is far too small to detect.

The mass changes and the associated energy changes in nuclear reactions are much greater than in chemical reactions. The energy released through the nuclear fission of only about a pound of uranium (Section 20.8) is equivalent to that released by combustion of 1500 tons of coal.

NUCLEAR BINDING ENERGIES

It was discovered in the 1930s that the masses of nuclei are always less than the masses of the individual nucleons of which they are composed. For example, the helium-4 nucleus has a mass of 4.00150 amu. The mass of a proton is 1.00728 amu, while that of a neutron is 1.00867 amu. Consequently, two protons and two neutrons have a total mass of 4.03190 amu:

Mass of two protons $= 2(1.00728 \text{ amu}) = 2.01456$ amu
Mass of two neutrons $= 2(1.00867 \text{ amu}) = \underline{2.01734 \text{ amu}}$
Total mass $= 4.03190$ amu

The individual nucleons weigh 0.03040 amu more than the helium-4 nucleus:

$$\text{Mass of two protons and two neutrons} = 4.03190 \text{ amu}$$
$$\text{Mass of } {}^{4}_{2}\text{He nucleus} = 4.00150 \text{ amu}$$
$$\text{Mass difference} = 0.03040 \text{ amu}$$

The mass difference between the helium-4 nucleus and the two protons and two neutrons is known as the **mass defect** of the nucleus. This mass is lost in the form of energy:

$$2{}^{1}_{1}\text{p} + 2{}^{1}_{0}\text{n} \longrightarrow {}^{4}_{2}\text{He} + \text{energy} \qquad\qquad [20.27]$$

The amount of energy produced in this reaction can be calculated from the mass difference:

$$\Delta E = c^2 \, \Delta m$$
$$= (3.00 \times 10^8 \text{ m/sec})^2(-0.0304 \text{ amu})$$
$$\times \left(\frac{1.00 \text{ g}}{6.02 \times 10^{23} \text{ amu}}\right)\left(\frac{1 \text{ kg}}{1000 \text{ g}}\right)$$
$$= -4.52 \times 10^{-12} \frac{\text{kg-m}^2}{\text{sec}^2} = -4.52 \times 10^{-12} \text{ J}$$

Formation of a mole of helium-4 nuclei in this fashion would produce a tremendous quantity of energy:

$$(6.02 \times 10^{23})(4.52 \times 10^{-12} \text{ J}) = 2.72 \times 10^{12} \text{ J}$$

If we turn Equation [20.27] around, we see that 4.52×10^{-12} J would be required to break a single helium-4 nucleus into protons and neutrons. The energy calculated from the mass defect is therefore a measure of the stability of the nucleus toward decomposition into individual nucleons. The energy required to decompose a nucleus into protons and neutrons is referred to as the **binding energy** of the nucleus.

The binding energy of any nucleus can be calculated from its mass and from the masses of the nucleons of which it is composed. Insight into the relative stabilities of nuclei is gained by comparing the binding energy per nucleon for different nuclei. Three nuclei (helium-4, iron-56, and uranium-238) are compared in Table 20.3. Similar calculations for other nuclei indicate that the binding energy per nucleon increases in magnitude to about 1.4×10^{-12} J with nuclei whose mass numbers are in the vicinity of iron-56. It then decreases slowly to about 1.2×10^{-12} J for very heavy nuclei. This trend is shown in Figure 20.9. These results

TABLE 20.3 Mass differences and binding energies for three nuclei

Nucleus	Mass of nucleus (amu)	Mass of individual nucleons (amu)	Mass difference (amu)	Binding energy (J)	Binding energy per nucleon (J)
${}^{4}_{2}\text{He}$	4.00150	4.03190	0.0304	4.52×10^{-12}	1.13×10^{-12}
${}^{56}_{26}\text{Fe}$	55.92066	56.44938	0.52872	7.90×10^{-11}	1.41×10^{-12}
${}^{238}_{92}\text{U}$	238.0003	239.9356	1.9353	2.89×10^{-10}	1.22×10^{-12}

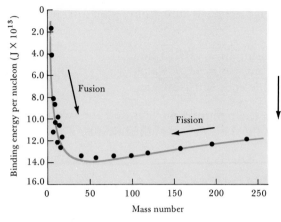

FIGURE 20.9 The average binding energy per nucleon increases to a maximum at a mass number of 50–60 and decreases slowly thereafter. As a result of these trends, fusion of light nuclei and fission of heavy nuclei are exothermic processes.

indicate that heavy nuclei will gain stability, and therefore give off energy, if they are fragmented. This process, known as fission, occurs in the atomic bomb and in nuclear power plants. Figure 20.9 also indicates that even greater amounts of energy should be available if very light nuclei are fused together. Such fusion reactions take place in the hydrogen bomb and are the essential energy-producing reactions in the sun. We shall look more closely at fission and fusion in Sections 20.8 and 20.9.

In calculating the energy changes that occur in a nuclear reaction we need to refer to tables of nuclear properties to determine the masses of the nuclear species involved. Generally the mass of a nuclide is expressed in tables as the mass of an *atom* containing the nucleus of interest. That is, the mass given includes the mass of the electrons surrounding the nucleus. Thus, for example, the mass of $^{56}_{26}Fe$ is given in a table of nuclides as 55.93494 amu. This includes the mass of 26 electrons. When the mass of the 26 electrons is subtracted we obtain the mass of the nucleus listed in Table 20.3. In calculating the mass change in a nuclear reaction, it is acceptable to employ the masses of the atoms containing the nuclides of interest, because the electrons are actually present in the nuclear reaction, and we are concerned with the overall mass change. However, we recognize that essentially all of the mass *change* is due to changes in nuclear binding energies.

SAMPLE EXERCISE 20.10

How much energy is lost or gained when a mole of cobalt-60 undergoes beta decay: $^{60}_{27}Co \longrightarrow {}^{0}_{-1}e + {}^{60}_{28}Ni$? The mass of the $^{60}_{27}Co$ atom is 59.9338 amu, that of a $^{60}_{28}Ni$ atom is 59.9308 amu, and that of an electron, $^{0}_{-1}e$, is 0.000549 amu.

Solution: The products of the nuclear reaction are a $^{60}_{28}Ni^+$ ion and an electron. The mass of these products is just the mass of the neutral $^{60}_{28}Ni$ atom. The mass lost in the reaction is 0.0030 g:

$$
\begin{array}{r}
59.9338 \text{ g} \\
-59.9308 \text{ g} \\
\hline
0.0030 \text{ g}
\end{array}
$$

The energy produced by the reaction can be calculated from this mass:

$$\Delta E = c^2 \, \Delta m$$

$$= (3.00 \times 10^8 \text{ m/sec})^2(-0.0030 \text{ g})\left(\frac{1 \text{ kg}}{1000 \text{ g}}\right)$$

$$= -2.7 \times 10^{11} \frac{\text{kg-m}^2}{\text{sec}^2} = -2.7 \times 10^{11} \text{ J}$$

By comparison, it takes only 9×10^5 J to break all the chemical bonds in a mole of water.

The increased pace of synthesis and use of radioisotopes has led to increased concern over the effects of radiation on matter, particularly in biological systems. It is therefore important for us to consider the health hazards associated with radioisotopes.

The biological damage caused by radiation is due to its ability to ionize and fragment molecules. The alpha, beta, or gamma rays emitted in the course of a nuclear decay process possess energies far in excess of ordinary chemical bond energies. As these forms of radiation move through matter they give up energy to the molecules they encounter, thereby leaving a trail of ions and molecular fragments. The substances produced in this fashion are very reactive. In a biological system they may be able to disrupt the normal operations of cells. The damage caused by a radiation source outside the body depends on the penetrating ability of the radiation. Gamma rays are particularly dangerous because they penetrate human tissue very effectively, just as do X rays. Consequently, their damage is not limited to the skin. In contrast, most alpha rays are stopped by skin, and beta rays are able to penetrate only about 1 cm beyond the surface of the skin. Hence, neither is as dangerous as gamma rays unless the radiation source somehow enters the body. Within the body, alpha rays are particularly dangerous because they leave a very dense trail of damaged molecules as they move through matter.

The chemical form of a radioisotope determines how easily it can enter an organism. It also determines whether the radioisotope will be retained in the body and, if so, where. Krypton-85 and strontium-90 illustrate these points. Krypton-85 is formed in nuclear fission and is released into the atmosphere during the reprocessing of nuclear fuels. Because krypton is chemically inert, no simple way of chemically immobilizing it has been devised. Once it is out in the atmosphere, the krypton-85 affects the skin and lungs. However, its lack of reactivity keeps it from getting elsewhere into the body or accumulating there. Strontium-90 is also formed by fission. Because it is an alkaline earth, strontium is able to replace calcium in its compounds. It is therefore able to enter bones where its radiation can cause bone cancer and leukemia.

Another important factor to consider is "biological concentration" (Section 17.6). As a substance passes from organism to organism through a food chain, it can become increasingly concentrated. For example, the concentration of strontium-90 may be very low in grass. However, a cow eats a great deal of it. Consequently, the concentration of strontium-90 becomes greater in the cow than in the grass. A person who then drinks milk taken from several cows could in turn end up with a concentration of strontium-90 that is higher than that found in the cows.

The effects of radiation on living systems can be classified as either somatic or genetic. Somatic damage is that which affects the organism during its own lifetime. Genetic damage is that which has a genetic effect; it harms offspring through damage to genes and chromosomes, the body's reproductive material. It is more difficult to study genetic effects than somatic ones because it may take several generations for genetic damage to become apparent. Somatic damage includes "burns," molecular disruptions similar to those produced by high temperatures. It also includes cancer. Cancer is brought about by damage to the growth-

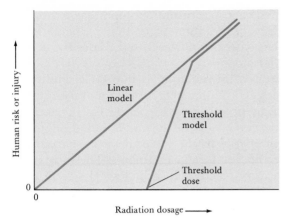

FIGURE 20.10 Comparison of two models of radiation damage. According to the linear model, any radiation dosage produces some finite risk of injury, even at low dosage. According to the threshold model, there is no risk of injury if the dosage is kept below a certain level called the threshold level.

regulation mechanism of cells, which causes them to reproduce in an uncontrolled manner. In general, the tissues that show the greatest damage from radiation are those that reproduce at a rapid rate, such as bone marrow, blood-forming tissues, and lymph nodes. Leukemia is probably the major cancer problem associated with radiation.

The clinical symptoms of acute (short-term) exposure to radiation include a decrease in the number of white blood cells, fatigue, nausea, and diarrhea. Sufficient exposure can result in death from blood disorders, gastrointestinal failure, and damage to the central nervous system. In light of these effects, it is important to determine whether there are any safe levels of exposure to radiation. What are the maximum levels of radiation that we should permit from various human activities? Unfortunately, we are hampered in our attempts to set realistic standards by our lack of understanding of the effects of chronic (long-term) exposure to radiation. Two models have been proposed for such exposure. According to one of these, the linear model, the effects of radiation are proportional to exposure, even down to low exposures. According to the other, the threshold model, no effects occur below a certain threshold of exposure. The two models are compared in Figure 20.10. Today most scientists concerned with radiation effects accept the linear hypothesis. This means that *any* amount of radiation causes some finite risk of injury. However, as we shall see shortly, it is important to keep in mind the relative magnitudes of various sources of radiation.

RADIATION DOSES

A gram of radium undergoes 3.7×10^{10} nuclear disintegrations in 1 sec. This number of disintegrations is known as a curie (Ci). The curie is the unit used in measuring nuclear activity. For example, a 5 millicurie sample of cobalt-60 undergoes $(5 \times 10^{-3})(3.7 \times 10^{10}) = 1.8 \times 10^8$ disintegrations per second.

The damage produced by radiation depends not only on the nuclear activity but also on the energy and penetrating power of the radiation. Therefore two other units, the rad and the rem, are commonly used to measure radiation doses. (A third unit, the roentgen, is essentially the same as the rad.) A rad (radiation *a*bsorbed *d*ose) is the amount of radiation that deposits 1×10^{-2} J of energy per kilogram of tissue. A rad of

TABLE 20.4 Effects of short-time exposures to radiation

Dose (rem)	Effect
0–25	No detectable clinical effects
25–50	Slight, temporary decrease in white blood cell counts
100–200	Nausea; marked decrease in white blood cells
500	Death of half the exposed population within 30 days after exposure

alpha rays can produce more damage than a rad of beta rays. Consequently, the rad is often multiplied by a factor that measures the relative biological damage caused by the radiation. This factor is known as the relative biological effectiveness of the radiation, abbreviated RBE. The RBE is approximately one for beta and gamma rays and ten for alpha rays. The exact value for the RBE varies with dose rate, total dose, and type of tissue affected. The product of the number of rads and the RBE of the radiation gives the effective dosage in **rems** (*r*oentgen *e*quivalent for *m*an):

$$\text{Number of rems} = (\text{number of rads})(\text{RBE}) \qquad [20.28]$$

The effects of some short-time exposures to radiation are given in Table 20.4. For most persons the effects of long-term exposure to low levels of radiation are more important. Each of us receives an average exposure of 0.1 to 0.2 rem each year as background radiation from natural sources, such as naturally occurring radioisotopes, and cosmic rays coming in from outer space. This amount may vary widely from one person to the next, depending on living situation. For example, persons who live in stone or brick houses normally receive more exposure than those who live in wooden houses. Those who live at higher elevation receive more cosmic ray exposure, because the earth's atmosphere absorbs cosmic rays to some extent.

In addition to natural sources, we all receive some exposure to radiation resulting from human activity. Table 20.5 shows radiation levels and the percent contributions of the more important sources to the average exposure of the U.S. population. It must be emphasized that these are merely averages; the relative contributions of the various sources will vary widely from one person to another. For example, a chest X ray

TABLE 20.5 Contributions to the average exposure of the U.S. population to ionizing radiation

Source	Percent contribution	Radiation level (millirem)
Natural background	52	100
X rays, other medical	43	80
Uranium mining, other technology-related sources	2	4
Fallout from nuclear weapons testing	3	6
Nuclear power plants	0.14	0.3
Consumer products— watch dials, color TV, etc.	0.02	0.04

involves a radiation dose of 20 to 40 millirem. Some people have never been X-rayed, others have had many exposures. Uranium miners clearly receive larger exposures than the general populace; similarly, those who live near a nuclear power plant may receive more exposure from this source than the average population.

Radiation exposure limits are presently set by the Environmental Protection Agency at 500 millirem each year for the general population and 5 rem for occupational exposure, exclusive of background radiation. As more scientists have become convinced that the linear hypothesis relating radiation damage to exposure is correct, there has been increasing pressure to set tougher standards for radiation exposure. As with many other aspects of human activity, the question at issue is one of risk versus benefit. Before we can make intelligent decisions we must have a deeper understanding of the risks than we do at present. In the next section we discuss a very controversial example of the risk–benefit debate, the operation of nuclear power plants and the associated problems of radioactive waste disposal.

20.8 Nuclear fission

Commercial nuclear power plants and the most common forms of nuclear weaponry depend for their operation on the process of nuclear fission. The first nuclear fission to be discovered was that of uranium-235. This nucleus, as well as those of uranium-233 and plutonium-239, undergoes fission when struck by slow-moving neutrons.* The fission process is illustrated in Figure 20.11. A heavy nucleus can split in many different ways, just as can a piece of glass. Two different ways that the uranium-235 nucleus splits are shown in Equations [20.29] and [20.30]:

$$\begin{array}{r}\nearrow\ {}^{137}_{52}\text{Te} + {}^{97}_{40}\text{Zr} + 2{}^{1}_{0}\text{n} \qquad [20.29]\\ {}^{1}_{0}\text{n} + {}^{235}_{92}\text{U} \\ \searrow\ {}^{142}_{56}\text{Ba} + {}^{91}_{36}\text{Kr} + 3{}^{1}_{0}\text{n} \qquad [20.30]\end{array}$$

Over 200 different isotopes of 35 different elements have been found among the fission products of uranium-235. In general, these are radioactive. As we have seen in the previous section, these radioactive products are hazardous.

On the average, 2.4 neutrons are produced by every fission of uranium-235. If one fission produces two neutrons, these two neutrons can cause two fissions. The four neutrons thereby released produce four fissions, and so forth as shown in Figure 20.12. The number of fissions and

*There are other heavy nuclei that can be induced to undergo fission. However, these three are the only ones of practical importance.

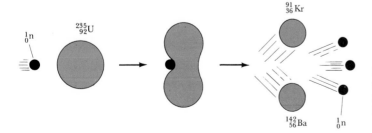

FIGURE 20.11 Schematic representation of the fission of uranium-235 showing one of its many fission patterns. In this process, 3.5×10^{-11} J of energy is produced.

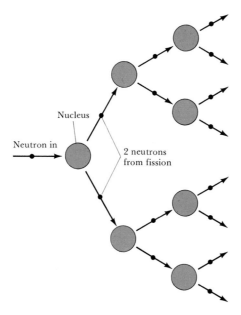

FIGURE 20.12 A chain fission reaction in which each fission produces two neutrons. The process leads to an accelerating rate of fission, with the number of fissions doubling at each stage.

their associated energies quickly escalate, and, if unchecked, the result is a violent explosion. Rections that multiply in this fashion are called branching chain reactions.

In order for a chain fission reaction to occur, the sample of fissionable material must have a minimum size. Otherwise, neutrons escape from the sample before they have an opportunity to strike a nucleus and cause fission. The chain stops if enough neutrons are lost. The reaction is then said to be subcritical. If the mass is large enough to maintain the chain reaction with a constant rate of fission, the reaction is said to be critical. This situation results if only one neutron from each fission is subsequently effective in producing another fission. If the mass is larger still, few of the neutrons produced are able to escape. The chain reaction then multiplies the number of fissions and the reaction is said to be supercritical. The effect of mass size on whether a reaction is subcritical, critical,

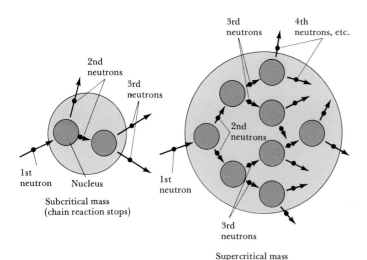

FIGURE 20.13 The chain reaction in a subcritical mass soon stops because neutrons are lost from the mass without causing fission. As the size of the mass increases, fewer neutrons are able to escape, and the chain reaction is able to expand.

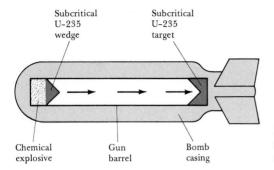

Subcritical
U-235
wedge

Subcritical
U-235
target

Chemical
explosive

Gun
barrel

Bomb
casing

FIGURE 20.14 One design used in atomic bombs. A conventional explosive is used to bring two subcritical masses together to form a supercritical mass.

or supercritical is illustrated in Figure 20.13. One of the ways that an atomic bomb is triggered to produce a supercritical mass is shown in Figure 20.14. As shown, two subcritical masses are brought together by use of chemical explosives to form a supercritical mass. We have seen that the basic design of such a bomb is quite simple. The fissionable materials are potentially available to any nation with a nuclear reactor. This simplicity has already resulted in the proliferation of atomic weapons. It is possible that such weapons could come into the hands of terrorist groups or groups intent on using the weapons to extort money from corporations or nations. Such events would add a frightening aspect to the nuclear age in which we live.

The fission of uranium-235 was first achieved in the late 1930s by Enrico Fermi and his colleagues in Rome and shortly thereafter by Otto Hahn and his co-workers in Berlin. Both groups were trying to produce transuranium elements. In 1938 Hahn identified barium among his reaction products. He was puzzled by this observation and questioned the identification because the presence of barium was so unexpected. He sent a detailed letter describing his experiments to Lise Meitner, a former co-worker. Meitner had been forced to leave Germany because of the anti-Semitism of the Third Reich, and she had settled in Sweden. She surmised that Hahn's experiment indicated that a new nuclear process was occurring in which the uranium-235 split. She called this nuclear fission. Meitner passed word of this discovery to her nephew, Otto Frisch, a physicist working at Niels Bohr's institute in Copenhagen. He repeated the experiment, verifying Hahn's observations and finding that tremendous energies were involved. In January 1939, Meitner and Frisch published a short article describing this new reaction. In March 1939, Leo Szilard and Walter Zinn at Columbia University discovered that more neutrons are produced than were used in each fission. As we have seen, this allows a branching chain reaction. News of these discoveries and an awareness of their potential use in explosive devices spread rapidly within the scientific community. Several scientists finally persuaded Albert Einstein, the most famous physicist of the time, to write a letter to President Roosevelt outlining the implications of these discoveries. Einstein's letter, written in August 1939, outlined the possible military applications of nuclear fission and emphasized the danger that weapons based on fission would pose if they were to be developed by the Nazis. Roosevelt judged it imperative that the United States investigate the possibility of such weapons. Late in 1941 the decision was made to build a bomb based on the fission reaction. An enormous research project, known as the "Manhattan Project," was initiated. This project led to the development of the atomic bomb and the dawning of the nuclear age.

NUCLEAR REACTORS

Nuclear fission produces the energy generated by nuclear power plants. The "fuel" of the nuclear reactor is therefore a fissionable substance such as uranium-235. Typically, uranium is enriched to about 3 percent ura-

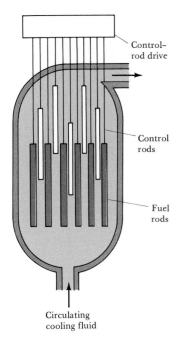

Control-
rod drive

Control
rods

Fuel
rods

Circulating
cooling fluid

FIGURE 20.15 Reactor core showing fuel elements, control rods, and cooling fluid.

nium-235 and then used in the form of UO$_2$ pellets. These enriched uranium-pellets are encased in zirconium or stainless-steel tubes. Rods composed of materials such as cadmium or boron are used to contol the fission process by absorbing neutrons. These **control rods** permit a sufficient flux of neutrons to keep the reaction chain self-sustaining but yet prevent the reactor core from overheating.* The reactor is started by using a neutron source; it is stopped by inserting the control rods more deeply into the reactor core, the site of the fission (Figure 20.15). The reactor core also contains a **moderator** that acts to slow down neutrons, so that they can be captured more readily by the fuel. Finally a **cooling liquid** circulates through the reactor core to carry off the heat generated by the nuclear fission. The cooling liquid can also serve as the neutron moderator.

The design of the power plant is basically the same as that of a power plant that burns fossil fuel (except that the burner is replaced by the reactor core). In both instances, steam is used to drive a turbine that is connected to an electrical generator. Because the steam must be condensed, additional cooling water is needed. This water is generally obtained from a large source such as a river or lake. The water is returned to its source at a higher temperature than when it was removed. Power plants are therefore a significant source of thermal pollution. The nuclear power plant design shown in Figure 20.16 is the one that is currently most popular. The primary coolant, which passes through the core, is in a closed system. Subsequent coolants never pass through the reactor core at all. This lessens the chance that radioactive products could escape the core. Additionally, the reactor is surrounded by a concrete shell to shield personnel and nearby residents from radiation.

*The reactor core cannot reach supercritical levels and explode with the violence of an atomic bomb because the concentration of uranium-235 is too low. However, if the core overheats, sufficient damage might be done to release radioactive materials into the environment.

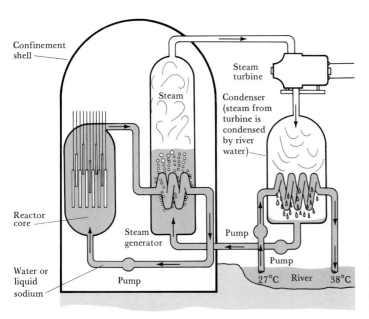

FIGURE 20.16 Basic design of a nuclear power plant. Heat produced by the reactor core is carried by a cooling fluid such as water or liquid sodium to a steam generator. The steam so produced is used to drive an electrical generator.

The nuclear reactor at Three Mile Island in Pennsylvania is basically of the type shown in Figure 20.16. On March 28, 1979, this reactor suffered a highly publicized accident that heightened concern for the safety of nuclear power plants. The accident began with failure of the pump that supplies water to the steam generator. The reactor overheated and shut down automatically. However, the water circulating through the reactor core became overheated. A valve that opened to relieve excess pressure failed to close properly, releasing highly radioactive water onto the floor of the containment structure. The low water level in the reactor caused it to overheat. An operator inadvertently shut off the emergency cooling system, resulting in even more overheating of the fuel rods. Some of the fuel rods burst or partially melted. Before the flow of cooling water was restored, a hydrogen gas bubble, formed by decomposition of water, appeared in the upper part of the reactor. This gas bubble, not anticipated in any of the emergency procedures, might have kept cooling water from reaching the fuel rods. In the absence of any cooling whatever, the rods might have overheated to the point of a meltdown, the so-called "China syndrome." In a meltdown, the molten nuclear fuel would drop through the floor of the containment structure and continue into the earth until it encountered a rock formation and cooled sufficiently to turn to a glassy mass. If the molten fuel struck water, a geyser of radioactive steam could result.

At Three Mile Island a major nuclear accident was averted with no major exposure of the surrounding population to radioactivity. (It was later estimated that persons living near the plant received at most perhaps 100 millirems.) The accident, however, illustrates the potential for disaster inherent in the operation of a complex facility where processes involving high levels of radioactivity are carried out.

Fission products accumulate as the reactor operates. These products lessen the efficiency of the reactor by capturing neutrons. Therefore, the reactor must be stopped periodically, so that the nuclear fuel can be reprocessed. When the fuel rods are removed from the reactor they are initially very radioactive. It was originally intended that they be stored for several months in pools at the reactor site to allow decay of short-lived radioactive nuclei. They were then to be transported in shielded containers to reprocessing plants where the fuel would be separated from the fission products. However, the reprocessing plants have been plagued with operational difficulties, and there is intense opposition to the transport of nuclear wastes on the nation's highways. Even if the transportation difficulties could be overcome, the high radioactivity of the spent fuel makes reprocessing a hazardous operation. At present the spent fuel rods are simply being kept in storage at the reactor site.

In reprocessing, the uranium is separated from nuclear waste products and repackaged into new fuel rods. The problem then becomes one of disposal of the leftover fission products. Storage poses a major problem because the fission products are extremely radioactive. It is estimated that 20 half-lives are required for their radioactivity to reach levels acceptable for biological exposure. Based on the 28.8-yr half-life of strontium-90, one of the longer-lived and most dangerous of the products, the wastes must be stored for 600 yr. If plutonium-239 is not removed, storage must be for longer periods; plutonium-239 has a half-life of 24,000 yr. It is advantageous, however, to remove plutonium-239 because it can be used as a fissionable fuel.

There has been and will continue to be a considerable amount of research devoted to disposal of radioactive wastes. At present the most attractive possibilities appear to be formation of a glass, ceramic, or synthetic rock from the wastes, as a means of immobilizing them. The solid materials could then be buried deep underground. Because the radioactivity will persist for a long time, there must be assurances that the solids will not crack from the heat generated by nuclear decay, possibly allowing radioactivity to find its way into underground water supplies. At present we simply do not have clear-cut answers to all the difficult and important questions relating to nuclear waste disposal.

Although the precise figures are kept secret, experts estimate that from 1000 to 3000 nuclear weapons are added to the U.S. nuclear stockpile each year. The radioactive waste from this military production in a single year far exceeds the total of all the wastes yet produced from operation of nuclear power plants. Production continues despite the fact that the United States already possesses about 30,000 nuclear bombs.

BREEDER REACTORS

Some scientists have estimated that it will take 40 yr or less to use all of the relatively low-cost uranium-235 available in this country. Because uranium-235 is very rare and because its supply can be exhausted so readily, methods of generating other fissionable materials are being actively investigated. Fissionable plutonium-239 and uranium-233 can be produced in nuclear reactors from nuclides that are far more abundant than uranium-235. The reactions involved are as follows:

$$^{238}_{92}U + ^{1}_{0}n \longrightarrow ^{239}_{92}U$$

$$^{239}_{92}U \longrightarrow ^{239}_{93}Np + ^{0}_{-1}e \qquad (t_{\frac{1}{2}} = 24 \text{ min})$$

$$^{239}_{93}Np \longrightarrow ^{239}_{94}Pu + ^{0}_{-1}e \qquad (t_{\frac{1}{2}} = 2.3 \text{ days})$$

$$^{232}_{90}Th + ^{1}_{0}n \longrightarrow ^{233}_{90}Th$$

$$^{233}_{90}Th \longrightarrow ^{233}_{91}Pa + ^{0}_{-1}e \qquad (t_{\frac{1}{2}} = 22 \text{ min})$$

$$^{233}_{91}Pa \longrightarrow ^{233}_{92}U + ^{0}_{-1}e \qquad (t_{\frac{1}{2}} = 27 \text{ days})$$

It is theoretically possible to build a reactor that both produces energy and converts uranium-238 or thorium-232 into fissionable fuel. We can envision a fission of a uranium-235 nucleus producing two neutrons, one causing another fission, the second initiating the change of uranium-238 into plutonium-239. Plutonium-239 is produced in ordinary reactors because of the presence of uranium-238. However, the hope is to be able to produce more fuel than the reactor uses. Reactors able to do this are still under development; they are called breeder reactors.

The design of breeder reactors poses many difficult technical problems. In addition, the breeder reactor program has become the subject of intense political debate. Because both breeder reactors and nuclear fuel reprocessing plants produce plutonium-239, any nation that acquired a breeder reactor or reprocessing technology would have in hand the raw materials for atomic weaponry. Fear of nuclear proliferation or the possibility of thefts of nuclear materials by terrorist groups has resulted in uncertainty as to whether the United States should proceed with breeder reactor development and fuel reprocessing. Development of breeder reactors is, however, progressing in several other countries.

20.9 **Nuclear fusion**

As shown in Section 20.6, energy is produced when light nuclei are fused to form heavier ones. Reactions of this type are responsible for the energy produced by the sun. Spectroscopic studies indicate that the sun is composed of 73 percent H, 26 percent He, and only 1 percent of all other elements, by mass. Among the several fusion processes that are believed to occur are the following:

$$\ce{^1_1H + ^1_1H \longrightarrow ^2_1H + ^0_1e} \tag{20.31}$$

$$\ce{^1_1H + ^2_1H \longrightarrow ^3_2He} \tag{20.32}$$

$$\ce{^3_2He + ^3_2He \longrightarrow ^4_2He + 2^1_1H} \tag{20.33}$$

$$\ce{^3_2H + ^1_1H \longrightarrow ^4_2He + ^0_1e} \tag{20.34}$$

Theories have been proposed for generation of other elements via fusion processes.

Fusion is appealing as an energy source because of the availability of light isotopes and because fusion products are generally not radioactive. It is therefore potentially a cleaner process than is fission. The problem is that high energies are required to overcome the repulsion between nuclei. The required energies are achieved by high temperatures. Fusion reactions are therefore also known as **thermonuclear reactions**. The lowest temperature required for any fusion is that needed to fuse $\ce{^2_1H}$ and $\ce{^3_1H}$ as shown in Equation [20.35]. This reaction requires a temperature of 40,000,000 K:

$$\ce{^2_1H + ^3_1H \longrightarrow ^4_2He + ^1_0n} \tag{20.35}$$

Such high temperatures have been achieved by using an atomic bomb to initiate the fusion process. This is done in the thermonuclear, or hydrogen, bomb. Clearly, this approach is unacceptable for controlled power generation.

Numerous problems must be overcome before fusion becomes a practical energy source. Besides the high temperatures necessary to initiate the reaction, there is the problem of confining the reaction. No known structural material is able to withstand the enormous temperatures necessary for fusion. Research has centered on the use of strong magnetic fields to contain the reaction. Use of powerful lasers to generate the temperatures required for fusion has also been the subject of much recent research. Although there is reason for some degree of optimism, it is impossible to tell if and when the tremendous technical difficulties involving nuclear fusion will be overcome. It is therefore not yet clear whether fusion will ever be a practical source of energy for humankind.

FOR REVIEW

Summary

Certain nuclei are radioactive. Most of these nuclei gain stability by emitting alpha particles ($\ce{^4_2He}$), beta particles ($\ce{^0_{-1}e}$), and/or gamma radiation ($\ce{^0_0\gamma}$). Some nuclei undergo decay by positron ($\ce{^0_1e}$) emission or by electron capture. The neutron-proton ratio is one factor determining nuclear stability. The presence of magic numbers of nucleons and an even number of protons and neutrons are also important. Nuclear transmutations can be induced by bombarding nuclei with charged particles using particle accelerators or with neutrons in a nuclear reactor.

Radioisotopes have characteristic rates of decay. These rates are generally expressed in terms of half-lives. The constant half-lives of nuclides permit their use in dating objects. The ease of detection of radioisotopes also permits their use as tracers, to follow elements through their reactions. Three methods of detection were discussed: use of photographic film, scintillation counters, and Geiger counters.

The energy produced in nuclear reactions is accompanied by measurable losses of mass in accordance with Einstein's relationship, $\Delta E = c^2 \Delta m$. The difference in mass between nuclei and the nucleons of which they are composed is known as the mass defect. The mass defect of a nuclide allows calculation of its nuclear binding energy, the energy required to separate the nucleus into individual nucleons. Examination of binding energies per nucleon reveals that energy is produced when heavy nuclei split (fission) and when light nuclei fuse (fusion).

The radioactivity of a sample is measured in curies. One curie corresponds to 3.7×10^{10} disintegrations per second. The amount of energy deposited in biological tissue by radiation is measured in terms of the rad; one rad corresponds to 1×10^{-2} J per kilogram of tissue. The rem is a more useful measure of the biological damage created by the deposited energy. We receive radiation from both naturally occurring and human sources in roughly equal amounts for the general population. The effects of long-term exposure to low levels of radiation are not completely understood, but the most commonly accepted hypothesis is that the extent of biological damage varies in direct proportion to the level of exposure (the linear hypothesis).

Uranium-235, uranium-233, and plutonium-239 undergo fission when they capture a neutron. The resulting nuclear reaction is a chain reaction. If the reaction maintains a constant rate it is said to be critical. If it slows it is said to be subcritical. In the atomic bomb, subcritical masses are brought together to form a mass that is supercritical. In nuclear reactors the fission is controlled to generate a constant power. The reactor core consists of fissionable fuel, control rods, moderator, and cooling fluid. The nuclear power plant resembles a conventional power plant except that the core replaces the fuel burner. In breeder reactors more nuclear fuel is produced than is used to generate energy. There is concern regarding the safety with which nuclear power plants can be operated. Beyond this, the reprocessing of spent fuel rods and disposal of highly radioactive nuclear wastes are unsolved problems.

Nuclear fusion requires high temperatures because nuclei must have large kinetic energies to overcome their mutual repulsions. It has not yet been possible to generate a controlled fusion process.

Learning goals

Having read and studied this chapter, you should be able to:

1 Write the nuclear symbols for protons, neutrons, electrons, alpha particles, and positrons.

2 Determine the effect of different types of decay on the proton-neutron ratio and predict the type of decay that a nucleus will undergo based on its position relative to the belt of stability.

3 Complete and balance nuclear equations, having been given all but one of the particles involved.

4 Use the half-life of a substance to predict the amount of radioisotope present after a given period of time.

5 Calculate half-life, age of an object, or the remaining amount of radioisotope, having been given any two of these pieces of information.

6 Explain how radioisotopes can be used in dating objects and as radiotracers.

7 Explain how radioactivity is detected, including a description of the basic design of a Geiger counter.

8 Use Einstein's relation, $\Delta E = c^2 \Delta m$, to calculate the energy change or the mass change of a reaction, having been given one of these quantities.

9 Calculate the binding energies of nuclei, having been given their masses and the masses of protons and neutrons.

10 Explain the roles played by the chemical behavior of an isotope and its mode of radioactivity in determining its ability to damage biological systems.

11 Define the units used to describe the level of radioactivity (curie) and to measure the effects of radiation on biological systems (rem and rad).

12 Describe various sources of radiation to which the general population is exposed and indicate the relative contributions of each.

13 Explain what fission and fusion are and what types of nuclei produce energy when undergoing these processes.

14 Describe the design of a nuclear power plant, including an explanation of the role of fuel elements, control rods, moderator, and cooling fluid.

Key terms

Among the more important terms and expressions used for the first time in this chapter are the following:

Alpha particles (Section 20.1) are identical to helium-4 nuclei, consisting of two protons and two neutrons, symbol ^4_2He.

Beta particles (Section 20.2) are energetic electrons emitted from the nucleus, symbol $^0_{-1}\text{e}$.

A **breeder reactor** (Section 20.8) is a nuclear-fission reactor that produces more fissionable fuel than it consumes.

The **binding energy** (Section 20.6) of a nucleus is the energy required to decompose that nucleus into nucleons; it is usually calculated from the **mass defect** of the nucleus.

A **chain reaction** (Section 20.8) is a series of reactions in which one reaction initiates the next.

A **critical mass** (Section 20.8) is the amount of fissionable material necessary to maintain a chain reaction. Smaller masses are said to be **subcritical**, whereas larger ones are **supercritical**.

A **curie** (Section 20.7) is a measure of radioactivity: 1 curie = 3.7×10^{10} nuclear disintegrations per second.

Electron capture (Section 20.2) is a mode of radioactive decay in which an inner-shell orbital electron is captured by the nucleus.

Fission (Section 20.8) is the splitting of a large nucleus into two intermediate-sized ones.

Fusion (Section 20.9) is the joining of small nuclei to form larger ones.

Gamma rays (Section 20.2) are energetic electromagnetic radiation emanating from the nucleus of a radioactive atom.

Half-life (Section 20.3) is the time required for half of a sample of a particular radioisotope to decay.

The **mass defect** (Section 20.6) of a nucleus is the difference between the mass of that nucleus and the total masses of the individual nucleons that it contains.

A **nuclear transmutation** (Section 20.1) is a conversion of one kind of atom to another.

A **positron** (Section 20.2) is a particle with the same mass as an electron but with a positive charge, symbol ^0_1e.

A **rad** (Section 20.7) is a measure of the energy absorbed from radiation by tissue or other biological material; 1 rad = transfer of 1×10^{-2} J of energy per kilogram of material.

A **radioisotope** (Section 20.1) is an isotope that is radioactive; that is, it is undergoing nuclear changes with emission of nuclear radiation.

A **rem** (Section 20.7) is a measure of the biological damage caused by radiation; rems = rads × RBE.

RBE (Section 20.7) (relative biological effectiveness) is an adjustment factor used to convert rads to rems; it accounts for differences in biological effects of different particles having the same energy.

EXERCISES

Nuclear reactions

20.1 Provide a short description or definition of each of the following terms: (a) nucleon; (b) mass number; (c) nuclear transmutation; (d) alpha ray; (e) positron; (f) gamma ray.

20.2 Indicate the number of protons and neutrons in each of the following nuclei: (a) oxygen-17; (b) $^{99}_{42}\text{Mo}$; (c) cesium-136; (d) ^{115}Ag.

20.3 Indicate the number of protons, neutrons, and nucleons in each of the following: (a) cerium-137; (b) ^{234}Pu; (c) cadmium-113; (d) $^{37}_{17}\text{Cl}$.

20.4 Write balanced nuclear equations for the following transformations: (a) ^{181}Hf undergoes beta decay; (b) radium-226 decays to a radon isotope; (c) lead-205 undergoes positron emission; (d) tungsten-179 undergoes orbital electron capture.

20.5 Write balanced nuclear equations for each of the following nuclear transformations: (a) zirconium-93 undergoes beta decay; (b) neptunium-233 undergoes alpha decay; (c) francium-218 is formed by decay of an actinium nuclide; (d) einsteinium-246 undergoes orbital electron capture.

20.6 Complete and balance the following nuclear equations by supplying the missing particle:

(a) $^{32}_{16}\text{S} + ^1_0\text{n} \longrightarrow ^1_1\text{H} + ?$

(b) $^7_4\text{Be} + ^0_{-1}\text{e}$ (orbital electron) $\longrightarrow ?$

(c) $? \longrightarrow ^{187}_{76}\text{Os} + ^0_{-1}\text{e}$

(d) $^{98}_{42}\text{Mo} + ^2_1\text{H} \longrightarrow ^1_0\text{n} + ?$

(e) $^{235}_{92}\text{U} + ^1_0\text{n} \longrightarrow ^{135}_{54}\text{Xe} + ? + 2^1_0\text{n}$

20.7 Complete and balance the following nuclear equations by supplying the missing particle:

(a) $^{252}_{98}\text{Cf} + ^{10}_5\text{B} \longrightarrow 3^1_0\text{n} + ?$

(b) $^2_1\text{H} + ^3_2\text{He} \longrightarrow ^4_2\text{He} + ?$

(c) $^1_1\text{H} + ^{11}_5\text{B} \longrightarrow 3?$

(d) $^{122}_{53}\text{I} \longrightarrow ^{122}_{54}\text{Xe} + ?$

(e) $^{59}_{26}\text{Fe} \longrightarrow ^0_{-1}\text{e} + ?$

20.8 Fermium-250, formed in a nuclear synthesis experiment, decays rather rapidly. The decay chain proceeds through several steps to form $^{234}_{92}\text{U}$, which has a very long half-life. Write the several reactions that lead from $^{250}_{100}\text{Fm}$ to $^{234}_{92}\text{U}$.

20.9 Write balanced equations for the following nuclear reactions: (a) $^{238}_{92}\text{U}(n,\gamma)^{239}_{92}\text{U}$; (b) $^{14}_{7}\text{N}(p,\alpha)^{11}_{6}\text{C}$; (c) $^{18}_{8}\text{O}(n,\beta)^{19}_{9}\text{F}$; (d) $^{59}_{26}\text{Fe}(\alpha,\beta)^{63}_{29}\text{Cu}$.

[20.10] The naturally occurring radioactive decay series that begins with $^{235}_{92}\text{U}$ stops with formation of the stable $^{207}_{82}\text{Pb}$. The decays proceed through a series of alpha particle and beta particle emissions. How many of each type of emission are involved in this series?

Nuclear stability

20.11 Which of the following nuclides would you expect to be radioactive: (a) $^{17}_{8}\text{O}$; (b) $^{176}_{74}\text{W}$; (c) $^{108}_{50}\text{Sn}$; (d) $^{92}_{40}\text{Zr}$; (e) $^{238}_{94}\text{Pu}$? Justify your choices.

20.12 In each of the following pairs, which nuclide would you expect to be the more abundant in nature: (a) $^{19}_{9}\text{F}$ or $^{18}_{9}\text{F}$ (b) $^{80}_{34}\text{Se}$ or $^{81}_{34}\text{Se}$ (c) $^{56}_{26}\text{Fe}$ or $^{57}_{26}\text{Fe}$ (d) $^{118}_{50}\text{Sn}$ or $^{118}_{51}\text{Sb}$? Justify your choice.

20.13 Which of the following nuclides is most likely to be a positron emitter: (a) $^{53}_{24}\text{Cr}$; (b) $^{51}_{25}\text{Mn}$; (c) $^{59}_{26}\text{Fe}$? Explain.

20.14 Indicate whether each of the following nuclides lie within the belt of stability in Figure 20.1: (a) $^{108}_{49}\text{In}$; (b) $^{102}_{47}\text{Ag}$; (c) $^{17}_{7}\text{N}$; (d) $^{210}_{86}\text{Rn}$. If they do not, describe a nuclear decay process that would alter the neutron-to-proton ratio in the direction of increased stability.

20.15 All of the following nuclides are radioactive and undergo either beta or positron emission: (a) $^{66}_{32}\text{Ge}$; (b) $^{105}_{45}\text{Rh}$; (c) $^{137}_{53}\text{I}$; (d) $^{133}_{58}\text{Ce}$. Indicate which nuclei undergo each of these types of emission.

20.16 We have seen that one possible mode of nuclear transformation is orbital electron capture. Which electron in a many-electron atom is most likely to be captured in such a process? How would you expect the likelihood of orbital electron capture to change with atomic number?

20.17 The neutron-to-proton ratio for maximum stability for nuclei with mass numbers in the range 205 to 220 is 1.52. Based on this fact, account for the following observations: (a) thallium-210 undergoes beta decay; (b) actinium-204 undergoes orbital electron capture; (c) bismuth-197 undergoes alpha particle emission.

20.18 Sulfur-35 is radioactive and undergoes beta decay. What differences would you expect in the chemical behavior of atoms containing sulfur-35 as compared with those containing sulfur-32, which is nonradioactive? Explain.

Half-life; dating

20.19 Germanium-66 decays by positron emission, with a half-life of 2.5 hr. Write the equation for the nuclear reaction. How much ^{66}Ge remains from a 25.0-mg sample after 10.0 hr?

20.20 The half-life of tritium (hydrogen-3) is 12.3 yr. If 48.0 mg of tritium is released from a nuclear power plant during the course of an accident, what mass of this nuclide remains after 12.3 yr? After 49.2 yr?

20.21 A sample of the synthetic nuclide curium-243 was prepared. After 1 yr the radioactivity of the sample had declined from 3012 disintegrations per second to 2921

disintegrations per second. What is the half-life of the decay process?

20.22 The half-life for decay of ^{139}Ba is 85 minutes. What mass of ^{139}Ba remains after 16 hr from a sample that originally contained 24.0 μg of the nuclide?

20.23 The half-life of ^{239}Pu is 24,000 yr. What fraction of the ^{239}Pu present in nuclear wastes generated today will be present in the year 3000?

20.24 An experiment was designed to determine whether an aquatic plant absorbed iodide ion from water. Iodine-131 ($t_{\frac{1}{2}} = 8.1$ days) was added as a tracer in the form of iodide ion to a tank containing the plants. The initial activity of a 1.00-μL sample of the water was determined to be 89 counts per minute. After 32 days the level of activity in a 1.00 μL sample was determined to be 5.7 counts per minute. Did the plants absorb iodide from the water?

20.25 A sample of strontium-89 has an initial activity of 4600 counts per minute on a device that measures the level of radioactivity. After exactly 30 days the activity has declined to 3130 counts per minute. What is the half-life for decay of strontium-89?

20.26 The half-life for the process $^{238}\text{U} \longrightarrow {}^{206}\text{Pb}$ is 4.5×10^9 yr. A mineral sample contains 50.0 mg of ^{238}U and 14.0 mg of ^{206}Pb. What is the age of the mineral?

20.27 A wooden artifact from a Chinese temple has a ^{14}C activity of 25.8 counts per minute as compared with an activity of 31.7 counts per minute for a standard of zero age. From the half-life for ^{14}C decay, 5.7×10^3 yr, determine the age of the artifact.

20.28 Potassium-40 decays to argon-40 with a half-life of 1.27×10^9 yr. What is the age of a rock in which the weight ratio of ^{40}Ar to ^{40}K is 3.6?

[20.29] The synthetic ratioisotope technetium-99, which decays by beta emission, is the most widely used isotope in nuclear medicine. The following data were collected on a sample of ^{99}Tc:

Disintegrations per minute	Time (hr)
180	0
130	2.5
104	5.0
77	7.5
59	10.0
46	12.5
24	17.5

Make a graph of these data similar to Figure 20.5 and determine the half-life. (You may wish to make a graph of the log of the disintegration rate versus time; a little rearranging of Equation [20.24] will produce an equation for a linear relation between $\log N_t$ and t; from the slope you can obtain $t_{\frac{1}{2}}$.)

[20.30] A 26.00-g sample of water containing tritium, ^3_1H, emits 1.50×10^3 beta particles per second. Tritium is a weak beta emitter, with a half-life of 12.26 yr. What fraction of all the hydrogen in the water sample is tritium? (Hint: Use Equations [20.20] and [20.22].)

Mass-energy relationships

20.31 Calculate the binding energy per nucleon for the following nuclei: (a) $^{12}_{6}C$ (atomic mass, 12.00000 amu); (b) $^{61}_{28}Ni$ (atomic mass, 60.93106 amu); (c) $^{206}_{82}Pb$ (atomic mass, 205.97447 amu).

20.32 Calculate the total binding energy and the binding energy per nucleon for each of the following nuclei: (a) $^{64}_{30}Zn$ (atomic mass, 63.92914); (b) $^{37}_{17}Cl$ (atomic mass, 36.96590 amu); (c) $^{4}_{2}He$ (atomic mass, 4.00260 amu).

20.33 How much energy must be supplied to break a single $^{6}_{3}Li$ nucleus into separated protons and neutrons if the nucleus has a mass of 6.01347 amu? What does this energy correspond to for a 1 mol of ^{6}Li nuclei?

20.34 What mass change occurs when 1 kg of crystalline silicon is converted to pure silica, $SiO_2(s)$? ΔH of the reaction is -910.8 kJ/mol of Si.

20.35 The solar radiation falling on earth amounts to 1.07×10^{16} kJ/min. What is the mass equivalence of the solar energy falling on earth in a 24-hr period? If the energy released in the reaction

$$^{235}_{92}U + ^{1}_{0}n \longrightarrow ^{141}_{56}Ba + ^{92}_{36}Kr + 3^{1}_{0}n$$

(^{235}U atomic mass, 235.0439 amu; ^{141}Ba atomic mass, 140.9140 amu; ^{92}Kr atomic mass, 91.9218 amu) is taken as typical of that occurring in a nuclear reactor, what mass of uranium-235 is required to equal 0.10 percent of the solar energy that falls on earth in 1 day?

20.36 Based on the following atomic mass values—($^{1}_{1}H$, 1.00782; $^{2}_{1}H$, 2.01410 amu; $^{3}_{1}H$, 3.01605 amu; $^{3}_{2}He$, 3.01603 amu; $^{4}_{2}He$, 4.00260 amu)—and the mass of the neutron given in the text, calculate the energy released in each of the following nuclear reactions, all of which are possibilities for a controlled fusion process:

(a) $^{2}_{1}H + ^{3}_{1}H \longrightarrow ^{4}_{2}He + ^{1}_{0}n$

(b) $^{2}_{1}H + ^{2}_{1}H \longrightarrow ^{3}_{2}He + ^{1}_{0}n$

(c) $^{2}_{1}H + ^{3}_{2}He \longrightarrow ^{4}_{2}He + ^{1}_{1}H$

20.37 It has been proposed that a nuclear reactor be built that derives its energy from fission of calcium-48. Is this a feasible project? Explain.

20.38 Which of the following nuclei is likely to have the largest mass defect per nucleon: (a) $^{59}_{27}Co$; (b) $^{11}_{5}B$; (c) $^{118}_{50}Sn$; (d) $^{243}_{96}Cm$? Explain your answer.

Effects and uses of radioisotopes

20.39 Describe or define the following: (a) rem; (b) curie; (c) moderator; (d) breeder reactor; (e) critical mass.

20.40 Which of the following radioactive nuclides are likely to concentrate in the food chain in the environment: (a) ^{90}Sr; (b) ^{3}H in the form of H_2; (c) ^{85}Kr; (d) ^{133}Ba.

20.41 Compare the characteristics of alpha, beta, and gamma rays as they pertain to the health hazards associated with radioactivity.

20.42 Plutonium-239 emits alpha particles that possess energies of about 5×10^8 kJ/mol. The half-life for the decay is about 24,000 yr. It has been said that the main danger from plutonium is inhalation of plutonium-containing dust. Why is such inhalation, rather than simply exposure to plutonium in the environment, the major concern?

20.43 Why is it not possible for the fuel in a nuclear reactor to blow up, as in an atomic bomb, when the reactor goes critical?

20.44 Explain how one might use radioactive ^{59}Fe (a beta emitter with $t_{\frac{1}{2}} = 46$ days) to determine the extent to which rabbits are able to convert a particular iron compound in their diet into blood hemoglobin, which contains an iron atom.

20.45 Explain how you might use radiotracer techniques to determine the extent to which nitrogen-containing fertilizer applied to farm soil finds its way into water supplies. Look in *Lange's Handbook of Chemistry*, or any other handbook containing a table of nuclear properties, to determine whether there is a nitrogen nuclide that would be suitable as a radiotracer in such a study.

20.46 Chlorine-36 is a convenient radiotracer. It is a weak beta emitter, with $t_{\frac{1}{2}} = 3 \times 10^5$ yr. Describe how you would use this radiotracer to carry out each of the following experiments: (a) determine whether trichloroacetic acid, CCl_3COOH, undergoes any ionization of its chlorines as chloride ion in aqueous solution; (b) demonstrate that the equilibrium between dissolved $BaCl_2$ and solid $BaCl_2$ in a saturated solution is a dynamic process; (c) determine the effects of soil pH on the uptake of chloride ion from the soil by soybeans.

[20.47] Tests on human subjects in Boston in 1965 and 1966, following the era of atomic bomb testing, revealed average quantities of about 2 picocuries of plutonium radioactivity in the average person. How many disintegrations per second does this level of activity imply? If each alpha particle deposits 8×10^{-13} J of energy and if the average person weighs 75 kg, calculate the number of rads of radiation dose in 1 yr from such a level of plutonium, and calculate also the number of rems.

20.48 The stresses placed on materials in nuclear reactor cores are very stringent, especially in breeder reactors. In terms of what we have learned in this chapter, what sorts of damage would you expect materials in the reactor core to undergo?

20.49 Why does it require such enormously high temperatures to initiate controlled fusion, whereas no such requirement exists for initiating the nuclear reaction in a fission reactor?

Additional exercises

20.50 Distinguish between the terms in each of the following pairs: (a) electron and positron; (b) binding energy and mass defect; (c) curie and rem; (d) gamma rays and beta rays.

20.51 Figure 20.3 shows the stepwise decay of uranium-238 to form the stable lead-206 nucleus. Write balanced nuclear equations for each step in this sequence.

20.52 Harmful chemicals are often destroyed by chemical treatment. For example, an acid can be neutralized by a base. Why can't chemical treatment be applied to destroy the fission products produced in a nuclear reactor?

20.53 A sample of cobalt-60 was purchased in 1968 for use as a source of beta rays in some biomedical experiments. The half-life for decay of ^{60}Co is 5.25 yr. What fraction of the activity present in the original sample remains in 1981?

[20.54] According to current regulations, the maximum permissible dose of strontium-90 in the body of an adult is 1 microcurie (1×10^{-6} Ci). Using the relationship

Rate = kN

calculate the number of atoms of strontium-90 this corresponds to. What mass of strontium-90 does this correspond to ($t_{\frac{1}{2}}$ for strontium-90 is 27.6 yr)?

20.55 During the past 30 yr nuclear scientists have synthesized approximately 1600 nuclei not known in nature. Many more might be discovered by using heavy-ion bombardment, which is possible only if high-energy instruments are used to accelerate the ions. Complete and balance the following reactions, which involve heavy-ion bombardments:

(a) $^{6}_{3}\text{Li} + ^{63}_{28}\text{Ni} \longrightarrow$?
(b) $^{48}_{20}\text{Ca} + ^{248}_{96}\text{Cm} \longrightarrow$?
(c) $^{88}_{38}\text{Sr} + ^{84}_{36}\text{Kr} \longrightarrow ^{116}_{46}\text{Pd} +$?
(d) $^{48}_{20}\text{Ca} + ^{238}_{92}\text{U} \longrightarrow ^{70}_{20}\text{Ca} + 4^{1}_{0}\text{n} +$ 2?

20.56 Suppose that the strontium-90 in the soil near the site of a nuclear bomb explosion gives rise to a radioactivity level of 7000 counts per second. How long will it be before the level drops to 1000 counts per second? ($t_{\frac{1}{2}}$ for strontium-90 is 28.8 yr.)

20.57 A portion of the sun's energy comes from the reaction

$4^{1}_{1}\text{H} \longrightarrow ^{4}_{2}\text{He} + 2^{0}_{+1}\text{e}$

Calculate the energy change in this reaction (see problem 20.36 and text for needed masses). This reaction requires a temperature of about 10^6 to 10^7 K. Why is such a high temperature required?

20.58 The 13 known nuclides of zinc range from $^{60}_{30}\text{Zn}$ to $^{72}_{30}\text{Zn}$. The naturally occurring nuclides have mass numbers 64, 66, 67, 68, and 70. What mode or modes of decay would you expect for the least massive radioactive nuclides of zinc? What mode for the most massive nuclides?

20.59 The sun radiates energy into space at the rate of 3.9×10^{26} J/sec. Calculate the rate of mass loss from the sun.

20.60 An ancient wooden object is found to have an activity of 9.6 disintegrations per minute per gram of carbon. By contrast, the carbon in a living tree undergoes 18.4 disintegrations per minute per gram of carbon. Based on the activity of carbon-14 in the object, calculate its age.

20.61 It has been suggested that strontium-90 deposited in the hot desert from nuclear testing will undergo radioactive decay more rapidly because it will be exposed to much higher average temperatures. Is this a reasonable suggestion?

[20.62] A radioactive decay series that begins with $^{237}_{93}\text{Np}$ ends with formation of the stable nuclide $^{209}_{83}\text{Bi}$. How many alpha particle emissions and how many beta particle emissions are involved in the sequence of radioactive decays?

[20.63] The half-life for decay of $^{230}_{90}\text{Th}$ is 8.0×10^4 yr. Because one cannot take the time to collect data for a graph such as that in Figure 20.5, how might one determine the half-life for such a long-lived isotope?

20.64 Rutherford was able to carry out the first nuclear transmutation reactions by bombarding nitrogen-14 nuclei with alpha particles. However, in the famous experiment on scattering of alpha particles by gold foil (Section 2.6), a nuclear transmutation reaction did not occur. What is the difference in the two experiments? What would one need to do to carry out a successful nuclear transmutation reaction involving gold nuclei and alpha particles?

20.65 Determine the product nucleus in the following cases: (a) $^{75}_{33}\text{As}(\alpha,\text{n})$_____ ; (b) $^{7}_{3}\text{Li}(\text{p},\text{n})$_____ ; (c) $^{31}_{15}\text{P}(^{2}_{1}\text{H},\text{p})$_____ .

20.66 Fusion energy is generally advanced as potentially a cleaner form of nuclear power than fission. It is anticipated, however, that the materials used to construct the inner components of a fusion reactor will become intensely radioactive and will need frequent replacement. What is the origin of this problem?

[20.67] Suppose you had a detection device that could count every decay from a radioactive sample of plutonium-239 ($t_{\frac{1}{2}}$ is 24,000 yr). How many counts per second would you obtain from a sample that contained 0.500 g of plutonium-239? (Hint: look at Equations [20.20] and [20.22].)

[20.68] When a positron is annihilated by combination with an electron, two photons of equal energy result. What are the wavelengths of these photons? Are they gamma ray photons?

Chemistry of
nonmetallic elements

For the most part, the previous chapters of this book have involved principles, such as rules for bonding, the laws of thermodynamics, the construction of electrochemical cells, and so forth. In the course of explaining these principles, we have described the chemical and physical properties of many substances. In this way, you have been exposed to a considerable amount of chemistry. However, at this point you might still find it difficult to make predictions of the chemical and physical characteristics of substances based on the chemical principles and the scattered facts you have been given. As an example, suppose that you are given a closed box labeled "Fluorine." What can you predict about the properties of the substance inside? Will it be a gas in a tank or a finely divided crystalline solid? Will it be a reactive substance or one that you can confidently expose to the air? What sorts of substances is it likely to react with? You could answer many questions on the basis of principles we have already discussed. For example, you might recall from Chapter 7 that fluorine exists as F_2 molecules; furthermore, you might deduce that F_2 is a gaseous substance because it is nonpolar and has weak intermolecular attractive forces. Remembering that fluorine is the most electronegative element, you could deduce that it is a very strong oxidizing agent; it should thus be very reactive. In short, you should now have some predictive abilities regarding the properties of chemical substances.

Just as important, you have the basis for understanding and remembering the many facts of chemistry as they come before you. In this chapter, we will consider the chemistry of several nonmetallic elements and their compounds. As we examine these substances, we will encounter certain facts about them that were discussed earlier. We will also frequently rely on principles introduced in earlier chapters to help us recognize and understand relationships among the physical and chemical properties of these substances. As you proceed through the chapter you will frequently encounter references to earlier sections in the text. Unless you are quite sure you remember the material thoroughly, it is probably a good idea to turn back to these sections for a review. Study of this chapter should serve to reinforce and deepen your understanding of the principles learned earlier; at the same time you will learn more about the so-called *descriptive* chemistry of the nonmetallic elements.

21.1 Comparison of metals and nonmetals

If you were asked to describe some of the properties of an element you might first find its position on the periodic table. We have seen that the elements can be classified into the broad categories of metals, semimetals (or metalloids), and nonmetals (Section 2.7). The dividing line between metals and nonmetals is shown on the periodic table on the inside front cover of the text. In this chapter we will be considering the properties of the elements of groups 5A, 6A, 7A, and 8A. We will see that the elements involved vary in character from the most nonmetallic to distinctly metallic. It will be useful for us at the very beginning then to consider the characteristics that distinguish a nonmetal from a metal.

Some of the distinguishing properties of metals and nonmetals are listed in Table 21.1. Metals in bulk form have a characteristic luster that we associate with a metal. Strongly metallic elements also exhibit good electrical and thermal conductivity and are malleable and ductile. By contrast, nonmetallic elements are not lustrous and are generally poor conductors of heat and electricity. Seven of the nonmetals exist as diatomic molecules. Included in this group are five gases (hydrogen, nitrogen, oxygen, fluorine, and chlorine), one liquid (bromine), and one volatile solid (iodine). The remaining nonmetals are solids that may be hard like diamond or soft like sulfur. These variations in properties are accounted for in terms of the bonding in the element, as discussed in Section 8.7.

Metals have structures and bonding arrangements that are substantially different from those of nonmetallic elements. We will consider the structures of metals in detail in Chapter 23. Here we need simply note that in metals each atom has a large number of nearest neighbors (characteristically 8 or 12), and the atoms are arranged in a close-packed arrangement or one that is nearly close-packed, such as body-centered cubic (Figure 11.15). In the metals the valence-shell electrons move readily throughout the solid, giving rise to the characteristic metallic properties of luster and good thermal and electrical conductivity.

TABLE 21.1   Characteristic properties of metallic and nonmetallic elements

Metallic elements	Nonmetallic elements
Distinguishing luster	Nonlustrous; various colors
Malleable and ductile as solids	Solids are usually brittle, may be hard or soft
Good thermal and electrical conductivity	Poor conductors of heat and electricity
Most metallic oxides are ionic solids; dissolve in water to form basic solutions	Most nonmetallic oxides are covalent compounds; dissolve in water to form acidic solutions
Exist in aqueous solution mainly as cations	Exist in aqueous solution mainly as anions or oxyanions

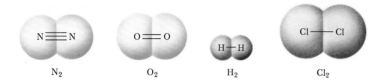

FIGURE 21.1 Examples of nonmetals composed of atoms having only a single nearest neighbor.

Although the rare gases, which are nonmetals, also crystallize in close-packed arrangements, the bonding between atoms is clearly different from that of metals. The low melting points of these rare-gas elements reflect the existence of weak van der Waals forces between atoms. In other nonmetals, the number of atoms that are bonded tightly together (nearest neighbors) is generally four or less. This gives rise to less densely packed arrangements than those possessed by metals. For example, nitrogen, oxygen, hydrogen, and the halogens are normally diatomic. Each atom therefore has only one nearest neighbor as seen in Figure 21.1. This relationship persists when these elements are solidified. In solid chlorine, for example, the atoms within the Cl_2 molecules are separated by 1.99 Å. The closest approach between nonbonded chlorine atoms is 2.79 Å. In the case of diamond, shown earlier in Figure 8.30, each carbon atom has four nearest neighbors. In all cases the bonds between nonmetal atoms are covalent, whereas the interactions between molecules are due to van der Waals forces. Furthermore, allotropism (Section 8.7) is common; carbon, for example, occurs in two crystalline modifications, diamond and graphite.

In Figure 21.2, the colored boxes indicate the so-called semimetals, elements that exhibit some properties of metals and some of nonmetals. Elements that border on the division between metals and nonmetals may have some characteristic metallic properties and not others. For example, antimony *looks* like a metal, but it is brittle rather than malleable and is a poor conductor of heat and electricity.

A traditional characteristic of metals is that the oxides of these elements are generally ionic solids that show basic properties in water. For

1A																	8A
H																	He
	2A											3A	4A	5A	6A	7A	
Li	Be											B	C	N	O	F	Ne
Na	Mg											Al	Si	P	S	Cl	Ar
		3B	4B	5B	6B	7B		8B		1B	2B						
K	Ca	Sc	Ti	V	Cr	Mn	Fe	Co	Ni	Cu	Zn	Ga	Ge	As	Se	Br	Kr
Rb	Sr	Y	Zr	Nb	Mo	Tc	Ru	Rh	Pd	Ag	Cd	In	Sn	Sb	Te	I	Xe
Cs	Ba	La	Hf	Ta	W	Re	Os	Ir	Pt	Au	Hg	Tl	Pb	Bi	Po	At	Rn

FIGURE 21.2 A portion of the periodic table. Colored boxes indicate the semimetals.

example, sodium oxide forms sodium hydroxide on addition to water, as illustrated in Equation [21.1]:

$$Na_2O(s) + H_2O(l) \longrightarrow 2Na^+(aq) + 2OH^-(aq) \qquad [21.1]$$

By contrast, the oxides of the nonmetallic elements are generally covalent compounds that form acidic solutions. Sulfur trioxide, for example, dissolves in water to form sulfuric acid, as illustrated in Equation [21.2]:

$$SO_3(g) + H_2O(l) \longrightarrow H^+(aq) + HSO_4^-(aq) \qquad [21.2]$$

Metal oxides such as Na_2O are called **basic anhydrides,** or basic oxides, whereas nonmetal oxides such as SO_3 are referred to as **acidic anhydrides,** or acidic oxides. (The term anhydride means "without water.") Many oxides are insoluble in water. Their acidic or basic character is then determined by whether they will dissolve in acids or bases. For example, Fe_2O_3 is insoluble in water and in basic solutions such as aqueous NaOH. However, it dissolves in acid, reacting as shown in Equation [21.3]:

$$Fe_2O_3(s) + 6H^+(aq) \longrightarrow 2Fe^{3+}(aq) + 3H_2O(l) \qquad [21.3]$$

It is therefore referred to as a basic oxide.

PERIODIC TRENDS

Much of the material in this chapter will be much clearer and easier to remember if you keep in mind several trends in physical and chemical properties that can be related to position in the periodic table. Some of the more important of these are illustrated in Figure 21.3. Recall that electronegativity increases as we move up in a given family, and increases from left to right across the table. As a result nonmetals have higher electronegativities than do metals. The tendency toward metallic character increases as we move downward in each group.

Because the nonmetallic elements have generally higher electronegativities, the compounds formed between strongly metallic and strongly

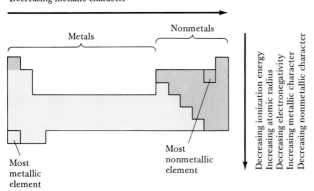

FIGURE 21.3 Trends in key properties of the elements as a function of position in the periodic table.

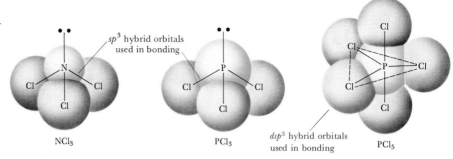

*sp*³ hybrid orbitals
used in bonding

NCl₃

PCl₃

*dsp*³ hybrid orbitals
used in bonding

PCl₅

FIGURE 21.4 Comparison of NCl₃, PCl₃, and PCl₅. Nitrogen is unable to form NCl₅ because it does not have *d* orbitals available for bonding.

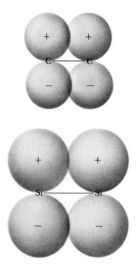

FIGURE 21.5 Comparison of *π*-bond formation by sideways overlap of *p* orbitals between two carbon atoms and between two silicon atoms. The distance between nuclei increases as we move from carbon to silicon because of the larger size of the silicon atom. The *p* orbitals do not overlap as effectively between two silicon atoms because of this greater separation.

nonmetallic elements tend to be ionic (for example, metal fluorides and metal oxides). In solutions metals generally appear as cations, whereas the nonmetallic elements appear as anions or oxyanions.

One important periodic trend that we have not explicitly pointed out in our earlier discussions will become evident as you proceed through this chapter. The first member of any group, or family, often differs from subsequent members of that group. One reason for this is the smaller size and greater electronegativity of the first member. Because the first member has only the 2*s* and the three 2*p* orbitals available for bonding, it can form a maximum of four bonds. Subsequent members of the family are able to use *d* orbitals in bonding in addition to *s* and *p* orbitals; they can therefore form more than four bonds. As an example, consider the chlorides of nitrogen and phosphorus, the first two members of group 5A. Nitrogen forms a maximum of three bonds with chlorine, NCl₃. Though phosphorus can also form the trichloride compound, PCl₃, it is able in addition to form five bonds with chlorine, PCl₅. These compounds are shown in Figure 21.4.

Another difference between the first member of any family and subsequent members of the same family is the greater ability of the former to form *π* bonds. We can understand this, in part, in terms of atomic size. As atoms increase in size, the sideways overlap of *p* orbitals, which form the strongest type of *π* bond, becomes less effective. This is shown in Figure 21.5. As an illustration of this effect, consider two differences in the chemistry of carbon and silicon, the first two members of group 4A. Carbon has two crystalline allotropes, diamond and graphite. In diamond there are σ bonds between carbon atoms but no *π* bonds. In graphite, *π* bonds result from sideways overlap of *p* orbitals (Section 8.7).

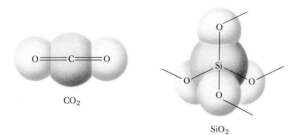

CO₂

SiO₂

FIGURE 21.6 Comparison of the structures of CO₂ and SiO₂; CO₂ has double bonds, whereas SiO₂ has only single bonds.

Silicon occurs only in the diamondlike crystal form. Silicon does not exhibit a graphitelike structure because of the low stability of π bonds between silicon atoms.

We see the same type of difference in the dioxides of these elements, Figure 21.6. In CO_2, carbon forms double bonds to oxygen, thereby achieving its valence by π bonding. In contrast, SiO_2 contains no double bonds. Instead, four oxygens are bonded to each silicon, forming an extended structure reminiscent of diamond.*

<table>
<tr><td>21.2 The noble-gas elements</td><td>

We have referred at several points in the text to the fact that the elements of group 8A are chemically unreactive. Indeed, most of our references to these elements have been in relation to their physical properties, as when we discussed intermolecular forces in Section 11.5. According to the Lewis theory of chemical bonding, the relative inertness of these elements is due to the formation of a completed octet of valence-shell electrons. The stability of such an arrangement is reflected in the high ionization energies of the group 8A elements (Section 6.5).

The group 8A elements are all gases; however, they condense and eventually freeze at low temperatures. The variation in melting point, boiling point, and enthalpy of vaporization among the group 8A elements (Table 11.4) is a measure of the variation in interatomic attractive forces. These attractions are due to the nonbonding dispersion force, as discussed in Section 11.5.

All the noble gases are components of the earth's atmosphere, except for radon, which exists only as a short-lived radioisotope. Only argon is relatively abundant (Table 10.1). Argon and the heavier noble gases are recovered from liquid air by fractional distillation. Argon is used as a blanketing atmosphere in electric light bulbs. The gas conducts heat away from the filament but does not react with it. It is also used as a protective atmosphere to prevent oxidation in welding and high-temperature metallurgical processes. Neon is used in electric signs; the gas is caused to radiate by passing an electric discharge through the tube.

Helium is, in many ways, the most important of the noble gases. Helium boils at 4.2 K under 1 atm pressure, the lowest boiling point of any substance. It is a very important liquid for the conduct of many experiments at very low temperatures. Because helium has such a low abundance in the atmosphere and boils at such a low temperature, recovery of the gas from the atmosphere would require an immense expenditure of energy. Helium is found in relatively high concentrations in many natural gas wells. Some of this helium is separated to meet current demands, and a little is kept for later use. Unfortunately, however, most of the helium is allowed to escape.

Because the noble gases are exceedingly stable, it is reasonable to expect that they will undergo reaction only under rather rigorous conditions. Furthermore, we might expect that the heavier noble gases would be most likely to undergo chemical transformation, because the ioniza-

</td></tr>
</table>

---

*The formula SiO_2 is consistent with this structure, because each oxygen is shared by two silicon atoms (not shown in Figure 21.6). For bookkeeping purposes, we may therefore count a half of each of the four oxygens that are bound to a given silicon atom as belonging to that silicon. We shall consider silicon-oxygen compounds in some detail in Chapter 22.

TABLE 21.2 Properties of xenon compounds

Compound	Oxidation state of Xe	Melting point (°C)	ΔH_f° (kJ/mol)[a]
XeF_2	+2	129	$-109(g)$
XeF_4	+4	117	$-218(g)$
XeF_6	+6	49	$-298(g)$
$XeOF_4$	+6	−41 to −28	$+146(l)$
XeO_3	+6	—[b]	$+402(s)$
XeO_2F_2	+6	31	$+145(s)$
XeO_4	+8	—[c]	—

[a]At 25°C, for the compound in the state indicated.
[b]A solid; decomposes at 40°C.
[c]A solid; decomposes at −40°C.

tion energies are lower for the heavier elements as illustrated in Figure 6.6. A lower ionization energy would suggest the possibility of losing an electron in formation of an ionic bond. In addition, since the group 8A elements already contain eight electrons in their valence shell (except, of course, for helium, which contains just two), formation of covalent bonds could require an expanded valence shell. We have seen (Section 7.7) that this occurs most readily with larger atoms.

The first noble-gas compound was prepared by Neil Bartlett in 1962. His work caused a sensation, because it undercut what had become a paradigm, the belief that the noble-gas elements were truly chemically inert. Bartlett's initial work involved xenon in combination with fluorine, the element we would expect to be most reactive. Since then several xenon compounds of fluorine and oxygen have been prepared. The properties of these substances are listed in Table 21.2. The three simple fluorides XeF_2, XeF_4, and XeF_6 are made by direct reaction of the elements. By varying the ratio of reactants and altering reaction conditions, one or the other of the three compounds can be obtained. The oxygen-containing compounds are formed when the fluorides are reacted with water, as in Equations [21.4] and [21.5]:

$$XeF_6(s) + H_2O(l) \longrightarrow XeOF_4(l) + 2HF(g) \tag{21.4}$$

$$XeF_6(s) + 3H_2O(l) \longrightarrow XeO_3(aq) + 6HF(aq) \tag{21.5}$$

SAMPLE EXERCISE 21.1

Describe the electronic and geometrical structure of $XeOF_4$.

Solution: We should first write a Lewis structure for the molecule. The total number of valence-shell electrons involved is 42; 8 from xenon, 7 each from four fluorines, and 6 from oxygen. This leads to the Lewis structure shown in Figure 21.7(a). We see that the Xe has 12 electrons in its valence shell. We thus expect an octahedral disposition of six electron pairs. One of these is taken up with the bond to oxygen and four

with the bonds to fluorine; the sixth position contains an unshared electron pair. Because oxygen is less electronegative than fluorine, the electron pair in the Xe—O bond should have a slightly larger volume requirement than the electron pairs in the Xe—F bonds. It is thus reasonable to expect that the Xe—O bond will be *trans* to (across from) the unshared electron pair, which has the larger volume requirement. The expected structure is shown in Figure 21.7(b).

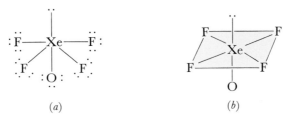

(a)　　　　　　　　　　(b)

FIGURE 21.7   The Lewis and geometrical structures of XeOF$_4$

The fact that the enthalpies of formation of the simple xenon fluorides are negative suggests that these compounds should be reasonably stable, and this is indeed found to be the case. They are, however, powerful fluorinating agents and must be handled in containers that do not readily react to form fluorides. Notice that the enthalpies of formation of the oxyfluorides and oxides of xenon are positive; these compounds are quite unstable.

As we might expect, formation of compounds by the other noble-gas elements occurs much less readily than in the case of xenon. Only one binary krypton compound, KrF_2, is known with certainty; it decomposes to the elements at $-10°C$.

21.3   The halogens

The elements of groups 7A, called the halogens, consist of the elements fluorine, chlorine, bromine, iodine, and astatine. These elements have played an important part in the development of chemistry as a science. Chlorine was first prepared by the Swedish chemist Karl Wilhelm Scheele in 1774, but it was not until 1810 that the English chemist Humphrey Davy identified it as an element. The discovery of iodine followed shortly thereafter, in 1811, and bromine was discovered in 1825. Compounds of fluorine were known for a long time, but it was not until 1886 that the French chemist Henri Moissan succeeded in preparing the very reactive free element.

The general valence-electron configuration of the halogens is ns^2np^5, where n may have values ranging from 2 through 6. When a halogen atom shares all seven of its electrons with a more electronegative atom, it is assigned the $+7$ oxidation state. The halogens also achieve a noble-gas configuration by gaining an electron, which results in a -1 oxidation state. Thus, we expect that the halogens might occur in oxidation states ranging from $+7$ at one extreme to -1 at the other. Oxidation states over this entire range are observed for all the halogens except fluorine; this most electronegative of all the elements is observed only in the 0 or -1 oxidation states.

Table 21.3 summarizes the occurrences of the halogens in nature. Both fluorine and chlorine are fairly abundant, but they are quite differently distributed. This happens because the salts of chlorine are generally quite soluble, whereas some of those of fluorine are not. Bromine is comparatively much less abundant than chlorine or fluorine, and iodine is much rarer still.

Chlorine, bromine, and iodine occur as the halides in seawater and in salt deposits. The concentration of iodine in these sources is generally very small. However, iodine is concentrated by certain seaweeds, which

TABLE 21.3   Occurrences of the halogens

Element	Occurrences
Fluorine	Fluorspar, CaF_2; Fluorapatite, $Ca_{10}F_2(PO_4)_6$; Cryolite, Na_3AlF_6 Biologically: teeth, bones
Chlorine	Seawater (0.55 M, chiefly as NaCl); salt beds, salt lakes (for example, Dead Sea, Great Salt Lake); NaCl deposits Biologically: gastric juice (as HCl(aq)), tissue fluids
Bromine	Seawater ($8.3 \times 10^{-4}\,M$); underground brines; salt beds, salt lakes Biologically: minor concentrations of bromide along with chlorine.
Iodine	Seawater ($4 \times 10^{-7}\,M$); seaweeds; $NaIO_3$ in minor amounts in nitrate deposits; oil-well brines Biologically: in human thyroid gland

are harvested, dried, and burned. Iodine is then extracted from the ashes. The element is also extracted commercially from oil-well brines in California. Fluorine occurs in the minerals fluorspar, cryolite, and fluorapatite. Only the first of these is an important commercial source of fluorine for the chemical industry. All isotopes of astatine are radioactive. The longest-lived isotope is astatine-210, which has a half-life of 8.3 hr and decays mainly by electron capture. Astatine was first synthesized by bombarding bismuth-209 with high-energy alpha particles as shown in Equation [21.6]:

$$^{209}_{83}\text{Bi} + ^{4}_{2}\text{He} \longrightarrow ^{211}_{85}\text{At} + 2^{1}_{0}\text{n} \qquad [21.6]$$

The fact that a cyclotron must be used to synthesize astatine makes it very expensive and limits its application and study.

In terms of commercial uses, chlorine is by far the most important halogen. About 10 billion kg of chlorine is produced in the United States each year. Hydrogen chloride production is about 2.5 billion kg annually. About half of this inorganic chlorine finds its way eventually into vinyl chloride, C_2H_3Cl, used in polyvinyl chloride (PVC) plastics manufacture, ethylene dichloride, $C_2H_2Cl_2$, an organic solvent, and other chlorine-containing organic compounds. Annual U.S. production of bromine is about 200 million kg; the much more expensive iodine is produced to the extent of about 4 million kg each year.

Before we begin a discussion of the chemical characteristics of the halogens, it will be useful to review key properties of the elements, summarized in Table 21.4. The symbol X represents any one of the halogens. Notice that all of the properties listed except the last two refer to the halogen atom. The last two refer to the diatomic molecules, X_2, with a single bond joining the atoms. You will recall from Section 8.7 that this is the stable form of the halogens as free elements.

Most of the properties listed in Table 21.4 vary in a regular fashion as a function of atomic number. Within each horizontal row of the periodic table each halogen has a high ionization energy, second only to the noble-gas element adjacent to it in the table. Similarly, each halogen has the highest electronegativity of all the elements in its horizontal row. Within the halogen family, atomic and ionic radii increase with increasing atomic number. Correspondingly, the ionization energy and electro-

TABLE 21.4 Some properties of the halogen atoms

Property	F	Cl	Br	I
Atomic radius (Å)	0.72	1.00	1.15	1.40
Ionic radius, X^- (Å)	1.33	1.81	1.96	2.17
First ionization energy (kJ/mol)	1.68×10^3	1.25×10^3	1.14×10^3	1.01×10^3
Electron affinity (kJ/mol)	-332	-349	-325	-295
Electronegativity	4.0	3.2	3.0	2.7
X—X bond energy (kJ/mol)	155	242	193	151
Reduction potential: $\frac{1}{2}X_2(aq) + e^- \longrightarrow X^-(aq)$	2.87	1.36	1.07	0.54

negativity steadily decrease. Under ordinary conditions the halogens exist as the diatomic molecules. At room temperature and 1 atm pressure I_2 is a solid, Br_2 is a liquid, and Cl_2 and F_2 are gases. Because of its high reactivity, F_2 is very difficult to work with. Certain metals such as copper and nickel can be used to contain F_2 because their surfaces form a protective coating of metal fluoride. Chlorine must also be handled with care. Because chlorine liquefies upon compression at room temperature, it is normally stored and handled in liquid form in steel containers. Chlorine and the heavier halogens are reactive, though less so than fluorine. They combine directly with most elements except the rare gases.

The comparatively low bond energy in F_2 accounts in part for the extreme reactivity of elemental fluorine. Notice from Table 21.4 that the potential for reduction of the halogens in aqueous medium is positive for all the elements and is exceptionally high for F_2. Fluorine gas readily oxidizes water according to the reaction

$$F_2(aq) + H_2O(l) \longrightarrow 2HF(aq) + \tfrac{1}{2}O_2(g) \qquad E° = +1.63 \text{ V} \qquad [21.7]$$

Fluorine cannot be prepared by electrolytic oxidation in water, because water itself is oxidized more readily than F^-, with formation of $O_2(g)$. In practice the element is formed by electrolytic oxidation of anhydrous HF. Because HF is not by itself a good conductor of electricity, a solution of KF in anhydrous HF is used. The KF reacts with HF to form a new salt, $K^+HF_2^-$, which acts as the current carrier in the liquid. (The HF_2^- ion is stable because of the very strong hydrogen bond between the two fluoride ions, as described in Section 11.5.) The overall cell reaction is

$$2KHF_2(l) \longrightarrow H_2(g) + F_2(g) + 2KF(l) \qquad [21.8]$$

Chlorine is also produced mainly by electrolysis of either molten sodium chloride or aqueous sodium chloride, as described in Section 19.6.

The reverse of the reduction equation listed in Table 21.4 represents oxidation of the halide ion to the free element. Because the potential for this reaction is not so negative for bromine or iodine, oxidation of bromide or iodide ions by chemical means is not too difficult. Bromine, for example, is produced commercially by oxidizing the aqueous bromide ion with chlorine gas, as described in Section 17.2. Similarly, iodine is obtained by chlorination of oil-well brines:

$$2I^-(aq) + Cl_2(g) \longrightarrow 2Cl^-(aq) + I_2(s) \qquad [21.9]$$

USES OF THE HALOGENS

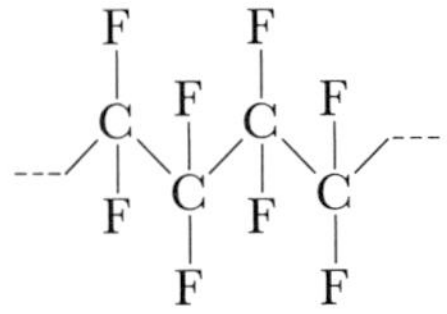

Fluorine has become an important industrial chemical. It is used, for example, to prepare fluorocarbons, very stable carbon-fluorine compounds. An example is CF_2Cl_2, known as Freon-12, which is used as a refrigerant and as a propellant for aerosol cans. As we noted in Section 10.4, the effects of these substances on the ozone layer have been under recent investigation. Fluorocarbons are also used as lubricants and in plastics; Teflon (Figure 21.8) is a polymeric fluorocarbon noted for its high thermal stability and lack of chemical reactivity.

FIGURE 21.8 Structure of Teflon, a fluorocarbon polymer.

Chlorine is used primarily as a bleach in the paper and textile industries. When Cl_2 dissolves in cold dilute base it forms Cl^- and the hypochlorite ion, ClO^-:

$$2OH^-(aq) + Cl_2(aq) \longrightarrow Cl^-(aq) + ClO^-(aq) + H_2O(l) \qquad [21.10]$$

Sodium hypochlorite, NaClO, is the active ingredient in Clorox bleach. Solid bleaching powder, which consists of Ca(ClO)Cl, is obtained by reaction of Cl_2 with a suspension of $Ca(OH)_2$ as shown in Equation [21.11]:

$$Ca(OH)_2(s) + Cl_2(aq) \longrightarrow Ca^{2+}(aq) + ClO^-(aq) + Cl^-(aq) + H_2O(l) \quad [21.11]$$

Ca(ClO)Cl is recovered as crystals upon evaporation of the solution. Chlorine is also used in water treatment to oxidize and thereby destroy bacteria. Recent studies have shown that such treatment of municipal sewage and drinking water produces small amounts (on the order of parts per billion) of chlorocarbon compounds. These compounds are claimed to be carcinogenic (cancer inducing) and are toxic to fish and other aquatic life. Whether they are harmful in the small amounts usually found in treated water is not known. Ozone, O_3, can be used as a substitute for Cl_2 in water treatment, but is more costly. Chlorine is also used in the manufacture of plastics and certain insecticides, such as DDT.

Bromine is used in the production of silver bromide used in photographic film. It is also used to prepare dibromoethane, $C_2H_4Br_2$, a gasoline additive used in gasoline that contains tetraethyl lead, $Pb(C_2H_5)_4$. The dibromoethane prevents formation of lead deposits in the engine by forming volatile $PbBr_2$, which is emitted in exhaust. Bromine-containing compounds also are used as fire retardants in clothing.

Iodine has not found as wide use as the other halogens. One familiar use is its addition, as KI, to salt to form iodized salt. Table salt contains about 0.02 percent potassium iodide by weight. Iodized salt is able to provide the small amount of iodine necessary in our diets; it is essential for the formation of thyroxin, a hormone secreted by the thyroid gland. Lack of iodine in the diet results in an enlarged thyroid gland, a condition called goiter. Iodine is also familiar as tincture of iodine, a solution of I_2 in alcohol that is used as an antiseptic.

TABLE 21.5 Properties of the hydrogen halides

Property	HF	HCl	HBr	HI
Molecular weight	20.01	36.46	80.92	127.91
Melting point (°C)	−83	−114	−87	−51
Boiling point (°C)	19.9	−85	−67	−35
Bond-dissociation energy (kJ/mol)	565	431	364	297
H—X distance (Å)	0.92	1.27	1.41	1.61
Solubility in H_2O (g/100 g H_2O, 10°C)	∞	78	210	234

THE HYDROGEN HALIDES

All of the halogens form stable diatomic molecules with hydrogen. These are very important compounds, in part because aqueous solutions of the hydrogen halides other than HF are strongly acidic. Table 21.5 lists some of the more important properties of the hydrogen halides.

Notice that the melting and boiling points of HF are abnormally high as compared with the other hydrogen halides. The cause of this unusual behavior is the strong hydrogen bonding that exists between HF molecules in the liquid state, as described in Section 11.5. Because HF is a small molecule and is capable of strong hydrogen-bonding interactions, it is completely miscible with water. The other hydrogen halides, though not capable of strong hydrogen-bonding interactions, are very soluble in water. We learned in Section 15.1 that this very high solubility is the result of reaction of the hydrogen halide with water, producing halide anions and the solvated proton, $H^+(aq)$.

It is noteworthy that as the H—X bond distance increases, the H—X bond-dissociation energy decreases. We can account for this decrease in terms of a lower polarity for the H—X bond based on electronegativity difference (Section 7.9) and a decreased overlap of the hydrogen $1s$ orbital with the valence p orbital of the halogen atom as the halogen atom grows larger.

The hydrogen halides can be formed by direct reaction of the elements. The reaction can be explosive in character, as in the reaction of H_2 with F_2 or when reaction between a mixture of H_2 and Cl_2 is initiated by a spark or a photon of sufficient energy. On the other hand, the reaction of either Br_2 or I_2 with hydrogen is much less vigorous, and the reaction mixtures must be heated to cause the system to approach equilibrium.

The most important means of formation of the hydrogen halides is through reaction of a salt of the halide with a strong, nonvolatile acid. Hydrogen fluoride and hydrogen chloride are prepared in this manner by reaction of a cheap, readily available salt with concentrated sulfuric acid:

$$CaF_2(s) + H_2SO_4(l) \longrightarrow 2HF(g) + CaSO_4(s) \qquad [21.12]$$

$$NaCl(s) + H_2SO_4(l) \longrightarrow HCl(g) + NaHSO_4(s) \qquad [21.13]$$

Because the hydrogen halide is the only volatile component in the mixture, it can be easily removed. The hydrogen halide is usually absorbed in water and marketed as the corresponding acid. In both cases the reaction mixtures must be heated to cause reaction to occur. The $NaHSO_4$ formed in producing HCl can be forced to react still one stage further by more extensive heating and the addition of more NaCl:

$$NaCl(s) + NaHSO_4(s) \longrightarrow HCl(g) + Na_2SO_4(s) \qquad [21.14]$$

This reaction provides an illustration of the acidic properties of the hydrogen sulfate ion, $HSO_4{}^-$.

Neither hydrogen bromide nor hydrogen iodide can be prepared by analogous reactions of salts with H_2SO_4, because HBr and HI undergo oxidation by H_2SO_4 at the higher temperatures of the reaction. The overall reactions are described by Equations [21.15] and [21.16]:

$$2NaBr(s) + 2H_2SO_4(l) \longrightarrow$$
$$Br_2(g) + SO_2(g) + Na_2SO_4(s) + 2H_2O(g) \qquad [21.15]$$

$$8NaI(s) + 9H_2SO_4(l) \longrightarrow$$
$$8NaHSO_4(s) + H_2S(g) + 4I_2(g) + 4H_2O(g) \qquad [21.16]$$

Notice that in the case of the bromide, part of the H_2SO_4 is reduced to SO_2, in which sulfur is in the $+4$ oxidation state. In the reaction with iodide, sulfur is reduced all the way to H_2S, in which sulfur is in the -2 oxidation state. This difference in products reflects the greater ease of oxidation of the iodide. The difficulties associated with use of H_2SO_4 can be avoided by using a nonvolatile acid that is a poorer oxidizing agent than H_2SO_4; concentrated phosphoric acid serves well.

SAMPLE EXERCISE 21.2

Write balanced equations for formation of HBr and HI from reaction of the appropriate sodium salt with phosphoric acid.

Solution: The formula for phosphoric acid is H_3PO_4. Let us assume that only one of the dissociable hydrogens of this acid undergoes reaction. (The ac-

tual number that reacts depends on reaction conditions.) The balanced equations are then:

$$NaBr(s) + H_3PO_4(l) \xrightarrow{\Delta} NaH_2PO_4(s) + HBr(g)$$

$$NaI(s) + H_3PO_4(l) \xrightarrow{\Delta} NaH_2PO_4(s) + HI(g)$$

The hydrogen halides are also formed when certain nonmetallic or metallic halides are hydrolyzed, as in these examples:

$$PBr_3(l) + 3H_2O(l) \longrightarrow H_3PO_3(l) + 3HBr(g) \qquad [21.17]$$

$$SeCl_4(s) + 3H_2O(l) \longrightarrow H_2SeO_3(s) + 4HCl(g) \qquad [21.18]$$

$$AlCl_3(s) + 3H_2O(l) \longrightarrow Al(OH)_3(s) + 3HCl(g) \qquad [21.19]$$

Because all of the hydrogen halides except HF form strongly acidic solutions in water, they exhibit the characteristic properties of all strong acids in their reactions in water. For example, they may react with an active metal to produce hydrogen gas or with a base in a neutralization reaction. In such reactions the halide ion is merely a spectator ion; it does not directly participate in the reaction. On the other hand, in oxidation-reduction reactions, the halide ion is oxidized, as in the example shown in Equation [21.20]:

$$14H^+(aq) + 6Cl^-(aq) + Cr_2O_7^{2-}(aq) \longrightarrow$$
$$2Cr^{3+}(aq) + 3Cl_2(g) + 7H_2O(l) \qquad [21.20]$$

The halide ion also plays a role in precipitation reactions, in which an insoluble metal halide is formed:

$$Pb^{2+}(aq) + 2Br^-(aq) \longrightarrow PbBr_2(s) \qquad [21.21]$$

In this reaction the solvated proton is a spectator ion.

Hydrofluoric acid is unusual among the aqueous hydrogen halides in that it reacts readily with silica, SiO_2, and with various silicates to form hexafluorosilicic acid, as in these examples:

$$SiO_2(s) + 6HF(aq) \longrightarrow H_2SiF_6(aq) + 2H_2O(l) \qquad [21.22]$$

$$CaSiO_3(s) + 8HF(aq) \longrightarrow H_2SiF_6(aq) + CaF_2(aq) + 3H_2O(l) \qquad [21.23]$$

Hydrofluoric acid must therefore be stored in wax or plastic bottles, because it reacts with ordinary glasses, which consist mostly of silicate structures (Section 22.3).

INTERHALOGEN COMPOUNDS

Because the halogens form diatomic molecules in their most stable state at ordinary temperatures and pressures, it is not surprising to discover that diatomic molecules consisting of two different halogen atoms exist. These compounds are the simplest example of interhalogens, that is, compounds formed between two different halogen elements. Some properties of the diatomic interhalogens are listed in Table 21.6. The table is incomplete because certain of the interhalogens are not stable; they undergo decomposition to form the diatomic halogen elements or to form more complex interhalogen compounds.

The higher interhalogen compounds have formulas of the form XX_3', XX_5', or XX_7', where X is chlorine, bromine, or iodine, and X′ is fluorine (the one exception is ICl_3, in which X′ is chlorine). From our earlier discussion of bonding and structure (Section 7.7), you will recall that in compounds of this kind the central atom has an expanded valence shell.

TABLE 21.6 Properties of interhalogen compounds, XX′

Compound	ClF	BrF	BrCl	IF	ICl	IBr
Molecular weight	54.6	98.9	115.4	145.9	162.4	206.8
Melting point (°C)	−155	−33	−54	—	27	42
Boiling point (°C)	−90	20	5	—	98	116
Bond distance (Å)	1.63	1.76	2.16	1.91	2.32	—
Dipole moment (D)	0.9	1.3	0.6	—	0.5	—
Bond-dissociation energy (kJ/mol)	253	237	218	278	208	175

Using the VSEPR model (Section 8.1), we can predict the geometrical structure of the compound. We can also describe the bonding about the central atom in terms of a hybrid orbital description (Section 8.2).

Account for the valence-shell electron distribution and geometrical structure in BrF_3. What hybrid orbital description is most suitable for the central atom in this molecule?

Solution: Bromine has seven valence-shell electrons. If the Br atom is singly bonded to three fluorine atoms, there are an additional three electrons from this source. According to the VSEPR model these ten electrons are disposed as five electron pairs about the central atom at the vertices of a trigonal bipyramid. Three of the electron pairs are used in bonding to fluorine, the other two are unshared electron pairs. These unshared pairs require a larger space, and so they are placed in the equatorial plane of the trigonal bipyramid:

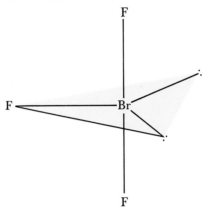

Because the unshared pairs push the bonding pairs back a little, the molecule should have the shape of a bent T. In terms of a hybridization description, we need to employ one of the bromine valence-shell d orbitals (the $4d$) in addition to the $4s$ and three $4p$ orbitals to provide the five atomic orbitals to contain the five electron pairs in the valence shell. Thus the appropriate hybrid orbital description is sp^3d, which results in orbitals directed toward the vertices of a trigonal bipyramid.

Notice that the central atom in these higher interhalogen compounds has valence-shell d orbitals available for bonding. Thus, the valence shell can be expanded beyond the octet. Secondly, the central atom in all cases is relatively large compared to the atoms grouped around it. Because fluorine is small and forms very strong bonds, it is ideally suited as the X' atom. Only when the central atom is very large, as in the case of iodine, can the larger chlorine atom form an interhalogen, in ICl_3. The importance of size is also seen in the fact that iodine is capable of forming IF_7, whereas with bromine a maximum of five fluorines can be fitted around the central atom, in BrF_5. Chlorine and fluorine can form the ClF_5 molecule, but only with great difficulty.

The interhalogen compounds are exceedingly reactive. They attack glass very readily and must be placed in special metal containers. They act as very active fluorinating agents, as in these examples:

$$2CoCl_2(s) + 2ClF_3(g) \longrightarrow 2CoF_3(s) + 3Cl_2(g) \qquad [21.24]$$

$$Se(s) + 3BrF_5(l) \longrightarrow SeF_6(l) + 3BrF_3(l) \qquad [21.25]$$

The *polyhalide ions* are closely related to the interhalogens. Many of these ions are stable as salts of the alkali metal ions—for example, KI_3,

$CsIBr_2$, $KICl_4$, and $KBrF_4$. Some of them, notably I_3^-, are also stable in aqueous solution.

OXYACIDS AND OXYANIONS

We have already had occasion to refer in the text to the oxyacids of the halogens in discussions of nomenclature, structure and bonding, and acid strengths. Table 21.7 summarizes the formulas of the known oxyacids, the naming system employed, and the oxidation state of the halogen. * The oxyacids are rather unstable; they generally decompose (sometimes explosively) when one attempts to isolate them. All of the oxyacids are strong oxidizing agents. The oxyanions, formed upon removal of the proton from the oxyacid (Section 15.9), are generally more stable than the oxyacids themselves. (Review the nomenclature of the oxyanions, Section 2.9.) Hypochlorite salts are used as bleaches and disinfectants because of the powerful oxidizing capabilities of the hypochlorite ion. Sodium chlorite, which can be isolated as the trihydrate, $NaClO_2 \cdot 3H_2O$, is used as a bleaching agent. Chlorite salts form potentially explosive mixtures with organic materials. Chlorate salts are similarly very reactive. For example, a mixture of potassium chlorate and sulfur may explode when struck. Potassium chlorate is used in making matches and fireworks. Perchloric acid and its salts are the most stable of the oxyacids and oxyanions. Dilute perchloric acid solutions are quite safe, and most perchlorate salts are stable except when heated with organic materials.

Iodic acid, HIO_3, can be isolated as colorless crystals after iodine is oxidized by an ozidizing agent such as nitric acid, chlorine, or chloric acid. An example of such a reaction is shown in Equation [21.26]:

$$8H^+(aq) + I_2(s) + 10NO_3^-(aq) \longrightarrow$$
$$2IO_3^-(aq) + 10NO_2(g) + 4H_2O(l) \qquad [21.26]$$

Further oxidation of iodic acids leads to periodic acid. We might expect the formula of this acid to be HIO_4, by analogy with the formula for perchloric acid. However, HIO_4 cannot be isolated. Instead, a substance of formula H_5IO_6, called paraperiodic acid, is obtained. Notice that this substance has the composition of HIO_4 with two added water molecules.

*Fluorine forms one oxyacid, HFO. Because the electronegativity of fluorine is greater than that of oxygen, we must consider fluorine to be a −1 oxidation state and oxygen to be in the 0 oxidation state.

TABLE 21.7   The oxyacids of the halogens

Oxidation state of halogen	Formula of acid			Type of name
	Cl	Br	I	
+1	HClO	HBrO	HIO	*Hypo*halous acid
+3	$HClO_2$	—	—	Hal*ous* acid
+5	$HClO_3$	$HBrO_3$	HIO_3	Hal*ic* acid
+7	$HClO_4$	$HBrO_4$	H_5IO_6	*Per*halic acid

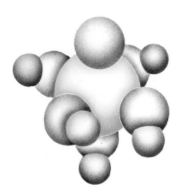

Because iodine is so large, there is room about the iodine for bonding to two additional atoms, forming a molecule in which the iodine is surrounded by six oxygen atoms, as illustrated in Figure 21.9. The protons are distributed so that there is one proton on each of five oxygens. The first two acid-dissociation constants for H_5IO_6 are $K_{a1} = 2.8 \times 10^{-2}$ and $K_{a2} = 5.3 \times 10^{-9}$.

The acid strengths of the oxyacids increase with increasing oxidation state of the central atom; the origin of this trend is discussed in Section 15.9. The stability of the oxyacids and of the corresponding oxyanions increases with increasing oxidation state of the central halogen atom. Because the halogens are relatively electronegative elements, we might expect that compounds in which the halogen has an increasingly positive oxidation number would be *less* stable. The origins of this unexpected trend are rather complicated and beyond the scope of our survey. However, the result is important, because it affects the chemical behavior of the oxyacids and oxyanions. The stabilities of the oxyacids and oxyanions are reflected in the reduction potentials. To illustrate, let's consider the reduction potentials for the oxychlorine species in acidic solution, which we can write in a compact form as shown in Figure 21.10. Note first of all that ClO_4^- and ClO_3^- are written as anions, whereas $HClO_2$ and $HClO$ are written as the acids. This follows from the fact that $HClO_2$ and $HClO$ are weak acids, whereas $HClO_3$ and $HClO_4$ are strong acids, and thus can be considered to be completely ionized, even in acidic solution. The voltage listed for each change in oxidation state is the standard reduction potential for that change. The potential corresponds to the complete half-cell reaction. For example, the half-reaction for reduction of $HClO_2$ to $HClO$ in acid solution has a standard potential of 1.64 V; the complete balanced half-reaction is written as follows:

$$HClO_2(aq) + 2H^+(aq) + 2e^- \longrightarrow$$
$$HClO(aq) + H_2O(l) \qquad E° = 1.64 \text{ V} \qquad [21.27]$$

You should be able to write a similar complete, balanced half-reaction for all the changes in oxidation state shown in Figure 21.10, using $H^+(aq)$ and $H_2O(l)$ to achieve a balanced equation.

Notice that all the reduction potentials given in Figure 21.10 are positive. This is in keeping with the strongly oxidizing character of the oxychlorides. Notice also that the potentials are greatest for reduction of the lower-oxidation-state species, $HClO_2$ and $HClO$. These acids are the most strongly oxidizing agents among the oxychlorine compounds.

$$ClO_4^- \xrightarrow{1.19 \text{ V}} ClO_3^- \xrightarrow{1.21 \text{ V}} HClO_2 \xrightarrow{1.64 \text{ V}} HClO \xrightarrow{1.63 \text{ V}} Cl_2 \xrightarrow{1.36 \text{ V}} Cl^-$$

(overarching: 1.38 V; lower connecting: 1.47 V)

FIGURE 21.10 Standard reduction potentials in acid solution for oxychlorine species. Note that Cl^-, ClO_4^-, and ClO_3^- are written as the anions because HCl, $HClO_4$, and $HClO_3$ are strong acids and are thus largely dissociated, even in acidic solution.

By using the standard potentials in Figure 21.10 it is possible to calculate the standard potentials for other half-reactions. For example, we can calculate the standard potential for the reduction of ClO_4^- to Cl^- using the potentials for reduction of ClO_4^- to ClO_3^- and ClO_3^- to Cl^-:

$$2e^- + 2H^+(aq) + ClO_4^-(aq) \longrightarrow ClO_3^-(aq) + H_2O(l) \qquad E° = 1.19 \text{ V} \qquad [21.28]$$

$$6e^- + 6H^+(aq) + ClO_3^-(aq) \longrightarrow Cl^-(aq) + 3H_2O(l) \qquad E° = 1.38 \text{ V} \qquad [21.29]$$

$$\overline{8e^- + 8H^+(aq) + ClO_4^-(aq) \longrightarrow Cl^-(aq) + 4H_2O(l)} \qquad\qquad\qquad\qquad [21.30]$$

Adding the two half-reactions together results in the desired half-reaction. However, the correct electrode potential cannot be obtained by merely adding the corresponding $E°$ values.

To understand why this is so, we must recall (Section 18.6) that when we add reactions (or half-reactions) to obtain a new reaction (or half-reaction), the free-energy change, ΔG, for the new reaction is the sum of the free-energy changes for the reactions that are added. We also know that the standard potential is related to the free-energy change by the equation $\Delta G° = -n\mathcal{F}E°$. Thus we have that

$$\Delta G° \text{ (reaction 21.30)} = \Delta G° \text{ (reaction 21.28)} + \Delta G \text{ (reaction 21.29)}$$

$$-n\mathcal{F}E° \text{ (reaction 21.30)} = -n\mathcal{F}E° \text{ (reaction 21.28)} - n\mathcal{F}E° \text{ (reaction 21.29)}$$

$$-8\mathcal{F}E° \text{ (reaction 21.30)} = -2\mathcal{F}E° \text{ (reaction 21.28)} - 6\mathcal{F}E° \text{ (reaction 21.29)}$$

$$= -2\mathcal{F}(1.19 \text{ V}) - 6\mathcal{F}(1.38 \text{ V})$$

$$E° \text{ (reaction 21.30)} = \frac{-2\mathcal{F}(1.19 \text{ V}) - 6\mathcal{F}(1.38 \text{ V})}{-8\mathcal{F}}$$

$$= \frac{10.66 \text{ V}}{-8}$$

$$= 1.33 \text{ V}$$

You may wonder why this procedure was not employed in Chapter 19, in which we added half-reactions and their corresponding $E°$ values. In Chapter 19 we added two half-reactions to give a balanced equation, in which the number of electrons gained in one half-reaction is just balanced by the number of electrons lost in the other half-reaction. When the half-reactions are added to give a complete balanced equation, n, the number of electrons, is the same for all reactions. It thus cancels out, and the result is the same whether we directly add $E°$ values or use the more complicated procedure of adding free energies, as described above. The important point to remember is that when half-reactions are added to give a new half-reaction in which there is a net gain or loss of electrons, the procedure described above must be used to obtain the correct half-reaction potential.

21.4 General characteristics of the group 6A elements

The elements of group 6A include oxygen, sulfur, selenium, tellurium, and polonium. In discussing the chemical properties of these elements we will find it useful to treat oxygen separately, because it differs substantially from the other elements of the group. We will not have much to say about polonium, an element produced by radioactive decay of radium. There are no stable isotopes of this element; as a result, it is found only in trace quantities in radium-containing minerals.

The group 6A elements possess the general outer electron configuration ns^2np^4, where n may have values ranging from 2 through 6. These elements thus may attain a noble-gas electron configuration by addition of two electrons, which results in a -2 oxidation state. Because the group 6A elements are nonmetals, this is a common oxidation state. Except for oxygen, however, the group 6A elements are also commonly found in positive oxidation states up to $+6$, which corresponds to shar-

TABLE 21.8 Some properties of the atoms of the group 6A elements

Property	O	S	Se	Te
Atomic radius (Å)	0.73	1.02	1.17	1.35
X^{2-} ionic radius (Å)	1.45	1.90	2.02	2.22
First ionization energy (kJ/mol)	1312	1004	946	870
Electron affinity (kJ/mol)	−141	−201	−195	−186
Electronegativity	3.4	2.6	2.6	2.1
Energy of X—X single bond (kJ/mol)	142[a]	266	172	126
Reduction potential to H_2X in acidic solution	1.23	0.14	−0.40	−0.72

[a] Based on O—O bond energy in H_2O_2.

ing of all six valence-shell electrons with atoms of a more electronegative element.

The group 6A elements vary considerably in their abundances. Oxygen is everywhere about us, in the atmosphere and in the earth's crust. Sulfur is generally abundant in the earth's crust, in a variety of forms as we shall see, but mainly in the form of sulfide ores. Selenium is rather scarce; it generally occurs as a minor constituent in sulfur-containing minerals. Tellurium is among the rarest of the elements and is less abundant than gold or platinum.

Some of the more important properties of the atoms of the group 6A elements are summarized in Table 21.8. The energy of the X—X single bond is estimated from data for the elements, except for oxygen. In this case, because the O—O bond in O_2 is not a single bond (Sections 8.6 and 8.7), the estimated O—O bond energy in hydrogen peroxide is employed. The reduction potential listed in the last line of the table refers to the reduction of the element in its standard state to form $H_2X(aq)$ in acidic solution. In most of the properties listed in Table 21.8 we again see a regular variation as a function of atomic number. Atomic and ionic radii increase; correspondingly, the ionization energy decreases, as we expect from the discussion in Section 6.5.

The electron affinities listed apply to the process shown in Equation [21.31]:

$$X(g) + e^- \longrightarrow X^-(g) \qquad [21.31]$$

This, of course, does not produce the commonly observed stable ionic form of these elements, X^{2-}, but it is the first step in its formation. It is interesting to note that addition of an electron to oxygen is less exothermic than addition of an electron to one of the other group 6A elements. The origin of this effect has to do with the relatively small size of the oxygen atom. Addition of a single extra electron to the smaller atom results in larger electron-electron repulsions, offsetting the gain in stability that results from the closer approach of the electron to the nucleus.

The decrease in ionization potential is the factor mainly responsible for the decrease in electronegativity in the series. In this connection it is interesting to note that sulfur and selenium are closely similar in many respects, whereas tellurium is substantially less electronegative. Note that the ease of reduction of the free element to form H_2X varies greatly throughout the series. Whereas oxygen is very readily reduced to the -2 oxidation state, the potential for reduction of tellurium is actually quite strongly negative. These observations indicate an increasingly metallic character in the group 6A elements with increasing atomic number. The physical properties of the elements show a corresponding trend. At one extreme we have oxygen, a diatomic molecule and sulfur, a yellow, non-conducting solid that melts at 114°C. At the other extreme, the stable form of tellurium has a bright luster and low electrical conductivity and melts at 452°C.

Several interesting comparisons can be made between the data listed in Table 21.8 and the corresponding values for the halogens listed in Table 21.4. Notice that the ionization energies and electron affinities of the halogens are generally higher. Correspondingly, the atomic radii of the halogens are smaller and their electronegativities are higher. The potentials for reduction of the free elements to the stable negative oxidation state are greater for the halogens, as expected. The X—X single bond energies for the corresponding members of the two families are not greatly different. For example, the S—S bond in S_8 has a bond energy of 226 kJ/mol as compared with 243 kJ/mol for the Cl—Cl bond in Cl_2. It is interesting that in both families of elements the X—X bond energy for the first member in each series is unusually low. With this background, let us now consider first the properties and chemical behavior of oxygen, then that of the other group 6A elements.

21.5 Oxygen

Oxygen, one of the most important of the elements, is vital to maintaining life as we know it, because it is essential to animal respiration. It is also essential in the combustion processes used to generate most of the energy used in our technological society. Oxygen plays an important role in the chemistries of most other elements and is found in combination with other elements in a great variety of compounds.

The electron configuration of oxygen is [He] $2s^2 2p^4$. Because it is the second most electronegative element, it exhibits principally negative oxidation states. The only examples of positive oxidation states occur in the oxygen-fluorine compounds OF_2 and O_2F_2.

Oxygen has two allotropes, diatomic oxygen, O_2, and triatomic oxygen, O_3 (ozone). When we speak of a molecule of oxygen, it is normally understood that we are speaking of O_2, the normal form of the element. The structure of ozone is shown in Figure 21.11.

OCCURRENCE AND ISOLATION

Oxygen is the most abundant element both in the earth's crust and in the human body. It constitutes 89 percent of water by weight. It also comprises about 50 percent of sand, clay, limestone, and igneous rocks that make up the bulk of the earth's crust. The principal source of the element is the atmosphere, which contains about 20 mole-percent O_2.

Oxygen is obtained for commercial use by the fractional distillation of liquefied air. The normal boiling point of O_2 is -183°C, whereas that

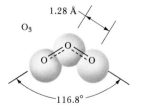

FIGURE 21.11 The structure of the ozone molecule.

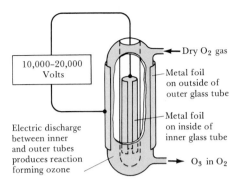

← Dry O_2 gas

Metal foil
on outside of
outer glass tube

Metal foil
on inside of
inner glass tube

Electric discharge
between inner
and outer tubes
produces reaction
forming ozone

→ O_3 in O_2

FIGURE 21.12 An apparatus for producing ozone from O_2.

of N_2, the other principal component of air, is $-196°C$. Where O_2 of higher purity is required, it can be obtained by electrolysis of water. Although water is inexpensive and plentiful, the cost of electricity makes this an expensive way of obtaining elemental oxygen. In choosing a commercial method of preparing a substance, cost is the most important factor. In contrast, in choosing a method for preparing small amounts of a substance in the laboratory, convenience becomes an important factor. The common laboratory preparation of O_2 involves the thermal decomposition of potassium chlorate, $KClO_3$, with MnO_2 added as a catalyst:

$$2KClO_3(s) \xrightarrow[\text{heat}]{MnO_2} 2KCl(s) + 3O_2(g) \qquad [21.32]$$

The experimental setup was shown earlier, in Figure 9.9.

Ozone can be prepared by passing electricity through dry O_2. An apparatus for accomplishing this is shown in Figure 21.12. The pungent odor of ozone gas can sometimes be detected around electrical equipment where there is a spark jump, and in the atmosphere during lightning storms. Ozone is an even more powerful oxidizing agent than O_2. However, it cannot be stored for long except at low temperatures; consequently, it must usually be used as it is generated. As we saw in Chapter 10, ozone is an important component of the upper atmosphere, in that it screens out ultraviolet radiation. In this way ozone protects the earth from the effects of these high-energy rays. However, ozone can do considerable damage in the lower atmosphere because it is such a powerful oxidizing agent. Because it causes damage to plants, animals, and structural materials, it is considered to be an air pollutant.

PROPERTIES AND COMPOUNDS

Oxygen in its normal elemental form, O_2, melts at $-218°C$ and boils at $-183°C$. At room temperature it is a colorless, odorless, and tasteless gas. It is paramagnetic (Section 8.5).

Oxygen is a chemically active substance. It combines either directly or indirectly with nearly all other elements to form oxides (O^{2-}) and occasionally peroxides (O_2^{2-}) and even superoxides (O_2^-). When the alkali metals are combusted in oxygen, the following oxides are the predominant products: Li_2O (oxide), Na_2O_2 (peroxide), KO_2 (superoxide), RbO_2 (superoxide), and CsO_2 (superoxide).

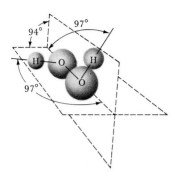

FIGURE 21.13 The structure of the hydrogen peroxide molecule.

Potassium superoxide is used as an oxygen source in masks worn for rescue work. Moisture in the breath causes the compound to decompose to form O_2 and KOH. The KOH so formed serves to remove CO_2 from the exhaled breath:

$$2KO_2(s) + 2H_2O(l) \longrightarrow O_2(g) + H_2O_2(l) + 2K^+(aq) + 2OH^-(aq)$$
$$2KOH(s) + CO_2(g) \longrightarrow H_2O(l) + K_2CO_3(s)$$

The most familiar peroxide is hydrogen peroxide, H_2O_2, whose structure is shown in Figure 21.13. One important commercial process for production of H_2O_2 involves electrolysis of a 50 percent aqueous solution of sulfuric acid. The products of the electrolysis are hydrogen and persulfate ion, $S_2O_8^{2-}$:

$$
\begin{array}{c}
2H^+(aq) + 2e^- \longrightarrow H_2(g) \\
2SO_4^{2-}(aq) \longrightarrow S_2O_8^{2-}(aq) + 2e^- \\
\hline
2H^+(aq) + 2SO_4^{2-}(aq) \longrightarrow H_2(g) + S_2O_8^{2-}(aq)
\end{array}
\qquad \text{[21.33]}
$$

Persulfate ion is the oxidation product; it possesses an O—O bond joining two SO_4^- units: $O_3SO\text{—}OSO_3^{2-}$. Distillation of the acidic solution containing persulfate ion results in decomposition to form H_2O_2 and sulfate ion. The net ionic equation is

$$2H_2O(l) + S_2O_8^{2-}(aq) \longrightarrow H_2O_2(aq) + 2H^+(aq) + 2SO_4^{2-}(aq) \qquad \text{[21.34]}$$

Pure hydrogen peroxide is a clear syrupy liquid, density 1.47 g/cm^3 at 0°C. It melts at −0.4°C, and its normal boiling point is 151°C. These properties are characteristic of a highly polar, strongly hydrogen-bonded liquid like water. Concentrated hydrogen peroxide is a dangerously reactive substance, because the decomposition to form water and oxygen gas is exothermic:

$$2H_2O_2(l) \longrightarrow 2H_2O(l) + O_2(g) \qquad \Delta H° = -196.0 \text{ kJ} \qquad \text{[21.35]}$$

The decomposition can occur with explosive violence if highly concentrated hydrogen peroxide comes in contact with substances that can catalyze the reaction. Hydrogen peroxide is marketed as a chemical reagent in aqueous solutions of up to about 30 percent by weight. A solution containing about 3 percent by weight H_2O_2 is commonly used as a mild antiseptic; somewhat more concentrated solutions are employed to bleach fabrics such as wool or silk.

Hydrogen peroxide is capable of acting as either an oxidizing or reducing agent. Equations [21.36] and [21.37] show the half-reactions for reaction in acid solution.

$$2H^+(aq) + H_2O_2(aq) + 2e^- \longrightarrow 2H_2O(l) \qquad E° = 1.77 \text{ V} \qquad \text{[21.36]}$$

$$H_2O_2(aq) \longrightarrow O_2(g) + 2H^+(aq) + 2e^- \qquad E° = -0.67 \text{ V} \qquad \text{[21.37]}$$

In basic solution, the corresponding standard electrode potentials are 0.87 V for reduction of H_2O_2 and 0.08 V for its oxidation.

In many old oil paints the white lead carbonate pigments have become discolored due to formation of black lead sulfide. Hydrogen peroxide has been used in restoration work, to convert the black sulfide to white lead sulfate. Both salts are insoluble in water. The reaction is as shown in Equation [21.38]:

$$PbS(s) + 4H_2O_2(aq) \longrightarrow PbSO_4(s) + 4H_2O(l) \qquad [21.38]$$

USES OF OXYGEN

Oxygen is the third most widely used industrial chemical, ranking behind sulfuric acid and lime, CaO. About 14 billion kg of the element are used annually. It is widely used as an oxidizing agent. Approximately half the oxygen produced is used in the steel industry, mainly to remove impurities from steel (Section 22.6). Oxygen is employed in medicine to quicken the life-sustaining oxidation processes. It is used together with acetylene, C_2H_2, in oxyacetylene welding. This use is based on the highly exothermic reaction between C_2H_2 and O_2, which results in temperatures in excess of 3000°C. This combustion reaction is shown in Equation [21.39]:

$$2C_2H_2(g) + 5O_2(g) \longrightarrow 4CO_2(g) + 2H_2O(g) \quad \Delta H° = -2510\,kJ \qquad [21.39]$$

THE OXYGEN CYCLE

Oxygen accounts for about one-fourth of the atoms in living matter. Because the number of oxygen atoms is fixed, as O_2 is removed from air through respiration and other processes, it needs to be replenished. The major nonliving sources of oxygen other than O_2 are CO_2 and H_2O. A simplified picture of the movement of oxygen in our environment is shown in Figure 21.14. This figure points out both how O_2 is removed from the atmosphere and how it is replenished. Oxygen, O_2, is reformed mainly from CO_2 through the process of photosynthesis. Energy is produced when O_2 is converted to CO_2; energy must therefore be supplied

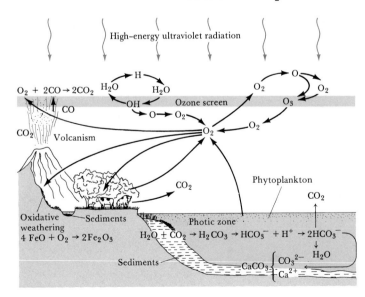

FIGURE 21.14 A simplified view of the oxygen cycle, showing some of the primary reactions involving oxygen in nature. The atmosphere, which contains O_2, is one of the primary sources of the element. Some O_2 is produced by radiation-induced dissociation of H_2O in the upper atmosphere. Some O_2 is produced by green plants from H_2O and CO_2 in the course of photosynthesis. Atmospheric CO_2, in turn, results from combustion reactions, animal respiration, and the dissociation of bicarbonate in water. The O_2 is used to produce ozone in the upper atmosphere, in oxidative weathering of rocks, in animal respiration, and in combustion reactions.

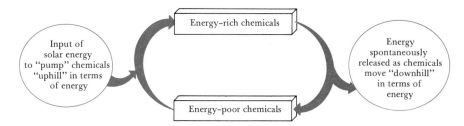

FIGURE 21.15 A schematic representation of the role of solar energy in the cycling of elements in nature.

to reform O_2 from CO_2. This energy is provided by the sun. Thus life on earth depends on chemical recycling made possible by solar energy. The situation is represented schematically in Figure 21.15.

21.6 Sulfur, selenium, and tellurium

The chemical behavior of sulfur, selenium, and tellurium is much different from that of oxygen. One notable difference is the occurrence of positive oxidation states up to $+6$, as in the ions SO_4^{2-}, SeO_4^{2-}, and $H_4TeO_6^{2-}$. In addition, the heavier elements exhibit expanded valence shells in many compounds, for example, in the hexafluorides SF_6, SeF_6, and TeF_6.

OCCURRENCES AND ISOLATION

Sulfur occurs in the elemental state in large underground deposits that serve as the principal source of the element. The Frasch process, illustrated in Figure 21.16, is used to obtain the element from these deposits. The method is based on the low melting point and low density of sulfur.

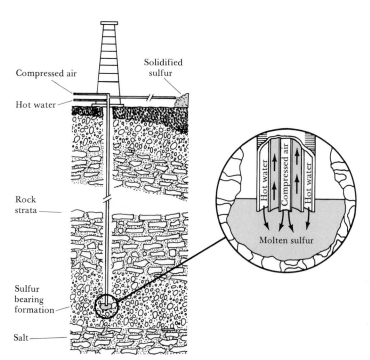

FIGURE 21.16 The mining of sulfur by the Frasch process. The process is named after Herman Frasch, who invented the process in the early 1890s. The process is particularly useful for recovering sulfur from deposits located under quicksand or water.

FIGURE 21.17 The production platform of the world's first offshore sulfur mine. The mine stands in 50 ft of water 7 mi off Grand Isle, Louisiana. (*Exxon Company, U.S.A.*)

Superheated water is forced into the deposit, where it melts the sulfur. Compressed air then forces the molten sulfur up a pipe that is concentric with the ones that introduce the hot water and compressed air into the deposit. The method is particularly useful for removing sulfur from beds that lie under water or quicksand. Figure 21.17 shows an offshore sulfur mine.

Sulfur also occurs widely as sulfide and sulfate minerals and as a minor component of coal and petroleum. The presence of sulfur in coal and petroleum poses a major problem. As we saw in Section 10.6, combustion of these "unclean" fuels leads to serious sulfur oxide pollution. Likewise, operations that use sulfide minerals as sources of metals liberate sulfur oxides. Much effort has therefore been directed at removing this sulfur, and these efforts have increased the availability of sulfur. The sale of this sulfur helps partially to offset the costs of the desulfurizing processes and equipment. However, sulfur obtained from sulfur deposits by the Frasch process is about 99.5 percent pure and can be used for most commercial processes without purification. It is therefore relatively cheap. Consequently, sulfur from desulfurizing processes must be sold at prices below its cost in order to compete on the market. Nevertheless, about half the sulfur used in the United States each year is produced by means other than the Frasch process.

Selenium and tellurium occur in rare minerals as Cu_2Se, $PbSe$, Ag_2Se, Cu_2Te, $PbTe$, Ag_2Te, and Au_2Te. They also occur as minor constituents in sulfide ores of copper, iron, nickel, and lead. From a commercial point of view, the most important sources of these elements are the copper ores. When these are roasted to form copper metal, the selenium and tellu-

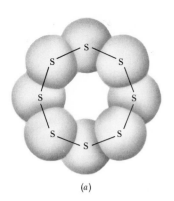

(a)

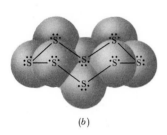

(b)

FIGURE 21.18 Top view (a) and side view (b) of the rhombic sulfur molecule.

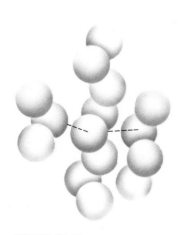

FIGURE 21.19 A portion of the structure of crystalline selenium or tellurium. The dashed lines represent weak bonding interactions between atoms in adjacent chains.

rium are retained to a large extent in the copper as the elements. When the copper is purified by electrolysis, as described in Section 19.6, the impurities such as selenium and tellurium along with precious metals such as gold and silver collect in the so-called anode sludge. If the anode sludge is treated with concentrated sulfuric acid at about 400°C, selenium is oxidized to selenium dioxide, which sublimes from the reaction mixture:

$$Se(s) + 2H_2SO_4(l) \xrightarrow{\Delta} SeO_2(g) + 2SO_2(g) + 2H_2O(g) \qquad [21.40]$$

The dioxide is recovered as a white powder from the upper furnace walls. It is then dissolved in dilute hydrochloric acid solution, in which it is quite soluble. Selenous acid, a weak acid, is formed:

$$SeO_2(s) + H_2O(l) \longrightarrow H_2SeO_3(aq) \qquad [21.41]$$

Sulfur dioxide is then bubbled into the solution, and an oxidation-reduction reaction occurs:

$$H_2O(l) + H_2SeO_3(aq) + 2SO_2(g) \longrightarrow Se(s) + 2H^+(aq) + 2HSO_4(aq) \quad [21.42]$$

Tellurium is recovered from the anode sludge by other procedures.

PROPERTIES AND USES

As we normally encounter it, sulfur is yellow, tasteless, and nearly odorless. It is insoluble in water and exists in several allotropic forms. The thermodynamically stable form at room temperature is rhombic sulfur, which consists of puckered S_8 rings, as shown in Figure 21.18. When heated above its melting point, at 113°C, sulfur undergoes a variety of changes. The molten sulfur first contains S_8 molecules and is fluid because the rings readily slip over each other. Further heating of this straw-colored liquid causes rings to break, the fragments joining to form very long molecules that can become entangled. The sulfur consequently becomes highly viscous. This change is marked by a color change to dark reddish-brown. Further heating breaks the chains and the viscosity again decreases.

Most of the 1.1×10^{10} kg of sulfur produced in the United States each year are used in the manufacture of sulfuric acid. Sulfur is also used in vulcanizing rubber, a process that toughens rubber by introducing cross-linking between polymer chains (Section 11.7).

The most stable allotropes of both selenium and tellurium are crystalline substances containing zigzag chains of atoms, as illustrated in Figure 21.19. Each atom of the chain, however, is close to atoms in adjacent chains, and it appears that there is some sharing of electron pairs between these atoms, as indicated by the dotted lines in Figure 21.19. The electrical conductivity of selenium is very small in the dark but increases greatly upon exposure to light. This property of the element is used in photoelectric cells and in light meters such as those used in cameras. Xerox copiers also depend on the photoconductivity of selenium for their operation.

Sulfur is widely distributed in biological systems. It is present in most proteins, as a component of the amino acids cysteine and methionine (Section 25.2). In contrast, selenium is rare in biological systems. It has only recently been established that there is a human nutritional requirement for the element. Selenium is present in trace quantities in most vegetables, especially spinach. The amount of selenium required for adequate nutrition is very small. In quantities much larger than this small nutritional requirement, the element is toxic. Tellurium does not have a known role in human nutrition. Its compounds are poisonous and, if they are volatile, usually have highly offensive odors.

OXIDES AND COMPOUNDS

Sulfur dioxide was first discovered by Joseph Priestley in 1774, when he heated mercury with concentrated sulfuric acid. The reaction which occurs is shown in Equation [21.43]:

$$Hg(l) + 2H_2SO_4(l) \longrightarrow HgSO_4(s) + SO_2(g) + 2H_2O(l) \qquad [21.43]$$

In the laboratory, SO_2 is prepared by the action of aqueous acid on a sulfite salt, as shown in Equation [21.44]:

$$2H^+(aq) + SO_3^{2-}(aq) \longrightarrow SO_2(g) + H_2O(l) \qquad [21.44]$$

Sulfur dioxide is formed when sulfur is combusted in air; it has a choking odor and is poisonous. The gas is particularly toxic to lower organisms such as fungi and is consequently used for sterilizing dried fruit. At 1 atm pressure and room temperature, SO_2 dissolves in water to the extent of 45 volumes of gas per volume of water, to produce a solution of about $1.6\,M$ concentration. The solution is acidic, and we describe it as $H_2SO_3(aq)$. Actually, there is evidence that much of the sulfur dioxide exists in solution as hydrated SO_2. When the saturated solution is cooled, crystals of the hydrate, $SO_2 \cdot 6H_2O$, can be recovered. It is convenient, however, to assume that all of the SO_2 that dissolves is in the form of the weak acid, H_2SO_3, which ionizes according to Equations [21.45] and [21.46]:

$$H_2SO_3(aq) \rightleftharpoons H^+(aq) + HSO_3^-(aq) \qquad K_{a1} = 1.7 \times 10^{-2}(25°C) \qquad [21.45]$$
$$HSO_3^-(aq) \rightleftharpoons H^+(aq) + SO_3^{2-}(aq) \qquad K_{a2} = 6.4 \times 10^{-8}(25°C) \qquad [21.46]$$

Aqueous sulfur dioxide can act as a reducing agent, for example, in the reaction with iodate ion in acidic solution:

$$2IO_3^-(aq) + 5H_2SO_3(aq) \longrightarrow$$
$$I_2(s) + 5HSO_4^-(aq) + H_2O(l) + 3H^+(aq) \qquad [21.47]$$

As we have seen, combustion of sulfur in air produces mainly SO_2, but small amounts of SO_3 are also formed. Sulfur trioxide is of great commercial importance because it is the anhydride of sulfuric acid. In the manufacture of sulfuric acid, SO_2 is first obtained by burning sulfur. The SO_2 is then oxidized to SO_3 using a catalyst such as V_2O_5 or platinum.

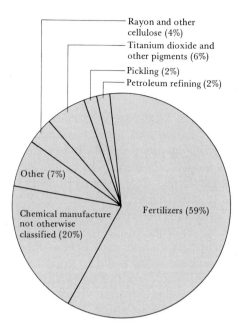

Rayon and other
cellulose (4%)

Titanium dioxide and
other pigments (6%)

Pickling (2%)

Petroleum refining (2%)

Other (7%)

Chemical manufacture
not otherwise
classified (20%)

Fertilizers (59%)

FIGURE 21.20 Sulfuric acid use in the United States.

The SO_3 is dissolved in H_2SO_4 because it does not dissolve quickly in water. The reaction is shown in Equation [21.48]. The $H_2S_2O_7$ formed in this reaction, called pyrosulfuric acid, is then added to water to form H_2SO_4 as shown in Equation [21.49]:

$$SO_3(g) + H_2SO_4(l) \longrightarrow H_2S_2O_7(l) \qquad [21.48]$$

$$H_2S_2O_7(l) + H_2O(l) \longrightarrow 2H_2SO_4(l) \qquad [21.49]$$

Commercial sulfuric acid is generally 98 percent H_2SO_4 and boils at 340°C. It is a colorless, oily liquid. Sulfuric acid has many useful properties. Most importantly, it is a strong acid, a good dehydrating agent,* and a moderately good oxidizing agent.

Year after year, the output of sulfuric acid has been the largest of any chemical produced in the United States. About 3.0×10^{10} kg are produced annually in this country. Sulfuric acid is employed in some way in almost all manufacturing. Consequently, its consumption is considered a standard measure of industrial activity. The primary uses for sulfuric acid are shown in Figure 21.20.

Only the first proton in sulfuric acid is completely ionized. The second proton ionizes only partially:

$$H_2SO_4(aq) \longrightarrow H^+(aq) + HSO_4^-(aq)$$

$$HSO_4^-(aq) \rightleftharpoons H^+(aq) + SO_4^{2-}(aq) \qquad K_a = 1.1 \times 10^{-2}$$

Consequently, sulfuric acid forms two series of compounds—sulfates and bisulfates (or hydrogen sulfates).

---

*Considerable heat is given off when sulfuric acid is diluted with water. Consequently, dilution must always be done carefully by pouring the acid into water to distribute the heat as uniformly as possible and to avoid spattering of the acid.

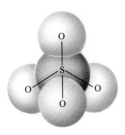

SO_4^{2-}

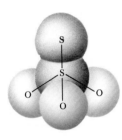

$S_2O_3^{2-}$

FIGURE 21.21 Comparison of the structures of the sulfate, SO_4^{2-}, and thiosulfate, $S_2O_3^{2-}$, ions.

Related to the sulfate ion is the thiosulfate ion, $S_2O_3^{2-}$, formed by boiling an alkaline solution of SO_3^{2-} with elemental sulfur, as shown in Equation [21.50]:

$$8SO_3^{2-}(aq) + S_8(s) \longrightarrow 8S_2O_3^{2-}(aq) \qquad [21.50]$$

The term "thio" indicates substitution of sulfur for oxygen. The structures of the sulfate and thiosulfate ions are compared in Figure 21.21. When acidified, the thiosulfate ion decomposes to form sulfur and H_2SO_3. The pentahydrated salt of sodium thiosulfate, $Na_2S_2O_3 \cdot 5H_2O$, is known as "hypo." It is used in photography to furnish thiosulfate ion as a complexing agent for silver ion. Photographic film consists of a suspension of microcrystals of AgBr in gelatin. In those portions of the film that are exposed to light the AgBr is "activated." This means that when the film is treated with a mild reducing agent (called the developer) the exposed AgBr is reduced. Thus black, metallic silver is formed in the film in concentrations proportional to the intensity of light exposure. The film is then treated with sodium thiosulfate solution to remove the remaining silver bromide, by forming the soluble silver thiosulfate complex:

$$AgBr(s) + 2S_2O_3^{2-}(aq) \rightleftharpoons Ag(S_2O_3)_2^{3-}(aq) + Br^-(aq) \qquad [21.51]$$

This step in the process is called "fixing." Thiosulfate ion is also used in quantitative analysis as an oxidizing agent for iodine:

$$2S_2O_3^{2-}(aq) + I_2(s) \longrightarrow 2I^-(aq) + S_4O_6^{2-}(aq) \qquad [21.52]$$

Both selenium and tellurium form dioxides upon burning the element in air or oxygen. Selenium dioxide is very soluble in water; it forms an acidic solution from which crystals of selenous acid, H_2SeO_3, can be isolated. Neutralization of selenous acid solutions yields salts of hydrogen selenite or selenite ion, for example, $NaHSeO_3$ or Na_2SeO_3. Tellurium dioxide is insoluble in water; thus the aqueous acid H_2TeO_3 is unknown. However, it is possible to prepare tellurite salts by dissolving TeO_2 in an aqueous base.

SAMPLE EXERCISE 21.4

In the vapor phase, SeO_2 exists as monomeric molecules. In the solid state it consists of zigzag chains involving Se—O—Se—O bonds. Write the Lewis structures of each form of SeO_2 and predict the bond angles about the selenium in each case.

Solution: In the gas phase, SeO_2 should have the Lewis structures

Using the VSEPR model (Section 8.1) we predict that the molecule will be bent, with an O—Se—O bond angle of somewhat less than 120°.

The Lewis structure for the solid state form is as follows:

According to the VSEPR model, the four valence-shell electron pairs about Se are arranged approximately at the vertices of a tetrahedron. The three Se—O bonds thus form a trigonal-pyramidal arrangement; the O—Se—O bond angles should be somewhat less than 109°. (It is found that the angles are not all the same; the O—Se—O angle involving the oxygens in the chain is 98°, the other two O—Se—O angles are 90°.)

Both selenium trioxide and tellurium trioxide are known. These are the anhydrides of selenic and telluric acids, but the acids are formed other than by dissolving the trioxides in water. Selenic acid, H_2SeO_4, can be formed by oxidation of aqueous SeO_2 with 30 percent hydrogen peroxide, Equation [21.53]:

$$H_2SeO_3(aq) + H_2O_2(aq) \longrightarrow HSeO_4^-(aq) + H_2O(l) + H^+(aq) \qquad [21.53]$$

Notice that we represent $H_2SeO_3(aq)$ in the un-ionized form; the acid-dissociation constants for H_2SeO_3 at 25°C are $K_{a1} = 2.3 \times 10^{-3}$ and $K_{a2} = 5.3 \times 10^{-9}$. The un-ionized acid is thus probably most representative of the species present in solution. On the other hand, selenic acid, H_2SeO_4, is a strong acid in its first dissociation constant; the second dissociation has an equilibrium constant $K_{a2} = 2.2 \times 10^{-2}$ at 25°C. Either SeO_4^{2-} or $HSeO_4^-$ might therefore be used to represent the most abundant species in acidic solution.

Telluric acid is formed when $TeO_2(s)$ is oxidized by various aqueous oxidizing agents. We might expect the formula for telluric acid to be H_2TeO_4, by analogy with the corresponding acid of sulfur or selenium. However, the acid which is recovered from solution as a white crystalline material is H_6TeO_6. (The formula can also be written as $Te(OH)_6$.) This substance is called orthotelluric acid. Notice the close similarity to iodine, which forms paraperiodic acid, H_5IO_6. In both cases, the large size of the central atom permits bonding with six surrounding oxygens. Orthotelluric acid is a weak acid, capable of dissociating up to two protons (at 25°C, $K_{a1} = 2.4 \times 10^{-8}$, $K_{a2} = 1.0 \times 10^{-11}$).

SULFIDES, SELENIDES, AND TELLURIDES

Sulfur forms compounds by direct combination with many elements. When the element is less electronegative than sulfur, sulfides, which contain S^{2-}, form. For example, iron(II) sulfide, FeS, forms by direct combination of iron and sulfur. Many metallic elements are found in the form of sulfide ores, e.g., PbS (galena) and HgS (cinnabar). A series of related ores, containing the S_2^{2-} ion (analogous to the peroxide ion) are known as *pyrites*. Iron pyrite, FeS_2, occurs as golden-yellow cubic crystals. Because it has been on occasion mistaken for gold by overeager gold miners, in some locales it is called "fool's gold."

One of the most important sulfides is hydrogen sulfide, H_2S. This substance is not normally produced by direct union of the elements because it is unstable at elevated temperatures and decomposes into the elements. It is normally prepared by action of dilute sulfuric acid on iron(II) sulfide as shown in Equation [21.54]:

$$FeS(s) + 2H^+(aq) \longrightarrow H_2S(aq) + Fe^{2+}(aq) \qquad [21.54]$$

A common laboratory source of H_2S is the reaction of thioacetamide with water, as shown in Equation [21.55]:

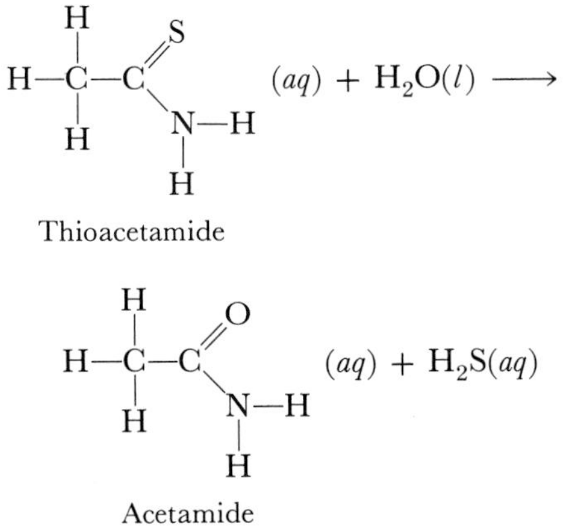

Thioacetamide

Acetamide

Hydrogen sulfide is often used in the laboratory for qualitative analysis of certain metal ions (Section 16.6).

One of hydrogen sulfide's most readily recognized properties is its odor; H_2S is largely responsible for the offensive odor of rotten eggs. Hydrogen sulfide is actually quite toxic; it has about the same level of toxicity as hydrogen cyanide, the gas that was used in California's gas chambers. The volatile hydrides H_2Se and H_2Te are similar to H_2S in many respects. Both compounds possess very offensive, lingering odors and are toxic. In aqueous solutions the acid strength increases in the order $H_2S < H_2Se < H_2Te$.

21.7 General characteristics of the group 5A elements

The elements of group 5A are nitrogen, phosphorus, arsenic, antimony, and bismuth. We will find it convenient to discuss the chemical properties of nitrogen and phosphorus each separately, and then to comment briefly on the chemical characteristics of the heavier elements of the group.

The group 5A elements possess the outer electron configuration ns^2np^3, where n may have values ranging from 2 to 6. A noble-gas configuration results from addition of three electrons to form the -3 oxidation state. Ionic compounds containing X^{3-} ions are not common, however, except for salts of the more active metals, for example, Na_3N. More commonly the group 5A element acquires an octet of electrons via covalent bonding. The oxidation number may range from -3 to $+5$, depending on the nature and number of the atoms to which it is bound.

Both nitrogen and phosphorus are abundant and important components of our environment. Nitrogen is, of course, the major component of the earth's atmosphere and is present in substantial quantities in biological systems. Phosphorus is found in minerals as phosphate; this element also is an important component of biological systems. Arsenic, antimony, and bismuth are much less abundant but are readily obtainable from accessible mineral sources. Bismuth holds an interesting place

TABLE 21.9 Properties of the atoms of group 5A elements

Property	N	P	As	Sb	Bi
Atomic radius (Å)	0.75	1.10	1.22	1.43	—
First ionization energy (kJ/mol)	1402	1012	947	834	703
Electron affinity (kJ/mol)	+6.8	−72	−77	−101	−106
Electronegativity	3.0	2.2	2.2	2.0	2.0
X—X single bond[a] energy (kJ/mol)	163	200	150	120	—
X≡X triple bond energy (kJ/mol)	941	480	380	295	192

[a]Approximate values only.

in chemistry. The only naturally occurring nuclide of this element, ^{209}Bi, has the highest atomic number of any nuclide that is stable with respect to radioactive decay. (This nuclide probably does decompose, but its half-life is estimated to be in excess of 10^{18} years.)

Some of the important properties of the atoms of the group 5A elements are listed in Table 21.9. The general pattern that emerges from these data is similar to what we have seen before; size and metallic character increase as atomic number increases within the group. Note also that in comparison with the corresponding elements of groups 6A and 7A, the atomic radii are larger, and ionization energies and electronegativities are lower.

The variation in properties among the elements of group 5A is more striking than that seen in the case of groups 6A or 7A. Nitrogen at the one extreme exists as a gaseous diatomic molecule; it is clearly nonmetallic in character. At the other extreme, bismuth is a reddish-white, metallic-looking substance that has most of the characteristics of a metal.

The electron affinities of the group 5A elements are especially interesting. In the ground state, the atoms of group 5A elements possess the electron configuration $ns^2np^1np^1np^1$. That is, according to Hund's rule, the valence-level p orbitals are exactly half-filled with electrons, with spins paired. Addition of an electron to this rather stable arrangement is not favored, and the electron affinity of nitrogen is in fact zero or slightly positive (Section 6.6). The electron affinities of the other group 5A elements are exothermic, but addition of an electron to any of the group 5A elements releases considerably less energy than the analogous process for the group 6A or 7A elements. The existence of a stable, half-filled valence shell is also responsible for the relatively high ionization energies of the group 5A elements. This is especially evident in the case of nitrogen, which has a higher ionization potential than oxygen.

The values listed for X—X single bond energies are not very reliable, because it is difficult to obtain such data from thermochemical experiments. However, there is no doubt about the general trend: an abnormally low value for the N—N single bond, a sharp increase at phosphorus, then a gradual decline among the other elements of the series. From observations of the group 5A elements in the gas phase (for all except N_2 this requires high temperatures), it is possible to estimate the X≡X triple bond energy, as listed in Table 21.9. Here we see a trend that is very much different than that for the X—X single bond. Nitrogen forms a much stronger bond than do the other elements, and there is a steady decline in the triple bond energy down through the group. These data help us to appreciate why nitrogen alone of the group 5A elements exists as a diatomic molecule in its stable state at 25°C. All the other elements exist in structural forms involving single bonds between the atoms.

21.8 Nitrogen

Nitrogen is the most abundant component of the earth's atmosphere. It occurs there as N_2, a colorless, odorless, and tasteless gas that boils at −196°C and freezes at −210°C. The atmosphere is 78 percent N_2 by volume. Nitrogen is separated from O_2 and the other less abundant components of the atmosphere by fractional distillation of liquid air. The N_2 molecule is quite unreactive because of the strong triple bond between nitrogen atoms. However, electrical discharges in the atmosphere can cause formation of nitric oxide, NO, as shown in Equation [21.56]:

$$N_2(g) + O_2(g) \longrightarrow 2NO(g) \qquad [21.56]$$

The reaction occurs because the intense heat and ionization generated in lightning causes disruption of the N_2 molecule. This simple reaction, which involves formation of a nitrogen-containing compound starting with N_2, is an example of nitrogen fixation. As discussed in Chapter 14, the Haber process is the primary commercial means available for the fixation of nitrogen. In this process N_2 from air and H_2, which is normally obtained from the CH_4 in natural gas,* are combined to form NH_3:

$$N_2(g) + 3H_2(g) \longrightarrow 2NH_3(g) \qquad [21.57]$$

Prior to the development of the Haber process, deposits of $NaNO_3$ in the Chilean desert formed the principal source of nitrogen used in the manufacture of nitrogen compounds. In nature, nitrogen is fixed by certain bacteria that occur in the root nodules of leguminous plants such as peas, beans, peanuts, and alfalfa, and during lightning, as described above.

The demand for fixed nitrogen has increased mainly because the element is required in maintaining soil fertility. Although we are immersed in an ocean of air that principally contains N_2, our supply of food is limited more by the availability of fixed nitrogen than by that of any other plant nutrient. Thus nitrogen is used primarily for the manufacture of nitrogen-containing fertilizers. It is also used in the manufacture of explosives, plastics, and many important chemicals.

*The H_2 can be obtained through the following series of reactions, which occur at elevated temperatures:

$$CH_4(g) + H_2O(g) \longrightarrow CO(g) + 3H_2(g)$$

$$2CH_4(g) + O_2(g) \longrightarrow 2CO(g) + 4H_2(g)$$

$$CO(g) + H_2O(g) \longrightarrow CO_2(g) + H_2(g)$$

Because CH_4 is used as the source of H_2, the fixation of N_2 is closely tied to the availability of natural gas. As gas prices increase, alternative sources of hydrogen become more attractive. The overall result, however, is that fertilizer costs, and thus food costs, increase.

When substances burn in ordinary air they normally react with oxygen, not with nitrogen. However, when active metals burn in air, as when a magnesium ribbon burns with an intensely hot, white flame, reaction with N_2 also occurs to form magnesium nitride. A similar reaction occurs with lithium:

$$6Li(s) + N_2(g) \longrightarrow 2Li_3N(s) \qquad [21.58]$$

$$3Mg(s) + N_2(g) \longrightarrow Mg_3N_2(s) \qquad [21.59]$$

These reactions are analogous to the reactions of the metal with oxygen to form metal oxides. The nitride ion, N^{3-}, so formed is a strong Brønsted base, just as is O^{2-}. It therefore forms ammonia when placed in contact with water:

$$Mg_3N_2(s) + 6H_2O(l) \longrightarrow 2NH_3(aq) + 3Mg(OH)_2(s) \qquad [21.60]$$

Ammonia is one of the most important compounds of nitrogen. It is a colorless, toxic gas that has a characteristic irritating odor. It boils at $-33°C$ and freezes at $-78°C$. In the laboratory, NH_3 is prepared by the action of NaOH on an ammonium salt as shown in Equation [21.61]:

$$NH_4Cl(aq) + NaOH(aq) \longrightarrow NH_3(g) + H_2O(l) + NaCl(aq) \qquad [21.61]$$

About 75 percent of the ammonia produced in this country is used for fertilizer.

Hydrazine, N_2H_4, bears the same relationship to ammonia that hydrogen peroxide does to water. As shown in Figure 21.22, the hydrazine molecule contains an N—N single bond. Pure hydrazine is an oily, colorless liquid with a melting point of 1.5°C and a boiling point of 113°C. The pure substance is highly dangerous; it explodes on heating and is highly reactive as a reducing agent. Normally it is employed as a reducing agent in aqueous solution, where it can be handled safely. For example, it reduces mercury(II) to the free element, as shown in Equation [21.62]:

$$2Hg^{2+}(aq) + N_2H_4(aq) \longrightarrow 2Hg(l) + N_2(g) + 4H^+(aq) \qquad [21.62]$$

Hydrazine is normally oxidized to N_2, as in this example.

The Lewis structure of hydroxylamine, NH_2OH, is shown in Figure 21.22. This compound may be considered a hybrid between hydrazine

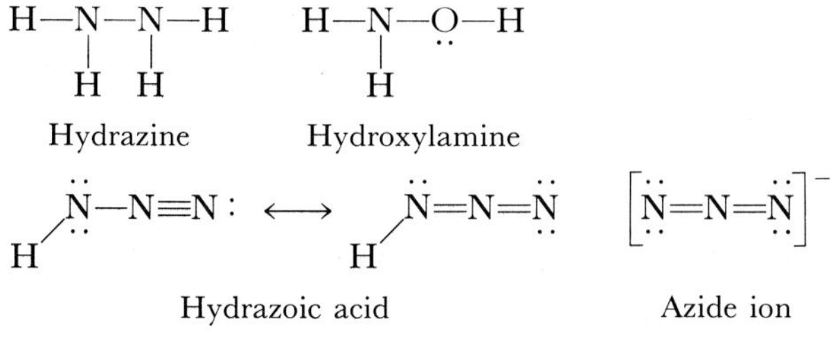

FIGURE 21.22 Lewis structures of some inorganic nitrogen compounds.

and hydrogen peroxide. Like both of these substances it is highly reactive and rather unstable as a pure substance. It normally acts as a reducing agent, with formation of N_2. However, it may be oxidized by some reagents to N_2O, or even NO_3^-.

SAMPLE EXERCISE 21.5

Hydroxylamine reduces copper (II) to the free metal in acidic aqueous medium. Write a balanced net ionic equation for the reaction, assuming that N_2 is the oxidation product.

Solution: The unbalanced equation is as follows:

$$Cu^{2+}(aq) + NH_2OH(aq) \longrightarrow Cu(s) + N_2(g)$$

Copper undergoes a change in oxidation number from $+2$ to 0. The nitrogen undergoes a change from -1 to 0. (We count OH as -1, each H as $+1$.) Thus, we require two NH_2OH molecules per copper ion. The other products on the right are $H_2O(l)$ and $H^+(aq)$. The balanced equation is thus:

$$Cu^{2+}(aq) + 2NH_2OH(aq) \longrightarrow$$
$$Cu(s) + N_2(g) + 2H_2O(l) + 2H^+(aq)$$

Hydrogen azide, HN_3, is the parent compound of a number of covalent and ionic azides. It can be obtained by reaction of sodium azide, NaN_3, with sulfuric acid. Hydrogen azide, also called hydrazoic acid, is a highly dangerous liquid that has a boiling point of $36°$ and decomposes explosively to the free elements. In water it is a weak acid ($K_a = 1.9 \times 10^{-5}$). The acid can be described in terms of two Lewis structures, as illustrated in Figure 21.22. The N—N bonds in HN_3 are not equivalent; the observed internal and terminal N—N distances are 1.24 Å and 1.13 Å, respectively. The azide ion, N_3^-, is isoelectronic with CO_2, and so we expect that it will have two N=N double bonds, as illustrated in Figure 21.22. The N=N distance in N_3^- is 1.15 Å. When azide salts of heavy metals are heated or struck, they decompose explosively to N_2 and the free metal. Lead azide is used as a primer in ammunition.

OXY COMPOUNDS OF NITROGEN

Only oxygen and fluorine have greater electronegativity values than nitrogen. Thus, only in compounds with these two elements will nitrogen exhibit positive oxidation states. In the oxides and oxyanions, nitrogen exhibits oxidation numbers ranging from $+1$ to $+5$.

Ammonia can be converted to nitric oxide by oxidation at elevated temperatures in the presence of platinum (Pt), which serves as a catalyst; the reaction is shown in Equation [21.63]:

$$4NH_3(g) + 5O_2(g) \xrightarrow[1000°C]{\substack{Pt \\ \text{catalyst}}} 4NO(g) + 6H_2O(g) \qquad [21.63]$$

In the absence of a catalyst, NH_3 is converted to N_2 instead of NO:

$$4NH_3(g) + 3O_2(g) \xrightarrow{1000°C} 2N_2(g) + 6H_2O(g) \qquad [21.64]$$

The catalytic conversion of NH_3 to NO is the commercial route to oxygen-containing compounds of nitrogen and is part of what is known as the Ostwald process.* Like the Haber process, this process was developed in Germany prior to World War I. It provided a means of converting NH_3 into nitric acid, which is used in making munitions.

Nitric oxide is one of three common oxides of nitrogen. The others are N_2O (nitrous oxide) and NO_2 (nitrogen dioxide). All three of these oxides are gases. Nitrous oxide is also known as laughing gas because a person becomes somewhat giddy after inhaling only a small amount of it. This colorless gas was the first substance used as a general anesthetic. It is used as the compressed gas propellant in several aerosols and foams such as in whipped cream. It can be prepared in the laboratory by carefully heating ammonium nitrate to about $200°C$:

$$NH_4NO_3(s) \longrightarrow N_2O(g) + 2H_2O(g) \qquad [21.65]$$

Nitric oxide is also a colorless gas, but unlike N_2O, it is slightly toxic. It can be prepared in the laboratory by reduction of dilute nitric acid, using copper or iron as a reducing agent:

$$3Cu(s) + 2NO_3^-(aq) + 8H^+(aq) \longrightarrow$$
$$3Cu^{2+}(aq) + 2NO(g) + 4H_2O(l) \qquad [21.66]$$

It is also produced by direct combination of N_2 and O_2 at elevated temperatures. As we saw in Section 10.5, this reaction is a significant source of nitrogen oxide air pollutants, which form during combustion reactions in air. However, the direct combination of N_2 and O_2 is not presently used for commercial production of NO because the yield is low; as noted in Section 14.6, the equilibrium constant K_c at 2400 K is only 0.05.

One of the important properties of NO is its ready ability to react with O_2, forming NO_2 when exposed to air:

$$2NO(g) + O_2(g) \longrightarrow 2NO_2(g) \qquad [21.67]$$

Nitrogen dioxide is a yellow-brown gas. It is poisonous and has a choking odor. At lower temperatures it reacts with itself to form the colorless N_2O_4 as shown in Equation [21.68]:

$$2NO_2(g) \rightleftharpoons N_2O_4(g) \qquad \Delta H° = -58 \text{ kJ} \qquad [21.68]$$

When dissolved in water, NO_2 forms HNO_3, nitric acid, as shown in Equation [21.69]:

$$H_2O(l) + 3NO_2(g) \longrightarrow 2H^+(aq) + 2NO_3^-(aq) + NO(g) \qquad [21.69]$$

Note that nitrogen in this reaction is both oxidized and reduced. The reduction product NO can be converted back into NO_2 by exposure to air and thereafter dissolved in water to prepare more HNO_3.

Nitric acid is a colorless, corrosive liquid. It is both a strong acid and a good oxidizing agent. The standard reduction potentials for conversion

---

*The Ostwald process is the process for converting ammonia to nitric acid utilizing the reactions summarized in Equations [21.63], [21.67], and [21.69].

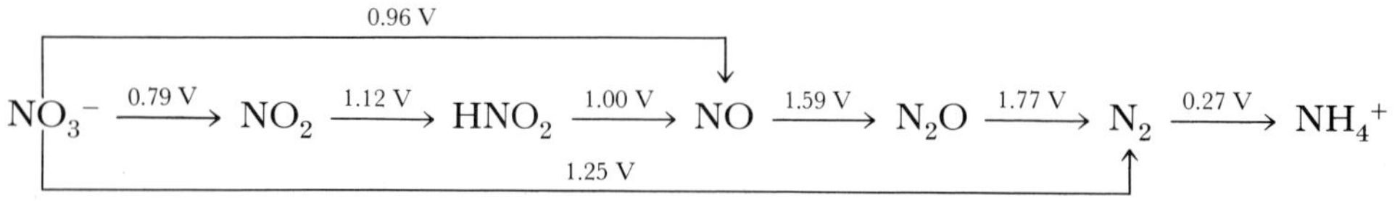

$$NO_3^- \xrightarrow{\text{0.79 V}} NO_2 \xrightarrow{\text{1.12 V}} HNO_2 \xrightarrow{\text{1.00 V}} NO \xrightarrow{\text{1.59 V}} N_2O \xrightarrow{\text{1.77 V}} N_2 \xrightarrow{\text{0.27 V}} NH_4^+$$

(with 0.96 V arc from NO_3^- to HNO_2, and 1.25 V arc below from NO_3^- to NO, and an arc from HNO_2 region to N_2)

FIGURE 21.23 Standard reduction potentials in acid solution for nitrogen-containing compounds in various oxidation states.

of nitrate into lower oxidation states in acid solution are shown in Figure 21.23. The notation in this figure is the same as that used in Figure 21.10, in connection with the chemistry of the oxychlorine compounds. The fact that the potentials in the diagram are large and positive is indication that the nitrogen oxides and anions derived from them are strong oxidizing agents. Nitric acid boils at 86°C and freezes at −42°C. It is used in the production of plastics, drugs, nitrate fertilizers, and explosives. The development of the Haber and Ostwald processes in Germany just prior to World War I permitted Germany to make munitions even though naval blockades prevented access to traditional sources of nitrates. Among the explosives made from nitric acid are nitroglycerin, trinitrotoluene (TNT), and nitrocellulose. The reaction of nitric acid with glycerin to form nitroglycerin is shown in Equation [21.70]:

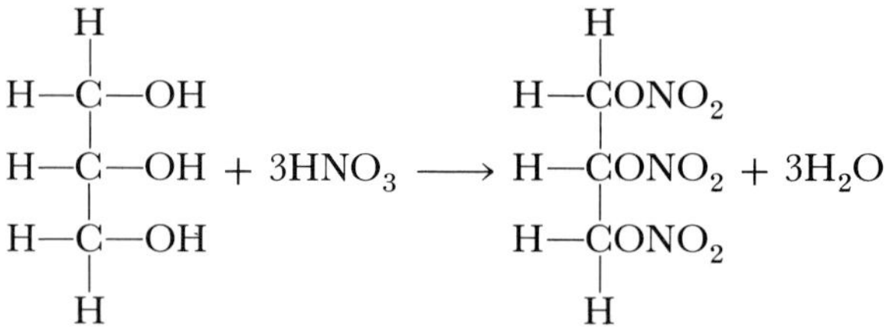

$$
\begin{array}{l}
\text{H} \\
| \\
\text{H—C—OH} \\
| \\
\text{H—C—OH} + 3HNO_3 \\
| \\
\text{H—C—OH} \\
| \\
\text{H}
\end{array}
\longrightarrow
\begin{array}{l}
\text{H} \\
| \\
\text{H—CONO}_2 \\
| \\
\text{H—CONO}_2 + 3H_2O \\
| \\
\text{H—CONO}_2 \\
| \\
\text{H}
\end{array}
\qquad [21.70]
$$

When nitroglycerin explodes, the reaction summarized in Equation [21.71] occurs:

$$4C_3H_5N_3O_9(l) \longrightarrow 6N_2(g) + 12CO_2(g) + 10H_2O(g) + O_2(g) \qquad [21.71]$$

A considerable amount of gaseous products form from the liquid. The sudden formation of these gases, together with their expansion resulting from the heat generated by the reaction, produces the explosion.

THE NITROGEN CYCLE IN NATURE

Plants are able to utilize several chemical forms of nitrogen, especially NH_3, NH_4^+, and NO_3^-. Common synthetic fertilizers therefore include liquid ammonia, ammonium nitrate, and urea. Ammonium nitrate, NH_4NO_3, is made by reaction of ammonia and nitric acid, as shown in Equation [21.72]:

$$
\begin{array}{l}
H^+(aq) + NH_3(aq) + NO_3^-(aq) \longrightarrow \\
\qquad NH_4^+(aq) + NO_3^-(aq) \longrightarrow NH_4NO_3(s)
\end{array}
\qquad [21.72]
$$

Urea is made by reaction of ammonia and carbon dioxide, Equation [21.73]:

$$2NH_3(aq) + CO_2(aq) \rightleftharpoons H_2N\overset{\overset{\displaystyle O}{\|}}{C}NH_2(aq) + H_2O(l) \qquad [21.73]$$

Urea slowly releases NH_3 as it reacts with water in the soil.

Plants use nitrogen to synthesize several nitrogen-containing compounds including proteins. Proteins are made from amino acids, which are compounds of the following type:

$$H-\overset{\overset{\displaystyle }{|}}{\underset{\underset{\displaystyle H}{|}}{N}}-\overset{\overset{\displaystyle R}{|}}{\underset{\underset{\displaystyle H}{|}}{C}}-\overset{\overset{\displaystyle O}{\|}}{C}-O-H$$

where R is H or any of a number of carbon-containing groups including CH_3. We shall consider these compounds in Chapter 25. Plant proteins are ingested by animals, are broken into amino acids, and are then either converted into animal protein or excreted as nitrogenous waste. Certain microorganisms are able to convert this waste back into N_2. Nitrogen is recycled in this fashion. As in the case of the cycling of oxygen, energy from the sun is required. A simplified picture of the nitrogen cycle is shown in Figure 21.24.

Large-scale cultivation of nitrogen-fixing legumes and industrial fixation have increased the quantity of fixed nitrogen in the biosphere. One effect of this intrusion into the nitrogen cycle has been increased water pollution. Much fixed nitrogen ends up as nitrates in the soil. These compounds are highly water soluble. They are therefore readily washed from the soil into water bodies. Only a portion of the added fertilizer ends up being used by the plants for which it was intended. Once in a lake, nitrates stimulate plant growth, encouraging, for example, rapid growth of algae. When these plants die, their decay consumes O_2 in the water. Without O_2, fish and other oxygen-dependent organisms die. Furthermore, when the oxygen is depleted, anaerobic decomposition begins.

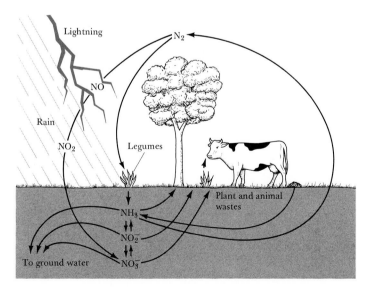

FIGURE 21.24 A simplified picture of the nitrogen cycle, showing some of the primary reactions involved in the utilization and formation of nitrogen in nature. The main reservoir of nitrogen is the atmosphere, which contains N_2. This nitrogen is fixed through the action of lightning and leguminous plants. The compounds of nitrogen reside in the soil as NH_3 (and NH_4^+), NO_2^-, and NO_3^-. All are water soluble and can be washed out of the soil by ground water. These nitrogen compounds are utilized by plants in their growth and are incorporated into animals that eat the plants. Animal waste and dead plants and animals are attacked by certain bacteria that free N_2, which escapes into the atmosphere, thereby completing the cycle.

This decomposition produces foul odors. The undecomposed plant matter falls to the bottom of the lake, thereby slowly turning it into a swamp and eventually a meadow. This process of fertilization of lakes, which leads to their death, is known as eutrophication. It takes thousands of years for this aging process to occur naturally. Through various activities, we have added nutrients to water and thereby succeeded in greatly accelerating this aging process. Nitrogen is not the only nutrient that promotes eutrophication. We shall discuss another important plant nutrient, phosphorus, in the next section.

21.9 Phosphorus

Phosphorus has an electron configuration of $[Ne]3s^2 3p^3$. Like nitrogen, it exhibits oxidation states ranging from -3 to $+5$. Because of its lower electronegativity, it is found more frequently in the positive oxidation states than is nitrogen. Furthermore, compounds in which phosphorus has the $+5$ oxidation state are not strong oxidizing agents, as are corresponding compounds of nitrogen. Compounds in which phosphorus has a -3 oxidation state are much stronger reducing agents than corresponding compounds of nitrogen.

OCCURRENCE, ISOLATION, AND PROPERTIES

Phosphorus occurs mainly in the form of phosphate minerals. The principal source of phosphorus is phosphate rock, which contains phosphate mainly in the form of $Ca_3(PO_4)_2$. Deposits of phosphate rock occur mainly in Florida, the western United States, North Africa, and parts of the USSR. The element is produced commercially by reduction of phosphate with coke* in the presence of SiO_2, Equation [21.74]:

$$2Ca_3(PO_4)_2(s) + 6SiO_2(s) + 10C(s) \xrightarrow{1500°C}$$
$$P_4(g) + 6CaSiO_3(l) + 10CO(g) \qquad [21.74]$$

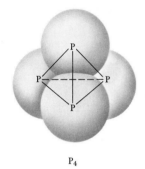

P_4

FIGURE 21.25 The structure of the P_4 molecule of white phosphorus.

The phosphorus produced in this fashion is the allotrope known as white phosphorus. This form distills from the reaction mixture as the reaction proceeds.

White phosphorus consists of P_4 tetrahedra as shown in Figure 21.25. As we noted in Section 8.7, the 60° bond angles in P_4 are unusually small for molecules. There must consequently be much strain in the bonding, a fact that is consistent with the high reactivity of white phosphorus. This allotrope bursts spontaneously into flames if exposed to air. It is a white, waxlike solid that melts at 44.2°C and boils at 280°C. When heated in the absence of air to about 400°C, it is converted to a more stable allotrope known as red phosphorus. This form does not ignite on contact

*Coke is a form of carbon made by heating coal in the absence of air.

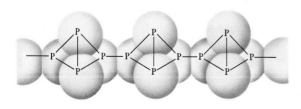

FIGURE 21.26 Red phosphorus is polymeric, possibly as shown; however, the details of its structure are not known.

TABLE 21.10 Properties of phosphorus-halogen compounds

Property	PF_3	PF_5	PCl_3	PCl_5	PBr_3	PBr_5	PI_3
Melting point (°C)	−160	−94	−112	167	−40	—	61
Boiling point (°C)	−95	−85	76	—[a]	173	106	120[b]
Dipole moment (D)	1.0	0	0.78	0	0.5	0	0

[a]Sublimes at 163°C at 1 atm pressure.
[b]At 15 mm Hg pressure.

with air. It is also considerably less poisonous than the white form. Red phosphorus appears to consist of a chain structure as shown in Figure 21.26.

PHOSPHORUS HALIDES

Phosphorus forms a wide range of compounds with the halogens. A few properties of the more important phosphorus-halogen compounds are listed in Table 21.10.

We have already discussed the Lewis structures and geometries of the phosphorus halide molecules at other places in the text (Section 7.5 and 7.6). The dipole moments (Section 8.2) listed in Table 21.10 reflect these geometries. The PX_3 compounds have a pyramidal shape (Figure 21.4), and the overall molecular dipole moment reflects the P—X bond polarity. We conclude that the bond polarities decrease in the order P—F > P—Cl > P—Br > P—I. This is the order expected on the basis of the electronegativity difference between phosphorus and the halogen. The PX_5 molecules possess trigonal-bipyramidal structures (Figure 21.4) with five electron pairs shared with the five X groups. The bond dipole moments of the five P—X bonds cancel to yield a net zero dipole moment in all cases.

Phosphorus trifluoride is prepared by reaction with another fluoride, as in Equation [21.75]:

$$PCl_3(l) + AsF_3(l) \longrightarrow PF_3(g) + AsCl_3(l) \qquad [21.75]$$

The driving force for this reaction is the greater strength of the phosphorus-fluorine bond as compared with the other bond energies involved (490 kJ/mol for P—F as compared with 406 kJ/mol for As—F, 326 kJ/mol for P—Cl, and 322 kJ/mol for As—Cl). Phosphorus pentafluoride can be prepared by direct combination of the elements, by reaction of PCl_5 with AsF_3, or by heating P_4O_{10} in a sealed tube with CaF_2:

$$4P_4O_{10}(l) + 15CaF_2(s) \longrightarrow 6PF_5(l) + 5Ca_3(PO_4)_2(s) \qquad [21.76]$$

Phosphorus trichloride is formed by passing a stream of dry chlorine gas over white or red phosphorus. In an excess of Cl_2, the PCl_3 reacts further to form PCl_5. Phosphorus tribromide is formed by reaction of liquid bromine on red phosphorus. Addition of still more bromine results in formation of PBr_5. However, this compound readily dissociates in the vapor state:

$$PBr_5(g) \rightleftharpoons PBr_3(g) + Br_2(g) \qquad [21.77]$$

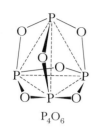

P_4O_6

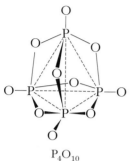

P_4O_{10}

FIGURE 21.27 The structures of P_4O_6 and P_4O_{10}.

Phosphorus triiodide can be formed by reacting white phosphorus and iodine in an appropriate solvent.

The phosphorus halides hydrolyze on contact with water. The reactions occur readily, and most of the phosphorus halides fume in air as a result of reaction with water vapor. In the presence of excess water, the products are the corresponding phosphorus oxyacid and hydrogen halide, as in the examples shown in Equations [21.78] and [21.79]:

$$PF_3(g) + 3H_2O(l) \longrightarrow H_3PO_3(aq) + 3HF(aq) \qquad [21.78]$$

$$PCl_5(l) + 4H_2O(l) \longrightarrow H_3PO_4(aq) + 5HCl(aq) \qquad [21.79]$$

When only a little water is used, the hydrogen halide can be obtained as a gas; these reactions thus provide convenient laboratory preparations of the hydrogen halides.

OXY COMPOUNDS OF PHOSPHORUS

Probably the most significant compounds of phosphorus are those in which the element is combined in some way with oxygen. Phosphorus(III) oxide, P_4O_6, is obtained by allowing white phosphorus to oxidize in a limited supply of oxygen. When oxidation takes place in the presence of excess oxygen, phosphorus(V) oxide, P_4O_{10}, forms. This compound is also readily formed by oxidation of P_4O_6. Phosphorus(III) oxide is often called phosphorus trioxide after its empirical formula; similarly, phosphorus(V) oxide is often called phosphorus pentoxide, and written as the empirical formula P_2O_5. These two oxides represent the two most common oxidation states for phosphorus, $+3$ and $+5$. The structural relationship between P_4O_6 and P_4O_{10} is shown in Figure 21.27. Notice the resemblance these molecules have to the P_4 molecule, Figure 21.25.

Phosphorus(V) oxide is the anhydride of phosphoric acid, H_3PO_4, a weak triprotic acid. In fact, P_4O_{10} has a very high affinity for water and is consequently used as a drying agent. Phosphorus(III) oxide is the anhydride of phosphorous acid, H_3PO_3, a weak diprotic acid. The structures of H_3PO_4 and H_3PO_3, are shown in Figure 21.28. The hydrogen atom that is attached directly to phosphorus in H_3PO_3 is not acidic.

One characteristic of phosphoric and phosphorous acids is their tendency to undergo condensation reactions when heated. A **condensation reaction** is one in which two or more molecules combine to form a larger

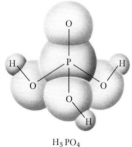

H_3PO_4

FIGURE 21.29 Structures of trimetaphosphoric acid and polymetaphosphoric acid.

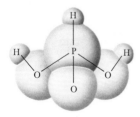

H_3PO_3

FIGURE 21.28 The structures of H_3PO_4 and H_3PO_3.

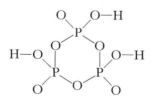

$(HPO_3)_3$
Trimetaphosphoric acid

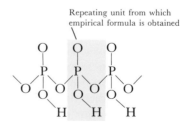

$(HPO_3)_n$
Polymetaphosphoric acid

molecule by eliminating a small molecule such as H_2O. Such a reaction in which two H_3PO_4 molecules are joined by the elimination of one H_2O molecule is shown in Equation [21.80]:

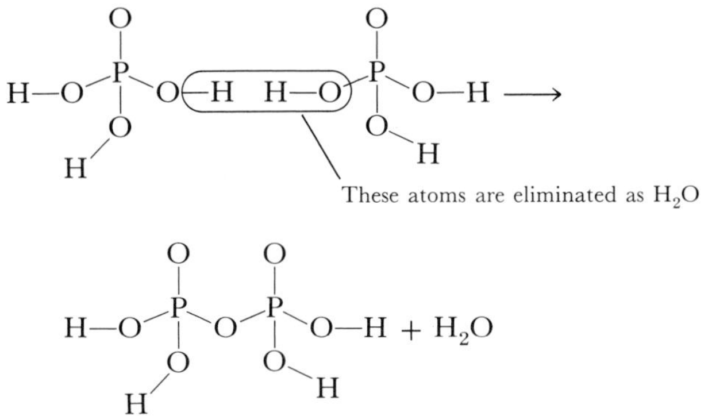

$$\begin{array}{c}
\text{(structural formula of two } H_3PO_4 \text{ molecules condensing)}
\end{array}$$

These atoms are eliminated as H_2O

$$\begin{array}{c}
\text{(structural formula of the product with } + H_2O\text{)} \qquad \text{[21.80]}
\end{array}$$

$$2H_3PO_4 \longrightarrow H_4P_2O_7 + H_2O$$

Further condensation produces phosphates having an empirical formula of HPO_3:

$$nH_3PO_4 \longrightarrow (HPO_3)_n + nH_2O \qquad \text{[21.81]}$$

Two phosphates having this empirical formula, one cyclic and the other polymeric, are shown in Figure 21.29. The three acids H_3PO_4, $H_4P_2O_7$, and $(HPO_3)_n$, all contain phosphorus in its $+5$ oxidation state, and all are therefore called phosphoric acids. To differentiate them, the prefixes *ortho-*, *pyro-*, and *meta-* are used: H_3PO_4 is orthophosphoric acid, $H_4P_2O_7$ is pyrophosphoric acid, and HPO_3 is metaphosphoric acid.

Phosphoric acid and its salts find their most important uses in detergents and fertilizers. The phosphates in detergents are often in the form of sodium tripolyphosphate, $Na_5P_3O_{10}$ (see Section 23.2). A typical detergent formulation contains 47 percent phosphate, 16 percent bleaches, perfumes, and abrasives, and 37 percent linear alkylsulfonate (LAS) surfactant such as that shown below:

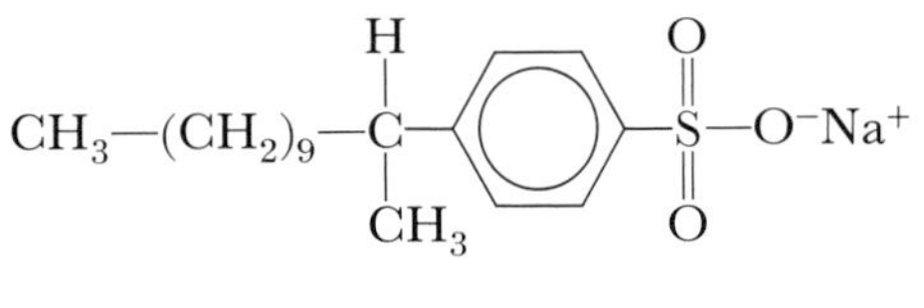

$$\begin{array}{c}
CH_3-(CH_2)_9-\overset{\overset{\displaystyle H}{|}}{\underset{\underset{\displaystyle CH_3}{|}}{C}}-\underset{}{\bigcirc}-\overset{\overset{\displaystyle O}{\|}}{\underset{\underset{\displaystyle O}{\|}}{S}}-O^-Na^+
\end{array}$$

(We have used the notation for the benzene ring described in Section 8.4.) The detergent action of such molecules has been described in Section 12.7. The phosphate ions form bonds with metal ions that contribute to the hardness of water. This keeps these ions from interfering with the action of the surfactant. The phosphates also keep the pH above 7 and thus prevent the surfactant molecules from becoming protonated (gaining an H^+ ion).

Most phosphate rock mined is converted to fertilizers. The mined phosphate rock contains large amounts of sand and clay. At a treatment plant, sand, clay, and organic materials are removed from the raw ore. The resultant concentrate is shipped to plants that convert it to phosphoric acid or water-soluble phosphate fertilizers. The $Ca_3(PO_4)_2$ in phosphate rock is insoluble ($K_{sp} = 2.0 \times 10^{-29}$). It is converted to a soluble form for use in fertilizers. This can be accomplished by treating the concentrated phosphate rock with sulfuric or phosphoric acid:

$$Ca_3(PO_4)_2(s) + 4H^+(aq) + 3SO_4^{2-}(aq) \longrightarrow$$
$$3CaSO_4(s) + 2H_2PO_4^-(aq) \qquad [21.82]$$

$$Ca_3(PO_4)_2(s) + 4H^+(aq) \longrightarrow 3Ca^{2+}(aq) + 2H_2PO_4^-(aq) \quad [21.83]$$

The mixture formed when ground phosphate rock is treated with sulfuric acid and then dried and pulverized is known as superphosphate. The $CaSO_4$ formed in this process is of little use in soil except when deficiencies in calcium or sulfur exist. It also dilutes the phosphorus, which is the nutrient of interest. If the phosphate rock is treated with phosphoric acid, the product contains no $CaSO_4$ and has a higher percentage of phosphorus. This product is known as triple superphosphate. Although the solubility of $Ca(H_2PO_4)_2$ allows it to be assimilated by plants, it also allows it to be washed from the soil and into water bodies, thereby contributing to eutrophication.

Phosphorus compounds are important in biological systems. The element occurs for example in phosphate groups in RNA and DNA, the molecules responsible for control of protein biosynthesis and transmission of genetic information. It also occurs in adenosine triphosphate (ATP), which stores energy within biological cells:

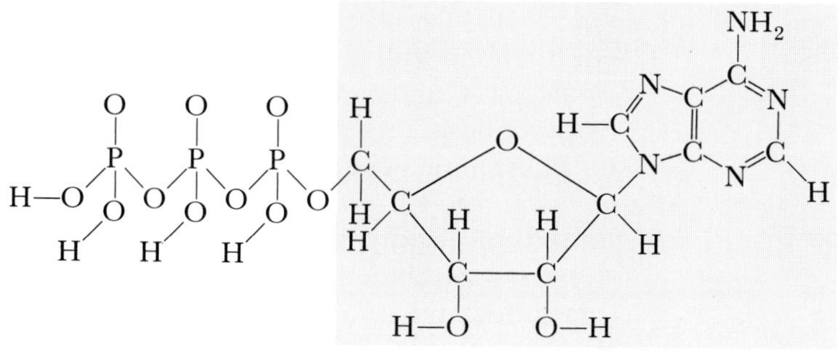

Adenosine

The P—O—P bond of the end phosphate group is broken by hydrolysis with water, forming adenosine diphosphate (ADP). This reaction produces 33 kJ of energy:

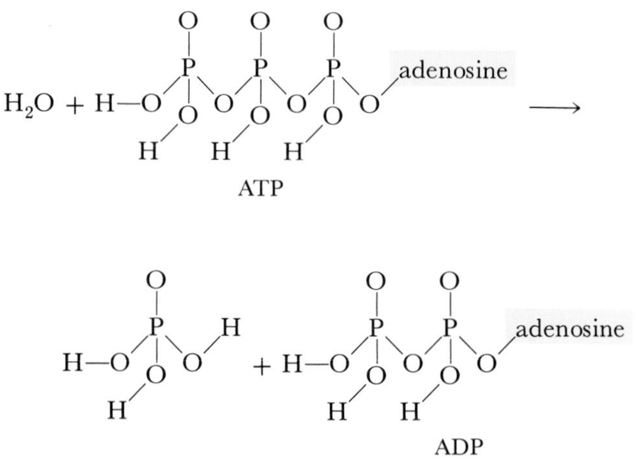

ATP

ADP

[21.84]

This energy is used to perform the mechanical work of muscle contraction and in many other biochemical reactions (see Section 15.7 and Figure 18.8). There is more on the biological properties of phosphorus-containing substances in Chapter 25.

21.10 Arsenic, antimony, and bismuth

Arsenic, antimony, and bismuth occur in nature in the form of sulfide ores, As_4S_4, As_2S_3, Sb_2S_3, and Bi_2S_3. In addition, the elements are found as minor components in ores of various metals such as Cu, Pb, Ag, and Hg. Arsenic and antimony exhibit allotropy similar to that of phosphorus. Both elements can be prepared as soft, yellow, nonmetallic solids by quickly cooling the high-temperature vapor of the element. In this allotropic form, analogous to the white allotrope of phosphorus, the elements are present as As_4 or Sb_4 tetrahedra. Heating or the action of light converts the substances into the gray, more metallic forms containing sheets of atoms. In its common form bismuth has a reddish-white, rather metallic appearance. The element forms alloys with many metallic elements. Alloys of lead, bismuth, and tin are used in the construction of plugs in sprinkler systems. Water is discharged when the fusible metal plug melts.

Arsenic and antimony resemble phosphorus in much of their chemical behavior. For example, both elements form halides of formula MX_3 and MX_5 with structures and chemical properties similar to the halides of phosphorus. The oxy compounds of these two elements are also rather similar to those of phosphorus, except that the higher oxidation state is not so easily attained. Thus, the product of burning arsenic in oxygen is As_4O_6, not As_4O_{10}. The higher oxide can be obtained by oxidation of As_4O_6 with a strong oxidizing agent such as nitric acid:

$$As_4O_6(s) + 4NO_3^-(aq) + 6H_2O(l) + 4H^+(aq) \longrightarrow$$
$$4H_3AsO_4(aq) + 4HNO_2(aq) \qquad [21.85]$$

In terms of the criteria discussed at the beginning of this chapter, bismuth is more logically considered a metal rather than a nonmetal. Bismuth usually appears in the $+3$ oxidation state; there is little tendency to attain the higher $+5$ oxidation state that is so common for phosphorus. The common oxide of bismuth is Bi_2O_3. This substance is insoluble in water or basic solution but is soluble in acidic solution. It thus is classified as a basic anhydride. As we have seen, the oxides of metals characteristically behave as basic anhydrides.

21.11 Carbon, silicon, and boron

At this point we have only three additional elements, carbon, silicon, and boron, to consider in our survey of the nonmetallic elements. We have already discussed the structures of carbon and silicon in Section 8.7. Carbon forms a large number of compounds with hydrogen, and we have referred to many of them in previous chapters. The systematic chemistry of these substances is the material of organic chemistry and biochemistry (Chapters 24 and 25). The most important inorganic compounds of carbon are the two oxides CO and CO_2. Both these substances have been mentioned at various places in the text—their structures in Chapters 7 and 8, and their importance in our environment in

Chapter 10. In addition, the acidic properties of aqueous solutions of CO_2 were discussed in Chapters 15 and 16.

You will recall from Section 21.1 that the oxides of carbon and silicon have very different characteristics. The most important compounds of silicon are the silicates, which may be considered to derive from silica, SiO_2. The chemical and structural properties of silicates, and their place in our environment, are described in Chapter 22.

Boron is the only element of group 3A that can be considered to be nonmetallic. The element has an extended network structure. The melting point of boron, 2300°C, is intermediate between that for carbon, 3550°C, and that for silicon, 1410°C. The electronic configuration of boron is $[He]2s^2 2p^1$. The element exhibits a valence of 3 in all its common chemical compounds. We have seen (Section 7.7) that the electron configuration about boron in the boron halides provides an exception to the octet rule, in that there are but six electrons in the boron valence shell. The boron halides are thus strong Lewis acids (Section 15.10).

Salts of the borohydride ion, BH_4^-, are widely used as reducing agents. This ion can be thought of as an isoelectronic analog of CH_4 or NH_4^+. The lower charge of the central atom in BH_4^- means that the hydrogens of the BH_4^- are "hydridic," that is, they carry a partial negative charge. Thus it is not surprising that borohydrides are good reducing agents, as illustrated by the reaction in Equation [21.86]:

$$3BH_4^-(aq) + 4IO_3^-(aq) \longrightarrow 3H_2BO_3^-(aq) + 4I^-(aq) + 3H_2O(l) \qquad [21.86]$$

The only important oxide of boron is boric oxide, B_2O_3. This substance is the anhydride of boric acid, which we may write as H_3BO_3 or $B(OH)_3$. Boric acid is a weak acid; so much so that solutions of boric acid are used as an eyewash. Upon warming to 100° orthoboric acid loses water by a condensation reaction similar to that described in Section 21.9. The product of the reaction, as shown in Equation [21.87], is metaboric acid, a polymeric substance of formula HBO_2.

$$H_3BO_3(s) \longrightarrow HBO_2(s) + H_2O(g) \qquad [21.87]$$

Heating of metaboric acid results in still more loss of water, as in Equations [21.88] and [21.89]:

$$4HBO_2(s) \longrightarrow H_2B_4O_7(s) + H_2O(l) \qquad [21.88]$$

$$H_2B_4O_7(s) \longrightarrow 2B_2O_3(s) + H_2O(g) \qquad [21.89]$$

The acid $H_2B_4O_7$ is called tetraboric acid. The sodium salt, $Na_2B_4O_7 \cdot 10H_2O$, called borax, occurs in dry lake deposits in California and can also be readily prepared from other borate minerals. Solutions of borax are alkaline; the substance is used in various laundry and cleaning products.

Summary

Metallic and nonmetallic elements are distinguished by several physical and chemical properties. Nonmetals lack luster, are not malleable or ductile, and are not good conductors of heat or electricity. In the structures of nonmetallic elements the atoms have relatively few near neighbors, and they are bonded to these neighbors via covalent bonds. Nonmetallic elements possess higher ionization energies and electronegativities than do metallic elements. The soluble oxides of the nonmetals generally produce acidic aqueous solutions; nonmetal oxides are thus said to be **acidic anhydrides.** By contrast, soluble metal oxides produce basic solutions and are referred to as **basic anhydrides.**

Among elements of a given family, size increases with increasing atomic number. Correspondingly, electronegativity and ionization energy decrease. Metallic character parallels electronegativity trends. Among the nonmetallic elements, the first member of each family differs dramatically from the other members; it forms a maximum of four bonds to other atoms (that is, it is confined to an octet of valence-shell electrons). Secondly, it exhibits a much greater tendency to form π bonds than the heavier elements in that family.

The noble-gas elements exhibit a very limited chemical behavior because of the exceptional stability of their electronic configuration. The xenon fluorides and oxides and KrF_2 are the only established examples of chemical reactivity among these elements.

The halogens occur as diatomic molecules. These elements possess the highest electronegativities of the elements in each row of the periodic table. All except fluorine exhibit oxidation states varying from -1 to $+7$. Fluorine, being the most electronegative element, is restricted to the oxidation states 0 and -1. The tendency to form the -1 oxidation state from the free element (that is, the oxidizing power of the element) decreases with increasing atomic number in the family. The halogens form compounds with one another, called **interhalogens.** In the higher interhalogens, XX'_n, the element X may be Cl, Br, or I, and X' is nearly always F; n may have values 3, 5, or 7.

The oxyacids and oxyanions of Cl, Br, and I are powerful oxidizing agents and are widely used as bleaches or in other oxidizing applications. Oxidizing strength in the oxyacids decreases with increasing oxidation state.

The group 6A elements range from the very abundant and strongly nonmetallic oxygen to the rare and rather metallic tellurium. The group 6A elements have generally lower electronegativities than do their neighboring halogen elements. Except for oxygen, oxidation states from -2 to $+6$ are known. Oxygen normally exhibits an oxidation number of -2 in its compounds, but in **peroxides,** containing an O—O bond, the oxidation state is -1. Oxygen is the most common and widely used oxidizing agent. In the allotropic form ozone, O_3, it is an even more powerful oxidizing agent.

Sulfur occurs in the elemental state in several allotropic forms; the most stable one consists of S_8 rings. The element is found in large underground deposits from which it is removed using the **Frasch process.** Sulfur exhibits oxidation states ranging from $+6$ to -2. Its most important compound is sulfuric acid, a strong acid; sulfuric acid is a good dehydrating agent and has a high boiling point. This acid is the most widely used industrial chemical.

Sulfur occurs widely in the form of sulfide ores, and these form important sources of several metals. Selenium and tellurium are chemically rather similar to sulfur, especially with respect to formation of oxides and oxyanions.

The group 5A elements exhibit a wide range of behavior, from strongly nonmetallic in the case of nitrogen to distinctly metallic in the case of bismuth. Nitrogen and phosphorus exhibit oxidation states ranging from $+5$ to -3. Because it is less electronegative than nitrogen, phosphorus is found in the positive oxidation states more frequently than is nitrogen. The primary source of nitrogen is the atmosphere, where it occurs as N_2 molecules. The most important commercial process for converting N_2 into compounds is the Haber process, which is used to manufacture ammonia. Another important commercial process is the **Ostwald process,** which is used to convert NH_3 to nitric acid, HNO_3. Nitric acid is both a strong acid and a good oxidizing agent. Nitrogen compounds are important fertilizers.

Phosphorus, which is also important in fertilizers, occurs in nature in certain phosphate minerals. The element exhibits several allotropes including one known as white phosphorus, a reactive form consisting of P_4 tetrahedra. Phosphorus forms compounds of formula PX_3 and PX_5 with the halogens. These undergo hydrolysis in water to produce the corresponding oxyacid of phosphorus and hydrogen halide. Phosphorus forms two oxides, P_4O_6 and P_4O_{10}.

Their corresponding acids, phosphorous acid and phosphoric acid, show a strong tendency to undergo condensation reactions when heated. Arsenic and antimony closely resemble phosphorus in chemical behavior. The tendency toward increasing metallic character with increasing atomic number in a family is evident in the properties of bismuth, which occurs most commonly in the $+3$ oxidation state.

Boron commonly exhibits a valence of $+3$, as in boric oxide, B_2O_3, and boric acid, H_3BO_3. The acid readily undergoes condensation reactions.

Learning goals

Having read and studied this chapter, you should be able to:

1 Identify an element as a metal, semimetal, or nonmetal, based on its position in the periodic table or its properties.

2 Give examples of how the first member in each family of nonmetallic elements differs from the other elements of the same family and account for these differences.

3 Describe the ways in which the structures of metallic and nonmetallic elements differ.

4 Explain what is meant by the terms basic anhydride and acidic anhydride and give examples of each.

5 Predict the relative electronegativities and metallic character of any two members of a periodic family or a horizontal row of the periodic table.

6 Predict the maximum and minimum oxidation state of any nonmetallic element discussed in the chapter. Give an example of a compound containing the element in each of those oxidation states.

7 Cite the most common occurrences of each nonmetallic element discussed in this chapter (for example, oxygen is found in the atmosphere, fluorine in CaF_2, and so forth).

8 Cite the most common molecular form of each nonmetallic element discussed in the chapter (for example, oxygen is found as O_2, selenium as chains of Se atoms).

9 Account for the fact that xenon forms several compounds with fluorine and oxygen, krypton forms only KrF_2, and no chemical reactivity is known for the lighter noble-gas elements.

10 Write the formulas of the known fluorides, oxyfluorides, and oxides of xenon and describe the relative stabilities of the oxides as compared with the fluorides.

11 Describe the electronic and geometrical structures of the known compounds of xenon.

12 Write balanced chemical equations describing at least one means of preparation of each halogen from naturally occurring sources.

13 Describe at least one important use of each halogen element.

14 Write a balanced chemical equation describing the preparation of each of the hydrogen halides.

15 Give examples of diatomic and higher interhalogen compounds and describe their electronic and geometrical structures.

16 Name the oxyacids or oxyanions of the halogens, given their formula, or vice versa.

17 Describe the variation in acid strength and oxidizing strength of the oxyacids of chlorine.

18 Describe the chemical and physical properties of hydrogen peroxide and its method of preparation.

19 Indicate the formulas of the common oxides of sulfur and the properties of their aqueous solutions.

20 Write balanced chemical equations for formation of sulfuric acid from sulfur and describe the important properties of the acid.

21 Compare the chemical behaviors of selenium and tellurium with that of sulfur, with respect to common oxidation states and formulas of oxides and oxyacids.

22 Cite at least one example each of a nitrogen compound in which nitrogen is in the -3, -1, $+1$, $+2$, $+3$, $+4$, or $+5$ oxidation states.

23 Write balanced chemical equations for formation of nitric acid via the Ostwald process, starting from NH_3.

24 Describe the preparation of elemental phosphorus from its ores, using balanced chemical equations.

25 Describe the formulas and structures of the stable halides and oxides of phosphorus.

26 Write balanced chemical equations for the reactions of the halides and oxides of phosphorus with water.

27 Describe a condensation reaction and give examples involving compounds of phosphorus or boron.

Key Terms

Among the more important terms and expressions used for the first time in this chapter are the following:

An acidic anhydride is an oxide that forms an acid when added to water; soluble nonmetal oxides are acidic anhydrides.

A **basic anhydride** is an oxide that forms a base when added to water; soluble metal oxides are basic anhydrides.

A **condensation reaction** is one in which two or more molecules combine to form larger ones by elimination of small molecules such as H_2O.

An **interhalogen** compound is one formed between two different halogen elements. Examples include IBr, BrF_3 and ICl_3.

The **Ostwald process** is used to make nitric acid from ammonia. The NH_3 is catalytically oxidized by O_2 to form NO; NO in air is oxidized to NO_2; HNO_3 is formed by dissolving NO_2 in water.

EXERCISES

Periodic trends

21.1 State two chemical and two physical characteristics that distinguish a nonmetallic element from a metal.

21.2 Using this chapter and material elsewhere in the text (use the index), compile a list of physical and chemical properties of the elements oxygen and zinc. How do these properties justify the classification of one as a nonmetal and the other as a metal?

21.3 In what general ways do the oxides of metallic elements differ from those of nonmetallic elements?

21.4 Illustrate the distinctive differences between metal and nonmetallic oxides, using CaO and SO_3 as examples. Write balanced chemical equations where appropriate.

21.5 Which element of group 5A would you expect to have the most metallic character? Describe two properties of the elements in the group that should relate to this degree of metallic character.

21.6 In each of the following pairs of substances, select the one that has the more polar bonds. Explain your choice in each case: (a) NF_3 and IF_3; (b) N_2O_3 and Bi_2O_3; (c) BN and BF_3; (d) HF and H_2Te; (e) CO_2 and SnO_2.

21.7 The highest fluoride compound formed by nitrogen is NF_3, whereas phosphorus and arsenic readily form PF_5 or AsF_5, respectively. Account for this difference.

21.8 Explain how the structures of the group 5A elements (Section 8.7) illustrate an important difference in characteristic bonding behavior between the lightest element in the family and all the others.

21.9 Give a brief explanation of each of the following observations: (a) AsH_3 is a stronger reducing agent than NH_3; (b) HNO_3 is a stronger oxidizing agent than H_3PO_4; (c) Cl_2 is capable of oxidizing sulfur in H_2S.

[21.10] The electronegativity of a given element can be regarded as varying with its oxidation state. How would you expect the electronegativity to change as a function of oxidation state? Although manganese (Mn) is completely different from chlorine in its properties as an element, the characteristics of MnO_4^- rather closely resemble those of ClO_4^-. Cite at least two instances of this similarity (you may need to look in a handbook; think in terms of acid-base properties, oxidation-reduction behavior, solubility, and so on) and discuss your observations in terms of the electronegativities of the central atom.

The halogens and noble gases

21.11 Write the formula for each of the following compounds and indicate the oxidation state of the halogen or rare-gas atom in each: (a) chloric acid; (b) bromine trifluoride; (c) xenon oxytetrafluoride; (d) periodic acid; (e) iodate ion; (f) potassium chlorite; (g) hypobromous acid; (h) potassium triiodide; (i) phosphorus(III) iodide.

21.12 Write a balanced chemical equation showing the commercial preparation of each halogen element.

21.13 Calculate the molarity of an aqueous hydrofluoric acid solution that is 50.0 percent HF by weight and has a density of $1.155 \, g/cm^3$.

21.14 Account for the variations in first ionization energy among the halogens in terms of the variations in atomic radii.

21.15 Why is it not possible to isolate elemental fluorine from an aqueous solution? In what respects does the process for isolation of fluorine resemble the process used in production of elemental chlorine in a Downs cell (Section 19.6)? In what respects do the processes differ?

21.16 Write a balanced chemical equation that describes a suitable means of preparation of each of the following substances: (a) HF; (b) I_2; (c) XeF_4; (d) $Ca(ClO)Cl$; (e) SF_6; (f) HIO_3.

21.17 It is observed that for a given oxidation state the acid strength of the oxyacid in aqueous solution decreases in the order chlorine > bromine > iodine. Account for this trend.

21.18 Write balanced net ionic equations for reaction of each of the following substances with water: (a) PBr_5; (b) IF_5; (c) $BrCl$; (d) Br_2; (e) ClO_2 (chloric acid is a product); (f) HI.

21.19 The interhalogen compound $BrF_3(l)$ reacts with antimony(V) fluoride in a Lewis acid-base reaction to form the salt $(BrF_2)(SbF_6)$. Write the Lewis structure for both the cation and anion in this substance and describe the likely structure of each.

21.20 Cite the most important reasons why iodine forms an IF_7 species, whereas for chlorine, ClF_3 is stable, ClF_5 is unstable, and ClF_7 is unknown.

21.21 Write balanced chemical equations for each of the following reactions (some of these are analogous but not identical to reactions shown in the chapter): (a) bromine forms hypobromite ion on addition to aqueous base; (b) bromine reacts with an aqueous solution of hydrogen peroxide, liberating O_2; (c) hydrogen bromide is produced upon heating calcium bromide with phosphoric acid; (d) hydrogen bromide is formed upon hydrolysis of aluminum bromide; (e) aqueous hydrogen fluoride reacts with solid calcium carbonate, forming water-insoluble calcium fluoride.

21.22 Write the complete, balanced half-reaction for reduction of ClO_3^- to Cl_2 and the complete balanced equation for reduction of ClO_3^- to Cl_2 by $Fe^{2+}(aq)$, which is oxidized to $Fe^{3+}(aq)$. Calculate the standard cell potential (see Appendix F and Figure 21.10).

21.23 Write the complete half-cell reactions and complete cell reaction for reduction of ClO_3^- to Cl_2 by chloride ion in aqueous solution. What is the standard cell potential?

[21.24] Write a balanced half-cell reaction for (a) reduction of $HClO_2$ to Cl_2; (b) oxidation of Cl^- to $HClO_2$. In each case calculate the standard potential using the reduction potentials shown in Figure 21.10.

The group 6A elements

21.25 Write the formula for each of the following compounds and indicate the oxidation state of the central group 6A element in each: (a) hydrogen sulfide; (b) selenous acid; (c) tellurium trioxide; (d) potassium hydrogen sulfite; (e) calcium peroxide; (f) aluminum(III) sulfide; (g) orthotelluric acid; (h) selenate ion; (i) thiosulfate ion.

21.26 Name the following: (a) K_2SeO_4; (b) H_2Te; (c) HSO_4^-; (d) O_3; (e) FeS_2; (f) SeF_6; (g) $NaHSO_4$.

21.27 Write a balanced chemical equation that shows the preparation of each of the following substances: (a) Se; (b) H_2Te; (c) H_2SeO_3; (d) S_8; (e) $H_2S_2O_7$; (f) $Na_2S_2O_3$.

21.28 What property of the atomic elements is mainly responsible for the variation in electronegativity seen among the group 6A elements?

21.29 Compare the electron affinities of the group 6A elements with those of their halogen neighbors. Why are the electron affinities smaller for the group 6A elements.

21.30 What volume of $SO_2(g)$ at $200°C$ and 10 mm Hg is produced from combustion of 600 g of H_2S?

21.31 Although addition of one electron to a group 6A element is in all cases exothermic (Table 21.8), addition of a second electron, to form $X^{2-}(g)$, is in all cases strongly endothermic (for example: $S^-(g) + e^- \longrightarrow S^{2-}(g)$, $+590$ kJ/mol). In light of this, why is sulfur found in solid compounds with metals not as S^-, but rather as S^{2-} (for example, Na_2S, not NaS)?

21.32 Write a balanced chemical equation for each of the following reactions: (a) selenium dioxide dissolves in water; (b) sodium superoxide reacts with water; (c) solid zinc sulfide reacts with hydrochloric acid; (d) sulfuric acid reacts with solid potassium iodide; (e) an aqueous solution of hydrogen peroxide decomposes.

21.33 Elemental sulfur in capable of reacting under suitable conditions with Fe, F_2, O_2, or H_2. Write balanced chemical equations to describe the reaction in each case. In which reactions is sulfur acting as a reducing agent and in which as an oxidizing agent?

21.34 One method proposed for removal of SO_2 from the flue gases of power plants involves reaction with aqueous H_2S. Elemental sulfur is the product. Write a balanced chemical equation for the reaction. What volume of H_2S at $27°C$ and 740 mm Hg would be required to remove the SO_2 formed by burning 1 ton of coal containing 3.5 percent S by weight? What mass of elemental sulfur is produced? Assume all reactions are 100 percent efficient.

21.35 Write Lewis structures for each of the following compounds and indicate the geometrical structure in each case: (a) $TeO_2(g)$; (b) SeO_4^{2-}; (c) S_2Cl_2; (d) chlorosulfonic acid, HSO_3Cl; (e) H_6TeO_6.

21.36 An aqueous solution of SO_2 acts as a reducing agent to reduce: (a) aqueous $KMnO_4$ to $MnSO_4(s)$; (b) acidic aqueous $K_2Cr_2O_7$ to aqueous Cr^{3+}; (c) aqueous mercury(I) nitrate to mercury metal. Write balanced chemical equations for these reactions.

21.37 In acidic solution the ferrocyanide ion, $Fe(CN)_6^{4-}$ is oxidized by H_2O_2 to the ferricyanide ion, $Fe(CN)_6^{3-}$. In basic solution, the ferricyanide ion is reduced by H_2O_2 to the ferrocyanide ion. Write balanced net ionic equations for both reactions.

The group 5A elements

21.38 Write formulas for each of the following compounds and indicate the oxidation state of the group 5A element in each: (a) nitrous acid; (b) potassium phosphite; (c) phosphorus(III) oxide; (d) hydrazine; (e) calcium dihydrogen phosphate; (f) potassium azide; (g) arsenic acid; (h) antimony(III) sulfide.

21.39 Write a balanced chemical equation or set of equations to represent commercial preparation of each of the following substances: (a) white phosphorus; (b) ammonia; (c) nitric acid; (d) superphosphate.

21.40 Write Lewis structures for each of the following species: (a) NH_4^+; (b) HNO_3; (c) NO_2; (d) P_4S_{10}; (e) HPO_4^{2-}; (f) $N_2H_5^+$.

21.41 Which of the following compounds would you expect to give rise to an acidic solution in water and which to a basic solution: (a) NH_4Br; (b) P_4O_6; (c) NO_2; (d) N_2H_4; (e) Na_3AsO_4?

21.42 Compare the X—X bond energies of the elements nitrogen, oxygen, and fluorine in their normal stable states; account for the variation observed.

21.43 Why does phosphorus not exist as a diatomic molecule, P_2, at room temperature, as does N_2?

21.44 Given the following heats of formation, calculate ΔH for the equilibrium $P_4(g) \rightleftharpoons 2P_2(g)$. $P_2(g)$, $\Delta H_f° = +146$ kJ/mol; $P_4(s)$, $\Delta H_f° = +17$ kJ/mol; ΔH (sublimation) of $P_4(s) = +112$ kJ/mol.

21.45 Compare the first ionization energies of phosphorus and sulfur (Tables 21.8 and 21.9) and account for the observed values in terms of electronic structures of the atoms.

21.46 Write balanced net ionic chemical equations for

each of the following reactions: (a) hydrazine is burned in excess fluorine gas (NF_3 is a product); (b) hydrazine reduces aqueous sodium selenite to selenium metal (N_2 is the oxidation product); (c) hydroxylamine is oxidized to N_2 by Cu^{2+} in aqueous solution (the copper is reduced to copper metal); (d) aqueous azide ion reacts with chlorine, forming N_2.

21.47 Write balanced net ionic equations for each of the following reactions: (a) dilute nitric acid reacts with zinc metal with formation of nitrous oxide; (b) concentrated nitric acid reacts with sulfur with formation of NO_2 as a product; (c) concentrated nitric acid oxidizes SO_2 with formation of nitric oxide; (d) urea reacts with water to form ammonia and CO_2.

21.48 A commercial lawn fertilizer contains 18 percent nitrogen by weight. All of this nitrogen is present as urea. What is the weight percentage of urea in the fertilizer?

21.49 Account for the fact that nitric acid is a strong acid, whereas phosphoric acid is weak.

21.50 Phosphorus pentachloride exists in one form in the solid state as an ionic lattice of PCl_4^+ and PCl_6^- ions. Draw the Lewis structures for these ions and predict their geometry. What set of hybrid orbitals is employed by the phosphorus in each case? Why should PCl_5 exist as an ionic substance in the solid state, whereas it is stable as the neutral molecule in the gas phase?

21.51 Write complete balanced half-cell equations for the half-cell reaction involving: (a) reduction of nitrate ion to NO in acidic solution; (b) oxidation of N_2O to NO in acidic solution. What is the standard potential in each case? (See Figure 21.23.)

[21.52] Write the complete balanced half-cell reaction for: (a) reduction of NO_2 to NO in acidic solution; (b) oxidation of NH_4^+ to NO_3^- in acidic solution. What is the standard potential in each case? (See Figure 21.23.)

Additional exercises

21.53 Xenon trioxide disproportionates in strongly alkaline solution to form the thermally stable perxenate ion, XeO_6^{4-}. Describe the bonding in this ion in terms of the hybridization of xenon valence-shell orbitals. What geometry should it have?

21.54 Chloride ion in aqueous solution is oxidized to $Cl_2(aq)$ by each of the following reagents: (a) $MnO_2(s)$; (b) $MnO_4^-(aq)$; (c) $Cr_2O_7^{2-}(aq)$. In each case, write a complete, balanced net ionic equation.

21.55 Offer a reasonable explanation of each of the following observations: (a) white phosphorus is quite volatile, whereas red phosphorus is not; (b) addition of SF_4 to water results in an acidic solution; (c) xenon hexafluoride is a stable compound, whereas krypton hexafluoride is unknown; (d) nitrogen exists abundantly in nature as the free element, whereas phosphorus is found only in its compounds.

21.56 What are the major factors responsible for the fact that xenon forms a stable series of fluorides, whereas no analogous chemical reactivity is known for argon?

21.57 What is the significance of the fact that the enthalpies of formation of the oxides of xenon are positive?

[21.58] Using the thermochemical data of Table 21.2 and Appendix D, calculate the average Xe—F bond energies in XeF_2, XeF_4, and XeF_6. What is the significance of the trend observed in these quantities?

21.59 Write balanced chemical equations to account for the following observations (there may not be closely similar reactions shown in the chapter; however, you should be able to make reasonable guesses at the likely products). (a) When burning sodium metal is immersed in a pure HCl atmosphere, it continues to burn. (b) Bubbling SO_2 gas through liquid bromine that is covered with a layer of water results in formation of a strongly acidic solution. Upon distillation, an aqueous HBr solution is collected. The remaining liquid is still strongly acidic. (c) When bromine is added to a basic solution containing potassium hypochlorite, insoluble potassium bromate is formed. (d) When bromic acid is reacted with SO_2, $Br_2(aq)$ is formed. (e) Uranium(VI) fluoride is formed by the action of ClF_3 on uranium(IV) chloride.

21.60 Write the formula for the anhydride of each of the following acids: (a) H_3PO_3; (b) $HClO_4$; (c) H_3AsO_4; (d) H_3BO_3; (e) H_2SO_4; (f) H_2CO_3.

21.61 The SF_5^- ion is formed when $SF_4(g)$ is reacted with fluoride salts containing large cations, for example, $CsF(s)$. Draw the Lewis structure for SF_5^- and predict its geometry.

21.62 Write balanced chemical equations for each of the following reactions (you may have to guess at one or more of the reaction products, but you should be able to make a reasonable guess based on your study of this chapter). (a) Selenous acid is reduced by hydrazine in aqueous solution to yield elemental selenium. (b) Heating orthotelluric acid to temperatures over 200°C yields the acid anhydride. (c) Hydrogen selenide can be prepared by reaction of aqueous acid solution on aluminum selenide. (d) When ozone reacts with aqueous iodide solution, O_2 and OH^- are among the products. (e) Lead(II) nitrate decomposes thermally at about 400°C to yield lead(II) oxide and NO_2 as products.

21.63 Calculate the mass of sodium iodide that must react with excess phosphoric acid to produce enough HI to form 2.50 L of 4.80 M solution.

21.64 What pressure of gas is formed when 0.654 g of XeO_3 decomposes completely to the free elements at 48°C in a 0.452-L volume?

21.65 In acidic aqueous solution, hydrogen sulfide reduces (a) Fe^{3+} to Fe^{2+}; (b) Br_2 to Br^-; (c) MnO_4^- to Mn^{2+}; (d) HNO_3 to NO_2. In all cases, under appropriate conditions, the product is elemental sulfur. Write a balanced net ionic equation for the reaction in each case.

21.66 Ammonium thiosulfate is frequently employed in photographic work in formulating hypo solutions. What mass of ammonium thiosulfate is needed to prepare 10.8 L of 0.28 M solution?

21.67 Hydrogen peroxide is capable of oxidizing (a) K_2S to S; (b) SO_2 to SO_4^{2-}; (c) NO_2^- to NO_3^-; (d) As_2O_3 to AsO_4^{3-}; (e) Fe^{2+} to Fe^{3+}. Write balanced net ionic equations for reactions in aqueous solution to represent each of these oxidations.

21.68 Hydrogen peroxide is capable of reducing (a) MnO_4^- to Mn^{2+}; (b) Cl_2 to Cl^-; (c) Ce^{4+} to Ce^{3+}; (d) O_3 to H_2O. Write balanced net ionic equations for reactions in aqueous acidic solution to illustrate each of these reductions.

21.69 A traditional laboratory preparation of nitrogen gas involves the decomposition of ammonium nitrite. Write a balanced equation for this reaction. Calculate the volume of nitrogen gas collected over water at 22°C and 1 atm pressure from decomposition of 3.26 g of NH_4NO_2.

21.70 Annual production of nitric acid in the U.S. is about 7.3×10^9 kg. What volume of ammonia measured at 45 atm and 25°C is required for production of this much HNO_3 via the Ostwald process, assuming an overall 92 percent conversion efficiency?

21.71 Write balanced chemical equations to represent reaction of each of the following substances with water: (a) phosphorus(V) oxide; (b) phosphorus(III) bromide; (c) arsenic(III) oxide; (d) lithium nitride; (e) nitrogen dioxide.

21.72 A mixture of three parts concentrated hydrochloric acid and one part nitric acid by volume is called *aqua regia*. This acid mixture dissolves gold according to the reaction

$$Au(s) + 4Cl^-(aq) + 3NO_3^-(aq) + 6H^+(aq) \longrightarrow$$
$$AuCl_4^-(aq) + 3NO_2(g) + 3H_2O(l)$$

What mass of NO_2 is produced by reaction of excess acid with 0.365 g of Au?

21.73 Why is the electron affinity of nitrogen (Table 21.9) so much more positive than that for the other elements of group 5A? Why is it more positive than that for carbon (-122 kJ/mol) or oxygen (-144 kJ/mol)?

[21.74] Thermodynamic data for $HNO_3(aq)$ at 25° are as follows: $\Delta H_f^\circ = -207.3$ kJ/mol; $\Delta G_f^\circ = -111.3$ kJ/mol. Calculate ΔH° and ΔG° for the reaction

$$\tfrac{1}{2}N_2(g) + \tfrac{1}{2}H_2O(l) + \tfrac{5}{4}O_2(g) \longrightarrow HNO_3(aq)$$

It has been suggested that if a suitable catalyst were present, the atmospheric gases could react to make the oceans a dilute nitric acid solution. In thermodynamic terms, is this a possible process?

21.75 Draw Lewis structures for the substances formed in each of the following condensation reactions: (a) a molecule of water is eliminated between two molecules of H_2SO_4; (b) three molecules of water are eliminated from three molecules of H_3PO_4 in forming a cyclic structure; (c) a molecule of water is eliminated from between two molecules of boric acid; (d) a chain structure is formed by elimination of one molecule of water per molecule of boric acid.

The lithosphere: geochemistry and metallurgy

In earlier chapters, we considered the properties of our atmospheric environment and of the earth's waters. In this chapter, we shall examine the characteristics of the lithosphere, the solid earth beneath our feet. The lithosphere provides us with most of the materials with which we feed, clothe, shelter, support, and entertain ourselves.

Although the bulk of the earth is solid, we have ready access only to a small area near the surface. The deepest well ever drilled is only 7.7 km deep, and the deepest mine descends only 3.4 km into the earth. In comparison, the earth has a radius of 6370 km. We shall see that many substances most useful to us are not especially plentiful in the portion of the lithosphere to which we have access. Furthermore, most of them occur in chemical forms that are not useful. Usually the compounds or elements that we desire must be separated from a large quantity of undesirable material and then chemically processed to render them useful. By consuming huge quantities of substances located close to the surface, we have literally changed the face of the earth (see Figure 22.1). Experts estimate that about 2.3×10^4 kg, or 25 tons, of materials are extracted from the lithosphere and processed annually to support each person in our society. The total quantity of materials mined is expected to increase because of rising populations and increasing per capita demand. In addition, because the richest sources of many substances are becoming exhausted, it will be necessary in the future to process larger volumes of lower quality raw materials. This means that the extraction of the compounds and elements we need will cost more in terms of energy and environmental impact.

22.1 The structure and composition of the earth

Although we have not penetrated very far into the earth, we have been able to construct a general picture of its structure and composition from indirect evidence. One line of evidence is based on comparison of the average density of the entire planet (5.5 g/cm^3) with the average density of rocks on its surface (2.8 g/cm^3). This comparison tells us that the innermost portions of the earth must be considerably more dense than the surface area. Further information is obtained by observing how earthquake waves are propagated through the earth. This type of study indicates that the earth has a layered arrangement consisting of four distinct regions: the crust, mantle, outer core, and inner core. These regions are shown in Figure 22.2.

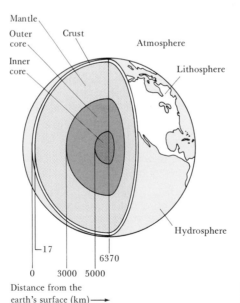

Distance from the
earth's surface (km) ⟶

FIGURE 22.2 Regions of the earth's surface and interior.

TABLE 22.1 The twelve most abundant elements in the earth's crust

Element	Percent by weight
Oxygen	50
Silicon	26
Aluminum	7.5
Iron	4.7
Calcium	3.4
Sodium	2.6
Potassium	2.4
Magnesium	1.9
Hydrogen	0.9
Titanium	0.6
Chlorine	0.2
Phosphorus	0.1

TABLE 22.2 Estimated annual world consumption of elements

Element	Annual consumption (kg)
C	10^{12}–10^{13}
Na, Fe	10^{11}–10^{12}
N, O, S, K, Ca	10^{10}–10^{11}
H, F, Mg, Al, P, Cl, Cr, Mn, Cu, Zn, Ba, Pb	10^{9}–10^{10}
B, Ti, Ni, Zr, Sn	10^{8}–10^{9}
Ar, Co, As, Mo, Sb, W, U	10^{7}–10^{8}
Li, V, Se, Sr, Nb, Ag, Cd, I, rare earths, Au, Hg, Bi	10^{6}–10^{7}
He, Be, Te, Ta	10^{5}–10^{6}

The outer portion, or **crust**, constitutes only about 0.4 percent of the total mass of the earth. It has an average thickness of 17 km, ranging in depth from 4 to 70 km. The term "crust" is a carry-over from the time when the entire interior of the planet was thought to be molten. The **mantle**, which lies immediately below the crust, is now believed to be almost entirely solid. It extends to a depth of about 3000 km and comprises 68.2 percent of the earth's mass. Although we have never penetrated into the mantle, it is believed that over 90 percent of it consists of four elements—magnesium, iron, silicon, and oxygen. Because the material in the mantle is at a higher pressure and temperature than that in the crust, the chemical forms in which these elements occur may differ considerably from those found in the crust. The innermost portions of the earth are known as the **inner** and **outer cores**. These portions are believed to contain about 80 percent iron. The outer core is thought to be molten, whereas the more dense inner core is thought to be solid.

Eighty-eight elements occur in the earth's crust. Twelve of these, listed in Table 22.1, make up 99.5 percent of the crust by weight.* However, if you were to analyze a shovelful of dirt from your backyard, you would not find it to have the composition listed in Table 22.1. Elements are not distributed uniformly throughout the crust.

It is interesting to compare the order of abundance of the elements with the order of estimated world consumption shown in Table 22.2. The ranking of elements in the two lists shows little correlation; some elements that are not very abundant are widely used. Consequently, the occurrence and distribution of concentrated deposits of certain elements in the lithosphere is important. Indeed, because a modern, high-technology society such as ours has need for a broad range of substances such as metals, the distribution of such deposits may become an important factor in international politics.

TABLE 22.3 Some common minerals

Mineral name	Chemical formula
Calcite	$CaCO_3$
Chalcopyrite	$CuFeS_2$
Cinnabar	HgS
Corundum	Al_2O_3
Fluorite	CaF_2
Galena	PbS
Gypsum	$CaSO_4 \cdot 2H_2O$
Halite	$NaCl$
Hematite	Fe_2O_3
Malachite	$Cu_2(CO_3)(OH)_2$
Pyrite	FeS_2
Perovskite	$CaTiO_3$
Quartz	SiO_2
Talc	$Mg_3(Si_4O_{10})(OH)_2$
Turquoise	$CuAl_6(PO_4)_4 \cdot (OH)_2 \cdot 4H_2O$
Wulfenite	$PbMoO_4$

22.2 Minerals

Most elements occur in nature in combination with other elements, that is, in compounds. Solid substances occurring in nature are referred to as **minerals.** Table 22.3 lists some common minerals. Some of these are pictured in Figure 22.3. Minerals are usually known by their common

*The top 1 km of the earth's crust has a total mass of approximately 10^{21} kg.

FIGURE 22.3 Three common minerals: (a) calcite; (b) fluorite; (c) wulfenite. Note the variety of crystal shapes.

names rather than by their chemical names. They are often named by the person who first described them. Although there is no systematic method of nomenclature, the name frequently ends in -*ite*. Minerals often have variable compositions owing to the substitution of one element for another within the solid (Section 22.5). Nevertheless, they have well-defined crystal structures. What we know as *rock* is merely an aggregate of different kinds of minerals.

Minerals can be classified into three groups: native elements, silicate minerals, and nonsilicate minerals. Examples are shown in Table 22.4. Metals found in elemental form include silver, gold, palladium, platinum, ruthenium, rhodium, osmium, and iridium. These metals, which are in families 8B and 1B of the periodic table, are known as noble metals because of their lack of reactivity. All of these metals have very high standard reduction potentials and are thus difficult to oxidize.

22.3 Silicate minerals

The silicate minerals are the most abundant group of minerals. It has been estimated that over 90 percent of the earth's crust consists of silicates, if quartz (SiO_2) is included. Suppose you pick up a piece of common granite rock* such as that shown in Figure 22.4 and determine its

*Granite is a mixture of small crystals of mica, quartz, and feldspars. We will discuss these minerals later in this section.

FIGURE 22.4 A piece of granite rock.

TABLE 22.4 Types of minerals

General category	Examples
Native elements	Cu, Ag, Au, Bi, Pt, Pd, S
Silicate minerals	$ZrSiO_4$, $Be_3AlSi_6O_{18}$
Nonsilicate minerals	
Oxides	Al_2O_3, Fe_2O_3, Fe_3O_4, Cu_2O, TiO_2
Hydroxides	$Mg(OH)_2$
Carbonates	$CaCO_3$, $MgCO_3$, $PbCO_3$, $ZnCO_3$
Sulfates	$BaSO_4$, $PbSO_4$, $CaSO_4$
Sulfides	Ag_2S, Cu_2S, HgS, ZnS
Halides	$NaCl$, $MgCl_2$
Phosphates	$Ca_3(PO_4)_2$

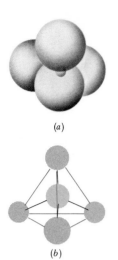

(a)

(b)

FIGURE 22.5 Two ways of representing a simple silicate tetrahedron, SiO_4^{4-}, such as that found in the mineral zircon, $ZrSiO_4$. (a) The small Si(IV) surrounded by four O^{2-} as it might appear if we could actually view the tetrahedron. (b) The geometric structure showing the orientation of the atoms.

elemental composition. You would find that it contains about 50 percent oxygen and 25 percent silicon. It also contains a surprising variety of important metals. Each 100 kg of granite contains approximately 8 kg of aluminum, 5 kg of iron, 90 g of manganese, 20 g of nickel, and 10 g of copper. Despite these figures, the silicate minerals are not presently economical sources of these or any other metals. The silicates are extremely stable chemical substances; a large amount of energy is required to remove metals from them. Nevertheless, they are important items of commerce and are used, for example, in the manufacture of cement and glass.

To understand why silicates are so stable, we need to examine their structures. The basic structural unit of the silicate minerals is a silicon atom bound in a tetrahedral fashion to four oxygens as shown in Figure 22.5.* This structural unit occurs as the orthosilicate ion, SiO_4^{4-}. However, this simple ion is found in very few minerals. Usually, the silicate tetrahedra are connected by sharing oxygen atoms, giving Si—O—Si bonds. When two silicate tetrahedra are so joined by sharing of an oxygen atom, the $Si_2O_7^{6-}$ ion shown in Figure 22.6 results. The situation is similar to the linking of phosphate tetrahedra, discussed in Section 21.9. By sharing oxygen atoms, silicate tetrahedra can be arranged in strands, sheets, or three-dimensional arrays. Two of the strandlike arrangements and the sheetlike arrangement are shown in Figure 22.7. Notice the empirical formula that results from each of these arrangements.

SAMPLE EXERCISE 22.1

Draw the Lewis structures, showing all valence electrons, for the SiO_4^{4-} and $Si_2O_7^{6-}$ ions. Indicate how the $Si_2O_7^{6-}$ ion is related to SiO_4^{4-}.

Solution: The total number of valence electrons in SiO_4^{4-} is 32 (remember we must count the 4 electrons that make the ionic charge -4). The total number of valence electrons in $Si_2O_7^{6-}$ is 56. When these electrons are placed about the atoms connected as shown in Figures 22.5 and 22.6, we obtain the following Lewis structures:

$$\left[\begin{array}{c} \ddot{O} \quad\quad \ddot{O} \\ \diagdown \;\; \diagup \\ Si \\ \diagup \;\; \diagdown \\ \ddot{O} \quad\quad \ddot{O} \end{array} \right]^{4-} \quad\quad \left[\begin{array}{c} \ddot{O} \quad\quad\quad \ddot{O} \\ \| \quad\quad\quad \| \\ \ddot{O}—Si—\ddot{O}—Si—\ddot{O} \\ \| \quad\quad\quad \| \\ \ddot{O} \quad\quad\quad \ddot{O} \end{array} \right]^{6-}$$

Notice that the bridging oxygen in $Si_2O_7^{6-}$ contains two rather than three unshared electron pairs. You can imagine the $Si_2O_7^{6-}$ ion to be formed by completely removing an O^{2-} from one SiO_4^{4-} ion, and then using an oxygen from a second SiO_4^{4-} ion to donate a pair of electrons into the vacancy created by removal of the O^{2-}.

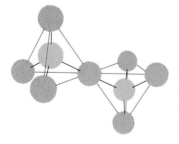

FIGURE 22.6 Geometric structure of the $Si_2O_7^{6-}$ ion, which is formed by the sharing of an oxygen atom by two silicon atoms. This ion occurs in the mineral hardystonite, $Ca_2Zn(Si_2O_7)$.

It is interesting how the cleavage properties of silicate minerals reflect the arrangement of the silicate ions. The asbestos minerals, which are fibrous, as shown in Figure 22.8, have either the double-strand chain structure or a rolled-sheet structure in which the sheets are rolled into strands. The fibrous texture of the asbestos minerals arises because the electrostatic bonds between strands are much weaker than are the cova-

*Under extreme conditions a different arrangement of oxygens about silicon might be formed. For example, under high pressures and high temperatures it is possible to prepare a silicate in which six oxygens surround each silicon. Such a structure has been found among the compounds in the impact craters of meteorites. It has been postulated that such silicates may also exist in the mantle of the earth.

(a)

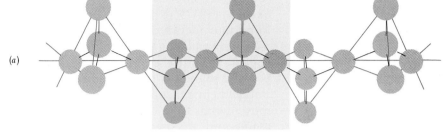

Repeating unit of chain

(b)

Repeating unit of chain

(c)

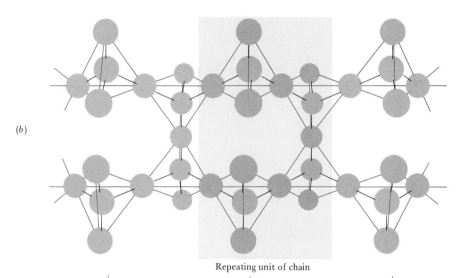

Repeating unit of sheet

FIGURE 22.7 Schematic representations of chain and sheet silicates formed by linking together silicate tetrahedra: (a) single-strand silicate, which has an empirical formula of SiO_3^{2-}, as in the mineral enstatite, $Mg(SiO_3)$; (b) double-strand silicate chain, which has an empirical formula of $Si_4O_{11}^{6-}$, as in the mineral tremolite, $Ca_2Mg_5(Si_4O_{11})_2(OH)_2$; (c) sheet silicate, which has an empirical formula $Si_2O_5^{2-}$, as in the mineral talc, $Mg_3(Si_2O_5)_2(OH)_2$. Note that in (a) the repeating unit is double the empirical formula.

FIGURE 22.8 Serpentine asbestos; note the fibrous character of this mineral.

lent bonds within the strands. Talc, $Mg_3Si_4O_{10}(OH)_2$, has a structure composed of planar sheets. The relatively weak forces between sheets allows them to slip over each other much as in the case of graphite. Talcum powder therefore has a slippery feel. Extension of the silicate tetrahedra in three dimensions in space, so that every oxygen atom is bridging between two silicons, gives quartz, SiO_2, which is shown in Figure 22.9. Because the silicate is locked together in a three-dimensional array, much like diamond, quartz is harder than the strand- or sheet-type silicates.

Asbestos is a general term applied to a group of silicates that are fibrous. These asbestos minerals are widely used, especially as thermal insulators. They are used, for example, to insulate furnaces and steam pipes and to make protective clothing for people exposed to high temperatures. Certain forms of asbestos pose a health hazard. Tiny fibers readily penetrate the tissues of the lungs and digestive tract, where they can apparently induce cancer.

ALUMINOSILICATES

In many silicate minerals Si^{4+} ions are replaced by Al^{3+} ions within the silicate tetrahedra. This replacement produces **aluminosilicates.** In order to maintain charge balance, an extra cation such as K^+ must accompany each of these substitutions. Muscovite, $KAl_2(AlSi_3O_{10})(OH)_2$,* a mica mineral, is an aluminosilicate. Replacement of a quarter of the silicon atoms in a sheet silicate, $Si_4O_{10}^{4-}$, with aluminum produces the $AlSi_3O_{10}^{5-}$ sheets found in this mineral. In both the silicate-sheet and aluminosilicate-sheet minerals, cations are located between the sheets to balance the charge. The electrostatic attraction between these cations and the sheets is greater for the aluminosilicate than for the silicate because of the former's greater negative charge. Thus while the sheets in the silicate mineral talc slide readily over each other, the sheets in the aluminosilicate mica do not. Nevertheless, mica does cleave readily into sheets, as shown in Figure 22.10.

FIGURE 22.9 Quartz crystals. (*American Museum of Natural History*)

*Aluminum is found in this mineral in two different environments. The first two Al^{3+} ions are located between the aluminosilicate sheets. The aluminum that is shown in parentheses is located within the sheets. It has replaced a silicon and is therefore located in an AlO_4 tetrahedron.

FIGURE 22.10 A piece of mica, showing its cleavage into thin sheets.

When aluminum replaces up to half of the silicon atoms in SiO_2, the feldspar minerals result (see Figure 22.11). The cations that compensate the extra negative charge accompanying this replacement are usually Na^+, K^+, or Ca^{2+} ions. The feldspars are the most abundant rock-forming silicates, comprising nearly 54 percent of the minerals in the earth's crust.

SAMPLE EXERCISE 22.2

The mineral anorthite is a feldspar mineral formed by replacing half of the silicon atoms in SiO_2 with aluminum and maintaining charge balance with Ca^{2+} ions. What is the simplest formula for this mineral?

Solution: If we write SiO_2 as Si_2O_4, the replacement of half of the Si^{4+} with Al^{3+} produces $AlSiO_4^-$. One Ca^{2+} therefore requires two $AlSiO_4^-$ units to maintain charge balance and the empirical formula of the mineral is $CaAl_2Si_2O_8$.

FIGURE 22.11 Orthoclase, $KAlSi_3O_8$, a feldspar mineral.

The clay minerals are hydrated aluminosilicates having sheet-type structures. They are thought to be formed during weathering when H_2O and CO_2 slowly attack feldspars. For example, the action of H_2O and CO_2 on the feldspar mineral anorthite to form the clay mineral known as kaolinite is given in Equation [22.1]:

$$CaAl_2Si_2O_8(s) + 3H_2O(l) + 2CO_2(aq) \longrightarrow$$
Anorthite

$$Al_2Si_2O_5(OH)_4(s) + Ca^{2+}(aq) + 2HCO_3^-(aq) \qquad \text{[22.1]}$$
Kaolinite

The ions released by such weathering reactions may be incorporated into plants and other minerals or washed into the sea. The clay minerals have small particle size and correspondingly large surface areas. They have the ability to absorb cations on their surfaces. This provides a means of storing plant nutrients such as K^+ that can then be made available to plants through equilibrium shifts:

$$K^+\text{—clay} \rightleftharpoons K^+(aq) + \text{clay} \qquad \text{[22.2]}$$

Some metal ions may exchange for others within the crystal lattice of the clay and thereby be stored in this fashion. Often the metal ions displace hydrogen ions from the OH groups on the surface of the clay particle:

$$M^+(aq) + H\text{—O—clay} \rightleftharpoons M\text{—O—clay} + H^+(aq) \qquad \text{[22.3]}$$

This situation gives rise to pH-dependent equilibria. Notice that the higher the concentration of $H^+(aq)$, the more the equilibrium is shifted to the left. If the soil is basic, the equilibrium lies to the right, and $M^+(aq)$ is not available to plants. Thus the pH of a soil plays an important role in determining its fertility (that is, its ability to supply plants with essential nutrients). The majority of plants grow best in soil whose pH is 6–7, that is, one that is slightly acidic. Basic or alkaline soils are common in areas where there is scanty rainfall or where soils drain poorly. However, it is more common for soils to be too acidic than for them to be too alkaline. The acidity of a soil is often lowered by adding lime, CaO, to adjust pH. This process is known as "liming." The CaO is a basic anhydride and therefore can react with $H^+(aq)$ as shown in Equation [22.4]:

$$CaO(s) + 2H^+(aq) \longrightarrow Ca^{2+}(aq) + H_2O(l) \qquad \text{[22.4]}$$

GLASS

As noted earlier, in Section 11.5, quartz melts at approximately 1600°C, forming a tacky liquid. In the course of melting, many silicon-oxygen bonds are broken. When the liquid is rapidly cooled, silicon-oxygen bonds are reformed before the atoms are able to arrange themselves in a regular fashion. An amorphous solid, known as quartz glass or silica

TABLE 22.5 Compositions, properties, and uses of various types of glass

Type of glass	Composition by weight	Properties and uses
Soda-lime	12% Na_2O, 12% CaO, 76% SiO_2	Window glass, bottles
Aluminosilicate	5% B_2O_3, 15% MgO, 15% CaO, 20% Al_2O_3, 55% SiO_2	High melting—used in cooking ware
Lead alkali	10% Na_2O, 20% PbO, 70% SiO_2	High refractive index—used in lenses, decorative glass
Borosilicate	5% Na_2O, 3% CaO, 16% B_2O_3, 76% SiO_2	Low coefficient of thermal expansion—used in laboratory ware, cooking utensils
Bioglass	24% Na_2O, 24% CaO, 6% P_2O_5, 46% SiO_2	Compatible with bone—used as coating on surgical implants

glass, therefore results (see Figure 11.28). Many different substances can be added to SiO_2 to cause it to melt at a lower temperature. The common glass used in windows and bottles is known as soda-lime glass. It contains CaO and Na_2O in addition to SiO_2 from sand. The CaO and Na_2O are produced by heating two inexpensive chemicals, limestone, $CaCO_3$, and soda ash, Na_2CO_3. These substances decompose at elevated temperatures as shown in Equations [22.5] and [22.6]:

$$CaCO_3(s) \longrightarrow CaO(s) + CO_2(g) \qquad [22.5]$$

$$Na_2CO_3(s) \longrightarrow Na_2O(s) + CO_2(g) \qquad [22.6]$$

Other substances can be added to soda-lime glass to produce color or to change the properties of the glass in various ways. For example, addition of CoO produces the deep blue color of "cobalt glass." Replacement of Na_2O by K_2O results in harder glass that has a higher melting point. Replacement of CaO by PbO results in a denser glass with a higher refractive index. Addition of nonmetal oxides such as B_2O_3 or P_2O_5, which form network structures related to the silicates, also causes a change in properties of the glass. For example, addition of B_2O_3 results in a glass with a higher melting point and a lower coefficient of thermal expansion. Such glasses, sold commercially under trade names such as Pyrex or Kimax, are used in applications where resistance to thermal shock is important, for example, in laboratory glassware. Table 22.5 lists the formulations and properties of several representative types of glass.

A recent popular development has been that of "photochromic" glasses that darken in the sun and return to their clear state in the dark. This glass contains a dispersion of $AgCl$ or $AgBr$. These substances are photosensitive, in that they decompose to silver and halogen atoms in the presence of light. The finely divided silver is black. The silver and halogen atoms are kept in close proximity by the glass matrix and reform $AgCl$ or $AgBr$ in the dark:

$$AgCl \underset{\text{dark reaction}}{\overset{\text{light reaction, } h\nu}{\rightleftharpoons}} Ag + Cl \qquad [22.7]$$

CEMENT

Portland cement, used as a binder in making concrete, is similar in structure to glass except that it is an aluminosilicate. It is made by

TABLE 22.6 Composition of Portland cement

Constituents[a]	Percentage range
CaO	60–64
SiO_2	18–26
Al_2O_3	4–12
Fe_2O_3	2–4
MgO	1–4
Other	2

[a]The cement is analyzed as if it were a mixture of oxides. However, it is not a simple mixture of these oxides but rather a complex mixture of aluminosilicates and oxides.

heating a powdered mixture of limestone, $CaCO_3$, sand, SiO_2, and clay in a kiln at about 1500°C. After the resulting mass is pulverized, a small amount of gypsum, $CaSO_4 \cdot 2H_2O$, is added. The composition of Portland cement is given in Table 22.6. The reactions involved in the setting of cement are complex, involving reaction with both water and CO_2. Concrete, like other materials containing Si—O bonds, is highly incompressible but lacks tensile strength.* The importance of cement and concrete in our society is indicated by the fact that the total economic value of the raw materials used annually in their manufacture is substantially higher than the value of the raw metallic ores that we process.

22.4 Nonsilicate minerals

The nonsilicate minerals, though less abundant than the silicates, are important as sources of metals. The greatest quantity of metals is obtained commercially from oxides, sulfides, and carbonates; most processes used for obtaining metals utilize oxides. Both sulfide and carbonate minerals are readily converted to oxides when heated in air. Equations [22.8] and [22.9] illustrate these conversions:

$$2ZnS(s) + 3O_2(g) \longrightarrow 2ZnO(s) + 2SO_2(g) \qquad [22.8]$$

$$PbCO_3(s) \longrightarrow PbO(s) + CO_2(g) \qquad [22.9]$$

We shall discuss the reduction of metal oxides in Section 22.6.

CALCIUM CARBONATE

The most abundant nonsilicate mineral is calcite, $CaCO_3$. This mineral is pictured in Figure 22.3. It is the principal mineral in the rock known as limestone. It is also the main constituent of marble, chalk, pearls, coral reefs, and the shells of marine animals such as clams and oysters. Although $CaCO_3$ is insoluble in pure water, it does dissolve in acidic solution. The reaction is given in Equation [22.10]:

$$CaCO_3(s) + 2H^+(aq) \longrightarrow Ca^{2+}(aq) + H_2O(l) + CO_2(g) \qquad [22.10]$$

Because water containing CO_2 is slightly acidic (Section 17.1), $CaCO_3$ dissolves slowly in this medium:

$$CaCO_3(s) + H_2O(l) + CO_2(aq) \rightleftharpoons Ca^{2+}(aq) + 2HCO_3^-(aq) \qquad [22.11]$$

This reaction occurs in nature when surface waters move underground through limestone deposits. If the dissolving limestone underlies a comparatively thin layer of earth, sinkholes like that shown in Figure 22.12 are produced. If the limestone deposit is deep enough underground, the dissolution of the limestone produces a cave; two well-known limestone caves are the Mammoth Cave in Kentucky and the Carlsbad Cavern in New Mexico. When a solution of $Ca(HCO_3)_2$ trickles through the cracks in a cave roof, the pressure release allows escape of CO_2 and evaporation of H_2O. Loss of these substances shifts the equilibrium of Equation

*Tensile strength measures the stress required to stretch a rod to the breaking point.

FIGURE 22.12 A sinkhole filled with water. (*U.S. Geological Survey*)

[22.11] to the left, forming deposits of $CaCO_3$. Those deposits that hang from the cave roof are known as stalactites, whereas those that rise from the cave floor are known as stalagmites. Figure 22.13 shows a limestone cave with prominent stalactites and stalagmites.

The chemical reaction between $CaCO_3$ and acidic solutions formed by CO_2 is responsible for the erosion of marble and limestone monuments (refer to Figure 10.7). The presence of SO_2 in polluted air accelerates this erosion as noted earlier (Section 10.5). To slow such erosion of monuments and thereby extend their lives, some have been treated with a mixture of $Ba(OH)_2$ and urea, $(NH_2)_2CO$. These two substances react with each other to form a layer of $BaCO_3$ on the monument's surface:

$$H_2N\overset{\overset{\displaystyle O}{\|}}{C}NH_2(aq) + H_2O(l) \longrightarrow 2NH_3(aq) + CO_2(aq) \qquad [22.12]$$

$$CO_2(aq) + Ba^{2+}(aq) + 2OH^-(aq) \longrightarrow BaCO_3(s) + H_2O(l) \qquad [22.13]$$

Barium carbonate ($K_{sp} = 5.1 \times 10^{-9}$) is about as insoluble as calcium carbonate ($K_{sp} = 2.8 \times 10^{-9}$). However, when it reacts with SO_2, it forms an even more insoluble compound, $BaSO_4$, as shown in Equation [22.14]:

$$2BaCO_3(s) + 2SO_2(aq) + O_2(aq) \longrightarrow 2BaSO_4(s) + 2CO_2(g) \qquad [22.14]$$

FIGURE 22.13 A limestone cave with stalagmite and stalactite formations. (*U.S. National Park Service*)

The solubility product, K_{sp}, of $BaSO_4$ is only 1.1×10^{-10}, whereas that of $CaSO_4$ is 9.1×10^{-6}.

One of the most important reactions of $CaCO_3$ is its decomposition into CaO and CO_2 at elevated temperatures, which was given earlier in Equation [22.5]. Over 2×10^{10} kg of calcium oxide, known as lime or quicklime, is used in the United States each year. Because calcium oxide reacts with water to form $Ca(OH)_2$, it is an important commercial base. In this chapter, we have already discussed the use of CaO in making glass and cement and in neutralizing acidic soils (Section 22.3); in Section 22.6 we shall consider its use in high-temperature reductions of metal ores. Calcium oxide is also used in making mortar, a mixture of sand, water, and CaO used in construction to bind bricks, blocks, and rocks together. The CaO reacts with water and CO_2 to form $CaCO_3$, which binds the sand in the mortar:

$$CaO(s) + H_2O(l) \longrightarrow Ca^{2+}(aq) + 2OH^-(aq) \qquad [22.15]$$

$$Ca^{2+}(aq) + 2OH^-(aq) + CO_2(aq) \longrightarrow CaCO_3(s) + H_2O(l) \qquad [22.16]$$

22.5 Close packing in minerals

In many minerals and other ionic compounds, the large ions in the lattice assume a close-packed arrangement identical to those discussed in Section 11.4. A close-packed array of spheres, whether hexagonal or cubic close packed, occupies 74 percent of the total volume taken up by the structure as a whole. The remaining 26 percent consists of holes between the spheres. The smaller ions in the lattice occupy these holes. For example, Fe_2O_3 consists of a cubic close-packed arrangement of oxide ions with Fe^{3+} ions in certain holes.

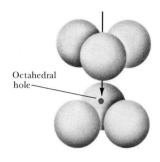

Octahedral hole

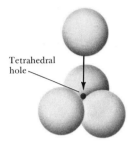

Tetrahedral hole

FIGURE 22.14 Views of the octahedral and tetrahedral holes that exist between two layers of close-packed spheres.

Examination of close-packed structures reveals that they possess two types of holes. These are shown in Figure 22.14. The first type is created by four large spheres arranged in a tetrahedral fashion. The second type is formed by six spheres arranged in an octahedral fashion. Using trigonometry, it can be shown that a **tetrahedral hole** is smaller than an **octahedral hole**. A small sphere that is 0.225 times the larger sphere in radius will fit perfectly into a tetrahedral hole. If the small sphere has a radius that is 0.414 times that of the larger spheres, it will just fit into an octahedral hole.

You may recall from our discussion of ionic radii in Section 7.3 that anions generally have larger radii than cations. It is useful then to think of the lattice as consisting of a close-packed arrangement of anions, with cations occupying one or another of the types of holes present. The size of the cation relative to that of the anion determines the type of hole that the cation occupies. The most stable arrangement is one that maximizes the number of cation-anion contacts, because these lead to electrostatic attractive forces. At the same time, however, the arrangement must also be one that prevents direct anion-anion contacts, because these contacts will generate electrostatic repulsive forces. We can examine the possibilities more closely by considering the situation in which a small cation fits just perfectly into a tetrahedral hole of a close-packed array of anions. As we have noted, this situation occurs when the ratio of the radius of the cation to that of the anion, r_c/r_a, equals 0.225. Under these conditions, the cation can just touch the four anions that surround it. Now consider what happens if the cation increases in size, so that $r_c/r_a > 0.225$. This condition forces the anions apart, thereby diminishing the destabilizing anion-anion contacts while maintaining the stabilizing cation-anion contacts. However, when the ratio of radii, r_c/r_a, reaches a value of 0.414, the cation is no longer most stable in a tetrahedral hole. Instead, the most stable location is the octahedral hole, which permits a greater number of stabilizing cation-anion contacts; in this location the cation can just touch six surrounding anions. Now consider what happens as the cation gets yet larger, so that $r_c/r_a > 0.414$. As before, the anions are forced apart, thereby lowering the destabilizing anion-anion contacts while maintaining the stabilizing cation-anion contacts. Lattices for which r_c/r_a falls in the range of 0.414 to 0.732 generally have cations that occupy octahedral holes. When r_c/r_a exceeds 0.732 the anions are no longer most stable in a close-packed array. Instead, they generally assume a simple cubic arrangement. This arrange-

TABLE 22.7 Radius ratios and cation location in a close-packed anion lattice

r_c/r_a[a]	Coordination number of cation	Arrangement of anions about the cation
0.225–0.414	4	Tetrahedral
0.414–0.732	6	Octahedral
0.732–1.000	8	Cubic

[a]If $r_c > r_a$, the ratio r_a/r_c gives the coordination number of the anion and the special arrangement of cations.

ment permits a cation to be in contact with eight anions, thereby increasing the number of stabilizing cation-anion contacts. These radius-ratio guidelines, which indicate the likely location of cations in a lattice composed of larger anions, are summarized in Table 22.7.

SAMPLE EXERCISE 22.3

The mineral hematite, Fe_2O_3, consists of a cubic close-packed array of oxide ions with Fe^{3+} ions occupying interstitial positions. Predict whether the iron ions are in octahedral or tetrahedral holes. (The radius of Fe^{3+} is 0.65 Å, whereas that of O^{2-} is 1.45 Å.)

Solution:

$$\frac{r_c}{r_a} = \frac{0.65 \text{ Å}}{1.45 \text{ Å}} = 0.45$$

Based on the radius ratio we would predict that Fe^{3+} ions would be located in octahedral holes. In fact, this is their location.

During the first 10 years of nuclear reactor operation, uranium metal was used as the fuel. However, fission produces two atoms for each original atom (Section 20.8). The fuel elements therefore expand, and the sheath intended to contain the fission fragments may burst under the pressure produced by the new atoms. It was then realized that use of UO_2 as the fuel greatly reduced this problem. In UO_2 the uranium atoms are arranged in a cubic close-packed array with oxygen atoms at the tetrahedral sites. The structure is relatively open, because the oxygen atoms move the uranium atoms apart slightly; in fact, the distance between uranium atoms is 42 percent greater in UO_2 than in uranium metal. Correspondingly, the octahedral holes are larger than in the metal and are available as sites for fission fragments to occupy.

Radius-ratio considerations have many important applications in geology. In minerals, it is not uncommon for one ion to substitute for another of similar size. For example, in the mineral olivine, Mg_2SiO_4, the Fe^{2+} ions ($r = 0.74$ Å) substitute freely for Mg^{2+} ions ($r = 0.65$ Å). Such substitution explains why this mineral and others are commonly **nonstoichiometric**, meaning that they deviate from ideal or simple chemical formulas. Substitution is most common when the radii differ by no more than 15 percent. Many commercially important ore deposits consist of minerals composed of two or more metallic elements occupying a common anionic lattice. For example, nickel sulfide and pyrite (S_2^{2-}) ores also commonly contain iron, because the ionic radius of Fe^{2+}, 0.61 Å, is near that of Ni^{2+}, 0.70 Å.

22.6 Metallurgy

We have already seen that silicate minerals are not useful sources of the various metals used in modern technology. Nevertheless, the chemical sources of these metals, such as sulfide or oxide minerals, are generally found in nature mixed with silicates. The deposits that contain the metal in economically exploitable quantities are known as **ores**. The undesirable materials present in the ore are referred to as **gangue** (pronounced "gang"). As worldwide demand for metals has increased, the mining of low-grade ores has increased also. When a particularly convenient or rich ore is closed off, either because the supply is exhausted or for reasons of international politics, alternate sources must be found. Often these are less rich and consequently more costly to work. In the face of these pressures, prices have been kept fairly low mainly by increases in mining and extraction efficiency. Prices have nevertheless increased, and this has made us aware of the need to recycle materials. It is likely that urban garbage and scrap piles will become economically feasible sources of metals in the future.

The availability and cost of a metal depends on a number of factors other than abundance. These include the occurrence of concentrated ore deposits and the ease of extraction of the metal from the ore. When an element possesses useful properties, it may be in demand even though it is not readily available. Demand stimulates search for extraction processes that can improve availability. As we have noted earlier (Section 19.6), aluminum was once costly and was exhibited as a rare metal, even though its compounds were readily available. Unfortunately, much aluminum is tied up in aluminosilicates; furthermore, the Al^{3+} ion is difficult to reduce. Aluminum is uniquely useful in many applications where low density and high electrical conductivity are important characteristics. In 1886, Charles M. Hall in the United States and Paul Heroult in France independently developed a new electrolysis procedure for producing aluminum from its oxide (Section 19.6). With the advent of this procedure, the price of aluminum fell sufficiently for the metal to find widespread use.

The science of extracting metals from their ores and preparing them for use is known as metallurgy. Metallurgical processes are conveniently divided into three types of operations: (1) preliminary treatment, which includes concentration of the ore; (2) reduction to obtain the metal in its elemental state; and (3) refining of the metal.

PRELIMINARY TREATMENT

In the processing of many ores, the metallic element is concentrated by removing the undesirable components and often by chemically modifying the mineral. The separation may be based on differences in either the physical or chemical properties of the mineral and the gangue. For example, gold minerals can be separated from the lighter gangue by shaking the crushed ore in a stream of water on an inclined table. In its simplest form, this separation technique is used by individuals panning for gold in streams.

Flotation is a process widely used on sulfide ores, especially those of copper, zinc, and lead. In this process, the finely crushed ore is mixed with oil, water, and a detergent or flotation agent. The mineral particles are wetted by the oil, and when air is blown through the mixture they float to the top of the water in the oil froth and are skimmed off as shown in Figure 22.15. The process is very economical and is widely used.

The Bayer method for obtaining Al_2O_3 from bauxite ore illustrates the use of chemical properties in separation. This procedure, which is discussed in Section 16.5, makes use of the amphoterism of aluminum. The extraction of magnesium from seawater, discussed in Section 17.2, provides another example of the use of chemical properties in separating substances.

Sometimes a component of an ore can be removed by leaching, that is, by allowing a solution to percolate through the ore and selectively dis-

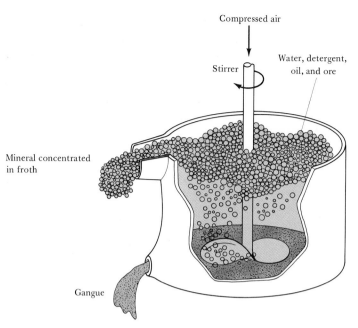

Compressed air

Stirrer

Water, detergent, oil, and ore

Mineral concentrated in froth

Gangue

FIGURE 22.15 A flotation tank.

solve certain components. For example, piles of silver and gold ore have been sprayed with dilute solutions of NaCN to leach silver and gold from them. These metals form very stable and soluble complex ions with cyanide as shown for silver in Equations [22.17] and [22.18]:

$$4Ag(s) + 8CN^-(aq) + O_2(aq) + 2H_2O(l) \longrightarrow$$
$$4Ag(CN)_2^-(aq) + 4OH^-(aq) \qquad [22.17]$$

$$Ag_2S(s) + 4CN^-(aq) \longrightarrow 2Ag(CN)_2^-(aq) + S^{2-}(aq) \qquad [22.18]$$

The ores are placed on slabs of blacktop or concrete or on plastic sheets; the solutions that have moved through the ore can then be readily captured and later treated to remove the gold and silver.

After preliminary removal of gangue, many ores are **roasted** in air. This drives off volatile impurities, burns off organic matter, and converts carbonates and sulfides into oxides as shown earlier in Equations [22.8] and [22.9].

REDUCTION

Metals can be obtained by either electrolytic reduction or chemical reduction of their compounds. Electrolytic reduction, discussed in Section 19.6, is used commercially to obtain the more active metals, such as sodium, magnesium, and aluminum. Less active metals, such as copper, iron, and zinc, are produced commercially using chemical reducing agents. Most of these less active metals are obtained by high-temperature reduction processes in which the metals are obtained in the molten state. These processes are known as **smelting operations.**

The reduction of iron illustrates several features of a typical smelting operation. A reactor designed to operate continuously is used. Such a device, known as a blast furnace, is shown in Figure 22.16. A mixture of coke, limestone, and crushed ore, usually containing Fe_2O_3, is added to

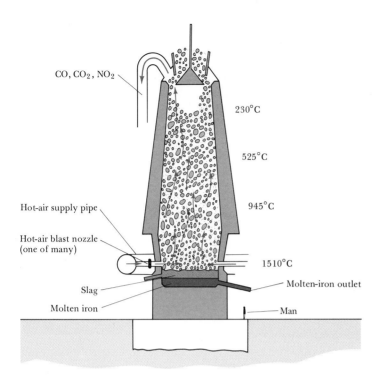

230°C

525°C

945°C

Hot-air supply pipe

Hot-air blast nozzle
(one of many)

1510°C

Slag

Molten iron

Molten-iron outlet

Man

FIGURE 22.16 A blast furnace used for reduction of iron ore. Notice the approximate temperatures in the various regions of the furnace.

the top of the furnace. (Coke is coal that has been heated in the absence of air to drive off volatile components.) Heated air, sometimes enriched with oxygen, is forced into the furnace at the bottom. About 2 tons of ore, 1 ton of coke, and 0.3 ton of limestone are required to produce 1 ton of iron. A single furnace may produce about 2000 tons of iron a day. The air coming into the furnace reacts with carbon to form CO; this liberates a considerable amount of heat and allows the furnace to reach a temperature of about 1500°C at the bottom. Reduction to metallic iron takes place as shown in Equations [22.19] and [22.20]:

$$Fe_2O_3(s) + 3C(s) \longrightarrow 2Fe(l) + 3CO(g) \qquad [22.19]$$

$$Fe_2O_3(s) + 3CO(g) \longrightarrow 2Fe(l) + 3CO_2(g) \qquad [22.20]$$

Carbon monoxide is produced not only by reaction between coke and oxygen, Equation [22.21], but also between coke and CO_2, Equation [22.22]:

$$2C(s) + O_2(g) \longrightarrow 2CO(g) \qquad [22.21]$$

$$C(s) + CO_2(g) \longrightarrow 2CO(g) \qquad [22.22]$$

Carbon dioxide is produced in the reduction of the iron oxide, Equation [22.20], as well as from decomposition of limestone. The limestone serves another important role in smelting operations. An ore typically still contains much gangue when it is reduced. The CaO formed by decomposition of limestone reacts with gangue to form a liquid known as slag. One of the important reactions is given in Equation [22.23]:

$$CaO(s) + SiO_2(s) \longrightarrow CaSiO_3(l) \qquad\qquad [22.23]$$

Gangue   Slag

In the smelting of iron, the slag floats on top of the molten iron, protecting it from oxidation by the incoming air. The slag and iron are removed periodically. The iron produced in the blast furnace, called **pig iron,** contains up to 5 percent carbon and as much as 2 percent other impurities such as silicon, phosphorus, and sulfur.

The thermodynamic basis for smelting operations using coke or CO as a reducing agent can be seen by reference to Figure 22.17. As this figure shows, metal oxides generally become less stable (ΔG_f more positive) as temperature increases. This follows from the fact that $\Delta G_f = \Delta H_f - T\Delta S_f$ (Section 18.5). The formation of an oxide from a metal and O_2, Equation [22.24],

$$2M(s) + O_2(g) \longrightarrow 2MO(s) \qquad [22.24]$$

generally represents a decrease in disorder, so that ΔS_f is negative. The factor $-T\Delta S_f$ therefore becomes increasingly positive with increasing temperature. In Figure 22.17 we notice that for CuO, ΔG_f becomes positive at approximately 900°C. Consequently, CuO spontaneously decomposes above this temperature. The negative slope of the CO line is noteworthy. For the formation of CO from carbon and oxygen, Equation [22.21], ΔS_f is positive. For a reduction of the type given in Equation [22.25],

$$MO(s) + C(s) \longrightarrow M(l) + CO(g) \qquad [22.25]$$

ΔG is negative, and the reaction is therefore spontaneous in all cases where the CO line is below the stability line for the metal oxide. To illustrate, we can use Figure 22.17 to estimate ΔG at 1500°C for the reaction

$$FeO(s) + C(s) \longrightarrow Fe(l) + CO(g)$$

For this reaction we have:

$$\Delta G = \Delta G_f(CO) - \Delta G_f(FeO)$$

From Figure 22.17, we obtain:

$$\Delta G_f(CO) = \frac{-550}{2} \text{ kJ/mol}$$

and

$$\Delta G_f(FeO) = \frac{-300}{2} \text{ kJ/mol}$$

Therefore,

$$\Delta G = -275 \text{ kJ/mol} + 150 \text{ kJ/mol}$$
$$= -125 \text{ kJ/mol} (-120 \text{ kJ/mol}$$
$$\text{to two significant figures)}$$

Notice that at 1500°C the stability line for CO in the figure is below the stability line for FeO. Correspondingly, ΔG is negative, and the reaction therefore is spontaneous.

Coke is a reasonably inexpensive reducing agent. There are occasions, however, when it is not satisfactory. It cannot be used when the reduction is not spontaneous (as for Al_2O_3) or when carbon impurities in the

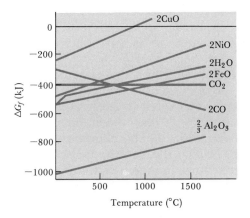

FIGURE 22.17 Stabilities of certain oxides (as measured by ΔG_f) shown as a function of temperature.

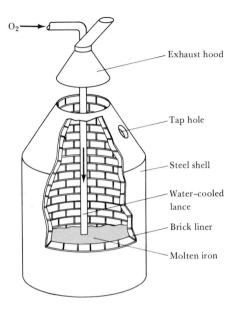

O$_2$ ⟶

— Exhaust hood

— Tap hole

— Steel shell

— Water–cooled
lance

— Brick liner

— Molten iron

FIGURE 22.18 Diagram of an oxygen-process furnace used to convert pig iron into steel. The furnace is generally mounted so that it can be tilted to pour steel out of the tap hole. On the average such furnaces are 8–16 ft in diameter and have a capacity of 30–300 tons.

metal are undesirable. Under these conditions, electrolytic reduction or chemical reduction using hydrogen or active metals such as sodium, magnesium, or zinc as reducing agents is employed. These reducing agents are more expensive than carbon. Equations [22.26] through [22.28] provide examples of chemical reduction using reducing agents other than carbon:

$$WO_3(s) + 3H_2(g) \longrightarrow W(s) + 3H_2O(g) \qquad [22.26]$$

$$2Ag(CN)_2^-(aq) + Zn(s) \longrightarrow 2Ag(s) + Zn(CN)_4^{2-}(aq) \qquad [22.27]$$

$$UF_4(g) + 2Mg(l) \longrightarrow U(s) + 2MgF_2(l) \qquad [22.28]$$

REFINING

Once the metal has been obtained from its ore, it may still be necessary to further purify it. Blister copper, obtained by smelting of copper ores, is about 99 percent pure. It contains small amounts of arsenic, antimony, silver, gold, and other metals. The electrical conductivity of copper is greatly decreased by certain impurities; an arsenic content of only 0.03 percent lowers the conductivity by 14 percent. Consequently, copper used for electrical applications must be very pure. It is refined electrolytically by making the blister copper the anode in an electrolytic cell, as discussed in Section 19.6.

Similarly, pig iron contains numerous impurities that cause it to be brittle and of low tensile strength; it therefore has few uses. Most pig iron is therefore converted to steel; over half is converted by means of an oxygen furnace. In this process, limestone is added to the molten iron to form a slag with the phosphorus and silicon; a high-pressure stream of oxygen is blown through the molten iron to burn off sulfur and excess carbon as shown in Figure 22.18. After this, small amounts of carbon and other materials may be added, depending on the type of steel desired. Small amounts of carbon increase the hardness and strength of the steel.

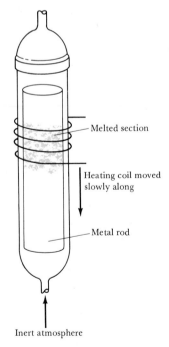

— Melted section

— Heating coil moved slowly along

— Metal rod

Inert atmosphere

FIGURE 22.19 A zone-refining apparatus.

Other purification or refining procedures include distillation, which can be used with mercury, and zone refining, which is used to purify silicon and germanium for use in semiconductors. In the zone-refining process a heating coil is passed slowly along a metal rod as shown in Figure 22.19; a narrow band of the element is thereby melted. As the molten area is swept slowly along the length of the rod the impurities concentrate in the molten region, following it to the end of the rod. The end in which the impurities are collected is cut off and the purified top portion retained.

22.7 The electronic structures and physical properties of metals

We have had occasion at several places in the text to refer to the differences between metals and nonmetals, with regard to both physical and chemical behavior. We discussed the characteristics of ionic and covalent bonds in some detail in Chapters 7 and 8. Let us now consider the distinct physical properties of metals, and then relate those properties to a model for metallic bonding.

Anyone who has handled a length of copper wire or an iron bolt or has observed a piece of freshly cut sodium knows that metals possess distinct physical properties. The three metals just mentioned differ from one another greatly in physical characteristics, and yet their properties are similar. In addition to their characteristic luster, metals are most distinctively characterized by their high electrical and heat conductivities. When an electrical potential is applied across a length of metal, current flows. This current flow occurs without any displacement of metal atoms. It is therefore due to the flow of electrons through the metal. Metallic conduction is characterized by a small resistance that increases with increase in temperature.

The characteristic feel of metals is due to their high heat conductivities. The electrical and thermal (heat) conductivities vary in the same manner from one metal to another. For example, silver and copper, which possess the highest electrical conductivities, also possess the highest thermal conductivities. This suggests that the two types of conductivity have the same origin.

A freshly prepared metal surface is characterized by its lustrous appearance. In most metals, light of all wavelengths is reflected. The color in metals such as gold and copper is due to absorption of light from the blue, or high-energy, region of the visible spectrum.

Most metals are malleable, which means that they can be hammered into sheets, and ductile, which means that they can be drawn into wires. These properties indicate that the atoms of the metallic lattice are capable of slipping with respect to one another. This is not a characteristic of ionic solids and the crystals of most covalent compounds. These types of solids are typically brittle and fracture easily along certain planes.

The metals form solid structures in which the atoms are arranged as close-packed spheres or in some similar packing arrangement. For example, copper possesses a close-packing arrangement called cubic close packing (Section 11.4), in which each copper atom is in contact with 12 other copper atoms. None of the metal atoms possess sufficient valence electrons to form a localized electron-pair bond with each of these neighbors. As another example, magnesium has only two valence electrons, yet

22 The lithosphere:
geochemistry and metallurgy

22.7 The electronic
structures and physical
properties of metals

it also is surrounded by 12 neighboring magnesium atoms. If each atom is to share its bonding electrons with all its neighbors, these electrons must be able to move from one bonding region to another.

In discussing the structures of molecules such as benzene, we saw that electrons can in some cases be delocalized, or distributed over several nuclear centers. This happens when the atomic orbitals on a single atom are able to interact with atomic orbitals on more than one other atom. In graphite (Section 8.7), the electrons are delocalized over an entire plane. It is useful to think of the bonding in metals in a similar way. The valence atomic orbitals on one metal atom overlap with those on several nearest neighbors. These in turn overlap with the atomic orbitals of still other atoms. We saw in Section 8.5 that overlap of atomic orbitals can lead to formation of molecular orbitals; the number of molecular orbitals formed is equal to the number of atomic orbitals that overlap.

In a metal, the number of atomic orbitals involved in a particular molecular orbital is very large, because the atomic orbitals each overlap with several others. Thus the number of molecular orbitals is also very large. Figure 22.20 shows schematically what happens as increasing numbers of metal atoms come together to form molecular orbitals. The difference in energy between the highest-energy and lowest-energy molecular orbitals is no greater than in ordinary covalent bonding, but the number of molecular orbitals in this energy range is very large. Thus, the interactions of all the valence orbitals of the metal atoms with the valence orbitals of adjacent metal atoms gives rise to a huge number of very closely spaced molecular orbitals that extend over the entire metal lattice. The energy separations between these metal orbitals is so tiny that for all practical purposes it is possible to think of them as forming a continuous *band* of allowed energy states, as shown in Figure 22.20. The electrons available do not completely fill this band. One can think of the energy band as a partially filled container for electrons. The incomplete filling of the allowed energy levels gives rise to characteristic metallic properties. The electrons near the top of the occupied levels require very little energy input to be "promoted" to a still higher unoccupied orbital. Under the influence of any source of excitation energy, such as an applied electrical potential or an input of thermal energy, electrons are excited to previously vacant levels and are thus freed to move through the lattice, giving rise to electrical or thermal conductivity.

The description we have given of the electronic structures of metals has been drastically simplified. In transition metals, in which s, p, and d

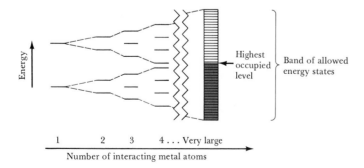

FIGURE 22.20 Schematic illustration of the interactions of atomic metal orbitals to form the delocalized orbitals of the metal lattice. The two atomic orbitals on each metal atom in this example might represent an s and p orbital. The main point is the formation of a very large number of molecular orbitals with very closely spaced energy levels. The number of electrons available does not completely fill these orbitals.

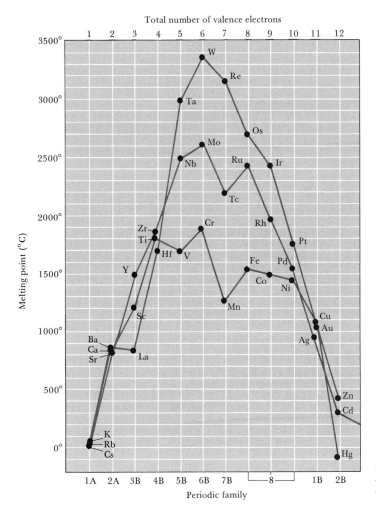

Total number of valence electrons

FIGURE 22.21 The melting points of metals as a function of their location in the periodic table and their number of valence electrons.

metal orbitals may overlap, the band structure is quite complex. In all these elements, however, the characteristic metallic properties of high electrical and thermal conductivities, luster, malleability, and ductility are found to varying degrees. The physical properties of metals depend on the structure of the metal lattice, the presence of irreglarities or lattice defects, and the strength of the bonds between metal atoms. Figure 22.21 shows the melting points of the metals through the transition series. Notice that the melting-point maximum in each period corresponds to outer electron configurations of $(n - 1)d^3(n)s^2$ or $(n - 1)d^4(n)s^2$, which are five or six electrons beyond the noble-gas arrangements. Curves for the boiling points, heats of fusion, heats of vaporization, hardnesses, and densities of metals have very similar appearances. We can qualitatively understand the shape of Figure 22.21 if we assume that only the outer s and d electrons are involved in metallic bonding. Because there are one s and five d orbitals of a given major quantum number, this means that each metal atom contributes a total of six atomic orbitals to the molecular orbital structure of the metal. Let us assume that the molecular orbitals formed by the s and d atomic orbitals all overlap, to make one continuous band of allowed energy states, as illustrated in Figure 22.20.

22 The lithosphere:
geochemistry and metallurgy

22.7 The electronic
structures and physical
properties of metals

TABLE 22.8 Some common alloys

Primary element	Name of alloy	Composition by weight	Properties	Uses
Bismuth	Wood's metal	50% Bi, 25% Pb, 2.5% Sn, 12.5% Cd	Low melting point (70°C)	Fuse plugs, automatic sprinklers
Copper	Yellow brass	67% Cu, 33% Zn	Ductile, takes polish	Hardware items
Iron	Stainless steel	80.6% Fe, 0.4% C, 18% Cr, 1% Ni	Resists corrosion	Tableware
Lead	Plumber's solder	67% Pb, 33% Sn	Low melting point (275°C)	Soldering joints
Silver	Sterling silver	92.5% Ag, 7.5% Cu	Bright surface	Tableware
Silver	Dental amalgam	70% Ag, 25% Pb, 3% Cu, 2% Hg	Easily worked	Dental fillings
Tin	Pewter	85% Sn, 6.8% Cu, 6% Bi, 1.7% Sb		Utensils

The lower half of these (those shaded in the figure) we can consider as bonding, the top half as antibonding. When we have six electrons per metal atom, there are just enough electrons to half fill the molecular orbitals in the band, giving rise to the maximum bonding interaction between atoms. On this basis the metal with six electrons should have the strongest bonds and consequently the highest melting point. Of course, this analysis is an oversimplification; other factors such as atomic radius, nuclear charge, and crystal form are involved.

ALLOYS

An **alloy** is a material that contains more than one element and has the characteristic properties of metals. The alloying of metals is of great importance, because it is one of the primary ways of modifying the properties of the pure metallic elements. For example, pure gold is too soft to be used in jewelry, whereas alloys of gold and copper are quite hard. Pure gold is 24 karat; the common alloy used in jewelry is 14 karat, meaning that it is 58 percent gold ($14/24 \times 100$). A gold alloy of this composition has suitable hardness to be used in jewelry. The alloy can be either yellow or white depending on the elements added. For the most part, pure metals are little used, and metals are consumed mainly in the form of alloys. Some further examples of alloys are given in Table 22.8.

Alloys can be classified as solution alloys, heterogeneous mixtures, and intermetallic compounds. The **solution alloys** are homogeneous mixtures with the components dispersed randomly and uniformly. Atoms of the solute can take positions normally occupied by the solvent, thereby forming a **substitutional alloy**, or they can occupy interstitial positions, thereby forming an **interstitial alloy**. These types are shown in Figure 22.22.

In interstitial alloys the interstitial atoms participate in bonding to neighboring atoms. The presence of the extra bonds provided by the interstitial atoms causes the metal lattice to become harder, stronger, and less ductile. For example, iron containing less than 3 percent carbon is much harder than pure iron and has a much higher tensile strength

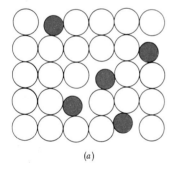

(a)

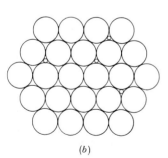

(b)

FIGURE 22.22 (a) Substitutional and (b) interstitial alloys. The open circles are the host metal, the colored circles the other component of the alloy.

and other desirable physical properties. So-called mild steels contain less than 0.2 percent carbon; they are malleable and ductile and are used to make cables, nails, and chains. Medium steels contain 0.2 to 0.6 percent carbon; they are tougher than the mild steels and are used to make girders and rails. High-carbon steel, used in cutlery, tools, and springs, contains 0.6 to 1.5 percent carbon. Other elements may be added to form alloy steels. One of the most common of these is stainless steel, which contains 0.4 percent carbon, 18 percent chromium, and 1 percent nickel. Alloys of this kind differ from ordinary chemical compounds in that the composition is not fixed. The ratio of nonmetallic to metallic element may vary over a wide range, imparting a variety of specific physical and chemical properties to the materials.

In heterogeneous alloys, the components are not dispersed uniformly. For example, in one form of steel known as pearlite there is a heterogeneous mixture of iron and Fe_3C known as cementite. We noted earlier that mixtures do not generally have characteristic melting points. However, in certain cases sharp and characteristic melting behavior can result. For example, pure lead melts at $237°C$, and pure tin melts at $232°C$. Addition of lead to tin causes a decrease in the melting temperature. The lowest melting temperature, $181°C$, is obtained for a mixture consisting of 64 percent tin. Such a lowest-melting composition is called a eutectic mixture.

Intermetallic alloys are homogeneous and have definite properties and composition. For example, copper and aluminum form a compound, $CuAl_2$, known as duralumin. Intermetallic compounds are rarely used as such, but are often found distributed through alloys just as cementite is found distributed through some steels.

FOR REVIEW

Summary

In this chapter, we have examined the structure and composition of our planet. The earth can be divided into four layers, the crust, mantle, outer core, and inner core. We have noted that we have access to only a small portion of the planet, largely its surface. The solid portion to which we have access is referred to as the lithosphere. The solid compounds that occur naturally in the lithosphere are called minerals. Rocks are solid mixtures of minerals.

We can divide minerals into three general types: native elements, silicate compounds, and nonsilicate compounds. Silicate minerals are the most abundant. These minerals are based on SiO_4 tetrahedra, which, through sharing oxygen atoms, are able to link together to form chains, sheets, and three-dimensional arrays. We examined how the macroscopic properties, such as cleavage, reflect the molecular-level arrangements of several silicate minerals. In many minerals, Si^{4+} ions are replaced by Al^{3+} ions, thus forming aluminosilicates such as the feldspar minerals. Silicates are important components of glass and cement, both of which were briefly discussed. However, the silicates are not presently economically feasible sources of most metals.

Nonsilicate minerals, on the other hand, are especially important as sources of metals. Calcium carbonate is the most abundant nonsilicate mineral in the lithosphere. Both $CaCO_3$ and CaO, obtained by heating $CaCO_3$, are useful in many processes discussed in this chapter.

Many ionic compounds can be viewed as consisting of a close-packed array of anions with cations located in octahedral or tetrahedral holes. The ratio of the cation radius to the anion radius is the most important factor in determining which type of hole a particular cation will occupy. Ions of similar size and the same charge can often substitute for one another in minerals.

Economically exploitable mixtures of minerals are known as **ores**. The undesirable portion of the ore is called the **gangue**. The process of extracting metals from their ores and modifying their properties is known as **metallurgy**. Three important types of processes involved in metallurgy are the **concentration** of an ore, its **reduction**, and the **refining** of the metal. Methods of concentration include **flotation**, **leaching**, and **roasting**. Most reductions involve carbon (coke), H_2, active metals, or electrolytic procedures. Methods of refining include electrolytic and chemical procedures, zone refining, and distillation. Much of our focus as we discussed reduction was on iron, which is obtained in a **smelting operation**.

In the solid state, metals form close-packed or similar types of structures. Thus, each atom has several nearest neighbors. The atomic orbitals of the metal atoms overlap to form molecular orbitals that extend over the metal lattice. Bonding in the metals maximizes with the elements that have available six outer *s* and *d* electrons per atom. **Alloys** possess the characteristic properties of metals but are composed of more than one element.

Learning goals

Having read and studied this chapter, you should be able to:

1 Describe the structure of the earth.
2 List the five most abundant elements in the earth's crust.
3 Describe the three major categories of minerals and give an example of each type.
4 Describe the structures possible for silicates and their empirical formulas (for example, silicate tetrahedra can combine through bridging oxygens to form a single-string silicate chain whose empirical formula is SiO_3^{2-}).
5 Draw the Lewis structures for the repeating units in the various types of silicate minerals.
6 Correlate the physical properties of certain silicate minerals, such as asbestos, with their structures.
7 Explain the changes in composition and properties that accompany substitution of Al^{3+} for Si^{4+} in a silicate.
8 Describe what is meant by a clay mineral.
9 Describe the role of clay minerals in soil fertility and the effects of soil pH.
10 Describe the composition of soda-lime glass.
11 Describe the manufacture of glass and cement.
12 Describe the behavior of $CaCO_3$ in acid solutions or at high temperatures.
13 Describe the octahedral and tetrahedral holes in a close-packed array of ions and use the radius-ratio rules to predict which type of hole a particular ion will occupy.
14 Give at least two examples each of concentration, reduction, and refining processes.
15 Describe the production of steel starting with Fe_2O_3 ore.
16 Describe the electronic structure and bonding in metals.
17 Describe the manner in which bonding in metals varies with the outer configuration of *ns* and $(n - 1)d$ electrons.
18 Describe what is meant by an alloy.
19 Name the various types of alloys and give an example of each.

Key terms

Among the more important terms and expressions used for the first time in this chapter are the following:

An **alloy** (Section 22.7) is a material that contains more than one element and has the characteristic properties of metals, such as luster and electrical conductivity.

In **aluminosilicate** minerals (Section 22.3), some of the Si^{4+} ions are replaced by Al^{3+}.

Clay minerals (Section 22.3) are hydrated aluminosilicates.

Glass (Section 22.3) is an amorphous solid formed by fusion of SiO_2, CaO, and Na_2O. Other oxides may also be used to form glasses with differing characteristics.

The **lithosphere** (Section 22.1) is the solid earth component of our environment.

Metallurgy (Section 22.6) is the science of extracting metals from their ores.

Minerals (Section 22.2) are naturally occurring forms of solid elements or compounds.

An **octahedral hole** (Section 22.5) is an opening in a close-packed array of atoms or ions created by six particles arranged in an octahedral fashion.

An **ore** (Section 22.6) is an economically feasible source of a metal.

A **rock** (Section 22.2) is an aggregate of minerals of different kinds.

Silicate minerals (Section 22.3) are those in which the structure is based on SiO_4 tetrahedra.

Smelting (Section 22.6) refers to processes in which metals are obtained in the molten state from ores via a high-temperature reduction process.

A **tetrahedral hole** (Section 22.5) is an opening in a close-packed array of atoms or ions created by four particles arranged in a tetrahedral fashion.

EXERCISES

Minerals and related minerals

22.1 The two most heavily utilized metallic elements listed in Table 22.2 are Na and Fe. Name the most important natural sources of these elements.

22.2 Describe the fundamental structural unit of all silicate minerals. How is this structural unit modified in aluminosilicate minerals?

22.3 Describe the chemical composition of each of the following: (a) lime; (b) gypsum; (c) limestone; (d) talc.

22.4 Distinguish (a) a mineral from an ore; (b) clay from glass; (c) calcite from limestone; (d) cement from rock.

22.5 What are the essential factors that must operate if a metallic element is to occur in nature as the free element?

22.6 Draw the Lewis structure for the cyclic $Si_4O_{12}{}^{8-}$ ion.

22.7 What empirical formula corresponds to each of the following structural arrangements: (a) isolated SiO_4 tetrahedra; (b) a chain structure of SiO_4 tetrahedra joined at corners to adjacent units; (c) a structure consisting of tetrahedra joined at corners to form a six-membered ring of alternating Si and O atoms?

22.8 Why does carbon not form a wide range of minerals analogous to the silicates, with carbon in place of silicon?

22.9 Which of the following metallic elements would you expect to be able to substitute for Si^{4+} in silicate minerals? Indicate in each case the expected oxidation state of the metallic element, and the reason for your answer: (a) aluminum; (b) palladium; (c) potassium; (d) titanium.

22.10 Provide a brief explanation for each of the following observations: (a) Mica is readily split into thin sheets. (b) In basic soils, metallic ions are not readily released for use by plants. (c) Mg^{2+} and Fe^{2+} readily substitute for one another in minerals. (d) Silicate sheet minerals are softer and more readily separated than aluminosilicate sheet minerals.

22.11 In the reaction shown in Equation 22.1 is the mineral anorthite acting as an acid or a base? Explain.

22.12 Using the weight percentages listed in Table 22.5, what is the mole ratio of boron to silicon in a typical borosilicate glass?

22.13 The lowest melting mixture of the three oxides Na_2O, SiO_2, and CaO, melting at $750°C$, has the composition of 21.3 percent Na_2O, 5.2 percent CaO, and 73.5 percent SiO_2, by weight. Calculate the relative molar quantities of Na, Ca, and Si in this mixture.

22.14 Write equations for the reactions that occur in each of the following cases: (a) CO_2 is bubbled through a solution of $Ba(OH)_2$; (b) $Ca(OH)_2$ is produced, starting with limestone; (c) solid $SrCO_3$ is heated.

22.15 Iron often occurs as the mineral siderite, $FeCO_3$, along with calcium carbonate in limestone deposits. When the $FeCO_3$ dissolves in ground water that passes through the limestone deposits, some dissolution of the $FeCO_3$ occurs to produce water hardness. The solubility of $FeCO_3$ at $25°C$ is 0.067 g/L of solution. Compare with the solubility calculated using K_{sp} from Appendix D. Account for the difference.

22.16 With the aid of balanced chemical equations, describe a procedure to protect a sculpture made from mortar against erosion due to acid rain.

22.17 Predict the type of hole in an oxide lattice that would be occupied by each of the following cations (radii in parentheses): (a) Li^+ (0.68 Å); (b) Ti^{4+} (0.53 Å); (c) Ni^{2+} (0.70 Å); (d) Ca^{2+} (0.94 Å).

22.18 The radius of the sulfide ion is 1.90 Å. Assuming that the sulfides can be viewed as close-packed structures of S^{2-} ions, predict the type of hole that each of the following cations will occupy: (a) Zn^{2+} (0.74 Å); (b) V^{5+} (0.54 Å); (c) Mg^{2+} (0.65 Å).

22.19 The ionic radius of Mn^{2+} is 0.80 Å. For which of the cations listed in problems 22.17 and 22.18 would you expect manganese to substitute fairly readily in minerals? Explain.

Metallurgy: structures and properties of metals

22.20 Describe the stages that occur in the overall processing of a raw material from the earth to obtain a pure metal.

22.21 Using the metal cobalt as an example, and assuming that it occurs as the ore CoS_2, write balanced chemical equations to demonstrate each of the following processes: (a) roasting (to form the oxide); (b) smelting; (c) electrolytic purification.

22.22 What role does each of the following materials play in the chemical processes that occur in a blast furnace: (a) coke; (b) air; (c) limestone. Write balanced chemical equations to illustrate your answers.

22.23 Copper is obtained in large quantities from ores containing chalcopyrite, $CuFeS_2$. What mass of copper metal is obtainable from 1.00 kg of this mineral, assuming 100 percent efficiency in the process?

22.24 Operation of the blast furnace is dependent on the fact that molten iron and molten slag are mutually insoluble. In terms of their compositions, and using the ideas developed in Chapters 11 and 12, explain why this mutual insolubility exists.

22.25 In refining of iron, limestone is added to the furnace to form a slag with the phosphorus and silicon present. Recognizing that the furnace is at a high temperature and is an oxygen-rich environment, write balanced chemical equations for formation of the slag. (*Note:* The slag may be a mixture and may consist of phosphates as well as of silicates.)

22.26 Write complete and balanced equations for each of the following processes: (a) reduction of Co_3O_4 with Mg; (b) roasting of NiS_2; (c) reduction of OsO_4 with hydrogen; (d) electrolytic formation of Cs from CsCl.

22.27 Write complete balanced chemical equations for

each of the following: (a) reduction of $HfCl_4$ with Na; (b) roasting of $HgCO_3$; (c) reduction of Re_2O_7 with hydrogen gas; (d) reduction of MnO_2 using charcoal.

22.28 Would you expect that a procedure similar to the one used in the oxygen furnace to refine iron could be used for refining of copper? Of magnesium? Explain.

22.29 Suppose you have four rods in the shape of a pencil composed of either sulfur, magnesium, pure iron, or high-carbon steel. Compare these four with respect to the following characteristics: (a) appearance; (b) flexibility; (c) melting point; (d) thermal conductivity.

22.30 Although the covalent radius of silver is considered to be 1.34 Å, the interatomic distance in silver metal is 2.88 Å. Explain this difference.

22.31 The enthalpy of sublimation of chromium metal at 25°C is 397 kJ/mol; the corresponding value for iodine at 25°C is 62 kJ/mol. Account for the large difference in these values.

22.32 Account for the observation that melting points and hardness increase in the order K < Ca < Sc among these three metallic elements.

22.33 Explain why diamond (structure in Figure 8.30) is an insulator, whereas silver, which crystallizes in the face-centered cubic lattice (Figure 11.15), has very high electrical conductivity.

22.34 Treatment of iron surfaces at high temperatures with ammonia leads to "nitriding," in which nitrogen atoms are incorporated into the metal lattice. What kind of alloy formation occurs in this case? How would you expect the nitriding to alter the properties of the metal?

22.35 Calculate the mole percentage of each metallic element present in: (a) brass; (b) plumber's solder; (c) the eutectic tin-lead mixture.

Additional exercises

22.36 Indicate which of the following statements is incorrect. Correct those which are incorrect: (a) The most stable type of lattice hole for a cation is one in which it is in contact with the maximum number of anions. (b) Limestone deposits dissolve when in contact with basic aqueous solution. (c) Addition of B_2O_3 to glass formulations leads to glass with a higher melting point and a lower coefficient of thermal expansion. (d) In aluminosilicate minerals, the Al^{3+} ions serve to balance the charges created by the SiO_4^{4-} units. (e) Nearly all metals are produced by electrolytic reduction methods.

22.37 What is the general chemical composition of asbestos minerals? How do the structures of the asbestos minerals give rise to their characteristic properties?

22.38 Propose a reasonable description of the structure in each of the following minerals: (a) albite, $NaAlSi_3O_8$; (b) leucite, $KAlSi_2O_6$; (c) zircon, $ZrSiO_4$; (d) sphene, $CaTiSiO_5$.

22.39 Iron often occurs along with calcite in limestone

deposits as the mineral siderite, $FeCO_3$. The $FeCO_3$ dissolves in slightly acidic water in the same mannner as calcium carbonate. When water containing $Fe^{2+}(aq)$ comes in contact with air, some oxidation occurs, with deposition of insoluble $Fe(OH)_3$. This gives rise to the rust color seen in sinks and other containers that continually come into contact with iron-containing water. Write balanced net ionic equations for the dissolution and precipitation reactions just described.

22.40 In the weathering of rocks, many metallic elements (for example, Ca^{2+}, Fe^{2+}, Mn^{2+}) pass into solution as the bicarbonate salts. Explain why this is so.

22.41 If cryolite, Na_3SiF_6, is viewed as a close-packed structure of fluoride ions, what types of interstitial holes do you predict the Al^{3+} and Na^+ ions will occupy (radii: F^-, 1.33 Å; Na^+, 0.98 Å; Al^{3+}, 0.45 Å)?

22.42 In the early days of iron mining in the United States, the iron oxide mined in the Lake Superior region was reduced to pig iron in blast furnaces located near the mines. The furnaces used charcoal made from hardwood as the reducing agent (transportation savings prompted this procedure). What mass of pig iron containing 3 percent carbon was produced from each kilogram of iron ore, assuming that it is mined as *magnetite,* Fe_3O_4, of 66 percent purity?

22.43 Using Figure 22.17, estimate ΔG at 1500°C for reduction of $NiO(s)$ to Ni, using carbon as reducing agent. For which product, CO or CO_2, is ΔG the more negative?

22.44 Explain the differences between CO_3^{2-} and SiO_3^{2-}.

22.45 Suppose a nation has good reserves of Fe_2O_3 ores but has no coal or petroleum sources. What means might be employed to obtain iron from the ore?

22.46 Zinc is purified by electrolytic refining. The electrolytic process is similar to that for copper (Section 19.6). A current of 25 amp is used, at a voltage of 3.5 V. Assuming that the electrolytic process is 80 percent efficient, how much electrical energy (kilowatt hours) is needed to produce 1.0×10^7 kg of pure zinc?

[22.47] Using the appropriate data from Appendix D, make a graph of the enthalpy of sublimation of the elements K through Zn as a function of atomic number. What characteristics of the metal does ΔH_s reflect? Account for the variation observed in ΔH_s as a function of atomic number.

22.48 What arguments can you give to refute the idea that bonding in metals consists of localized, shared electron-pair bonds between metal atoms?

22.49 What is the percentage of gold in 18 karat gold?

22.50 Account for the fact that although copper is found in several places in the form of the native metal, aluminum is never found as the free element.

22.51 Write balanced chemical reactions for processes that might be used to recover free zinc metal from each of the following ores: (a) ZnS; (b) $ZnCO_3$.

23

Chemistry of metals: coordination compounds

In earlier chapters of this text we noted that metallic elements are characterized by a tendency to lose electrons in their chemical reactions. For this reason, positively charged metal ions play a primary role in the chemical behavior of metals. Of course, metal ions do not exist in isolation. In the first place they are accompanied by anions that serve to maintain charge balance. In addition, the metal ion acts as a Lewis acid (Section 15.10). Neutral molecules or anions with unshared pairs of electrons may be bound to the metal center. On several occasions we have discussed compounds in which a metal ion is surrounded by a group of anions or neutral molecules. Examples include $Ag(CN)_2^-$, discussed in connection with metallurgy in Section 22.6; hemoglobin, discussed in connection with the oxygen-carrying capacities of the blood in Section 10.5; $Cu(CN)_4^{2-}$ and $Ag(NH_3)_2^+$, encountered in our discussion of equilibria in Section 16.5. Such species are known as complex ions or merely complexes. Compounds containing them are called coordination compounds.

23.1 The structure of complexes

The molecules or ions that surround a metal ion in a complex are known as ligands (from the Latin word *ligare*, meaning "to bind"). Ligands are normally either anions or polar molecules. Furthermore, they have at least one unshared pair of valence electrons, as illustrated in the following examples:

$$: \overset{\displaystyle ..}{\underset{\displaystyle H}{O}} - H \qquad : \overset{\displaystyle H}{\underset{\displaystyle H}{N}} - H \qquad : \overset{\displaystyle ..}{\underset{\displaystyle ..}{Cl}} : ^- \qquad : C \equiv N : ^-$$

In some instances, the bonding between a metal and its ligands can be understood to result from electrostatic attraction between positive ions and either negative ions or the negative end of a dipole. Correspondingly, the ability of metals to form complexes normally increases as the positive charge of the metal increases and as its size decreases. The alkali metals such as Na^+ and K^+ do not readily form complexes. On the other hand, the dipositive and tripositive transition-metal ions excel in forming complexes. In fact, transition-metal ions often form complexes more readily than their size and charge alone would suggest. For example, on the basis of size alone we might expect that Al^{3+} ($r = 0.45$ Å) would form complexes more readily than would the larger Cr^{3+} ($r = 0.62$ Å). However, Cr^{3+} forms much more stable complexes than does Al^{3+}. Thus the bonding in these complexes cannot be explained merely on the basis of electrostatic interaction between metal and ligand. It is therefore convenient to view the bonding from a covalent viewpoint. Because metal ions have empty valence orbitals, they can act as Lewis acids (electron-pair acceptors). Because ligands have unshared pairs of electrons, they are able to function as Lewis bases (electron-pair donors). The bond between metal and ligand can therefore be pictured as resulting from the sharing of a pair of electrons that was originally on the ligand:

$$Ag^+(aq) + 2 : \overset{\displaystyle H}{\underset{\displaystyle H}{N}} - H(aq) \longrightarrow \left[H - \overset{\displaystyle H}{\underset{\displaystyle H}{N}} : Ag : \overset{\displaystyle H}{\underset{\displaystyle H}{N}} - H \right]^+ (aq) \qquad [23.1]$$

We shall examine the bonding in complexes more closely in Section 23.8.

In forming a complex, the ligands are said to coordinate to the metal or to complex the metal. The central metal and the ligands bound to it constitute the coordination sphere. In writing the chemical formula for a coordination compound, we use square brackets to set off the groups within the coordination sphere from other parts of the compound. For example, the formula $[Cu(NH_3)_4]SO_4$ represents a coordination compound consisting of the $Cu(NH_3)_4^{2+}$ ion and the SO_4^{2-} ion. The four ammonia groups in the compound are bound directly to the copper(II) ion.

The charge of a complex is the sum of the charges on the central metal and on its surrounding ligands. In $[Cu(NH_3)_4]SO_4$ we can deduce the charge on the complex if we first recognize SO_4 as being the sulfate ion and therefore having a -2 charge. Because the compound is neutral, the complex ion must have a $+2$ charge, $Cu(NH_3)_4^{2+}$. The oxidation number of the copper must be $+2$ because the NH_3 groups are neutral:

$$\underset{\displaystyle Cu(NH_3)_4{}^{2+}}{+2 \; + \; 4(0) = +2}$$

SAMPLE EXERCISE 23.1

What is the oxidation number of the central metal in $[Co(NH_3)_5Cl](NO_3)_2$?

Solution: The NO_3 group is the nitrate anion and has a -1 charge, NO_3^-. The NH_3 ligands are neutral; the Cl is a coordinated chloride and therefore has a -1 charge. The sum of all the charges must be zero:

$$x + 5(0) + (-1) + 2(-1) = 0$$

$$[Co(NH_3)_5Cl](NO_3)_2$$

The charge on the cobalt, x, must therefore be $+3$.

SAMPLE EXERCISE 23.2

Given that a complex ion contains a chromium(III) bound to four water molecules and two chloride ions, write its formula.

Solution: The metal has a $+3$ oxidation number, water is neutral, and chloride has a -1 charge:

$$+3 + 4(0) + 2(-1) = +1$$

$$Cr(H_2O)_4Cl_2$$

Therefore the charge on the ion is $+1$, $[Cr(H_2O)_4Cl_2]^+$.

COORDINATION NUMBERS AND GEOMETRIES

The atom in a ligand that is bound directly to the metal is known as the donor atom. For example, nitrogen is the donor atom in the $Ag(NH_3)_2^+$ complex shown in Equation [23.1]. The number of donor atoms attached to a metal is known as its **coordination number**. In $Ag(NH_3)_2^+$, silver has a coordination number of two; in $Cr(H_2O)_4Cl_2^+$, chromium has a coordination number of six.

Some metal ions exhibit constant coordination numbers. For example, the coordination number of chromium(III) and cobalt(III) is invariably six, whereas that of platinum(II) is always four. However, the coordination numbers of most metal ions vary with the ligand. The most common coordination numbers are four and six.

The coordination number of a metal ion is often influenced by the relative sizes of the metal ion and the surrounding ligands. As the ligand gets larger, fewer can coordinate to the metal. This explains why iron is able to coordinate to six fluorides in FeF_6^{4-}, but to only four chlorides in $FeCl_4^-$. Ligands that transfer substantial negative charge to the metal also produce reduced coordination numbers. For example, six neutral ammonia molecules can coordinate to nickel(II) forming $Ni(NH_3)_6^{2+}$; however, only four negatively charged chlorides coordinate, $NiCl_4^{2-}$.

Four-coordinate complexes have two common geometries, tetrahedral and square planar, as shown in Figure 23.1. The tetrahedral geometry is the more common of the two and is especially common among the non-transition metals. The square-planar geometry is characteristic of transition-metal ions with eight d electrons in the valence shell, for example, platinum(II) and gold(III); it is also found in some copper(II) complexes.

Six-coordinate complexes have an octahedral geometry, as shown in Figure 23.2(a). Notice that the octahedron can be represented as a planar square with ligands above and below the plane, as in Figure 23.2(b). In this representation, the ligands along the vertical axis are geometrically equivalent to those in the plane.

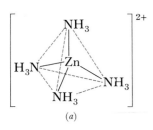

FIGURE 23.1 The structures of (a) $Zn(NH_3)_4^{2+}$ and (b) $Pt(NH_3)_4^{2+}$, illustrating the tetrahedral and square-planar geometries, respectively. These are the two common geometries for complexes in which the metal ion has a coordination number of four. The dashed lines shown in the figure are not bonds; they are included merely to assist in visualizing the shape of the metal complex.

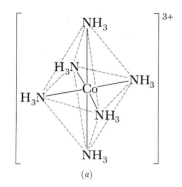

(a)

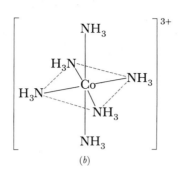

(b)

FIGURE 23.2 Two representations of an octahedral coordination sphere, the common geometric arrangement for complexes in which the metal ion has a coordination number of six.

23.2 Chelates

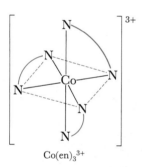

Co(en)₃³⁺

FIGURE 23.3 The Co(en)₃³⁺ ion showing how each bidentate ethylenediamine ligand is able to occupy two positions in the coordination sphere.

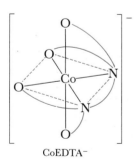

CoEDTA⁻

FIGURE 23.4 The CoEDTA⁻ ion showing how the ethylenediaminetetraacetate ion is able to wrap around a metal ion, occupying six positions in the coordination sphere.

The ligands that we have discussed so far, such as NH_3 and Cl^-, are known as monodentate ligands (from the Latin meaning "one-toothed"). These ligands possess a single donor atom. Some ligands have two or more donor atoms situated so that they can simultaneously coordinate to a metal ion. They are called polydentate ligands ("many-toothed"); because they appear to grasp the metal between two or more donor atoms, they are also known as chelating agents (from the Greek word *chele*, "claw"). One such ligand is ethylenediamine:

$$H_2\ddot{N} \quad \overset{CH_2-CH_2}{\diagup \diagdown} \quad \ddot{N}H_2$$

This ligand, which is abbreviated en, has two nitrogen atoms (shown in color) that have unshared pairs of electrons. These donor atoms are sufficiently far apart that the ligand can wrap around a metal ion with the two nitrogen atoms simultaneously complexing to the metal. The $Co(en)_3^{3+}$ ion, which contains three ethylenediamine ligands bound to the octahedral coordination sphere of cobalt(III), is shown in Figure 23.3. Notice that the ethylenediamine has been written in a shorthand notation as two nitrogen atoms connected by a line.

Ethylenediamine is an example of a bidentate ligand. Other common bidentate ligands include oxalate, $C_2O_4^{2-}$, and carbonate, CO_3^{2-} (the donor atoms are shown in color):

$$\begin{matrix} \ddot{O} & \ddot{O} \\ \| & \| \\ C & -C \\ \ddot{O} & \ddot{O} \end{matrix}^{2-} \qquad \begin{matrix} \ddot{O} \\ \| \\ C \\ \diagup \diagdown \\ \ddot{O} \quad \ddot{O} \end{matrix}^{2-}$$

Another common polydentate ligand is the ethylenediaminetetraacetate ion:

$$\begin{matrix} :O: & & & :O: \\ \| & & & \| \\ {}^-:\ddot{O}-CCH_2 & & & CH_2C-\ddot{O}:^- \\ & \diagdown & & \diagup \\ & NCH_2CH_2N & \\ & \diagup & & \diagdown \\ {}^-:\ddot{O}-CCH_2 & & & CH_2C-\ddot{O}:^- \\ \| & & & \| \\ :O: & & & :O: \end{matrix} \quad \equiv \quad \begin{matrix} {}^-O & & O^- \\ \diagdown & & \diagup \\ & N\diagdown N & \\ \diagup & & \diagdown \\ {}^-O & & O^- \end{matrix}$$

EDTA⁴⁻

This ion, abbreviated EDTA^{4-}, has six donor atoms. It can wrap around a metal ion using all six of these donor atoms as shown in Figure 23.4.

In general, chelating agents form more stable complexes than do related monodentate ligands. This is illustrated by the formation constants for $Ni(NH_3)_6^{2+}$ and $Ni(en)_3^{2+}$, shown in Equations [23.2] and [23.3]:

$$Ni(H_2O)_6^{2+}(aq) + 6NH_3(aq) \rightleftharpoons$$
$$Ni(NH_3)_6^{2+}(aq) + 6H_2O(l) \qquad K_f = 4 \times 10^8 \qquad [23.2]$$

$$Ni(H_2O)_6^{2+}(aq) + 3en(aq) \rightleftharpoons$$
$$Ni(en)_3^{2+}(aq) + 6H_2O(l) \qquad K_f = 2 \times 10^{18} \qquad [23.3]$$

Although the donor atom is nitrogen in both instances, $Ni(en)_3^{2+}$ has a stability constant nearly 10^{10} times larger than $Ni(NH_3)_6^{2+}$.

Complexing agents often can be added to a solution to prevent one or more of the customary reactions of a metal ion without actually removing it from the solution. For example, a metal ion that interferes with a chemical analysis can often be complexed and its interference thereby removed. In a sense the complexing agent hides the metal ion and is therefore referred to as a sequestering agent. Because chelates perform this role more effectively than do monodentate ligands, the term sequestering agent is usually reserved for chelates.

One of the most common applications of sequestering agents is in complexing cations in natural waters to keep them from interfering with the action of soap or detergent molecules. As noted in Section 17.5, Mg^{2+} and Ca^{2+} ions react with soap to form a precipitate commonly known as soap scum. Although these and other metal ions do not precipitate detergent molecules, they do complex with them, thereby interfering with their cleansing action.* Phosphates are effective and cheap sequestering agents for these ions. The most important phosphate used for this purpose is sodium tripolyphosphate:

Complexing agents also enhance the solubility of metal salts. For example, AgBr, the photosensitive material in photographic film, is insoluble in water, but dissolves in the presence of thiosulfate ion, $S_2O_3^{2-}$:

$$AgBr(s) \rightleftharpoons Ag^+(aq) + Br^-(aq)$$
$$K_{sp} = 7.7 \times 10^{-13}$$
$$Ag^+(aq) + 2S_2O_3^{2-}(aq) \rightleftharpoons Ag(S_2O_3)_2^{3-}(aq)$$
$$K_f = 1.6 \times 10^{13}$$

The thiosulfate can be visualized as shifting the first equilibrium to the right by complexing the Ag^+. Sodium thiosulfate decahydrate, $Na_2S_2O_3 \cdot 10H_2O$, known as hypo, is used in black-and-white photography to dissolve unexposed and undeveloped AgBr from the photographic film (Section 21.6).

METALS AND CHELATES IN LIVING SYSTEMS

Living systems consist mainly of hydrogen, oxygen, carbon, and nitrogen. In fact, more than 99 percent of the atoms required by living cells are one of these four elements. Nevertheless, many other elements are known to be essential for life. Those presently known to be essential are shown in Figure 23.5. Six of the essential elements are transition metals—iron, copper, zinc, manganese, cobalt, and molybdenum. These elements owe their roles in living systems mainly to their ability to form complexes with a variety of donor groups present in biological systems. Many enzymes, the body's catalysts, require metal ions to function. We shall take a close look at enzymes in Chapter 25.

*The action of soaps and detergents was discussed earlier, in Section 12.7; the formula of a typical detergent molecule is shown in Section 21.9.

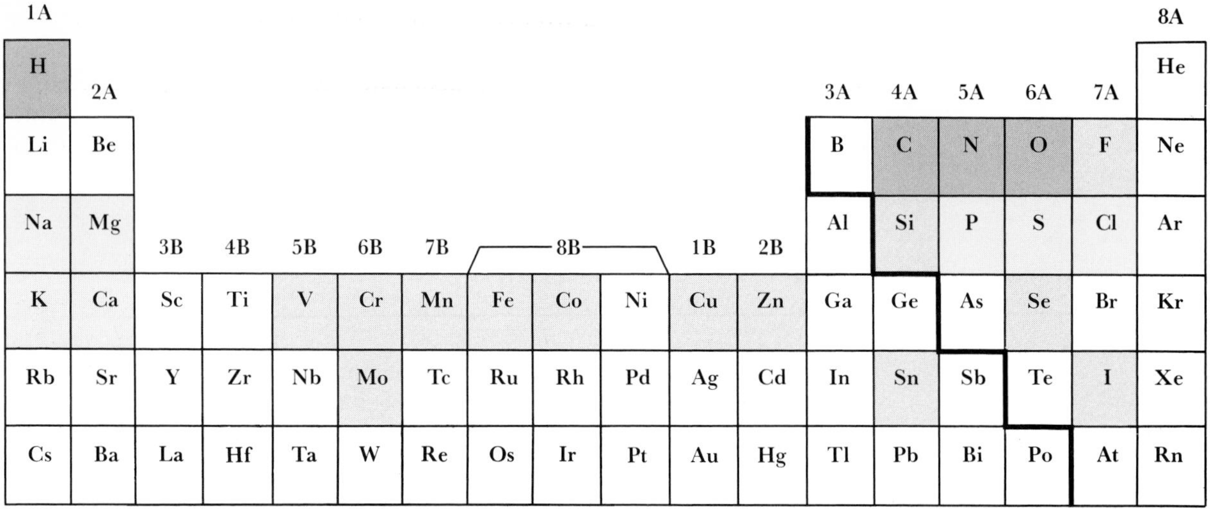

1A	2A	3B	4B	5B	6B	7B	8B	8B	8B	1B	2B	3A	4A	5A	6A	7A	8A
H																	He
Li	Be											B	C	N	O	F	Ne
Na	Mg											Al	Si	P	S	Cl	Ar
K	Ca	Sc	Ti	V	Cr	Mn	Fe	Co	Ni	Cu	Zn	Ga	Ge	As	Se	Br	Kr
Rb	Sr	Y	Zr	Nb	Mo	Tc	Ru	Rh	Pd	Ag	Cd	In	Sn	Sb	Te	I	Xe
Cs	Ba	La	Hf	Ta	W	Re	Os	Ir	Pt	Au	Hg	Tl	Pb	Bi	Po	At	Rn

FIGURE 23.5 The elements that are essential for life are indicated by the shaded areas. The dark color indicates the four most abundant elements in living systems (hydrogen, carbon, nitrogen, and oxygen). The light color indicates the seven next most common elements. The gray shading indicates the elements needed in only trace amounts.

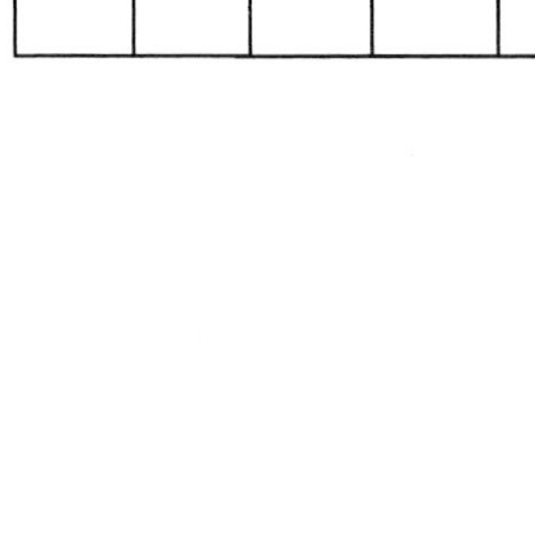

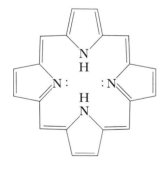

FIGURE 23.6 The structure of the porphine molecule. This molecule forms a tetradentate ligand with the loss of the two protons bound to nitrogen atoms. Porphine is the basic structure of porphyrins, compounds whose complexes play a variety of important roles in nature.

Although our bodies require only small quantities of certain metals, a deficiency of the element can lead to serious illness. For example, it was recently discovered that a deficiency of manganese in the diet can lead to convulsive disorders. This is an especially important consideration during pregnancy. Some epilepsy patients have been helped by addition of manganese to their diets.

Among the most important chelating agents in nature are those derived from the porphine molecule shown in Figure 23.6. This molecule can coordinate to a metal using the four nitrogen atoms as donors. Upon coordination to a metal, the two H^+ shown bonded to nitrogen are displaced. Complexes derived from porphine are called **porphyrins**. Different porphyrins contain different metals and have different substituent groups attached to the carbon atoms at the ligand's periphery. Two of the most important porphyrins are heme, which contains iron(II), and chlorophyll, which contains magnesium(II). We discussed heme earlier in Section 10.5. Hemoglobin contains four heme subunits as shown in Figure 10.10. The iron is coordinated to the four nitrogen atoms of the porphyrin and also to a nitrogen atom from the protein that composes the bulk of the hemoglobin molecule. The sixth position around the iron is occupied either by O_2 (in oxyhemoglobin, the red form) or by water (in deoxyhemoglobin, the blue form). Oxyhemoglobin is shown in Figure 23.7. As noted in Section 10.5, some groups such as CO act as poisons because of their ability to bind to iron more strongly than O_2 can.

When we have an insufficient quantity of iron in our diet, we suffer from iron-deficiency anemia. The lack of iron leads to a reduction in the amount of hemoglobin; we develop what advertisements have referred to as "iron-poor blood." Without hemoglobin to transport oxygen, our body's cells are unable to produce energy. Therefore, the symptoms of anemia include weakness and drowsiness.

Plants also need iron. Iron is part of a plant enzyme that participates in the production of chlorophyll, the green pigment essential to photosynthesis. Plants suffering from a deficiency in iron develop a condition known as iron chlorosis. The effect is usually first noticed in young leaves, which have a yellow coloration. Often the plant develops chlorosis because the soil conditions interfere with the availability of the iron. For example, iron may be present in the soil, but only in insoluble forms unavailable to the plant.

Iron may occur in soil either as iron(II) or iron(III), depending on the ability of oxygen to penetrate the soil and oxidize the iron. Although both Fe^{2+} and Fe^{3+} can be used by plants, the Fe^{3+} is much less soluble than Fe^{2+} in normal soils, which usually have pH's in the vicinity of 7. For example, in typical soil, Fe^{3+} precipitates as $Fe(OH)_3$ when the pH exceeds 3, whereas Fe^{2+} precipitates as $Fe(OH)_2$ when the pH exceeds 6. Therefore plants growing in alkaline soils readily suffer from iron deficiency. Simple addition of iron salts to the soil does not correct this condition, because the iron is merely precipitated. Current practice in agriculture is to add the iron in a complexed form, such as $FeEDTA^{2-}$. The chelated iron will not precipitate and is therefore available to the plant; once in the plant, the iron is removed as needed.

Some plants, such as certain strains of soybeans, secrete their own chelates to solubilize the needed iron. Similarly, mosses and lichens growing on rocks generate chelating agents to extract the metals that they need.

All living cells need iron. One mechanism that our body uses to fight invading bacteria is to withhold iron needed by the bacteria. The bacteria use powerful chelating agents to obtain their required iron. Our body, in turn, keeps its iron tightly complexed. Thus there is a confrontation between the chelating agents of our body and those of the microbial invaders. It has been found that the ability of bacteria to synthesize their chelates is suppressed at elevated temperatures. Consequently, fever is part of the body's attempt to overcome the invading bacteria.

Chelates and chelating agents have also been used as drugs. For example, they have been used to destroy bacteria by depriving them of essential metals. In this fashion the drug mimics the body's natural defenses described above. Chelating agents are also used to remove metals such as Hg^{2+}, Pb^{2+}, and Cd^{2+}, which are detrimental to health. For example, one method of treating lead poisoning is to administer $Na_2[CaEDTA]$. The EDTA chelates the lead, allowing its removal from the body in urine.

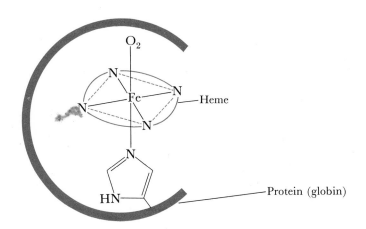

FIGURE 23.7 A schematic representation of oxyhemoglobin, showing one of the four heme units in the molecule. The iron is bound to four nitrogen atoms of the porphyrin, to a nitrogen from the surrounding protein, and to an O_2 molecule.

23.3 Nomenclature

Before we go too far in our discussions of complexes, it is useful to describe how these substances are named. When complexes were first discovered and few were known, they were named after the chemist who originally prepared them. A few of these names persist; for example, $NH_4[Cr(NH_3)_2(NCS)_4]$ is known as Reinecke's salt. As the number of known complexes grew, they began to be named by color. For example, $[Co(NH_3)_5Cl]Cl_2$, whose formula was then written $CoCl_3 \cdot 5NH_3$, was known as purpureocobaltic chloride after its purple color. Once the structures of complexes were more fully understood, it became possible to name them in a more systematic manner. Before we give the rules for naming complexes, let's consider two examples. We can thereby get a little better idea of how the nomenclature rules apply. The two complexes are:

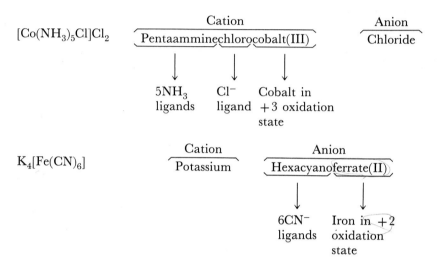

The rules of nomenclature are as follows:*

1   *In naming salts, the name of the cation is given before the name of the anion.* Thus in $[Co(NH_3)_5Cl]Cl_2$ we name the $Co(NH_3)_5Cl^{2+}$ and then Cl^-.

2   *Within a complex ion or molecule, the ligands are named before the metal. Ligands are listed in alphabetical order, regardless of charge on the ligand. Prefixes that give the number of ligands are not considered part of the ligand name in determining alphabetical order.)* Thus, in the $Co(NH_3)_5Cl^{2+}$ ion, we name the ammonia ligands first, then the chloride, then the metal: pentaamminechlorocobalt(III). Note, however, that in writing the formula the metal is listed first.

3   *Anionic ligands end in the letter o, whereas neutral ones ordinarily bear the name of the molecule.* Some common ligands and their names are listed in Table 23.1. Special names are given to H_2O (aqua) and NH_3 (ammine). Thus the terms chloro and ammine occur in the name for $[Co(NH_3)_5Cl]Cl_2$.

TABLE 23.1   Some common ligands

Ligand	Ligand name
Azide, N_3^-	Azido
Bromide, Br^-	Bromo
Chloride, Cl^-	Chloro
Cyanide, CN^-	Cyano
Hydroxide, OH^-	Hydroxo
Carbonate, CO_3^{2-}	Carbonato
Oxalate, $C_2O_4^{2-}$	Oxalato
Ammonia, NH_3	Ammine
Ethylenediamine, en	Ethylenediamine
Water, H_2O	Aqua

*The rules of nomenclature are approved by the International Union of Pure and Applied Chemistry and are subject to periodic revision. Some rules are clearly more important than others. For example, the order in which ligands are named (rule 2) is not as important as assigning each ligand its correct name.

4 *A Greek prefix (for example,* di-, tri-, tetra-, penta-, *and* hexa-) *is used to indicate the number of each kind of ligand when more than one is present.* Thus in the name for $Co(NH_3)_5Cl^{2+}$ we have pentaammine, indicating five NH_3 ligands. *If the name of the ligand itself contains a Greek prefix such as* mono- *or* di-, *then the ligand is enclosed in parentheses and alternate prefixes* (bis-, tris-, tetrakis-, pentakis-, *and* hexakis-) *used.* For example, the name for $[Co(en)_3]Cl_3$ is tris(ethylenediamine)cobalt(III) chloride.

5 *If the complex is an anion, its name ends in* -ate. For example, in $K_4[Fe(CN)_6]$ the anion is called the hexacyanoferrate(II) ion. The suffix *-ate* is often added to the Latin stem as in this example.

6 *The oxidation number of the metal is given in parentheses in Roman numerals following the name of the metal.* For example, the Roman numeral III is used to indicate the $+3$ oxidation state of cobalt in $Co(NH_3)_5Cl^{2+}$.

SAMPLE EXERCISE 23.3

Give the name of the following compounds: (a) $[Cr(H_2O)_4Cl_2]Cl$; (b) $K_4[Ni(CN)_4]$.

Solution: (a) We begin with the four waters, which are indicated as tetraaqua. Then there are two chloride ions, indicated as dichloro. The oxidation state of Cr is $+3$:

$$+3 + 4(0) + 2(-1) + (-1) = 0$$
$$[Cr(H_2O)_4Cl_2]Cl$$

Thus we have chromium(III). Finally, the anion is chloride. Putting these parts together we have the compound's name: tetraaquadichlorochromium(III) chloride.

(b) The complex has four CN^-, which we indicate as tetracyano. The oxidation state of the nickel is zero:

$$4(+1) + 0 + 4(-1) = 0$$
$$K_4[Ni(CN)_4]$$

Because the complex is an anion, the metal is indicated as nickelate(0). Putting these parts together and naming the cation first we have: potassium tetracyanonickelate(0).

Given the name of a coordination compound, you should be able to write out the formula. Remember that cations are listed before anions and that the metal and ligands of the coordination sphere are enclosed within brackets.

SAMPLE EXERCISE 23.4

Write the formula for bis(ethylenediamine)difluorocobalt(III) perchlorate.

Solution: The complex cation contains two fluorides, two ethylenediamines, and a cobalt with a $+3$ oxidation number. Knowing this we can determine the charge on the complex:

$$+3 + 2(0) + 2(-1) = +1$$
$$Co(en)_2F_2$$

The perchlorate anion has a single negative charge, ClO_4^-. Therefore, only one is needed to balance the charge on the complex cation. The formula is thus $[Co(en)_2F_2]ClO_4$.

23.4 Isomerism

It sometimes happens that two or more compounds have the same formula but differ in one or more physical or chemical properties, such as color, solubility, or rate of reaction with some reagent. Such compounds,

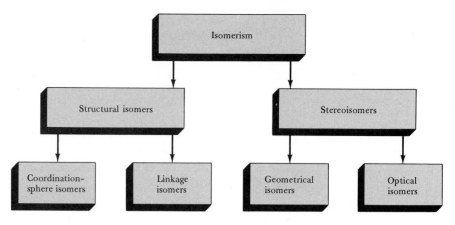

FIGURE 23.8 Forms of isomerism in coordination compounds.

which have the same collection of atoms arranged in different ways, are called isomers. Several types of isomerism are possible, as outlined in Figure 23.8.

STRUCTURAL ISOMERISM

Structural isomers differ in the bonding arrangements of the atoms; that is, they are chemically distinct. Many different types of structural isomerism are known in coordination chemistry. The two listed in Figure 23.8 are given as examples. Linkage isomerism is a relatively rare but interesting type that arises when a particular ligand is capable of coordinating to a metal in two different ways. For example, the nitrite ion, NO_2^-, can coordinate through either a nitrogen or an oxygen atom, as shown in Figure 23.9. When it coordinates through the nitrogen atom the NO_2^- ligand is called *nitro;* when it coordinates through the oxygen atom, it is referred to as *nitrito.* The isomers shown in Figure 23.9 differ in their chemical and physical properties. For example, the N-bonded isomer is yellow, while the O-bonded isomer is red. Other ligands that are capable of coordinating through either of two donor atoms include thiocyanate, SCN^-, whose potential donor atoms are N and S.

Coordination-sphere isomers differ in the ligands that are directly bonded to the metal, as opposed to being outside the coordination sphere in the solid lattice. For example, the compound $CrCl_3(H_2O)_6$ exists in three forms: $[Cr(H_2O)_6]Cl_3$, a violet compound, $[Cr(H_2O)_5Cl]Cl_2 \cdot H_2O$, a green compound, and $[Cr(H_2O)_4Cl_2]Cl \cdot 2H_2O$, also a green compound. In the second and third compounds the water has been displaced from the coordination sphere by chloride ions and occupies a site in the solid lattice.

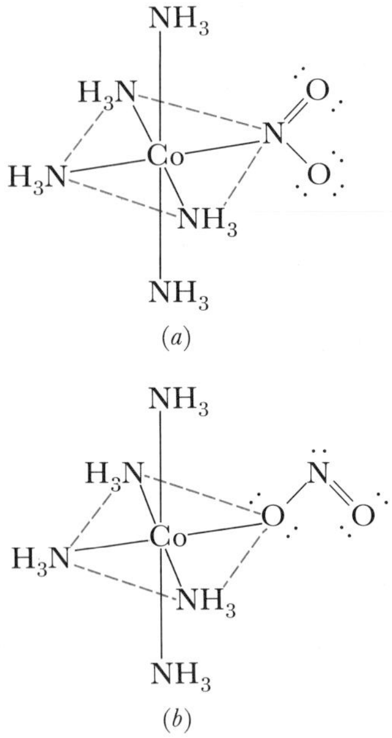

STEREOISOMERISM

The most important form of isomerism is known as stereoisomerism. Stereoisomers have the same chemical bonds but different spatial arrangements. For example, in $Pt(NH_3)_2Cl_2$, the chloro ligands can be arranged either adjacent or opposite to each other, as shown in Figure 23.10. This particular form of stereoisomerism, in which the arrange-

FIGURE 23.9 (a) The yellow N-bound and (b) the red O-bound isomers of $Co(NH_3)_5NO_2^{2+}$.

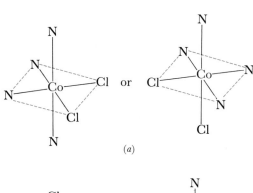

FIGURE 23.10 (a) The *cis* and (b) the *trans* geometric isomers of the square-planar $Pt(NH_3)_2Cl_2$.

ments of donor atoms around the coordination sphere are different, is known as geometric or *cis-trans* isomerism. The isomer with like groups close together is called the *cis* isomer, whereas the one with like groups far apart is called the *trans* isomer.

Because all of the corners of a tetrahedron are adjacent to one another, *cis-trans* isomerism is not observed in the case of tetrahedral complexes. However, it is found in octahedral ones.

SAMPLE EXERCISE 23.5

How many geometric isomers are there for $Co(NH_3)_4Cl_2^+$?

Solution: To answer this question, draw several octahedra with ligands in different locations. A systematic approach would be to keep one chloride in the same location each time. Examination of these structures indicates that there are two unique geometric isomers: one with adjacent Cl^- ligands (the *cis* isomer) and one with Cl^- ligands across the metal from each other (the *trans* isomer). Two different orientations of both of these isomers are shown in Figure 23.11. It is easy to overestimate the number of geometric isomers; sometimes different orientations of a single isomer are incorrectly thought to be different isomers. Therefore, you should keep in mind that if two structures can be rotated so that they are equivalent, they are not isomers of one another. The problem of identifying isomers is compounded by the difficulty we often have in visualizing three-dimensional molecules from their two-dimensional representations. It is easier to determine the number of isomers if we are working with three-dimensional models.

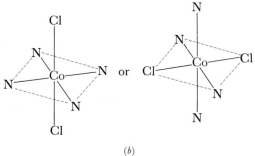

FIGURE 23.11 (a) The *cis* and (b) the *trans* geometric isomers of the octahedral $Co(NH_3)_4Cl_2^+$ ion. (The symbol N represents the coordinated NH_3 group.)

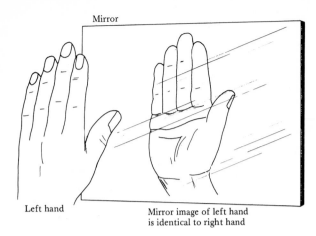

Left hand

Mirror image of left hand
is identical to right hand

FIGURE 23.12 Our hands are nonsuperimposable mirror images of each other.

A second type of stereoisomerism is known as optical isomerism. Optical isomers are nonsuperimposable mirror images of one another. They bear the same resemblance to one another that our left hand bears to our right hand. If you look at your left hand in a mirror, as shown in Figure 23.12, the image is identical to your right hand. Furthermore, your two hands are not superimposable on one another. A good example of a complex that exhibits this type of isomerism is the $Co(en)_3^{3+}$ ion. The two isomers of $Co(en)_3^{3+}$ and their mirror-image relationship to one another are shown in Figure 23.13. Just as there is no way that we can twist or turn our right hand to make it look identical to our left, so also there is no way to rotate one of these isomers to make it identical to the other. If we had models of each that we could handle, perhaps we could more easily satisfy ourselves of this fact. Molecules or ions that have nonsuperimposable mirror images are said to be chiral (pronounced KY-rul). Enzymes, the body's catalysts, are among the most highly chiral molecules known. As noted in Section 23.2, many enzymes contain complexed metal ions.

SAMPLE EXERCISE 23.6

Does either *cis*- or *trans*-$Co(en)_2Cl_2^+$ have optical isomers?

Solution: To answer this question you should draw out both the *cis* and *trans* isomers of $Co(en)_2Cl_2^+$, and then their mirror images. Note that the mirror image of the *trans* isomer is identical to the original. Consequently *trans*-$Co(en)_2Cl_2^+$ has no optical isomer. However, the mirror image of *cis*-$Co(en)_2Cl_2^+$ is not identical to the original. Consequently, there are optical isomers for this complex.

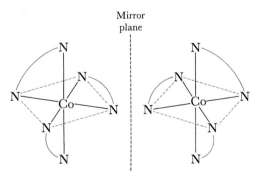

FIGURE 23.13 The two optical isomers of $Co(en)_3^{3+}$; notice that the ions are nonsuperimposable mirror images of each other.

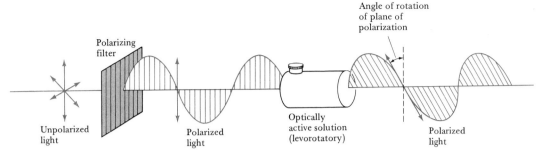

Unpolarized light · Polarizing filter · Polarized light · Optically active solution (levorotatory) · Angle of rotation of plane of polarization · Polarized light

FIGURE 23.14 The effect of an optically active solution on the plane of polarization of plane-polarized light. The unpolarized light is passed through a polarizing filter. The resultant polarized light thereafter passes through a solution containing a levorotatory optical isomer. As a result, the plane of polarization of the light is rotated to the left relative to an observer looking toward the light source.

Most of the physical and chemical properties of optical isomers are identical. The properties of two optical isomers differ only if they are in a chiral environment—that is, one in which there is a sense of right- and left-handedness. For example, in the presence of a chiral enzyme, the reaction of one optical isomer might be catalyzed, whereas the other isomer remains totally unreacted. Consequently, one optical isomer may produce a specific physiological effect within our body, whereas its mirror image produces a different effect or none at all.

Optical isomers are usually distinguished from each other by their interaction with plane-polarized light. If light is polarized, for example, by passage through a sheet of Polaroid film, the light waves are vibrating in a single plane, as shown in Figure 23.14. If the polarized light is passed through a solution containing one optical isomer, the plane of polarization is rotated either to the right (clockwise) or to the left (counterclockwise). The isomer that rotates the plane of polarization to the right is said to be dextrorotatory; it is labeled the dextro, or *d*, isomer (Latin *dexter*, "right"). Its mirror image will rotate the plane of polarization to the left; it is said to be levorotatory and is labeled the levo, or *l*, isomer (Latin *laevus*, "left"). Because of their effect on plane-polarized light, chiral molecules are said to be optically active.

When a substance with optical isomers is prepared in the laboratory, the chemical environment during the synthesis is not usually chiral. Consequently, equal amounts of the two isomers are obtained; the mixture is said to be racemic. A racemic mixture will not rotate polarized light because the rotatory effects of the two isomers cancel each other. In order to separate the isomers from the racemic mixture, they must be placed in a chiral environment. For example, one optical isomer of the chiral tartrate anion,* $C_4H_4O_6^{2-}$, can be used to separate a racemic mixture of $[Co(en)_3]Cl_3$. If *d*-tartrate is added to an aqueous solution of the $[Co(en)_3]Cl_3$, *d*-$[Co(en)_3](d$-$C_4H_4O_6)Cl$ will precipitate leaving *l*-$[Co(en)_3]^{3+}$ in solution.

*When sodium ammonium tartrate, $NaNH_4C_4H_4O_6$, is crystallized from solution, the two isomers form separate crystals whose shapes are mirror images of each other. In 1848, Louis Pasteur achieved the first separation of a racemic mixture into optical isomers; using a microscope he picked the "right-handed" crystals of this compound from the "left-handed" ones.

If we were to examine a number of different complexes in solution, we would observe that some exchange ligands at an extremely rapid rate, whereas others do so quite slowly. For example, addition of ammonia to an aqueous solution of $CuSO_4$ produces an essentially instantaneous color change as the pale blue $Cu(H_2O)_4^{2+}$ ion is converted to the deep blue $Cu(NH_3)_4^{2+}$ ion. When this solution is acidified, the pale blue color is regenerated again at a rapid rate:

$$Cu(NH_3)_4^{2+}(aq) + 4H_2O(l) + 4H^+(aq) \longrightarrow$$
$$Cu(H_2O)_4^{2+}(aq) + 4NH_4^+(aq) \qquad [23.4]$$

In contrast, $Co(NH_3)_6^{3+}$ is more difficult to prepare than is $Cu(NH_3)_4^{2+}$. However, once it has been formed and is then placed in an acidic solution, reaction to form NH_4^+ takes several days. This tells us that the coordinated NH_3 groups are not readily removed from the metal. Complexes like $Cu(NH_3)_4^{2+}$ that undergo rapid ligand exchange are said to be **labile;** those like $Co(NH_3)_6^{3+}$ that undergo slow ligand exchange are said to be **inert.** The distinction between labile and inert complexes applies to how rapidly equilibrium is attained and not to the position of the equilibrium. For example, although $Co(NH_3)_6^{3+}$ is inert in acidified aqueous solutions, the equilibrium constant indicates that the complex is not thermodynamically stable under these conditions:

$$Co(NH_3)_6^{3+}(aq) + 6H_3O^+(aq) \rightleftharpoons$$
$$Co(H_2O)_6^{3+}(aq) + 6NH_4^+(aq) \qquad K_f \simeq 10^{20} \qquad [23.5]$$

The kinetic inertness of $Co(NH_3)_6^{3+}$ can be attributed to a high activation energy for the reaction.

Cobalt(III) is one of few metal ions to consistently form inert complexes; others include chromium(III), platinum(IV), and platinum(II). Complexes of these ions maintain their identity in solution long enough to permit study of their structures and properties. They were therefore among the first complexes studied. Much of our understanding of structure and isomerism comes from studies of these complexes.

23.6 Structure and isomerism: a historical perspective

Many early systematic studies of coordination compounds involved complexes of cobalt(III), chromium(III), platinum(II), and platinum(IV) with ammonia as a ligand. One of the earliest reports of the preparation of an ammine complex dates from 1798 when a chemist by the name of Tassaert accidentally prepared an ammonia complex of cobalt. The compound was found to have the empirical formula $CoN_6H_{18}Cl_3$. Tassaert wrote the formula as $CoCl_3 \cdot 6NH_3$, suggesting that the compound was analogous to hydrated salts like $CoCl_2 \cdot 6H_2O$.

By 1890, many ammine complexes had been prepared, and a great deal of information about them had been gathered by a number of different investigators. By this time chemists had begun to wonder about how the atoms in these complexes were connected to each other and about the possible effect of these arrangements on the properties of the complexes. Among the observations that any successful theory would have to account for were the electrical conductivity of the complexes and

TABLE 23.2 Properties of some ammonia complexes of cobalt(III)

Original formulation	Color	Ions per formula unit	Cl⁻ ions in solution per formula unit	Modern formulation
$CoCl_3 \cdot 6NH_3$	Orange	4	3	$[Co(NH_3)_6]Cl_3$
$CoCl_3 \cdot 5NH_3$	Purple	3	2	$[Co(NH_3)_5Cl]Cl_2$
$CoCl_3 \cdot 4NH_3$	Green	2	1	$trans\text{-}[Co(NH_3)_4Cl_2]Cl$
$CoCl_3 \cdot 4NH_3$	Violet	2	1	$cis\text{-}[Co(NH_3)_4Cl_2]Cl$

their behavior toward $AgNO_3$. Recall from our previous discussions (Sections 12.5 and 19.6) that solutions of ionic substances conduct electrical current. The ease with which the solution conducts current is referred to as the conductivity. The conductivity of solutions of ionic substances increases with the total concentration of ions and is also greater for ions of higher charge. By comparing the conductivities of solutions of coordination compounds with those of simple salts, it was possible to determine the number and types of ions present in the solution. For example, the molar conductivity of an aqueous solution of $CoCl_3 \cdot 5NH_3$ is about the same as that of $CaCl_2$ and other $1:2$ electrolytes. We can thus conclude that an aqueous solution of $CoCl_3 \cdot 5NH_3$ produces three ions, one that carries a $+2$ charge and two carrying negative charges. The number of free chloride ions present in the solution was determined by treating solutions with $AgNO_3$. When cold, freshly prepared solutions of $CoCl_3 \cdot 5NH_3$ were treated with $AgNO_3$, 2 mol of AgCl precipitated for each mole of complex; one chloride in the compound did not precipitate. These results are summarized in Table 23.2.

In 1893, a 26-year-old Swiss chemist, Alfred Werner, proposed a theory that successfully explained these facts and became the basis for our subsequent understanding of metal complexes. Werner's first basic postulate was that metals exhibit both primary and secondary valences. We now refer to these as the metal's oxidation state and coordination number, respectively. This postulate had no theoretical basis; it predated Lewis's theory of covalent bonding by 23 years. However, it allowed Werner to explain many experimental facts. Werner postulated a primary valence of three and a secondary valence of six for cobalt(III). He therefore wrote the formula for $CoCl_3 \cdot 5NH_3$ as $[Co(NH_3)_5Cl]Cl_2$. The ligands within the brackets satisfied cobalt's secondary valence of six; the three chlorides satisfied the primary valence of three. Werner proposed that the chlorides within the coordination sphere of cobalt(III) are bound so tightly that they are unavailable to contribute to the compound's conductivity or to react with $AgNO_3$. Thus $CoCl_3 \cdot 5NH_3$ consists of a $Co(NH_3)_5Cl^{2+}$ ion and two Cl^- ions.

Werner also sought to deduce the arrangement of ligands around the central metal. He postulated that cobalt(III) complexes exhibit an octahedral geometry and sought to verify this postulate by comparing the number of observed isomers with the number expected for various geometries. For example, $[Co(NH_3)_4Cl_2]^+$ should exhibit two geometric isomers if it was octahedral. When he first postulated an octahedral geometry for cobalt(III), only one isomer of $[Co(NH_3)_4Cl_2]^+$ was known, the green *trans* isomer. In 1907, after considerable effort, Werner succeeded in isolating the violet *cis* isomer. Even before that, however, he had succeeded in isolating other *cis* and *trans* isomers of cobalt(III) complexes. The occurrence of two isomers was consistent with his postulate of octahedral geometry. Another result consistent with octahedral geometry was the demonstration that $Co(en)_3^{3+}$ and certain other complexes were optically active. In 1913 Werner was awarded the Nobel Prize in chemistry for his outstanding research work in the field of coordination chemistry.

SAMPLE EXERCISE 23.7

Suggest the structure for $CoCl_2 \cdot 6H_2O$.

Solution: By analogy to the ammonia complexes of cobalt(III), we might write the formula of this compound as $[Co(H_2O)_6]Cl_2$. Indeed, experimental evidence indicates that the water molecules are attached to the metal as ligands. Hydrated metal salts generally have water coordinated to the metal. However, water can also be hydrogen bonded to the anion, particularly to oxyanions. For example, the familiar $CuSO_4 \cdot 5H_2O$ has four water molecules coordinated to Cu^{2+}, and one to SO_4^{2-}.

23.7 Magnetism and color

Werner's theory helps us to understand many properties of complexes, including isomerism and conductivity. However, Werner's theory must be extended before we can use it to explain certain other properties such as the magnetic behavior and color of transition-metal complexes. These two properties have played an important role in the further development of concepts about metal-ligand bonding. Therefore, let's briefly discuss these properties before taking a closer look at the bonding in transition-metal complexes.

MAGNETISM

In our earlier discussions of chemical bonding in Section 8.6, it was noted that substances containing unpaired electrons are paramagnetic; they are drawn into a magnetic field. The magnitude of the paramagnetism is related to the number of unpaired electrons.* Substances without unpaired electrons are diamagnetic; they are slightly repelled by a magnetic field. Therefore, one way to determine the number of unpaired electrons is to measure the effect of a magnetic field on the sample, as shown in Figure 23.15. The mass of the sample is first measured in the absence of a magnetic field and then in its presence. If the sample appears to weigh more in the magnetic field, it is being drawn into the field and is consequently paramagnetic. If the sample appears to weigh less in the magnetic field, it is being pushed upward, out of the field; it must consequently be diamagnetic. The point of interest in the complexes of the transition metals is that the number of unpaired electrons associated with a particular metal ion depends on the surrounding ligands. For example, there are no unpaired electrons in $Co(NH_3)_6^{3+}$, but there are four in CoF_6^{3-}; yet both complexes contain cobalt(III). Any successful bonding theory must explain this observation.

---

*The behavior of a sample in a magnetic field is related to its magnetic moment, μ. The magnetic moment of transition-metal complexes is proportional to the quantity $\sqrt{n(n + 2)}$, where n is the number of unpaired electrons.

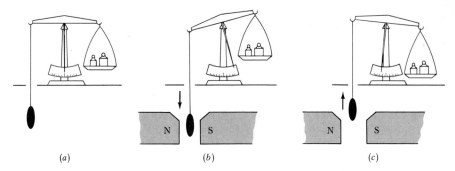

FIGURE 23.15 (a) A sample is weighed in the absence of a magnetic field. If it is attracted into a magnetic field (b), it is paramagnetic. If it is slightly repelled by a magnetic field (c), it is diamagnetic.

COLOR

Another interesting characteristic of transition-metal complexes is the variety of colors that they exhibit. The colors of some complexes of cobalt(III) were shown earlier in Table 23.2. From that list, we see that color can change as the ligands surrounding the metal ion change. If we were to examine a more extensive list of substances, we would see that color also depends on the metal and its oxidation state.

Before we can attempt to explain the origin of these colors, we need to review our earlier discussion of light and introduce some new concepts. Recall first that visible light consists of electromagnetic radiation whose wavelength, λ, ranges from 400 nm to 700 nm (Figure 5.3). The energy of this radiation is inversely proportional to its wavelength, as discussed earlier, in Section 5.2:

$$E = h\nu = h(c/\lambda) \tag{23.6}$$

The visible spectrum is shown in Figure 23.16. The colors of the spectrum, indicated by their first letters, are shown in order of increasing energy. Notice that these letters spell out what appears to be a man's name: Roy G. Biv.

When a sample absorbs light, what we see is the sum of the remaining colors that strike our eyes. If a sample absorbs all wavelengths of visible light, none reaches our eyes from that sample. Consequently, it appears black. If the sample absorbs no light, it is white or colorless. If it absorbs all but orange, the sample appears orange. Each of these situations is

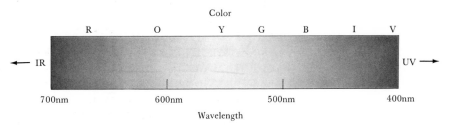

FIGURE 23.16 The visible spectrum showing the relation between color and wavelength (R = red, O = orange, Y = yellow, G = green, B = blue, I = indigo, and V = violet). Notice that the colors spell out a name: Roy G. Biv.

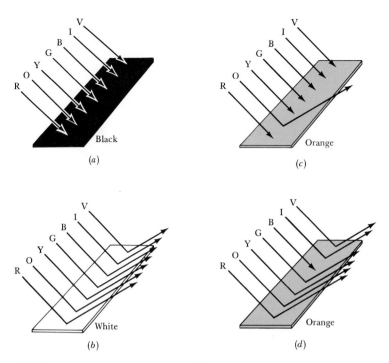

FIGURE 23.17 (a) An object is black if it absorbs all colors of light. (b) An object is white if it reflects all colors of light. (c) An object is orange if it reflects only this color and absorbs all others. (d) An object is also orange if it reflects all colors except blue, the complementary color of orange.

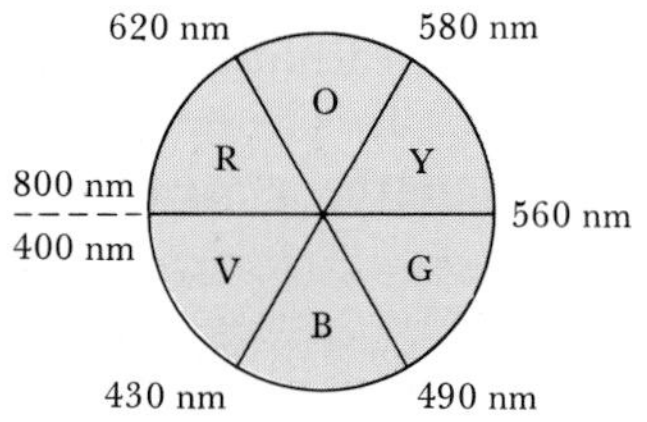

FIGURE 23.18 An artist's color wheel, showing the colors that are complementary to one another and the wavelength range of each color.

shown in Figure 23.17. That figure shows one further situation; we also perceive an orange color when visible light of all colors except blue strikes our eyes. In a complementary fashion, if the sample absorbed only orange, it would appear blue; blue and orange are said to be complementary colors. Complementary colors can be determined with the aid of the artist's color wheel, shown in Figure 23.18. The colors that are complementary to one another, like orange and blue, are across the wheel from each other.

SAMPLE EXERCISE 23.8

The complex ion *trans*-$[Co(NH_3)_4Cl_2]^+$ absorbs light primarily in the red region of the visible spectrum (the most intense absorption is at 640 nm). What is the color of the complex?

Solution: Because the complex absorbs red light, its color will be the complementary color of red. From Figure 23.18, we see that this is green.

The amount of light absorbed by a sample as a function of wavelength is known as its absorption spectrum. The visible absorption spectrum of a transparent sample, such as a solution of *trans*-$[Co(NH_3)_4Cl_2]^+$, can be determined as shown in Figure 23.19. The spectrum of $Ti(H_2O)_6^{3+}$, which we shall discuss in the next section, is shown in Figure 23.20. The absorption maximum of $Ti(H_2O)_6^{3+}$ is at 510 nm. Because the sample absorbs most strongly in the green and yellow regions of the visible spectrum, it appears purple.

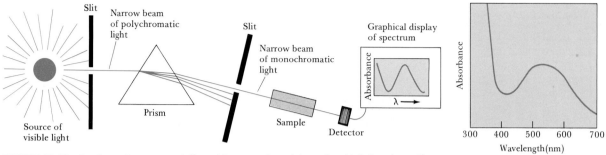

FIGURE 23.19 A schematic representation of the experimental way in which the absorption spectrum of a solution is determined. The prism is rotated so that different wavelengths of light pass through the sample. The detector measures the amount of light reaching it, and this information can be displayed as the absorption at each wavelength.

FIGURE 23.20 The visible absorption spectrum of the $Ti(H_2O)_6^{3+}$ ion.

23.8 Crystal-field theory

Although the ability to form complexes is common to all metal ions, the most numerous and interesting complexes are formed by the transition elements. It has been recognized for a long time that the magnetic properties and colors of transition metal complexes are related to the presence of d electrons in metal orbitals. In this section we will consider a model for bonding in transition metal complexes, called the crystal-field theory, that accounts very well for the observed properties of these interesting substances. *

We have already noted that the ability of a metal ion to attract ligands such as water around itself can be viewed as a Lewis acid-base interaction (Section 15.10). The base, that is, the ligand, can be considered to donate a pair of electrons into a suitable empty hybrid orbital on the metal, as shown in Figure 23.21. However, we can assume that much of the attractive interaction between the metal ion and the surrounding ligands is due to the electrostatic forces between the positive charge on the metal and negative charges on the ligands. If the ligand is ionic, as in the case of Cl^- or SCN^-, the electrostatic interaction occurs between the positive charge on the metal center and the negative charge on the ligand. When the ligand is neutral, as in the case of H_2O or NH_3, the negative ends of these polar molecules, containing an unshared electron

*The name "crystal field" arose because the theory was first developed to explain the properties of solid, crystalline materials such as ruby. The same theoretical model applies to complexes in solution.

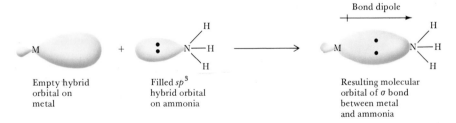

FIGURE 23.21 A representation of the metal-ligand bond in a complex as a Lewis acid-base interaction. The ligand, which acts as a Lewis base, donates charge to the metal via a metal hybrid orbital. The bond that results is strongly polar, with some covalent character. For many purposes it is sufficient to assume that the metal-ligand interaction is entirely electrostatic in character, as is done in the crystal-field model.

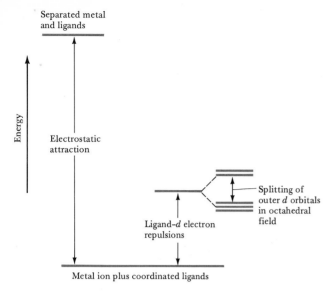

Energy

Separated metal
and ligands

Electrostatic
attraction

Ligand–d electron
repulsions

Splitting of
outer d orbitals
in octahedral
field

Metal ion plus coordinated ligands

FIGURE 23.22 In the crystal-field model, the bonding between metal ion and donor atoms is considered to be largely electrostatic. The energy of the metal ion plus coordinated ligands is less than that of the separated metal ion plus ligands because of the electrostatic attraction between metal ion and ligands. At the same time the energies of the metal d electrons are increased by the repulsive interaction between these electrons and the ligands. However, because of their distributions in space, the electrons that occupy the d_{z^2} and $d_{x^2-y^2}$ orbitals are more strongly repelled by the ligands than the electrons occupying the d_{xz}, d_{yz}, and d_{xy} orbitals. This difference in repulsive interactions gives rise to the splitting of the metal d orbital energies shown on the right and is referred to as the crystal-field splitting.

pair, are directed toward the metal. In this case the attractive interaction is of the ion-dipole type (Section 11.5). In either case, the result is the same; the ligands are attracted strongly toward the metal center. At the same time, however, the ligands repel one another, because they possess the same charge, or because the negative ends of the molecular dipoles are directed toward the same center. In any metal complex, then, there is a balance between the attractive forces between metal and ligands and the ligand-ligand repulsive interactions. The most stable complex results when the geometrical arrangement of the ligands around the metal is such as to maximize metal-ligand attractions while minimizing the ligand-ligand repulsions. The assembly of metal ion and ligands is lower in energy than the fully separated charges, as illustrated on the left in Figure 23.22.

In a six-coordinate complex the ligand-ligand repulsions cause the ligands to approach along the x, y, and z axes, as shown in Figure 23.23. With the physical arrangement of ligands and metal ion shown in this figure as our starting point, let us now consider what happens to the energies of electrons in the metal d orbitals as the ligands approach the metal ion. Keep in mind that the d electrons are the outermost electrons of the metal ion. We know that the overall energy of the metal ion plus ligands will be lower (that is, more stable) when the ligands are drawn toward the metal center. At the same time, however, there is a repulsive interaction between the outermost electrons on the metal and the negative charges on the ligands. This interaction is called the crystal field. As a result, if we consider just the energies of the d electrons on the metal ion, we find that these have *increased*. This increase in energy is shown in the center part of Figure 23.22. But the d orbitals of the metal ion do not all behave in the same way under the influence of the ligand field. To see why this is so, recall the shapes of the five d orbitals, illustrated in Figure 23.24. In the isolated metal ion, these five orbitals are equivalent in energy. However, as the ligands approach the metal ion, the $d_{x^2-y^2}$ and d_{z^2} orbitals, which are directed *along* the x, y, and z axes, are more strongly

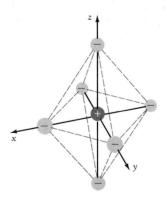

FIGURE 23.23 An octahedral array of negative charges surrounding a positive charge.

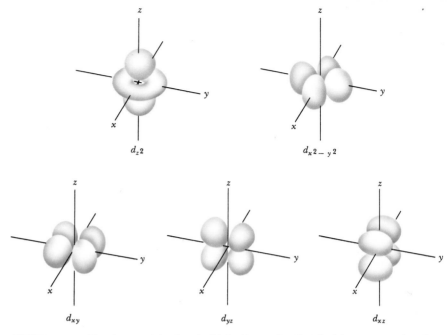

FIGURE 23.24 The shapes of the five d orbitals. Remember that the lobes represent regions in which the electrons occupying an orbital are most likely to be found.

repelled by the ligands than the d_{xy}, d_{xz}, and d_{yz} orbitals. These latter are directed *between* the axes along which the ligands approach. Thus, an energy separation, or splitting, occurs. The $d_{x^2-y^2}$ and d_{z^2} orbitals are raised in energy, and the d_{xz}, d_{yz}, and d_{xy} orbitals are lowered. This energy splitting is illustrated on the right side of Figure 23.22. In the material that follows we will concentrate on just the splitting of the d orbital energies by the ligand field. This is illustrated in a slightly different form in Figure 23.25.

Let's examine how the crystal-field model accounts for the observed colors in transition metal complexes. The energy gap between the d orbitals, labeled Δ, is of the same order of magnitude as the energy of a photon of visible light. It is therefore possible for a transition-metal complex to absorb visible light, which thereby excites an electron from the lower energy d orbitals into the higher energy ones. The $Ti(H_2O)_6^{3+}$ ion provides a simple example, because titanium(III) has only one 3d elec-

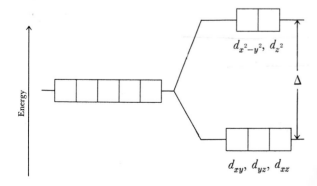

FIGURE 23.25 The energies of the d orbitals in an octahedral crystal field.

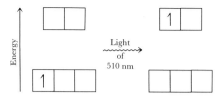

FIGURE 23.26 The 3*d* electron of Ti(H₂O)₆³⁺ is excited from the lower-energy *d* orbitals to the higher-energy ones when irradiated with light of 510 nm wavelength.

tron. As shown in Figure 23.20, $Ti(H_2O)_6^{3+}$ has a single absorption peak in the visible region of the spectrum. The maximum absorption is at 510 nm (3.9×10^{-19} J/molecule). Light of this wavelength causes the *d* electron to move from the lower energy set of *d* orbitals into the higher energy set, as shown in Figure 23.26.

SAMPLE EXERCISE 23.9

Al^{3+}, Zn^{2+}, and Co^{2+} ions are placed in octahedral environments. Which can absorb visible light and thereby exhibit color?

Solution: The Al^{3+} ion has an electron configuration of [Ne]. Because it has no outer *d* electrons, it is colorless.

The Zn^{2+} ion has an electron configuration of [Ar]$3d^{10}$. In this case all of the 3*d* orbitals are filled.

There is no room in the d_{z^2} and $d_{x^2-y^2}$ orbitals to accept an electron from a lower energy d_{xy}, d_{yz}, or d_{xz} orbital. The complex is therefore colorless.

The Co^{2+} ion has an electron configuration of [Ar]$3d^7$. In this case there is room for movement of a *d* electron from the lower energy d_{xy}, d_{yz}, and d_{xz} into the higher energy d_{z^2} and $d_{x^2-y^2}$ orbitals. The complex is therefore colored.

Gemstones such as ruby and emerald owe their color to the presence of trace amounts of transition-metal ions. For example, replacement of a fraction of the aluminum in the colorless mineral corundum, Al_2O_3, produces several different gems: chromium forms ruby, manganese forms amethyst, and iron forms topaz. Sapphire, which occurs in a variety of colors but is most often blue, contains titanium and cobalt. Several other gems are produced by replacing a trace of aluminum in the colorless mineral beryl, $Be_3Al_2Si_6O_{18}$, with transition-metal ions. For example, emerald contains chromium, whereas aquamarine contains iron.

The magnitude of the energy gap, Δ, and consequently the color of a complex, depend on both the metal and the surrounding ligands. For example, $Fe(H_2O)_6^{3+}$ is yellow, $Cr(H_2O)_6^{3+}$ is violet, and $Cr(NH_3)_6^{3+}$ is yellow. Ligands can be arranged in order of their abilities to increase the the energy gap, Δ. The following is an abbreviated list of common ligands arranged in order of increasing Δ:

$$Cl^- < F^- < H_2O < NH_3 < en < NO_2^- < CN^-$$

This list is known as the **spectrochemical series.**

Ligands that lie on the low end of the spectrochemical series are termed **weak field ligands.** Those that lie on the high end are termed **strong field ligands.** Figure 23.27 shows schematically what happens to the crystal-field splitting when the ligand is varied in a series of chromium(III) complexes. (This is a good place to remind you that when a transition metal is ionized, the valence *s* electrons are removed first. Thus, the outer electron configuration for chromium is [Ar]$4s^13d^5$; that

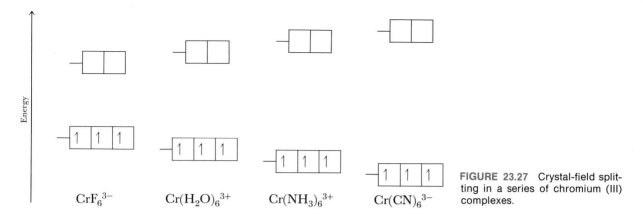

FIGURE 23.27 Crystal-field splitting in a series of chromium (III) complexes.

for Cr^{3+} is $[Ar]3d^3$.) Notice that as the field exerted by the six surrounding ligands increases, the splitting of the metal d orbitals increases. Because the absorption spectrum is related to this energy separation, these complexes vary in color.

The crystal-field model helps us understand the magnetic properties and some important chemical properties of the transition metal ions. In discussing these it is useful to use the idea of a crystal-field stabilization energy. Figure 23.28 shows a modification of Figure 23.25, in which the crystal-field splitting energy, Δ, is divided between the orbitals of lower energy and those of higher energy. An electron located in one of the lower energy orbitals is *stabilized* relative to the average energy by an amount 0.4 Δ. This is called the crystal-field stabilization energy, CFSE. On the other hand, an electron placed in one of the higher energy orbitals is *destabilized*, by an amount 0.6 Δ. It thus has a negative crystal-field stabilization energy. With the spectrochemical series in mind, let's consider the CoF_6^{3-} and $Co(CN)_6^{3-}$ ions. The F^- ion, on the low end of the spectrochemical series, is a weak-field ligand. The CN^-, on the high end of the series, is a strong-field ligand. It produces a larger energy gap than does the F^- ion. The splitting of the d orbital energies in the complexes is compared in Figure 23.29. A count of the electrons in cobalt(III) tells us that we have six electrons to place in the $3d$ orbitals. If no other factors were involved, these six electrons would go into the $3d$ orbitals one at a time, with parallel spin, until five such electrons have been added. The remaining electron would then be paired up in one of the $3d$ orbitals. The CFSE for this arrangement would be $4(0.4\ \Delta) - 2(0.6\ \Delta) = 0.4\ \Delta$. Because the two sets of $3d$ orbitals differ in energy, there is an energy

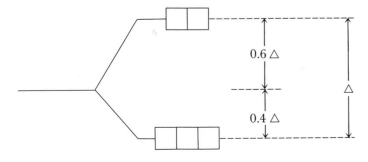

FIGURE 23.28 Crystal-field stabilization energy (CFSE) of 0.4 Δ and destabilization energy of 0.6 Δ for electrons in d orbitals split by an octahedral crystal field. Keep in mind that the actual magnitude of Δ depends on the metal ion, its charge, and on the ligands.

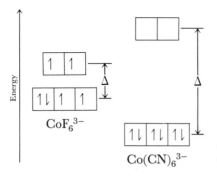

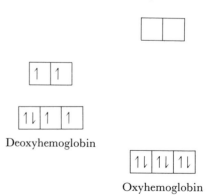

FIGURE 23.29 The population of *d* orbitals in the high-spin CoF_6^{3-} ion (small Δ) and low-spin $Co(CN)_6^{3-}$ ion (large Δ).

penalty for adding an electron to one of the higher-energy orbitals instead of pairing it up with one of those in the lower-energy set. Instead of gaining a crystal field stabilization energy of 0.4 Δ, we lose 0.6 Δ of energy relative to the average energy of the *d* electrons. We thus have two conflicting energy considerations at work. If the crystal-field splitting is not too large, the electrons follow Hund's rule (Section 6.3) and remain unpaired as much as possible. This situation occurs in CoF_6^{3-}. However, in the case of $Co(CN)_6^{3-}$, the separation between the *d* orbital energies is so large that the electrons pair up in the lower-energy set until all three of these orbitals are filled, as shown in Figure 23.29. In this case the CFSE is 6(0.4 Δ) = 2.4 Δ. In this complex the crystal-field stabilization energy more than compensates for the unfavorable spin-pairing energy. Thus, the CoF_6^{3-} complex is high spin; that is, the electrons are arranged so that they remain unpaired as much as possible. On the other hand, the electron arrangement found in the $Co(CN)_6^{3-}$ complex is termed low spin. These two different electronic arrangements can be readily distinguished by measuring the properties of the two substances in a magnetic field, as described above.

SAMPLE EXERCISE 23.10

Deoxyhemoglobin, a complex of iron(II), is blue and has four unpaired electrons. Coordination of an O_2 molecule to deoxyhemoglobin forms oxyhemoglobin, which is red and diamagnetic. Assuming that the diagram for octahedral crystal-field splitting applies, explain the relation between the color and magnetic properties in these complexes.

Solution: Deoxyhemoglobin appears blue because it absorbs orange light; this light is on the low-energy side of the visible spectrum (Figure 23.16). Consequently, the energy gap, Δ, between the *d* orbitals must not be large. In contrast, oxyhemoglobin is red because it absorbs green light; this light is on the high-energy end of the spectrum. Consequently, Δ must be larger in this case than in the case of deoxyhemoglobin.

The electron configuration of iron(II) is [Ar]$3d^6$.

The population of the *d* electrons in the two cases is shown in the following diagrams:

Deoxyhemoglobin

Oxyhemoglobin

The fact that deoxyhemoglobin is a high-spin complex, whereas oxyhemoglobin is a low-spin complex therefore correlates with their colors.

TABLE 23.3 Crystal-field stabilization energies (CFSE) of the divalent metal ions in high-spin $M(H_2O)_6^{2+}$ complexes

Ion	Number of $3d$ electrons	Electrons in lower-energy orbitals	Electrons in higher-energy orbitals	CFSE (Δ)
Ca^{2+}	0	0	0	0
Sc^{2+}	1	1	0	0.4
Ti^{2+}	2	2	0	0.8
V^{2+}	3	3	0	1.2
Cr^{2+}	4	3	1	0.6
Mn^{2+}	5	3	2	0
Fe^{2+}	6	4	2	0.4
Co^{2+}	7	5	2	0.8
Ni^{2+}	8	6	2	1.2
Cu^{2+}	9	6	3	0.6
Zn^{2+}	10	6	4	0

The idea of a crystal-field stabilization energy, CFSE, can be tested by considering the enthalpies of hydration of the divalent metal ions of the transition elements. The enthalpy of hydration, ΔH_h, of a metal ion is the heat evolved in the process

$$M^{2+}(g) + \text{water} \longrightarrow M(H_2O)_6^{2+}(aq) \qquad [23.7]$$

The heat of such a process can be estimated quite well from various thermochemical data. The $3d$ electron configurations of the divalent ions are listed in Table 23.3. Because H_2O is a relatively weak field ligand, all the hexaaqua complexes of the metal ions listed are high-spin, octahedral complexes of the form $M(H_2O)_6^{2+}$. Thus the $3d$ electrons must occupy one or the other of the two sets of orbitals that are split in energy by the crystal field of the coordinated water molecules. Table 23.3 lists the number of electrons in the lower- and higher-energy sets for each metal ion. The total CFSE for the metal complex is the sum of the CFSE terms for each electron.

If it were not for the influence of the CFSE, we would expect ΔH_h to vary in a smooth way with the metal-ion radius. The smaller the metal ion, the greater is the interaction between ligands and metal ion, and thus the more negative is ΔH_h. In passing from Ca^{2+} to Zn^{2+}, ΔH_h should become steadily more negative because of the steady decrease in ionic radius in proceeding across the transition-element series (Sections 6.7 and 7.3). Superimposed on this general trend, however, is the effect of the CFSE. Figure 23.30 shows the variation in ΔH_h as a function of atomic number for the divalent metal ions listed in Table 23.3. For the three ions that have no CFSE, Ca^{2+}, Mn^{2+}, and Zn^{2+}, the ΔH_h values lie near a smooth curve. For the other transition-metal ions, the enthalpies are larger than expected from radius considerations alone. The extent of departure from the line is related to the magnitude of the total CFSE for the complex, as listed in Table 23.3. Notice that the discrepancies are quite large, on the order of ordinary bond energies. These data provide strong support for the crystal-field model. The idea of a crystal-field stabilization energy can be used to explain other properties of transi-

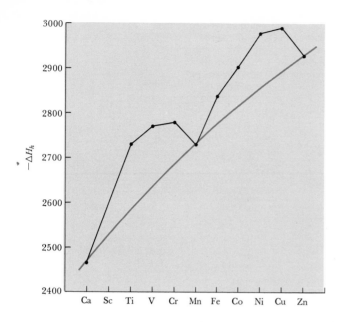

FIGURE 23.30 Enthalpies of hydration of the divalent transition metal ions. The smooth curve connecting the data for Ca^{2+}, Mn^{2+} and Zn^{2+} represents the expected variation in the absence of crystal field stabilization effects. (Sc^{2+} is not known.)

tion-metal ions, such as the lattice energies of transition-metal ionic solids and the relative stabilities of different oxidation states.

Thus far we have considered the crystal-field model only for complexes of octahedral geometry. When there are only four ligands about the metal, the geometry is tetrahedral, except for the special case of metal ions with a d^8 electron configuration, which we will discuss in a moment. The crystal-field splitting of the metal d orbitals in tetrahedral complexes differs from that in octahedral complexes. Four equivalent ligands can interact with a central metal ion most effectively by approaching along the vertices of a tetrahedron. (Figure 22.14 offers a good comparison of the octahedral and tetrahedral geometries.) It turns out, and this is not easy to explain in just a few sentences, that the splitting of the metal d orbitals in a tetrahedral crystal field is just the opposite of that for the octahedral case. That is, three of the metal d orbitals are raised in energy, and the other two are lowered, as illustrated in Figure 23.31. Because there are only four ligands instead of six as in the octahedral case, the crystal-field splitting is much smaller for tetrahedral complexes. Calculations show that for the same metal ion and ligand set, the crystal-field splitting for a tetrahedral complex is only $\frac{4}{9}$ as large as for the octahedral complex. For this reason, all tetrahedral complexes are high spin; the crystal field is never large enough to overcome the spin-pairing energies.

Square-planar complexes, in which four ligands are arranged about the metal ion in a plane, represent a common geometrical form. One can think of the square-planar complex as formed by removing two ligands from along the vertical z axis of the octahedral complex. As this happens the four ligands in the plane are drawn in more tightly. The changes that occur in the energy levels of the d orbitals are illustrated in Figure 23.32.

Square-planar complexes are characteristic of metal ions with a d^8 electron configuration. They are nearly always low spin; that is, the eight d electrons are spin-paired to form a diamagnetic complex. Such an electronic arrangement is particularly common among the heavier metals such as Pd, Pt, Ir, and Au.

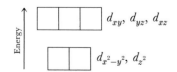

Energy →

$d_{xy},\ d_{yz},\ d_{xz}$

$d_{x^2-y^2},\ d_{z^2}$

FIGURE 23.31 The energies of the d orbitals in a tetrahedral crystal field.

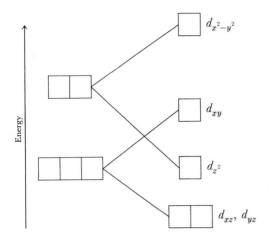

FIGURE 23.32 The effect on the relative energies of the d orbitals of removing the two negative charges from the z axis of an octahedral complex. When the charges are completely removed, the square-planar geometry results.

SAMPLE EXERCISE 23.11

Four-coordinate nickel(II) complexes exhibit both square-planar and tetrahedral geometries. The tetrahedral ones such as $NiCl_4{}^{2-}$ are paramagnetic; the square-planar ones such as $Ni(CN)_4{}^{2-}$ are diamagnetic. Show how the d electrons of nickel(II) populate the d orbitals in the appropriate crystal-field-splitting diagram in each of these cases.

Solution: Nickel(II) has an electron configuration of $[Ar]3d^8$. The population of the d electrons in the two geometries is given at right:

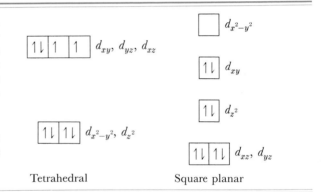

We have seen that the crystal-field model provides a basis for explaining many features of transition-metal complexes. In fact, it can be used to explain many observations in addition to those we have discussed. However, many lines of evidence show that the bonding between transition-metal ions and ligands must have some covalent character. Molecular-orbital theory (Sections 8.5 and 8.6) can also be used to describe the bonding in complexes. However, the application of molecular-orbital theory to coordination compounds is beyond the scope of our discussion. The crystal-field model, though not entirely accurate in all details, provides an adequate and useful description.

One of the simplest examples of a square-planar metal complex, *cis*-diamminedichloroplatinum(II), Figure 23.10(*a*), is a potent anticancer drug that is now used in treatment of cancer of the ovaries, testes, prostate, and other organs. When used along with other chemotherapeutic drugs, the platinum complex has led to complete, long-term remissions in 60 to 70 percent of cases treated. The anticancer activity of *cis*-Pt(NH₃)₂Cl₂ was discovered by accident by Professor Barnett Rosenberg, of Michigan State University, and his co-workers, when they noted that bacterial cell division in growth media containing platinum electrodes was inhibited near the electrodes. The causative agent proved to be *cis*-Pt(NH₃)₂Cl₂.

Summary

Coordination compounds or complexes contain metal ions bonded to several surrounding anions or molecules known as ligands. The metal ion and its ligands comprise the coordination sphere of the complex. The atom of the ligand that bonds to the metal ion is known as the donor atom. The number of donor atoms attached to the metal ion is known as the coordination number of the metal ion. The common coordination numbers are four and six; the common coordination geometries are tetrahedral, square planar, and octahedral.

If a ligand has several donor atoms that can coordinate simultaneously to the metal, it is said to be polydentate and is referred to as a chelating agent. Two common examples are ethylenediamine (en), which is potentially bidentate, and ethylenediaminetetraacetate (EDTA^{4-}), which is potentially hexadentate. Many biologically important molecules such as the porphyrins are complexes of chelating agents.

Isomerism is common among coordination compounds. Structural isomerism involves differences in the bonding arrangements of the ligands. One simple form of structural isomerism, known as linkage isomerism, occurs when a ligand is capable of coordinating to a metal through either of two donor atoms. Coordination-sphere isomerism occurs when two compounds with the same overall formula contain different ligands in the coordination sphere. Stereoisomerism involves complexes with the same chemical bonding arrangements but with differing spatial arrangements of ligands. The most common forms are geometric and optical isomerism. Geometric isomers differ from one another in the relative locations of donor atoms in the coordination sphere; the most common are cis-trans isomers. Optical isomers differ from one another in that they are non-superimposable mirror images of one another. Isomers differ from one another in their chemical and physical properties; however, optical isomers differ only in the presence of a chiral environment. Most often optical isomers are distinguished from one another by their interactions with plane-polarized light; solutions of one isomer rotate the plane of polarization to the right, whereas solutions of its mirror image rotate the plane to the left. The isomer that rotates the plane of polarization to the right is said to be dextrorotatory, whereas its isomer is levorotatory. Chiral molecules are said to be optically active. A 50–50 mixture of two optical isomers does not rotate plane-polarized light and is said to be racemic.

Many of the early studies that served as a basis for our current understanding of complexes involved complexes of chromium(III), cobalt(III), platinum(II), and platinum(IV). Complexes of these metal ions are inert, in that they undergo ligand exchange at a slow rate. Complexes undergoing rapid exchange are said to be labile.

Studies of the magnetic properties and colors of transition-metal complexes have played an important role in formulation of bonding theories for these compounds. The crystal-field theory successfully accounts for many properties of coordination compounds. In this model the interaction between metal ion and ligand is viewed as electrostatic. The ligands produce an electric field that causes a splitting in the energies of the metal d orbitals. In the spectrochemical series the ligands are listed in order of their ability to split the d orbital energies in octahedral complexes.

Electrons in the lower-energy d orbitals experience a stabilization relative to the average energy of the d orbitals, called the crystal-field stabilization energy. In strong-field complexes the splitting of d orbital energies is large enough to overcome spin-pairing energies, and d electrons preferentially pair up in the lower-energy orbitals. Such complexes are referred to as low spin. When the ligands exert a relatively weak crystal field the electrons occupy the higher-energy d orbitals in preference to pairing up in the lower-energy set, and the complexes are high spin.

The crystal-field model is applicable also to tetrahedral and square-planar complexes. However, the ordering of the d orbital energies is different than in octahedral complexes.

Learning goals

Having read and studied this chapter, you should be able to:

1 Determine either the charge of a complex ion, having been given the oxidation state of the metal, or the oxidation state, having been given the charge of the complex. (You will need to recognize the common ligands and their charges.)

2 Describe, with the aid of drawings, the common geometries of complexes. (You will need to recognize whether the common ligands are functioning as monodentate or polydentate ligands.)

3 Name coordination compounds, having been given their formulas, or write their formulas, having been given their names.

4 Describe the common types of isomerism and distinguish between structural isomerism and stereoisomerism.

5 Determine the possible number of stereoisomers for a complex, having been given its composition.

6 Distinguish between inert and labile complexes.

7 Explain how the conductivity, precipitation reactions, and isomerism of complexes are used to infer their structures.

8 Explain how the magnetic properties of a compound can be measured and used to infer the number of unpaired electrons.

9 Explain how the colors of substances are related to their absorption and reflection of incident light.

10 Explain how the electrostatic interaction between ligands and metal d orbitals in an octahedral complex results in a splitting of energy levels.

11 Explain the significance of the spectrochemical series.

12 Explain what is meant by the term crystal-field stabilization energy.

13 Account for the tendency of electrons to pair in strong-field, low-spin complexes.

14 Account for the larger-than-expected enthalpies of hydration among the divalent metal ions of the transition metals.

15 Sketch a representation of the d orbital energy levels in a tetrahedral complex and explain the reason for a smaller crystal-field splitting in this geometry as compared with octahedral complexes.

16 Sketch the d orbital energy levels in a square-planar complex.

Key terms

Among the more important terms and expressions used for the first time in this chapter are the following:

A **chelating agent** (Section 23.3) is a polydentate ligand that is capable of occupying two or more sites in the coordination sphere.

Chiral (Section 23.4) means having a nonsuperimposable mirror image; for example, we might refer to a molecule as being chiral.

A *cis* geometrical arrangement (Section 23.4) refers to one with like groups adjacent to each other.

The **coordination number** (Section 23.1) of an atom is the number of adjacent atoms to which it is directly bonded; in a complex, the coordination number of the metal ion is the number of donor atoms to which it is bonded.

The **coordination sphere** (Section 23.1) of a complex is the volume enclosing the metal ion and its surrounding ligands.

Coordination-sphere isomers (Section 23.4) are structural isomers of coordination compounds in which the ligands within the coordination sphere differ.

Crystal-field stabilization energy (Section 23.8) is the stabilization, as compared with the average d orbital energy, that results when an electron is placed in a lower-energy d orbital in an octahedral complex.

The **crystal-field theory** (Section 23.8) accounts for the colors and magnetic and other properties of transition-metal complexes in terms of the splitting of the energies of metal-ion d orbitals by the electrostatic interaction with the ligands.

The term **dextrorotatory**, or merely **dextro** or d (Section 23.4), is used to label a chiral molecule that rotates the plane of polarization of plane-polarized light to the right (clockwise).

A **diamagnetic substance** (Section 23.7) has no unpaired electrons and is therefore weakly repelled by a magnetic field.

The **donor atom** (Section 23.1) of a ligand is the one that bonds to the metal.

Geometric isomers (Section 23.4) have different spatial arrangements of donor atoms around the coordination sphere.

A **high-spin** complex (Section 23.8) has the same number of unpaired electrons as does the isolated metal ion.

An **inert** complex (Section 23.5) exchanges ligands at a slow rate.

Isomers (Section 23.4) are compounds whose molecules have the same overall composition but different structures.

A **labile** complex (Section 23.3) exchanges ligands at a rapid rate.

The term **levorotatory**, or merely **levo** or l (Section 23.4), is used to label a chiral molecule that rotates the plane of polarization of plane-polarized light to the left (counterclockwise).

A **ligand** (Section 23.1) is an ion or molecule that coordinates to a single central metal atom or to an ion to form a complex.

Linkage isomers (Section 23.4) are structural isomers of coordination compounds in which a ligand differs in its mode of attachment to a metal ion.

A **low-spin** complex (Section 23.8) has fewer unpaired electrons than does the isolated metal ion.

A **monodentate ligand** (Section 23.2) is one that binds to the metal ion via a single donor atom. It occupies one position in the coordination sphere.

Optical isomers (Section 23.4) are stereoisomers in which the two forms of the compound are non-superimposable mirror images.

A **paramagnetic substance** (Section 23.7) is one that is attracted into a magnetic field; paramagnetic compounds contain unpaired electrons.

A polydentate ligand (Section 23.2) is one in which two or more donor atoms can coordinate to the same metal ion. In a bidentate ligand the number of coordinating atoms bound to the metal is two.

The spectrochemical series (Section 23.8) is a list of ligands arranged in order of their abilities to split the d orbital energies (using the terminology of the crystal-field model).

Stereoisomers (Section 23.4) are compounds possessing the same formula and bonding arrangement but differing in the spatial arrangements of the atoms.

Structural isomers (Section 23.4) are compounds possessing the same formula but differing in the bonding arrangements of the atoms.

The *trans* geometrical arrangement (Section 23.4) refers to one with like groups opposite each other.

EXERCISES

Structure and nomenclature

23.1 Provide a brief definition or description of each of the following terms: (a) coordination sphere; (b) ligand; (c) coordination number; (d) tridentate ligand.

23.2 Indicate the coordination number about the central metal in each of the following compounds: (a) $[Zn(NH_3)_4]Cl_2$; (b) $[Co(NH_3)_3Cl_3]$; (c) $[Cr(en)_2Cl_2]^+$; (d) $K_2[FeCl_4]$; (e) $Na_2[TaF_7]$.

23.3 Choosing from among the metallic ions Cd^{2+}, Pt^{2+}, and Co^{3+}, and using NH_3 and Br^- as ligands, sketch two examples each of different metal complexes, not isomers, with tetrahedral, square-planar, and octahedral geometry.

23.4 Indicate the oxidation number of the central metal in each of the following coordination compounds: (a) $K_3[Fe(CN)_6]$; (b) $Na_3[Cr(C_2O_4)_3]$; (c) $[Pt(NH_3)_4Cl_2]Cl_2$; (d) $[Cr(NH_3)_4Br_2]Br$.

23.5 Write the formulas for all the possible six-coordinate complexes of nickel(II) containing H_2O and/or ethylenediamine (en) as ligands.

23.6 Sketch the structure of each of the following: (a) $[Zn(H_2O)_4]^{2+}$; (b) $[AuCl_4]^-$; (c) *cis*-$[PtH(Cl)(PH_3)_2]$; (d) *cis*-$[Rh(en)_2(CN)_2]^+$; (e) $[PdCl_2(C_2O_4)]^{2-}$.

23.7 Sketch the structures of each of the following complexes: (a) *trans*-$[Cr(NH_3)_4Cl_2]^+$; (b) $[Co(C_2O_4)_3]^{3-}$; (c) $[Cr(C_2O_4)Br_4]^-$; (d) *cis*-$[Pt(en)_2(CN)_2]^{2+}$.

23.8 Name each of the coordination compounds or complex ions listed in problems 23.6 and 23.7.

23.9 Name each of the following compounds: (a) $Cs_2[Ni(CN)_4]$; (b) $[Pt(NH_3)_4][Co(NH_3)_2Cl_4]_2$; (c) $Na_2[Zn(gly)_2]$ (gly represents the anion of the amino acid glycine, $H_2NCH_2COO^-$); (d) $[Fe(H_2O)_4SO_4]Cl$; (e) $K_2[OsCl_5(ONO)]$; (f) $[Co(en)(NH_3)(OH)_3]$.

23.10 Write the formulas for each of the following compounds, being sure to use brackets as appropriate to indicate the coordination sphere: (a) tetraamminenickel(II) perchlorate; (b) sodium hexanitritocobaltate(III); (c) diaquabis(ethylenediamine)chromium(III) nitrate; (d) chlorobis(ethylenediamine)thiocyanatocobalt(III)tetrachlorocadmate(II); (e) potassium hexacyanonickelate(III); (f) sulfatobis(ethylenediamine)iron(II); (g) pentaammineazidocobalt(III)sulfate.

23.11 Using Br^- and NH_3 as ligands, (a) give the formula of a six-coordinate palladium(IV) complex that would be a nonelectrolyte in aqueous solution; (b) give the coordination compound of Cr(III) that has about the same solution electrical conductivity as KBr; (c) give an octahedral complex of V(III) containing four NH_3 groups.

23.12 Polydentate ligands can vary in the number of coordination positions they occupy. In each of the following, indicate the most probable number of coordination positions occupied by the polydentate ligand present: (a) $[Cr(NH_3)_4SO_4]Cl$; (b) $[Cr(EDTA)(H_2O)_2]^-$; (c) $[Zn(en)_2](NO_3)_2$; (d) $[Co(NH_3)_4CO_3]ClO_4$.

23.13 From the following list of metal complexes and simple salts, pair up each complex with a simple salt that is likely to have about the same electrical conductivity in solution: (a) $Pt(NH_3)_3Cl_4$; (b) $Co(NH_3)_6Cl_3$; (c) K_2PtCl_6; (d) KBr; (e) $Ca(NO_3)_2$; (f) $ScBr_3$.

Isomerism

23.14 Two forms of a platinum complex having identical chemical formulas have different colors and different solubilities in various solvents. Neither substance forms an electrically conducting solution when dissolved in water. On the basis of this limited information, what types of isomerism are possibly involved? Which types are definitely excluded by the data?

23.15 By writing formulas or drawing structures related to any one of the following complexes, illustrate (a) geometrical isomerism; (b) optical isomerism; (c) linkage isomerism; (d) coordination sphere isomerism. The complexes are: $[Co(NH_3)_4CO_3]C_2O_4$; $[Pt(NH_3)_2(SCN)_2]$; $[Cr(C_2O_4)_3]^{3-}$.

23.16 Sketch all possible isomeric structures (there might be only one) for each of the following complex ions or coordination compounds: (a) $[Zn(en)_2]^{2+}$; (b) $[Cr(C_2O_4)(NH_3)_2Cl_2]^-$; (c) $[PtCl_4(SCN)_2]^{2-}$; (d) $[OsCl_3(NH_3)_3]$.

[23.17] Diethylenetriamine,

$$H_2NCH_2CH_2\overset{..}{N}HCH_2CH_2\overset{..}{N}H_2$$

(dien), can act as a tridentate ligand. Indicate all the possible isomeric forms and the types of isomerism involved for each of the following coordination compounds or complex ions: (a) $[Zn(dien)Cl]^+$; (b) $[Co(dien)_2]^{3+}$; (c) $[Cr(dien)Cl_3]$; (d) $[Cr(dien)Cl_2(NO_2)]$.

23.18 Sketch the stereoisomers of $Pt(en)(NO_2)_2Br_2$.

23.19 The compound $Co(NH_3)_5(SO_4)Br$ exists in two forms, one red and one violet. Both forms dissociate in solution to form two ions. Solutions of the red compound form a precipitate of AgBr on addition of $AgNO_3$ solution, but no precipitate of $BaSO_4$ on addition of $BaCl_2$ solution. For the violet compound just the reverse occurs. From this evidence indicate the structures of the complex ions in each case and give the correct name of each compound.

23.20 Write the formulas for, and properly name, two possible coordination-sphere isomers with formula $Co(NH_3)_4Cl_2(OH)$.

[23.21] Write formulas showing the coordination spheres, and give the correct names, for two different substances with formula $PtCu(NH_3)_4Cl_4$.

Color; magnetism; crystal-field theory

23.22 What is the observed color of a coordination compound that absorbs radiation of 530 nm wavelength?

23.23 Give the number of outer d electrons associated with the central metal in each of the following complexes: (a) $[PtCl_4]^{2-}$; (b) $[Cu(CN)_4]^-$; (c) $[Cr(NH_3)_3Cl_3]$; (d) $[V(H_2O)_3(OH)_3]^+$.

23.24 Three isomeric complexes are found to be colored red, green, and yellow, respectively. Assuming that each complex has just one major absorption band in the visible region of the spectrum, which complex absorbs at highest energy? Which at lowest?

23.25 What characteristic of the d orbitals is responsible for their splitting into two groups in an octahedral field of ligands?

23.26 For each of the following metals, write the electronic configurations of the metal atom and the $3+$ ion; draw the crystal-field energy-level diagram for the d orbitals of an octahedral complex and show the placement of the d electrons in each case, assuming a strong-field complex. (a) Cr; (b) Ru; (c) Ni.

23.27 Classify the following complexes as either high spin or low spin: (a) $[Mn(NH_3)_6]^{3+}$ (two unpaired electrons); (b) $[Rh(CN)_6]^{3+}$ (no unpaired electrons); (c) $[Co(C_2O_4)_3]^{4-}$ (three unpaired electrons).

23.28 The value of Δ for the $Fe(H_2O)_6^{2+}$ complex is 120 kJ/mol. Calculate the expected wavelength of the absorption corresponding to promotion of an electron from the lower-energy to the higher-energy d orbital set in this complex. Should the complex absorb in the visible range? (Hint: You may need to review Sample Exercise 5.2; remember to divide by Avogadro's number.)

23.29 How would you go about determining whether there are unpaired electrons in the compound $K_2Ni(SCN)_4$? If the compound proved to have two unpaired electrons per molecule, how would you account for this in terms of crystal-field theory?

23.30 In each of the following pairs of complexes, indicate which you would expect to absorb light at higher energy: (a) $[CoF_6]^{4-}$, $[Co(NH_3)_6]^{2+}$; (b) $[FeCl_4]^-$, $[FeCl_4]^{2-}$; (c) $[V(H_2O)_6]^{3+}$, $[V(NO_2)_6]^{3-}$.

23.31 Draw the crystal-field energy-level diagrams and show the placement of electrons for each of the following

complexes: (a) $[ZrCl_6]^{4-}$; (b) $[CoF_6]^{3-}$ (a high-spin complex); (c) $[Mn(H_2O)_6]^{3+}$ (a high-spin complex); (d) $[CoCl_4]^{2-}$; (e) $[OsCl_6]^{2-}$; (f) $[Pd(CN)_4]^{2-}$.

23.32 The formation constants for complexes (Section 16.5) often provide interesting evidence of the relative stabilities of metal complexes. In each of the following comparisons, indicate a possible origin of the relative values of K_f, in terms of the electronic configuration or charge on the metal ion. (a) $[Co(en)_3]^{3+}$, $K_f = 5 \times 10^{48}$, and $[Co(en)_3]^{2+}$, $K_f = 7 \times 10^{13}$; (b) $[Fe(CN)_6]^{3-}$, $K_f = 7.7 \times 10^{43}$, and $[Fe(CN)_6]^{4-}$, $K_f = 7.7 \times 10^{36}$; (c) $[Fe(en)_3]^{2+}$, $K_f = 5 \times 10^9$, and $[Co(en)_3]^{3+}$, $K_f = 5 \times 10^{48}$.

[23.33] Suppose that a transition-metal ion were located in a lattice in which it was in contact with just two nearby anions, located along the $+$ and $-$ axis directions. Diagram the splitting of the metal d orbitals that would result from such a crystal field. Assuming a strong field, how many unpaired electrons would you expect for a metal ion with six d electrons?

Additional exercises

23.34 Distinguish between the following terms: (a) structural isomer and stereoisomer; (b) labile and inert complex; (c) high-spin and low-spin complex; (d) chiral complex and *cis*-isomer; (e) spectrochemical series and crystal-field stabilization energy.

23.35 Based on the molar conductance values listed below for the series of platinum(IV) complexes, write the formula for each complex so as to show which ligands are in the coordination sphere of the metal.

Complex	Molar conductance $(ohm^{-1})^a$ of 0.05 M solution
$Pt(NH_3)_6Cl_4$	523
$Pt(NH_3)_4Cl_4$	228
$Pt(NH_3)_3Cl_4$	97
$Pt(NH_3)_2Cl_4$	0
$KPt(NH_3)Cl_5$	108

^{a}The ohm is the unit of resistance; conductance is the inverse of resistance.

23.36 In Werner's early studies, he observed that when the complex $[Co(NH_3)_4Br_2]Br$ was placed in water, the electrical conductivity of a 0.05 M solution changed from an initial value of 191 ohm^{-1} to a final value of 374 ohm^{-1} over a period of an hour or so. Suggest an explanation of the observed results. Write a balanced chemical equation to describe the reaction.

23.37 The SCN^- ligand is capable of linkage isomerism. When it is S-bonded it is termed thiocyanato; when it is N-bonded it is termed isothiocyanato. Draw the structures for, and name, *all* the possible isomers of diamminedithiocyanatoplatinum(II).

23.38 Solutions containing the $[Co(H_2O)_6]^{2+}$ ion absorb at about 520 nm; those containing the $[CoCl_4]^{2-}$ ion absorb at about 690 nm. What colors do you expect for the solutions? Why is the absorption maximum for the $[CoCl_4]^{2-}$ solution at a longer wavelength than for the $[Co(H_2O)_6]^{2+}$ ion?

23.39 Draw out the d orbital energy-level diagram for each of the following complexes and indicate the most likely placement of d electrons: (a) $[AuCl_4]^-$; (b) $[V(C_2O_4)_3]^{2-}$; (c) $[NiBr_4]^{2-}$ (tetrahedral); (d) $[Fe(CN)_6]^{3-}$.

23.40 The complex ion $[Fe(NH_3)_6]^{2+}$ contains four unpaired electrons, whereas $[Co(NH_3)_6]^{3+}$ contains none. Account for this difference in terms of the crystal-field model.

23.41 Acetylacetonate anion forms very stable complexes with many metallic ions, in which it acts as a bidentate ligand. Suppose that one of the CH_3 groups on the ligand is replaced by a CF_3 group, as shown below.

$$\left[\begin{array}{c} \overset{\displaystyle H}{\underset{\displaystyle |}{C}} \\ CF_3-C\overset{\displaystyle \diagup \quad \diagdown}{}C-CH_3 \\ :\ddot{O}: \quad :\ddot{O}: \end{array} \right]^-$$

Trifluoromethyl acetylacetonate
(tfac)

Sketch all the possible isomers for the tris complex of tfac with cobalt(III).

23.42 What changes would you expect in the absorption spectrum of octahedral V(III) complexes as the ligand is varied in the order H_2O, NH_3, CN^-?

[23.43] Write balanced chemical equations to represent the following observations. (In some instances, the complex involved has been discussed previously at some point in the text.) (a) Solid silver bromide dissolves in an excess of aqueous sodium thiosulfate solution. (b) The green complex $[Cr(en)_2Cl_2]Cl$ on treatment with water over a long time converts to a brown-orange complex. Reaction of $AgNO_3$ with a molar solution of the product results in precipitation of 3 mol of AgCl. (Write *two* reactions.) (c) Insoluble zinc hydroxide dissolves in excess aqueous ammonia. (d) A pink solution of $Co(NO_3)_2$ turns deep blue on addition of concentrated hydrochloric acid.

23.44 The absorption maxima in the visible-absorption spectra of several complexes are as follows: (a) $Co(NH_3)_6{}^{3+}$, 470 nm; (b) *trans*-$Co(NH_3)_4(NO_2)_2{}^+$, 440 nm; (c) *cis*-$Co(NH_3)_4(H_2O)_2{}^{3+}$, 510 nm; (d) *cis*-$Co(en)_2Cl_2{}^+$, 535 nm. Predict the color of each complex.

23.45 The anion of the amino acid glycine,

$$H_2NCH_2\overset{\displaystyle O}{\overset{\displaystyle \|}{C}}\!-\!O^-$$

symbol gly, is capable of acting as a bidentate ligand, coordinating to the metal through the nitrogen and $-O^-$ atoms. How many isomers are possible for: (a) $[Zn(gly)_2]$ (tetrahedral); (b) $[Pt(gly)_2]$ (square planar); (c) $[Co(gly)_3]$. Sketch all the possible isomers. Use $N\frown O$ to represent the ligand.

[23.46] In many metal-oxide minerals, transition-metal ions may be located in either the octahedral or tetrahedral holes in the oxide close-packed lattice (see Section 22.5, especially Figure 22.14). Electronic as well as radius-ratio factors may determine which site is occupied. Recognizing that the crystal-field splitting is only about half as large in a tetrahedral hole as compared with an octahedral hole and taking account of the crystal-field stabilization energy, for which of the following ions would the preference for an octahedral site be greatest: Mn^{2+}; Cr^{3+}; Co^{2+}? Explain your answer.

23.47 From the following values of Δ for each complex, calculate the total CFSE: (a) $[CrF_6]^{3-}$, 182 kJ/mol; (b) $[Cr(NH_3)_6]^{3+}$, 258 kJ/mol; (c) $[MoCl_6]^{3-}$, 230 kJ/mol; (d) $[Rh(CN)_6]^{3-}$, 545 kJ/mol.

[23.48] The red color of ruby is due to the presence of Cr(III) ions in octahedral sites in the close-packed oxide lattice. Draw the crystal-field-splitting diagram for Cr(III) in this environment. Suppose the ruby crystal is subjected to high pressure. What do you predict for the variation in the wavelength of absorption of the ruby as a function of pressure? Explain.

23.49 A palladium complex formed from a solution containing bromide ion and pyridine, C_5H_5N (a good donor ligand toward metal ions), is found on elemental analysis to contain 37.6 percent bromine, 28.3 percent carbon, 6.60 percent nitrogen, and 2.37 percent hydrogen. The compound is slightly soluble in several organic solvents; its solutions in water or alcohol do not conduct electricity. It is found experimentally to have a zero dipole moment. Write the chemical formula for the complex and indicate its probable structure.

23.50 Experimental evidence indicates that the ionic radius of the iron(II) ion is smaller when iron is in the low-spin state than when it is in the high-spin state. Suggest a reason why this should be so.

24

Organic compounds

Organic chemistry deals mainly with the chemical characteristics of compounds in which carbon is a principal element. There are no precise boundaries between organic chemistry and other areas of chemical science. The concept of a separate area of chemistry that could be thought of as organic developed out of the vitalist theory, which held that substances that make up living matter are fundamentally different from those that form inanimate matter. In 1828, Friedrich Wöhler, a German chemist, reacted potassium cyanate, $KOCN$, with ammonium chloride, NH_4Cl; much to his surprise he obtained urea, H_2NCONH_2, a well-known substance that had been isolated from the urine of mammals. Following Wöhler, many other organic substances were prepared from inorganic starting materials, and the vitalist theory gradually disappeared. Nevertheless organic chemistry continued to develop as a distinct area of chemistry. This may be explained partly by the fact that the raw materials for much of organic chemistry—oil, coal, wood, animal matter, and so forth—are of plant or animal origin. Secondly, the chemical characteristics of organic substances could be classified and systematized. The "rules" for how organic substances behave seemed to fit into a pattern that made for a classification system that was separate from that used for inorganic materials. Even though the large body of orderly chemical information and theory called organic chemistry is treated as a separate topic, we should not forget that the same laws of thermodynamics, bonding rules, and so forth govern both inorganic and organic chemical processes.

In this chapter, we shall present a brief view of some of the elementary aspects of organic chemistry. We can do no more than hint at the magnitude of the subject. It has been estimated that there are now more than a million known organic substances. Each year several thousand new organic substances are discovered in nature or synthesized in the laboratory. These huge numbers might lead one to think that learning the subject of organic chemistry is hopelessly difficult. However, certain arrangements of atoms and groups of atoms, called functional groups, occur repeatedly in organic substances. These arrangements lead to particular chemical characteristics that are very similar among the compounds containing that functional group. Thus, by learning the characteristic chemical properties of these functional groups, you can understand the chemical characteristics of many organic substances.

24.1 The hydrocarbons

Hydrocarbons contain only two elements, carbon and hydrogen. With such a limited range of composition, you might suppose that there would be little variety in the chemical properties of the hydrocarbons. However, such is not the case. The key structural feature of hydrocarbons, and for that matter of most other organic substances, is the presence of stable carbon-carbon bonds. Carbon alone among the elements is able to form stable, extended chains of atoms bonded through single, double, or triple bonds. No other element is capable of forming similar structures.

The hydrocarbons can be divided into four groups, the **alkanes, alkenes, alkynes,** and **aromatic hydrocarbons.** We have already encountered at least one member from each of these series. Figure 24.1 shows the

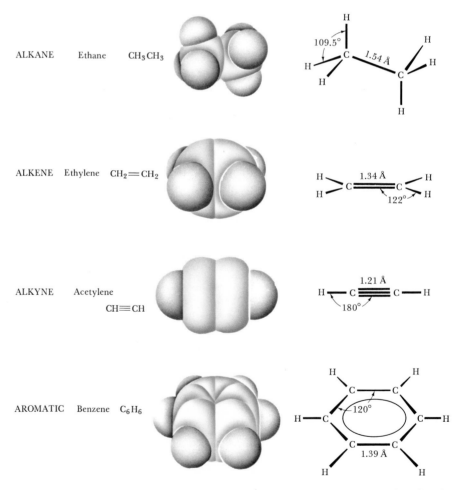

ALKANE   Ethane CH_3CH_3

ALKENE   Ethylene $CH_2{=}CH_2$

ALKYNE   Acetylene

                  $CH{\equiv}CH$

AROMATIC   Benzene C_6H_6

FIGURE 24.1 Names, geometrical structures, and molecular formulas of examples of each type of hydrocarbon.

name, molecular formula, and geometrical structure of the simplest member that contains a carbon-carbon bond in each series.

The alkanes consist of carbon atoms bonded either to hydrogen or to other carbon atoms by four single bonds. Depending on its place in the alkane structure, a carbon atom may be bonded to three hydrogens and one carbon, two hydrogens and two carbons, a hydrogen and three carbons, or four carbons. The alkenes are hydrocarbons with one or more carbon-carbon double bonds. The simplest member of the alkene series is ethylene; you might wish to review the electronic structure of ethylene, discussed in Section 8.4. In the alkynes, there is at least one carbon-carbon triple bond, as in the simplest member of the series, acetylene (Section 8.4). In the aromatic hydrocarbons the carbon atoms are connected in a planar ring structure, joined by both σ and π bonds between carbon atoms. Benzene is the best-known example of an aromatic hydrocarbon. Other examples are illustrated in Figure 8.15. The nonaromatic hydrocarbons, that is, the alkanes, alkenes, and alkynes, are referred to as aliphatic compounds to distinguish them from aromatic substances.

The members of the different series of hydrocarbons exhibit different chemical behaviors, as we shall see shortly. However, the hydrocarbons are very similar in many ways. Because carbon and hydrogen are not greatly different in electronegativity (2.5 for carbon, 2.2 for hydrogen), the C—H bond is not very polar. Hydrocarbons are formed entirely from C—H bonds and bonds between carbon atoms; this means that hydrocarbon molecules are relatively nonpolar. Thus, they are very much unlike water; hydrocarbons are almost completely insoluble in water. Those that are liquids are good solvents toward nonpolar molecules, but poor solvents toward ionic substances, such as sodium chloride, or polar substances, such as NH_3.

THE ALKANES

Table 24.1 lists several of the simplest alkanes. Many of these substances are familiar because of their widespread use. Methane is a major component of natural gas and is used for home heating and in gas stoves and hot-water heaters. Propane is the major component of bottled, or LP, gas used for home heating, cooking, and so forth in areas where natural gas is not available. Butane is used in the disposable lighters sold in drug stores

TABLE 24.1   First several members of the straight-chain alkane series

Molecular formula	Condensed structural formula	Name	Boiling point (°C)
CH_4	CH_4	Methane	−161
C_2H_6	CH_3CH_3	Ethane	−89
C_3H_8	$CH_3CH_2CH_3$	Propane	−44
C_4H_{10}	$CH_3CH_2CH_2CH_3$	Butane	−0.5
C_5H_{12}	$CH_3CH_2CH_2CH_2CH_3$	Pentane	36
C_6H_{14}	$CH_3CH_2CH_2CH_2CH_2CH_3$	Hexane	68
C_7H_{16}	$CH_3CH_2CH_2CH_2CH_2CH_2CH_3$	Heptane	98
C_8H_{18}	$CH_3CH_2CH_2CH_2CH_2CH_2CH_2CH_3$	Octane	125
C_9H_{20}	$CH_3CH_2CH_2CH_2CH_2CH_2CH_2CH_2CH_3$	Nonane	151
$C_{10}H_{22}$	$CH_3CH_2CH_2CH_2CH_2CH_2CH_2CH_2CH_2CH_3$	Decane	174

and in the fuel cannisters for gas camping stoves and lanterns. Alkanes with from five to twelve carbon atoms are found in gasoline.

The formulas for the alkanes given in Table 24.1 are written in a notation called the condensed structural formula. This notation reveals the way in which atoms are bonded to one another, but does not require drawing in all the bonds. For example, the Lewis structure and condensed structural formula for butane, C_4H_{10}, are:

$$H-\overset{\displaystyle H}{\underset{\displaystyle H}{\overset{|}{\underset{|}{C}}}}-\overset{\displaystyle H}{\underset{\displaystyle H}{\overset{|}{\underset{|}{C}}}}-\overset{\displaystyle H}{\underset{\displaystyle H}{\overset{|}{\underset{|}{C}}}}-\overset{\displaystyle H}{\underset{\displaystyle H}{\overset{|}{\underset{|}{C}}}}-H \qquad\qquad CH_3CH_2CH_2CH_3$$

Lewis structure   Condensed structural formula

We shall frequently use either Lewis structures or condensed structural formulas to represent organic compounds. You should practice drawing the structural formulas from the condensed ones. To aid you in this task, notice that each carbon atom in an alkane has four single bonds, while each hydrogen atom forms one single bond.

Notice that each succeeding compound in the series listed in Table 24.1 is related to the one before it by the addition of a CH_2 unit. A series such as that shown in Table 24.1 is known as a homologous series. The general formula for all the compounds listed in the table is C_nH_{2n+2}, where n is the number of carbon atoms. One of the characteristics of a homologous series is that all the compounds of the series can be described by the same general formula. We shall see several other examples of homologous series as we proceed.

The alkanes listed in Table 24.1 are called straight-chain alkanes, because all the carbon atoms are joined in a continuous chain. However, for alkanes consisting of four or more carbon atoms, other arrangements of the carbon atoms, consisting of branched chains, are possible. Figure 24.2 shows the Lewis structures and condensed structural formulas for all the possible structures of alkanes containing four or five carbon atoms. Notice that the two possible forms of butane have the same molecular formula, C_4H_{10}. Similarly, the three possible forms of pentane have the same molecular formula, C_5H_{12}. Compounds with the same molecular formula, but with different structures, are called isomers. The isomers of a given alkane differ slightly from one another in physical properties. By way of illustration, the melting and boiling points (°C) of the isomers of butane and pentane are given in Figure 24.2. The number of possible isomers increases rapidly with the number of carbon atoms in the alkane. For example, there are 18 possible isomers of octane, C_8H_{18}, and 75 possible isomers of decane, $C_{10}H_{22}$.

NOMENCLATURE OF ALKANES

The first names given to the structural isomers shown in Figure 24.2 are the so-called common names. The straight-chain isomer is referred to as the normal isomer, abbreviated by the prefix n-. The isomer in which one CH_3 group is branched off the major chain is labeled the iso-isomer; for example, isobutane. However, as the number of isomers grows, it becomes impossible to find a suitable prefix to denote each isomer. The

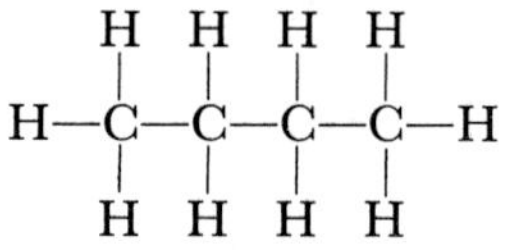

$CH_3CH_2CH_2CH_3$
n-Butane
(butane)
m.p. −135°C
b.p. −0.5°C

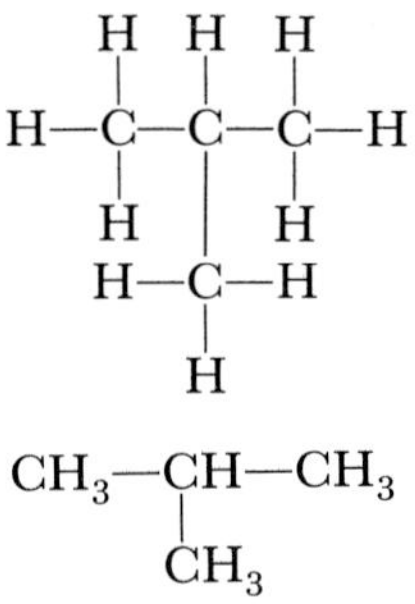

Isobutane
(2-methylpropane)
m.p. −145°C
b.p. −10°C

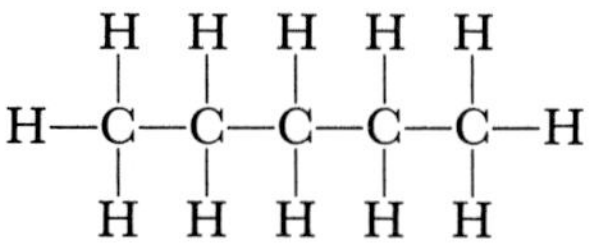

$CH_3CH_2CH_2CH_2CH_3$
n-Pentane
(pentane)
m.p. −130°C
b.p. +36°C

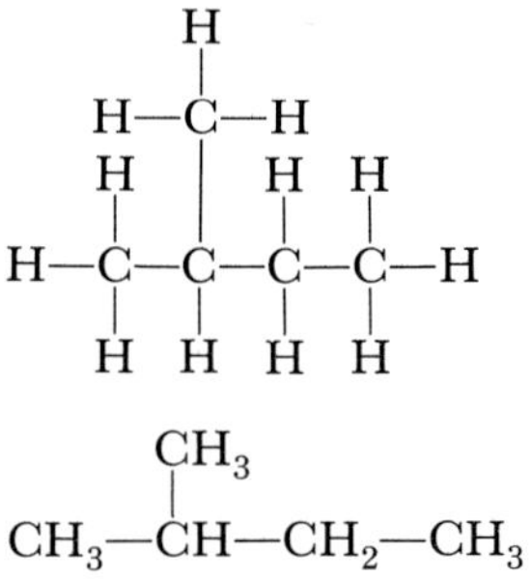

Isopentane
(2-methylbutane)
m.p. −160°C
b.p. +28°C

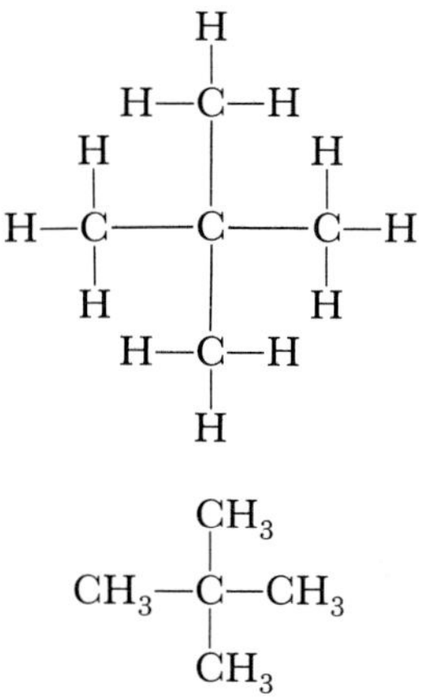

Neopentane
(2,2-dimethylpropane)
m.p. −20°C
b.p. +9°C

FIGURE 24.2 Possible structures, names, and melting and boiling points of alkanes of formula C_4H_{10} and C_5H_{12}.

need for a systematic means of naming organic compounds was recognized early in the history of organic chemistry. In 1892 an organization called the International Union of Chemistry met in Geneva, Switzerland, to formulate rules for systematic naming of organic substances. Since that time the task of keeping the rules for naming compounds up to date has fallen to the International Union of Pure and Applied Chemistry (IUPAC). It is interesting to note that through two devastating world wars and major social upheavals, the work of IUPAC has continued. Chemists everywhere, regardless of their nationality or political affiliation, subscribe to a common system for naming compounds.

The IUPAC names for the isomers of butane and pentane are the second ones given for each compound in Figure 24.2. The following rules summarize the procedures used to arrive at these names. We shall see that a similar approach is taken to write the names for other organic compounds.

1 Each compound is named for the longest continuous chain of carbon atoms present. For example, the longest chain of carbon atoms in isobutane is three (Figure 24.2). Consequently, this compound is

TABLE 24.2 Names and condensed structural formulas for several alkyl groups

Group	Name
CH_3-	Methyl
CH_3CH_2-	Ethyl
$CH_3CH_2CH_2-$	n-Propyl
$CH_3CH_2CH_2CH_2-$	n-Butyl
$\begin{array}{c} CH_3 \\ \mid \\ HC- \\ \mid \\ CH_3 \end{array}$	Isopropyl
$\begin{array}{c} CH_3 \\ \mid \\ CH_3-C- \\ \mid \\ CH3 \end{array}$	t-Butyl

named as a derivative of propane, which has three carbon atoms; in the IUPAC system it is called 2-methylpropane.

2 A CH_3 group that branches off the main hydrocarbon chain is called a methyl group. In general, a group that is formed by removing a hydrogen atom from an alkane is called an **alkyl group.** The names for alkyl groups are derived by dropping the *-ane* ending from the name of the parent alkane and adding *-yl.* For example, the methyl group, CH_3, is derived from methane, CH_4; likewise, the ethyl group, C_2H_5, is derived from ethane, C_2H_6. Table 24.2 lists several of the more common alkyl groups.

3 The location of an alkyl group along a carbon-atom chain is indicated by numbering the carbon atoms along the chain. Thus the name 2-methylpropane indicates the presence of a methyl (CH_3) group on the second carbon atom of a propane (three-carbon) chain. In general, the chain is numbered from the end that gives the lowest numbers for the alkyl positions.

4 If there is more than one substituent group of a certain type along the chain, the number of groups of that type is indicated by a prefix: *di-* (two), *tri-* (three), *tetra-* (four), *penta-* (five), and so forth. Thus the IUPAC name for neopentane (Figure 24.2) is 2,2-dimethylpropane. Dimethyl indicates the presence of two methyl groups; the 2,2- prefix indicates that both are on the second carbon atom of the propane chain.

SAMPLE EXERCISE 24.1

Name the following alkane:

$$CH_3-CH-CH_3$$
$$CH_3-CH-CH_2$$
$$\qquad\qquad\quad CH_3$$

Solution: To name this compound properly, you must first find the longest continuous chain of carbon atoms. This chain, extending from the upper left CH_3 group to the lower right CH_3 group, is five carbon atoms long:

$$\overset{①}{CH_3}-\overset{②}{CH}-CH_3$$
$$CH_3-\overset{③}{CH}-\overset{④}{CH_2}$$
$$\qquad\qquad\qquad \overset{⑤}{CH_3}$$

The compound is thus named as a derivative of pentane. We could number the carbon atoms starting from either end. However, the IUPAC rules state that the numbering should be done so that the numbers of those carbons which bear side chains are as low as possible. This means that we should start numbering with the upper carbon. There is a methyl group on carbon number two, and one on carbon number three. The compound is thus called, 2,3-dimethylpentane.

SAMPLE EXERCISE 24.2

Write the condensed structural formula for 2-methyl-3-ethylpentane.

Solution: The longest chain of continuous carbon atoms in this compound is five. We can therefore begin by writing out a string of five C atoms:

$$C-C-C-C-C$$

We next place a methyl group on the second carbon,

and an ethyl group on the middle carbon atom of the chain. Hydrogens are then added to all the other carbon atoms to make their covalencies equal to four:

$$\begin{array}{c} CH_3 \\ \mid \\ CH_3-CH-CH-CH_2CH_3 \\ \mid \\ CH_2CH_3 \end{array}$$

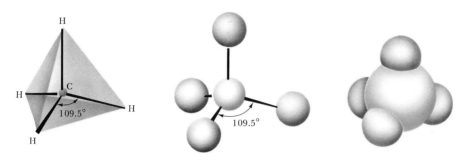

FIGURE 24.3 Representations of the three-dimensional arrangement of bonds about carbon in alkanes.

STRUCTURES OF ALKANES

The Lewis structures or condensed structural formulas for alkanes do not tell us anything about the three-dimensional structures of these substances. As we would predict from the VSEPR model (Section 8.1), the geometry about each carbon atom in an alkane is tetrahedral; that is, the four groups attached to each carbon are located at the vertices of tetrahedron. The three-dimensional structures can be represented as shown for methane in Figure 24.3. The bonding may be described as involving sp^3 hybridized orbitals on the carbon, as discussed earlier, in Section 8.2.

One of the characteristics of the carbon-carbon single bond is that rotation about this bond is relatively free. You might imagine grasping the top left methyl group in Figure 24.4, which shows the structure of propane, and twisting it relative to the rest of the structure. Motion of this sort occurs very rapidly in alkanes at room temperature. Thus, a long-chain alkane is constantly undergoing motions that cause it to change its shape, something like a length of chain that is being shaken.

One possible structural form for alkanes is that in which the carbon chain forms a ring, or cycle. Alkanes with this form of structure are called **cycloalkanes.** A few examples of cycloalkanes are shown in Figure 24.5. The cycloalkane structures are sometimes drawn as simple polygons, as illustrated in Figure 24.5. In this shorthand notation, each corner of the polygon represents a CH_2 group. This method of representation is similar to that used for aromatic rings, as illustrated in Figure 8.15. In the case of the aromatic structures, each corner represents a CH group.

Carbon rings containing less than six carbon atoms are strained, because the C—C—C bond angle in the smaller rings must be less than the

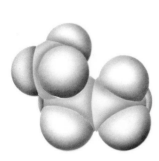

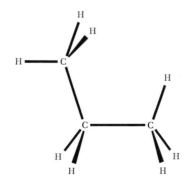

FIGURE 24.4 Three-dimensional models for propane, C_3H_8.

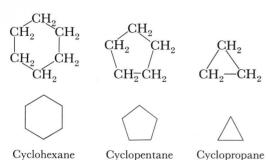

Cyclohexane   Cyclopentane   Cyclopropane   **FIGURE 24.5** Condensed structural formulas of three cycloalkanes.

109.5° tetrahedral angle. The amount of strain increases as the rings get smaller. In cyclopropane, which has the shape of an equilateral triangle, the angle is only 60°; this molecule is therefore much more reactive than either propane, its straight-chain analogue, or cyclohexane, which has no ring strain. Note that the empirical formula for the cycloalkanes is C_nH_{2n}, which differs from that for the straight-chain alkanes. The cycloalkanes thus form a separate homologous series.

ALKENES

The alkenes are close relatives of the alkanes. They differ in that there is at least one carbon-carbon double bond in the molecule. Alkenes are sometimes referred to as olefins. The presence of a double bond results in two fewer hydrogens than would be present in an alkane. Because the alkenes possess fewer hydrogens than are needed to form the alkane, they are said to be unsaturated. As we shall see a little later, the presence of the double bond confers considerably more chemical reactivity on the alkenes than is found in the alkanes. The simplest alkene is C_2H_4, called ethene, or ethylene. The next member of the series is CH_3—CH=CH_2, called propene or propylene. When there are more than three carbon atoms in the molecule, there are several possibilities for forming isomers. For example, the possible alkenes with four carbon atoms, and with molecular formula C_4H_8, are shown in Figure 24.6. The first compound shown has a branched chain; the other three all have continuous chains of four carbon atoms. In naming alkenes, the compound is named for the length of the longest continuous chain of carbon atoms, modifying the ending of the name as listed in Table 24.1 from -*ane* to -*ene*. The location of the double bond is indicated by a prefix number that designates the number of the carbon atom that is part of the double bond and is nearest an end of the chain. Thus, the compound on the left in Figure 24.6 is called a propene, because the longest continuous chain length is three

2-Methylpropene   1-Butene   *cis*-2-Butene   *trans*-2-Butene
b.p. −7°C   b.p. −6°C   b.p. 4°C   b.p. 1°C

FIGURE 24.6 Structures, names, and boiling points of the alkenes with molecular formula C_4H_8.

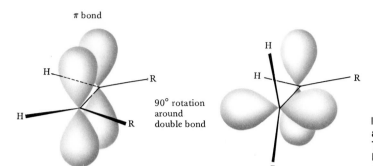

π bond

90° rotation around double bond

FIGURE 24.7 Schematic illustration of rotation about a carbon-carbon double bond in an alkene. The overlap of the p orbitals that form the π bond is lost in the rotation. For this reason, rotation about the carbon-carbon double bonds does not occur readily.

carbons. The location of the double bond is obvious in this compound. The placement of the methyl group is designated by a prefix numeral; thus, the name is 2-methylpropene. The other three compounds are named as butenes, as shown.

Cis- and *trans*-2-butene are geometrical isomers; that is, they are compounds that have the same molecular formula and the same groups bonded to one another, but that differ in the spatial arrangement of those groups. In the *cis* form, the two methyl groups are on the same side of the double bond; in the *trans* form they are on opposite sides. We have already met examples of geometrical isomerism, in the chemistry of transition metal coordination compounds (Section 23.4). Geometrical isomers possess distinct physical properties and may even differ significantly in their chemical behavior in certain circumstances.

Recall from our earlier discussion of the geometry about carbon (Section 8.4) that the double bond between two carbon atoms consists of a σ and π part. Figure 24.7 shows the geometrical arrangement as in a *cis* alkene. The bonding arrangement around each carbon is planar; that is, the carbon-carbon bond axis and the bonds to the other two groups, either hydrogen or carbon, are all in a plane. It is easy to see from Figure 24.7 that rotation about the carbon-carbon double bond will not be easy. Such a rotation would cause the *p* orbitals which form the π bond to lose their overlap, and the π bond would be destroyed. Because rotation about the carbon-carbon bond is difficult the *cis* and *trans* isomers can be separated and studied. If the interconversion were easy, we would always have an equilibrium mixture of the two forms rather than the pure isomers.

SAMPLE EXERCISE 24.3

Name the following compound:

$$CH_3CH_2CH_2-\overset{\overset{\displaystyle CH_3}{|}}{CH}\diagdown \underset{\underset{\displaystyle H}{|}}{C}=\underset{\underset{\displaystyle H}{|}}{C}\diagup \overset{\displaystyle CH_3}{}$$

Solution: The longest continuous chain of carbon atoms in this compound is seven in length. Because it possesses a double bond it is an alkene. Thus, the

compound is a heptene. The double bond begins at carbon atom number two; thus the name is 2-heptene. Continuing the numbering along the chain, a methyl group is bound at carbon atom number four. Thus, the compound is 4-methyl-2-heptene. Finally, we note that the geometrical configuration at the double-bond is *cis;* that is, the alkyl groups are bonded to the double bond on the same side. Thus, the full name is 4-methyl-*cis*-2-heptene.

In substances containing two alkene functional groups, each must be located by a number. The ending of the name is altered to identify the number of functional groups: diene (two double bonds), triene (three double bonds), and so forth. For example, 1,4-pentadiene is $CH_2\!\!=\!\!CH\!-\!CH_2\!-\!CH\!\!=\!\!CH_2$.

ALKYNES

The alkynes are yet another series of unsaturated hydrocarbons, in which there is one or more carbon-carbon triple bonds between carbon atoms. The empirical formula for simple alkynes is C_nH_{2n-2}. The simplest alkyne, acetylene, is a highly reactive molecule. When acetylene is burned in a stream of oxygen in the so-called oxyacetylene torch, the flame reaches a very high temperature, about 3200 K (Section 21.4). The oxyacetylene torch is widely used in welding, where high temperatures are required. Alkynes in general are highly reactive molecules. Because of their higher reactivity they are not as widely distributed in nature as the alkenes; however, they are important intermediates in many industrial processes.

The alkynes are named by identifying the longest continuous chain in the molecule containing the triple bond, and modifying the ending of the name as listed in Table 24.1 from *-ane* to *-yne,* as shown in the following sample exercise.

SAMPLE EXERCISE 24.4

Name the following compounds:

(a) $CH_3CH_2CH_2\!-\!C\!\!\equiv\!\!C\!-\!CH_3$

(b) $CH_3CH_2CH_2\underset{\displaystyle CH_2CH_2CH_3}{CH}\!-\!C\!\!\equiv\!\!CH$

Solution: In (a) the longest chain of carbon atoms is six. There are no side chains. The triple bond begins at carbon atom number two (remember we always arrange the numbering so that the smallest possible number is assigned to the carbon containing the multiple bond). Thus, the name is 2-hexyne.

In (b) the longest continuous chain of carbon atoms is seven; but because this chain does not contain the triple bond we do not count it as derived from heptane. The longest chain containing the triple bond is six, so this compound is named as a derivative of hexyne, 3-propyl-1-hexyne.

AROMATIC HYDROCARBONS

The aromatic hydrocarbons are a large and important class of hydrocarbons. The simplest member of the series is benzene (see Figure 24.1), with molecular formula C_6H_6. As we have already noted, benzene is a planar, highly symmetrical molecule. The molecular formula for benzene suggests a high degree of unsaturation. One might thus expect that benzene would be highly reactive, and that it might resemble the unsaturated hydrocarbons in reactivity. In fact, however, benzene is not at all similar to alkenes or alkynes in chemical behavior. The great stability of benzene and the other aromatic hydrocarbons as compared with alkenes and alkynes is due to stabilization of the π electrons through delocalization in the π orbitals (Section 8.4).

We can obtain an estimate of the stabilization of the π electrons in benzene by comparing the energy involved in adding hydrogen to benzene to form a saturated compound, as compared with that involved in hydrogenating simple alkenes. The hydrogenation of benzene to form cyclohexane can be represented as

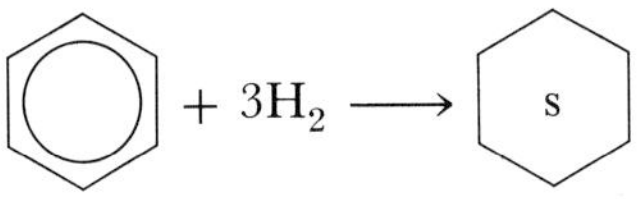

$$\text{(benzene)} + 3H_2 \longrightarrow \text{(cyclohexane, s)} \qquad \Delta H° = -208 \text{ kJ/mol} \qquad [24.1]$$

(The s in the ring on the right indicates that it is a cycloalkane, with CH_2 groups at each corner.) The enthalpy change in this reaction is -208 kJ/mol. The heat of hydrogenation of the cyclic alkene cyclohexene, is -120 kJ/mol:

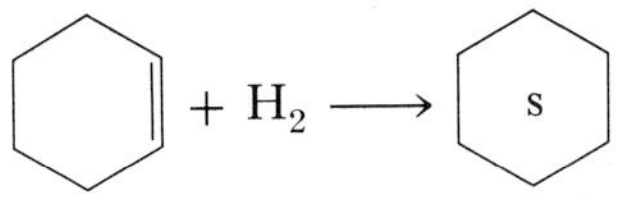

$$\text{(cyclohexene)} + H_2 \longrightarrow \text{(s)} \qquad \Delta H° = -120 \text{ kJ/mol} \qquad [24.2]$$

Cyclohexene

Similarly, the heat released on hydrogenating 1,4-cyclohexadiene is -232 kJ/mol:

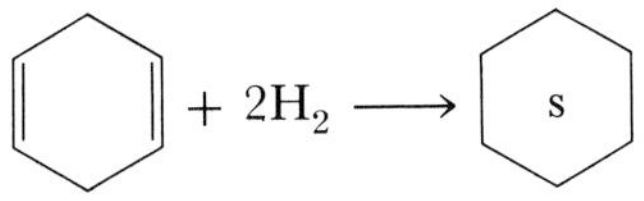

$$\text{(1,4-cyclohexadiene)} + 2H_2 \longrightarrow \text{(s)} \qquad \Delta H° = -232 \text{ kJ/mol} \qquad [24.3]$$

1,4-Cyclo-
hexadiene

From these last two reactions it would appear that the heat of hydrogenating a double bond is about 116 kJ/mol for each bond. There is the equivalent of three double bonds in benzene. Thus we might expect that the heat of hydrogenating benzene would be about three times -116, or -348 kJ/mol, if benzene behaved as though it were "cyclohexatriene," that is, if it behaved as though it were three double bonds in a ring. Instead, the heat released is much less than this, indicating that benzene is more stable than would be expected for three double bonds. The difference of 140 kJ/mol between -348 kJ/mol and the observed heat of hydrogenation, -208 kJ/mol, can be ascribed to stabilization of the π electrons through delocalization in the π orbitals that extend around the ring.

There is no widely used systematic nomenclature for naming aromatic rings. Each ring system is given a common name; several aromatic compounds are shown in Figure 24.8. The aromatic rings are represented by hexagons with a circle inscribed inside to denote aromatic character. Each corner represents a carbon atom. Each carbon is bound to three other atoms—either three carbons or two carbons and a hydrogen. The hydrogen atoms are not shown. In naming derivatives of the aromatic hydrocarbons, it is often necessary to indicate the position in the aromatic ring at which some side chain or other group is located. For this purpose the numbering shown in Figure 24.8 is used.

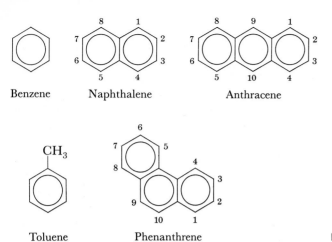

Benzene   Naphthalene   Anthracene

Toluene
(Methylbenzene)   Phenanthrene

FIGURE 24.8 Structures, names, and numbering system of several aromatic compounds.

24.2 Petroleum

Petroleum is a complex mixture of organic compounds, mainly hydrocarbons, with smaller quantities of other organic compounds containing nitrogen, oxygen, or sulfur. Petroleum is formed over a period of millions of years by the decomposition of marine plants and animals.

The usual first step in the refining, or processing, of petroleum is separation of the crude oil into fractions on the basis of boiling points. The fractions commonly taken are shown in Table 24.3. As you might expect, the fractions that boil at higher temperatures are made up of molecules with larger numbers of carbon atoms per molecule. The fractions collected in the initial separation may require further processing to yield a usable product. For example, modifications must be made to the straight-run gasoline obtained from fractionation of petroleum to render it suitable for use as a fuel in automobile engines. Similarly, the fuel oil fraction may need additional processing to remove sulfur before it is suitable for use in an electrical power station or a home heating system. At present, the most commercially important single product from petroleum refining is gasoline.

TABLE 24.3 Hydrocarbon fractions from petroleum

Fraction	Size range of molecules	Boiling-point range (°C)	Uses
Gas	C_1–C_5	−160 to 30	Gaseous fuel, production of H_2
Straight-run gasoline	C_5–C_{12}	30 to 200	Motor fuel
Kerosene, fuel oil	C_{12}–C_{18}	180 to 400	Diesel fuel, furnace fuel, cracking
Lubricants	C_{16} and up	350 and up	Lubricants
Paraffins	C_{20} and up	Low-melting solids	Candles, matches
Asphalt	C_{36}	Gummy residues	Surfacing roads, fuel

Gasoline is a mixture of volatile hydrocarbons. Depending on the source of the crude oil, it may contain varying amounts of cyclic alkanes and aromatic hydrocarbons in addition to alkanes. Straight-run gasoline consists mainly of straight-chain hydrocarbons, which in general are not very suitable for use as fuel in an automobile engine. In an automobile engine, a mixture of air and gasoline vapor is ignited by the spark plug at the moment when the gas mixture inside the cylinder has been compressed by the piston. The burning of the gasoline should create a strong, smooth expansion of gas in the cylinder, forcing the piston outward and imparting force along the drive shaft of the engine. If the gas burns too rapidly, the piston receives a single hard slam rather than a strong, smooth push. The result is a "knocking" or pinging sound; the efficiency with which the energy of gasoline combustion is converted to power is reduced.

Gasolines are rated according to octane number. Gasolines with high octane numbers burn more slowly and smoothly and thus are more effective fuels, especially in engines in which the gas-air mixture is highly compressed. It happens that the more highly branched alkanes have higher octane numbers than the straight-chain compounds; some examples are shown in Table 24.4. The octane number of gasoline is obtained by comparing its knocking characteristics with those of "isooctane" (2,2,4-trimethylpentane) and heptane. Isooctane is assigned an octane number of 100, whereas heptane is assigned 0. Gasoline with the same knocking characteristics as a mixture of 95 percent isooctane and 5 percent heptane would be rated as 95-octane.

Because straight-run gasoline contains mostly straight-chain hydrocarbons, it has a low octane number. It is therefore subjected to a process called cracking to convert the straight-chain compounds into more desirable branched-chain molecules. Cracking is also used to convert some of the less volatile kerosene and fuel-oil fractions into compounds with

TABLE 24.4 Octane numbers of some C_7 and C_8 hydrocarbons

Name	Condensed structural formula	Octane number
n-Heptane	$CH_3CH_2CH_2CH_2CH_2CH_2CH_3$	0
2-Methylhexane	$CH_3CH_2CH_2CH_2{-}CH{-}CH_3$ CH_3	40
Methylcyclohexane	(structure)	75
2,3-Dimethylpentane	CH_3 $CH_3CH_2{-}CH{-}CH{-}CH_3$ CH_3	90
2,2,4-Trimethylpentane ("isooctane")	CH_3 $CH_3{-}CH{-}CH_2{-}C{-}CH_3$ CH_3 CH_3	100

lower molecular weights that are suitable for use as automobile fuel. In the cracking process, the hydrocarbons are mixed with a catalyst and heated to 400–500°C. The catalysts used are naturally occurring clay minerals, or synthetic Al_2O_3–SiO_2 mixtures. In addition to forming molecules more suitable for gasoline, cracking results in the formation of hydrocarbons of lower molecular weight, such as ethylene and propene. These are used in a variety of processes to form plastics and other chemicals.

The octane number of a given blend of hydrocarbons can be improved by adding an **antiknock agent,** a substance that helps control the burning rate of the gasoline. The most widely used substances for this purpose are tetraethyl lead, $(CH_3CH_2)_4Pb$, and tetramethyl lead, $(CH_3)_4Pb$. The premium gasolines contain 2 or 3 mL of one of these lead compounds per gallon, with a resultant increase of 10 to 15 in octane rating. Although alkyl lead compounds are undoubtedly effective in improving gasoline performance, their use in gasolines has been drastically curtailed because of the environmental hazards associated with lead. This metal is highly toxic, and there is good evidence that the lead released from automobile exhausts is a general health hazard. Although other substances have been tried as antiknock agents in gasolines, none of these has proved to be an effective and inexpensive antiknock agent that is environmentally safe. The 1975 and later model cars are designed to operate with unleaded gasolines. The gasolines blended for these cars are made up of more highly branched components and more aromatic components, because these have relatively high octane ratings.

24.3 Reactions of hydrocarbons

The hydrocarbons are capable of undergoing a variety of reactions with other substances. Many of these reactions are of considerable importance in the chemical industry, because they lead to useful products or to substances that can in turn be converted to useful products.

FIGURE 24.9 Drilling for oil on the North Slope in Alaska. (*Courtesy of Exxon Corp.*)

OXIDATION

The most common oxidation reactions of hydrocarbons are those that result in complete oxidation to form carbon dioxide and water:

$$CH_3CH_2CH_3 + 5O_2 \longrightarrow 3CO_2 + 4H_2O \qquad [24.4]$$

$$CH_3CH{=}CH_2 + \tfrac{9}{2}O_2 \longrightarrow 3CO_2 + 3H_2O \qquad [24.5]$$

$$CH_3C{\equiv}CH + 4O_2 \longrightarrow 3CO_2 + 2H_2O \qquad [24.6]$$

$$C_6H_6 + \tfrac{15}{2}O_2 \longrightarrow 6CO_2 + 3H_2O \qquad [24.7]$$

These reactions are all highly exothermic. Combustion of the hydrocarbons results in release of energy that can be used to propel an auto or an airplane, to generate steam in the boiler of an electric power plant, or to heat a building. The tremendous demand for petroleum to meet the world's energy needs has led to the tapping of oil wells in such forbidding places as the North Sea and the North Slope of the continent in Alaska, facing the Arctic Ocean (Figure 24.9).

Controlled oxidation of hydrocarbons yields organic substances that contain oxygen in addition to carbon and hydrogen. These products of controlled oxidations will be considered later when we discuss hydrocarbon derivatives.

ADDITION REACTIONS

Under appropriate conditions, it is possible to add atoms or groups of atoms to alkenes, alkynes, or aromatic hydrocarbons by disrupting the carbon-carbon multiple bonds. For example, ethylene reacts with bromine to form 1,2-dibromoethane, as in Equation [24.8]:

$$H_2C{=}CH_2 + Br_2 \longrightarrow \underset{\underset{Br \quad Br}{|\qquad|}}{H_2C{-}CH_2} \qquad [24.8]$$

The pair of electrons that form the π bonds in ethylene are uncoupled and are used to form two new bonds to the two bromine atoms. The σ bond between the carbon atoms remains.

Addition of halogens to alkynes also occurs readily, as in the example shown in Equation [24.9]:

$$CH_3C{\equiv}CH + 2Cl_2 \longrightarrow \underset{\underset{Cl \quad Cl}{|\qquad|}}{\overset{\overset{Cl \quad Cl}{|\qquad|}}{CH_3{-}C{-}CH}} \qquad [24.9]$$

1,1,2,2-Tetrachloropropane

Reaction of alkenes or alkynes with hydrogen halides also results in addition:

$$CH_3CH{=}CH_2 + HBr \longrightarrow \underset{\underset{Br}{|}}{CH_3CH{-}CH_3} \qquad [24.10]$$

Notice that this reaction might have proceeded to give another product, in which the bromine atom is on the end carbon. It turns out that when a hydrogen halide adds to an alkene, the more electronegative halogen atom always ends up on the carbon atom of the double bond that has the fewer hydrogen atoms. This rule, known as Markovnikoff's rule, was first formulated by the Russian chemist V. V. Markovnikoff.

SAMPLE EXERCISE 24.5

Predict the product of reaction of HCl with the following compound:

$$CH_3CH_2\underset{\underset{CH_3}{|}}{\overset{\overset{CH_3}{|}}{C}}{=}CH$$

Solution: The double-bonded carbon atom on the right has one hydrogen on it, the other has none. According to Markovnikoff's rule, the chloride of HCl will end up on the carbon atom on the left:

$$CH_3CH_2-\underset{\underset{CH_3}{|}}{\overset{\overset{Cl}{|}}{C}}-CH_2 \quad \overset{CH_3}{}$$

Markovnikoff's rule also applies to the addition of unsymmetrical reagents to alkynes. For example, reaction of HBr with 1-butyne yields 2,2-dibromobutane:

$$CH_3CH_2C{\equiv}CH + 2HBr \longrightarrow CH_3CH_2-\underset{\underset{Br}{|}}{\overset{\overset{Br}{|}}{C}}-CH_3 \qquad [24.11]$$

Using an acid such as H_2SO_4 as catalyst, it is possible to add H_2O to a double bond. The products of such reactions are alcohols; that is, compounds containing an OH group bonded to carbon. Markovnikoff's rule applies here as well; we consider the water molecule as being polarized as follows: H^+-OH^-. Thus, the OH group adds to the carbon atom with fewer hydrogens:

$$CH_3CH{=}CH_2 + HOH \xrightarrow{H_2SO_4} CH_3-\underset{\underset{}{}}{\overset{\overset{OH}{|}}{CH}}-CH_3 \qquad [24.12]$$

Propene   2-Propanol

Addition of H_2 to an alkene converts it into an alkane. This reaction, referred to as hydrogenation, does not occur readily under ordinary temperature and pressure conditions. One of the reasons for the lack of reactivity of H_2 toward an alkene is the high bond energy of the H_2 bond. To promote the reaction it is necessary to use a catalyst that assists in the rupture of the H—H bond. The most widely used catalysts are heterogeneous and consist of finely divided metals on which H_2 is adsorbed. The action of these heterogeneous catalysts in reaction of H_2 with an alkene is described in detail in Section 13.6. Molecular hydrogen also reacts with alkynes in the presence of a catalyst to yield alkanes, as in the example of Equation [24.13].

$$CH_3C{\equiv}CH + 2H_2 \xrightarrow{Ni} CH_3CH_2CH_3 \qquad [24.13]$$

Addition reactions of aromatic hydrocarbons proceed much less readily than such reactions of alkenes and alkynes. For example, benzene does not react at all with Cl_2 or Br_2 under ordinary conditions. However, if conditions are sufficiently rigorous, such addition reactions may be forced to occur.

ADDITION POLYMERIZATION

One of the most important addition reactions of alkenes is **addition polymerization.** In a polymerization reaction, small molecules called **monomers** react with one another to form long, chainlike molecules of high molecular weight called **polymers.** For example, 2-methylpropene (common name, isobutylene) reacts with itself in the presence of a small amount of acid catalyst to form polyisobutylene, which may contain more than a thousand isobutylene units:

$$nCH_2{=}\overset{\displaystyle CH_3}{\overset{|}{\underset{\displaystyle CH_3}{\underset{|}{C}}}} \xrightarrow{acid} {-}CH_2{-}\overset{\displaystyle CH_3}{\overset{|}{\underset{\displaystyle CH_3}{\underset{|}{C}}}}{-}CH_2{-}\overset{\displaystyle CH_3}{\overset{|}{\underset{\displaystyle CH_3}{\underset{|}{C}}}}{-}CH_2{-}\overset{\displaystyle CH_3}{\overset{|}{\underset{\displaystyle CH_3}{\underset{|}{C}}}}{-}CH_2{-}\overset{\displaystyle CH_3}{\overset{|}{\underset{\displaystyle CH_3}{\underset{|}{C}}}}{-}\ldots \qquad [24.14]$$

Butylene　　　Polyisobutylene

The product on the right in this reaction is a rubbery, rather gummy material that can be further treated to form a wide variety of useful products. In its finished form, it is known as butyl rubber. Many other alkenes can be similarly polymerized to form the variety of useful materials summarized in Table 24.5.

Other substances besides acids can catalyze the polymerization of molecules such as isobutylene. One of the most important types of catalyst is the **free radical.** A free radical is a species with an odd number of electrons. There is thus an atom in the radical that does not have its full complement of electrons. As a result, the radical is quite reactive. The simplest examples of free radicals are single atoms such as a chlorine atom, which is one electron short of completing its valence shell. Free

TABLE 24.5 Alkenes that undergo addition polymerization

Monomer formula	Monomer name	Polymer name	Uses
$CH_2{=}CH_2$	Ethylene	Polyethylene	Coating for milk cartons, wire insulation, plastic bags
$CF_2{=}CF_2$	Tetrafluoroethylene	Teflon	Insulation, bearings, frying-pan surfaces
$CH_2{=}\underset{\displaystyle Cl}{\underset{\mid}{C}}H$	Vinyl chloride	Polyvinyl chloride (PVC)	Phonograph records, rainwear, piping
$CH_2{=}\underset{\displaystyle CN}{\underset{\mid}{C}}H$	Acrylonitrile	Polyacrylonitrile (Orlon)	Rug fibers
$CH_2{=}\underset{\displaystyle C_6H_5}{\underset{\mid}{C}}H$	Vinyl benzene (styrene)	Polystyrene	TV lead-in wire, combs

radicals can be formed by rupture of chemical bonds. For example, peroxides, which contain O—O bonds, decompose on heating, with rupture of the O—O bond. The result is a pair of free radicals:

$$R-\overset{..}{\underset{..}{O}}-\overset{..}{\underset{..}{O}}-R \rightleftharpoons 2R-\overset{..}{\underset{..}{O}}\cdot \qquad [24.15]$$

In this equation R could be any of a number of different organic groups. By placing a small mount of a peroxide in contact with an alkene and then heating it, radicals that can catalyze addition reactions are generated. For example, consider the polymerization of ethylene initiated (that is, started) by a radical formed from a peroxide:

$$R-\overset{..}{\underset{..}{O}}-\overset{..}{\underset{..}{O}}-R \rightleftharpoons 2R-\overset{..}{\underset{..}{O}}\cdot$$

$$R-\overset{..}{\underset{..}{O}}\cdot + H_2C{=}CH_2 \longrightarrow H-\overset{\overset{\displaystyle H}{|}}{\underset{\underset{\displaystyle OR}{|}}{C}}-\overset{\overset{\displaystyle H}{|}}{\underset{\underset{\displaystyle H}{|}}{C}}\cdot \qquad [24.16]$$

In Equation [24.16], the R—O radical reacts with a molecule of ethylene, forming a bond to one of the carbons. For this to occur, the pair of electrons forming the π bond in ethylene must have uncoupled. One of them is involved in the bond to the O—R group, the other is on the second carbon. Thus, the product of this reaction is itself a free radical. This species goes on to react with a second molecule of ethylene:

$$H-\overset{\overset{\displaystyle H}{|}}{\underset{\underset{\displaystyle OR}{|}}{C}}-\overset{\overset{\displaystyle H}{|}}{\underset{\underset{\displaystyle H}{|}}{C}}\cdot + H_2C{=}CH_2 \longrightarrow RO-\overset{\overset{\displaystyle H}{|}}{\underset{\underset{\displaystyle H}{|}}{C}}-\overset{\overset{\displaystyle H}{|}}{\underset{\underset{\displaystyle H}{|}}{C}}-\overset{\overset{\displaystyle H}{|}}{\underset{\underset{\displaystyle H}{|}}{C}}-\overset{\overset{\displaystyle H}{|}}{\underset{\underset{\displaystyle H}{|}}{C}}\cdot \qquad [24.17]$$

By means of such successive reactions, the polymer grows in length. A reaction of this type is known as a chain reaction, because it is self-propagating. However, chain-termination reactions might occur. If two radicals come into contact, their unpaired spins might couple, and no further reaction would occur. In forming polymers by such free-radical reactions, we must control conditions carefully so that a polymer chain with the desired average length results.

In addition to acids and free radicals, other substances may serve as catalysts for addition polymerization reactions of alkenes. The choice of catalyst is important, because it influences various characteristics of the resulting polymer. For example, polyethylene (see Section 11.7), formed by polymerizing ethylene, can be a pliable material with a low melting point or a much stiffer substance with a higher melting point, depending on the nature of the catalyst.

SUBSTITUTION REACTIONS

In a substitution reaction of a hydrocarbon, one or more hydrogen atoms are replaced by other atoms or groups. Substitution reactions are difficult to carry out with aliphatic compounds. One of the most important substitution reactions of alkanes involves replacement of hydrogen by a halogen atom. The chlorination of an alkane is a photo-initiated reaction. That is, it requires the use of light, which causes dissociation of the Cl_2 molecule, forming reactive chlorine atoms. A chlorine atom then attacks the alkane, removing a hydrogen and forming HCl and an alkyl radical. The alkyl radical then attacks a Cl_2 molecule, forming an alkyl halide and a chlorine atom:

$$Cl_2 \xrightarrow{\text{light}} 2Cl\cdot \qquad\qquad [24.18]$$

$$Cl\cdot + CH_3CH_3 \longrightarrow HCl + CH_3CH_2\cdot \qquad\qquad [24.19]$$

$$CH_3CH_2\cdot + Cl_2 \longrightarrow CH_3CH_2Cl + Cl\cdot \qquad\qquad [24.20]$$

The chlorine atom formed in Equation [24.20] reacts with more ethane, continuing the cycle. Thus, for each quantum of light absorbed by a chlorine molecule, many molecules of ethyl chloride may be formed. This reaction provides an example of a **radical chain** process. One of the disadvantages of radical chain reactions such as this is that they are not very selective. As the concentration of ethyl chloride builds up in the reaction, chlorine atoms may abstract hydrogen atoms from it, so that eventually dichloroethane and even more highly chlorinated molecules may be formed. Thus, several products are formed in the reaction, and these must be carefully separated by distillation or some other separation procedure.

Substitution into alkenes or alkynes is difficult to carry out, because the presence of the reactive double or triple bond usually leads to addition reactions rather than substitution. By contrast, substitution into aromatic hydrocarbons is relatively easy. For example, when benzene is warmed in a mixture of nitric and sulfuric acid, hydrogen is replaced by the nitro group, NO_2:

$$\text{(benzene)} + HNO_3 \xrightarrow{H_2SO_4} \text{(nitrobenzene)} + H_2O \qquad\qquad [24.21]$$

More vigorous treatment results in substitution of a second nitro group into the molecule:

$$\text{(nitrobenzene)} + HNO_3 \xrightarrow{H_2SO_4} \text{(dinitrobenzene)} + H_2O \qquad\qquad [24.22]$$

There are three possible isomers of benzene with two nitro groups attached. These three isomers are named *ortho-*, *meta-*, and *para-*dinitrobenzene:

ortho-Dinitrobenzene	*meta*-Dinitrobenzene	*para*-Dinitrobenzene
m.p. 118°C	m.p. 90°C	m.p. 174°C

Only the *meta* isomer is formed in the reaction of nitric acid with nitro-benzene.

Bromination of benzene is carried out using $FeBr_3$ as a catalyst:

$$+ Br_2 \xrightarrow{FeBr_3} \qquad + HBr \qquad [24.23]$$

In a similar reaction, called the **Friedel-Crafts reaction**, alkyl groups can be substituted onto an aromatic ring by reaction of an alkyl halide with an aromatic compound in the presence of $AlCl_3$:

$$+ CH_3CH_2Cl \xrightarrow{AlCl_3} \qquad + HCl \qquad [24.24]$$

The substitution reactions of aromatic compounds occur via attack of a positively charged reagent on the ring. Substances such as H_2SO_4, $FeCl_3$, or $AlCl_3$ serve as catalysts by generating the positively charged species. For example, the bromination of benzene proceeds via the following steps:

$$FeBr_3 + Br_2 \rightleftharpoons FeBr_4{}^- + Br^+ \qquad [24.25]$$

$$+ \overset{\curvearrowleft}{Br}^+ \longrightarrow \qquad [24.26]$$

$$\xrightarrow{FeBr_4{}^-} \qquad + FeBr_3 + HBr \qquad [24.27]$$

The Br^+ ion, called the **bromonium ion,** has only six electrons in its valence shell. Using a pair of π electrons from the benzene, it forms a single bond to a carbon atom. In doing so, it leaves the adjacent carbon atom with only six electrons in its valence shell. Such a species is not very stable. Loss of a proton from the carbon atom to which the bromine is attached converts the molecule to a more stable one, bromobenzene. The overall result is that bromine has replaced hydrogen on the benzene ring.

24.4 Hydrocarbon derivatives

We noted in the introduction to this chapter that certain groups or arrangements of atoms impart characteristic behavior to an organic molecule. These groups are called functional groups. Two of these groups have already been discussed, the double and triple carbon-carbon bonds, both of which impart considerable chemical reactivity to a hydrocarbon. The other functional groups contain elements other than carbon and hydrogen, most often oxygen, nitrogen, or a halogen. Compounds containing these elements are generally considered to be hydrocarbon derivatives; they may be viewed as being derived from a hydrocarbon by replacing one or more of the hydrogens with functional groups. The compound can then be considered to consist of two parts: a hydrocarbon fragment such as an alkyl group (often designated R) and one or more functional groups.

Although a hydrocarbon derivative may consist largely of carbon and hydrogen, the hydrocarbon portion of the molecule may be essentially unchanged in various chemical reactions. The chemical properties are generally determined by the functional group. Now let's take a brief look at some of these functional groups and some of the properties they impart to organic compounds.

ALCOHOLS

Alcohols are hydrocarbon derivatives in which one or more hydrogens of a parent hydrocarbon have been replaced by a hydroxyl or alcohol functional group, OH. Figure 24.10 shows the structural formulas and names of several alcohols. Note that the accepted name for an alcohol ends in *-ol*. The simple alcohols are named by changing the last letter in the name of the corresponding alkane to *-ol*—for example, ethan*e* becomes ethan*ol*. Where necessary, the location of the OH group is designated by an appropriate prefix numeral that indicates the number of the carbon atom bearing the OH group, as shown in the examples in Figure 24.10.

Aliphatic alcohols are classified according to the number of carbon groups bonded to the carbon that contains the OH group. If there is only one other carbon atom, as with ethanol or 1-propanol, the alcohol is termed a **primary alcohol**. If there are two other carbon groups attached, as in 2-propanol, the alcohol is **secondary**. If there are three other carbons, as in *t*-butanol, the alcohol is **tertiary**.

Because the OH group is quite polar, the presence of the hydroxyl group confers considerable polarity on the hydrocarbon molecule. Table 24.6 lists the solubilities of several alcohols in water at room temperature. In the alcohols having low molecular weights, the presence of the OH group plays an important role in determining physical properties. However, as the length of the hydrocarbon chain increases, the OH group

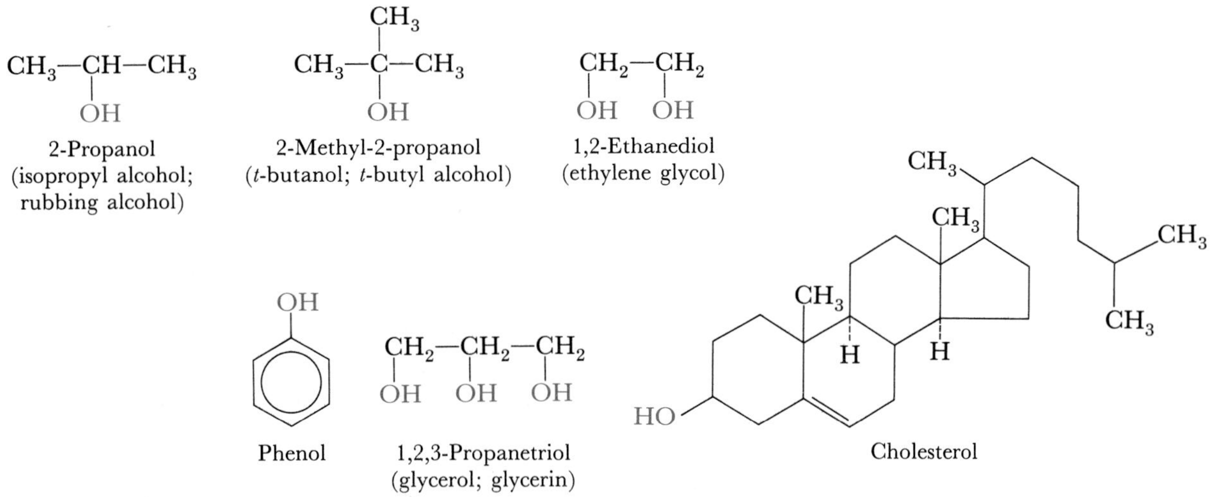

FIGURE 24.10 Structural formulas of several important alcohols.

TABLE 24.6 Solubilities of several straight-chain alcohols in water

Alcohol	Boiling point (°C)	Solubility in water at 25° (g/100 g H_2O)
Methanol	65	Miscible
Ethanol	78	Miscible
1-Propanol	97	Miscible
1-Butanol	117	9
Cyclohexanol	161	5.6
1-Hexanol	158	0.6

becomes less important in determining overall behavior (see also Section 12.3).

The simplest alcohol, methanol, has many important industrial uses, and it is produced on a large scale. Carbon monoxide and hydrogen are heated together under pressure in the presence of a metal oxide catalyst:

$$CO(g) + 2H_2(g) \xrightarrow[400°C]{200-300 \text{ atm}} CH_3OH(g) \qquad [24.28]$$

Methanol has a very high octane rating of about 110 when it is blended with gasoline. There has been considerable discussion of using it as a fuel extender by mixing it with gasolines. Methanol boils at 65°C, and freezes at −98°. It is therefore suitable in terms of volatility for use as an automotive fuel. It could probably be manufactured for about twice the present cost of gasoline (on a per mile basis), if the manufacture were carried out on a larger scale than at present. To offset its higher cost, methanol has the advantage of requiring less pollution-control equipment. A disadvantage of methanol for fuel use is that it is miscible with water in all proportions; it thus might attract water into methanol-gasoline mixtures. Secondly, methanol is quite toxic. However, gasoline is also a toxic substance, and only moderate additional precautions would need to be taken if methanol were added to gasoline.

Ethanol, C_2H_5OH, is a product of the fermentation of carbohydrates such as sugar or starch. Bacterial cultures such as yeast work in the absence of air to convert carbohydrates into a mixture of ethanol and CO_2, as shown in Equation [24.29]. In the process, they derive the energy necessary for growth:

$$C_6H_{12}O_6(aq) \xrightarrow{\text{yeast}} 2C_2H_5OH(aq) + 2CO_2(g) \qquad [24.29]$$

This naturally occurring reaction is carried out under carefully controlled conditions to produce beer, wine, and other beverages in which ethanol is the active ingredient. It is often said that ethanol is the least toxic of the straight-chain alcohols. Although this is true in the strictest sense, the combination of ethanol and automobiles produces far more human fatalities each year than any other chemical agent.

Many alcohols are formed industrially by addition of the elements of water to an olefin. For example, isopropyl alcohol is formed by hydration of propene using sulfuric acid as a catalyst:

$$CH_3CH{=}CH_2 + H_2O \xrightarrow{H_2SO_4} CH_3{-}\underset{\underset{OH}{|}}{CH}{-}CH_3 \qquad [24.30]$$

Ethylene glycol, the major ingredient in automobile antifreeze, is formed in a two-stage reaction. First ethylene is reacted with hypochlorous acid:

$$CH_2\!\!=\!\!CH_2 + HClO \longrightarrow \underset{\underset{OH \quad Cl}{|\quad\;\;|}}{CH_2\!-\!CH_2} \qquad [24.31]$$

Chlorine is removed from the product of this reaction by displacing the chloride ion with hydroxide ion:

$$\underset{\underset{OH \quad Cl}{|\quad\;\;|}}{CH_2\!-\!CH_2} + OH^- \longrightarrow \underset{\underset{OH \quad OH}{|\quad\;\;|}}{CH_2\!-\!CH_2} + Cl^- \qquad [24.32]$$

A reaction of this sort, in which hydroxide ion displaces another functional group is called **base hydrolysis.**

Phenol is the simplest example of a compound with an OH group attached to an aromatic ring. One of the most striking effects of the aromatic group is the greatly increased acidity of the proton. Phenol is about a million times more acidic in water than a typical aliphatic alcohol such as ethanol. Even so, it is not a very strong acid (K_a, 1.3×10^{-10}). Phenol is used industrially in the making of several kinds of plastics and in the preparation of dyes.

Cholesterol, shown in Figure 24.10, is an example of a biochemically important alcohol. Notice that the OH group forms only a small component of this rather large molecule. As a result, cholesterol is not very soluble in water (0.26 g per 100 mL H_2O). Cholesterol is a normal component of our bodies. However, when present in excessive amounts it may precipitate from solution. It precipitates in the gall bladder to form crystalline lumps called gall stones. It may also precipitate against the walls of veins and arteries and thus contribute to high blood pressure and other cardiovascular problems. The amount of cholesterol in our blood is determined not only by how much cholesterol we eat, but by total dietary intake. There is evidence that excessive caloric intake leads the body to synthesize excessive cholesterol.

ETHERS

Ethers can be formed from two molecules of alcohol by splitting out a molecule of water. The reaction is thus a dehydration process; it is catalyzed by sulfuric acid, which takes up water to remove it from the system:

$$CH_3CH_2\!-\!OH + H\!-\!OCH_2CH_3 \xrightarrow{H_2SO_4}$$
$$CH_3CH_2\!-\!O\!-\!CH_2CH_3 + H_2O \qquad [24.33]$$

A reaction in which water is split out from two substances is called a **condensation reaction.** An inorganic example of a condensation reaction was presented earlier (Section 21.9) in our discussion of the chemistry of phosphates. We shall encounter several additional examples in this chapter and in Chapter 25.

Ethers are used as solvents; both dimethyl ether and tetrahydrofuran are common solvents for organic reactions.

$$CH_3-O-CH_3$$

$$\begin{array}{c} CH_2-CH_2 \\ | \qquad | \\ CH_2 \quad CH_2 \\ \diagdown O \diagup \end{array}$$

Dimethyl ether   Tetrahydrofuran (THF)

OXIDATION OF ALCOHOLS; ALDEHYDES AND KETONES

It is fairly easy to oxidize alcohols. Complete oxidation results in formation of CO_2 and H_2O, as in the burning of methanol:

$$CH_3OH(g) + \tfrac{3}{2}O_2(g) \longrightarrow CO_2(g) + 2H_2O(g) \qquad \text{[24.34]}$$

$$\Delta H^\circ = -676 \text{ kJ/mol}$$

Controlled oxidation to form other organic substances that may have value is carried out by using various oxidizing agents such as air, hydrogen peroxide (H_2O_2), ozone (O_3), or potassium dichromate ($K_2Cr_2O_7$). We shall not concern ourselves with balancing most of these oxidation-reduction equations; instead we shall simply show the source of oxygen as an O in parentheses.

More careful oxidation of alcohols produces aldehydes and ketones. Both of these classes of compounds contain the carbonyl group, C=O. In aldehydes the carbonyl group has at least one hydrogen atom attached as in the following examples:

$$\begin{array}{cc} \overset{\displaystyle O}{\underset{\displaystyle \|}{CH_3C}}H & \overset{\displaystyle O}{\underset{\displaystyle \|}{HC}}H \\ \text{Acetaldehyde} & \text{Formaldehyde} \end{array}$$

In ketones, the carbonyl group occurs at the interior of a carbon chain and is therefore flanked by carbon atoms:

$$\begin{array}{cc} \overset{\displaystyle O}{\underset{\displaystyle \|}{CH_3-C-CH_3}} & \overset{\displaystyle O}{\underset{\displaystyle \|}{CH_3-C-CH_2CH_3}} \\ \text{Acetone} & \text{Methyl ethyl ketone} \end{array}$$

Aldehydes are produced by oxidation of primary alcohols. For example, acetaldehyde is formed by oxidation of ethanol:

$$CH_3CH_2OH + (O) \longrightarrow \overset{\displaystyle O}{\underset{\displaystyle \|}{CH_3C}}H + H_2O \qquad \text{[24.35]}$$
$$\text{Ethanol} \qquad\qquad\qquad \text{Acetaldehyde}$$

Ketones are produced by oxidation of secondary alcohols. For example, acetone is formed in large quantities by oxidation of isopropyl alcohol:

$$\overset{\displaystyle OH}{\underset{\displaystyle |}{CH_3-CH-CH_3}} + (O) \longrightarrow \overset{\displaystyle O}{\underset{\displaystyle \|}{CH_3-C-CH_3}} + H_2O \qquad \text{[24.36]}$$
$$\qquad\qquad\qquad\qquad\qquad \text{Acetone}$$

Ketones are less reactive than are aldehydes because they do not possess the reactive C—H bond attached to the carbonyl carbon. Ketones are used extensively as solvents. Acetone, which boils at 56°C, is the most widely used. The carbonyl functional group imparts polarity to the solvent. Acetone is completely miscible with water, and yet it dissolves a wide range of organic substances. Methyl ethyl ketone, $CH_3COCH_2CH_3$, which boils at 80°C, is also used industrially as a solvent.

24.5 Carboxylic acids and their derivatives

When a primary alcohol is oxidized with 2 mol of oxygen atoms per mole of alcohol, the reaction does not stop with the formation of an aldehyde, but rather proceeds further to form a carboxylic acid. For example, such oxidation of ethanol produces acetic acid:

$$CH_3CH_2OH + 2(O) \longrightarrow CH_3\overset{\displaystyle O}{\overset{\|}{C}}{-}OH + H_2O \qquad [24.37]$$

This reaction is the one responsible for causing wine to turn sour, producing vinegar.

Carboxylic acids are widely distributed in nature and are important compounds in many industrial chemical processes. Figure 24.11 shows the structural formulas of several substances containing one or more carboxylic acid functional groups. Note that these substances are not named in a systematic way. The names of many acids are based on their historical origins. For example, formic acid was first prepared by extraction from ants; its name is therefore derived from the Latin word *formica,* meaning "ant." Oxalic acid was first identified as a constituent of the *oxalis* family of plants.

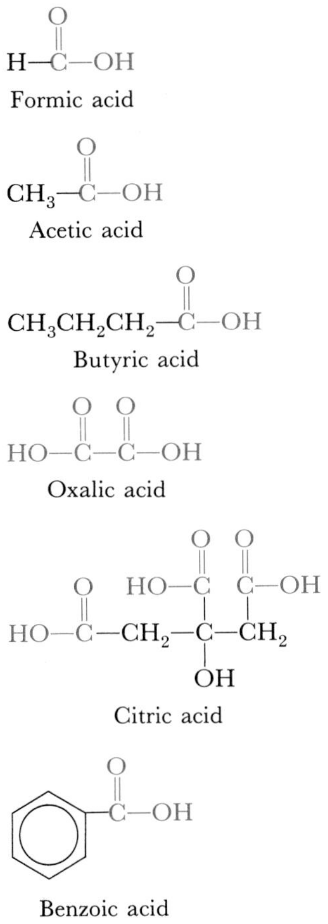

Carboxylic acids are important in the manufacture of polymers used to make fibers, films, and paints. Among the compounds having low molecular weights, acetic acid is an industrially important compound. A relatively new method for manufacture of acetic acid involves reaction of methanol with carbon monoxide, in the presence of a rhodium catalyst:

$$CH_3OH + CO \xrightarrow{\text{catalyst}} CH_3\overset{\displaystyle O}{\overset{\|}{C}}{-}OH \qquad [24.38]$$

Notice that this reaction involves, in effect, the insertion of a carbon monoxide molecule between the CH_3 and OH groups. A reaction of this kind is called **carbonylation.**

Carboxylic acids may react with alcohols, with splitting out of water, to form **esters:**

$$CH_3\overset{\displaystyle O}{\overset{\|}{C}}{-}OH + HO{-}CH_2CH_3 \longrightarrow$$
Acetic acid   Ethanol

$$CH_3\overset{\displaystyle O}{\overset{\|}{C}}{-}O{-}CH_2CH_3 + H_2O \qquad [24.39]$$
Ethyl acetate

$$H{-}\overset{\displaystyle O}{\overset{\|}{C}}{-}OH$$
Formic acid

$$CH_3{-}\overset{\displaystyle O}{\overset{\|}{C}}{-}OH$$
Acetic acid

$$CH_3CH_2CH_2{-}\overset{\displaystyle O}{\overset{\|}{C}}{-}OH$$
Butyric acid

$$HO{-}\overset{\displaystyle O}{\overset{\|}{C}}{-}\overset{\displaystyle O}{\overset{\|}{C}}{-}OH$$
Oxalic acid

$$HO{-}\overset{\displaystyle O}{\overset{\|}{C}}{-}CH_2{-}\underset{\underset{\displaystyle OH}{|}}{\overset{\overset{\displaystyle HO{-}\overset{O}{\overset{\|}{C}}\ \ \overset{O}{\overset{\|}{C}}{-}OH}{|}}{C}}{-}CH_2$$
Citric acid

$$\underset{\text{Benzoic acid}}{\bigcirc\!\!-\!\!\overset{\displaystyle O}{\overset{\|}{C}}{-}OH}$$

FIGURE 24.11 Structural formulas of several familiar carboxylic acids.

761

This is another example of a condensation reaction. The reaction product, the ester, is named by using first the group from which the alcohol is derived and then the group from which the acid is derived.

SAMPLE EXERCISE 24.6

Name the following esters:

(a)

$$\text{C}_6\text{H}_5 - \overset{\displaystyle O}{\overset{\displaystyle \|}{\text{C}}} - \text{OCH}_2\text{CH}_3$$

(b) $\text{CH}_3\text{CH}_3\text{CH}_2 - \overset{\displaystyle O}{\overset{\displaystyle \|}{\text{C}}} - \text{O} - \text{C}_6\text{H}_5$

Solution: In (a) the ester is derived from ethanol and benzoic acid. Its name is therefore ethyl benzoate. In (b) the ester is derived from phenol and butyric acid. The residue from the phenol, C_6H_5, is called the phenyl group. The ester is therefore named phenyl butyrate.

SAMPLE EXERCISE 24.7

Beginning with ethene as the only carbon-containing compound, suggest a series of reactions by which ethyl acetate might be prepared.

Solution: As shown in Equation [24.39], ethyl acetate can be prepared by a condensation reaction between acetic acid and ethanol. The ethanol can be prepared from ethene by H_2SO_4-catalyzed hydration:

$$\text{CH}_2{=}\text{CH}_2 + \text{H}_2\text{O} \xrightarrow{\text{H}_2\text{SO}_4} \text{CH}_3\text{CH}_2\text{OH}$$

This process is analogous to that given in Equation [24.30] for the hydration of propene. A portion of the ethanol thus produced can be oxidized to acetic acid as shown in Equation [24.37].

Esters generally have very pleasant odors; they are largely responsible for the pleasant aromas of fruit. Ethyl butyrate, for example, is responsible for the odor of apples. When esters are treated with acid or base in aqueous solution, they are hydrolyzed; that is, the molecule is split into its alcohol and acid components:

$$\underset{\text{Methyl propionate}}{\text{CH}_3\text{CH}_2 - \overset{\displaystyle O}{\overset{\displaystyle \|}{\text{C}}} - \text{O} - \text{CH}_3} + \text{Na}^+ + \text{OH}^- \longrightarrow$$

$$\underset{\text{Sodium propionate}}{\text{CH}_3\text{CH}_2 - \overset{\displaystyle O}{\overset{\displaystyle \|}{\text{C}}} - \text{O}^-} + \text{Na}^+ + \underset{\text{Methanol}}{\text{CH}_3\text{OH}} \qquad [24.40]$$

In this example, the hydrolysis was carried out in basic medium. The products of the reaction are the sodium salt of the carboxylic acid and the alcohol.

Amides are rather closely related to carboxylic acids and esters. In a simple amide, the OH group of the carboxylic acid is replaced by an NH_2 group, as in these examples:

$$CH_3\overset{\displaystyle O}{\overset{\|}{C}}-NH_2 \qquad \underset{}{\bigcirc}-\overset{\displaystyle O}{\overset{\|}{C}}-NH_2$$

Acetamide   Benzamide

The nitrogen may have one or more organic groups attached in place of the hydrogens. The amide linkage,

$$R-\overset{\displaystyle O}{\overset{\|}{C}}-\underset{\displaystyle H}{\overset{\displaystyle }{N}}-R'$$

where R and R′ are organic groups, is the key functional group in the structures of proteins, about which we shall have more to say in the next chapter. This same linkage is also very important in many synthetic polymeric materials formed by condensation polymerization.

CONDENSATION POLYMERS

A condensation polymer is formed by condensation reactions between molecules in which the functional groups are arranged so as to give long-chain, polymeric products. The most important condensation polymers are the nylons. A substance with two carboxylic acid functional groups is reacted with a substance with two amine functional groups, to yield a polymer. For example, nylon-66 is formed by heating a six-carbon diacid with a six-carbon diamine:

$$\text{---COOH} \quad H_2N-(CH_2)_6-NH_2 \quad HOOC-(CH_2)_4-COOH \quad H_2N-(CH_2)_6-NH_2 \quad HOOC-(CH_2)_4-COOH \quad H_2N\text{---}$$

$$\downarrow \text{heat}$$

$$\text{---}\overset{O}{\overset{\|}{C}}-NH-(CH_2)_6-NH-\overset{O}{\overset{\|}{C}}-(CH_2)_4-\overset{O}{\overset{\|}{C}}-NH-(CH_2)_6-NH-\overset{O}{\overset{\|}{C}}-(CH_2)_4-\overset{O}{\overset{\|}{C}}-NH\text{---} \qquad [24.41]$$

The polymer is formed by loss of water from between each pair of reacting functional groups. The resulting polymer is labeled 66 because there are six carbon atoms in the sections between each NH group along the chain.

Condensation polymers can also result from formation of ester linkages; that is, by condensation of carboxylic acids with alcohols. In forming a polymer, it is important to use difunctional acids and alcohols. By varying the acid and alcohol functions, various kinds of polyester materials can be formed. The familiar polyester Dacron, from which so much of our clothing is produced, is a condensation polymer of ethylene glycol and terephthalic acid:

$$\text{---HOCH}_2CH_2OH \quad HO\overset{O}{\overset{\|}{C}}-\bigcirc-\overset{O}{\overset{\|}{C}}OH \quad HOCH_2CH_2OH \quad HO\overset{O}{\overset{\|}{C}}-\bigcirc-\overset{O}{\overset{\|}{C}}OH\text{---} \xrightarrow{\text{heat}}$$

Ethylene glycol   Terephthalic acid

$$\text{---OCH}_2CH_2O\overset{O}{\overset{\|}{C}}-\bigcirc-\overset{O}{\overset{\|}{C}}OCH_2CH_2O\overset{O}{\overset{\|}{C}}-\bigcirc-\overset{O}{\overset{\|}{C}}OCH_2CH_2O\overset{O}{\overset{\|}{C}}-\bigcirc-\overset{O}{\overset{\|}{C}}O\text{---} \qquad [24.42]$$

Dacron

The polymer formed under typical reaction conditions may have 80 to 100 units per molecule. This material can be melted without decomposition. The molten polymer is forced through tiny holes, and then it cools and solidifies in the form of fibers. As the tiny fibers are stretched, the long molecules are forced to lie in a more or less parallel arrangement. The drawn or stretched fibers are then spun into a yarn.

FOR REVIEW

Summary

In this chapter we have studied the chemical and physical characteristics of simple organic substances. Hydrocarbons are composed of only carbon and hydrogen. There are four major classes of hydrocarbons. The alkanes are composed of only carbon-hydrogen and carbon-carbon single bonds. The alkenes contain one or more carbon-carbon double bonds. Alkynes contain one or more carbon-carbon triple bonds. Aromatic hydrocarbons contain cyclic arrangements of carbon atoms bonded through both σ and π bonds. Isomers are substances that possess the same molecular formula but that differ in some other respect. Isomers may differ in the bonding arrangements of atoms within the molecule or in the geometrical arrangements of groups. Geometrical isomerism is possible in the alkenes because of restricted rotation about the C=C double bond.

The major sources of hydrocarbons are fossil fuels such as coal and oil. Crude oil is separated by distillation into several fractions according to variations in volatility. The most important fraction is gasoline, used as automotive fuel. Less volatile fractions are cracked; that is, they are heated with catalysts to convert them to more volatile substances that have lower molecular weights and are suitable for gasoline. In the same process, straight-chain hydrocarbons are caused to rearrange to the more desirable branched-chain isomers, which have higher octane numbers.

Combustion of hydrocarbons is a highly exothermic process. The chief use of hydrocarbons is as a source of heat energy via combustion. The unsaturated hydrocarbons, the alkenes and alkynes, readily undergo addition reactions to the carbon-carbon multiple bonds. By contrast, addition reactions to aromatic hydrocarbons are difficult to carry out. Addition polymerization of alkenes is a source of many useful synthetic materials.

Substitution reactions are those in which one or more hydrogens of a hydrocarbon are replaced by some other atom or group. The substitution reactions of aliphatic (nonaromatic) hydrocarbons are not easily carried out. However, substitution reactions of aromatic hydrocarbons are easily carried out in the presence of acid catalysts.

The chemistry of hydrocarbon derivatives is often dominated by the nature of their functional groups. The functional groups we have considered are summarized here:

R—O—H   Alcohol

$\overset{\displaystyle O}{\underset{\displaystyle \|}{R—C—H}}$ Aldehyde

$\overset{}{C=C}$ Alkene

—C≡C—   Alkyne

$\overset{\displaystyle O}{\underset{\displaystyle \|}{R—C—N}}$ Amide

$\overset{\displaystyle O}{\underset{\displaystyle \|}{R—C—O—H}}$ Carboxylic Acid

$\overset{\displaystyle O}{\underset{\displaystyle \|}{R—C—O—R'}}$ Ester

R—O—R'   Ether

$\overset{\displaystyle O}{\underset{\displaystyle \|}{R—C—R'}}$ Ketone

(Remember that R and R' represent some hydrocarbon group—for example, methyl, CH_3, or phenyl, C_6H_5.)

Alcohols are hydrocarbon derivatives containing one or more OH groups. Ethers are related compounds that are formed by a condensation reaction of two molecules of alcohol, with splitting out of water. Oxidation of primary alcohols can lead to formation of aldehydes; further oxidation of the aldehydes produces carboxylic acids. Oxidation of secondary alcohols leads to formation of ketones.

Carboxylic acids can form esters by a condensation reaction with alcohols, or they can form amides by condensation reaction with amines. Condensation polymers are formed by condensation reactions involving molecules with functional groups on both ends.

Learning goals

Having read and studied this chapter, you should be able to:

1 List the four groups of hydrocarbons and draw the structural formula of an example from each group.

2 List the names of the first ten members of the alkane series of hydrocarbons.

3 Write the structural formula of an alkane, alkene, or alkyne, given its systematic (IUPAC) name.

4 Name an alkane, alkene, or alkyne, given its structural formula.

5 Give an example of geometrical isomerism in alkenes.

6 Explain why aromatic hydrocarbons do not readily undergo addition reactions as do the alkenes and alkynes.

7 List and describe the fractions obtained in petroleum refining.

8 Explain the general relationship between octane number and structure in alkanes and describe the methods used to increase the octane numbers of straight-run gasolines.

9 Give examples of addition reactions of alkenes and alkynes, showing the structural formulas of reactants and products.

10 Describe Markovnikoff's rule and apply it to the addition reactions of alkenes and alkynes.

11 Give an example of addition polymerization of an alkene.

12 Write the steps in the photo-initiated chlorination of an alkane.

13 Give two or three examples of substitution of an aromatic compound.

14 Write structural formulas of molecules that are examples of an alcohol, an ether, an aldehyde, a ketone, a carboxylic acid, an ester, and an amide.

15 Describe the general character of condensation reactions and give an example of condensation polymerization.

Key terms

Among the more important terms and expressions used for the first time in this chapter are the following:

Addition polymerization (Section 24.3) is a reaction in which alkenes add together end-to-end by opening of the carbon-carbon double bond, to form long polymeric chain molecules.

An **addition reaction** (Section 24.3) is one in which a reagent adds to the two carbon atoms of a carbon-carbon multiple bond.

Alcohols (Section 24.4) are hydrocarbons in which one or more hydrogens have been replaced by a hydroxyl group, OH.

Aliphatic hydrocarbons (Section 24.1) are those that are not aromatic; that is, the alkanes, cycloalkanes, alkenes, and alkynes.

Alkanes (Section 24.1) are compounds of carbon and hydrogen containing only single carbon-carbon bonds.

Alkenes (Section 24.1) are hydrocarbons containing one or more carbon-carbon double bonds.

Alkynes (Section 24.1) are hydrocarbons containing one or more carbon-carbon triple bonds.

An **antiknock agent** (Section 24.2) is a substance added to automotive fuel to increase its octane number. This agent slows the rate of burning of fuel in the cylinder.

Aromatic hydrocarbons (Section 24.1) are hydrocarbon compounds that contain a planar, cyclic arrangement of carbon atoms linked by both σ and π bonds.

Base hydrolysis (Section 24.4) is a chemical reaction involving the replacement of a negatively charged functional group by aqueous hydroxide ion.

The **carbonyl** (Section 24.5) functional group, $C=O$, is a characteristic feature of several organic functional groups such as ketones or carboxylic acids.

Carbonylation (Section 24.5) is a reaction in which the carbonyl functional group is introduced into a molecule.

A **condensation polymer** (Section 24.5) is a substance having a high molecular weight that is formed by elimination of water from between molecules.

Cycloalkanes (Section 24.1) are saturated hydrocarbons of general formula C_nH_{2n}, in which the carbon atoms form a closed ring.

A **free radical** (Section 24.3) is a species with an odd number of electrons.

A **functional group** (introduction, Section 24.4) is an atom or group of atoms that has characteristic chemical properties.

Geometrical isomers (Section 24.1) have the same molecular formulas and functional groups, but differ in the geometrical arrangements of atoms. For example, *cis*- and *trans*-2-butenes are geometrical isomers.

In **homologous series** (Section 24.1), compounds contain common structural elements, but differ in the number of atoms making up the molecule. Members of a homologous series have the same general formula. For example, the aliphatic saturated alcohols have the general formula $C_nH_{2n+2}O$.

A **hydrocarbon** (Section 24.1) is a compound composed of carbon and hydrogen. Substituted hydrocarbons may contain functional groups composed of atoms other than carbon and hydrogen.

Markovnikoff's rule (Section 24.3) states that when a reagent of general formula XY adds to an alkene, the more negatively charged species, X or Y, ends up on the carbon with the fewer attached hydrogen atoms. For example, on addition of HCl to propene, Cl ends up on the middle carbon atom.

An olefin (Section 24.1) is a compound containing one or more carbon-carbon double bonds.

In a **substitution reaction** (Section 24.3), one atom or functional group bonded to a larger molecule is replaced by another.

Unsaturated hydrocarbons (Section 24.1) are nonaromatic compounds of carbon and hydrogen that contain one or more carbon-carbon multiple bonds.

EXERCISES

Hydrocarbon structures and nomenclature

24.1 Write the molecular formulas of an alkane, alkene, alkyne, and aromatic hydrocarbon that in each case contains six carbon atoms.

24.2 Write the structural formulas for all the alkanes with the molecular formula C_6H_{14}. Name each compound.

24.3 Write the structural formula for each of the following hydrocarbons: (a) $CH_2{=}CHCH_3$; (b) $CH_3CH_2CH_2CH_2CH_3$; (c) $CH_3C{\equiv}CCH_3$; (d) $(CH_3)_2CHCH_2CH(CH_3)_2$; (e) $CH_2{=}CHCH_2CH{=}CH_2$.

24.4 Draw the condensed structural formulas of as many aliphatic compounds as you can think of that have the molecular formula C_5H_8.

24.5 What are the characteristic bond angles (a) about carbon in an alkane; (b) about the carbon-carbon double bond in an alkene; (c) about the carbon-carbon triple bond in an alkyne?

24.6 Write the condensed structural formula for each of the following compounds: (a) 5-methyl-*trans*-2-heptene; (b) 3-chloropropyne; (c) *ortho*-dichlorobenzene; (d) 2,2,4,4-tetramethylpentane; (e) 2-methyl-3-ethylhexane; (f) 2-methyl-6-chloro-3-heptyne; (g) 1,5-dimethylnaphthalene; (h) methylcyclobutane.

24.7 Write the structural formula for each of the following compounds: (a) 2,2-dimethylpentane; (b) 2,3-dimethylhexane; (c) *cis*-2-hexene; (d) methylcyclopentane; (e) 2-chlorobutane; (f) 1,2-dibromobenzene.

24.8 Name each of the following compounds:

(a)
$$\begin{array}{c} CH_3CH_2 \quad\; CH_3 \\ \quad\; | \qquad\; | \\ CH_2CHCHCH_3 \\ \quad\; | \\ \quad\; CH_3 \end{array}$$

(b)
$$\begin{array}{c} \qquad\qquad CH_3 \\ \qquad\qquad\; | \\ CH_3CH_2\!\!\diagdown \qquad CH_2CHCH_3 \\ \qquad\quad C{=}C \\ \qquad H \qquad\quad H \end{array}$$

(c) (naphthalene ring with Cl, Cl)

(d)
$$\begin{array}{c} \qquad\qquad CH_3 \\ \qquad\qquad\; | \\ CH{\equiv}CCH_2CCH_3 \\ \qquad\qquad\; | \\ \qquad\qquad CH_3 \end{array}$$

(e)
$$\begin{array}{c} CH_3CHCH_2CH_3 \\ \;\;| \\ Br \end{array}$$

(f) $CH_3CH{=}CHCH_2CH{=}CHCH_3$

(g)
$$\begin{array}{c} \;\; Br \\ \;\; | \\ CH_3CHCHCH_2CH_2CH_3 \end{array}$$
(benzene ring)

24.9 Indicate whether each of the following molecules is capable of *cis-trans* geometric isomerism; for those that are, draw the structures of the isomers: (a) 2,3-dichlorobutane; (b) 2,3-dichloro-2-butene; (c) 1,3-dimethylbenzene; (d) 1,4-dimethyl-2-pentyne.

24.10 Draw the structural formulas for all possible geometric isomers for 2,6-octadiene.

24.11 Draw the structures of all isomers of (a) dichlorobenzene; (b) trichlorobenzene; (c) chlorobromomethylbenzene.

Petroleum

24.12 Indicate briefly the meaning of each of the following terms: (a) cracking; (b) isomerization; (c) antiknock agent; (d) octane number; (e) branched-chain alkane; (f) refining.

24.13 What is the octane number of a mixture of 20 percent *n*-heptane, 30 percent methylcyclohexane, and 50 percent isooctane? (Refer to Table 24.4.)

24.14 Describe two ways in which the octane number of a gasoline consisting of alkanes can be increased.

24.15 In places where the average temperature varies widely with the season, gasoline used during the winter contains a greater proportion of hydrocarbons with lower molecular weights than does gasoline used during the summer. Suggest a reason for this.

Reactions of hydrocarbons

24.16 Give an example, in the form of a balanced equation, of each of the following chemical reactions: (a) substitution reaction of an alkane; (b) oxidation of an alkene; (c) addition reaction of an alkyne; (d) substitution reaction of an aromatic hydrocarbon.

24.17 In each of the following pairs, indicate which molecule is the more reactive and give a reason for the greater reactivity: (a) butane and cyclobutane; (b) cyclohexane and cyclohexene; (c) benzene and 1-hexene; (d) 2-hexyne and 2-hexene.

24.18 Using condensed structural formulas, write a balanced chemical equation for each of the following reactions: (a) hydrogenation of 1-butene; (b) addition of H_2O to *cis*-2-butene using H_2SO_4 as a catalyst; (c) complete oxidation of cyclobutane; (d) addition polymerization of vinyl chloride (see Table 24.5).

24.19 Using Markovnikoff's rule, predict the product formed when HCl is reacted with each of the following compounds:

(a) $CH_3CH{=}CH_2$

(b) 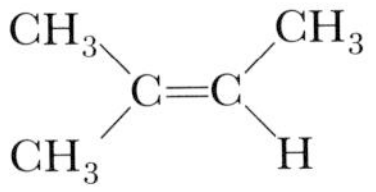
$$
\begin{array}{c}
CH_3 \\
\diagdown \\

\end{array}
C{=}C
\begin{array}{c}
CH_3 \\
 \\
H
\end{array}
$$
with CH_3 below left

(c) $CH_3C{\equiv}CH$

24.20 Why do addition reactions occur more readily with alkenes and alkynes than with aromatic hydrocarbons?

24.21 Predict the product or products formed in each of the following reactions:

(a) *cis*-$CH_3CH{=}CHCH_3 + H_2 \xrightarrow{Ni}$

(b) $CH_3CH_2C{\equiv}CH + HI \longrightarrow$

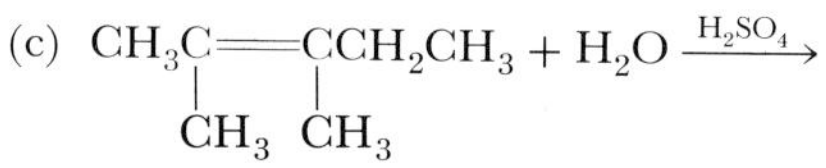

(c) $CH_3C{=}CCH_2CH_3 + H_2O \xrightarrow{H_2SO_4}$ with CH_3 and CH_3 below

(d) $CH_2{=}CHCH_3 + Br_2 \longrightarrow$

24.22 When cyclopropane is treated with HI, 1-iodopropane is formed. A similar type of reaction does not occur with cyclopentane or cyclohexane. How do you account for the reactivity of cyclopropane?

24.23 Using Appendix D, calculate the heat of combustion per mole for (a) ethane; (b) ethene; (c) ethyne. In each case assume gaseous products.

24.24 The heat of combustion of decahydronaphthalene, $C_{10}H_{18}$, is -6286 kJ/mol. The heat of combustion of naphthalene, $C_{10}H_8$, is -5157 kJ/mol. (In each case $CO_2(g)$ and $H_2O(l)$ are the products.) Using these data and data in Appendix D, calculate the heat of hydrogenation of naphthalene. Does this value provide any evidence for aromatic character in naphthalene?

24.25 Suggest a method of preparing ethylbenzene starting with benzene and ethylene as the only organic reagents.

Hydrocarbon derivatives

24.26 Identify the functional groups in each of the following compounds:

(a) $CH_3CCH_2CH_3$ with $\underset{O}{\|}$

(b) $CH_3C{=}O$ with $\underset{OH}{|}$

(c) $CH_2CH_2CH_3$ with HO below

(d) $CH_3OCCH_2CH_3$ with $\underset{O}{\|}$

(e) $HCCH_2CH_3$ with $\underset{O}{\|}$

(f) $CH_3C{=}O$ with CH_3 above and $|$

(g) NH_2CCH_3 with $\underset{O}{\|}$

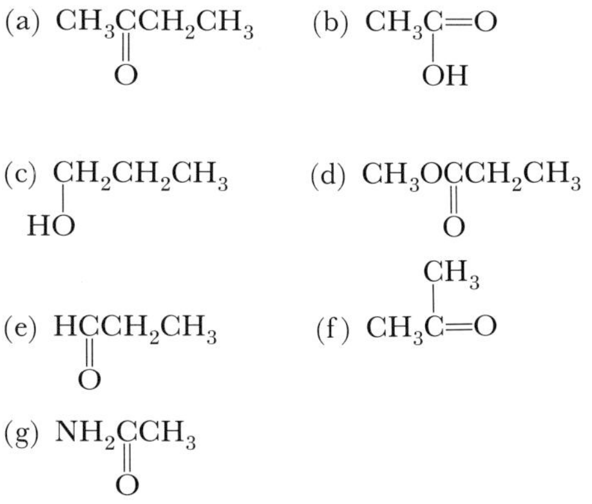

24.27 What are the bond angles expected around each carbon atom in acetone, whose structural formula is shown below?

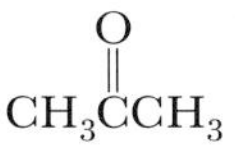

$$CH_3\overset{O}{\overset{\|}{C}}CH_3$$

24.28 Alcohols of low molecular weight are moderately soluble in water, whereas ethers of about the same molecular weight are not. Explain.

24.29 What products (if any) occur when each of the following compounds is mildly oxidized:

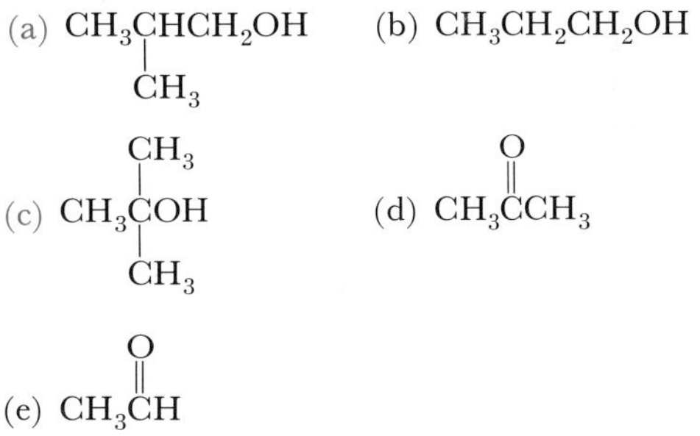

(a) CH_3CHCH_2OH with CH_3 below

(b) $CH_3CH_2CH_2OH$

(c) CH_3COH with CH_3 above and CH_3 below

(d) CH_3CCH_3 with $\overset{O}{\|}$

(e) CH_3CH with $\overset{O}{\|}$

24.30 Draw the structures of the esters formed from (a) acetic acid and 2-propanol; (b) acetic acid and 1-propanol; (c) formic acid and ethanol.

24.31 Name each of the compounds in question 24.29.

24.32 Write structural formulas for each of the following compounds: (a) 2-butanol; (b) 1,2-ethanediol; (c) methylformate; (d) diethyl ketone; (e) diethyl ether.

24.33 The IUPAC name for a carboxylic acid is based on the name of the hydrocarbon with the same number of carbon atoms. The ending *-oic* is appended as in ethanoic acid, which is the IUPAC name for acetic acid,

$$CH_3\overset{\overset{\displaystyle O}{\|}}{C}OH$$

Give the IUPAC name for each of the following acids:

(a) $H\overset{\overset{\displaystyle O}{\|}}{C}OH$

(b) $CH_3CH_2CH_2\overset{\overset{\displaystyle O}{\|}}{C}OH$

(c) $CH_3CH_2\underset{\underset{\displaystyle CH_3}{|}}{C}HCH_2\overset{\overset{\displaystyle O}{\|}}{C}OH$

24.34 Aldehydes and ketones can be named in a systematic way by counting the number of carbon atoms (including the carbonyl carbon) that they contain. The name of the aldehyde or ketone is based on the hydrocarbon with the same number of carbon atoms. The endings *-al*, for aldehyde, or *-one*, for ketone, are added as appropriate. Draw the structural formulas for the following aldehydes or ketones: (a) propanal; (b) 2-pentanone; (c) 3-methyl-2-butanone; (d) 2-methylbutanal.

24.35 Write balanced chemical equations, with organic substances given by their condensed formulas, describing the preparation of (a) acetone from 1-propene; (b) propanoic acid (a carboxylic acid with three carbon atoms) from 1-propanol; (c) sodium acetate from ethene; (d) methyl acetate from methyl alcohol and acetic acid; (e) methyl ethyl ketone from 2-butanol.

24.36 Write a balanced chemical equation to describe what happens when (a) methyl acetate is treated with sodium hydroxide; (b) methanol is treated with sulfuric acid.

Additional exercises

24.37 Draw the structural formulas of two molecules with the formula C_4H_6.

24.38 What is the molecular formula of (a) an alkane with 20 carbon atoms; (b) an alkene with 18 carbon atoms; (c) an alkyne with 12 carbon atoms?

24.39 Draw the Lewis structures for the *cis* and *trans* isomers of 2-pentene. Does cyclopentane exhibit *cis-trans* isomerism? Explain.

24.40 Classify each of the following substances as alkane, alkene, or alkyne (assuming that none are cyclic hydrocarbons): (a) C_5H_{12}; (b) C_5H_8; (c) C_6H_{12}; (d) C_8H_{18}.

24.41 Give the IUPAC name for each of the following molecules:

(a) $CH_3\underset{\underset{\displaystyle OH}{|}}{C}HCH_3$ (b) CH_3OCH_3

(c) $CH_2\text{=}CHCH_2CH_2OH$

(d) $CH_3\underset{\underset{\displaystyle CH_2CH_3}{|}}{C}\text{=}CHCH_2CH_3$ (e) $CH_3C\text{≡}C\text{—}\underset{\underset{\displaystyle CH_3}{|}}{C}H\;\overset{\overset{\displaystyle CH_3}{}}{}$

24.42 Describe how you would prepare (a) 2-bromo-2-chloropropane from propyne; (b) 1,2-dichloroethane from ethene; (c) a condensation polymer from $HOCH_2CH_2OH$ and

$$HO\overset{\overset{\displaystyle O}{\|}}{C}CH_2CH_2\overset{\overset{\displaystyle O}{\|}}{C}OH$$

(d) sodium benzoate from methylbenzoate, (e) polypropylene from propene; (f) 1-ethylnaphthalene from naphthalene and ethene.

24.43 Write the formulas for all of the structural isomers of C_3H_8O.

24.44 Explain the fact that *trans*-1,2-dichloroethene has no dipole moment while *cis*-1,2-dichloroethene has a dipole moment.

24.45 Would you expect cyclohexyne to be a stable compound? Explain.

24.46 Identify all of the functional groups in each of the following molecules:

(a) $CH_2\text{=}CH\text{—}O\text{—}CH\text{=}CH_2$ (an anesthetic)

(b)   (acetylsalicyclic acid, aspirin)

(c)   (testosterone, a male sex hormone)

24.47 Write the structural formulas for each of the following: (a) an ether with the formula C_3H_8O; (b) an aldehyde with the formula C_3H_6O; (c) a ketone with the formula C_3H_6O; (d) a secondary alkyl fluoride with the formula C_3H_7F; (e) a primary alcohol with the formula $C_4H_{10}O$; (f) a carboxylic acid with the formula $C_3H_6O_2$; (g) an ester with the formula $C_3H_6O_2$.

24.48 Give the condensed formulas for the carboxylic acid and the alcohol that each of the following esters is formed from:

(a)   (b)

24.49 A 256-mg sample of an organic compound is combusted producing 512 mg of CO_2 and 209 mg of H_2O. At 127°C, 155 mg of this substance occupies a volume of 0.100 L with a pressure of 1.16 atm. What is the molecular formula of the compound? Suggest a possible structure for the compound and suggest some chemical test for the functional group.

[24.50] Suggest a process that would cause the breakup of a condensation polymer by essentially reversing the condensation reaction.

[24.51] An unknown organic compound is found on analysis to have the empirical formula $C_5H_{12}O$. It is slightly soluble in water. Upon careful oxidation it is converted into a compound of empirical formula $C_5H_{10}O$, which behaves chemically like a ketone. Indicate two or more reasonable structures for the unknown.

[24.52] Consider two hydrocarbons, A and B, whose molecular formulas are both C_4H_8. The first compound, A, is converted to $C_4H_{10}O$ when treated with water and sulfuric acid. Mild oxidation of this product produces C_4H_8O, which is an excellent solvent. Treatment with $K_2Cr_2O_7$ does not produce any further change in the C_4H_8O. In contrast, compound B gives a different $C_4H_{10}O$ substance upon treatment with sulfuric acid and water. Upon treating this $C_4H_{10}O$ substance with $K_2Cr_2O_7$, no oxidation occurs. Identify the two compounds, A and B, and trace all the reactions described that begin with these compounds, writing the names and structural formulas for all the organic compounds involved.

[24.53] An unknown substance is found to contain only carbon and hydrogen. It is a liquid that boils at 49°C at 1 atm pressure. Upon analysis it is found to contain 85.7 percent carbon and 14.3 percent hydrogen. At 100°C and 735 mm Hg, the vapor of the unknown has a density of 2.21 g/L. When it is dissolved in hexane solution and bromine water added, no reaction occurs. Suggest the identity of the unknown compound.

24.54 Bromination of butane requires irradiation. Write a series of reaction steps that can account for the bromination of butane under irradiation conditions.

25

The biosphere

In earlier chapters of this text, we have considered various aspects of the physical world in which we live. We have discussed chemistry of the atmosphere (Chapter 10), of natural waters (Chapter 17), and of the solid earth (Chapter 22). In this final chapter we consider the biosphere, which is defined as that part of the earth in which living organisms are formed and live out their life cycles. The biosphere is not distinct from the atmosphere, natural waters, or the solid earth; rather, it is an integral part of it in those places where conditions permit life to exist. For living organisms to be sustained, there must be a supply of available energy, because the growth and maintenance of organisms requires energy. Secondly, there must be an adequate supply of water, because organisms are composed largely of water and use it for exchange of materials with their environment.

25.1 Energy requirements of organisms

Living organisms require energy for their maintenance, growth, and reproduction. The ultimate source of this energy is the sun. However, as living matter proliferated on earth, and as organisms became more and more specialized, many of them developed the capacity for obtaining energy indirectly, by utilizing the energy stored in other organisms. Thus, for example, our bodies have essentially no capacity for directly utilizing solar energy. We consume animal and plant materials to obtain substances that our bodies can utilize as energy sources.

There are two distinct reasons why living systems need energy. First, organisms rely on substances readily available in their surroundings for synthesis of compounds needed for their existence. Most of these reactions are endothermic. To make such reactions proceed, energy must be supplied from an outside source. Secondly, living organisms are highly organized. The complexity of all the substances that make up even the simplest of single-cell organisms and the relationships between all the many chemical processes occurring are truly amazing. In thermodynamic terms, this means that living systems are very low in entropy as compared with the raw materials from which they are formed. The orderliness that is characteristic of living systems is attained by expenditure of energy.

Recall from the discussion in Section 18.5 that the free-energy change, ΔG, is related to both the enthalpy and the entropy changes that occur in a process:

$$\Delta G = \Delta H - T\Delta S$$

If the entropy change in a process that builds up a living organism is negative (in other words, if a more highly ordered state results), the contribution to ΔG is positive. This means that the process becomes less spontaneous. Thus, both the enthalpy and the entropy changes that result in the formation, maintenance, and reproduction of living systems are such that the overall process is nonspontaneous. To overcome the positive values for ΔG associated with the essential processes, living systems must be coupled to some outside source of energy that can be converted into a form useful for driving the biochemical processes. The ultimate source of this needed energy is the sun.

The major means for conversion of solar energy into forms that can be utilized by living organisms is photosynthesis. The photosynthetic reaction that occurs in the leaves of plants is conversion of carbon dioxide and water to carbohydrate, with release of oxygen:

$$6CO_2 + 6H_2O \xrightarrow{48\ h\nu} C_6H_{12}O_6 + 6O_2 \qquad [25.1]$$

Note that formation of a mole of sugar, $C_6H_{12}O_6$, requires the absorption and utilization of 48 photons. This needed radiant energy comes from the visible region of the spectrum (Figure 5.3). The photons are absorbed by photosynthetic pigments in the leaves of plants. The key pigments are the chlorophylls; the structure of the most abundant chlorophyll, called chlorophyll-a, is shown in Figure 25.1. Chlorophyll is a coordination compound; it contains a Mg^{2+} ion bound to four nitrogen atoms arranged around the metal in a planar array. The nitrogen atoms are part of a porphyrin ring (Section 23.2). Notice that there is a series of

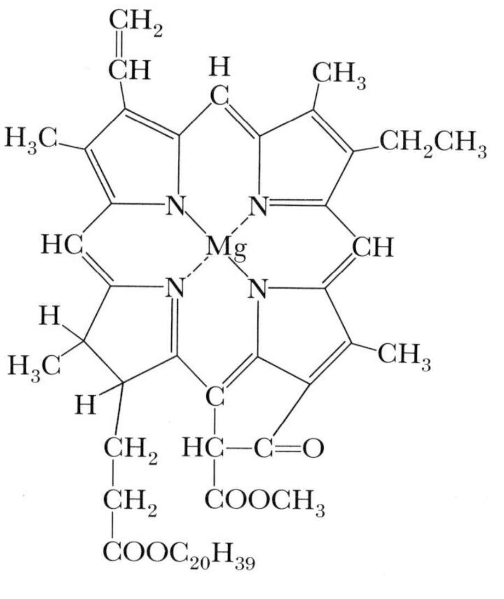

FIGURE 25.1 The structure of chlorophyll-a. All chlorophyll molecules are essentially alike; they differ only in details of the side chains.

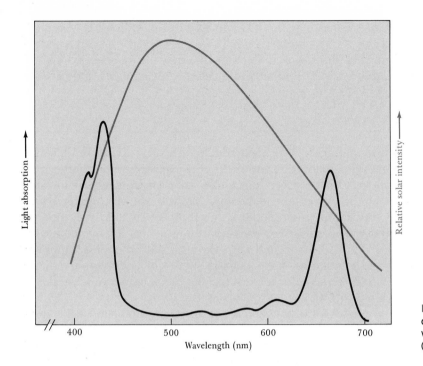

alternating double bonds in the ring that surrounds the metal ion. This system of alternating, or conjugated, double bonds gives rise to the strong absorptions of chlorophyll in the visible region of the spectrum. Figure 25.2 shows the absorption spectrum of chlorophyll as compared with the distribution of solar energy at the earth's surface. Chlorophyll is green in color because it absorbs red light (maximum absorption at 655 nm) and blue light (maximum absorption at 430 nm) and transmits green light.

The solar energy absorbed by chlorophyll is converted in a complex series of steps into chemical energy. This stored energy is then used to drive the reaction in Equation [25.1] to the right, a direction in which it is highly endothermic. Thus, plant photosynthesis is nature's solar-energy-conversion machine; all living systems on earth are dependent on it for continued existence. A field of corn in the Midwest at the height of its growing season converts several percent of all the incident solar radiation into plant matter. It has been suggested that by cultivating about 6 percent of the land surface in the United States and using optimal growing conditions, we could obtain sufficient energy to satisfy all the energy needs of modern society.

Let us now turn our attention to the materials from which living organisms are formed. At some level of concentration, in some organism somewhere, almost every element seems to have a role to play in the biosphere. However, the elements of major importance in terms of their abundances in living systems are carbon, hydrogen, oxygen, nitrogen, phosphorus, and sulfur. Many other elements, including many metallic elements, are present in lesser quantities (refer back to Figure 23.5).

Much of the material present in living systems is in the form of macromolecules, polymers of high molecular weight. These biopolymers can be classified into three broad groups: proteins, carbohydrates, and nucleic acids. The proteins and carbohydrates, along with a class of molecules called fats and oils, are the major sources of energy in the food supply of animals. In addition, polymeric carbohydrates are the major construction material that gives form to plant systems, whereas proteins perform a similar role in animal systems. The nucleic acids are the storehouse of information that determines the form of a living system and that regulates its reproduction and development.

25.2 Proteins

Proteins are macromolecular substances present in all living cells. They serve as the major structural component in animal tissues; they are a major component of skin, cartilage, nails, and the skeletal muscles. Enzymes, the catalysts for the biochemical reactions occurring in all living systems, are proteins. Proteins transport vital substances within the body. For examle, hemoglobin, which carries O_2 from the lungs to cells, is a protein. The antibodies that serve as a defense mechanism against undesirable substances within the organism are also composed of proteins.

The molecular weights of proteins vary from about 10,000 to over 50 million. Simple proteins are composed entirely of amino acids; other proteins, called conjugated proteins, are made up of simple proteins bonded to other kinds of biochemical structures. The most abundant elements present in proteins are carbon (50 to 55 percent), hydrogen (7 percent), oxygen (23 percent), and nitrogen (16 percent). Sulfur is present in most proteins to the extent of about 1 or 2 percent; phosphorus either is absent or is present to only a very slight extent.

Simple proteins are linear polymers of amino acids. The characteristic polymeric linkage in proteins is the amide linkage, introduced in Section 24.5:

$$R-\overset{\displaystyle O}{\overset{\displaystyle \|}{C}}-\underset{\displaystyle \underset{\textstyle H}{|}}{N}-R'$$

The particular functional group is called the peptide linkage when it is formed from amino acids. Proteins are sometimes referred to as polypeptides. Usually, however, the term polypeptide refers to a polymer of amino acids that has a molecular weight of less than 10,000. To see how the peptide linkage might serve to form a polymer we must investigate the structures of the amino acids, substances from which proteins are formed.

AMINO ACIDS

Twenty amino acids are found to occur commonly in various proteins. All of these are α-amino acids; that is, the amino group is located on the carbon atom immediately adjacent to the carboxylic acid functional group. The general formula for an amino acid is as follows:

$$R-\overset{\displaystyle \overset{\textstyle H}{|}}{\underset{\displaystyle \underset{\textstyle NH_2}{|}}{C}}-\overset{\displaystyle O}{\overset{\displaystyle \|}{C}}-OH$$

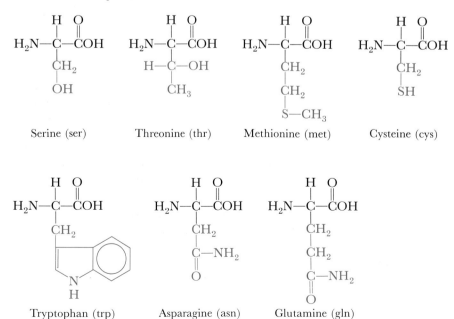

Amino acids with hydrocarbon side chains

Glycine (gly)   Alanine (ala)   Valine (val)   Leucine (leu)

Isoleucine (ile)   Phenylalanine (phe)   Proline (pro)

Amino acids with polar, neutral side chains

Serine (ser)   Threonine (thr)   Methionine (met)   Cysteine (cys)

Tryptophan (trp)   Asparagine (asn)   Glutamine (gln)

FIGURE 25.3 (cont.)

Amino acids with acidic or basic side chain functional groups

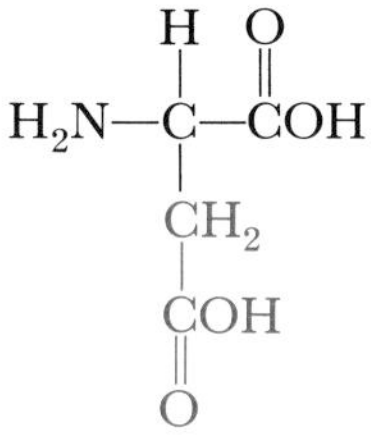

Aspartic acid (asp) 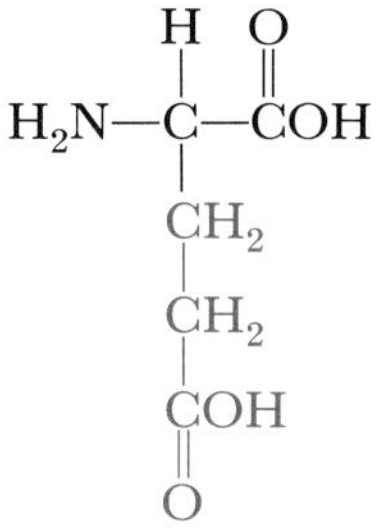   Glutamic acid (glu) 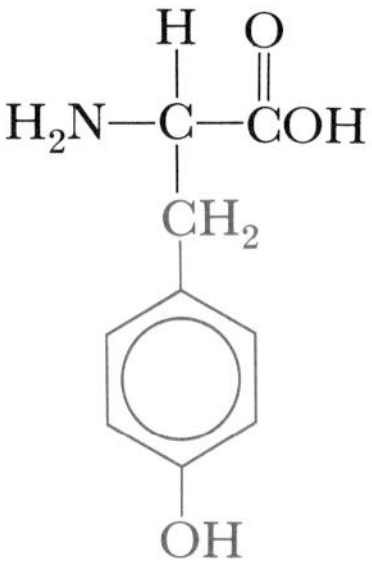   Tyrosine (tyr)

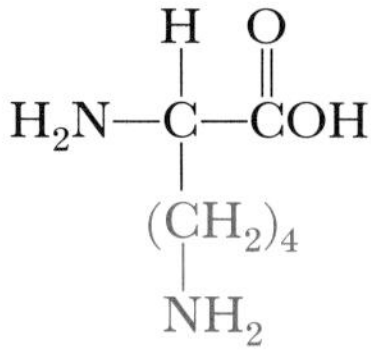

Lysine (lys) 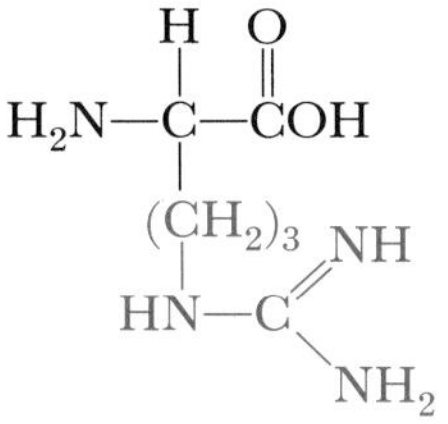   Arginine (arg) 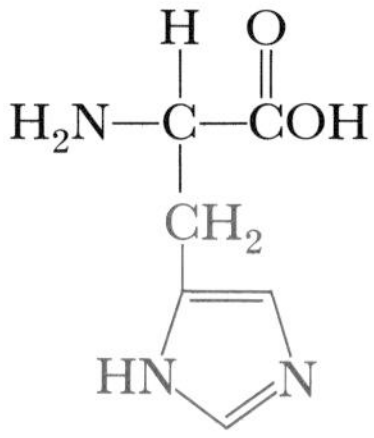   Histidine (his)

The amino acids differ in the nature of the R group. Figure 25.3 shows the structural formulas of the 20 amino acids found in most proteins. Although certain of the amino acids are more common than others, most large proteins contain most of the amino acids.

You can see from the condensed structural formula shown above that the α-carbon atom, which bears both the amino and carboxylic acid groups, has four different groups attached to it.* Any molecule containing a carbon with four different attached groups will be **chiral.** As we pointed out in Section 23.4, a chiral molecule is one that is not superimposable with its mirror image. Let us consider a particular amino acid, say alanine, and suppose that we have two such molecules that are mirror images of one another, as in Figure 25.4. This figure also shows two familiar objects that are mirror images of one another, a pair of hands. Your right and left hands are not superimposable. That is, if there were some way that you could hollow out your left hand and place it face down on a surface, there is no way that you could slide your right hand, also face down, into the left. (That is, of course, why you can't wear your right-hand glove on your left hand.) In the same way, one of the mirror-image forms of alanine cannot be superimposed on the other.

The two molecules of a chiral substance that are mirror images of one another are called **enantiomers.** Because two enantiomers are not exactly

*The sole exception is glycine, for which R = H. For this amino acid, there are two H atoms on the α-carbon atom.

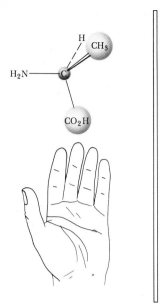

the same, they are isomers of one another. This type of isomerism is called **configurational isomerism,** or **optical isomerism.** The two enanti-omers of a pair are sometimes labeled *R-* (from the Latin *rectus,* "right") or *S-* (from the Latin *sinister,* "left") to distinguish them. Another widely used notation for distinguishing the enantiomers uses the prefix labels D- (from the Latin *dexter,* "right") or L- (from the Latin *laevus,* "left"). The enantiomers of a chiral substance possess the same physical properties, such as solubility, melting point, and so forth. Their chemical behavior toward ordinary chemical reagents is also the same. However, they differ in chemical reactivity toward other chiral molecules. It is a striking fact that all the amino acids in nature are of the *S-,* or L-, configuration at the carbon center (except glycine, which is not chiral). Only amino acids of this specific configuration at the chiral carbon center are biologically effective in forming polypeptides and proteins in most organisms; pep-tide linkages are formed in cells under such specific conditions that enan-tiomeric molecules can be distinguished.

Chiral substances differ from one another in their effect on the plane of polarized light. We saw an example of this in Section 23.4, which dealt with the isomerism of coordination compounds. Suppose a beam of po-larized light is passed through a solution containing a chiral substance such as alanine, using the arrangement shown in Figure 23.14. A solution of one of the enantiomers causes the plane of polarization to be rotated in one direction; a solution of the other enantiomer has the opposite effect. In each case the amount of rotation is proportional to the concentration of the solution and the length of the cell containing it. A solution con-taining equal concentrations of the two enantiomers, called a **racemic mixture,** causes no net rotation. Enantiomers of a chiral substance are often called **optical isomers** because of their effect on polarized light.

The relative amounts of the various amino acids in a protein vary with the nature and function of the protein material. Amino acids that have hydrocarbon side chains predominate in the insoluble, fiberlike proteins such as silk, wool, and collagen (a tissue-supporting protein). The amino acids with polar side chains are relatively more abundant in the water-soluble proteins. In addition, the polar groups often play an important role in determining the overall structure of the protein molecule, about which we shall have more to say later. Amino acids with acid or base side chains also help to increase water solubility. In addition, the functional groups on the side chains can act as sources of acid or base character in reactions catalyzed by acids or bases.

PROTEIN STRUCTURE

We have noted that the characteristic functional group in proteins is the amide, or peptide, linkage. This linkage is formed by a condensation reaction (Sections 24.4 and 24.5) between two amino acid molecules. As an example, alanine and glycine might form the dipeptide glycylalanine:

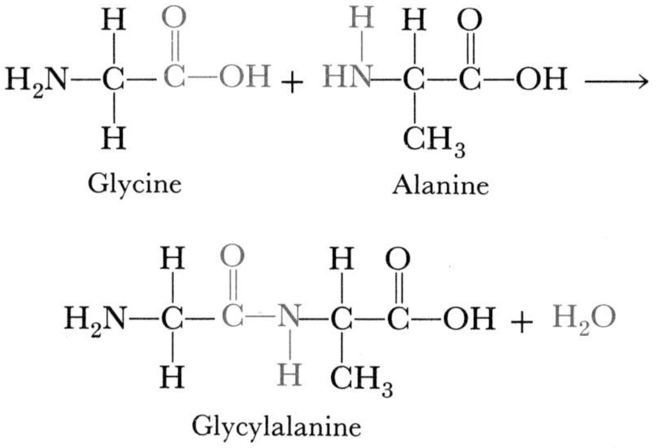

$$H_2N-\underset{\underset{H}{|}}{\overset{\overset{H}{|}}{C}}-\overset{\overset{O}{\|}}{C}-OH + HN-\underset{\underset{CH_3}{|}}{\overset{\overset{H}{|}}{C}}-\overset{\overset{O}{\|}}{C}-OH \longrightarrow$$

Glycine   Alanine

$$H_2N-\underset{\underset{H}{|}}{\overset{\overset{H}{|}}{C}}-\overset{\overset{O}{\|}}{C}-\underset{\underset{H}{|}}{N}-\underset{\underset{CH_3}{|}}{\overset{\overset{H}{|}}{C}}-\overset{\overset{O}{\|}}{C}-OH + H_2O \qquad [25.2]$$

Glycylalanine

Notice that the acid that furnishes the carboxyl group is named first, with a *-yl* ending; then the amino acid furnishing the amino group.

SAMPLE EXERCISE 25.1

Draw the structural formula for histidylglycine.

Solution: The name for this dipeptide tells us that the amino group of glycine and the carboxylic acid function of the histidine are involved in formation of the peptide bond. The structure is therefore:

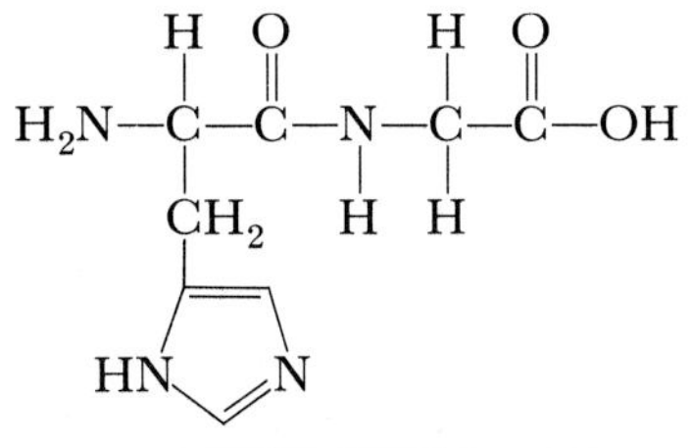

$$H_2N-\underset{\underset{CH_2}{|}}{\overset{\overset{H}{|}}{C}}-\overset{\overset{O}{\|}}{C}-\underset{\underset{H}{|}}{N}-\underset{\underset{H}{|}}{\overset{\overset{H}{|}}{C}}-\overset{\overset{O}{\|}}{C}-OH$$

Because 20 amino acids are commonly found in proteins and because the protein chain may consist of hundreds of amino acids, the number of possible arrangements of amino acids is huge beyond imagination. Nature makes use of only certain of these combinations, but even so the number of different proteins is very large.

The arrangement, or sequence, of amino acids along the chain determines the **primary structure** of a protein. This primary structure gives the protein its unique identity. A change of even one amino acid can alter the biochemical characteristics of the protein. For example, sickle-cell anemia is a genetic disorder resulting from a single misplacement in a protein chain in hemoglobin. The chain that is affected contains 146 amino acids. The first seven in the chain are valine, histidine, leucine, threonine, proline, glutamic acid, and glutamic acid. In a person suffering from sickle-cell anemia, the sixth amino acid is valine instead of glutamic acid. This one substitution of an amino acid with a hydrocarbon side chain for one that has an acidic functional group in the side chain alters the solubility properties of the hemoglobin, and normal blood flow is impeded (see also Section 12.8).

Proteins in living organisms are not simply long, flexible chains with more or less random shape. Rather, the chain coils or stretches in particular ways that are essential to the proper functioning of the protein. This aspect of the protein structure is called the **secondary structure.** One of the most important and common secondary structure arrangements is the **α-helix,** first proposed by Linus Pauling and R. B. Corey. The helix arrangement is shown in schematic form in Figure 25.5. Imagine winding a long protein chain in a helical fashion around a long cylinder. The helix is held in position by hydrogen-bond interactions between an N—H bond and the oxygen of a carbonyl function located directly above or below. The pitch of the helix and the diameter of the cylinder must be such that (1) no bond angles are strained and (2) the N—H and C=O functional groups on adjacent turns are in proper position for hydrogen bonding. An arrangement of this kind is possible for some amino acids along the chain, but not for others. Large protein molecules may contain segments of chain that have the α-helical arrangement interspersed with sections in which the chain is in a random coil. The overall shape of the protein, determined by all the bends, kinks, and sections of a rodlike α-helical structure, is called the **tertiary structure.** You might suppose that a long, complex protein molecule could adopt almost any shape under a given set of conditions, but this is not the case. Certain foldings of the protein chain lead to lower-energy (more stable) arrangements than do other folding patterns. For example, a protein dissolved in aqueous solution folds in such a way that the nonpolar hydrocarbon portions are tucked within the molecule, whereas the more polar acidic and basic side-chain functional groups are projected into the solution.

When proteins are heated above the temperatures characteristic of living organisms, or when they are subjected to unusual acid or base conditions, they begin to lose their particular tertiary and secondary structure. As this happens the protein loses its biological activity; it is said to be **denatured.** When denaturation occurs under very mild conditions, it is often reversible. That is, a return to normal conditions results in a return of biological activity. On the other hand, denaturation may also be irreversible; chemical changes that permanently alter the character of the protein may occur. As an example, the protein material in the white of an egg is irreversibly denatured when the egg is placed for a time in boiling water.

With this brief introduction to proteins, let us now consider the characteristics of a very important group of proteins, the enzymes.

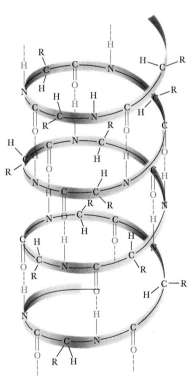

FIGURE 25.5 The α-helix structure for a protein. The symbol R represents any one of the several side chains shown in Figure 25.3.

The human body is characterized by an extremely complex system of interrelated chemical reactions. All of these reactions must occur at carefully controlled rates, so that thousands of individual chemical components are maintained at proper concentration levels and the system is able to respond as needed to demands made upon it. Every one of the many thousands of chemical reactions occurring in a biochemical system can be represented by an ordinary chemical equation. Indeed, many of the same reactions can be carried out under ordinary laboratory conditions. What is extraordinary about biochemical systems is that so many reactions occur with great rapidity at moderate temperatures. We cited in Chapter 13 the example of sugar, which is quite rapidly oxidized in the body at 37°C to carbon dioxide and water. In the laboratory, sugar does not react with oxygen at room temperature at a measurable rate. It must be heated to rather high temperature, for example, in a burner flame, before it burns.

The oxidation of sugar in the body occurs by means of a series of more than two dozen biochemically catalyzed steps. The catalyst in each step is called an enzyme. Enzymes are formed in living cells and are protein in nature. The molecular weights of enzymes range from a low of perhaps 10,000 to a high of about 1 million. The enzyme may consist of just a single protein chain or may involve several chains that are loosely held together.

The reaction that the enzyme catalyzes occurs at a specific site on the protein; this is called the active site. The substances that undergo reaction at this site are called substrates. Besides the substrate, other substances, called cofactors or coenzymes, may be required in the enzyme-catalyzed reaction. For example, the enzyme may require the presence of Mg^{2+} or some other metal ion, or it may require the presence and participation of a small organic molecule.

In many enzyme-catalyzed reactions, the coenzyme is one of the vitamins. An enzyme without its required coenzyme is called an apoenzyme. The combination of apoenzyme and coenzyme is called the holoenzyme:

$$\text{Apoenzyme} + \text{coenzyme} \longrightarrow \text{holoenzyme}$$

Enzymes possess certain characteristics not common to other types of catalysts. In the first place, they have a rather special sensitivity to temperature. Experimental studies have shown that the activity of a particular enzyme maximizes at around the normal temperature of the organism in which the enzyme is found. Figure 25.6 illustrates a typical activity-versus-temperature curve. Often it happens that when the temperature is raised above the usual operating temperature of the enzyme, the activity temporarily increases, but then subsequently decreases. The secondary and tertiary structures of the protein chain, on which the activity of the active site depends, are the result of many weak forces that induce the chain to take up that particular arrangement. Heating causes the protein chain to come undone, as it were; the enzyme is denatured and loses its activity completely.

A second respect in which enzymes are special is that they often show a sharp change in activity with change in the acidity or basicity of the

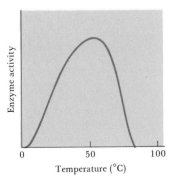

FIGURE 25.6 Enzyme activity as a function of temperature. At the higher temperature at which the activity of the enzyme drops to zero, the protein is denatured.

solution. This suggests that acid-base reactions are important in the catalyzed reaction. Thirdly, enzymes may be very specific in catalyzing exclusively a particular kind of reaction and no other, or even in catalyzing a particular reaction for only one compound and no other. The degree of specificity of enzymes varies widely. For example, enzymes called carboxypeptidases are rather general in their action.* They catalyze the hydrolysis of polypeptides into amino acids:

$$\cdots \overset{\overset{\displaystyle H}{|}}{\underset{\underset{\displaystyle R'}{|}}{C}} - \overset{\overset{\displaystyle O}{\|}}{C} - \overset{\overset{\displaystyle H}{|}}{\underset{\underset{\displaystyle H}{|}}{N}} - \overset{\overset{\displaystyle H}{|}}{\underset{\underset{\displaystyle R}{|}}{C}} - \overset{\overset{\displaystyle O}{\|}}{C} - OH + H_2O \xrightarrow{\text{Carboxypeptidase}}$$

$$\cdots \overset{\overset{\displaystyle H}{|}}{\underset{\underset{\displaystyle R'}{|}}{C}} - \overset{\overset{\displaystyle O}{\|}}{C} - OH + H_2N - \overset{\overset{\displaystyle H}{|}}{\underset{\underset{\displaystyle R}{|}}{C}} - \overset{\overset{\displaystyle O}{\|}}{C} - OH \qquad [25.3]$$

This equation shows only the end member of a peptide chain; the carboxypeptidase attacks only the amide group at the end of the chain. However, it is active regardless of the nature of the particular side chains R and R'. Although carboxypeptidases catalyze the hydrolysis of peptides, they are not at all active in the hydrolysis of fats; an entirely separate set of enzymes is responsible for the latter reaction. The high degree of specificity that enzymes possess is necessary to maintain some degree of independence of all the reactions occurring in a complex organism.

Finally, many enzymes differ from ordinary nonbiochemical catalysts in that they are enormously more efficient. The number of individual, catalyzed-reaction events occurring per second at a particular active site, called the **turnover number,** is generally in the range 10^3/sec to 10^7/sec. Such large turnover numbers are indicative of reactions with very low activation energies.

*The names of most enzymes end in -*ase*. The -*ase* ending is attached to the name of the substrate on which the enzyme acts, or to the name of the type of reaction that the enzyme catalyzes. Thus, *peptidases* are enzymes that act on peptides; *lipase* is an enzyme that acts on lipids; a *transmethylase* is an enzyme that catalyzes transfer of a methyl group.

SAMPLE EXERCISE 25.2

Solutions of dilute (about 3 percent) hydrogen peroxide are stable for long periods of time, especially if a small amount of a stabilizer is added. However, when such a solution is poured onto an open cut or wound, the hydrogen peroxide decomposes very rapidly with evolution of oxygen and formation of water:

$$2H_2O_2(aq) \longrightarrow O_2(g) + 2H_2O(l)$$

How can you account for this result?

Solution: Because the reaction occurs so rapidly on contact with the open wound, some chemical species present at the wound must catalyze the decomposition of hydrogen peroxide. That substance is an enzyme called peroxidase. Peroxidase is present to ensure that hydrogen peroxide does not accumulate in cells, because it could interfere with many other cell reactions. Peroxidase is an efficient enzyme, with a very high turnover number.

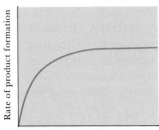

Rate of product formation

Concentration of substrate S

FIGURE 25.7 The rate of product formation as a function of concentration in a typical enzyme-catalyzed reaction. The rate of product formation is proportional to substrate concentration in the region of low concentration. At high substrate concentrations, the active sites on the enzyme are all complexed; further addition of substrate does not affect reaction rate.

Biochemists have been trying for a long time to discover how and why enzymes are so fantastically efficient and selective. During the past 20 years, the development of new experimental techniques for studying the structures of molecules and for following the rates of very rapid reactions has led to a much better understanding of how enzymes work.

The high turnover numbers observed for enzymes suggest that the substrate molecules cannot be very tightly bound to the enzyme; if they were, they might block the active site. Reaction would then be slow, because the active site is not quickly cleared out. Most enzyme systems that have been studied behave as though there were an equilibrium between the substrate (S) and the active site (E); we can write this equilibrium as an equation:

$$E + S \underset{k_{-1}}{\overset{k_1}{\rightleftharpoons}} ES \qquad [25.4]$$

The symbol ES represents a species in which the substrate is attached in some way to the enzyme. This **enzyme-substrate complex,** as it is called, then reacts to give product (P) and free active site:

$$ES \overset{k_2}{\longrightarrow} E + P \qquad [25.5]$$

In this picture, it is assumed that the substrate molecules may come off and on the active site very rapidly compared with the rate at which they undergo reaction to form products P. This means that the equilibrium described by Equation [25.4] is rapidly established between enzyme and substrate. It is also supposed that the equilibrium is such that most of the enzyme sites are not occupied by S when the substrate is present at normal concentrations. Now suppose that we study and graph (Figure 25.7) the rate of the enzyme-catalyzed reaction as a function of increasing concentration of substrate S. When S is present in low concentration, most of the enzyme active sites are not in use. Increasing the concentration of S thus shifts the equilibrium in Equation [25.4] to the right, so that a larger number of enzyme-substrate complexes are formed. This in turn increases the overall rate of the reaction, because that rate depends on the concentration of ES, $[ES]$:

$$\text{Rate} = \frac{\Delta[P]}{\Delta t} = k_2[ES] \qquad [25.6]$$

However, with still further increases in the concentration of S, a sizable fraction of the active sites are occupied. Adding still more S thus does not result in the same degree of increase in $[ES]$, and the rate does not increase as much. Eventually, with still higher concentration of S, *all* the active sites are effectively occupied. Further increases in S cannot result in faster reaction, because all available active sites are in use.

Many enzyme systems have been found to obey the sort of behavior illustrated in Figure 25.7. This strongly indicates that an enzyme-substrate complex is involved in the action of enzymes.

THE LOCK-AND-KEY MODEL

One of the earliest models for the enzyme-substrate interaction was the **lock-and-key model** illustrated in Figure 25.8. The substrate is pictured

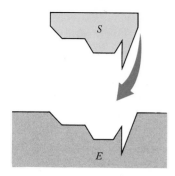

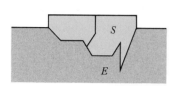

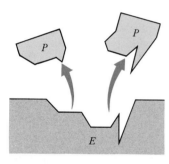

FIGURE 25.8 Lock-and-key model for formation of the enzyme-substrate complex.

as fitting neatly into a special place on the protein (the active site) tailored more or less specifically for that particular substrate. The catalyzed reaction occurs while the substrate is bound to the enzyme, and the reaction products then depart. Clearly, this picture of enzyme action has much in common with the models for heterogeneous catalysis that were discussed in Section 13.7. The difference is that an element of greater specificity is brought into the picture.

The lock-and-key model for enzyme action goes a long way toward explaining many aspects of enzyme action, but it cannot be the whole story. As the substrate molecules are drawn into the active site, they are somehow activated so that they are capable of extremely rapid reaction. This activation may result from the withdrawal or donation of electron density at a particular bond by the enzyme. In addition, in the process of fitting into the active site, the molecule may be distorted by the portions of the enzyme protein chain near the active site and thus made more reactive. If coenzymes are involved, the enzyme may promote reaction by binding both the coenzyme and substrate to the same site or to adjacent sites, thus keeping the reactants in close proximity. Enzymes, like all proteins, contain many acidic and basic sites along the chain, and these may be involved in acid-base reactions with the substrate. It is clear that the enzyme-coenzyme-substrate interactions can be very complex and are often difficult to unravel.

One of the more interesting corollaries of the lock-and-key model for enzyme action is that certain molecules should be capable of inhibiting the enzyme. Suppose that a certain molecule is capable of fitting the active site on the enzyme but is not reactive for one reason or another. If it is present in the solution along with substrate, this molecule competes with substrate for binding at the active sites. The substrate is thus prevented from forming the necessary enzyme-substrate complex, and the rate of product formation decreases. Metals that are highly toxic—for example, lead and mercury—probably operate as enzyme inhibitors. The heavy metal ions are especially strongly bound to sulfur-containing groups on protein side chains. By complexing very strongly to these sites on proteins, they interfere with the normal reactions of enzymes.

SAMPLE EXERCISE 25.3

Many bacteria employ *para*-aminobenzoic acid (PABA) in a vital enzyme-catalyzed metabolic reaction. Sulfa drugs such as sulfanilamide (*para*-aminobenzenesulfonamide) kill bacteria. Explain this in terms of the concept of the enzyme-substrate complex.

Solution: The structural formulas reveal that PABA and sulfanilamide are very similar in shape and in the locations of polar functional groups. The sulfa drug acts as an inhibitor by binding at the active site in the enzyme and blocking access by PABA. Without this needed ingredient, the metabolic processes of the bacteria cease.

$$H_2N-\!\!\!\bigcirc\!\!\!-\overset{\displaystyle O}{\underset{\displaystyle O}{S}}-NH_2$$

Sulfanilamide
(*para*-aminobenzenesulfonamide)

$$H_2N-\!\!\!\bigcirc\!\!\!-\overset{\displaystyle O}{C}-OH$$

para-Aminobenzoic acid (PABA)

25.4 Carbohydrates

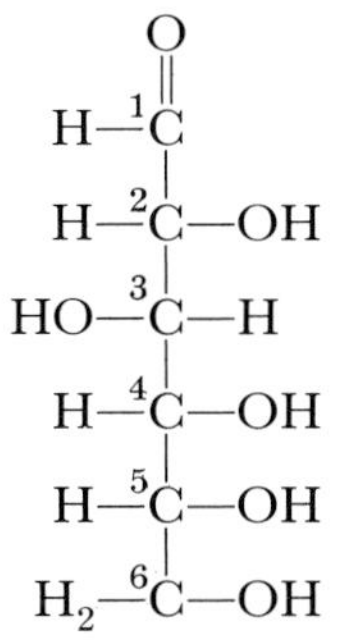

The carbohydrates are a very important class of naturally occurring substances found in both plant and animal matter. The name **carbohydrate** (hydrate of carbon) comes from the empirical formulas for most substances in this class; they can be written as $C_x(H_2O)_y$. For example, **glucose,** the most abundant carbohydrate, has molecular formula $C_6H_{12}O_6$, or $C_6(H_2O)_6$. The carbohydrates are not really hydrates of carbon; rather, they are polyhydroxy aldehydes and ketones. The structural formula for glucose can be written as shown in Figure 25.9.

Notice that four of the carbon atoms in this molecule, numbered 2, 3, 4, and 5, are chiral. Thus, there are many configurational isomers of glucose. Several naturally occurring sugars differ from glucose only in the configuration at one of these four chiral carbon atoms. The fact that these sugars have different biological properties is another example of the extreme specificity of biochemical systems. Many sugars are optically active substances; their solutions cause the plane of polarization of polarized light to be rotated, as illustrated in Figure 23.14.

When glucose is placed in solution, the aldehyde functional group at one end of the molecule reacts with the hydroxy group at the other end to form a cyclic structure, called a **hemiacetal:**

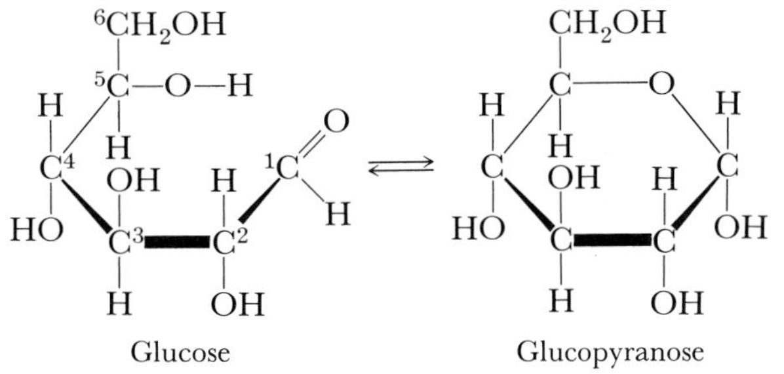

FIGURE 25.9 Structure of glucose.

Glucose   Glucopyranose   [25.7]

Note that in forming the ring structure in Equation [25.7], the OH group on carbon 1 can be on the same side of the ring as the OH group of carbon 2 (the α form) or on the opposite side (the β form). This seemingly minor difference is very important in distinguishing starch from cellulose, as we shall see shortly.

Six-membered cyclic hemiacetal structures are called **pyranoses.** The name glucopyranose indicates that the sugar glucose forms a six-membered ring. Sugars may also form five-membered rings called **furanoses.**

Some sugars are ketones rather than aldehydes. One important example is **fructose,** shown here in its straight-chain (linear) form as well as the five-membered cyclic form called fructofuranose:

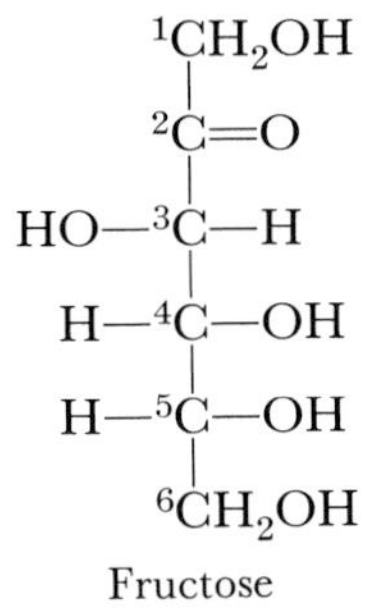

Fructose

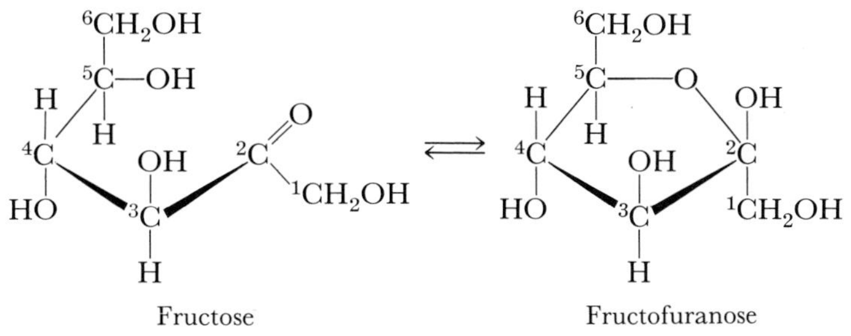

Fructose   Fructofuranose   [25.8]

Both glucose and fructose are examples of simple sugars, also called **monosaccharides.** Two or more such units can be joined together to form **polysaccharides.**

SAMPLE EXERCISE 25.4

Draw the structural formula for fructopyranose.

Solution: The name of the substance for which we are to draw the structure tells us that it involves a six-membered ring. Thus, beginning with fructose, for which the linear formula was given above, we complete a six-membered ring by closure with the OH group on the terminal carbon atom:

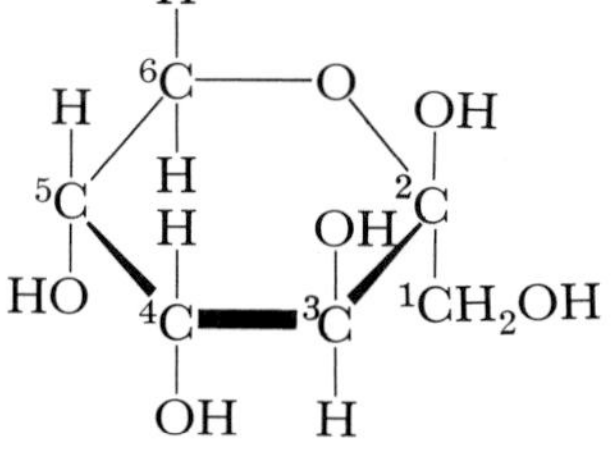

Notice that we might have drawn this structure so that the OH group of carbon atom 2 was on the opposite side of the ring as compared with the OH group of carbon 3. Thus there are two forms of fructopyranose.

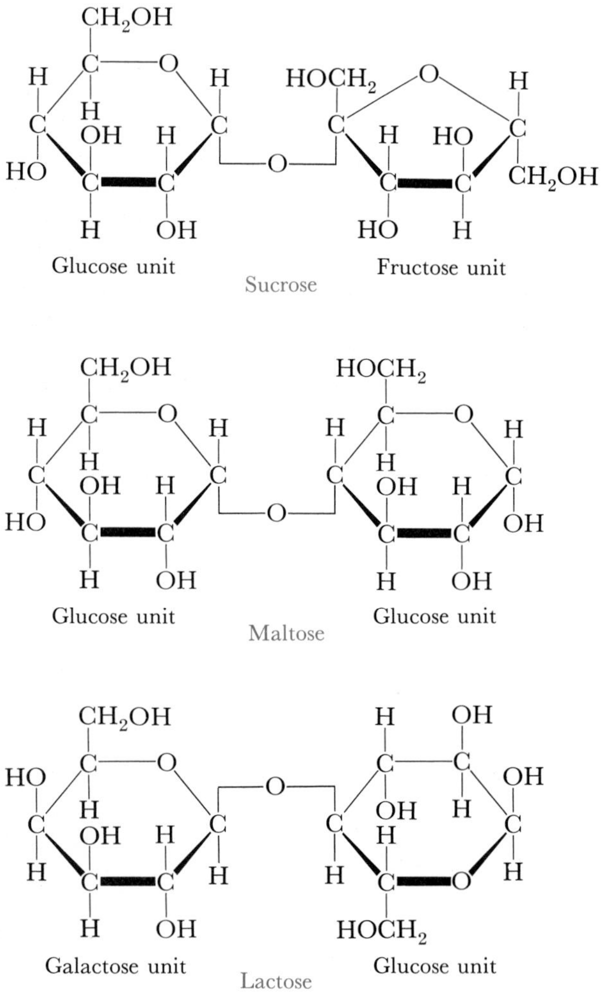

Glucose unit   Fructose unit

Sucrose

Glucose unit   Glucose unit

Maltose

Galactose unit   Glucose unit

Lactose

FIGURE 25.10 Structures of three disaccharides.

DISACCHARIDES

Disaccharides are formed by the joining together of two monosaccharide units. The monosaccharides are linked by a condensation reaction, in which a molecule of water is split out from between two hydroxyl groups, one on each sugar. Because there are several hydroxyl groups, there are several ways in which disaccharides can be formed. The structures of three common disaccharides, **sucrose** (table sugar), **maltose** (malt sugar), and **lactose** (milk sugar) are shown in Figure 25.10. The word "sugar" makes us think of sweetness; all sugars are sweet, but they differ in the degree of sweetness we perceive when we taste them. Sucrose is about six times sweeter than lactose, about three times sweeter than maltose, slightly sweeter than glucose, but only about half as sweet as fructose. Disaccharides can be hydrolyzed; that is, they can be reacted with water in the presence of an acid catalyst to form the monosaccharides. When sucrose is hydrolyzed, the mixture of glucose and fructose that forms, called *invert sugar,* * is sweeter to the taste than the original sucrose. The sweet syrup present in canned fruits and candies is largely invert sugar formed from hydrolysis of added sucrose.

POLYSACCHARIDES

Polysaccharides are made up of several monosaccharide units joined together by a bonding arrangement similar to those shown for disaccharides in Figure 25.10. The most important polysaccharides are starch, glycogen, and cellulose, which are formed from repeating glucose units.

Starch is not a pure substance; the term refers to a group of polysaccharides found in plants. Starches serve as a major method of food storage in plant seeds and tubers. Corn, potatoes, wheat, and rice all contain substantial amounts of starch. These plant products serve as major sources of needed food energy for humans. Enzymes within the digestive system catalyze the hydrolysis of starch to glucose.

Some starch molecules are linear chains, whereas others are branched. All contain the type of chain structure illustrated in Figure 25.11. It is particularly important to note the geometrical arrangement in the joining of one ring to another; the glucose units are in the α form.

---

*The term "invert sugar" comes from the fact that the rotation of the plane of polarized light by the glucose-fructose mixture is in the opposite direction, or inverted, from that of the sucrose solution.

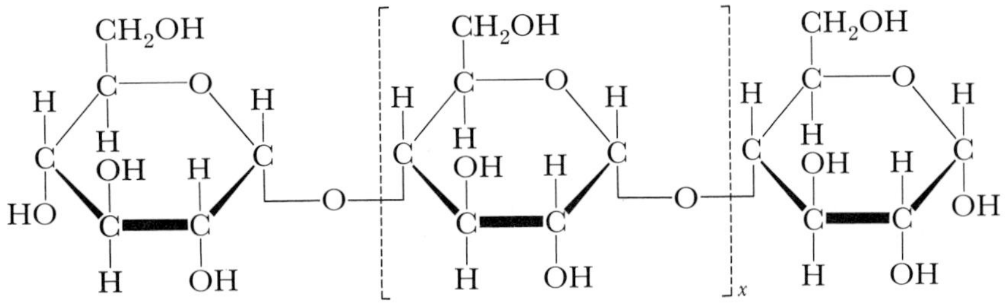

FIGURE 25.11 Structure of starch molecules.

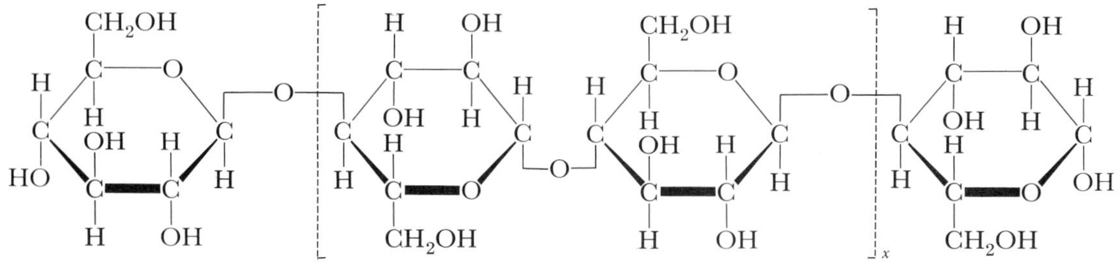

FIGURE 25.12 Structure of cellulose.

Glycogen is a starchlike substance synthesized in the body. Glycogen molecules vary in molecular weight from about 5000 to more than 5 million. Glycogen acts as a kind of energy bank in the body. It is concentrated in the muscles and liver. In muscles, it serves as an immediate source of energy; in the liver, it serves as a storage place for glucose and helps to maintain a constant glucose level in the blood.

Cellulose forms the major structural unit of plants. Wood is about 50 percent cellulose; cotton fibers are almost entirely cellulose. Cellulose consists of a straight chain of glucose units, with molecular weights averaging more than 500,000. The structure of cellulose is shown in Figure 25.12. At first glance, this structure looks very similar to that of starch. However, the differences in the arrangements of bonds that join the glucose units are important. Notice that glucose in cellulose is in the β form. Enzymes that readily hydrolyze starches do not hydrolyze cellulose. Thus, you might chew up and swallow a pound of cellulose and receive no caloric value from it whatsoever, even though the heat of combustion per unit weight is essentially the same for both cellulose and starch. A pound of starch, on the other hand, would represent a substantial caloric intake. The difference is that the starch is hydrolyzed to glucose, which is eventually oxidized with release of energy. Cellulose, on the other hand, is not hydrolyzed by any enzymes present in the body, and so it passes through the body unchanged. Many bacteria contain enzymes, called cellulases, that hydrolyze cellulose. These bacteria are present in the digestive systems of grazing animals, such as cattle, that utilize cellulose for food.

25.5 Fats and oils

Both plant and animal systems need means of storing energy in various chemical forms so that this energy supply can be tapped as needed later on. In plant seeds, the stored energy is used to promote rapid growth after germination. In animals that hibernate during cold periods, the stored energy is needed when other sources of food are absent or scarce. One of the most important classes of compounds used for energy storage are the **fats** and **oils.** These substances are esters of long-chain carboxylic acids with 1,2,3-trihydroxypropane (glycerol) and are called **triglycerides.** Their general structural formula is shown in Figure 25.13. The groups R_1, R_2, and R_3 can be alike or different. The triglycerides are found as fat deposits in animals and as oils in nuts and seeds. Those that are liquid at room temperature are generally referred to as oils; those that are solid are called fats.

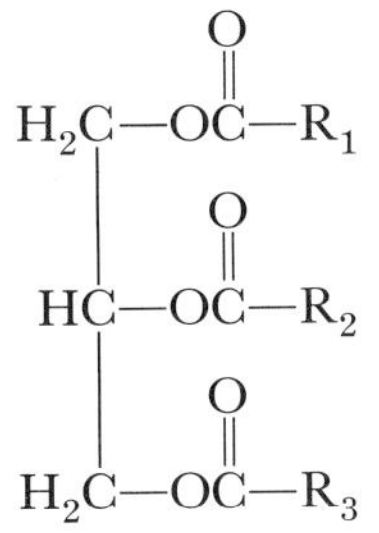

FIGURE 25.13 Structure of triglycerides. The groups R_1, R_2, and R_3 are hydrocarbon chains of from 3 to 21 carbons. The chains may be entirely saturated or may contain one or more *cis* olefin groups.

TABLE 25.1 Name, source, and structure of several fatty acids

Formula or structure[a]	Name	Source
$CH_3(CH_2)_{10}COOH$	Lauric acid	Coconuts
$CH_3(CH_2)_{12}COOH$	Myristic acid	Butter
$CH_3(CH_2)_{16}COOH$	Stearic acid	Animal fats
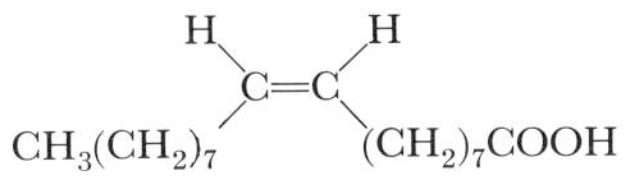	Oleic acid	Corn oil
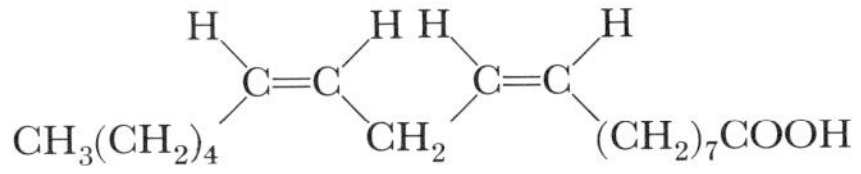	Linoleic acid	Vegetable oils
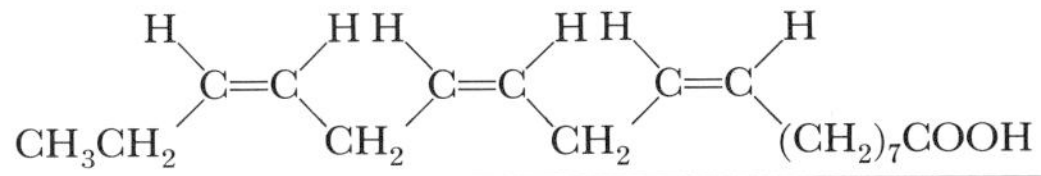	Linolenic acid	Linseed oil

[a]Notice that the *cis* configuration is adopted at the double bonds in the unsaturated fatty acids.

Hydrolysis of a fat or oil yields glycerol and the carboxylic acids of the R_1, R_2, and R_3 chains. The acids are commonly referred to as fatty acids. Most commonly the fatty acids contain from 12 to 22 carbons. It is an interesting fact that nearly all fatty acids contain an even number of carbon atoms, including the carboxylic acid carbon. Several of the more common fatty acids are listed in Table 25.1. The oils, which are obtained mainly from plant products (corn, peanuts, and soybeans), are formed primarily from unsaturated fatty acids. On the other hand, animal fats (beef, butter, and pork) contain mainly saturated fatty acids.

Some of the fat extracted from cottonseed, corn, or soybean oil is utilized as liquid cooking oil. Other plant-derived fats are hydrogenated to convert the carbon-carbon double bonds in the acid chains to carbon-carbon single bonds. In the process, the liquid fats are converted to solids; they are used to produce oleomargarine, peanut butter, vegetable shortening, and other similar food products. As an example, trilinolein, a major constituent of cottonseed oil, is hydrogenated to form tristearin, which is solid at room temperature:

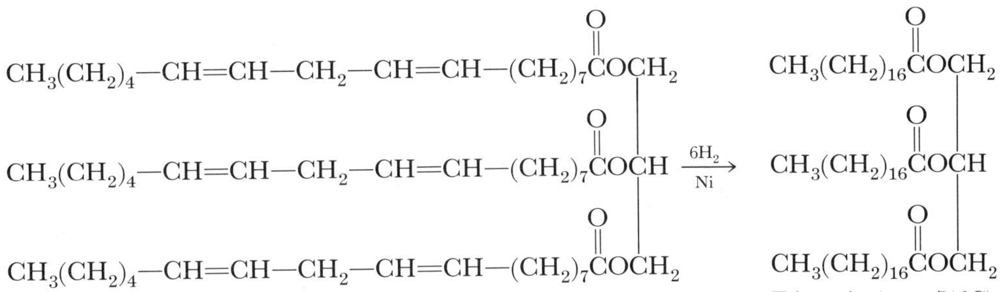

[25.9]

Fats and oils are an important source of energy in our food supply. In the body they undergo hydrolysis to form glycerol and carboxylic acids. The hydrolysis is promoted by enzymes called **lipases:**

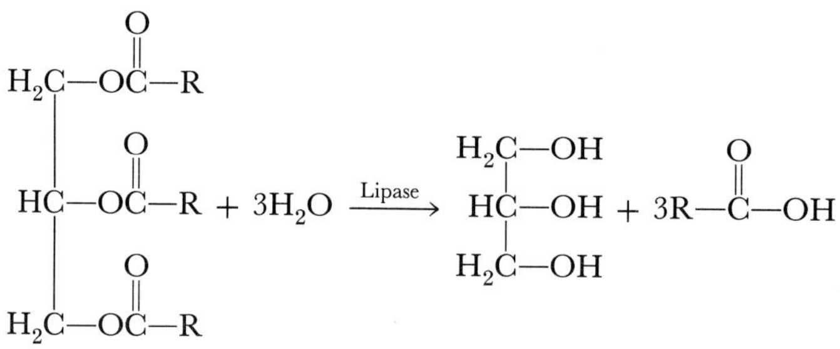

$$H_2C-O\overset{\overset{\displaystyle O}{\|}}{C}-R$$

$$HC-O\overset{\overset{\displaystyle O}{\|}}{C}-R + 3H_2O \xrightarrow{\text{Lipase}} \begin{array}{l} H_2C-OH \\ HC-OH \\ H_2C-OH \end{array} + 3R-\overset{\overset{\displaystyle O}{\|}}{C}-OH \qquad [25.10]$$

$$H_2C-O\overset{\overset{\displaystyle O}{\|}}{C}-R$$

This hydrolysis reaction, which takes place in aqueous solution, is hampered by the fact that fats and oils are essentially insoluble in water. As a result, not much hydrolysis occurs in the stomach. The gall bladder excretes compounds called bile salts into the small intestine to assist in the hydrolysis. The bile salts break up the larger droplets of fats into an emulsion (a suspension of very small droplets), so that hydrolysis can proceed more rapidly.

25.6 Nucleic acids

The nucleic acids are a class of biopolymers present in nearly all cells. They are classified into two groups—**deoxyribonucleic acids (DNA)** and **ribonucleic acids (RNA)**. DNA molecules are very large; their molecular weights may range from 6 to 16 million. RNA molecules are much smaller, with molecular weights in the range from 20,000 to 40,000. DNA is found primarily in the nucleus of the cell; RNA is found outside the nucleus, in the surrounding fluid called the cytoplasm.

Both DNA and RNA are polymers of a basic repeating unit called a **nucleotide.** To understand the structure of the polymer, we must first examine the structure of the nucleotide units. The nucleotide consists of three parts: (1) a phosphoric acid molecule, (2) a sugar in the furanose (five-membered ring) form, and (3) a nitrogen-containing organic base with a ring structure analogous to the ring structures of aromatic molecules. The sugar molecule found in RNA is ribose, shown in Figure 25.14. In DNA the sugar is deoxyribose, which differs from ribose only in the substitution of hydrogen for an —OH group on one carbon, as illustrated in Figure 25.14. In DNA the organic base may be any one of the four shown in Figure 25.15. An example of a complete nucleotide (deoxyadenylic acid) formed from these three components is illustrated in Figure 25.16. Substitution of one of the other bases for adenine in this structure produces one of the other three nucleotides that make up DNA.

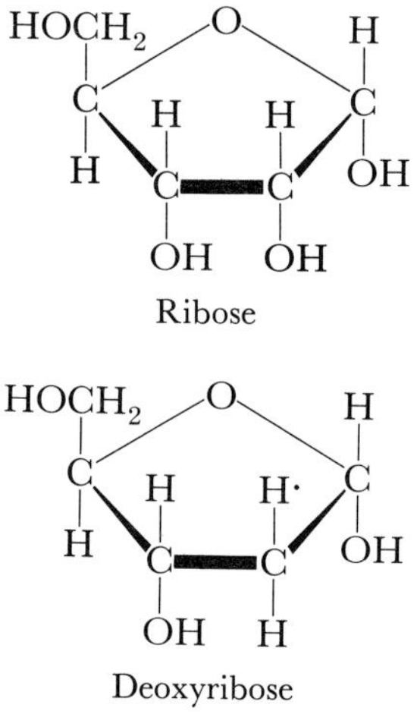

Ribose

Deoxyribose

FIGURE 25.14 Ribose and deoxyribose, the two sugars found in RNA and DNA, respectively.

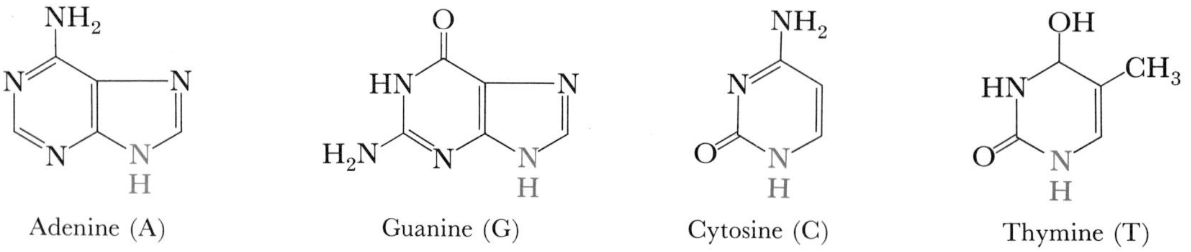

Adenine (A)

Guanine (G)

Cytosine (C)

Thymine (T)

FIGURE 25.15 The four organic bases present in DNA. In DNA each base is attached to a ribose molecule through a bond from the nitrogen shown in color.

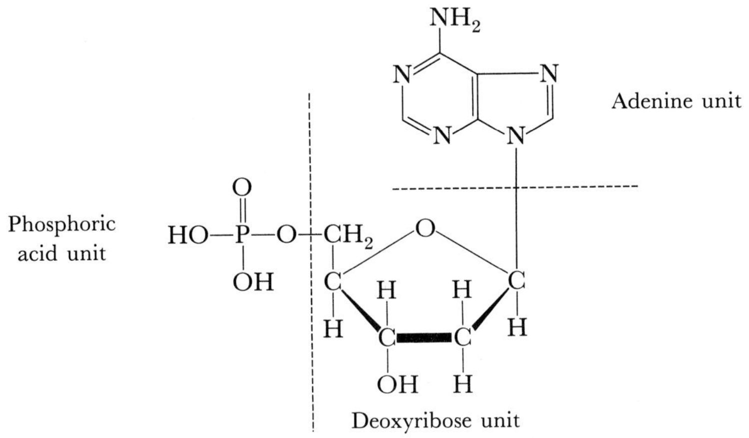

FIGURE 25.16 Structure of deoxyadenylic acid, a nucleotide formed from phosphoric acid, deoxyribose, and an organic base, adenine.

Just as a polypeptide is formed by a condensation reaction between amino acids, creating a peptide bond, a polynucleotide is formed by a condensation reaction creating a phosphate ester bond. The condensation reaction results in elimination of water from between an OH group of the phosphoric acid and an OH group of the sugar. The formation of the polymeric chain is shown schematically in Figure 25.17. The organic

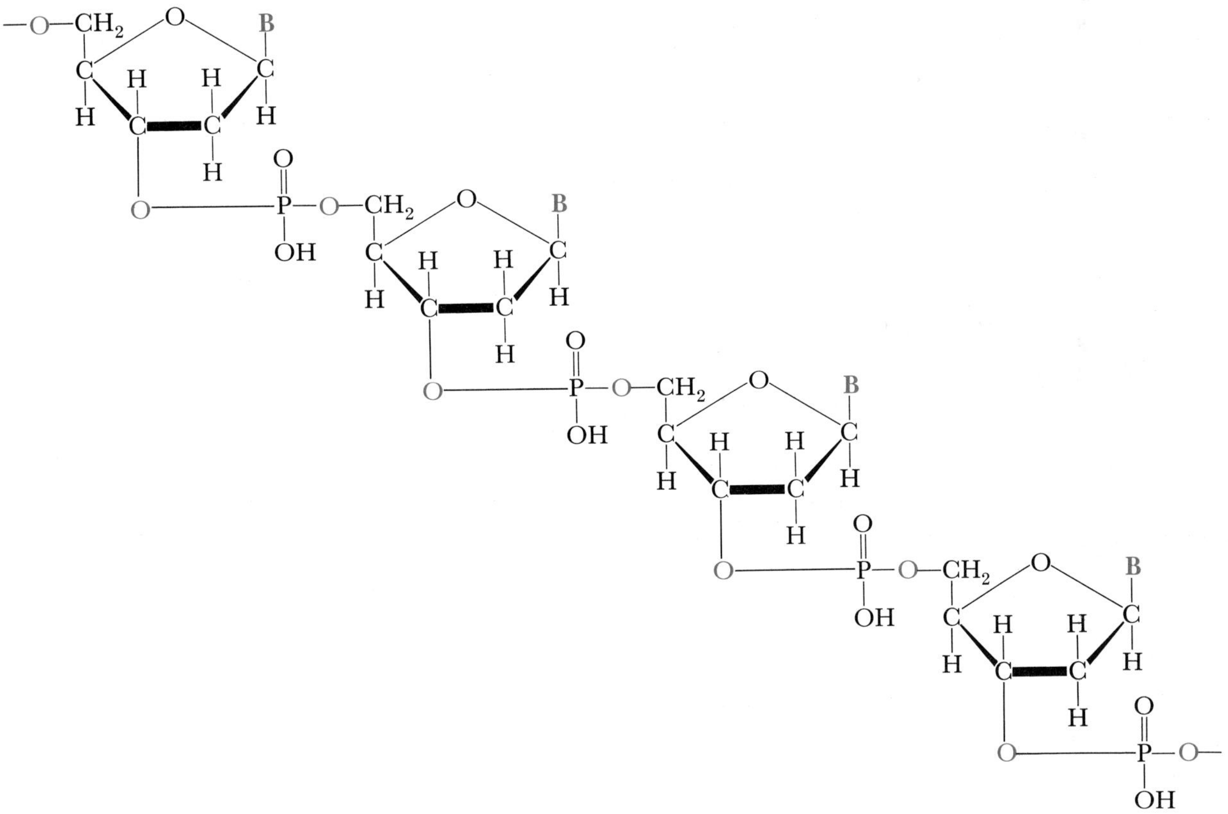

FIGURE 25.17 Structure of a polynucleotide. The symbol B represents any one of the four bases shown in Figure 25.15. Because the sugar in each nucleotide is deoxyribose, this polynucleotide is of the form found in DNA.

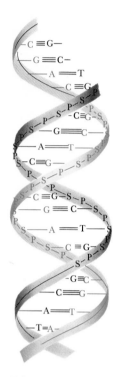

FIGURE 25.18 Schematic illustration of the double-stranded helical structure for DNA. The hydrogen-bond interactions between complementary base pairs is represented as illustrated in Figure 25.19.

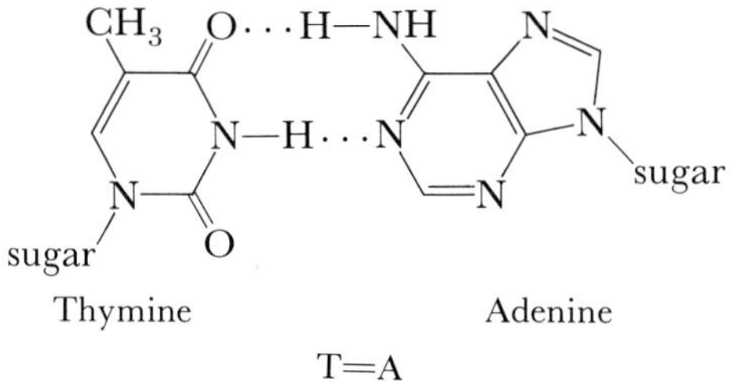

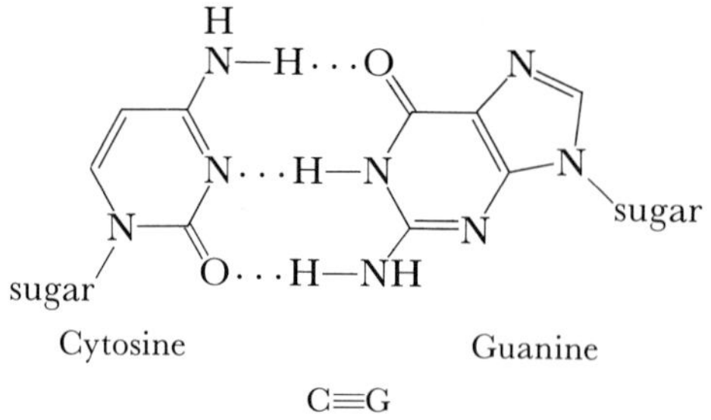

bases in this illustration are simply represented as B; they might be the same or different bases.

DNA molecules consist of two linear strands of the sort shown in Figure 25.17, which are wound together in the form of a **double helix.** It is very confusing to look at a model of double-stranded DNA in which all the atoms are represented. However, we can see from a more schematic illustration, Figure 25.18, how the DNA molecule is constructed. Remember that the polymeric strand itself consists of alternating sugar and phosphate groups, represented as —S— and —P—, respectively. The various bases are attached to the polymeric strand at each sugar unit. The key to the double-helix structure for DNA is the formation of hydrogen bonds between bases on the two chains. When adenine and thymine are located opposite one another along the chain, they form a strong hydrogen bond as shown in Figure 25.19. Similarly, guanine and cytosine form especially strong hydrogen bonds with one another. This means that the two strands of DNA are complementary; opposite each adenine on one chain there is always a thymine; opposite each guanine there is a cytosine. The double-helix structure for DNA, first proposed by James Watson and Francis Crick in 1953, has proved to be the key to understanding protein synthesis in cells, the means by which viral particles infect cells, the means by which genetic information is transmitted in reproduction of cells, and many other problems of central importance in molecular biology. Those themes are beyond the scope of this book; however, if you are interested in study in the area of life sciences, you will learn a good deal about such matters in other courses.

Thymine   Adenine

T=A

Cytosine   Guanine

C≡G

FIGURE 25.19 Hydrogen bonding between complementary base pairs. The hydrogen bonds shown here are responsible for formation of the double-stranded helical structure for DNA, as shown in Figure 25.18.

Summary

In this chapter, we have taken a brief look at some of the major constituents of the biosphere—that part of the physical world in which organisms live out their life cycles. The maintenance of life requires a source of energy as well as appropriate environmental conditions. The ultimate source of the needed energy is the sun. Plants convert solar energy to chemical energy in the process called photosynthesis. Solar energy is absorbed by a plant pigment called chlorophyll and then utilized to form carbohydrate and O_2 from CO_2 and H_2O.

Proteins are polymers of amino acids. They form a major structural element in animal systems. Enzymes, the catalysts of biochemical reactions, are protein in nature. All naturally occurring proteins are formed from some 20 amino acids. The amino acids are chiral substances; that is, they are capable of existing as nonsuperimposable mirror-image isomers called enantiomers. Usually, only one of the enantiomers is found to be biologically active. Protein structure is determined by the sequence of amino acids in the chain, the coiling or stretching of the chain, and the overall shape of the complete molecule. All these aspects of protein structure are important in determining its biological activity. Heating or other forms of treatment may inactivate, or denature, the protein.

In this chapter, we also have considered the action of enzymes—their specificity, a model for how they may operate, and the ways in which they are sometimes inhibited.

Carbohydrates, which are formed from polyhydroxy aldehydes and ketones, are the major structural constituent of plants and a source of energy in both plants and animals. The three most important groups of carbohydrates are starch, which is found in plants, glycogen, which is found in mammals, and cellulose, which is also found in plants. All are polysaccharides; that is, they are polymers of a simple sugar, glucose.

Fats and oils, along with proteins and carbohydrates, form the important sources of energy in our food supply. The fats and oils are esters of long-chain acids with glycerol, 1,2,3-trihydroxypropane; they are often called triglycerides. The major sources of these substances are animal fats and the oils of plant seeds, such as corn, peanuts, cottonseeds, and soybeans.

The nucleic acids are biopolymers of high molecular weight that carry the genetic information necessary for cell reproduction. In addition, the nucleic acids control cell development through control of protein synthesis. The nucleic acids consist of a polymeric backbone of alternating phosphate and ribose sugar groups, with organic bases attached to the sugar molecules. The DNA polymer is a double-stranded helix that is held together by hydrogen bonding between matching organic bases situated across from one another on the two strands.

Learning goals

Having read and studied this chapter, you should be able to:

1 Describe two distinct reasons why energy is required by living organisms.

2 List the several functions of proteins in living systems.

3 Define the terms chiral and enantiomer and draw the enantiomer of a given chiral molecule.

4 Write the reaction for formation of a peptide bond between two amino acids.

5 Explain the structures of proteins in terms of primary, secondary, and tertiary structure.

6 Explain the following characteristics of enzymes in terms of what is known about proteins: specificity; temperature dependence of enzyme activity, as shown in Figure 25.6; inhibition.

7 Explain the lock-and-key mechanism for enzyme action.

8 Distinguish the cyclic hemiacetal and linear forms of a sugar and describe the difference between the pyranose and furanose forms of sugar.

9 Describe the manner in which monosaccharides are joined together to form polysaccharides.

10 Enumerate the major groups of polysaccharides and indicate their sources and general functions.

11 Describe the structures of fats and oils and list the sources of these substances.

12 Draw the structures of any of the nucleotides that make up the polynucleotide DNA.

13 Describe the nature of the polymeric unit of polynucleotides.

14 Describe the double-stranded structure of DNA and explain the principle that determines the relationship between bases in the two strands.

Key terms

Among the more important terms and expressions used for the first time, or in a new context, in this chapter are the following:

The active site (Section 25.3) is the specific site in

an enzyme at which the enzyme-catalyzed reaction occurs.

An **amino acid** (Section 25.2) is a carboxylic acid that contains an amino (—NH$_2$) group attached to the carbon atom adjacent to the carboxylic acid (CO$_2$H) functional group. Twenty different amino acids are important in living systems.

The **biosphere** (introduction) is that part of the earth in which living organisms can exist and live out their life cycles.

Carbohydrates (Section 25.4) are a class of substances formed from polyhydroxy aldehydes or ketones.

Cellulose (Section 25.4) is a polysaccharide of glucose; it is the major structural element in plant matter.

A **chiral** molecule (Section 25.2) is one that is not superimposable on its mirror image.

Chlorophyll (Section 25.1), found in plant leaves, is a plant pigment that plays a major role in conversion of solar energy to chemical energy in photosynthesis.

A **coenzyme,** or **cofactor** (Section 25.3), is a substance that is needed along with the enzyme if an enzyme-catalyzed reaction is to occur.

Configurational isomers (Section 25.2) are molecules that differ only by being nonsuperimposable mirror images of one another. Such isomers are often called **optical isomers** because their solutions affect the plane of polarized light differently.

Denaturation (Section 25.2) is the loss of biological activity in a protein because of disruption of its tertiary structure by heating, by the action of acids or bases, or by other influence.

Deoxyribonucleic acid, DNA (Section 25.6), is a polynucleotide in which the sugar component is deoxyribose (in the furanose-ring form).

The **double-helix** structure for DNA (Section 25.6) involves the winding of two DNA polynucleotide chains together in a helical arrangement. The two strands of the double helix are complementary, in that the organic bases on the two strands are paired for optimal hydrogen-bond interaction.

Enantiomers (Section 25.2) are two mirror-image molecules of a chiral substance. The enantiomers are nonsuperimposable.

Enzymes (Section 25.3) are proteins that act as catalysts in biochemical reactions.

Fats and oils (Section 25.5) are esters of long-chain carboxylic acids and the alcohol 1,2,3-trihydroxypropane (glycerol). These esters are also called **triglycerides.**

A **furanose** (Section 25.4) is a cyclic sugar molecule in the form of a five-membered ring.

Glucose (Section 25.4), a polyhydroxy aldehyde

of formula CH$_2$OH(CHOH)$_4$CHO, is the most important of the monosaccharides.

Glycogen (Section 25.4) is the general name given to a group of polysaccharides of glucose that are synthesized in mammals and used as a means of carbohydrate energy storage.

In the **α-helix** structure (Section 25.2) for a protein, the protein is coiled in the form of a helix, with hydrogen bonds between C=O and N—H groups on adjacent turns.

A **hemiacetal** (Section 25.4) is a bonding arrangement formed by reaction of an aldehyde functional group with an alcohol. The hemiacetal linkage is formed when aldehyde sugars form the cyclic structure.

Lipases (Section 25.5) are enzymes that catalyze the hydrolysis of fats.

In the **lock-and-key model** for enzyme action (Section 25.3), the substrate molecule is pictured as fitting rather specifically into the active site on the enzyme. It is assumed that in being bound to the active site the substrate is somehow activated for reaction.

Macromolecules (Section 25.1) are polymeric molecules of high molecular weight. In referring to substances of biochemical origin, the term biopolymers is often used.

A **monosaccharide** (Section 25.4) is a simple sugar, most commonly containing six carbon atoms. The joining together of monosaccharide units by a condensation reaction results in formation of polysaccharides.

Nucleic acids (Section 25.6) are high-molecular-weight polymers of nucleotides.

A **nucleotide** (Section 25.6) is formed from a molecule of phosphoric acid, a sugar molecule, and an organic base. Nucleotides form liner polymers called DNA and RNA that are involved in protein synthesis and cell reproduction.

Photosynthesis (Section 25.1) is the process occurring in plant leaves by which light energy is used to convert CO$_2$ and water to carbohydrates and oxygen.

Polarized light (Section 25.2) is electromagnetic radiation in which the waves that form the light move in a single plane.

The **primary structure** of a protein (Section 25.2) refers to the sequence of amino acids along the protein chain.

Proteins (Section 25.2) are biopolymers formed from amino acids.

A **pryranose** (Section 25.4) is a cyclic sugar molecule in the form of a six-membered ring.

Ribonucleic acid, RNA (Section 25.6), is a polynucleotide in which ribose (in the furanose-ring form) is the sugar component.

The **secondary structure** of a protein (Section 25.2) refers to the manner in which the protein is coiled or stretched.

Starch (Section 25.4) is the general name given to a group of polysaccharides that act as energy-storage substances in plants.

The **tertiary structure** of a protein (Section 25.2) refers to the overall shape of the molecule—specifically, the manner in which sections of the chain fold back upon themselves or intertwine.

Turnover number (Section 25.3) refers to the number of individual reaction events per unit time at a particular active site in an enzyme.

EXERCISES

Energy requirements

25.1 Explain the relationship between entropy and the energy needs of organisms.

25.2 Name at least four processes occurring in an organism (such as yourself) that require energy. These may occur at the molecular level or may involve the complete organism.

[**25.3**] ΔG_{298}° for oxidation of glucose in solution is -2878 kJ:

$$C_6H_{12}O_6(aq) + 6O_2(g) \longrightarrow$$
$$6CO_2(g) + 6H_2O(l) \qquad \Delta G_{298}^{\circ} = -2878 \text{ kJ}$$

(a) What is ΔG_{298}° for photosynthesis? Is this reaction spontaneous under standard conditions? (b) What is ΔG_{298}° for photosynthesis if $P_{O_2} = 0.02$ atm, $P_{CO_2} = 3.1 \times 10^{-4}$ atm, and $C_6H_{12}O_6 = 1.0 \times 10^{-3} M$ (assume that H_2O can be omitted from the reaction quotient; refer to Section 18.6, if necessary).

25.4 What is the role of pigments in photosynthesis? What characteristics would you expect their absorption spectra to have?

25.5 A tree might convert about 50 g of CO_2 per day into carbohydrate at its greatest rate of photosynthesis. How many grams of oxygen does the tree produce in one day at this rate? How many liters of O_2 (at STP) does this correspond to?

25.6 Assume that 1×10^{14} kg of carbon are photosynthetically fixed as glucose each year on the earth's surface. If the total solar energy falling on the earth's surface is 4×10^{21} kJ/yr, and 2878 kJ are required to produce a mole of glucose, calculate the percentage of the solar energy that is absorbed annually by photosynthetic processes.

Proteins

25.7 Name at least four distinct roles for proteins in animal organisms.

25.8 (a) What is an α-amino acid? (b) How do amino acids react to form proteins?

25.9 How do the side chains (R groups) of amino acids affect their behavior?

25.10 Draw the dipeptides formed by condensation reaction between glycine and valine.

25.11 Write a chemical equation for the formation of aspartylcysteine from the constituent amino acids.

25.12 Draw the structure of the tripeptide trp-gly-ser.

25.13 What amino acids would be obtained by hydrolysis of the following tripeptide?

25.14 In what form would you expect glycine to exist in basic solution? In acidic solution?

25.15 How many different tripeptides could one make from the amino acids glycine, valine, and alanine?

25.16 Draw the two enantiometric forms of aspartic acid.

25.17 Explain why glycine is not optically active.

25.18 Describe what is meant by the terms primary, secondary, and tertiary structures of proteins.

25.19 Describe the role of hydrogen bonding in determining the α-helix structure of a protein.

25.20 What is meant by the term denaturation? Describe how increased temperature could lead to denaturation.

Enzymes

25.21 Provide a definition of each of the following terms: (a) enzyme; (b) apoenzyme; (c) denaturation; (d) active site; (e) holoenzyme; (f) specificity; (g) peptidase; (h) turnover number.

25.22 In terms of the lock-and-key model, what characteristics must an enzyme inhibitor possess?

25.23 The names of most enzymes end in -*ase*. The -*ase* ending is attached to the name of the substrate on which the enzyme acts, or to the type of reaction it catalyzes. Match the following enzyme names and reactions:

1 esterase	(a) removal of carboxyl groups from compounds
2 decarboxylase	
3 urease	(b) hydrolysis of peptide linkages
4 transmethylase	
5 peptidase	(c) transfers a methyl group
	(d) formation of ester linkages
	(e) hydrolysis of urea

25.24 Normally the rates of enzyme-catalyzed reactions increase linearly with increase in substrate concentration. What does this tell us about the position of equilibrium in Equation [25.4]? Explain.

25.25 Protein enzymes are sold commercially for use as meat tenderizers and for stain removal (for example, blood stains) from cloth. What kind of enzymes do you expect that these are? Explain how they work.

25.26 One of the many remarkable enzymes in the human body is carbonic anhydrase, which catalyzes the release of dissolved carbon dioxide from the blood into the air of our lungs. If it were not for this enzyme, the body could not rid itself rapidly enough of the CO_2 accumulated by cell metabolism. The enzyme has a molecular weight of 30,000, contains one atom of zinc per protein molecule, and catalyzes the dehydration (the release to air) of up to 10^7 CO_2 molecules per second. Which components of this description correspond to the terms holoenzyme, apoenzyme, cofactor, and turnover number?

Carbohydrates

25.27 What is the difference between α-glucose and β-glucose? Show the condensation of two glucose molecules to form a disaccharide with α-linkages; with β-linkages.

25.28 Identify each of the following as an α and β form of a hemiacetal:

25.29 The structural formula for the linear form of galactose is shown here. Draw the structure of the pyranose form of this sugar.

25.30 Which carbon atoms in galactose (see problem 25.29) are chiral?

25.31 Glucose is a hexose that can exist in solution as a pyranose; fructose, which is also a hexose, forms a furanose. Explain what is meant by the terms hexose, pyranose, and furanose. Why does glucose form a pyranose, whereas fructose forms a furanose?

Fats and oils

25.32 Draw the structural formula for the triglyceride of oleic acid.

25.33 The reaction between glycerol and a fatty acid is called esterification. Write the chemical equation for the esterification of a mole of glycerol with 3 mol of lauric acid.

25.34 The hydrolysis of fats and oils is catalyzed by bases. Many household cleaners (such as aqueous ammonia solutions and Drano, which contains NaOH) are basic. Write the chemical equation for the hydrolysis of the triglyceride of myristic acid.

25.23 Write a balanced chemical equation for complete hydrogenation of trilinolein. If partial hydrogenation were to occur, with uptake of only 4 mol of H_2 per mole of fat, draw the structural formulas of two possible products.

25.36 Fats and oils belong to a general class of compounds called lipids, which are distinguished by the fact that they are soluble in organic solvents. Discuss, in terms of their comparative structures, why fats belong to this class, whereas carbohydrates such as glucose and sucrose do not.

Nucleic acids

25.37 Describe a nucleotide. Draw the structural formula for deoxycytidine monophosphate in which cytosine is the organic base.

25.38 Write a balanced chemical equation for the condensation reaction between a mole of deoxyribose and a mole of phosphoric acid.

25.39 Imagine a single DNA strand containing a section with the following base sequence: A, C, T, C, G, A. What is the base sequence of the complementary strand?

25.40 When samples of double-stranded DNA are analyzed, the quantity of adenine present equals that of thymine. Similarly, the quantity of guanine equals that of cytosine. Explain the significance of these observations.

Additional exercises

25.41 In a temperate climate 1.0 m² of leaf area absorbs about 2.0×10^4 kJ of energy per day. About 1.2 percent of this energy is used in photosynthesis. (a) Calculate the leaf area required to convert 10,000 kJ per day into plant matter. This is the approximate energy requirement of a person doing an average quantity of work (Section 4.8). (b) In terms of what you know about the composition of plants, explain why more area than this would be required to provide a 10,000-kJ daily diet for a person.

25.42 In each of the following substances, locate the chiral carbon atoms if any:

(c)
$$
\begin{array}{c}
\quad\ \ O \quad\ CH_3 \\
\quad\ \ \| \qquad | \\
HOCCHCHC_2H_5 \\
\quad\ \ \ \ | \\
\quad\ \ \ \ NH_2
\end{array}
$$

25.43 Predict the products of the hydrolysis of each of the following compounds:

(a)
$$
\begin{array}{c}
\quad\ \ O \qquad\ O \\
\quad\ \ \| \qquad\ \ \| \\
CH_3CHCNHCH_2COH \\
\quad\ | \\
\quad\ NH_2
\end{array}
$$

(b)
$$
\begin{array}{c}
\quad O \\
\quad \| \\
C_{17}H_{33}COCH_2 \\
\quad\ \ O \\
\quad\ \ \| \\
C_{17}H_{33}COCH \\
\quad\ \ O \\
\quad\ \ \| \\
C_{17}H_{33}COCH_2
\end{array}
$$

(c)

(d)

25.44 Draw the condensed structural formula of each of the following tripeptides: (a) val-gly-thr; (b) pro-ser-ala.

[25.45] Explain why the peptide bond is planar.

25.46 Glutathione is a tripeptide found in most living cells. Partial hydrolysis yields cys-gly and glu-cys. What structures are possible for glutathione?

25.47 Phenylketonuria (PKU) is a disease caused by the lack in some individuals of an enzyme, phenylalanine hydroxylase. This enzyme catalyzes the conversion of phenylalanine to tyrosine (both amino acids). The disease can lead to severe mental retardation. (a) Write the condensed structural formulas for phenylalanine and tyrosine; (b) suggest the origin for the name given to the enzyme.

25.48 (a) Describe in qualitative terms how an enzyme

works. (b) What is meant by enzyme substrate? Enzyme inhibition?

[25.49] Ingestion of large amounts of alcohol by humans causes some enzymes to lose their active sites. Suggest a possible mechanism at the molecular level.

25.50 The popular flavor enhancer MSG (monosodium glutamate) is the monosodium salt of glutamic acid. (a) What is its condensed structural formula? (b) Only the L-isomer is effective. Is this surprising?

25.51 The standard free energy of formation of glycine(s) is -369 kJ/mol, whereas that of glycylglycine(s) is -488 kJ/mol. What is $\Delta G°$ for condensation of glycine to form glycylglycine?

25.52 Give the chain structural formulas for ribose and deoxyribose (see Figure 25.14).

[25.53] The enzyme *invertase* catalyzes the conversion of sucrose, a disaccharide, to invert sugar. When the concentration of invertase is $3 \times 10^{-7} M$, and the concentration of sucrose is $0.01 M$, invert sugar is formed at the rate of $2 \times 10^{-4} M/\mathrm{sec}$. When the sucrose concentration is doubled, the rate of formation of invert sugar is doubled also. Assuming that the enzyme-substrate model is operative, is the fraction of enzyme tied up as complex large or small? Explain. Addition of innositol, another sugar, causes a decrease in rate of formation of invert sugar. Suggest a mechanism by which this occurs.

25.54 Give a specific example of each of the following: (a) a disaccharide; (b) a sugar present in nucleic acids; (c) a sugar present in human blood serum; (d) a polysaccharide.

[25.55] The standard free energy of formation for aqueous solutions of glucose is -917.2 kJ/mol, whereas that of glycogen is -662.3 kJ/mol of glucose units. Derive a general expression for $\Delta G°$ for the formation of a glycogen molecule that contains n units of glucose.

25.56 Define, in your own words, the terms condensation and hydrolysis. Why are such reactions so important in biochemistry?

25.57 Write a complementary nucleic acid strand for the following strand, using the concept of complementary base pairing: TATGCA.

25.58 The monoanion of adenosine monophosphate (AMP) is an intermediate in phosphate metabolism:

$$
\begin{array}{c}
\quad\ \ \ O^- \\
\quad\ \ \ | \\
A{-}O{-}P{-}OH = AMP{-}OH^- \\
\quad\ \ \ \| \\
\quad\ \ \ O
\end{array}
$$

where A = adenosine. If the pK_a for this anion is 7.21, what is the ratio of [AMP—OH$^-$] to [AMP—O^{2-}] in blood at pH = 7.40?

Appendices

Appendix A

Mathematical operations

A.1 EXPONENTIAL NOTATION

The numbers used in chemistry are often either extremely large or extremely small. Such numbers are conveniently expressed in the form

$$N \times 10^n$$

where N is a number between 1 and 10, and n is the exponent. Some examples of this exponential notation are as follows:

1,200,000 is 1.2×10^6 (read "one point two times ten to the sixth power")

0.000604 is 6.04×10^{-4} (read "six point oh four times ten to the negative fourth power")

    A positive exponent, as in our first example, tells us how many times a number must be multiplied by 10 to give the long form of the number:

$$1.2 \times 10^6 = 1.2 \times 10 \times 10 \times 10 \times 10 \times 10 \times 10 \quad \text{(six tens)}$$
$$= 1,200,000$$

It is also convenient to think of the positive exponent as the number of places the decimal point must be moved to the *left* to give a number greater than 1 and less than 10: if we begin with 3450 and move the decimal point three places to left, we end up with 3.45×10^3.

In a related fashion, a negative exponent tells us how many times we must divide a number by 10 to give the long form of the number:

$$6.04 \times 10^{-4} = \frac{6.04}{10 \times 10 \times 10 \times 10}$$

$$= 0.000604$$

It is convenient to think of the negative exponent as the number of places the decimal point must be moved to the *right* to give a number greater than 1 but less than 10: if we begin with 0.0048 and move the decimal point three places to right, we end up with 4.8×10^{-3}.

In the system of exponential notation, with each shift of the decimal point one place to the right, the exponent *decreases* by 1:

$$4.8 \times 10^{-3} = 48 \times 10^{-4}$$

Similarly, with each shift of the decimal point one place to the left, the exponent *increases* by 1:

$$4.8 \times 10^{-3} = 0.48 \times 10^{-2}$$

In working with exponents, it is important to know that $10^0 = 1$. The following rules are useful for carrying exponents through calculations:

1 **Addition and subtraction** In order to add or subtract numbers expressed in exponential notation, the powers of 10 must be the same:

$$(5.22 \times 10^4) + (3.21 \times 10^2) = (522 \times 10^2) + (3.21 \times 10^2)$$
$$= 525 \times 10^2$$
$$= 5.25 \times 10^4$$
$$(6.25 \times 10^{-2}) - (5.77 \times 10^{-3}) = (6.25 \times 10^{-2}) - (0.577 \times 10^{-2})$$
$$= 5.67 \times 10^{-2}$$

2 **Multiplication and division** When numbers expressed in exponential notation are multiplied, the exponents are added; when numbers expressed in exponential notation are divided, the exponent of the divisor is subtracted from the exponent of the dividend:

$$(5.4 \times 10^2)(2.1 \times 10^3) = (5.4)(2.1) \times 10^{2+3}$$
$$= 11 \times 10^5$$
$$= 1.1 \times 10^6$$
$$(1.2 \times 10^5)(3.22 \times 10^{-3}) = (1.2)(3.22) \times 10^{5-3}$$
$$= 3.8 \times 10^2$$

$$\frac{3.2 \times 10^5}{6.5 \times 10^2} = \frac{3.2}{6.5} \times 10^{5-2} = 0.49 \times 10^3$$

$$= 4.9 \times 10^2$$

$$\frac{5.7 \times 10^7}{8.5 \times 10^{-2}} = \frac{5.7}{8.5} \times 10^{7-(-2)} = 0.67 \times 10^9$$

$$= 6.7 \times 10^8$$

3 **Powers and roots** When numbers expressed in exponential notation are raised to a power, the exponents are multiplied by the power; when the roots of numbers expressed in exponential notation are taken, the exponents are divided by the root:

$$(1.2 \times 10^5)^3 = 1.2^3 \times 10^{5\times3}$$
$$= 1.7 \times 10^{15}$$
$$\sqrt[3]{2.5 \times 10^6} = \sqrt[3]{2.5} \times 10^{6/3}$$
$$= 1.3 \times 10^2$$

A.2 LOGARITHMS

The common, or base-10, logarithm (abbreviated log) of any number is the power to which 10 must be raised to equal the number. For example, the common logarithm of 1000 (written log 1000) is 3, because raising 10 to the third power gives $1000:10^3 = 1000$. Further examples are:

$$\log 10^5 = 5$$
$$\log 1 = 0$$
$$\log 10^{-2} = -2$$

In these examples, the logarithm can be obtained by simple inspection. However, it is not possible to obtain the logarithm of a number like 3.2 by inspection. The logarithms of such numbers can be obtained from a log table such as that given in Table 1. This table can be used to find the logs of numbers from 1 to 10. To do so, we locate the first digit of the number in the first vertical column. We then move horizontally to the column headed by the second digit of the number. In this way we find that the log of 3.2 is 505. Because decimals are omitted from the table the log of 3.2 is actually 0.505. Further examples are:

$$\log 2.50 = 0.398$$
$$\log 2.55 = 0.406$$

TABLE 1 Three-place common logarithms

	0	1	2	3	4	5	6	7	8	9
1	000	041	079	114	146	176	204	230	255	279
2	301	322	342	362	380	398	415	431	447	462
3	477	491	505	519	532	544	556	568	580	591
4	602	613	623	634	644	653	663	672	681	690
5	699	708	716	724	732	740	748	756	763	771
6	778	785	792	799	806	813	820	826	833	839
7	845	851	857	863	869	875	881	887	892	898
8	903	909	914	919	924	929	935	940	945	949
9	954	959	964	969	973	978	982	987	991	996

In using Table 1 to find the log of 2.55, we must estimate, or interpolate, the value from the logs given for 2.5 and 2.6. Appendix B is a table of four-place logarithms from which the logarithm of 2.55 can be read directly as 0.4065. (In using Appendix B, you should mentally insert a decimal point between the two-digit numbers that form the first column.)

The logarithm of a number that is less than 1 or greater than 10 can be determined by writing the number first in standard exponential notation as the following examples show:

$$\log 450 = \log (4.50 \times 10^2)$$
$$= \log 4.50 + \log 10^2$$
$$= 0.653 + 2 = 2.653$$

$$\log 0.0673 = \log (6.73 \times 10^{-2})$$
$$= \log 6.73 + \log 10^{-2}$$
$$= 0.828 - 2$$
$$= -1.172$$

Check these examples yourself, using Appendix B.

Because logarithms are exponents, mathematical operations involving logarithms follow the rules for the use of exponents:

1 Multiplication and division:

$$\log ab = \log a + \log b$$

$$\log \frac{a}{b} = \log a - \log b$$

2 Powers and roots:

$$\log a^n = n(\log a)$$

$$\log a^{1/n} = \left(\frac{1}{n}\right)(\log a)$$

Obtaining antilogarithms   The process of finding a number given its logarithm is known as obtaining an antilogarithm. It is the reverse of taking a logarithm, as the following examples show:

1 Find the number whose logarithm is 5.322.

$$\text{antilog } 5.322 = \text{antilog } 0.322 \times \text{antilog } 5$$
$$= 2.10 \times 10^5$$

2 Find the number whose logarithm is -2.133.

$$\text{antilog } (-2.133) = \text{antilog } 0.867 \times \text{antilog } (-3)$$
$$= 7.37 \times 10^{-3}$$

pH problems   In general chemistry, logarithms are used most frequently in working pH problems. The pH is defined as $-\log [H^+]$, as discussed in Section 15.3. The following sample exercise illustrates this application.

SAMPLE EXERCISE 1

(a) What is the pH of a solution whose hydrogen-ion concentration is 0.015 M? (b) If the pH of a solution is 3.80, what is its hydrogen-ion concentration?

Solution:

(a) $pH = -\log [H^+]$
$= -\log (0.015)$
$= -\log (1.5 \times 10^{-2})$
$= -\log 1.5 - \log (10^{-2})$
$= -0.18 + 2 = 1.82$

(b) $pH = -\log [H^+] = 3.80$
$\log [H^+] = -3.80$
$[H^+] = $ antilog (-3.80)
$= $ antilog 0.20
$\times$ antilog (-4)
$= 1.6 \times 10^{-4} M$

Natural logarithms Natural, or base e, logarithms (abbreviated ln) are the power to which e, which has the value 2.71828 . . . , must be raised to equal a number. The relation between common and natural logarithms is as follows:

$\ln a = 2.303 \log a$

A.3 QUADRATIC EQUATIONS

An algebraic equation of the form $ax^2 + bx + c = 0$ is called a quadratic equation. The two solutions to such an equation are given by the quadratic formula:

$$x = \frac{-b \pm \sqrt{b^2 - 4ac}}{2a}$$

SAMPLE EXERCISE 2

Find x if $2x^2 + 4x = 1$.

Solution: To solve the given equation for x, we must first put it in the form

$ax^2 + bx + c = 0$

and then use the quadratic formula. If

$2x^2 + 4x = 1$

then

$2x^2 + 4x - 1 = 0$

Using the quadratic formula, where $a = 2$, $b = 4$, and $c = -1$, we have

$x = \frac{-4 \pm \sqrt{(4)(4) - 4(2)(-1)}}{2(2)}$

$= \frac{-4 \pm \sqrt{16 + 8}}{4} = \frac{-4 \pm \sqrt{24}}{4}$

$= \frac{-4 \pm 4.899}{4}$

The two solutions are

$x = \frac{0.899}{4} = 0.225$

$x = \frac{-8.899}{4} = -2.225$

Often in chemical problems the negative solution has no physical meaning, and only the positive answer is used.

A.4 GRAPHS

Often the clearest way to represent the interrelationship between two variables is to graph them. Usually the variable that is being experimentally varied, called the independent variable, is shown along the horizontal axis (x-axis). The variable that responds to the change in the independent variable, called the dependent variable, is then shown along the vertical axis (y-axis). For example, consider an experiment in which we vary the temperature of an enclosed gas and measure its pressure. The independent variable is temperature, whereas the dependent variable is pressure. The data shown in Table 2 could be obtained by means of this experiment. These data are shown graphically in Figure 1. The relationship between temperature and pressure is linear. The equation for any straight-line graph has the form

$$y = ax + b$$

where a is the slope of the line and b is the intercept with the y-axis. In the case of Figure 1, we could say that the relationship between temperature and pressure takes the form

$$P = aT + b$$

where P is pressure in atm, and T is temperature in °C. As shown on Figure 1, the slope is 4.10×10^{-4} atm/°C, and the intercept, the point where the line crosses the y-axis, is 0.112 atm. Therefore the equation for the line is:

$$P = \left(4.10 \times 10^{-4}\,\frac{\text{atm}}{\text{°C}}\right) T + 0.112 \text{ atm}$$

TABLE 2 Interrelation between pressure and temperature

Temperature (°C)	Pressure (atm)
20.0	0.120
30.0	0.124
40.0	0.128
50.0	0.132

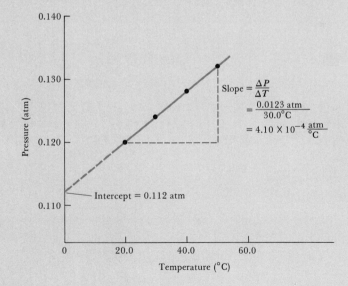

FIGURE 1

	0	1	2	3	4	5	6	7	8	9
10	0000	0043	0086	0128	0170	0212	0253	0294	0334	0374
11	0414	0453	0492	0531	0569	0607	0645	0682	0719	0755
12	0792	0828	0864	0899	0934	0969	1004	1038	1072	1106
13	1139	1173	1206	1239	1271	1303	1335	1367	1399	1430
14	1461	1492	1523	1553	1584	1614	1644	1673	1703	1732
15	1761	1790	1818	1847	1875	1903	1931	1959	1987	2014
16	2041	2068	2095	2122	2148	2175	2201	2227	2253	2279
17	2304	2330	2355	2380	2405	2430	2455	2480	2504	2529
18	2553	2577	2601	2625	2648	2672	2695	2718	2742	2765
19	2788	2810	2833	2856	2878	2900	2923	2945	2967	2989
20	3010	3032	3054	3075	3096	3118	3139	3160	3181	3201
21	3222	3243	3263	3284	3304	3324	3345	3365	3385	3404
22	3424	3444	3464	3483	3502	3522	3541	3560	3579	3598
23	3617	3636	3655	3674	3692	3711	3729	3747	3766	3784
24	3802	3820	3838	3856	3874	3892	3909	3927	3945	3962
25	3979	3997	4014	4031	4048	4065	4082	4099	4116	4133
26	4150	4166	4183	4200	4216	4232	4249	4265	4281	4298
27	4314	4330	4346	4362	4378	4393	4409	4425	4440	4456
28	4472	4487	4502	4518	4533	4548	4564	4579	4594	4609
29	4624	4639	4654	4669	4683	4698	4713	4728	4742	4757
30	4771	4786	4800	4814	4829	4843	4857	4871	4886	4900
31	4914	4928	4942	4955	4969	4983	4997	5011	5024	5083
32	5051	5065	5079	5092	5105	5119	5132	5145	5159	5172
33	5185	5198	5211	5224	5237	5250	5263	5276	5289	5302
34	5315	5328	5340	5353	5366	5378	5391	5403	5416	5428
35	5441	5453	5465	5478	5490	5502	5514	5527	5539	5551
36	5563	5575	5587	5599	5611	5623	5635	5647	5658	5670
37	5682	5694	5705	5717	5729	5740	5752	5763	5775	5786
38	5798	5809	5821	5832	5843	5855	5866	5877	5888	5899
39	5911	5922	5933	5944	5955	5966	5977	5988	5999	6010
40	6021	6031	6042	6053	6064	6075	6085	6096	6107	6117
41	6128	6138	6149	6160	6170	6180	6191	6201	6212	6222
42	6232	6243	6253	6263	6274	6284	6294	6304	6314	6325
43	6335	6345	6355	6365	6375	6385	6395	6405	6415	6425
44	6435	6444	6454	6464	6474	6484	6493	6503	6513	6522
45	6532	6542	6551	6561	6571	6580	6590	6599	6609	6618
46	6628	6637	6646	6656	6665	6675	6684	6693	6702	6712
47	6721	6730	6739	6749	6758	6767	6776	6785	6794	6803
48	6812	6821	6830	6839	6848	6857	6866	6875	6884	6893
49	6902	6911	6920	6928	6937	6946	6955	6964	6972	6981
50	6990	6998	7007	7016	7024	7033	7042	7050	7059	7067
51	7076	7084	7093	7101	7110	7118	7126	7135	7143	7152
52	7160	7168	7177	7185	7193	7202	7210	7218	7226	7235
53	7243	7251	7259	7267	7275	7284	7292	7300	7308	7316
54	7324	7332	7340	7348	7356	7364	7372	7380	7388	7396
55	7404	7412	7419	7427	7435	7443	7451	7459	7466	7474
56	7482	7490	7497	7505	7513	7520	7528	7536	7543	7551
57	7559	7566	7574	7582	7589	7597	7604	7612	7619	7627
58	7634	7642	7649	7657	7664	7672	7679	7686	7694	7701
59	7709	7716	7723	7731	7738	7745	7752	7760	7767	7774

	0	1	2	3	4	5	6	7	8	9
60	7782	7789	7796	7803	7810	7818	7825	7832	7839	7846
61	7853	7860	7868	7875	7882	7889	7896	7903	7910	7917
62	7924	7931	7938	7945	7952	7959	7966	7973	7980	7987
63	7993	8000	8007	8014	8021	8028	8035	8041	8048	8055
64	8062	8069	8075	8082	8089	8096	8102	8109	8116	8122
65	8129	8136	8142	8149	8156	8162	8169	8176	8182	8189
66	8195	8202	8209	8215	8222	8228	8235	8241	8248	8254
67	8261	8267	8274	8280	8287	8293	8299	8306	8312	8319
68	8325	8331	8338	8344	8351	8357	8363	8370	8376	8382
69	8388	8395	8401	8407	8414	8420	8426	8432	8439	8445
70	8451	8457	8463	8470	8476	8482	8488	8494	8500	8506
71	8513	8519	8525	8531	8537	8543	8549	8555	8561	8567
72	8573	8579	8585	8591	8597	8603	8609	8615	8621	8627
73	8633	8639	8645	8651	8657	8663	8669	8675	8681	8686
74	8692	8698	8704	8710	8716	8722	8727	8733	8739	8745
75	8751	8756	8762	8768	8774	8779	8785	8791	8797	8802
76	8808	8814	8820	8825	8831	8837	8842	8848	8854	8859
77	8865	8871	8876	8882	8887	8893	8899	8904	8910	8915
78	8921	8927	8932	8938	8943	8949	8954	8960	8965	8971
79	8976	8982	8987	8993	8998	9004	9009	9015	9020	9025
80	9031	9036	9042	9047	9053	9058	9063	9069	9074	9079
81	9085	9090	9096	9101	9106	9112	9117	9122	9128	9133
82	9138	9143	9149	9154	9159	9165	9170	9175	9180	9186
83	9191	9196	9201	9206	9212	9217	9222	9227	9232	9238
84	9243	9248	9253	9258	9263	9269	9274	9279	9284	9289
85	9294	9299	9304	9309	9315	9320	9325	9330	9335	9340
86	9345	9350	9355	9360	9365	9370	9375	9380	9385	9390
87	9395	9400	9405	9410	9415	9420	9425	9430	9435	9440
88	9445	9450	9455	9460	9465	9469	9474	9479	9484	9489
89	9494	9499	9504	9509	9513	9518	9523	9628	9533	9538
90	9542	9547	9552	9557	9562	9566	9571	9576	9581	9586
91	9590	9595	9600	9605	9609	9614	9619	9624	9628	9633
92	9638	9643	9647	9652	9657	9661	9666	9671	9675	9680
93	9685	9689	9694	9699	9703	9708	9713	9717	9722	9727
94	9731	9736	9741	9745	9750	9754	9759	9763	9768	9773
95	9777	9782	9786	9791	9795	9800	9805	9809	9814	9818
96	9823	9827	9832	9836	9841	9845	9850	9854	9859	9863
97	9868	9872	9877	9881	9886	9890	9894	9899	9903	9908
98	9912	9917	9921	9926	9930	9934	9939	9943	9948	9952
99	9956	9961	9965	9969	9974	9978	9983	9987	9991	9996

Appendix C
Properties of water

Density:
0.99987 g/cm^3 at 0 °C
1.00000 g/cm^3 at 4 °C
0.99707 g/cm^3 at 25 °C
0.95838 g/cm^3 at 100 °C

Heat of fusion: 6.02 kJ/mol at 0 °C

Heat of vaporization:
44.94 kJ/mol at 0 °C
44.02 kJ/mol at 25 °C
40.67 kJ/mol at 100 °C

Specific heat:
Ice (-3 °C)—2.092 J/°C-g
Water at 14.5 °C—4.184 J/°C-g
Steam (100 °C)—1.841 J/°C-g

Vapor pressure (mm Hg):

T(°C)	VP	T(°C)	VP	T(°C)	VP	T(°C)	VP
0	4.58	21	18.65	35	42.2	92	567.0
5	6.54	22	19.83	40	55.3	94	610.9
10	9.21	23	21.07	45	71.9	96	657.6
12	10.52	24	22.38	50	92.5	98	707.3
14	11.99	25	23.76	55	118.0	100	760.0
16	13.63	26	25.21	60	149.4	102	815.9
17	14.53	27	26.74	65	187.5	104	875.1
18	15.48	28	28.35	70	233.7	106	937.9
19	16.48	29	30.04	80	355.1	108	1004.4
20	17.54	30	31.82	90	525.8	110	1074.6

Appendix D
Thermodynamic
quantities for selected
substances at 25°C

Substance	ΔH_f°(kJ/mol)	ΔG_f°(kJ/mol)	S°(J/mol-K)
Al(s)	0.00	0.00	28.32
Al$_2$O$_3$(s)	−1669.8	−1576.5	51.00
Ag$^+$(aq)	105.90	77.11	73.93
AgCl(s)	−127.0	−109.70	96.11
Ba(s)	0.00	0.00	63.2
Br(g)	111.8	82.38	174.9
Br$^-$(aq)	−120.9	−102.8	80.71
Br$_2$(g)	30.71	3.14	245.3
Br$_2$(l)	0.00	0.00	152.3
C(g)	718.4	672.9	158.0
C(diamond)	1.88	2.84	2.43
C(graphite)	0.00	0.00	5.69
CCl$_4$(g)	−106.7	−64.0	309.4
CCl$_4$(l)	−139.3	−68.6	214.4
CF$_4$(g)	−679.9	−635.1	262.3
CO(g)	−110.5	−137.3	197.9
CO$_2$(g)	−393.5	−394.4	213.6
CH$_4$(g)	−74.8	−50.8	186.3
C$_2$H$_2$(g)	226.7	209.2	200.8
C$_2$H$_4$(g)	52.30	68.11	219.4
C$_2$H$_6$(g)	−84.68	−32.89	229.5
C$_3$H$_8$(g)	−103.85	−23.47	269.9
CH$_3$OH(g)	−201.2	−161.9	237.6
CH$_3$OH(l)	−238.6	−166.23	126.8
C$_2$H$_5$OH(l)	−277.7	−174.76	160.7
CH$_3$COOH(l)	−487.0	−392.4	159.8
C$_6$H$_6$(l)	49.0	124.5	172.8
C$_6$H$_6$(g)	82.9	129.7	269.2
Ca(s)	0	0	41.4
Ca(g)	179.3	145.5	154.8
CaCO$_3$(calcite)	−1207.1	−1128.76	92.88
CaO(s)	−635.5	−604.17	39.75
Ca(OH)$_2$(s)	−986.2	−898.5	83.4
Cl$_2$(g)	0.00	0.00	222.96
Co(s)	0.00	0.00	28.4
Co(g)	439	393	179
Cr(s)	0.00	0.00	23.6
Cr(g)	397.5	352.6	174.2
Cu(s)	0.00	0.00	33.30
Cu(g)	338.4	298.6	166.3
F(g)	80.0	61.9	158.7
F$_2$(g)	0.00	0.00	202.7
Fe(s)	0.00	0.00	27.15
Fe^{2+}(aq)	−87.86	−84.93	113.4
Fe^{3+}(aq)	−47.69	−10.54	293.3
FeCl$_3$(s)	−405	—	—
Fe$_2$O$_3$(s)	−822.16	−740.98	89.96
Fe$_3$O$_4$(s)	−1117.1	−1014.2	146.4
H(g)	217.94	203.26	114.60
H$^+$(aq)	0.00	0.00	0.00
HBr(g)	−36.23	−53.22	198.49
HCl(g)	−92.30	−95.27	186.69
HF(g)	−268.61	−270.70	173.51

Substance	ΔH_f°(kJ/mol)	ΔG_f°(kJ/mol)	S°(J/mol-K)
HI(g)	25.94	1.30	206.3
H_2(g)	0.00	0.00	130.58
H_2O(g)	−241.8	−228.61	188.7
H_2O(l)	−285.85	−236.81	69.96
H_2O_2(g)	−136.10	−105.48	232.9
H_2O_2(l)	−187.8	−120.4	109.6
H_2S(g)	−20.17	−33.01	205.6
Hg(g)	60.83	31.76	174.89
Hg(l)	0.00	0.00	77.40
I(g)	106.60	70.16	180.66
I_2(s)	0.00	0.00	116.73
I_2(g)	62.25	19.37	260.57
K(g)	89.99	61.17	160.2
KCl(s)	−435.9	−408.3	82.7
$KClO_3$(s)	−391.2	−289.9	143.0
$KClO_3$(aq)	−349.5	−284.9	265.7
KNO_3(s)	−492.70	−393.13	288.1
Mg(s)	0.00	0.00	32.51
$MgCl_2$(s)	−641.6	−592.1	89.6
Mn(s)	0	0	32.0
Mn(g)	280.7	238.5	173.6
MnO_2(s)	−519.6	−464.8	53.14
NH_3(g)	−46.19	−16.66	192.5
NH_4CN(s)	0.00	—	—
NH_4Cl(s)	−314.4	−203.0	94.6
NH_4NO_3(s)	−365.6	−184.0	151
NO(g)	90.37	86.71	210.62
NO_2(g)	33.84	51.84	240.45
NOCl(g)	52.6	66.3	264
N_2(g)	0.00	0.00	191.50
N_2O(g)	81.6	103.59	220.0
N_2O_4(g)	9.66	98.28	304.3
Na(g)	107.7	77.3	51.5
NaBr(aq)	−360.6	−364.7	141
NaCl(s)	−410.9	−384.0	72.33
NaCl(aq)	−407.1	−393.0	115.5
$NaHCO_3$(s)	−947.7	−851.8	102.1
Na_2CO_3(s)	−1130.9	−1047.7	136.0
$NaNO_3$(aq)	−446.2	−372.4	207
Ni(s)	0	0	29.9
Ni(g)	429.7	384.5	182.1
O(g)	247.5	230.1	161.0
O_2(g)	0.00	0.00	205.0
O_3(g)	142.3	163.4	237.6
OH^-(aq)	−230.0	−157.3	−10.7
P_4(g)	54.8	24.3	280
PCl_3(l)	−319.6	−272.4	217
PH_3(g)	23.0	25.5	210
$POCl_3$(g)	−542.2	−502.5	325
$POCl_3$(l)	−597.0	−520.9	222
P_4O_6(s)	−1640.1	—	—
P_4O_{10}(s, hexagonal)	−2940.1	−2675.2	228.9
$PbBr_2$(s)	−277.4	−260.7	161

Substance	ΔH_f°(kJ/mol)	ΔG_f°(kJ/mol)	S°(J/mol-K)
$Pb(NO_3)_2(s)$	−451.9	—	—
$Pb(NO_3)_2(aq)$	−421.3	—	—
$Rb(g)$	85.8	55.8	170.0
$RbCl(s)$	−430.5	−412.0	92
$RbClO_3(s)$	−392.4	−292.0	152
$S(s,\ \text{rhombic})$	0.00	0.00	31.88
$SO_2(g)$	−296.9	−300.4	248.5
$SO_3(g)$	−395.2	−370.4	256.2
$SOCl_2(l)$	−245.6	—	—
$Sc(s)$	0	0	34.6
$Sc(g)$	377.8	336.1	174.7
$Si(g)$	368.2	323.9	167.8
$SiCl_4(l)$	−640.1	−572.8	239.3
$Ti(g)$	468	422	180.3
$V(s)$	0	0	28.9
$V(g)$	514.2	453.1	182.2
$Zn(s)$	0.00	0.00	41.63
$Zn(g)$	130.7	95.2	160.9
$ZnO(s)$	−348.0	−318.2	43.9

Appendix E

Aqueous-equilibrium constants

E.1 DISSOCIATION CONSTANTS FOR ACIDS AT 25°C

Name	Formula	K_{a1}	K_{a2}	K_{a3}
Acetic	$HC_2H_3O_2$	1.8×10^{-5}		
Ascorbic	$HC_6H_7O_6$	8.0×10^{-5}		
Arsenic	H_3AsO_4	5.6×10^{-3}	1.0×10^{-7}	3.0×10^{-12}
Arsenous	H_3AsO_3	6×10^{-10}		
Benzoic	$HC_7H_5O_2$	6.5×10^{-5}		
Boric	H_3BO_3	5.8×10^{-10}		
Carbonic	H_2CO_3	4.3×10^{-7}	5.6×10^{-11}	
Chloroacetic	$HC_2H_2O_2Cl$	1.4×10^{-3}		
Cyanic	$HCNO$	3.5×10^{-4}		
Citric	$H_3C_6H_5O_7$	7.4×10^{-4}	1.7×10^{-5}	4.0×10^{-7}
Formic	$HCHO_2$	1.8×10^{-4}		
Hydrazoic	HN_3	1.9×10^{-5}		
Hydrocyanic	HCN	4.9×10^{-10}		
Hydrofluoric	HF	6.8×10^{-4}		
Hydrogen chromate	$HCrO_4^-$	3.0×10^{-7}		
Hydrogen peroxide	H_2O_2	2.4×10^{-12}		
Hydrogen selenate	$HSeO_4^-$	2.2×10^{-2}		
Hydrogen sulfate	HSO_4^-	1.2×10^{-2}		
Hydrogen sulfide	H_2S	5.7×10^{-8}	1.3×10^{-13}	
Hypobromous	$HBrO$	2×10^{-9}		
Hypochlorous	$HClO$	3.0×10^{-8}		
Iodic	HIO_3	1.7×10^{-1}		
Lactic	$HC_3H_5O_3$	1.4×10^{-4}		
Malonic	$H_2C_3H_2O_4$	1.5×10^{-3}	2.0×10^{-6}	
Nitrous	HNO_2	4.5×10^{-4}		
Oxalic	$H_2C_2O_4$	5.9×10^{-2}	6.4×10^{-5}	
Phenol	HC_6H_5O	1.3×10^{-10}		
Phosphoric	H_3PO_4	7.5×10^{-3}	6.2×10^{-8}	4.2×10^{-13}
para-Periodic	H_5IO_6	2.8×10^{-2}	5.3×10^{-9}	
Propionic	$HC_3H_5O_2$	1.3×10^{-5}		
Pyrophosphoric	$H_4P_2O_7$	3.0×10^{-2}	4.4×10^{-3}	
Selenous	H_2SeO_3	2.3×10^{-3}	5.3×10^{-9}	
Sulfuric	H_2SO_4	strong acid	1.2×10^{-2}	
Sulfurous	H_2SO_3	1.7×10^{-2}	6.4×10^{-8}	
Tartaric	$H_2C_4H_4O_6$	1.0×10^{-3}	4.6×10^{-5}	

E.2 DISSOCIATION CONSTANTS FOR BASES AT 25°C

Name	Formula	K_b
Ammonia	NH_3	1.8×10^{-5}
Aniline	$C_6H_5NH_2$	4.3×10^{-10}
Dimethylamine	$(CH_3)_2NH$	5.4×10^{-4}
Ethylamine	$C_2H_5NH_2$	6.4×10^{-4}
Hydrazine	H_2NNH_2	1.3×10^{-6}
Hydroxylamine	$HONH_2$	1.1×10^{-8}
Methylamine	CH_3NH_2	4.4×10^{-4}
Pyridine	C_5H_5N	1.7×10^{-9}
Trimethylamine	$(CH_3)_3N$	6.4×10^{-5}

E.3 SOLUBILITY-PRODUCT CONSTANTS FOR COMPOUNDS AT 25°C

Name	Formula	K_{sp}
Barium carbonate	$BaCO_3$	5.1×10^{-9}
Barium chromate	$BaCrO_4$	1.2×10^{-10}
Barium fluoride	BaF_2	1.0×10^{-6}
Barium hydroxide	$Ba(OH)_2$	5×10^{-3}
Barium oxalate	BaC_2O_4	1.6×10^{-7}
Barium phosphate	$Ba_3(PO_4)_2$	3.4×10^{-23}
Barium sulfate	$BaSO_4$	1.1×10^{-10}
Cadmium carbonate	$CdCO_3$	5.2×10^{-12}
Cadmium hydroxide	$Cd(OH)_2$	2.5×10^{-14}
Cadmium sulfide	CdS	8.0×10^{-27}
Calcium carbonate	$CaCO_3$	2.8×10^{-9}
Calcium chromate	$CaCrO_4$	7.1×10^{-4}
Calcium fluoride	CaF_2	3.9×10^{-11}
Calcium hydroxide	$Ca(OH)_2$	5.5×10^{-6}
Calcium phosphate	$Ca_3(PO_4)_2$	2.0×10^{-29}
Calcium sulfate	$CaSO_4$	9.1×10^{-6}
Cerium(III) fluoride	CeF_3	8×10^{-16}
Chromium(III) fluoride	CrF_3	6.6×10^{-11}
Chromium(III) hydroxide	$Cr(OH)_3$	6.3×10^{-31}
Cobalt(II) carbonate	$CoCO_3$	1.4×10^{-13}
Cobalt(II) hydroxide	$Co(OH)_2$	1.6×10^{-15}
Cobalt(III) hydroxide	$Co(OH)_3$	1.6×10^{-44}
α-Cobalt(II) sulfide[a]	CoS	4.0×10^{-21}
Copper(I) bromide	$CuBr$	5.3×10^{-9}
Copper(I) chloride	$CuCl$	1.2×10^{-6}
Copper(I) sulfide	Cu_2S	2.5×10^{-48}
Copper(II) carbonate	$CuCO_3$	1.4×10^{-10}
Copper(II) chromate	$CuCrO_4$	3.6×10^{-6}
Copper(II) hydroxide	$Cu(OH)_2$	2.2×10^{-20}
Copper(II) phosphate	$Cu_3(PO_4)_2$	1.3×10^{-37}
Copper(II) sulfide	CuS	6.3×10^{-36}
Gold(I) chloride	$AuCl$	2.0×10^{-13}
Gold(III) chloride	$AuCl_3$	3.2×10^{-25}
Iron(II) carbonate	$FeCO_3$	3.2×10^{-11}
Iron(II) hydroxide	$Fe(OH)_2$	8.0×10^{-16}
Iron(II) sulfide	FeS	6.3×10^{-18}
Iron(III) hydroxide	$Fe(OH)_3$	4×10^{-38}
Lanthanum fluoride	LaF_3	7×10^{-17}
Lanthanum iodate	$La(IO_3)_3$	6.1×10^{-12}
Lead carbonate	$PbCO_3$	7.4×10^{-14}
Lead chloride	$PbCl_2$	1.6×10^{-5}
Lead chromate	$PbCrO_4$	2.8×10^{-13}
Lead fluoride	PbF_2	2.7×10^{-8}
Lead hydroxide	$Pb(OH)_2$	1.2×10^{-15}
Lead sulfate	$PbSO_4$	1.6×10^{-8}
Lead sulfide	PbS	8.0×10^{-28}
Magnesium hydroxide	$Mg(OH)_2$	1.8×10^{-11}
Magnesium oxalate	MgC_2O_4	8.6×10^{-5}
Manganese carbonate	$MnCO_3$	1.8×10^{-11}

[a] Some substances exist in more than one crystalline form; the prefix indicates the particular form for which K_{sp} is listed.

Name	Formula	K_{sp}
Manganese hydroxide	$Mn(OH)_2$	1.9×10^{-13}
Manganese(II) sulfide	MnS	1.0×10^{-13}
Mercury(I) chloride	Hg_2Cl_2	1.3×10^{-18}
Mercury(I) oxalate	$Hg_2C_2O_4$	2.0×10^{-13}
Mercury(I) sulfide	Hg_2S	1.0×10^{-47}
Mercury(II) hydroxide	$Hg(OH)_2$	3.0×10^{-26}
Mercury(II) sulfide	HgS	4×10^{-53}
Nickel carbonate	$NiCO_3$	6.6×10^{-9}
Nickel hydroxide	$Ni(OH)_2$	1.6×10^{-14}
Nickel oxalate	NiC_2O_4	4×10^{-10}
α-Nickel sulfide[a]	NiS	3.2×10^{-19}
Silver arsenate	Ag_3AsO_4	1.0×10^{-22}
Silver bromide	$AgBr$	5.0×10^{-13}
Silver carbonate	Ag_2CO_3	8.1×10^{-12}
Silver chloride	$AgCl$	1.8×10^{-10}
Silver chromate	Ag_2CrO_4	1.1×10^{-12}
Silver cyanide	$AgCN$	1.2×10^{-16}
Silver iodide	AgI	8.3×10^{-17}
Silver sulfate	Ag_2SO_4	1.4×10^{-5}
Strontium carbonate	$SrCO_3$	1.1×10^{-10}
Tin(II) hydroxide	$Sn(OH)_2$	1.4×10^{-28}
Tin(II) sulfide	SnS	1.0×10^{-25}
Zinc carbonate	$ZnCO_3$	1.4×10^{-11}
Zinc hydroxide	$Zn(OH)_2$	1.2×10^{-17}
Zinc oxalate	ZnC_2O_4	2.7×10^{-8}
α-Zinc sulfide[a]	ZnS	1.1×10^{-21}

[a] Some substances exist in more than one crystalline form; the prefix indicates the particular form for which K_{sp} is listed.

Half-reaction	$E°(V)$
$Ag^+(aq) + e^- \rightleftharpoons Ag(s)$	$+0.799$
$AgBr(s) + e^- \rightleftharpoons Ag(s) + Br^-(aq)$	$+0.095$
$AgCl(s) + e^- \rightleftharpoons Ag(s) + Cl^-(aq)$	$+0.222$
$Ag(CN)_2^-(aq) + e^- \rightleftharpoons Ag(s) + 2CN^-(aq)$	-0.31
$Ag_2CrO_4(s) + 2e^- \rightleftharpoons 2Ag(s) + CrO_4^{2-}(aq)$	$+0.446$
$AgI(s) + e^- \rightleftharpoons Ag(s) + I^-(aq)$	-0.151
$Ag(S_2O_3)_2^{3-} + e^- \rightleftharpoons Ag(s) + 2S_2O_3^{2-}(aq)$	$+0.01$
$Al^{3+}(aq) + 3e^- \rightleftharpoons Al(s)$	-1.66
$H_3AsO_4(aq) + 2H^+(aq) + 2e^- \rightleftharpoons H_3AsO_3(aq) + H_2O(l)$	$+0.559$
$Ba^{2+}(aq) + 2e^- \rightleftharpoons Ba(s)$	-2.90
$BiO^+(aq) + 2H^+(aq) + 3e^- \rightleftharpoons Bi(s) + H_2O(l)$	$+0.32$
$Br_2(l) + 2e^- \rightleftharpoons 2Br^-(aq)$	$+1.065$
$BrO_3^-(aq) + 6H^+(aq) + 5e^- \rightleftharpoons \frac{1}{2}Br_2(l) + 3H_2O(l)$	$+1.52$
$Ca^{2+}(aq) + 2e^- \rightleftharpoons Ca(s)$	-2.87
$2CO_2(g) + 2H^+(aq) + 2e^- \rightleftharpoons H_2C_2O_4(aq)$	-0.49
$Cd^{2+}(aq) + 2e^- \rightleftharpoons Cd(s)$	-0.403
$Ce^{4+}(aq) + e^- \rightleftharpoons Ce^{3+}(aq)$	$+1.61$
$Cl_2(g) + 2e^- \rightleftharpoons 2Cl^-(aq)$	$+1.359$
$HClO(aq) + H^+(aq) + e^- \rightleftharpoons \frac{1}{2}Cl_2(g) + H_2O(l)$	$+1.63$
$ClO_3^-(aq) + 6H^+(aq) + 5e^- \rightleftharpoons \frac{1}{2}Cl_2(g) + 3H_2O(l)$	$+1.47$
$Co^{2+}(aq) + 2e^- \rightleftharpoons Co(s)$	-0.277
$Co^{3+}(aq) + e^- \rightleftharpoons Co^{2+}(aq)$	$+1.842$
$Cr^{3+}(aq) + 3e^- \rightleftharpoons Cr(s)$	-0.74
$Cr_2O_7^{2-}(aq) + 14H^+(aq) + 6e^- \rightleftharpoons 2Cr^{3+}(aq) + 7H_2O(l)$	$+1.33$
$CrO_4^{2-}(aq) + 4H_2O(l) + 3e^- \rightleftharpoons Cr(OH)_3(s) + 5OH^-(aq)$	-0.13
$Cu^{2+}(aq) + 2e^- \rightleftharpoons Cu(s)$	$+0.337$
$Cu^{2+}(aq) + e^- \rightleftharpoons Cu^+(aq)$	$+0.153$
$Cu^+(aq) + e^- \rightleftharpoons Cu(s)$	$+0.521$
$CuI(s) + e^- \rightleftharpoons Cu(s) + I^-(aq)$	-0.185
$F_2(g) + 2e^- \rightleftharpoons 2F^-(aq)$	$+2.87$
$Fe^{2+}(aq) + 2e^- \rightleftharpoons Fe(s)$	-0.440
$Fe^{3+}(aq) + e^- \rightleftharpoons Fe^{2+}(aq)$	$+0.771$
$Fe(CN)_6^{3-}(aq) + e^- \rightleftharpoons Fe(CN)_6^{4-}(aq)$	$+0.36$
$2H^+(aq) + 2e^- \rightleftharpoons H_2(g)$	0.000
$H_2O(l) + 2e^- \rightleftharpoons H_2(g) + 2OH^-(aq)$	-0.83
$HO_2^-(aq) + H_2O(l) + 2e^- \rightleftharpoons 3OH^-(aq)$	$+0.88$
$H_2O_2(aq) + 2H^+(aq) + 2e^- \rightleftharpoons 2H_2O(l)$	$+1.776$
$Hg_2^{2+}(aq) + 2e^- \rightleftharpoons 2Hg(l)$	$+0.789$
$2Hg^{2+}(aq) + 2e^- \rightleftharpoons Hg_2^{2+}(aq)$	$+0.920$
$Hg^{2+}(aq) + 2e^- \rightleftharpoons Hg(l)$	$+0.854$
$I_2(s) + 2e^- \rightleftharpoons 2I^-(aq)$	$+0.536$
$IO_3^-(aq) + 6H^+(aq) + 5e^- \rightleftharpoons \frac{1}{2}I_2(s) + 3H_2O(l)$	$+1.195$
$K^+(aq) + e^- \rightleftharpoons K(s)$	-2.925
$Li^+(aq) + e^- \rightleftharpoons Li(s)$	-3.05
$Mg^{2+}(aq) + 2e^- \rightleftharpoons Mg(s)$	-2.37

Half-reaction	$E°$(V)
$Mn^{2+}(aq) + 2e^- \rightleftarrows Mn(s)$	-1.18
$MnO_2(s) + 4H^+(aq) + 2e^- \rightleftarrows Mn^{2+}(aq) + 2H_2O(l)$	$+1.23$
$MnO_4^-(aq) + 8H^+(aq) + 5e^- \rightleftarrows Mn^{2+}(aq) + 4H_2O(l)$	$+1.51$
$MnO_4^-(aq) + 2H_2O(l) + 3e^- \rightleftarrows MnO_2(s) + 4OH^-(aq)$	$+0.59$
$HNO_2(aq) + H^+(aq) + e^- \rightleftarrows NO(g) + H_2O(l)$	$+1.00$
$NO_3^-(aq) + 4H^+(aq) + 3e^- \rightleftarrows NO(g) + 2H_2O(l)$	$+0.96$
$Na^+(aq) + e^- \rightleftarrows Na(s)$	-2.71
$Ni^{2+}(aq) + 2e^- \rightleftarrows Ni(s)$	-0.28
$O_2(g) + 4H^+(aq) + 4e^- \rightleftarrows 2H_2O(l)$	$+1.23$
$O_2(g) + 2H_2O(l) + 4e^- \rightleftarrows 4OH^-(aq)$	$+0.40$
$O_2(g) + 2H^+(aq) + 2e^- \rightleftarrows H_2O_2(aq)$	$+0.68$
$O_3(g) + 2H^+(aq) + 2e^- \rightleftarrows O_2(g) + H_2O(l)$	$+2.07$
$Pb^{2+}(aq) + 2e^- \rightleftarrows Pb(s)$	-0.126
$PbO_2(s) + HSO_4^-(aq) + 3H^+(aq) + 2e^- \rightleftarrows PbSO_4(s) + 2H_2O(l)$	$+1.685$
$PbSO_4(s) + H^+(aq) + 2e^- \rightleftarrows Pb(s) + HSO_4^{2-}(aq)$	-0.356
$PtCl_4^{2-}(aq) + 2e^- \rightleftarrows Pt(s) + 4Cl^-(aq)$	$+0.73$
$S(s) + 2H^+(aq) + 2e^- \rightleftarrows H_2S(g)$	$+0.141$
$H_2SO_3(aq) + 4H^+(aq) + 4e^- \rightleftarrows S(s) + 3H_2O(l)$	$+0.45$
$HSO_4^-(aq) + 3H^+(aq) + 2e^- \rightleftarrows H_2SO_3(aq) + H_2O(l)$	$+0.17$
$Sn^{2+}(aq) + 2e^- \rightleftarrows Sn(s)$	-0.136
$Sn^{4+}(aq) + 2e^- \rightleftarrows Sn^{2+}(aq)$	$+0.154$
$Zn^{2+}(aq) + 2e^- \rightleftarrows Zn(s)$	-0.763

Answers to selected exercises

CHAPTER 1

1.3 It should be Coulomb's law, because it simply summarizes experimental observations regarding the forces between charged objects. A theory would advance an *explanation* for why the charged objects behave as they do. **1.6** (a) meters, m; (b) cubic meters, m^3; (c) kilograms, kg; (d) square meters, m^2; (e) second, sec, or simply s. **1.8** (a) 51.8 cm^3; (b) 5.18×10^{-3} cm^3; (c) 5.18×10^{-5} cm^3; (d) 0.518 cm^3; (e) 5.18×10^{-9} cm^3; (f) 5.18×10^{-6} cm^3. **1.9** (a) 650 cm or 6.50×10^2 cm; (b) 3.3×10^7 mg; (c) 1.2×10^4 msec; (d) 0.043 g; (e) 2.35×10^3 g; (f) 2.25×10^4 μm. **1.12** 3.08×10^9 cm/nparsec **1.15** (a) four; (b) three; (c) three; (d) one; (e) two; (f) two, three, or four—can't be sure without knowing the context; (g) six. **1.18** (a) 1.245×10^3; (b) 6.50×10^4; (c) 5.975×10^4; (d) 4.56×10^{-3}. **1.19** (a) 0.55; (b) 3.5×10^3; (c) 5.74×10^{-9}; (d) 2×10^{-6}; (e) 1.5; (f) 1.82×10^3. **1.21** (a) 1.78 m; (b) 2.20×10^3 kg; (c) 3282 ft; (d) 43.4 L. **1.24** (a) 86.62 dm^3; (b) 5.24 L; (c) 10^6 m^2; (d) 0.386 mi^2. **1.27** (a) 80 chains; (b) 5.027 m; (c) 1 mi^2 = 6400 $chain^2$; (d) 1 yd^3 = 5.13×10^{-3} rod^3. **1.28** (a) 100 km/hr; (b) 28 m/sec; (c) \$5.40/kg, 11.7 Mark/kg; (d) 0.35 L. **1.29** 238,850 mi = 3.842×10^5 km (where we have used four significant figures in converting from meters to yards); 27.32 day = 2.360×10^6 sec; 1081 mi = 1739 km; 1.62×10^{23} lb = 7.35×10^{22} kg. **1.32** 9.8066 m/sec^2. **1.35** (a) 19.4°C; (b) −22°C; (c) 628 K; (d) −123°C; (e) 322 K; (f) 227 K. **1.38** \$0.82/kg. **1.40** 360 kg. **1.42** weight—extensive; color—intensive; density—intensive; volume—extensive; temperature—intensive; melting point—intensive; ease of corrosion—intensive. **1.45** (a) 5 ft 3 in. = 1.60 m, 110 lb = 49.9 kg, 122 lb = 55.3 kg; (b) 5 ft 9 in. = 1.75 m, 154 lb = 69.8 kg, 2.5 oz = 71 g. **1.46** 15 g CO. **1.48** (a) 4.5×10^3 kg/m^3; (b) 2.48×10^{-2} m^2; (c) 2.89×10^{-3} kg-m/sec^2; (d) 6.19×10^{-4} m^3/sec. **1.51** 9.55 km/L. **1.53** (a) r = 1.82 cm; (b) 2.60 cm; (c) 1.61 cm.

CHAPTER 2

2.1 (a) mercury; (b) Sulfur melts at 112°C, phenol (carbolic acid) melts at 41°C, and naphthalene (used as moth crystals) melts at 80°C. (c) Naphthalene or *para*-dichlorobenzene, used as moth crystals, sublimes slowly into the vapor state when allowed to stand. Solid CO_2, "dry ice," sublimes without melting. **2.3** (a) A battery fluid is a solution of sulfuric acid in water. (b) The gold bars should be pure substance. (c) Sand is a heterogeneous mixture of different minerals. (d) Seven-up is a solution, that is, a homogeneous mixture, of sugar and other substances, including carbon dioxide. (e) A Bufferin or Anacin tablet is a heterogeneous mixture of two or more substances. **2.4** (a) Burning of the magnesium ribbon is a chemical process; it involves oxidation of magnesium to magnesium oxide. (b) physical process; (c) physical process (however, some small amount of chemical change may occur; eventually the filament burns out); (d) physical process; (e) physical process; (f) chemical process. **2.7** Physical properties: silver-white color; soft; good conductor of electricity; boils at 883°C; vapor is violet-colored. Chemical properties: prepared by electrolysis of molten sodium chloride; tarnishes rapidly in air; burns on heating in air or in bromine vapor. **2.9** In the early days of the atomic theory, the evidence for an element was that the substance was incapable of being chemically broken down into still simpler substances. Thus the evidence was always negative in character. Only when means were devised for analyzing materials carefully could the elemental nature of a substance be fully established. There was also much reasoning by analogy. A substance with a bright, shiny, metallic appearance was likely to be a metallic element. **2.11** The product of the reaction of hydrogen and bromine is a substance with a fixed ratio of one substance to the other because the *molecules* of the product each contain the same relative numbers of hydrogen and bromine atoms. The composition of the product is fixed; the identity of the product depends on the fact that the composition of each molecule is the same. If there is an excess of one gas over the other when the two are reacted, molecules of the component present in excess remain unreacted. These observations illustrate the *law of constant composition*. **2.14** Calculate the ratio mass of oxygen/mass of nitrogen: A = 1.14; B = 2.28; C = 1.705. Dividing through by the smallest of these, we obtain A = 1.00; B = 2.00; C = 1.50. To convert these to whole numbers, multiply by 2: A = 2.00; B = 4.00; C = 3.00. These data tell us that the ratios of oxygen to nitrogen in the three compounds are related to one another as small whole numbers. In other words, while we don't know yet, without doing more calculating, just what the ratio of oxygen to nitrogen *atoms* in compound A is, we do know that in compound B there is twice as much oxygen relative to nitrogen, and in compound C there is 1.5 times as much. **2.17** The tracks are long because the alpha particles encounter only empty space or the outer electrons of atoms in their path. Through many contacts with the electrons of atoms, the particles eventually slow down and are stopped. Occasionally an alpha particle makes a direct hit on a small but highly charged nucleus of an oxygen or nitrogen atom. These rare encounters cause the alpha particles to be deflected. **2.19** Proton fraction of mass = 0.429; neutron fraction of mass = 0.571; electron fraction of mass = 2.3×10^{-4}. **2.21** 1×10^{18} g/cm^3. **2.22** (a) must contain the same number of protons, neutrons, and electrons; (b) must contain the same number of protons and electrons; (c) must contain the same total number of protons plus neutrons. **2.23** (a) 6 protons, 7 neutrons, 6 electrons; (b) 25 protons, 30 neutrons, 25 electrons; (c) 42 protons, 55 neutrons, 42 electrons. **2.26** (a) Mn, a metal; (b) Se, a nonmetal; (c) Br, a nonmetal; (d) Zn, a metal; (e) Cr, a metal; (f) Ge, a metalloid (or semimetal); (g) S, a nonmetal. **2.28** (a) calcium, Ca, and strontium, Sr; (b) arsenic, As, and antimony, Sb. **2.30** (a) P_2O_5; (b) CH_2; (c) C_3H_7; (d) SiO_2; (e) $Na_2B_4O_7$; (f) C_2O_2H. **2.33** (a) Ca; (b) Kr; (c) Br; (d) Zn. **2.35** (a) H_2CO; (b) CH; (c) CHCl; (d) C_5H_7N; (e) P_2S_5; (f) HgCl. Acetic acid, benzene, dichloroethylene, and nicotine are organic substances. **2.36** (a) chromium(II) or chromous ion; (b) strontium ion (we generally don't write strontium-(II) because strontium always appears as the $2+$ ion in its compounds); (c) bromide ion; (d) sulfide ion; (e) cesium ion; (f) titanium(III) ion. **2.39** (a) VBr_3; (b) $Zn(HCO_3)_2$; ZnO, CO_2, H_2O; (c) HF, SiO_2, SiF_4, H_2O; (d) SO_2, H_2O, H_2SO_3, H_2SO_4; (e) PH_3; (f) $HClO_4$, Cd, $Cd(ClO_4)_2$. **2.42** (a) strontium nitrate; (b) tin(IV) bromide (or stannic bromide); (c) silver nitrate; (d) potassium cyanide; (e) aluminum hydroxide; (f) sodium hydrogen sulfate or sodium bisulfate; (g) ammonium sulfate; (h) iodic acid; (i) copper(I) bromide or cuprous bromide. **2.44** (a) Homogeneous indicates the same throughout; heterogeneous indicates variation throughout a material. (b) A gas expands so as uniformly to occupy all of the volume containing it. A liquid is of fixed volume, but it adapts its shape to that of the container. Gases are highly compressible, liquids are not. (c) Atomic number is a measure of the number of protons in the atomic nucleus; the mass number is a measure of the total number of protons plus neutrons. (d) A chemical property is a characteristic of a substance as it undergoes chemical change; that is, as it converts into one or more other substances. A physical property, for example, the melting point of a solid, is exhibited by a substance while it retains its chemical properties and composition. (e) Ca refers to the neutral calcium atom; Ca^{2+} refers to the calcium ion, with two fewer electrons than the neutral atom. (f) Hydrochloric acid is HCl; chloric acid is $HClO_3$, an oxyacid. (g) Iron(II) and iron(III) are two forms of the element iron. In iron(II) the atom has lost two electrons; in iron(III) it has lost three. (h) Sodium carbonate is Na_2CO_3; sodium bicarbonate is $NaHCO_3$. (i) H_2O is water; H_2O_2 is hydrogen peroxide. (j) Metals and nonmetals have distinct physical and chemical properties. Metals are located at the left side of the periodic table; the nonmetals are located in the upper right side. (k) H represents the element hydrogen, with one proton in the nucleus and one external electron per atom. He represents helium, with two protons per nucleus and two electrons in each atom. (1) Chloride ion is Cl^-; chlorate ion is ClO_3^-, an oxyanion. **2.48** (a) lithium, Li; (b) aluminum, Al; (c) neon, Ne.

2.49

Symbol	$^{12}_{6}C$	$^{17}_{8}O^{2-}$	$^{25}_{12}Mg$	$^{23}_{11}Na^+$	$^{18}_{8}O^{2-}$
Protons	6	8	12	11	8
Neutrons	6	9	13	12	10
Electrons	6	10	12	10	10
Net charge	0	2−	0	1+	2−

2.53 (a) $CdCl_2$; (b) CdS; (c) ZnO. **2.56** Elements are I_2, S_8, O_3; compounds are CO_2, NH_3, H_2O_2. **2.58** A family of elements is a group of elements located in a vertical column on the periodic table. Elements related in this way possess similar chemical and physical properties because they have similar arrangements of electrons in the outermost shells.

CHAPTER 3

3.2 Equation (a) is consistent with the law of conservation of mass. In (c) there should be $2H_2O$ on the right. **3.3** (a) $2Al(s) + 3Cl_2(g) \longrightarrow 2AlCl_3(s)$; (b) $P_2O_3(s) + 3H_2O(l) \longrightarrow 2H_3PO_3(aq)$; (c) $Ca(OH)_2(aq) + 2HBr(aq) \longrightarrow CaBr_2(aq) + 2H_2O(l)$; (d) $16Al(s) + 3S_8(s) \longrightarrow 8Al_2S_3(s)$; (e) $Mg_2C_3(s) + 4H_2O(l) \longrightarrow 2Mg(OH)_2 (aq) + C_3H_4(g)$. **3.5** (a) $2PH_3(g) + 4O_2(g) \longrightarrow P_2O_5 (s) + 3H_2O(g)$; (b) $Ba(s) + 2CH_3OH (l) \longrightarrow H_2(g) + Ba(OCH_3)_2(alcohol, alc)$; (c) $B_2S_3(s) + 6H_2O(l) \longrightarrow 2H_3BO_3(aq) + 3H_2S(g)$; (d) $Cu(s) + 2H_2SO_4(l) \longrightarrow CuSO_4(s) + SO_2(g) + 2H_2O(l)$; (e) $2NH_3(g) + 2Na(l) \longrightarrow H_2(g) + 2NaNH_2(s)$. **3.7** (a) $2H_2(g)$; (c) $Cu(s)$; (e) $H_2SO_4(l)$. **3.10** (a) $SOCl_2$, FW = 118.98; (b) $Tl_2(C_2O_4)$, FW = 496.80; (c) $C_9H_6O_2$, FW = 146.14; (d) $C_{16}H_{22}O_8 \cdot 2H_2O$, FW = 378.4; (e) XeF_4, FW = 207.29. **3.12** 1.00×10^{24} atoms C; weight percentage C = 59.9%. **3.14** $Cl_2(g) + 3F_2(g) \longrightarrow 2ClF_3(g)$. **3.17** Atomic weight of element X is 85 g/mol; the element is rubidium, Rb. **3.18** Berzelius's value for the atomic weight of Zn is

129.0, a factor of two too high. His error lay in the wrong assumed formula for the oxide, which is ZnO, not ZnO_2 **3.22** (a) FW = 44.09; (b) 64.06; (c) 32.12; (d) 65.14; (e) 65.99. **3.24** Average mass of a neon atom is 20.17 amu. **3.28** 1.97×10^{16} molecules C_2H_3Cl/L. **3.30** (a) 1.0×10^{22} molecules I_2; (b) 5.2×10^{-5} g Fe; (c) 2.1×10^{22} Ag atoms; (d) 8.1×10^{-3} g $C_6H_{12}O_6$; (e) 2.08 mol H_2O. **3.33** Formula of the hydrate is $Sr(OH)_2 \cdot 8H_2O$. **3.36** Empirical formula is CH_3O_2. **3.39** (a) 6.45×10^{-2} mol P_2O_5; (b) 12.6 g H_3PO_4. **3.41** (a) 0.0385 mol SiH_4; (b) 2.77 g H_2O; (c) Water is the limiting reagent; 6.24 mg SiH_4 can be formed. **3.43** 3.762 g Al_4C_3; 13.93 g $AlCl_3$. **3.45** KBr is the limiting reagent; 3.87 g AgBr is formed. **3.47** (a) 0.112 M; (b) 0.035 M; (c) 4.54×10^{-6} M; (d) 1.2×10^{-3} M (1.2 mM); **3.50** (a) 0.14 L solution; (b) 0.359 L solution; (c) 0.059 L = 59 mL solution; (d) 0.200 L solution. **3.52** 0.0759 L Na_2CrO_4 solution. **3.54** 2.60×10^{-3} g $ZnCl_2$; the weight percentage is $9.05 \times 10^{-2}\%$. **3.56** (a) $Li_3N(s) + 3H_2O(l) \longrightarrow NH_3(g) + 3LiOH(aq)$; (c) $PBr_3(l) + 3H_2O(l) \longrightarrow H_3PO_3(aq) + 3HBr(aq)$; (e) $2CCl_4(g) + O_2(g) \longrightarrow 2CCl_2O(g) + 2Cl_2(g)$. **3.59** 12.011 amu. **3.60** weight percentage H is 2.46%; weight percentage S is 39.06%; weight percentage O is 58.48%. **3.63** Empirical formula is Cr_2O_3. **3.65** Simplest formula is CH_2. **3.67** Empirical formula is C_2HF_3ClBr. **3.69** When reaction is complete, essentially all O_2 is consumed; 0.22 g H_2 and 29.2 g H_2O are present. **3.71** 7.27×10^{-6} g $Fe(CO)_5$. **3.75** 0.0637 g As; weight percentage As is 5.22%. **3.77** (a) 0.595 M; (b) 0.0462 M; (c) 0.382 M; (d) total molarity of KBr = 0.41 M. **3.79** (a) Dissolve 5.9 g KOH with stirring in a small amount of water. Add to a 300-mL volumetric flask, with rinsing. Add water to bring volume to exactly 300 mL. (A graduated cylinder would also be accurate enough for two-place precision.) (b) Weigh out 85.0 g $AgNO_3$; dissolve in a few hundred mL water. Add to a 1-L volumetric flask; rinse a couple of times, add rinses to flask also. Dilute with water, while stirring, to bring volume to 1-L mark. **3.82** 3.39 M H_2SO_4, solution. **3.84** 2.8 L solution required.

CHAPTER 4

4.4 kinetic energy = 2.91×10^4 BTU. **4.5** At the time of entry, $E_k = 7.7 \times 10^6$ J; at the time of striking the earth, $E_k = 1.4 \times 10^3$ J. The loss in kinetic energy has appeared as heat as the meteorite makes its way through the atmosphere. **4.7** 4.30×10^4 J. **4.9** (a) exothermic; (b) endothermic; (c) endothermic; (d) exothermic; (e) exothermic. **4.10** (a) ΔH positive—heat must be added to evaporate the liquid; (b) ΔH negative—heat is released in the combustion; (c) ΔH positive—the solution absorbs heat from surroundings as NH_4Cl dissolves; (d) ΔH positive—heat must be added to break up the stable HCl molecules. **4.12** -56.0 kJ. **4.14** Addition of the following equations gives us the equation we need:

$$2H_2(g) + 2F_2(g) \longrightarrow 4HF(g) \qquad \Delta H = 2(-537 \text{ kJ})$$
$$2C(s) + 4F_2(g) \longrightarrow 2CF_4(g) \qquad \Delta H = 2(-680 \text{ kJ})$$
$$C_2H_4(g) \longrightarrow 2C(s) + 2H_2(g)$$
$$\Delta H = -52.3 \text{ kJ}$$

$$\overline{C_2H_4(g) + 6F_2(g) \longrightarrow 2CF_4(g) + 4HF(g)}$$
$$\Delta H = -2486 \text{ kJ}$$

4.16 In evaluating ΔH_{rxn}, be sure you have chosen the correct physical state of each reactant and product. (a) -38.9 kJ; (b) -36.4 kJ; (c) -1372.4 kJ; (d) -804.0 kJ. **4.18** -54.9 kJ. **4.19** for nitroethane, 18.9 kJ/cm^3; for ethanol, 23.4 kJ/cm^3; for diethyl ether, 26.3 kJ/cm^3. **4.22** Heat capacity of molybdenum is 0.27 J/°C-g. **4.25** Total heat capacity of solution is 422 J/°C; heat of solution of 1.38 g PCl_3 is -2.72 kJ; the molar heat of solution is -271 kJ/mol PCl_3. **4.27** Apple is 13% carbohydrate. **4.28** Food value is 2500 Calories/lb peanut brittle; about 275 Calories in a 50-g bar; 470 g of peanut brittle per day would be required. **4.30** Complete metabolism of 10 g of $C_3H_8O_3$ would yield 180 kJ, or 43 nutritional Calories. **4.33** 7.3×10^6 tons. **4.36** 3.1×10^{13} kJ. **4.38** The lighter car, traveling at the higher speed, possesses the larger kinetic energy, 5.4×10^6 kg-km^2/hr^2. This converts to 4.2×10^5 kg-m^2/sec^2, or 4.2×10^5 J, or 1.0×10^5 cal. The heavier, slower car has a kinetic energy of 4.0×10^6 kg-km^2/hr^2, or 3.1×10^5 J, or 7.4×10^4 cal. **4.41** (a) $\Delta H_{rxn} = +36.8$ kJ, endothermic process; (b) $\Delta H_{rxn} = -1117$ kJ, exothermic process; (c) $\Delta H_{rxn} = +464.8$ kJ, endothermic process; (d) $\Delta H_{rxn} = -178.1$ kJ, exothermic process. **4.42** Add the first and fourth equations, double and add the third equation, then double and add the second equation. Overall $\Delta H = -830$ kJ. **4.45** $2C(s) + 4H_2(g) + O_2(g) \longrightarrow 2CH_3OH(g)$
$$\Delta H = -402.6 \text{ kJ}$$
$$CH_3OH(l) \longrightarrow CH_3OH(g) \qquad \Delta H = 37.4 \text{ kJ}$$
Reverse and double the second reaction, add to first to obtain desired reaction, with overall $\Delta H = -477.4$ kJ. **4.47** 3.82°C rise. **4.49** 2.2×10^{10} kJ/hr; 8.4×10^6 g S/hr; 7.8×10^7 kJ/hr results from combustion of S to SO_2. **4.50** 7.0×10^{18} kJ; 64 yr; 7.1×10^{17} g CO_2.

CHAPTER 5

5.1 In order of increasing wavelength: X rays, green light, red light, microwaves, FM radio signal. **5.2** (a) $\lambda = c/\nu = (3.00 \times 10^8 \text{ m/sec})/(5.00 \times 10^{13}/\text{sec}) = 6.00 \times 10^{-6}$ m = 6.00 μm; (b) 2.00×10^{25}/sec; (c) 1.8×10^{10} m. **5.5** (a) 9.47×10^{15} m, 5.89×10^{12} mi; (b) 3.09×10^{13} km. **5.7** (a) 3.31×10^{-20} J; (b) 9.94×10^{-19} J; (c) 1.96×10^{14}/sec. **5.10** 3.61×10^{-19} J/photon; 40 photons. **5.12** $\lambda = 297$ nm. Doubling the intensity of radiation doubles the number of electrons ejected. No change in the kinetic energy of the ejected electrons. **5.15** (a) Energy is emitted because the electron moves to a lower-energy (more stable) orbital. (b) absorbed; (c) absorbed. **5.16** $\Delta E = 1.94 \times 10^{-18}$ J. **5.19** Ionization energy for Li^{2+} is larger than for H because of the larger nuclear charge. **5.23** (a) momentum, $mv = 6.31 \times 10^{-24}$ kg-m/sec; (b) $\lambda = 1.05 \times 10^{-10}$ m; (c) $v = 1.99 \times 10^3$ m/sec. **5.27** The incorrect designations are $2d$, $1p$, and $3f$. **5.30** $1s < 2s = 2p < 3s = 3d < 4s = 4f$. **5.34** (a) The principal quantum number (n) determines the radial extent of the orbital; it also determines the energy of the orbital. (b) The azimuthal quantum number (l) determines the angular shape of the orbital. (c) The magnetic quantum number (m_l) determines the direction in which an orbital is oriented. **5.36** (a) $\nu = 2.19 \times 10^{14}$/sec; (b) $E(535 \text{ nm})/E(0.15 \text{ nm}) = 0.15 \text{ nm}/535 \text{ nm} = 2.8 \times$

10^{-4} times as many calories from the longer wavelength photon. **5.39** (a) 3.27×10^{18} photons; (b) 2.01×10^{18} photons. **5.42** (a) $E = 3.82 \times 10^{-19}$ J. On a molar basis, 230 kJ/mol. (b) $E = 5.52 \times 10^{-19}$ J; kinetic energy is 1.70×10^{-19} J/electron. **5.45** 2.2×10^{3} sec. **5.46** $E = 7.42 \times 10^{-28}$ J. **5.49** The emissions from the neon lamp arise from transitions of the electrons in neon atoms that have been excited by an electric discharge. The electrons may exist only in allowed states of definite energy. When they undergo a transition from one allowed state to another of lower energy, a photon of energy corresponding to the difference in energies of the two allowed states is emitted. **5.52** (a) one; (b) five; (c) one; (d) three. **5.53** $v = 2.19 \times 10^{6}$ m/sec. **5.55** The *change* in major quantum number is the only factor that determines the energy change. The change is determined by $(1/n_1^2 - 1/n_2^2)$. The $2p \longrightarrow 3s$ transition requires the lowest-energy photon. **5.57** (a) $n = 1$, $l = 0$, $m_l = 0$; (b) $n = 2$, $l = 1$, $m_l = 1, 0, -1$; (c) $n = 3$, $l = 2$, $m_l = 2, 1, 0, -1, -2$.

CHAPTER 6

6.1 (a) $3p$; (b) $2s$; (d) $3s$; (e) $3d$. **6.3** In a hydrogen atom the orbital energy depends only on n. In a many-electron atom the energy depends on both n and l. **6.5** The $3p_x$ and $3p_y$ orbitals are degenerate; that is, they have the same energy. **6.7** (a) For $n = 3$, $l = 1$, m_l may have values $+1$, 0, -1, and m_s may have values $+\frac{1}{2}$ and $-\frac{1}{2}$. Hence, a total of six electrons. (b) 10; (c) 2; (d) 32; (e) 9. These nine are $n = 3$, $l = 2$, with $m_l = 2, 1, 0, -1, -2$; $l = 1$ with $m_l = 1, 0, -1$; $l = 0$ with $m_l = 0$. There is just one electron in each orbital, with $m_s = +\frac{1}{2}$. **6.10** (a) $[Ar]4s^1$; (b) $[Ne]3s^23p^2$; (c) $[Ar]4s^23d^{10}4p^4$; (d) $[Ar]4s^23d^5$; (e) $[Xe]6s^25d^1$.

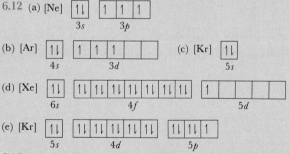

6.14 (a) ground state of Be; (b) excited state of Li (ground state would be $1s^22s^1$); (c) excited state of Al; (d) ground state of Ti; (e) excited state of Na. **6.16** (a) alkaline earths—ns^2 ($n = 2 \ldots 7$); (b) group 1B—$ns^1(n - 1)d^{10}$ ($n = 4, 5, 6$); (c) group 5A—ns^2np^3 ($n = 2 \ldots 6$); (d) halogens—ns^2np^5 ($n = 2 \ldots 6$); (e) noble gases—ns^2np^6 ($n = 2 \ldots 6$); (f) group 5B—$ns^2(n - 1)d^3$ ($n = 4, 5, 6$). **6.18** (a) Atomic size decreases from left to right in a horizontal row of the table. (b) ionization energy increases; (c) increases. As one moves down a family in the table, atomic size generally increases, ionization energy decreases, and electron affinity remains relatively constant. **6.19** (a) Na; (b) Li; (c) P; (d) Si. **6.23** The third electron removed from Mg comes from the $2p$ subshell. The $2p$ electrons experience a much larger effective nuclear charge because they do not completely shield one another from the nucleus. The $3s$

electrons, on the other hand, were well shielded by the $2p$ electrons, and thus they experienced a much smaller effective nuclear charge. **6.25** (a) $2K(s) + H_2O(l) \longrightarrow 2KOH(aq) + H_2(g)$; (b) $Li_2O(s) + H_2O(l) \longrightarrow 2LiOH(aq)$; (c) $4Li(s) + O_2(g) \longrightarrow 2Li_2O(s)$; (d) $2Na(l) + H_2(g) \longrightarrow 2NaH(s)$; (e) $2Li(s) + I_2(g) \longrightarrow 2LiI(s)$. **6.28** Reversals of atomic weights occur at Ar, 39.948–K, 39.098; Co, 58.9332–Ni, 58.70; Te, 127.60–I, 126.9045; Th, 232.038–Pa, 231.0359; U, 238.03–Np, 237.0482. These "violations" occur because the average number of neutrons in the nuclei of the element with one less proton is sufficient to outweigh the smaller total mass of protons. The modern version of the periodic law is that the chemical and physical properties of the elements are periodic functions of their atomic *number*. **6.31** (a) $2s$; (b) $2p$; (c) $3d$. **6.33** Use $E = -R_HZ^2/n^2$. On a molar basis, for H, $E = -1312$ kJ; for He^+, $E = -5248$ kJ; for Li^{2+}, $E = -1.181 \times 10^4$ kJ. **6.34** Separation of atoms into two beams could occur with H or Li, but not with He or Be, because they have no unpaired electrons. **6.37** If there were three possible values for m_s, this would suggest that it would require three electrons to fill each orbital. Thus for carbon,

$$\boxed{1 \to \downarrow} \quad \boxed{1 \to \downarrow}$$
$$1s \qquad\qquad 2s$$

6.38 The g orbitals correspond to $l = 4$. Thus the lowest value which n could have would be 5. There are nine g orbitals for a given n. No known elements in their ground states have electrons in g orbitals. **6.41** (a) C possesses two unpaired electrons; (b) none; (c) five; (d) two. **6.43** (a) chlorine; (b) alkaline earth metals (Mg, Ca, etc.); (c) Co, Rh (actually Rh is not exactly this, but is $5s^14d^8$, as shown in Table 6.2). Note that Ir, a member of the same family, is not included because it possesses $4f$ electrons in addition to the s and d (Table 6.2). **6.44** (a) K; (b) He; (c) N; (d) He; (e) F or B; (f) Cl.

CHAPTER 7

7.1 (a) $\cdot \text{Ca} \cdot$ (b) $\cdot \overset{\cdot\cdot}{\text{Se}} \cdot$ (c) $: \overset{\cdot\cdot}{\text{Br}} \cdot$ (d) $\cdot \overset{\cdot}{\text{B}} \cdot$ **7.2** Atoms or ions with octets are (b) O^{2-}; (c) Ne; (d) Ca^{2+}. **7.5** (a) $SrBr_2$; (b) BeO; (c) Al_2S_3; (d) ZnF_2; (e) K_2Se; (f) ZnO. **7.6** (a) Mn outer electron configuration is $4s^23d^5$; Mn^{4+} is $3d^3$; (b) $3s^23p^6$; (c) $3d^{10}$; (d) none; (e) $3d^3$; (f) $6s^24f^{14}5d^{10}$. **7.8** (a) ions in MgO have higher charge; (b) larger lattice energy because smaller oxide ion allows closer approach of ions; (c) chloride ion is smaller than bromide ion—larger lattice energy because of closer approach of ions. **7.11** (a) $CaBr_2$; (b) $Al_2(SO_4)_3$; (c) $(NH_4)_3PO_4$; (d) K_3N; (e) TiF_4. **7.12** (a) Cs^+; (b) Cl^-; (c) Fe^{2+}; (d) S^{2-}; (e) Zn; (f) K^+. **7.14** (a) $Mg^{2+} < Ca^{2+} < Sr^{2+}$; (b) $Ca^{2+} < Na^+ < K^+ < Cl^-$; (c) $F < O < O^{2-}$. **7.17** (a) covalent; (b) ionic; (c) covalent.

7.18 (a)
$$\begin{array}{c} \text{H} \\ | \\ \text{H}-\text{Si}-\text{H} \\ | \\ \text{H} \end{array}$$
(b) $\text{H}-\overset{\cdot\cdot}{\underset{\cdot\cdot}{\text{S}}}-\text{H}$ (c) $: \text{C} \equiv \text{O} :$

(d) $: \text{N} \equiv \text{N} :$ (e) $\text{H}-\overset{\cdot\cdot}{\underset{\cdot\cdot}{\text{O}}}-\overset{\cdot\cdot}{\underset{\cdot\cdot}{\text{O}}}-\text{H}$ (f) $\overset{\cdot\cdot}{\underset{\cdot\cdot}{\text{S}}} = \text{C} = \overset{\cdot\cdot}{\underset{\cdot\cdot}{\text{S}}}$

7.19 In all these problems, first determine the number of valence shell electrons to be placed. Be sure to consider the charge on the ion.

(a) $[:\!\ddot{C}l\!-\!\ddot{O}\!:]^-$ (b) $[:\!\ddot{O}\!-\!S\!-\!\ddot{O}\!:]^{2-}$ with $:\ddot{O}:$ below   (d) $[:\!C\!\equiv\!N\!:]^-$

7.21 (a) resonance structures of SO_3:

$$\ddot{O}\!\!=\!\!S(\ddot{O})(\ddot{O}) \longleftrightarrow S(\ddot{O})(\ddot{O})\!\!=\!\!\ddot{O} \longleftrightarrow S(\ddot{O})(\ddot{O})\!\!=\!\!\ddot{O}$$

(b) resonance structures of CO_3^{2-}:

$$\left[\ddot{O}\!\!=\!\!C(\ddot{O})(\ddot{O})\right]^{2-} \longleftrightarrow \left[C(\ddot{O})(\ddot{O})\!\!=\!\!\ddot{O}\right]^{2-} \longleftrightarrow \left[C(\ddot{O})(\ddot{O})\!\!=\!\!\ddot{O}\right]^{2-}$$

7.24 (a) SF_4 structure F—S—F with F,F; (c) B—Cl with Cl, Cl (BCl$_3$); (e) Cl—P with Cl,Cl,Cl,Cl (PCl$_5$)

(g) $:\!-\!Xe$ with F, F (XeF$_2$)

7.27 (a) C—F bond is stronger—larger electronegativity difference; (b) C=O bond is stronger—higher bond order; (c) S—S bond is stronger; (d) O—H bond—larger electronegativity difference. **7.28** (a) $\Delta H = -D(O{-}H) - D(C{-}I) + D(C{-}O) + D(H{-}I) = -463 - 240 + 358 + 299 = -46$ kJ **7.30** (a) P < S < O; (b) Mg < Al < Si; (c) S < Br < Cl; (d) Si < C < N. **7.31** (a) Cl—I is polar—Cl is the more electronegative atom; (b) P—P is nonpolar; (c) C—N is polar—N is the more electronegative atom; (d) F—F is nonpolar; (e) O—H is polar—O is the more electronegative atom. **7.36** (a) N, +4; O, −2; (c) H, +1; S, +4; O, −2. **7.37** (a) N may range from −3 (Na_3N) to +5 (HNO_3); (b) Cl may range from −1 (Cl^-) to +7 (ClO_4^-). **7.38** (a) C is oxidized, O is reduced; (b) S is oxidized, N is reduced. **7.40** (a) nitrogen trifluoride or nitrogen(III) fluoride; (b) dichromium hexaoxide or chromium(VI) oxide; (c) niobium pentafluoride or niobium(V) fluoride; (d) selenium dioxide or selenium(IV) oxide; (e) diphosphorus tetrachloride or phosphorus(II) chloride; (f) titanium trifluoride or titanium(III) fluoride. **7.43** (a) XeF_4; (b) Cr_2O_3; (c) CuBr; (d) N_2O; (e) Ga_2S_3. **7.44** (a) $SrO(s) + H_2O(l) \longrightarrow Sr(OH)_2(s)$ (or $Sr(OH)_2(aq)$ as product; solubility of $Sr(OH)_2$ is 18 g/L at room temperature); (b) $SO_3(g) + 2KOH(aq) \longrightarrow K_2SO_4(aq) + H_2O(l)$ or $SO_3(g) + KOH(aq) \longrightarrow KHSO_4(aq)$; (c) $P_4O_{10}(s) + 6H_2O(l) \longrightarrow 4H_3PO_4(aq)$; (d) $2HCl(aq) + CoO(s) \longrightarrow CoCl_2(aq) + H_2O(l)$; (e) $Ga_2O_3(s) + 6NaOH(aq) \longrightarrow 2Na_3GaO_3(aq) + 3H_2O(l)$. **7.47** (a) MgO, high lattice energy; (b) SiO_2, extended network solid. **7.51** (a) should double energy of ionic interaction; (b) should double energy of ionic interaction; (c) should quadruple energy of ionic interaction; (d) should reduce ionic energy to one-half.

7.53 -3.88×10^{-18} J/ion pair; 2.33×10^6 J/mol ion pairs. **7.56** Effective nuclear charge increases in moving from left to right, and ionic radii decrease. These quantities vary in the opposite senses, because as the effective nuclear charge increases, the electrons are drawn more tightly toward the nucleus, thus reducing the atomic or ionic radius. **7.57** As^{3-}, Se^{2-}, Br^-, Rb^+, Sr^{2+}, Y^{3+}, and Zr^{4+} are all isoelectronic with Kr. **7.60** Mn^{2+}, $[Ar]3d^5$; Mn^{4+}, $[Ar]3d^3$; Mn^{7+}, $[Ar]$. Oxides are MnO, MnO_2, and Mn_2O_7.

7.61 (a) $\ddot{N}\!\!=\!\!N\!\!=\!\!\ddot{O} \longleftrightarrow :\!N\!\!\equiv\!\!N\!\!-\!\!\ddot{O}:$ (b) $H\!-\!\underset{H}{\overset{H}{C}}\!-\!\underset{H}{\overset{H}{C}}\!-\!\ddot{O}\!-\!H$

(c) $\underset{H}{\overset{H}{>}}C\!\!=\!\!\ddot{O}$ (d) $H\!-\!\underset{H}{\ddot{N}}\!-\!\overset{:\ddot{O}:}{\underset{}{C}}\!-\!\underset{H}{\ddot{N}}\!-\!H$

7.64 Write the Lewis structures. The S—O bond order in SO_3^{2-} is 1, whereas it is 1.5 in SO_2. The S—O bond distance should be shorter and bond energy greater in SO_2 than in SO_3^{2-}. **7.71** (a) $\Delta H = -289$ kJ

CHAPTER 8

8.1 (a) 180°; (b) 120°; (c) 109.5°; (d) 90°. **8.2** (a) tetrahedral; (b) trigonal pyramid; (c) planar about C, 120° bond angles; (d) see-saw, as in SF_4, Table 8.3; (e) linear, 180° bond angles; (f) planar trigonal. **8.6** (a) BF_3 (3 electron pairs versus 4 in NF_3); (b) NH_3 (smaller lone pair repulsions than in H_2O); (c) C_2H_2 (two bonds to adjacent atoms versus three in C_2H_4); (d) H_2O (F is more electronegative than H, bonding pairs have smaller volume requirement). **8.9** (a) $CHCl_3$ is polar; (b) CF_4 is nonpolar; (c) nonpolar; (d) polar; (e) polar (however, the dipole moment of CO is surprisingly small, only 0.12 debye); (f) polar. **8.12** $\mu = 6.78$ D. **8.13** (a) 180°; (b) 120°; (c) 109.5°; (d) 90°; (e) 90° angles between axial and equatorial, 120° angles in equatorial plane; (f) 90°; (g) 90°. **8.14** (a) sp^3, no π bonds; (b) sp^2, one π bond; (c) sp, two π bonds; (d) sp^2 about carbon, one π bond. **8.16** (a) the bond angles about the oxygen-bearing carbon are 120°; about the second carbon, 109°; (b) about the oxygen-bearing carbon, sp^2; about the second carbon, sp^3; (c) seven σ bonds; (d) one π bond.

8.20

	C_2	C_2^{2-}
σ_{2p}^*	☐	☐
π_{2p}^*	☐ ☐	☐ ☐
σ_{2p}	☐	⇅
π_{2p}	⇅ ⇅	⇅ ⇅
σ_{2s}^*	⇅	⇅
σ_{2s}	⇅	⇅

The C_2^{2-} ion contains two electrons more than C_2; these are located in the σ_{2p} orbital, a bonding orbital. The bond order in C_2 is 2; in C_2^{2-} it is 3. We thus expect the C—C bond energy in C_2^{2-} to be greater and the C—C distance to be shorter than in C_2. Both species are expected to be diamagnetic. Note that C_2^{2-} is isoelectronic with N_2. **8.22** A major factor in stabilizing the bonding molecular orbital is that electrons can reside in a region of lower potential energy than in the separated atomic orbitals. The bonding molecular orbital is concentrated between the two nuclear centers, thus affording electron-nuclear attractive interactions with two positive charges. This more than overcomes any additional electron-electron repulsions. **8.24** There are no unpaired electrons in N_2, so all electrons in the molecule are paired up in some way. The pairing results in the course of bond formation. The N—N distance in N_2 is remarkably short (Table 8.7) and the N—N bond energy is especially large; both observations suggest a very strong bonding interaction. **8.26** The nonmetallic elements of the third row of the periodic table and beyond do not form strong homonuclear π bonds. On the other hand, their σ bonds are relatively strong. Thus phosphorus is more stable as an extended sheetlike structure involving P—P σ bonds between phosphorus and three neighboring phosphorus atoms. **8.28** The three electron pairs that determine the geometry about the central oxygen in O_3 consist of one unshared electron pair and two bonding pairs in the O—O bonds. The volume requirement of the unshared pair is greater than that of the two shared pairs, so the O—O bonds are pushed together to make the O—O—O bond angle less than 120°. **8.32** In BF_3 there are three shared electron pairs in the B valence orbitals. The molecule is planar with 120° F—B—F bond angles. When the adduct with NH_3 is formed, the boron acquires another shared electron pair. In F_3BNH_3 the boron has four shared electron pairs, and the bond angles are approximately tetrahedral. **8.34** (a) Ordinarily no, but there are a few exceptions, e.g., O_3 has a small dipole moment (0.53 D). All homonuclear diatomic molecules have zero dipole moment. (b) A molecule may have polar bonds but overall molecular dipole moment of zero when the bond dipoles add to zero by virtue of their directions (e.g., CF_4). **8.36** When the two Cl atoms are on the same side of the planar ethylene structure, the C—Cl bond dipoles do not completely cancel. When they are on opposite sides (the *trans* form) they do exactly cancel. **8.38** (a) two sp hybrid orbitals; linear; (b) four $4sp^3$ orbitals; tetrahedral; (c) six d^2sp^3 hybrid orbitals; octahedral. **8.40** (a) sp^3—no orbitals available for π bond formation; (b) sp^2—one p orbital available for π bond formation; (c) sp—two p orbitals available for π bond formation.

8.46 (a) H—N̈—N̈—H

with H atoms below each N

(b) The geometry about each nitrogen is an approximate trigonal pyramid. Thus, the molecule is shaped something like this:

There is a rotation about the N—N bond, so each end of

the molecule can have varying orientation relative to the other end. (c) The nitrogen may be considered to employ sp^3 hybrid orbitals, just as in ammonia. In fact, hydrazine can be viewed as two ammonia molecules joined by removing a hydrogen atom from each and making the N—N bond in their place. (d) All of the covalent bonds in hydrazine are σ in character. **8.48** Empirical formula is $AlCl_3$; molecular formula is Al_2Cl_6. The structure involves bridging chlorines between two aluminum atoms. **8.52** O_2^+, bond order 2.5; O_2, bond order 2.0; O_2^-, bond order 1.5; O_2^{2-}, bond order 1.0. Bond length increases in the series from O_2^+ to O_2^{2-}

CHAPTER 9

9.3 temperature (Kelvin); pressure (atm, mm Hg, kilopascals); volume (liter, cm^3); quantity (number of moles). **9.4** force = $(1.033 \text{ kg})(9.81 \text{ m/sec}^2)$ = 10.13 kg-m/sec² = 10.13 N; pressure = 1.013×10^5 N/m² = 101.3 kPa. **9.7** An equation of state is a relationship between all the variables that are needed to specify the state, or condition, of a system. For an ideal gas these variables are pressure, temperature, volume, and quantity of gas. **9.8** density = 2.91 g/L. **9.10** 1.14 tanks needed. **9.13** pressure = 1.98 atm. **9.14** pressure = 0.175 atm. **9.16** pressure = 192 kPa. **9.18** (a) pressure = 6.77 atm; (b) volume = 6.67×10^4 L; (c) pressure = 0.0212 atm. **9.21** P_{H_2} = 6.9×10^{-3} atm; P_t = 3.55×10^{-2} atm. **9.23** Ratio of pressures is 12.6. **9.25** 1.8×10^3 g H_2O. **9.27** H_2O is the limiting reagent; 0.181 L NH_3 is formed. **9.29** volume of O_2 = 3.69 L. **9.32** u_{H_2}/u_{N_2} = 3.72. **9.33** (a) The rms speed increases as $T^{1/2}$ (Equation [9.18]); thus it increases by 1.414; (b) doubles; (c) decreases. **9.35** (a) same; (b) more collisions per unit time for N_2; (c) the N_2 distribution curve is spread out more toward higher speeds; (d) N_2 effuses 2.28 times faster. **9.37** (a) He; (b) N_2; (c) O_2. **9.39** (a) UF_6; (b) $SiCl_4$; (c) SO_2. **9.40** At STP the molar volume is 22.4 L. $b/4$ = 8.0×10^{-3} L. Thus the volume of the Ar atoms is 0.036% of total volume at 1 atm pressure, 3.6% at 100 atm pressure. **9.43** (a) 760 mm Hg; (b) 1 mm Hg; (c) 59 mm Hg; (d) 646.9 mm Hg; (e) 120.9 kPa. **9.45** (a) 2.34 atm; (b) 8.22×10^{-10} atm; (c) 0.359 atm; (d) 112 kPa. **9.46** 3.83 L. **9.48** pressure = 4.1×10^{-2} mm Hg. **9.50** temperature = 450 K (two significant figures). **9.52** Mol. wt. = 83.5 g/mol. **9.55** (a) 62.2 g/mol; (b) 730 mm Hg, 1.15×10^3 cm³; (c) 0.596 g/L (use the average molecular weight, 21.97 g/mol). **9.58** (a) r_{He}/r_{Ar} = 3.16; (b) r_{CO}/r_{CO_2} = 1.25. **9.60** r_1/r_2 = 1.018 for CO; r_1/r_2 = 1.011 for CO_2. The degree of separation would be greater for CO than for CO_2.

CHAPTER 10

10.2 tropopause (220 K); stratopause (270 K); mesopause (180 K). **10.4** He partial pressure is 3.98×10^{-3} mm Hg; CH_4 partial pressure is 2×10^{-3} mm Hg. **10.5** 569 nm **10.7** Temperature is about 220 K; n/V = 1.0×10^{14} molecules/cm³. **10.10** 127 nm **10.12** (a) $O_2(g) + h\nu \longrightarrow 2O(g)$; (b) $N_2^+(g) + e^- \longrightarrow N(g) + N(g)$; (c) $O^+(g) + N_2(g) \longrightarrow NO^+(g) + N(g)$. **10.14** (a) $O_2(g) \longrightarrow O_2^+(g) + e^-$, ΔE = 1205 kJ; $NO^+(g) + e^- \longrightarrow NO(g)$, ΔE = −890 kJ; $NO^+(g) + O_2(g) \longrightarrow NO(g) + O_2^+(g)$, ΔE = 315 kJ. (c) $O_2^+(g) + e^- \longrightarrow O_2(g)$, ΔE = −1205 kJ; $O_2(g) \longrightarrow O(g) + O(g)$,

$\Delta E = 495$ kJ; $O_2^+(g) + e^- \longrightarrow O(g) + O(g)$, $\Delta E = -710$ kJ. **10.16** For oxygen to combine with another O atom to form O_2 it must have present a third body to carry off excess energy. This is much more likely to occur at 50 km than at 120 km because the atmosphere is much denser at the lower elevation. **10.19** $NO(g) + O_3(g) \longrightarrow NO_2(g) + O_2(g)$, $\Delta H = 33.8 - 142.3 - 90.4 = -198.9$ kJ; $NO_2(g) + O(g) \longrightarrow O_2(g) + NO(g)$, $\Delta H = 90.4 - 247.5 - 33.8 = -190.9$ kJ. **10.22** The $CFCl_3$ molecule is approximately tetrahedral. The Cl—C—Cl bond angles should be slightly larger than tetrahedral, the Cl—C—F angles slightly smaller. **10.24** $CF_2Cl_2(g) + h\nu \longrightarrow CF_2Cl(g) + Cl(g)$; $CF_2Cl(g) + h\nu \longrightarrow CF_2(g) + Cl(g)$; $Cl(g) + O_3(g) \longrightarrow ClO(g) + O_2(g)$; $ClO(g) + O(g) \longrightarrow Cl(g) + O_2(g)$. **10.26** $n/V = 6.0 \times 10^{18}$ molecules/m³. **10.29** 5.5×10^{11} g SO_2 **10.31** It may be oxidized following excitation by sunlight; upon adsorption onto dust and other particulates in the atmosphere; upon dissolving in rain droplets. The product of the oxidation, SO_3, is soluble in water to form H_2SO_4 solutions. This substance is the major component of acid rain. **10.33** 2.2×10^5 g $CaCO_3$. **10.37** Iron-containing hemoglobin in the blood is the binding site for CO; it is also the binding site for O_2 in the lungs. Because CO has a larger binding constant, low levels of CO can result in substantial reduction of the blood hemoglobin available for binding of O_2. **10.41** Solar dissociation of O_2 is a major factor. In addition there is some gravitational separation at the highest altitudes, which results in relatively more H_2 and He. **10.43** (a) $N_2(g) + h\nu \longrightarrow N_2^+(g) + e^-$; $N_2^+(g) + e^- \longrightarrow N(g) + N(g)$; (b) $O^+(g) + N_2(g) \longrightarrow NO^+(g) + N(g)$; (c) $NO^+(g) + e^- \longrightarrow N(g) + O(g)$; (d) $N_2^+(g) + O(g) \longrightarrow N(g) + NO^+(g)$. **10.45** A catalyst is a substance that affects the speed of a chemical process without itself being consumed in the process. Cl acts as a catalyst in the two-step process as follows: $Cl(g) + O_3(g) \longrightarrow ClO(g) + O_2(g)$; $ClO(g) + O(g) \longrightarrow Cl(g) + O_2(g)$. The sum of these two reactions is just $O_3(g) + O(g) \longrightarrow 2O_2(g)$; Cl is recycled continuously. **10.47** The ionization energies of the metal atoms are much lower than those of any of the atomic or molecular ions present at 120 km; that is, O_2^+, O^+, NO^+. Thus any of these react with a metal atom as follows: $M + NO^+ \longrightarrow M^+ + NO$. **10.49** 3.9×10^5 g sulfate/mi². **10.51** The origin of acid rain is mainly SO_2. It is oxidized by one means or another to SO_3, which forms sulfuric acid, H_2SO_4, on dissolving in water. Nitric acid, HNO_3, may also contribute.

CHAPTER 11

11.2 The solid is slightly denser because the molecules are packed more regularly in the solid state. Thus there is less "free volume," or space between the molecules, than there is in the liquid. Water is one of the rare exceptions to the rule. **11.6** Liquid ethyl chloride at room temperature is far above its boiling point at 1 atm pressure. When the liquid boils, heat is required to overcome the intermolecular forces between molecules. This heat is extracted from the surface that contacts the liquid. **11.7** $\Delta H = 50.87$ kJ/mol. **11.10** (a) no change; (b) increases with increasing temperature; (c) higher intermolecular forces mean lower vapor pressure; (d) no change. **11.12** diethyl ether, about 23°C;

ethyl alcohol, about 68°C; water, 90°C. **11.15** As chain length increases, the intermolecular forces increase. Furthermore, the molecules are longer and thus more readily entangled. **11.18** (a) decreases; (b) increase; (c) increase; (d) increase; (e) increases; (f) increases; (g) increases; (h) unaffected. **11.20** You should have estimated about 354°C. **11.22** The slope of the graph of log v.p. versus $1/T$ (absolute temperature) is about 2.57×10^3, from which we calculate ΔH_s of 49.1 kJ/mol. **11.24** Well-defined X-ray patterns might be seen for (a) sugar; (b) KBr; (d) iron; (e) ice. **11.25** $\theta = 17.9°$, 38.0°, 67.4°. **11.27** The unit cell is 7.98×10^{-23} cm³ in volume; each side is 4.30×10^{-8} cm. **11.30** The area bounded by the lines represents the unit cell of the two-dimensional array. **11.32** Twenty cannonballs are required. The stacking is of the form ABCA; this is cubic close packing. **11.36** The ionic substance $FeCl_2$ is most likely to exist as a solid; F_2 is most likely to exist as a gas. **11.38** (a) London dispersion forces; (b) ion-ion attraction; (c) dipole-dipole; (d) hydrogen bonding. **11.40** $C_6H_5OH > CH_3NH_2 > H_2S$. **11.42** Effective ionic radius can be estimated from a graph as about $(2.14 - 0.68) = 1.46$ Å. **11.44** (a) Most important intermolecular forces in CH_3OH—H_2O in order of decreasing importance are hydrogen bonding, dipole-dipole interaction, and dispersion force interaction. (c) London dispersion forces. **11.47** The dispersion force contribution in CH_2Br_2 will be larger than for CH_2Cl_2. Dipole-dipole contribution in CH_2Cl_2 should be greater than in CH_2Br_2. By contrast, the dipole-dipole contribution is more important in CH_2F_2 because it is more polar and less polarizable than CH_2Cl_2. **11.50** (a) $CuBr_2$; (b) SiO_2; (c) Cr; (d) CaF_2. **11.51** C_2H_3Cl. It should be amorphous in character, because the long chains do not pack in a very regular way. **11.53** London dispersion forces should be much greater in hexachlorobenzene than in benzene because of the presence of the six large, polarizable chlorines in place of hydrogen. The two molecules have the same general shape. **11.56** The heat added at 80°C is required to convert liquid benzene to benzene vapor. The amount of heat required per mole is the heat of vaporization of liquid benzene at 80°C. **11.58** $\Delta H = 80.4$ kJ. **11.61** In the face-centered unit cell there are four atoms; in the body-centered unit cell there are two atoms of one kind if the atoms at the corners are the same as at the center. **11.63** Because the molecules possess greater average kinetic energies at the higher temperature, less energy is required in the process of vaporization to disrupt the hydrogen-bonding structure that remains. **11.67** The major source of intermolecular attractive energy in glycerol is hydrogen bonding. As temperature is raised the increasing kinetic energies of the molecules result in more disruption of hydrogen bonds. Glycerol is viscous because the hydrogen bonds keep molecules from moving readily with respect to one another. Viscosity decreases rapidly with increasing temperature because the added kinetic energy of the molecules causes disruption of the hydrogen-bonding network. **11.70** $\theta = 12.7°$, 26.0°, 41.1°. **11.75** These molecules are planar-trigonal in shape and nonpolar. The increasing critical temperature simply reflects increasing intermolecular attractive forces of the London dispersion type. **11.77** A, solid; B, liquid and gas in equilibrium; C, solid, liquid, and gas in equilibrium (the triple point); D, gas.

12.1 (a) 0.0916 mol $Ca(NO_3)_2$; (b) 2.27×10^3 mol HBr; (c) 4.35×10^{-2} mol NaCl. 12.3 (i) wt. fraction = 0.010, mole fraction NaCl = 3.2×10^{-3}, $m = 0.181\,m$ NaCl; (ii) wt. fraction = 7.81×10^{-3}, mole fraction KBr = 3.05×10^{-3}, $m = 0.0663\,m$ KBr; (iii) wt. fraction = 0.351, mole fraction solute 0.136, $m = 8.72\,m$ $C_2H_6O_2$; (iv) wt. fraction NaCl = 0.027, wt. fraction KBr = 0.041, mole fraction NaCl = 8.9×10^{-3}, mole fraction KBr = 6.5×10^{-3}, m (NaCl) = 0.50, m (KBr) = 0.37, total molality = 0.87. 12.6 (a) 1.62 g $La(NO_3)_3$; (b) 19.1 g $La(NO_3)_3$; (c) 1.28 g $La(NO_3)_3$. 12.9 (a) Normality here is twice molarity; thus a 0.36 M solution is 0.72 N, a 0.42 N solution is 0.21 M. (b) Three protons are transferred by H_3PO_4. Thus, a 0.105 N solution is 0.0350 M. 12.11 $4.8 \times 10^{-5}\,m$, $3.8 \times 10^{-5}\,M$. 12.12 (a) ion-dipole; (b) London dispersion force; (c) F—H···O hydrogen bonding (d) dipole-dipole (both molecules are polar). 12.15 (a) Ethyl alcohol is polar; it can solvate ions much as water does. Ethyl alcohol also engages in hydrogen bonding in the liquid state, just as water does. (b) The oxygen atoms of dioxane can act as electron-pair-donor atoms toward the H—O bonds of water molecules. No such interaction is possible for cyclohexane. (c) Ethyl alcohol, though polar, has a nonpolar portion compatible with solute molecules of low polarity. Chloroform is not very polar, and thus can mix with other substances of low polarity. 12.20 (a) miscible—hydrogen-bonding interactions; (b) nonmiscible; (c) nonmiscible; (d) miscible. 12.22 $2.88 \times 10^{-4}\,M$ 12.24 He solubility is $1.8 \times 10^{-4}\,M$, N_2 solubility is $3.0 \times 10^{-4}\,M$. 12.28 (a) nonelectrolyte; (b) nonelectrolyte; (c) weak electrolyte; (d) strong electrolyte; (e) strong electrolyte; (f) weak electrolyte (methyl amine is a close relative to ammonia, NH_3); (g) weak electrolyte; (h) nonelectrolyte; (i) strong electrolyte. The acids are HF, H_2SO_4, HClO. The bases are CH_3NH_2 and LiOH. $Ba(NO_3)_2$ is a salt. 12.30 (a) molecular form; (b) as Ca^{2+} and Cl^- ions; (c) as Ca^{2+} and Cl^- ions (The difference between (b) and (c) is that in molten NaCl the Ca^{2+} ions are surrounded by Cl^- ions. In H_2O they are surrounded by solvating water molecules, as in Figure 12.2.); (d) as separated Na^+ and OH^- ions; (e) as the hydrated H^+ ion, written as $H^+(aq)$, and the separated $ClO_4^-(aq)$ ion. 12.33 (a) $K^+(aq) + OH^-(aq) + H^+(aq) + ClO_4^-(aq)$ $\longrightarrow$ $H_2O(l) + K^+(aq) + ClO_4^-(aq)$—Net: $OH^-(aq) +$ $H^+(aq) \longrightarrow H_2O(l)$; (b) $Mg^{2+}(aq) + 2Cl^-(aq) +$ $2K^+(aq) + 2OH^-(aq) \longrightarrow Mg(OH)_2(s) + 2K^+(aq) +$ $2Cl^-(aq)$—Net: $Mg^{2+}(aq) + 2OH^-(aq) \longrightarrow Mg(OH)_2(s)$; (c) $BaCO_3(s) + 2H^+(aq) + 2ClO_4^-(aq) \longrightarrow Ba^{2+}(aq)$ $+ 2ClO_4^-(aq) + H_2O(l) + CO_2(g)$—Net: $BaCO_3(s) +$ $2H^+(aq) \longrightarrow Ba^{2+}(aq) + H_2O(l) + CO_2(g)$. 12.35 (a) $BaSO_3(s) + 2HBr(aq) \longrightarrow BaBr_2(aq) + H_2O(l) +$ $SO_2(g)$; $BaSO_3(s) + 2H^+(aq) + 2Br^-(aq) \longrightarrow$ $Ba^{2+}(aq) + 2Br^-(aq) + H_2O(l) + SO_2(g)$; $BaSO_3(s) +$ $2H^+(aq) \longrightarrow Ba^{2+}(aq) + H_2O(l) + SO_2(g)$. 12.37 Addition of acid should cause dissolution of $SrCO_3$, with evolution of CO_2; of $MnSO_3$, with evolution of SO_2; of CaF_2, with formation of aqueous HF, a moderately weak electrolyte. 12.38 Properties (c) and (e) are colligative properties. 12.40 There are 0.351 mol of $C_{12}H_{22}O_{11}$ in 120 g of the substance. We require 0.351 mol of *solute particles* in each of

the other solutes. (a) 20.9 g KBr; (b) 18.6 g $SrCl_2$; (c) 22.4 g HI. 12.43 420 g CH_3OH. 12.44 In order of decreasing boiling point, 0.030 m glycerine < 0.030 m benzoic acid < 0.020 m KBr < 0.018 m $MgCl_2$. 12.47 From the osmotic pressure, $M = 0.117\,M$. Assume as an approximation that molality and molarity are the same; MW = 340 g/mol. 12.49 (a) hydrophobic; (b) hydrophilic; (c) hydrophobic; (d) hydrophilic. 12.51 The polar ends of the soap molecules would be directed inward, with the long, nonpolar hydrocarbon chains projecting into the oil. 12.54 1.76 kg H_2SO_4/L; wt. percentage H_2SO_4 = 96%; mole fraction H_2SO_4 is 0.80. 12.56 2.2×10^8 kg toluene. 12.58 It is most important to ensure that equilibrium is attained. Stir an excess of KNO_3 with ethyl alcohol at a temperature above 25°C for some time, then reduce temperature to 25°C. Stir at 25°C for a time, allow excess solid to settle, withdraw a known volume of the solution and measure the amount of KNO_3 it contains. Resume stirring, repeat sample withdrawal at a later time, until successive experiments yield a constant concentration of KNO_3 in the alcohol. Repeat experiments at a higher temperature, say 35°C. If the solubility is higher at the higher temperature, the solution process is endothermic; if it is lower, the solution process is exothermic. 12.60 (a) A decrease in randomness or disorder results from the formation of an ordered ionic lattice. (b) The disorder increases as the CCl_4 molecules expand into the larger volume of the gas state. (c) Melting causes the sulfur molecules to move with respect to one another in a random fashion. It thus results in an increased disorder or randomness. 12.62 $2.0 \times 10^{-6}\,M$. 12.66 $Cu(CN)_2(s) + 2H^+(aq) \longrightarrow$ $Cu^{2+}(aq) + 2HCN(aq)$. 12.68 solution is 0.0178 M; $\pi =$ 0.43 atm. 12.70 0.511 mol $C_3H_9O_3$/kg H_2O; freezing point = -0.95°C; boiling point = 100.26°C. 12.72 Molecular weight is 256 g/mol; molecular formula is S_8. 12.74 Molecular formula is $C_9H_{13}O_3N$.

CHAPTER 13

13.1 In each case the time interval is followed by the average rate, in units of M/min. 0–79, 2.3×10^{-3}; 79–158, 1.9×10^{-3}; 158–316, 1.4×10^{-3}; 316–632, 0.95×10^{-3}. 13.2 Determine the slope at each temperature, then divide the slope by the concentration of CH_3NC at that time. At $t = 0$, rate = 0.101 mm/sec, rate constant is 2.0×10^{-4}/sec. At $t = 5000$ sec, rate = 0.0348 mm/sec, rate constant = 1.9×10^{-4}/sec. At $t = 10,000$ sec, rate = 0.0138 mm/sec, rate constant = 2.1×10^{-4}/sec. Note that the estimated first-order rate constants are reasonably constant, even though the CH_3NC pressure changes substantially. 13.4 (a) $-\Delta[H_2]/\Delta t = \frac{3}{2}\Delta[NH_3]/\Delta t$; $-\Delta[N_2]/\Delta t = \frac{1}{2}\Delta[NH_3]/\Delta t$; (b) $-\Delta[NOCl]/\Delta t =$ $\Delta[NO]/\Delta t = 2\Delta[Cl_2]/\Delta t$; (c) $-\Delta[HI]/\Delta t =$ $-\Delta[CH_3I]/\Delta t = \Delta[CH_4]/\Delta t = \Delta[I_2]/\Delta t$. 13.7 (a) first order in A, first order in B, second order overall; (b) second order in A, zero order in B, second order overall; (c) first order in A, second order in B, third order overall. 13.9

Rate law	Effect of doubling [A]	Effect of doubling [B]
$k[A][B]$	Double rate	Double rate
$k[A]^2$	Fourfold increase	No effect
$k[A][B]^2$	Double rate	Fourfold increase

13.11 (a) rate $= k[B]^2$; (b) reaction is zero order with respect to A, second order with respect to B; (c) $k = 7.0 \times 10^{-3}/M$-sec. **13.13** $2.8 \times 10^{-4}/$sec. **13.15** $k = 2.19 \times 10^{-5}/$sec. **13.18** (a) 0.115 mol N_2O_5; (b) $t = 330$ sec. **13.21** (a) Reaction (b) would be fastest, (c) slowest. The reverse reaction would be fastest in (c), slowest in (b). **13.23** $E_a = 47$ kJ/mol. **13.25** (a) $k_{308}/k_{298} = 2.51$; (b) $k_{383}/k_{373} = 1.80$. **13.27** (a) $k_{373} = 0.231/$sec; (b) $k_{373} = 53.5/$sec. **13.29** (a) $NOBr_2$ is the intermediate. (b) For the first step, rate $= \Delta[NOBr_2]/\Delta t = k[NO][Br_2]$; for the second step, rate $= -\Delta[NOBr_2]/\Delta t = k[NOBr_2][NO]$. (c) If the first step is slow, the rate law for the overall reaction is just the rate law for the first step written in (b). (d) If the rate law is $k[NO]^2[Br_2]$, it is evident that the second step in the two-step reaction is slow. **13.32** Heterogeneous catalysts operate by providing a surface on which reaction can occur. The amount of surface per unit mass of catalyst varies with the method of preparation. In general, large surface areas are desirable. Also, the chemical activity of the surface depends on the method of preparation of the solid, and how it has been handled. **13.34** (a) The factors are concentrations of reactants, temperature, and addition of a catalyst. (b) Increases in concentrations of reactants result in more frequent collisions per unit time, thus in a greater probability of reaction. An increase in temperature results in an increase in average kinetic energy of the reactants, so that a larger fraction may possess sufficient energy to form product. A catalyst lowers the activation energy for reaction, so that a larger fraction of the encounters between reactants can lead to products. **13.38** (a) Doubling [A] changes [A] and rate, but not k or [B]. (b) Addition of a catalyst changes rate and k. (c) $[A]^2$ increases by fourfold, [B] decreases by fourfold, rate and k are unchanged. (d) Increase in temperature increases k and rate. **13.40** Reactions often proceed through several steps. The rate at which product is formed will depend on which step in the overall reaction is slowest. There is no unique relation between the "mechanism," or detailed path taken by the reactants, and the overall balanced equation for the reaction. **13.43** (a) $k = 4.28 \times 10^{-4}/$sec; (b) [urea] $= 0.059\, M$. **13.46** $-\Delta[H_2S]/\Delta t = 5.8 \times 10^{-14}\, M/$sec. **13.50** $E_a = 24.4$ kJ/mol. **13.51** $k_{773} = 9.82 \times 10^{-2}/M$-sec.

CHAPTER 14

14.1 (a) $K = [SO_3]^2/[SO_2]^2\,[O_2]$; (b) $K = [H_2O]/[H_2]$; (c) $K = [CO][H_2O]/[CO_2][H_2]$; (d) $K = [N_2]^2\,[H_2O]^6/[NH_3]^4[O_2]^3$; (e) $K = [N_2][H_2O]^2$. **14.3** $K_c = [H_2][I_2]/[HI]^2 = (4.79 \times 10^{-4})(4.79 \times 10^{-4})/(3.53 \times 10^{-3})^2 = 1.82 \times 10^{-2}$. **14.6** $K_p = K_c(RT)^{\Delta n}$. Here $\Delta n = 0$, so $K_p = K_c = 0.11$. **14.10** (a) The number of moles of NO reacting is $0.100 - 0.070 = 0.030$. Thus the quantity of H_2 remaining is $0.050 - 0.030 = 0.020$ mol. By reaction of 0.030 mol of NO we produce 0.030 mol of H_2O, and half as many, 0.015 mol, of N_2. We had initially 0.100 mol H_2O. (b) $K_c = (0.015)(0.130)^2/(0.070)^2(0.020)^2 = 129$. **14.12** (a) $K_c = 6.42 \times 10^{-2}$, (b) $K_p = 1.98$; (c) $P = 0.968$ atm. **14.14** K_p is 4.51×10^{-5}. Compare the reaction quotient Q in each case with this value. (a) $Q = 2.7 \times 10^{-6}$. Since $Q < K_p$, reaction shifts to the right to attain equilibrium. (b) $Q > K_p$. Reaction must shift to the left to attain equilib-

rium. *Some* N_2 must be present to attain equilibrium. (c) $Q = K_p$. Reaction system is at equilibrium. (d) $Q > K_p$. Reaction proceeds to the left to attain equilibrium. **14.16** $[O_2] = 61\, M$. **14.18** 3.07×10^{-3} mol Br. **14.20** $P = 1.2 \times 10^{-2}$ atm. **14.22** $[NO] = 3.95 \times 10^{-4}\, M$, $[O_2] = 0.125\, M$, $[N_2] = 0.125\, M$. **14.25** (a) shift equilibrium to the right; more CO will be formed; (b) no effect on equilibrium, so long as *any* C(s) is present; (c) equilibrium shifted to the right, in direction in which reaction is endothermic; (d) equilibrium shifts to the left; (e) no effect; (f) a shift to the right occurs; that is, more $CO_2(g)$ reacts with C(s) to form CO(g). **14.28** The dissolving of $CuSO_4(s)$ is an endothermic process. As temperature is increased the equilibrium shifts in the direction that absorbs heat, that is, in the direction of greater solubility of the salt. **14.30** $K = 1.48 \times 10^{-39}$. **14.35** Reactions (a), (d), and (e) are homogeneous; (b) and (c) are heterogeneous. **14.38** $K_c = 3.74 \times 10^{-3}$. **14.40** $K_c = 4.4 \times 10^{-2}$; $K_p = 4.0$. **14.43** (a) at equilibrium, $[NO] = 1.78 \times 10^{-2}\, M$, $[O_2] = 2.89 \times 10^{-2}\, M$, $[NO_2] = 2.2 \times 10^{-3}\, M$; (b) $K_c = 0.53$. **14.47** (a) to the left, to partially reduce the CO_2 concentration; (b) to the right, to form more CO_2; (c) to the left, to compensate for the lower pressure; (d) to the right, in the direction of fewer moles of gas; (e) shift to the left, the direction in which the reaction is endothermic. **14.48** Increase in temperature causes a decrease in the value of K for reactions such as this that are exothermic in the forward direction. **14.51** $\Delta H = -128.1$ kJ; the process is exothermic; thus the equilibrium constant decreases with increasing temperature. Extent of reaction will increase with an increase in total pressure. **14.53** (a) $Q = 2.0 \times 10^7$. (b) Since $Q < K$, reaction will proceed to the right to attain equilibrium. That is, AgCl(s) will form until the concentrations of $Ag^+(aq)$ and $Cl^-(aq)$ are sufficiently low to meet the condition $Q = K$. **14.54** $K_c = 1.04 \times 10^{-3}$; at equilibrium [Br] $= 3.2 \times 10^{-2}\, M$. The fraction of Br_2 that dissociates is $0.016/1 = 0.016$. **14.58** $Q = 8.3 \times 10^{-6}$. Since $Q > K_p$, the system shifts to the left to attain equilibrium. Thus a catalyst that promoted the attainment of equilibrium would result in a lower CO content in the exhaust.

CHAPTER 15

15.2 The proton is bound to one or more water molecules, forming species such as H_3O^+ and $H_9O_4^+$ (Figure 15.3). The state of the proton is essentially independent of the anion from which it is derived, so long as ionization is complete or nearly so. This is the case for the strong acids HCl and HI. **15.5** (a) CN^-; (b) HSO_4^-; (c) $C_2H_3O_2^-$; (d) NH_3; (e) NH_2^-; (f) SO_3^{2-}; (g) O^{2-}. **15.7** (a) $NH_4^+(aq) + OH^-(aq) \rightleftharpoons NH_3(aq) + H_2O(l)$. In this reaction NH_4^+ is the acid, NH_3 the conjugate base; OH^- is the base, H_2O the conjugate acid. (b) $HCN(aq) + H_2O(l) \rightleftharpoons H_3O^+(aq) + CN^-(aq)$. HCN is the acid, CN^- the conjugate base; H_2O is the base, H_3O^+ the conjugate acid. **15.8** (a) F^-; (b) S^{2-}; (c) NO_2^-; (d) OH^-. **15.10** (a) basic; (b) acidic; (c) neutral; (d) acidic; (e) basic; (f) acidic; (g) basic. **15.12** (a) 1.30; (b) 4.16. **15.13** (a) $3.63 \times 10^{-3}\, M$; (b) $1.0 \times 10^{-3}\, M$; (c) $1.35 \times 10^{-13}\, M$. **15.15** (a) pOH $= 3.19$; (b) $[OH^-] = 5 \times 10^{-5}\, M$; (c) pOH $= 8.2$; (d) pH $= 11.38$; (e) $pK_a = 3.23$; (f) $K_a = 1.1 \times 10^{-6}$. **15.17** (a) 12.0; (b) 1.26; (c) 12.60; (d) 4.43. **15.21** A strong acid in water is a substance that transfers a proton to the solvent in a reaction that proceeds essentially to completion. In the proton-transfer reaction of a weak acid, the reaction proceeds only partly toward completion. **15.24** Phenol is the weakest acid, ascorbic acid is the strongest. **15.26** (a) $[H^+] = 4.4 \times 10^{-4} M$. **15.29** pH = 2.99. **15.31** $K_a = 4.20 \times 10^{-4}$. **15.33** HSO_3^- is the strongest acid, HPO_4^{2-} is the weakest. **15.35** (a) $K_b = [CH_3NH_3^+] \times [OH^-]/[CH_3NH_2]$, (b) $K_b = [HCN][OH^-]/[CN^-]$. **15.36** (a) pH = 8.61; (b) pH = 11.44 (use the quadratic formula here); (c) pH = 11.13. **15.38** (a) $K_b = 5.6 \times 10^{-11}$; (b) $K_b = 2.38 \times 10^{-2}$. **15.40** pH = 5.48. **15.43** percentage ionized = 1.3×10^{-2}. **15.45** (a) neutral; (b) acidic; (c) acidic; (d) neutral or slightly acidic; (e) basic; (f) neutral. **15.48** (a) pH = 10.10; (b) pH = 4.89; (c) pH = 10.84. **15.50** (a) solution is acidic; (b) neutral. **15.54** (a) H_2SO_4; (b) H_2SO_4; (c) HClO; (d) H_3PO_4; (e) H_2SO_3. **15.56** (a) $BF_3(g)$ is the Lewis acid, $F^-(aq)$ is the Lewis base. (c) $HF(aq)$ is the Lewis acid, H_2O is the Lewis base. **15.58** (a) ZnI_2; (b) $FeCl_3$; (c) NH_4Cl; (d) $FeCl_3$. **15.59** The ratio of the hydrogen ion concentrations is 100. **15.60** HBr, strong acid; KOH, strong base; H_3PO_4, weak acid; CH_3NH_2, weak base; NH_4^+, weak acid; CN^-, weak base; Rb_2O, strong base. **15.62** The reaction rate, which depends on $[H^+]$, will decrease in the order 0.1 M $H_3PO_4 > 0.1$ M formic acid > 0.1 M HCN. **15.64** (a) 5.00×10^{-4} mol HCl is required. (b) 2.8×10^{-2} mol $HC_2H_3O_2$. (c) 1.25×10^{-3} mol HF (you must add $[H^+]$ to the calculated [HF]). **15.67** $NH_2^- = 5.0 \times 10^{-30} M$. **15.69** The equilibrium is shifted to the right by addition of OH^-, which acts to remove H^+. Thus the concentration of the conjugate base form increases at the expense of the acid form; the blue color of the anion is observed in NaOH solution. **15.71** (a) NH_3 cannot be a proton donor in water. If there were a base strong enough to remove a proton from NH_3, that base would react preferentially with the water, since H_2O is a stronger proton donor than NH_3. (b) NH_2^- is a strong base in water: $NH_2^-(aq) + H_2O(l) \rightleftharpoons NH_3(aq) + OH^-(aq)$. (c) pH = 12.18. **15.75** (a) Use the full quadratic form for the first ionization. $[H^+] = 2.67 \times 10^{-2} M$, $[HC_3H_2O_4^-] = 2.67 \times 10^{-2} M$, $[C_3H_2O_4^{2-}] = 2.0 \times 10^{-6} M$; pH = 1.57. **15.77** (a) pH = 9.45; (b) pH = 5.45.

CHAPTER 16

16.1 pH = 4.36. **16.2** (a) pH = 3.17; (b) pH = 5.04; (c) $[C_3H_5O_2^-] = 2.8 \times 10^{-2} M$. **16.4** (a) $C_2H_3O_2^-(aq) + H^+(aq) \rightleftharpoons HC_2H_3O_2(aq)$; (b) $K = 1/K_a = 5.6 \times 10^4$; (c) $[H^+] = [HC_2H_3O_2] = [Na^+] = 0.050 M$; $[ClO_4^-] = 0.10 M$, $[C_2H_3O_2^-] = 1.8 \times 10^{-5} M$. **16.10** The greatest buffer capacity is possessed by solution (c), because the number of moles of acid or conjugate base per unit volume available to react with added base or acid is largest. Solution (b) has the least capacity as a buffer. **16.11** $[HCO_3^-]/[H_2CO_3] = 11$. **16.13** The initial pH of the buffer is 3.74. (a) pH = 3.73; (b) pH = 3.65; (c) pH = 3.75; (d) pH = 3.83. **16.15** (a) The quantity of base required to reach the equivalence point is the same in the two titrations. (b) Initial pH is higher in titration of the weak acid. (c) pH is higher at the equivalence point in titration of the weak acid. (d) pH in excess base is essen-

tially the same in the two titrations. (e) In titrating a weak acid one needs an indicator that changes color at a higher pH than for the strong acid titration. **16.16** (a) 0.040 L NaOH solution; (b) 36.1 mL NaOH solution. **16.18** (a) initially, pH = 1.00; (b) pH = 1.48; (c) pH = 4.00; (d) pH = 7.00; (e) pH = 10.0; (f) pH = 12.30. **16.22** (a) $K_{sp} = [Ag^+][Br^-]$; (c) $K_{sp} = [Cd^{2+}][IO_3^-]^2$. **16.23** $K_{sp} = 2.3 \times 10^{-11}$. **16.24** (a) $K_{sp} = 1.0 \times 10^{-8}$. **16.25** $[Ni^{2+}] = [CO_3^{2-}] = 8.1 \times 10^{-5} M$. **16.29** (a) Molar solubility is $1.3 \times 10^{-4} M$; (b) Molar solubility of $PbSO_4 = [Pb^{2+}] = 1.6 \times 10^{-6} M$; (c) Molar solubility = $[SO_4^{2-}] = 1.6 \times 10^{-6} M$. **16.31** After mixing, the product $Q = [Pb^{2+}][Cl^-]^2 > K_{sp}$, so a precipitate forms. **16.35** The salts soluble in acid solution are those for which the anion is the conjugate base of a weak acid. These include (a) CdS, (d) $Co(OH)_2$, and (f) $MnCO_3$. $PbSO_4$ may be slightly more soluble in strongly acid solution; SO_4^{2-} is a very weak base. **16.36** (a) pH = 1.0; (b) pH = 6.0; (c) pH = 9.0. **16.39** The amphoteric metal hydroxides are those that are insoluble in neutral water but soluble in the presence of either excess acid or excess base. The amphoteric ones are (b) $Cr(OH)_3$ and (e) $Al(OH)_3$. **16.43** The first two experiments eliminate groups 1 and 2. The fact that no insoluble carbonates are formed eliminates group 4. The ions that might be present are those of groups 3 and 5. **16.48** (a) $C_2H_3O_2^-(aq) + H^+(aq) \rightleftharpoons HC_2H_3O_2(aq)$; (b) pH = 5.22. **16.50** The equilibrium constant for the reaction $Mg(OH)_2(s) + 2H^+(aq) \rightleftharpoons Mg^{2+}(aq) + 2H_2O(l)$ is the product of K's for the solubility equilibrium and the neutralization reaction; $K = (1.8 \times 10^{-11})(1.0 \times 10^{28}) = 1.8 \times 10^{17}$. Clearly the reaction proceeds far to the right in acidic solution. **16.52** $pK_a = 6.60$. **16.55** pH = 7.00. **16.58** (a) pH = 10.25; (c) pH = 10.64. **16.60** (a) pH = 3.75; (b) pH = 3.65; (c) pH = 3.85. **16.63** (a) pH = 5.97; (b) pH = 8.39; (c) pH = 7.0. **16.68** (a) 0.53 g MgC_2O_4; (b) 4.8×10^{-2} g MgC_2O_4. **16.74** The pH is 9.86; since $Q = [Ca^{2+}][OH^-]^2 = 6.4 \times 10^{-10}$ is smaller than K_{sp}, no precipitation will occur. **16.78** pH = 7.82. **16.79** The $[OH^-]$ that results from dissolving of $Fe(OH)_3$ in water is negligible compared to that formed by ionization of water itself. We must therefore use the value $[OH^-] = 10^{-7} M$ in calculating the solubility of $Fe(OH)_3$ in water. $[Fe^{3+}] = 4 \times 10^{-17} M$.

CHAPTER 17

17.1 There are 4.7×10^{22} g salts in the entire world ocean. The amount added each year is only one part in 10^7 of this, so the composition of seawater cannot change rapidly due to this source. **17.3** $2.3 \times 10^{-6} M$ PO_4^{3-}. **17.5** Seawater is a solution, in which the vapor pressure of water (the solvent) is reduced by the presence of solutes, in accordance with Raoult's law (Section 12.7). (b) The effect of interionic attractive forces, and the slightly alkaline character of seawater (pH ~ 8) combine to shift the $H_2CO_3 \rightleftharpoons HCO_3^- \rightleftharpoons CO_3^{2-}$ equilibria in seawater farther to the right. The result is that CO_3^{2-} is more abundant in seawater than in fresh water, and HCO_3^- is correspondingly lower in concentration. (c) Near the surface nutrients such as nitrate are used up in forming phytoplankton. At lower depths dead plant and animal matter decompose via oxidation to yield nitrate ion. **17.6** 1.5×10^{13} L seawater must be processed. **17.9** 4.8×10^6 g CaO. **17.10** $\pi = 5.8$ atm. **17.17** (a) Products of aerobic decomposition are mainly CO_2, H_2O, NO_3^-, SO_2, SO_4^{2-}, and phosphates, e.g., $H_2PO_4^-$. (b) Anaerobic decomposition leads to hydrocarbons (e.g., CH_4) and other hydrides such as NH_3, PH_3, H_2S, and H_2O. **17.18** 25.6 ppm O_2. **17.23** 3.4×10^3 gal H_2O. **17.25** The culprit is cadmium, closely related chemically to zinc and exceedingly toxic. **17.28** Parts per million refers to the mass of a solute component per million mass units of the total solution. **17.33** 3.3×10^8 L H_2O. **17.37** Addition of CaO produces a basic medium via the following reactions: $CaO(s) + H_2O(l) \rightleftharpoons Ca^{2+}(aq) + 2OH^-(aq)$, $H_2PO_4^-(aq) + 2OH^-(aq) \rightleftharpoons PO_4^{3-}(aq) + 2H_2O(l)$. Precipitation of $Ca_3(PO_4)_2$, $K_{sp} = 2 \times 10^{-29}$, results in removal of phosphate. **17.41** 1.1×10^7 g $C_{108}H_{226}N_{16}O_{109}$-P/0.1 km^3 water.

CHAPTER 18

18.2 (a) 1 mol He at 25°C—more kinetic energy of motion; (b) 2 mol H at 25°C—more particles, thus more randomness in the possible arrangements; (c) 1 mol $H_2O(g)$ at 100°C—more volume in gas phase, thus more randomness in locations of H_2O molecules; (d) 1 mol $HCl(g)$ at 25°C—larger volume for gaseous molecules, thus more randomness in locations of molecules; (e) 1 mol $C_2H_6(g)$ at 25°C—larger molecule, thus more ways in which energy may be distributed in vibrations within the molecule. **18.5** (a) ΔS positive—larger volume; (b) ΔS positive—larger volume for each component, increased randomness in mixing of two different substances; (c) ΔS negative—entropy of $AgCl(s)$ is much smaller than that of the separated aqueous ions, which can move freely throughout the solution. **18.6** (a) ΔS positive—a gas is formed; (b) ΔS positive—a gas is formed; (c) ΔS negative—a gas is consumed; (d) ΔS positive—the number of gaseous particles is doubled; (e) ΔS positive—in aqueous NaCl the separated ions are able to move throughout the solution volume; (f) ΔS positive—dilution means that the ions in $HCl(aq)$ are free to move throughout a larger volume. **18.7** (a) $\Delta S° = 237.6 - 197.9 - 2(130.6) = -221.5$ J/K-mol. (Note that a smaller number of gas molecules is formed in the products as compared with the reactants; thus ΔS is negative.) **18.10** Spontaneity is determined by the change in entropy, which measures the change in randomness of the system, and by the enthalpy change, which measures the change in potential energy. Formation of $H_2O(l)$ from $H_2(g)$ and $O_2(g)$ represents a large decrease in potential energy (the products are more stable, or lower in energy). The enthalpy change here is more important than the entropy decrease. **18.13** (a) $\Delta G° = -394.4 - (-318.2) - (-137.3) = 61.1$ kJ/mol **18.15** (a) $\Delta G° = 2(-370.4) - 2(-300.4) = -140.0$ kJ; $\Delta H° = 2(-395.2) - 2(-296.9) = -196.6$ kJ; $\Delta S° = 2(256.2) - 2(248.5) - 205.0 = -189.6$ J/K. **18.18** (a) $\Delta H° = -90.7$ kJ, $\Delta S° = -221.5$ J/K. The reaction has a negative ΔG at lower temperature, and positive ΔG at higher temperature. Synthesis of CH_3OH from CO and H_2 is more spontaneous (proceeds farther to the right) at lower temperature, but a catalyst is needed to make the reaction proceed at a reasonable rate. **18.20** (a) an increase in N_2 pressure causes

the reaction to shift toward more product. ΔG for the reaction mixture becomes more negative. (b) $\Delta G = -53.85$ kJ. **18.23** (a) $\Delta G° = -193.1$ kJ, $K = 7.8 \times 10^{33}$; (b) $\Delta G° = 2.60$ kJ, $K = 0.35$. **18.26** $\Delta G° = 60.7$ kJ. **18.32** The denaturation process is endothermic; thus ΔH is positive. The entropy change is also positive, because the denatured protein is less regular and ordered than the original protein. ΔG is positive at lower temperature but becomes more negative as temperature is increased. **18.34** The entropy of a pure crystalline solid at 0 K is defined as zero. As temperature increases above 0 K the energy the system gains from its surroundings is distributed among the various forms of energy the system can have. Entropy is a measure of the degree to which this energy is randomized among the various energy states. This measure is inherently positive with respect to the reference state. **18.35** (a) the shuffled deck; (b) the distribution of toys throughout the house; (c) unassembled TV parts. **18.37** (a) This process occurs spontaneously to yield a saturated $MgCl_2$ solution. When $[MgCl_2] = 6\,M$, $\Delta G = 0$ and equilibrium is attained. (b) The reverse reaction is spontaneous. (c) Reaction occurs spontaneously. **18.40** (a) ΔH positive, ΔS positive, $\Delta G = 0$; (b) ΔH positive, ΔS positive, ΔG negative; (c) ΔH positive, ΔS positive, ΔG positive. **18.43** (a) This suggests that ΔH is positive. That is, the process is endothermic. (b) If ΔG is negative when ΔH is positive, ΔS must be positive, so that the $-T\Delta S$ term in $\Delta G = \Delta H - T\Delta S$ is larger than the ΔH term. $\Delta S = (\Delta G - \Delta H)/333$. **18.48** $\Delta G°$ is zero when $\Delta H° - T\Delta S°$ is zero, so $T = \Delta H°/\Delta S°$. For the reaction $O_2(g) \longrightarrow 2O(g)$, $\Delta H° = 495$ kJ, $\Delta S° = 117$ J/K. Thus $T = 4230$ K. Of course, it is only an estimate to use the ΔH and ΔS values for 298 K at such a high temperature, so this is only a rough estimate of T. **18.50** For the aerobic reaction $\Delta G° = -2875$ kJ; for the anaerobic reaction $\Delta G° = -226$ kJ. The equilibrium constant for the first reaction is the larger, 3×10^{504}. This first reaction is capable of supplying more energy for the accomplishment of work, through coupling with other reactions.

CHAPTER 19

19.1 (a) $SO_4^{2-}(aq) + 4H^+(aq) + 2e^- \longrightarrow SO_2(g) + 2H_2O(l)$, reduction; (d) $O_2(g) + 2H_2O(l) + 4e^- \longrightarrow 4OH^-(aq)$, reduction. **19.3** (a) The half-reactions are: $2[Fe(s) \longrightarrow Fe^{3+}(aq) + 3e^-]$ and $3[SO_4^{2-}(aq) + 4H^+(aq) + 2e^- \longrightarrow SO_2(g) + 2H_2O(l)]$. The balanced equation is as follows: $2Fe(s) + 12H^+(aq) + 3SO_4^{2-} \longrightarrow 2Fe^{3+}(aq) + 3SO_2(g) + 6H_2O(l)$. Fe is oxidized, SO_4^{2-} is reduced. (b) $Cl_2(g) + 2OH^-(aq) \longrightarrow Cl^-(aq) + ClO^-(aq) + H_2O(l)$. Here Cl_2 is both oxidized and reduced. **19.4** (a) $2Cr_2O_7^{2-}(aq) + 3CH_3OH(aq) + 16H^+(aq) \longrightarrow 4Cr^{3+}(aq) + 3HCO_2H(aq) + 11H_2O(l)$. (b) $2MnO_4^-(aq) + 6I^-(aq) + 4H_2O(l) \longrightarrow 2MnO_2(s) + 3I_2(s) + 8OH^-(aq)$ **19.7** Your cell should look much like the one in Figure 19.2, with $Fe(s)$ and $Fe^{2+}(aq)$ replacing Zn and $Zn^{2+}(aq)$, respectively. The standard cell potential is $0.44 + 0.34 = 0.78$ V. **19.10** $2Cr(s) + 3MnO_2(s) + 12H^+(aq) \longrightarrow 2Cr^{3+}(aq) + 3Mn^{2+}(aq) + 6H_2O(l)$, $E° = 1.97$ V. **19.11** Na^+ and SO_4^{2-} are not at all readily oxidized. Cl^- can be oxidized both electrochemically and with strong oxidizing agents. Cl_2 is oxidized to form species such as $ClO_3^-(aq)$, but a strong oxidizing agent is required, as you can tell from the $E°$ values listed in Appendix F. **19.13** (a) $Cu^{2+}(aq) < O_2(g) < Cr_2O_7^{2-}(aq) < Cl_2(aq) < H_2O_2(aq)$. **19.15** (a) $E° = +1.23$ V $+ (-1.36$ V$) = -0.13$ V, nonspontaneous; (b) $E° = +0.74$ V $+ (-1.66$ V$) = -0.92$ V, nonspontaneous. **19.17** (a) $E° = 0.323$ V, $\Delta G° = -62.3$ kJ; (b) $E° = 2.041$ V, $\Delta G° = -393.9$ kJ. **19.21** We have that $E = E° - (0.0591/2)\log([Cl^-]^2/[I^-]^2)$. Thus increasing $[I^-]$ makes the log term smaller; E becomes more positive. The effect of increasing $[Cl^-]$ is exactly the opposite. **19.22** $E° = 1.562$ V; $E = 1.60$ V **19.26** (a) A major advantage is that there is no need to separate the anode and cathode compartments via a salt bridge or similar device. Mechanical ruggedness is improved by the use of solids as reactants. (b) The concentration of a solid is effectively constant, so long as there is any solid present to contact the electrolyte. Thus the voltage remains more or less constant as the cell discharges. **19.31** (a) $Mg(l) + Cl_2(g)$; (b) $Cu(s) + O_2(g)$; (c) $Ag(s) + O_2(g)$; (d) $H_2(g) + Br_2(g)$. **19.36** (a) 5 $\mathfrak{F}$; (c) 1 $\mathfrak{F}$. **19.37** 0.348 g Cd. **19.41** 0.0373 mol Ni is deposited. After electrolysis $[Ni^{2+}] = 0.351$ M, $[H^+] = 0.298$ M, and $[SO_4^{2-}]$ remains unchanged. **19.43** (a) 2.98×10^3 amp. **19.46** Zn is oxidized in preference to iron, since the oxidation potential for Zn is much more positive than that for Fe. Chromium also acts as a sacrificial anode when in contact with iron, since $E°$ for Cr oxidation is more positive than for Fe. **19.50** (a) $I_2(s) + H_3AsO_3(aq) + H_2O(l) \longrightarrow 2I^-(aq) + H_3AsO_4(aq) + 2H^+(aq)$. (f) $2CrO_4^{2-}(aq) + H_2O(l) + 3HSnO_2^-(aq) \longrightarrow 2CrO_2^-(aq) + 3HSnO_3^-(aq) + 2OH^-(aq)$. **19.51** (a) $5C_6H_{12}O_6(aq) + 24NO_3^-(aq) + 24H^+(aq) \longrightarrow 30CO_2(g) + 12N_2(g) + 42H_2O(l)$. **19.53** The balanced equation is $I_2(s) + 10NO_3^-(aq) + 8H^+(aq) \longrightarrow 2IO_3^-(aq) + 10NO_2(g) + 4H_2O(l)$. 13.1 g NO_2 is formed. **19.55** (a) Iron serves as the anode. (b) $Fe(s) + Sn^{2+}(aq) \longrightarrow Fe^{2+}(aq) + Sn(s)$; $E° = +0.304$ V. **19.57** For the spontaneous reaction zinc is oxidized to Zn^{2+}, Ag_2O is reduced to Ag. (b) Electrons come from the anode, the Zn electrode, during cell operation. Thus, the anode is negatively charged. (c) The overall reaction and cell voltage are: $Zn(s) + Ag_2O(s) + H_2O(l) \longrightarrow Zn^{2+}(aq) + 2Ag(s) + 2OH^-(aq)$, $E° = +1.10$ V. **19.59** (a) E_{ox} must be between -0.54 and -0.80 V. Thus the potential choices are $Fe^{2+}(aq)$, $H_2O_2(aq)$ in acidic solution, and $MnO_2(s)$ in basic solution. (b) E_{red} must be between 1.36 and 2.87 V. $MnO_4^-(aq)$ in acidic solution is the only choice listed in the table. **19.62** $E = 0.051$ V. The anode is the electrode immersed in the less concentrated solution, 5×10^{-3} M. **19.63** (a) $E° = 0.14$ V, $\Delta G° = -n\mathfrak{F}E° = -27$ kJ, $K = 5.5 \times 10^4$; (b) $E° = 0.440$ V, $\Delta G° = -84.9$ kJ, $K = 8 \times 10^{14}$. **19.66** The cell shown in Figure 19.2 is rechargeable. Application of an external voltage of appropriate sign and magnitude could cause the cell reaction to run "backward," because all components of the half-cell reactions remain in the system and because of the large overvoltage for reducing H^+. **19.69** (a) 0.103 g Ag; (b) 8.1×10^4 sec or 22.5 hr. **19.71** 1.2 amp. **19.73** 2651 sec, about 45 min. **19.75** Maximum work obtainable is 4.3×10^3 kJ. **19.80** 1.37×10^6 coul.

20.2 (a) 8 protons, 9 neutrons; (b) 42 protons, 57 neutrons; (c) 55 protons, 81 neutrons; (d) 47 protons, 68 neutrons. **20.4** (a) $^{181}_{72}Hf \longrightarrow ^{0}_{-1}e + ^{181}_{73}Ta$; (b) $^{226}_{88}Ra \longrightarrow ^{222}_{86}Rn + ^4_2He$; (c) $^{205}_{82}Pb \longrightarrow ^{0}_{1}e + ^{205}_{81}Tl$; (d) $^{179}_{74}W + ^{0}_{-1}e \longrightarrow ^{179}_{73}Ta$. **20.6** (a) $^{32}_{16}S + ^1_0n \longrightarrow ^1_1H + ^{32}_{15}P$; (b) $^7_4Be + ^{0}_{-1}e$ (orbital electron) $\longrightarrow ^7_3Li$; (c) $^{187}_{75}Re \longrightarrow ^{187}_{76}Os + ^{0}_{-1}e$; (d) $^{98}_{42}Mo + ^2_1H \longrightarrow ^1_0n + ^{99}_{43}Tc$; (e) $^{235}_{92}U + ^1_0n \longrightarrow ^{135}_{54}Xe + ^{99}_{38}Sr + 2^1_0n$. **20.8** $^{250}_{100}Fm \longrightarrow ^{246}_{98}Cf + \alpha$; $^{246}_{98}Cf \longrightarrow ^{242}_{96}Cm + \alpha$; $^{242}_{96}Cm \longrightarrow ^{238}_{94}Pu + \alpha$; $^{238}_{94}Pu \longrightarrow ^{234}_{92}U + \alpha$. (In these equations we employ α as shorthand for ^4_2He.) **20.11** Radioactive nuclei are: (b), low neutron/proton ratio; (c), low neutron/proton ratio; and (e), high atomic number. **20.14** (a) No—low neutron/proton ratio. Should be a positron emitter. (b) No—low neutron/proton ratio. Should be a positron emitter, or possibly undergo orbital electron capture. (c) No—high neutron/proton ratio. Should be a beta emitter. (d) No—high atomic number. Should be an alpha emitter. **20.17** (a) Thallium-210 has a high neutron/proton ratio. Beta emission in effect converts a neutron into a proton, thus lowering the neutron/proton ratio. (b) $^{204}_{89}Ac$ has a low neutron/proton ratio. Orbital electron capture converts a proton into a neutron, thus increases the ratio. (c) $^{197}_{83}Bi$ has a low neutron/proton ratio. Alpha particle emission lowers both neutron and proton numbers and lowers the atomic number toward a value for which a lower neutron/proton ratio is stable. **20.19** $^{66}_{32}Ge \longrightarrow ^{66}_{31}Ga + ^{0}_{1}e$. Ten hours represents four half-lives. The amount of ^{66}Ge remaining is $(\frac{1}{2})^4 \times 25.0 = 1.56$ mg. **20.21** $t_{\frac{1}{2}} = 22.6$ yr. **20.23** About 97% of the ^{239}Pu will remain in the year A.D. 3000. **20.26** $t = 1.8 \times 10^9$ yr. **20.30** First calculate k, then use Equation [20.20] to obtain N. Calculate the number of hydrogens in 26.0 g H_2O. The mole fraction of ^3_1H in the sample is 4.8×10^{-13}. **20.31** (a) $\Delta E = 1.23 \times 10^{-12}$ J; (b) $\Delta E = 1.41 \times 10^{-12}$ J; (c) $\Delta E = 1.26 \times 10^{-12}$ J. **20.33** For 1 mol of ^6Li nuclei, $\Delta E = 3.09 \times 10^{12}$ J. **20.36** (a) $\Delta E = 1.70 \times 10^{12}$ J/mol; (b) $\Delta E = 3.15 \times 10^{11}$ J/mol; (c) $\Delta E = 1.77 \times 10^{12}$ J/mol. **20.38** The binding energy per nucleon is greatest for nuclei of mass number around 50 (Figure 20.8). Thus $^{59}_{27}Co$ should possess the greatest mass defect per nucleon. **20.40** Both ^{90}Sr and ^{133}Ba are likely to be incorporated into the food chain as replacements for calcium or perhaps zinc. Neither H_2 nor Kr will be concentrated into living systems. **20.42** Alpha-emitting substances are dangerous only upon inhalation or ingestion because of the short range of the alpha particle. Plutonium is not readily excreted once inside the body, so it remains to cause radiation damage over a long time. **20.46** (a) Add ^{36}Cl to water as a chloride salt. Dissolve ordinary CCl_3COOH. After a time distill the volatile materials away from the salt; CCl_3COOH is volatile and will distill with water. Count the radioactivity in the volatile material. If chlorine exchange has occurred, there will be radioactivity present in the volatile material. **20.53** $N_t/N_0 = 0.180$. **20.55** (a) $^6_3Li + ^{63}_{31}Ni \longrightarrow ^{69}_{31}Ga$; (b) $^{48}_{20}Ca + ^{248}_{96}Cm \longrightarrow ^{296}_{116}X$; (c) $^{88}_{38}Sr + ^{84}_{36}Kr \longrightarrow ^{116}_{46}Pd + ^{56}_{28}Ni$; (d) $^{48}_{20}Ca + ^{238}_{92}U \longrightarrow ^{70}_{20}Ca + 4^1_0n + 2^{106}_{46}Pd$. **20.58** The most massive nuclides will possess the highest neutron/proton ratios and thus will be most likely to decay by beta emission. The least massive nuclides will decay by positron emission or orbital electron capture. **20.61** The energies of nuclear states are so large that merely changing the temperature by less than 100 K will have no observable effect on nuclear decay rates. **20.62** There are seven alpha emissions and four beta emissions. **20.65** (a) $^{78}_{35}Br$; (b) $^{7}_{4}Be$; (c) $^{32}_{15}P$. **20.68** Each photon has wavelength of 2.4×10^{-3} nm. This lies at the short-wavelength (high-energy) side of the range of gamma ray wavelengths (see Figure 5.3).

CHAPTER 21

21.3 Metal oxides are generally ionic solids with relatively high melting points, whereas nonmetal oxides are often gases, liquids, or perhaps low-melting solids. The oxides of metals, when they are soluble in water, generally form basic solutions; metal oxides tend to be more soluble in acidic solutions. Nonmetal oxides are generally soluble in water; they dissolve to form acidic solutions. **21.5** Bismuth should be the most metallic in character. Metallic character would be manifested in electrical conductivity, in the structure of the solid (large number of near-neighbor atoms), in the tendency of the oxide to form a basic solution on dissolving in water, and in a tendency to form ionic compounds with the more electronegative elements. **21.6** (a) IF_3; (b) Bi_2O_3; (c) BF_3; (d) HF; (e) SnO_2. **21.11** (a) $HClO_3$ $(+5)$; (b) BrF_3 (Br is $+3$, F is -1); (c) $XeOF_4$ (Xe is $+6$, F is -1); (d) H_5IO_6 $(+7)$; (e) IO_3^- $(+5)$; (f) $KClO_2$ $(+3)$; (g) $HBrO$ $(+1)$; (h) KI_3 (two I are 0, one is -1; the average oxidation state of I is $-\frac{1}{3}$); (i) PI_3 (I is -1). **21.13** 28.9 M. **21.16** (a) $CaF_2(s) + H_2SO_4(l) \longrightarrow 2HF(g) + CaSO_4(s)$; (b) $2I^-(aq) + Cl_2(g) \longrightarrow I_2(s) + 2Cl^-(aq)$; (c) $Xe(g) + 2F_2(g) \longrightarrow XeF_4(s)$; (d) $Ca^{2+}(aq) + 2OH^-(aq) + Cl_2(aq) \longrightarrow ClO^-(aq) + Cl^-(aq) + H_2O(l) + Ca^{2+}(aq)$; (e) $S_8(s) + 24F_2(g) \longrightarrow 8SF_6(g)$ or $S_8(s) + 24BrF_5(l) \longrightarrow 8SF_6(g) + 24BrF_3(l)$; (f) $8H^+(aq) + I_2(s) + 10NO_3^-(aq) \longrightarrow 2IO_3^-(aq) + 10NO_2(g) + 4H_2O(l)$.

21.19 $\left[:\ddot{F}-\ddot{Br}-\ddot{F}:\right]^+$ $\left[\begin{array}{c} F \quad F \\ F-Sb-F \\ F \quad F \end{array}\right]^-$

From VSEPR theory we predict a bent structure for the cation and an octahedral geometry for the anion. (For clarity we have omitted the unshared pairs from the fluorines in the anion.) **21.21** (a) $Br_2(l) + 2OH^-(aq) \longrightarrow BrO^-(aq) + Br^-(aq) + H_2O(l)$; (b) $Br_2(aq) + H_2O_2(aq) \longrightarrow O_2(g) + 2H^+(aq) + 2Br^-(aq)$; (c) $3CaBr_2(s) + 2H_3PO_4(l) \longrightarrow Ca_3(PO_4)_2(s) + 6HBr(g)$; (d) $AlBr_3(s) + 3H_2O(l) \longrightarrow 3HBr(g) + Al(OH)_3(s)$; (e) $2HF(aq) + CaCO_3(s) \longrightarrow CaF_2(s) + H_2O(l) + CO_2(g)$. **21.22** The balanced equation is $2ClO_3^-(aq) + 10Fe^{2+}(aq) + 12H^+(aq) \longrightarrow Cl_2(g) + 10Fe^{3+}(aq) + 6H_2O(l)$, $E° = +1.47 + (-0.771) = +0.70$ V. **21.25** (a) H_2S (-2); (b) H_2SeO_3 $(+4)$; (c) TeO_3 $(+6)$; (d) $KHSO_3$ $(+4)$; (e) CaO_2 (-1); (f) Al_2S_3 (-2); (g) H_6TeO_6 $(+6)$; (h) SeO_4^{2-} $(+6)$; (i) $S_2O_3^{2-}$.

(The average oxidation state of the two sulfurs is $+2$. However, they are not equivalent; one is effectively in the $+4$ oxidation state, the other in the zero oxidation state.) **21.28** Electronegativity is related to both the ionization energy and electron affinity (Section 7.9). The ionization energy varies much more throughout the group 6A elements than does the electron affinity. **21.30** $V = 5.2 \times 10^4$ L SO_2. **21.32** (a) $SeO_2(s) + H_2O(l) \longrightarrow H_2SeO_3(aq)$; (b) $2NaO_2(s) + 2H_2O(l) \longrightarrow O_2(g) + H_2O_2(aq) + 2Na^+(aq) + 2OH^-(aq)$; (c) $ZnS(s) + 2H^+(aq) \longrightarrow Zn^{2+}(aq) + H_2S(g)$; (d) $8KI(s) + 9H_2SO_4(l) \longrightarrow 8KHSO_4(s) + H_2S(g) + 4I_2(g) + 4H_2O(g)$; (e) $2H_2O_2(aq) \longrightarrow 2H_2O(l) + O_2(g)$. **21.34** 2.0×10^3 mol H_2S per ton of coal, with a volume of 5.1×10^4 L at 300 K, 740 mm Hg pressure. 9.6×10^4 g S. **21.36** (a) $2MnO_4^-(aq) + 5H_2SO_3(aq) \longrightarrow 2MnSO_4(s) + 3SO_4^{2-}(aq) + 3H_2O(l) + 4H^+(aq)$; (b) $Cr_2O_7^{2-}(aq) + 3H_2SO_3(aq) + 2H^+(aq) \longrightarrow 2Cr^{3+}(aq) + 3SO_4^{2-}(aq) + 4H_2O(l)$; (c) $Hg_2^{2+}(aq) + H_2SO_3(aq) + H_2O(l) \longrightarrow 2Hg(l) + SO_4^{2-}(aq) + 4H^+(aq)$. **21.39** (a) $2Ca_3(PO_4)_2(s) + 6SiO_2(s) + 10C(s) \xrightarrow{\Delta} P_4(g) + 6CaSiO_3(l) + 10CO(g)$; (b) $N_2(g) + 3H_2(g) \longrightarrow 2NH_3(g)$; (c) $4NH_3(g) + 5O_2(g) \xrightarrow{\Delta}_{cat} 4NO(g) + 6H_2O(g)$; $2NO(g) + O_2(g) \longrightarrow 2NO_2(g)$; $3NO_2(g) + H_2O(l) \longrightarrow 2H^+(aq) + 2NO_3^-(aq) + NO(g)$; (d) $Ca_3(PO_4)_2(s) + 4H^+(aq) + 3SO_4^{2-}(aq) \longrightarrow 3CaSO_4(s) + 2H_2PO_4^-(aq)$. **21.41** (a) acidic ($NH_4^+$ is the acid); (b) acidic (forms $H_3PO_3(aq)$); (c) acidic (see Equation [21.69]); (d) basic ($N_2H_4(aq) + H_2O(l) \longrightarrow N_2H_5^+(aq) + OH^-(aq)$); (e) basic (anion hydrolysis). **21.44** $\Delta H = 163$ kJ/mol P_4. **21.46** (a) $N_2H_4(g) + 5F_2(g) \longrightarrow 2NF_3(g) + 4HF(g)$; (b) $SeO_3^{2-}(aq) + N_2H_4(aq) \longrightarrow Se(s) + N_2(g) + H_2O(l) + 2OH^-(aq)$; (c) $Cu^{2+}(aq) + 2NH_2OH(aq) \longrightarrow Cu(s) + N_2(g) + 2H_2O(l) + 2H^+(aq)$; (d) $2N_3^-(aq) + Cl_2(g) \longrightarrow 3N_2(g) + 2Cl^-(aq)$. **21.48** 39% weight percentage.

21.50 $\left[\begin{array}{c} :\ddot{Cl}: \\ :\ddot{Cl}\cdots P \cdots \ddot{Cl}: \\ :\ddot{Cl}: \end{array}\right]^+$ $\left[\begin{array}{c} Cl \quad Cl \\ Cl-P-Cl \\ Cl \quad Cl \end{array}\right]^-$

(Each chlorine in PCl_6^- has three unshared electron pairs, not shown here for clarity.) In PCl_4^+, phosphorus employs an sp^3 hybrid set. In PCl_6^-, phosphorus employs an sp^3d^2 hybrid set. The ionic form is stabilized in the solid state by the lattice energy gained in forming the ions. **21.51** (a) $NO_3^-(aq) + 4H^+(aq) + 3e^- \longrightarrow NO(g) + 2H_2O(l)$, $E° = 0.96$ V; (b) $N_2O(aq) + H_2O(l) \longrightarrow 2NO(aq) + 2H^+(aq) + 2e^-$, $E° = -1.59$ V. **21.56** Xenon has a considerably lower ionization energy than argon. Thus the electrons of the octet about xenon can be promoted to higher-energy orbitals (e.g., the $5d$) to allow formation of bonds. In addition, argon is smaller than xenon; there is less room about the atom to accommodate surrounding groups of atoms. **21.59** (a) $2Na(l) + 2HCl(g) \longrightarrow 2NaCl(s) + H_2(g)$; (b) $H_2SO_3(aq) + Br_2(aq) + H_2O(l) \longrightarrow HSO_4^-(aq) + 2Br^-(aq) + 3H^+(aq)$; (c) $2K^+(aq) + 2OH^-(aq) + Br_2(l) + 5ClO^-(aq) \longrightarrow 5Cl^-(aq) + 2KBrO_3(s) + H_2O(l)$; (d) $2BrO_3^-(aq) + 5H_2SO_3(aq) \longrightarrow Br_2(l) + 5HSO_4^-(aq) + 3H^+(aq) + H_2O(l)$; (e)

$UCl_4(s) + 4ClF_3(g) \longrightarrow 3UF_4(g) + 8Cl_2(g)$. **21.60** (a) P_4O_6; (b) Cl_2O_7; (c) As_4O_{10}; (d) B_2O_3; (e) SO_3; (f) CO_2. **21.63** 1.8 kg NaI. **21.66** 448 g $(NH_4)_2S_2O_3$. **21.67** Let us make the assumption that in all cases except (e) the reactions occur in basic solution. (a) $H_2O_2(aq) + S^{2-}(aq) \longrightarrow 2OH^-(aq) + S(s)$; (b) $SO_2(g) + 2OH^-(aq) + H_2O_2(aq) \longrightarrow SO_4^{2-}(aq) + 2H_2O(l)$; (c) $NO_2^-(aq) + H_2O_2(aq) \longrightarrow NO_3^-(aq) + H_2O(l)$; (d) $As_2O_3(s) + 2H_2O_2(aq) + 6OH^-(aq) \longrightarrow 2AsO_4^{3-}(aq) + 5H_2O(l)$; (e) $2Fe^{2+}(aq) + H_2O_2(aq) + 2H^+(aq) \longrightarrow 2Fe^{3+}(aq) + 2H_2O(l)$. **21.70** Refer to Equations [21.69], [21.67], and [21.63] in working through the problem. If we assume that the NO formed is recirculated so as to be consumed with 0.92 overall efficiency, $V = 7.0 \times 10^{10}$ L.

21.75 (a), (b), (c), (d) [structural formulas]

CHAPTER 22

22.1 For sodium the most important mineral is NaCl. For iron the most important minerals from a commercial viewpoint are Fe_2O_3 and Fe_3O_4. Iron pyrite, FeS_2, is abundant, but not commercially utilized. **22.3** (a) CaO; (b) $CaSO_4 \cdot 2H_2O$; (c) $CaCO_3$; (d) $Mg_3Si_4O_{10}(OH)_2$. **22.7** (a) SiO_4^{4-}; (b) SiO_3^{2-}; (c) SiO_3^{2-}. **22.9** (a) Yes; as Al^{3+}. (b) No. Pd forms +2 and +4 oxidation states, but the stable geometrical environments are square-planar and octahedral, respectively. (c) No. K^+ is too large and has too low a charge. (d) Yes; as +4. Ti^{4+} should be similar to Si^{4+}. **22.12** moles B/moles Si = 0.36. **22.14** (a) $Ba^{2+}(aq) + 2OH^-(aq) + CO_2(g) \longrightarrow BaCO_3(s) + H_2O(l)$; (b) $CaCO_3(s) \xrightarrow{\Delta} CaO(s) + CO_2(g)$; $CaO(s) + H_2O(l) \longrightarrow Ca(OH)_2(s)$; (c) $SrCO_3(s) \xrightarrow{\Delta} SrO(s) + CO_2(g)$. **22.15** The solubility equilibrium $FeCO_3(s) \rightleftharpoons Fe^{2+}(aq) + CO_3^{2-}(aq)$ is shifted to the right because of the hydrolysis of the CO_3^{2-} ion. Ground water is saturated with CO_2, and is thus slightly acidic: $CO_3^{2-}(aq) + H_2CO_3(aq) \longrightarrow 2HCO_3^-(aq)$. **22.17** (a) octahedral; (b) tetrahedral; (c) octahedral; (d) octahedral. **22.21** (a) $2CoS_2(s) + 5O_2(g) \longrightarrow 2CoO(s) + 4SO_2(g)$; (b) $2CoO(s) + C(s) \xrightarrow{\Delta} 2Co(l) + CO_2(g)$; (c) $Co(s) \longrightarrow Co^{2+}(aq) + 2e^-$; $Co^{2+}(aq) + 2e^- \longrightarrow Co(s)$. **22.23** 347 g Cu. **22.26** (a) $Co_3O_4(s) + 4Mg(s) \longrightarrow 3Co(s) + 4MgO(s)$; (b) $2NiS_2(s) + 5O_2(g) \longrightarrow 2NiO(s) + 4SO_2(g)$; (c) $OsO_4(g) + 4H_2(g) \longrightarrow Os(s) + 4H_2O(g)$; (d) $2Cs^+(l) + 2e^- \longrightarrow 2Cs(l)$; $2Cl^-(l) \longrightarrow Cl_2(g) + 2e^-$. **22.31** In sublimation of Cr, the metallic bonds between Cr atoms need to be broken. In subliming I_2 only the intermolecular forces between I_2 molecules need to be overcome. **22.35** (a) mole % Cu = 68; mole % Zn = 32; (b) mole % Pb = 53; mole % Sn = 47; (c) mole % Sn = 76; mole % Pb = 24. **22.38** (a) A three-dimensional SiO_2 lattice in which half the Si^{4+} ions

are replaced by Al^{3+}. (b) A three-dimensional SiO_2 lattice in which one-third of the Si^{4+} ions are replaced by Al^{3+} ions. The K^+ ions are present in interstitial positions. (c) Probably best viewed as a SiO_2 lattice in which half the Si^{4+} are replaced by Zr^{4+} ions. Both cations should be in tetrahedral locations. (d) A derivative of $Si_2O_5^{2-}$ in which half the Si^{4+} are replaced by Ti^{4+}. Ca^{2+} is present for charge balance. **22.41** Na^+ in octahedral holes, possibly cubic; Al^{3+} ions in tetrahedral holes. **22.42** 493 g pig iron. **22.46** 3.0×10^{13} coul; 3.6×10^7 kwh. **22.49** 75%. **22.51** (a) $2ZnS(s) + 3O_2(g) \longrightarrow 2ZnO(s) + 2SO_2(g)$; (b) $ZnCO_3(s) \xrightarrow{\Delta} ZnO(s) + CO_2(g)$. Then, $ZnO(s) + C(s) \xrightarrow{\Delta} Zn(l) + CO(g)$. Zinc is usually further refined by electrolysis, after dissolving in aqueous acid.

CHAPTER 23

23.2 (a) four; (b) six; (c) six; (d) four; (e) seven. **23.4** (a) +3; (b) +3; (c) +4; (d) +3. **23.9** (a) cesium tetracyanonickelate(II); (b) tetraammineplatinum(II) diamminetetrachlorocobaltate(III). **23.10** (a) $[Ni(NH_3)_4]$-$(ClO_4)_2$; (d) $[Co(en)_2Cl(SCN)]_2[CdCl_4]$. **23.12** (a) SO_4^{2-} is bidentate; (b) EDTA is tetradentate; (c) en is bidentate; (d) CO_3^{2-} is bidentate. **23.17** (a) one isomer only

(b) [two structural formulas of Co complexes, each labeled $3+$]

(c) There are two geometrical isomers. (d) There are four geometrical isomers; one is chiral, and so it has an optical isomer. There is also the possibility of linkage isomerism of the NO_2 group in each isomer. **23.20** $[Co(NH_3)_4Cl_2]OH$; tetraamminedichlorocobalt(III) hydroxide, $[Co(NH_3)_4Cl(OH)]Cl$; tetraamminechlorohydroxocobalt(III) chloride. **23.22** purple, rose, magenta—something near a red. **23.23** (b) 8. **23.27** (a) a low-spin $3d^4$; (b) low spin; (c) high spin. **23.30** (a) $[Co(NH_3)_6]^{2+}$. NH_3 is a stronger field ligand. (b) $[FeCl_4]^-$. Splitting is larger for the more highly charged ion. (c) $[V(NO_2)_6]^{3-}$. Nitrite is higher in the spectrochemical series.

23.31 (b) [orbital diagrams]

23.32 (a) The larger formation constant for $[Co(en)_3]^{3+}$ is due to the higher charge on the central metal ion. **23.36** Reaction is probably $[Co(NH_3)_4Br_2]^+(aq) + H_2O(l) \longrightarrow [Co(NH_3)_4(H_2O)Br]^{2+}(aq) + Br^-(aq)$. **23.40** The crystal-field splitting is larger in $Co(NH_3)_6^{3+}$ than in $Fe(NH_3)_6^{2+}$ because of the higher central metal ion charge. Thus the cobalt complex is low spin, the iron complex is high spin. **23.42** Wavelength of absorption due to a $3d$ electronic transition should decrease in the order $V(H_2O)_6^{3+} > V(NH_3)_6^{3+} > V(CN)_6^{3-}$. This is so because the crystal-field splitting is increasing in this order. **23.43** (b) $[Cr(en)_2Cl_2]^+(aq) + H_2O(l) \longrightarrow [Cr(en)_2$-

$(H_2O)Cl]^{2+}(aq) + Cl^-(aq)$ and $[Cr(en)_2(H_2O)Cl]^{2+}(aq) + H_2O(l) \longrightarrow [Cr(en)_2(H_2O)_2]^{3+} + Cl^-(aq)$. **23.47** (a) A d^3 complex; the CFSE is $1.2\ \Delta$, $= 218$ kJ.

CHAPTER 24

24.1 alkane, C_6H_{14}; alkene, C_6H_{12}; alkyne, C_6H_{10}; aromatic, C_6H_6.

24.4 $CH_3CH_2CH_2C\equiv CH$ $CH_3CH_2C\equiv CCH_3$

$CH_3CH=C=CHCH_3$ $CH_3CHC\equiv CH$ (with CH_3 below)

$CH_2=CHCH_2CH=CH_2$ $CH_2=CHC=CH_2$ (with CH_3 below)

$CH_2=CHCH=CHCH_3$ $CH_2=C=CCH_3$ (with CH_3 below)

$CH_2=C=CHCH_2CH_3$ CH_2—$CHCH_3$ / $CH=CH$

$CH=CCH_3$ / CH_2—CH_2 ring structure with $CH=CH$, CH_2, CH_2, CH_2 $CH=CH$ / CH / CH_2CH_3

CH_2—$C=CH_2$ / CH_2—CH_2 CH_2 / CH_2—$CHCH=CH_2$

24.6 (a)

H, $CH_2CHCH_2CH_3$ / C=C / CH_3, CH_3, H

(b) $HC\equiv CCH_2Cl$ (c) benzene ring with two Cl substituents

24.8 (a) 2,3-dimethylhexane; (b) 6-methyl-*cis*-3-heptene; (c) 2,3-dichloronaphthalene; (d) 4,4-dimethyl-1-pentyne; (e) 2-bromobutane; (f) 2,5-heptadiene; (g) 2-phenyl-3-bromohexane.

24.9 (a) no;

(b)

CH_3, CH_3 / C=C / Cl, Cl *cis* CH_3, Cl / C=C / Cl, CH_3 *trans*

(c) no; (d) no. **24.13** Octane number is 72.5. **24.16** (a) $CH_3CH_2CH_2CH_2CH_3(g) + Cl_2(g) \xrightarrow{hv} CH_3CH_2CH_2CHClCH_3(g) + HCl(g)$; (b) $2CH_3CH=CH_2(g) + 9O_2(g) \longrightarrow 6CO_2(g) + 6H_2O(g)$; (c) $CH_3CH_2C\equiv CH + Br_2 \longrightarrow CH_3CH_2CBr=CHBr$; (d) $C_6H_6(l) + Br_2(l) \xrightarrow{FeBr_3} C_6H_5Br(l) + HBr(g)$ or $C_6H_6(l) + HNO_3(l) \xrightarrow{H_2SO_4} C_6H_5NO_2(l) + H_2O(l)$. **24.18** (a) $CH_2=CHCH_2CH_3 + H_2 \longrightarrow CH_2CH_2CH_2CH_3$; (b) $CH_3CH=CHCH_3 + H_2O \xrightarrow{H_2SO_4} CH_3CH(OH)CH_2CH_3$; (c) $CH_2CH_2CH_2CH_2(g) + 6O_2(g) \longrightarrow 4CO_2(g) + 4H_2O(g)$.

(d) $nCH_2=CHCl \longrightarrow$ —CH_2—CH—CH_2—CH— (with Cl on CH groups, n repeat)

24.19 (a) CH_3CHCH_3 (with Cl below)   (b) $(CH_3)_2CCH_2CH_3$ (with Cl below)

(c) CH_3CCH_3 (with Cl above and Cl below) or $CH_3C=CH_2$ (with Cl below)

24.21 (a) $CH_3CH_2CH_2CH_3$ (b) $CH_3CH_2CI=CH_2$

(c) CH_3CH—C—CH_2CH_3 (with OH above, CH_3 CH_3 below)

(d) $CH_2BrCHBrCH_3$

24.23 (a) $\Delta H° = -1427.7$ kJ. **24.26** (a) ketone; (b) carboxylic acid; (c) alcohol; (d) ester; (e) aldehyde; (f) ketone; (g) amide.

24.29 (a) CH_3CHCH (with O double bond above, CH_3 below)   (c) no possibility for mild oxidation.

24.30 (a) CH_3C—OCH (with O above, CH_3 and CH_3 at right/below)   (b) $CH_3COCH_2CH_2CH_3$ (with O above)

(c) $HCOCH_2CH_3$ (with O above)

24.32 (a) $CH_3CH_2CHCH_3$ (with OH below)   (b) $HOCH_2CH_2OH$

(c) $HCOCH_3$ (with O above)   (d) $CH_3CH_2CCH_2CH_3$ (with O above)   (e) $CH_3CH_2OCH_2CH_3$

24.34 (a) CH_3CH_2CH (with O above)   (b) $CH_3CH_2CH_2CCH_3$ (with O above)

24.38 (a) $C_{20}H_{42}$; (b) $C_{18}H_{36}$; (c) $C_{12}H_{22}$. **24.42** (a) $CH_3C\equiv CH + HCl \longrightarrow CH_3CCl=CH_2$;

$CH_3CCl=CH_2 + HBr \longrightarrow CH_3$—C—$CH_3$ (with Br above, Cl below)

(b) $CH_2=CH_2 + Cl_2 \longrightarrow CH_2ClCH_2Cl$

(c) product is $\cdots CH_2OCCH_2CH_2COCH_2CH_2OC \cdots$ (with O double bonds above)

24.45 Cyclohexyne would not be stable because the 180° angles about the carbon in an alkyne are inconsistent with the requirements of the six-membered ring structure.

24.47 (a) $CH_3OCH_2CH_3$ (c) $CH_3\overset{\overset{\displaystyle O}{\|}}{C}CH_3$ (e) $CH_3CH_2CH_2CH_2OH$

24.49 Empirical formula is C_2H_4O. Use Equation [9.11]; $MW = 43.8$ g/mol. Thus empirical formula is also the molecular formula; probably acetaldehyde. 24.51 Compound is a secondary alcohol; there are three possibilities:

$$CH_3\underset{\underset{\displaystyle OH}{|}}{C}HCH_2CH_2CH_3 \qquad CH_3CH_2\underset{\underset{\displaystyle OH}{|}}{C}HCH_2CH_3 \qquad CH_3\underset{\underset{\displaystyle OH}{|}}{C}HCH(CH_3)_2$$

24.52 A is either $CH_3CH=CHCH_3$ or $CH_2=CHCH_2CH_3$. B is $(CH_3)_2C=CH_2$.

CHAPTER 25

25.2 All growth processes require energy. In plants, for example, transpiration of water requires energy. In animals, the actions of muscles, maintenance of body temperature, and so forth, all require energy. 25.5 36.4 g O_2; this corresponds to 25.4 L at STP. 25.8 (a) An α-amino acid contains an $-NH_2$ function attached to the carbon that is bound to the carbon of the carboxylic acid function. (b) In protein formation there is a condensation reaction between the amino group of one amino acid and the carboxylic acid group of another. 25.10 Two dipeptides are possible:

$$H_2N-\overset{\overset{\displaystyle H}{|}}{\underset{\underset{\displaystyle H}{|}}{C}}-\overset{\overset{\displaystyle O}{\|}}{C}-N-\overset{\overset{\displaystyle H}{|}}{\underset{\underset{\displaystyle CH(CH_3)_2}{|}}{C}}-\overset{\overset{\displaystyle O}{\|}}{C}-OH \qquad H_2N-\overset{\overset{\displaystyle H}{|}}{\underset{\underset{\displaystyle CH(CH_3)_2}{|}}{C}}-\overset{\overset{\displaystyle O}{\|}}{C}-N-\overset{\overset{\displaystyle H}{|}}{\underset{\underset{\displaystyle H}{|}}{C}}-\overset{\overset{\displaystyle O}{\|}}{C}-OH$$

Glycylvaline   Valinylglycine

25.13 alanine, serine. 25.14 In basic solution as the anion, $H_2NCH_2CO_2^-$; in acidic solution as the cation, $H_3NCH_2CO_2H^+$. 25.17 In glycine the α-carbon atom has two hydrogens. It therefore is not chiral in the sense illustrated in Figure 25.4. 25.20 Denaturation, a change in the tertiary or secondary structure of the protein, can drastically alter its chemical behavior. It can be brought about by a change in solvent or by an increase in temperature. 25.23 1, (d); 2, (a); 3, (e); 4, (c); 5, (b). 25.25 These enzymes are peptidases; their function is to hydrolyze the amide functions along the polypeptide chain. 25.27 Glucose exists in solution as a cyclic structure in which the aldehyde function on carbon 1 reacts with the OH group of carbon 5 to form a so-called hemiacetal. Carbon 1 carries an OH group in the hemiacetal form; in α-glucose this OH group is on the same side of the ring as an OH group on adjacent carbon atom 2 (Figure 25.7). In the β-form the OH groups are on opposite sides.

25.29

Galactose

25.32 cis-$CH_3(CH_2)_7CH=CH(CH_2)_7\overset{\overset{\displaystyle O}{\|}}{C}OCH_2$

cis-$CH_3(CH_2)_7CH=CH(CH_2)_7\overset{\overset{\displaystyle O}{\|}}{C}OCH_2$

cis-$CH_3(CH_2)_7CH=CH(CH_2)_7\overset{\overset{\displaystyle O}{\|}}{C}OCH_2$

25.34 $C_3H_5(OC(O)C_{13}H_{27})_3(l) + 3Na^+(aq) + 3OH^-(aq) \longrightarrow 3C_{13}H_{27}C(O)O^-(aq) + 3Na^+(aq) + C_3H_5(OH)_3(aq)$. 25.37 A nucleotide is a molecule consisting of a nitrogen-containing aromatic compound, a sugar in the furanose form, and a phosphoric acid molecule. 25.39 Complementary strand is TGAGCT 25.42 (a) None. (b) The carbon bearing the secondary OH is chiral. (c) The carbon bearing the CH_3 group and the carbon bearing the NH_2 group are both chiral. 25.46 The only possible structure is glu-cys-gly. 25.48 (a) An enzyme is thought to work by providing a kind of template on which reactants can come together in a certain arrangement to undergo reaction. The so-called lock-and-key model is illustrated in Figure 25.8. (b) A substrate is any substance that undergoes the reaction catalyzed by a particular enzyme. Enzyme inhibition occurs when the active site is blocked by some reagent which is not a substrate but which is able to interact with the active site to keep out substances that are substrates. 25.51 The reaction is $2NH_2CH_2CO_2H(aq) \longrightarrow H_2NCH_2CONHCH_2CO_2H(aq) + H_2O(l)$, $\Delta G = -35.8$ kJ. 25.53 The fact that the rate doubles with a doubling of the concentration of sugar tells us that the fraction of enzyme tied up in the form of an enzyme-substrate complex is small. A doubling of the substrate concentration leads to a doubling of the concentration of enzyme-substrate complex, because only a small fraction of enzyme molecules is tied up. The behavior of innositol suggests that it acts as a competitor with sucrose for binding at the active sites of the enzyme system. Such a competition results in a lower effective concentration of active sites for binding of sucrose, and thus results in a lower reaction rate. 25.58 $AMPOH^-(aq) \rightleftharpoons AMPO^{2-}(aq) + H^+(aq)$. When pH $= 7.40$, $[AMPOH^-]/[AMPO^{2-}] = 0.65$.

Index

Italic page numbers indicate illustrations or structures.

Physical and chemical constants

Atomic mass unit	$1 \text{ amu} = 1.66057 \times 10^{-27} \text{ kg}$
	$6.022169 \times 10^{23} \text{ amu} = 1 \text{ g}$
Avogadro's number	$N = 6.022045 \times 10^{23}/\text{mol}$
Boltzmann's constant	$k = 1.38066 \times 10^{-23} \text{ J/K}$
Electron rest mass	$m_e = 5.48580 \times 10^{-4} \text{ amu}$
	$= 9.10953 \times 10^{-28} \text{ g}$
Electronic charge	$e = 1.6022 \times 10^{-19} \text{ coul}$
Faraday's constant	$\mathcal{F} = Ne = 9.6485 \times 10^4 \text{ coul/mol}$
Gas constant	$R = Nk = 8.3144 \text{ J/K-mol}$
	$= 0.082057 \text{ L-atm/K-mol}$
Neutron rest mass	$m_n = 1.00866 \text{ amu}$
	$= 1.67495 \times 10^{-24} \text{ g}$
Pi	$\pi = 3.1415926536$
Planck's constant	$h = 6.6262 \times 10^{-34} \text{ J-sec}$
Proton rest mass	$m_p = 1.00728 \text{ amu}$
	$= 1.67265 \times 10^{-24} \text{ g}$
Speed of light (in vacuum)	$c = 2.997925 \times 10^8 \text{ m/sec}$